Principles of Electronic Communication Systems

Third Edition

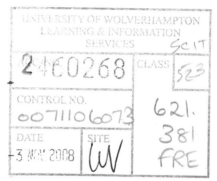

Louis E. Frenzel

Electronic Design Magazine

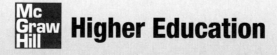 **Higher Education**

Boston   Burr Ridge, IL   Dubuque, IA   Madison, WI   New York   San Francisco   St. Louis
Bangkok   Bogotá   Caracas   Kuala Lumpur   Lisbon   London   Madrid   Mexico City
Milan   Montreal   New Delhi   Santiago   Seoul   Singapore   Sydney   Taipei   Toronto

Higher Education

PRINCIPLES OF ELECTRONIC COMMUNICATION SYSTEM, THIRD EDITION

Published by McGraw-Hill, a business unit of The McGraw-Hill Companies, Inc., 1221 Avenue of the Americas, New York, NY 10020. Copyright © 2008 by The McGraw-Hill Companies, Inc. All rights reserved. No part of this publication may be reproduced or distributed in any form or by any means, or stored in a database or retrieval system, without the prior written consent of The McGraw-Hill Companies, Inc., including, but not limited to, in any network or other electronic storage or transmission, or broadcast for distance learning.

Some ancillaries, including electronic and print components, may not be available to customers outside the United States.

This book is printed on acid-free paper.

1 2 3 4 5 6 7 8 9 0 CCI/CCI 0 9 8 7

ISBN 978–0–07–310704–2
MHID 0–07–310704–2

Publisher: *David T. Culverwell*
Senior Sponsoring Editor: *Thomas Casson*
Managing Developmental Editor: *Jonathan Plant*
Marketing Manager: *James Connely*
Project Manager: *Joyce Watters*
Senior Production Supervisor: *Laura Fuller*
Lead Media Project Manager: *Audrey A. Reiter*
Designer: *Laurie B. Janssen*
(USE) Cover Image: *©GettyImages/Digital Vision, Global Interface DV207*
Senior Photo Research Coordinator: *John Leland*
Photo Research: *LouAnn K. Wilson*
Compositor: *Techbooks*
Typeface: *10/12 Times Roman*
Printer: *Courier Kendallville, Inc.*

The credits section for this book begins on page 919 and is considered an extension of the copyright page.

Library of Congress Cataloging-in-Publication Data

Frenzel, Louis. E.
   Principles of electronic communication systems / Louis E. Frenzel. —3rd ed.
      p. cm.
   ISBN 978-0-07-310704-2 — ISBN 0-07-310704-2
      1. Telecommunications—Textbooks. 2. Wireless communication systems—Textbooks.
   I. Title.

   TK5101.F664 2008
   621.382—dc22

                                                              2006052501

Contents

Preface

This new third edition of *Principles of Electronic Communication Systems* is fully revised and updated to make it one of the most current textbooks available on wireless, networking, and other communications technologies. Because the field of electronic communications changes so fast, it is a never-ending challenge to keep a textbook up to date. While principles do not change, their emphasis and relevance do as technology evolves. Furthermore, students need not only a firm grounding in the fundamentals but also an essential understanding of the real world components, circuits, equipment, and systems in everyday use. This latest edition attempts to balance the principles with an overview of the latest techniques.

One of the major goals of this latest revision is to increase the emphasis on the *system level understanding* of wireless, networking, and other communications technologies. Because of the heavy integration of communications circuits today, the engineer and the technician now work more with printed circuit boards, modules, plug-in cards, and equipment rather than component level circuits. As a result, older obsolete circuits have been removed from this text and replaced with more integrated circuits and block diagram level analysis. Modern communications engineers and technicians work with specifications and standards and spend their time testing, measuring, installing, and troubleshooting. This edition moves in that direction. Detailed circuit analysis is still included in selected areas where it proves useful in understanding the concepts and issues in current equipment.

In the past, a course in communications was considered an option in many electronic programs. Today, communications is the largest sector of the electronics field with the most employees and the largest equipment sales annually. In addition, wireless, networking or other communications technologies are now contained in almost every electronic product. This makes a knowledge and understanding of communication a must rather than an option for every student. Without at least one course in communications, the student may graduate with an incomplete view of the products and systems so common today. This book can provide the background to meet the needs of such a general course.

As the Communications and Networking Editor for Electronic Design Magazine (Penton Media) and editor of the Wireless System Design Update online newsletter, I witness daily the continuous changes in the components, circuits, equipment, systems, and applications of modern communications. As I research the field, interview engineers and executives, and attend the many conferences for the articles and columns I write, I have come to see the growing importance of communications in all of our lives. I have tried to bring that perspective to this latest edition where the most recent techniques and technologies are explained. That perspective coupled with the feedback and insight from some of you who teach this subject has resulted in a text that best fits the 21st century student.

New to this Edition

Here is a chapter-by-chapter summary of revisions and additions to this new edition.

Chapter 1   Significant update of the applications section.

Chapter 2   Revised and updated section on filters.

Chapters 3–6   General editing and updating of circuits.

Chapter 7	Previously chapter 8 on Digital Communications Techniques. Extensive update of the section on data conversion to include new ADC and DAC circuits and expanded specifications section. DSP section has also been updated.
Chapter 8	Previously chapter 7 on Radio Transmitters. Expanded coverage of the I/Q architecture for digital data transmission. Addition of broadband linear power amplifiers using feedforward and adaptive predistortion techniques. Addition of ISM band IC transmitters. The section on vacuum tube power amplifiers has been removed but will be available on line if anyone needs it.
Chapter 9	Expanded coverage of receiver sensitivity and signal to noise ratio, its importance and calculation. Increased coverage of the software-defined radio (SDR) and introduction to cognitive radio. Updated section on receiver circuits and transceivers. Description of a typical wireless LAN transceiver chip.
Chapter 10	Addition of code division multiple access, the Radio Data System and SCA subsystems in FM radios. Elimination of the older no-longer-used PAM telemetry system coverage. A new section on time and frequency division duplexing.
Chapter 11	Expanded coverage of digital modulation and spectral efficiency. Addition of an explanation of how different digital modulation schemes affect the bit error rate (BER) in communications systems. Comparisons based on BER vs. carrier to noise radio (C/N) are added. Updated sections on spread spectrum and OFDM. A new section on convolutional and turbo coding.
Chapter 12	Previous chapter 12 Computer Networking has been revised into a new chapter called Introduction to Networking and LANs. The coverage has been expanded and updated to include things like mesh networking fundamentals, the latest Ethernet standards including Power over Ethernet (PoE), and improved explanation of LAN equipment.
Chapter 13	Minor revisions and updates.
Chapter 14	Improved explanation of the near and far fields. Introduction to the automatic antenna tuner.
Chapter 15	A new chapter focusing on the Internet, chapter 9 includes the Internet material from the previous chapter 12 but with extensive new material. Detailed explanation of how information travels via the Internet. Addition of descriptions of Internet core technologies like ATM, Frame Relay, and Sonet. Considerably expanded discussion of the TCP/IP protocol. Expanded explanation of routers including line cards and switch fabrics. Introduction of a new section on storage area networks (SANs) and their transmission technologies including Fibre Channel and iSCSI. A new section on Internet security including encryption and authentication.
Chapter 16	Extensively revised and updated. New material on microwave antennas including phased arrays, beam forming arrays, adaptive antennas, and the smaller ceramic and PC board antennas like the loop, meander line, and inverted-F. The concepts of diversity and multiple input multiple output (MIMO) are added.
Chapter 17	Revised and updated. New materials include a section on Very Small Aperture Terminals and expanded coverage of GPS.
Chapter 18	Elimination of the section on paging. Updated section on cordless phones. New section on voice over Internet protocol (VoIP) digital telephones.

Chapter 19	New section on MSA optical transceiver modules, types and specifications. Expanded section on electronic dispersion compensation. New section on passive optical networks (PONs) used in fiber to the home (FTTH) broadband systems.
Chapter 20	This is a new chapter on Cell Phone Technologies. It covers all major analog and digital cell phone standards and systems and frequency allocations. GSM, GPRS, and EDGE TDM systems are covered as well as both cdma2000 and WCDMA systems. Typical chips are reviewed. Fourth generation systems are introduced.
Chapter 21	A new chapter on wireless technologies. Coverage includes wireless LAN (802.11a/b/g/n), Bluetooth, ZigBee, Ultra wideband (UWB), WiMAX, RFID, near field communications (NFC), ISM band short range radios, and infrared wireless. Coverage of personal area networks and mesh systems is included.
Chapter 22	Communications Tests and Measurement chapter is revised and updated. A new section on the widely used boundary scan and JTAG test system for chips and boards has been added.
Chapter 23	Television has been dropped from the book, but the chapter has been revised and updated, and placed on the Online Learning Center website for those who choose to assign it. It now includes new digital television information, new cable standards, and mobile (cell phone video) television standards.

In a large book such as this, it's difficult to give every one what he or she wants. Some want more depth others greater breadth. I tried to strike a balance between the two. As always, I am always eager to hear from those of you who use the book and welcome your suggestions for the next edition.

Learning Features

Principles of Electronic Communication Systems third edition has been completely redesigned to give it a more attractive and accessible page layout. To guide readers and provide an integrated learning approach, each chapter contains the following features:

- Chapter Objectives
- Key Terms
- Pioneers of Electronics articles
- Good to Know margin features
- Examples with solutions
- Chapter Summary
- Questions
- Problems
- Critical Thinking

Student Resources

Laboratory & Activities Manual

A major change with this third edition is the availability of a new laboratory manual. The Lab Manual developed for the second edition will be retained for those of you who use it. This new *Laboratory & Activities Manual* provides more actual hands-on hardware experiments with modern circuits and components. While many circuits are still explored, the attempt is to push toward more systems-level experiments. Building a practical, affordable but meaningful lab is one of the more difficult parts of creating a college course in communications. This new manual provides practice in the principles by using the latest components and methods. Affordable and readily available components and equipment have been used to make it easy for professors to put together a communications lab that validates and complements the text.

Many of the exercises in the *Laboratory & Activities Manual* involve web access and search to build the student's ability to use the vast resources of the Internet and World Wide Web. The practical engineers and technicians of today have become experts at finding relevant information and answers to their questions and solutions to their problems this way. While practicing this essential skill of any communications engineer or technician knowledge, the student will be able to expand his or her knowledge of any of the subjects in this book, either to dig deeper into the theory and practice or get the latest update information on chips and other products.

Online Learning Center ("OLC") website, www.mhhe.com/frenzel3e

This text-specific site includes a number of student-oriented resources, including:

- Chapter outlines and summaries.
- MultiSim version 9 Primer, for those who want to get up and running with this popular simulation software. The section is written to provide communications examples and applications.
- MultiSim circuit files for communications electronics.
- Web Links to industrial and educational sites of interest.
- Link to the Work-Ready Electronics; these activities, created by the MATEC research center, show the practical skills needed in various areas of interest—including communications—in the context of modern industry.

Instructor Resources

Instructor Productivity Center CD–ROM

This CD includes the following resources for adopters of the text:

- Answers and solutions to all text problems.
- Answers and information for the Lab & Activity Manual.
- Electronic test banks with a mix of questions for each text chapter.
- PowerPoint presentations for all chapters of the text.

Online Learning Center ("OLC") website, www.mhhe.com/frenzel3e

The OLC contains student resources, plus the following instructor resources:

- Answers and solutions to the text problems and lab activities, under password protection.
- PowerPoint presentations for each chapter online.
- Additional quiz questions for each chapter, which can be assigned or used for student self-study.
- Blackboard and WebCT cartridges for use with these popular classroom management systems.

Classroom Performance System (CPS) from eInstruction is available for adopters; its "clicker" system provides a vehicle for in-class quizzing and concept reinforcement, and classroom management.

Acknowledgements

While producing a new edition of a book does not involve the same effort as writing a new book, this latest revision was a major project. My special thanks to Managing Developmental Editor Jonathan Plant, and Publisher Thomas Casson for their continued support and encouragement to make this happen. It has been a pleasure to work with you both.

And my appreciation also goes out to those professors who reviewed the book and offered your feedback, criticism and suggestions. Thanks for taking the time to provide that valuable input. I have implemented virtually all of your recommendations. I especially appreciate the extensive input from Walt Curry of the United States Naval Academy, most of which I have included. The following reviewers looked over the manuscript in various stages, and provided a wealth of good suggestions for the new edition:

Heng Chan
  Mohawk College (ON)
Captain Walter N. Currier Jr.
  United States Naval Academy (MD)
William C. Donaldson
  Wake Techical College (NC)
Robbie Edens
  ECPI College of Technology (SC)
Terry Fleischman
  Fox Valley Technical College (WI)
Richard Fornes
  Johnson College (PA)
G. J. Gerard
  Gateway Community College (CT)
Georges C. Livanos
  Humber College (ON)

Robert J. Lovelace
  East Mississippi Junior College (MS)
Robert Most
  Ferris State University (MI)
Tom N. Neal Jr.
  Griffin Technical College (GA)
Phillip C. Purvis
  George C. Wallace Community College (AL)
Pravin M. Raghuwanshi
  DeVry University (NJ)
William Salice
  ECPI College of Technology (VA)
Randy Winzer
  Pittsburg Sate University (KS)

With the latest input from industry and the suggestions from those who use the book, this edition should come closer than ever to being an ideal text for teaching current day communications electronics.

Lou Frenzel
Austin, Texas
2006

Guided Tour

Learning Features

Many new learning features have been incorporated into the seventh edition of *Electronic Principles*. These learning features, found throughout the chapters, include:

chapter

2

The Fundamentals of Electronics: A Review

To understand communication electronics as presented in this book, you need a knowledge of certain basic principles of electronics, including the fundamentals of alternating-current (ac) and direct-current (dc) circuits, semiconductor operation and characteristics, and basic electronic circuit operation (amplifiers, oscillators, power supplies, and digital logic circuits). Some of the basics are particularly critical to understanding the chapters that follow. These include the expression of gain and loss in decibels, *LC* tuned circuits, resonance and filters, and Fourier theory. The purpose of this chapter is to briefly review all these subjects. If you have studied the material before, it will simply serve as a review and reference. If, because of your own schedule or the school's curriculum, you have not previously covered this material, use this chapter to learn the necessary information before you continue.

Objectives

After completing this chapter, you will be able to:

- Calculate voltage, current, gain, and attenuation in decibels and apply these formulas in applications involving cascaded circuits.
- Explain the relationship between *Q*, resonant frequency, and bandwidth.
- Describe the basic configuration of the different types of filters that are used in communication networks and compare and contrast active filters with passive filters.
- Explain how using switched capacitor filters enhances selectivity.
- Explain the benefits and operation of crystal, ceramic, and SAW filters.
- Calculate bandwidth by using Fourier analysis.

30

Chapter Introduction

Each chapter begins with a brief introduction setting the stage for what the student is about to learn.

Chapter Objectives

Chapter Objectives provide a concise statement of expected learning outcomes.

Examples

Each chapter contains worked-out Examples that demonstrate important concepts or circuit operations, including circuit analysis, applications, troubleshooting, and basic design.

Good To Know

Good To Know statements, found in margins, provide interesting added insights to topics being presented

$$dBm = 10 \log \frac{P_{out}(W)}{0.001(W)}$$

Here P_{out} is the output power, or some power value you want to compare to 1 mW, and 0.001 is 1 mW expressed in watts.

The output of a 1-W amplifier expressed in dBm is, e.g.,

$$dBm = 10 \log \frac{1}{0.001} = 10 \log 1000 = 10(3) = 30 \text{ dBm}$$

Sometimes the output of a circuit or device is given in dBm. For example, if a microphone has an output of -50 dBm, the actual output power can be computed as follows:

$$-50 \text{ dBm} = 10 \log \frac{P_{out}}{0.001}$$

$$\frac{-50 \text{ dBm}}{10} = \log \frac{P_{out}}{0.001}$$

Therefore

$$\frac{P_{out}}{0.001} = 10^{-50 \text{ dBm}/10} = 10^{-5} = 0.00001$$

$$P_{out} = 0.001 \times 0.00001 = 10^{-3} \times 10^{-5} = 10^{-8} \text{ W} = 10 \times 10^{-9} = 10 \text{ nW}$$

Example 2-10

A power amplifier has an input of 90 mV across 10 kΩ. The output is 7.8 V across an 8-Ω speaker. What is the power gain, in decibels? You must compute the input and output power levels first.

GOOD TO KNOW

From the standpoint of sound measurement, 0 dB is the least perceptible sound (hearing threshold), and 120 dB equals the pain threshold of sound. This list shows intensity levels for common sounds. (Tippens, *Physics*, 6th ed., Glencoe/ McGraw-Hill, 2001, p. 497)

Sound	Intensity level, dB
Hearing threshold	0
Rustling leaves	10
Whisper	20
Quiet radio	40
Normal conversation	65
Busy street corner	80
Subway car	100
Pain threshold	120
Jet engine	140–160

Pioneers of Electronics

Students can use summaries when reviewing for examinations, or just to make sure they haven't missed any key concepts. are listed to help solidly learning outcomes.

Chapter Review

Students can use summaries when reviewing for examinations, or just to make sure they haven't missed any key concepts. Important circuit derivations and definition are listed to help solidly learning outcomes.

Problems

Students obtain back by Problems that immediately follow most Examples. Answers to these problems are found at the end of each chapter.

Critical Thinking

A wide variety of questions and problems are found at the end of each chapter; over 30% are new or revised in this edition. Those include circuit analysis, trouble shooting, critical thinking, and job interview questions.

Figure 1-14 The electromagnetic spectrum used in electronic communication.

Name	Frequency	Wavelength
Extremely low frequencies (ELFs)	30–300 Hz	10^7–10^6 m
Voice frequencies (VFs)	300–3000 Hz	10^6–10^5 m
Very low frequencies (VLFs)	3–30 kHz	10^5–10^4 m
Low frequencies (LFs)	30–300 kHz	10^4–10^3 m
Medium frequencies (MFs)	300 kHz–3 MHz	10^3–10^2 m
High frequencies (HFs)	3–30 MHz	10^2–10^1 m
Very high frequencies (VHFs)	30–300 MHz	10^1–1 m
Ultra high frequencies (UHFs)	300 MHz–3 GHz	1–10^{-1} m
Super high frequencies (SHFs)	3–30 GHz	10^{-1}–10^{-2} m
Extremely high frequencies (EHFs)	30–300 GHz	10^{-2}–10^{-3} m
Infrared	—	0.7–10 µm
The visible spectrum (light)	—	0.4–0.8 µm

Units of Measure and Abbreviations:

kHz = 1000 Hz
MHz = 1000 kHz = 1×10^6 = 1,000,000 Hz
GHz = 1000 MHz = 1×10^6 = 1,000,000 kHz
= 1×10^9 = 1,000,000,000 Hz
m = meter
μm = micrometer = $\dfrac{1}{1,000,000}$ m = 1×10^{-6} m

Prefixes representing powers of 10 are often used to express frequencies. The most frequently used prefixes are as follows:

$$k = kilo = 1000 = 10^3$$
$$M = mega = 1,000,000 = 10^6$$
$$G = giga = 1,000,000,000 = 10^9$$
$$T = tera = 1,000,000,000,000 = 10^{12}$$

Thus, 1000 Hz = 1 kHz (kilohertz). A frequency of 9,000,000 Hz is more commonly expressed as 9 MHz (megahertz). A signal with a frequency of 15,700,000,000 Hz is written 15.7 GHz (gigahertz).

PIONEERS OF ELECTRONICS

In 1887 German physicist Heinrich Hertz was the first to demonstrate the effect of electromagnetic radiation through space. The distance of transmission was only a few feet, but this transmission proved that radio waves could travel from one place to another without the need for any connecting wires. Hertz also proved that radio waves, although invisible, travel at the same velocity as light waves. (Grob/Schultz, *Basic Electronics*, 9th ed., Glencoe/McGraw-Hill, 2003, p. 4)

CHAPTER REVIEW

Summary

All electronic communication systems consist of three basic components: a transmitter, a communication channel (medium), and a receiver. Messages are converted to electrical signals and sent over electrical or fiber-optic cable or free space to a receiver. Attenuation (weakening) and noise can interfere with transmission.

Electronic communication is classified as (1) one-way (simplex) or two-way (full duplex or half duplex) transmissions and (2) analog or digital signals. Analog signals are smoothly varying, continuous signals. Digital signals are discrete, two-state (on/off) codes. Electronic signals are often changed from analog to digital and vice versa. Before transmission, electronic signals are known as baseband signals.

Amplitude and frequency modulation make an information signal compatible with the channel over which it is to be sent, modifying the carrier wave by changing its amplitude, frequency, or phase angle and sending it to an antenna for transmission, a process known as broadband communication. Frequency-division and time-division multiplexing allow more than one signal at a time to be transmitted over the same medium.

All electronic signals that radiate into space are part of the electromagnetic spectrum; their location on the spectrum is determined by frequency. Most information signals to be transmitted occur at lower frequencies and modulate a carrier wave of a higher frequency.

How much information a given signal can carry depends in part on its bandwidth. Available space for transmitting signals is limited, and signals transmitting on the same frequency or on overlapping frequencies interfere with one another. Research efforts are being devoted to developing use of higher-frequency signals and minimizing the bandwidth required.

Spectrum usage is regulated by governments, in the United States by the FCC and NTIA, and by equivalent agencies in other governments. Standards for communication systems state specifically how the information is transmitted and received. Standards are set by independent organizations such as ANSI, EIA, ETSI, IEEE, ITU, IETF, and TIA.

The four major electronic specialties are computers, communication, industrial control, and instrumentation. There are many job opportunities in the field of electronic communication.

Questions

1. In what century did electronic communication begin?
2. Name the four main elements of a communication system, and draw a diagram that shows their relationship.
3. List five types of media used for communication, and state which three are the most commonly used.
4. Name the device used to convert an information signal to a signal compatible with the medium over which it is being transmitted.
5. What piece of equipment acquires a signal from a communication medium and recovers the original information signal?

Introduction to Electronic Communication   27

Problems

1. Calculate the frequency of signals with wavelengths of 40 m, 5 m, and 8 cm.
2. In what frequency range does the common ac power line frequency fall?
3. What is the primary use of the SHF and EHF ranges? ◆

◆ *Answers to Selected Problems follow Chapt. 22.*

Critical Thinking

1. Name three ways that a higher-frequency signal called the carrier can be varied to transmit the intelligence.
2. Name two common household remote-control units, and state the type of media and frequency ranges used for each.
3. How is radio astronomy used to locate and map stars and other heavenly bodies?
4. In what segment of the communication field are you interested in working, and why?
5. Assume that all the electromagnetic spectrum from ELF through microwaves was fully occupied. Explain some ways that communication capability could be added.
6. What is the speed of light in feet per microsecond? In inches per nanosecond? In meters per second?
7. Make a general statement comparing the speed of light with the speed of sound. Give an example of how the principles mentioned might be demonstrated.
8. List five real-life communication applications not specifically mentioned in this chapter.
9. "Invent" five new communication methods, wired or wireless, that you think would be practical.
10. Assume that you have a wireless application you would like to design, build, and sell as a commercial product. You have selected a target frequency in the UHF range. How would you decide what frequency to use, and how would you get permission to use it?
11. Make an exhaustive list of all the electronic communication products that you own, have access to at home or in the office, and/or use on a regular basis.
12. You have probably seen or heard of a simple communication system made of two paper cups and a long piece of string. How could such a simple system work?

Introduction to Electronic Communication

Objectives

After completing this chapter, you will be able to:

- Explain the functions of the three main parts of an electronic communication system.
- Describe the system used to classify different types of electronic communication and list examples of each type.
- Discuss the role of modulation and multiplexing in facilitating signal transmission.
- Define the electromagnetic spectrum and explain why the nature of electronic communication makes it necessary to regulate the electromagnetic spectrum.
- Explain the relationship between frequency range and bandwidth and give the frequency ranges for spectrum uses ranging from voice to ultra-high-frequency television.
- List the major branches of the field of electronic communication and describe the qualifications necessary for different jobs.

Figure 1-1 Milestones in the history of electronic communication.

When?	Where or Who?	What?
1837	Samuel Morse	Invention of the telegraph (patented in 1844).
1843	Alexander Bain	Invention of facsimile.
1866	United States and England	The first trans-Atlantic telegraph cable laid.
1876	Alexander Bell	Invention of the telephone.
1877	Thomas Edison	Invention of the phonograph.
1879	George Eastman	Invention of photography.
1887	Heinrich Hertz (German)	Discovery of radio waves.
1887	Guglielmo Marconi (Italian)	Demonstration of "wireless" communications by radio waves.
1901	Marconi (Italian)	First trans-Atlantic radio contact made.
1903	John Fleming	Invention of the two-electrode vacuum tube rectifier.
1906	Reginald Fessenden	Invention of amplitude modulation; first electronic voice communication demonstrated.
1906	Lee de Forest	Invention of the triode vacuum tube.
1914	Hiram P. Maxim	Founding of American Radio Relay League, the first amateur radio organization.
1920	KDKA Pittsburgh	First radio broadcast.
1923	Vladimir Zworykin	Invention and demonstration of television.
1933–1939	Edwin Armstrong	Invention of the superheterodyne receiver and frequency modulation.
1939	United States	First use of two-way radio (walkie-talkies).
1940–1945	Britain, United States	Invention and perfection of radar (World War II).
1948	John von Neumann and others	Creation of the first stored program electronic digital computer.
1948	Bell Laboratories	Invention of transistor.
1953	RCA/NBC	First color TV broadcast.
1958–1959	Jack Kilby (Texas Instruments) and Robert Noyce (Fairchild)	Invention of integrated circuits.
1958–1962	United States	First communication satellite tested.
1961	United States	Citizens band radio first used.
1975	United States	First personal computers.
1977	United States	First use of fiber-optic cable.
1983	United States	Cellular telephone networks.
1990s	United States	Adoption and growth of computer networking, including local-area networks (LANs). Global Positioning System (GPS) for satellite navigation. The Internet and World Wide Web.
2000–present	Worldwide	Third-generation digital cell phones, wireless local-area networks, digital broadcast radio, and 40-Gbps fiber-optic communication.

1-1 The Significance of Human Communication

Communication is the process of exchanging information. People communicate to convey their thoughts, ideas, and feelings to others. The process of communication is inherent to all human life and includes verbal, nonverbal (body language), print, and electronic processes.

Two of the main barriers to human communication are language and distance. Language barriers arise between persons of different cultures or nationalities.

Communicating over long distances is another problem. Communication between early human beings was limited to face-to-face encounters. Long-distance communication was first accomplished by sending simple signals such as drumbeats, horn blasts, and smoke signals and later by waving signal flags (semaphores). When messages were relayed from one location to another, even greater distances could be covered.

The distance over which communication could be sent was extended by the written word. For many years, long-distance communication was limited to the sending of verbal or written messages by human runner, horseback, ship, and later trains.

Human communication took a dramatic leap forward in the late nineteenth century, when electricity was discovered and its many applications were explored. The telegraph was invented in 1844 and the telephone in 1876. Radio was discovered in 1887 and demonstrated in 1895. Figure 1-1 is a timetable listing important milestones in the history of electronic communication.

Well-known forms of electronic communication, such as the telephone, radio, TV, and the Internet, have increased our ability to share information. The way we do things and the success of our work and personal lives are directly related to how well we communicate. It has been said that the emphasis in our society has now shifted from that of manufacturing and mass production of goods to the accumulation, packaging, and exchange of information. Ours is an information society, and a key part of it is communication. Without electronic communication, we could not access and apply the available information in a timely way.

This book is about electronic communication, and how electrical and electronic principles, components, circuits, equipment, and systems facilitate and improve our ability to communicate. Rapid communication is critical in our very fast-paced world. It is also addictive. Once we adopt and get used to any form of electronic communication, we become hooked on its benefits. In fact, we cannot imagine conducting our lives or our businesses without it. Just imagine our world without the telephone, radio, fax, television, cell phones, or computer networking.

> ## GOOD TO KNOW
> Fax machines have been in use since the 1930s. These early machines were primarily utilized by news services to transmit photographs by using free space or radio waves rather than telephone lines.

1-2 Communication Systems

All electronic communication systems have a transmitter, a communication channel or medium, and a receiver. These basic components are shown in Fig. 1-2. The process of communication begins when a human being generates some kind of message, data, or other intelligence that must be received by others. A message may also be generated by a computer or electronic current. In *electronic communication systems,* the message is referred to as *information,* or an intelligence signal. This message, in the form of an electronic signal, is fed to the transmitter, which then transmits the message over the communication channel. The message is picked up by the receiver and relayed to another human. Along the way, noise is added in the communication channel and in the receiver. *Noise* is the general term applied to any phenomenon that degrades or interferes with the transmitted information.

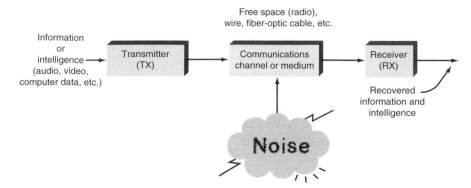

Figure 1-2 A general model of all communication systems.

Transmitter

The first step in sending a message is to convert it into electronic form suitable for transmission. For voice messages, a microphone is used to translate the sound into an electronic *audio* signal. For TV, a camera converts the light information in the scene to a video signal. In computer systems, the message is typed on a keyboard and converted to binary codes that can be stored in memory or transmitted serially. Transducers convert physical characteristics (temperature, pressure, light intensity, and so on) into electrical signals.

The *transmitter* itself is a collection of electronic components and circuits designed to convert the electrical signal to a signal suitable for transmission over a given communication medium. Transmitters are made up of oscillators, amplifiers, tuned circuits and filters, modulators, frequency mixers, frequency synthesizers, and other circuits. The original intelligence signal usually modulates a higher-frequency carrier sine wave generated by the transmitter, and the combination is raised in amplitude by power amplifiers, resulting in a signal that is compatible with the selected transmission medium.

Communication Channel

The *communication channel* is the medium by which the electronic signal is sent from one place to another. Many different types of media are used in communication systems, including wire conductors, fiber-optic cable, and free space.

Electrical Conductors. In its simplest form, the medium may simply be a pair of wires that carry a voice signal from a microphone to a headset. It may be a coaxial cable such as that used to carry cable TV signals. Or it may be a twisted-pair cable used in a local-area network (LAN) for personal computers.

Optical Media. The communication medium may also be a fiber-optic cable or "light pipe" that carries the message on a light wave. These are widely used today to carry long-distance calls and all Internet communications. The information is converted to digital form that can be used to turn a laser diode off and on at high speeds. Alternatively, audio or video analog signals can be used to vary the amplitude of the light.

Free Space. When free space is the medium, the resulting system is known as radio. Also known as *wireless*, radio is the broad general term applied to any form of wireless communication from one point to another. Radio makes use of the electromagnetic spectrum. Intelligence signals are converted to electric and magnetic fields that propagate nearly instantaneously through space over long distances. Communication by visible or infrared light also occurs in free space.

Audio

Transmitter

Communication channel

Wireless radio

Other Types of Media. Although the most widely used media are conducting cables and free space (radio), other types of media are used in special communication systems. For example, in sonar, water is used as the medium. Passive sonar "listens" for underwater sounds with sensitive hydrophones. Active sonar uses an echo-reflecting technique similar to that used in radar for determining how far away objects under water are and in what direction they are moving.

The earth itself can be used as a communication medium, because it conducts electricity and can also carry low-frequency sound waves.

Alternating-current (ac) power lines, the electrical conductors that carry the power to operate virtually all our electrical and electronic devices, can also be used as communication channels. The signals to be transmitted are simply superimposed on or added to the power line voltage. This is known as *carrier current transmission*. It is used for some types of voice intercoms, for remote control of electrical equipment, and in some LANs.

Carrier current transmission

Receivers

A *receiver* is a collection of electronic components and circuits that accepts the transmitted message from the channel and converts it back to a form understandable by humans. Receivers contain amplifiers, oscillators, mixers, tuned circuits and filters, and a demodulator or detector that recovers the original intelligence signal from the modulated carrier. The output is the original signal, which is then read out or displayed. It may be a voice signal sent to a speaker, a video signal that is fed to a cathode-ray tube for display, or binary data that is received by a computer and then printed out or displayed on a video monitor.

Receiver

Transceivers

Most electronic communication is two-way, and so both parties must have both a transmitter and a receiver. As a result, most communication equipment incorporates circuits that both send and receive. These units are commonly referred to as *transceivers*. All the transmitter and receiver circuits are packaged within a single housing and usually share some common circuits such as the power supply. Telephones, fax machines, handheld CB radios, cellular telephones, and computer modems are examples of transceivers.

Transceiver

Attenuation

Attenuation

Signal *attenuation,* or degradation, is inevitable no matter what the medium of transmission. Attenuation is proportional to the square of the distance between the transmitter and receiver. Media are also frequency-selective, in that a given medium will act as a low-pass filter to a transmitted signal, distorting digital pulses in addition to greatly reducing signal amplitude over long distances. Thus considerable signal amplification, in both the transmitter and the receiver, is required for successful transmission. Any medium also slows signal propagation to a speed slower than the speed of light.

Noise

Noise is mentioned here because it is the bane of all electronic communications. Its effect is experienced in the receiver part of any communications system. For that reason, we cover noise at that more appropriate time in Chapter 9. While some noise can be filtered out, the general way to minimize noise is to use components that contribute less noise and to lower their temperatures. The measure of noise is usually expressed in terms of the signal-to-noise (S/N) ratio (SNR), which is the signal power divided by the noise power and can be stated numerically or in terms of decibels (dB). Obviously, a very high SNR is preferred for best performance.

GOOD TO KNOW

Solar flares can send out storms of ionized radiation that can last for a day or more. The extra ionization in the atmosphere can interfere with communication by adding noise. It can also interfere because the ionized particles can damage or even disable communication satellites. The most serious X-class flares can cause planetwide radio blackouts.

1-3 Types of Electronic Communication

Electronic communications are classified according to whether they are (1) one-way (simplex) or two-way (full duplex or half duplex) transmissions and (2) analog or digital signals.

Simplex

Simplex communication

The simplest way in which electronic communication is conducted is one-way communications, normally referred to as *simplex communication.* Examples are shown in Fig. 1-3. The most common forms of simplex communication are radio and TV broadcasting. Another example of one-way communication is transmission via a paging system to a personal receiver (beeper).

Figure 1-3   Simplex communication.

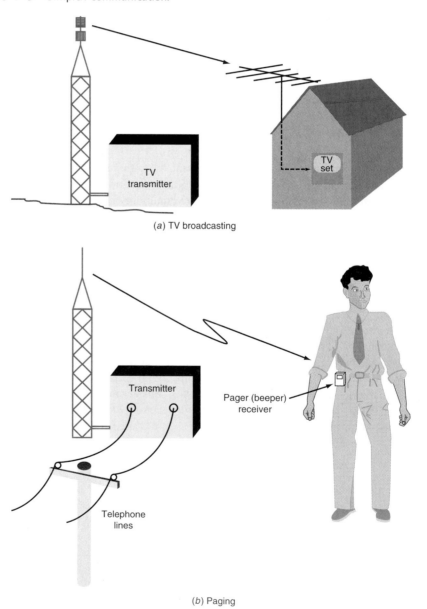

(a) TV broadcasting

(b) Paging

Chapter 1

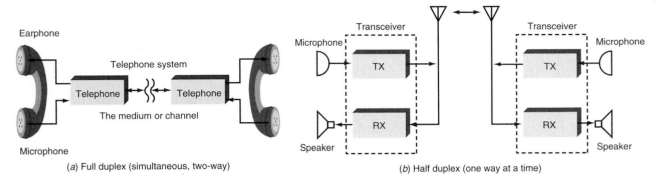

(*a*) Full duplex (simultaneous, two-way)

(*b*) Half duplex (one way at a time)

Full Duplex

The bulk of electronic communication is two-way, or *duplex communication.* Typical duplex applications are shown in Fig. 1-4. For example, people communicating with one another over the telephone can talk and listen simultaneously, as Fig. 1-4(*a*) illustrates. This is called *full duplex communication.*

Duplex communication

Full duplex communication

Half Duplex

The form of two-way communication in which only one party transmits at a time is known as *half duplex communication* [see Fig. 1-4(*b*)]. The communication is two-way, but the direction alternates: the communicating parties take turns transmitting and receiving. Most radio transmissions, such as those used in the military, fire, police, aircraft, marine, and other services, are half duplex communication. Citizens band (CB), Family Radio, and amateur radio communication are also half duplex.

Half duplex communication

Analog Signals

An *analog signal* is a smoothly and continuously varying voltage or current. Some typical analog signals are shown in Fig. 1-5. A sine wave is a single-frequency analog signal. Voice and video voltages are analog signals that vary in accordance with the sound or light variations that are analogous to the information being transmitted.

Analog signal

Digital Signals

Digital signals, in contrast to analog signals, do not vary continuously, but change in steps or in discrete increments. Most digital signals use binary or two-state codes. Some

Digital signal

Figure 1-5 Analog signals. (*a*) Sine wave "tone." (*b*) Voice. (*c*) Video (TV) signal.

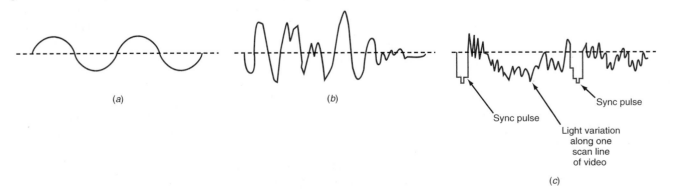

(*a*)

(*b*)

Sync pulse

Sync pulse

Light variation
along one
scan line
of video

(*c*)

Figure 1-6 Digital signals. (*a*) Telegraph (Morse code). (*b*) Continuous-wave (CW) code. (*c*) Serial binary code.

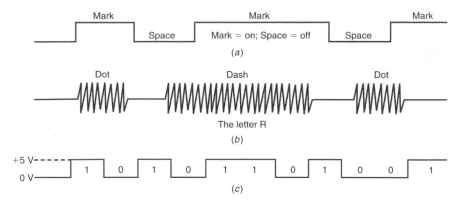

examples are shown in Fig. 1-6. The earliest forms of both wire and radio communication used a type of on/off digital code. The telegraph used Morse code, with its system of short and long signals (dots and dashes) to designate letters and numbers. See Fig. 1-6(*a*). In radio telegraphy, also known as continuous-wave (CW) transmission, a sine wave signal is turned off and on for short or long durations to represent the dots and dashes. Refer to Fig. 1-6(*b*).

Data used in computers is also digital. Binary codes representing numbers, letters, and special symbols are transmitted serially by wire, radio, or optical medium. The most commonly used digital code in communications is the *American Standard Code for Information Interchange* (*ASCII*, pronounced "ask key"). Figure 1-6(*c*) shows a serial binary code.

Many transmissions are of signals that originate in digital form, e.g., telegraphy messages or computer data, but that must be converted to analog form to match the transmission medium. An example is the transmission of digital data over the telephone network, which was designed to handle analog voice signals only. If the digital data is converted to analog signals, such as tones in the audio frequency range, it can be transmitted over the telephone network.

Analog signals can also be transmitted digitally. It is very common today to take voice or video analog signals and digitize them with an analog-to-digital (A/D) converter. The data can then be transmitted efficiently in digital form and processed by computers and other digital circuits.

1-4 Modulation and Multiplexing

Modulation and multiplexing are electronic techniques for transmitting information efficiently from one place to another. *Modulation* makes the information signal more compatible with the medium, and *multiplexing* allows more than one signal to be transmitted concurrently over a single medium. Modulation and multiplexing techniques are basic to electronic communication. Once you have mastered the fundamentals of these techniques, you will easily understand how most modern communication systems work.

Baseband Transmission

Before it can be transmitted, the information or intelligence must be converted to an electronic signal compatible with the medium. For example, a microphone changes voice signals (sound waves) into an analog voltage of varying frequency and amplitude. This signal is then passed over wires to a speaker or headphones. This is the way the telephone system works.

A video camera generates an analog signal that represents the light variations along one scan line of the picture. This analog signal is usually transmitted over a coaxial cable. Binary

ASCII

Modulation

Multiplexing

GOOD TO KNOW

Multiplexing has been used in the music industry to create stereo sound. In stereo radio, two signals are transmitted and received— one for the right and one for the left channel of sound. (For more information on multiplexing, see Chap. 10.)

data is generated by a keyboard attached to a computer. The computer stores the data and processes it in some way. The data is then transmitted on cables to peripherals such as a printer or to other computers over a LAN. Regardless of whether the original information or intelligence signals are analog or digital, they are all referred to as baseband signals.

In a communication system, baseband information signals can be sent directly and unmodified over the medium or can be used to modulate a carrier for transmission over the medium. Putting the original voice, video, or digital signals directly into the medium is referred to as *baseband transmission.* For example, in many telephone and intercom systems, it is the voice itself that is placed on the wires and transmitted over some distance to the receiver. In some computer networks, the digital signals are applied directly to coaxial or twisted-pair cables for transmission to another computer.

Baseband transmission

In many instances, baseband signals are incompatible with the medium. Although it is theoretically possible to transmit voice signals directly by radio, realistically it is impractical. As a result, the baseband information signal, be it audio, video, or data, is normally used to modulate a high-frequency signal called a *carrier.* The higher-frequency carriers radiate into space more efficiently than the baseband signals themselves. Such wireless signals consist of both electric and magnetic fields. These electromagnetic signals, which are able to travel through space for long distances, are also referred to as *radio-frequency (RF) waves,* or just radio waves.

Carrier

Radio-frequency (RF) wave

Broadband Transmission

Modulation is the process of having a baseband voice, video, or digital signal modify another, higher-frequency signal, the carrier. The process is illustrated in Fig. 1-7. The information or intelligence to be sent is said to be *impressed* upon the carrier. The carrier is usually a sine wave generated by an oscillator. The carrier is fed to a circuit called a modulator along with the baseband intelligence signal. The intelligence signal changes the carrier in a unique way. The modulated carrier is amplified and sent to the antenna for transmission. This process is called *broadband transmission.*

Broadband transmission

Consider the common mathematical expression for a sine wave:

$$v = V_p \sin (2\pi ft + \theta) \qquad \text{or} \qquad v = V_p \sin (\omega t + \theta)$$

where v = instantaneous value of sine wave voltage
$\quad V_p$ = peak value of sine wave
$\quad f$ = frequency, Hz
$\quad \omega$ = angular velocity = $2\pi f$
$\quad t$ = time, s
$\quad \omega t = 2\pi ft$ = angle, rad ($360° = 2\pi$ rad)
$\quad \theta$ = phase angle

Figure 1-7 Modulation at the transmitter.

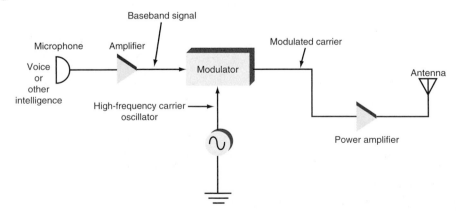

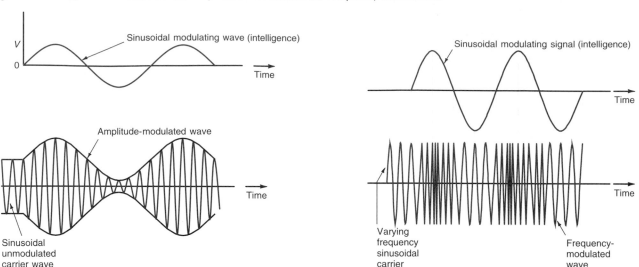

Amplitude modulation (AM)

Frequency modulation (FM)

Phase modulation (PM)

Frequency-shift keying (FSK)

Phase-shift keying (PSK)

Modems

The three ways to make the baseband signal change the carrier sine wave are to vary its amplitude, vary its frequency, or vary its phase angle. The two most common methods of modulation are *amplitude modulation (AM)* and *frequency modulation (FM)*. In AM, the baseband information signal called the modulating signal varies the amplitude of the higher-frequency carrier signal, as shown in Fig. 1-8(*a*). It changes the V_p part of the equation. In FM, the information signal varies the frequency of the carrier, as shown in Fig. 1-8(*b*). The carrier amplitude remains constant. FM varies the value of f in the first angle term inside the parentheses. Varying the phase angle produces *phase modulation (PM)*. Here, the second term inside the parentheses (θ) is made to vary by the intelligence signal. Phase modulation produces frequency modulation; therefore, the PM signal is similar in appearance to a frequency-modulated carrier. Two common examples of transmitting digital data by modulation are given in Fig. 1-9. In Fig. 1-9(*a*), the data is converted to frequency-varying tones. This is called *frequency-shift keying (FSK)*. In Fig. 1-9(*b*), the data introduces a 180°-phase shift. This is called *phase-shift keying (PSK)*. Devices called *modems (mo*dulator-*dem*odulator) translate the data from digital to analog and back again. Both FM and PM are forms of angle modulation.

At the receiver, the carrier with the intelligence signal is amplified and then demodulated to extract the original baseband signal. Another name for the demodulation process is detection. (See Fig. 1-10.)

Figure 1-9 Transmitting binary data in analog form. (*a*) FSK. (*b*) PSK.

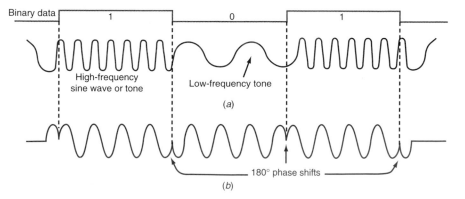

Figure 1–10 Recovering the intelligence signal at the receiver.

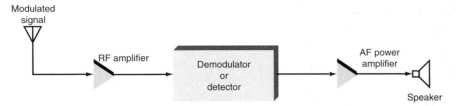

Multiplexing

The use of modulation also permits another technique, known as multiplexing, to be used. Multiplexing is the process of allowing two or more signals to share the same medium or channel; see Fig. 1-11. A multiplexer converts the individual baseband signals to a composite signal that is used to modulate a carrier in the transmitter. At the receiver, the composite signal is recovered at the demodulator, then sent to a demultiplexer where the individual baseband signals are regenerated (see Fig. 1-12).

There are three basic types of multiplexing: frequency division, time division, and code dimension. In *frequency-division multiplexing,* the intelligence signals modulate subcarriers that are then added together, and the composite signal is used to modulate the carrier. In optical networking, wavelength division multiplexing (WDM) is equivalent to frequency-division multiplexing for optical signal. In *time-division multiplexing,* the multiple intelligence signals are sequentially sampled, and a small piece of each is used to modulate the carrier. If the information signals are sampled fast enough, sufficient details are transmitted that at the receiving end the signal can be reconstructed with great accuracy. In code-division multiplexing, the signals to be transmitted are converted to digital data that is then uniquely coded with a faster binary code. The signals modulate a carrier on the same frequency. All use the same communications channel simultaneously. The unique coding is used at the receiver to select the desired signal.

Frequency-division multiplexing

Time-division multiplexing

Figure 1–11 Multiplexing at the transmitter.

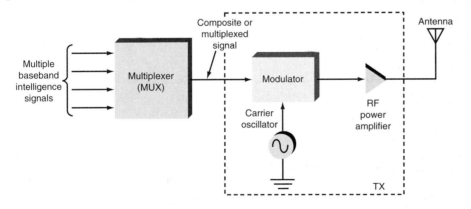

Figure 1–12 Demultiplexing at the receiver.

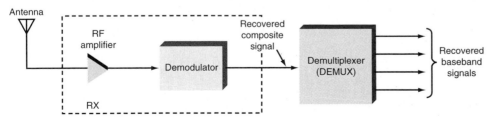

1-5 The Electromagnetic Spectrum

Electromagnetic waves are signals that oscillate; i.e., the amplitudes of the electric and magnetic fields vary at a specific rate. The field intensities fluctuate up and down, and the polarity reverses a given number of times per second. The electromagnetic waves vary sinusoidally. Their frequency is measured in cycles per second (cps) or hertz (Hz). These oscillations may occur at a very low frequency or at an extremely high frequency. The range of electromagnetic signals encompassing all frequencies is referred to as the *electromagnetic spectrum.*

Electromagnetic spectrum

All electrical and electronic signals that radiate into free space fall into the electromagnetic spectrum. Not included are signals carried by cables. Signals carried by cable may share the same frequencies of similar signals in the spectrum, but they are not radio signals. Figure 1-13 shows the entire electromagnetic spectrum, giving both frequency and wavelength. Within the middle ranges are located the most commonly used radio frequencies for two-way communication, TV, cell phones, wireless LANs, radar, and other applications. At the upper end of the spectrum are infrared and visible light. Figure 1-14 is a listing of the generally recognized segments in the spectrum used for electronic communication.

Frequency and Wavelength

A given signal is located on the frequency spectrum according to its frequency and wavelength.

Frequency

Frequency. *Frequency* is the number of times a particular phenomenon occurs in a given period of time. In electronics, frequency is the number of cycles of a repetitive wave that occurs in a given time period. A cycle consists of two voltage polarity reversals, current reversals, or electromagnetic field oscillations. The cycles repeat, forming a continuous but repetitive wave. Frequency is measured in cycles per second (cps). In electronics, the unit of frequency is the hertz, named for the German physicist Heinrich Hertz, who was a pioneer in the field of electromagnetics. One cycle per second is equal to one hertz, abbreviated (Hz). Therefore, 440 cps = 440 Hz.

Figure 1-15(*a*) shows a sine wave variation of voltage. One positive alternation and one negative alternation form a cycle. If 2500 cycles occur in 1 s, the frequency is 2500 Hz.

Figure 1-13 The electromagnetic spectrum.

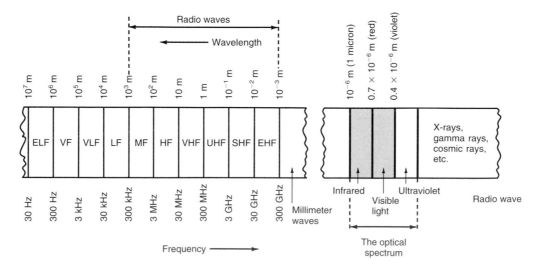

Figure 1-14 The electromagnetic spectrum used in electronic communication.

Name	Frequency	Wavelength
Extremely low frequencies (ELFs)	30–300 Hz	$10^7 - 10^6$ m
Voice frequencies (VFs)	300–3000 Hz	$10^6 - 10^5$ m
Very low frequencies (VLFs)	3–30 kHz	$10^5 - 10^4$ m
Low frequencies (LFs)	30–300 kHz	$10^4 - 10^3$ m
Medium frequencies (MFs)	300 kHz–3 MHz	$10^3 - 10^2$ m
High frequencies (HFs)	3–30 MHz	$10^2 - 10^1$ m
Very high frequencies (VHFs)	30–300 MHz	$10^1 - 1$ m
Ultra high frequencies (UHFs)	300 MHz–3 GHz	$1 - 10^{-1}$ m
Super high frequencies (SHFs)	3–30 GHz	$10^{-1} - 10^{-2}$ m
Extremely high frequencies (EHFs)	30–300 GHz	$10^{-2} - 10^{-3}$ m
Infrared	—	0.7–10 μm
The visible spectrum (light)	—	0.4–0.8 μm

Units of Measure and Abbreviations:
kHz = 1000 Hz
MHz = 1000 kHz = 1×10^6 = 1,000,000 Hz
GHz = 1000 MHz = 1×10^6 = 1,000,000 kHz
= 1×10^9 = 1,000,000,000 Hz
m = meter
μm = micrometer = $\dfrac{1}{1,000,000}$ m = 1×10^{-6} m

Prefixes representing powers of 10 are often used to express frequencies. The most frequently used prefixes are as follows:

$$k = kilo = 1000 = 10^3$$

$$M = mega = 1,000,000 = 10^6$$

$$G = giga = 1,000,000,000 = 10^9$$

$$T = tera = 1,000,000,000,000 = 10^{12}$$

Thus, 1000 Hz = 1 kHz (kilohertz). A frequency of 9,000,000 Hz is more commonly expressed as 9 MHz (megahertz). A signal with a frequency of 15,700,000,000 Hz is written as 15.7 GHz (gigahertz).

Wavelength. *Wavelength* is the distance occupied by one cycle of a wave, and it is usually expressed in meters. One meter (m) is equal to 39.37 in (just over 3 ft, or

Wavelength

Figure 1-15 Frequency and wavelength. (*a*) One cycle. (*b*) One wavelength.

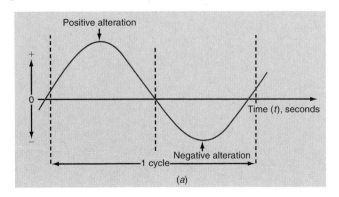

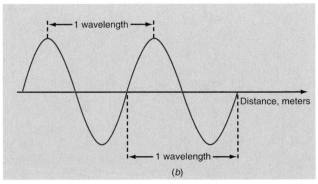

Introduction to Electronic Communication **13**

1 yd). Wavelength is measured between identical points on succeeding cycles of a wave, as Fig. 1-15(b) shows. If the signal is an electromagnetic wave, one wavelength is the distance that one cycle occupies in free space. It is the distance between adjacent peaks or valleys of the electric and magnetic fields making up the wave.

Wavelength is also the distance traveled by an electromagnetic wave during the time of one cycle. Electromagnetic waves travel at the speed of light, or 299,792,800 m/s. The speed of light and radio waves in a vacuum or in air is usually rounded off to 300,000,000 m/s (3×10^8 m/s), or 186,000 mi/s. The speed of transmission in media such as a cable is less.

The wavelength of a signal, which is represented by the Greek letter λ (lambda), is computed by dividing the speed of light by the frequency f of the wave in hertz: $\lambda = 300,000,000/f$. For example, the wavelength of a 4,000,000-Hz signal is

$$\lambda = 300,000,000/4,000,000 = 75 \text{ m}$$

If the frequency is expressed in megahertz, the formula can be simplified to $\lambda(\text{m}) = 300/f(\text{MHz})$ or $\lambda(\text{ft}) = 984\,f(\text{MHz})$.

The 4,000,000-Hz signal can be expressed as 4 MHz. Therefore $\lambda = 300/4 = 75$ m.

A wavelength of 0.697 m, as in the second equation in Example 1-1, is known as a *very high frequency signal wavelength*. Very high frequency wavelengths are sometimes expressed in centimeters (cm). Since 1 m equals 100 cm, we can express the wavelength of 0.697 m in Example 1-1 as 69.7, or about 70 cm.

Very high frequency signal wavelength

Example 1-1

Find the wavelengths of (*a*) a 150-MHz, (*b*) a 430-MHz, (*c*) an 8-MHz, and (*d*) a 750-kHz signal.

a. $\lambda = \dfrac{300,000,000}{150,000,000} = \dfrac{300}{150} = 2$ m

b. $\lambda = \dfrac{300}{430} = 0.697$ m

c. $\lambda = \dfrac{300}{8} = 37.5$ m

d. For Hz (750 kHz = 750,000 Hz):

$$\lambda = \dfrac{300,000,000}{750,000} = 400 \text{ m}$$

For MHz (750 kHz = 0.75 MHz):

$$\lambda = \dfrac{300}{0.75} = 400 \text{ m}$$

If the wavelength of a signal is known or can be measured, the frequency of the signal can be calculated by rearranging the basic formula $f = 300/\lambda$. Here, f is in megahertz and λ is in meters. As an example, a signal with a wavelength of 14.29 m has a frequency of $f = 300/14.29 = 21$ MHz.

Example 1-2

A signal with a wavelength of 1.5 m has a frequency of

$$f = \frac{300}{1.5} = 200 \text{ MHz}$$

Example 1-3

A signal travels a distance of 75 ft in the time it takes to complete 1 cycle. What is its frequency?

$$1 \text{ m} = 3.28 \text{ ft}$$

$$\frac{75 \text{ ft}}{3.28} = 22.86 \text{ m}$$

$$f = \frac{300}{22.86} = 13.12 \text{ MHz}$$

Example 1-4

The maximum peaks of an electromagnetic wave are separated by a distance of 8 in. What is the frequency in megahertz? In gigahertz?

$$1 \text{ m} = 39.37 \text{ in}$$

$$8 \text{ in} = \frac{8}{39.37} = 0.203 \text{ m}$$

$$f = \frac{300}{0.203} = 1477.8 \text{ MHz}$$

$$\frac{1477.8}{10^3} = 1.4778 \text{ GHz}$$

Frequency Ranges from 30 Hz to 300 GHz

For the purpose of classification, the electromagnetic frequency spectrum is divided into segments, as shown in Fig. 1-13. The signal characteristics and applications for each segment are discussed in the following paragraphs.

Extremely Low Frequencies. *Extremely low frequencies (ELFs)* are in the 30- to 300-Hz range. These include ac power line frequencies (50 and 60 Hz are common), as well as those frequencies in the low end of the human audio range.

Extremely low frequency (ELF)

Introduction to Electronic Communication

Voice frequency (VF)

Voice Frequencies. *Voice frequencies (VFs)* are in the range of 300 to 3000 Hz. This is the normal range of human speech. Although human hearing extends from approximately 20 to 20,000 Hz, most intelligible sound occurs in the VF range.

Very low frequency (VLF)

Very Low Frequencies. *Very low frequencies (VLFs)* extend from 9 kHz to 30 kHz and include the higher end of the human hearing range up to about 15 or 20 kHz. Many musical instruments make sounds in this range as well as in the ELF and VF ranges. The VLF range is also used in some government and military communication. For example, VLF radio transmission is used by the navy to communicate with submarines.

Low frequency (LF)

Subcarrier

Low Frequencies. *Low frequencies (LFs)* are in the 30- to 300-kHz range. The primary communication services using this range are in aeronautical and marine navigation. Frequencies in this range are also used as *subcarriers,* signals that are modulated by the baseband information. Usually, two or more subcarriers are added, and the combination is used to modulate the final high-frequency carrier.

Medium frequency (MF)

Medium Frequencies. *Medium frequencies (MFs)* are in the 300- to 3000-kHz (0.3- to 3.0-MHz) range. The major application of frequencies in this range is AM radio broadcasting (535 to 1605 kHz). Other applications in this range are various marine and aeronautical communication.

High frequency (HF)

High Frequencies. *High frequencies (HFs)* are in the 3- to 30-MHz range. These are the frequencies generally known as short waves. All kinds of simplex broadcasting and half duplex two-way radio communication take place in this range. Broadcasts from Voice of America and the British Broadcasting Company occur in this range. Government and military services use these frequencies for two-way communication. An example is diplomatic communication between embassies. Amateur radio and CB communication also occur in this part of the spectrum.

Very high frequency (VHF)

Very High Frequencies. *Very high frequencies (VHFs)* encompass the 30- to 300-MHz range. This popular frequency range is used by many services, including mobile radio, marine and aeronautical communication, FM radio broadcasting (88 to 108 MHz), and television channels 2 through 13. Radio amateurs also have numerous bands in this frequency range.

Ultrahigh frequency (UHF)

Ultrahigh Frequencies. *Ultrahigh frequencies (UHFs)* encompass the 300- to 3000-MHz range. This, too, is a widely used portion of the frequency spectrum. It includes the UHF TV channels 14 through 67, and it is used for land mobile communication and services such as cellular telephones as well as for military communication. Some radar and navigation services occupy this portion of the frequency spectrum, and radio amateurs also have bands in this range.

Microwave

Superhigh frequency (SHF)

Microwaves and SHFs. Frequencies between the 1000-MHz (1-GHz) and 30-GHz range are called *microwaves.* Microwave ovens usually operate at 2.45 GHz. *Superhigh frequencies (SHFs)* are in the 3- to 30-GHz range. These microwave frequencies are widely used for satellite communication and radar. Wireless local-area networks (LANs) also occupy this region.

Extremely high frequency (EHF)

Millimeter wave

Extremely High Frequencies. *Extremely high frequencies (EHFs)* extend from 30 to 300 GHz. Electromagnetic signals with frequencies higher than 30 GHz are referred to as *millimeter waves.* Equipment used to generate and receive signals in this range is extremely complex and expensive but there is growing use of this range for satellite communication telephony, computer data, and some specialized radar.

Frequencies Between 300 GHz and the Optical Spectrum. This portion of the spectrum is virtually uninhabited. It is a cross between RF and optical. Lack of hardware and components prevents its use.

With three sons serving abroad during World War II, the Rubis family of Muse, Pennsylvania, listened to President Roosevelt's Sunday radio address in 1943.

The Optical Spectrum

Right above the millimeter wave region is what is called the *optical spectrum,* the region occupied by light waves. There are three different types of light waves: infrared, visible, and ultraviolet.

Optical spectrum

Infrared. The *infrared region* is sandwiched between the highest radio frequencies (i.e., millimeter waves) and the visible portion of the electromagnetic spectrum. Infrared occupies the range between approximately 0.1 millimeter (mm) and 700 nanometers (nm), or 100 to 0.7 micrometer (μm). One micrometer is one-millionth of a meter. Infrared wavelengths are often given in micrometers or nanometers.

Infrared region

Infrared radiation is generally associated with heat. Infrared is produced by light-bulbs, our bodies, and any physical equipment that generates heat. Infrared signals can also be generated by special types of light-emitting diodes (LEDs) and lasers.

Infrared signals are used for various special kinds of communication. For example, infrared is used in astronomy to detect stars and other physical bodies in the universe, and for guidance in weapons systems, where the heat radiated from airplanes or missiles can be picked up by infrared detectors and used to guide missiles to targets. Infrared is also used in most new TV remote-control units where special coded signals are transmitted by an infrared LED to the TV receiver for the purpose of changing channels, setting the volume, and performing other functions. Infrared is the basis for some of the newer wireless LANs and all fiber-optic communication.

Infrared signals have many of the same properties as signals in the visible spectrum. Optical devices such as lenses and mirrors are often used to process and manipulate infrared signals, and infrared light is the signal usually propagated over fiber-optic cables.

The Visible Spectrum. Just above the infrared region is the *visible spectrum* we ordinarily refer to as *light.* Light is a special type of electromagnetic radiation that has a wavelength in the 0.4- to 0.8-μm range (400 to 800 nm). Light wavelengths are usually expressed in terms of angstroms (Å). An angstrom is one ten-thousandth of a micrometer; for example, 1 Å = 10^{-10} m. The visible range is approximately 8000 Å

Visible spectrum

Light

Ultraviolet light (UV)

Bandwidth (BW)

(red) to 4000 Å (violet). Red is low-frequency or long-wavelength light, whereas violet is high-frequency or short-wavelength light.

Light is used for various kinds of communication. Light waves can be modulated and transmitted through glass fibers, just as electric signals can be transmitted over wires. The great advantage of light wave signals is that their very high frequency gives them the ability to handle a tremendous amount of information. That is, the bandwidth of the baseband signals can be very wide.

Light signals can also be transmitted through free space. Various types of communication systems have been created using a laser that generates a light beam at a specific visible frequency. Lasers generate an extremely narrow beam of light which is easily modulated with voice, video, and data information.

Ultraviolet. *Ultraviolet light (UV)* covers the range from about 4 to 400 nm. Ultraviolet generated by the sun is what causes sunburn. Ultraviolet is also generated by mercury vapor lights and some other types of lights such as fluorescent lamps and sun lamps. Ultraviolet is not used for communication; its primary use is medical.

Beyond the visible region are the X-rays, gamma rays, and cosmic rays. These are all forms of electromagnetic radiation, but they do not figure into communication systems and are not covered here.

1-6 Bandwidth

Bandwidth (BW) is that portion of the electromagnetic spectrum occupied by a signal. It is also the frequency range over which a receiver or other electronic circuit operates. More specifically, bandwidth is the difference between the upper and lower frequency limits of the signal or the equipment operation range. Figure 1-16 shows the bandwidth of the voice frequency range from 300 to 3000 Hz. The upper frequency is f_2 and the lower frequency is f_1. The bandwidth, then, is

$$BW = f_2 - f_1$$

Example 1-5

A commonly used frequency range is 902 to 928 MHz. What is the width of this band?

$$f_1 = 902 \text{ MHz} \qquad f_2 = 928 \text{ MHz}$$

$$BW = f_2 - f_1 = 928 - 902 = 26 \text{ MHz}$$

Figure 1-16 Bandwidth is the frequency range over which equipment operates or that portion of the spectrum occupied by the signal. This is the voice frequency bandwidth.

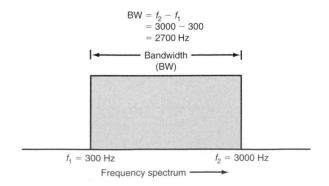

Example 1-6

A television signal occupies a 6-MHz bandwidth. If the low-frequency limit of channel 2 is 54 MHz, what is the upper-frequency limit?

$$BW = 54\,MHz \qquad f_1 = 6\,MHz$$

$$BW = f_1 - f_2$$

$$f_2 = BW + f_1 = 6 + 54 = 60\,MHz$$

Channel Bandwidth

When information is modulated onto a carrier somewhere in the electromagnetic spectrum, the resulting signal occupies a small portion of the spectrum surrounding the carrier frequency. The modulation process causes other signals, called *sidebands,* to be generated at frequencies above and below the carrier frequency by an amount equal to the modulating frequency. For example, in AM broadcasting, audio signals up to 5 kHz can be transmitted. If the carrier frequency is 1000 kHz, or 1 MHz, and the modulating frequency is 5 kHz, sidebands will be produced at $1000 - 5 = 995$ kHz and at $1000 + 5 = 1005$ kHz. In other words, the modulation process generates other signals that take up spectrum space. It is not just the carrier at 1000 kHz that is transmitted. Thus the term bandwidth refers to the range of frequencies that contain the information. The term *channel bandwidth* refers to the range of frequencies required to transmit the desired information.

Sideband

Channel bandwidth

The bandwidth of the AM signal described above is the difference between the highest and lowest transmitting frequencies: $BW = 1005$ kHz $- 995$ kHz $= 10$ kHz. In this case, the channel bandwidth is 10 kHz. An AM broadcast signal, therefore, takes up a 10-kHz piece of the spectrum.

Signals transmitting on the same frequency or on overlapping frequencies do, of course, interfere with one another. Thus a limited number of signals can be transmitted in the frequency spectrum. As communication activities have grown over the years, there has been a continuous demand for more frequency channels over which communication can be transmitted. This has caused a push for the development of equipment that operates at the higher frequencies. Prior to World War II, frequencies above 1 GHz were virtually unused, since there were no electronic components suitable for generating signals at those frequencies. But technological developments over the years have given us many microwave components such as klystrons, magnetrons, and traveling-wave tubes, and today transistors, integrated circuits, and other semiconductor devices that work in the microwave range.

More Room at the Top

The benefit of using the higher frequencies for communication carriers is that a signal of a given bandwidth represents a smaller percentage of the spectrum at the higher frequencies than at the lower frequencies. For example, at 1000 kHz, the 10-kHz-wide AM signal discussed earlier represents 1 percent of the spectrum:

$$\% \text{ of spectrum} = \frac{10\,kHz}{1000\,KHz} \times 100$$

$$= 1\%$$

But at 1 GHz, or 1,000,000 kHz, it represents only one-thousandth of 1 percent:

$$\% \text{ of spectrum} = \frac{10\,kHz}{1,000,000\,kHz} \times 100$$

$$= 0.001\%$$

National Telecommunications and Information Administration (NTIA)

International Telecommunications Union (ITU)

Standards

In practice, this means that there are many more 10-kHz channels at the higher frequencies than at the lower frequencies. In other words, there is more spectrum space for information signals at the higher frequencies.

The higher frequencies also permit wider-bandwidth signals to be used. A TV signal, e.g., occupies a bandwidth of 6 MHz. Such a signal cannot be used to modulate a carrier in the MF or HF ranges because it would use up all the available spectrum space. Television signals are transmitted in the VHF and UHF portions of the spectrum, where sufficient space is available.

Today, virtually the entire frequency spectrum between approximately 30 kHz and 300 MHz has been spoken for. Some open areas and portions of the spectrum are not heavily used, but for the most part, the spectrum is filled with communication activities of all kinds generated from all over the world. There is tremendous competition for these frequencies, not only between companies, individuals, and government services in individual carriers but also between the different nations of the world. The electromagnetic spectrum is one of our most precious natural resources. Because of this, communication engineering is devoted to making the best use of that finite spectrum. A considerable amount of effort goes into developing communication techniques that will minimize the bandwidth required to transmit given information and thus conserve spectrum space. This provides more room for additional communication channels and gives other services or users an opportunity to take advantage of it. Many of the techniques discussed later in this book evolved in an effort to minimize transmission bandwidth.

Spectrum Management

Governments of the United States and other countries recognized early on that the frequency spectrum was a valuable and finite natural resource and so set up agencies to control spectrum use. In the United States, Congress passed the Communications Act of 1934. This act and its various amendments established regulations for the use of spectrum space. It also established the Federal Communications Commission (FCC), a regulatory body whose function is to allocate spectrum space, issue licenses, set standards, and police the airwaves. The Telecommunications Act of 1996 has also greatly influenced the use of spectrum. The FCC controls all telephone and radio communications in this country and, in general, regulates all electromagnetic emissions. The *National Telecommunications and Information Administration (NTIA)* performs a similar function for government and military services. Other countries have similar organizations.

The *International Telecommunications Union (ITU),* an agency of the United Nations that is headquartered in Geneva, Switzerland, comprises 189 member countries that meet at regular intervals to promote cooperation and negotiate national interests. Typical of these meetings are the World Administrative Radio Conferences, held approximately every two years. Various committees of the ITU set standards for various areas within the communication field. The ITU brings together the various countries to discuss how the frequency spectrum is to be divided up and shared. Because many of the signals generated in the spectrum do not carry for long distances, countries can use these frequencies simultaneously without interference. On the other hand, some ranges of the frequency spectrum can literally carry signals around the world. As a result, countries must negotiate with one another to coordinate usage of various portions of the high-frequency spectrum to prevent mutual interference.

Standards

Standards are specifications and guidelines that companies and individuals follow to ensure compatibility between transmitting and receiving equipment in communication systems. Although the concepts of communication are simple, there are obviously many ways to send and receive information. A variety of methods are used to modulate, multiplex, and otherwise process the information to be transmitted. If each system used different methods created at the whim of the designing engineer, the systems would be incompatible with one another and no communication could take place. In the real world, standards are set and followed so that when equipment is designed and built, compatibility is ensured. The term used to describe the ability of equipment from one manufacturer to work compatibly with that of another is *interoperability*.

Interoperability

Standards are detailed outlines of principles of operation, blueprints for construction, and methods of measurement that define communication equipment. Some of the specifications covered are modulation methods, frequency of operation, multiplexing methods, word length and bit formats, data transmission speeds, line coding methods, and cable and connector types. These standards are set and maintained by numerous nonprofit organizations around the world. Committees made up of individuals from industry and academia meet to establish and agree upon the standards, which are then published for others to use. Other committees review, revise, and enhance the standards over time, as needs change.

In working in the communication field, you will regularly encounter many different standards. For example, there are standards for long-distance telephone transmission, digital cell phones, local-area networks, and computer modems. Listed below are organizations that maintain standards for communication systems. For more details, go to the corresponding website.

American National Standards Institute (ANSI)—www.ansi.org

Electronic Industries Alliance (EIA)—www.eia.org

European Telecommunications Standards Institute (ETSI)—www.etsi.org

Institute of Electrical and Electronics Engineers (IEEE)—www.ieee.org

International Telecommunications Union (ITU)—www.itu.org

Internet Engineering Task Force (IETF)—www.ietf.org

Telecommunications Institute of America (TIA)—www.tiaonline.org

Federal Communications Commission

The U.S. Federal Communications Commission (FCC) handles both government and consumer issues. Its special focus is on six top concerns.

Homeland security. Protecting telecommunication, broadcast, and other communication infrastructure.

Broadband. Encouraging the rapid availability to users of high-speed, switched, broadband telecommunication for voice, data, graphics, and video.

Digital television. Working with the industry to accelerate the transition to DTV.

FCC reform. Attempting to be as efficient a bureau as possible. Recently, FCC added a media group and a division on wireline competition.

Media ownership. Promoting media ownership that fosters diversity, localism, and competition in the marketplace.

Spectrum policy. Supporting innovation and the efficient, flexible use of the spectrum, including issues such as interference protection, effective safety communication, and international spectrum policies.

1-7 A Survey of Communication Applications

The applications of electronic techniques to communication are so common and pervasive that you are already familiar with most of them. You use the telephone, listen to the radio, and watch TV. You also use other forms of electronic communication such as cellular telephones, ham radios, CB and Family radios, pagers, electronic mail, and remote-control garage door openers. Figure 1-17 lists all the various major applications of electronic communication.

Figure 1-17 Applications of electronic communication.

SIMPLEX (ONE-WAY)

1. *AM and FM radio broadcasting.* Stations broadcast music, news, weather reports, and programs for entertainment and information. It includes shortwave.

2. *Digital radio.* There is both satellite and terrestrial. Radio programming is transmitted in digital format.

3. *TV broadcasting.* Stations broadcast entertainment, informational, and educational programs by radio.

4. *Digital television (DTV).* Radio transmission of television programming is performed by digital methods, both satellite and terrestrial, e.g., high-definition television (HDTV) and Internet Protocol Television (IPTV).

5. *Cable television.* Movies, sports events, and other programs are distributed to subscribers by fiber-optic and coaxial cable.

6. *Facsimile.* Printed visual material is transmitted over telephone lines. A facsimile, or fax, machine scans a document and converts it to electronic signals that are sent over the telephone system for reproduction in printed form by another fax machine. Faxes can also be sent from a computer.

7. *Wireless remote control.* This category includes a device that controls any remote item by radio or infrared. Examples are missiles, satellites, robots, toys, and other vehicles or remote plants or stations. A remote keyless entry device, garage door opener, and the remote control on your TV set are other examples.

8. *Paging services.* A radio system is used to page individuals, usually in connection with their work. Persons carry tiny battery-powered receivers that can pick up signals from a local paging station that receives telephone requests to notify individuals.

9. *Navigation and direction-finding services.* Special stations transmit signals that can be picked up by receivers for the purpose of identifying exact location (latitude and longitude) or determining direction and/or distance from a station. Such systems employ both land-based and satellite stations. The services are used primarily by boats and ships or airplanes, although systems for cars and trucks are being developed. The Global Positioning System (GPS) which uses 24 satellites is the most widely used.

10. *Telemetry.* Measurements are transmitted over a long distance. Telemetry systems use sensors to determine physical conditions (temperature, pressure, flow rate, voltages, frequency, etc.) at a remote location. The sensors modulate a carrier signal that is sent by wire or radio to a remote receiver which stores and/or displays the data for analysis. Examples are satellites, rockets, pipelines, plants, and factories.

11. *Radio astronomy.* Radio signals, including infrared, are emitted by virtually all heavenly bodies such as stars and planets. With the use of large directional antennas and sensitive high-gain receivers, these signals may be picked up and used to plot star locations and study the universe. Radio astronomy is an alternative and supplement to traditional optical astronomy.

(continues on next page)

Figure 1-17 (*continued*)

12. *Surveillance.* Surveillance means discreet monitoring or "spying." Electronic techniques are widely used by police forces, governments, the military, business and industry, and others to gather information for the purpose of gaining some competitive advantage. Techniques include phone taps, tiny wireless "bugs," clandestine listening stations, and reconnaissance airplanes and satellites.

13. *Music services.* Continuous background music is transmitted for doctors' offices, stores, elevators, and so on by local FM radio stations on special high-frequency subcarriers that cannot be picked up by conventional FM receivers.

14. *Internet radio and video.* Music and video are delivered on a computer via the Internet.

DUPLEX (TWO-WAY)

15. *Telephones.* One-on-one verbal communication is transmitted over the vast worldwide telephone networks employing wire, fiber optics, radio, and satellites.

   a. Cordless telephones provide short-distance wireless communication for cord-free convenience.

   b. Cell phones provide worldwide wireless communications via handsets and base stations and the wired telephone system. In addition to voice communications, cell phones facilitate e-mail, Internet access, instant message service, video, and games.

   c. Internet telephones, known as voice over the Internet protocol (VoIP) phones, use high-speed broadband services (cable, DSL, wireless, fiber) over the Internet to provide digital voice communications.

   d. Satellite phones use low-earth-orbit satellites to give worldwide voice service from any remote location on earth.

16. *Two-way radio.* Commercial, industrial, and government communication is transmitted between vehicles, handheld units, and base stations. Examples include police, fire, taxi, forestry service, trucking companies, aircraft, marine, military, and government.

17. *Radar.* This special form of communication makes use of reflected microwave signals for the purpose of detecting ships, planes, and missiles and for determining their range, direction, and speed. Most radar is used in military applications, but civilian aircraft and marine services also use it. Police use radar in speed detection and enforcement.

18. *Sonar.* In underwater communication, audible baseband signals use water as the transmission medium. Submarines and ships use sonar to detect the presence of enemy submarines. Passive sonar uses audio receivers to pick up water, propeller, and other noises. Active sonar is like an underwater radar with which reflections from a transmitted ultrasonic pulse are used to determine the direction, range, and speed of an underwater target.

19. *Amateur radio.* This is a hobby for individuals interested in radio communication. Individuals may become licensed "hams" to build and operate two-way radio equipment for personal communication with other hams.

20. *Citizens radio.* Citizens band (CB) radio is a special service that any individual may use for personal communication with others. Most CB radios are used in trucks and cars for exchanging information about traffic conditions, speed traps, and emergencies.

21. *Family Radio Service.* This is a two-way personal communication with handheld units over short distances (< 2 mi).

22. *The Internet.* Worldwide interconnections via fiber-optic networks, telecommunications companies, cable TV companies, Internet service providers, and others provide World Wide Web (WWW) access to millions of websites and pages and electronic mail (e-mail).

23. *Wide-Area Networks (WANs).* Worldwide fiber-optic networks provide long-distance telephone and Internet services.

24. *Metropolitan-area networks (MANs).* Networks of computers transmit over a specific geographic area such as a college campus, company facility, or city. Normally they are implemented with fiber-optic cable, but may also be coaxial cable or wireless.

25. *Local-area networks (LANs).* Wired (or wireless) interconnections of personal computers (PCs), laptops, servers, or mainframe computers within an office or building for the purpose of e-mail, Internet access, or the sharing of mass storage, peripherals, data, and software.

1-8 Jobs and Careers in the Communication Industry

The electronics industry is roughly divided into four major specializations. The largest in terms of people employed and the dollar value of equipment purchased is the communications field, closely followed by the computer field. The industrial control and instrumentation fields are considerably smaller. Hundreds of thousands of employees are in the communication field, and billions of dollars' worth of equipment is purchased each year. The growth rate varies from year to year depending upon the economy, technological developments, and other factors. But, as in most areas in electronics, the communication field has grown steadily over the years, creating a relatively constant opportunity for employment. If your interests lie in communication, you will be glad to know that there are many opportunities for long-term jobs and careers. The next section outlines the types of jobs available and the major kinds of employers.

Types of Jobs

The two major types of technical positions available in the communication field are engineer and technician.

Engineer

Engineers. *Engineers* design communication equipment and systems. They have bachelor's (B.S.E.E.), master's (M.S.E.E.), or doctoral (Ph.D.) degrees in electrical engineering, giving them a strong science and mathematics background combined with specialized education in electronic circuits and equipment. Engineers work from specifications and create new equipment or systems which are then manufactured.

Many engineers have a bachelor's degree in electronics technology from a technical college or university. Some typical degree titles are bachelor of technology (B.T.), bachelor of engineering technology (B.E.T.), and bachelor of science in engineering technology (B.S.E.T.).

Bachelor of technology programs are sometimes extensions of two-year associate degree programs. In the two additional years required for a bachelor of technology degree, the student takes more complex electronics courses along with additional science, math, and humanities courses. The main difference between the B.T. graduate and the engineering graduate is that the technologist usually takes courses that are more practical and hands-on than engineering courses. Holders of B.T. degrees can generally design electronic equipment and systems but do not typically have the depth of background in analytical mathematics or science that is required for complex design jobs. However, B.T. graduates are generally employed as engineers. Although many do design work, others are employed in engineering positions in manufacturing and field service rather than design.

Some engineers specialize in design; others work in manufacturing, testing, quality control, and management, among other areas. Engineers may also serve as field service personnel, installing and maintaining complex equipment and systems. If your interest lies in the design of communication equipment, then an engineering position may be for you.

Although a degree in electrical engineering is generally the minimum entrance requirement for engineers' jobs in most organizations, people with other educational backgrounds (e.g., physics and math) do become engineers. Technicians who obtain sufficient additional education and appropriate experience may go on to become engineers.

Technician

Technicians. *Technicians* have some kind of postsecondary education in electronics, from a vocational or technical school, a community college, or a technical institute. Many technicians are educated in military training programs. Most technicians have an average of two years of formal post–high school education and an associate degree. Common degrees are associate in arts (A.A.), associate in science (A.S.) or associate of science in engineering technology or electronic engineering technology (A.S.E.T. or A.S.E.E.T.), and associate in applied science (A.A.S.). The A.A.S. degrees tend to cover more occupational and job-related subjects; the A.A. and A.S. degrees are more general and are designed to provide a foundation for transfer to a bachelor's degree program. Technicians with an associate degree from a community college can usually transfer to a bachelor of technology program and complete the bachelor's degree in another two years. However, associate degree holders are usually not able to transfer to an engineering degree program but must literally start over if the engineering career path is chosen.

Technicians are most often employed in service jobs. The work typically involves equipment installation, troubleshooting and repair, testing and measuring, maintenance and adjustment, or operation. Technicians in such positions are sometimes called *field service technicians, field service engineers,* or *customer representatives.*

Technicians can also be involved in engineering. Engineers may use one or more technicians to assist in the design of equipment. They build and troubleshoot prototypes and in many cases actually participate in equipment design. A great deal of the work involves testing and measuring. In this capacity, the technician is known as an *engineering technician, lab technician, engineering assistant,* or *associate engineer.*

Technicians are also employed in manufacturing. They may be involved in the actual building and assembling of the equipment, but more typically are concerned with final testing and measurement of the finished products. Other positions involve quality control or repair of defective units.

Other Positions. There are many jobs in the communication industry other than those of engineer or technician. For example, there are many outstanding jobs in technical sales. Selling complex electronic communication equipment often requires a strong technical education and background. The work may involve determining customer needs and related equipment specifications, writing technical proposals, making sales presentations to customers, and attending shows and exhibits where equipment is sold. The pay potential in sales is generally much higher than in the engineering or service positions.

Another position is that of technical writer. Technical writers generate the technical documentation for communication equipment and systems, producing installation and service manuals, maintenance procedures, and customer operations manuals. This important task requires considerable depth of education and experience.

Finally, there is the position of trainer. Engineers and technicians are often used to train other engineers and technicians or customers. With the high degree of complexity that exists in communication equipment today, there is a major need for training. Many individuals find education and training positions to be very desirable and satisfying. The work typically involves developing curriculum and programs, generating the necessary training manuals and presentation materials, creating online training, and conducting classroom training sessions in-house or at a customer site.

Major Employers

The overall structure of the communication electronics industry is shown in Fig. 1-18. The four major segments of the industry are manufacturers, resellers, service organizations, and end users.

Manufacturers. It all begins, of course, with customer needs. Manufacturers translate customer needs into products, purchasing components and materials from other electronics companies to use in creating the products. Engineers design the products, and

Technical product sales is big business. Product personalization for cell phones offers choices such as faceplate color, ringtones, and other accessories like hands-free headsets.

Figure 1-18 Structure of the communication electronics industry.

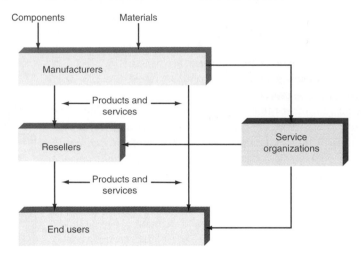

manufacturing produces them. There are jobs for engineers, technicians, salespeople, field service personnel, technical writers, and trainers.

Resellers. Manufacturers who do not sell products directly to end users sell the products to reselling organizations, which in turn sell them to the end user. For example, a manufacturer of marine communication equipment may not sell directly to a boat owner but instead to a regional distributor or marine electronics store or shop. This shop not only sells the equipment but also takes care of installation, service, and repairs. A cellular telephone or fax machine manufacturer also typically sells to a distributor or dealer who takes care of sales and service. Most of the jobs available in the reselling segment of the industry are in sales, service, and training.

Service Organizations. These companies usually perform some kind of service, such as repair, installation, or maintenance. One example is an avionics company that does installation or service work on electronic equipment for private planes. Another is a systems integrator, a company that designs and assembles a piece of communication equipment or more often an entire system by using the products of other companies. Systems integrators put together systems to meet special needs and customize existing systems for particular jobs.

End Users. The end user is the ultimate customer—and a major employer. Today, almost every person and organization is

Most communication technicians perform installation, maintenance, and troubleshooting.

an end user of communication equipment. The major categories of end users in the communication field are

Telephone companies

Radio users—mobile, marine, aircraft, etc.

Radio and TV broadcast stations and cable TV companies

Business and industry users of satellites, networks, etc.

Transportation companies (airlines, shipping, railroads)

Government and military

Personal and hobby

Consumers

There are an enormous number of communication jobs with end users. Most are of the service type: installation, repair, maintenance, and operation of equipment.

CHAPTER REVIEW

Summary

All electronic communication systems consist of three basic components: a transmitter, a communication channel (medium), and a receiver. Messages are converted to electrical signals and sent over electrical or fiber-optic cable or free space to a receiver. Attenuation (weakening) and noise can interfere with transmission.

Electronic communication is classified as (1) one-way (simplex) or two-way (full duplex or half duplex) transmissions and (2) analog or digital signals. Analog signals are smoothly varying, continuous signals. Digital signals are discrete, two-state (on/off) codes. Electronic signals are often changed from analog to digital and vice versa. Before transmission, electronic signals are known as baseband signals.

Amplitude and frequency modulation make an information signal compatible with the channel over which it is to be sent, modifying the carrier wave by changing its amplitude, frequency, or phase angle and sending it to an antenna for transmission, a process known as broadband communication. Frequency-division and time-division multiplexing allow more than one signal at a time to be transmitted over the same medium.

All electronic signals that radiate into space are part of the electromagnetic spectrum; their location on the spectrum is determined by frequency. Most information signals to be transmitted occur at lower frequencies and modulate a carrier wave of a higher frequency.

How much information a given signal can carry depends in part on its bandwidth. Available space for transmitting signals is limited, and signals transmitting on the same frequency or on overlapping frequencies interfere with one another. Research efforts are being devoted to developing use of higher-frequency signals and minimizing the bandwidth required.

Spectrum usage is regulated by governments, in the United States by the FCC and NTIA, and by equivalent agencies in other governments. Standards for communication systems state specifically how the information is transmitted and received. Standards are set by independent organizations such as ANSI, EIA, ETSI, IEEE, ITU, IETF, and TIA.

The four major electronic specialties are computers, communication, industrial control, and instrumentation. There are many job opportunities in the field of electronic communication.

Questions

1. In what century did electronic communication begin?
2. Name the four main elements of a communication system, and draw a diagram that shows their relationship.
3. List five types of media used for communication, and state which three are the most commonly used.
4. Name the device used to convert an information signal to a signal compatible with the medium over which it is being transmitted.
5. What piece of equipment acquires a signal from a communication medium and recovers the original information signal?

6. What is a transceiver?
7. What are two ways in which a communication medium can affect a signal?
8. What is another name for *communication medium?*
9. What is the name given to undesirable interference that is added to a signal being transmitted?
10. Name three common sources of interference.
11. What is the name given to the original information or intelligence signals that are transmitted directly via a communication medium?
12. Name the two forms in which intelligence signals can exist.
13. What is the name given to one-way communication? Give three examples.
14. What is the name given to simultaneous two-way communication? Give three examples.
15. What is the term used to describe two-way communication in which each party takes turns transmitting? Give three examples.
16. What type of electronic signals are continuously varying voice and video signals?
17. What are on/off intelligence signals called?
18. How are voice and video signals transmitted digitally?
19. What terms are often used to refer to original voice, video, or data signals?
20. What technique must sometimes be used to make an information signal compatible with the medium over which it is being transmitted?
21. What is the process of recovering an original signal called?
22. What is a broadband signal?
23. Name the process used to transmit two or more baseband signals simultaneously over a common medium.
24. Name the technique used to extract multiple intelligence signals that have been transmitted simultaneously over a single communication channel.
25. What is the name given to signals that travel through free space for long distances?
26. What does a radio wave consist of?
27. Calculate the wavelength of signals with frequencies of 1.5 kHz, 18 MHz, and 22 GHz in miles, feet, and centimeters, respectively.
28. Why are audio signals not transmitted directly by electromagnetic waves?
29. What is the human hearing frequency range?
30. What is the approximate frequency range of the human voice?
31. Do radio transmissions occur in the VLF and LF ranges?
32. What is the frequency range of AM radio broadcast stations?
33. What is the name given to radio signals in the high-frequency range?

34. In what segment of the spectrum do TV channels 2 to 13, and FM broadcasting, appear?
35. List five major uses of the UHF band.
36. What are frequencies above 1 GHz called?
37. What are the frequencies just above the EHF range called?
38. What is a micrometer, and what is it used to measure?
39. Name the three segments of the optical frequency spectrum.
40. What is a common source of infrared signals?
41. What is the approximate spectrum range of infrared signals?
42. Define the term *angstrom* and explain how it is used.
43. What is the wavelength range of visible light?
44. Light signals use which two channels or media for electronic communication?
45. Name two methods of transmitting visual data over a telephone network.
46. What is the name given to the signaling of individuals at remote locations by radio?
47. What term is used to describe the process of making measurements at a distance?
48. List four ways radio is used in the telephone system.
49. What principle is used in radar?
50. What is underwater radar called? Give two examples.
51. What is the name of a popular radio communication hobby?
52. What device enables computers to exchange digital data over the telephone network?
53. What do you call the systems of interconnections of PCs and other computers in offices or buildings?
54. What is a generic synonym for radio?
55. Name the three main types of technical positions available in the communication field.
56. What is the main job of an engineer?
57. What is the primary degree for an engineer?
58. What is the primary degree for a technician?
59. Name a type of technical degree in engineering other than engineer or technician.
60. Can the holder of an associate of technology degree transfer the credits to an engineering degree program?
61. What types of work does a technician ordinarily do?
62. List three other types of jobs in the field of electronic communication that do not involve engineering or technician's work.
63. What are the four main segments of the communication industry? Explain briefly the function of each.
64. Why are standards important?
65. What types of characteristics do communication standards define?

Problems

1. Calculate the frequency of signals with wavelengths of 40 m, 5 m, and 8 cm. ◆
2. In what frequency range does the common ac power line frequency fall?
3. What is the primary use of the SHF and EHF ranges? ◆

◆ *Answers to Selected Problems follow Chap. 22.*

Critical Thinking

1. Name three ways that a higher-frequency signal called the carrier can be varied to transmit the intelligence.
2. Name two common household remote-control units, and state the type of media and frequency ranges used for each.
3. How is radio astronomy used to locate and map stars and other heavenly bodies?
4. In what segment of the communication field are you interested in working, and why?
5. Assume that all the electromagnetic spectrum from ELF through microwaves was fully occupied. Explain some ways that communication capability could be added.
6. What is the speed of light in feet per microsecond? In inches per nanosecond? In meters per second?
7. Make a general statement comparing the speed of light with the speed of sound. Give an example of how the principles mentioned might be demonstrated.
8. List five real-life communication applications not specifically mentioned in this chapter.
9. "Invent" five new communication methods, wired or wireless, that you think would be practical.
10. Assume that you have a wireless application you would like to design, build, and sell as a commercial product. You have selected a target frequency in the UHF range. How would you decide what frequency to use, and how would you get permission to use it?
11. Make an exhaustive list of all the electronic communication products that you own, have access to at home or in the office, and/or use on a regular basis.
12. You have probably seen or heard of a simple communication system made of two paper cups and a long piece of string. How could such a simple system work?

The Fundamentals of Electronics: A Review

To understand communication electronics as presented in this book, you need a knowledge of certain basic principles of electronics, including the fundamentals of alternating-current (ac) and direct-current (dc) circuits, semiconductor operation and characteristics, and basic electronic circuit operation (amplifiers, oscillators, power supplies, and digital logic circuits). Some of the basics are particularly critical to understanding the chapters that follow. These include the expression of gain and loss in decibels, *LC* tuned circuits, resonance and filters, and Fourier theory. The purpose of this chapter is to briefly review all these subjects. If you have studied the material before, it will simply serve as a review and reference. If, because of your own schedule or the school's curriculum, you have not previously covered this material, use this chapter to learn the necessary information before you continue.

Objectives

After completing this chapter, you will be able to:

- Calculate voltage, current, gain, and attenuation in decibels and apply these formulas in applications involving cascaded circuits.
- Explain the relationship between *Q*, resonant frequency, and bandwidth.
- Describe the basic configuration of the different types of filters that are used in communication networks and compare and contrast active filters with passive filters.
- Explain how using switched capacitor filters enhances selectivity.
- Explain the benefits and operation of crystal, ceramic, and SAW filters.
- Calculate bandwidth by using Fourier analysis.

2-1 Gain, Attenuation, and Decibels

Most electronic circuits in communication are used to process signals, i.e., to manipulate signals to produce a desired result. All signal processing circuits involve either gain or attenuation.

Gain

Gain means amplification. If a signal is applied to a circuit such as the amplifier shown in Fig. 2-1 and the output of the circuit has a greater amplitude than the input signal, the circuit has gain. Gain is simply the ratio of the output to the input. For input (V_{in}) and output (V_{out}) voltages, voltage gain A_V is expressed as follows:

$$A_V = \frac{\text{output}}{\text{input}} = \frac{V_{\text{out}}}{V_{\text{in}}}$$

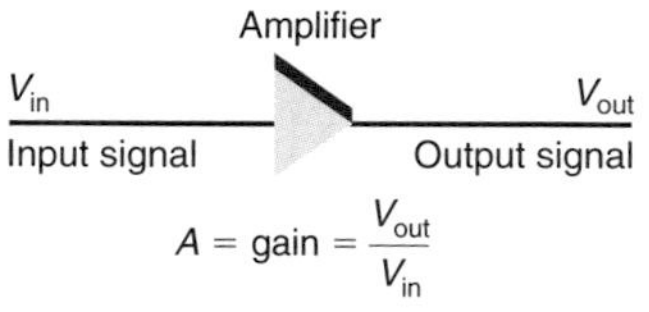

The number obtained by dividing the output by the input shows how much larger the output is than the input. For example, if the input is 150 μV and the output is 75 mV, the gain is $A_V = (75 \times 10^{-3})/(150 \times 10^{-6}) = 500$.

The formula can be rearranged to obtain the input or the output, given the other two variables: $V_{\text{out}} = V_{\text{in}} \times A_V$ and $V_{\text{in}} = V_{\text{out}}/A_V$.

If the output is 0.6 V and the gain is 240, the input is $V_{\text{in}} = 0.6/240 = 2.5 \times 10^{-3} = 2.5$ mV.

Figure 2-1   An amplifier has gain.

Amplifier

V_{in} V_{out}
Input signal   Output signal

$$A = \text{gain} = \frac{V_{\text{out}}}{V_{\text{in}}}$$

Example 2-1

What is the voltage gain of an amplifier that produces an output of 750 mV for a 30-μV input?

$$A_V = \frac{V_{\text{out}}}{V_{\text{in}}} = \frac{750 \times 10^{-3}}{30 \times 10^{-6}} = 25,000$$

Since most amplifiers are also power amplifiers, the same procedure can be used to calculate power gain A_P:

$$A_P = \frac{P_{\text{out}}}{P_{\text{in}}}$$

where P_{in} is the power input and P_{out} is the power output.

Example 2-2

The power output of an amplifier is 6 watts (W). The power gain is 80. What is the input power?

$$A_P = \frac{P_{\text{out}}}{P_{\text{in}}} \qquad \text{therefore} \qquad P_{\text{in}} = \frac{P_{\text{out}}}{A_P}$$

$$P_{\text{in}} = \frac{6}{80} = 0.075 \text{ W} = 75 \text{ mW}$$

$V_{in} = 1$ mV   5 mV   15 mV $V_{out} = 60$ mV

$A_1 = 5$ $A_2 = 3$ $A_3 = 4$

$A_T = A_1 \times A_2 \times A_3 = 5 \times 3 \times 4 = 60$

When two or more stages of amplification or other forms of signal processing are cascaded, the overall gain of the combination is the product of the individual circuit gains. Figure 2-2 shows three amplifiers connected one after the other so that the output of one is the input to the next. The voltage gains of the individual circuits are marked. To find the total gain of this circuit, simply multiply the individual circuit gains: $A_T = A_1 \times A_2 \times A_3 = 5 \times 3 \times 4 = 60$.

If an input signal of 1 mV is applied to the first amplifier, the output of the third amplifier will be 60 mV. The outputs of the individual amplifiers depend upon their individual gains. The output voltage from each amplifier is shown in Fig. 2-2.

Example 2-3

Three cascaded amplifiers have power gains of 5, 2, and 17. The input power is 40 mW. What is the output power?

$$A_P = A_1 \times A_2 \times A_3 = 5 \times 2 \times 17 = 170$$

$$A_P = \frac{P_{out}}{P_{in}} \quad \text{therefore} \quad P_{out} = A_P P_{in}$$

$$P_{out} = 170(40 \times 10^{-3}) = 6.8 \text{ W}$$

Example 2-4

A two-stage amplifier has an input power of 25 μW and an output power of 1.5 mW. One stage has a gain of 3. What is the gain of the second stage?

$$A_P = \frac{P_{out}}{P_{in}} = \frac{1.5 \times 10^{-3}}{25 \times 10^{-6}} = 60$$

$$A_P = A_1 \times A_2$$

If $A_1 = 3$, then $60 = 3 \times A_2$ and $A_2 = 60/3 = 20$.

Figure 2-3 A voltage divider introduces attenuation.

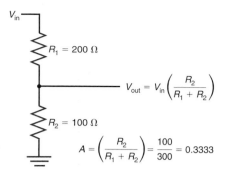

Attenuation

Attenuation refers to a loss introduced by a circuit or component. Many electronic circuits, sometimes called stages, reduce the amplitude of a signal rather than increase it. If the output signal is lower in amplitude than the input, the circuit has loss, or attenuation. Like gain, attenuation is simply the ratio of the output to the input. The letter A is used to represent attenuation as well as gain:

$$\text{Attenuation } A = \frac{\text{output}}{\text{input}} = \frac{V_{out}}{V_{in}}$$

Circuits that introduce attenuation have a gain that is less than 1. In other words, the output is some fraction of the input.

An example of a simple circuit with attenuation is a voltage divider such as that shown in Fig. 2-3. The output voltage is the input voltage multiplied by a ratio based on the resistor values. With the resistor values shown, the gain or attenuation factor of the circuit is $A = R_2/(R_1 + R_2) = 100/(200 + 100) = 100/300 = 0.3333$. If a signal of 10 V is applied to the attenuator, the output is $V_{out} = V_{in}A = 10(0.3333) = 3.333$ V.

When several circuits with attenuation are cascaded, the total attenuation is, again, the product of the individual attenuations. The circuit in Fig. 2-4 is an example. The attenuation factors for each circuit are shown. The overall attenuation is

$$A_T = A_1 \times A_2 \times A_3$$

With the values shown in Fig. 2-4, the overall attenuation is

$$A_T = 0.2 \times 0.9 \times 0.06 = 0.0108$$

Given an input of 3 V, the output voltage is

$$V_{out} = A_T V_{in} = 0.0108(3) = 0.0324 = 32.4 \text{ mV}$$

It is common in communication systems and equipment to cascade circuits and components that have gain and attenuation. For example, loss introduced by a circuit can be

Figure 2-4 Total attenuation is the product of individual attenuations of each cascaded circuit.

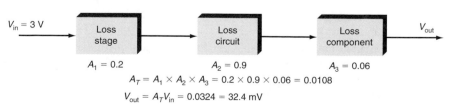

Figure 2-5 Gain exactly offsets the attenuation.

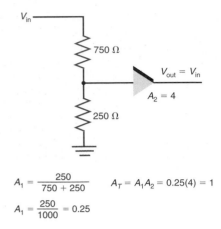

$$A_1 = \frac{250}{750 + 250} \qquad A_T = A_1 A_2 = 0.25(4) = 1$$

$$A_1 = \frac{250}{1000} = 0.25$$

compensated for by adding a stage of amplification that offsets it. An example of this is shown in Fig. 2-5. Here the voltage divider introduces a 4-to-1 voltage loss, or an attenuation of 0.25. To offset this, it is followed with an amplifier whose gain is 4. The overall gain or attenuation of the circuit is simply the product of the attenuation and gain factors. In this case, the overall gain is $A_T = A_1 A_2 = 0.25(4) = 1$.

Another example is shown in Fig. 2-6, which shows two attenuation circuits and two amplifier circuits. The individual gain and attenuation factors are given. The overall circuit gain is $A_T = A_1 A_2 A_3 A_4 = (0.1)(10)(0.3)(15) = 4.5$.

For an input voltage of 1.5 V, the output voltage at each circuit is shown in Fig. 2-6.

In this example, the overall circuit has a net gain. But in some instances, the overall circuit or system may have a net loss. In any case, the overall gain or loss is obtained by multiplying the individual gain and attenuation factors.

Example 2-5

A voltage divider such as that shown in Fig. 2-5 has values of $R_1 = 10\ k\Omega$ and $R_2 = 470\ \Omega$.

a. What is the attenuation?

$$A_1 = \frac{R_2}{R_1 + R_2} = \frac{470}{10,470} \qquad A_1 = 0.045$$

b. What amplifier gain would you need to offset the loss for an overall gain of 1?

$$A_T = A_1 A_2$$

where A_1 is the attenuation and A_2 is the amplifier gain.

$$1 = 0.045A_2 \qquad A_2 = \frac{1}{0.045} = 22.3$$

Note: To find the gain that will offset the loss for unity gain, just take the reciprocal of attenuation: $A_2 = 1/A_1$.

Figure 2-6 The total gain is the product of the individual stage gains and attenuations.

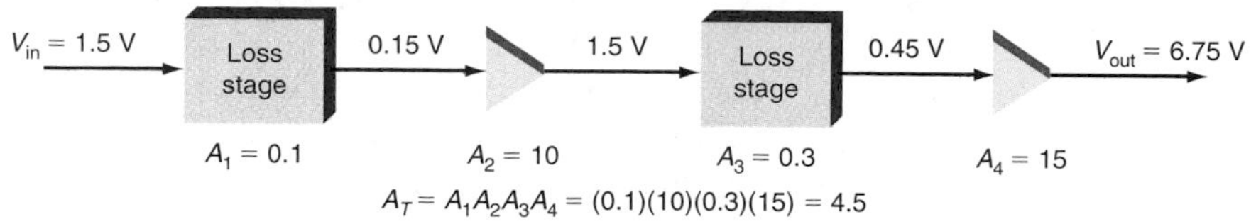

$$A_T = A_1 A_2 A_3 A_4 = (0.1)(10)(0.3)(15) = 4.5$$

Example 2-6

An amplifier has a gain of 45,000, which is too much for the application. With an input voltage of 20 μV, what attenuation factor is needed to keep the output voltage from exceeding 100 mV? Let A_1 = amplifier gain = 45,000; A_2 = attenuation factor; A_T = total gain.

$$A_T = \frac{V_{out}}{V_{in}} = \frac{100 \times 10^{-3}}{20 \times 10^{-6}} = 5000$$

$$A_T = A_1 A_2 \qquad \text{therefore} \qquad A_2 = \frac{A_T}{A_1} = \frac{5000}{45,000} = 0.1111$$

Decibels

The gain or loss of a circuit is usually expressed in *decibels (dB)*, a unit of measurement that was originally created as a way of expressing the hearing response of the human ear to various sound levels. A decibel is one-tenth of a bel.

Decibel (dB)

When gain and attenuation are both converted to decibels, the overall gain or attenuation of an electronic circuit can be computed by simply adding the individual gains or attenuations, expressed in decibels.

It is common for electronic circuits and systems to have extremely high gains or attenuations, often in excess of 1 million. Converting these factors to decibels and using logarithms result in smaller gain and attenuation figures, which are easier to use.

Decibel Calculations. The formulas for computing the decibel gain or loss of a circuit are

$$dB = 20 \log \frac{V_{out}}{V_{in}} \tag{1}$$

$$dB = 20 \log \frac{I_{out}}{I_{in}} \tag{2}$$

$$dB = 10 \log \frac{P_{out}}{P_{in}} \tag{3}$$

Formula (1) is used for expressing the voltage gain or attenuation of a circuit; formula (2), for current gain or attenuation. The ratio of the output voltage or current to the input voltage or current is determined as usual. The base-10 or common log of the input/output ratio is then obtained and multiplied by 20. The resulting number is the gain or attenuation in decibels.

Formula (3) is used to compute power gain or attenuation. The ratio of the power output to the power input is computed, and then its logarithm is multiplied by 10.

Example 2-7

a. An amplifier has an input of 3 mV and an output of 5 V. What is the gain in decibels?

$$dB = 20 \log \frac{5}{0.003} = 20 \log 1666.67 = 20(3.22) = 64.4$$

b. A filter has a power input of 50 mW and an output of 2 mW. What is the gain or attenuation?

$$dB = 10 \log \frac{2}{50} = 10 \log 0.04 = 10(-1.398) = -13.98$$

Note that when the circuit has gain, the decibel figure is positive. If the gain is less than 1, which means that there is an attenuation, the decibel figure is negative.

Now, to calculate the overall gain or attenuation of a circuit or system, you simply add the decibel gain and attenuation factors of each circuit. An example is shown in Fig. 2-7, where there are two gain stages and an attenuation block. The overall gain of this circuit is

$$A_T = A_1 + A_2 + A_3 = 15 - 20 + 35 = 30 \text{ dB}$$

Decibels are widely used in the expression of gain and attenuation in communication circuits. The table on the next page shows some common gain and attenuation factors and their corresponding decibel figures.

Ratios less than 1 give negative decibel values, indicating attenuation. Note that a 2:1 ratio represents a 3-dB power gain or a 6-dB voltage gain.

Antilog margin note: Antilog

Antilogs. To calculate the input or output voltage or power, given the decibel gain or attenuation and the output or input, the *antilog* is used. The antilog is the number obtained when the base is raised to the logarithm which is the exponent:

$$dB = 10 \log \frac{P_{out}}{P_{in}} \qquad \text{and} \qquad \frac{dB}{10} = \log \frac{P_{out}}{P_{in}}$$

and

$$\frac{P_{out}}{P_{in}} = \text{antilog} \frac{dB}{10} = \log^{-1} \frac{dB}{10}$$

The antilog is simply the base 10 raised to the dB/10 power.

Figure 2-7 Total gain or attenuation is the algebraic sum of the individual stage gains in decibels.

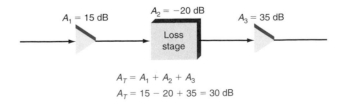

$A_1 = 15 \text{ dB}$ $A_2 = -20 \text{ dB}$ $A_3 = 35 \text{ dB}$

Loss stage

$$A_T = A_1 + A_2 + A_3$$
$$A_T = 15 - 20 + 35 = 30 \text{ dB}$$

dB GAIN OR ATTENUATION		
Ratio (Power or Voltage)	Power	Voltage
0.000001	−60	−120
0.00001	−50	−100
0.0001	−40	−80
0.001	−30	−60
0.01	−20	−40
0.1	−10	−20
0.5	−3	−6
1	0	0
2	3	6
10	10	20
100	20	40
1000	30	60
10,000	40	80
100,000	50	100

Remember that the logarithm y of a number N is the power to which the base 10 must be raised to get the number.

$$N = 10^y \quad \text{and} \quad y = \log N$$

Since

$$\text{dB} = 10 \log \frac{P_{out}}{P_{in}}$$

$$\frac{\text{dB}}{10} = \log \frac{P_{out}}{P_{in}}$$

Therefore

$$\frac{P_{out}}{P_{in}} = 10^{\text{dB}/10} = \log^{-1} \frac{\text{dB}}{10}$$

The antilog is readily calculated on a scientific calculator. To find the antilog for a common or base-10 logarithm, you normally press the (Inv) or (2nd) function key on the calculator and then the (log) key. Sometimes the log key is marked with 10^x, which is the antilog. The antilog with base e is found in a similar way, by using the (Inv) or (2nd) function on the (ln) key. It is sometimes marked e^x, which is the same as the antilog.

Example 2-8

A power amplifier with a 40-dB gain has an output power of 100 W. What is the input power?

$$dB = 10 \log \frac{P_{out}}{P_{in}} \qquad antilog = \log^{-1}$$

$$\frac{dB}{10} = \log \frac{P_{out}}{P_{in}}$$

$$\frac{40}{10} = \log \frac{P_{out}}{P_{in}}$$

$$4 = \log \frac{P_{out}}{P_{in}}$$

$$antilog\ 4 = antilog \left(\log \frac{P_{out}}{P_{in}} \right)$$

$$\log^{-1} 4 = \frac{P_{out}}{P_{in}}$$

$$\frac{P_{out}}{P_{in}} = 10^4 = 10{,}000$$

$$P_{in} = \frac{P_{out}}{10{,}000} = \frac{100}{10{,}000} = 0.01\ W = 10\ mW$$

Example 2-9

An amplifier has a gain of 60 dB. If the input voltage is 50 μV, what is the output voltage?

Since

$$dB = 20 \log \frac{V_{out}}{V_{in}}$$

$$\frac{dB}{20} = \log \frac{V_{out}}{V_{in}}$$

Therefore

$$\frac{V_{out}}{V_{in}} = \log^{-1} \frac{dB}{20} = 10^{dB/20}$$

$$\frac{V_{out}}{V_{in}} = 10^{60/20} = 10^3$$

$$\frac{V_{out}}{V_{in}} = 10^3 = 1000$$

$$V_{out} = 1000 V_{in} = 1000(50 \times 10^{-6}) = 0.05\ V = 50\ mV$$

dBm. When the gain or attenuation of a circuit is expressed in decibels, implicit is a comparison between two values, the output and the input. When the ratio is computed, the units of voltage or power are canceled, making the ratio a dimensionless, or relative, figure. When you see a decibel value, you really do not know the actual voltage or power values. In some cases, this is not a problem; in others, it is useful or necessary to know the actual values involved. When an absolute value is needed, you can use a *reference value* to compare any other value.

An often used reference level in communication is 1 mW. When a decibel value is computed by comparing a power value to 1 mW, the result is a value called the *dBm*. It is computed with the standard power decibel formula with 1 mW as the denominator of the ratio:

$$dBm = 10 \log \frac{P_{out}(W)}{0.001(W)}$$

Here P_{out} is the output power, or some power value you want to compare to 1 mW, and 0.001 is 1 mW expressed in watts.

The output of a 1-W amplifier expressed in dBm is, e.g.,

$$dBm = 10 \log \frac{1}{0.001} = 10 \log 1000 = 10(3) = 30 \text{ dBm}$$

Sometimes the output of a circuit or device is given in dBm. For example, if a microphone has an output of -50 dBm, the actual output power can be computed as follows:

$$-50 \text{ dBm} = 10 \log \frac{P_{out}}{0.001}$$

$$\frac{-50 \text{ dBm}}{10} = \log \frac{P_{out}}{0.001}$$

Therefore

$$\frac{P_{out}}{0.001} = 10^{-50 \text{ dBm}/10} = 10^{-5} = 0.00001$$

$$P_{out} = 0.001 \times 0.00001 = 10^{-3} \times 10^{-5} = 10^{-8} \text{ W} = 10 \times 10^{-9} = 10 \text{ nW}$$

GOOD TO KNOW

From the standpoint of sound measurement, 0 dB is the least perceptible sound (hearing threshold), and 120 dB equals the pain threshold of sound. This list shows intensity levels for common sounds. (Tippens, *Physics,* 6th ed., Glencoe/ McGraw-Hill, 2001, p. 497)

Sound	Intensity level, dB
Hearing threshold	0
Rustling leaves	10
Whisper	20
Quiet radio	40
Normal conversation	65
Busy street corner	80
Subway car	100
Pain threshold	120
Jet engine	140–160

Example 2-10

A power amplifier has an input of 90 mV across 10 kΩ. The output is 7.8 V across an 8-Ω speaker. What is the power gain, in decibels? You must compute the input and output power levels first.

$$P = \frac{V^2}{R}$$

$$P_{in} = \frac{(90 \times 10^{-3})^2}{10^4} = 8.1 \times 10^{-7} \text{ W}$$

$$P_{out} = \frac{(7.8)^2}{8} = 7.605 \text{ W}$$

$$A_P = \frac{P_{out}}{P_{in}} = \frac{7.605}{8.1 \times 10^{-7}} = 9.39 \times 10^6$$

$$A_P(\text{dB}) = 10 \log A_P = 10 \log 9.39 \times 10^6 = 69.7 \text{ dB}$$

dBc. This is a decibel gain attenuation figure where the reference is the carrier. The carrier is the base communication signal, a sine wave that is modulated. Often the amplitude's sidebands, spurious or interfering signals, are referenced to the carrier. For example, if the spurious signal is 1 mW compared to the 10-W carrier, the dBc is

$$\text{dBc} = 10 \log \frac{P_\text{signal}}{P_\text{carrier}}$$

$$\text{dBc} = 10 \log \frac{0.001}{10} = 10(-4) = -40$$

Example 2-11

An amplifier has a power gain of 28 dB. The input power is 36 mW. What is the output power?

$$\frac{P_\text{out}}{P_\text{in}} = 10^{\text{dB}/10} = 10^{2.8} = 630.96$$

$$P_\text{out} = 630.96 P_\text{in} = 630.96(36 \times 10^{-3}) = 22.71 \text{ W}$$

Example 2-12

A circuit consists of two amplifiers with gains of 6.8 and 14.3 dB and two filters with attenuations of -16.4 and -2.9 dB. If the output voltage is 800 mV, what is the input voltage?

$$A_T = A_1 + A_2 + A_3 + A_4 = 6.8 + 14.3 - 16.4 - 2.9 = 1.8 \text{ dB}$$

$$A_T = \frac{V_\text{out}}{V_\text{in}} = 10^{\text{dB}/20} = 10^{1.8/20} = 10^{0.09}$$

$$\frac{V_\text{out}}{V_\text{in}} = 10^{0.09} = 1.23$$

$$V_\text{in} = \frac{V_\text{out}}{1.23} = \frac{800}{1.23} = 650.4 \text{ mV}$$

Example 2-13

Express $P_\text{out} = 12.3$ dBm in watts.

$$\frac{P_\text{out}}{0.001} = 10^{\text{dBm}/10} = 10^{12.3/10} = 10^{1.23} = 17$$

$$P_\text{out} = 0.001 \times 17 = 17 \text{ mW}$$

2-2 Tuned Circuits

Virtually all communication equipment contains *tuned circuits,* circuits made up of inductors and capacitors that resonate at specific frequencies. In this section you will review how to calculate the reactance, resonant frequency, impedance, Q, and bandwidth of series and parallel resonance circuits.

Reactive Components

All tuned circuits and many filters are made up of inductive and capacitive elements, including discrete components such as coils and capacitors and the stray and distributed inductance and capacitance that appear in all electronic circuits. Both coils and capacitors offer an opposition to alternating-current flow known as *reactance,* which is expressed in ohms (abbreviated Ω). Like resistance, reactance is an opposition that directly affects the amount of current in a circuit. In addition, reactive effects produce a phase shift between the currents and voltages in a circuit. Capacitance causes the current to lead the applied voltage, whereas inductance causes the current to lag the applied voltage. Coils and capacitors used together form tuned, or resonant, circuits.

Capacitors. A *capacitor* used in an ac circuit continually charges and discharges. A capacitor tends to oppose voltage changes across it. This translates to an opposition to alternating current known as *capacitive reactance X_C.*

The reactance of a capacitor is inversely proportional to the value of capacitance C and operating frequency f. It is given by the familiar expression

$$X_C = \frac{1}{2\pi f C}$$

The reactance of a 100-pF capacitor at 2 MHz is

Chip capacitors.

$$X_C = \frac{1}{6.28(2 \times 10^6)(100 \times 10^{-12})} = 796.2 \ \Omega$$

The formula can also be used to calculate either frequency or capacitance depending on the application. These formulas are

$$f = \frac{1}{2\pi X_C C} \qquad \text{and} \qquad C = \frac{1}{2\pi f X_C}$$

The wire leads of a capacitor have resistance and inductance, and the dielectric has leakage which appears as a resistance value in parallel with the capacitor. These characteristics, which are illustrated in Fig. 2-8, are sometimes referred to as *residuals* or parasitics. The series resistance and inductance are very small, and the leakage resistance

> **GOOD TO KNOW**
>
> Stray and distributed capacitances and inductances can greatly alter the operation and performance of a circuit.

Reactance

Capacitor

Capacitive reactance

Residual

Figure 2-8 What a capacitor looks like at high frequencies.

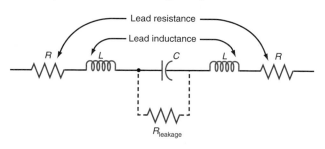

Stray (or distributed) capacitance

Inductor (coil or choke)

Inductance

Inductive reactance

is very high, so these factors can be ignored at low frequencies. At radio frequencies, however, these residuals become noticeable, and the capacitor functions as a complex *RLC* circuit. Most of these effects can be greatly minimized by keeping the capacitor leads very short. This problem is mostly eliminated by using the newer chip capacitors, which have no leads as such.

Capacitance is generally added to a circuit by a capacitor of a specific value, but capacitance can occur between any two conductors separated by an insulator. For example, there is capacitance between the parallel wires in a cable, between a wire and a metal chassis, and between parallel adjacent copper patterns on a printed-circuit board. These are known as *stray,* or *distributed, capacitances.* Stray capacitances are typically small, but they cannot be ignored, especially at the high frequencies used in communication. Stray and distributed capacitances can significantly affect the performance of a circuit.

Inductors. An *inductor,* also called a *coil* or *choke,* is simply a winding of multiple turns of wire. When current is passed through a coil, a magnetic field is produced around the coil. If the applied voltage and current are varying, the magnetic field alternately expands and collapses. This causes a voltage to be self-induced into the coil winding, which has the effect of opposing current changes in the coil. This effect is known as *inductance.*

The basic unit of inductance is the henry (H). Inductance is directly affected by the physical characteristics of the coil, including the number of turns of wire in the inductor, the spacing of the turns, the length of the coil, the diameter of the coil, and the type of magnetic core material. Practical inductance values are in the millihenry ($mH = 10^{-3}$ H), microhenry ($\mu H = 10^{-6}$ H), and nanohenry ($nH = 10^{-9}$ H) regions.

Figure 2-9 shows several different types of inductor coils.

- Figure 2-9(*a*) is an inductor made of a heavy, self-supporting wire coil.
- In Fig. 2-9(*b*) the inductor is formed of a copper spiral that is etched right onto the board itself.
- In Fig. 2-9(*c*) the coil is wound on an insulating form containing a powdered iron or ferrite core in the center, to increase its inductance.
- Figure 2-9(*d*) shows another common type of inductor, one using turns of wire on a toroidal or doughnut-shaped form.
- Figure 2-9(*e*) shows an inductor made by placing a small ferrite bead over a wire; the bead effectively increases the wire's small inductance.
- Figure 2-9(*f*) shows a chip inductor. It is typically no more than $\frac{1}{8}$ to $\frac{1}{4}$ in long. A coil is contained within the body, and the unit is soldered to the circuit board with the end connections. These devices look exactly like chip resistors and capacitors.

In a dc circuit, an inductor will have little or no effect. Only the ohmic resistance of the wire affects current flow. However, when the current changes, such as during the time the power is turned off or on, the coil will oppose these changes in current.

When an inductor is used in an ac circuit, this opposition becomes continuous and constant and is known as *inductive reactance.* Inductive reactance X_L is expressed in ohms and is calculated by using the expression

$$X_L = 2\pi fL$$

For example, the inductive reactance of a 40-μH coil at 18 MHz is

$$X_L = 6.28(18 \times 10^6)(40 \times 10^{-6}) = 4522 \ \Omega$$

In addition to the resistance of the wire in an inductor, there is stray capacitance between the turns of the coil. See Fig. 2-10(*a*). The overall effect is as if a small capacitor were connected in parallel with the coil, as shown in Fig. 2-10(*b*). This is the

Figure 2-9 Types of inductors. (*a*) Heavy self-supporting wire coil. (*b*) Inductor made as copper pattern. (*c*) Insulating form. (*d*) Toroidal inductor. (*e*) Ferrite bead inductor. (*f*) Chip inductor.

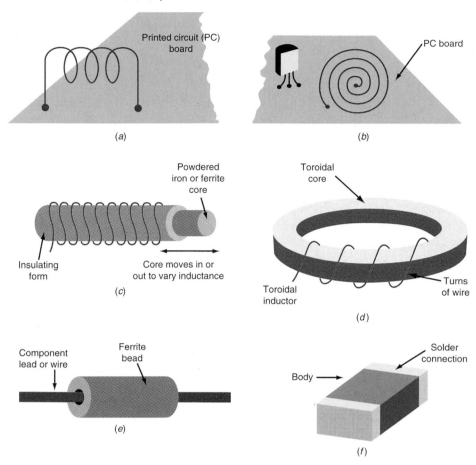

equivalent circuit of an inductor at high frequencies. At low frequencies, capacitance may be ignored, but at radio frequencies, it is sufficiently large to affect circuit operation. The coil then functions not as a pure inductor, but as a complex *RLC* circuit with a self-resonating frequency.

Any wire or conductor exhibits a characteristic inductance. The longer the wire, the greater the inductance. Although the inductance of a straight wire is only a fraction of

Figure 2-10 Equivalent circuit of an inductor at high frequencies. (*a*) Stray capacitance between turns. (*b*) Equivalent circuit of an inductor at high frequencies.

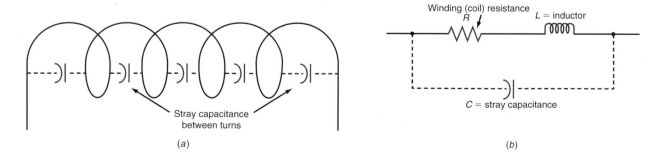

a microhenry, at very high frequencies, the reactance can be significant. For this reason, it is important to keep all lead lengths short in interconnecting components in *RF* circuits. This is especially true of capacitor and transistor leads, since stray or distributed inductance can significantly affect the performance and characteristics of a circuit.

Quality factor *Q*

Another important characteristic of an inductor is its *quality factor Q*, the ratio of inductive power to resistive power:

$$Q = \frac{I^2 X_L}{I^2 R} = \frac{X_L}{R}$$

This is the ratio of the power returned to the circuit to the power actually dissipated by the coil resistance. For example, the Q of a 3-μH inductor with a total resistance of 45 Ω at 90 MHz is calculated as follows:

$$Q = \frac{2\pi f L}{R} = \frac{6.28(90 \times 10^6)(3 \times 10^{-6})}{45} = \frac{1695.6}{45} = 37.68$$

Resistor

Resistors. At low frequencies, a standard low-wattage color-coded *resistor* offers nearly pure resistance, but at high frequencies its leads have considerable inductance, and stray capacitance between the leads causes the resistor to act as a complex *RLC* circuit, as shown in Fig. 2-11. To minimize the inductive and capacitive effects, the leads are kept very short in radio applications.

The tiny resistor chips used in surface-mount construction of the electronic circuits preferred for radio equipment have practically no leads except for the metallic end pieces soldered to the printed-circuit board. They have virtually no lead inductance and little stray capacitance.

Many resistors are made from a carbon-composition material in powdered form sealed inside a tiny housing to which leads are attached. The type and amount of carbon material determine the value of these resistors. They contribute noise to the circuit in which they are used. The noise is caused by thermal effects and the granular nature of the resistance material. The noise contributed by such resistors in an amplifier used to amplify very low level radio signals may be so high as to obliterate the desired signal.

To overcome this problem, film resistors were developed. They are made by depositing a carbon or metal film in spiral form on a ceramic form. The size of the spiral and the kind of metal film determine the resistance value. Carbon film resistors are quieter than carbon-composition resistors, and metal film resistors are quieter than carbon film resistors. Metal film resistors should be used in amplifier circuits that must deal with very low level RF signals. Most surface-mount resistors are of the metallic film type.

Skin Effect. The resistance of any wire conductor, whether it is a resistor or capacitor lead or the wire in an inductor, is primarily determined by the ohmic resistance of the wire itself. However, other factors influence it. The most significant one is *skin effect,* the tendency of electrons flowing in a conductor to flow near and on the outer surface

Skin effect

Figure 2-11   Equivalent circuit of a resistor at high (radio) frequencies.

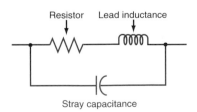

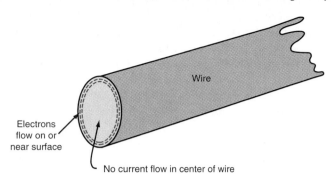

of the conductor frequencies in the VHF, UHF, and microwave regions (Fig. 2-12). This has the effect of greatly decreasing the total cross-sectional area of the conductor, thus increasing its resistance and significantly affecting the performance of the circuit in which the conductor is used. For example, skin effect lowers the Q of an inductor at the higher frequencies, causing unexpected and undesirable effects. Thus many high-frequency coils, particularly those in high-powered transmitters, are made with copper tubing. Since current does not flow in the center of the conductor, but only on the surface, tubing provides the most efficient conductor. Very thin conductors, such as a copper pattern on a printed-circuit board, are also used. Often these conductors are silver- or gold-plated to further reduce their resistance.

Tuned Circuits and Resonance

A tuned circuit is made up of inductance and capacitance and resonates at a specific frequency, the resonant frequency. In general, the terms *tuned circuit* and *resonant circuit* are used interchangeably. Because tuned circuits are frequency-selective, they respond best at their resonant frequency and at a narrow range of frequencies around the resonant frequency.

Tuned (resonant) circuit

Series Resonant Circuits. A *series resonant circuit* is made up of inductance, capacitance, and resistance, as shown in Fig. 2-13. Such circuits are often referred to as *LCR circuits* or *RLC circuits*. The inductive and capacitive reactances depend upon the frequency of the applied voltage. Resonance occurs when the inductive and capacitive reactances are equal. A plot of reactance versus frequency is shown in Fig. 2-14, where f_r is the resonant frequency.

Series resonant circuit

***LCR* circuit**

***RLC* circuit**

Figure 2-13   Series *RLC* circuit.

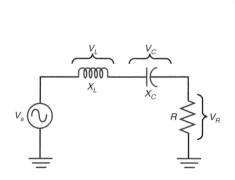

Figure 2-14   Variation of reactance with frequency.

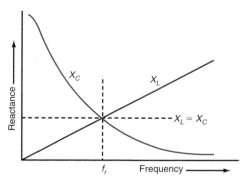

The total impedance of the circuit is given by the expression

$$Z = \sqrt{R^2 + (X_L - X_C)^2}$$

When X_L equals X_C, they cancel each other, leaving only the resistance of the circuit to oppose the current. At resonance, the total circuit impedance is simply the value of all series resistances in the circuit. This includes the resistance of the coil and the resistance of the component leads, as well as any physical resistor in the circuit.

The resonant frequency can be expressed in terms of inductance and capacitance. A formula for resonant frequency can be easily derived. First, express X_L and X_C as an equivalence: $X_L = X_C$. Since

$$X_L = 2\pi f_r L \qquad \text{and} \qquad X_C = \frac{1}{2\pi f_r C}$$

we have

$$2\pi f_r L = \frac{1}{2\pi f_r C}$$

Solving for f_r gives

$$f_r = \frac{1}{2\pi\sqrt{LC}}$$

In this formula, the frequency is in hertz, the inductance is in henrys, and the capacitance is in farads.

Example 2-14

What is the resonant frequency of a 2.7-pF capacitor and a 33-nH inductor?

$$f_r = \frac{1}{2\pi\sqrt{LC}} = \frac{1}{6.28\sqrt{33 \times 10^{-9} \times 2.7 \times 10^{-12}}}$$
$$= 5.33 \times 10^8 \text{ Hz or 533 MHz}$$

It is often necessary to calculate capacitance or inductance, given one of those values and the resonant frequency. The basic resonant frequency formula can be rearranged to solve for either inductance and capacitance as follows:

$$L = \frac{1}{4\pi^2 f^2 C} \qquad \text{and} \qquad C = \frac{1}{4\pi^2 f^2 L}$$

For example, the capacitance that will resonate at a frequency of 18 MHz with a 12-μH inductor is determined as follows:

$$C = \frac{1}{4\pi^2 f_r^2 L} = \frac{1}{39.478(18 \times 10^6)^2(12 \times 10^{-6})}$$

$$= \frac{1}{39.478(3.24 \times 10^{14})(12 \times 10^{-6})} = 6.5 \times 10^{-12} \text{ F or 6.5 pF}$$

Example 2-15

What value of inductance will resonate with a 12-pF capacitor at 49 MHz?

$$L = \frac{1}{4\pi^2 f_r^2 C} = \frac{1}{39.478(49 \times 10^6)^2(12 \times 10^{-12})}$$

$$= 8.79 \times 10^{-7} \text{ H or } 879 \text{ nH}$$

As indicated earlier, the basic definition of resonance in a series tuned circuit is the point at which X_L equals X_C. With this condition, only the resistance of the circuit impedes the current. The total circuit impedance at resonance is $Z = R$. For this reason, resonance in a series tuned circuit can also be defined as the point at which the circuit impedance is lowest and the circuit current is highest. Since the circuit is resistive at resonance, the current is in phase with the applied voltage. Above the resonant frequency, the inductive reactance is higher than the capacitive reactance and the inductor voltage drop is greater than the capacitor voltage drop. Therefore, the circuit is inductive and the current will lag the applied voltage. Below resonance, the capacitive reactance is higher than the inductive reactance; the net reactance is capacitive, thereby producing a leading current in the circuit. The capacitor voltage drop is higher than the inductor voltage drop.

The response of a series resonant circuit is illustrated in Fig. 2-15, which is a plot of the frequency and phase shift of the current in the circuit with respect to frequency.

At very low frequencies, the capacitive reactance is much greater than the inductive reactance; therefore the current in the circuit is very low because of the high impedance. In addition, because the circuit is predominantly capacitive, the current leads the voltage by nearly 90°. As the frequency increases, X_C goes down and X_L goes up. The amount of leading phase shift decreases. As the values of reactances approach one another, the current begins to rise. When X_L equals X_C, their effects cancel and the impedance in the circuit is just that of the resistance. This produces a current peak, where the current is in phase with the voltage (0°). As the frequency continues to rise,

Figure 2-15 Frequency and phase response curves of a series resonant circuit.

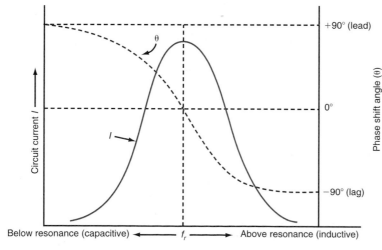

Figure 2-16 Bandwidth of a series resonant circuit.

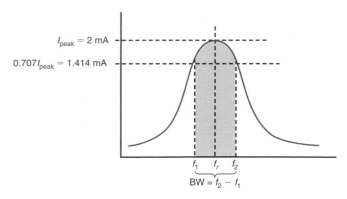

X_L becomes greater than X_C. The impedance of the circuit increases and the current decreases. With the circuit predominantly inductive, the current lags the applied voltage. If the output voltage were being taken from across the resistor in Fig. 2-13, the response curve and phase angle of the voltage would correspond to those in Fig. 2-15. As Fig. 2-15 shows, the current is highest in a region centered on the resonant frequency. The narrow frequency range over which the current is highest is called the *bandwidth*. This area is illustrated in Fig. 2-16.

The upper and lower boundaries of the bandwidth are defined by two cutoff frequencies designated f_1 and f_2. These cutoff frequencies occur where the current amplitude is 70.7 percent of the peak current. In the figure, the peak circuit current is 2 mA, and the current at both the lower (f_1) and upper (f_2) cutoff frequency is 0.707 of 2 mA, or 1.414 mA.

Current levels at which the response is down 70.7 percent are called the *half-power points* because the power at the cutoff frequencies is one-half the power peak of the curve.

$$P = I^2R = (0.707\,I_{peak})^2R = 0.5\,I_{peak}^2R$$

The bandwidth BW of the tuned circuit is defined as the difference between the upper and lower cutoff frequencies:

$$BW = f_2 - f_1$$

For example, assuming a resonant frequency of 75 kHz and upper and lower cutoff frequencies of 76.5 and 73.5 kHz, respectively, the bandwidth is BW = 76.5 − 73.5 = 3 kHz.

The bandwidth of a resonant circuit is determined by the Q of the circuit. Recall that the Q of an inductor is the ratio of the inductive reactance to the circuit resistance. This holds true for a series resonant circuit, where Q is the ratio of the inductive reactance to the total circuit resistance, which includes the resistance of the inductor plus any additional series resistance:

$$Q = \frac{X_L}{R_T}$$

Recall that bandwidth is then computed as

$$BW = \frac{f_r}{Q}$$

If the Q of a circuit resonant at 18 MHz is 50, then the bandwidth is BW = 18/50 = 0.36 MHz = 360 kHz.

Example 2-16

What is the bandwidth of a resonant circuit with a frequency of 28 MHz and a Q of 70?

$$BW = \frac{f_r}{Q} = \frac{28 \times 10^6}{70} = 400{,}000 \text{ Hz} = 400 \text{ kHz}$$

The formula can be rearranged to compute Q, given the frequency and the bandwidth:

$$Q = \frac{f_r}{BW}$$

Thus the Q of the circuit whose bandwidth was computed previously is $Q = 75 \text{ kHz}/3\text{kHz} = 25$.

Since the bandwidth is approximately centered on the resonant frequency, f_1 is the same distance from f_r as f_2 is from f_r. This fact allows you to calculate the resonant frequency by knowing only the cutoff frequencies:

$$f_r = \sqrt{f_1 \times f_2}$$

For example, if $f_1 = 175 \text{ kHz}$ and $f_2 = 178 \text{ kHz}$, the resonant frequency is

$$f_r = \sqrt{175 \times 10^3 \times 178 \times 10^3} = 176.5 \text{ kHz}$$

For a linear frequency scale, you can calculate the center or resonant frequency by using an average of the cutoff frequencies.

$$f_r = \frac{f_1 + f_2}{2}$$

If the circuit Q is very high (> 100), then the response curve is approximately symmetric around the resonant frequency. The cutoff frequencies will then be roughly equidistant from the resonant frequency by the amount of BW/2. Thus the cutoff frequencies can be calculated if the bandwidth and the resonant frequency are known:

$$f_1 = f_r - \frac{BW}{2} \quad \text{and} \quad f_2 = f_r + \frac{BW}{2}$$

For instance, if the resonant frequency is 49 MHz (49,000 kHz) and the bandwidth is 10 kHz, then the cutoff frequencies will be

$$f_1 = 49{,}000 \text{ kHz} - \frac{10k}{2} = 49{,}000 \text{ kHz} - 5 \text{ kHz} = 48{,}995 \text{ kHz}$$

$$f_2 = 49{,}000 \text{ kHz} + 5 \text{ kHz} = 49{,}005 \text{ kHz}$$

Keep in mind that although this procedure is an approximation, it is useful in many applications.

The bandwidth of a resonant circuit defines its *selectivity,* i.e., how the circuit responds to varying frequencies. If the response is to produce a high current only over a narrow range of frequencies, a narrow bandwidth, the circuit is said to be highly selective. If the current is high over a broader range of frequencies, i.e., the bandwidth is wider, the circuit is less selective. In general, circuits with high selectivity and narrow bandwidths are more desirable. However, the actual selectivity and bandwidth of a circuit must be optimized for each application.

Selectivity

The relationship between circuit resistance Q and bandwidth is extremely important. The bandwidth of a circuit is inversely proportional to Q. The higher Q is, the smaller the bandwidth. Low Qs produce wide bandwidths or less selectivity. In turn, Q is a function of the circuit resistance. A low resistance produces a high Q, a narrow bandwidth, and a highly selective circuit. A high circuit resistance produces a low Q, wide bandwidth, and poor selectivity. In most communication circuits, circuit Qs are at least 10 and typically higher. In most cases, Q is controlled directly by the resistance of the inductor. Figure 2-17 shows the effect of different values of Q on bandwidth.

Example 2-17

The upper and lower cutoff frequencies of a resonant circuit are found to be 8.07 and 7.93 MHz. Calculate (a) the bandwidth, (b) the approximate resonant frequency, and (c) Q.

a. $\text{BW} = f_2 - f_1 = 8.07 \text{ MHz} - 7.93 \text{ MHz} = 0.14 \text{ MHz} = 140 \text{ kHz}$

b. $f_r = \sqrt{f_1 f_2} = \sqrt{(8.07 \times 10^6)(7.93 \times 10^6)} = 8 \text{ MHz}$

c. $Q = \dfrac{f_r}{\text{BW}} = \dfrac{8 \times 10^6}{140 \times 10^3} = 57.14$

Example 2-18

What are the approximate 3-dB down frequencies of a resonant circuit with a Q of 200 at 16 MHz?

$$\text{BW} = \frac{f_r}{Q} = \frac{16 \times 10^6}{200} = 80,000 \text{ Hz} = 80 \text{ kHz}$$

$$f_1 = f_r - \frac{\text{BW}}{2} = 16,000,000 - \frac{80,000}{2} = 15.96 \text{ MHz}$$

$$f_2 = f_r + \frac{\text{BW}}{2} = 16,000,000 + \frac{80,000}{2} = 16.04 \text{ MHz}$$

Resonance produces an interesting but useful phenomenon in a series RLC circuit. Consider the circuit in Fig. 2-18(a). At resonance, assume $X_L = X_C = 500 \ \Omega$. The total circuit resistance is 10 Ω. The Q of the circuit is then

$$Q = \frac{X_L}{R} = \frac{500}{10} = 50$$

If the applied or source voltage V_s is 2 V, the circuit current at resonance will be

$$I = \frac{V_s}{R} = \frac{2}{10} = 0.2 \text{ A}$$

Figure 2-17 The effect of Q on bandwidth and selectivity in a resonant circuit.

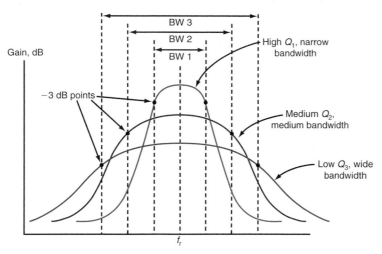

Figure 2-18 Resonant step-up voltage in a series resonant circuit.

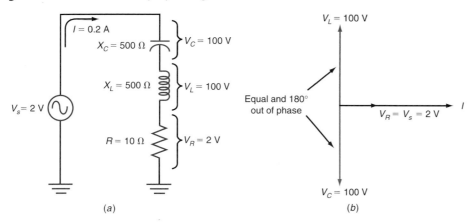

(a)   (b)

When the reactances, the resistances, and the current are known, the voltage drops across each component can be computed:

$$V_L = IX_L = 0.2(500) = 100 \text{ V}$$

$$V_C = IX_C = 0.2(500) = 100 \text{ V}$$

$$V_R = IR = 0.2(10) = 2 \text{ V}$$

As you can see, the voltage drops across the inductor and capacitor are significantly higher than the applied voltage. This is known as the *resonant step-up voltage*. Although the sum of the voltage drops around the series circuit is still equal to the source voltage, at resonance the voltage across the inductor leads the current by 90° and the voltage across the capacitor lags the current by 90° [see Fig. 2-18(b)]. Therefore the inductive and reactive voltages are equal but 180° out of phase. As a result, when added, they cancel each other, leaving a total reactive voltage of 0. This means that the entire applied voltage appears across the circuit resistance.

The resonant step-up voltage across the coil or capacitor can be easily computed by multiplying the input or source voltage by Q:

$$V_L = V_C = QV_s$$

In the example in Fig. 2-18, $V_L = 50(2) = 100$ V.

Resonant step-up voltage

This interesting and useful phenomenon means that small applied voltages can essentially be stepped up to a higher voltage—a form of simple amplification without active circuits that is widely applied in communication circuits.

Example 2-19

A series resonant circuit has a Q of 150 at 3.5 MHz. The applied voltage is 3 μV. What is the voltage across the capacitor?

$$V_C = QV_s = 150(3 \times 10^{-6}) = 450 \times 10^{-6} = 450 \ \mu\text{V}$$

Parallel resonant circuit

Parallel Resonant Circuits. A *parallel resonant circuit* is formed when the inductor and capacitor are connected in parallel with the applied voltage, as shown in Fig. 2-19(*a*). In general, resonance in a parallel tuned circuit can also be defined as the point at which the inductive and capacitive reactances are equal. The resonant frequency is therefore calculated by the resonant frequency formula given earlier. If we assume lossless components in the circuit (no resistance), then the current in the inductor equals the current in the capacitor:

$$I_L = I_C$$

Although the currents are equal, they are 180° out of phase, as the phasor diagram in Fig. 2-19(*b*) shows. The current in the inductor lags the applied voltage by 90°, and the current in the capacitor leads the applied voltage by 90°, for a total of 180°.

Now, by applying Kirchhoff's current law to the circuit, the sum of the individual branch currents equals the total current drawn from the source. With the inductive and capacitive currents equal and out of phase, their sum is 0. Thus, at resonance, a parallel tuned circuit appears to have infinite resistance, draws no current from the source and thus has infinite impedance, and acts as an open circuit. However, there is a high circulating current between the inductor and capacitor. Energy is being stored and transferred between the inductor and capacitor. Because such a circuit acts as a kind of storage vessel for electric energy, it is often

Tank circuit

Tank current

referred to as a *tank circuit* and the circulating current is referred to as the *tank current*.

In a practical resonant circuit where the components do have losses (resistance), the circuit still behaves as described above. Typically, we can assume that the capacitor has practically zero losses and the inductor contains a resistance, as illustrated in Fig. 2-20(*a*). At resonance, where $X_L = X_C$, the impedance of the inductive branch of the circuit is higher than the impedance of the capacitive branch because of the coil resistance. The capacitive current is slightly higher than the inductive current. Even if the reactances are

Figure 2-19 Parallel resonant circuit currents. (*a*) Parallel resonant circuit. (*b*) Current relationships in parallel resonant circuit.

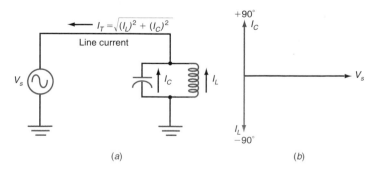

(*a*)   (*b*)

Figure 2-20 A practical parallel resonant circuit. (*a*) Practical parallel resonant circuit with coil resistance R_W. (*b*) Phase relationships.

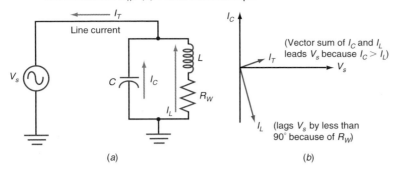

(*a*)   (*b*)

equal, the branch currents will be unequal and therefore there will be some net current flow in the supply line. The source current will lead the supply voltage, as shown in Fig. 2-20(*b*). Nevertheless, the inductive and capacitive currents in most cases will cancel because they are approximately equal and of opposite phase, and consequently the line or source current will be significantly lower than the individual branch currents. The result is a very high resistive impedance, approximately equal to

$$Z = \frac{V_s}{I_T}$$

The circuit in Fig. 2-20(*a*) is not easy to analyze. One way to simplify the mathematics involved is to convert the circuit to an equivalent circuit in which the coil resistance is translated to a parallel resistance that gives the same overall results, as shown in Fig. 2-21. The equivalent inductance L_{eq} and resistance R_{eq} are calculated with the formulas

$$L_{eq} = \frac{L(Q^2 + 1)}{Q^2} \qquad \text{and} \qquad R_{eq} = R_W(Q^2 + 1)$$

and Q is determined by the formula

$$Q = \frac{X_L}{R_W}$$

where R_W is the coil winding resistance.

If Q is high, usually more than 10, L_{eq} is approximately equal to the actual inductance value L. The total impedance of the circuit at resonance is equal to the equivalent parallel resistance:

$$Z = R_{eq}$$

Figure 2-21 An equivalent circuit makes parallel resonant circuits easier to analyze.

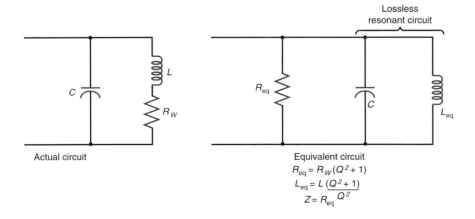

Actual circuit

Equivalent circuit
$R_{eq} = R_W(Q^2 + 1)$
$L_{eq} = L\dfrac{(Q^2 + 1)}{Q^2}$
$Z = R_{eq}$

Example 2-20

What is the impedance of a parallel LC circuit with a resonant frequency of 52 MHz and a Q of 12? $L = 0.15 \, \mu H$.

$$Q = \frac{X_L}{R_W}$$

$$X_L = 2\pi f L = 6.28(52 \times 10^6)(0.15 \times 10^{-6}) = 49 \, \Omega$$

$$R_W = \frac{X_L}{Q} = \frac{49}{12} = 4.1 \, \Omega$$

$$Z = R_{eq} = R_W(Q^2 + 1) = 4.1(12^2 + 1) = 4.1(145) = 592 \, \Omega$$

If the Q of the parallel resonant circuit is greater than 10, the following simplified formula can be used to calculate the resistive impedance at resonance:

$$Z = \frac{L}{CR_W}$$

The value of R_W is the winding resistance of the coil.

Example 2-21

Calculate the impedance of the circuit given in Example 2-20 by using the formula $Z = L/CR$.

$$f_r = 52 \text{ MHz} \qquad R_W = 4.1 \, \Omega \qquad L = 0.15 \, \mu H$$

$$C = \frac{1}{4\pi^2 f_r^2 L} = \frac{1}{39.478(52 \times 10^6)^2(0.15 \times 10^{-6})}$$

$$= 6.245 \times 10^{-11}$$

$$Z = \frac{L}{CR_W} = \frac{0.15 \times 10^{-6}}{(62.35 \times 10^{-12})(4.1)} = 586 \, \Omega$$

This is close to the previously computed value of 592 Ω. The formula $Z = L/CR_W$ is an approximation.

GOOD TO KNOW

The bandwidth of a circuit is inversely proportional to the circuit Q. The higher the Q, the smaller the bandwidth. Low Q values produce wide bandwidths or less selectivity.

A frequency and phase response curve of a parallel resonant circuit is shown in Fig. 2-22. Below the resonant frequency, X_L is less than X_C; thus the inductive current is greater than the capacitive current, and the circuit appears inductive. The line current lags the applied voltage. Above the resonant frequency, X_C is less than X_L; thus the capacitive current is more than the inductive current, and the circuit appears capacitive. Therefore, the line current leads the applied voltage.

At the resonant frequency, the impedance of the circuit peaks. This means that the line current at that time is at its minimum. At resonance, the circuit appears to have a very high resistance, and the small line current is in phase with the applied voltage.

Figure 2-22 Response of a parallel resonant circuit.

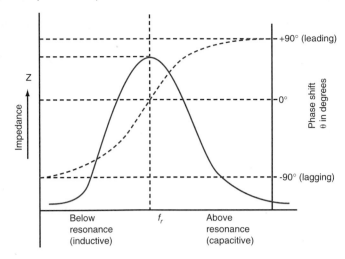

Note that the Q of a parallel circuit, which was previously expressed as $Q = X_L/R_W$, can also be computed with the expression

$$Q = \frac{R_P}{X_L}$$

where R_P is the equivalent parallel resistance, R_{eq} in parallel with any other parallel resistance, and X_L is the inductive reactance of the equivalent inductance L_{eq}.

You can set the bandwidth of a parallel tuned circuit by controlling Q. The Q can be determined by connecting an external resistor across the circuit. This has the effect of lowering R_P and increasing the bandwidth.

Example 2-22

What value of parallel resistor is needed to set the bandwidth of a parallel tuned circuit to 1 MHz? Assume $X_L = 300\ \Omega$, $R_W = 10\ \Omega$, and $f_r = 10$ MHz.

$$Q = \frac{X_L}{R_W} = \frac{300}{10} = 30$$

$$R_P = R_W(Q^2 + 1) = 10(30^2 + 1) = 10(901) = 9010\ \Omega$$

(equivalent resistance of the parallel circuit at resonance)

$$\text{BW} = \frac{f_r}{Q}$$

$$Q = \frac{f_r}{\text{BW}} = \frac{10\ \text{MHz}}{1\ \text{MHz}} = 10\ (Q\text{ needed for 1-MHz bandwidth})$$

$$R_{Pnew} = QX_L = 10(300) = 3000\ \Omega$$

(this is the total resistance of the circuit R_{Pnew} made up of the original R_P and an externally connected resistor R_{ext})

$$R_{Pnew} = \frac{R_P R_{ext}}{R_P + R_{ext}}$$

$$R_{ext} = \frac{R_{Pnew}R_P}{R_P - R_{Pnew}} = \frac{9010(3000)}{9010 - 3000} = 4497.5\ \Omega$$

2-3 Filters

A *filter* is a frequency-selective circuit. Filters are designed to pass some frequencies and reject others. The series and parallel resonant circuits reviewed in Section 2-2 are examples of filters.

There are numerous ways to implement filter circuits. Simple filters created by using resistors and capacitors or inductors and capacitors are called *passive filters* because they use passive components that do not amplify. In communication work, many filters are of the passive *LC* variety, although many other types are used. Some special types of filters are active filters that use *RC* networks with feedback in op amp circuits, switched capacitor filters, crystal and ceramic filters, surface acoustic wave (SAW) filters, and digital filters implemented with digital signal processing (DSP) techniques.

The five basic kinds of filter circuits are as follows:

Low-pass filter. Passes frequencies below a critical frequency called the *cutoff frequency* and greatly attenuates those above the cutoff frequency.

High-pass filter. Passes frequencies above the cutoff but rejects those below it.

Bandpass filter. Passes frequencies over a narrow range between lower and upper cutoff frequencies.

Band-reject filter. Rejects or stops frequencies over a narrow range but allows frequencies above and below to pass.

All-pass filter. Passes all frequencies equally well over its design range but has a fixed or predictable phase shift characteristic.

RC Filters

A low-pass filter allows the lower-frequency components of the applied voltage to develop output voltage across the load resistance, whereas the higher-frequency components are attenuated, or reduced, in the output.

A high-pass filter does the opposite, allowing the higher-frequency components of the applied voltage to develop voltage across the output load resistance.

The case of an *RC* coupling circuit is an example of a high-pass filter because the ac component of the input voltage is developed across *R* and the dc voltage is blocked by the series capacitor. Furthermore, with higher frequencies in the ac component, more ac voltage is coupled.

Any low-pass or high-pass filter can be thought of as a frequency-dependent voltage divider, because the amount of output voltage is a function of frequency.

RC filters use combinations of resistors and capacitors to achieve the desired response. Most *RC* filters are of the low-pass or high-pass type. Some band-reject or notch filters are also made with *RC* circuits. Bandpass filters can be made by combining low-pass and high-pass *RC* sections, but this is rarely done.

Low-Pass Filter. A *low-pass filter* is a circuit that introduces no attenuation at frequencies below the cutoff frequency but completely eliminates all signals with frequencies above the cutoff. Low-pass filters are sometimes referred to as *high cut filters*.

The ideal response curve for a low-pass filter is shown in Fig. 2-23. This response curve cannot be realized in practice. In practical circuits, instead of a sharp transition at the cutoff frequency, there is a more gradual transition between little or no attenuation and maximum attenuation.

The simplest form of low-pass filter is the *RC* circuit shown in Fig. 2-24(a). The circuit forms a simple voltage divider with one frequency-sensitive component, in this case the capacitor. At very low frequencies, the capacitor has very high reactance compared to the resistance and therefore the attenuation is minimum. As the frequency increases, the capacitive reactance decreases. When the reactance becomes smaller than the resistance, the attenuation increases rapidly. The frequency response of the basic

Figure 2-23 Ideal response curve of a low-pass filter.

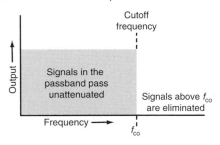

Figure 2-24 RC low-pass filter. (a) Circuit. (b) Low-pass filter.

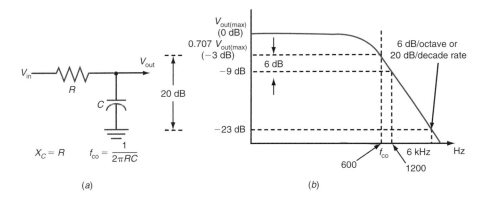

$$X_C = R \qquad f_{co} = \frac{1}{2\pi RC}$$

(a)   (b)

circuit is illustrated in Fig. 2-24(b). The cutoff frequency of this filter is that point where R and X_C are equal. The cutoff frequency, also known as the critical frequency, is determined by the expression

$$X_C = R$$

$$\frac{1}{2\pi f_c} = R$$

$$f_{co} = \frac{1}{2\pi RC}$$

For example, if $R = 4.7$ kΩ and $C = 560$ pF, the cutoff frequency is

$$f_{co} = \frac{1}{2\pi(4700)(560 \times 10^{-12})} = 60,469 \text{ Hz} \quad \text{or} \quad 60.5 \text{ kHz}$$

Example 2-23

What is the cutoff frequency of a single-section RC low-pass filter with $R = 8.2$ kΩ and $C = 0.0033$ μF?

$$f_{co} = \frac{1}{2\pi RC} = \frac{1}{2\pi(8.2 \times 10^3)(0.0033 \times 10^{-6})}$$

$$f_{co} = 5881.56 \text{ Hz} \quad \text{or} \quad 5.88 \text{ kHz}$$

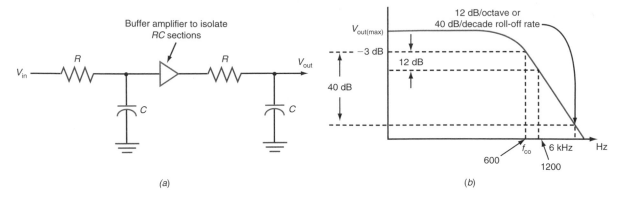

(*a*)

(*b*)

At the cutoff frequency, the output amplitude is 70.7 percent of the input amplitude at lower frequencies. This is the so-called 3-dB down point. In other words, this filter has a voltage gain of −3 dB at the cutoff frequency. At frequencies above the cutoff frequency, the amplitude decreases at a linear rate of 6 dB per octave or 20 dB per decade. An *octave* is defined as a doubling or halving of frequency, and a *decade* represents a one-tenth or times-10 relationship. Assume that a filter has a cutoff of 600 Hz. If the frequency doubles to 1200 Hz, the attenuation will increase by 6 dB, or from 3 dB at cutoff to 9 dB at 1200 Hz. If the frequency increased by a factor of 10 from 600 Hz to 6 kHz, the attenuation would increase by a factor of 20 dB from 3 dB at cutoff to 23 dB at 6 kHz.

Octave

Decade

If a faster rate of attenuation is required, two *RC* sections set to the same cutoff frequency can be used. Such a circuit is shown in Fig. 2-25(*a*). With this circuit, the rate of attenuation is 12 dB per octave or 40 dB per decade. Two identical *RC* circuits are used, but an isolation or buffer amplifier such as an emitter-follower (gain ≈ 1) is used between them to prevent the second section from loading the first. Cascading two *RC* sections without the isolation will give an attenuation rate less than the theoretically ideal 12-dB octave because of the loading effects.

With a steeper attenuation curve, the circuit is said to be more selective. The disadvantage of cascading such sections is that higher attenuation makes the output signal considerably smaller. This signal attenuation in the passband of the filter is called *insertion loss.*

Insertion loss

A low-pass filter can also be implemented with an inductor and a resistor, as shown in Fig. 2-26. The response curve for this *RL* filter is the same as that shown in Fig. 2-24(*b*). The cutoff frequency is determined by using the formula

$$f_{\text{co}} = \frac{R}{2\pi L}$$

Figure 2-26 A low-pass filter implemented with an inductor.

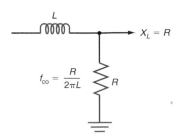

Figure 2-27 Frequency response curve of a high-pass filter. (*a*) Ideal. (*b*) Practical.

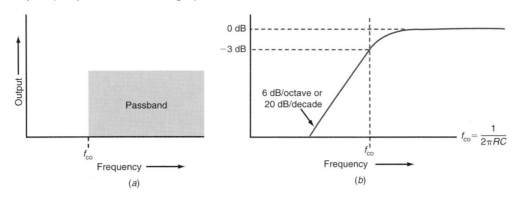

$$f_{co} = \frac{1}{2\pi RC}$$

(a)

(b)

The *RL* low-pass filters are not as widely used as *RC* filters because inductors are usually larger, heavier, and more expensive than capacitors. Inductors also have greater loss than capacitors because of their inherent winding resistance.

High-Pass Filter. A *high-pass filter* passes frequencies above the cutoff frequency with little or no attenuation but greatly attenuates those signals below the cutoff. The ideal high-pass response curve is shown in Fig. 2-27(*a*). Approximations to the ideal response curve shown in Fig. 2-27(*b*) can be obtained with a variety of *RC* and *LC* filters.

High-pass filter

The basic *RC* high-pass filter is shown in Fig. 2-28(*a*). Again, it is nothing more than a voltage divider with the capacitor serving as the frequency-sensitive component in a voltage divider. At low frequencies, X_C is very high. When X_C is much higher than *R*, the voltage divider effect provides high attenuation of the low-frequency signals. As the frequency increases, the capacitive reactance decreases. When the capacitive reactance is equal to or less than the resistance, the voltage divider gives very little attenuation. Therefore, high frequencies pass relatively unattenuated.

The cutoff frequency for this filter is the same as that for the low-pass circuit and is derived from setting X_C equal to *R* and solving for frequency:

$$f_{co} = \frac{1}{2\pi RC}$$

The roll-off rate is 6 dB per octave or 20 dB per decade.

A high-pass filter can also be implemented with a coil and a resistor, as shown in Fig. 2-28(*b*). The cutoff frequency is

$$f_{co} = \frac{R}{2\pi L}$$

The response curve for this filter is the same as that shown in Fig. 2-27(*b*). The rate of attenuation is 6 dB per octave or 20 dB per decade, as was the case with the low-pass filter. Again, improved attenuation can be obtained by cascading filter sections.

Figure 2-28 (*a*) RC high-pass filter. (*b*) RL high-pass filter.

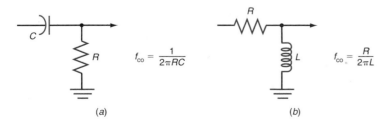

(a)

(b)

The Fundamentals of Electronics: A Review

Example 2-24

What is the closest standard EIA resistor value that will produce a cutoff frequency of 3.4 kHz with a 0.047-μF capacitor in a high-pass RC filter?

$$f_{co} = \frac{1}{2\pi RC}$$

$$R = \frac{1}{2\pi f_{co} C} = \frac{1}{2\pi(3.4 \times 10^3)(0.047 \times 10^{-6})} = 996\ \Omega$$

The closest standard values are 910 and 1000 Ω, with 1000 being the closest.

RC **Notch Filter.** *Notch filters* are also referred to as *bandstop* or *band-reject* filters. Band-reject filters are used to greatly attenuate a narrow range of frequencies around a center point. Notch filters accomplish the same purpose, but for a single frequency.

A simple notch filter that is implemented with resistors and capacitors as shown in Fig. 2-29(*a*) is called a *parallel-T* or *twin-T* notch filter. This filter is a variation of a bridge circuit. Recall that in a bridge circuit the output is zero if the bridge is balanced. If the component values are precisely matched, the circuit will be in balance and produce an attenuation of an input signal at the design frequency as high as 30 to 40 dB. A typical response curve is shown in Fig. 2-29(*b*).

The center notch frequency is computed with the formula

$$f_{notch} = \frac{1}{2\pi RC}$$

For example, if the values of resistance and capacitance are 100 kΩ and 0.02 μF, the notch frequency is

$$f_{notch} = \frac{1}{6.28(10^5)(0.02 \times 10^{-6})} = 79.6\ \text{Hz}$$

Twin-T notch filters are used primarily at low frequencies, audio and below. A common use is to eliminate 60-Hz power line hum from audio circuits and low-frequency medical equipment amplifiers. The key to high attenuation at the notch frequency is

Figure 2-29 *RC* notch filter.

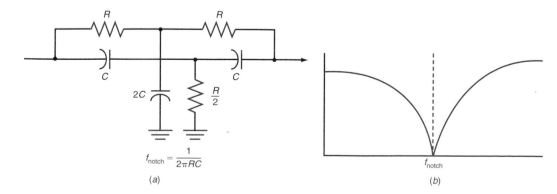

(*a*)   (*b*)

Example 2-25

What values of capacitors would you use in an RC twin-T notch filter to remove 120 Hz if $R = 220\ \text{k}\Omega$?

$$f_{\text{notch}} = \frac{1}{2\pi RC}$$

$$C = \frac{1}{2\pi f_{\text{notch}}R} = \frac{1}{6.28(120)(220 \times 10^3)}$$

$$C = 6.03 \times 10^{-9} = 6.03\ \text{nF} \quad \text{or} \quad 0.006\ \mu\text{F}$$

$$2C = 0.012\ \mu\text{F}$$

precise component values. The resistor and capacitor values must be matched to achieve high attenuation.

LC Filters

RC filters are used primarily at the lower frequencies. They are very common at audio frequencies but are rarely used above about 100 kHz. At radio frequencies, their passband attenuation is just too great, and the cutoff slope is too gradual. It is more common to see LC filters made with inductors and capacitors. Inductors for lower frequencies are large, bulky, and expensive, but those used at higher frequencies are very small, light, and inexpensive. Over the years, a multitude of filter types have been developed. Filter design methods have also changed over the years, thanks to computer design.

Filter Terminology. When working with filters, you will hear a variety of terms to describe the operation and characteristics of filters. The following definitions will help you understand filter specifications and operation.

1. **Passband.** This is the frequency range over which the filter passes signals. It is the frequency range between the cutoff frequencies or between the cutoff frequency and zero (for low-pass) or between the cutoff frequency and infinity (for high-pass). Passband

2. **Stop band.** This is the range of frequencies outside the passband, i.e., the range of frequencies that is greatly attenuated by the filter. Frequencies in this range are rejected. Stop band

3. **Attenuation.** This is the amount by which undesired frequencies in the stop band are reduced. It can be expressed as a power ratio or voltage ratio of the output to the input. Attenuation is usually given in decibels. Attenuation

4. **Insertion loss.** Insertion loss is the loss the filter introduces to the signals in the passband. Passive filters introduce attenuation because of the resistive losses in the components. Insertion loss is typically given in decibels. Insertion loss

5. **Impedance.** Impedance is the resistive value of the load and source terminations of the filter. Filters are usually designed for specific driving source and load impedances that must be present for proper operation. Impedance

6. **Ripple.** Amplitude variation with frequency in the passband, or the repetitive rise and fall of the signal level in the passband of some types of filters, is known as ripple. It is usually stated in decibels. There may also be ripple in the stop bandwidth in some types of filters. Ripple

Figure 2-30 Shape factor.

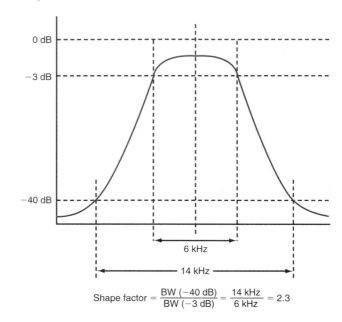

$$\text{Shape factor} = \frac{\text{BW }(-40\text{ dB})}{\text{BW }(-3\text{ dB})} = \frac{14\text{ kHz}}{6\text{ kHz}} = 2.3$$

Shape factor

7. *Shape factor.* Shape factor, also known as bandwidth ratio, is the ratio of the stop bandwidth to the pass bandwidth of a bandpass filter. It compares the bandwidth at minimum attenuation, usually at the -3-dB points or cutoff frequencies, to that of maximum attenuation and thus gives a relative indication of attenuation rate or selectivity. The smaller the ratio, the greater the selectivity. The ideal is a ratio of 1, which in general cannot be obtained with practical filters. The filter in Fig. 2-30 has a bandwidth of 6 kHz at the -3-dB attenuation point and a bandwidth of 14 kHz at the -40-dB attenuation point. The shape factor then is 14 kHz/6 kHz = 2.333. The points of comparison vary with different filters and manufacturers. The points of comparison may be at the 6-dB down and 60-dB down points or at any other designated two levels.

Pole

8. *Pole.* A pole is a frequency at which there is a high impedance in the circuit. It is also used to describe one RC section of a filter. A simple low-pass RC filter such as that in Fig. 2-24(a) has one pole. The two-section filter in Fig. 2-25 has two poles. For LC low- and high-pass filters, the number of poles is equal to the number of reactive components in the filter. For bandpass and band-reject filters, the number of poles is generally assumed to be one-half the number of reactive components used.

Zero

9. *Zero.* This term refers to a frequency at which there is zero impedance in the circuit.

Envelope delay

10. *Envelope delay.* Also known as time delay, envelope delay is the time it takes for a specific point on an input waveform to pass through the filter.

Roll-off

11. *Roll-off.* Also called the attenuation rate, roll-off is the rate of change of amplitude with frequency in a filter. The faster the roll-off, or the higher the attenuation rate, the more selective the filter is, i.e., the better able it is to differentiate between two closely spaced signals, one desired and the other not.

Any of the four basic filter types can be easily implemented with inductors and capacitors. Such filters can be made for frequencies up to about several hundred megahertz before the component values get too small to be practical. At frequencies above this frequency, special filters made with microstrip techniques on printed-circuit boards, surface acoustic wave filters, and cavity resonators are common. Because two types of reactances are used, inductive combined with capacitive, the roll-off rate of attenuation

Figure 2-31 Low-pass filter configurations and response. (*a*) L section. (*b*) T section. (*c*) π section. (*d*) Response curve.

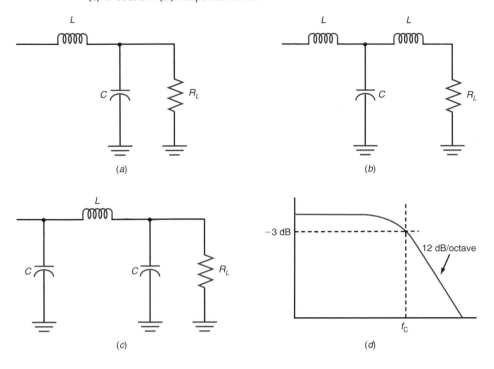

(a)

(b)

(c)

(d)

Figure 2-32 Low-pass Butterworth filter attenuation curves beyond the cutoff frequency f_c.

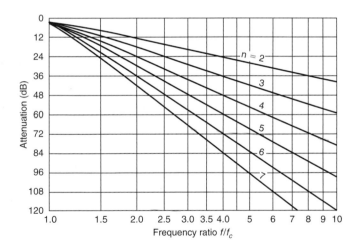

is greater with *LC* filters than with *RC* filters. The inductors make such filters larger and more expensive, but the need for better selectivity makes them necessary.

Low- and High–Pass *LC* Filters. Figure 2-31 shows the basic low-pass filter configurations. The basic two-pole circuit in Fig. 2-31 (*a*) provides an attenuation rate of 12 dB per octave or 20 dB per decade. These sections may be cascaded to provide an even greater roll-off rate. The chart in Fig. 2-32 shows the attenuation rates for low-pass filters with two through seven poles. The horizontal axis f/f_c is the ratio of any given frequency in ratio to the filter cutoff frequency f_c. The value *n* is the number of poles in the filter. Assume a cutoff frequency of 20 MHz. The ratio for a frequency of 40 MHz would be 40/20 = 2. This represents a doubling of the frequency,

Figure 2-33 High-pass filters. (*a*) L section. (*b*) T section (*b*) π section.

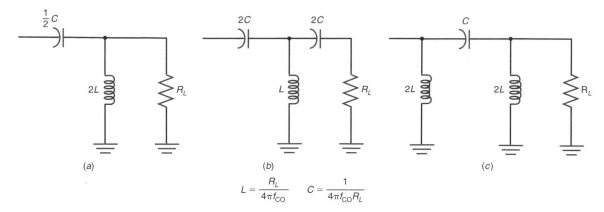

$$L = \frac{R_L}{4\pi f_{CO}} \qquad C = \frac{1}{4\pi f_{CO} R_L}$$

or one octave. The attenuation on the curve with two poles is 12 dB. The π and T filters in Fig. 2-31 (*b*) and (*c*) with three poles give an attenuation rate of 18 dB for a 2:1 frequency ratio. Figure 2-33 shows the basic high-pass filter configurations. A curve similar to that in Fig. 2-32 is also used to determine attenuation for filters with multiple poles. Cascading these sections provides a greater attenuation rate. Those filter configurations using the least number of inductors are preferred for lower cost and less space.

Types of Filters

The major types of *LC* filters in use are named after the person who discovered and developed the analysis and design method for each filter. The most widely used filters are Butterworth, Chebyshev, Cauer (elliptical), and Bessel. Each can be implemented by using the basic low- and high-pass configurations shown previously. The different response curves are achieved by selecting the component values during the design.

Butterworth filter

Butterworth. The *Butterworth* filter effect has maximum flatness in response in the passband and a uniform attenuation with frequency. The attenuation rate just outside the passband is not as great as can be achieved with other types of filters. See Fig. 2-34 for an example of a low-pass Butterworth filter.

Chebyshev filter

Chebyshev. *Chebyshev* (or Tchebyschev) filters have extremely good selectivity; i.e., their attenuation rate or roll-off is high, much higher than that of the Butterworth filter (see Fig. 2-34). The attenuation just outside the passband is also very high—again, better than that of the Butterworth. The main problem with the Chebyshev filter is that it has ripple in the passband, as is evident from the figure. The response is not flat or constant, as it is with the Butterworth filter. This may be a disadvantage in some applications.

Cauer (elliptical) filter

Cauer (Elliptical). *Cauer* filters produce an even greater attenuation or roll-off rate than do Chebyshev filters and greater attenuation out of the passband. However, they do this with an even higher ripple in the passband as well as outside of the passband.

Bessel (Thomson) filter

Bessel. Also called *Thomson filters, Bessel* circuits provide the desired frequency response (i.e., low-pass, bandpass, etc.) but have a constant time delay in the passband.

Figure 2-34 Butterworth, elliptical, Bessel, and Chebyshev response curves.

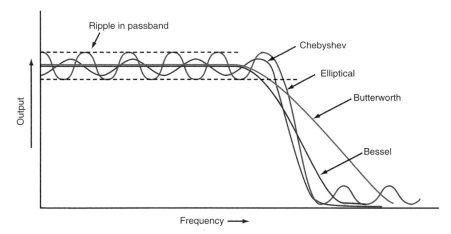

Bessel filters have what is known as a *flat group delay:* as the signal frequency varies in the passband, the phase shift or time delay it introduces is constant. In some applications, constant group delay is necessary to prevent distortion of the signals in the passband due to varying phase shifts with frequency. Filters that must pass pulses or wideband modulation are examples. To achieve this desired response, the Bessel filter has lower attenuation just outside the passband.

Mechanical Filters. An older but still useful filter is the mechanical filter. This type of filter uses resonant vibrations of mechanical disks to provide the selectivity. The signal to be filtered is applied to a coil that interacts with a permanent magnet to produce vibrations in the rod connected to a sequence of seven or eight disks whose dimensions determine the center frequency of the filter. The disks vibrate only near their resonant frequency, producing movement in another rod connected to an output coil. This coil works with another permanent magnet to generate an electrical output. Mechanical filters are designed to work in the 200- to 500-kHz range and have very high Qs. Their performance is comparable to that of crystal filters.

Whatever the type, passive filters are usually designed and built with discrete components although they may also be put into integrated-circuit form. A number of filter design software packages are available to simplify and speed up the design process. The design of *LC* filters is specialized and complex and beyond the scope of this text. However, filters can be purchased as components. These filters are predesigned and packaged in small sealed housings with only input, output, and ground terminals and can be used just as integrated circuits are. A wide range of frequencies, response characteristics, and attenuation rates can be obtained.

Bandpass Filters. A *bandpass filter* is one that allows a narrow range of frequencies around a center frequency f_c to pass with minimum attenuation but rejects frequencies above and below this range. The ideal response curve of a bandpass filter is shown in Fig. 2-35(*a*). It has both upper and lower cutoff frequencies f_2 and f_1, as indicated. The bandwidth of this filter is the difference between the upper and lower cutoff frequencies, or BW $= f_2 - f_1$. Frequencies above and below the cutoff frequencies are eliminated.

The ideal response curve is not obtainable with practical circuits, but close approximations can be obtained. A practical bandpass filter response curve is shown in Fig. 2-35(*b*). The simple series and parallel resonant circuits described in the previous section have a response curve like that in the figure and make good bandpass filters. The cutoff frequencies are those at which the output voltage is down 0.707 percent from the peak output value. These are the 3-dB attenuation points.

Bandpass filter

Figure 2-35 Response curves of a bandpass filter. (*a*) Ideal. (*b*) Practical.

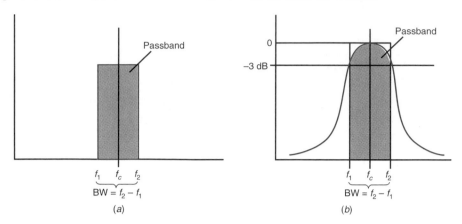

Two types of bandpass filters are shown in Fig. 2-36. In Fig. 2-36(*a*), a series resonant circuit is connected in series with an output resistor, forming a voltage divider. At frequencies above and below the resonant frequency, either the inductive or the capacitive reactance will be high compared to the output resistance. Therefore, the output amplitude will be very low. However, at the resonant frequency, the inductive and capacitive reactances cancel, leaving only the small resistance of the inductor. Thus most of the input voltage appears across the larger output resistance. The response curve for this circuit is shown in Fig. 2-35(*b*). Remember that the bandwidth of such a circuit is a function of the resonant frequency and Q: BW = f_c/Q.

A parallel resonant bandpass filter is shown in Fig. 2-36(*b*). Again, a voltage divider is formed with resistor R and the tuned circuit. This time the output is taken from across the parallel resonant circuit. At frequencies above and below the center resonant frequency, the impedance of the parallel tuned circuit is low compared to that of the resistance. Therefore, the output voltage is very low. Frequencies above and below the center frequency are greatly attenuated. At the resonant frequency, the reactances are equal and the impedance of the parallel tuned circuit is very high compared to that of the resistance. Therefore, most of the input voltage appears across the tuned circuit. The response curve is similar to that shown in Fig. 2-35(*b*).

Improved selectivity with steeper "skirts" on the curve can be obtained by cascading several bandpass sections. Several ways to do this are shown in Fig. 2-37. As sections are cascaded, the bandwidth becomes narrower and the response curve becomes steeper. An example is shown in Fig. 2-38. As indicated earlier, using multiple filter sections greatly improves the selectivity but increases the passband attenuation (insertion loss), which must be offset by added gain.

Figure 2-36 Simple bandpass filters.

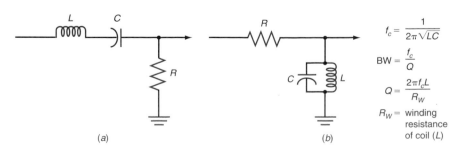

Figure 2-37 Some common bandpass filter circuits.

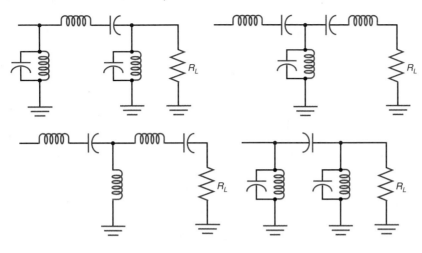

Figure 2-38 How cascading filter sections narrow the bandwidth and improve selectivity.

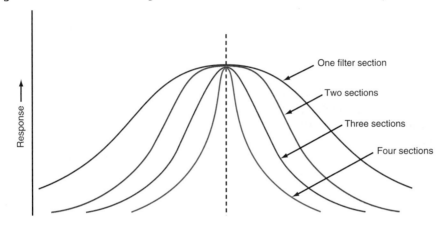

Figure 2-39 *LC* tuned bandstop filters. (*a*) Shunt. (*b*) Series. (*c*) Response curve.

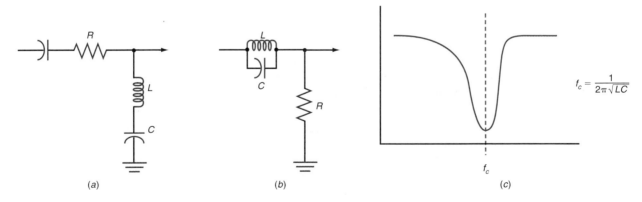

$$f_c = \frac{1}{2\pi\sqrt{LC}}$$

Figure 2-40 Bridge-T notch filter.

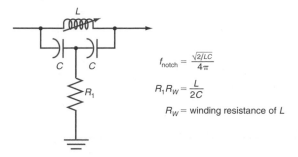

$$f_{notch} = \frac{\sqrt{2/LC}}{4\pi}$$

$$R_1 R_W = \frac{L}{2C}$$

R_W = winding resistance of L

Band–Reject Filters. *Band-reject filters,* also known as *bandstop filters,* reject a narrow band of frequencies around a center or notch frequency. Two typical *LC* bandstop filters are shown in Fig. 2-39. In Fig. 2-39(*a*), the series *LC* resonant circuit forms a voltage divider with input resistor *R*. At frequencies above and below the center rejection or notch frequency, the *LC* circuit impedance is high compared to that of the resistance. Therefore, signals at frequencies above and below center frequency are passed with minimum attenuation. At the center frequency, the tuned circuit resonates, leaving only the small resistance of the inductor. This forms a voltage divider with the input resistor. Since the impedance at resonance is very low compared to the resistor, the output signal is very low in amplitude. A typical response curve is shown in Fig. 2-39(*c*).

A parallel version of this circuit is shown in Fig. 2-39(*b*), where the parallel resonant circuit is connected in series with a resistor from which the output is taken. At frequencies above and below the resonant frequency, the impedance of the parallel circuit is very low; there is, therefore, little signal attenuation, and most of the input voltage appears across the output resistor. At the resonant frequency, the parallel *LC* circuit has an extremely high resistive impedance compared to the output resistance, and so minimum voltage appears at the center frequency. *LC* filters used in this way are often referred to as *traps*.

Another bridge-type notch filter is the *bridge-T filter* shown in Fig. 2-40. This filter, which is widely used in RF circuits, uses inductors and capacitors and thus has a steeper response curve than the *RC* twin-T notch filter. Since *L* is variable, the notch is tunable.

Figure 2-41 shows common symbols used to represent *RC* and *LC* filters or any other type of filter in system block diagrams or schematics.

Active Filters

Active filters are frequency-selective circuits that incorporate *RC* networks and amplifiers with feedback to produce low-pass, high-pass, bandpass, and bandstop performance. These filters can replace standard passive *LC* filters in many applications. They offer the following advantages over standard passive *LC* filters.

Figure 2-41 Block diagram or schematic symbols for filters.

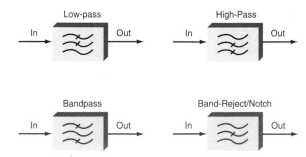

1. **Gain.** Because active filters use amplifiers, they can be designed to amplify as well as filter, thus offsetting any insertion loss.

2. **No inductors.** Inductors are usually larger, heavier, and more expensive than capacitors and have greater losses. Active filters use only resistors and capacitors.

3. **Easy to tune.** Because selected resistors can be made variable, the filter cutoff frequency, center frequency, gain, Q, and bandwidth are adjustable.

4. **Isolation.** The amplifiers provide very high isolation between cascaded circuits because of the amplifier circuitry, thereby decreasing interaction between filter sections.

5. **Easier impedance matching.** Impedance matching is not as critical as with LC filters.

Figure 2-42 shows two types of low-pass active filters and two types of high-pass active filters. Note that these active filters use op amps to provide the gain. The voltage divider, made up of R_1 and R_2, sets the circuit gain in the circuits of Fig. 2-42(a) and (c) as in any noninverting op amp. The gain is set by R_3 and/or R_1 in Fig. 2-42(b) and by C_3 and/or C_1 in Fig. 2-42(d). All circuits have what is called a *second-order response,* which means that they provide the same filtering action as a two-pole LC filter. The roll-off rate is 12 dB per octave, or 40 dB per decade. Multiple filters can be cascaded to provide faster roll-off rates.

Second-order response

Figure 2-42   Types of active filters. (a) Low-pass. (b) Low-pass. (c) High-pass. (d) High-pass.

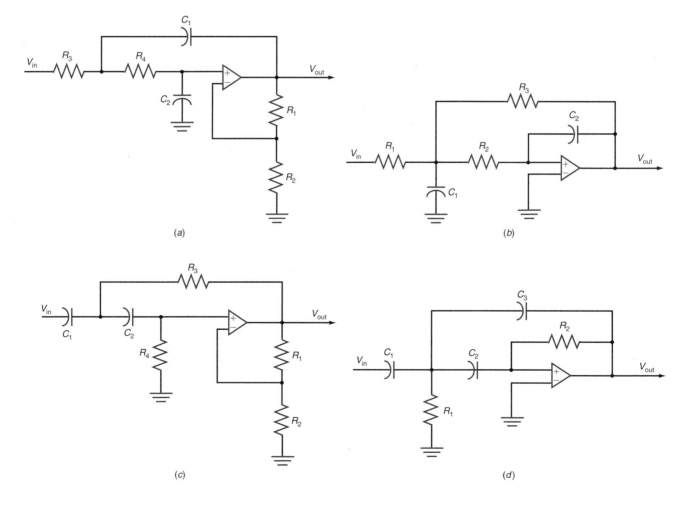

Two active bandpass filters and a notch filter are shown in Fig. 2-43. In Fig. 2-43(a), both *RC* low-pass and high-pass sections are combined with feedback to give a bandpass result. In Fig. 2-43(b), a twin-T *RC* notch filter is used with negative feedback to provide a bandpass result. A notch filter using a twin-T is illustrated in Fig. 2-43(c). The feedback makes the response sharper than that with a standard passive twin-T.

A special form of active filter is the variable-state filter, which can simultaneously provide low-pass, high-pass, and bandpass operation from one circuit. The basic circuit is shown in Fig. 2-44(a). It uses op amps and *RC* networks and a feedback arrangement. Op amps 2 and 3 are connected as integrators or low-pass filters. Op amp 1 is connected as a summing amplifier that adds the input signal to the feedback signals from op amps 2 and 3. Note the outputs at each op amp. The center and cutoff frequencies are set by the integrator feedback capacitors and the value of R_f. And R_q and R_g set the Q and gain of the circuit. The circuit can be made tunable by simultaneously varying the values of R_f.

A variation of the variable-state filter is the biquad filter shown in Fig. 2-44(b). It, too, uses two op amp integrators and a summing amplifier. Again, low-pass, high-pass, and bandpass characteristics are obtained simultaneously. However, the primary use of the biquad filter is bandpass filtering. Again, the center or cutoff frequencies are set by the value of the integrator feedback capacitors and the value of R_f. And R_b sets the filter bandwidth, and R_g sets the circuit gain.

Active filters are made with integrated-circuit (IC) op amps and discrete *RC* networks. They can be designed to have any of the responses discussed earlier, such as

Figure 2-43 Active bandpass and notch filters. (*a*) Bandpass. (*b*) Bandpass. (*c*) High-*Q* notch.

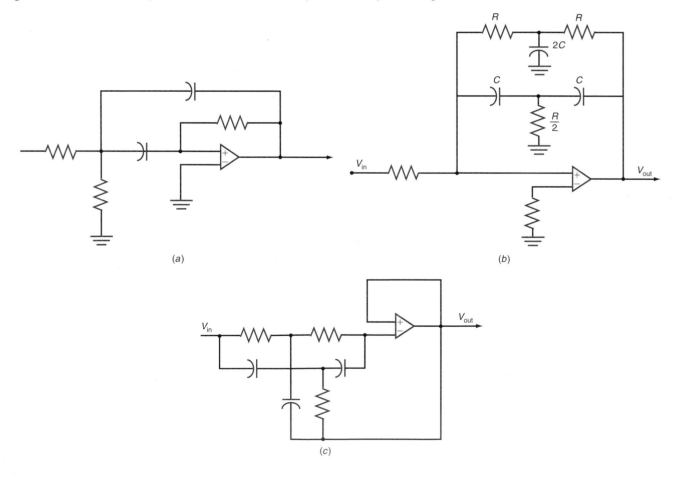

Figure 2-44 Multifunction active filters. (*a*) Variable-state filter. (*b*) Biquad filter.

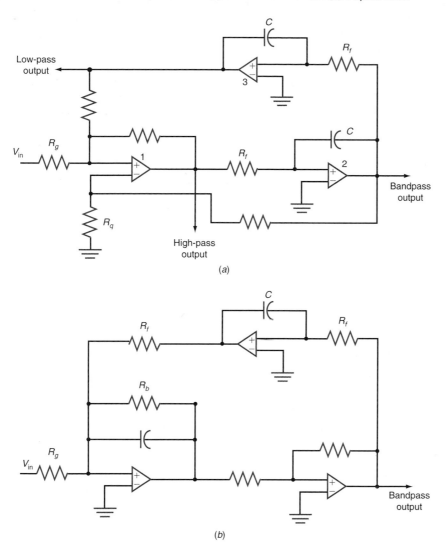

(a)

(b)

Butterworth and Chebyshev, and they are easily cascaded to provide even greater selectivity. Active filters are also available as complete packaged components. The primary disadvantage of active filters is that their upper frequency of operation is limited by the frequency response of the op amps and the practical sizes of resistors and capacitors. Most active filters are used at frequencies below 1 MHz, and most active circuits operate in the audio range and slightly above. However, today op amps with frequency ranges up to one microwave (>1 GHz) mated with chip resistors and capacitors have made RC active filters practical for applications up to the RF range.

Crystal and Ceramic Filters

The selectivity of a filter is limited primarily by the Q of the circuits, which is generally the Q of the inductors used. With LC circuits, it is difficult to achieve Q values over 200. In fact, most LC circuit Qs are in the range of 10 to 100, and as a result, the roll-off rate is limited. In some applications, however, it is necessary to select one desired signal, distinguishing it from a nearby undesired signal (see Fig. 2-45). A conventional filter has a slow roll-off rate, and the undesired signal is

Figure 2-45 How selectivity affects the ability to discriminate between signals.

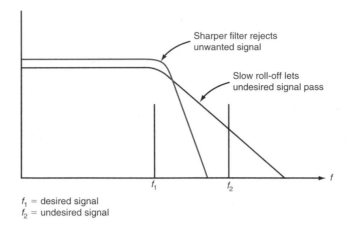

f_1 = desired signal
f_2 = undesired signal

not, therefore, fully attenuated. The way to gain greater selectivity and higher Q, so that the undesirable signal will be almost completely rejected, is to use filters that are made of thin slivers of quartz crystal or certain types of ceramic materials. These materials exhibit what is called *piezoelectricity*. When they are physically bent or otherwise distorted, they develop a voltage across the faces of the crystal. Alternatively, if an ac voltage is applied across the crystal or ceramic, the material vibrates at a very precise frequency, a frequency that is determined by the thickness, shape, and size of the crystal as well as the angle of cut of the crystal faces. In general, the thinner the crystal or ceramic element, the higher the frequency of oscillation.

Crystals and ceramic elements are widely used in oscillators to set the frequency of operation to some precise value, which is held despite temperature and voltage variations that may occur in the circuit.

Crystals and ceramic elements can also be used as circuit elements to form filters, specifically bandpass filters. The equivalent circuit of a crystal or ceramic device is a tuned circuit with a Q of 10,000 to 1,000,000, permitting highly selective filters to be built.

Crystal Filters. *Crystal filters* are made from the same type of quartz crystals normally used in crystal oscillators. When a voltage is applied across a crystal, it vibrates at a specific resonant frequency, which is a function of the size, thickness, and direction of cut of the crystal. Crystals can be cut and ground for almost any frequency in the 100-kHz to 100-MHz range. The frequency of vibration of crystal is extremely stable, and crystals are therefore widely used to supply signals on exact frequencies with good stability.

The equivalent circuit and schematic symbol of a quartz crystal are shown in Fig. 2-46. The crystal acts as a resonant LC circuit. The series LCR part of the equivalent circuit represents the crystal itself, whereas the parallel capacitance C_P is the capacitance of the metal mounting plates with the crystal as the dielectric.

Figure 2-47 shows the impedance variations of the crystal as a function of frequency. At frequencies below the crystal's resonant frequency, the circuit appears capacitive and has a high impedance. However, at some frequency, the reactances of the equivalent inductance L and the series capacitance C_S are equal, and the circuit resonates. The series circuit is resonant when $X_L = X_{C_S}$. At this series resonant frequency f_S, the circuit is resistive. The resistance of the crystal is extremely low, giving the circuit an extremely high Q. Values of Q in the 10,000 to 1,000,000 range are common. This makes the crystal a highly selective series resonant circuit.

Piezoelectricity

Crystal filter

Chapter 2

Figure 2-46 Quartz crystal. (a) Equivalent circuit. (b) Schematic symbol.

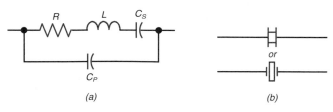

(a)   (b)

Figure 2-47 Impedance variation with frequency of a quartz crystal.

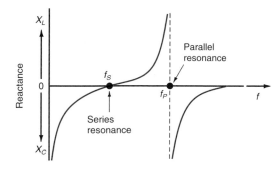

If the frequency of the signal applied to the crystal is above f_S, the crystal appears inductive. At some higher frequency, the reactance of the parallel capacitance C_P equals the reactance of the net inductance. When this occurs, a parallel resonant circuit is formed. At this parallel resonant frequency f_P, the impedance of the circuit is resistive but extremely high.

Because the crystal has both series and parallel resonant frequencies which are close together, it makes an ideal component for use in filters. By combining crystals with selected series and parallel resonant points, highly selective filters with any desired bandpass can be constructed.

The most commonly used crystal filter is the full crystal lattice shown in Fig. 2-48. It is a bandpass filter. Note that transformers are used to provide the input to the filter and to extract the output. Crystals Y_1 and Y_2 resonate at one frequency, and crystals Y_3 and Y_4 resonate at another frequency. The difference between the two crystal frequencies determines the bandwidth of the filter. The 3-dB down bandwidth is approximately 1.5 times the crystal frequency spacing. For example, if the Y_1 to Y_2 frequency is 9 MHz and the Y_3 to Y_4 frequency is 9.002 MHz, the difference is $9.002 - 9.000 = 0.002$ MHz $= 2$ kHz. The 3-dB bandwidth is, then, 1.5×2 kHz $= 3$ kHz.

The crystals are also chosen so that the parallel resonant frequency of Y_3 to Y_4 equals the series resonant frequency of Y_1 to Y_2. The series resonant frequency of Y_3 to Y_4 is equal to the parallel resonant frequency of Y_1 to Y_2. The result is a passband with

Figure 2-48 Crystal lattice filter.

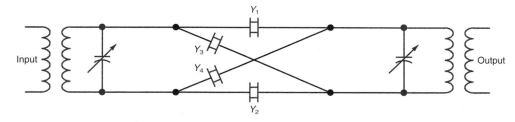

Figure 2-49 Crystal ladder filter.

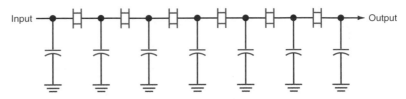

extremely steep attenuation. Signals outside the passband are rejected as much as 50 to 60 dB below those inside the passband. Such a filter can easily discriminate between very closely spaced desired and undesired signals.

Another type of crystal filter is the ladder filter shown in Fig. 2-49, which is also a bandpass filter. All the crystals in this filter are cut for exactly the same frequency. The number of crystals used and the values of the shunt capacitors set the bandwidth. At least six crystals must usually be cascaded to achieve the kind of selectivity needed in communication applications.

Ceramic Filters. Ceramic is a manufactured crystallike compound that has the same piezoelectric qualities as quartz. Ceramic disks can be made so that they vibrate at a fixed frequency, thereby providing filtering actions. *Ceramic filters* are very small and inexpensive and are, therefore, widely used in transmitters and receivers. Although the Q of ceramic does not have as high an upper limit as that of quartz, it is typically several thousand, which is very high compared to the Q obtainable with LC filters. Typical ceramic filters are of the bandpass type with center frequencies of 455 kHz and 10.7 MHz. These are available in different bandwidths depending upon the application. Such ceramic filters are widely used in communication receivers.

A schematic diagram of a ceramic filter is shown in Fig. 2-50. For proper operation, the filter must be driven from a generator with an output impedance of R_g and be terminated with a load of R_L. The values of R_g and R_L are usually 1.5 or 2 kΩ.

Surface Acoustic Wave Filters. A special form of a crystal filter is the *surface acoustic wave (SAW) filter*. This fixed tuned bandpass filter is designed to provide the exact selectivity required by a given application. Figure 2-51 shows the schematic design

Figure 2-50 Schematic symbol for a ceramic filter.

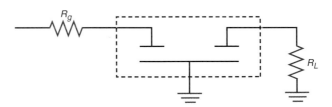

Figure 2-51 A surface acoustic wave filter.

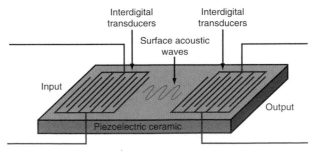

of a SAW filter. SAW filters are made on a piezoelectric ceramic substrate such as lithium niobate. A pattern of interdigital fingers on the surface converts the signals into acoustic waves that travel across the filter surface. By controlling the shapes, sizes, and spacings of the interdigital fingers, the response can be tailored to any application. Interdigital fingers at the output convert the acoustic waves back to electrical signals.

SAW filters are normally bandpass filters used at very high radio frequencies where selectivity is difficult to obtain. Their common useful range is from 10 MHz to 3 GHz. They have a low shape factor, giving them exceedingly good selectivity at such high frequencies. They do have a significant insertion loss, usually in the 10- to 35-dB range, which must be overcome with an accompanying amplifier. SAW filters are widely used in modern TV receivers, radar receivers, wireless LANs, and cell phones.

Switched Capacitor Filters

Switched capacitor filters (SCFs) are active IC filters made of op amps, capacitors, and transistor switches. Also known as *analog sampled data filters* or *commutating filters,* these devices are usually implemented with MOS or CMOS circuits. They can be designed to operate as high-pass, low-pass, bandpass, or bandstop filters. The primary advantage of SCFs is that they provide a way to make tuned or selective circuits in an IC without the use of discrete inductors, capacitors, or resistors.

Switched capacitor filters are made of op amps, MOSFET switches, and capacitors. All components are fully integrated on a single chip, making external discrete components unnecessary. The secret to the SCF is that all resistors are replaced by capacitors that are switched by MOSFET switches. Resistors are more difficult to make in IC form and take up far more space on the chip than transistors and capacitors. With switched capacitors, it is possible to make complex active filters on a single chip. Other advantages are selectibility of filter type, full adjustability of the cutoff or center frequency, and full adjustability of bandwidth. One filter circuit can be used for many different applications and can be set to a wide range of frequencies and bandwidths.

Switched Integrators. The basic building block of SCFs is the classic op amp integrator, as shown in Fig. 2-52(a). The input is applied through a resistor and the feedback is provided by a capacitor. With this arrangement, the output is a function of the integral of the input:

$$V_{out} = -\frac{1}{RC} \int V_{in} \, dt$$

With ac signals, the circuit essentially functions as a low-pass filter with a gain of $1/RC$.

Switched capacitor filter (SCF) or analog sampled data or commutating filter

Switched integrator

Figure 2-52 IC integrators. (a) Conventional integrator. (b) Switched capacitor integrator.

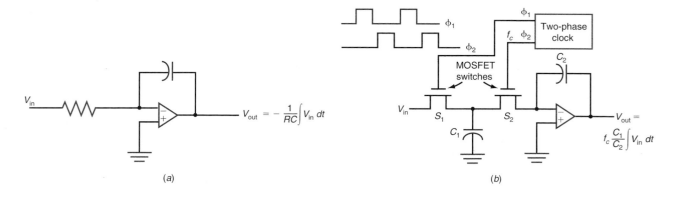

(a)

(b)

To work over a wide range of frequencies, the integrator RC values must be changed. Making low and high resistor and capacitor values in IC form is difficult. However, this problem can be solved by replacing the input resistor with a switched capacitor, as shown in Fig. 2-52(b). The MOSFET switches are driven by a clock generator whose frequency is typically 50 to 100 times the maximum frequency of the ac signal to be filtered. The resistance of a MOSFET switch when on is usually less than 1000 Ω. When the switch is off, its resistance is many megohms.

The clock puts out two phases, designated ϕ_1 and ϕ_2, that drive the MOSFET switches. When S_1 is on, S_2 is off and vice versa. The switches are of the break-before-make type, which means that one switch opens before the other is closed. When S_1 is closed, the charge on the capacitor follows the input signal. Since the clock period and time duration that the switch is on are very short compared to the input signal variation, a brief "sample" of the input voltage remains stored on C_1 and S_1 turns off.

Now S_2 turns on. The charge on capacitor C_1 is applied to the summing junction of the op amp. It discharges, causing a current to flow in the feedback capacitor C_2. The resulting output voltage is proportional to the integral of the input. But this time, the gain of the integrator is

$$f\left(\frac{C_1}{C_2}\right)$$

where f is the clock frequency. Capacitor C_1, which is switched at a clock frequency of f with period T, is equivalent to a resistor value of $R = T/C_1$.

The beauty of this arrangement is that it is not necessary to make resistors on the IC chip. Instead, capacitors and MOSFET switches, which are smaller than resistors, are used. Further, since the gain is a function of the ratio of C_1 to C_2, the exact capacitor values are less important than their ratio. It is much easier to control the ratio of matched pairs of capacitors than it is to make precise values of capacitance.

By combining several such switching integrators, it is possible to create low-pass, high-pass, bandpass, and band-reject filters of the Butterworth, Chebyshev, elliptical, and Bessel type with almost any desired selectivity. The center frequency or cutoff frequency of the filter is set by the value of the clock frequency. This means that the filter can be tuned on the fly by varying the clock frequency.

A unique but sometimes undesirable characteristic of an SCF is that the output signal is really a stepped approximation of the input signal. Because of the switching action of the MOSFETs and the charging and discharging of the capacitors, the signal takes on a stepped digital form. The higher the clock frequency compared to the frequency of the input signal, the smaller this effect. The signal can be smoothed back into its original state by passing it through a simple RC low-pass filter whose cutoff frequency is set to just above the maximum signal frequency.

A variety of SCFs are available in IC form. Both dedicated single-purpose or universal SCFs can be purchased for less than $2 in bulk. One of the most popular is the MF10 made by National Semiconductor. It is a universal SCF that can be set for low-pass, high-pass, bandpass, or band-reject operation. It can be used for center or cutoff frequencies up to about 20 kHz. The clock frequency is about 50 to 100 times the operating frequency.

Commutating filter

Commutating Filters. An interesting variation of a switched capacitor filter is the *commutating filter* shown in Fig. 2-53. It is made of discrete resistors and capacitors with MOSFET switches driven by a counter and decoder. The circuit appears to be a low-pass RC filter, but the switching action makes the circuit function as a bandpass filter. The operating frequency f_{out} is related to the clock frequency f_c and the number N of switches and capacitors used.

$$f_c = N f_{\text{out}} \qquad \text{and} \qquad f_{\text{out}} = \frac{f_c}{N}$$

Figure 2-53 A commutating SCF.

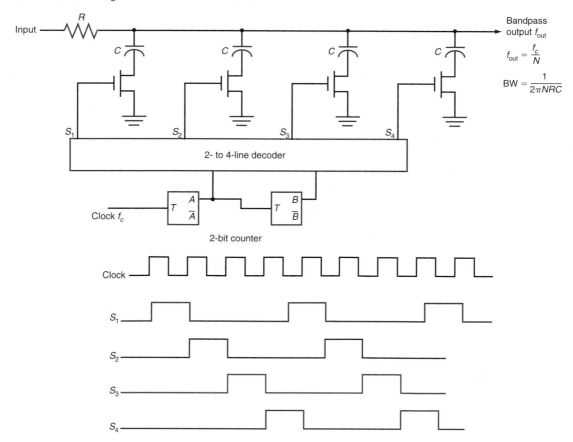

The bandwidth of the circuit is related to the RC values and number of capacitors and switches used as follows:

$$BW = \frac{1}{2\pi NRC}$$

For the filter in Fig. 2-53, the bandwidth is $BW = 1/(8\pi RC)$.

Very high Q and narrow bandwidth can be obtained, and varying the resistor value makes the bandwidth adjustable.

The operating waveforms in Fig. 2-53 show that each capacitor is switched on and off sequentially so that only one capacitor is connected to the circuit at a time. A sample of the input voltage is stored as a charge on each capacitor as it is connected to the input. The capacitor voltage is the average of the voltage variation during the time the switch connects the capacitor to the circuit.

Figure 2-54(a) shows typical input and output waveforms, assuming a sine wave input. The output is a stepped approximation of the input because of the sampling action of the switched capacitors. The steps are large, but their size can be reduced by simply using a greater number of switches and capacitors. Increasing the number of capacitors from four to eight, as in Fig. 2-54(b), makes the steps smaller, and thus the output more closely approximates the input. The steps can be eliminated or greatly minimized by passing the output through a simple RC low-pass filter whose cutoff is set to the center frequency value or slightly higher.

One characteristic of the commutating filter is that it is sensitive to the harmonics of the center frequency for which it is designed. Signals whose frequency is some integer multiple of the center frequency of the filter are also passed by the filter, although

Figure 2-54 Input and output for commutating filter. (*a*) Four-capacitor filter. (*b*) Eight-capacitor filter.

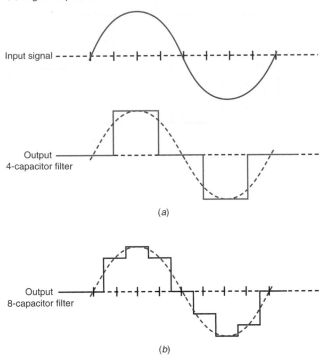

Input signal

Output
4-capacitor filter

(*a*)

Output
8-capacitor filter

(*b*)

Figure 2-55 Comb filter response of a commutating filter.

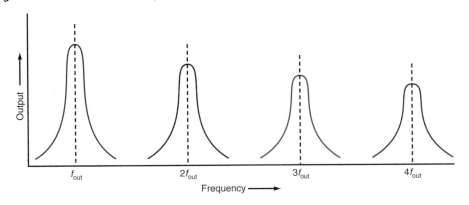

Output

f_{out} $2f_{out}$ $3f_{out}$ $4f_{out}$

Frequency

Comb response

at a somewhat lower amplitude. The response of the filter, called a *comb response,* is shown in Fig. 2-55. If such performance is undesirable, the higher frequencies can be eliminated with a conventional *RC* or *LC* low-pass filter connected to the output.

2-4 Fourier Theory

The mathematical analysis of the modulation and multiplexing methods used in communication systems assumes sine wave carriers and information signals. This simplifies the analysis and makes operation predictable. However, in the real world, not all information signals are sinusoidal. Information signals are typically more complex voice and video signals that are essentially composites of sine waves of many frequencies and amplitudes. Information signals can take on an infinite number of shapes, including rectangular waves (i.e., digital pulses), triangular waves, sawtooth waves, and other non-sinusoidal forms. Such signals require that a non–sine wave approach be taken to determine the characteristics and performance of any communication circuit or system. One of the methods used to do this is *Fourier analysis,* which provides a means of accurately analyzing the content of most complex nonsinusoidal signals. Although Fourier analysis requires the use of calculus and advanced mathematical techniques beyond the scope of this text, its practical applications to communication electronics are relatively straightforward.

Fourier analysis

Basic Concepts

Figure 2-56(*a*) shows a basic sine wave with its most important dimensions and the equation expressing it. A basic cosine wave is illustrated in Fig. 2-56(*b*). Note that the cosine wave has the same shape as a sine wave but leads the sine wave by 90°. A *harmonic* is a sine wave whose frequency is some integer multiple of a fundamental sine wave. For example, the third harmonic of a 2-kHz sine wave is a sine wave of 6 kHz. Figure 2-57 shows the first four harmonics of a fundamental sine wave.

Harmonic

What the Fourier theory tells us is that we can take a nonsinusoidal waveform and break it down into individual harmonically related sine wave or cosine wave components. The classic example of this is a *square wave,* which is a rectangular signal with equal-duration positive and negative alternations. In the ac square wave in Fig. 2-58, this means that t_1 is equal to t_2. Another way of saying this is that the square wave has a 50 percent *duty cycle D,* the ratio of the duration of the positive alteration t_1 to the period T expressed as a percentage:

Square wave

Duty cycle *D*

$$D = \frac{t_1}{T} \times 100$$

Fourier analysis tells us that a square wave is made up of a sine wave at the fundamental frequency of the square wave plus an infinite number of odd harmonics. For example, if the fundamental frequency of the square wave is 1 kHz, the square wave can be synthesized by adding the 1-kHz sine wave and harmonic sine waves of 3 kHz, 5 kHz, 7 kHz, 9 kHz, etc.

Figure 2-59 shows how this is done. The sine waves must be of the correct amplitude and phase relationship to one another. The fundamental sine wave in this case has a value of 20 V peak to peak (a 10-V peak). When the sine wave values are added instantaneously,

Figure 2-56 Sine and cosine waves.

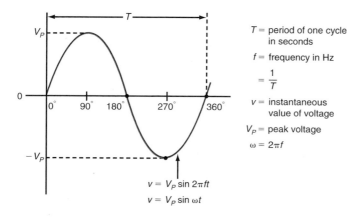

T = period of one cycle in seconds

f = frequency in Hz

$= \dfrac{1}{T}$

v = instantaneous value of voltage

V_P = peak voltage

$\omega = 2\pi f$

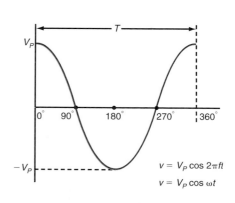

The Fundamentals of Electronics: A Review

Figure 2-57 A sine wave and its harmonics.

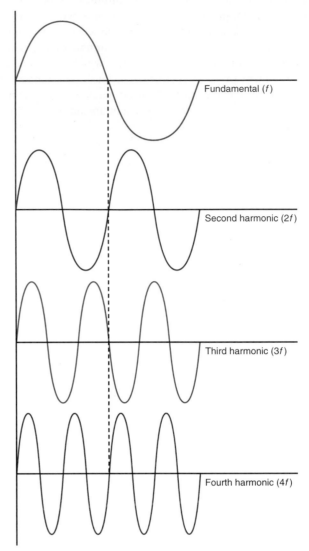

Fundamental (*f*)

Second harmonic (2*f*)

Third harmonic (3*f*)

Fourth harmonic (4*f*)

the result approaches a square wave. In Fig. 2-59(*a*), the fundamental and third harmonic are added. Note the shape of the composite wave with the third and fifth harmonics added, as in Fig. 2-59(*b*). The more higher harmonics that are added, the more the composite wave looks like a perfect square wave. Figure 2-60 shows how the composite wave would look with 20 odd harmonics added to the fundamental. The results very closely approximate a square wave.

Figure 2-58 A square wave.

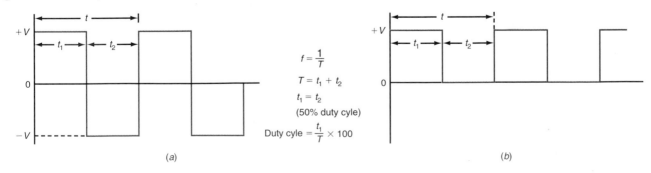

$$f = \frac{1}{T}$$

$$T = t_1 + t_2$$

$$t_1 = t_2$$

(50% duty cyle)

$$\text{Duty cyle} = \frac{t_1}{T} \times 100$$

(a)

(b)

Chapter 2

Figure 2-59   A square wave is made up of a fundamental sine wave and an infinite number of odd harmonics.

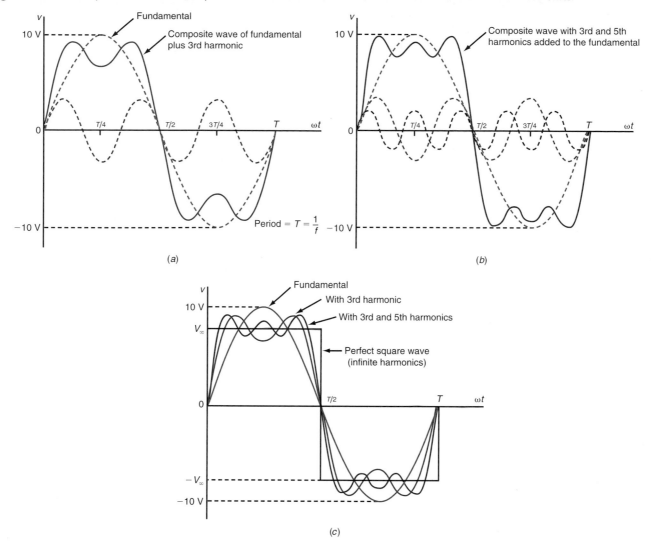

(a)

(b)

(c)

Figure 2-60   Square wave made up of 20 odd harmonics added to the fundamental.

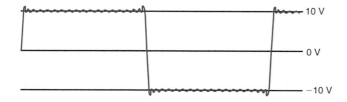

The implication of this is that a square wave should be analyzed as a collection of harmonically related sine waves rather than a single square wave entity. This is confirmed by performing a Fourier mathematical analysis on the square wave. The result is the following equation, which expresses voltage as a function of time:

$$f(t) = \frac{4V}{\pi}\left[\sin 2\pi\left(\frac{1}{T}\right)t + \frac{1}{3}\sin 2\pi\left(\frac{3}{T}\right)t + \frac{1}{5}\sin 2\pi\left(\frac{5}{T}\right)t + \frac{1}{7}\sin 2\pi\left(\frac{7}{T}\right)t + \cdots\right]$$

where the factor $4V/\pi$ is a multiplier for all sine terms and V is the square wave peak voltage. The first term is the fundamental sine wave, and the succeeding terms are the third, fifth, seventh, etc., harmonics. Note that the terms also have an amplitude factor. In this case, the amplitude is also a function of the harmonic. For example, the third harmonic has an amplitude that is one-third of the fundamental amplitude, and so on. The expression could be rewritten with $f = 1/T$. If the square wave is direct current rather than alternating current, as shown in Fig. 2-58(b), the Fourier expression has a dc component:

$$f(t) = \frac{V}{2} + \frac{4V}{\pi} \left(\sin 2\pi ft + \frac{1}{3} \sin 2\pi 3ft + \frac{1}{5} \sin 2\pi 5ft + \frac{1}{7} \sin 2\pi 7ft + \cdots \right)$$

In this equation, $V/2$ is the dc component, the average value of the square wave. It is also the baseline upon which the fundamental and harmonic sine waves ride.

A general formula for the Fourier equation of a waveform is

$$f(t) = \frac{V}{2} + \frac{4V}{\pi n} \sum_{n=1}^{\infty} (\sin 2\pi nft)$$

where n is odd. The dc component, if one is present in the waveform, is $V/2$.

By using calculus and other mathematical techniques, the waveform is defined, analyzed, and expressed as a summation of sine and/or cosine terms, as illustrated by the expression for the square wave above. Figure 2-61 gives the Fourier expressions for some of the most common nonsinusoidal waveforms.

Example 2-26

An ac square wave has a peak voltage of 3 V and a frequency of 48 kHz. Find (a) the frequency of the fifth harmonic and (b) the rms value of the fifth harmonic. Use the formula in Fig. 2-61(a).

a. 5 × 48 kHz = 240 kHz

b. Isolate the expression for the fifth harmonic in the formula, which is $\frac{1}{5} \sin 2\pi (5/T)t$. Multiply by the amplitude factor $4V/\pi$. The peak value of the fifth harmonic V_P is

$$V_P = \frac{4V}{\pi} \left(\frac{1}{5} \right) = \frac{4(3)}{5\pi} = 0.76$$

rms = 0.707 × peak value
$V_{rms} = 0.707 V_P = 0.707(0.76) = 0.537$ V

The triangular wave in Fig. 2-61(b) exhibits the fundamental and odd harmonics, but it is made up of cosine waves rather than sine waves. The sawtooth wave in Fig. 2-61(c) contains the fundamental plus all odd and even harmonics. Figure 2-61(d) and (e) shows half sine pulses like those seen at the output of half and full wave rectifiers. Both have an average dc component, as would be expected. The half wave signal is made up of even harmonics only, whereas the full wave signal has both odd and even harmonics. Figure 2-61(f) shows the Fourier expression for a dc square wave where the average dc component is Vt_0/T.

Time Domain Versus Frequency Domain

Time domain

Most of the signals and waveforms that we discuss and analyze are expressed in the *time domain*. That is, they are variations of voltage, current, or power with respect to time.

Figure 2-61 Common nonsinusoidal waves and their Fourier equations. (a) Square wave. (b) Triangle wave. (c) Sawtooth. (d) Half cosine wave. (e) Full cosine wave. (f) Rectangular pulse.

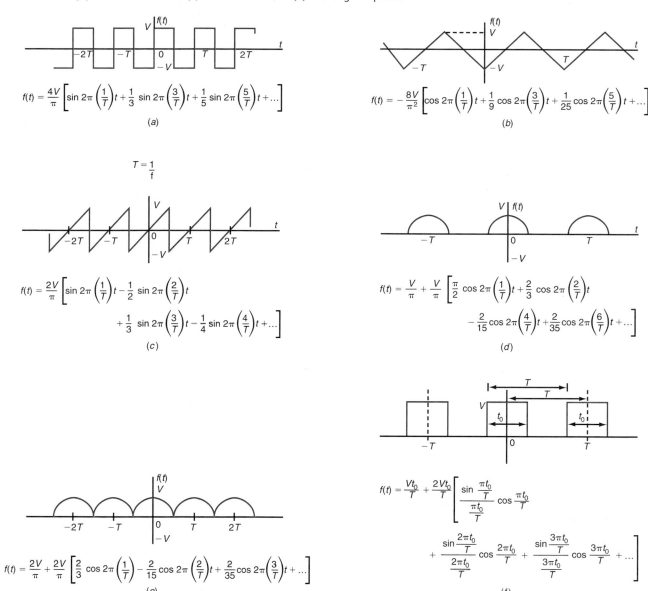

$$f(t) = \frac{4V}{\pi}\left[\sin 2\pi\left(\frac{1}{T}\right)t + \frac{1}{3}\sin 2\pi\left(\frac{3}{T}\right)t + \frac{1}{5}\sin 2\pi\left(\frac{5}{T}\right)t + \ldots\right]$$

(a)

$$f(t) = -\frac{8V}{\pi^2}\left[\cos 2\pi\left(\frac{1}{T}\right)t + \frac{1}{9}\cos 2\pi\left(\frac{3}{T}\right)t + \frac{1}{25}\cos 2\pi\left(\frac{5}{T}\right)t + \ldots\right]$$

(b)

$$T = \frac{1}{f}$$

$$f(t) = \frac{2V}{\pi}\left[\sin 2\pi\left(\frac{1}{T}\right)t - \frac{1}{2}\sin 2\pi\left(\frac{2}{T}\right)t + \frac{1}{3}\sin 2\pi\left(\frac{3}{T}\right)t - \frac{1}{4}\sin 2\pi\left(\frac{4}{T}\right)t + \ldots\right]$$

(c)

$$f(t) = \frac{V}{\pi} + \frac{V}{\pi}\left[\frac{\pi}{2}\cos 2\pi\left(\frac{1}{T}\right)t + \frac{2}{3}\cos 2\pi\left(\frac{2}{T}\right)t - \frac{2}{15}\cos 2\pi\left(\frac{4}{T}\right)t + \frac{2}{35}\cos 2\pi\left(\frac{6}{T}\right)t + \ldots\right]$$

(d)

$$f(t) = \frac{2V}{\pi} + \frac{2V}{\pi}\left[\frac{2}{3}\cos 2\pi\left(\frac{1}{T}\right) - \frac{2}{15}\cos 2\pi\left(\frac{2}{T}\right)t + \frac{2}{35}\cos 2\pi\left(\frac{3}{T}\right)t + \ldots\right]$$

(e)

$$f(t) = \frac{Vt_0}{T} + \frac{2Vt_0}{T}\left[\frac{\sin\frac{\pi t_0}{T}}{\frac{\pi t_0}{T}}\cos\frac{\pi t_0}{T}\right.$$
$$\left. + \frac{\sin\frac{2\pi t_0}{T}}{\frac{2\pi t_0}{T}}\cos\frac{2\pi t_0}{T} + \frac{\sin\frac{3\pi t_0}{T}}{\frac{3\pi t_0}{T}}\cos\frac{3\pi t_0}{T} + \ldots\right]$$

(f)

All the signals shown in the previous illustrations are examples of time-domain waveforms. Their mathematical expressions contain the variable time t, indicating that they are a time-variant quantity.

Fourier theory gives us a new and different way to express and illustrate complex signals. Here, complex signals containing many sine and/or cosine components are expressed as sine or cosine wave amplitudes at different frequencies. In other words, a graph of a particular signal is a plot of sine and/or cosine component amplitudes with respect to frequency.

A typical frequency-domain plot of the square wave is shown in Fig. 2-62(a). Note that the straight lines represent the sine wave amplitudes of the fundamental and harmonics, and these are plotted on a horizontal frequency axis. Such a frequency-domain plot can be made directly from the Fourier expression by simply using the frequencies of the fundamentals and harmonics and their amplitudes. Frequency-domain plots for some of the other common nonsinusoidal waves are also shown in Fig. 2-62.

Figure 2-62 The frequency-domain plots of common nonsinusoidal waves. (*a*) Square wave. (*b*) Sawtooth. (*c*) Triangle. (*d*) Half cosine wave.

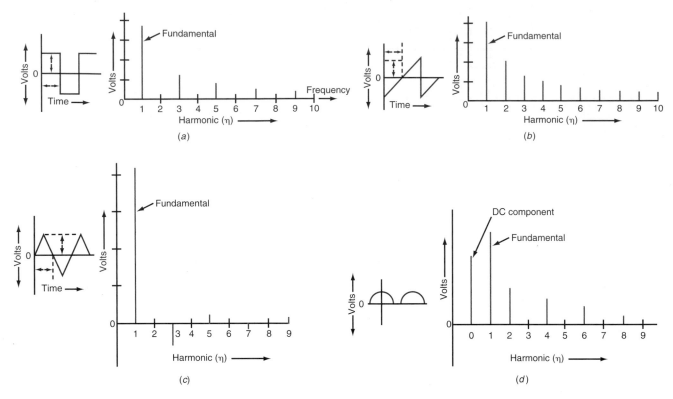

Note that the triangle wave in Fig. 2-62(*c*) is made up of the fundamental and odd harmonics. The third harmonic is shown as a line below the axis, which indicates a 180° phase shift in the cosine wave making it up.

Figure 2-63 shows how the time and frequency domains are related. The square wave discussed earlier is used as an example. The result is a three-axis three-dimensional view.

Figure 2-63 The relationship between time and frequency domains.

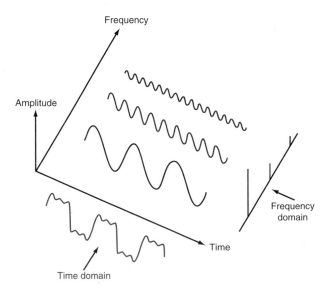

Figure 2-64 Converting a square wave to sine wave by filtering out all the harmonics.

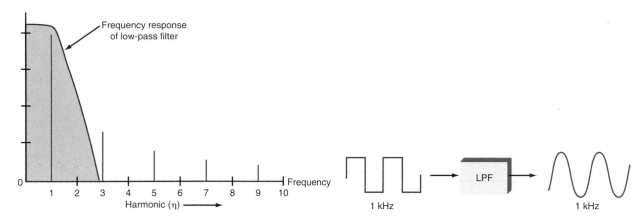

Signals and waveforms in communication applications are expressed by using both time-domain and frequency-domain plots, but in many cases the frequency-domain plot is far more useful. This is particularly true in the analysis of complex signal waveforms as well as the many modulation and multiplexing methods used in communication.

Test instruments for displaying signals in both time and frequency domains are readily available. You are already familiar with the oscilloscope, which displays the voltage amplitude of a signal with respect to a horizontal time axis.

The test instrument for producing a frequency-domain display is the *spectrum analyzer.* Like the oscilloscope, the spectrum analyzer uses a cathode-ray tube for display, but the horizontal sweep axis is calibrated in hertz and the vertical axis is calibrated in volts or power units or decibels.

Spectrum analyzer

The Importance of Fourier Theory

Fourier analysis allows us to determine not only the sine wave components in any complex signal but also how much bandwidth a particular signal occupies. Although a sine or cosine wave at a single frequency theoretically occupies no bandwidth, complex signals obviously take up more spectrum space. For example, a 1-MHz square wave with harmonics up to the eleventh occupies a bandwidth of 11 MHz. If this signal is to pass unattenuated and undistorted, then all harmonics must be passed.

An example is shown in Fig. 2-64. If a 1-kHz square wave is passed through a low-pass filter with a cutoff frequency just above 1 kHz, all the harmonics beyond the third harmonic are greatly attenuated or, for the most part, filtered out completely. The result is that the output of the low-pass filter is simply the fundamental sine wave at the square wave frequency.

If the low-pass filter were set to cut off at a frequency above the third harmonic, then the output of the filter would consist of a fundamental sine wave and the third harmonic. Such a waveshape was shown in Fig. 2-59(*a*). As you can see, when the higher harmonics are not all passed, the original signal is greatly distorted. This is why it is important for communication circuits and systems to have a bandwidth wide enough to accommodate all the harmonic components within the signal waveform to be processed.

Figure 2-65 shows an example in which a 1-kHz square wave is passed through a bandpass filter set to the third harmonic, resulting in a 3-kHz sine wave output. In this case the filter used is sharp enough to select out the desired component.

Pulse Spectrum

Pulse spectrum

The Fourier analysis of binary pulses is especially useful in communication, for it gives a way to analyze the bandwidth needed to transmit such pulses. Although theoretically

Figure 2-65   Selecting the third harmonic with a bandpass filter.

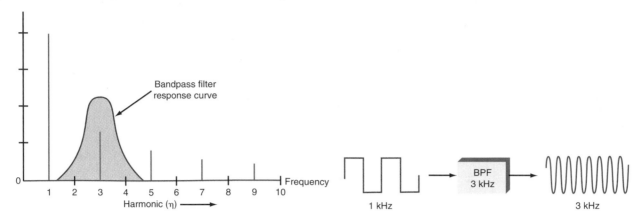

the system must pass all the harmonics in the pulses, in reality, relatively few must be passed to preserve the shape of the pulse. In addition, the pulse train in data communication rarely consists of square waves with a 50 percent duty cycle. Instead, the pulses are rectangular and exhibit varying duty cycles, from very low to very high. [The Fourier response of such pulses is given in Fig. 2-61(f).]

Look back at Fig. 2-61(f). The period of the pulse train is T and the pulse width is t_0. The duty cycle is t_0/T. The pulse train consists of dc pulses with an average dc value of Vt_0/T. In terms of Fourier analysis, the pulse train is made up of a fundamental and all even and odd harmonics. The special case of this waveform is where the duty cycle is 50 percent; in that case all the even harmonics drop out. But with any other duty cycle the waveform is made up of both odd and even harmonics. Since this is a series of dc pulses, the average dc value is Vt_0/T.

A frequency-domain graph of harmonic amplitudes plotted with respect to frequency is shown in Fig. 2-66. The horizontal axis is frequency-plotted in increments of the pulse repetition frequency f, where $f = 1/T$ and T is the period. The first component is the average dc component at zero frequency Vt_0/T, where V is the peak voltage value of the pulse.

Now, note the amplitudes of the fundamental and harmonics. Remember that each vertical line represents the peak value of the sine wave components of the pulse train. Some of the higher harmonics are negative; that simply means that their phase is reversed.

Figure 2-66   Frequency domain of a rectangular pulse train.

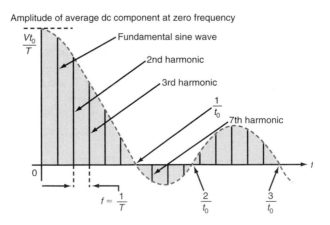

The dashed line in Fig. 2-66, the outline of the peaks of the individual components, is what is known as the *envelope* of the frequency spectrum. The equation for the envelope curve has the general form $(\sin x)/x$, where $x = \dfrac{n\pi t_0}{T}$ and t_0 is the pulse width. This is known as the *sinc* function. In Fig. 2-66, the sinc function crosses the horizontal axis several times. These times can be computed and are indicated in the figure. Note that they are some multiple of $1/t_0$.

Envelope

Sinc function

The sinc function drawn on a frequency-domain curve is used in predicting the harmonic content of a pulse train and thus the bandwidth necessary to pass the wave. For example, in Fig. 2-66, as the frequency of the pulse train gets higher, the period T gets shorter and the spacing between the harmonics gets wider. This moves the curve out to the right. And as the pulse duration t_0 gets shorter, which means that the duty cycle gets shorter, the first zero crossing of the envelope moves farther to the right. The practical significance of this is that higher-frequency pulses with shorter pulse durations have more harmonics with greater amplitudes, and thus a wider bandwidth is needed to pass the wave with minimum distortion. For data communication applications, it is generally assumed that a bandwidth equal to the first zero crossing of the envelope is the minimum that is sufficient to pass enough harmonics for reasonable waveshape:

$$BW = \frac{1}{t_0}$$

Example 2-27

A dc pulse train like that in Fig. 2-61(f) has a peak voltage value of 5 V, a frequency of 4 MHz, and a duty cycle of 30 percent.

a. What is the average dc value? [$V_{avg} = Vt_0/T$. Use the formula given in Fig. 2-61(f).]

$$\text{Duty cycle} = \frac{t_0}{T} = 30\% \quad \text{or} \quad 0.30$$

$$T = \frac{1}{f} = \frac{1}{4 \times 10^6} = 2.5 \times 10^{-7}\,\text{s} = 250 \times 10^{-9}\,\text{s}$$

$$T = 250\,\text{ns}$$

$$t_0 = \text{duty cycle} \times T = 0.3 \times 250 = 75\,\text{ns}$$

$$V_{avg} = \frac{Vt_0}{T} = V \times \text{duty cycle} = 5 \times 0.3 = 1.5\,\text{V}$$

b. What is the minimum bandwidth necessary to pass this signal without excessive distortions?

$$\text{Minimum bandwidth BW} = \frac{1}{t_0} = \frac{1}{75 \times 10^{-9}}$$

$$= 0.013333 \times 10^9 = 13.333 \times 10^6$$

$$= 13.333\,\text{MHz}$$

Most of the higher-amplitude harmonics and thus the most significant part of the signal power are contained within the larger area between zero frequency and the $1/t_0$ point on the curve.

The Relationship Between Rise Time and Bandwidth

Because a rectangular wave such as a square wave theoretically contains an infinite number of harmonics, we can use a square wave as the basis for determining the bandwidth of a signal. If the processing circuit should pass all or an infinite number of harmonics, the rise and fall times of the square wave will be zero. As the bandwidth is decreased by rolling off or filtering out the higher frequencies, the higher harmonics are greatly attenuated. The effect this has on the square wave is that the rise and fall times of the waveform become finite and increase as more and more of the higher harmonics are filtered out. The more restricted the bandwidth, the fewer the harmonics passed and the greater the rise and fall times. The ultimate restriction is where all the harmonics are filtered out, leaving only the fundamental sine wave (Fig. 2-64).

The concept of rise and fall times is illustrated in Fig. 2-67. The rise time t_r is the time it takes the pulse voltage to rise from its 10 percent value to its 90 percent value. The fall time t_f is the time it takes the voltage to drop from the 90 percent value to the 10 percent value. Pulse width t_0 is usually measured at the 50 percent amplitude points on the leading (rise) and trailing (fall) edges of the pulse.

A simple mathematical expression relating the rise time of a rectangular wave and the bandwidth of a circuit required to pass the wave without distortion is

$$BW = \frac{0.35}{t_r}$$

Example 2-28

A pulse train has a rise time of 6 ns. What is the minimum bandwidth to pass this pulse train faithfully?

$$BW = \frac{0.35}{t_r} \qquad t_r = 6 \text{ ns} = 0.006 \ \mu s$$

$$\text{Minimum BW} = \frac{0.35}{0.006} = 58.3 \text{ MHz}$$

This is the bandwidth of the circuit required to pass a signal containing the highest-frequency component in a square wave with a rise time of t_r. In this expression, the bandwidth is really the upper 3-dB down cutoff frequency of the circuit given in megahertz. The rise time of the output square wave is given in microseconds. For example, if the square wave output of an amplifier has a rise time of 10 ns (0.01 μs), the bandwidth of the circuit must be at least BW = 0.35/0.01 = 35 MHz.

Rearranging the formula, you can calculate the rise time of an output signal from the circuit whose bandwidth is given: $t_r = 0.35/\text{BW}$. For example, a circuit with a 50-MHz bandwidth will pass a square wave with a minimum rise time of $t_r = 0.35/50 = 0.007 \ \mu s = 7$ ns.

This simple relationship permits you to quickly determine the approximate bandwidth of a circuit needed to pass a rectangular waveform with a given rise time. This relationship is widely used to express the frequency response of the vertical amplifier in an oscilloscope. Oscilloscope specifications often give only a rise time figure for the

Figure 2-67   Rise and fall times of a pulse.

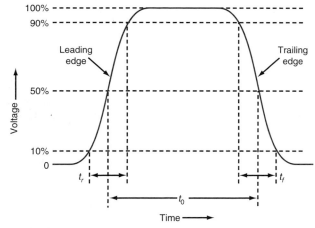

t_r = rise time
t_f = fall time
t_0 = pulse width (duration)

vertical amplifier. An oscilloscope with a 60-MHz bandwidth would pass rectangular waveforms with rise times as short as $t_r = 0.35/60 = 0.00583\ \mu s = 5.83$ ns.

Example 2-29

A circuit has a bandwidth of 200 kHz. What is the fastest rise time this circuit will pass?

$$t_r(\mu s) = \frac{0.35}{f(\text{MHz})} \qquad \text{and} \qquad 200\ \text{kHz} = 0.2\ \text{MHz}$$

$$t_r = \frac{0.35}{0.2} = 1.75\ \mu s$$

Similarly, an oscilloscope whose vertical amplifier is rated at 2 ns (0.002 μs) has a bandwidth or upper cutoff frequency of BW = 0.35/0.002 = 175 MHz. What this means is that the vertical amplifier of the oscilloscope has a bandwidth adequate to pass a sufficient number of harmonics so that the resulting rectangular wave has a rise time of 2 ns. This does not indicate the rise time of the input square wave itself. To take this into account, you use the formula

$$t_r = 1.1\sqrt{t_{ri}^2 + t_{ra}^2}$$

where t_{ri} = rise time of input square wave
  t_{ra} = rise time of amplifier
  t_r = composite rise time of amplifier output

The expression can be expanded to include the effect of additional stages of amplification by simply adding the squares of the individual rise times to the above expression before taking the square root of it.

Example 2-30

An oscilloscope has a bandwidth of 60 MHz. The input square wave has a rise time of 15 ns. What is the rise time of the displayed square wave?

$$t_{ra} \text{ (oscilloscope)} = \frac{0.35}{60} = 0.005833 \ \mu s = 5.833 \text{ ns}$$

$$t_{ri} = 15 \text{ ns}$$

$$t_{ra} \text{ (composite)} = 1.1 \ \sqrt{t_{ri}^2 + t_{ra}^2} = 1.1 \ \sqrt{(15)^2 + (5.833)^2}$$

$$= 1.1 \ \sqrt{259} = 17.7 \text{ ns}$$

Keep in mind that the bandwidth or upper cut-off frequency derived from the rise time formula on the previous page passes only the harmonics needed to support the rise time. There are harmonics beyond this bandwidth that also contribute to unwanted emissions and noise.

The spectrum analyzer shows a frequency-domain plot of electronic signals. It is the key test instrument in designing, analyzing, and troubleshooting communication equipment.

CHAPTER REVIEW

Summary

It is common in communication systems to cascade components that have gain and loss so that loss can be offset by adding a gain stage, and vice versa. The formulas for voltage, current, and power gain and loss are commonly expressed in terms of decibels, or in dBm. Tuned parallel or series circuits are made up of inductors and capacitors that resonate at specific frequencies. Both coils and capacitors offer an opposition to alternating current known as reactance. Like resistance, it is an opposition that directly affects the amount of current in the circuit. Another reactive effect is capacitance. Combining resistance, inductance, and capacitance produces a total opposition of the combined components known as impedance.

A filter is a frequency-selective circuit designed to pass some frequencies and reject others. There are both active and passive filters. The five basic kinds of filter circuits are low-pass, high-pass, bandpass, band-reject, and all-pass. The type of filter material (e.g., crystal or ceramic) affects selectivity.

Filters are made with resistor-capacitor (*RC*) or inductor-capacitor (*LC*) networks. *RC* filters are used at frequencies below 100 kHz, and *LC* filters are used at frequencies to several hundred megahertz. Crystal, ceramic, and surface acoustic wave (SAW) filters are used to provide high selectivity at high frequencies from 1 MHz to 5 GHz. Switched capacitor filters and active filters offer a way to get very high selectivity without inductors in integrated circuits.

The Fourier theory provides a way to analyze complex (nonsinusoidal) signals to determine their harmonic frequency content. Fourier theory allows you to determine the bandwidth needed to pass a rectangular wave of given frequency and duty cycle. The bandwidth of a pulse signal is related to its rise time.

Questions

1. What happens to capacitive reactance as the frequency of operation increases?
2. As frequency decreases, how does the reactance of a coil vary?
3. What is skin effect, and how does it affect the Q of a coil?
4. What happens to a wire when a ferrite bead is placed around it?
5. What is the name given to the widely used coil form that is shaped like a doughnut?
6. Describe the current and impedance in a series *RLC* circuit at resonance.
7. Describe the current and impedance in a parallel *RLC* circuit at resonance.
8. State in your own words the relationship between Q and the bandwidth of a tuned circuit.
9. What kind of filter is used to select a single signal frequency from many signals?
10. What kind of filter would you use to get rid of an annoying 120-Hz hum?
11. What does selectivity mean?
12. State the Fourier theory in your own words.
13. Define the terms *time domain* and *frequency domain*.
14. Write the first four odd harmonics of 800 Hz.
15. What waveform is made up of even harmonics only? What waveform is made up of odd harmonics only?
16. Why is a nonsinusoidal signal distorted when it passes through a filter?

Problems

1. What is the gain of an amplifier with an output of 1.5 V and an input of 30 μV? ◆
2. What is the attenuation of a voltage divider like that in Fig. 2-3, where R_1 is 3.3 kΩ and R_2 is 5.1 kΩ?
3. What is the overall gain or attenuation of the combination formed by cascading the circuits described in Problems 1 and 2? ◆
4. Three amplifiers with gains of 15, 22, and 7 are cascaded; the input voltage is 120 μV. What are the overall gain and the output voltages of each stage?
5. A piece of communication equipment has two stages of amplification with gains of 40 and 60 and two loss stages with attenuation factors of 0.03 and 0.075. The output voltage is 2.2 V. What are the overall gain (or attenuation) and the input voltage? ◆
6. Find the voltage gain or attenuation, in decibels, for each of the circuits described in Problems 1 through 5.
7. A power amplifier has an output of 200 W and an input of 8 W. What is the power gain in decibels? ◆

8. A power amplifier has a gain of 55 dB. The input power is 600 mW. What is the output power?
9. An amplifier has an output of 5 W. What is its gain in dBm? ◆
10. A communication system has five stages, with gains and attenuations of 12, −45, 68, −31, and 9 dB. What is the overall gain?
11. What is the reactance of a 7-pF capacitor at 2 GHz?
12. What value of capacitance is required to produce 50 Ω of reactance at 450 MHz?
13. Calculate the inductive reactance of a 0.9-μH coil at 800 MHz.
14. At what frequency will a 2-μH inductor have a reactance of 300 Ω?
15. A 2.5-μH inductor has a resistance of 23 Ω. At a frequency of 35 MHz, what is its Q?
16. What is the resonant frequency of a 0.55-μH coil with a capacitance of 22 pF?
17. What is the value of inductance that will resonate with an 80-pF capacitor at 18 MHz?

18. What is the bandwidth of a parallel resonant circuit that has an inductance of 33 μH with a resistance of 14 Ω and a capacitance of 48 pF?
19. A series resonant circuit has upper and lower cutoff frequencies of 72.9 and 70.5 MHz. What is its bandwidth?
20. A resonant circuit has a peak output voltage of 4.5 mV. What is the voltage output at the upper and lower cutoff frequencies?
21. What circuit Q is required to give a bandwidth of 36 MHz at a frequency of 4 GHz?
22. Find the impedance of a parallel resonant circuit with $L = 60\ \mu$H, $R_W = 7$ Ω, and $C = 22$ pF.
23. Write the first four terms of the Fourier equation of a sawtooth wave that has a peak-to-peak amplitude of 5 V and a frequency of 100 kHz.
24. An oscilloscope has a rise time of 8 ns. What is the highest-frequency sine wave that the scope can display?
25. A low-pass filter has a cutoff frequency of 24 MHz. What is the fastest rise time that a rectangular wave that will pass through the filter can have?

◆ *Answers to Selected Problems follow Chap. 22.*

Critical Thinking

1. Explain how capacitance and inductance can exist in a circuit without lumped capacitors and inductor components being present.
2. How can the voltage across the coil or capacitor in a series resonant circuit be greater than the source voltage at resonance?
3. What type of filter would you use to prevent the harmonics generated by a transmitter from reaching the antenna?
4. What kind of filter would you use on a TV set to prevent a signal from a CB radio on 27 MHz from interfering with a TV signal on channel 2 at 54 MHz?
5. Explain why it is possible to reduce the effective Q of a parallel resonant circuit by connecting a resistor in parallel with it.
6. A parallel resonant circuit has an inductance of 800 nH, a winding resistance of 3 Ω, and a capacitance of 15 pF.

Calculate (*a*) resonant frequency, (*b*) Q, (*c*) bandwidth, (*d*) impedance at resonance.
7. For the previous circuit, what would the bandwidth be if you connected a 33-kΩ resistor in parallel with the tuned circuit?
8. What value of capacitor would you need to produce a high-pass filter with a cutoff frequency of 48 kHz with a resistor value of 2.2 kΩ?
9. What is the minimum bandwidth needed to pass a periodic pulse train whose frequency is 28.8 kHz and duty cycle is 20 percent? 50 percent?
10. Refer to Fig. 2-61. Examine the various waveforms and Fourier expressions. What circuit do you think might make a good but simple frequency doubler?

Amplitude Modulation Fundamentals

In the modulation process, the baseband voice, video, or digital signal modifies another, higher-frequency signal called the *carrier*, which is usually a sine wave. A sine wave carrier can be modified by the intelligence signal through amplitude modulation, frequency modulation, or phase modulation. The focus of this chapter is *amplitude modulation (AM)*.

Carrier

Amplitude modulation (AM)

Objectives

After completing this chapter, you will be able to:

- Calculate the modulation index and percentage of modulation of an AM signal, given the amplitudes of the carrier and modulating signals.
- Define overmodulation and explain how to alleviate its effects.
- Explain how the power in an AM signal is distributed between the carrier and the sideband, and then compute the carrier and sideband powers, given the percentage of modulation.
- Compute sideband frequencies, given carrier and modulating signal frequencies.
- Compare time-domain, frequency-domain, and phasor representations of an AM signal.
- Explain what is meant by the terms *DSB* and *SSB* and state the main advantages of an SSB signal over a conventional AM signal.
- Calculate peak envelope power (PEP), given signal voltages and load impedances.

3-1 AM Concepts

As the name suggests, in AM, the information signal varies the amplitude of the carrier sine wave. The instantaneous value of the carrier amplitude changes in accordance with the amplitude and frequency variations of the modulating signal. Figure 3-1 shows a single-frequency sine wave intelligence signal modulating a higher-frequency carrier. The carrier frequency remains constant during the modulation process, but its amplitude varies in accordance with the modulating signal. An increase in the amplitude of the modulating signal causes the amplitude of the carrier to increase. Both the positive and the negative peaks of the carrier wave vary with the modulating signal. An increase or a decrease in the amplitude of the modulating signal causes a corresponding increase or decrease in both the positive and the negative peaks of the carrier amplitude.

An imaginary line connecting the positive peaks and negative peaks of the carrier waveform (the dashed line in Fig. 3-1) gives the exact shape of the modulating information signal. This imaginary line on the carrier waveform is known as the *envelope*.

Envelope

Because complex waveforms such as that shown in Fig. 3-1 are difficult to draw, they are often simplified by representing the high-frequency carrier wave as many equally spaced vertical lines whose amplitudes vary in accordance with a modulating signal, as in Fig. 3-2. This method of representation is used throughout this book.

The signals illustrated in Figs. 3-1 and 3-2 show the variation of the carrier amplitude with respect to time and are said to be in the time domain. Time-domain signals—voltage or current variations that occur over time—are displayed on the screen of an oscilloscope.

Using trigonometric functions, we can express the sine wave carrier with the simple expression

$$v_c = V_c \sin 2\pi f_c t$$

In this expression, v_c represents the instantaneous value of the carrier sine wave voltage at some specific time in the cycle; V_c represents the peak value of the constant unmodulated carrier sine wave as measured between zero and the maximum amplitude of either the positive-going or the negative-going alternations (Fig. 3-1); f_c is the frequency of the carrier sine wave; and t is a particular point in time during the carrier cycle.

A sine wave modulating signal can be expressed with a similar formula

$$v_m = V_m \sin 2\pi f_m t$$

where v_m = instantaneous value of information signal
V_m = peak amplitude of information signal
f_m = frequency of modulating signal

Figure 3-1   Amplitude modulation. (*a*) The modulating or information signal. (*b*) The modulated carrier.

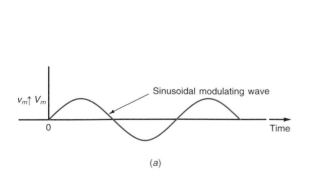

(a)

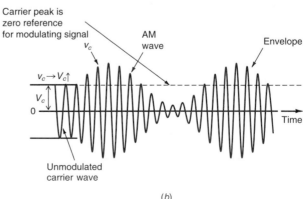

(b)

Chapter 3

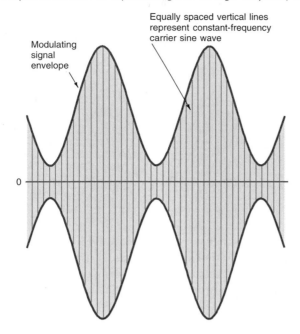

In Fig. 3-1, the modulating signal uses the peak value of the carrier rather than zero as its reference point. The envelope of the modulating signal varies above and below the peak carrier amplitude. That is, the zero reference line of the modulating signal coincides with the peak value of the unmodulated carrier. Because of this, the relative amplitudes of the carrier and modulating signal are important. In general, the amplitude of the modulating signal should be less than the amplitude of the carrier. When the amplitude of the modulating signal is greater than the amplitude of the carrier, distortion will occur, causing incorrect information to be transmitted. In amplitude modulation, it is particularly important that the peak value of the modulating signal be less than the peak value of the carrier. Mathematically,

$$V_m < V_c$$

Values for the carrier signal and the modulating signal can be used in a formula to express the complete modulated wave. First, keep in mind that the peak value of the carrier is the reference point for the modulating signal; the value of the modulating signal is added to or subtracted from the peak value of the carrier. The instantaneous value of either the top or the bottom voltage envelope v_1 can be computed by using the equation

$$v_1 = V_c + v_m = V_c + V_m \sin 2\pi f_m t$$

which expresses the fact that the instantaneous value of the modulating signal algebraically adds to the peak value of the carrier. Thus we can write the instantaneous value of the complete modulated wave v_2 by substituting v_1 for the peak value of carrier voltage V_c as follows:

$$v_2 = v_1 \sin 2\pi f_c t$$

Now substituting the previously derived expression for v_1 and expanding, we get the following:

$$v_2 = (V_c + V_m \sin 2\pi f_m t) \sin 2\pi f_c t = V_c \sin 2\pi f_c t + (V_m \sin 2\pi f_m t)(\sin 2\pi f_c t)$$

GOOD TO KNOW

In this text, radian measure will be used for all angles unless otherwise indicated. One radian is approximately 57.3°.

GOOD TO KNOW

If the amplitude of the modulating signal is greater than the amplitude of the carrier, distortion will occur.

Amplitude Modulation Fundamentals

Figure 3-3 Amplitude modulator showing input and output signals.

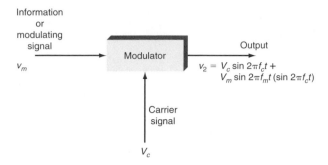

where v_2 is the instantaneous value of the AM wave (or v_{AM}), $V_c \sin 2\pi f_c t$ is the carrier waveform, and $(V_m \sin 2\pi f_m t)(\sin 2\pi f_c t)$ is the carrier waveform multiplied by the modulating signal waveform. It is the second part of the expression that is characteristic of AM. A circuit must be able to produce mathematical multiplication of the carrier and modulating signals in order for AM to occur. The AM wave is the product of the carrier and modulating signals.

Modulator

The circuit used for producing AM is called a *modulator.* Its two inputs, the carrier and the modulating signal, and the resulting outputs are shown in Fig. 3-3. Amplitude modulators compute the product of the carrier and modulating signals. Circuits that compute the product of two analog signals are also known as analog multipliers, mixers, converters, product detectors, and phase detectors. A circuit that changes a lower-frequency baseband or intelligence signal to a higher-frequency signal is usually called a modulator. A circuit used to recover the original intelligence signal from an AM wave is known as a detector or demodulator. Mixing and detection applications are discussed in detail in later chapters.

3-2 Modulation Index and Percentage of Modulation

As stated previously, for undistorted AM to occur, the modulating signal voltage V_m must be less than the carrier voltage V_c. Therefore the relationship between the amplitude of the modulating signal and the amplitude of the carrier signal is important. This relationship, known as the *modulation index m* (also called the modulating factor or coefficient, or the degree of modulation), is the ratio

Modulation index *m*

$$m = \frac{V_m}{V_c}$$

These are the peak values of the signals, and the carrier voltage is the unmodulated value.

Percentage of modulation

Multiplying the modulation index by 100 gives the *percentage of modulation.* For example, if the carrier voltage is 9 V and the modulating signal voltage is 7.5 V, the modulation factor is 0.8333 and the percentage of modulation is $0.833 \times 100 = 83.33$.

Overmodulation and Distortion

The modulation index should be a number between 0 and 1. If the amplitude of the modulating voltage is higher than the carrier voltage, m will be greater than 1, causing

Figure 3-4 Distortion of the envelope caused by overmodulation where the modulating signal amplitude V_m is greater than the carrier signal V_c.

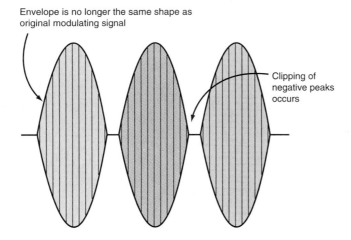

Envelope is no longer the same shape as original modulating signal

Clipping of negative peaks occurs

distortion of the modulated waveform. If the distortion is great enough, the intelligence signal becomes unintelligible. Distortion of voice transmissions produces garbled, harsh, or unnatural sounds in the speaker. Distortion of video signals produces a scrambled and inaccurate picture on a TV screen.

Simple distortion is illustrated in Fig. 3-4. Here a sine wave information signal is modulating a sine wave carrier, but the modulating voltage is much greater than the carrier voltage, resulting in a condition called *overmodulation*. As you can see, the waveform is flattened at the zero line. The received signal will produce an output waveform in the shape of the envelope, which in this case is a sine wave whose negative peaks have been clipped off. If the amplitude of the modulating signal is less than the carrier amplitude, no distortion will occur. The ideal condition for AM is when $V_m = V_c$, or $m = 1$, which gives 100 percent modulation. This results in the greatest output power at the transmitter and the greatest output voltage at the receiver, with no distortion.

Preventing overmodulation is tricky. For example, at different times during voice transmission voices will go from low amplitude to high amplitude. Normally, the amplitude of the modulating signal is adjusted so that only the voice peaks produce 100 percent modulation. This prevents overmodulation and distortion. Automatic circuits called *compression circuits* solve this problem by amplifying the lower-level signals and suppressing or compressing the higher-level signals. The result is a higher average power output level without overmodulation.

Distortion caused by overmodulation also produces adjacent channel interference. Distortion produces a nonsinusoidal information signal. According to Fourier theory, any nonsinusoidal signal can be treated as a fundamental sine wave at the frequency of the information signal plus harmonics. Obviously, these harmonics also modulate the carrier and can cause interference with other signals on channels adjacent to the carrier.

Distortion

Overmodulation

> **GOOD TO KNOW**
>
> Distortion caused by overmodulation also produces adjacent channel interference.

Compression circuit

Percentage of Modulation

The modulation index can be determined by measuring the actual values of the modulation voltage and the carrier voltage and computing the ratio. However, it is more common

Figure 3–5 An AM wave showing peaks (V_{max}) and troughs (V_{min}).

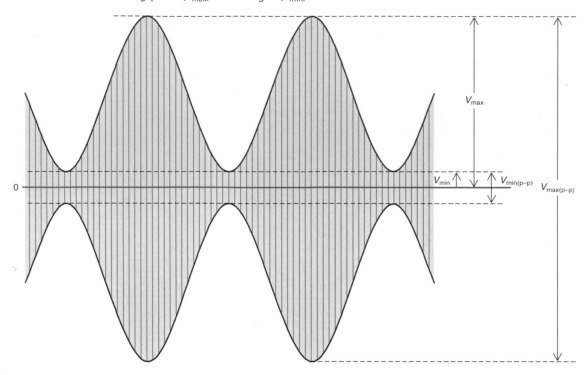

to compute the modulation index from measurements taken on the composite modulated wave itself. When the AM signal is displayed on an oscilloscope, the modulation index can be computed from V_{max} and V_{min}, as shown in Fig. 3-5. The peak value of the modulating signal V_m is one-half the difference of the peak and trough values:

$$V_m = \frac{V_{max} - V_{min}}{2}$$

As shown in Fig. 3-5, V_{max} is the peak value of the signal during modulation, and V_{min} is the lowest value, or trough, of the modulated wave. The V_{max} is one-half the peak-to-peak value of the AM signal, or $V_{max(p-p)}/2$. Subtracting V_{min} from V_{max} produces the peak-to-peak value of the modulating signal. One-half of that, of course, is simply the peak value.

The peak value of the carrier signal V_c is the average of the V_{max} and V_{min} values:

$$V_c = \frac{V_{max} + V_{min}}{2}$$

The modulation index is

$$m = \frac{V_{max} - V_{min}}{V_{max} + V_{min}}$$

The values for $V_{max(p-p)}$ and $V_{min(p-p)}$ can be read directly from an oscilloscope screen and plugged directly into the formula to compute the modulation index.

The amount, or depth, of AM is more commonly expressed as the percentage of modulation rather than as a fractional value. In Example 3-1, the percentage of modulation is $100 \times m$, or 66.2 percent. The maximum amount of modulation without signal distortion, of course, is 100 percent, where V_c and V_m are equal. At this time, $V_{min} = 0$ and $V_{max} = 2V_m$, where V_m is the peak value of the modulating signal.

Example 3-1

Suppose that on an AM signal, the $V_{max(p-p)}$ value read from the graticule on the oscilloscope screen is 5.9 divisions and $V_{min(p-p)}$ is 1.2 divisions.

a. What is the modulation index?

$$m\frac{V_{max} - V_{min}}{V_{max} + V_{min}} = \frac{5.9 - 1.2}{5.9 + 1.2} = \frac{4.7}{7.1} = 0.662$$

b. Calculate V_c, V_m, and m if the vertical scale is 2 V per division. (*Hint:* Sketch the signal.)

$$V_c = \frac{V_{max} + V_{min}}{2} = \frac{5.9 + 1.2}{2} = \frac{7.1}{2} = 3.55 @ \frac{2 \text{ V}}{\text{div}}$$

$$V_c = 3.55 \times 2 \text{ V} = 7.1 \text{ V}$$

$$V_m = \frac{V_{max} - V_{min}}{2} = \frac{5.9 - 1.2}{2} = \frac{4.7}{2}$$

$$= 2.35 @ \frac{2 \text{ V}}{\text{div}}$$

$$V_m = 2.35 \times 2 \text{ V} = 4.7 \text{ V}$$

$$m = \frac{V_m}{V_c} = \frac{4.7}{7.1} = 0.662$$

3-3 Sidebands and the Frequency Domain

Whenever a carrier is modulated by an information signal, new signals at different frequencies are generated as part of the process. These new frequencies, which are called *side frequencies,* or *sidebands,* occur in the frequency spectrum directly above and directly below the carrier frequency. More specifically, the sidebands occur at frequencies that are the sum and difference of the carrier and modulating frequencies. When signals of more than one frequency make up a waveform, it is often better to show the AM signal in the frequency domain rather than in the time domain.

Sideband

Sideband Calculations

When only a single-frequency sine wave modulating signal is used, the modulation process generates two sidebands. If the modulating signal is a complex wave, such as voice or video, a whole range of frequencies modulate the carrier, and thus a whole range of sidebands are generated.

The upper sideband f_{USB} and lower sideband f_{LSB} are computed as

$$f_{USB} = f_c + f_m \quad \text{and} \quad f_{LSB} = f_c - f_m$$

where f_c is the carrier frequency and f_m is the modulating frequency.

The existence of sidebands can be demonstrated mathematically, starting with the equation for an AM signal described previously:

$$v_{AM} = V_c \sin 2\pi f_c t + (V_m \sin 2\pi f_m t)(\sin 2\pi f_c t)$$

By using the trigonometric identity that says that the product of two sine waves is

$$\sin A \sin B = \frac{\cos (A - B)}{2} - \frac{\cos (A + B)}{2}$$

and substituting this identity into the expression a modulated wave, the instantaneous amplitude of the signal becomes

$$v_{AM} = V_c \sin 2\pi f_c t + \frac{V_m}{2} \cos 2\pi t(f_c - f_m) - \frac{V_m}{2} \cos 2\pi t(f_c + f_m)$$

where the first term is the carrier; the second term, containing the difference $f_c - f_m$, is the lower sideband; and the third term, containing the sum $f_c + f_m$, is the upper sideband.

For example, assume that a 400-Hz tone modulates a 300-kHz carrier. The upper and lower sidebands are

$$f_{USB} = 300{,}000 + 400 = 300{,}400 \text{ Hz} \quad \text{or} \quad 300.4 \text{ kHz}$$

$$f_{LSB} = 300{,}000 - 400 = 299{,}600 \text{ Hz} \quad \text{or} \quad 299.6 \text{ kHz}$$

Observing an AM signal on an oscilloscope, you can see the amplitude variations of the carrier with respect to time. This time-domain display gives no obvious or outward indication of the existence of the sidebands, although the modulation process does indeed produce them, as the equation above shows. An AM signal is really a composite signal formed from several components: the carrier sine wave is added to the upper and lower sidebands, as the equation indicates. This is illustrated graphically in Fig. 3-6.

Figure 3-6   The AM wave is the algebraic sum of the carrier and upper and lower sideband sine waves. (*a*) Intelligence or modulating signal. (*b*) Lower sideband. (*c*) Carrier. (*d*) Upper sideband. (*e*) Composite AM wave.

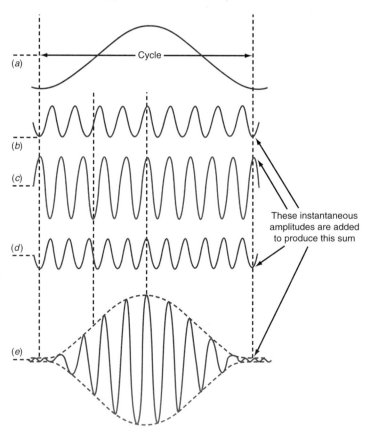

Figure 3-7 A frequency-domain display of an AM signal (voltage).

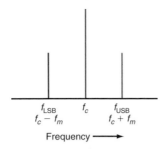

$$f_{LSB} \quad\quad f_c \quad\quad f_{USB}$$
$$f_c - f_m \quad\quad\quad\quad f_c + f_m$$

Frequency ──────▶

Adding these signals together algebraically at every instantaneous point along the time axis and plotting the result yield the AM wave shown in the figure. It is a sine wave at the carrier frequency whose amplitude varies as determined by the modulating signal.

Frequency-Domain Representation of AM

Another method of showing the sideband signals is to plot the carrier and sideband amplitudes with respect to frequency, as in Fig. 3-7. Here the horizontal axis represents frequency, and the vertical axis represents the amplitudes of the signals. The signals may be voltage, current, or power amplitudes and may be given in peak or rms values. A plot of signal amplitude versus frequency is referred to as a *frequency-domain display*. A test instrument known as a *spectrum analyzer* is used to display the frequency domain of a signal.

Figure 3-8 shows the relationship between the time- and frequency-domain displays of an AM signal. The time and frequency axes are perpendicular to each other. The amplitudes shown in the frequency-domain display are the peak values of the carrier and sideband sine waves.

Whenever the modulating signal is more complex than a single sine wave tone, multiple upper and lower sidebands are produced by the AM process. For example, a voice

Frequency-domain display

Spectrum analyzer

Figure 3-8 The relationship between the time and frequency domains.

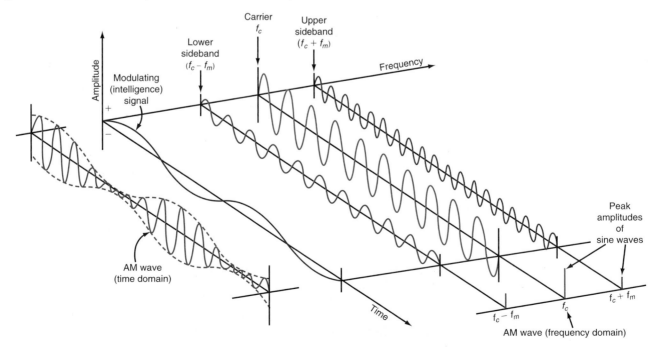

Figure 3-9 The upper and lower sidebands of a voice modulator AM signal.

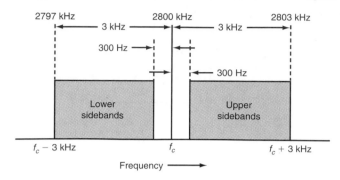

signal consists of many sine wave components of different frequencies mixed together. Recall that voice frequencies occur in the 300- to 3000-Hz range. Therefore, voice signals produce a range of frequencies above and below the carrier frequency, as shown in Fig. 3-9. These sidebands take up spectrum space. The total bandwidth of an AM signal is calculated by computing the maximum and minimum sideband frequencies. This is done by finding the sum and difference of the carrier frequency and maximum modulating frequency (3000 Hz, or 3 kHz, in Fig. 3-9). For example, if the carrier frequency is 2.8 MHz (2800 kHz), then the maximum and minimum sideband frequencies are

$$f_{USB} = 2800 + 3 = 2803 \text{ kHz} \quad \text{and} \quad f_{LSB} = 2800 - 3 = 2797 \text{ kHz}$$

The total bandwidth is simply the difference between the upper and lower sideband frequencies:

$$BW = f_{USB} - f_{LSB} = 2803 - 2797 = 6 \text{ kHz}$$

As it turns out, the bandwidth of an AM signal is twice the highest frequency in the modulating signal: $BW = 2f_m$, where f_m is the maximum modulating frequency. In the case of a voice signal whose maximum frequency is 3 kHz, the total bandwidth is simply

$$BW = 2(3 \text{ kHz}) = 6 \text{ kHz}$$

Example 3-2

A standard AM broadcast station is allowed to transmit modulating frequencies up to 5 kHz. If the AM station is transmitting on a frequency of 980 kHz, compute the maximum and minimum upper and lower sidebands and the total bandwidth occupied by the AM station.

$$f_{USB} = 980 + 5 = 985 \text{ kHz}$$
$$f_{LSB} = 980 - 5 = 975 \text{ kHz}$$
$$BW = f_{USB} - f_{LSB} = 985 - 975 = 10 \text{ kHz} \quad \text{or}$$
$$BW = 2(5 \text{ kHz}) = 10 \text{ kHz}$$

As Example 3-2 indicates, an AM broadcast station has a total bandwidth of 10 kHz. In addition, AM broadcast stations are spaced every 10 kHz across the spectrum from 540 to 1600 kHz. This is illustrated in Fig. 3-10. The sidebands from the first AM broadcast frequency extend down to 535 kHz and up to 545 kHz, forming a 10-kHz channel for the signal. The highest channel frequency is 1600 kHz, with sidebands extending

Figure 3-10   Frequency spectrum of AM broadcast band.

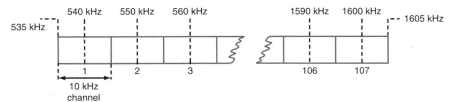

from 1595 up to 1605 kHz. There are a total of 107 10-kHz-wide channels for AM radio stations.

Pulse Modulation

When complex signals such as pulses or rectangular waves modulate a carrier, a broad spectrum of sidebands are produced. According to Fourier theory, complex signals such as square waves, triangular waves, sawtooth waves, and distorted sine waves are simply made up of a fundamental sine wave and numerous harmonic signals at different amplitudes. Assume that a carrier is amplitude-modulated by a square wave which is made up of a fundamental sine wave and all odd harmonics. A modulating square wave will produce sidebands at frequencies based upon the fundamental sine wave as well as at the third, fifth, seventh, etc., harmonics, resulting in a frequency-domain plot like that shown in Fig. 3-11. As can be seen, pulses generate extremely wide-bandwidth signals. In order for a square wave to be transmitted and faithfully received without distortion or degradation, all the most significant sidebands must be passed by the antennas and the transmitting and receiving circuits.

Figure 3-12 shows the AM wave resulting when a square wave modulates a sine wave carrier. In Fig. 3-12(*a*), the percentage of modulation is 50; in Fig. 3-12(*b*), it is 100. In this case, when the square wave goes negative, it drives the carrier amplitude to zero. Amplitude modulation by square waves or rectangular binary pulses is referred to as *amplitude-shift keying (ASK)*. ASK is used in some types of data communication when binary information is to be transmitted.

Another crude type of amplitude modulation can be achieved by simply turning the carrier off and on. An example is the transmitting of Morse code by using dots and dashes.

Figure 3-11   Frequency spectrum of an AM signal modulated by a square wave.

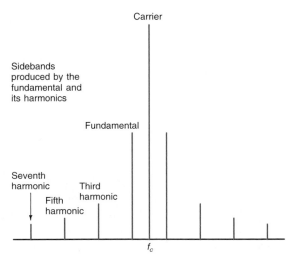

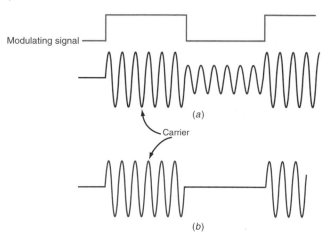

A dot is a short burst of carrier, and a dash is a longer burst of carrier. Figure 3-13 shows the transmission of the letter P, which is dot-dash-dash-dot (pronounced "dit-dah-dah-dit"). The time duration of a dash is 3 times the length of a dot, and the spacing between dots and dashes is one dot time. Code transmissions such as this are usually called *continuous-wave (CW) transmissions.* This kind of transmission is also referred to as *ON/OFF keying (OOK).* Despite the fact that only the carrier is being transmitted, sidebands are generated by such ON/OFF signals. The sidebands result from the frequency or repetition rate of the pulses themselves plus their harmonics.

As indicated earlier, the distortion of an analog signal by overmodulation also generates harmonics. For example, the spectrum produced by a 500-Hz sine wave modulating a carrier of 1 MHz is shown in Fig. 3-14(*a*). The total bandwidth of the signal is 1 kHz. However, if the modulating signal is distorted, the second, third, fourth, and higher harmonics are generated. These harmonics also modulate the carrier, producing many more sidebands, as illustrated in Fig. 3-14(*b*). Assume that the distortion is such that the harmonic amplitudes beyond the fourth harmonic are insignificant (usually less than 1 percent); then the total bandwidth of the resulting signal is about 4 kHz instead of the 1-kHz bandwidth that would result without overmodulation and distortion. The harmonics can overlap into adjacent channels, where other signals may be present and interfere with them. Such harmonic sideband interference is sometimes called *splatter* because of the way it sounds at the receiver. Overmodulation and splatter are easily eliminated simply by reducing the level of the modulating signal by using gain control or in some cases by using amplitude-limiting or compression circuits.

Figure 3-13 Sending the letter P by Morse code. An example of ON/OFF keying (OOK).

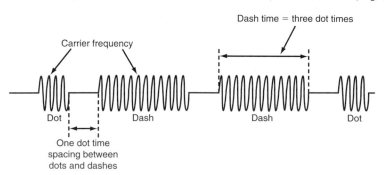

Continuous-wave (CW) transmission

ON/OFF keying (OOK)

Splatter

Figure 3-14 The effect of overmodulation and distortion on AM signal bandwidth. (*a*) Sine wave of 500 Hz modulating a 1-MHz carrier. (*b*) Distorted 500-Hz sine wave with significant second, third, and fourth harmonics.

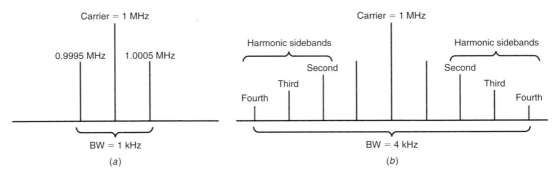

3-4 AM Power

In radio transmission, the AM signal is amplified by a power amplifier and fed to the antenna with a characteristic impedance that is ideally, but not necessarily, almost pure resistance. The AM signal is really a composite of several signal voltages, namely, the carrier and the two sidebands, and each of these signals produces power in the antenna. The total transmitted power P_T is simply the sum of the carrier power P_c and the power in the two sidebands P_{USB} and P_{LSB}:

$$P_T = P_c + P_{LSB} + P_{USB}$$

You can see how the power in an AM signal is distributed and calculated by going back to the original AM equation:

$$v_{AM} = V_c \sin 2\pi f_c t + \frac{V_m}{2} \cos 2\pi t (f_c - f_m) - \frac{V_m}{2} \cos 2\pi t (f_c + f_m)$$

where the first term is the carrier, the second term is the lower sideband, and the third term is the upper sideband.

Now, remember that V_c and V_m are peak values of the carrier and modulating sine waves, respectively. For power calculations, rms values must be used for the voltages. We can convert from peak to rms by dividing the peak value by $\sqrt{2}$ or multiplying by 0.707. The rms carrier and sideband voltages are then

$$v_{AM} = \frac{V_c}{\sqrt{2}} \sin 2\pi f_c t + \frac{V_m}{2\sqrt{2}} \cos 2\pi t (f_c - f_m) - \frac{V_m}{2\sqrt{2}} \cos 2\pi t (f_c + f_m)$$

The power in the carrier and sidebands can be calculated by using the power formula $P = V^2/R$, where P is the output power, V is the rms output voltage, and R is the resistive part of the load impedance, which is usually an antenna. We just need to use the coefficients on the sine and cosine terms above in the power formula:

$$P_T = \frac{(V_c/\sqrt{2})^2}{R} + \frac{(V_m/2\sqrt{2})^2}{R} + \frac{(V_m/2\sqrt{2})^2}{R} = \frac{V_c^2}{2R} + \frac{V_m^2}{8R} + \frac{V_m^2}{8R}$$

Remembering that we can express the modulating signal V_m in terms of the carrier V_c by using the expression given earlier for the modulation index $m = V_m/V_c$; we can write

$$V_m = mV_c$$

If we express the sideband powers in terms of the carrier power, the total power becomes

$$P_T = \frac{(V_c)^2}{2R} + \frac{(mV_c)^2}{8R} + \frac{(mV_c)^2}{8R} = \frac{V_c^2}{2R} + \frac{m^2V_c^2}{8R} + \frac{m^2V_c^2}{8R}$$

Since the term $V_c^2/2R$ is equal to the rms carrier power P_c, it can be factored out, giving

$$P_T = \frac{V_c^2}{2R}\left(1 + \frac{m^2}{4} + \frac{m^2}{4}\right)$$

Finally, we get a handy formula for computing the total power in an AM signal when the carrier power and the percentage of modulation are known:

$$P_T = P_c\left(1 + \frac{m^2}{2}\right)$$

For example, if the carrier of an AM transmitter is 1000 W and it is modulated 100 percent ($m = 1$), the total AM power is

$$P_T = 1000\left(1 + \frac{1^2}{2}\right) = 1500 \text{ W}$$

Of the total power, 1000 W of it is in the carrier. That leaves 500 W in both sidebands. Since the sidebands are equal in size, each sideband has 250 W.

For a 100 percent modulated AM transmitter, the total sideband power is always one-half that of the carrier power. A 50-kW transmitter carrier that is 100 percent modulated will have a sideband power of 25 kW, with 12.5 kW in each sideband. The total power for the AM signal is the sum of the carrier and sideband power, or 75 kW.

When the percentage of modulation is less than the optimum 100, there is much less power in the sidebands. For example, for a 70 percent modulated 250-W carrier, the total power in the composite AM signal is

$$P_T = 250\left(1 + \frac{0.7^2}{2}\right) = 250(1 + 0.245) = 311.25 \text{ W}$$

Of the total, 250 W is in the carrier, leaving $311.25 - 250 = 61.25$ W in the sidebands. There is 61.25/2 or 30.625 W in each sideband.

Example 3-3

An AM transmitter has a carrier power of 30 W. The percentage of modulation is 85 percent. Calculate (a) the total power and (b) the power in one sideband.

a. $P_T = P_c\left(1 + \dfrac{m^2}{2}\right) = 30\left[1 + \dfrac{(0.85)^2}{2}\right] = 30\left(1 + \dfrac{0.7225}{2}\right)$

$P_T = 30(1.36125) = 40.8$ W

b. P_{SB} (both) $= P_T - P_c = 40.8 - 30 = 10.8$ W

P_{SB} (one) $= \dfrac{P_{SB}}{2} = \dfrac{10.8}{2} = 5.4$ W

In the real world, it is difficult to determine AM power by measuring the output voltage and calculating the power with the expression $P = V^2/R$. However, it is easy to measure the current in the load. For example, you can use an RF ammeter connected in series with an antenna to observe antenna current. When the antenna impedance is known, the output power is easily calculated by using the formula

$$P_T = I_T^2 R$$

where $I_T = I_c\sqrt{(1 + m^2/2)}$. Here I_c is the unmodulated carrier current in the load, and m is the modulation index. For example, the total output power of an 85 percent modulated AM transmitter, whose unmodulated carrier current into a 50-Ω antenna load impedance is 10 A, is

$$I_T = 10\sqrt{\left(1 + \frac{0.85^2}{2}\right)} = 10\sqrt{1.36125} = 11.67 \text{ A}$$

$$P_T = 11.67^2(50) = 136.2(50) = 6809 \text{ W}$$

One way to find the percentage of modulation is to measure both the modulated and the unmodulated antenna currents. Then, by algebraically rearranging the formula above, m can be calculated directly:

$$m = \sqrt{2\left[\left(\frac{I_T}{I_c}\right)^2 - 1\right]}$$

Suppose that the unmodulated antenna current is 2.2 A. That is the current produced by the carrier only, or I_c. Now, if the modulated antenna current is 2.6 A, the modulation index is

$$m = \sqrt{2\left[\left(\frac{2.6}{2.2}\right)^2 - 1\right]} = \sqrt{2[(1.18)^2 - 1]} = \sqrt{0.7934} = 0.89$$

The percentage of modulation is 89.

As you can see, the power in the sidebands depends on the value of the modulation index. The greater the percentage of modulation, the higher the sideband power and the higher the total power transmitted. Of course, maximum power appears in the sidebands when the carrier is 100 percent modulated. The power in each sideband P_{SB} is given by

$$P_{SB} = P_{LSB} = P_{USB} = \frac{P_c m^2}{4}$$

An example of a time-domain display of an AM signal (power) is as follows.

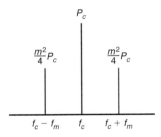

Assuming 100 percent modulation where the modulation factor $m = 1$, the power in each sideband is one-fourth, or 25 percent, of the carrier power. Since there are two sidebands, their power together represents 50 percent of the carrier power. For example, if the carrier power is 100 W, then at 100 percent modulation, 50 W will appear in the sidebands, 25 W in each. The total transmitted power, then, is the sum of the carrier and sideband powers, or 150 W. The goal in AM is to keep the percentage of modulation as high as possible without overmodulation so that maximum sideband power is transmitted.

The carrier power represents two-thirds of the total transmitted power. Assuming 100-W carrier power and a total power of 150 W, the carrier power percentage is $100/150 = 0.667$, or 66.7 percent. The sideband power percentage is thus $50/150 = 0.333$, or 33.3 percent.

The carrier itself conveys no information. The carrier can be transmitted and received, but unless modulation occurs, no information will be transmitted. When modulation occurs, sidebands are produced. It is easy to conclude, therefore, that all the transmitted information is contained within the sidebands. Only one-third of the total transmitted power is allotted to the sidebands, and the remaining two-thirds is literally wasted on the carrier.

At lower percentages of modulation, the power in the sidebands is even less. For example, assuming a carrier power of 500 W and a modulation of 70 percent, the power in each sideband is

$$P_{SB} = \frac{P_c m^2}{4} = \frac{500(0.7)^2}{4} = \frac{500(0.49)}{4} = 61.25 \text{ W}$$

and the total sideband power is 122.5 W. The carrier power, of course, remains unchanged at 500 W.

As stated previously, complex voice and video signals vary over a wide amplitude and frequency range, and 100 percent modulation occurs only on the peaks of the modulating signal. For this reason, the average sideband power is considerably lower than the ideal 50 percent that would be produced by 100 percent modulation. With less sideband power transmitted, the received signal is weaker and communication is less reliable.

Example 3-4

An antenna has an impedance of 40 Ω. An unmodulated AM signal produces a current of 4.8 A. The modulation is 90 percent. Calculate (*a*) the carrier power, (*b*) the total power, and (*c*) the sideband power.

a. $P_c = I^2 R = (4.8)^2(40) = (23.04)(40) = 921.6 \text{ W}$

b. $I_T = I_c \sqrt{1 + \frac{m^2}{2}} = 4.8 \sqrt{1 + \frac{(0.9)^2}{2}} = 4.8 \sqrt{1 + \frac{0.81}{2}}$

$I_T = 4.8 \sqrt{1.405} = 5.7 \text{ A}$

$P_T = I_T^2 R = (5.7)^2(40) = 32.49(40) = 1295 \text{ W}$

c. $P_{SB} = P_T - P_c = 1295 - 921.6 = 373.4 \text{ W (186.7 W each sideband)}$

Example 3-5

The transmitter in Example 3-4 experiences an antenna current change from 4.8 A unmodulated to 5.1 A. What is the percentage of modulation?

$$m = \sqrt{2\left[\left(\frac{I_T}{I_c}\right)^2 - 1\right]}$$

$$= \sqrt{2\left[\left(\frac{5.1}{4.8}\right)^2 - 1\right]}$$

$$= \sqrt{2[(1.0625)^2 - 1]}$$

$$= \sqrt{2(1.13 - 1)}$$

$$= \sqrt{2(0.13)}$$

$$= \sqrt{0.26}$$

$$m = 0.51$$

The percentage of modulation is 51.

Example 3-6

What is the power in one sideband of the transmitter in Example 3-4?

$$P_{SB} = m^2 \frac{P_c}{4} = \frac{(0.9)^2(921.6)}{4} = \frac{746.5}{4} = 186.6 \text{ W}$$

Despite its inefficiency, AM is still widely used because it is simple and effective. It is used in AM radio broadcasting, CB radio, TV broadcasting, and aircraft tower communication. Some simple control radios use ASK because of its simplicity. Examples are garage door openers and remote keyless entry devices on cars. AM is also widely used in combination with phase modulation to produce quadrature amplitude modulation (QAM) which facilitates high-speed data transmissions in modems, cable TV, and some wireless applications.

3-5 Single-Sideband Modulation

Single-sideband modulation

In amplitude modulation, two-thirds of the transmitted power is in the carrier, which itself conveys no information. The real information is contained within the sidebands. One way to improve the efficiency of amplitude modulation is to suppress the carrier and eliminate one sideband. The result is a single-sideband (SSB) signal. SSB is a form of AM that offers unique benefits in some types of electronic communication.

DSB Signals

The first step in generating an SSB signal is to suppress the carrier, leaving the upper and lower sidebands. This type of signal is referred to as a *double-sideband suppressed carrier (DSSC or DSB)* signal. The benefit, of course, is that no power is wasted on the carrier. Double-sideband suppressed carrier modulation is simply a special case of AM with no carrier.

Double-sideband suppressed carrier (DSSC or DSB)

A typical DSB signal is shown in Fig. 3-15. This signal, the algebraic sum of the two sinusoidal sidebands, is the signal produced when a carrier is modulated by a single-tone sine wave information signal. The carrier is suppressed, and the time-domain DSB signal is a sine wave at the carrier frequency, varying in amplitude as shown. Note that the envelope of this waveform is not the same as that of the modulating signal, as

Figure 3-15 A time-domain display of a DSB AM signal.

Time-domain display

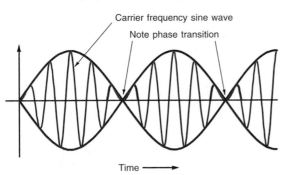

Carrier frequency sine wave

Note phase transition

Time ⟶

Figure 3-16 A frequency-domain display of DSB signal.

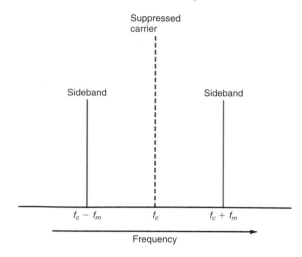

it is in a pure AM signal with carrier. A unique characteristic of the DSB signal is the phase transitions that occur at the lower-amplitude portions of the wave. In Fig. 3-15, note that there are two adjacent positive-going half-cycles at the null points in the wave. That is one way to tell from an oscilloscope display whether the signal shown is a true DSB signal.

A *frequency-domain display* of a DSB signal is given in Fig. 3-16. As shown, the spectrum space occupied by a DSB signal is the same as that for a conventional AM signal.

Double-sideband suppressed carrier signals are generated by a circuit called a *balanced modulator.* The purpose of the balanced modulator is to produce the sum and difference frequencies but to cancel or balance out the carrier. Balanced modulators are covered in detail in Chap. 4.

Despite the fact that elimination of the carrier in DSB AM saves considerable power, DSB is not widely used because the signal is difficult to demodulate (recover) at the receiver. One important application for DSB, however, is the transmission of the color information in a TV signal.

SSB Signals

In DSB transmission, since the sidebands are the sum and difference of the carrier and modulating signals, the information is contained in both sidebands. As it turns out, there is no reason to transmit both sidebands in order to convey the information. One side-band can be suppressed; the remaining sideband is called a *single-sideband suppressed carrier (SSSC* or *SSB)* signal. SSB signals offer four major benefits.

1. The primary benefit of an SSB signal is that the spectrum space it occupies is only one-half that of AM and DSB signals. This greatly conserves spectrum space and allows more signals to be transmitted in the same frequency range.

2. All the power previously devoted to the carrier and the other sideband can be chan-neled into the single sideband, producing a stronger signal that should carry farther and be more reliably received at greater distances. Alternatively, SSB transmitters can be made smaller and lighter than an equivalent AM or DSB transmitter because less circuitry and power are used.

3. Because SSB signals occupy a narrower bandwidth, the amount of noise in the sig-nal is reduced.

4. There is less selective fading of an SSB signal over long distances. An AM signal is really multiple signals, at least a carrier and two sidebands. These are on different

Figure 3-17 An SSB signal produced by a 2-kHz sine wave modulating a 14.3-MHz sine wave carrier.

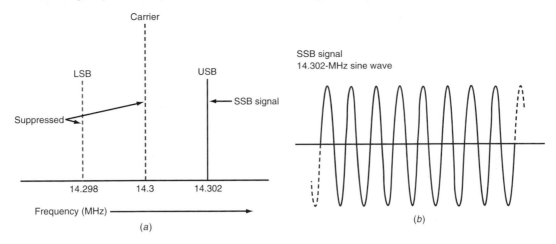

(a)

(b)

frequencies, so they are affected in slightly different ways by the ionosphere and upper atmosphere, which have a great influence on radio signals of less than about 50 MHz. The carrier and sidebands may arrive at the receiver at slightly different times, causing a phase shift that can, in turn, cause them to add in such a way as to cancel one another rather than add up to the original AM signal. Such cancellation, or *selective fading*, is not a problem with SSB since only one sideband is being transmitted.

Selective fading

An SSB signal has some unusual characteristics. First, when no information or modulating signal is present, no RF signal is transmitted. In a standard AM transmitter, the carrier is still transmitted even though it may not be modulated. This is the condition that might occur during a voice pause on an AM broadcast. But since there is no carrier transmitted in an SSB system, no signals are present if the information signal is zero. Sidebands are generated only during the modulation process, e.g., when someone speaks into a microphone. This explains why SSB is so much more efficient than AM.

Figure 3-17 shows the frequency- and time-domain displays of an SSB signal produced when a steady 2-kHz sine wave tone modulates a 14.3-MHz carrier. Amplitude modulation would produce sidebands of 14.298 and 14.302 MHz. In SSB, only one sideband is used. Figure 3-17(a) shows that only the upper sideband is generated. The RF signal is simply a constant-power 14.302-MHz sine wave. A time-domain display of this SSB signal is shown in Fig. 3-17(b).

Of course, most information signals transmitted by SSB are not pure sine waves. A more common modulation signal is voice, with its varying frequency and amplitude content. The voice signal creates a complex RF SSB signal that varies in frequency and amplitude over the narrow spectrum defined by the voice signal bandwidth. The waveform at the output of the SSB modulator has the same shape as the baseband waveform, but it is shifted in frequency.

Disadvantages of DSB and SSB

The main disadvantage of DSB and SSB signals is that they are harder to recover, or demodulate, at the receiver. Demodulation depends upon the carrier being present. If the carrier is not present, then it must be regenerated at the receiver and reinserted into the signal. To faithfully recover the intelligence signal, the reinserted carrier must have the same phase and frequency as those of the original carrier. This is a difficult requirement. When SSB is used for voice transmission, the reinserted carrier can be made variable in frequency so that it can be adjusted manually while listening to recover an intelligible signal. This is not possible with some kinds of data signals.

Pilot carrier

To solve this problem, a low-level carrier signal is sometimes transmitted along with the two sidebands in DSB or a single sideband in SSB. Because the carrier has a low power level, the essential benefits of SSB are retained, but a weak carrier is received so that it can be amplified and reinserted to recover the original information. Such a low-level carrier is referred to as a *pilot carrier.* This technique is used in FM stereo transmissions as well as in the transmission of the color information in a TV picture.

Signal Power Considerations

In conventional AM, the transmitted power is distributed among the carrier and two sidebands. For example, given a carrier power of 400 W with 100 percent modulation, each sideband will contain 100 W of power and the total power transmitted will be 600 W. The effective transmission power is the combined power in the sidebands, or 200 W.

An SSB transmitter sends no carrier, so the carrier power is zero. A given SSB transmitter will have the same communication effectiveness as a conventional AM unit running much more power. For example, a 10-W SSB transmitter offers the performance capabilities of an AM transmitter running a total of 40 W, since they both show 10 W of power in one sideband. The power advantage of SSB over AM is 4:1.

Peak envelope power (PEP)

In SSB, the transmitter output is expressed in terms of *peak envelope power (PEP),* the maximum power produced on voice amplitude peaks. PEP is computed by the equation $P = V^2/R$. For example, assume that a voice signal produces a 360-V, peak-to-peak signal across a 50-Ω load. The rms voltage is 0.707 times the peak value, and the peak value is one-half the peak-to-peak voltage. In this example, the rms voltage is $0.707(360/2) = 127.26$ V.

The peak envelope power is then

$$\text{PEP} = V_{\text{rms}}{}^2/R = \frac{(127.26)^2}{50} = 324 \text{ W}$$

The PEP input power is simply the dc input power of the transmitter's final amplifier stage at the instant of the voice envelope peak. It is the final amplifier stage dc supply voltage multiplied by the maximum amplifier current that occurs at the peak, or

$$\text{PEP} = V_s I_{\text{max}}$$

where V_s = amplifier supply voltage
I_{max} = current peak

For example, a 450-V supply with a peak current of 0.8 A produces a PEP of $450(0.8) = 360$ W.

Note that voice amplitude peaks are produced only when very loud sounds are generated during certain speech patterns or when some word or sound is emphasized. During normal speech levels, the input and output power levels are much less than the PEP level. The average power is typically only one-fourth to one-third of the PEP value with typical human speech:

$$P_{\text{avg}} = \frac{\text{PEP}}{3} \quad \text{or} \quad P_{\text{avg}} = \frac{\text{PEP}}{4}$$

With a PEP of 240 W, the average power is only 60 to 80 W. Typical SSB transmitters are designed to handle only the average power level on a continuous basis, not the PEP.

The transmitted sideband will, of course, change in frequency and amplitude as a complex voice signal is applied. This sideband will occupy the same bandwidth as one sideband in a fully modulated AM signal with carrier.

Incidentally, it does not matter whether the upper or lower sideband is used, since the information is contained in either. A filter is typically used to remove the unwanted sideband.

Example 3-7

An SSB transmitter produces a peak-to-peak voltage of 178 V across a 75-Ω antenna load. What is the PEP?

$$V_p = \frac{V_{p-p}}{2} = \frac{178}{2} = 89 \text{ V}$$

$$V_{rms} = 0.707 \qquad V_p = 0.707(89) = 62.9 \text{ V}$$

$$P = \frac{V^2}{R} = \frac{(62.9)^2}{75} = 52.8 \text{ W}$$

$$\text{PEP} = 52.8 \text{ W}$$

Example 3-8

An SSB transmitter has a 24-V dc power supply. On voice peaks the current achieves a maximum of 9.3 A.

a. What is the PEP?

$$\text{PEP} = V_s I_m = 24(9.3) = 223.2 \text{ W}$$

b. What is the average power of the transmitter?

$$P_{avg} = \frac{\text{PEP}}{3} = \frac{223.2}{3} = 74.4 \text{ W}$$

$$P_{avg} = \frac{\text{PEP}}{4} = \frac{223.2}{4} = 55.8 \text{ W}$$

$$P_{avg} = 55.8 \text{ to } 74.4 \text{ W}$$

Applications of DSB and SSB

Both DSB and SSB techniques are widely used in communication. SSB signals are still used in some two-way radios. Two-way SSB communication is used in marine applications, in the military, and by hobbyists known as radio amateurs (hams). DSB signals are used in FM and TV broadcasting to transmit two-channel stereo signals and to transmit the color information for a TV picture.

An unusual form of AM is that used in TV broadcasting. A TV signal consists of the picture (video) signal and the audio signal, which have different carrier frequencies. The audio carrier is frequency-modulated, but the video information amplitude-modulates the picture carrier. The picture carrier is transmitted, but one sideband is partially suppressed.

Video information typically contains frequencies as high as 4.2 MHz. A fully amplitude-modulated TV signal would then occupy 2(4.2) = 8.4 MHz. This is an excessive amount of bandwidth that is wasteful of spectrum space because not all of it is required to reliably transmit a TV signal. To reduce the bandwidth to the 6-MHz

Figure 3-18 Vestigial sideband transmission of a TV picture signal.

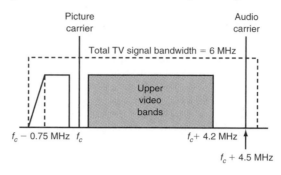

maximum allowed by the FCC for TV signals, a portion of the lower sideband of the TV signal is suppressed, leaving only a small part, or vestige, of the lower sideband. This arrangement, known as a *vestigial sideband (VSB) signal,* is illustrated in Fig. 3-18. Video signals above 0.75 MHz (750 kHz) are suppressed in the lower (vestigial) sideband, and all video frequencies are transmitted in the upper sideband.

Vestigial sideband (VSB) signal

The newer high-definition or digital TV also uses VSB but with multilevel digital modulation called VSB.

3-6 Classification of Radio Emissions

Figure 3-19 shows the codes used to designate the many types of signals that can be transmitted by radio and wire. The basic code is made up of a capital letter and a number, and lowercase subscript letters are used for more specific definitions. For example, a basic AM voice signal such as that heard on the AM broadcast band or on a CB or aircraft radio has the code A3. All the variations of AM using voice or video intelligence have the A3 designation, but subscript letters are used to distinguish them. Examples of codes designating signals described in this chapter are as follows:

DSB two sidebands, full carrier = A3

DSB two sidebands, suppressed carrier = $A3_b$

SSB single sideband, suppressed carrier = $A3_j$

SSB single sideband, 10 percent pilot carrier = $A3_a$

Vestigial sideband TV = $A3_c$

OOK and ASK = A1

Note that there are special designations for fax and pulse transmissions, and that the number 9 covers any special modulation or techniques not covered elsewhere. When a number precedes the letter code, the number refers to bandwidth in kilohertz. For example, the designation 10A3 refers to a 10-kHz bandwidth voice AM signal. The designation $20A3_h$ refers to an AM SSB signal with full carrier and message frequency to 20 kHz.

Another system used to describe a signal is given in Fig. 3-20. It is similar to the method just described, but with some variations. This is the definition used by the standards organization International Telecommunications Union (ITU). Some examples are

A3F	amplitude-modulated analog TV
J3E	SSB voice
F2D	FSK data
G7E	phase-modulated voice, multiple signals

Figure 3-19 Radio emission code designations.

Letter	A	Amplitude modulation
	F	Frequency modulation
	P	Phase modulation
Number	0	Carrier ON only, no message (radio beacon)
	1	Carrier ON/OFF, no message (Morse code, radar)
	2	Carrier ON, keyed tone ON/OFF (code)
	3	Telephony, message as voice or music
	4	Fax, nonmoving graphics (slow-scan TV)
	5	Vestigial sideband (commercial TV)
	6	Four-frequency diplex telegraphy
	7	Multiple sidebands each with different message
	8	
	9	General (all others)
Subscripts		
	None	Double sideband, full carrier
	a	Single sideband, reduced carrier
	b	Double sideband, no carrier
	c	Vestigial sideband
	d	Carrier pulses only, pulse amplitude modulation (PAM)
	e	Carrier pulses only, pulse width modulation (PWM)
	f	Carrier pulses only, pulse position modulation (PPM)
	g	Quantized pulses, digital video
	h	Single sideband, full carrier
	j	Single sideband, no carrier

Figure 3-20 ITU emissions designations.

Type of Modulation
- N   Unmodulated carrier
- A   Amplitude modulation
- J   Single sideband
- F   Frequency modulation
- G   Phase modulation
- P   Series of pulses, no modulation

Type of Modulating Signals
- 0   None
- 1   Digital, single channel, no modulation
- 2   Digital, single channel, with modulation
- 3   Analog, single channel
- 7   Digital, two or more channels
- 8   Analog, two or more channels
- 9   Analog plus digital

Type of Intelligence Signal
- N   None
- A   Telegraphy, human
- B   Telegraphy, machine
- C   Fax
- D   Data, telemetry, control signals
- E   Telephony (human voice)
- F   Video, TV
- W   Some combination of any of the above

Amplitude Modulation Fundamentals

CHAPTER REVIEW

Summary

In amplitude modulation, an increase or a decrease in the amplitude of the modulating signal causes a corresponding increase or decrease in both the positive and the negative peaks of the carrier amplitude. Interconnecting the adjacent positive or negative peaks of the carrier waveform yields the shape of the modulating information signal, known as the envelope.

Using trigonometric functions, we can form mathematical expressions for the carrier and the modulating signal, and we combine these to create a formula for the complete modulated wave. Modulators (circuits that produce amplitude modulation) compute the product of the carrier and modulating signals.

The relationship between the amplitudes of the modulating signal and the carrier is expressed as the modulation index m, a number between 0 and 1. If the amplitude of the modulating voltage is higher than the carrier voltage $m > 1$, then distortion, or overmodulation, will result.

When a carrier is modulated by an information signal, new signals at different frequencies are generated. These side frequencies, or sidebands, occur in the frequency spectrum directly above and below the carrier frequency. An AM signal is a composite of several signal voltages, the carrier, and the two sidebands, each of which produces power in the antenna.

Total transmitted power is the sum of the carrier power and the power in the two sidebands.

AM signals can be expressed through time-domain displays or frequency-domain displays.

In AM transmission, two-thirds of the transmitted power appears in the carrier, which itself conveys no information. One way to overcome this wasteful effect is to suppress the carrier. When the carrier is initially suppressed, both the upper and the lower sidebands are left, leaving a double-sideband suppressed (DSSC or DSB) signal. Because both sidebands are not necessary to transmit the desired information, one of the remaining sidebands can be suppressed, leaving a single-sideband (SSB) signal. SSB signals offer important benefits: they conserve spectrum space, produce strong signals, reduce noise, and result in less fading over long distances.

In SSB, the transmitter output is expressed as peak envelope power (PEP), the maximum power produced on voice amplitude peaks.

Both DSB and SSB techniques are widely used in communication. Two-way SSB communication is used in marine applications, in the military, and by hams. In some TV applications, to reduce the signal bandwidth to the 6-MHz maximum allowed by the FCC for TV signals, a vestigial sideband signal is used to suppress the lower sideband of the TV signal.

Questions

1. Define modulation.
2. Explain why modulation is necessary or desirable.
3. Name the circuit that causes one signal to modulate another, and give the names of the two signals applied to this circuit.
4. In AM, how does the carrier vary in accordance with the information signal?
5. True or false? The carrier frequency is usually lower than the modulating frequency.
6. What is the outline of the peaks of the carrier signal called, and what shape does it have?
7. What are voltages that vary over time called?
8. Write the trigonometric expression for a sine wave carrier signal.
9. True or false? The carrier frequency remains constant during AM.
10. What mathematical operation does an amplitude modulator perform?
11. What is the ideal relationship between the modulating signal voltage V_m and the carrier voltage V_c?
12. What is the modulation index called when it is expressed as a percentage?
13. Explain the effects of a modulation percentage greater than 100.
14. What is the name given to the new signals generated by the modulation process?
15. What is the name of the type of signal that is displayed on an oscilloscope?
16. What is the type of signal whose amplitude components are displayed with respect to frequency called, and on what instrument is this signal displayed?
17. Explain why complex nonsinusoidal and distorted signals produce a greater bandwidth AM signal than a simple sine wave signal of the same frequency.
18. What three signals can be added to give an AM wave?
19. What is the name given to an AM signal whose carrier is modulated by binary pulses?
20. What is the value of phasor representation of AM signals?
21. True or false? The modulating signal appears in the output spectrum of an AM signal.
22. What percentage of the total power in an AM signal is in the carrier? One sideband? Both sidebands?
23. Does the carrier of an AM signal contain any information? Explain.

24. What is the name of a signal that has both sidebands but no carrier?
25. What is the name of the circuit used to eliminate the carrier in DSB/SSB transmissions?
26. What is the minimum bandwidth AM signal that can be transmitted and still convey all the necessary intelligence?
27. State the four main benefits of SSB over conventional AM.
28. Name two applications for SSB and two applications for DSB.
29. Name the type of AM used in TV picture transmission. Why is it used? Draw the frequency-domain spectrum of the TV signal.
30. Using Figs. 3-19 and 3-20, write the designations for a pulse-amplitude-modulated radio signal and an amplitude-modulated (V_{SB}) analog fax signal.
31. Explain the bandwidth requirements of a voice signal of 2 kHz and a binary data signal with a rate of 2 kHz.

Problems

1. Give the formula for modulation index and explain its terms. ◆
2. An AM wave displayed on an oscilloscope has values of V_{max} = 4.8 and V_{min} = 2.5 as read from the graticule. What is the percentage of modulation?
3. What is the ideal percentage of modulation for maximum amplitude of information transmission? ◆
4. To achieve 75 percent modulation of a carrier of V_c = 50 V, what amplitude of the modulating signal V_m is needed?
5. The maximum peak-to-peak value of an AM wave is 45 V. The peak-to-peak value of the modulating signal is 20 V. What is the percentage of modulation? ◆
6. What is the mathematical relationship of the carrier and modulating signal voltages when overmodulation occurs?
7. An AM radio transmitter operating on 3.9 MHz is modulated by frequencies up to 4 kHz. What are the maximum upper and lower side frequencies? What is the total bandwidth of the AM signal? ◆
8. What is the bandwidth of an AM signal whose carrier is 2.1 MHz modulated by a 1.5-kHz square wave with significant harmonics up to the fifth? Calculate all the upper and lower sidebands produced.
9. How much power appears in one sideband of an AM signal of a 5-kW transmitter modulated by 80 percent? ◆
10. What is the total power supplied by an AM transmitter with a carrier power of 2500 W and modulation of 77 percent?

11. An AM signal has a 12-W carrier and 1.5 W in each sideband. What is the percentage of modulation?
12. An AM transmitter puts a carrier of 6 A into an antenna whose resistance is 52 Ω. The transmitter is modulated by 60 percent. What is the total output power?
13. The antenna current produced by an unmodulated carrier is 2.4 A into an antenna with a resistance of 75 Ω. When amplitude-modulated, the antenna current rises to 2.7 A. What is the percentage of modulation?
14. A ham transmitter has a carrier power of 750 W. How much power is added to the signal when the transmitter is 100 percent modulated?
15. An SSB transmitter has a power supply voltage of 250 V. On voice peaks, the final amplifier draws a current of 3.3 A. What is the input PEP?
16. The peak-to-peak output voltage of 675 V appears across a 52-Ω antenna on voice peaks in an SSB transmitter. What is the output PEP?
17. What is the average output power of an SSB transmitter rated at 100-W PEP?
18. An SSB transmitter with a carrier of 2.3 MHz is modulated by an intelligence signal in the 150-Hz to 4.2-kHz range. Calculate the frequency range of the lower sideband.

◆ *Answers to Selected Problems follow Chap. 22.*

Critical Thinking

1. Can intelligence be sent without a carrier? If so, how?
2. How is the output power of an SSB transmitter expressed?
3. A subcarrier of 70 kHz is amplitude-modulated by tones of 2.1 and 6.8 kHz. The resulting AM signal is then used to amplitude-modulate a carrier of 12.5 MHz. Calculate all sideband frequencies in the composite signal, and draw a frequency-domain display of the signal. Assume 100 percent modulation. What is the bandwidth occupied by the complete signal?
4. Explain how you could transmit two independent baseband information signals by using SSB on a common carrier frequency.
5. An AM signal with 100 percent modulation has an upper sideband power of 32 W. What is the carrier power?
6. Can an information signal have a higher frequency than that of the carrier signal? What would happen if a 1-kHz signal amplitude-modulated a 1-kHz carrier signal?

chapter 4

Amplitude Modulator and Demodulator Circuits

Dozens of modulator circuits have been developed that cause the carrier amplitude to be varied in accordance with the modulating information signal. There are circuits to produce AM, DSB, and SSB at low or high power levels. This chapter examines some of the more common and widely used discrete-component and integrated-circuit (IC) amplitude modulators. Also covered are demodulator circuits for AM, DSB, and SSB.

Objectives

After completing this chapter, you will be able to:

- Explain the relationship of the basic equation for an AM signal to the production of amplitude modulation, mixing, and frequency conversion by a diode or other nonlinear frequency component or circuit.
- Describe the operation of diode modulator circuits and diode detector circuits.
- Compare the advantages and disadvantages of low- and high-level modulation.
- Explain how the performance of a basic diode detector is enhanced by using full wave rectifier circuits.
- Define synchronous detection and explain the role of clippers in synchronous detector circuits.
- State the function of balanced modulators and describe the differences between lattice modulators and IC modulator circuits.
- Draw the basic components of both filter-type and phase-shift-type circuits for generation of SSB signals.

4-1 Basic Principles of Amplitude Modulation

Examining the basic equation for an AM signal, introduced in Chap. 3, gives us several clues as to how AM can be generated. The equation is

$$v_{AM} = V_c \sin 2\pi f_c t + (V_m \sin 2\pi f_m t)(\sin 2\pi f_c t)$$

where the first term is the sine wave carrier and second term is the product of the sine wave carrier and modulating signals. (Remember that v_{AM} is the instantaneous value of the amplitude modulation voltage.) The modulation index m is the ratio of the modulating signal amplitude to the carrier amplitude, or $m = V_m/V_c$, and so $V_m = mV_c$. Then substituting this for V_m in the basic equation yields $v_{AM} = V_c \sin 2\pi f_c t + (mV_c \sin 2\pi f_m t)(\sin 2\pi f_c t)$. Factoring gives $v_{AM} = V_c \sin 2\pi f_c t(1 + m \sin 2\pi f_m t)$.

AM in the Time Domain

When we look at the expression for v_{AM}, it is clear that we need a circuit that can multiply the carrier by the modulating signal and then add the carrier. A block diagram of such a circuit is shown in Fig. 4-1. One way to do this is to develop a circuit whose gain (or attenuation) is a function of $1 + m \sin 2\pi f_m t$. If we call that gain A, the expression for the AM signal becomes

$$v_{AM} = A(v_c)$$

where A is the gain or attenuation factor. Figure 4-2 shows simple circuits based on this expression. In Fig. 4-2(a), A is a gain greater than 1 provided by an amplifier. In Fig. 4-2(b), the carrier is attenuated by a voltage divider. The gain in this case is less than 1 and is therefore an attenuation factor. The carrier is multiplied by a fixed fraction A.

Now, if the gain of the amplifier or the attenuation of the voltage divider can be varied in accordance with the modulating signal plus 1, AM will be produced. In Fig. 4-2(a) the modulating signal would be used to increase or decrease the gain of the amplifier as the amplitude of the intelligence changed. In Fig. 4-2(b), the modulating

Figure 4-1   Block diagram of a circuit to produce AM.

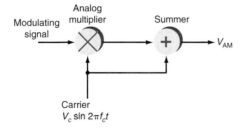

Figure 4-2   Multiplying the carrier by a fixed gain A.

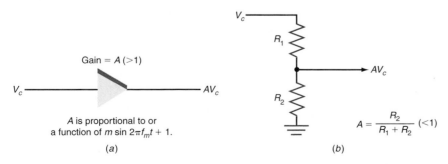

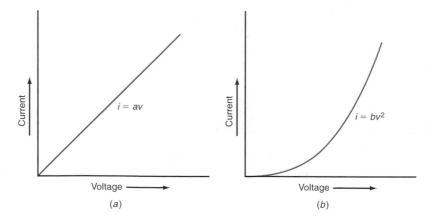

signal could be made to vary one of the resistances in the voltage divider, creating a varying attenuation factor. A variety of popular circuits permit gain or attenuation to be varied dynamically with another signal, producing AM.

AM in the Frequency Domain

Another way to generate the product of the carrier and modulating signal is to apply both signals to a nonlinear component or circuit, ideally one that generates a square-law function. A linear component or circuit is one in which the current is a linear function of the voltage [see Fig. 4-3(*a*)]. A resistor or linearly biased transistor is an example of a linear device. The current in the device increases in direct proportion to increases in voltage. The steepness or slope of the line is determined by the coefficient *a* in the expression $i = av$.

A nonlinear circuit is one in which the current is not directly proportional to the voltage. A common nonlinear component is a diode which has the nonlinear parabolic response shown in Fig. 4-3(*b*), where increasing the voltage increases the current but not in a straight line. Instead, the current variation is a square-law function. A *square-law function* is one that varies in proportion to the square of the input signals. A diode gives a good approximation of a square-law response. Bipolar and field-effect transistors (FETs) can also be biased to give a square-law response. An FET gives a near perfect square-law response, whereas diodes and bipolar transistors, which contain higher-order components, only approximate the square-law function.

The current variation in a typical semiconductor diode can be approximated by the equation ·

$$i = av + bv^2$$

where av is a linear component of the current equal to the applied voltage multiplied by the coefficient *a* (usually a dc bias) and bv^2 is second-order or square-law component of the current. Diodes and transistors also have higher-order terms, such as cv^3 and dv^4; however, these are smaller and often negligible and so are neglected in an analysis.

To produce AM, the carrier and modulating signals are added and applied to the nonlinear device. A simple way to do this is to connect the carrier and modulating sources in series and apply them to the diode circuit, as in Fig. 4-4. The voltage applied to the diode is then

$$v = v_c + v_m$$

The diode current in the resistor is

$$i = a(v_c + v_m) + b(v_c + v_m)^2$$

Expanding, we get

$$i = a(v_c + v_m) + b(v_c^2 + 2v_cv_m + v_m^2)$$

Square-law function

Figure 4-4   A square-law circuit for producing AM.

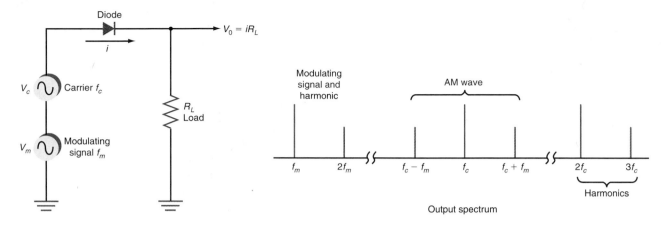

Output spectrum

Substituting the trigonometric expressions for the carrier and modulating signals, we let $v_c \sin 2\pi f_c t = v_c \sin \omega_c t$, where $\omega = 2\pi f_c$, and $v_m = \sin 2\pi f_m t = v_m \sin \omega_m t$, where $\omega_m = 2\pi f_m$. Then

$$i = aV_c \sin \omega_c t + aV_m \sin \omega_m t + bV_c^2 \sin^2 \omega_c t + 2bV_c V_m \sin \omega_c t \sin \omega_m t \\ + bv_m^2 \sin^2 \omega_m t$$

Next, substituting the trigonometric identity $\sin^2 A = 0.5(1 - \cos 2A)$ into the preceding expression gives the expression for the current in the load resistor in Fig. 4-4:

$$i = av_c \sin \omega_c t + av_m \sin \omega_m t + 0.5bv_c^2(1 - \cos 2\omega_c t) \\ + 2bv_c v_m \sin \omega_c t \sin \omega_m t + 0.5bv_m^2(1 - \cos \omega_m t)$$

The first term is the carrier sine wave, which is a key part of the AM wave; the second term is the modulating signal sine wave. Normally, this is not part of the AM wave. It is substantially lower in frequency than the carrier, so it is easily filtered out. The fourth term, the product of the carrier and modulating signal sine waves, defines the AM wave. If we make the trigonometric substitutions explained in Chap. 3, we obtain two additional terms—the sum and difference frequency sine waves, which are, of course, the upper and lower sidebands. The third term $\cos 2\omega_c t$ is a sine wave at 2 times the frequency of the carrier, i.e., the second harmonic of the carrier. The term $\cos 2\omega_m t$ is the second harmonic of the modulating sine wave. These components are undesirable, but are relatively easy to filter out. Diodes and transistors whose function is not a pure square-law function produce third-, fourth-, and higher-order harmonics, which are sometimes referred to as *intermodulation products* and which are also easy to filter out.

Intermodulation product

Figure 4-4 shows both the circuit and the output spectrum for a simple diode modulator. The output waveform is shown in Fig. 4-5. This waveform is a normal AM wave to which the modulating signal has been added.

If a parallel resonant circuit is substituted for the resistor in Fig. 4-4, the modulator circuit shown in Fig. 4-6 results. This circuit is resonant at the carrier frequency and has a bandwidth wide enough to pass the sidebands but narrow enough to filter out the modulating signal as well as the second- and higher-order harmonics of the carrier. The result is an AM wave across the tuned circuit.

This analysis applies not only to AM but also to frequency translation devices such as mixers, product detectors, phase detectors, balanced modulators, and other heterodyning circuits. In fact, it applies to any device or circuit that has a square-law function. It explains how sum and difference frequencies are formed and also explains why most mixing and modulation in any nonlinear circuit are accompanied by undesirable components such as harmonics and intermodulation products.

Figure 4-5 AM signal containing not only the carrier and sidebands but also the modulating signal.

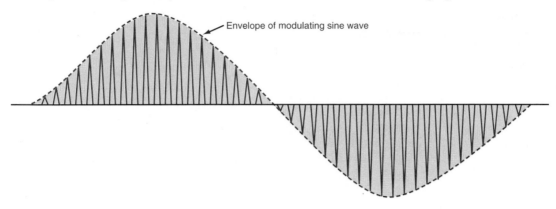

Envelope of modulating sine wave

Figure 4-6 The tuned circuit filters out the modulating signal and carrier harmonics, leaving only the carrier and sidebands.

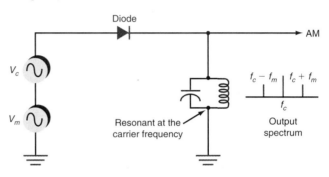

Diode

AM

V_c

V_m

Resonant at the carrier frequency

$f_c - f_m$ | $f_c + f_m$

f_c

Output spectrum

4-2 Amplitude Modulators

Amplitude modulators are generally one of two types: low level or high level. Low-level modulators generate AM with small signals and thus must be amplified considerably if they are to be transmitted. High-level modulators produce AM at high power levels, usually in the final amplifier stage of a transmitter. Although the discrete component circuits discussed in the following sections are still used to a limited extent, keep in mind that today most amplitude modulators and demodulators are in integrated-circuit form.

Low-level AM

Diode modulator

Low-Level AM

Diode Modulator. One of the simplest amplitude modulators is the *diode modulator* described in Sec. 4-1. The practical implementation shown in Fig. 4-7 consists of a resistive mixing network, a diode rectifier, and an *LC* tuned circuit. The carrier is applied to one input resistor and the modulating signal to the other. The mixed signals appear across R_3. This network causes the two signals to be linearly mixed, i.e., algebraically added. If both the carrier and the modulating signal are sine waves, the waveform resulting at the junction of the two resistors will be like that shown in Fig. 4-8(*c*), where the carrier wave is riding on the modulating signal. This signal is not AM. Modulation is a multiplication process, not an addition process.

The composite waveform is applied to a diode rectifier. The diode is connected so that it is forward-biased by the positive-going half-cycles of the input wave. During the negative portions of the wave, the diode is cut off and no signal passes. The current through the diode is a series of positive-going pulses whose amplitude varies in proportion to the amplitude of the modulating signal (see Fig. 4-8(*d*)].

Figure 4-7 Amplitude modulation with a diode.

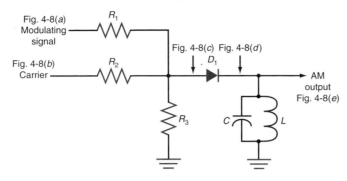

Figure 4-8 Waveforms in the diode modulator. (*a*) Modulating signal. (*b*) Carrier. (*c*) Linearly mixed modulating signal and carrier. (*d*) Positive-going signal after diode D_1. (*e*) AM output signal.

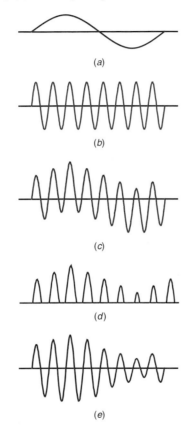

(*a*)

(*b*)

(*c*)

(*d*)

(*e*)

These positive-going pulses are applied to the parallel tuned circuit made up of L and C, which are resonant at the carrier frequency. Each time the diode conducts, a pulse of current flows through the tuned circuit. The coil and capacitor repeatedly exchange energy, causing an oscillation, or "ringing," at the resonant frequency. The oscillation of the tuned circuit creates one negative half-cycle for every positive input pulse. High-amplitude positive pulses cause the tuned circuit to produce high-amplitude negative pulses. Low-amplitude positive pulses produce corresponding low-amplitude negative pulses. The resulting waveform across the tuned circuit is an AM signal, as Fig. 4-8(*e*) illustrates. The Q of the tuned circuit should be high enough to eliminate the harmonics and produce a clean sine wave and to filter out the modulating signal, and low enough that its bandwidth accommodates the sidebands generated.

Figure 4-9   Simple transistor modulator.

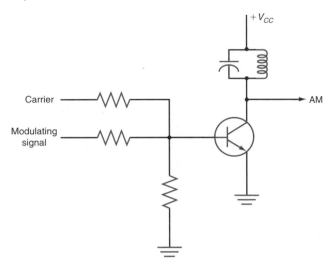

This signal produces high-quality AM, but the amplitudes of the signals are critical to proper operation. Because the nonlinear portion of the diode's characteristic curve occurs only at low voltage levels, signal levels must be low, less than a volt, to produce AM. At higher voltages, the diode current response is nearly linear. The circuit works best with millivolt-level signals.

Transistor Modulator. An improved version of the circuit just described is shown in Fig. 4-9. Because it uses a transistor instead of the diode, the circuit has gain. The emitter-base junction is a diode and a nonlinear device. Modulation occurs as described previously, except that the base current controls a larger collector current, and therefore the circuit amplifies. Rectification occurs because of the emitter-base junction. This causes larger half-sine pulses of current in the tuned circuit. The tuned circuit oscillates (rings) to generate the missing half-cycle. The output is a classic AM wave.

PIN Diode Modulator. Variable attenuator circuits for producing AM are shown in Fig. 4-10. These circuits use PIN diodes to produce AM at VHF, UHF, and microwave frequencies. The PIN diodes are a special type of silicon junction diode designed for use at frequencies above approximately 100 MHz. When forward-biased, these diodes act as variable resistors. The resistance of the diode varies linearly with the amount of current flowing through it. A high current produces a low resistance, whereas a low current produces a high resistance. As the modulating signal varies the forward-bias current through the PIN diode, AM is produced.

In Fig. 4-10(a), two PIN diodes are connected back to back and are forward-biased by a fixed negative dc voltage. The modulating signal is applied to the diodes through capacitor C_1. This ac modulating signal rides on the dc bias, adding to and subtracting from it and thus varying the resistance of the PIN diodes. These diodes appear in series with the carrier oscillator and the load. A positive-going modulating signal reduces the bias on the PIN diodes, causing their resistance to go up. This causes a reduction in amplitude of the carrier across the load. A negative-going modulating signal adds to the forward bias, causing the resistance of the diodes to go down, thereby increasing the carrier amplitude.

A variation of the PIN diode modulator circuit is shown in Fig. 4-10(b), where the diodes are arranged in a π network. This configuration is used when it is necessary to maintain a constant circuit impedance even under modulation.

In both circuits of Fig. 4-10, the PIN diodes form a variable attenuator circuit whose attenuation varies with the amplitude of the modulating signal. Such modulator circuits

Transistor modulator

PIN diode

GOOD TO KNOW

PIN modulators are widely used because they are one of the few methods available for producing AM at microwave frequencies.

Figure 4-10 High-frequency amplitude modulators using PIN diodes.

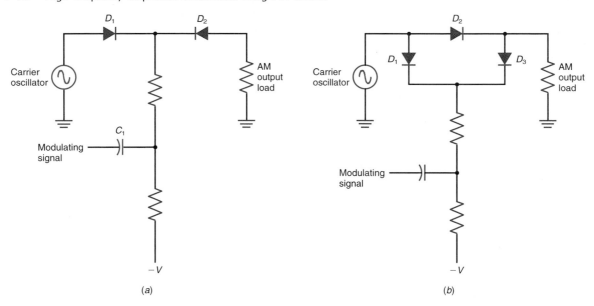

(a)   (b)

introduce a considerable amount of loss and must therefore be followed by amplifiers to increase the AM signal to a usable level. Despite this disadvantage, PIN modulators are widely used because they are one of the few methods available for producing AM at microwave frequencies.

Differential Amplifier. A *differential amplifier modulator* makes an excellent amplitude modulator. A typical circuit is shown in Fig. 4-11(a). Transistors Q_1 and Q_2 form the differential pair, and Q_3 is a constant-current source. Transistor Q_3 supplies a fixed emitter current I_E to Q_1 and Q_2, one-half of which flows in each transistor. The output is developed across the collector resistors R_1 and R_2. — Differential amplifier modulator

The output is a function of the difference between inputs V_1 and V_2; that is, $V_{out} = A(V_2 - V_1)$, where A is the circuit gain. The amplifier can also be operated with a single input. When this is done, the other input is grounded or set to zero. In Fig. 4-11(a), if V_1 is zero, the output is $V_{out} = A(V_2)$. If V_2 is zero, the output is $V_{out} = A(-V_1) = -AV_1$. This means that the circuit inverts V_1.

The output voltage can be taken between the two collectors, producing a *balanced,* or *differential,* output. The output can also be taken from the output of either collector to ground, producing a single-ended output. The two outputs are 180° out of phase with each other. If the balanced output is used, the output voltage across the load is twice the single-ended output voltage.

No special biasing circuits are needed, since the correct value of collector current is supplied directly by the constant-current source Q_3 in Fig. 4-11(a). Resistors R_3, R_4, and R_5, along with V_{EE}, bias the constant-current source Q_3. With no inputs applied, the current in Q_1 equals the current in Q_2, which is $I_E/2$. The balanced output at this time is zero. The circuit formed by R_1 and Q_1 and R_2 and Q_2 is a *bridge circuit.* When no inputs — Bridge circuit are applied, R_1 equals R_2, and Q_1 and Q_2 conduct equally. Therefore, the bridge is balanced and the output between the collectors is zero.

Now, if an input signal V_1 is applied to Q_1, the conduction of Q_1 and Q_2 is affected. Increasing the voltage at the base of Q_1 increases the collector current in Q_1 and decreases the collector current in Q_2 by an equal amount, so that the two currents sum to I_E. Decreasing the input voltage on the base of Q_1 decreases the collector current in Q_1 but increases it in Q_2. The sum of the emitter currents is always equal to the current supplied by Q_3.

Figure 4-11 (*a*) Basic differential amplifier. (*b*) Differential amplifier modulator.

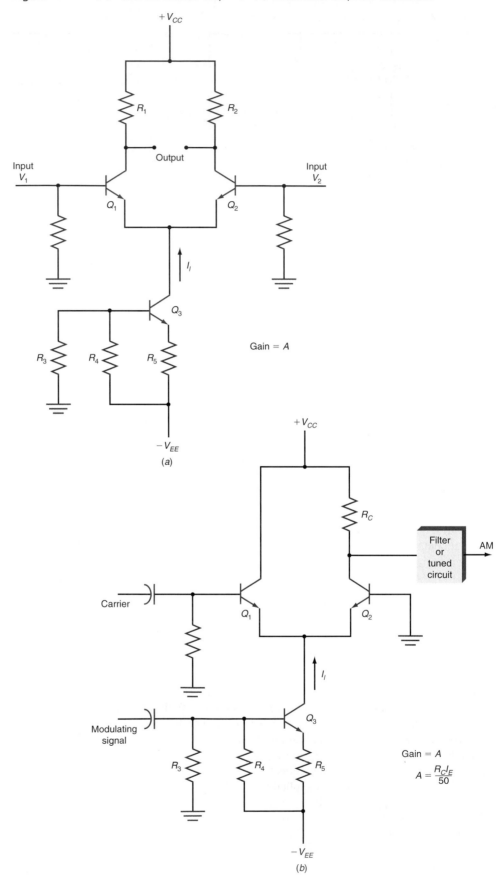

$$A = \frac{R_C I_E}{50}$$

The gain of a differential amplifier is a function of the emitter current and the value of the collector resistors. An approximation of the gain is given by the expression $A = R_C I_E / 50$. This is the single-ended gain, where the output is taken from one of the collectors with respect to ground. If the output is taken between the collectors, the gain is 2 times the above value.

Resistor R_C is the collector resistor value in ohms, and I_E is the emitter current in milliamperes. If $R_C = R_1 = R_2 = 4.7 \text{ k}\Omega$ and $I_E = 1.5 \text{ mA}$, the gain will be about $A = 4700(1.5)/50 = 7050/50 = 141$.

In most differential amplifiers, both R_C and I_E are fixed, providing a constant gain. But as the formula above shows, the gain is directly proportional to the emitter current. Thus if the emitter current can be varied in accordance with the modulating signal, the circuit will produce AM. This is easily done by changing the circuit only slightly, as in Fig. 4-11(b). The carrier is applied to the base of Q_1, and the base of Q_2 is grounded. The output, taken from the collector of Q_2, is single-ended. Since the output from Q_1 is not used, its collector resistor can be omitted with no effect on the circuit. The modulating signal is applied to the base of the constant-current source Q_3. As the intelligence signal varies, it varies the emitter current. This changes the gain of the circuit, amplifying the carrier by an amount determined by the modulating signal amplitude. The result is AM in the output.

This circuit, like the basic diode modulator, has the modulating signal in the output in addition to the carrier and sidebands. The modulating signal can be removed by using a simple high-pass filter on the output, since the carrier and sideband frequencies are usually much higher than that of the modulating signal. A bandpass filter centered on the carrier with sufficient bandwidth to pass the sidebands can also be used. A parallel tuned circuit in the collector of Q_2 replacing R_C can be used.

The differential amplifier makes an excellent amplitude modulator. It has a high gain and good linearity, and it can be modulated 100 percent. And if high-frequency transistors or a high-frequency IC differential amplifier is used, this circuit can be used to produce low-level modulation at frequencies well into the hundreds of megahertz. MOSFETs may be used in place of the bipolar transistors to produce a similar result in ICs.

Amplifying Low-Level AM Signals. In low-level modulator circuits such as those discussed above, the signals are generated at very low voltage and power amplitudes. The voltage is typically less than 1 V, and the power is in milliwatts. In systems using low-level modulation, the AM signal is applied to one or more linear amplifiers, as shown in Fig. 4-12, to increase its power level without distorting the signal. These amplifier

Figure 4-12 Low-level modulation systems use linear power amplifiers to increase the AM signal level before transmission.

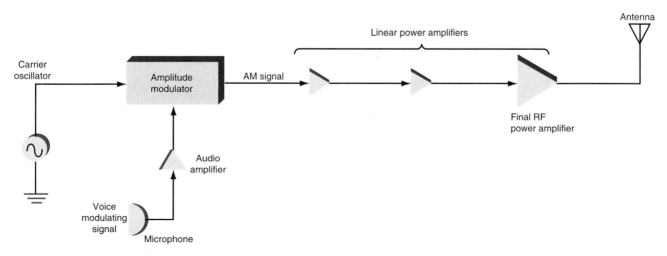

Figure 4-13 A high-level collector modulator.

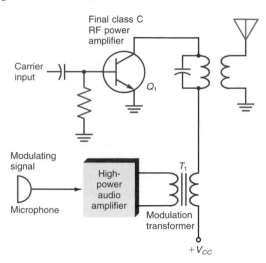

circuits—class A, class AB, or class B—raise the level of the signal to the desired power level before the AM signal is fed to the antenna.

High–Level AM

In high-level AM, the modulator varies the voltage and power in the final RF amplifier stage of the transmitter. The result is high efficiency in the RF amplifier and overall high-quality performance.

Collector Modulator. One example of a high-level modulator circuit is the *collector modulator* shown in Fig. 4-13. The output stage of the transmitter is a high-power class C amplifier. Class C amplifiers conduct for only a portion of the positive half-cycle of their input signal. The collector current pulses cause the tuned circuit to oscillate (ring) at the desired output frequency. The tuned circuit, therefore, reproduces the negative portion of the carrier signal (see Chap. 7 for more details).

The modulator is a linear power amplifier that takes the low-level modulating signal and amplifies it to a high-power level. The modulating output signal is coupled through modulation transformer T_1 to the class C amplifier. The secondary winding of the modulation transformer is connected in series with the collector supply voltage V_{CC} of the class C amplifier.

With a zero-modulation input signal, there is zero-modulation voltage across the secondary of T_1, the collector supply voltage is applied directly to the class C amplifier, and the output carrier is a steady sine wave.

When the modulating signal occurs, the ac voltage of the modulating signal across the secondary of the modulation transformer is added to and subtracted from the dc collector supply voltage. This varying supply voltage is then applied to the class C amplifier, causing the amplitude of the current pulses through transistor Q_1 to vary. As a result, the amplitude of the carrier sine wave varies in accordance with the modulated signal. When the modulation signal goes positive, it adds to the collector supply voltage, thereby increasing its value and causing higher current pulses and a higher-amplitude carrier. When the modulating signal goes negative, it subtracts from the collector supply voltage, decreasing it. For that reason, the class C amplifier current pulses are smaller, resulting in a lower-amplitude carrier output.

For 100 percent modulation, the peak of the modulating signal across the secondary of T_1 must be equal to the supply voltage. When the positive peak occurs, the voltage applied to the collector is twice the collector supply voltage. When the modulating signal goes negative, it subtracts from the collector supply voltage. When the negative peak is equal to the supply voltage, the effective voltage applied to the collector of Q_1 is zero, producing zero carrier output. This is illustrated in Fig. 4-14.

In practice, 100 percent modulation cannot be achieved with the high-level collector modulator circuit shown in Fig. 4-13 because of the transistor's nonlinear response to small signals. To overcome this problem, the amplifier driving the final class C amplifier is collector-modulated simultaneously.

High-level modulation produces the best type of AM, but it requires an extremely high-power modulator circuit. In fact, for 100 percent modulation, the power supplied by the modulator must be equal to one-half the total class C amplifier input power. If the class C amplifier has an input power of 1000 W, the modulator must be able to deliver one-half this amount, or 500 W.

Example 4–1

An AM transmitter uses high-level modulation of the final RF power amplifier, which has a dc supply voltage V_{CC} of 48 V with a total current I of 3.5 A. The efficiency is 70 percent.

a. What is the RF input power to the final stage?

DC input power $= P_i = V_{CC}I$ $P = 48 \times 3.5 = 168$ W

b. How much AF power is required for 100 percent modulation? (*Hint:* For 100 percent modulation, AF modulating power P_m is one-half the input power.)

$$P_m = \frac{P_i}{2} = \frac{168}{2} = 84 \text{ W}$$

c. What is the carrier output power?

$$\% \text{ efficiency} = \frac{P_{\text{out}}}{P_{\text{in}}} \times 100$$

$$P_{\text{out}} = \frac{\% \text{ efficiency} \times P_{\text{in}}}{100} = \frac{70(168)}{100} = 117.6 \text{ W}$$

d. What is the power in one sideband for 67 percent modulation?

$$P_s = \text{sideband power}$$

$$P_s = \frac{P_c(m^2)}{4}$$

$$m = \text{modulation percentage } (\%) = 0.67$$

$$P_c = 168$$

$$P_s = \frac{168(0.67)^2}{4} = 18.85 \text{ W}$$

e. What is the maximum and minimum dc supply voltage swing with 100 percent modulation? (See Fig. 4-14.)

$$\text{Minimum swing} = 0$$

$$\text{Supply voltage } v_{CC} = 48 \text{ V}$$

$$\text{Maximum swing } 2 \times V_{CC} = 2 \times 48 = 96 \text{ V}$$

Figure 4-14   For 100 percent modulation the peak of the modulating signal must be equal to V_{CC}.

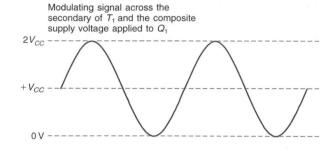

Modulating signal across the
secondary of T_1 and the composite
supply voltage applied to Q_1

$2V_{CC}$

$+V_{CC}$

0 V

Figure 4-15   Series modulation. Transistors may also be MOSFETs with appropriate biasing.

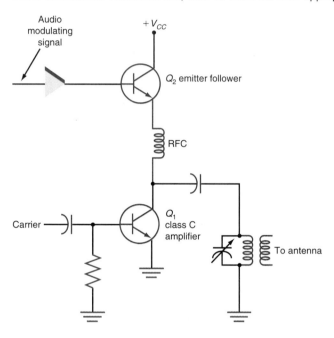

Audio
modulating
signal

$+V_{CC}$

Q_2 emitter follower

RFC

Carrier

Q_1
class C
amplifier

To antenna

Series modulator

Series Modulator. A major disadvantage of collector modulators is the need for a modulation transformer that connects the audio amplifier to the class C amplifier in the transmitter. The higher the power, the larger and more expensive the transformer. For very high power applications, the transformer is eliminated and the modulation is accomplished at a lower level with one of the many modulator circuits described in previous sections. The resulting AM signal is amplified by a high-power linear amplifier. This arrangement is not preferred because linear RF amplifiers are less efficient than class C amplifiers.

One approach is to use a transistorized version of a collector modulator in which a transistor is used to replace the transformer, as in Fig. 4-15. This series modulator replaces the transformer with an emitter follower. The modulating signal is applied to the emitter follower Q_2, which is an audio power amplifier. Note that the emitter follower appears in series with the collector supply voltage $+V_{CC}$. This causes the amplified audio modulating signal to vary the collector supply voltage to the class C amplifier Q_1, as illustrated in Fig. 4-14. And Q_2 simply varies the supply voltage to Q_1. If the modulating signal goes positive, the supply voltage to Q_1 increases; thus the carrier amplitude increases in proportion to the modulating signal. If the modulating signal goes negative, the supply voltage to Q_1 decreases, thereby decreasing the carrier amplitude in proportion to the modulating signal. For 100 percent modulation, the emitter follower can reduce the supply voltage to zero on maximum negative peaks.

Using this high-level modulating scheme eliminates the need for a large, heavy, and expensive transformer, and considerably improves frequency response. However, it is very inefficient. The emitter-follower modulator must dissipate as much power as the class C RF amplifier. For example, assume a collector supply voltage of 24 V and a collector current of 0.5 A. With no modulating signal applied, the percentage of modulation is 0. The emitter follower is biased so that the base and the emitter are at a dc voltage of about one-half the supply voltage, or in this example 12 V. The collector supply voltage on the class C amplifier is 12 V, and the input power is therefore

$$P_{in} = V_{CC}I_c = 12(0.5) = 6 \text{ W}$$

To produce 100 percent modulation, the collector voltage on Q_1 must double, as must the collector current. This occurs on positive peaks of the audio input, as described above. At this time most of the audio signal appears at the emitter of Q_1; very little of the signal appears between the emitter and collector of Q_2, and so at 100 percent modulation, Q_2 dissipates very little power.

When the audio input is at its negative peak, the voltage at the emitter of Q_2 is reduced to 12 V. This means that the rest of the supply voltage, or another 12 V, appears between the emitter and collector of Q_2. Since Q_2 must also be able to dissipate 6 W, it has to be a very large power transistor. The efficiency drops to less than 50 percent. With a modulation transformer, the efficiency is much greater, in some cases as high as 80 percent.

This arrangement is not practical for very high power AM, but it does make an effective higher-level modulator for power levels below about 100 W.

4-3 Amplitude Demodulators

Demodulators, or *detectors,* are circuits that accept modulated signals and recover the original modulating information. The demodulator circuit is the key circuit in any radio receiver. In fact, demodulator circuits can be used alone as simple radio receivers.

Demodulator (detector)

Diode Detectors

The simplest and most widely used amplitude demodulator is the *diode detector* (see Fig. 4-16). As shown, the AM signal is usually transformer-coupled and applied to a basic half wave rectifier circuit consisting of D_1 and R_1. The diode conducts when the positive half cycles of the AM signals occur. During the negative half cycles, the diode is reverse-biased and no current flows through it. As a result, the voltage across R_1 is a series of positive pulses whose amplitude varies with the modulating signal. A capacitor C_1 is connected across resistor R_1, effectively filtering out the carrier and thus recovering the original modulating signal.

Diode detector

One way to look at the operation of a diode detector is to analyze its operation in the time domain. The waveforms in Fig. 4-17 illustrate this. On each positive alternation of the AM signal, the capacitor charges quickly to the peak value of the pulses passed

Figure 4-16 A diode detector AM demodulator.

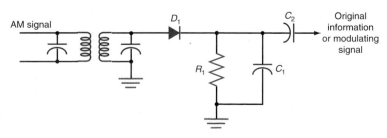

Figure 4-17 Diode detector waveforms.

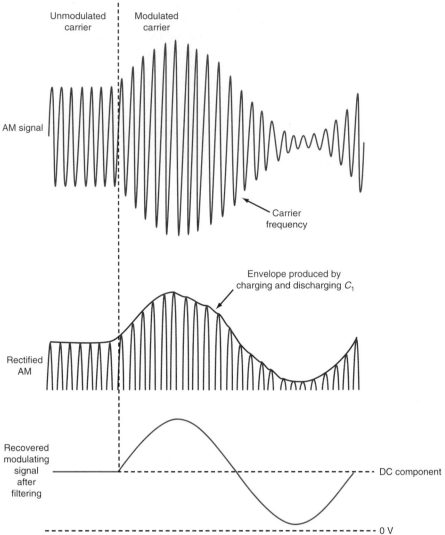

Unmodulated carrier | Modulated carrier

AM signal

Carrier frequency

Envelope produced by charging and discharging C_1

Rectified AM

Recovered modulating signal after filtering

DC component

0 V

by the diode. When the pulse voltage drops to zero, the capacitor discharges into resistor R_1. The time constant of C_1 and R_1 is chosen to be long compared to the period of the carrier. As a result, the capacitor discharges only slightly during the time that the diode is not conducting. When the next pulse comes along, the capacitor again charges to its peak value. When the diode cuts off, the capacitor again discharges a small amount into the resistor. The resulting waveform across the capacitor is a close approximation to the original modulating signal.

Because the capacitor charges and discharges, the recovered signal has a small amount of ripple on it, causing distortion of the modulating signal. However, because the carrier frequency is usually many times higher than the modulating frequency, these ripple variations are barely noticeable.

Because the diode detector recovers the envelope of the AM signal, which is the original modulating signal, the circuit is sometimes referred to as an *envelope detector*. Distortion of the original signal can occur if the time constant of the load resistor R_1 and the shunt filter capacitor C_1 is too long or too short. If the time constant is too long, the capacitor discharge will be too slow to follow the faster changes in the modulating signal. This is referred to as *diagonal distortion*. If the time constant is too short, the capacitor will discharge too fast and the carrier will not be sufficiently filtered out. The dc

Envelope detector

Diagonal distortion

Figure 4-18 Output spectrum of a diode detector.

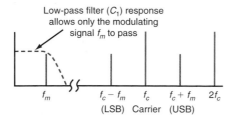

Low-pass filter (C_1) response
allows only the modulating
signal f_m to pass

f_m $f_c - f_m$ f_c $f_c + f_m$ $2f_c$
(LSB)   Carrier   (USB)

component in the output is removed with a series coupling or blocking capacitor, C_2 in Fig. 4-16, which is connected to an amplifier.

Another way to view the operation of the diode detector is in the frequency domain. In this case, the diode is regarded as a nonlinear device to which are applied multiple signals where modulation will take place. The multiple signals are the carrier and sidebands, which make up the input AM signal to be demodulated. The components of the AM signal are the carrier f_c, the upper sideband $f_c + f_m$, and the lower sideband $f_c - f_m$. The diode detector circuit combines these signals, creating the sum and difference signals:

$$f_c + (f_c + f_m) = 2f_c + f_m$$
$$f_c - (f_c + f_m) = -f_m$$
$$f_c + (f_c - f_m) = 2f_c - f_m$$
$$f_c - (f_c - f_m) = f_m$$

All these components appear in the output. Since the carrier frequency is very much higher than that of the modulating signal, the carrier signal can easily be filtered out with a simple low-pass filter. In a diode detector, this low-pass filter is just capacitor C_1 across load resistor R_1. Removing the carrier leaves only the original modulating signal. The frequency spectrum of a diode detector is illustrated in Fig. 4-18. The low-pass filter, C_1 in Fig. 4-16, removes all but the desired original modulating signal.

Crystal Radio Receivers

The crystal component of the *crystal radio receivers* that were widely used in the past is simply a diode. In Fig. 4-19 the diode detector circuit of Fig. 4-16 is redrawn, showing an antenna connection and headphones. A long wire antenna picks up the radio signal, which is inductively coupled to the secondary winding of T_1, which forms a series resonant circuit with C_1. Note that the secondary is not a parallel circuit, because the voltage induced into the secondary winding appears as a voltage source in series with the coil and capacitor. The variable capacitor C_1 is used to select a station. At resonance, the voltage across the capacitor is stepped up by a factor equal to the Q of the tuned circuit. This resonant voltage rise is a form of amplification. This higher-voltage signal

Crystal radio receiver

Figure 4-19 A crystal radio receiver.

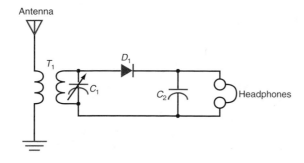

is applied to the diode. The diode detector D_1 and its filter C_2 recover the original modulating information, which causes current flow in the headphones. The headphones serve as the load resistance, and capacitor C_2 removes the carrier. The result is a simple radio receiver; reception is very weak because no active amplification is provided. Typically, a germanium diode is used because its voltage threshold is lower than that of a silicon diode and permits reception of weaker signals. Crystal radio receivers can easily be built to receive standard AM broadcasts.

Synchronous Detection

Synchronous detectors use an internal clock signal at the carrier frequency in the receiver to switch the AM signal off and on, producing rectification similar to that in a standard diode detector (see Fig. 4-20.) The AM signal is applied to a series switch that is opened and closed synchronously with the carrier signal. The switch is usually a diode or transistor that is turned on or off by an internally generated clock signal equal in frequency to and in phase with the carrier frequency. The switch in Fig. 4-20 is turned on by the clock signal during the positive half-cycles of the AM signal, which therefore appears across the load resistor. During the negative half-cycles of the AM signal, the clock turns the switch off, so no signal reaches the load or filter capacitor. The capacitor filters out the carrier.

A full wave synchronous detector is shown in Fig. 4-21. The AM signal is applied to both inverting and noninverting amplifiers. The internally generated carrier signal operates two switches A and B. The clock turns A on and B off or turns B on and A off. This arrangement simulates an electronic single-pole, double-throw (SPDT) switch. During positive half-cycles of the AM signal, the A switch feeds the noninverted AM output of positive half-cycles to the load. During the negative half-cycles of the input, the B switch connects the output of the inverter to the load. The negative half-cycles are inverted, becoming positive, and the signal appears across the load. The result is full wave rectification of the signal.

The key to making the synchronous detector work is to ensure that the signal producing the switching action is perfectly in phase with the received AM carrier. An internally generated carrier signal from, say, an oscillator will not work. Even though the frequency

Figure 4–20   Concept of a synchronous detector.

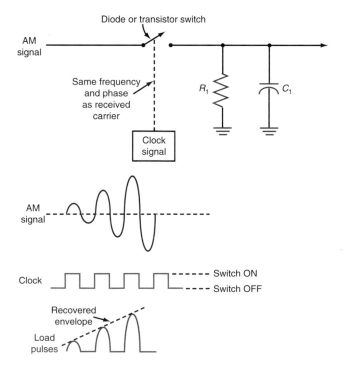

Figure 4-21 A full-wave synchronous detector.

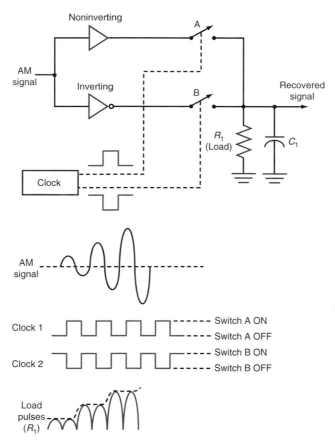

and phase of the switching signal might be close to those of the carrier, they would not be perfectly equal. However, there are a number of techniques, collectively referred to as *carrier recovery circuits,* that can be used to generate a switching signal that has the correct frequency and phase relationship to the carrier.

Carrier recovery circuit

A practical synchronous detector is shown in Fig. 4-22. A center-tapped transformer provides the two equal but inverted signals. The carrier signal is applied to the center tap. Note that one diode is connected oppositely from the way it would be if used in a full wave rectifier. These diodes are used as switches, which are turned off and on by the clock, which is used as the bias voltage. The carrier is usually a square wave derived by clipping and amplifying the AM signal. When the clock is positive, diode D_1 is

Figure 4-22 A practical synchronous detector.

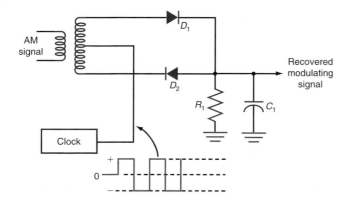

Figure 4-23 A simple carrier recovery circuit.

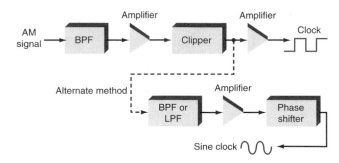

forward-biased. It acts as a short circuit and connects the AM signal to the load resistor. Positive half-cycles appear across the load.

When the clock goes negative, D_2 is forward-biased. During this time, the negative cycles of the AM signal are occurring, which makes the lower output of the secondary winding positive. With D_2 conducting, the positive half-cycles are passed to the load, and the circuit performs full wave rectification. As before, the capacitor across the load filters out the carrier, leaving the original modulating signal across the load.

The circuit shown in Fig. 4-23 is one way to supply the carrier to the synchronous detector. The AM signal to be demodulated is applied to a highly selective bandpass filter that picks out the carrier and suppresses the sidebands, thus removing most of the amplitude variations. This signal is amplified and applied to a clipper or limiter that removes any remaining amplitude variations from the signal, leaving only the carrier. The clipper circuit typically converts the sine wave carrier into a square wave that is amplified and thus becomes the clock signal. In some synchronous detectors, the clipped carrier is put through another bandpass filter to get rid of the square wave harmonics and generate a pure sine wave carrier. This signal is then amplified and used as the clock. A small phase shifter may be introduced to correct for any phase differences that occur during the carrier recovery process. The resulting carrier signal is exactly the same frequency and phase as those of the original carrier, as it is indeed derived from it. The output of this circuit is applied to the synchronous detector. Some synchronous detectors use a phase-locked loop to generate the clock, which is locked to the incoming carrier.

Synchronous detectors are also referred to as *coherent detectors,* and were known in the past as homodyne detectors. Their main advantage over standard diode detectors is that they have less distortion and a better signal-to-noise ratio. They are also less prone to *selective fading,* a phenomenon in which distortion is caused by the weakening of a sideband on the carrier during transmission.

Selective fading

4-4 Balanced Modulators

Balanced modulator

A *balanced modulator* is a circuit that generates a DSB signal, suppressing the carrier and leaving only the sum and difference frequencies at the output. The output of a balanced modulator can be further processed by filters or phase-shifting circuitry to eliminate one of the sidebands, resulting in an SSB signal.

Lattice Modulators

Lattice modulator (diode ring)

One of the most popular and widely used balanced modulators is the diode ring or *lattice modulator* in Fig. 4-24, consisting of an input transformer T_1, an output transformer T_2, and four diodes connected in a bridge circuit. The carrier signal is applied to the center taps of the input and output transformers, and the modulating signal is applied to the input transformer T_1. The output appears across the secondary of the output

Figure 4-24 Lattice-type balanced modulator.

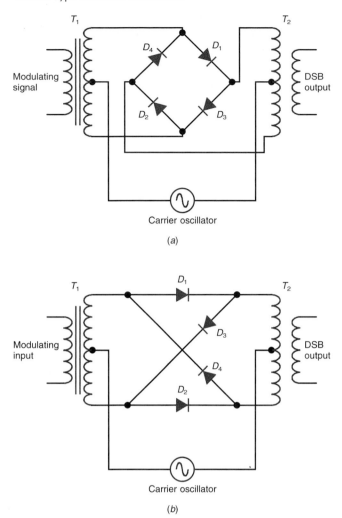

(a)

(b)

transformer T_2. The connections in Fig. 4-24(a) are the same as those in Fig. 4-24(b), but the operation of the circuit is perhaps more easily visualized as represented in part (b).

The operation of the lattice modulator is relatively simple. The carrier sine wave, which is usually considerably higher in frequency and amplitude than the modulating signal, is used as a source of forward and reverse bias for the diodes. The carrier turns the diodes off and on at a high rate of speed, and the diodes act as switches that connect the modulating signal at the secondary of T_1 to the primary of T_2.

Figures 4-25 and 4-26 show how lattice modulators operate. Assume that the modulating input is zero. When the polarity of the carrier is positive, as illustrated in Fig. 4-26(a), diodes D_1 and D_2 are forward-biased. At this time, D_3 and D_4 are reverse-biased and act as open circuits. As you can see, current divides equally in the upper and lower portions of the primary winding of T_2. The current in the upper part of the winding produces a magnetic field that is equal and opposite to the magnetic field produced by the current in the lower half of the secondary. The magnetic fields thus cancel each other out. No output is induced in the secondary, and the carrier is effectively suppressed.

When the polarity of the carrier reverses, as shown in Fig. 4-26(b), diodes D_1 and D_2 are reverse-biased and diodes D_3 and D_4 conduct. Again, the current flows in the secondary winding of T_1 and the primary winding of T_2. The equal and opposite magnetic fields produced in T_2 cancel each other out. The carrier is effectively balanced out, and its output is zero. The degree of carrier suppression depends on the degree of precision with which the transformers are made and the placement of the center tap: the goal is

Amplitude Modulator and Demodulator Circuits

Figure 4-25 Operation of the lattice modulator.

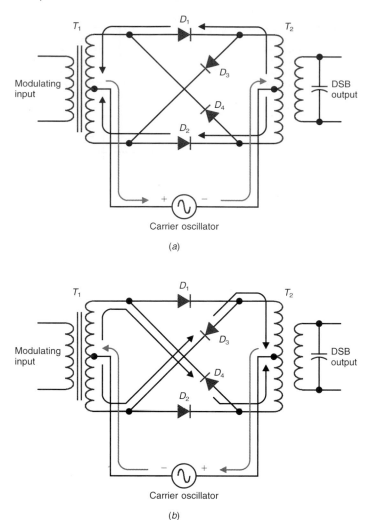

(a)

(b)

exactly equal upper and lower currents and perfect magnetic field cancellation. The degree of carrier attenuation also depends upon the diodes. The greatest carrier suppression occurs when the diode characteristics are perfectly matched. A carrier suppression of 40 dB is achievable with well-balanced components.

Now assume that a low-frequency sine wave is applied to the primary of T_1 as the modulating signal. The modulating signal appears across the secondary of T_1. The diode switches connect the secondary of T_1 to the primary of T_2 at different times depending upon the carrier polarity. When the carrier polarity is as shown in Fig. 4-26(a), diodes D_1 and D_2 conduct and act as closed switches. At this time, D_3 and D_4 are reverse-biased and are effectively not in the circuit. As a result, the modulating signal at the secondary of T_1 is applied to the primary of T_2 through D_1 and D_2.

When the carrier polarity reverses, D_1 and D_2 cut off and D_3 and D_4 conduct. Again, a portion of the modulating signal at the secondary of T_1 is applied to the primary of T_2, but this time the leads have been effectively reversed because of the connections of D_3 and D_4. The result is a 180° phase reversal. With this connection, if the modulating signal is positive, the output will be negative, and vice versa.

In Fig. 4-26, the carrier is operating at a considerably higher frequency than the modulating signal. Therefore, the diodes switch off and on at a high rate of speed, causing portions of the modulating signal to be passed through the diodes at different times. The DSB signal appearing across the primary of T_2 is illustrated in Fig. 4-26(c). The steep

Figure 4-26 Waveforms in the lattice-type balanced modulator. (*a*) Carrier. (*b*) Modulating signal. (*c*) DSB signal—primary T_2. (*d*) DSB output.

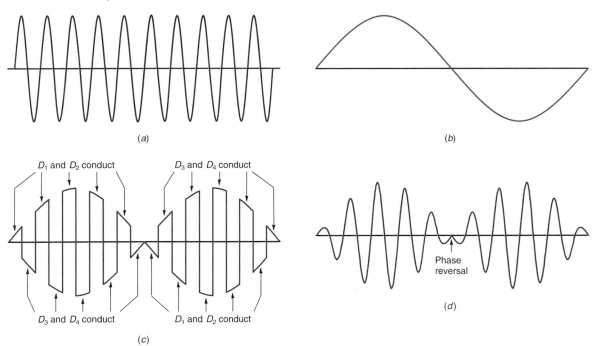

(*a*)

(*b*)

D_1 and D_2 conduct D_3 and D_4 conduct

D_3 and D_4 conduct D_1 and D_2 conduct

(*c*)

Phase
reversal

(*d*)

rise and fall of the waveform are caused by the rapid switching of the diodes. Because of the switching action the waveform contains harmonics of the carrier. Ordinarily, the secondary of T_2 is a resonant circuit as shown, and therefore the high-frequency harmonic content is filtered out, leaving a DSB signal like that shown in Fig. 4-26(*d*).

There are several important things to notice about this signal. First, the output waveform occurs at the carrier frequency. This is true even though the carrier has been removed. If two sine waves occurring at the sideband frequencies are added algebraically, the result is a sine wave signal at the carrier frequency with the amplitude variation shown in Fig. 4-26(*c*) or (*d*). Observe that the envelope of the output signal is *not* the shape of the modulating signal. Note also the phase reversal of the signal in the very center of the waveform, which is one indication that the signal being observed is a true DSB signal.

Although lattice modulators can be constructed of discrete components, they are usually available in a single module containing the transformers and diodes in a sealed package. The unit can be used as an individual component. The transformers are carefully balanced, and matched hot-carrier diodes are used to provide a wide operating frequency range and superior carrier suppression.

The diode lattice modulator shown in Fig. 4-25 uses one low-frequency iron-core transformer for the modulating signal and an air-core transformer for the RF output. This is an inconvenient arrangement because the low-frequency transformer is large and expensive. More commonly, two RF transformers are used, as shown in Fig. 4-27, where the modulating signal is applied to the center taps of the RF transformers. The operation of this circuit is similar to that of other lattice modulators.

IC Balanced Modulators

Another widely used balanced modulator circuit uses differential amplifiers. A typical example, the popular *1496/1596 IC balanced modulator,* is seen in Fig. 4-28. This circuit can work at carrier frequencies up to approximately 100 MHz and can achieve a carrier suppression of 50 to 65 dB. The pin numbers shown on the inputs and outputs of the IC are those for a standard 14-pin dual in-line package (DIP) IC. The device is also available in a 10-lead metal can.

GOOD TO KNOW

In DSB and SSB, the carrier that was suppressed at the DSB and SSB transmitter must be reinserted at the receiver to recover the intelligence.

1496/1596 IC balanced modulator

Figure 4-27 A modified version of the lattice modulator not requiring an iron-core transformer for the low-frequency modulating signal.

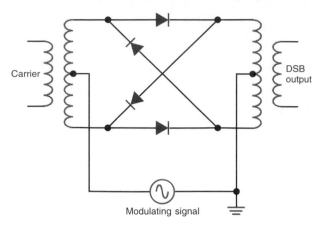

Figure 4-28 Integrated-circuit balanced modulator.

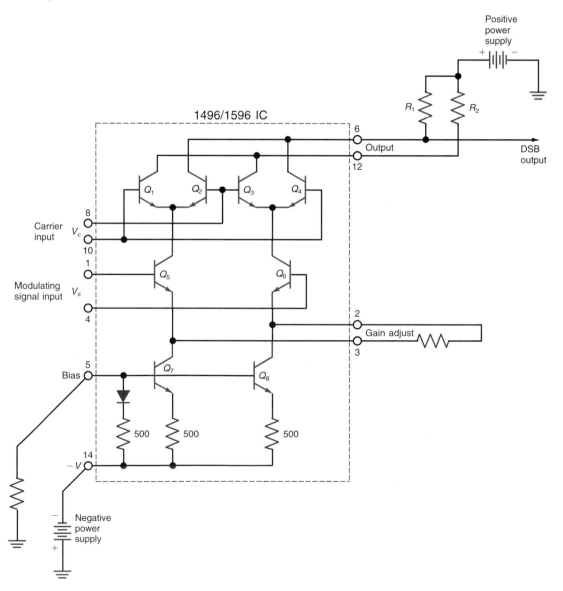

In Fig. 4-28, transistors Q_7 and Q_8 are constant-current sources that are biased with a single external resistor and the negative supply. They supply equal values of current to the two differential amplifiers. One differential amplifier is made up of Q_1, Q_2, and Q_5, and the other of Q_3, Q_4, and Q_6. The modulating signal is applied to the bases of Q_5 and Q_6. These transistors are connected in the current paths to the differential transistors and vary the amplitude of the current in accordance with the modulating signal. The current in Q_5 is 180° out of phase with the current in Q_6. As the current in Q_5 increases, the current through Q_6 decreases, and vice versa.

The differential transistors Q_1 through Q_4, which are controlled by the carrier, operate as switches. When the carrier input is such that the lower input terminal is positive with respect to the upper input terminal, transistors Q_1 and Q_4 conduct and act as closed switches and Q_2 and Q_3 are cut off. When the polarity of the carrier signal reverses, Q_1 and Q_4 are cut off and Q_2 and Q_3 conduct, acting as closed switches. These differential transistors, therefore, serve the same switching purpose as the diodes in the lattice modulator circuit discussed previously. They switch the modulating signal off and on at the carrier rate.

Assume that a high-frequency carrier wave is applied to switching transistors Q_1 and Q_4 and that a low-frequency sine wave is applied to the modulating signal input at Q_5 and Q_6. Assume that the modulating signal is positive-going so that the current through Q_5 increases while the current through Q_6 decreases. When the carrier polarity is positive, Q_1 and Q_4 conduct. As the current through Q_5 increases, the current through Q_1 and R_2 increases proportionately; therefore, the output voltage at the collector of Q_1 goes in a negative direction. As the current through Q_6 decreases, the current through Q_4 and R_1 decreases. Thus the output voltage at the collector of Q_4 increases. When the carrier polarity reverses, Q_2 and Q_3 conduct. The increasing current of Q_5 is passed through Q_2 and R_1 and therefore the output voltage begins to decrease. The decreasing current through Q_6 is now passed through Q_3 and R_2, causing the output voltage to increase. The result of the carrier switching off and on and the modulating signal varying as indicated produces the classical DSB output signal described before [see Fig. 4-26(c)]. The signal at R_1 is the same as the signal at R_2, but the two are 180° out of phase.

Figure 4-29 shows the 1496 connected to operate as a DSB or AM. The additional components are included in the circuit in Fig. 4-28 to provide for single-ended rather than balanced inputs to the carrier, modulating signal inputs, and a way to fine-tune the carrier balance. The potentiometer on pins 1 and 4 allows tuning for minimum carrier output, compensates for minor imbalances in the internal balanced modulator circuits, and corrects for parts tolerances in the resistors, thus giving maximum carrier suppression. The carrier suppression can be adjusted to at least 50 dB under most conditions and as high as 65 dB at low frequencies.

Applications for 1496/1596 ICs. The 1496 IC is one of the most versatile circuits available for communication applications. In addition to its use as balanced modulator, it can be reconfigured to perform as an amplitude modulator or as a synchronous detector.

In Fig. 4-29, the 1-kΩ resistors bias the differential amplifiers into the linear region so that they amplify the input carrier. The modulating signal is applied to the series emitter transistors Q_5 and Q_6. An adjustable network using a 50-kΩ potentiometer allows control of the amount of modulating signal that is applied to each internal pair of differential amplifiers. If the potentiometer is set near the center, the carrier balances out and the circuit functions as a balanced modulator. When the potentiometer is fine-tuned to the center position, the carrier is suppressed and the output is DSB AM.

If the potentiometer is offset one way or another, one pair of differential amplifiers receives little or no carrier amplification and the other pair gets all or most of the carrier. The circuit becomes a version of the differential amplifier modulator shown in Fig. 4-11(b). This circuit works quite nicely, but has very low input impedances. The carrier and modulating signal input impedances are equal to the input resistor values

Figure 4-29 AM modulator made with 1496 IC.

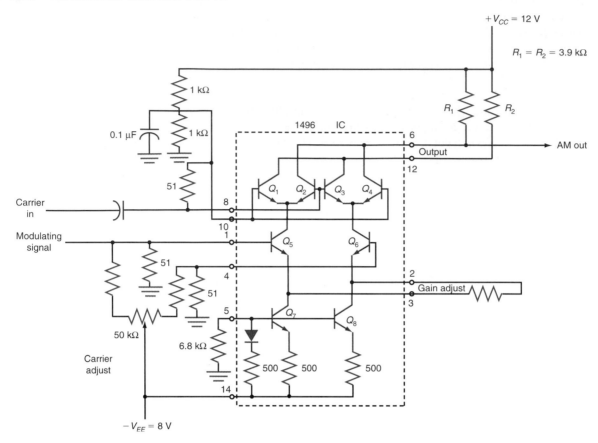

of 51 Ω. This means that the carrier and modulating signal sources must come from circuits with low output impedances, such as emitter followers or op amps.

Figure 4-30 shows the 1496 connected as a synchronous detector for AM. The AM signal is applied to the series emitter transistors Q_5 and Q_6, thus varying the emitter currents in the differential amplifiers, which in this case are used as switches to turn the AM signal off and on at the right time. The carrier must be in phase with the AM signal.

In this circuit, the carrier can be derived from the AM signal itself. In fact, connecting the AM signal to both inputs works if the AM signal is high enough in amplitude. When the amplitude is high enough, the AM signal drives the differential amplifier transistors Q_1 through Q_4 into cutoff and saturation, thereby removing any amplitude variations. Since the carrier is derived from the AM signal, it is in perfect phase to provide high-quality demodulation. The carrier variations are filtered from the output by an *RC* low-pass filter, leaving the recovered intelligence signal.

Analog Multiplier. Another type of IC that can be used as a balanced modulator is the *analog multiplier*. Analog multipliers are often used to generate DSB signals. The primary difference between an IC balanced modulator and an analog multiplier is that the balanced modulator is a switching circuit. The carrier, which may be a rectangular wave, causes the differential amplifier transistors to turn off and on to switch the modulating signal. The analog multiplier uses differential amplifiers, but they operate in the linear mode. The carrier must be a sine wave, and the analog multiplier produces the true product of two analog inputs.

IC Devices. In large-scale integrated circuits in which complete receivers are put on a single silicon chip, the circuits described here are applicable. However, the circuitry is more likely to be implemented with MOSFETs instead of bipolar transistors.

Analog multiplier

IC devices

Figure 4-30   Synchronous AM detector using a 1496.

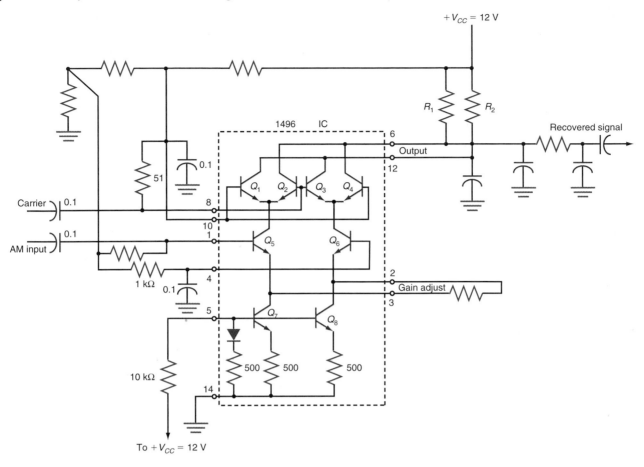

4-5 SSB Circuits

Generating SSB Signals: The Filter Method

The simplest and most widely used method of generating SSB signals is the filter method. Figure 4-31 shows a general block diagram of an SSB transmitter using the filter method. The modulating signal, usually voice from a microphone, is applied to the audio amplifier, the output of which is fed to one input of a balanced modulator. A crystal oscillator provides the carrier signal, which is also applied to the balanced modulator. The output of the balanced modulator is a *double-sideband (DSB)* signal. An SSB signal is produced by passing the DSB signal through a highly selective bandpass filter that selects either the upper or lower sideband.

SSB circuit

Double-sideband (DSB)

The primary requirement of the filter is, of course, that it pass only the desired sideband. Filters are usually designed with a bandwidth of approximately 2.5 to 3 kHz, making them wide enough to pass only standard voice frequencies. The sides of the filter response curve are extremely steep, providing for excellent selectivity. Filters are fixed tuned devices; i.e., the frequencies they can pass are not alterable. Therefore, the carrier oscillator frequency must be chosen so that the sidebands fall within the filter bandpass. Many commercially available filters are tuned to the 455-kHz, 3.35-MHz, or 9-MHz frequency ranges, although other frequencies are also used. Digital signal processing (DSP) filters are also used in modern equipment.

With the filter method, it is necessary to select either the upper or the lower sideband. Since the same information is contained in both sidebands, it generally makes no difference which one is selected, provided that the same sideband is used in both

Figure 4-31 An SSB transmitter using the filter method.

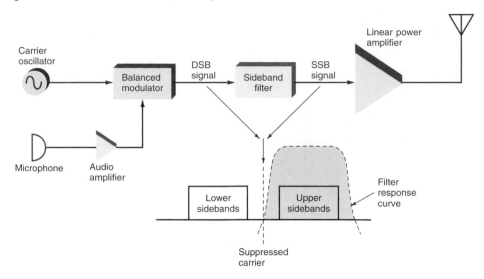

transmitter and receiver. However, the choice of the upper or lower sideband as a standard varies from service to service, and it is necessary to know which has been used to properly receive an SSB signal.

There are two methods of sideband selection. Many transmitters simply contain two filters, one that will pass the upper sideband and another that will pass the lower sideband, and a switch is used to select the desired sideband [Fig. 4-32(a)]. An alternative method is to provide two carrier oscillator frequencies. Two crystals change the carrier

Figure 4-32 Methods of selecting the upper or lower sideband. (a) Two filters. (b) Two carrier frequencies.

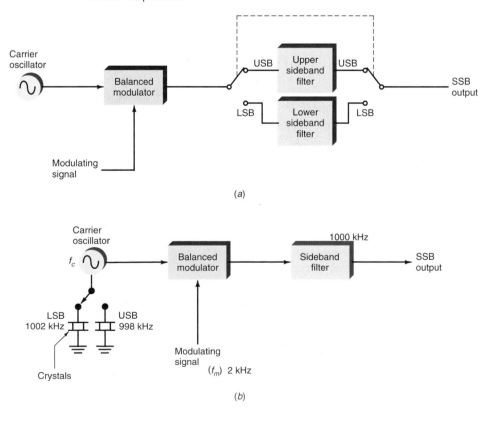

oscillator frequency to force either the upper sideband or the lower sideband to appear in the filter bandpass (see Fig. 4-32(b)].

As an example, assume that a bandpass filter is fixed at 1000 kHz and the modulating signal f_m is 2 kHz. The balanced modulator generates the sum and difference frequencies. Therefore, the carrier frequency f_c must be chosen so that the USB or LSB is at 1000 kHz. The balanced modulator outputs are USB $= f_c + f_m$ and LSB $= f_c - f_m$.

To set the USB at 1000 kHz, the carrier must be $f_c + f_m = 1000, f_c + 2 = 1000$, and $f_c = 1000 - 2 = 998$ kHz. To set the LSB at 1000 kHz, the carrier must be $f_c - f_m = 1000, f_c - 2 = 1000$, and $f_c = 1000 + 2 = 1002$ kHz.

Crystal filters, which are low in cost and relatively simple to design, are by far the most commonly used filters in SSB transmitters. Their very high Q provides extremely good selectivity. Ceramic filters are used in some designs. Typical center frequencies are 455 kHz and 10.7 MHz. DSP filters are also used in contemporary designs.

Example 4-2

An SSB transmitter using the filter method of Fig. 4-31 operates at a frequency of 4.2 MHz. The voice frequency range is 300 to 3400 Hz.

a. Calculate the upper and lower sideband ranges.

Upper sideband

Lower limit $f_{LL} = f_c + 300 = 4,200,000 + 300 = 4,200,300$ Hz

Upper limit $f_{UL} = f_c + 3400 = 4,200,000 + 3400$

$$= 4,203,400 \text{ Hz}$$

Range, USB $= 4,200,300$ to $4,203,400$ Hz

Lower sideband

Lower limit $f_{LL} = f_c - 300 = 4,200,000 - 300 = 4,199,700$ Hz

Upper limit $f_{UL} = f_c - 3400 = 4,200,000 - 3400$

$$= 4,196,600 \text{ Hz}$$

Range, LSB $= 4,196,000$ to $4,199,700$ Hz

b. What should be the approximate center frequency of a bandpass filter to select the lower sideband? The equation for the center frequency of the lower sideband f_{LSB} is

$$f_{LSB} = \sqrt{f_{LL} \, f_{UL}} = \sqrt{4,196,660 \times 4,199,700} = 4,198,149.7 \text{ Hz}$$

An approximation is

$$f_{LSB} = \frac{f_{LL} + f_{UL}}{2} = \frac{4,196,600 + 4,199,700}{2} = 4,198,150 \text{ Hz}$$

Generating SSB Signals: Phasing

The phasing method of SSB generation uses a phase-shift technique that causes one of the sidebands to be canceled out. A block diagram of a phasing-type SSB generator is shown in Fig. 4-33. It uses two balanced modulators, which effectively eliminate the carrier. The carrier oscillator is applied directly to the upper balanced modulator along with the audio modulating signal. The carrier and modulating signal are then both shifted in phase by 90° and applied to the second, lower, balanced modulator. The phase-shifting

Figure 4-33 An SSB generator using the phasing method.

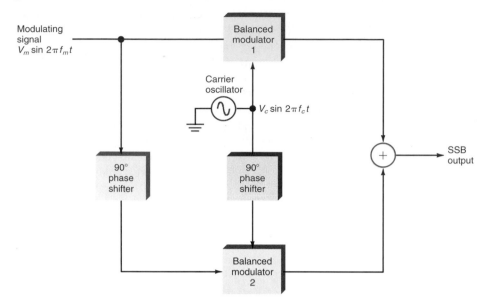

action causes one sideband to be canceled out when the two balanced modulator outputs are added to produce the output.

The carrier signal is $V_c \sin 2\pi f_c t$. The modulating signal is $V_m \sin 2\pi f_m t$. Balanced modulator 1 produces the product of these two signals: $(V_m \sin 2\pi f_m t)\,(V_c \sin 2\pi f_c t)$. Applying a common trigonometric identity

$$\sin A \sin B = 0.5[\cos(A - B) - \cos(A + B)]$$

we have

$$(V_m \sin 2\pi f_m t)\,(V_c \sin 2\pi f_c t) = 0.5 V_m V_c [\cos(2\pi f_c - 2\pi f_m)t - \cos(2\pi f_c + 2\pi f_m)t]$$

Note that these are the sum and difference frequencies or the upper and lower sidebands.

It is important to remember that a cosine wave is simply a sine wave shifted by 90°; that is, it has exactly the same shape as a sine wave, but it occurs 90° earlier in time. A cosine wave *leads* a sine wave by 90°, and a sine wave *lags* a cosine wave by 90°.

The 90° phase shifters in Fig. 4-33 create cosine waves of the carrier and modulating signals that are multiplied in balanced modulator 2 to produce $(V_m \cos 2\pi f_m t) \times (V_c \cos 2\pi f_c t)$. Applying another common trigonometric identity

$$\cos A \cos B = 0.5[\cos(A - B) + \cos(A + B)]$$

we have

$$(V_m \cos 2\pi f_m t)(V_c \cos 2\pi f_c t) = 0.5 V_m V_c [\cos(2\pi f_c - 2\pi f_m)t + \cos(2\pi f_c + 2\pi f_m)t]$$

When you add the sine expression given previously to the cosine expression just above, the sum frequencies cancel and the difference frequencies add, producing only the lower sideband $\cos[(2\pi f_c - 2\pi f_m)t]$.

Carrier Phase Shift. A phase shifter is usually an *RC* network that causes the output to either lead or lag the input by 90°. Many different kinds of circuits have been devised for producing this phase shift. A simple *RF* phase shifter consisting of two *RC* sections, each set to produce a phase shift of 45°, is shown in Fig. 4-34. The section made up of R_1 and C_1 produces an output that lags the input by 45°. The section made up of C_2 and R_2 produces a phase shift that leads the input by 45°. The total phase shift between the two outputs is 90°. One output goes to balanced modulator 1, and the other goes to balanced modulator 2.

Figure 4-34   A single-frequency 90° phase shifter.

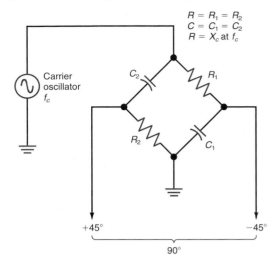

$$R = R_1 = R_2$$
$$C = C_1 = C_2$$
$$R = X_c \text{ at } f_c$$

Since a phasing-type SSB generator can be made with IC balanced modulators such as the 1496 and since these can be driven by a square wave carrier frequency signal, a digital phase shifter can be used to provide the two carrier signals that are 90° out of phase. Figure 4-35 shows two D-type flip-flops connected as a simple shift register with feedback from the complement output of the B flip-flop to the D input of the A flip-flop. Also JK flip-flops could be used. It is assumed that the flip-flops trigger or change state on the negative-going edge of the clock signal. The clock signal is set to a frequency exactly 4 times higher than the carrier frequency. With this arrangement, each flip-flop produces a 50 percent duty cycle square wave at the carrier frequency, and the two signals are exactly 90° out of phase with each other. These signals drive the differential amplifier switches in the 1496 balanced modulators, and this phase relationship is maintained regardless of the clock or carrier frequency. TTL flip-flops can be used at frequencies up to about 50 MHz. For higher frequencies, in excess of 100 MHz, emitter

Figure 4-35   A digital phase shifter.

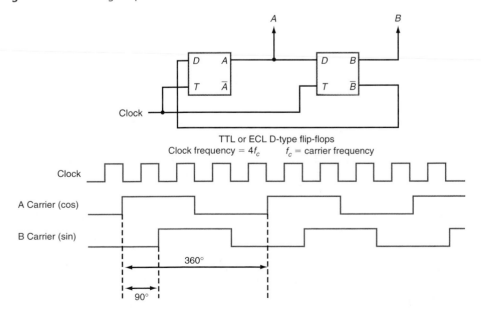

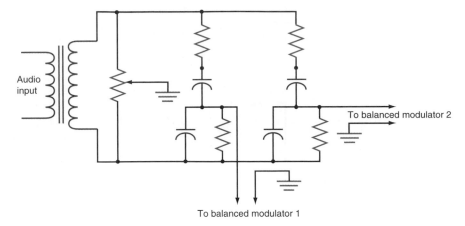

coupled logic (ECL) flip-flops can be used. In CMOS integrated circuits, this technique is useful to frequencies up to 10 GHz.

Audio Phase Shift. The most difficult part of creating a phasing-type SSB generator is to design a circuit that maintains a constant 90° phase shift over a wide range of audio modulating frequencies. (Keep in mind that a phase shift is simply a time shift between sine waves of the same frequency.) An *RC* network produces a specific amount of phase shift at only one frequency because the capacitive reactance varies with frequency. In the carrier phase shifter, this is not a problem, since the carrier is maintained at a constant frequency. However, the modulating signal is usually a band of frequencies, typically in the audio range from 300 to 3000 Hz.

One of the circuits commonly used to produce a 90° phase shift over a wide bandwidth is shown in Fig. 4-36. The phase-shift difference between the output to modulator 1 and the output to modulator 2 is 90° ± 1.5° over the 300- to 3000-Hz range. Resistor and capacitor values must be carefully selected to ensure phase-shift accuracy, since inaccuracies cause incomplete cancellation of the undesired sideband.

A wideband audio phase shifter that uses an op amp in an active filter arrangement is shown in Fig. 4-37. Careful selection of components will ensure that the phase shift of the output will be close to 90° over the audio frequency range of 300 to 3000 Hz. Greater precision of phase shift can be obtained by using multiple stages, with each stage having different component values and therefore a different phase-shift value. The phase shifts in the multiple stages produce a total shift of 90°.

Figure 4-37 An active phase shifter.

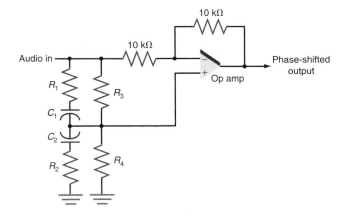

The phasing method can be used to select either the upper or the lower sideband. This is done by changing the phase shift of either the audio or the carrier signals to the balanced modulator inputs. For example, applying the direct audio signal to balanced modulator 2 in Fig. 4-33 and the 90° phase-shifted signal to balanced modulator 1 will cause the upper sideband to be selected instead of the lower sideband. The phase relationship of the carrier can also be switched to make this change.

The output of the phasing generator is a low-level SSB signal. The degree of suppression of the carrier depends on the configuration and precision of the balanced modulators, and the precision of the phase shifting determines the degree of suppression of the unwanted sideband. The design of phasing-type SSB generators is critical if complete suppression of the undesired sideband is to be achieved. The SSB output is then applied to linear RF amplifiers, where its power level is increased before being applied to the transmitting antenna.

DSB and SSB Demodulation

To recover the intelligence in a DSB or SSB signal, the carrier that was suppressed at the receiver must be reinserted. Assume, e.g., that a 3-kHz sine wave tone is transmitted by modulating a 1000-kHz carrier. With SSB transmission of the upper sideband, the transmitted signal is $1000 + 3 = 1003$ kHz. Now at the receiver, the SSB signal (the 1003-kHz USB) is used to modulate a carrier of 1000 kHz. See Fig. 4-38(a). If a balanced modulator is used, the 1000-kHz carrier is suppressed, but the sum and difference signals are generated. The balanced modulator is called a *product detector* because it is used to recover the modulating signal rather than generate a carrier that will transmit it. The sum and difference frequencies produced are

Product detector

$$\text{Sum:} \quad 1003 + 1000 = 2003 \text{ kHz}$$
$$\text{Difference:} \quad 1003 - 1000 = 3 \text{ kHz}$$

The difference is, of course, the original intelligence or modulating signal. The sum, the 2003-kHz signal, has no importance or meaning. Since the two output frequencies of the balanced modulator are so far apart, the higher undesired frequency is easily filtered out by a low-pass filter that keeps the 3-kHz signal but suppresses everything above it.

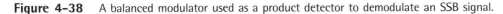

Figure 4-38 A balanced modulator used as a product detector to demodulate an SSB signal.

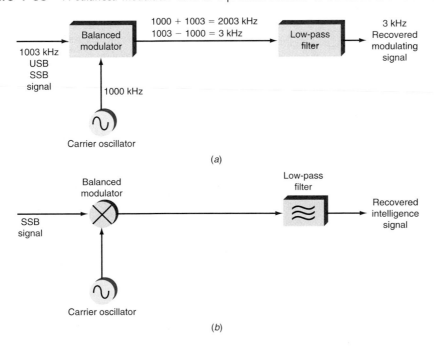

Any balanced modulator can be used as a product detector to demodulate SSB signals. Many special product detector circuits have been developed over the years. Lattice modulators or ICs such as the 1496 both make good product detectors. All that needs to be done is to connect a low-pass filter on the output to get rid of the undesired high-frequency signal while passing the desired difference signal. Figure 4-38(*b*) shows a widely accepted convention for representing balanced modulator circuits. Note the special symbols used for the balanced modulator and low-pass filter.

CHAPTER REVIEW

Summary

One type of AM circuit varies the gain of the amplifier or the attenuation of the voltage divider according to the modulating signal plus 1. Another applies the product of the carrier and modulating signals to a nonlinear component or circuit. A parallel tuned circuit resonant at the carrier frequency, with a bandwidth wide enough to filter out the modulating signal as well as the second and higher harmonics of the carrier, can be used to produce an AM wave.

Low-level AM can be produced by many types of circuits. In high-level modulation, the modulator varies the voltage and power in the transmitter's final RF amplifier stage.

Demodulator (detector) circuits accept a modulated signal and recover the original modulating information. The basic AM detector is a half wave rectifier. Synchronous detectors use an internal clock signal to switch the AM signal off and on, producing rectification.

A balanced modulator is a circuit that generates a DSB signal. The diode ring or lattice modulator is a widely used balanced modulator.

Filters used to generate SSB signals must have high selectivity. Crystal filters are the most common but DSP is becoming more widely used.

Product detectors, which are circuits for demodulating or detecting DSB or SSB signals, generate the mathematical product of the SSB signal and the carrier.

Questions

1. What mathematical operation does an amplitude modulator perform?
2. A device that produces amplitude modulation must have what type of response curve?
3. Describe the two basic ways in which amplitude modulator circuits generate AM.
4. What type of semiconductor device gives a near-perfect square-law response?
5. Which four signals and frequencies appear at the output of a low-level diode modulator?
6. What component does a PIN diode appear to be when it is used in an amplitude modulator?
7. Name the primary application of PIN diodes as amplitude modulators.
8. What kind of amplifier must be used to boost the power of a low-level AM signal?
9. How does a differential amplifier modulator work?
10. To what stage of a transmitter does the modulator connect in a high-level AM transmitter?
11. What is the simplest and most common technique for demodulating an AM signal?

12. What is the most critical component value in a diode detector circuit? Explain.
13. What is the basic component in a synchronous detector? What operates this component?
14. What signals does a balanced modulator generate? Eliminate?
15. What type of balanced modulator uses transformers and diodes?
16. What is the most commonly used filter in a filter-type SSB generator?
17. What is the most difficult part of producing SSB for voice signals by using the phasing methods?
18. Which type of balanced modulator gives the greatest carrier suppression?
19. What is the name of the circuit used to demodulate an SSB signal?
20. What signal must be present in an SSB demodulator besides the signal to be detected?

Problems

1. A collector modulated transmitter has a supply voltage of 48 V and an average collector current of 600 mA. What is the input power to the transmitter? How much modulating signal power is needed to produce 100 percent modulation? ◆
2. An SSB generator has a 9-MHz carrier and is used to pass voice frequencies in the 300- to 3300-Hz range. The lower sideband is selected. What is the approximate center frequency of the filter needed to pass the lower sideband?
3. A 1496 IC balanced modulator has a carrier-level input of 200 mV. The amount of suppression achieved is 60 dB. How much carrier voltage appears at the output? ◆

◆ *Answers to Selected Problems follow Chap. 22.*

Critical Thinking

1. State the relative advantages and disadvantages of synchronous detectors versus other types of amplitude demodulators.
2. Could a balanced modulator be used as a synchronous detector? Why or why not?
3. An SSB signal is generated by modulating a 5-MHz carrier with a 400-Hz sine tone. At the receiver, the carrier is reinserted during demodulation, but its frequency is 5.00015 MHz rather than exactly 5 MHz. How does this affect the recovered signal? How would a voice signal be affected by a carrier that is not exactly the same as the original?

5

Fundamentals of Frequency Modulation

A sine wave carrier can be modulated by varying its amplitude, frequency, or phase shift. The basic equation for a carrier wave is

$$v = V_c \sin(2\pi ft \pm \theta)$$

where V_c = peak amplitude, f = frequency, and θ = phase angle

Impressing an information signal on a carrier by changing its frequency produces FM. Varying the amount of phase shift that a carrier experiences is known as phase modulation (PM). Varying the phase shift of a carrier also produces FM. FM and PM are collectively referred to as *angle modulation.* Since FM is generally superior in performance to AM, it is widely used in many areas of communication electronics.

Angle modulation

Objectives

After completing this chapter, you will be able to:

- Compare and contrast frequency modulation and phase modulation.
- Calculate the modulation index given the maximum deviation and the maximum modulating frequency and use the modulation index and Bessel coefficients to determine the number of significant sidebands in an FM signal.
- Calculate the bandwidth of an FM signal by using two methods and explain the difference between the two.
- Explain how preemphasis is used to solve the problem of the interference of high-frequency components by noise.
- List the advantages and disadvantages of FM as compared to AM.
- Give the reasons for FM's superior immunity to noise.

5-1 Basic Principles of Frequency Modulation

In FM, the carrier amplitude remains constant and the carrier frequency is changed by the modulating signal. As the amplitude of the information signal varies, the carrier frequency shifts proportionately. As the modulating signal amplitude increases, the carrier frequency increases. If the amplitude of the modulating signal decreases, the carrier frequency decreases. The reverse relationship can also be implemented. A decreasing modulating signal increases the carrier frequency above its center value, whereas an increasing modulating signal decreases the carrier frequency below its center value. As the modulating signal amplitude varies, the carrier frequency varies above and below its normal center, or *resting,* frequency with no modulation. The amount of change in carrier frequency produced by the modulating signal is known as the *frequency deviation* f_d. Maximum frequency deviation occurs at the maximum amplitude of the modulating signal.

The frequency of the modulating signal determines the frequency deviation rate, or how many times per second the carrier frequency deviates above and below its center frequency. If the modulating signal is a 500-Hz sine wave, the carrier frequency shifts above and below the center frequency 500 times per second.

An FM signal is illustrated in Fig. 5-1(c). Normally the carrier [Fig. 5-1(a)] is a sine wave, but it is shown as a triangular wave here to simplify the illustration. With no modulating signal applied, the carrier frequency is a constant-amplitude sine wave at its normal resting frequency.

The modulating information signal [Fig. 5-1(b)] is a low-frequency sine wave. As the sine wave goes positive, the frequency of the carrier increases proportionately. The highest frequency occurs at the peak amplitude of the modulating signal. As the modulating signal amplitude decreases, the carrier frequency decreases. When the modulating signal is at zero amplitude, the carrier is at its center frequency point.

When the modulating signal goes negative, the carrier frequency decreases. It continues to decrease until the peak of the negative half-cycle of the modulating sine wave is reached. Then as the modulating signal increases toward zero, the carrier frequency again increases. This phenomenon is illustrated in Fig. 5-1(c), where the carrier sine waves seem to be first compressed and then stretched by the modulating signal.

Assume a carrier frequency of 150 MHz. If the peak amplitude of the modulating signal causes a maximum frequency shift of 30 kHz, the carrier frequency will deviate up to 150.03 MHz and down to 149.97 MHz. The total frequency deviation is 150.03 − 149.97 = 0.06 MHz = 60 kHz. In practice, however, the frequency deviation is expressed as the amount of frequency shift of the carrier above or below the center frequency. Thus the frequency deviation for the 150-MHz carrier frequency is represented as ± 30 kHz. This means that the modulating signal varies the carrier above and below its center frequency by

Frequency modulation

Frequency deviation f_d

GOOD TO KNOW

The frequency of the modulating signal determines the frequency deviation rate, or how many times per second the carrier frequency deviates above and below its center frequency.

Example 5-1

A transmitter operates on a frequency of 915 MHz. The maximum FM deviation is ± 12.5 kHz. What are the maximum and minimum frequencies that occur during modulation?

$$915 \text{ MHz} = 915{,}000 \text{ kHz}$$

$$\text{Maximum deviation} = 915{,}000 + 12.5 = 915{,}012.5 \text{ kHz}$$

$$\text{Minimum deviation} = 915{,}000 - 12.5 = 914{,}987.5 \text{ kHz}$$

Figure 5-1 FM and PM signals. The carrier is drawn as a triangular wave for simplicity, but in practice it is a sine wave. (*a*) Carrier. (*b*) Modulating signal. (*c*) FM signal. (*d*) PM signal.

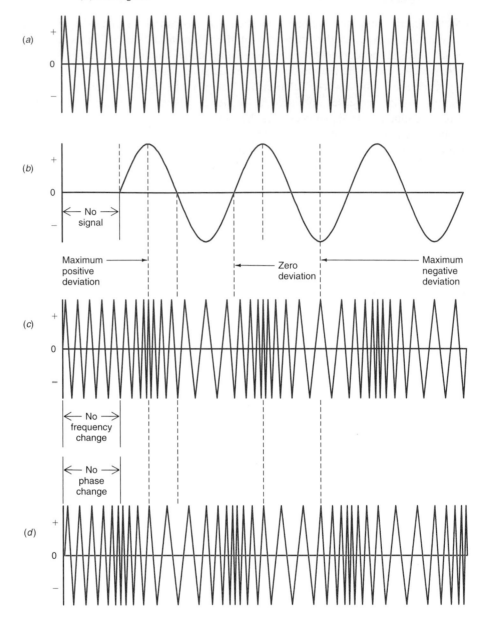

30 kHz. Note that the frequency of the modulating signal has no effect on the *amount* of deviation, which is strictly a function of the amplitude of the modulating signal.

Frequently, the modulating signal is a pulse train or series of rectangular waves, e.g., serial binary data. When the modulating signal has only two amplitudes, the carrier frequency, instead of having an infinite number of values, as it would have with a continuously varying (analog) signal, has only two values. This phenomenon is illustrated in Fig. 5-2. For example, when the modulating signal is a binary 0, the carrier frequency is the center frequency value. When the modulating signal is a binary 1, the carrier frequency abruptly changes to a higher frequency level. The amount of the shift depends on the amplitude of the binary signal. This kind of modulation, called *frequency-shift keying (FSK)*, is widely used in the transmission of binary data in digital cell phones and in some types of low-speed computer modems.

Frequency-shift keying

Figure 5-2 Frequency-modulating of a carrier with binary data produces FSK.

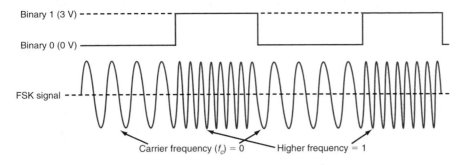

5-2 Principles of Phase Modulation

When the amount of phase shift of a constant-frequency carrier is varied in accordance with a modulating signal, the resulting output is a *phase modulation (PM)* signal [see Fig. 5-1(*d*)]. Imagine a modulator circuit whose basic function is to produce a *phase shift,* i.e., a time separation between two sine waves of the same frequency. Assume that a phase shifter can be built that will cause the amount of phase shift to vary with the amplitude of the modulating signal. The greater the amplitude of the modulating signal, the greater the phase shift. Assume further that positive alternations of the modulating signal produce a lagging phase shift and negative signals produce a leading phase shift.

> **Phase modulation (PM)**

If a constant-amplitude, constant-frequency carrier sine wave is applied to the phase shifter whose phase shift is varied by the intelligence signal, the output of the phase shifter is a PM wave. As the modulating signal goes positive, the amount of phase lag, and thus the delay of the carrier output, increases with the amplitude of the modulating signal. The result at the output is the same as if the constant-frequency carrier signal had been stretched out, or had its frequency lowered. When the modulating signal goes negative, the phase shift becomes leading. This causes the carrier sine wave to be effectively speeded up, or compressed. The result is the same as if the carrier frequency had been increased.

Note that it is the dynamic nature of the modulating signal that causes the frequency variation at the output of the phase shifter: FM is produced only as long as the phase shift is varying. To understand this better, look at the modulating signal shown in Fig. 5-3(*a*), which is a triangular wave whose positive and negative peaks have been clipped off at a fixed amplitude. During time t_0, the signal is zero, so the carrier is at its center frequency.

Applying this modulating signal to a frequency modulator produces the FM signal shown in Fig. 5-3(*b*). During the time the waveform is rising (t_1), the frequency increases. During the time the positive amplitude is constant (t_2), the FM output frequency is constant. During the time the amplitude decreases and goes negative (t_3), the frequency decreases. During the constant-amplitude negative alternation (t_4), the frequency remains constant, at a lower frequency. During t_5, the frequency increases.

Now, refer to the PM signal in Fig. 5-3(*c*). During increases or decreases in amplitude (t_1, t_3, and t_5), a varying frequency is produced. However, during the constant-amplitude positive and negative peaks, no frequency change takes place. The output of the phase modulator is simply the carrier frequency which has been shifted in phase. This clearly illustrates that when a modulating signal is applied to a phase modulator, the output frequency changes only during the time that the amplitude of the modulating signal is varying.

The maximum frequency deviation produced by a phase modulator occurs during the time that the modulating signal is changing at its most rapid rate. For a sine wave modulating signal, the rate of change of the modulating signal is greatest when the modulating wave changes from plus to minus or from minus to plus. As Fig. 5-3(*c*) shows,

> **GOOD TO KNOW**
>
> The maximum frequency deviation produced by a phase modulator occurs when the modulating signal is changing most quickly. For a sine wave modulating signal, that time is when the modulating wave changes from plus to minus or from minus to plus.

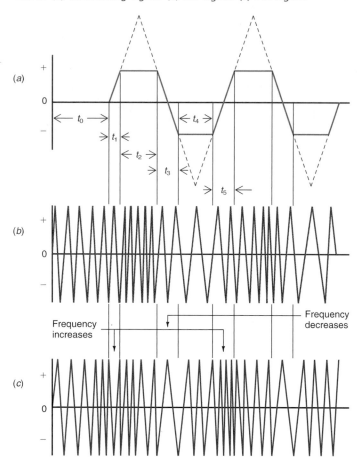

Figure 5-3 A frequency shift occurs in PM only when the modulating signal amplitude varies. (*a*) Modulating signal. (*b*) FM signal. (*c*) PM signal.

the maximum rate of change of modulating voltage occurs exactly at the zero crossing points. In contrast, note that in an FM wave the maximum deviation occurs at the peak positive and negative amplitude of the modulating voltage. Thus, although a phase modulator does indeed produce FM, maximum deviation occurs at different points of the modulating signal.

In PM, the amount of carrier deviation is proportional to the rate of change of the modulating signal, i.e., the calculus derivative. With a sine wave modulating signal, the PM carrier appears to be frequency-modulated by the cosine of the modulating signal. Remember that the cosine occurs 90° earlier (leads) than the sine.

Since the frequency deviation in PM is proportional to the rate of change in the modulating signal, the frequency deviation is proportional to the modulating signal frequency as well as its amplitude. This effect is compensated for prior to modulation.

Relationship Between the Modulating Signal and Carrier Deviation

In FM, the frequency deviation is directly proportional to the amplitude of the modulating signal. The maximum deviation occurs at the peak positive and negative amplitudes of the modulating signal. In PM, the frequency deviation is also directly proportional to the amplitude of the modulating signal. The maximum amount of leading or lagging phase shift occurs at the peak amplitudes of the modulating signal. This effect, for both FM and PM, is illustrated in Fig. 5-4(*a*).

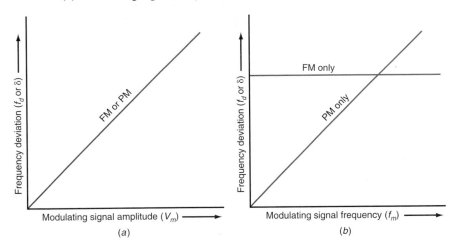

(a)   (b)

Now look at Fig. 5-4(b), which shows that the frequency deviation of an FM signal is constant for any value of modulating frequency. Only the amplitude of the modulating signal determines the amount of deviation. But look at how the deviation varies in a PM signal with different modulating signal frequencies. The higher the modulating signal frequency, the shorter its period and the faster the voltage changes. Higher modulating voltages result in greater phase shift, and this, in turn, produces greater frequency deviation. However, higher modulating frequencies produce a faster rate of change of the modulating voltage and thus greater frequency deviation. In PM, then, the carrier frequency deviation is proportional to both the modulating frequency (slope of modulating voltage) and the amplitude. In FM, frequency deviation is proportional only to the amplitude of the modulating signal, regardless of its frequency.

Converting PM to FM

To make PM compatible with FM, the deviation produced by frequency variations in the modulating signal must be compensated for. This can be done by passing the intelligence signal through a low-pass RC network, as illustrated in Fig. 5-5. This low-pass filter,

Figure 5-5 Using a low-pass filter to roll off the audio modulating signal amplitude with frequency.

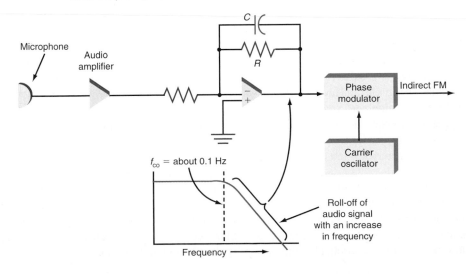

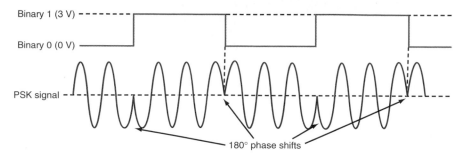

Figure 5-6 Phase modulation of a carrier by binary data produces PSK.

Binary 1 (3 V)

Binary 0 (0 V)

PSK signal

180° phase shifts

called a *frequency-correcting network, predistorter,* or *1/f filter,* causes the higher modulating frequencies to be attenuated. Although the higher modulating frequencies produce a greater rate of change and thus a greater frequency deviation, this is offset by the lower amplitude of the modulating signal, which produces less phase shift and thus less frequency deviation. The predistorter compensates for the excess frequency deviation caused by higher modulating frequencies. The result is an output that is the same as an FM signal. The FM produced by a phase modulator is called *indirect FM.*

Phase-Shift Keying

PM is also used with binary signals, as Fig. 5-6 shows. When the binary modulating signal is 0 V, or binary 0, the PM signal is simply the carrier frequency. When a binary 1 voltage level occurs, the modulator, which is a phase shifter, simply changes the phase of the carrier, not its frequency. In Fig. 5-6 the phase shift is 180°. Each time the signal changes from 0 to 1 or 1 to 0, there is a 180° phase shift. The PM signal is still the carrier frequency, but the phase has been changed with respect to the original carrier with a binary 0 input.

The process of phase-modulating a carrier with binary data is called *phase-shift keying (PSK)* or *binary phase-shift keying (BPSK).* The PSK signal shown in Fig. 5-6 uses a 180° phase shift from a reference, but other phase-shift values can be used, for example, 45°, 90°, 135°, or 225°. The important thing to remember is that no frequency variation occurs. The PSK signal has a constant frequency, but the phase of the signal from some reference changes as the binary modulating signal occurs.

5-3 Modulation Index and Sidebands

Any modulation process produces *sidebands.* When a constant-frequency sine wave modulates a carrier, two side frequencies are produced. The side frequencies are the sum and difference of the carrier and the modulating frequency. In FM and PM, as in AM, sum and difference sideband frequencies are produced. In addition, a large number of pairs of upper and lower sidebands are generated. As a result, the spectrum of an FM or a PM signal is usually wider than that of an equivalent AM signal. It is also possible to generate a special narrowband FM signal whose bandwidth is only slightly wider than that of an AM signal.

Figure 5-7 shows the frequency spectrum of a typical FM signal produced by modulating a carrier with a single-frequency sine wave. Note that the sidebands are spaced from the carrier f_c and from one another by a frequency equal to the modulating frequency f_m. If the modulating frequency is 1 kHz, the first pair of sidebands is above and below the carrier by 1000 Hz. The second pair of sidebands is above and below the

Figure 5-7 Frequency spectrum of an FM signal. Note that the carrier and sideband amplitudes shown are just examples. The amplitudes depend upon the modulation index m_f.

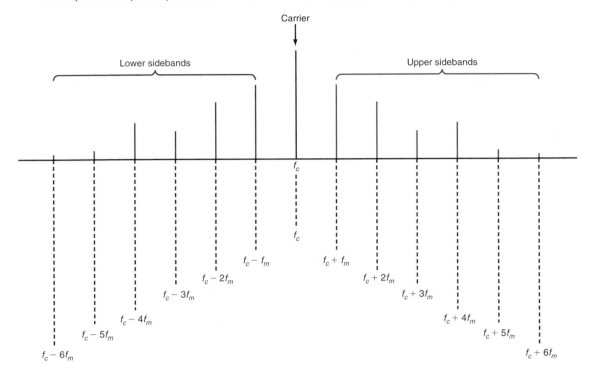

carrier by 2×1000 Hz = 2000 Hz, or 2 kHz, and so on. Note also that the amplitudes of the sidebands vary. If each sideband is assumed to be a sine wave, with a frequency and an amplitude as indicated in Fig. 5-7, and all the sine waves are added, then the FM signal producing them will be created.

As the amplitude of the modulating signal varies, the frequency deviation changes. The number of sidebands produced, and their amplitude and spacing, depend on the frequency deviation and modulating frequency. Keep in mind that an FM signal has a constant amplitude. Since an FM signal is a summation of the sideband frequencies, the sideband amplitudes must vary with frequency deviation and modulating frequency if their sum is to produce a constant-amplitude but variable-frequency FM signal.

Theoretically, the FM process produces an infinite number of upper and lower sidebands and, therefore, a theoretically infinitely large bandwidth. However, in practice, only those sidebands with the largest amplitudes are significant in carrying the information. Typically any sideband whose amplitude is less than 1 percent of the unmodulated carrier is considered insignificant. Thus FM is readily passed by circuits or communication media with finite bandwidth. Despite this, the bandwidth of an FM signal is usually much wider than that of an AM signal with the same modulating signal.

Modulation Index

The ratio of the frequency deviation to the modulating frequency is known as the *modulation index* m_f:

Modulation index

$$m_f = \frac{f_d}{f_m}$$

where f_d is the frequency deviation and f_m is the modulating frequency. Sometimes the lowercase Greek letter delta (δ) is used instead of f_d to represent deviation; then $m_f = \delta/f_m$. For example, if the maximum frequency deviation of the carrier is ±12 kHz and the maximum modulating frequency is 2.5 kHz, the modulating index is $m_f = 12/2.5 = 4.8$.

In most communication systems using FM, maximum limits are put on both the frequency deviation and the modulating frequency. For example, in standard FM broadcasting, the maximum permitted frequency deviation is 75 kHz and the maximum permitted modulating frequency is 15 kHz. This produces a modulation index of $m_f = 75/15 = 5$.

Deviation ratio

When the maximum allowable frequency deviation and the maximum modulating frequency are used in computing the modulation index, m_f is known as the *deviation ratio*.

Example 5-2

What is the deviation ratio of TV sound if the maximum deviation is 25 kHz and the maximum modulating frequency is 15 kHz?

$$m_f = \frac{f_d}{f_m} = \frac{25}{15} = 1.667$$

Bessel Functions

Bessel function

Given the modulation index, the number and amplitudes of the significant sidebands can be determined by solving the basic equation of an FM signal. The FM equation, whose derivation is beyond the scope of this book, is $v_{FM} = V_c \sin\left[2\pi f_c t + m_f \sin\left(2\pi f_m t\right)\right]$, where v_{FM} is the instantaneous value of the FM signal and m_f is the modulation index. The term whose coefficient is m_f is the phase angle of the carrier. Note that this equation expresses the phase angle in terms of the sine wave modulating signal. This equation is solved with a complex mathematical process known as *Bessel functions*. It is not necessary to show this solution, but the result is as follows:

$$\begin{aligned}
v_{FM} = V_c\{&J_0(\sin \omega_c t) + J_1[\sin (\omega_c + \omega_m)t - \sin(\omega_c - \omega_m)t] \\
&+ J_2[\sin(\omega_c + 2\omega_m)t + \sin(\omega_c - 2\omega_m)t] \\
&+ J_3[\sin(\omega_c + 3\omega_m)t - \sin(\omega_c - 3\omega_m)t] \\
&+ J_4[\sin(\omega_c + 4\omega_m)t + \sin(\omega_c - 4\omega_m)t] \\
&+ J_5[\sin \cdots] + \cdots\}
\end{aligned}$$

where $\omega_c = 2\pi f_c$ = carrier frequency
$\omega_m = 2\pi f_m$ = modulating signal frequency
V_c = peak value of unmodulated carrier

The FM wave is expressed as a composite of sine waves of different frequencies and amplitudes that, when added, give an FM time-domain signal. The first term is the carrier with an amplitude given by a J_n coefficient, in this case J_0. The next term represents a pair of upper and lower side frequencies equal to the sum and difference of the carrier and modulating signal frequency. The amplitude of these side frequencies is J_1. The next term is another pair of side frequencies equal to the carrier ±2 times the modulating signal frequency. The other terms represent additional side frequencies spaced from one another by an amount equal to the modulating signal frequency.

The amplitudes of the sidebands are determined by the J_n coefficients, which are, in turn, determined by the value of the modulation index. These amplitude coefficients are computed by using the expression

$$J_n(m_f) = \left(\frac{m_f}{2^n n!}\right)^n \left[1 - \frac{(m_f)^2}{2(2n+2)} + \frac{(m_f)^4}{2 \cdot 4(2n+2)(2n+4)} - \frac{(m_f)^6}{2 \cdot 4 \cdot 6(2n+2)(2n+4)(2n+6)} + \cdots\right]$$

where ! = factorial
n = sideband number (1, 2, 3, etc.)
n = 0 is the carrier
$m_f = \dfrac{f_d}{f_m}$ = frequency deviation

In practice, you do not have to know or calculate these coefficients, since tables giving them are widely available. The Bessel coefficients for a range of modulation indexes are given in Fig. 5-8. The leftmost column gives the modulation index m_f. The remaining columns indicate the relative amplitudes of the carrier and the various pairs of sidebands. Any sideband with a relative carrier amplitude of less than 1 percent (0.01) has been eliminated. Note that some of the carrier and sideband amplitudes have negative signs. This means that the signal represented by that amplitude is simply shifted in phase 180° (phase inversion).

Figure 5-9 shows the curves that are generated by plotting the data in Fig. 5-8. The carrier and sideband amplitudes and polarities are plotted on the vertical axis; the modulation index is plotted on the horizontal axis. As the figures illustrate, the carrier amplitude J_0 varies with the modulation index. In FM, the carrier amplitude and the amplitudes of the sidebands change as the modulating signal frequency and deviation change. In AM, the carrier amplitude remains constant.

Note that at several points in Figs. 5-8 and 5-9, at modulation indexes of about 2.4, 5.5, and 8.7, the carrier amplitude J_0 actually drops to zero. At those points, all the signal power is completely distributed throughout the sidebands. And as can be seen in Fig. 5-9, the sidebands also go to zero at certain values of the modulation index.

> **GOOD TO KNOW**
>
> The symbol ! means factorial. This tells you to multiply all integers from 1 through the number to which the symbol is attached. For example, 5! means $1 \times 2 \times 3 \times 4 \times 5 = 120$.

Figure 5-8 Carrier and sideband amplitudes for different modulation indexes of FM signals based on the Bessel functions.

Modulation Index	Carrier	Sidebands (Pairs)															
		1st	2d	3d	4th	5th	6th	7th	8th	9th	10th	11th	12th	13th	14th	15th	16th
0.00	1.00	—	—	—	—	—	—	—	—	—	—	—	—	—	—	—	—
0.25	0.98	0.12	—	—	—	—	—	—	—	—	—	—	—	—	—	—	—
0.5	0.94	0.24	0.03	—	—	—	—	—	—	—	—	—	—	—	—	—	—
1.0	0.77	0.44	0.11	0.02	—	—	—	—	—	—	—	—	—	—	—	—	—
1.5	0.51	0.56	0.23	0.06	0.01	—	—	—	—	—	—	—	—	—	—	—	—
2.0	0.22	0.58	0.35	0.13	0.03	—	—	—	—	—	—	—	—	—	—	—	—
2.5	−0.05	0.50	0.45	0.22	0.07	0.02	—	—	—	—	—	—	—	—	—	—	—
3.0	−0.26	0.34	0.49	0.31	0.13	0.04	0.01	—	—	—	—	—	—	—	—	—	—
4.0	−0.40	−0.07	0.36	0.43	0.28	0.13	0.05	0.02	—	—	—	—	—	—	—	—	—
5.0	−0.18	−0.33	0.05	0.36	0.39	0.26	0.13	0.05	0.02	—	—	—	—	—	—	—	—
6.0	0.15	−0.28	−0.24	0.11	0.36	0.36	0.25	0.13	0.06	0.02	—	—	—	—	—	—	—
7.0	0.30	0.00	−0.30	−0.17	0.16	0.35	0.34	0.23	0.13	0.06	0.02	—	—	—	—	—	—
8.0	0.17	0.23	−0.11	−0.29	−0.10	0.19	0.34	0.32	0.22	0.13	0.06	0.03	—	—	—	—	—
9.0	−0.09	0.24	0.14	−0.18	−0.27	−0.06	0.20	0.33	0.30	0.21	0.12	0.06	0.03	0.01	—	—	—
10.0	−0.25	0.04	0.25	0.06	−0.22	−0.23	−0.01	0.22	0.31	0.29	0.20	0.12	0.06	0.03	0.01	—	—
12.0	−0.05	−0.22	−0.08	0.20	0.18	−0.07	−0.24	−0.17	0.05	0.23	0.30	0.27	0.20	0.12	0.07	0.03	0.01
15.0	−0.01	0.21	0.04	0.19	−0.12	0.13	0.21	0.03	−0.17	−0.22	−0.09	0.10	0.24	0.28	0.25	0.18	0.12

Figure 5-9 Plot of the Bessel function data from Fig. 5-8.

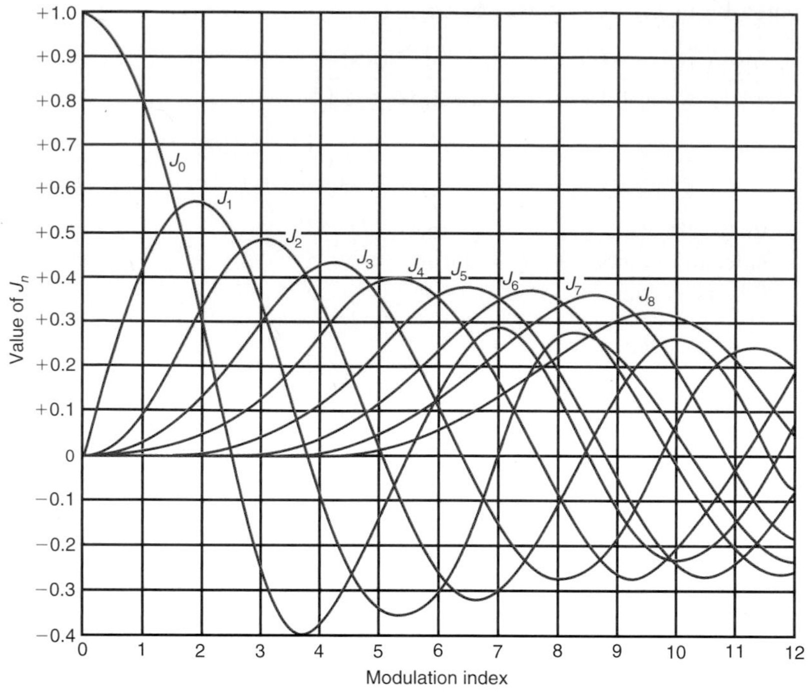

Example 5-3

What is the maximum modulating frequency that can be used to achieve a modulation index of 2.2 with a deviation of 7.48 kHz?

$$f_m = \frac{f_d}{m_f} = \frac{7480}{2.2} = 3400 \text{ Hz} = 3.4 \text{ kHz}$$

Figure 5-10 shows several examples of an FM signal spectrum with different modulation indexes. Compare the examples to the entries in Fig. 5-8. The unmodulated carrier in Fig. 5-10(a) has a relative amplitude of 1.0. With no modulation, all the power is in the carrier. With modulation, the carrier amplitude decreases while the amplitudes of the various sidebands increase.

In Fig. 5-10(d), the modulation index is 0.25. This is a special case of FM in which the modulation process produces only a single pair of significant sidebands like those produced by AM. With a modulation index of 0.25, the FM signal occupies no more spectrum space than an AM signal. This type of FM is called *narrowband FM,* or *NBFM.* The formal definition of NBFM is any FM system in which the modulation index is less than $\pi/2 = 1.57$, or $m_f < \pi/2$. However, for true NBFM with only a single pair of sidebands, m_f must be much less than $\pi/2$. Values of m_f in the 0.2 to 0.25 range will give true NBFM. Common FM mobile radios use a maximum deviation of 5 kHz, with a maximum voice frequency of 3 kHz, giving a modulation index of $m_f = 5 \text{ kHz}/3 \text{ kHz} = 1.667$. Although these systems do not fall within the formal definition of NBFM, they are nonetheless regarded as narrowband transmissions.

Narrowband FM (NBFM)

Figure 5-10

Examples of FM signal spectra. (*a*) Modulation index of 0 (no modulation or sidebands). (*b*) Modulation index of 1. (*c*) Modulation index of 2. (*d*) Modulation index of 0.25 (NBFM).

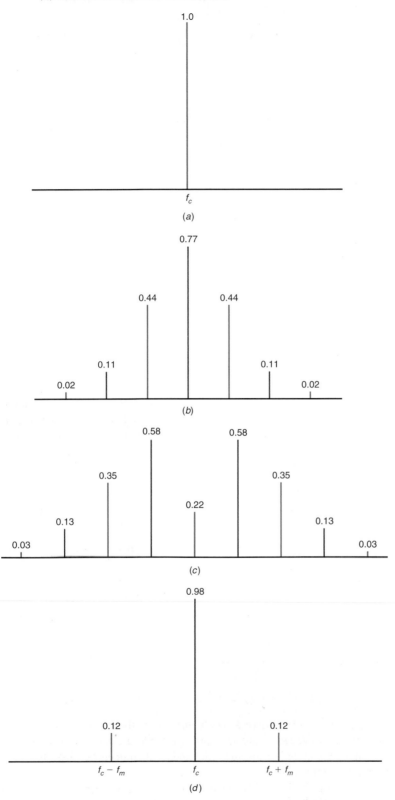

Example 5-4

State the amplitudes of the carrier and first four sidebands of an FM signal with a modulation index of 4. (Use Figs. 5-8 and 5-9.)

$$J_0 = -0.4$$
$$J_1 = -0.07$$
$$J_2 = 0.36$$
$$J_3 = 0.43$$
$$J_4 = 0.28$$

The primary purpose of NBFM is to conserve spectrum space, and NBFM is widely used in radio communication. Note, however, that NBFM conserves spectrum space at the expense of the signal-to-noise ratio.

FM Signal Bandwidth

As stated previously, the higher the modulation index in FM, the greater the number of significant sidebands and the wider the bandwidth of the signal. When spectrum conservation is necessary, the bandwidth of an FM signal can be deliberately restricted by putting an upper limit on the modulation index.

The total bandwidth of an FM signal can be determined by knowing the modulation index and using Fig. 5-8. For example, assume that the highest modulating frequency of a signal is 3 kHz and the maximum deviation is 6 kHz. This gives a modulation index of $m_f = 6$ kHz/3 kHz $= 2$. Referring to Fig. 5-8, you can see that this produces four significant pairs of sidebands. The bandwidth can then be determined with the simple formula

$$BW = 2f_m N$$

where N is the number of significant sidebands in the signal. According to this formula, the bandwidth of our FM signal is

$$BW = 2(3 \text{ kHz})(4) = 24 \text{ kHz}$$

In general terms, an FM signal with a modulation index of 2 and a highest modulating frequency of 3 kHz will occupy a 24-kHz bandwidth.

Carson's rule

Another way to determine the bandwidth of an FM signal is to use *Carson's rule*. This rule recognizes only the power in the most significant sidebands with amplitudes greater than 2 percent of the carrier (0.02 or higher in Fig. 5-8). This rule is

$$BW = 2[f_{d(\max)} + f_{m(\max)}]$$

According to Carson's rule, the bandwidth of the FM signal in the previous example would be

$$BW = 2(6 \text{ kHz} + 3 \text{ kHz}) = 2(9 \text{ kHz}) = 18 \text{ kHz}$$

Carson's rule will always give a bandwidth lower than that calculated with the formula $BW = 2f_m N$. However, it has been proved that if a circuit or system has the bandwidth calculated by Carson's rule, the sidebands will indeed be passed well enough to ensure full intelligibility of the signal.

So far, all the examples of FM have assumed a single-frequency sine wave modulating signal. However, as you know, most modulating signals are not pure sine waves, but complex waves made up of many different frequencies. When the modulating signal is a pulse or binary wave train, the carrier is modulated by the equivalent signal, which is a mix of a fundamental sine wave and all the relevant harmonics, as determined by Fourier theory. For example, if the modulating signal is a square wave, the fundamental sine wave and all the odd harmonics modulate the carrier. Each harmonic produces multiple pairs of sidebands depending on the modulation index. As you can imagine, FM by a square or rectangular wave generates many sidebands and produces a signal with an enormous bandwidth. The circuits or systems that will carry, process, or pass such a signal must have the appropriate bandwidth so as not to distort the signal. In most equipment that transmits digital or binary data by FSK, the binary signal is filtered to remove higher-level harmonics prior to modulation. This reduces the bandwidth required for transmission.

Example 5-5

What is the maximum bandwidth of an FM signal with a deviation of 30 kHz and a maximum modulating signal of 5 kHz as determined by (*a*) Fig. 5-8 and (*b*) Carson's rule?

a. $m_f = \dfrac{f_d}{f_m} = \dfrac{30 \text{ kHz}}{5 \text{ kHz}} = 6$

Figure 5-8 shows nine significant sidebands spaced 5 kHz apart for $m_f = 6$.
$$\text{BW} = 2f_m N = 2(5 \text{ kHz})\, 9 = 90 \text{ kHz}$$

b. $\text{BW} = 2[f_{d\,(\text{max})} + f_{m\,(\text{max})}]$

$= 2(30 \text{ kHz} + 5 \text{ kHz})$

$= 2(35 \text{ kHz})$

$\text{BW} = 70 \text{ kHz}$

5-4 Noise Suppression Effects of FM

Noise is interference generated by lightning, motors, automotive ignition systems, and any power line switching that produces transient signals. Such noise is typically narrow spikes of voltage with very high frequencies. They add to a signal and interfere with it. The potential effect of such noise on an FM signal is shown in Fig. 5-11. If the noise signals were strong enough, they could completely obliterate the information signal.

FM signals, however, have a constant modulated carrier amplitude, and FM receivers contain limiter circuits that deliberately restrict the amplitude of the received signal. Any

Noise

Figure 5-11 An FM signal with noise.

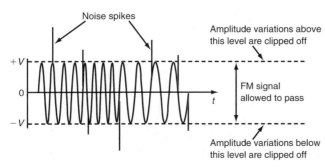

amplitude variations occurring on the FM signal are effectively clipped off, as shown in Fig. 5-11. This does not affect the information content of the FM signal, since it is contained solely within the frequency variations of the carrier. Because of the clipping action of the limiter circuits, noise is almost completely eliminated. Even if the peaks of the FM signal itself are clipped or flattened and the resulting signal is distorted, no information is lost. In fact, one of the primary benefits of FM over AM is its superior noise immunity. The process of demodulating or recovering an FM signal actually suppresses noise and improves the signal-to-noise ratio.

Noise and Phase Shift

The noise amplitude added to an FM signal introduces a small frequency variation, or phase shift, which changes or distorts the signal. Figure 5-12 shows how this works. The carrier signal is represented by a fixed-length (amplitude) phasor S. The noise is usually a short-duration pulse containing many frequencies at many amplitudes and phases according to Fourier theory. To simplify the analysis, however, we assume a single high-frequency noise signal varying in phase. In Fig. 5-12(a), this noise signal is represented as a rotating phasor N. The composite signal of the carrier and the noise, labeled C, is a phasor whose amplitude is the phasor sum of the signal and noise and a phase angle shifted from the carrier by an amount ϕ. If you imagine the noise phasor rotating, you can also imagine the composite signal varying in amplitude and phase angle with respect to the carrier.

The maximum phase shift occurs when the noise and signal phasors are at a right angle to each other, as illustrated in Fig. 5-12(b). This angle can be computed with the arcsine or inverse sine according to the formula

$$\phi = \sin^{-1}\frac{N}{S}$$

It is possible to determine just how much of a frequency shift a particular phase shift produces by using the formula

$$\delta = \phi(f_m)$$

where δ = frequency deviation produced by noise
ϕ = phase shift, rad
f_m = frequency of modulating signal

Assume that the signal-to-noise ratio (S/N) is 3 : 1 and the modulating signal frequency is 800 Hz. The phase shift is then $\phi = \sin^{-1}(N/S) = \sin^{-1}(1/3) = \sin^{-1}0.3333 = 19.47°$.

Figure 5-12   How noise introduces a phase shift.

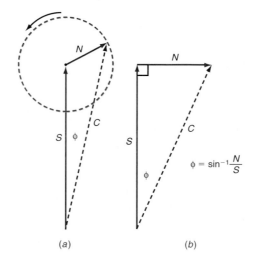

(a)   (b)

Since there is 57.3° per radian, this angle is $\phi = 19.47/57.3 = 0.34$ rad. The frequency deviation produced by this brief phase shift can be calculated as

$$\delta = 0.34(800) = 271.8 \text{ Hz}$$

Just how badly a particular phase shift will distort a signal depends on several factors. Looking at the formula for deviation, you can deduce that the worst-case phase shift and frequency deviation will occur at the highest modulating signal frequency. The overall effect of the shift depends upon the maximum allowed frequency shift for the application. If very high deviations are allowed, i.e., if there is a high modulation index, the shift can be small and inconsequential. If the total allowed deviation is small, then the noise-induced deviation can be severe. Remember that the noise interference is of very short duration; thus the phase shift is momentary, and intelligibility is rarely severely impaired. With heavy noise, human speech might be temporarily garbled, but so much that it could not be understood.

Assume that the maximum allowed deviation is 5 kHz in the example above. The ratio of the shift produced by the noise to the maximum allowed deviation is

$$\frac{\text{Frequency deviation produced by noise}}{\text{Maximum allowed deviation}} = \frac{271.8}{5000} = 0.0544$$

This is only a bit more than a 5 percent shift. The 5-kHz deviation represents the maximum modulating signal amplitude. The 271.8-Hz shift is the noise amplitude. Therefore this ratio is the noise-to-signal ratio N/S. The reciprocal of this value gives you the FM signal-to-noise ratio:

$$\frac{S}{N} = \frac{1}{N/S} = \frac{1}{0.0544} = 18.4$$

For FM, a 3:1 input S/N translates to an 18.4:1 output S/N.

Example 5-6

The input to an FM receiver has an S/N of 2.8. The modulating frequency is 1.5 kHz. The maximum permitted deviation is 4 kHz. What are (a) the frequency deviation caused by the noise and (b) the improved output S/N?

a. $\phi = \sin^{-1} \dfrac{N}{S} = \sin^{-1} \dfrac{1}{2.8} = \sin^{-1} 0.3571 = 20.92°$ or 0.3652 rad

$\delta = \phi(f_m) = (0.3652)(1.5 \text{ kHz}) = 547.8 \text{ Hz}$

b. $\dfrac{N}{S} = \dfrac{\text{frequency deviation produced by noise}}{\text{maximum allowed deviation}} = \dfrac{547.8}{4000}$

$\dfrac{N}{S} = 0.13695$

$\dfrac{S}{N} = \dfrac{1}{N/S} = 7.3$

Preemphasis

Noise *can* interfere with an FM signal, and particularly with the high-frequency components of the modulating signal. Since noise is primarily sharp spikes of energy, it contains a lot of harmonics and other high-frequency components. These frequencies can be larger in amplitude than the high-frequency content of the modulating signal, causing frequency distortion that can make the signal unintelligible.

Most of the content of a modulating signal, particularly voice, is at low frequencies. In voice communication systems, the bandwidth of the signal is limited to about 3 kHz, which permits acceptable intelligibility. In contrast, musical instruments typically generate signals at low frequencies but contain many high-frequency harmonics that give them their unique sound and must be passed if that sound is to be preserved. Thus a wide bandwidth is needed in high-fidelity systems. Since the high-frequency components are usually at a very low level, noise can obliterate them.

Preemphasis

To overcome this problem, most FM systems use a technique known as *preemphasis* that helps offset high-frequency noise interference. At the transmitter, the modulating signal is passed through a simple network that amplifies the high-frequency components more than the low-frequency components. The simplest form of such a circuit is a simple high-pass filter of the type shown in Fig. 5-13(a). Specifications

Figure 5-13 Preemphasis and deemphasis. (*a*) Preemphasis circuit. (*b*) Preemphasis curve. (*c*) Deemphasis circuit. (*d*) Deemphasis curve. (*e*) Combined frequency response.

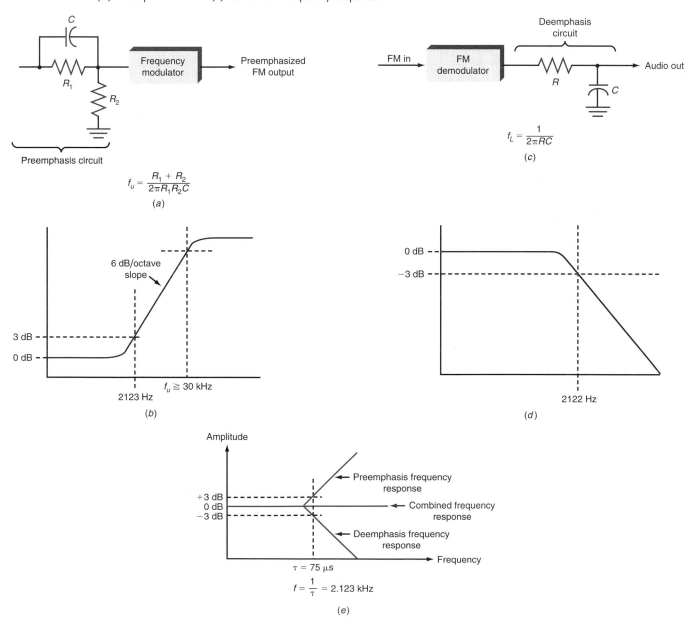

dictate a time constant t of 75 μs, where $t = RC$. Any combination of resistor and capacitor (or resistor and inductor) giving this time constant will work.

$$f_L = \frac{1}{2\pi RC} = \frac{1}{2\pi t} = \frac{1}{2\pi(75\ \mu s)} = 2123\ Hz$$

Such a circuit has a cutoff frequency of 2122 Hz; frequencies higher than 2122 Hz will be linearly enhanced. The output amplitude increases with frequency at a rate of 6 dB per octave. The preemphasis circuit increases the energy content of the higher-frequency signals so that they become stronger than the high-frequency noise components. This improves the signal-to-noise ratio and increases intelligibility and fidelity.

The preemphasis circuit also has an upper break frequency f_u, at which the signal enhancement flattens out [see Fig. 5-13(b)], which is computed with the formula

$$f_u = \frac{R_1 + R_2}{2\pi R_1 R_2 C}$$

The value of f_u is usually set well beyond the audio range and is typically greater than 30 kHz.

To return the frequency response to its normal, "flat" level, a *deemphasis circuit*, a simple low-pass filter with a time constant of 75 μs, is used at the receiver [see Fig. 5-13(c)]. Signals above its cutoff frequency of 2123 Hz are attenuated at the rate of 6 dB per octave. The response curve is shown in Fig. 5-13(d). As a result, the preemphasis at the transmitter is exactly offset by the deemphasis circuit in the receiver, providing a flat frequency response. The combined effect of preemphasis and deemphasis is to increase the signal-to-noise ratio for the high-frequency components during transmission so that they will be stronger and not masked by noise. Figure 5-13(e) shows the overall effect of preemphasis and deemphasis.

Deemphasis circuit

5-5 Frequency Modulation Versus Amplitude Modulation

Advantages of FM

In general, FM is considered to be superior to AM. Although both AM and FM signals can be used to transmit information from one place to another, FM typically offers some significant benefits over AM.

Noise Immunity. The main benefit of FM over AM is its superior immunity to noise, made possible by the clipper limiter circuits in the receiver, which effectively strip off all the noise variations, leaving a constant-amplitude FM signal. Although clipping does not result in total recovery in all cases, FM can nevertheless tolerate a much higher noise level than AM for a given carrier amplitude. This is also true for phase-shift-induced distortion.

Capture Effect. Another major benefit of FM is that interfering signals on the same frequency are effectively rejected. Because of the amplitude limiters and the demodulating methods used by FM receivers, a phenomenon known as the *capture effect* takes place when two or more FM signals occur simultaneously on the same frequency. If one signal is more than twice the amplitude of the other, the stronger signal captures the channel, totally eliminating the weaker signal. With modern receiver circuitry, a difference in signal amplitudes of only 1 dB is usually sufficient to produce the capture effect. In contrast, when two AM signals occupy the same frequency, both signals are generally heard, regardless of their relative signal strengths. When one AM signal is significantly stronger than another, naturally the stronger signal is intelligible; however, the weaker signal is not eliminated and can still be heard in the background. When the signal

Capture effect

strengths of given AM signals are nearly the same, they will interfere with each other, making both nearly unintelligible.

Although the capture effect prevents the weaker of two FM signals from being heard, when two stations are broadcasting signals of approximately the same amplitude, first one may be captured and then the other. This can happen, e.g., when a driver moving along a highway is listening to a clear broadcast on a particular frequency. At some point, the driver may suddenly hear the other broadcast, completely losing the first, and then, just as suddenly, hear the original broadcast again. Which one dominates depends on where the car is and on the relative signal strengths of the two signals.

Transmitter Efficiency. A third advantage of FM over AM involves efficiency. Recall that AM can be produced by both low-level and high-level modulation techniques. The most efficient is high-level modulation in which a class C amplifier is used as the final RF power stage and is modulated by a high-power modulation amplifier. The AM transmitter must produce both very high RF and modulating signal power. In addition, at very high power levels, large-modulation amplifiers are impractical. Under such conditions, low-level modulation must be used if the AM information is to be preserved without distortion. The AM signal is generated at a lower level and then amplified with linear amplifiers to produce the final RF signal. Linear amplifiers are either class A or class B and are far less efficient than class C amplifiers.

FM signals have a constant amplitude, and it is therefore not necessary to use linear amplifiers to increase their power level. In fact, FM signals are always generated at a lower level and then amplified by a series of class C amplifiers to increase their power. The result is greater use of available power because of the high level of efficiency of class C amplifiers. Even more efficient class D, E, or F amplifiers are also used in FM or PM equipment.

Disadvantages of FM

Excessive Spectrum Use. Perhaps the greatest disadvantage of FM is that it simply uses too much spectrum space. The bandwidth of an FM signal is, in general, considerably wider than that of an AM signal transmitting similar information. Although it is possible to keep the modulation index low to minimize bandwidth, reducing the modulation index also reduces the noise immunity of an FM signal. In commercial two-way FM radio systems, the maximum allowed deviation is 5 kHz, with a maximum modulating frequency of 3 kHz. This produces a deviation ratio of 5/3 = 1.67. Deviation ratios as low as 0.25 are possible, although they result in signals that are much less desirable than wideband FM signals. Both of these deviation ratios are classified as narrowband FM.

Since FM occupies so much bandwidth, it is typically used only in those portions of the spectrum where adequate bandwidth is available, i.e., at very high frequencies. In fact, it is rarely used below frequencies of 30 MHz. Most FM communication work is done at the VHF, UHF, and microwave frequencies.

Circuit Complexity. One major disadvantage of FM in the past involved the complexity of the circuits used for frequency modulation and demodulation in comparison with the simple circuits used for amplitude modulation and demodulation. Today, this disadvantage has almost disappeared because of the use of integrated circuits. Although the ICs used in FM transmission are still complex, they require very little effort to use and their price is just as low as those of comparable AM circuits.

Since the trend in electronic communication is toward higher and higher frequencies and because ICs are so cheap and easy to use, FM and PM have become by far the most widely used modulation method in electronic communication today.

FM and AM Applications

Here are some of the major applications for AM and FM.

Application	Type of Modulation
AM broadcast radio	AM
FM broadcast radio	FM
FM stereo multiplex sound	DSB (AM) and FM
TV sound	FM
TV picture (video)	AM, VSB
TV color signals	Quadrature DSB (AM)
Cellular telephone	FM, FSK, PSK
Cordless telephone	FM, PSK
Fax machine	FM, QAM (AM plus PSK)
Aircraft radio	AM
Marine radio	FM and SSB (AM)
Mobile and handheld radio	FM
Citizens band radio	AM and SSB (AM)
Amateur radio	FM and SSB (AM)
Computer modems	FSK, PSK, QAM (AM plus PSK)
Garage door opener	OOK
TV remote control	OOK
VCR	FM
Family Radio service	FM

CHAPTER REVIEW

Summary

Frequency deviation in FM is proportional only to the amplitude of the modulating signal regardless of its frequency. In FM, the frequency of the modulating signal determines how many times per second the carrier frequency deviates above and below its nominal center frequency. In frequency-shift keying (FSK), as in transmission of serial binary data, the modulating signal is a pulse train or series of rectangular waves. In PM, the amount of phase shift of a constant-frequency carrier is varied in accordance with a modulating signal, and the carrier frequency deviation is proportional to both the modulating frequency and the amplitude. Since FM is produced by PM, often PM is referred to as indirect FM.

Transmission of binary data by phase modulation is known as phase-shift keying (PSK).

To make PM compatible with FM, the deviation produced by frequency variations in the modulating signal must be compensated for. The ratio between the maximum permitted frequency deviation and the maximum permitted modulating frequency is the modulation index, or the deviation ratio. Given the modulation index, the number of significant sidebands, and the sideband amplitude coefficents as determined by Bessel functions, the basic equation of an FM signal can be solved. The FM wave is expressed as a composite of sine waves of different frequencies and amplitudes that when added will give an FM time-domain signal.

The bandwidth of an FM signal can be calculated by using the modulation index and Bessel functions or by Carson's rule.

One of the primary benefits of FM over AM is its superior immunity to noise. FM receivers contain limiter circuits that restrict the amplitude of the received signal, clipping off any amplitude variations and almost completely eliminating noise. Certain kinds of high-frequency components in the modulating signal can, however, interfere with FM transmission. To overcome this problem, most FM systems use a technique known as preemphasis. At the transmitter, the modulating signal is passed through a simple network that amplifies the high-frequency components more than the low-frequency components.

In FM, the stronger of two signals on the same frequency will reject the weaker, through a phenomenon known as the capture effect. In contrast, when one AM signal is significantly stronger than the other, the weaker signal can be heard in the background. A final advantage of FM over AM is efficiency of transmission. FM signals are always generated at a low level and then amplified by a series of highly efficient class C, D, E, or F amplifiers.

Questions

1. What is the general name given to both FM and PM?
2. State the effect on the amplitude of the carrier during FM or PM.
3. What are the name and mathematical expressions for the amount that the carrier varies from its unmodulated center frequency during modulation?
4. State how the frequency of a carrier varies in an FM system when the modulating signal amplitude and frequency change.
5. State how the frequency of a carrier varies in a PM system when the modulating signal amplitude and frequency change.
6. When does maximum frequency deviation occur in an FM signal? A PM signal?
7. State the conditions that must exist for a phase modulator to produce FM.
8. What do you call FM produced by PM techniques?
9. State the nature of the output of a phase modulator during the time when the modulating signal voltage is constant.
10. What is the name given to the process of frequency modulation of a carrier by binary data?
11. What is the name given to the process of phase modulation of a carrier by binary data?
12. How must the nature of the modulating signal be modified to produce FM by PM techniques?
13. What is the difference between the modulation index and the deviation ratio?
14. Define narrowband FM. What criterion is used to indicate NBFM?
15. What is the name of the mathematical equation used to solve for the number and amplitude of sidebands in an FM signal?
16. What is the meaning of a negative sign on the sideband value in Fig. 5-8?
17. Name two ways that noise affects an FM signal.
18. How is the noise on an FM signal minimized at the receiver?
19. What is the primary advantage of FM over AM?
20. List two additional advantages of FM over AM.
21. What is the nature of the noise that usually accompanies a radio signal?
22. In what ways is an FM transmitter more efficient than a low-level AM transmitter? Explain.
23. What is the main disadvantage of FM over AM? State two ways in which this disadvantage can be overcome.
24. What type of power amplifier is used to amplify FM signals? Low-level AM signals?
25. What is the name of the receiver circuit that eliminates noise?
26. What is the capture effect and what causes it?
27. What is the nature of the modulating signals that are most negatively affected by noise on an FM signal?
28. Describe the process of preemphasis. How does it improve communication performance in the presence of noise? Where is it performed, at the transmitter or receiver?
29. What is the basic circuit used to produce preemphasis?
30. Describe the process of deemphasis. Where is it performed, at the transmitter or receiver?
31. What type of circuit is used to accomplish deemphasis?
32. What is the cutoff frequency of preemphasis and deemphasis circuits?
33. List four major applications for FM.

Problems

1. A 162-MHz carrier is deviated by 12 kHz by a 2-kHz modulating signal. What is the modulation index? ◆
2. The maximum deviation of an FM carrier with a 2.5-kHz signal is 4 kHz. What is the deviation ratio?
3. For Problems 1 and 2, compute the bandwidth occupied by the signal, by using the conventional method and Carson's rule. Sketch the spectrum of each signal, showing all significant sidebands and their exact amplitudes.
4. For a single-frequency sine wave modulating signal of 3 kHz with a carrier frequency of 36 MHz, what is the spacing between sidebands?
5. What are the relative amplitudes of the fourth pair of sidebands for an FM signal with a deviation ratio of 8? ◆
6. At approximately what modulation index does the amplitude of the first pair of sidebands go to zero? Use Fig. 5-8 or 5-9 to find the lowest modulation index that gives this result.
7. An available channel for FM transmission is 30 kHz wide. The maximum allowable modulating signal frequency is 3.5 kHz. What deviation ratio should be used? ◆
8. The signal-to-noise ratio in an FM system is 4:1. The maximum allowed deviation is 4 kHz. How much frequency deviation is introduced by the phase shift caused by the noise when the modulating frequency is 650 Hz? What is the real signal-to-noise ratio?
9. A deemphasis circuit has a capacitor value of 0.02 μF. What value of resistor is needed? Give the closest standard EIA value. ◆
10. Use Carson's rule to determine the bandwidth of an FM channel when the maximum deviation allowed is 5 kHz at frequencies up to 3.333 kHz. Sketch the spectrum, showing carrier and sideband values.

◆ *Answers to Selected Problems follow Chap. 22.*

Critical Thinking

1. The AM broadcast band consists of 107 channels for stations 10 kHz wide. The maximum permitted modulating frequency is 5 kHz. Could FM be used on this band? If so, explain what would be necessary to make it happen.
2. A carrier of 49 MHz is frequency-modulated by a 1.5-kHz square wave. The modulation index is 0.25. Sketch the spectrum of the resulting signal. (Assume that only harmonics less than the sixth are passed by the system.)
3. The FM radio broadcast band is allocated the frequency spectrum from 88 to 108 MHz. There are 100 channels spaced 200 kHz apart. The first channel center frequency is 88.1 MHz; the last, or 100th, channel center frequency is 107.9 MHz. Each 200-kHz channel has a 150-kHz modulation bandwidth with 25-kHz "guard bands" on either side of it to minimize the effects of overmodulation (overdeviation). The FM broadcast band permits a maximum deviation of ± 75 kHz and a maximum modulating frequency of 15 kHz.
   a. Draw the frequency spectrum of the channel centered on 99.9 MHz, showing all relevant frequencies.
   b. Draw the frequency spectrum of the FM band, showing details of the three lowest-frequency channels and the three highest-frequency channels.
   c. Determine the bandwidth of the FM signal by using the deviation ratio and the Bessel table.
   d. Determine the bandwidth of the FM signal by using Carson's rule.
   e. Which of the above bandwidth calculations best fits the available channel bandwidth?
4. A 450-MHz radio transmitter uses FM with a maximum allowed deviation of 6 kHz and a maximum modulating frequency of 3.5 kHz. What is the minimum bandwidth required? Use Fig. 5-14 to determine the approximate amplitudes of the carrier and first three significant sidebands.
5. Assume that you could transmit digital data over the FM broadcast band radio station. The maximum allowed bandwidth is 200 kHz. The maximum deviation is 75 kHz, and the deviation ratio is 5. Assuming that you wished to preserve up to the third harmonic, what is the highest-frequency square wave you could transmit?

Figure 5-14

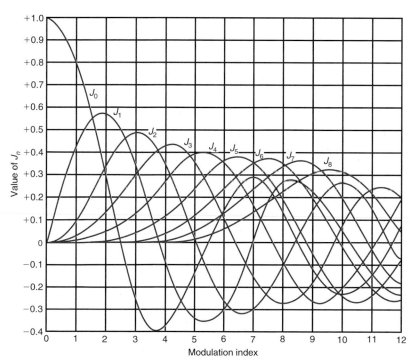

chapter 6

FM Circuits

Many different circuits have been devised to produce FM and PM signals. There are two different types of frequency modulator circuits, direct circuits and circuits that produce FM indirectly by phase modulation techniques. Direct FM circuits make use of techniques for varying the frequency of the carrier oscillator in accordance with the modulating signal. Indirect modulators produce FM via a phase shifter after the carrier oscillator stage. Frequency demodulator or detector circuits convert the FM signal back to the original modulating signal.

Objectives

After completing this chapter, you will be able to:

- Compare and contrast FM using crystal oscillator circuits with FM using varactors.
- Explain the general principles of phase modulator circuits and list the basic techniques for achieving phase shift.
- Calculate the total frequency deviation of an FM transmitter given the original oscillator frequency and the frequency multiplication factor.
- Describe the operation of slope detectors, pulse-averaging discriminators, and quadrature detectors.
- Draw a block diagram of a phase-locked loop (PLL), state what each component does, explain the operation of the circuit, and define the capture range and the lock range of a PLL.
- Explain the operation of a PLL as a frequency demodulator.

6-1 Frequency Modulators

A *frequency modulator* is a circuit that varies *carrier frequency* in accordance with the modulating signal. The carrier is generated by either an *LC* or a crystal oscillator circuit, and so a way must be found to change the frequency of oscillation. In an *LC* oscillator, the carrier frequency is fixed by the values of the inductance and capacitance in a tuned circuit, and the carrier frequency can therefore be changed by varying either inductance or capacitance. The idea is to find a circuit or component that converts a modulating voltage to a corresponding change in capacitance or inductance.

When the carrier is generated by a crystal oscillator, the frequency is fixed by the crystal. However, keep in mind that the equivalent circuit of a crystal is an *LCR* circuit with both series and parallel resonant points. Connecting an external capacitor to the crystal allows minor variations in operating frequency to be obtained. Again, the objective is to find a circuit or component whose capacitance will change in response to the modulating signal. The component most frequently used for this purpose is a *varactor*. Also known as a voltage variable capacitor, variable capacitance diode, or varicap, this device is basically a semiconductor junction diode operated in a reverse-bias mode.

Frequency modulator

Carrier frequency

Varactor

Varactor Operation

A junction diode is created when P- and N-type semiconductors are formed during the manufacturing process. Some electrons in the N-type material drift over into the P-type material and neutralize the holes there [see Fig. 6-1(*a*)], forming a thin area called the *depletion region,* where there are no free carriers, holes, or electrons.

Depletion region

Figure 6-1 Depletion region in a junction diode.

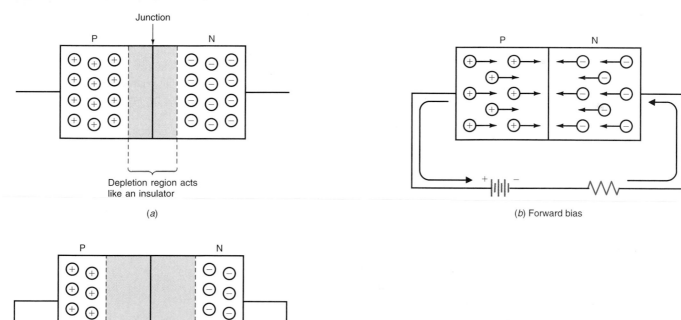

(*a*)

(*b*) Forward bias

(*c*) Reverse bias

FM Circuits

175

Figure 6-2 Schematic symbols of a varactor diode.

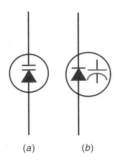

(a)   (b)

This region acts as a thin insulator that prevents current from flowing through the device.

If a forward bias is applied to the diode, it will conduct. The external potential forces the holes and electrons toward the junction, where they combine and cause a continuous current inside the diode as well as externally. The depletion layer simply disappears [see Fig. 6-1(b)]. If an external reverse bias is applied to the diode, as in Fig. 6-1(c), no current will flow. The bias increases the width of the depletion layer, with the amount of increase depending on the amount of the reverse bias. The higher the reverse bias, the wider the depletion layer and the less chance for current flow.

A reverse-biased junction diode acts as a small capacitor. The P- and N-type materials act as the two plates of the capacitor, and the depletion region acts as the dielectric. With all the active current carriers (electrons and holes) neutralized in the depletion region, it functions just as an insulating material. The width of the depletion layer determines the width of the dielectric and, therefore, the amount of capacitance. If the reverse bias is high, the depletion region will be wide and the dielectric will cause the plates of the capacitor to be widely spaced, producing a low capacitance. Decreasing the amount of reverse bias narrows the depletion region; the plates of the capacitor are effectively closer together, producing a higher capacitance.

All junction diodes exhibit variable capacitance as the reverse bias is changed. However, varactors are designed to optimize this particular characteristic, so that the capacitance variations are as wide and linear as possible. The symbols used to represent varactor diodes are shown in Fig. 6-2.

Varactors are made with a wide range of capacitance values, most units having a nominal capacitance in the 1- to 200-pF range. The capacitance variation range can be as high as 12:1. Figure 6-3 shows the curve for a typical diode. A maximum

Figure 6-3 Capacitance versus reverse junction voltage for a typical varactor.

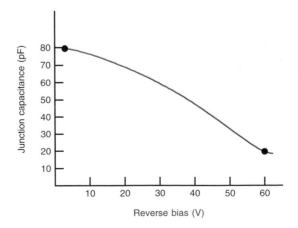

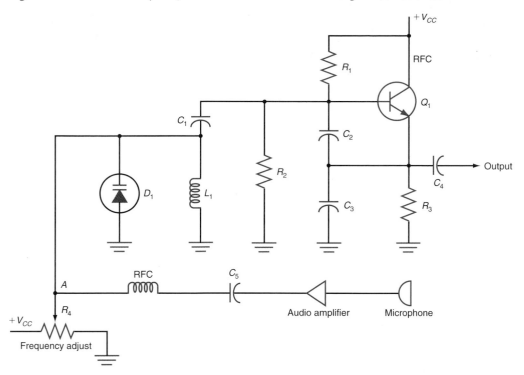

capacitance of 80 pF is obtained at 1 V. With 60 V applied, the capacitance drops to 20 pF, a 4 : 1 range. The operating range is usually restricted to the linear center portion of the curve.

Varactor Modulators

Varactor modulator

Figure 6-4, a carrier oscillator for a transmitter, shows the basic concept of a varactor frequency modulator. The capacitance of varactor diode D_1 and L_1 form the parallel tuned circuit of the oscillator. The value of C_1 is made very large at the operating frequency so that its reactance is very low. As a result, C_1 connects the tuned circuit to the oscillator circuit. Also C_1 blocks the dc bias on the base of Q_1 from being shorted to ground through L_1. The values of L_1 and D_1 fix the center carrier frequency.

The capacitance of D_1 is controlled in two ways, through a fixed dc bias and by the modulating signal. In Fig. 6-4, the bias on D_1 is set by the voltage divider potentiometer R_4. Varying R_4 allows the center carrier frequency to be adjusted over a narrow range. The modulating signal is applied through C_5 and the radio frequency choke (RFC); C_5 is a blocking capacitor that keeps the dc varactor bias out of the modulating-signal circuits. The reactance of the RFC is high at the carrier frequency to prevent the carrier signal from getting back into the audio modulating-signal circuits.

The modulating signal derived from the microphone is amplified and applied to the modulator. As the modulating signal varies, it adds to and subtracts from the fixed-bias voltage. Thus the effective voltage applied to D_1 causes its capacitance to vary. This, in turn, produces the desired deviation of the carrier frequency. A positive-going signal at point A adds to the reverse bias, decreasing the capacitance and increasing the carrier frequency. A negative-going signal at A subtracts from the bias, increasing the capacitance and decreasing the carrier frequency.

Example 6-1

The value of capacitance of a varactor at the center of its linear range is 40 pF. This varactor will be in parallel with a fixed 20-pF capacitor. What value of inductance should be used to resonate this combination to 5.5 MHz in an oscillator? Total capacitance $C_T = 40 + 20 = 60$ pF.

$$f_0 = 5.5 \text{ MHz} = \frac{1}{2\pi\sqrt{LC_T}}$$

$$L = \frac{1}{(2\pi f)^2 C_T} = \frac{1}{(6.28 \times 5.5 \times 10^6)^2 \times 60 \times 10^{-12}}$$

$$L = 13.97 \times 10^{-6} \text{ H} \quad \text{or} \quad 14 \ \mu\text{H}$$

The main problem with the circuit in Fig. 6-4 is that most *LC* oscillators are simply not stable enough to provide a carrier signal. Even with high-quality components and optimal design, the frequency of *LC* oscillators will vary because of temperature changes, variations in circuit voltage, and other factors. Such instabilities cannot be tolerated in most modern electronic communication systems, where a transmitter must stay on frequency as precisely as possible. The *LC* oscillators simply are not stable enough to meet the stringent requirements imposed by the FCC. As a result, crystal oscillators are normally used to set carrier frequency. Not only do crystal oscillators provide a highly accurate carrier frequency, but also their frequency stability is superior over a wide temperature range.

Frequency–Modulating a Crystal Oscillator

It is possible to vary the frequency of a crystal oscillator by changing the value of capacitance in series or in parallel with the crystal. Figure 6-5 shows a typical crystal oscillator. When a small value of capacitance is connected in series with the crystal, the crystal

Figure 6-5 Frequency modulation of a crystal oscillator with a VVC.

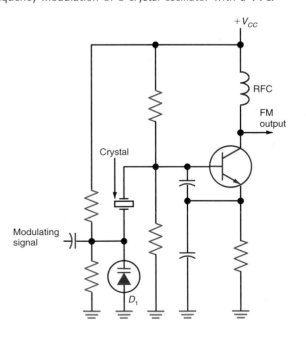

Figure 6-6 How frequency multipliers increase carrier frequency and deviation.

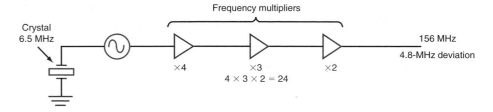

frequency can be "pulled" slightly from its natural resonant frequency. By making the series capacitance a varactor diode, frequency modulation of the crystal oscillator can be achieved. The modulating signal is applied to the varactor diode D_1, which changes the oscillator frequency.

It is important to note that only a very small frequency deviation is possible with frequency-modulated crystal oscillators. Rarely can the frequency of a crystal oscillator be changed more than several hundred hertz from the nominal crystal value. The resulting deviation may be less than the total deviation desired. For example, to achieve a total frequency shift of 75 kHz, which is necessary in commercial FM broadcasting, other techniques must be used. In NBFM communication systems, the narrower deviations are acceptable.

Although it is possible to achieve a deviation of only several hundred cycles from the crystal oscillator frequency, the total deviation can be increased by using frequency multiplier circuits after the carrier oscillator. A *frequency multiplier circuit* is one whose output frequency is some integer multiple of the input frequency. A frequency multiplier that multiplies a frequency by 2 is called a *doubler;* a frequency multiplier circuit that multiplies an input frequency by 3 is called a *tripler;* and so on. Frequency multipliers can also be cascaded.

When the FM signal is applied to a frequency multiplier, both the carrier frequency of operation and the amount of deviation are increased. Typical frequency multipliers can increase the carrier oscillator frequency by 24 to 32 times. Figure 6-6 shows how frequency multipliers increase carrier frequency and deviation. The desired output frequency from the FM transmitter in the figure is 156 MHz, and the desired maximum frequency deviation is 5 kHz. The carrier is generated by a 6.5-MHz crystal oscillator, which is followed by frequency multiplier circuits that increase the frequency by a factor of 24 (6.5 MHz $\times$ 24 = 156 MHz). Frequency modulation of the crystal oscillator by the varactor produces a maximum deviation of only 200 Hz. When multiplied by a factor of 24 in the frequency multiplier circuits, this deviation is increased to 200 $\times$ 24 = 4800 Hz, or 4.8 kHz, which is close to the desired deviation. Frequency multiplier circuits are discussed in greater detail in Chap. 8.

Frequency multiplier circuit

Doubler

Tripler

Voltage-Controlled Oscillators

Oscillators whose frequencies are controlled by an external input voltage are generally referred to as *voltage-controlled oscillators (VCOs). Voltage-controlled crystal oscillators* are generally referred to as *VXOs.* Although some VCOs are used primarily in FM, they are also used in other applications where voltage-to-frequency conversion is required. As you will see, their most common application is in phase-locked loops, discussed later in this chapter.

Although VCOs for VHF, UHF, and microwaves are still implemented with discrete components, more and more they are being integrated on a single chip of silicon along with other transmitter or receiver circuits. An example of such a VCO is shown in Fig. 6-7. This circuit uses silicon-germanium (SiGe) bipolar transistor to achieve an operating frequency centered near 10 GHz. The oscillator uses cross-coupled transistors Q_1 and Q_2 in a multivibrator or flip-flop type of design. The signal is a sine wave whose frequency is set by the collector inductances and varactor capacitances. The modulating

Voltage-controlled oscillator (VCO)

Voltage-controlled crystal oscillator (VXO)

Figure 6-7 A 10-GHz SiGe integrated VCO.

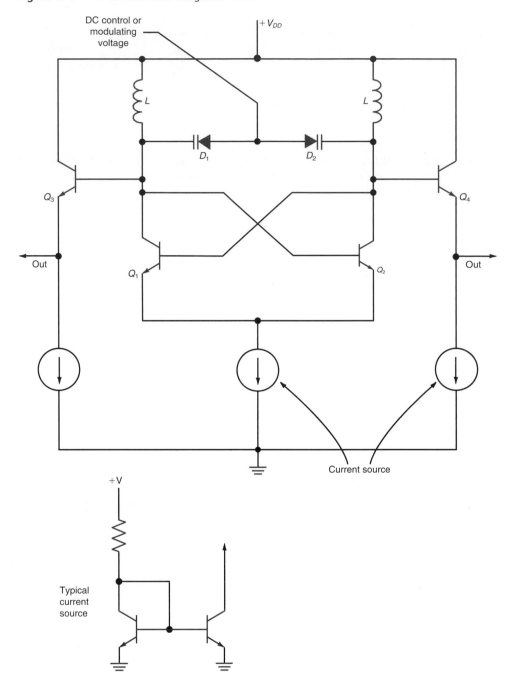

voltage, usually a binary signal to produce FSK, is applied to the junction of D_1 and D_2. Two complementary outputs are available from the emitter followers Q_3 and Q_4. In this circuit, the inductors are actually tiny spirals of aluminum (or copper) inside the chip, with inductance in the 500- to 900-pH range. The varactors are reverse-biased diodes that function as variable capacitors. The tuning range is from 9.953 to 10.66 GHz.

A CMOS version of the VCO is shown in Fig. 6-8. This circuit also uses a cross-coupled LC resonant circuit design and operates in the 2.4- to 2.5-GHz range. Variations of it are used in Bluetooth transceivers and wireless LAN applications. (See Chap. 20.)

There are also many different types of lower-frequency VCOs in common use, including IC VCOs using RC multivibrator-type oscillators whose frequency can be

Figure 6-8   A CMOS VCO for a 2.4-GHz FSK.

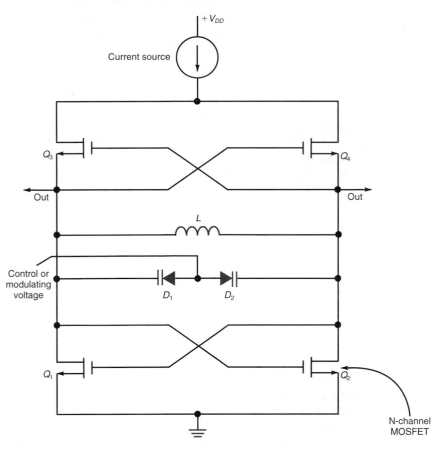

Figure 6-9   Frequency modulation with an IC VCO. (*a*) Block diagram with an IC VCO.
(*b*) Basic frequency modulator using the NE566 VCO.

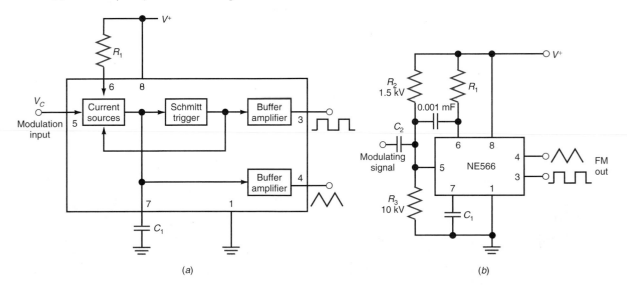

(*a*)   (*b*)

controlled over a wide range by an ac or dc input voltage. These VCOs typically have
an operating range of less than 1 Hz to approximately 1 MHz. The output is either a
square or a triangular wave rather than a sine wave.

Figure 6-9(*a*) is a block diagram of one widely used *IC VCO*, the popular *NE566*.
External resistor R_1 at pin 6 sets the value of current produced by the internal current

NE566 IC VCO

sources. The current sources linearly charge and discharge external capacitor C_1 at pin 7. An external voltage V_C applied at pin 5 is used to vary the amount of current produced by the current sources. The *Schmitt trigger circuit* is a level detector that controls the current source by switching between charging and discharging when the capacitor charges or discharges to a specific voltage level. A linear sawtooth of voltage is developed across the capacitor by the current source. This is buffered by an amplifier and made available at pin 4. The Schmitt trigger output is a square wave at the same frequency available at pin 3. If a sine wave output is desired, the triangular wave is usually filtered with a tuned circuit resonant to the desired carrier frequency.

A complete frequency modulator circuit using the NE566 is shown in Fig. 6-9(*b*). The current sources are biased with a voltage divider made up of R_2 and R_3. The modulating signal is applied through C_2 to the voltage divider at pin 5. The 0.001-μF capacitor between pins 5 and 6 is used to prevent unwanted oscillations. The center carrier frequency of the circuit is set by the values of R_1 and C_1. Carrier frequencies up to 1 MHz may be used with this IC. If higher frequencies and deviations are necessary, the outputs can be filtered or used to drive other circuits, such as a frequency multiplier. The modulating signal can vary the carrier frequency over nearly a 10:1 range, making very large deviations possible. The deviation is linear with respect to the input amplitude over the entire range.

Reactance Modulators

Another way to produce direct FM is to use a *reactance modulator.* This circuit uses a transistor amplifier that acts as either a variable capacitor or an inductor. When the circuit is connected across the tuned circuit of an oscillator, the oscillator frequency can be varied by applying the modulating signal to the amplifier.

Figure 6-10 shows a standard reactance modulator, which is basically a common-emitter class A amplifier. Resistors R_1 and R_2 form a voltage divider to bias the transistor into the linear region. Resistor R_3 is an emitter bias resistor that is bypassed with capacitor C_3. An RF choke (RFC$_2$), rather than a collector resistor, is used to provide a high-impedance load at the operating frequency. The collector of the transistor is connected to the tuned circuit in the carrier oscillator. Capacitor C_4 has a very low impedance at the oscillator frequency. Its main purpose is to keep the direct current from the collector of Q_1 from being shorted to ground through the oscillator coil L_0. The reactance modulator circuit is connected directly across the parallel tuned circuit that sets the oscillator frequency.

Figure 6-10   A reactance modulator.

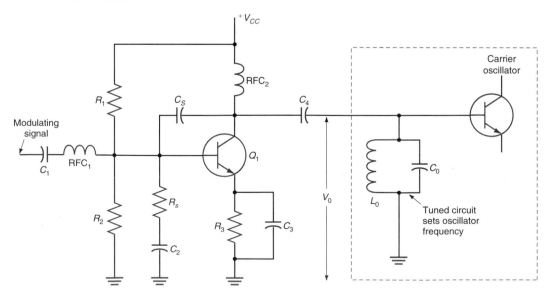

The oscillator signal from the tuned circuit V_0 is connected back to an RC phase-shift circuit made up of C_s and R_s. Capacitor C_2 in series with R_s has a very low impedance at the operating frequency, so it does not affect the phase shift. However, it does prevent R_s from disturbing the dc bias on Q_1. The value of C_s is chosen so that its reactance at the oscillator frequency is about 10 or more times the value of R_s. If the reactance is much greater than the resistance, the circuit acts predominantly capacitively, and therefore the current through the capacitor and R_s leads the applied voltage by about 90°. This means that the voltage across R_s that is applied to the base of Q_1 leads the voltage from the oscillator. Since the collector current in Q_1 is in phase with the base current, which, in turn, is in phase with the base voltage, the collector current in Q_1 leads the oscillator voltage V_0 by 90°. Of course, any circuit whose current leads its applied voltage by 90° looks capacitive to the source voltage. This means that the reactance modulator looks like a capacitor to the oscillator tuned circuit.

The modulating signal is applied to the modulator circuit through C_1 and RFC$_1$. The RFC helps keep the RF signal from the oscillator out of the audio circuits from which the modulating signal is derived. The audio modulating signal varies the base voltage and current of Q_1 according to the intelligence to be transmitted, and the collector current varies proportionally. As the collector current amplitude varies, the phase-shift angle changes with respect to the oscillator voltage, which is interpreted by the oscillator as a change in the capacitance. Thus, as the modulating signal changes, the effective capacitance of the circuit varies and the oscillator frequency varies accordingly. An increase in capacitance lowers the frequency, and a decrease in capacitance increases the frequency. The circuit produces direct FM.

If the positions of R_s and C_s in the circuit of Fig. 6-10 are reversed, the current in the phase shifter still leads the oscillator voltage by 90°. However, voltage from across the capacitor is now applied to the base of the transistor, and that voltage lags the oscillator voltage by 90°. With this configuration, the reactance modulator acts as an inductor. The equivalent inductance changes as the modulating signal is applied. Again, the oscillator frequency varies in proportion to the amplitude of the intelligence signal amplitude.

Reactance modulator circuits can produce frequency deviation over a wide range. They are highly linear, and so distortion is minimal. These circuits can also be implemented with field-effect transistors (FETs) instead of the NPN bipolar component shown in Fig. 6-10. Despite these advantages, reactance modulators are now almost entirely obsolete.

6-2 Phase Modulators

Most modern FM transmitters use some form of phase modulation to produce indirect FM. The reason for using PM instead of direct FM is that the carrier oscillator can be optimized for frequency accuracy and stability. Crystal oscillators or crystal-controlled frequency synthesizers can be used to set the carrier frequency accurately and maintain solid stability.

The output of the carrier oscillator is fed to a phase modulator where the phase shift is made to vary in accordance with the modulating signal. Since phase variations produce frequency variations, indirect FM is the result.

Some phase modulators are based upon the phase shift produced by an RC or LC tuned circuit. It should be pointed out that simple phase shifters of this type do not produce linear response over a large range of phase shift. The total allowable phase shift must be restricted to maximize linearity, and multipliers must be used to achieve the desired deviation. The simplest phase shifters are RC networks like those shown in Fig. 6-11(a) and (b). Depending on the values of R and C, the output of the phase shifter can be set to any phase angle between 0 and 90°. In (a), the output leads the input by some angle between 0 and 90°. For example, when X_c equals R, the phase shift is 45°. The phase shift is computed by using the formula

$$\phi = \tan^{-1} \frac{X_C}{R}$$

Phase modulator

GOOD TO KNOW

Simple phase shifters do not produce a linear response over a large range of phase shift. To compensate for this, restrict the total allowable phase shift to maximize linearity. Multipliers must also be used to achieve the desired deviation.

Figure 6-11 *RC* phase-shifter basics.

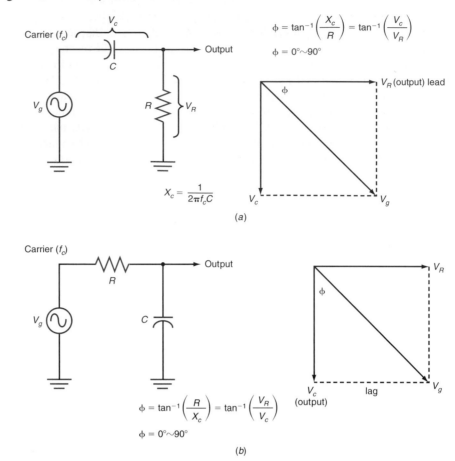

(a)

(b)

A low-pass *RC* filter can also be used, as shown in Fig. 6-11(*b*). Here the output is taken from across the capacitor, so it lags the input voltage by some angle between 0 and 90°. The phase angle is computed by using the formula

$$\phi = \tan^{-1} \frac{R}{X_C}$$

Varactor phase modulator

Varactor Phase Modulators

A simple phase-shift circuit can be used as a phase modulator if the resistance or capacitance can be made to vary with the modulating signal. One way to do this is to replace the capacitor shown in the circuit of Fig. 6-11(*b*) with a varactor. The resulting phase-shift circuit is shown in Fig. 6-12.

In this circuit, the modulating signal causes the capacitance of the varactor to change. If the modulating signal amplitude at the output of amplifier *A* becomes more positive, it adds to the varactor reverse bias from R_1 and R_2, causing the capacitance to decrease. This causes the reactance to increase; thus the circuit produces less phase shift and less deviation. A more negative modulating signal from *A* subtracts from the reverse bias on the varactor diode, increasing the capacitance or decreasing the capacitive reactance. This increases the amount of phase shift and the deviation.

With this arrangement, there is an inverse relationship between the modulating signal polarity and the direction of the frequency deviation. This is the opposite of the desired variation. To correct this condition, an inverting amplifier *A* can be inserted between the modulating signal source and the input to the modulator. Then when the modulating signal goes positive, the inverter output and modulator input go negative and the deviation

Figure 6-12 A varactor phase modulator.

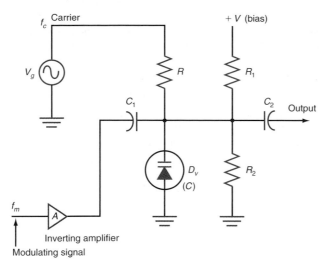

increases. In Fig. 6-12 C_1 and C_2 are dc blocking capacitors and have very low reactance at the carrier frequency. The phase shift produced is lagging, and as in any phase modulator, the output amplitude and phase vary with a change in the modulating signal amplitude.

Transistor Phase Modulators

Figure 6-13 shows a bipolar transistor used as a variable resistor to create a phase modulator. FETs may also be used. The circuit is simply a standard common emitter class A amplifier biased into the linear region by resistors R_1 and R_2. With no modulation, the collector current is 1.22 mA. The voltage from the collector to ground is 6.28 V. The transistor from collector to ground is acting as a resistor, and has a resistance value of $R = V_C/I_C = 6.28/1.22 \times 10^{-3} = 5147\ \Omega$. This resistance forms part of the phase shifter, with $C_1 = 22$ pF. With a carrier frequency of 1.4 MHz, the capacitive reactance is $X_C = 1/2\pi f_c C_1 = 1/6.28(1.4 \times 10^6)(22 \times 10^{-12}) = 5170\ \Omega$. The phase shift produced by the circuit is $\phi = \tan^{-1}(5170/5147) = \tan^{-1} 1.004 = 45°$.

Now assume that a modulating signal is applied to the circuit through dc blocking capacitor C_2. The RFC keeps the RF out of the audio modulating circuits. If the

Figure 6-13 A transistor phase shifter.

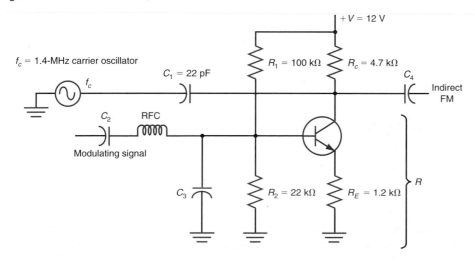

signal goes positive, it increases the base current; this increases the collector current which, in turn, increases the voltage across the collector and decreases the voltage between the collector and ground. If the collector current rises by 0.5 mA to 1.72 mA, the voltage at the collector drops to 3.9 V. The transistor now represents a resistance value of $R = 3.9/0.00172 = 2267\ \Omega$. The new value of phase shift with C_1 is now $\phi = \tan^{-1}(5170/2267) = \tan^{-1} 2.28 = 66.3°$.

If the input signal goes negative, the base current decreases and the collector current decreases proportionally. If the collector current goes down by 0.5 mA, to 0.72 mA, the new output voltage at the collector is 8.6 V. The new value of resistance represented by the transistor is $R = 8.6/0.00072 = 11{,}944\ \Omega$. The new phase shift is $\phi = \tan^{-1}(5170/11944) = 23.4°$. The total phase shift produced by the circuit is $66.3 - 23.4 = 42.9°$.

A phase shift of 43° can also be represented as $\pm 21.5°$. Expressed in radians, this is a total shift of $43/57.3 = 0.75$ rad or ± 0.375 rad.

Example 6-2

A transmitter must operate at a frequency of 168.96 MHz with a deviation of ± 5 kHz. It uses three frequency multipliers—a doubler, a tripler, and a quadrupler. Phase modulation is used. Calculate (a) the frequency of the carrier crystal oscillator and (b) the phase shift $\Delta\phi$ required to produce the necessary deviation at a 2.8-kHz modulation frequency.

a. The frequency multiplier produces a total multiplication of $2 \times 3 \times 4 = 24$. The crystal oscillator frequency is multiplied by 24 to obtain the final output frequency of 168.96 MHz. Thus the crystal oscillator frequency is

$$f_0 = \frac{168.96}{24} = 7.04 \text{ MHz}$$

b. The frequency multipliers multiply the deviation by the same factor. To achieve a deviation of ± 5 kHz, the phase modulator must produce a deviation of $f_d = 5 \text{ kHz}/24 = \pm 208.33$ Hz. The deviation is computed with $f_d = \Delta\phi f_m$; $f_m = 2.8$ kHz.

$$\Delta\phi = \frac{f_d}{f_m} = \frac{208.33}{2800} = \pm 0.0744 \text{ rad}$$

Converting to degrees gives

$$0.0744(57.3°) = \pm 4.263°$$

The total phase shift is

$$\pm 4.263° = 2 \times 4.263° = 8.526°$$

A simple formula for determining the amount of frequency deviation f_d represented by a specific phase angle is

$$f_d = \Delta\phi f_m$$

where $\Delta\phi$ = change in phase angle, rad
$\quad f_m$ = modulating signal frequency

Assume that the lowest modulating frequency for the circuit with the shift of 0.75 rad is 300 Hz. The deviation is $f_d = 0.75(300) = 225$ Hz or ± 112.5 Hz. Since this is PM, the actual deviation is also proportional to the frequency of the modulating signal. With the same maximum deviation of 0.75 rad, if the modulating frequency is 3 kHz (3000 Hz), the deviation is $f_d = 0.75(3000) = 2250$ Hz or ± 1125 Hz.

To eliminate this effect and to generate real FM, the audio input frequency must be applied to a low-pass filter to roll off the signal amplitude at the higher frequencies. In Fig. 6-13, this is the function of C_3, which, with the output impedance of the driving audio amplifier, creates a low-pass filter.

Example 6-3

For the transmitter in Example 6-2, a phase shifter like that in Fig. 6-11 is used, where C is a varactor and $R = 1$ kΩ. Assume that the total phase-shift range is centered on 45°. Calculate the two capacitance values required to achieve the total deviation.

The phase range is centered on 45°, or $45° \pm 4.263° = 40.737°$ and $49.263°$. The total phase range is $49.263 - 40.737 = 8.526°$. If $\phi = \tan^{-1}(R/X_C)$, then $\tan \phi = R/X_C$.

$$X_C = \frac{R}{\tan \phi} = \frac{1000}{\tan 40.737} = 1161 \ \Omega$$

$$C = \frac{1}{2\pi f X_C} = \frac{1}{6.28 \times 7.04 \times 10^6 \times 1161} = 19.48 \text{ pF}$$

$$X_C = \frac{R}{\tan \phi} = \frac{1000}{\tan 49.263} = 861 \ \Omega$$

$$C = \frac{1}{2\pi f X_C} = \frac{1}{6.28 \times 7.04 \times 10^6 \times 861} = 26.26 \text{ pF}$$

To achieve the desired deviation, the voice signal must bias the varactor to vary over the 19.48- to 26.26-pF range.

Tuned–Circuit Phase Modulators

Most phase modulators are capable of producing only a small amount of phase shift—the total being essentially limited to $\pm 20°$ because of the narrow range of linearity of the transistor or varactor. The limited phase shift produces, in turn, a limited frequency shift. One technique for solving this problem is to use a parallel tuned circuit to produce the phase shift. At resonance, a parallel resonant circuit acts as a very high-value resistor. Off resonance, the circuit acts inductively or capacitively and, as a result, produces a phase shift between its current and applied voltage.

Figure 6-14 shows the basic impedance response curve and the phase variation of a parallel resonant circuit. At the resonant frequency f_r, the inductive and capacitive reactances are equal and their effects cancel each other. The result is an extremely high resistive impedance at f_r. The circuit acts resistively at this point, and the phase angle between the current and the applied voltage is therefore zero.

At frequencies below resonance, X_L decreases and X_C increases. This causes the circuit to act as an inductor, and the current lags the applied voltage. Above resonance, X_L increases and X_C decreases. This causes the circuit to act as a capacitor, and the current leads the applied voltage. If the Q of the resonant circuit is relatively high, the phase shift will be quite pronounced, as shown in Fig. 6-14. The same effect is achieved if the frequency is constant and either L or C is varied. A relatively small change in L

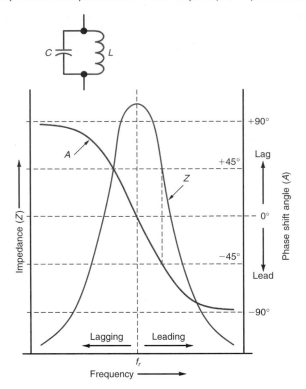

or C can produce a significant phase shift. The idea, then, is to cause the inductance or capacitance to vary with the modulating voltage, producing a phase shift.

One of the variety of circuits that have been developed based on this technique is illustrated in Fig. 6-15. In this type of circuit, the parallel tuned circuit is usually part of the output circuit of an RF amplifier driven by the carrier oscillator. A varactor diode D_1 is connected in parallel with the tuned circuit, thereby providing a capacitance change with the modulating signal. The voltage divider, made up of R_1 and R_2, sets the reverse bias on D_1. And C_2 acts as a dc blocking capacitor, preventing bias from being applied to the tuned circuit. Its value is very large, so it is essentially an ac short circuit at the carrier frequency, and the capacitance of D_1 controls the resonant frequency.

The modulating signal is first passed through a low-pass network, made up of R_3 and C_3, which provides the amplitude compensation necessary to produce FM. The

Figure 6-15 One form of phase modulator.

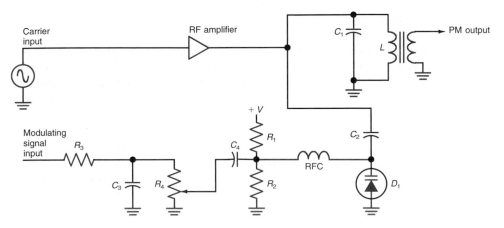

modulating signal appears across potentiometer R_4, allowing the desired amount of modulating signal to be tapped off and applied to the phase-shift circuit. Potentiometer R_4 acts as a deviation control. The higher the modulating voltage, the greater the frequency deviation. The modulating signal is applied to the varactor diode through capacitor C_4. With zero modulating voltage, the value of the capacitance of D_1, along with capacitor C_1 and inductor L, sets the resonant frequency of the tuned circuit. Capacitor C_2 is a dc blocking capacitor with a near-zero impedance at the carrier frequency. The indirect FM output across L is inductively coupled to the output.

When the modulating signal goes negative, it subtracts from the reverse bias of D_1. This increases the capacitance of the circuit and lowers the reactance, making the circuit appear capacitive. Thus a leading phase shift is produced. The parallel LC circuit looks like a capacitor to the output resistance of the carrier amplifier, so the output lags the input. A positive-going modulating voltage decreases the capacitance; the tuned circuit then becomes inductive, producing a lagging phase shift. The LC circuit looks like an inductor to the output resistance of the carrier amplifier, so the output leads the input. The result at the output is a relatively wide phase shift, which, in turn, produces excellent linear frequency deviation.

Phase modulators are relatively easy to implement, but they have two main disadvantages. First, the amount of phase shift they produce and the resulting frequency deviation are relatively low. For that reason, the carrier is usually generated at a lower frequency, and frequency multipliers are used to increase the carrier frequency and the amount of frequency deviation. Second, all the phase-shift circuits described above, including the tuned-circuit phase shifter, produce amplitude variations as well as phase changes. When the value of one of the components is changed, the phase shifts but the output amplitude changes as well. Both of these problems are solved by feeding the output of the phase modulator to class C amplifiers used as frequency multipliers. These amplifiers eliminate amplitude variations at the same time as they increase the carrier frequency and deviation to the desired final values. (Figure 6-6 illustrated how frequency multipliers increase both carrier frequency and deviation.)

6-3 Frequency Demodulators

Frequency demodulator

Any circuit that will convert a frequency variation in the carrier back to a proportional voltage variation can be used to demodulate or detect FM signals. Circuits used to recover the original modulating signal from an FM transmission are called demodulators, detectors, or discriminators.

Slope Detectors

The simplest frequency demodulator, the *slope detector,* makes use of a tuned circuit and a diode detector to convert frequency variations to voltage variations. The basic circuit is shown in Fig. 6-16(a). This has the same configuration as the basic AM diode detector described in Chap. 4, although it is tuned differently.

Slope detector

The FM signal is applied to transformer T_1 made up of L_1 and L_2. Together L_2 and C_1 form a series resonant circuit. Remember that the signal voltage induced into L_2 appears in series with L_2 and C_1 and the output voltage is taken from across C_1. The response curve of this tuned circuit is shown in Fig. 6-16(b). Note that at the resonant frequency f_r the voltage across C_1 peaks. At lower or higher frequencies, the voltage falls off.

To use the circuit to detect or recover FM, the circuit is tuned so that the center or carrier frequency of the FM signals is approximately centered on the leading edge of the response curve, as shown in Fig. 6-16(b). As the carrier frequency varies above and below its center frequency, the tuned circuit responds as shown in the figure. If the frequency goes lower than the carrier frequency, the output voltage across C_1 decreases. If the frequency goes higher, the output across C_1 goes higher. Thus the ac voltage across C_1 is proportional to the frequency of the FM signal. The voltage across C_1 is rectified into

Figure 6-16 Slope detector operation.

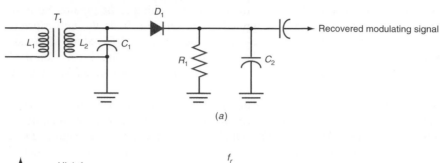

(a)

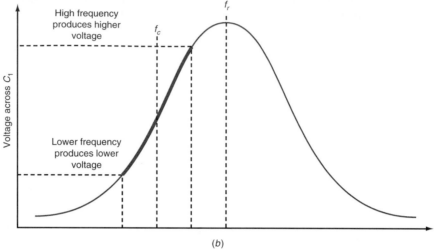

(b)

dc pulses that appear across the load R_1. These are filtered into a varying dc signal that is an exact reproduction of the original modulating signal.

The main difficulty with slope detectors lies in tuning them so that the FM signal is correctly centered on the leading edge of the tuned circuit. In addition, the tuned circuit does not have a perfectly linear response. It is approximately linear over a narrow range, as Fig. 6-16(b) shows, but for wide deviations, amplitude distortion occurs because of the nonlinearity.

The slope detector is never used in practice, but it does show the principle of FM demodulation, i.e., converting a frequency variation to a voltage variation. Numerous practical designs based upon these principles have been developed. These include the Foster-Seeley discriminator and the ratio detector, neither of which is used in modern equipment.

Pulse–Averaging Discriminators

Pulse-averaging discriminator

A simplified block diagram of a *pulse-averaging discriminator* is illustrated in Fig. 6-17. The FM signal is applied to a zero-crossing detector or a clipper-limiter which generates a binary voltage-level change each time the FM signal varies from

Figure 6-17 Pulse-averaging discriminator.

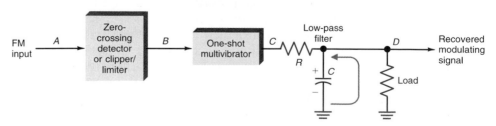

Figure 6-18 (a) FM input. (b) Output of zero-crossing detector. (c) Output of one shot. (d) Output of discriminator (original modulating signal).

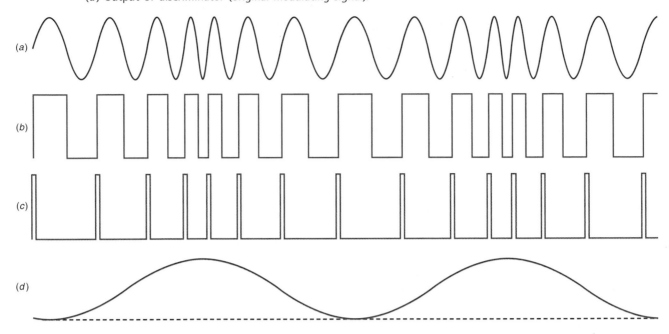

minus to plus or from plus to minus. The result is a rectangular wave containing all the frequency variations of the original signal but without amplitude variations. The FM square wave is then applied to a one-shot (monostable) multivibrator which generates a fixed-amplitude, fixed-width dc pulse on the leading edge of each FM cycle. The duration of the one shot is set so it is less than one-half the period of the highest frequency expected during maximum deviation. The one-shot output pulses are then fed to a simple *RC* low-pass filter which averages the dc pulses to recover the original modulating signal.

The waveforms for the pulse-averaging discriminator are illustrated in Fig. 6-18. At low frequencies, the one-shot pulses are widely spaced; at higher frequencies, they occur very close together. When these pulses are applied to the averaging filter, a dc output voltage is developed, the amplitude of which is directly proportional to the frequency deviation.

When a one-shot pulse occurs, the capacitor in the filter charges to the amplitude of the pulse. When the pulse turns off, the capacitor discharges into the load. If the *RC* time constant is high, the charge on the capacitor does not decrease much. When the time interval between pulses is long, however, the capacitor loses some of its charge into the load so the average dc output is low. When the pulses occur rapidly, the capacitor has little time between pulses to discharge; the average voltage across it therefore remains higher. As the figure shows, the filter output voltage varies in amplitude with the frequency deviation. The original modulating signal is developed across the filter output. The filter components are carefully selected to minimize the ripple caused by the charging and discharging of the capacitor while at the same time providing the necessary high-frequency response for the original modulating signal.

Some pulse-averaging discriminators generate a pulse every half-cycle or at every zero crossing instead of every one cycle of the input. With a greater number of pulses to average, the output signal is easier to filter and contains less ripple.

The pulse-averaging discriminator is a very high-quality frequency demodulator. In the past, its use was limited to expensive telemetry and industrial control applications. Today, with the availability of low-cost ICs, the pulse-averaging discriminator is easily implemented and is used in many electronic products.

FM Circuits   **191**

Figure 6-19 A quadrature FM detector.

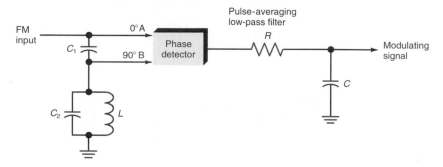

Quadrature Detectors

Quadrature detector

The *quadrature detector* is probably the single most widely used FM demodulator. Its primary application is in TV audio demodulation, although it is also used in some FM radio systems. The quadrature detector uses a phase-shift circuit to produce a phase shift of 90° at the unmodulated carrier frequency. The most commonly used phase-shift arrangement is shown in Fig. 6-19. The frequency-modulated signal is applied through a very small capacitor (C_1) to the parallel tuned circuit, which is adjusted to resonate at the center carrier frequency. At resonance, the tuned circuit appears as a high value of pure resistance. The small capacitor has a very high reactance compared to the tuned circuit impedance. Thus the output across the tuned circuit at the carrier frequency is very close to 90° and leads the input. When frequency modulation occurs, the carrier frequency deviates above and below the resonant frequency of the tuned circuit, resulting in an increasing or a decreasing amount of phase shift between the input and the output.

The two quadrature signals are then fed to a phase detector circuit. The most commonly used phase detector is a balanced modulator using differential amplifiers like those discussed in Chap. 4. The output of the phase detector is a series of pulses whose width varies with the amount of phase shift between the two signals. These signals are averaged in an RC low-pass filter to recreate the original modulating signal.

Normally the sinusoidal FM input signals to the phase detector are at a very high level and drive the differential amplifiers in the phase detector into cutoff and saturation. The differential transistors act as switches, so the output is a series of pulses. No limiter is needed if the input signal is large enough. The duration of the output pulse is determined by the amount of phase shift. The phase detector can be regarded as an AND gate whose output is ON only when the two input pulses are ON and is OFF if either one or both of the inputs are OFF.

Figure 6-20 shows the typical waveforms involved in a quadrature detector. When there is no modulation, the two input signals are exactly 90° out of phase and therefore provide an output pulse with the indicated width. When the FM signal frequency increases, the amount of phase shift decreases, resulting in a wider output pulse. The wider pulses averaged by the RC filter produce a higher average output voltage, which corresponds to the higher amplitude required to produce the higher carrier frequency. When the signal frequency decreases, greater phase shift and narrower output pulses occur. The narrower pulses, when averaged, produce a lower average output voltage, which corresponds to the original lower-amplitude modulating signal.

Quadrature detectors are usually built into other ICs, such as intermediate-frequency amplifiers and complete receiver ICs, which are discussed in Chap. 9.

Phase-Locked Loops

Phase-locked loop (PLL)

A *phase-locked loop* (PLL) is a frequency- or phase-sensitive feedback control circuit used in frequency demodulation, frequency synthesizers, and various filtering and signal detection applications. All phase-locked loops have the three basic elements, shown in Fig. 6-21.

Figure 6-20 Waveforms in the quadrature detector.

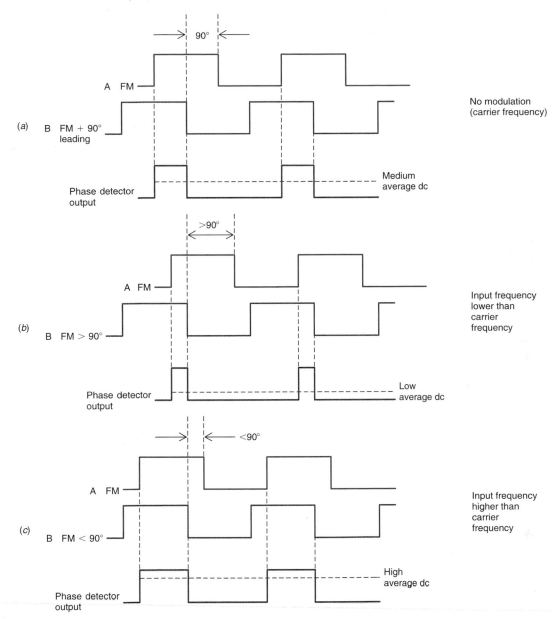

(a) A FM

B FM + 90°
leading

Phase detector
output

No modulation
(carrier frequency)

Medium
average dc

90°

(b) A FM

B FM > 90°

Phase detector
output

Input frequency
lower than
carrier
frequency

Low
average dc

>90°

(c) A FM

B FM < 90°

Phase detector
output

Input frequency
higher than
carrier
frequency

High
average dc

<90°

Figure 6-21 Block diagram of a PLL.

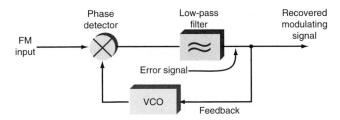

1. A phase detector is used to compare the FM input, sometimes referred to as the *reference signal*, to the output of a voltage-controlled oscillator (VCO).
2. The VCO frequency is varied by the dc output voltage from a low-pass filter.
3. The low-pass filter smoothes the output of the phase detector into a control voltage that varies the frequency of the VCO.

The primary job of the phase detector is to compare the two input signals and generate an output signal that, when filtered, will control the VCO. If there is a phase or frequency difference between the FM input and VCO signals, the phase detector output varies in proportion to the difference. The filtered output adjusts the VCO frequency in an attempt to correct for the original frequency or phase difference. This dc control voltage, called the *error signal,* is also the feedback in this circuit.

When no input signal is applied, the phase detector and low-pass filter outputs are zero. The VCO then operates at what is called the *free-running frequency,* its normal operating frequency as determined by internal frequency-determining components. When an input signal close to the frequency of the VCO is applied, the phase detector compares the VCO free-running frequency to the input frequency and produces an output voltage proportional to the frequency difference. Most PLL phase detectors operate just as the one discussed in the section on quadrature detectors. The phase detector output is a series of pulses that vary in width in accordance with the amount of phase shift or frequency difference that exists between the two inputs. The output pulses are then filtered into a dc voltage that is applied to the VCO. This dc voltage is such that it forces the VCO frequency to move in a direction that reduces the dc error voltage. The error voltage forces the VCO frequency to change in the direction that reduces the amount of phase or frequency difference between the VCO and the input. At some point, the error voltage causes the VCO frequency to equal the input frequency; when this happens, the PLL is said to be in a *locked* condition. Although the input and VCO frequencies are equal, there is a phase difference between them, usually exactly 90°, which produces the dc output voltage that will cause the VCO to produce the frequency that keeps the circuit locked.

If the input frequency changes, the phase detector and low-pass filter produce a new value of dc control voltage that forces the VCO output frequency to change until it is equal to the new input frequency. Any variation in input frequency is matched by a VCO frequency change, so the circuit remains locked. The VCO in a PLL is, therefore, capable of *tracking* the input frequency over a wide range. The range of frequencies over which a PLL can track an input signal and remain locked is known as the *lock range.* The lock range is usually a band of frequencies above and below the free-running frequency of the VCO. If the input signal frequency is out of the lock range, the PLL will not lock. When this occurs, the VCO output frequency jumps to its free-running frequency.

If an input frequency within the lock range is applied to the PLL, the circuit immediately adjusts itself into a locked condition. The phase detector determines the phase difference between the free-running and input frequencies of the VCO and generates the error signal that forces the VCO to equal the input frequency. This action is referred to as *capturing* an input signal. Once the input signal is captured, the PLL remains locked and will track any changes in the input signal as long as the frequency is within the lock range. The range of frequencies over which a PLL will capture an input signal, known as the *capture range,* is much narrower than the lock range, but, like the lock range, is generally centered on the free-running frequency of the VCO (see Fig. 6-22).

Figure 6-22 Capture and lock ranges of a PLL.

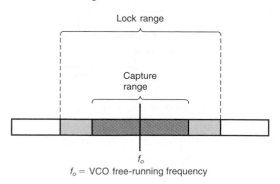

f_o = VCO free-running frequency

The characteristic that causes the PLL to capture signals within a certain frequency range causes it to act as a bandpass filter. Phase-locked loops are often used in signal conditioning applications, where it is desirable to pass signals only in a certain range and to reject signals outside of that range. The PLL is highly effective in eliminating the noise and interference on a signal.

The ability of a PLL to respond to input frequency variations makes it useful in FM applications. The PLL's tracking action means that the VCO can operate as a frequency modulator that produces exactly the same FM signal as the input. In order for this to happen, however, the VCO input must be identical to the original modulating signal. The VCO output follows the FM input signal because the error voltage produced by the phase detector and low-pass filter forces the VCO to track it. Thus the VCO output must be identical to the input signal if the PLL is to remain locked. The error signal must be identical to the original modulating signal of the FM input. The low-pass filter cutoff frequency is designed in such a way that it is capable of passing the original modulating signal.

The ability of a PLL to provide frequency selectivity and filtering gives it a signal-to-noise ratio superior to that of any other type of FM detector. The linearity of the VCO ensures low distortion and a highly accurate reproduction of the original modulating signal. Although PLLs are complex, they are easy to apply because they are readily available in low-cost IC form.

Figure 6-23 is a block diagram of a popular and widely used IC PLL, the 565. The 565 is connected as an FM demodulator. The 565 circuitry is shown inside the dashed lines; all components outside the dashed lines are discrete components. The numbers on the connections are the pin numbers on the 565 IC, which is housed in a standard 14-pin dual-in-line package (DIP). The circuit is powered by ± 12-V power supplies.

The low-pass filter is made up of a 3.6-kΩ resistor inside the 565 that terminates at pin 7. A 0.1-μF external capacitor C_2 completes the filter. Note that the recovered original modulating signal is taken from the filter output. The free-running frequency of the VCO (f_0) is set by external components R_1 and C_1 according to the formula $f_0 = 1.2/4R_1C_1 = 1.2/4(2700)(0.01 \times 10^{-6}) = 11,111$ Hz or 11.11 kHz.

PLL demodulator

Figure 6-23 A PLL FM demodulator using the 565 IC.

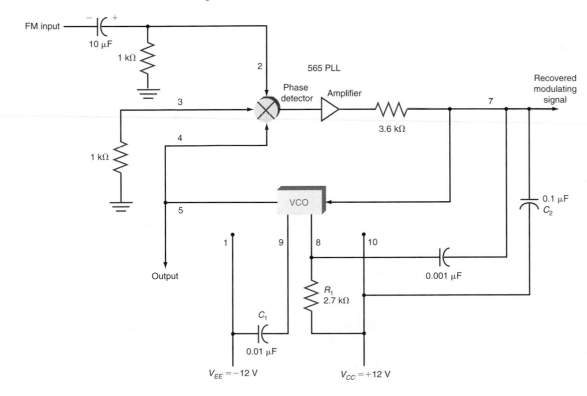

The lock range f_L can be computed with an expression supplied by the manufacturer for this circuit $f_L = 16 f_0 / V_S$, where V_S is the total supply voltage. In the circuit of Fig. 6-23, V_S is the sum of the two 12-V supplies, or 24 V, so the total lock range centered on the free-running frequency is $f_L = 16(11.11 \times 10^3)/24 = 7406.7$ Hz, or ± 3703.3 Hz.

With this circuit, it is assumed that the unmodulated carrier frequency is the same as the free-running frequency, 11.11 kHz. Of course, it is possible to set this type of circuit to any other desired center frequency simply by changing the values of R_1 and C_1. The upper frequency limitation for the 565 IC is 500 kHz.

No one likes static on the radio. But until Edwin Armstrong (1890–1954) came along, people just lived with it. Listening to AM radio was an exercise in patience, but no one considered any other method viable. The public learned to live with static caused by storms, vacuum cleaners, and oil burners.

Armstrong took the idea of frequency modulation (conceived by others around 1904) and between 1928 and 1933 turned it into clear FM radio. Armstrong figured out that narrowband FM (which was where FM research had focused until Armstrong) was the wrong path, and that broadband FM would work.

Today it is hard to imagine a world without FM. It was used for mobile military radio (for tanks during World War II), for FM radar, and the audio for TV. By 1940 there were 40 FM broadcast stations, and FM was used widely in buses, in taxis, and by firefighter and police.

In addition to numerous other contributions to technology, Armstrong developed a system for multiplexing FM radio so that more than one program could be broadcast concurrently on the same wavelength.

Although Armstrong did not live to see it, by the 1960s there were almost 2000 U.S. FM stations. Almost all radio sets are FM, all microwave links are FM, and FM is used in communicating in space. Armstrong's old 1936 station and tower were eventually turned into a pager and repeater site for New York City.

Today FM is being replaced with digital methods. The first cell phones were FM, but quickly most manufacturers went to digital. Radio is headed in that direction, too, with the new satellite digital radios; TV is headed for digital; satellite TVs are digital, as are the new digital high-definition TVs. Before long, virtually everything will be digital! Strangely enough, some digital will be transmitted by a digital version of FM called frequency-shift keying (FSK).

CHAPTER REVIEW

Summary

Frequency-modulating circuits are designed to convert a modulating voltage to a corresponding change in capacitance or inductance. The varactor is often used in this application. It is basically a semiconductor junction diode operated in a reverse-bias mode.

Because *LC* oscillators are not stable enough to meet the stringent frequency stability requirements imposed by the FCC, crystal oscillators, or oscillators combined with frequency-multiplier circuits, are used to set the carrier frequency.

Simple phase-shift circuits can be used as phase modulators if the resistance or capacitance can be made to vary with the modulating signal. This can be achieved by adding a varactor phase modulator to the circuit.

Any circuit that will convert a frequency variation in the carrier back to a proportional voltage variation can be used for frequency demodulation. The most widely used FM demodulators today are quadrature detectors and phase-locked loops.

All PLLs have three basic elements: a phase detector, a voltage-controlled oscillator (VCO), and a low-pass filter. The range of frequencies over which the PLL will track an input signal and remain locked is known as the lock range, and the range of frequencies over which the PLL will capture an input signal is known as the capture range.

Questions

1. What parts of the varactor act as the plates of a capacitor?
2. How does capacitance vary with applied voltage?
3. Do varactors operate with forward or reverse bias?
4. What is the main reason why *LC* oscillators are not used in transmitters today?
5. Can the most widely used type of carrier oscillator be frequency-modulated by a varactor?
6. How would the reactance modulator in Fig. 6-10 act if capacitor C_S were replaced by an inductor?
7. What is the main advantage of using a phase modulator rather than a direct frequency modulator?
8. What is the term for frequency modulation produced by PM?
9. What is the advantage of a parallel tuned circuit as a phase shifter over a simple *RC* circuit?
10. What components in Fig. 6-15 compensate for the greater frequency deviation at the higher modulating signal frequencies?
11. What are the primary applications for quadrature detectors?
12. What two IC demodulators use the concept of averaging pulses in a low-pass filter to recover the original modulating signal?
13. Which is probably the best FM demodulator of all those discussed in this chapter?
14. What is a capture range? What is a lock range?
15. What frequency does the VCO assume when the input is outside the capture range?
16. What type of circuit does a PLL look like over its lock range?

Problems

1. A parallel tuned circuit in an oscillator consists of a 40-μH inductor in parallel with a 330-pF capacitor. A varactor with a capacitance of 50 pF is connected in parallel with the circuit. What is the resonant frequency of the tuned circuit and the oscillator operating frequency? ◆
2. If the varactor capacitance of the circuit in Prob. 1 is decreased to 25 pF, (a) how does the frequency change and (b) what is the new resonant frequency?
3. A phase modulator produces a maximum phase shift of 45°. The modulating frequency range is 300 to 4000 Hz. What is the maximum frequency deviation possible? ◆
4. The FM input to a PLL demodulator has an unmodulated center frequency of 10.7 MHz. (a) To what frequency must the VCO be set? (b) From which circuit is the recovered modulating signal taken?
5. A 565 IC PLL has an external resistor R_1 of 1.2 kΩ and a capacitor C_1 of 560 pF. The power supply is 10 V. (a) What is the free-running frequency? (b) What is the total lock range? ◆
6. A varactor phase modulator like the one in Fig. 6-12 has a resistance value of 3.3 kΩ. The capacitance of the varactor at the center unmodulated frequency is 40 pF and the carrier frequency is 1 MHz. (a) What is the phase shift? (b) If the modulating signal changes the varactor capacitance to 55 pF, what is the new phase shift? (c) If the modulating signal frequency is 400 Hz, what is the approximate frequency deviation represented by this phase shift?

◆ *Answers to Selected Problems follow Chap. 22.*

Critical Thinking

1. What circuit must be used ahead of the Foster-Seeley discriminator in order for it to work properly? Explain.
2. Name the three key components of a phase-locked loop and write a brief explanation of how each component works.
3. What happens to an FM signal that has been passed through a tuned circuit that is too narrow, resulting in the higher upper and lower sidebands to be eliminated? What would the output of a demodulator processing this signal look like compared to that of the original modulating signal?
4. A direct-frequency (DF) modulated crystal oscillator has a frequency of 9.3 MHz. The varactor produces a maximum deviation of 250 Hz. The oscillator is followed by two triplers, a doubler, and a quadrupler. What are the final output frequency and the deviation?
5. Refer to Fig. 6-4. To decrease oscillator frequency, would you adjust potentiometer R_4 closer to $+V_{CC}$ or closer to ground?
6. Refer to Fig. 6-15. If R_1 became open, would the circuit still operate? Explain.

Digital Communication Techniques

Since the mid-1970s, digital methods of transmitting data have slowly but surely replaced the older, more conventional analog ones. Today, thanks to the availability of fast, low-cost analog-to-digital (A/D) and digital-to-analog (D/A) converters and high-speed digital signal processors, most electronic communications is digital.

This chapter begins with the reasons for using digital transmission. Then the concepts and operation of A/D and D/A converters are summarized. Next, pulse modulation techniques are described, and the chapter concludes with an introduction to digital signal processing (DSP), techniques.

Objectives

After completing this chapter, you will be able to:

- Give a step-by-step account of the transmission of analog signals using digital techniques.
- Explain how quantizing error occurs, describe the techniques used to minimize it, and calculate the minimum sampling rate given the upper frequency limit of the analog signal to be converted.
- List the advantages and disadvantages of the three most common types of analog-to-digital converters.
- Explain why pulse-code modulation has superseded pulse-amplitude modulation (PAM), pulse-width modulation (PWM), and pulse-position modulation (PPM).
- Draw and fully label a block diagram of a digital signal-processing (DSP) circuit.

7-1 Digital Transmission of Data

Data

The term *data* refers to information to be communicated. Data is in digital form if it comes from a computer. If the information is in the form of voice, video, or some other analog signal, it can be converted to digital form before it is transmitted.

Digital communication was originally limited to the transmission of data between computers. Numerous large and small networks have been formed to support communication between computers, e.g., local-area networks (LANs), which permit PCs to communicate (see Chap. 12). The use of modems to allow PCs and larger computers to communicate via the telephone system is another example. Now, because analog signals can be readily and inexpensively converted to digital and vice versa, data communication techniques can be used to transmit voice, video, and other analog signals in digital form.

There are three primary reasons for the growth of digital communication systems. First, the increased use of computers has made it necessary to find a way for computers to communicate and exchange data. Second, digital transmission methods offer some major benefits over analog communication techniques. Third, the telephone system, the largest and most widely used communication system, has been converting from analog methods to digital over the years. These reasons are discussed in the following sections.

Proliferation of Computers

Since PCs were introduced during the 1970s, their numbers have increased by several orders of magnitude so that hundreds of millions of PCs are now in use worldwide. Most white-collar workers have PCs on their desks. PCs have speeded up and simplified our work and increased our productivity.

At the same time, the need has arisen for users to share or exchange data and programs. At a simple level, data can be transferred by handing a diskette to another person or by mail. But it is far easier if computers can communicate directly. When computers are connected via some communication medium and programmed appropriately, they can exchange data or share programs and peripherals as well as other resources. The result is increased convenience and usefulness of computers.

Some common examples of computer data communication are as follows:

File transfer

1. *File transfer.* The transfer of files, records, or whole databases such as company accounting records, customer orders from a sales office to the main office, or bank data transfers.

Electronic mail (e-mail)

2. *Electronic mail (e-mail).* Communication between and among individuals by means of a computer. Users send messages to one another as they would letters or memos in printed form.

3. *Computer-peripheral links.* The use of data communication techniques to send data between a main computer and peripherals. Data is keyed into the computer, sent to and from floppy and hard disk drives for data storage or retrieval, and then sent to a printer.

4. *Internet access.* Tapping into remote databases, websites, or other sources of information by way of the telephone system or a cable TV network. Online services and the Internet are examples.

Local-area network (LAN)

5. *Local-area networks (LANs).* Groups of PCs in an office or company that are connected in order to share data and other resources. This is the fastest-growing segment of data communication.

Noncomputer Uses of Digital Communication

Among the noncomputer applications of digital techniques is remote control, for example:

1. **TV remote control.** Binary signals generated by pushbuttons modulate an infrared light beam for the purpose of changing channels or volume.

2. **Garage door opener.** Depressing a button on a remote control unit generates a unique binary code that modulates a VHF or UHF radio transmitter for the purpose of opening or closing a garage door.

3. **Carrier current controls.** These systems generate binary codes that modulate a carrier signal that is then superimposed on a 60-Hz power signal. The ac power lines in a house or building become the transmission medium for control signals from one room or area to another. An example is the X-10 system widely used in homes for the remote control of lights and appliances and powerline networking.

4. **Radio control of models.** Hobbyists build model airplanes, boats, and cars and control them remotely by radio transmission of binary control codes.

5. **Remote keyless entry.** Wireless door locks and other accessories are used in automobiles.

Benefits of Digital Communication

Transmitting information by digital means offers several important advantages over analog methods, as discussed in the next section and in Sec. 7-5.

Noise Immunity. When a signal is sent over a medium or channel, noise is invariably added to the signal. The S/N ratio decreases, and the signal becomes harder to recover. Noise, which is a voltage of randomly varying amplitude and frequency, easily corrupts analog signals. Signals of insufficient amplitude can be completely obliterated by noise. Some improvement can be achieved with preemphasis circuits at the transmitter and deemphasis circuits at the receiver, and other similar techniques. If the signal is analog FM, the noise can be clipped off at the receiver so that the signal can be more readily recovered, but phase modulation of the signal by the noise will still degrade quality.

Digital signals, which are usually binary, are more immune to noise than analog signals because the noise amplitude must be much higher than the signal amplitude to make a binary 1 look like a binary 0 or vice versa. If the binary amplitudes for binary 0 and 1 are sufficiently large, the receiving circuitry can easily distinguish between the 0 and 1 levels even with a significant amount of noise (see Fig. 7-1).

Noise immunity

Figure 7-1 (*a*) Noise on a binary signal. (*b*) Clean binary signal after regeneration.

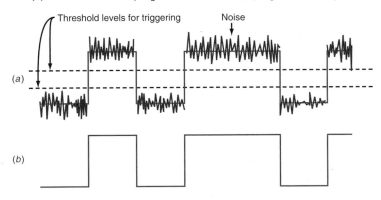

At the receiver, circuits can be set up so that the noise is clipped off. A threshold circuit made with a line receiver circuit, an op amp comparator, or a Schmitt trigger will trigger above or below the thresholds to which it is set. If the thresholds are set carefully, only the logic levels will trigger the circuit. Thus a clean output pulse will be generated by the circuit. This process is called *signal regeneration.*

Digital signals, like analog signals, experience distortion and attenuation when transmitted over a cable or by radio. The cable acts as a low-pass filter and thus filters out the higher harmonics in a pulse signal, causing the signal to be rounded and distorted. When a signal is transmitted by radio, its amplitude is seriously reduced. However, digital signals can be transmitted over long distances if the signal is regenerated along the way to restore the amplitude lost in the medium and to overcome the noise added in the process. When the signal reaches its destination, it has almost exactly the same shape as the original. Consequently, with digital transmission, the error rate is minimal.

Error Detection and Correction. With digital communication, transmission *errors* can usually be detected and even corrected. If an error occurs because of a very high noise level, it can be detected by special circuitry. The receiver recognizes that an error is contained in the transmission, and data can be retransmitted. A variety of techniques have been developed to find errors in binary transmissions; some of them are discussed in Chap. 11. In addition, elaborate error detection schemes have been developed so that the type of error and its location can be identified. This kind of information makes it possible to correct errors before the data is used at the receiver.

Compatibility with Time–Division Multiplexing. Digital data communication is adaptable to time-division multiplexing schemes. *Multiplexing* is the process of transmitting two or more signals simultaneously on a single communication channel or medium. There are two types of multiplexing: frequency-division multiplexing, an analog technique using modulation methods, and time-division multiplexing, a digital technique. These techniques are discussed further in Chap. 10.

Digital ICs. A further benefit of digital techniques is that digital ICs are smaller and easier to make than linear ICs so therefore can be more complex and provide a processing capability greater than what can be accomplished with analog ICs.

Digital Signal Processing (DSP). DSP is the processing of analog signals by digital methods. This involves converting an analog signal to digital and then processing with a fast digital computer. Processing means filtering, equalization, phase shifting, mixing, and other traditionally analog methods. Processing also includes data compression techniques that enhance the speed of data transmission and reduce the digital data storage capacity required for some applications. Even modulation and demodulation can be accomplished by DSP. The processing is accomplished by executing unique mathematical algorithms on the computer. The digital signal is then converted back to analog form. DSP permits significant improvements in processing over equivalent analog techniques. But best of all, it permits types of processing that were never available in analog form.

Finally, processing also involves storage of data. Analog data is difficult to store. But digital data is routinely stored in computers by using a variety of well-proven digital storage methods and equipment such as RAM; ROM; flash, floppy, and hard disk drives; optical drives; and tape units.

Disadvantages of Digital Communication

There are some disadvantages to digital communication. The most important is the bandwidth size required by a digital signal. With binary techniques, the bandwidth of a signal can be 2 or more times greater than it would be with analog methods. Also, digital communication circuits are usually more complex than analog circuits. However,

Signal regeneration

Error

Multiplexing

Digital signal processing (DSP)

GOOD TO KNOW

With binary techniques, the bandwidth of a signal may be 2 or more times greater than it would be with analog methods.

although more circuitry is needed to do the same job, the circuits are usually in IC form, are inexpensive, and do not require much expertise or attention on the part of the user.

7-2 Parallel and Serial Transmission

There are two ways to move binary bits from one place to another: transmit all bits of a word simultaneously or send only 1 bit at a time. These methods are referred to, respectively, as parallel transfer and serial transfer.

Parallel Transfer

In *parallel* data transfers, all the bits of a code word are transferred simultaneously (see Fig. 7-2). The binary word to be transmitted is usually loaded into a register containing one flip-flop for each bit. Each flip-flop output is connected to a wire to carry that bit to the receiving circuit, which is usually also a storage register. As can be seen in Fig. 7-2, in parallel data transmission, there is one wire for each bit of information to be transmitted. This means that a multiwire cable must be used. Multiple parallel lines that carry binary data are usually referred to as a *data bus*. All eight lines are referenced to a common ground wire.

Parallel transfer

Data bus

Parallel data transmission is extremely fast because all the bits of the data word are transferred simultaneously. The speed of parallel transfer depends on the propagation delay in the sending and receiving of logic circuits and any time delay introduced by the cable. Such data transfers can occur in only a few nanoseconds in many applications.

Parallel data transmission is not practical for long-distance communication. To transfer an 8-bit data word from one place to another, eight separate communication channels are needed, one for each bit. Although multiwire cables can be used over limited distances (usually no more than a few feet), for long-distance data communication they are impractical because of cost and signal attenuation. And, of course, parallel data transmission by radio would be even more complex and expensive, because one transmitter and receiver would be required for each bit.

Figure 7-2   Parallel data transmission.

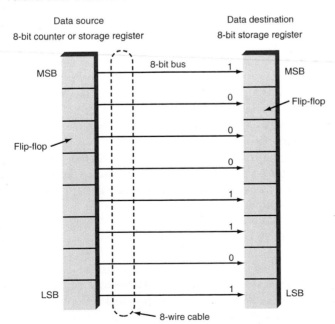

Figure 7-3 Serial data transmission.

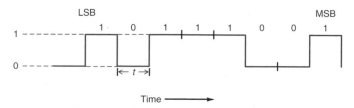

Over the years, data transfer rates over parallel buses have continued to increase. For example, in personal computers bus transfer rates have increased from 33 to 66 to 133 MHz and now to 400 MHz and beyond. However, to achieve these speeds, the bus line lengths have had to be considerably shortened. The capacitance and inductance of the bus lines severely distort the pulse signals. Furthermore, cross talk between the lines also limits the speed. Reducing the line length decreases the inductance and capacitance, permitting higher speeds. To achieve speeds up to 400 MHz, bus lengths must be limited to only a few inches. To achieve higher rates, serial data transfers are being used.

Serial Transfer

Data transfers in communication systems are made serially; each bit of a word is transmitted one after another (see Fig. 7-3). This figure shows the code 10011101 being transmitted 1 bit at a time. The least significant bit (LSB) is transmitted first, and the most significant bit (MSB) last. The MSB is on the right, indicating that it was transmitted later in time than the LSB. Each bit is transmitted for a fixed interval of time t. The voltage levels representing each bit appear on a single data line (with respect to ground) one after another until the entire word has been transmitted. For example, the bit interval may be 10 μs, which means that the voltage level for each bit in the word appears for 10 μs. It would therefore take 80 μs to transmit an 8-bit word.

Serial–Parallel Conversion

Because both parallel and serial transmission occur in computers and other equipment, there must be techniques for converting between parallel and serial and vice versa. Such data conversions are usually taken care of by shift registers (see Fig. 7-4).

A *shift register* is a sequential logic circuit made up of a number of flip-flops connected in cascade. The flip-flops are capable of storing a multibit binary word, which is usually loaded in parallel into the transmitting register. When a clock pulse (CP) is applied to the flip-flops, the bits of the word are shifted from one flip-flop to another in sequence. The last (right-hand) flip-flop in the transmitting register ultimately stores each bit in sequence as it is shifted out.

The serial data word is then transmitted over the communication link and is received by another shift register. The bits of the word are shifted into the flip-flops one at a time until the entire word is contained within the register. The flip-flop outputs can then be observed and the data stored in them transferred in parallel to other circuits. These serial-parallel data transfers take place inside the interface circuits and are referred to as *serializer/deserializer (SERDES) devices.*

Serial data can typically be transmitted faster over longer distances than parallel data. If a two-wire transmission line, rather than multiple interconnecting wires, is used, speeds over 2 GHz can be achieved over a serial link up to several feet long. If the serial data is converted to infrared light pulses, fiber-optic cable can be used. Serial data rates up to 40 GHz can be achieved at distances of many kilometers. Serial buses are now replacing parallel buses in computers, storage systems, and telecommunication equipment where very high speeds are required.

Serial-parallel conversion

Shift register

Serializer/deserializer (SERDES) devices

Figure 7-4 Parallel-to-serial and serial-to-parallel data transfers with shift registers.

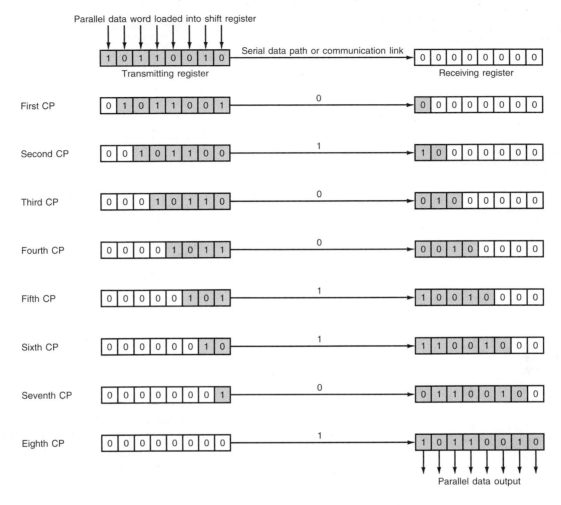

For example, suppose that you must transmit data at a rate of 400 Mbytes/s. In a parallel system you could transmit 4 bytes at a time on a 32-bit 100-MHz parallel bus. Bus length would be limited to a few inches.

You could also do this serially. Remember that 400 Mbytes/s is the same as 8 × 400 Mbytes/s or 3.2 gigabits per second (Gbps) or 3.2 GHz. This rate is easily obtained serially for several feet with a copper transmission line or up to many kilometers with a fiber-optic cable.

Delta Modulation. *Delta modulation* is a special form of analog-to-digital conversion that results in a continuous serial data signal being transmitted. The delta modulator looks at a sample of the analog input signal, compares it to a previous sample, and then transmits a 0 or a 1 if the sample is less than or more than the previous sample.

Delta modulation

7-3 Data Conversion

The key to digital communication is to convert data in analog form to digital form. Special circuits are available to do this. Once it is in digital form, the data can be processed or stored. Data must usually be reconverted to analog form for final consumption by the user; e.g., voice and video must be in analog form. *Data conversion* is the subject of this section.

Data conversion

Figure 7-5 A/D converter.

Analog signal input → Analog-to-digital converter (A/D converter) → 8-bit binary word output

Figure 7-6 D/A converter.

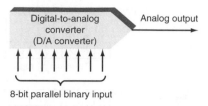

Digital-to-analog converter (D/A converter) → Analog output

8-bit parallel binary input

Basic Principles of Data Conversion

Analog-to-digital (A/D) conversion

Translating an analog signal to a digital signal is called *analog-to-digital (A/D) conversion, digitizing a signal,* or *encoding.* The device used to perform this translation is known as an *analog-to-digital (A/D) converter* or *ADC.* A modern A/D converter is usually a single-chip IC that takes an analog signal and generates a parallel or serial binary output (see Fig. 7-5).

Digital-to-analog (D/A) conversion

The opposite process is called *digital-to-analog (D/A) conversion.* The circuit used to perform this is called a *digital-to-analog (D/A) converter* (or *DAC*) or a *decoder.* The input to a D/A converter is usually a parallel binary number, and the output is a proportional analog voltage level. Like the A/D converter, a D/A converter is usually a single-chip IC (see Fig. 7-6) or a part of a large IC.

A/D Conversion. An analog signal is a smooth or continuous voltage or current variation (see Fig. 7-7). It could be a voice signal, a video waveform, or a voltage representing a variation of some other physical characteristic such as temperature. Through A/D conversion these continuously variable signals are changed to a series of binary numbers.

A/D conversion is a process of sampling or measuring the analog signal at regular time intervals. At the times indicated by the vertical dashed lines in Fig. 7-7, the instantaneous value of the analog signal is measured and a proportional binary number is generated to represent that sample. As a result, the continuous analog signal is translated to a series of discrete binary numbers representing samples.

A key factor in the sampling process is the frequency of sampling f, which is the reciprocal of the sampling interval t shown in Fig. 7-7. To retain the high-frequency information in the analog signal, a sufficient number of samples must be taken so that the waveform is adequately represented. It has been found that the minimum sampling frequency is twice the highest analog frequency content of the signal. For example, if the analog signal contains a maximum frequency variation of 3000 Hz, the

Figure 7-7 Sampling an analog signal.

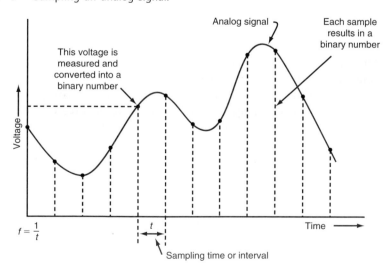

This voltage is measured and converted into a binary number

Analog signal

Each sample results in a binary number

$f = \dfrac{1}{t}$

Voltage

Time

Sampling time or interval

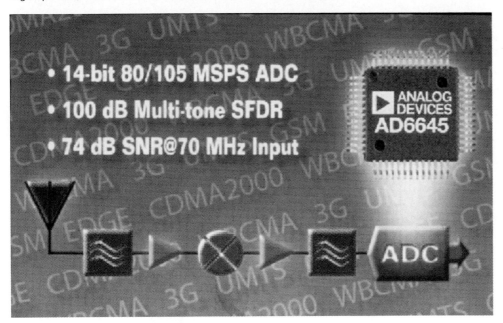

- **14-bit 80/105 MSPS ADC**
- **100 dB Multi-tone SFDR**
- **74 dB SNR@70 MHz Input**

ANALOG DEVICES
AD6645

ADC

analog wave must be sampled at a rate of at least twice this, or 6000 Hz. This minimum sampling frequency is known as the *Nyquist frequency* f_N. (And $f_N \geq 2f_m$, where f_m is the frequency of the input signal.) For bandwidth limited signals with upper and lower limits of f_2 and f_1, the Nyquist sampling rate is just twice the bandwidth or $2(f_2 - f_1)$.

Although theoretically the highest frequency component can be adequately represented by a sampling rate of twice the highest frequency, in practice the sampling rate is much higher than the Nyquist minimum, typically 2.5 to 3 times more. The actual sampling rate depends on the application as well as factors such as cost, complexity, channel bandwidth, and availability of practical circuits.

Assume, e.g., that the output of an FM radio is to be digitized. The maximum frequency of the audio in an FM broadcast is 15 kHz. To ensure that the highest frequency is represented, the sampling rate must be twice the highest frequency: $f = 2 \times 15 \text{ kHz} = 30 \text{ kHz}$. But in practice, the sampling rate is made higher, that is, 3 to 10 times higher, or $3 \times 15 \text{ kHz} = 45 \text{ kHz}$ to $10 \times 15 \text{ kHz} = 150 \text{ kHz}$. The sampling rate for compact disk players that store music signals with frequencies up to about 20 kHz is 44.1 kHz or 48 kHz.

Another important factor in the conversion process is that, because the analog signal is smooth and continuous, it represents an infinite number of actual voltage values. In a practical A/D converter, it is not possible to convert all analog samples to a precise proportional binary number. Instead, the A/D converter is capable of representing only a finite number of voltage values over a specific range. The samples are converted to a binary number whose value is close to the actual sample value. For example, an 8-bit binary number can represent only 256 states, which may be the converted values from an analog waveform having an infinite number of positive and negative values between $+1$ V and -1 V.

The physical nature of an A/D converter is such that it divides a voltage range into discrete increments, each of which is then represented by a binary number. The analog voltage measured during the sampling process is assigned to the increment of voltage closest to it. For example, assume that an A/D converter produces 4 output bits. With 4 bits, 2^4 or 16 voltage levels can be represented. For simplicity, assume an analog voltage range of 0 to 15 V. The A/D converter divides the voltage range as shown in Fig. 7-8. The binary number represented by each increment is indicated. Note that although there are 16 levels, there are only 15 increments. The number of levels is 2^N and the number of increments is $2^N - 1$, where N is the number of bits.

Nyquist frequency

GOOD TO KNOW

MSPS means millions of samples per second. SFDR is spurious free dynamic range. SNR refers to signal-to-noise ratio.

Figure 7-8 The A/D converter divides the input voltage range into discrete voltage increments.

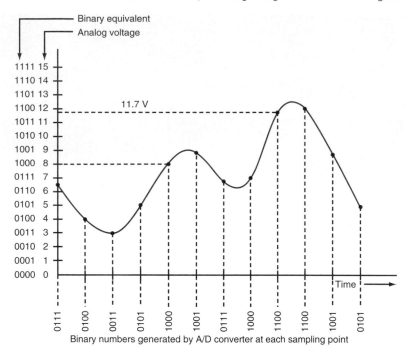

Binary numbers generated by A/D converter at each sampling point

Now assume that the A/D converter samples the analog input and measures a voltage of 0 V. The A/D converter will produce a binary number as close as possible to this value, in this case 0000. If the analog input is 8 V, the A/D converter generates the binary number 1000. But what happens if the analog input is 11.7 V, as shown in Fig. 7-8? The A/D converter produces the binary number 1011, whose decimal equivalent is 11. In fact, any value of analog voltage between 11 and 12 V will produce this binary value.

As you can see, there is some error associated with the conversion process. This is

referred to as *quantizing error*.

The quantizing error can be reduced, of course, by simply dividing the analog voltage range into a larger number of smaller voltage increments. To represent more voltage increments, a greater number of bits must be used. For example, using 12 bits instead of 10 allows the analog voltage range to produce 2^{12} or 4096 voltage increments. This more finely divides the analog voltage range and thus permits the A/D converter to output a proportional binary number closer to the actual analog value. The greater the number of bits, the greater the number of increments over the analog range and the smaller the quantizing error.

The maximum amount of error can be computed by dividing the voltage range over which the A/D converter operates by the number of increments. Assume a 10-bit A/D converter, with 10 bits, $2^{10} = 1024$ voltage levels, or $1024 - 1 = 1023$ increments. Assume that the input voltage range is from 0 to 6 V. The minimum voltage step increment then is $6/1023 = 5.86 \times 10^{-3} = 5.865$ mV.

As you can see, each increment has a range of less than 6 mV. This is the maximum error that can occur; the average error is one-half that value. The maximum error is said to be $\pm 1/2$ LSB or one-half the LSB increment value.

The quantizing error can also be considered as a type of random or white noise. This noise limits the dynamic range of an A/D converter since it makes low-level signals difficult or impossible to convert. An approximate value of this noise is

$$V_n = \frac{q}{\sqrt{12}}$$

where V_n is the rms noise voltage and q is the weight of the LSB. This approximation is valid only over the bandwidth from direct current to $f_s/2$ (called the Nyquist bandwidth).

Figure 7-9 A D/A converter produces a stepped approximation of the original signal.

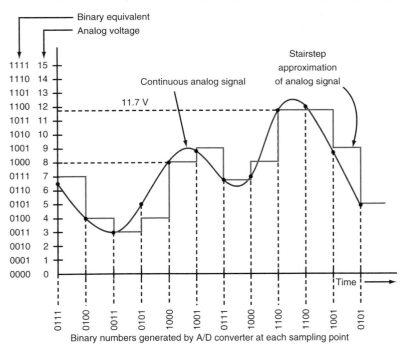

The input signal is in this range. Using the 10-bit example above, the LSB is 5.865 mV. The rms noise voltage then is

$$V_n = \frac{q}{\sqrt{12}} = \frac{0.005865}{3.464} = 0.0017 \text{ V} \quad \text{or} \quad 1.7 \text{ mV}$$

The signal to be digitized should be 2 or more times this noise level to ensure a reasonably error-free conversion.

It can be shown that the overall quantization noise can be reduced by oversampling, i.e., sampling the signal at a rate that is many times the Nyquist sampling rate of 2 times the highest signal frequency. Oversampling reduces the quantizing noise by a factor equal to the square root of the oversampling ratio, which is $f_s/2f_m$.

D/A Conversion. To retain an analog signal converted to digital, some form of binary memory must be used. The multiple binary numbers representing each of the samples can be stored in random access memory (RAM), on disk, or on magnetic tape. Once they are in this form, the samples can be processed and used as data by a microcomputer which can perform mathematical and logical manipulations. This is called digital signal processing (DSP) and is discussed in Sec. 7-5.

At some point it is usually desirable to translate the multiple binary numbers back to the equivalent analog voltage. This is the job of the D/A converter, which receives the binary numbers sequentially and produces a proportional analog voltage at the output. Because the input binary numbers represent specific voltage levels, the output of the D/A converter has a stairstep characteristic. Figure 7-9 shows the process of converting the 4-bit binary numbers obtained in the conversion of the waveform in Fig. 7-8. If these binary numbers are fed to a D/A converter, the output is a stairstep voltage as shown. Since the steps are very large, the resulting voltage is only an approximation to the actual analog signal. However, the stairsteps can be filtered out by passing the D/A converter output through a low-pass filter with an appropriate cutoff frequency.

Example 7-1

An information signal to be transmitted digitally is a rectangular wave with a period of 71.4 μs. It has been determined that the wave will be adequately passed if the bandwidth includes the fourth harmonic. Calculate (*a*) the signal frequency, (*b*) the fourth harmonic, and (*c*) the minimum sampling frequency (Nyquist rate).

a. $f = \dfrac{1}{t} = \dfrac{1}{71.4 \times 10^{-6}} = 14{,}006 \text{ Hz} \cong 14 \text{ kHz}$

b. $f_{\text{4th harmonic}} = 4 \times 14 \text{ kHz} = 56 \text{ kHz}$

c. Minimum sampling rate $= 2 \times 56 \text{ kHz} = 112 \text{ kHz}$

If the binary words contain a larger number of bits, the analog voltage range is divided into smaller increments and the output step increments will be smaller. This leads to a closer approximation to the original analog signal.

Pulse-amplitude modulation (PAM)

Aliasing. Whenever an analog waveform is sampled, a form of modulation called *pulse-amplitude modulation (PAM)* takes place. The modulator is a gating circuit that momentarily allows a portion of the analog wave to pass through, producing a pulse for a fixed time duration and at an amplitude equal to the signal value at that time. The result is a series of pulses as shown in Fig. 7-10. These pulses are passed to the A/D converter, where each is converted to a proportional binary value. PAM is discussed in greater detail later in this chapter, but for now we need to analyze the process to see how it affects the A/D conversion process.

Recall from Chap. 3 that amplitude modulation is the process of multiplying the carrier by the modulating signal. In this case, the carrier or sampling signal is a series of narrow pulses that can be described by a Fourier series:

$$v_c = D + 2D\left(\frac{\sin \pi D}{\pi D} \cos \omega_s t + \frac{\sin 2\pi D}{2\pi D} \cos 2\omega_s t + \frac{\sin 3\pi D}{3\pi D} \cos 3\omega_s t + \cdots\right)$$

Here, v_c is the instantaneous carrier voltage, and D is the duty cycle, which is the ratio of the pulse duration t to the pulse period T, or $D = t/T$. The term ω_s is $2\pi f_s$, where f_s is the pulse-sampling frequency. Note that the pulses have a dc component (the D term) plus sine-cosine waves representing the fundamental frequency and its odd and even harmonics.

Figure 7-10 Sampling and analog signal to produce pulse-amplitude modulation.

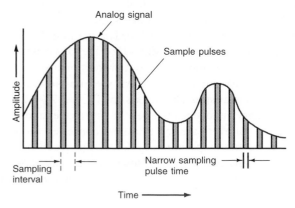

Figure 7-11 Spectrum of PAM signal.

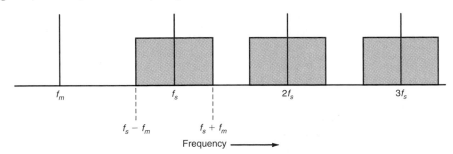

When you multiply this by the analog modulating or information signal to be digitized, you get a messy-looking equation that is remarkably easy to decipher. Assume that the analog wave to be digitized is a sine wave at a frequency of f_m or $V_m \sin(2f_m t)$. When you multiply that by the Fourier equation describing the carrier or sampling pulses, you get

$$v = V_m D \sin \omega_m t + 2V_m D \left(\frac{\sin \pi D}{\pi D} \sin \omega_m t \cos \omega_s t + V_m \frac{\sin 2\pi D}{2\pi D} \sin \omega_m t \cos \omega_s t \right.$$
$$\left. + V_m \frac{\sin 3\pi D}{3\pi D} \sin \omega_m t \cos \omega_s t + \cdots \right)$$

The first term in this equation is the original information sine signal. If we put this complex signal through a low-pass filter set to a frequency a bit above the modulating signal, all the pulses will be filtered out, leaving only the desired information signal.

Looking at the complex signal again, you should see the familiar AM equations showing the product of sine and cosine waves. If you remember from Chap. 3, these sine-cosine expressions convert to sum and difference frequencies that form the sidebands in AM. Well, the same thing happens here. The modulating signal forms sidebands with the sampling frequency f_s or $f_s + f_m$ and $f_s - f_m$. Furthermore, sidebands are also formed with all the harmonics of the carrier or sampling frequency ($2f_s \pm f_m, 3f_s \pm f_m, 4f_s \pm f_m$, etc.). The resulting output is best shown in the frequency domain as in Fig. 7-11. We are not generally concerned about all the higher harmonics and their sidebands, for they will ultimately get filtered out. But we do need to look at the sidebands formed with the fundamental sine wave.

All is well with this arrangement as long as the frequency (f_s) or carrier is 2 or more times the highest frequency in the modulating or information signal. However, if the sampling frequency is not high enough, then a problem called *aliasing* arises. Aliasing causes a new signal near the original to be created. This signal has a frequency of $f_s - f_m$. When the sampled signal is eventually converted back to analog by a D/A converter, the output will be the alias $f_s - f_m$, not the original signal f_m. Figure 7-12(*a*) shows the spectrum, and Fig. 7-12(*b*) shows the original analog signal and the recovered alias signal.

Assume a desired input signal of 2 kHz. The minimum sampling or Nyquist frequency is 4 kHz. But what if the sampling rate is only 2.5 kHz? This results in an alias signal of 2.5 kHz − 2 kHz = 0.5 kHz or 500 Hz. It is this alias signal that will be recovered by a D/A converter, not the desired signal of 2 kHz.

To eliminate this problem, a low-pass filter called an *antialiasing filter* is usually placed between the modulating signal source and the A/D converter input to ensure that no signal with a frequency greater than one-half the sampling frequency is passed. This filter must have extremely good selectivity. The roll-off rate of a common *RC* or *LC* low-pass filter is too gradual. Most antialiasing filters use multiple-stage *LC* filters, an *RC* active filter, or high-order switched capacitor filters to give the steep roll-off required to eliminate any aliasing. The filter cutoff is usually set just slightly above the highest-frequency content of the input signal.

Aliasing

Antialiasing filter

Digital Communication Techniques   **211**

Figure 7-12   Aliasing.

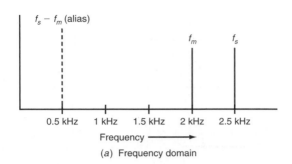

(a) Frequency domain

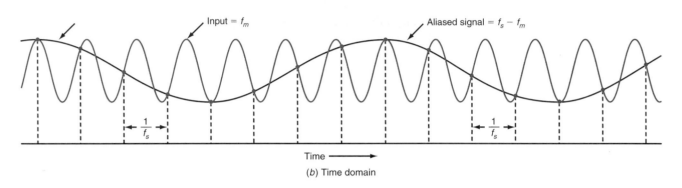

(b) Time domain

D/A Converters

There are many ways to convert digital codes to proportional analog voltages. However, the most popular methods are the *R-2R*, string, and weighted current source converters. These are available in integrated-circuit form and are also integrated into other larger systems on a chip (SoC).

R-2R Converter. The *R-2R* converter consists of four major sections, as shown in Fig. 7-13 and described in the next paragraphs.

Reference regulator

Reference Regulators. The precise *reference voltage regulator*, a zener diode, receives the dc supply voltage as an input and translates it to a highly precise reference voltage. This voltage is passed through a resistor that establishes the maximum input

Figure 7-13   Major components of a D/A converter.

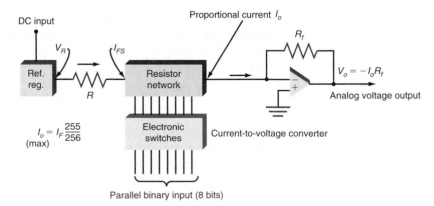

current to the resistor network and sets the precision of the circuit. The current is called the *full-scale current,* or I_{FS}:

$$I_{FS} = \frac{V_R}{R_R}$$

where V_R = reference voltage
R_R = reference resistor

Resistor Networks. The precision *resistor network* is connected in a unique configuration. The voltage from the reference is applied to this resistor network, which converts the reference voltage to a current proportional to the binary input. The output of the resistor network is a current that is directly proportional to the binary input value and the full-scale reference current. Its maximum value is computed as follows:

Resistor network

$$I_O = \frac{I_{FS}(2^N - 1)}{2^N}$$

For an 8-bit D/A converter, $N = 8$.

Some modern D/A converters use a capacitor network instead of the resistor network to perform the conversion from a binary number to a proportional current.

Output Amplifiers. The proportional current is then converted by an op amp to a proportional voltage. The output of the resistive network is connected to the summing junction of the op amp. The output voltage of the op amp is equal to the output current of the resistor network multiplied by the feedback resistor value. If the appropriate value of feedback resistance is selected, the output voltage can be scaled to any desired value. The op amp inverts the polarity of the signal:

$$V_O = -I_O R_f$$

Electronic Switches. The resistor network is modified by a set of electronic switches that can be either current or voltage switches and are usually implemented with diodes or transistors. These switches are controlled by the parallel binary input bits from a counter, a register, or a microcomputer output port. The switches turn on or off to configure the resistor network.

All the components shown in Fig. 7-13 are usually integrated onto a single IC chip. The only exception may be the amp, which is often an external circuit.

D/A converters of this type are available in a variety of configurations and can convert 8-, 10-, 12-, 14-, and 16-bit binary words.

The implementation of D/A converter circuitry varies widely. One of the most popular configurations is shown in detail in Fig. 7-14. Only 4 bits is shown, to simplify the drawing. Of particular interest is the resistor network, which uses only two values of resistance and thus is known as an *R-2R* ladder network. More complex networks have been devised but use a wider range of resistor values that are difficult to make with precise values in IC form. In Fig. 7-14 the switches are shown as mechanical devices, whereas in reality they are transistor switches controlled by the binary input. Many newer D/A converters and A/D converters use a capacitive network instead of the *R-2R* network.

String DAC. The string DAC gets its name from the fact that it is made up of a series string of equal-value resistors forming a voltage divider. See Fig. 7-15. This voltage divider divides the input reference voltage into equal steps of voltage proportional to the binary input. There are 2^N resistors in the string, where N is the number of input bits that determines resolution. In Fig. 7-12 the resolution is $2^3 = 8$, so eight resistors are used. Higher resolutions of 10 and 12 bits are available in this configuration. If the input reference is 10 V, the resolution is $10/2^3 = 10/8 = 1.25$ V. The output varies in increments of 1.25 V from 0 to 8.75 V.

Digital Communication Techniques

213

Figure 7-14 D/A converter with *R-2R* ladder network.

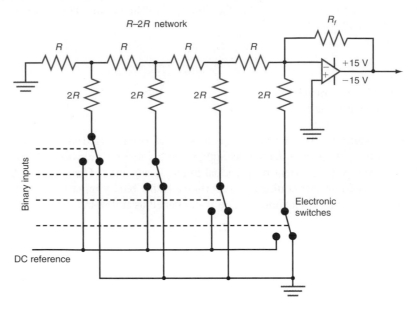

Figure 7-15 A string DAC.

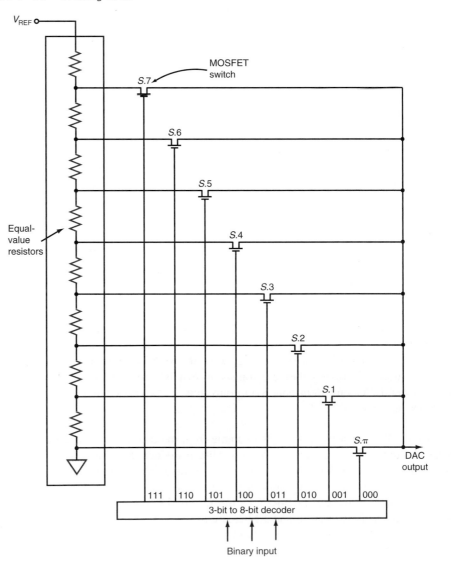

Figure 7-16 Weighted current source DAC.

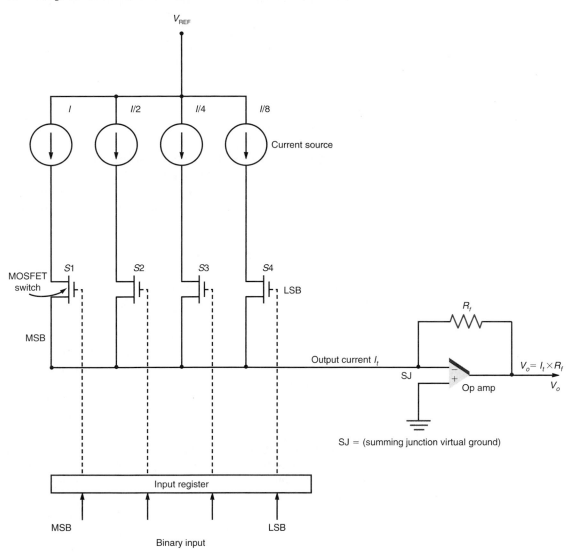

The output voltage is determined by a set of enhancement mode MOSFET switches controlled by a standard binary decoder. With 3 bits in, there are eight decoder outputs, each driving one MOSFET switch. If the input code is 000, switch $S0$ is turned on and the output is ground or 0 V. All other MOSFETs are off at this time. If the input code is 111, then $S7$ is turned on and the output voltage is 8.75 V. The output is a voltage and may be further conditioned by an op amp with gain and a lower output impedance as needed by the application.

Weighted Current Source DAC. A popular configuration for very high-speed DACs is the weighted current source DAC shown in Fig. 7-16. The current sources supply a fixed current that is determined by the external reference voltage. Each current source supplies a binary weighted value of I, $I/2$, $I/4$, $I/8$, etc. The current sources are made up of some combination of resistors, MOSFETs, or in some cases bipolar transistors. The switches are usually fast enhancement mode MOSFETs, but bipolar transistors are used in some models. The parallel binary input is usually stored in an input register, and the register outputs turn the switches off and on as dictated by the binary value.

The current source outputs are added at the summing junction of an op amp. The output voltage V_O is the sum of the currents I_t multiplied by the feedback resistor R_f.

$$V_O = I_t R_f$$

In Fig. 7-16, with 4 bits of resolution, there are $2^N = 2^4 = 16$ increments of current. Assume $I = 100$ μA. If the input binary number is 0101, then switches $S2$ and $S4$ are closed and the current is $50 + 12.5 = 62.5$ μA. With a 10-kΩ feedback resistor, the output voltage would be $62.5 \times 10^{-6} \times 10 \times 10^3 = 0.625$ V.

Current source DACs are used for very fast conversions and are available in resolutions of 8, 10, 12, and 14 bits.

D/A Converter Specifications. Three important specifications are associated with D/A converters: resolution, error, and settling time.

Resolution is the smallest increment of voltage that the D/A converter produces over its output voltage range. Resolution is directly related to the number of input bits. It is computed by dividing the reference voltage V_R by the number of output steps $2^N - 1$. There is 1 fewer increment than the number of binary states.

For a 10-V reference and an 8-bit D/A converter the resolution is $10/(2^8 - 1) = 10/255 = 0.039$ V $= 39$ mV.

For high-precision applications, D/A converters with larger input words should be used. D/A converters with 8 and 12 bits are the most common, but D/A converters with 10, 14, 16, 20, and 24 bits are available.

Error is expressed as a percentage of the maximum, or full-scale, output voltage, which is the reference voltage value. Typical error figures are less than ± 0.1 percent. This error should be less than one-half the minimum increment. The smallest increment of an 8-bit D/A converter with a 10-V reference is 0.039 V, or 39 mV. Expressed as a percentage, this is $0.039 \div 10 = 0.0039 \times 100 = 0.39$ percent. One-half of this is 0.195 percent. With a 10-V reference, this represents a voltage of $0.00195 \times 10 = 0.0195$ V, or 19.5 mV. A stated error of 0.1 percent of full scale is $0.001 \times 10 = 0.01$ V, or 10 mV.

Settling time is the amount of time it takes for the output voltage of a D/A converter to stabilize to within a specific voltage range after a change in binary input. Refer to Fig. 7-17. When a binary input change occurs, a finite amount of time is needed for the electronic switches to turn on and off and for any circuit capacitance to charge or discharge. During the change, the output rings and overshoots, and contains transients from the switching action. The output is thus not an accurate representation of the binary input; it is not usable until it settles down.

Settling time is the time it takes for the D/A converter's output to settle to within $\pm \frac{1}{2}$ LSB change. In the case of the 8-bit D/A converter described earlier, when the output voltage settles to less than one-half the minimum voltage change of 39 mV, or 19.5 mV, the output can be considered stable. Typical settling times are in the 100-ns range. This specification is important because it determines the maximum speed of operation of the circuit, called the conversion time. A 100-ns settling time translates to a frequency of $1/100 \times 10^{-9} = 10$ MHz. Operations faster than this result in output errors.

Monotonicity is another DAC specification. A DAC is monotonic if the output increases in one increment of resolution voltage for each increment of binary number input. In very high-resolution DACs with very small increments, it is possible that circuit inaccuracies could literally result in a decrease in output voltage for a binary increase. This is often caused by poorly trimmed and matched resistors or current sources in the DAC.

Another specification is dc operating voltage and current. Older DACs operated from $+5$ V, but most of the newer ones operate from 3.3 or 2.5 V. A current consumption figure is also usually given.

Another consideration is how many DACs there are per chip. ICs with two, four, and eight DACs per chip are available. In the multi-DAC chips, the binary input is serial. Serial input is an option on most DACs today as the serial input greatly reduces the number of pins devoted to input signals. A 16-bit parallel DAC has, of course, 16 input pins. The same DAC with a serial input has only one input pin. Typical serial input formats

Resolution

Error

Settling time

GOOD TO KNOW

Settling time is usually equal to the time it takes for the output of the D/A converter to settle to $\pm \frac{1}{2}$ least significant bit (LSB) change.

Figure 7-17 Settling time.

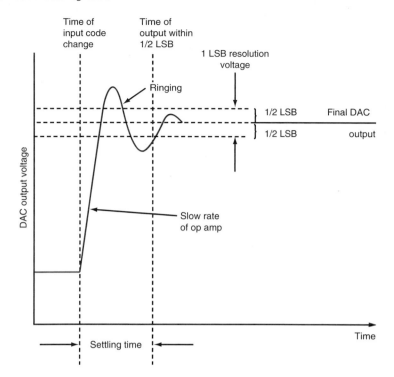

are the serial peripheral interface (SPI) or the I^2C interface common on most embedded controllers and microprocessors.

The binary input voltage is also a specification. Older DACs used a TTL or CMOS compatible +5 V input while newer chips use lower input signal voltages of 1.8, 2.5, or 3.3 V. High-speed DACs often use *current mode logic (CML)* inputs or low-voltage differential signaling (LVDS) levels with a differential signal swing of only a few hundred millivolts.

The reference voltage is typically from 1 to 5 V from a temperature-compensated zener diode which is usually on the DAC chip.

A/D Converters

A/D conversion begins with the process of sampling, which is usually carried out by a sample-and-hold (S/H) circuit. The S/H circuit takes a precise measurement of the analog voltage at specified intervals. The A/D converter then converts this instantaneous value of voltage and translates it to a binary number.

S/H Circuits. A *sample-and-hold (S/H) circuit,* also called a *track/store circuit,* accepts the analog input signal and passes it through, unchanged, during its sampling mode. In the hold mode, the amplifier remembers or memorizes a particular voltage level at the instant of sampling. The output of the S/H amplifier is a fixed dc level whose amplitude is the value at the sampling time.

Figure 7-18 is a simplified drawing of an S/H amplifier. The main element is a high-gain dc differential op amp. The amplifier is connected as a follower with 100 percent feedback. Any signal applied to the noninverting (+) input is passed through unaffected. The amplifier has unity gain and no inversion.

A storage capacitor is connected across the very high input impedance of the amplifier. The input signal is applied to the storage capacitor and the amplifier input through a MOSFET gate. An enhancement mode MOSFET that acts as an on/off switch is normally used. As long as the control signal to the gate of the MOSFET is held high, the input signal will be connected to the op amp input and capacitor. When the gate is high,

Sample-and-hold (S/H) circuit (track/store circuit)

Figure 7-18   An S/H amplifier.

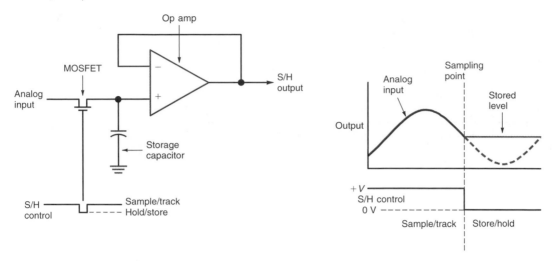

the transistor turns on and acts as a very low-value resistor, connecting the input signal to the amplifier. The charge on the capacitor follows the input signal. This is the sample or track mode for the amplifier. The op amp output is equal to the input.

When the S/H control signal goes low, the transistor is cut off, but the charge on the capacitor remains. The very high input impedance of the amplifier allows the capacitor to retain the charge for a relatively long time. The output of the S/H amplifier, then, is the voltage value of the input signal at the instant of sampling, i.e., the point at which the S/H control pulse switches from high (sample) to low (hold). The op amp output voltage is applied to the A/D converter for conversion to a proportional binary number.

The primary benefit of an S/H amplifier is that it stores the analog voltage during the sampling interval. In some high-frequency signals, the analog voltage may increase or decrease during the sampling interval; this is undesirable because it confuses the A/D converter and introduces what is referred to as aperture error. The S/H amplifier, however, stores the voltage on the capacitor; with the voltage constant during the sampling interval, quantizing is accurate.

There are many ways to translate an analog voltage to a binary number. The next sections describe the most common ones.

Successive–Approximations Converters. This converter contains an 8-bit *successive-approximations register (SAR)*, as shown in Fig. 7-19. Special logic in the register causes each bit to be turned on one at a time from MSB to LSB until the closest binary value is stored in the register. The clock input signal sets the rate of turning the bits off and on.

Assume that the SAR is initially reset to zero. When the conversion is started, the MSB is turned on, producing 10 000 000 at the output and causing the D/A converter output to go to half-scale. The D/A converter output is applied to the op amp, which applies it to the comparator along with the analog input. If the D/A converter output is greater than the input, the comparator signals the SAR to turn off the MSB. The next MSB is turned on. The D/A converter output goes to the proportional analog value, which is again compared to the input. If the D/A converter output is still greater than the input, the bit will be turned off; if the D/A converter output is less than the input, the bit will be left at binary 1.

The next MSB is then turned on, and another comparison is made. The process continues until all 8 bits have been turned on or off and eight comparisons have been made. The output is a proportional 8-bit binary number. With a clock frequency of 200 kHz, the clock period is $1/200 \times 10^3 = 5 \ \mu s$. Each bit decision is made during the clock period. For eight comparisons at 5 μs each, the total conversion time is $8 \times 5 = 40 \ \mu s$.

Successive-approximations converter

Successive-approximations register (SAR)

Figure 7-19   Successive-approximations A/D converter.

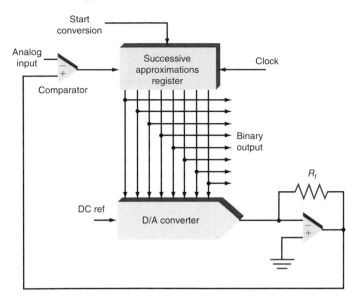

Successive-approximations converters are fast and consistent. They are available with conversion times from about 0.25 to 200 μs, and 8-, 10-, 12-, and 16-bit versions are available. Conversion times are also expressed as megasamples per second (MSPS). Successive-approximations converters with speeds to 5 MSPS are available. Most IC A/D converters in use are the successive-approximations type.

Instead of using a D/A converter with an R-$2R$ network, many newer successive-approximations converters use capacitors instead of resistors in the weighting network. The most difficult part of making an integrated-circuit (IC) A/D or D/A converter is the resistor network. It can be made with laser-trimmed thin-film resistors, but these require very expensive processing steps in making the IC. Resistors also take up more space on an IC than any other component. In A/D converters, the R-$2R$ network takes up probably 10 or more times the space of all the rest of the circuitry. To eliminate these problems, a capacitor network can be used to replace the resistor network. Capacitors are easy to make and take up little space.

The basic concept of a capacitive network is shown in Fig. 7-20. This is a simple 3-bit D/A converter. Note that the capacitors have binary weights of C, $C/2$, and $C/4$. The total capcitance of all capacitors in parallel is $2C$. The actual capacitor values are irrelevant since the capacitor ratios determine the outcome of the conversion. This fact also makes it easy to make the IC, since precise capacitor values are not needed. Only the ratio must be carefully controlled, and this is easier to do when making the IC than with laser trimming resistors. The switches in this diagram represent MOSFET switches in the actual circuit. A 3-bit successive-approximations register operates the switches labeled S_1 through S_4.

To start the conversion, switches S_C and S_{in} are closed, and switches S_1 through S_4 connect V_i to the capacitors, which are in parallel at this time. The comparator is shorted out temporarily. The input analog signal to be sampled and converted V_i is applied to all capacitors, causing each to charge up to the current signal value. Next, the S_C and S_{in} switches are opened, storing the current value of the signal on the capacitors. Since the capacitors store the input value at sampling time, no separate S/H circuit is needed. The successive-approximations register and related circuitry switch the reference voltage V_{REF} to the various capacitors in a specific sequence, and the comparator looks at the resulting voltage at each step and makes a decision about whether a 0 or 1 results from the comparison at each step. For example, in the first step S_1 connects V_{REF} to capacitor C, and all other capacitors are switched to ground via S_2

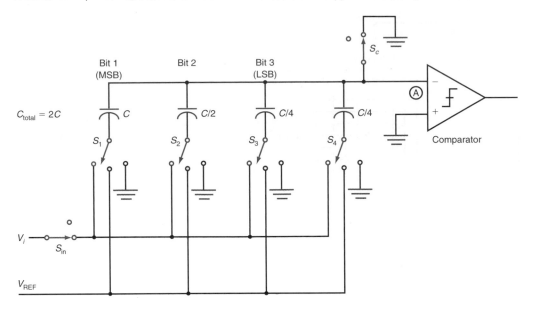

to S_4. Capacitor C forms a voltage divider with all the other capacitors in parallel. The comparator looks at the junction of the capacitors (node A) and then outputs a 0 or 1 depending upon the voltage. If the voltage at the junction is greater than the comparator threshold (usually one-half of the supply voltage), a 0 bit appears at the comparator output and is also stored in an output register. If the voltage at the junction is less than the threshold, a 1 bit appears at the comparator output and is stored in the output register. If a 1 bit occurs, capacitor C remains connected to V_{REF} throughout the remainder of the conversion.

The process continues by connecting the $C/2$ capacitor from ground to V_{REF}, and the comparison again takes place and another output bit is generated. This process continues until all capacitor voltages have been compared. During this process the initial charges on the capacitors are redistributed according to the value of the input voltage. The binary output appears in the successive-approximations register.

The circuit is easily expanded with more capacitors to produce a greater number of output bits. Both positive and negative reference voltages can be used to accommodate a bipolar input signal.

A switched-capacitor network makes the A/D converter very small. It can then be easily integrated into other circuits. A typical case is an A/D converter integrated into a microcontroller chip with memory.

Flash converter

Flash Converters. A *flash converter* takes an entirely different approach to the A/D conversion process. It uses a large resistive voltage divider and multiple analog comparators. The number of comparators required is equal to $2^N - 1$, where N is the number of desired output bits. A 3-bit A/D converter requires $2^3 - 1 = 8 - 1 = 7$ comparators (see Fig. 7-21).

The resistive voltage divider divides the dc reference voltage range into a number of equal increments. Each tap on the voltage divider is connected to a separate analog comparator. All the other comparator inputs are connected together and driven by the analog input voltage. Some comparators will be on and others will be off, depending on the actual value of input voltage. The comparators operate in such a way that if the analog input is greater than the reference voltage at the divider tap, the comparator output will be binary 1. For example, if the analog input voltage in Fig. 7-21 is 4.5 V, the outputs of comparators 4, 5, 6, and 7 will be binary 1. The other comparator outputs will

Figure 7-21 A flash converter.

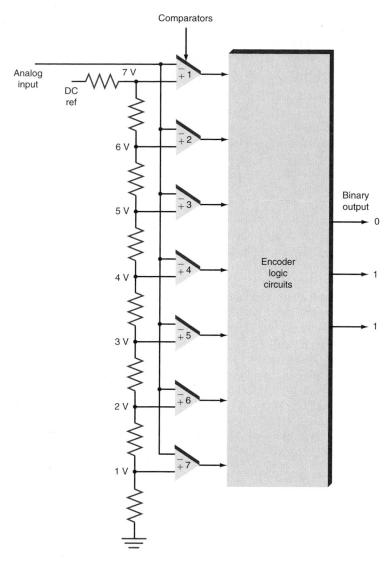

be binary 0. The encoder logic, which is a special combinational logic circuit, converts the 7-bit input from the comparators to a 3-bit binary output.

Successive-approximations converters generate their output voltage after the circuits go through their decision-making process. The flash converter, on the other hand, produces a binary output almost instantaneously. Counters do not have to be incremented, and a sequence of bits in a register does not have to be turned on and off. Instead, the flash converter produces an output as fast as the comparators can switch and the signals can be translated to binary levels by the logic circuits. Comparator switching and logic propagation delays are extremely short. Flash converters, therefore, are the fastest type of A/D converter. Conversion speeds of less than 100 ns are typical, and speeds of less than 0.5 ns are possible. Flash speeds are given in MSPS or gigasamples per second (GSPS) or 10^9 samples per second. Flash A/D converters are complicated and expensive because of the large number of analog comparators required for large binary numbers. The total number of comparators required is based upon the power of 2. An 8-bit flash converter has $2^8 - 1 = 255$ comparator circuits. Obviously, ICs requiring this many components are large and difficult to make. They also consume much more power than a digital circuit because the comparators are linear circuits. Yet for high-speed conversions, they are the best choice. With the high speed they can achieve, high-frequency signals such as video signals can be easily digitized. Flash converters are available with output word lengths of 6, 8, and 10.

Example 7-2

The voltage range of an A/D converter that uses 14-bit numbers is -6 to $+6$ V. Find (*a*) the number of discrete levels (binary codes) that are represented, (*b*) the number of voltage increments used to divide the total voltage range, and (*c*) the resolution of digitization expressed as the smallest voltage increment.

a. $2^N = 2^{14} = 16{,}384$

b. $2^N - 1 = 16{,}384 - 1 = 16{,}383$

c. The total voltage range is -6 to $+6$ V, or 12 V; thus,

$$\text{Resolution} = \frac{12}{16{,}383} = 0.7325 \text{ mV} \quad \text{or} \quad 732.5 \text{ } \mu\text{V}$$

Pipelined Converters. A pipelined converter is one that uses two or more low-resolution flash converters to achieve higher speed and higher resolution than successive-approximations converters but less than a full flash converter. High-resolution flash converters with more than 8 bits are essentially impractical because the large number of comparators required makes the power consumption very high. However, it is possible to use several flash converters with a smaller bit count to achieve very high conversion speeds and higher resolution. An example is the two-stage 8-bit pipeline converter shown in Fig. 7-22. The sampled analog input signal from a sample/hold (S/H) amplifier is applied to a 4-bit flash converter that generates the 4 most significant bits. These bits are applied to a 4-bit DAC and converted back to analog. The DAC output signal is then subtracted from the original analog input signal in a differential amplifier. The residual analog signal represents the least significant part of the signal. It is amplified and applied to a second 4-bit flash converter. Its output represents the 4 least significant bits of the output.

Figure 7-22   A two-stage 8-bit pipelined converter.

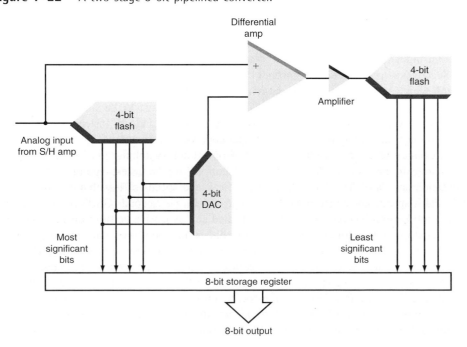

Chapter 7

With just two 4-bit flash converters, only 30 comparators are needed to achieve an 8-bit resolution. Otherwise, it would take 255 comparators, as indicated earlier. The tradeoff here is lower speed. A pipelined converter is obviously slower because it has to undergo a two-step conversion, one in each flash converter. However, the overall result is still very fast, much faster than that of any successive-approximations converter.

This principle can be extended to three, four, or more pipeline stages to achieve resolutions of 12, 14, and 16-bits. Speeds as high as 500 MSPS are possible with this arrangement.

ADC Specifications

The key ADC specifications are resolution, dynamic range, signal-to-noise ratio, effective number of bits, and spurious free dynamic range.

Resolution is related to the number of bits. Resolution indicated the smallest input voltage that is recognized by the converter and is the reference voltage V_{REF} divided by 2^N, where N is the number of output bits. ADCs with resolutions of 8, 10, 12, 14, 16, 18, 20, 22, and 24 are used in a wide range of applications.

Dynamic range is a measure of the range of input voltages that can be converted. It is expressed as the ratio of the maximum input voltage to the minimum recognizable voltage and converted to decibels. In any ADC, the minimum input voltage is simply the value of the LSB voltage or 1. The maximum input is simply related to the maximum output code or $2^N - 1$, where N is the number of bits. Therefore you can express the dynamic range with the expression

$$dB = \frac{20 \log 2^N - 1}{1} \quad \text{or just} \quad 20 \log(2^N - 1)$$

The dynamic range of a 12-bit converter then is

$$dB = 20 \log (2^{12} - 1) = 20 \log (4096 - 1) = 20 \log 4095 = 72.24 \text{ dB}$$

The greater the decibel value, the better.

The signal-to-noise *(S/N)* ratio (SNR) plays a major part in the performance of an ADC. This is the ratio of the actual input signal voltage to the total noise in the system. The noise comes from a combination of clock-related noise, power supply ripple, external signal coupling, and quantizing noise. The clock noise can be minimized by locating clock wiring away from the ADC and minimizing jitter on the clock signal. Good bypassing on the power supply should take care of most ripple noise. Then shielding the converter will reduce signals coupled in by inductive or capacitive coupling. Quantizing noise is another matter altogether. It is the result of the conversion process itself and cannot be reduced beyond a certain point.

Quantizing noise is an actual voltage that manifests itself as noise added to the analog input signal as the result of the error produced in converting an analog signal to its closest digital value. You can see this error if you plot it over the input voltage range, as shown in Fig. 7-23. This figure is a plot showing the input voltage and the related output code in a simple 3-bit ADC. The 1-LSB resolution is $V_R/2^N$. Below the plot is the noise or error voltage. When the ADC input voltage is exactly equal to the voltage represented by each output code, the error is zero. But as the voltage difference between the actual input voltage and the voltage represented by the code becomes greater, the error voltage increases. The result is a sawtoothlike error voltage that in effect becomes noise added to the input signal. Luckily, the maximum noise peak is just 1 LSB, but that can reduce the conversion accuracy depending upon the input signal level. Quantization noise can be reduced by using a converter with a greater number of bits, as this reduces the maximum noise represented by the LSB value.

Another way to show quantizing noise is given in Fig. 7-24. If you could take the binary output of the ADC and convert it back to analog in a DAC and then show a frequency-domain plot of the result, this is what you would see. The noise, which is

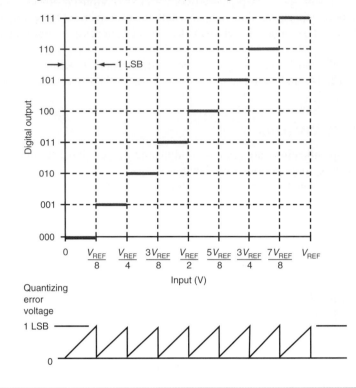

Figure 7-23 Quantizing noise is the error resulting from the difference between the input signal level and the available quantizing levels.

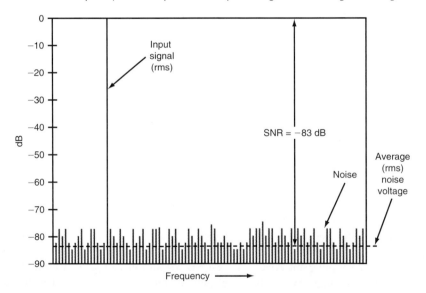

Figure 7-24 A frequency-domain plot of the quantizing noise and signal voltages.

mostly quantizing noise, has multiple frequency components over a wide frequency range. The large vertical line represents the analog input signal voltage being converted. This plot also shows the signal-to-noise ratio in decibels. The rms value of the signal voltage and the average rms value of the noise are used in computing the decibel value of the SNR.

Spurious free dynamic range (SFDR)

A related specification is *Spurious free dynamic range (SFDR)*. See Fig. 7-25. It is the ratio of the rms signal voltage to the voltage value of the highest "spur" expressed in decibels. A spur is any spurious or unwanted signal that may result from intermodulation

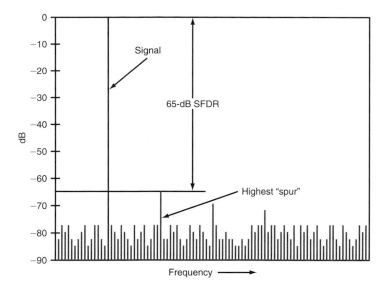

distortion. This is the formation of signals that are the result of mixing or modulation action caused by any nonlinear characteristic of the converter circuitry, an amplifier, or related circuit or component. The spurs are sums or differences between the various signals present and their harmonics.

As you might suspect, any noise, harmonics, or spurious signals all add together and basically reduce the resolution of an ADC. Often the combined noise level is greater than the LSB value, so only those more significant bits really define the signal amplitude. This effect is expressed by a measure known as the *effective number of bits (ENOB)*. ENOB is computed with the expression

$$\text{ENOB} = \frac{\text{SINAD} - 1.76}{6.02}$$

SINAD is the ratio of the signal amplitude to all the noise plus harmonic distortion in the circuit. SINAD in a totally noise- and distortion-free ADC is $6.02N + 1.76$, where N is the number of bits of resolution. This is the best SINAD figure possible, and it will be less in a practical converter.

Example 7-3

1. Calculate the SINAD for a 12-bit converter.

2. Calculate the ENOB for a converter with a SINAD of 78 dB.

Solution

1. SINAD = 6.02(12) + 1.76 = 74 dB

2. ENOB = (78 − 1.76)/6.02 = 12.66 bits, or just 12 bits.

In Fig. 7-26, the analog signal is sampled by an S/H circuit, as in most other forms of A/D converter. The sample is also applied to a comparator as in other A/D converter circuits. The other input to the comparator comes from a D/A converter driven by an up/down counter. The counter counts up (increments) or down (decrements) depending

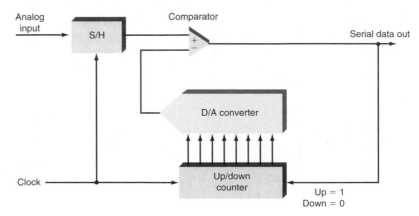

Figure 7-26 Delta modulator.

on the output state of the comparator. The comparator output is also the serial data signal representing the analog value.

Assume that the counter is initially at zero. This means that the D/A converter output will be zero. The analog input and S/H output are at some nonzero value, causing the comparator output to be binary 1. A binary 1 output sets the counter to count up. The clock increments the counter, causing the D/A converter output to rise a step at a time. As long as the D/A converter output is less than the analog input value, the comparator output will be binary 1, the counter will continue to count up, and the D/A converter output will rise a step at a time. When the D/A converter output exceeds the analog input by one increment, the comparator output switches to binary 0. Figure 7-27 shows the various signals in the circuit.

If the analog input decreases, the comparator output will be binary 0. The comparator compares the current analog sample to the previous sample that appears at the D/A converter output. The D/A converter output is always one clock period behind. If the analog signal continues to decrease, the comparator output remains binary 0, as Fig. 7-27 shows. If the analog signal is constant, it will not change with each sample; therefore, the comparator just switches between 0 and 1.

Basically, a delta modulator is a 1-bit A/D converter. It does not transmit the absolute value of a sample. Instead, it transmits a 0 or 1, indicating whether the new sample is

Figure 7-27 Waveforms in a delta modulator A/D converter.

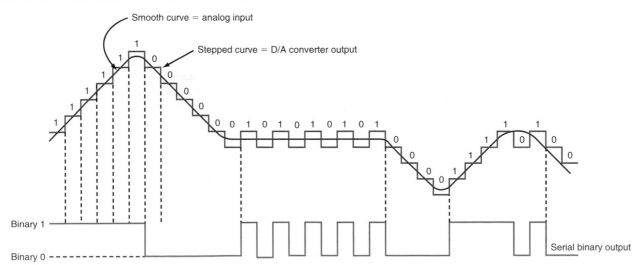

Figure 7-28 A delta demodulator.

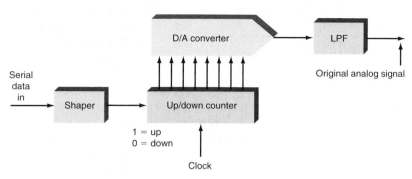

higher or lower than the previous sample. The resolution of the D/A converter establishes the minimum value of the step.

A delta demodulator is shown in Fig. 7-28. It is really a type of D/A converter. The serial data signal controls an up/down counter. A clock steps the counter, which drives a D/A converter. The D/A converter reproduces the stepped approximation shown in Fig. 7-27. A *low-pass filter (LPF)* on the output of the D/A converter removes the steps and smoothes the wave to its original form.

The delta modulator was once used in some early digital telephone systems; today it is no longer widely used.

Sigma–Delta Converter. A variation of the delta converter is the *sigma-delta* $(\Sigma\Delta)$ *converter*. Also known as a delta-sigma or charge balance converter, this circuit provides extreme precision, wide dynamic range, and low noise compared to other converters. It is available with output word lengths of 18, 20, 22, and 24 bits. These converters are widely used in digital audio applications, for example, CD, DVD, and MP3 players, as well as in industrial and geophysical applications where low-speed sensor data is to be captured and digitized. They are not designed for high speed, nor are they adaptable to applications in which many separate channels must be multiplexed into one.

The $\Sigma\Delta$ converter is what is known as an *oversampling converter*. It uses a clock or sampling frequency that is many times the minimum Nyquist rate required for other types of converters. Conversion rates are typically 64 to 128 times or more the highest frequency in the analog input signal. For example, assume a music signal with harmonics up to 24 kHz. A successive-approximations converter would have to sample this at a rate of 2 or more times (more than 48 kHz) to avoid aliasing and loss of data. A $\Sigma\Delta$ converter would use a clock or sampling rate in the 1.5- to 3-MHz range. Sampling rates of several hundred times the highest input signal have been used. The reason for this is that the quantizing noise is reduced by a factor equal to the square root of the oversampling ratio. The higher the sampling frequency, the lower the noise and as a result the wider the dynamic range. The oversampling techniques used in the sigma-delta converter essentially translate the noise to a higher frequency that can be easily filtered out by a low-pass filter. With a lower noise level, lower input levels can be converted, giving the converter extra dynamic range. Remember, the dynamic range is the difference between the lowest and the highest signal voltage levels that the converter can resolve, expressed in decibels. Of course, the other benefit of this technique is that aliasing is no longer a big problem. Often only a simple *RC* low-pass filter is needed to provide adequate protection from aliasing effects.

Figure 7-29 shows the basic $\Sigma\Delta$ circuit. The input is applied to a differential amplifier that subtracts the output voltage of a 1-bit D/A converter from the input signal. This D/A converter is driven by the comparator output. If the output is a binary 1, the D/A

Sigma-delta $(\Sigma\Delta)$ converter

Oversampling converter

Digital Communication Techniques **227**

Figure 7–29 A sigma-delta ($\Sigma\Delta$) converter.

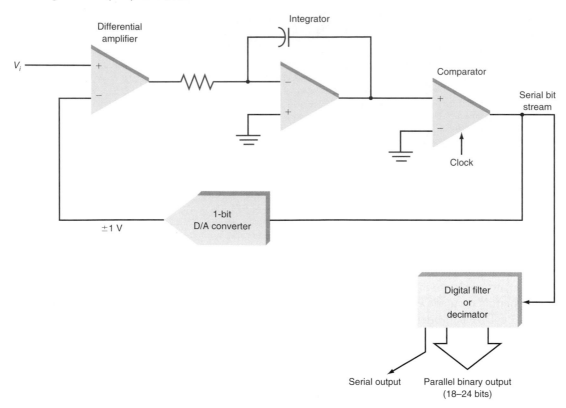

converter outputs $+1$ V. If the comparator output is binary 0, the D/A converter outputs a -1 V. This sets the input voltage range to $+1$ V.

The output of the differential amplifier is averaged in an integrator. The integrator output is compared to ground (0 V) in the comparator. The comparator is clocked by an external clock oscillator so that the comparator produces one output bit decision for each clock cycle. The resulting bit stream of binary 0s and 1s represents the varying analog input signal. This serial bit stream is fed to a digital filter or decimator that produces the final binary output word.

As the input signal is applied, the $\Sigma\Delta$ converter produces a serial bit stream output that represents the average value of the input. The closed-loop circuit causes the input signal to be compared to the D/A converter output every clock cycle, resulting in a comparator decision that may or may not change the bit value or the D/A converter output. If the input signal is increasing, the D/A converter will continually output binary 1s so that the average in the integrator increases. If the input signal decreases, the comparator switches to binary 0, forcing the D/A converter output to -1 V. What happens is that the D/A converter output, averaged over many cycles, produces an output that equals the input voltage. The closed loop continually tries to force the differential amplifier output to zero.

To clarify, consider the D/A converter output that switches between $+1$ and -1 V. If the output is all binary 1s or $+1$-V pulses, the average value at the D/A converter output is just $+1$. If the D/A converter input is all binary 0s, then a series of -1-V pulses occur, making the output average -1 V over many cycles. Now assume that the D/A converter input is a series of alternating binary 0s and 1s. The D/A converter output is $+1$ V for one cycle and -1 V for the next cycle. The average over time is zero. Now you can see that with more binary 1s at the D/A converter input, the average output will rise above zero. With a stream of more binary 0s, the average will become negative. The density of the 0s or 1s determines the average output value over time. The output of the comparator, then, is a bit stream that represents the average of the input value. It is a continuous nonbinary output.

This serial bit stream is not very useful as is. Therefore, it is passed through a digital filter called a *decimator*. This filter uses digital signal processing (DSP) techniques that are beyond the scope of this book. But the overall effect of the filter is to digitally average the serial bit stream and to produce multibit sequential output words that are in effect a rolling average of the input. The filter or decimator produces binary outputs at some fraction of the clock rate. The overall result is as if the input signal were sampled at a much lower rate but with a very high-resolution converter. The true binary output words may be in serial or parallel form.

Decimator

7-4 Pulse Modulation

Pulse modulation is the process of changing a binary pulse signal to represent the information to be transmitted. The primary benefits of transmitting information by binary techniques arise from the great noise tolerance and the ability to regenerate the degraded signal. Any noise that gets added to the binary signal along the way is usually clipped off. Further, any distortion of the signal can be eliminated by reshaping the signal with a Schmitt trigger, comparator, or similar circuit. If information can be transmitted on a carrier consisting of binary pulses, these aspects of binary techniques can be used to improve the quality of communications. Pulse modulation techniques were developed to take advantage of these qualities. The information signal, usually analog, is used to modify a binary (on/off) or pulsed carrier in some way.

Pulse modulation

With pulse modulation the carrier is not transmitted continuously but in short bursts whose duration and amplitude correspond to the modulation. The duty cycle of the carrier is usually made short so that the carrier is off for a longer time than the bursts. This arrangement allows the *average* carrier power to remain low, even when high peak powers are involved. For a given average power, the peak power pulses can travel a longer distance and more effectively overcome any noise in the system.

There are four basic forms of pulse modulation: *Pulse-amplitude modulation (PAM), pulse-width modulation (PWM), pulse-position modulation (PPM),* and *pulse-code modulation (PCM).*

Pulse-amplitude modulation (PAM)

Comparing Pulse-Modulation Methods

Figure 7-30 shows an analog modulating signal and the various waveforms produced by PAM, PWM, and PPM modulators. In all three cases, the analog signal is sampled, as it would be in A/D conversion. The sampling points are shown on the analog waveform. The sampling time interval t is constant and subject to the Nyquist conditions described earlier. The sampling rate of the analog signal must be at least 2 times the highest frequency component of the analog wave.

The PAM signal in Fig. 7-30 is a series of constant-width pulses whose amplitudes vary in accordance with the analog signal. The pulses are usually narrow compared to the period of sampling; this means that the duty cycle is low. The PWM signal is binary in amplitude (has only two levels). The width or duration of the pulses varies according to the amplitude of the analog signal: At low analog voltages, the pulses are narrow; at the higher amplitudes, the pulses get wider. In PPM, the pulses change position according to the amplitude of the analog signal. The pulses are very narrow. These pulse signals may be transmitted in a baseband form, but in most applications they modulate a high-frequency radio carrier. They turn the carrier on and off in accordance with their shape.

Of the four types of pulse modulation, PAM is the simplest and least expensive to implement. On the other hand, because the pulses vary in amplitude, they are far more susceptible to noise, and clipping techniques to eliminate noise cannot be used because they would also remove the modulation. PWM and PPM are binary and therefore clipping can be used to reduce the noise level.

Digital Communication Techniques   **229**

Figure 7-30   Types of pulse modulation.

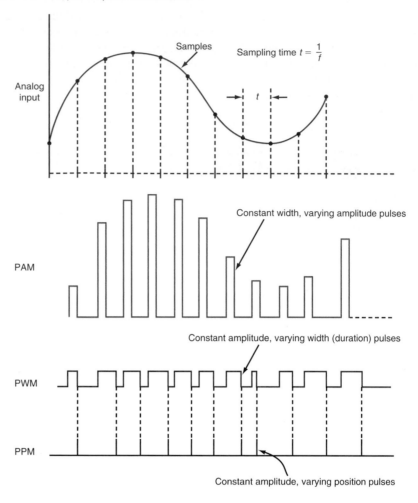

Although the techniques of pulse modulation have been known for decades, their development surged in the late 1950s and 1960s as a result of military missile development and the space program. Pulse modulation techniques were widely used in telemetry systems. *Telemetry,* a system of monitoring and measuring at a distance, allows scientists and engineers to monitor physical characteristics such as temperature, speed, acceleration, and pressure in a remote missile or spacecraft. Pulse modulation techniques are also used for remote-control purposes, e.g., in model airplanes, boats, and cars. Pulse-width modulation (PWM) methods are also used in switch mode power supplies (dc-dc convertors, regulators, etc.) as well as in class D audio switching power amplifiers.

Today pulse-modulation techniques have been largely superseded by more advanced digital techniques such as pulse-code modulation (PCM), in which actual binary numbers representing the digital data are transmitted.

Pulse-Code Modulation

Pulse-code modulation (PAM)

The most widely used technique for digitizing information signals for electronic data transmission is *pulse-code modulation (PCM)*. PCM signals are serial digital data. There are two ways to generate them. The more common is to use an S/H circuit and traditional A/D converter to sample and convert the analog signal to a sequence of binary words, convert the parallel binary words to serial form, and transmit the data serially, 1 bit at a time. The second way is to use the delta modulator described earlier.

Figure 7-31 Basic PCM system.

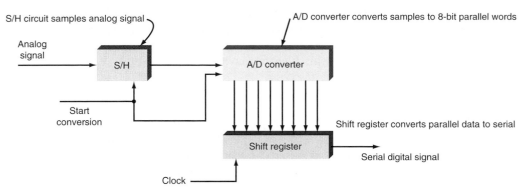

Traditional PCM. In traditional PCM, the analog signal is sampled and converted to a sequence of parallel binary words by an A/D converter. The parallel binary output word is converted to a serial signal by a shift register (see Fig. 7-31). Each time a sample is taken, an 8-bit word is generated by the A/D converter. This word must be transmitted serially before another sample is taken and another binary word is generated. The clock and start conversion signals are synchronized so that the resulting output signal is a continuous train of binary words.

Figure 7-32 shows the timing signals. The start conversion signal triggers the S/H to hold the sampled value and starts the A/D converter. Once the conversion is complete, the parallel word from the A/D converter is transferred to the shift register. The clock pulses start shifting the data out 1 bit at a time. When one 8-bit word has been transmitted, another conversion is initiated and the next word is transmitted. In Fig. 7-32, the first word sent is 01010101; the second word is 00110011.

At the receiving end of the system, the serial data is shifted into a shift register (see Fig. 7-33). The clock signal is derived from the data to ensure exact synchronization with

Figure 7-32 Timing signals for PCM.

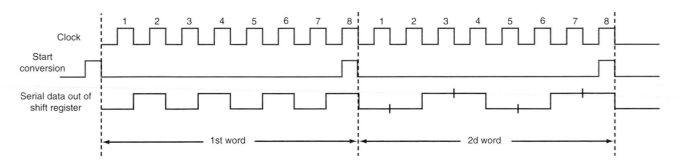

Figure 7-33 PCM to analog translation at the receiver.

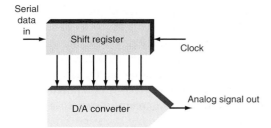

the transmitted data. (The process of clock recovery is discussed in Chap. 11.) Once one 8-bit word is in the register, the D/A converter converts it to a proportional analog output. Thus the analog signal is reconstructed one sample at a time as each binary word representing a sample is converted to the corresponding analog value. The D/A converter output is a stepped approximation of the original signal. This signal may be passed through a low-pass filter to smooth out the steps.

Companding. *Companding* is a process of signal compression and expansion that is used to overcome problems of distortion and noise in the transmission of audio signals.

The range of voice amplitude levels in the telephone system is approximately 1000 : 1. In other words, the largest-amplitude voice peak is approximately 1000 times the smallest voice signal or 1000 : 1, representing a 60-dB range. If a quantizer with 1000 increments were used, very high-quality analog signal representation would be achieved. For example, an A/D converter with a 10-bit word can represent 1024 individual levels. A 10-bit A/D converter would provide excellent signal representation. If the maximum peak audio voltage were 1 V, the smallest voltage increment would be 1/1023 of this, or 0.9775 mV.

As it turns out, it is not necessary to use that many quantizing levels for voice, and in most practical PCM systems, a 7- or 8-bit A/D converter is used for quantizing. One popular format is to use an 8-bit code, where 7 bits represents 128 amplitude levels and the eighth bit designates polarity $(0 = +, 1 = -)$. Overall, this provides 255 levels; one-half are positive, and one-half negative.

Although the analog voltage range of the typical voice signal is approximately 1000 : 1, lower-level signals predominate. Most conversations take place at a low level, and the human ear is most sensitive in the low-amplitude range. Thus the upper end of the quantizing scale is not often used.

Since most signals are low-level, quantizing error is relatively large. That is, small increments of quantization become a large percentage of the lower-level signal. This is a small amount of the peak amplitude value, of course, but this fact is irrelevant when the signals are low in amplitude. The increased quantizing error can produce garbled or distorted sound.

In addition to their potential for increasing quantizing error, low-level signals are susceptible to noise. Noise represents random spikes or voltage impulses added to the signal. The result is static that interferes with the low-level signals and makes intelligibility difficult.

Companding is the most common means of overcoming the problems of quantizing error and noise. At the transmitting end of the system, the voice signal to be transmitted is compressed; i.e., its dynamic range is decreased. The lower-level signals are emphasized, and the higher-level signals are deemphasized.

At the receiving end, the recovered signal is fed to an expander circuit that does the opposite, deemphasizing the lower-level signals and emphasizing the higher-level signals, thereby returning the transmitted signal to its original condition. Companding greatly improves the quality of the transmitted signal.

Originally, companding circuits were analog, and the concept is most easily understood when described in analog terms. One type of analog compression circuit is a nonlinear amplifier that amplifies lower-level signals more than it does upper-level signals. Figure 7-34 illustrates the companding process. The curve shows the relationship between the input and output of the compander. At the lower input voltages, the gain of the amplifier is high and produces high output voltages. As the input voltage increases, the curve begins to flatten, producing proportionately lower gain. The nonlinear curve compresses the upper-level signals while bringing the lower-level signals to a higher amplitude. Such compression greatly reduces the dynamic range of the audio signal. Compression reduces the customary ratio of 1000 : 1 to approximately 60 : 1. The degree of compression can be controlled by careful design of the gain characteristics of the compression amplifier, in which case the 60-dB voice range can be reduced to more like 36 dB.

In addition to minimizing quantizing error and the effects of noise, compression lowers the dynamic range so that fewer binary bits are required to digitize the audio signal. A 64 : 1 voltage ratio could be easily implemented with a 6-bit A/D converter, but in practice, a 7-bit A/D converter is used.

Figure 7-34 Compression and expansion curves.

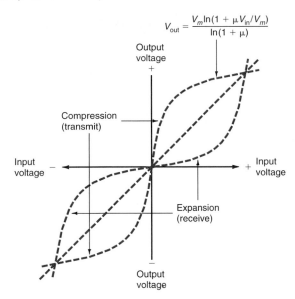

Two basic types of companding are used in telephone systems: the μ-*law* (pronounced "mu law") *compander* and the *A-law compander*. The two companders differ slightly in their compression and expansion curves. The μ-law compander is used in telephone systems in the United States and Japan, and the A-law compander is used in European telephone networks. The two are incompatible, but conversion circuits have been developed to convert μ-law to A-law and vice versa. According to international telecommunication regulations, users of μ-law companders are responsible for the conversions. The voltage formulas for both are as follows:

μ-law: $$V_{out} = \frac{V_m \ln\left(1 + \mu V_{in}/V_m\right)}{\ln\left(1 + \mu\right)}$$

A-law: $$V_{out} = \frac{1 + \ln\left(A V_{in}/V_m\right)}{1 + \ln A}$$

where V_{out} = output voltage
V_m = maximum possible input voltage
V_{in} = instantaneous value of input voltage

The value of μ is usually 255; A is usually 87.6.

μ-law compander

A-law compander

Example 7-4

The input voltage of a compander with a maximum voltage range of 1 V and a μ of 255 is 0.25. What are the output voltage and gain?

$$V_{out} = \frac{V_m \ln\left(1 + \mu V_{in}/V_m\right)}{\ln\left(1 + \mu\right)}$$

$$V_{out} = \frac{1 \ln\left[1 + 255(0.25)/1\right]}{\ln\left(1 + 255\right)} = \frac{\ln 64.75}{\ln 256} = \frac{4.17}{5.55} = 0.75 \text{ V}$$

$$\text{Gain} = \frac{V_{out}}{V_{in}} = \frac{0.75}{0.25} = 3$$

Example 7-5

The input to the compander of Example 7-4 is 0.8 V. What are the output voltage and gain?

$$V_{\text{out}} = \frac{V_m \ln\left(1 + \mu V_{\text{in}}/V_m\right)}{\ln\left(1 + \mu\right)}$$

$$V_{\text{out}} = \frac{1 \ln\left[1 + 255(0.8)\right]/1}{\ln\left(1 + 255\right)} = \frac{\ln 205}{\ln 256} = \frac{5.32}{5.55} = 1.02 \text{ V}$$

$$\text{Gain} = \frac{V_{\text{out}}}{V_{\text{in}}} = \frac{0.96}{0.8} = 1.2$$

As the examples show, the gain of a compander is higher at the lower input voltages than at the higher input voltages.

Older companding circuits used analog methods such as the nonlinear amplifiers described earlier. Today, most companding is digital. One method is to use a nonlinear A/D converter. These converters provide a greater number of quantizing steps at the lower levels than at the higher levels, providing compression. On the receiving end, a matching nonlinear D/A converter is used to provide the opposite compensating expansion effect. Compression can also be accomplished by digitizing the signal in a linear ADC and then using an appropriate algorithm to compute the companded digital output in an embedded microcontroller.

Codecs and Vocoders. Both ends of the communication link in telephone systems have transmitting and receiving capability. All A/D and D/A conversion and related functions such as serial-to-parallel and parallel-to-serial conversion as well as companding are usually taken care of by a single large-scale IC chip known as a *codec* or *vocoder*. One codec is used at each end of the communication channel. Codecs are usually combined with digital multiplexers and demultiplexers; clock and synchronizing circuits complete the system. These elements are discussed in Chap. 10.

Codec

Vocoder

Figure 7-35 is a simplified block diagram of a codec. The analog input is sampled by the S/H amplifier at an 8-kHz rate. The samples are quantized by the successive-approximations type of A/D converter. Compression is done digitally in the A/D converter.

Figure 7-35   Simplified block diagram of an IC codec.

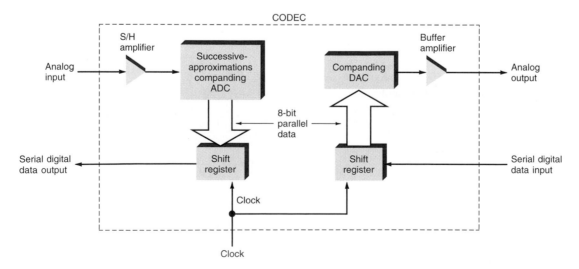

The parallel A/D converter output is sent to a shift register to create the serial data output, which usually goes to one input of a digital multiplexer.

The serial digital input is generally derived from a digital demultiplexer. The clock shifts the binary words representing the voice into a shift register for serial-to-parallel conversion. The 8-bit parallel word is sent to the D/A converter, which has built-in digital expansion. The analog output is then buffered, and it may be filtered externally. Most vocoders are made with a complementary metal-oxide semiconductor (CMOS) circuits and are part of large chips used in telephone (wired and cellular) systems.

7-5 Digital Signal Processing

As previous chapters have emphasized, communication involves a great deal of signal processing. To carry out communication, analog signals must be processed in some way; e.g., they may be amplified or attenuated. Often they must be filtered to remove undesirable frequency components. They must be shifted in phase and modulated or demodulated. Or they may have to be mixed, compared, or analyzed to determine their frequency components. Thousands of circuits have been devised to process analog signals, and many have been described in this book.

Although analog signals are still widely processed by analog circuits, increasingly they are being converted to digital for transmission and processing. As described earlier in this chapter, there are several important advantages to transmitting and using data in digital form. One advantage is that signals can now be manipulated by *digital signal processing (DSP)*.

Digital signal processing (DSP)

The Basis of DSP

DSP is the use of a fast digital computer to perform processing on digital signals. Any digital computer with sufficient speed and memory can be used for DSP. The superfast 32-bit *reduced-instruction-set computing (RISC)* processors are especially adept at DSP. However, DSP is most often implemented with processors developed specifically for this application because they differ in organization and operation from traditional microprocessors.

The basic DSP technique is shown in Fig. 7-36. An analog signal to be processed is fed to an A/D converter, where it is converted to a sequence of binary numbers which

Figure 7-36 Concept of digital signal processing.

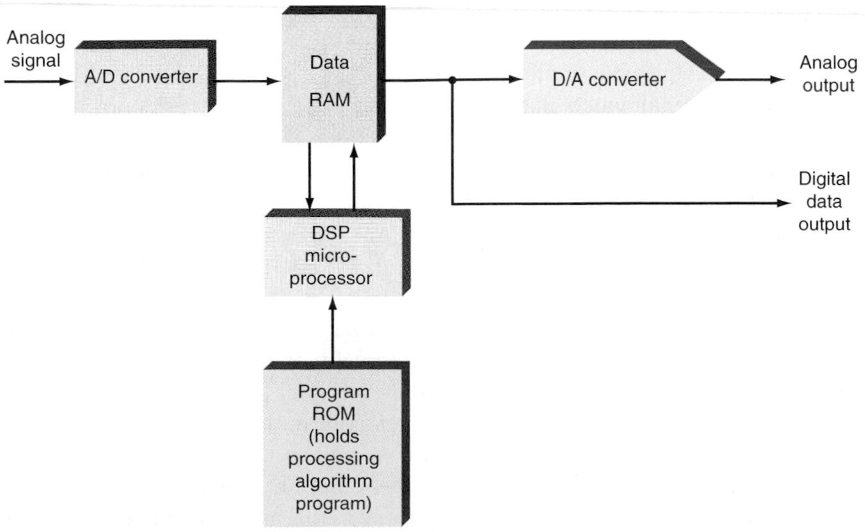

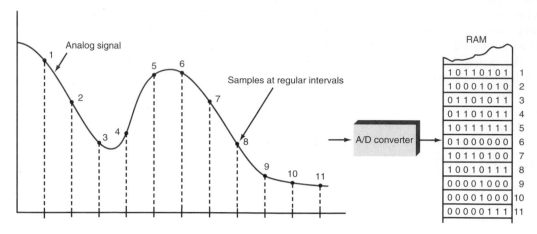

are stored in a read/write random-access memory (RAM). (See Fig. 7-37.) A program, usually stored in a read-only memory (ROM), performs mathematical and other manipulations on the data. Most digital processing involves complex mathematical algorithms that are executed in real time; i.e., the output is produced simultaneously with the occurrence of the input. With real-time processing, the processor must be extremely fast so that it can perform all the computations on the samples before the next sample comes along.

The processing results in another set of data words which are also stored in RAM. They can then be used or transmitted in digital form, or they may be fed to a D/A converter where they are converted back to an analog signal. The output analog signal then looks as though it has undergone processing by an equivalent analog circuit.

Almost any processing operation that can be done with analog circuits can also be done with DSP. The most common is filtering, but equalization, companding, phase shifting, mixing, modulation, and demodulation can also be programmed on a DSP computer.

DSP Processors

When DSP was first developed during the 1960s, only the largest and fastest mainframe computers were able to handle it, and even then, in some applications real-time processing could not be achieved. As computers got faster, more sophisticated processing could be performed, and in real time. However, only the most demanding of applications could afford a fast mainframe or minicomputer. For example, NASA used DSP to process and enhance the digital video data from remote exploratory spacecraft such as the *Voyager*, which passed by Mars and Jupiter. Oil companies used DSP in the 1960s and 1970s to process geological data to determine whether oil deposits were present in structures beneath the earth.

With the appearance of fast 16- and 32-bit microprocessors the use of DSP became practical for many applications, and finally in the 1980s special microprocessors optimized for DSP were developed.

Most computers and microprocessors use an organization known as the Von Neumann architecture. Physicist John Von Neumann is generally credited with creating the stored program concept that is the basis of operation of all digital computers. Binary words representing computer instructions are stored sequentially in a memory to form a program. The instructions are fetched and executed one at a time at high speed. The program usually processes data in the form of binary numbers that are stored in the same memory. The key feature of the Von Neumann arrangement is that both instructions and data are stored in a common memory space. That memory space may be read/write RAM or ROM or some combination thereof. But the important point is that there is only one path between the memory and the CPU, and therefore only one data or instruction word can be accessed at a time. This has the effect of greatly limiting execution speed. This short-coming is generally referred to as the *Von Neumann bottleneck*.

Von Neumann bottleneck

DSP microprocessors work in a similar way, but they use a variation called the *Harvard architecture*. In a Harvard architecture microprocessor, there are two memories, a program or instruction memory, usually a ROM, and a data memory, which is a RAM. Also, there are two data paths into and out of the CPU between the memories. Because both instructions and data can be accessed simultaneously, very high-speed operation is possible.

Harvard architecture

DSP microprocessors are designed to perform the math operations common to DSP. Most DSP is a combination of multiplication and addition or accumulation operations on the data words developed by the A/D converter and stored in RAM. DSP processors carry out addition and multiplication faster than any other type of CPU, and most combine these operations in a single instruction for even greater speed. DSP CPUs contain two or more *multiply and accumulate (MAC)* processors.

> **GOOD TO KNOW**
>
> DSP microprocessors are designed to operate at the highest speeds possible. Speeds up to 100 MHz are not uncommon.

DSP microprocessors are designed to operate at the highest speeds possible. Clock speeds over 100 MHz are common, and DSP chips with clock rates as high as 1 GHz are now being used. Some DSP processors are available as just the CPU chip, but others combine the CPU with data RAM and a program ROM on chip. Some even include the A/D and D/A converter circuits. If the desired processing program is written and stored in ROM, a complete single-chip DSP circuit can be created for customized analog signal processing by digital techniques. Many conventional embedded (core) processors such as the ARM, MIPS, and Power PC now have special DSP instructions such as the MAC operation built in.

Finally, some DSP circuits are embedded or dedicated. Instead of being programmed in a general-purpose DSP, they are made of logic hardwired to perform only the desired filtering or other function. *Complex programmable logic devices (CPLDs)* and *field-programmable logic arrays (FPGAs)* are widely used to implement custom DSP.

DSP Applications

Filtering. The most common DSP application is *filtering*. A DSP processor can be programmed to perform bandpass, low-pass, high-pass, and band-reject filter operations. With DSP, the filters can have characteristics far superior to those of equivalent analog filters: Selectivity can be better, and the passband or reject band can be customized to the application. Further, the phase response of the filter can be controlled more easily than with analog filters.

Filtering

Compression. Data *compression* is a process that reduces the number of binary words needed to represent a given analog signal. It is often necessary to convert a video analog signal to digital for storage and processing. Digitizing a video signal with an A/D converter produces an immense amount of binary data. If the video signal contains frequencies up to 4 MHz, the A/D converter must sample at 8 MHz or faster. Assuming a sampling rate of 8 MHz with an 8-bit A/D converter, 8 Mbytes/s will be produced. Digitizing 1 min of video is equal to 60 s × 8 Mbytes, or 480 Mbytes of data. This amount of data exceeds the capacity of most computer RAM, although a larger hard disk could store this data. In terms of data communication, it would take a great deal of time to transmit this amount of data serially.

Compression

To solve this problem, the data is compressed. Numerous algorithms have been developed to compress data. Examples are MPEG-2 and MPEG-4, widely used in digital photography and video. The data is examined for redundancy and other characteristics, and a new group of data, based upon various mathematical operations, is created. Data can be compressed by a factor of up to 100; in other words, the compressed data is 1/100 its original size. With compression, 480 Mbytes of data becomes 4.8 Mbytes. This is still a lot, but it is now within the capabilities of RAM and disk storage components. Audio data is also compressed. An example is MP3, the algorithm used in portable music players.

A DSP chip does the compression on the data received from the A/D converter. The compressed version of the data is then stored or transmitted. In the case of data communication, compression greatly reduces the time needed to transmit data.

When the data is needed, it must be decompressed. A reverse-calculation DSP algorithm is used to reconstruct the original data. Again, a special DSP chip is used for this purpose.

Spectrum analysis

Discrete Fourier transform (DFT)

Spectrum Analysis. *Spectrum analysis* is the process of examining a signal to determine its frequency content. Recall that all nonsinusoidal signals are a combination of a fundamental sine wave to which have been added harmonic sine waves of different frequency, amplitude, and phase. An algorithm known as the *discrete Fourier transform (DFT)* can be used in a DSP processor to analyze the frequency content of an input signal. The analog input signal is converted to a block of digital data, which is then processed by the DFT program. The result is a frequency-domain output that indicates the content of the signal in terms of sine wave frequencies, amplitudes, and phases.

The DFT is a complex program that is long and time-consuming to run. In general, computers are not fast enough to perform DFT in real time as the signal occurs. Therefore, a special version of the algorithm has been developed to speed up the calculation.

Fast Fourier transform (FFT)

Known as the *fast Fourier transform (FFT)*, it permits real-time signal spectrum analysis.

Other Applications. As mentioned, DSP can do almost everything analog circuits can do, e.g., phase shifting, equalization, and signal averaging. *Signal averaging* is the process of sampling a recurring analog signal which is transmitted in the presence of noise. If the signal is repeatedly converted to digital and the mathematical average of the samples is taken, the signal-to-noise ratio is greatly improved. Since the noise is random, an average of it tends to be zero. The signal, which is constant and unchanging, averages into a noise-free version of itself.

DSP can also be used for signal synthesis. Waveforms of any shape or characteristics can be stored as digital bit patterns in a memory. Then when it is necessary to generate a signal with a specific shape, the bit pattern is called up and transmitted to the DAC, which generates the analog version. This type of technique is used in voice and music synthesis.

Modulation, mixing, and demodulation are also easy to implement in DSP.

DSP is widely used in fax machines, CD players, modems, all cell phones, and a variety of other common electronic products. Its use in communication is increasing as DSP processors become even faster. Some fast DSP processors have been used to perform all the normal communication receiver functions from the IF stages through signal

Software-defined radios (SDRs)

recovery. All-digital or *Software-defined radios (SDRs)* are a reality now.

How DSP Works

The advanced mathematical techniques used in DSP are beyond the scope of this book and certainly beyond the knowledge required of electronics technicians in their jobs. For the most part, it is sufficient to know that the techniques exist. However, without getting bogged down in the math, it is possible to give some insight into the workings of a DSP circuit. For example, it is relatively easy to visualize the digitizing of an analog signal into a block of sequential binary words representing the amplitudes of the samples, and then to imagine that the binary words representing the analog signal are stored in a RAM (see Fig. 7-37). Once the signal is in digital form, it can be processed in many different ways. Two common applications are filtering and spectrum analysis.

Finite impulse response (FIR) filter or nonrecursive filter

Filter Applications. One of the most popular DSP filters is called a *finite impulse response (FIR) filter*. It is also called a *nonrecursive filter*. (A nonrecursive filter is one

Figure 7-38   A block diagram showing the processing algorithm of a nonrecursive FIR filter.

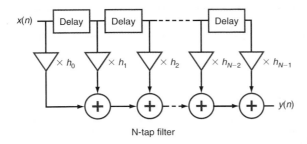

N-tap filter

whose output is only a function of the sum of products of the current input samples.) A program can be written to create a low-pass, high-pass, bandpass, or band-reject filter of the FIR type. The algorithm of such a filter has the mathematical form $Y = \Sigma a_i \, b_i$. In this expression, Y is the binary output, which is the summation (Σ) of the products of a and b. The terms a and b represent the binary samples, and i is the number of the sample. Usually these samples are multiplied by coefficients appropriate to the type of filter, and the results summed.

Figure 7-38 is a graphical representation of what goes on inside the filter. The term $X(n)$, where n is the number of the sample, represents the input data samples from RAM. The boxes labeled Delay represent delay lines. (A *delay line* is a circuit that delays a signal or sample by some constant time interval.) In reality, nothing is being delayed. Rather, the circuit generates samples that occur one after another at a fixed time interval equal to the sampling time, which is a function of the A/D converter clock frequency. In effect, the outputs of the delay boxes in Fig. 7-38 are the sequential samples which occur one after another at the sampling rate which is equivalent to a series of delays.

Delay line

Note that the samples are multiplied by some constant represented by the term h_n. These constants, or coefficients, are determined by the algorithm and the type of filter desired. After the samples have been multiplied by the appropriate coefficient, they are summed. The first two samples are added, this sum is added to the next-multiplied sample, that sum is added to the next sample, and so on. The result is the output Y, which is a value made up of the sum of products of the other samples. The DSP solves the equation: $y(n) = h_0 x_0 + h_1 x_1 + h_2 x_2 + \cdots$. The x samples come from the A/D converter. The h values are constants or coefficients that define the function (filtering in this case) to be performed. Designing DSP software is essentially figuring out what the constants should be. These new data samples are also stored in RAM. This block of new data is sent to the D/A converter at whose output the filtered analog signal appears. The decimation circuit in the $\Sigma\Delta$ converter discussed earlier is a type of FIR filter.

Another type of DSP filter is the *infinite impulse response (IIR) filter,* a recursive filter that uses feedback: Each new output sample is calculated by using both the current output and past samples (inputs).

Infinite impulse response (IIR) filter

DIT/FFT. As indicated earlier, a DSP processor can perform spectrum analysis by using the discrete or fast Fourier transform (FFT). Figure 7-39 illustrates the processing that takes place with FFT. It is called a *decimation in time (DIT).* The $x(n)$ values at the input are the samples, which are processed in three stages. In the first stage, a so-called butterfly operation is performed on pairs of samples. Some of the samples are multiplied by a constant and then added. At the second stage, some of the outputs are multiplied by constants, and new pairs of sums, called *groups,* are formed. Then a similar process is performed to create the final outputs, called *stages*. These outputs are converted to new values that can be plotted in the frequency domain.

Decimation in time (DIT)

Group

Stage

Figure 7-39   The fast Fourier transform decimation in time.

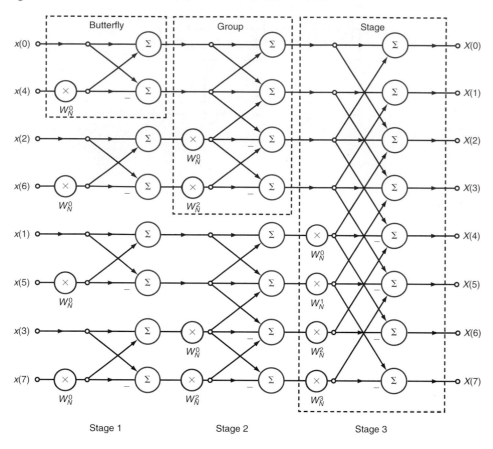

Figure 7-40   Output plot of an FFT spectrum analysis.

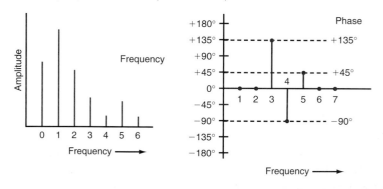

In the graph in Fig. 7-40, the horizontal axis in the upper plot is frequency, and the vertical axis is the amplitude of the dc and ac sine wave components that make up the sampled wave. A 0 frequency component is represented by a vertical line indicating the dc component of a signal. The 1 indicates the amplitude of the fundamental sine wave making up the signal. The other values, at 2, 3, 4, and so on, are the amplitudes of the harmonics. In the lower plot, the phase angle of the sine waves is given for each harmonic. A negative value indicates phase inversion of the sine wave (180°).

CHAPTER REVIEW

Summary

Transmitting data by using digital techniques offers a number of advantages over analog processing: high immunity to noise, excellent error detection and error correction capabilities, compatibility with time-division multiplexing techniques, and the use of digital signal processing (DSP) circuits.

In parallel data transfers, all the bits of a code word are transferred simultaneously. In serial data transfers, each bit of the word is transmitted in sequence. The conversion between parallel and serial and serial and parallel is accomplished by using shift registers.

Before analog signals can be transmitted digitally, they must be converted to digital signals by analog-to-digital (A/D) conversion, in which the signal is translated into a series of discrete binary numbers representing samples. A sufficient number of samples must be taken to retain the high-frequency information in the analog signal.

Modern A/D converters are usually single-chip ICs that take an analog signal and generate a parallel binary output. Since A/D converters can represent only a finite number of voltage values over a specific range, the samples are converted to binary numbers whose values are close to the actual sample values. The error associated with the conversion process, the quantizing error, can be reduced by dividing the analog voltage range into a larger number of smaller voltage increments. A process of signal compression and expansion known as companding is used to overcome the problems of quantizing error and noise.

D/A converters receive the binary signals sequentially and produce a proportional analog voltage at the output. D/A converters have four major sections: a regulator, a resistance or capacitor network, an output amplifier, and electronic switches. The three most important specifications associated with D/A converters are resolution, error, and settling time.

The most common A/D conversion circuits are the successive-approximations converter, the flash converter, the pipelined converter, and the sigma-delta converter. For high-speed conversions, flash converters are the circuits of choice, offering conversion speeds to several gigahertz.

In pulse modulation, the information signal, usually analog, is used to modify a binary or pulsed carrier in some way. There are three basic forms of pulse modulation: pulse-amplitude modulation (PAM), pulse-width modulation (PWM), and pulse-position modulation (PPM). Today, these techniques have been almost entirely superseded by the more sophisticated and effective pulse-code modulation (PCM).

In digital signal processing (DSP), very fast, specially designed computers control the conversion process. The analog signal to be processed is fed to an A/D converter where it is converted to a series of binary numbers, stored in RAM, and executed in real time. Programs for filtering, equalization, companding, phase shifting, modulation, and so on are written for DSP computers.

Questions

1. Name the four primary benefits of using digital techniques in communication. Which of these is probably the most important?
2. What is data conversion? Name two basic types.
3. What is the name given to the process of measuring the value of an analog signal at some point in time?
4. What is the name given to the process of assigning a specific binary number to an instantaneous value on an analog signal?
5. What is another name commonly used for A/D conversion?
6. Describe the nature of the signals and information obtained when an analog signal is converted to digital form.
7. Describe the nature of the output waveform obtained from a D/A converter.
8. Name the four major components in a D/A converter.
9. Define aliasing and explain its effect in an A/D converter.
10. What types of circuits are commonly used to translate the current output from a D/A converter to a voltage output?
11. Name three types of A/D converters and state which is the most widely used.
12. What A/D converter circuit sequentially turns the bits of the output on one at a time in sequence from MSB to LSB in seeking a voltage level equal to the input voltage level?
13. What is the fastest type of A/D converter? Briefly describe the method of conversion used.
14. What type of A/D converter generates a serial output data signal directly from the conversion process?
15. What circuit is normally used to perform serial-to-parallel and parallel-to-serial data conversion? What is the abbreviation for this process?
16. What circuit performs the sampling operation prior to A/D conversion, and why is it so important?
17. Where are sigma-delta converters used? Why?

18. What process converts an analog signal to sequential binary numbers and transmits them serially?

19. What is the name given to the process of compressing the dynamic range of an analog signal at the transmitter and expanding it later at the receiver?

20. What is the general mathematical shape of a companding curve?

21. Name the three basic types of pulse modulation. Which type is not binary?

22. Name the DAC that produces a voltage output.

23. What type of DAC is used for very high-speed conversions?

24. True or false? ADC outputs or DAC inputs may be either parallel or serial.

25. What type of ADC is faster than a successive-approximations converter but slower than a flash converter?

26. Which type of ADC gives the best resolution?

27. Why are capacitor D/A converters preferred over R-$2R$ D/A converters?

28. What does oversampling mean? What converter uses this technique? Why is it used?

29. How is aliasing prevented?

30. Name two common noncommunication applications for PWM.

31. Describe briefly the techniques known as digital signal processing (DSP).

32. What type of circuit performs DSP?

33. Briefly describe the basic mathematical process used in the implementation of DSP.

34. Give the names for the basic architecture of non-DSP microprocessors and for the architecture normally used in DSP microprocessors. Briefly describe the difference between the two.

35. Name five common processing operations that take place with DSP. What is probably the most commonly implemented DSP application?

36. Briefly describe the nature of the output of a DSP processor that performs the discrete Fourier transform or the fast Fourier transform.

37. Name the two types of filters implemented with DSP and explain how they differ.

38. What useful function is performed by an FFT computation?

Problems

1. A video signal contains light variations that change at a frequency as high as 3.5 MHz. What is the minimum sampling frequency for A/D conversion? ◆

2. A D/A converter has a 12-bit binary input. The output analog voltage range is 0 to 5 V. How many discrete output voltage increments are there and what is the smallest voltage increment?

3. Compute the alias created by sampling a 5-kHz signal at 8 kHz. ◆

4. Calculate the quantizing noise on a 14-bit A/D converter with a voltage range up to 3 V.

5. What is the SINAD for a 15-bit ADC?

6. Calculate the ENOB for a converter with a SINAD of 83 dB.

◆ *Answers to Selected Problems follow Chap. 22.*

Critical Thinking

1. List three major types of communication services that are not yet digital but could eventually be, and explain how digital techniques could be applied to those applications.

2. Explain how an all-digital receiver would process the signal of an analog AM radio broadcast signal.

3. What type of A/D converter would work best for video signals with a frequency content up to 5 MHz? Why?

4. Under what conditions can serial data transfers be faster than parallel data transfers?

chapter 8

Radio Transmitters

A radio transmitter takes the information to be communicated and converts it to an electronic signal compatible with the communication medium. Typically this process involves carrier generation, modulation, and power amplification. The signal is then fed by wire, coaxial cable, or waveguide to an antenna that launches it into free space. This chapter covers transmitter configurations and the circuits commonly used in radio transmitters, including oscillators, amplifiers, frequency multipliers, and impedance-matching networks.

Objectives

After completing this chapter, you will be able to:

- Calculate the frequency tolerance of crystal oscillators in percent and in parts per million (ppm).
- Discuss the operation of phase-locked loop (PLL) and direct digital synthesis (DDS) frequency synthesizers and explain how the output frequency is changed.
- Calculate the output frequency of a transmitter given the oscillator frequency and the number and types of multipliers.
- Explain the biasing and operation of class A, AB, and C power amplifiers using transistors.
- Define *neutralization* and explain how it is implemented.
- Discuss the operation and benefits of class D, E, and F switching amplifiers and explain why they are more efficient.
- Explain the basic design of L, π, and T-type *LC* circuits and discuss how they are used for impedance matching.
- Explain the use of transformers and baluns in impedance matching.

8-1 Transmitter Fundamentals

Transmitter

The *transmitter* is the electronic unit that accepts the information signal to be transmitted and converts it to an RF signal capable of being transmitted over long distances. Every transmitter has four basic requirements.

1. It must generate a carrier signal of the correct frequency at a desired point in the spectrum.
2. It must provide some form of modulation that causes the information signal to modify the carrier signal.
3. It must provide sufficient power amplification to ensure that the signal level is high enough to carry over the desired distance.
4. It must provide circuits that match the impedance of the power amplifier to that of the antenna for maximum transfer of power.

Transmitter Configurations

Continuous-wave (CW) transmission

The simplest transmitter is a single-transistor oscillator connected directly to an antenna. The oscillator generates the carrier and can be switched off and on by a telegraph key to produce the dots and dashes of the International Morse code. Information transmitted in this way is referred to as *continuous-wave (CW) transmission* Such a transmitter is rarely used today, for the Morse code is nearly extinct and the oscillator power is too low for reliable communication. Nowadays transmitters such as this are built only by amateur (ham) radio operators for what is called *QRP* or *low-power operation* for personal hobby communication.

The CW transmitter can be greatly improved by simply adding a power amplifier to it, as illustrated in Fig. 8-1. The oscillator is still keyed off and on to produce dots and dashes, and the amplifier increases the power level of the signal. The result is a stronger signal that carries farther and produces more reliable transmission.

The basic oscillator-amplifier combination shown in Fig. 8-1 is the basis for virtually all radio transmitters. Many other circuits are added depending on the type of modulation used, the power level, and other considerations.

High-Level AM Transmitters. Figure 8-2 shows an AM transmitter using high-level modulation. An oscillator, in most applications a crystal oscillator, generates the final carrier frequency. The carrier signal is then fed to a buffer amplifier whose primary purpose is to isolate the oscillator from the remaining power amplifier stages. The buffer amplifier usually operates at the class A level and provides a modest increase in power output. The main purpose of the buffer amplifier is simply to prevent load

Figure 8-1 A more powerful CW transmitter.

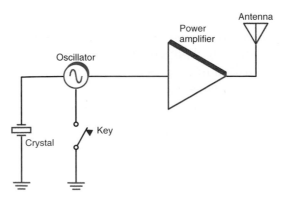

Figure 8-2 An AM transmitter using high-level collector modulation.

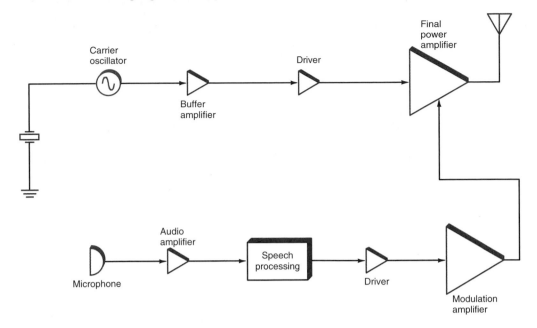

changes in the power amplifier stages or in the antenna from causing frequency variations in the oscillator.

The signal from the buffer amplifier is applied to a class C driver amplifier designed to provide an intermediate level of power amplification. The purpose of this circuit is to generate sufficient output power to drive the final power amplifier stage. The *final power amplifier,* normally just referred to as *the final,* also operates at the class C level at very high power. The actual amount of power depends on the application. For example, in a CB transmitter, the power input is only 5 W. However, AM radio stations operate at much higher powers—say, 250, 500, 1000, 5000, or 50,000 W—and the video transmitter at a TV station operates at even higher power levels. Cell phone base stations operate at the 30- to 40-W level.

All the RF circuits in the transmitter are usually solid-state; i.e., they are implemented with either bipolar transistors or metal-oxide semiconductor field-effect transistors (MOSFETs). Although bipolar transistors are by far the most common type, the use of MOSFETs is increasing because they are now capable of handling high power at high frequencies. Transistors are also typically used in the final as long as the power level does not exceed several hundred watts. Individual RF power transistors can handle up to about 300 W. Many of these can be connected in parallel or in push-pull configurations to increase the power-handling capability to many kilowatts. For higher power levels, vacuum tubes are still used in some transmitters, but rarely in new designs. Vacuum tubes function into the VHF and UHF ranges, with power levels of 1 kW or more.

Now, assume that the AM transmitter shown in Fig. 8-2 is a voice transmitter. The input from the microphone is applied to a low-level class A audio amplifier, which boosts the small signal from the microphone to a higher voltage level. (One or more stages of amplification could be used.) The voice signal is then fed to some form of *speech-processing* (filtering and amplitude control) circuit. The filtering ensures that only voice frequencies in a certain range are passed, which helps to minimize the bandwidth occupied by the signal. Most communication transmitters limit the voice frequency to the 300- to 3000-Hz range, which is adequate for intelligible communication. However, AM broadcast stations offer higher fidelity and allow frequencies up to 5 kHz to be used. In practice, many AM stations modulate with frequencies up to 7.5 kHz, and even 10 kHz, since the FCC uses alternate channel assignments within a given region and the outer sidebands are very weak, so no adjacent channel interference occurs.

Final power amplifier

Speech processing

Speech processors also contain a circuit used to hold the amplitude to some maximum level. High-amplitude signals are compressed and lower-amplitude signals are given more amplification. The result is that overmodulation is prevented, yet the transmitter operates as close to 100 percent modulation as possible. This reduces the possibility of signal distortion and harmonics, which produce wider sidebands that can cause adjacent channel interference, but maintains the highest possible output power in the sidebands.

After the speech processor, a driver amplifier is used to increase the power level of the signal so that it is capable of driving the high-power modulation amplifier. In the AM transmitter of Fig. 8-2, high-level or collector modulation (plate modulation in a tube) is used. As stated previously, the power output of the modulation amplifier must be one-half the input power of the RF amplifier. The high-power modulation amplifier usually operates with a class AB or class B push-pull configuration to achieve these power levels.

Low-Level FM Transmitters.

In low-level modulation, modulation is performed on the carrier at low power levels, and the signal is then amplified by power amplifiers. This arrangement works for both AM and FM. FM transmitters using this method are far more common than low-level AM transmitters.

Figure 8-3 shows the typical configuration for an FM or PM transmitter. The indirect method of FM generation is used. A stable crystal oscillator is used to generate the carrier signal, and a buffer amplifier is used to isolate it from the remainder of the circuitry. The carrier signal is then applied to a phase modulator such as those discussed in Chap. 6. The voice input is amplified and processed to limit the frequency range and prevent overdeviation. The output of the modulator is the desired FM signal.

Most FM transmitters are used in the VHF and UHF range. Because crystals are not available for generating those frequencies directly, the carrier is usually generated at a frequency considerably lower than the final output frequency. To achieve the desired output frequency, one or more frequency multiplier stages are used. A frequency multiplier is a class C amplifier whose output frequency is some integer multiple of the input frequency. Most frequency multipliers increase the frequency by a factor of 2, 3, 4, or 5. Because they are class C amplifiers, most frequency multipliers also provide a modest amount of power amplification.

Not only does the frequency multiplier increase the carrier frequency to the desired output frequency, but also it multiplies the frequency deviation produced by the modulator. Many frequency and phase modulators generate only a small frequency shift, much lower than the desired final deviation. The design of the transmitter must be such that the frequency multipliers will provide the correct amount of multiplication not only for the carrier frequency, but also for the modulation deviation. After the frequency multiplier stage, a class C driver amplifier is used to increase the power level sufficiently to operate the final power amplifier, which also operates at the class C level.

Figure 8-3   A typical FM transmitter using indirect FM with a phase modulator.

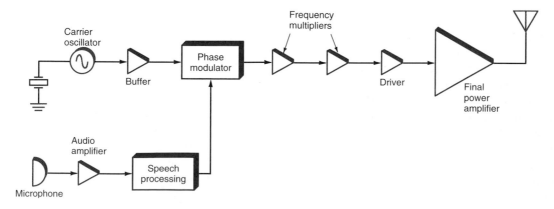

Figure 8-4 An SSB transmitter.

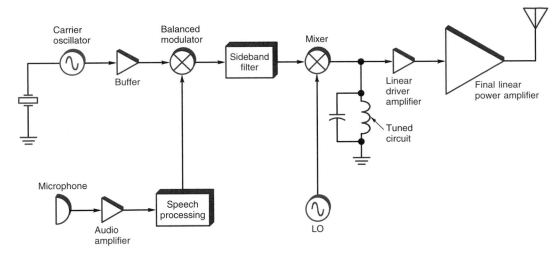

Most FM communication transmitters operate at relatively low power levels, typically less than 100 W. All the circuits, even in the VHF and UHF range, use transistors. For power levels beyond several hundred watts, vacuum tubes must be used. The final amplifier stages in FM broadcast transmitters typically use large vacuum tube class C amplifiers. In FM transmitters operating in the microwave range, klystrons, magnetrons, and traveling-wave tubes are used to provide the final power amplification.

SSB Transmitters. A typical *single-sideband (SSB) transmitter* is shown in Fig. 8-4. An oscillator signal generates the carrier, which is then fed to the buffer amplifier. The buffer amplifier supplies the carrier input signal to the balanced modulator. The audio amplifier and speech-processing circuits described previously provide the other input to the balanced modulator. The balanced modulator output—a DSB signal—is then fed to a sideband filter which selects either the upper or lower sideband. Following this, the SSB signal is fed to a mixer circuit, which is used to convert the signal to its final operating frequency. Mixer circuits, which operate as simple amplitude modulators, are used to convert a lower frequency to a higher one or a higher frequency to a lower one. (Mixers are discussed more fully in Chap. 9.)

Typically, the SSB signal is generated at a low RF. This makes the balanced modulator and filter circuits simpler and easier to design. The mixer translates the SSB signal to a higher desired frequency. The other input to the mixer is derived from a local oscillator set at a frequency that, when mixed with the SSB signal, produces the desired operating frequency. The mixer can be set up so that the tuned circuit at its output selects either the sum or the difference frequency. The oscillator frequency must be set to provide the desired output frequency. For fixed-channel operation, crystals can be used in this local oscillator. However, in some equipment, such as that used by hams, a *variable frequency oscillator (VFO)* is used to provide continuous tuning over a desired range. In most modern communication equipment, a frequency synthesizer is used to set the final output frequency.

The output of the mixer in Fig. 8-4 is the desired final carrier frequency containing the SSB modulation. It is then fed to linear driver and power amplifiers to increase the power level as required. Class C amplifiers distort the signal and therefore cannot be used to transmit SSB or low-level AM of any kind, including DSB. Class A or AB linear amplifiers must be used to retain the information content in the AM signal.

Most modern digital radios such as cell phones use DSP to produce the modulation and related processing of the data to be transmitted. Refer to Fig. 8-5. The serial data representing the data to be transmitted is sent to the DSP which then generates two data streams that are then converted to RF for transmission. The data paths from

Single-sideband (SSB) transmitter

Variable frequency oscillator (VFO)

Figure 8-5 Modern digital transmitter.

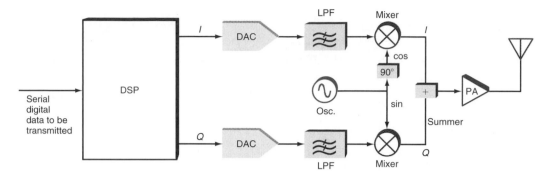

the DSP chip are sent to DACs where they are translated to equivalent analog signals. The analog signals are filtered in a *low-pass filter (LPF)* and then applied to mixers that will up-convert them to the final output frequency. The mixers receive their second inputs from an oscillator or a frequency synthesizer that selects the operating frequency. Note that the oscillator signals are in quadrature; i.e., one is shifted 90° from the other. One is a sine wave, and the other is a cosine wave. The upper signal is referred to as the in-phase (*I*) signal and the other as the quadrature (*Q*) signal. The output signals from the mixers are then added, and the result is amplified and transmitted by the power amplifier (PA). Two quadrature signals are needed at the receiver to recover the signal and demodulate it in a DSP chip. You will learn more about this technique in a later chapter.

8-2 Carrier Generators

The starting point for all transmitters is carrier generation. Once generated, the carrier can be modulated, processed in various ways, amplified, and finally transmitted. The source of most carriers in modern transmitters is a crystal oscillator. PLL frequency synthesizers in which a crystal oscillator is the basic stabilizing reference are used in applications requiring multiple channels of operation.

Crystal Oscillators

Most radio transmitters are licensed by the FCC either directly or indirectly to operate not only within a specific frequency band but also on predefined frequencies or channels. Deviating from the assigned frequency by even a small amount can cause interference with signals on adjacent channels. Therefore the transmitter carrier generator must be very precise, operating on the exact frequency assigned, often within very close tolerances. In some radio services, the frequency of operation must be within 0.001 percent of the assigned frequency. In addition, the transmitter must remain on the assigned frequency. It must not drift off or wander from its assigned value despite the many operating conditions, such as wide temperature variations and changes in power supply voltage, that affect frequency. The only oscillator capable of meeting the precision and stability demanded by the FCC is a *crystal oscillator*.

A *crystal* is a piece of quartz that has been cut and ground into a thin, flat wafer and mounted between two metal plates. When the crystal is excited by an ac signal across its plates, it vibrates. This action is referred to as the *piezoelectric effect*. The frequency of vibration is determined primarily by the thickness of the crystal. Other factors influencing frequency are the cut of the crystal, i.e., the place and angle of cut made in the base quartz rock from which the crystal was derived, and the size of the

> **GOOD TO KNOW**
>
> The only oscillator capable of maintaining the frequency precision and stability demanded by the FCC is a crystal oscillator. In fact, the FCC requires that a crystal oscillator be used in all transmitters.

crystal wafer. Crystals frequencies range from as low as 30 kHz to as high as 150 MHz. As the crystal vibrates or oscillates, it maintains a very constant frequency. Once a crystal has been cut or ground to a particular frequency, it will not change to any great extent even with wide voltage or temperature variations. Even greater stability can be achieved by mounting the crystal in sealed, temperature-controlled chambers known as *crystal ovens*. These devices maintain an absolute constant temperature, ensuring a stable output frequency.

As you saw in Chap. 4, the crystal acts as an *LC* tuned circuit. It can emulate a series or parallel *LC* circuit with a *Q* as high as 30,000. The crystal is simply substituted for the coil and capacitor in a conventional oscillator circuit. The end result is a very precise, stable oscillator. The precision, or stability, of a crystal is usually expressed in parts per million (ppm). For example, to say that a crystal with a frequency of 1 MHz has a precision of 100 ppm means that the frequency of the crystal can vary from 999,900 to 1,000,100 Hz. Most crystals have tolerance and stability values in the 10- to 1000-ppm range. Expressed as a percentage, the precision is $(100/1,000,000) \times 100 = 0.0001 \times 100 = 0.01$ percent.

You can also use ratio and proportion to figure the frequency variation for a crystal with a given precision. For example, a 24-MHz crystal with a stability of ± 50 ppm has a maximum frequency variation Δf of $(50/1,000,000) \times 24,000,000$. Thus $\Delta f = 50(24,000,000)/1,000,000 = 24 \times 50 = 1200$ Hz or ± 1200 Hz.

Example 8-1

What are the maximum and minimum frequencies of a 16-MHz crystal with a stability of 200 ppm?

The frequency can vary as much as 200 Hz for every 1 MHz of frequency or $200 \times 16 = 3200$ Hz.

The possible frequency range is

$$16,000,000 - 3200 = 15,996,800 \text{ Hz}$$

$$16,000,000 + 3200 = 16,003,200 \text{ Hz}$$

Expressed as a percentage, this stability is $(3200/16,000,000) \times 100 = 0.0002 \times 100 = 0.02$ percent.

In other words, the actual frequency may be different from the designated frequency by as much as 50 Hz for every 1 MHz of designated frequency, or $24 \times 50 = 1200$ Hz.

A precision value given as a percentage can be converted to a ppm value as follows. Assume that a 10-MHz crystal has a precision percentage of ± 0.001 percent; 0.001 percent of 10,000,000 is $0.00001 \times 10,000,000 = 100$ Hz. Thus

$$\text{ppm}/1,000,000 = 100/10,000,000$$

$$\text{ppm} = 100(1,000,000)/10,000,000 = 10 \text{ ppm}$$

However, the simplest way to convert from percentage to ppm is to convert the percentage value to its decimal form by dividing by 100, or moving the decimal point two places to the left, and then multiplying by 10^6, or moving the decimal point six places to the right. For example, the ppm stability of a 5-MHz crystal with a precision of 0.005 percent is found as follows. First, put 0.005 percent in decimal form: 0.005 percent = 0.00005. Next, multiply by 1 million:

$$0.00005 \times 1,000,000 = 50 \text{ ppm}$$

Example 8-2

A radio transmitter uses a crystal oscillator with a frequency of 14.9 MHz and a frequency multiplier chain with factors of 2, 3, and 3. The crystal has a stability of ± 300 ppm.

a. Calculate the transmitter output frequency.

Total frequency multiplication factor $= 2 \times 3 \times 3 = 18$

Transmitter output frequency $= 14.9\ \text{MHz} \times 18$

$= 268.2\ \text{MHz}$

b. Calculate the maximum and minimum frequencies that the transmitter is likely to achieve if the crystal drifts to its maximum extreme.

$$\pm 300\ \text{ppm} = \frac{300}{1,000,000} \times 100 = \pm 0.03\%$$

This variation is multiplied by the frequency multiplier chain, yielding ± 0.03 percent $\times\ 18 = \pm 0.54$ percent. Now, $268.2\ \text{MHz} \times 0.0054 = 1.45\ \text{MHz}$. Thus the frequency of the transmitter output is $268.2 \pm 1.45\ \text{MHz}$. The upper limit is

$$268.2 + 1.45 = 269.65\ \text{MHz}$$

The lower limit is

$$268.2 - 1.45 = 266.75\ \text{MHz}$$

Typical Crystal Oscillator Circuits. The most common crystal oscillator is a *Colpitts type,* in which the feedback is derived from the capacitive voltage divider made up of C_1 and C_2. An emitter-follower version is shown in Fig. 8-6. Again, the feedback comes from the capacitor voltage divider C_1–C_2. The output is taken from the emitter, which is untuned.

Occasionally you will see a capacitor in series or in parallel with the crystal (not both), as shown in Fig. 8-6. These capacitors can be used to make minor adjustments in

Colpitts oscillator

Figure 8-6 An emitter-follower crystal oscillator.

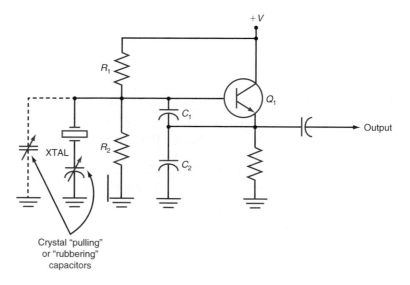

Crystal "pulling"
or "rubbering"
capacitors

Figure 8-7 The Pierce crystal oscillator using an FET.

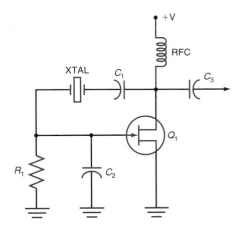

the crystal frequency. As discussed previously, it is not possible to affect large frequency changes with series or shunt capacitors, but they can be used to make fine adjustments. The capacitors are called crystal *pulling* capacitors, and the whole process of fine-tuning a crystal is sometimes referred to as *rubbering*. When the pulling capacitor is a varactor, FM or FSK can be produced. The analog or binary modulating signal varies the varactor capacitance that, in turn, shifts the crystal frequency.

Field-effect transistors (FETs) also make good crystal oscillators. Figure 8-7 shows a FET used in the popular *Pierce oscillator* configuration. Most crystal oscillators are some variation of these basic types. They operate as class A linear amplifiers and generate a clean sine wave output signal.

Rubbering

Field-effect transistor (FET)
Pierce oscillator

Overtone Oscillators. The main problem with crystals is that their upper frequency operation is limited. The higher the frequency, the thinner the crystal must be to oscillate at that frequency. At an upper limit of about 50 MHz, the crystal is so fragile that it becomes impractical to use. However, over the years, operating frequencies have continued to move upward as a result of the quest for more frequency space and greater channel capacity, and the FCC has continued to demand the same stability and precision that are required at the lower frequencies. One way to achieve VHF, UHF, and even microwave frequencies using crystals is by employing frequency multiplier circuits, as described earlier. The carrier oscillator operates on a frequency less than 50 MHz, and multipliers raise that frequency to the desired level. For example, if the desired operating frequency is 163.2 MHz and the frequency multipliers multiply by a factor of 24, the crystal frequency must be 163.2/24 = 6.8 MHz.

Another way to achieve crystal precision and stability at frequencies above 50 MHz is to use *overtone crystals*. An overtone crystal is cut in a special way so that it optimizes its oscillation at an overtone of the basic crystal frequency. An overtone is like a harmonic as it is usually some multiple of the fundamental vibration frequency. However, the term *harmonic* is usually applied to electric signals, and the term *overtone* refers to higher mechanical vibration frequencies. Like a harmonic, an overtone is usually some integer multiple of the base vibration frequency. However, most overtones are slightly more or slightly less than the integer value. In a crystal, the second harmonic is the first overtone, the third harmonic is the second overtone, and so on. For example, a crystal with a fundamental frequency of 20 MHz would have a second harmonic or first overtone of 40 MHz, and a third harmonic or second overtone of 60 MHz.

The term *overtone* is often used as a synonym for harmonic. Most manufacturers refer to their third overtone crystals as *third harmonic crystals.*

The odd overtones are far greater in amplitude than the even overtones. Most overtone crystals oscillate reliably at the third or fifth overtone of the frequency at which the crystal is originally ground. There are also seventh-overtone crystals. Overtone crystals can be

Overtone crystal

Harmonic
Overtone

GOOD TO KNOW

Overtones refer to multiples of the harmonic frequency. The second harmonic is the first overtone, the third harmonic is the second overtone, and so on.

Figure 8-8 An overtone crystal oscillator.

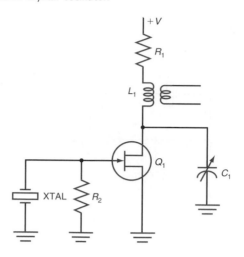

obtained with frequencies up to about 250 MHz. A typical overtone crystal oscillator is shown in Fig. 8-8. With this design, a crystal cut for a frequency of, say, 16.8 MHz and optimized for overtone service will have a third-overtone oscillation at $3 \times 16.8 = 50.4$ MHz. The tuned output circuit made up of L_1 and C_1 will be resonant at 50.4 MHz.

Crystal Switching. If a transmitter must operate on more than one frequency, as is often the case, but crystal precision and stability are required, multiple crystals can be used and the desired one switched in. The most straightforward way to do this is to use a mechanical rotary switch. This arrangement works fine at the lower frequencies if the crystals are located close to the switch. The connections between the crystals and the switch and oscillator must be kept short to minimize stray inductance and capacitance, which can affect the feedback and the frequency of operation. At higher frequencies this approach is unacceptable because of excessive distributed stray inductance and capacitance.

One approach to crystal switching, using diode switches, is shown in Fig. 8-9. The mechanical switch is used to apply a dc bias voltage to the diodes to select the desired frequency. Note that a silicon switching diode is connected in series with each crystal. With the switch set to channel A, diode D_1 is forward-biased by the dc voltage applied by the switch. The diode conducts, acting as a very low-value resistor. The diode connects crystal X_1 to ground. The other diode is cut off because no forward bias is applied to it. The RFCs and the capacitors keep the RF signal out of the dc bias circuits.

The diode switching arrangement is fast and reliable and overcomes the problem of long connecting wires between the crystal, switch, and oscillator circuit. The diodes are mounted near the crystals, which in turn are close to the oscillator components, usually

Figure 8-9 Using diodes to switch crystals.

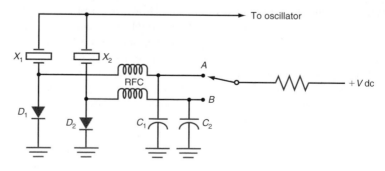

on a printed-circuit board. The switch can be located any distance away. Since the switch is switching direct current and not high-frequency alternating current at the crystal itself, the length of the wires between the switch and diodes is not a factor.

Frequency Synthesizers

Frequency synthesizers are variable-frequency generators that provide the frequency stability of crystal oscillators but with the convenience of incremental tuning over a broad frequency range. Frequency synthesizers usually provide an output signal that varies in fixed frequency increments over a wide range. In a transmitter, a frequency synthesizer provides basic carrier generation for channelized operation. Frequency synthesizers are also used in receivers as local oscillators and perform the receiver tuning function.

Using frequency synthesizers overcomes certain cost and size disadvantages associated with crystals. Assume, e.g., that a transmitter must operate on 50 channels. Crystal stability is required. The most direct approach is simply to use one crystal per frequency and add a large switch. Although such an arrangement works, it has major disadvantages. Crystals are expensive, ranging from $1 to $10 each, and even at the lowest price, 50 crystals may cost more than all the rest of the parts in the transmitter. The same 50 crystals would also take up a great deal of space, possibly occupying more than 10 times the volume of all the rest of the transmitter parts. With a frequency synthesizer, only one crystal is needed, and the requisite number of channels can be generated by using a few tiny ICs.

Over the years, many techniques have been developed for implementing frequency synthesizers with frequency multipliers and mixers. Today, however, most frequency synthesizers use some variation of the *phase-locked loop (PLL)*. A newer technique called *digital signal synthesis (DSS)* is becoming more popular as integrated-circuit technology has made high-frequency generation practical.

Phase-Locked Loop Synthesizers

An elementary frequency synthesizer based on a PLL is shown in Fig. 8-10. Like all phase-locked loops, it consists of a phase detector, a low-pass filter, and a VCO. The input to the phase detector is a reference oscillator. The reference oscillator is normally crystal-controlled to provide high-frequency stability. The frequency of the reference oscillator sets the increments in which the frequency may be changed. Note that the VCO output is not connected directly back to the phase detector, but applied to a frequency divider first. A *frequency divider* is a circuit whose output frequency is some integer submultiple of the input frequency. A divide-by-10 frequency synthesizer produces an output frequency that is one-tenth of the input frequency. Frequency dividers can be easily implemented with digital circuits to provide any integer value of frequency division.

In the PLL in Fig. 8-10, the reference oscillator is set to 100 kHz (0.1 MHz). Assume that the frequency divider is initially set for a division of 10. For a PLL to become locked or synchronized, the second input to the phase detector must be equal in frequency to the reference frequency; for this PLL to be locked, the frequency divider

Frequency synthesizer

Phase-locked loop (PLL)

Digital signal synthesis (DSS)

Phase-locked loop synthesizer

Frequency divider

Figure 8-10   Basic PLL frequency synthesizer.

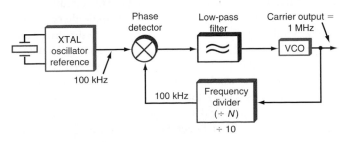

Figure 8-11 VHF/UHF frequency synthesizer.

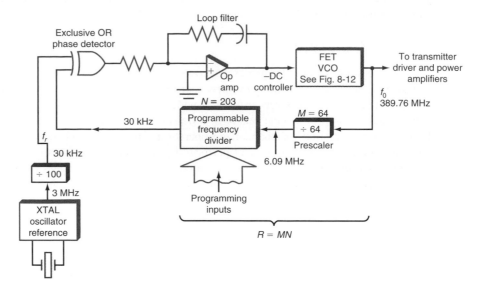

output must be 100 kHz. The VCO output has to be 10 times higher than this, or 1 MHz. One way to look at this circuit is as a frequency *multiplier:* The 100-kHz input is multiplied by 10 to produce the 1-MHz output. In the design of the synthesizer, the VCO frequency is set to 1 MHz so that when it is divided, it will provide the 100-kHz input signal required by the phase detector for the locked condition. The synthesizer output is the output of the VCO. What has been created, then, is a 1-MHz signal source. Because the PLL is locked to the crystal reference source, the VCO output frequency has the same stability as that of the crystal oscillator. The PLL will track any frequency variations, but the crystal is very stable and the VCO output is as stable as that of the crystal reference oscillator.

To make the frequency synthesizer more useful, some means must be provided to vary its output frequency. This is done by varying the frequency division ratio. Through various switching techniques, the flip-flops in a frequency divider can be arranged to provide any desired frequency division ratio. The frequency division ratio is normally designed to be manually changed in some way. For example, rotary-switch-controlled logic circuits may provide the correct configuration, or a thumb-wheel switch may be used. Some designs actually incorporate a keyboard on which the desired frequency division ratio can be keyed in. In the most sophisticated circuits, a microprocessor generates the correct frequency division ratio and provides a direct frequency readout display.

Varying the frequency division ratio changes the output frequency. For example, in the circuit in Fig. 8-10, if the frequency division ratio is changed from 10 to 11, the VCO output frequency must change to 1.1 MHz. The output of the divider then remains at 100 kHz (1,100,000/11 = 100,000), as necessary to maintain a locked condition. Each incremental change in frequency division ratio produces an output frequency change of 0.1 MHz. This is how the frequency increment is set by the reference oscillator.

A more complex PLL synthesizer, a circuit that generates VHF and UHF frequencies over the 100- to 500-MHz range, is shown in Fig. 8-11. This circuit uses a FET oscillator to generate the carrier frequency directly. No frequency multipliers are needed. The output of the frequency synthesizer can be connected directly to the driver and power amplifiers in the transmitter. This synthesizer has an output frequency in the 390-MHz range, and the frequency can be varied in 30-kHz increments above and below that frequency.

The VCO circuit for the synthesizer in Fig. 8-11 is shown in Fig. 8-12. The frequency of this LC oscillator is set by the values of L_1, C_1, C_2 and the capacitances of the varactor diodes D_1 and D_2, C_a and C_b, respectively. The dc voltage applied to the varactors changes the frequency. Two varactors are connected back to back, and thus the

Figure 8-12 VHF/UHF range VCO.

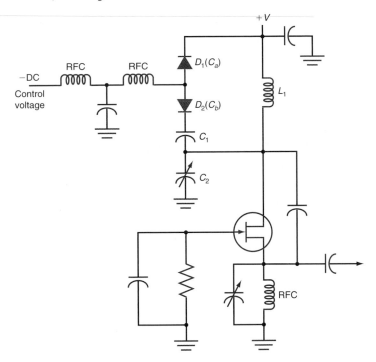

total effective capacitance of the pair is less than either individual capacitance. Specifically, it is equal to the series capacitance C_S, where $C_S = C_aC_b/(C_a + C_b)$. If D_1 and D_2 are identical, $C_S = C_a/2$. A negative voltage with respect to ground is required to reverse-bias the diodes. Increasing the negative voltage increases the reverse bias and decreases the capacitance. This, in turn, increases the oscillator frequency.

Using two varactors allows the oscillator to produce higher RF voltages without the problem of the varactors becoming forward-biased. If a varactor, which is a diode, becomes forward-biased, it is no longer a capacitor. High voltages in the tank circuit of the oscillator can sometimes exceed the bias voltage level and cause forward conduction. When forward conduction occurs, rectification takes place, producing a dc voltage that changes the dc tuning voltage from the phase detector and loop filter. The result is called *phase noise*. With two capacitors in series, the voltage required to forward-bias the combination is double that of one varactor. An additional benefit is that two varactors in series produce a more linear variation of capacitance with voltage than one diode. The dc frequency control voltage is, of course, derived by filtering the phase detector output with the low-pass loop filter.

In most PLLs the phase detector is a digital circuit rather than a linear circuit, since the inputs to the phase detector are usually digital. Remember, one input comes from the output of the feedback frequency divider chain, which is certainly digital, and the other comes from the reference oscillator. In some designs, the reference oscillator frequency is also divided down by a digital frequency divider to achieve the desired frequency step increment. This is the case in Fig. 8-11. Since the synthesizer frequency can be stepped in increments of 30 kHz, the reference input to the phase detector must be 30 kHz. This is derived from a stable 3-MHz crystal oscillator and a frequency divider of 100.

The design shown in Fig. 8-11 uses an exclusive-OR gate as a phase detector. Recall that an exclusive-OR (XOR) gate generates a binary 1 output only if the two inputs are complementary; otherwise, it produces a binary 0 output.

Figure 8-13 shows how the XOR phase detector works: Remember that the inputs to a phase detector must have the same frequency. This circuit requires that the inputs have a 50 percent duty cycle. The phase relationship between the two signals determines

Figure 8-13 Operation of XOR phase detector.

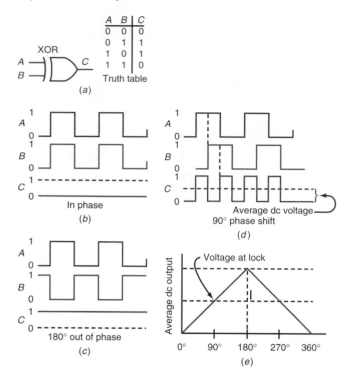

the output of the phase detector. If the two inputs are exactly in phase with each other, the XOR output will be zero, as Fig. 8-13(b) shows. If the two inputs are 180° out of phase with each other, the XOR output will be a constant binary 1 [see Fig. 8-13(c)]. Any other phase relationship will produce output pulses at twice the input frequency. The duty cycle of these pulses indicates the amount of phase shift. A small phase shift produces narrow pulses; a larger phase shift produces wider pulses. Figure 8-13(d) shows a 90° phase shift.

The output pulses are fed to the loop filter (Fig. 8-11), an op amp with a capacitor in the feedback path which makes it into a low-pass filter. This filter averages the phase detector pulses into a constant dc voltage that biases the VCO varactors. The average dc voltage is proportional to the duty cycle, which is the ratio of the binary 1 pulse time to the period of the signal. Narrow pulses (low duty cycle) produce a low average dc voltage, and wide pulses (high duty cycle) produce a high average dc voltage. Figure 8-13(e) shows how the average dc voltage varies with phase shift. Most PLLs lock in at a phase difference of 90°. Then, as the frequency of the VCO changes because of drift or because of changes in the frequency divider ratio, the input to the phase detector from the feedback divider changes, varying the duty cycle. This changes the dc voltage from the loop filter and forces a change of the VCO frequency to compensate for the original change. Note that the XOR produces a positive dc average voltage, but the op amp used in the loop filter inverts this to a negative dc voltage, as required by the VCO.

The output frequency of the synthesizer f_o and the phase detector reference frequency f_r are related to the overall divider ratio R as follows:

$$R = \frac{f_o}{f_r} \qquad f_o = Rf_r \qquad \text{or} \qquad f_r = \frac{f_o}{R}$$

In our example, the reference input to the phase detector f_r must be 30 kHz to match the feedback from the VCO output f_o. Assume a VCO output frequency of 389.76 MHz. A frequency divider reduces this amount to 30 kHz. The overall division ratio is $R = f_o/f_r = 389,760,000/30,000 = 12,992$.

Frequency dividers are usually designed to change the division ratio in integer increments. Programmable or presettable digital counter and divider ICs of the TTL or CMOS variety are available for this purpose. They can be programmed by applying an external binary code from thumbwheel switches, a keypad, a ROM, or a microprocessor.

In some very high-frequency PLL synthesizers, a special frequency divider called a *prescaler* is used between the high-output frequency of the VCO and the programmable part of the divider. The prescaler could be one or more *emitter-coupled logic (ECL)* flip-flops or a low-ratio CMOS frequency divider that can operate at frequencies up to 1 to 2 GHz. Refer again to Fig. 8-11. The prescaler divides by a ratio of $M = 64$ to reduce the 389.76-MHz output of the VCO to 6.09 MHz, which is well within the range of most programmable frequency dividers. Since we need an overall division ratio of $R = 12,992$ and a factor of $M = 64$ is in the prescaler, the programmable portion of the feedback divider N can be computed. The total division factor is $R = MN = 12,992$. Rearranging, we have $N = R/M = 12,992/64 = 203$.

Prescaler

Now, to see how the synthesizer changes output frequencies when the division ratio is changed, assume that the programmable part of the divider is changed by one increment, to $N = 204$. For the PLL to remain in a locked state, the phase detector input must remain at 30 kHz. This means that the VCO output frequency must change. The new frequency division ratio is $204 \times 64 = 13,056$. Multiplying this by 30 kHz yields the new VCO output frequency $f_o = 30,000 \times 13,056 = 391,680,000$ Hz $= 391.68$ MHz. Instead of the desired 30-kHz increment, the VCO output varied by $391,680,000 - 389,760,000 = 1,920,000$ Hz, or a step of 1.92 MHz. This was caused by the prescaler. For a 30-kHz step to be achieved, the feedback divider should have changed its ratio from 12,992 to 12,993. Since the prescaler is fixed with a division factor of 64, the smallest increment step is 64 times the reference frequency, or $64 \times 30,000 = 1,920,000$ Hz. The prescaler solves the problem of having a divider with a high enough frequency capability to handle the VCO output, but forces the use of programmable dividers for only a portion of the total divider ratio. Because of the prescaler, the divider ratio is not stepped in integer increments but in increments of 64. Circuit designers can either live with this or find another solution.

One possible solution is to reduce the reference frequency by a factor of 64. In the example, the reference frequency would become 30 kHz/64 $= 468.75$ Hz. To achieve this frequency at the other input of the phase detector, an additional division factor of 64 must be included in the programmable divider, making it $N = 203 \times 64 = 12,992$. Assuming the original output frequency of 389.76 MHz, the overall divider ratio is $R = MN = 12,992(64) = 831,488$. This makes the output of the programmable divider equal to the reference frequency, or $f_r = 389,760,000/831,488 = 468.75$ Hz.

This solution is logical, but it has several disadvantages. First, it increases cost and complexity by requiring two more divide-by-64 ICs in the reference and feedback paths. Second, the lower the operating frequency of the phase detector, the more difficult it is to filter the output into direct current. Further, the low-frequency response of the filter slows the process of achieving lock. When a change in the divider ratio is made, the VCO frequency must change. It takes a finite amount of time for the filter to develop the necessary value of corrective voltage to shift the VCO frequency. The lower the phase detector frequency, the greater this lock delay time. It has been determined that the lowest acceptable frequency is about 1 kHz, and even this is too low in some applications. At 1 kHz, the change in VCO frequency is very slow as the filter capacitor changes its charge in response to the different duty-cycle pulses of the phase detector. With a 468.75-Hz phase detector frequency, the loop response becomes even slower. For more rapid frequency changes, a much higher frequency must be used. For spread spectrum and in some satellite applications, the frequency must change in a few microseconds or less, requiring an extremely high reference frequency.

To solve this problem, designers of high-frequency PLL synthesizers created special IC frequency dividers, such as the one diagrammed in Fig. 8-14. This is known as a *fractional N divider PLL*. The VCO output is applied to a special variable-modulus prescaler

Fractional N divider

Figure 8-14 Using a variable-modulus prescaler in a PLL frequency divider.

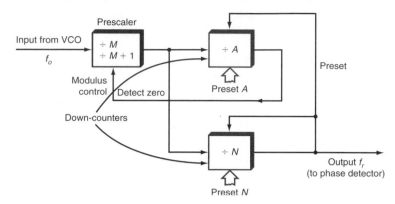

divider. It is made of emitter-coupled logic or CMOS circuits. It is designed to have two divider ratios, M and $M + 1$. Some commonly available ratio pairs are 10/11, 64/65, and 128/129. Let's assume the use of a 64/65 counter. The actual divider ratio is determined by the modulus control input. If this input is binary 0, the prescaler divides by M, or 64; if this input is binary 1, the prescaler divides by $M + 1$, or 65. As Fig. 8-14 shows, the modulus control receives its input from an output of counter A. Counters A and N are programmable down-counters used as frequency dividers. The divider ratios are preset into the counters each time a full divider cycle is achieved. These ratios are such that $N > A$. The count input to each counter comes from the output of the variable-modulus prescaler.

A divider cycle begins by presetting the down-counters to A and N and setting the prescaler to $M + 1 = 65$. The input frequency from the VCO is f_o. The input to the down-counters is $f_o/65$. Both counters begin down-counting. Since A is a shorter counter than N, A will decrement to 0 first. When it does, its detect-0 output goes high, changing the modulus of the prescaler from 65 to 64. The N counter initially counts down by a factor of A, but continues to down-count with an input of $f_o/64$. When it reaches 0, both down-counters are preset again, the dual modulo prescaler is changed back to a divider ratio of 65, and the cycle starts over.

The total division ratio R of the complete divider in Fig. 8-14 is $R = MN + A$. If $M = 64$, $N = 203$, and $A = 8$, the total divider ratio is $R = 64(203) + 8 = 12,992 + 8 = 13,000$. The output frequency is $f_o = Rf_r = 13,000(30,000) = 390,000,000 = 390$ MHz.

Any divider ratio in the desired range can be obtained by selecting the appropriate preset values for A and N. Further, this divider steps the divider ratio one integer at a time so that the step increment in the output frequency is 30 kHz, as desired.

As an example, assume that N is set to 207 and A is set to 51. The total divider ratio is $R = MN + A = 64(207) + 51 = 13,248 + 51 = 13,299$. The new output frequency is $f_o = 13,299(30,000) = 398,970,000 = 398.97$ MHz.

If the A value is changed by 1, raising it to 52, the new divide ratio is $R = MN + A = 64(207) + 52 = 13,248 + 52 = 13,300$. The new frequency is $f_o = 13,300(3,000) = 399,000,000 = 399$ MHz. Note that with an increment change in A of 1, R changed by 1 and the final output frequency increased by a 30-kHz (0.03-MHz) increment, from 398.97 to 399 MHz.

The preset values for N and A can be supplied by almost any parallel digital source but are usually supplied by a microprocessor or are stored in a ROM. Although this type of circuit is complex, it achieves the desired results of stepping the output frequency in increments equal to the reference input to the phase detector and allowing the reference frequency to remain high so that the change delay in the output frequency is shorter.

Example 8-3

A frequency synthesizer has a crystal reference oscillator of 10 MHz followed by a divider with a factor of 100. The variable-modulus prescaler has $M = 31/32$. The A and N down-counters have factors of 63 and 285, respectively. What is the synthesizer output frequency?

The reference input signal to the phase detector is

$$\frac{10 \text{ MHz}}{100} = 0.1 \text{ MHz} = 100 \text{ kHz}$$

The total divider factor R is

$$R = MN + A = 32(285) + 63 = 9183$$

The output of this divider must be 100 kHz to match the 100-kHz reference signal to achieve lock. Therefore, the input to the divider, the output of the VCO, is R times 100 kHz, or

$$f_o = 9183(0.1 \text{ MHz}) = 918.3 \text{ MHz}$$

Example 8-4

Demonstrate that the step change in output frequency for the synthesizer in Example 8-3 is equal to the phase detector reference range, or 0.1 MHz.

Changing the A factor one increment to 64 and recalculating the output yield

$$R = 32(285) + 64 = 9184$$

$$f_o = 9184(0.1 \text{ MHz}) = 918.4 \text{ MHz}$$

The increment is $918.4 - 918.3 = 0.1$ MHz.

Direct Digital Synthesis

A newer form of frequency synthesis is known as *direct digital synthesis (DDS)*. A DDS synthesizer generates a sine wave output digitally. The output frequency can be varied in increments depending upon a binary value supplied to the unit by a counter, a register, or an embedded microcontroller.

The basic concept of the DDS synthesizer is illustrated in Fig. 8-15. A read-only memory (ROM) is programmed with the binary representation of a sine wave. These are the values that would be generated by an analog-to-digital (A/D) converter if an analog sine wave were digitized and stored in the memory. If these binary values are fed to a digital-to-analog (D/A) converter, the output of the D/A converter will be a stepped approximation of the sine wave. A low-pass filter (LPF) is used to remove the high-frequency content near the clock frequency, thereby smoothing the ac output into a nearly perfect sine wave.

To operate this circuit, a binary counter is used to supply the address word to the ROM. A clock signal steps the counter that supplies a sequentially increasing address to ROM. The binary numbers stored in ROM are applied to the D/A converter, and the stepped sine wave is generated. The frequency of the clock determines the frequency of the sine wave.

Direct digital synthesis

GOOD TO KNOW

High-frequency content near the clock freqency of the D/A converter must be removed from the waveform. A low-pass filter is used to accomplish this, and a smooth sine wave results.

Figure 8-15 Basic concept of a DDS frequency source.

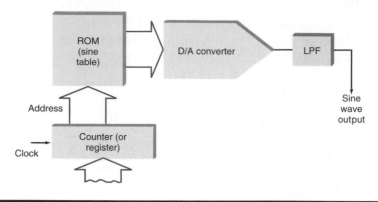

Figure 8-16 Shifting a sine wave to direct current.

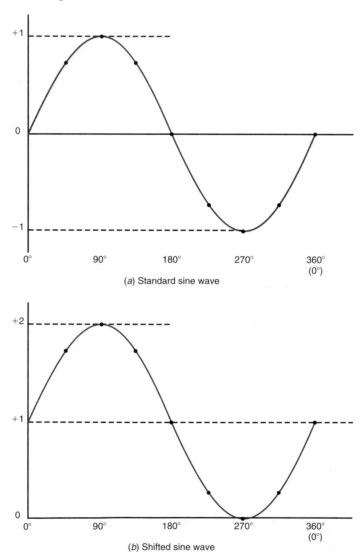

(a) Standard sine wave

(b) Shifted sine wave

To illustrate this concept, assume a 16-word ROM in which each storage location has a 4-bit address. The addresses are supplied by a 4-bit binary counter that counts from 0000 through 1111 and recycles. Stored in ROM are binary numbers representing values that are the sine of particular angles of the sine wave to be generated. Since a sine wave is 360° in length, and since the 4-bit counter produces

16 addresses or increments, the binary values represent the sine values at 360/16 = 22.5° increments.

Assume further that these sine values are represented with 8 bits of precision. The 8-bit binary sine values are fed to the D/A converter, where they are converted to a proportional voltage. If the D/A converter is a simple unit capable of a dc output voltage only, it cannot produce a negative value of voltage as required by a sine wave. Therefore, we will add to the sine value stored in ROM an offset value that will produce a sine wave output, but shifted so that it is all positive. For example, if we wish to produce a sine wave with a 1-V peak value, the sine wave would vary from 0 to +1, then back to 0, from 0 to −1, and then back to 0, as shown in Fig. 8-16(*a*). We add a binary 1 to the waveform so that the output of the D/A converter will appear as shown in Fig. 8-16(*b*). The D/A converter output will be 0 at the peak negative value of the sine wave. This value of 1 is added to each of the sine values stored in ROM. Figure 8-17 shows the ROM address, the phase angle, sine value, and the sine value plus 1.

If the counter starts counting at zero, the sine values will be sequentially accessed from ROM and fed to the D/A converter, which produces a stepped approximation of the sine wave. The resulting waveform (red) for one complete count of the counter is shown in Fig. 8-18. If the clock continues to count, the counter will recycle and the sine wave output cycle will be repeated.

An important point to note is that this frequency synthesizer produces one complete sine wave cycle for every 16 clock pulses. The reason for this is that we used 16 sine values to create the one cycle of the sine wave in ROM. To get a more accurate representation of the sine wave, we could have used more bits. For example, if we had used an 8-bit counter with 256 states, the sine values would be spaced every 360/256 = 1.4°, giving a highly accurate representation of the sine wave. Because of this relationship, the output frequency of the sine wave f_0 = the clock frequency $f_{\text{clk}}/2^N$, where N is equal to the number of address bits in ROM.

If a clock frequency of 1 MHz were used with our 4-bit counter, the sine wave output frequency would be

$$f_0 = 1{,}000{,}000/2^4 = 1{,}000{,}000/16 = 62{,}500 \text{ Hz}$$

The stepped approximation of the sine wave is then applied to a low-pass filter where the high-frequency components are removed, leaving a low-distortion sine wave.

The only way to change the frequency in this synthesizer is to change the clock frequency. This arrangement does not make much sense in view of the fact that we want our

GOOD TO KNOW

The output frequency f_0 equals the clock frequency f_{clk} divided by 2^N, where N is the number of address bits in ROM.

Figure 8-17 Address and sine values for a 4-bit DDS.

Address	Angle (degrees)	Sine	Sine + 1
0000	90	1	2
0001	112.5	0.924	1.924
0010	135	0.707	1.707
0011	157.5	0.383	1.383
0100	180	0	1
0101	202.5	−0.383	0.617
0110	225	−0.707	0.293
0111	247.5	−0.924	0.076
1000	270	−1	0
1001	292.5	−0.924	0.076
1010	315	−0.707	0.293
1011	337.5	−0.383	0.617
1100	360	0	1
1101	22.5	0.383	1.383
1110	45	0.707	1.707
1111	67.5	0.924	1.924
0000	90	1	2

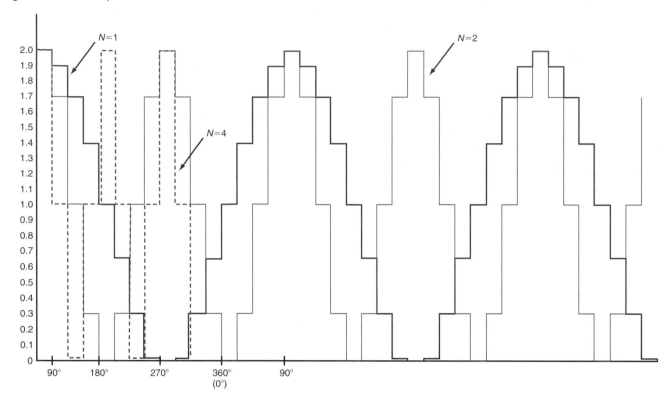

Figure 8-18 Output waveforms of a 4-bit DDS.

synthesizer output to have crystal oscillator precision and stability. To achieve this, the clock oscillator must be crystal-controlled. The question then becomes, How can you modify this circuit to maintain a constant clock frequency and also change the frequency digitally?

The most commonly used method to vary the synthesizer output frequency is to replace the counter with a register whose content will be used as the ROM address but also one that can be readily changed. For example, it could be loaded with an address from an external microcontroller. However, in most DDS circuits, this register is used in conjunction with a binary adder, as shown in Fig. 8-19. The output of the address register is applied to the adder along with a constant binary input value. This constant value can also be changed. The output of the adder is fed back into the register. The combination of the register and adder is generally referred to as an *accumulator*. This circuit is arranged so that upon the occurrence of each clock pulse, the constant *C* is added to the previous value of the register content and the sum is re-stored in the address register. The constant value comes from the phase increment register, which in turn gets it from an embedded microcontroller or other source.

To show how this circuit works, assume that we are using a 4-bit accumulator register and the same ROM described previously. Assume also that we will set the constant value to 1. For this reason, each time a clock pulse occurs, a 1 is added to the content of the register. With the register initially set to 0000, the first clock pulse will cause the register to increment to 1. On the next clock pulse the register will increment to 2, and so on. As a result, this arrangement acts just as the binary counter described earlier.

Now assume that the constant value is 2. This means that for each clock pulse, the register value will be incremented by 2. Starting at 0000, the register contents would be 0, 2, 4, 6, and so on. Looking at the sine value table in Fig. 8-17, you can see that the values output to the D/A converter also describe the sine wave, but the sine wave is being generated at a more rapid rate. Instead of having eight amplitude values represent the peak-to-peak value of the sine wave, only four values are used. Refer to Fig. 8-18, which illustrates what the output looks like (blue curve). The output is, of course, a stepped approximation of a sine wave, but during the complete cycle of the

Accumulator

Figure 8-19 Complete DDS block diagram.

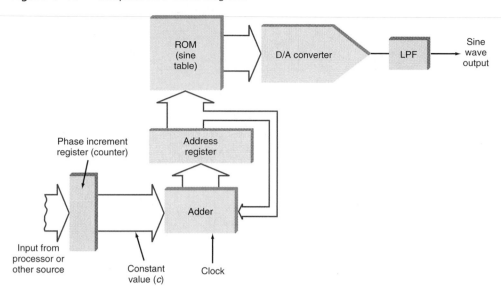

counter from 0000 through 1111, two cycles of the output sine wave occur. The output has fewer steps and is a cruder representation. With an adequate low-pass filter, the output will be a sine wave whose frequency is twice that generated by the circuit with a constant input of 1.

The frequency of the sine wave can further be adjusted by changing the constant value added to the accumulator. Setting the constant to 3 will produce an output frequency that is 3 times that produced by the original circuit. A constant value of 4 produces a frequency 4 times the original frequency.

With this arrangement we can now express the output sine wave frequency with the formula

$$f_0 = \frac{Cf_{\text{clk}}}{2^N}$$

The higher the constant value C, the fewer the samples used to reconstruct the output sine wave. When the constant is set to 4, every fourth value in Fig. 8-17 will be sent to the D/A converter, generating the dashed waveform in Fig. 8-18. Its frequency is 4 times the original. This corresponds to two samples per cycle, which is the least number that can be used and still generate an accurate output frequency. Recall the Nyquist criterion which says that to adequately reproduce a sine wave, it must be sampled a minimum of 2 times per cycle to reproduce it accurately in a D/A converter.

To make the DDS effective, then, the total number of sine samples stored in ROM must be a very large value. Practical circuits use a minimum of 12 address bits, giving 4096 sine samples. Even a larger number of samples can be used.

The DDS synthesizer described earlier offers some advantages over a PLL synthesizer. First, if a sufficient number of bits of resolution in ROM word size and the accumulator size are provided, the frequency can be varied in very fine increments. And because the clock is crystal-controlled, the resulting sine wave output will have the accuracy and precision of the crystal clock.

A second benefit is that the frequency of the DDS synthesizer can usually be changed much faster than that of a PLL synthesizer. Remember that to change the PLL synthesizer frequency, a new frequency-division factor must be entered into the frequency divider. Once this is done, it takes a finite amount of time for the feedback loop to detect the error and settle into the new locked condition. The storage time of the loop low-pass filter considerably delays the frequency change. This is not a problem in the DDS synthesizer, which can change frequencies within nanoseconds.

GOOD TO KNOW

For stepped-up constants added to the 4-bit accumulator register, the output can be calculated based on the constant C multiplied by the clock frequency f_{clk} divided by 2^N, where N is the number of bits in the register.

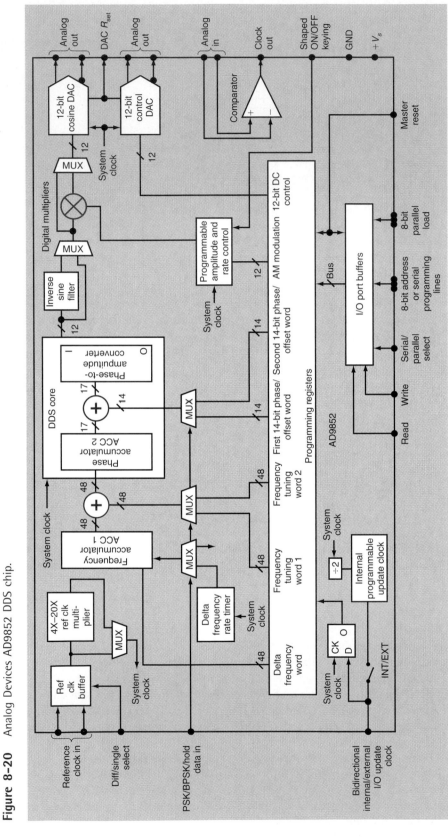

Figure 8-20 Analog Devices AD9852 DDS chip.

A downside of the DDS synthesizer is that it is difficult to make one with very high output frequencies. The output frequency is limited by the speed of the available D/A converter and digital logic circuitry. With today's components, it is possible to produce a DDS synthesizer with an output frequency as high as 200 MHz. Further developments in IC technology will increase that in the future. For applications requiring higher frequencies, the phase-locked loop (PLL) is still the best alternative.

DDS synthesizers are available from several IC companies. The entire DDS circuitry is contained on a chip. The clock circuit is usually contained within the chip, and its frequency is set by an external crystal. Parallel binary input lines are provided to set the constant value required to change the frequency. A 12-bit D/A converter is typical. An example of such a chip is the Analog Devices AD9852, shown in Fig. 8-20. The on-chip clock is derived from a PLL used as a frequency multiplier that can be set to multiply by any integer value between 4 and 20. With a maximum of 20, a clock frequency of 300 MHz is generated. To achieve this frequency, the external reference clock input must be $300/20 = 15$ MHz. With a 300-MHz clock, the synthesizer can generate sine waves up to 150 MHz.

The outputs come from two 12-bit DACs that produce both the sine and the cosine waves simultaneously. A 48-bit frequency word is used to step the frequency in 2^{48} increments. A 17-bit phase accumulator lets you shift the phase in 2^{17} increments.

This chip also has circuitry that lets you modulate the sine wave outputs. AM, FM, FSK, PM, and BPSK can be implemented.

8-3 Power Amplifiers

The three basic types of power amplifiers used in transmitters are linear, class C, and switching.

Linear amplifiers provide an output signal that is an identical, enlarged replica of the input. Their output is directly proportional to their input, and they therefore faithfully reproduce an input, but at a higher power level. Most audio amplifiers are linear. Linear RF amplifiers are used to increase the power level of variable-amplitude RF signals such as low-level AM or SSB signals. Linear amplifiers are class A, AB, or B. The class of an amplifier indicates how it is biased.

Class A amplifiers are biased so that they conduct continuously. The bias is set so that the input varies the collector (or drain) current over a linear region of the transistor's characteristics. Thus its output is an amplified linear reproduction of the input. Usually we say that the class A amplifier conducts for 360° of an input sine wave.

Class B amplifiers are biased at cutoff so that no collector current flows with zero input. The transistor conducts on only one-half, or 180°, of the sine wave input. This means that only one-half of the sine wave is amplified. Normally, two class B amplifiers are connected in a push-pull arrangement so that both the positive and negative alternations of the input are amplified.

Class AB linear amplifiers are biased near cutoff with some continuous collector current flow. They conduct for more than 180° but less than 360° of the input. They too are used primarily in push-pull amplifiers and provide better linearity than class B amplifiers, but with less efficiency.

Class A amplifiers are linear but not very efficient. For that reason, they make poor power amplifiers. As a result, they are used primarily as small-signal voltage amplifiers or for low-power amplifications. The buffer amplifiers described previously are class A amplifiers.

Class B amplifiers are more efficient than class A amplifiers, because current flows for only a portion of the input signal, and they make good power amplifiers. However, they distort an input signal because they conduct for only one-half of the cycle. Therefore, special techniques are often used to eliminate or compensate for the distortion. For example, operating class B amplifiers in a push-pull configuration minimizes the distortion.

Class C amplifiers conduct for even less than one-half of the sine wave input cycle, making them very efficient. The resulting highly distorted current pulse is used to ring

Linear amplifier

Class A amplifier

Class B amplifier

Class AB amplifier

GOOD TO KNOW

Most audio amplifiers are linear and are therefore class A or AB.

Class C amplifier

Figure 8-21   A linear (class A) RF buffer amplifier.

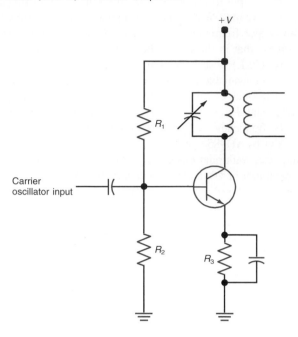

a tuned circuit to create a continuous sine wave output. Class C amplifiers cannot be used to amplify varying-amplitude signals. They will clip off or otherwise distort an AM or SSB signal. However, FM signals do not vary in amplitude and can therefore be amplified with more efficient nonlinear class C amplifiers. This type of amplifier also makes a good frequency multiplier as harmonics are generated in the amplification process.

Switching amplifiers act like on/off or digital switches. They effectively generate a square wave output. Such a distorted output is undesirable; however, by using high-Q tuned circuits in the output, the harmonics generated as part of the switching process can be easily filtered out. The on/off switching action is highly efficient because current flows during only one-half of the input cycle, and when it does, the voltage drop across the transistor is very low, resulting in low power dissipation. Switching amplifiers are designated class D, E, F, and S.

Linear Amplifiers

Linear amplifiers are used primarily in AM and SSB transmitters, and both low- and high-power versions are used. Some examples follow.

Class A Buffers. A simple class A buffer amplifier is shown in Fig. 8-21. This type of amplifier is used between the carrier oscillator and the final power amplifier to isolate the oscillator from the power amplifier load, which can change the oscillator frequency. It also provides a modest power increase to provide the driving power required by the final amplifier. Such circuits usually provide milliwatts of power and rarely more than 1 W. The carrier oscillator signal is capacitively coupled to the input. The bias is derived from R_1, R_2, and R_3. The emitter resistor R_3 is bypassed to provide maximum gain. The collector is tuned with a resonant LC circuit at the operating frequency. An inductively coupled secondary loop transfers power to the next stage.

High–Power Linear Amplifiers. A high-power class A linear amplifier is shown in Fig. 8-22. A power MOSFET may also be used in this circuit with a few modifications. Base bias is supplied by a constant-current circuit that is temperature-compensated. The RF input from a 50-Ω source is connected to the base via an impedance-matching circuit made up of C_1, C_2, and L_1. The output is matched to a 50-Ω load by the impedance-matching

Figure 8-22   A high-power class A linear RF amplifier.

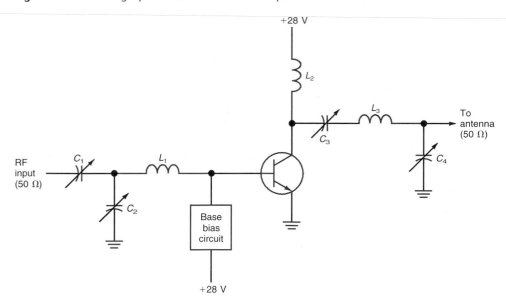

network made up of L_2, L_3, C_3, and C_4. When connected to a proper heat sink, the transistor can generate up to 100 W of power up to about 200 MHz. The amplifier is designed for a specific frequency that is set by the input and output tuned circuits. Class A amplifiers have a maximum efficiency of 50 percent. Thus only 50 percent of the dc power is converted to RF, with the remaining 50 percent being dissipated in the transistor. For 100-W RF output, the transistor dissipates 100 W. Efficiencies of less than 50 percent are typical.

Commonly available RF power transistors have an upper power limit of several hundred watts. To produce more power, two or more devices can be connected in parallel, in a push-pull configuration, or in some combination. Power levels of up to several thousand watts are possible with these arrangements.

Class B Push–Pull Amplifiers. A class B linear power amplifier using push-pull is shown in Fig. 8-23. The RF driving signal is applied to Q_1 and Q_2 through input transformer T_1. It provides impedance-matching and base drive signals to Q_1 and Q_2 that are 180° out of phase. An output transformer T_2 couples the power to the antenna or load. Bias is provided by R_1 and D_1.

For class B operation, Q_1 and Q_2 must be biased right at the cutoff point. The emitter-base junction of a transistor will not conduct until about 0.6 to 0.8 V of forward bias is applied because of the built-in potential barrier. This effect causes the transistors to be naturally biased beyond cutoff, not right at it. A forward-biased silicon diode D_1 has about 0.7 V across it, and this is used to put Q_1 and Q_2 right on the conduction threshold.

On the positive half-cycle of the RF input, the base of Q_1 is positive and the base of Q_2 is negative. The Q_2 is cut off, but Q_1 conducts, linearly amplifying the positive half-cycle. Collector current flows in the upper half of T_2, which induces an output voltage in the secondary. On the negative half-cycle of the RF input, the base of Q_1 is negative, so it is cut off. The base of Q_2 is positive, so Q_2 amplifies the negative half-cycle. Current flows in Q_2 and the lower half of T_2, completing a full cycle. The power is split between the two transistors.

The circuit in Fig. 8-23 is an untuned broadband circuit that can amplify signals over a broad frequency range, typically from 2 to 30 MHz. A low-power AM or SSB signal is generated at the desired frequency and then applied to this power amplifier before being sent to the antenna. With push-pull circuits, power levels of up to 1 kW are possible.

Figure 8-24 shows another push-pull RF power amplifier. It uses two power MOSFETs, can produce an output up to 1 kW over the 10- to 90-MHz range, and has

Push-pull amplifier

Figure 8-23 A push-pull class B power amplifier.

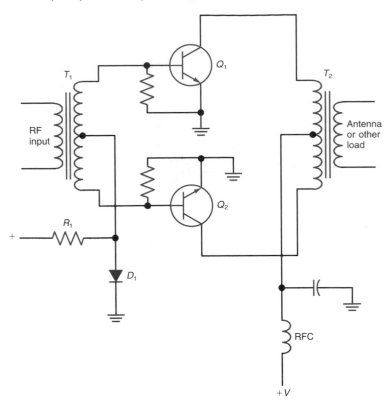

Figure 8-24 A 1-kW push-pull RF power amplifier using MOSFETs.

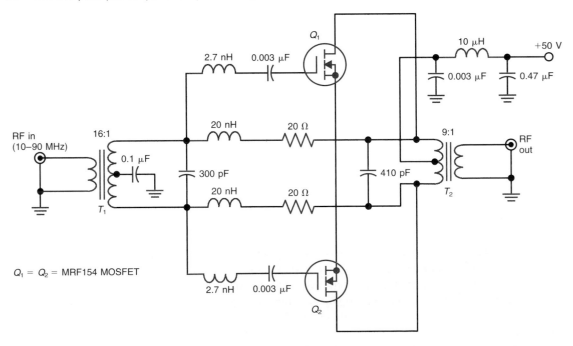

a 12-dB power gain. The RF input driving power must be 63 W to produce the full 1-kW output. Toroidal transformers T_1 and T_2 are used at the input and output for impedance matching. They provide broadband operation over the 10- to 90-MHz range without tuning. The 20-nH chokes and 20-Ω resistors form neutralization circuits that provide out-of-phase feedback from output to input to prevent self-oscillation.

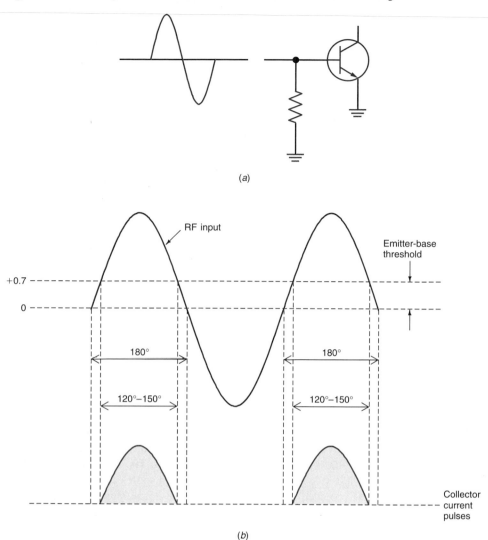

(a)

(b)

Class C Amplifiers

The key circuit in most AM and FM transmitters is the class C amplifier. These amplifiers are used for power amplification in the form of drivers, frequency multipliers, and final amplifiers. Class C amplifiers are biased, so they conduct for less than 180° of the input. A class C amplifier typically has a conduction angle of 90 to 150°. Current flows through it in short pulses, and a resonant tuned circuit is used for complete signal amplification.

Biasing Methods. Figure 8-25(a) shows one way of biasing a class C amplifier. The base of the transistor is simply connected to ground through a resistor. No external bias voltage is applied. An RF signal to be amplified is applied directly to the base. The transistor conducts on the positive half-cycles of the input wave and is cut off on the negative half-cycles. Although this sounds like a class B configuration, that is not the case. Recall that the emitter-base junction of a bipolar transistor has a forward voltage threshold of approximately 0.7 V. In other words, the emitter-base junction does not really conduct until the base is more positive than the emitter by 0.7 V. Because of this, the transistor has an inherent built-in reverse bias. When the input signal is applied, the collector current does not flow until the base is positive by 0.7 V. This is illustrated in Fig. 8-25(b). The result is that collector current flows through the transistor in positive pulses for less than the full 180° of the positive ac alternation.

Figure 8-26 Methods of biasing a class C amplifier. (*a*) Signal bias. (*b*) External bias. (*c*) Self-bias.

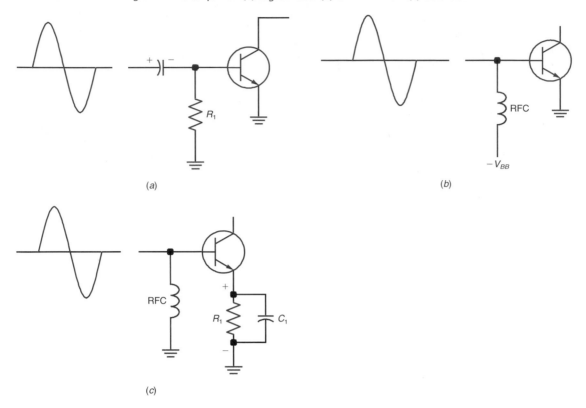

(*a*)

(*b*)

(*c*)

Signal bias

Self-bias method

In many low-power driver and multiplier stages, no special biasing provisions other than the inherent emitter-base junction voltage are required. The resistor between base and ground simply provides a load for the driving circuit. In some cases, a narrower conduction angle than that provided by the circuit in Fig. 8-25(*a*) must be used. In such cases, some form of bias must be applied. A simple way of supplying bias is with the *RC* network shown in Fig. 8-26(*a*). Here the signal to be amplified is applied through capacitor C_1. When the emitter-base junction conducts on the positive half-cycle, C_1 charges to the peak of the applied voltage less the forward drop across the emitter-base junction. On the negative half-cycle of the input, the emitter-base junction is reverse-biased, so the transistor does not conduct. During this time, however, capacitor C_1 discharges through R_1, producing a negative voltage across R_1 which serves as a reverse bias on the transistor. By properly adjusting the time constant of R_1 and C_1, an average dc reverse-bias voltage can be established. The applied voltage causes the transistor to conduct, but only on the peaks. The higher the average dc bias voltage, the narrower the conduction angle and the shorter the duration of the collector current pulses. This method is referred to as *signal bias*.

Of course, negative bias can also be supplied to a class C amplifier from a fixed dc supply voltage, as shown in Fig. 8-26(*b*). After the desired conduction angle is established, the value of the reverse voltage can be determined and applied to the base through the RFC. The incoming signal is then coupled to the base, causing the transistor to conduct on only the peaks of the positive input alternations. This is called *external bias* and requires a separate negative dc supply.

Another biasing method is shown in Fig. 8-26(*c*). As in the circuit shown in Fig. 8-26(*a*), the bias is derived from the signal. This arrangement is known as the *self-bias method*. When current flows in the transistor, a voltage is developed across R_1. Capacitor C_1 is charged and holds the voltage constant. This makes the emitter more positive than the base, which has the same effect as a negative voltage on the base. A strong input

Figure 8-27 Class C amplifier operation.

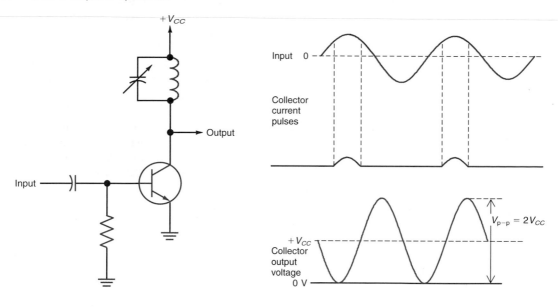

signal is required for proper operation. These circuits also work with an enhancement mode MOSFET.

Tuned Output Circuits. All class C amplifiers have some form of tuned circuit connected in the collector, as shown in Fig. 8-27. The primary purpose of this tuned circuit is to form the complete ac sine wave output. A parallel tuned circuit rings, or oscillates, at its resonant frequency whenever it receives a dc pulse. The pulse charges the capacitor, which, in turn, discharges into the inductor. The magnetic field in the inductor increases and then collapses, inducing a voltage which then recharges the capacitor in the opposite direction. This exchange of energy between the inductor and the capacitor, called the *flywheel effect,* produces a damped sine wave at the resonant frequency. If the resonant circuit receives a pulse of current every half-cycle, the voltage across the tuned circuit is a constant-amplitude sine wave at the resonant frequency. Even though the current flows through the transistor in short pulses, the class C amplifier output is a continuous sine wave.

Another way to look at the operation of a class C amplifier is to view the transistor as supplying a highly distorted pulse of power to the tuned circuit. According to Fourier theory, this distorted signal contains a fundamental sine wave plus both odd and even harmonics. The tuned circuit acts as a bandpass filter to select the fundamental sine wave contained in the distorted composite signal.

The tuned circuit in the collector is also used to filter out unwanted harmonics. The short pulses in a class C amplifier are made up of second, third, fourth, fifth, etc., harmonics. In a high-power transmitter, signals are radiated at these harmonic frequencies as well as at the fundamental resonant frequency. Such harmonic radiation can cause out-of-band interference, and the tuned circuit acts as a selective filter to eliminate these higher-order harmonics. If the Q of the tuned circuit is made high enough, the harmonics will be adequately suppressed.

The Q of the tuned circuit in the class C amplifier should be selected so that it provides adequate attenuation of the harmonics but also has sufficient bandwidth to pass the sidebands produced by the modulation process. Remember that the bandwidth and Q of a tuned circuit are related by the expression

$$\text{BW} = \frac{f_r}{Q} \qquad Q = \frac{f_r}{\text{BW}}$$

Tuned output circuit

If the Q of the tuned circuit is too high, then the bandwidth will be very narrow and some of the higher-frequency sidebands will be eliminated. This causes a form of frequency distortion called *sideband clipping* and may make some signals unintelligible or will at least limit the fidelity of reproduction.

One of the main reasons why class C amplifiers are preferred for RF power amplification over class A and class B amplifiers is their high efficiency. Remember, efficiency is the ratio of the output power to the input power. If all the generated power, the input power, is converted to output power, the efficiency is 100 percent. This doesn't happen in the real world because of losses. But in a class C amplifier more of the total power generated is applied to the load. Because the current flows for less than 180° of the ac input cycle, the average current in the transistor is fairly low; i.e., the power dissipated by the device is low. A class C amplifier functions almost as a transistor switch that is off for over 180° of the input cycle. The switch conducts for approximately 90 to 150° of the input cycle. During the time that it conducts, its emitter-to-collector resistance is low. Even though the peak current may be high, the total power dissipation is much less than that in class A and class B circuits. For this reason, more of the dc power is converted to RF energy and passed on to the load, usually an antenna. The efficiency of most class C amplifiers is in the 60 to 85 percent range.

The input power in a class C amplifier is the average power consumed by the circuit, which is simply the product of the supply voltage and the average collector current, or

$$P_{in} = V_{CC}(I_C)$$

For example, if the supply voltage is 13.5 V and the average dc collector current is 0.7 A, the input power is $P_{in} = 13.5(0.7) = 9.45$ W.

The output power is the power actually transmitted to the load. The amount of power depends upon the efficiency of the amplifier. The output power can be computed with the familiar power expression

$$P_{out} = \frac{V^2}{R_L}$$

where V is the RF output voltage at the collector of the amplifier and R_L is the load impedance. When a class C amplifier is set up and operating properly, the peak-to-peak RF output voltage is 2 times the supply voltage, or $2V_{CC}$ (see Fig. 8-27).

Frequency multiplier

Frequency Multipliers. Any class C amplifier is capable of performing frequency multiplication if the tuned circuit in the collector resonates at some integer multiple of the input frequency. For example, a frequency doubler can be constructed by simply connecting a parallel tuned circuit in the collector of a class C amplifier that resonates at twice the input frequency. When the collector current pulse occurs, it excites or rings the tuned circuit at twice the input frequency. A current pulse flows for every other cycle of the input. A tripler circuit is constructed in exactly the same way, except that the tuned circuit resonates at 3 times the input frequency, receiving one input pulse for every 3 cycles of oscillation it produces (see Fig. 8-28).

Multipliers can be constructed to increase the input frequency by any integer factor up to approximately 10. As the multiplication factor gets higher, the power output of the multiplier decreases. For most practical applications, the best result is obtained with multipliers of 2 and 3.

Another way to look at the operation of a class C frequency multiplier is to remember that the nonsinusoidal current pulse is rich in harmonics. Each time the pulse occurs, the second, third, fourth, fifth, and higher harmonics are generated. The purpose of the tuned circuit in the collector is to act as a filter to select the desired harmonic.

In many applications, a multiplication factor greater than that achievable with a single multiplier stage is required. In such cases, two or more multipliers are cascaded. Figure 8-29 shows two multiplier examples. In the first case, multipliers of 2 and 3 are cascaded to produce an overall multiplication of 6. In the second, three multipliers

GOOD TO KNOW

Although multipliers can be constructed to increase the input frequency by any integer up to approximately 10, the best results are obtained with multipliers of 2 and 3.

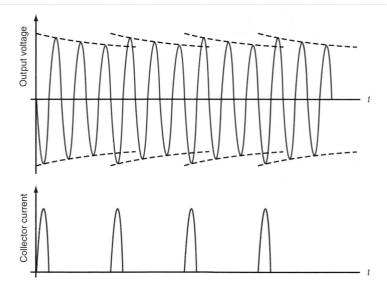

Figure 8-29 Frequency multiplication with class C amplifiers.

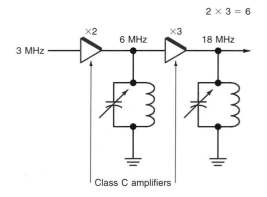

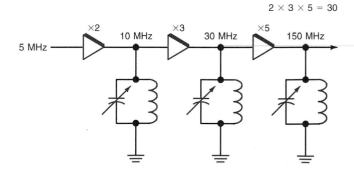

provide an overall multiplication of 30. The total multiplication factor is the product of
the multiplication factors of the individual stages.

Neutralization

One problem that all RF amplifiers have, both linear and class C amplifiers, is *self-
oscillation*. When some of the output voltage finds its way back to the input of the

Self-oscillation

amplifier with the correct amplitude and phase, the amplifier oscillates, sometimes at its tuned frequency and in some cases at a much higher frequency. When the circuit oscillates at a higher frequency unrelated to the tuned frequency, the oscillation is referred to as *parasitic oscillation*. In both cases, the oscillation is undesirable, and it either prevents amplification from taking place or, in the case of parasitic oscillation, reduces the power amplification and introduces distortion of the signal.

Self-oscillation at the tuned frequency in an amplifier is the result of positive feedback that occurs because of the interelement capacitance of the amplifying device, whether it is a bipolar transistor, FET, or vacuum tube. In a bipolar transistor this is the collector-to-base capacitance C_{bc} as shown in Fig. 8-30(a). Transistor amplifiers are biased so that the emitter-base junction is forward-biased and the base-collector junction is reversed-biased. As discussed previously, a reversed-biased diode or transistor junction acts as a capacitor. This small capacitance permits output from the collector to be fed back to the base. Depending on the frequency of the signal, the value of the capacitance, and the values of stray inductances and capacitances in the circuit, the signal fed back may be in phase with the input signal and high enough in amplitude to cause oscillation.

This interelement capacitance cannot be eliminated; therefore its effect must be compensated for, or neutralized. In the *neutralization* process, another signal, equal in amplitude to the signal fed back through C_{bc} and 180° out of phase with it, is fed back. The result is that the two signals cancel.

Several methods of neutralization are shown in Fig. 8-30. In Fig. 8-30(a), a signal of equal and opposite phase is provided by the inductor L_n. Capacitor C_1 is a high-value capacitor used strictly for dc blocking to prevent collector voltage from being applied to the base. So that its value can be set to make its reactance equal to the reactance of C_{bc} at the oscillation frequency, L_n is made adjustable. As a result, C_{bc} and L_n form a parallel resonant circuit that acts as a very high-value resistor at the resonant frequency. The result is effective cancellation of the positive feedback. The type of neutralization shown in Fig. 8-30(b) uses a tapped collector coil and a neutralization capacitor C_n. The two equal halves of the collector inductance, the junction capacitance C_{bc}, and C_n, form a bridge circuit [Fig. 8-30(c)]. When C_n is adjusted to be equal to C_{bc}, the bridge is balanced and no feedback signal V_f occurs. A variation of this is shown in Fig. 8-30(d), where a center-tapped base input inductor is used.

Parasitic oscillations are usually eliminated by connecting a low value of resistor in the collector or base lead. A value of 10 to 22 Ω is typical. Parasitic oscillations can also be eliminated by putting one or more ferrite beads over the collector or base leads. Another practice is to wind a small inductor on a resistor, creating a parallel R_L circuit which is then placed in the collector or base leads.

Switching Power Amplifiers

As stated previously, the primary problem with RF power amplifiers is their inefficiency and high power dissipation. To generate RF power to transfer to the antenna, the amplifier must dissipate a considerable amount of power itself. For example, a class A power amplifier using a transistor conducts continuously. It is a linear amplifier whose conduction varies as the signal changes. Because of the continuous conduction, the class A amplifier generates a considerable amount of power that is not transferred to the load. No more than 50 percent of the total power consumed by the amplifier can be transferred to the load. Because of the high-power dissipation, the output power of a class A amplifier is generally limited. For that reason, class A amplifiers are normally used only in low-power transmitter stages.

To produce greater power output, class B amplifiers are used. Each transistor conducts for 180° of the carrier signal. Two transistors are used in a push-pull arrangement to form a complete carrier sine wave. Since each transistor is conducting for only 180° of any carrier cycle, the amount of power it dissipates is considerably less,

Figure 8-30 Neutralization circuits. (*a*) Canceling the effect of C_{bc} with an equivalent inductance L_n. (*b*) Neutralization using a tapped collector coil and a neutralization capacitor C_n. (*c*) Equivalent circuit of part *b*. (*d*) Neutralization with a tapped input inductor.

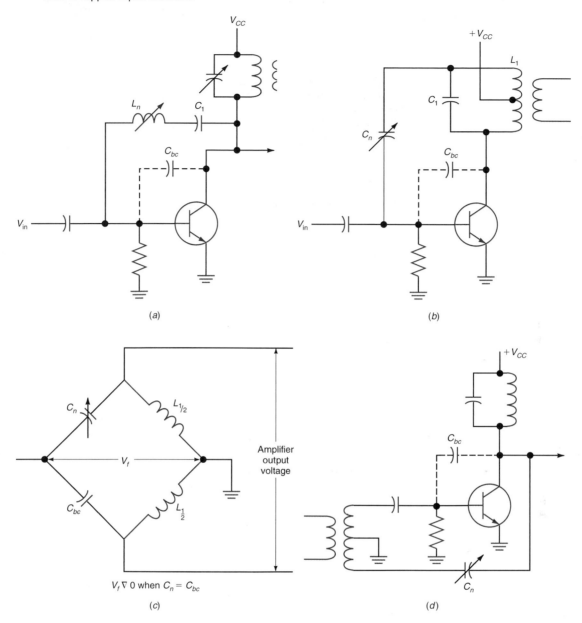

(a)

(b)

$V_f \nabla 0$ when $C_n = C_{bc}$

(c)

(d)

and efficiencies of 70 to 75 percent are possible. Class C power amplifiers are even more efficient, since they conduct for less than 180° of the carrier signal, relying on the tuned circuit in the plate or collector to supply power to the load when they are not conducting. With current flowing for less than 180° of the cycle, class C amplifiers dissipate less power and can, therefore, transfer more power to the load. Efficiencies as high as about 85 percent can be achieved, and class C amplifiers are therefore the most widely used type in power amplifiers when the type of modulation permits.

Another way to achieve high efficiencies in power amplifiers is to use a switching amplifier. A *switching amplifier* is a transistor that is used as a switch and is either conducting or nonconducting. Both bipolar transistors and enhancement mode MOSFETs are widely used in switching-amplifier applications. A bipolar transistor as a switch is

Switching amplifier

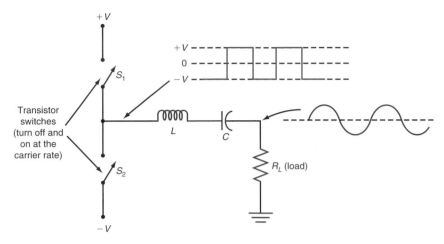

Figure 8-31 Basic configuration of a class D amplifier.

either cut off or saturated. When it is cut off, no power is dissipated. When it is saturated, current flow is maximum, but the emitter-collector voltage is extremely low, usually less than 1 V. As a result, power dissipation is extremely low.

When enhancement mode MOSFETs are used, the transistor is either cut off or turned on. In the cutoff state, no current flows, so no power is dissipated. When the transistor is conducting, its on resistance between source and drain is usually very low—again, no more than several ohms and typically far less than 1 Ω. As a result, power dissipation is extremely low even with high currents.

The use of switching power amplifiers permits efficiencies of over 90 percent. The current variations in a switching power amplifier are square waves and thus harmonics are generated. However, these are relatively easy to filter out by the use of tuned circuits and filters between the power amplifier and the antenna.

The three basic types of switching power amplifiers, class D, class E, and class S, were originally developed for high-power audio applications. But with the availability of high-power, high-frequency switching transistors, they are now widely used in radio transmitter design.

Class D Amplifiers. A class D amplifier uses a pair of transistors to produce a square wave current in a tuned circuit. Figure 8-31 shows the basic configuration of a *class D amplifier.* Two switches are used to apply both positive and negative dc voltages to a load through the tuned circuit. When switch S_1 is closed, S_2 is open; when S_2 is closed, S_1 is open. When S_1 is closed, a positive dc voltage is applied to the load. When S_2 is closed, a negative dc voltage is applied to the load. Thus the tuned circuit and load receive an ac square wave at the input.

The series resonant circuit has a very high Q. It is resonant at the carrier frequency. Since the input waveform is a square wave, it consists of a fundamental sine wave and odd harmonics. Because of the high Q of the tuned circuit, the odd harmonics are filtered out, leaving a fundamental sine wave across the load. With ideal switches, that is, no leakage current in the off state and no on resistance when conducting, the theoretical efficiency is 100 percent.

Figure 8-32 shows a class D amplifier implemented with enhancement mode MOSFETs. The carrier is applied to the MOSFET gates 180° out of phase by use of a transformer with a center-tapped secondary. When the input to the gate of Q_1 is positive, the input to the gate of Q_2 is negative. Thus Q_1 conducts and Q_2 is cut off. On the next half-cycle of the input, the gate to Q_2 goes positive and the gate of Q_1 goes negative. The Q_2 conducts, applying a negative pulse to the tuned circuit. Recall that enhancement mode MOSFETs are normally nonconducting until a gate voltage higher than a specific threshold value is applied,

Class D amplifier

Figure 8-32 A class D amplifier made with enhancement mode MOSFETs.

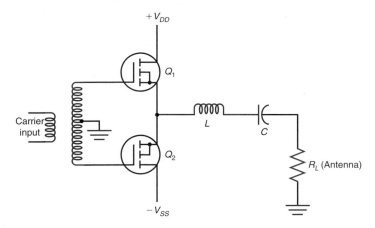

at which time the MOSFET conducts. The on resistance is very low. In practice, efficiencies of up to 90 percent can be achieved by using a circuit like that in Fig. 8-32.

Class E and F Amplifiers.

In *class E amplifiers,* only a single transistor is used. Both bipolar and MOSFETs can be used, although the MOSFET is preferred because of its low drive requirements. Figure 8-33 shows a typical class E RF amplifier. The carrier, which may initially be a sine wave, is applied to a shaping circuit which effectively converts it to a square wave. The carrier is usually frequency-modulated. The square wave carrier signal is then applied to the base of the class E bipolar power amplifier. The Q_1 switches off and on at the carrier rate. The signal at the collector is a square wave which is applied to a low-pass filter and tuned impedance-matching circuit made up of C_1, C_2, and L_1. The odd harmonics are filtered out, leaving a fundamental sine wave which is applied to the antenna. A high level of efficiency is achieved with this arrangement.

A *class F amplifier* is a variation of the class E amplifier. It contains an additional resonant network in the collector or drain circuit. This circuit, a lumped LC or even a tuned transmission line at microwave frequencies, is resonant at the second or third harmonic of the operating frequency. The result is a waveform at the collector (drain) that more closely resembles a square wave. The steeper waveform produces faster transistor switching and better efficiency.

Class S Amplifiers.

Class S amplifiers, which use switching techniques but with a scheme of pulse-width modulation, are found primarily in audio applications but have also been used in low- and medium-frequency RF amplifiers such as those used in AM broadcast transmitters. The low-level audio signal to be amplified is applied to a circuit

Class E amplifier

Class F amplifier

Class S amplifier

Figure 8-33 A class E RF amplifier.

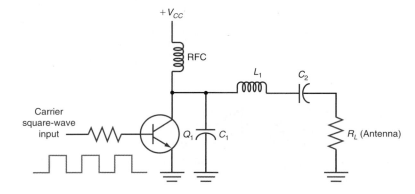

called a *pulse-width modulator.* A carrier signal at a frequency 5 to 10 times the highest audio frequency to be amplified is also applied to the pulse-width modulator. At the output of the modulator is a series of constant-amplitude pulses whose pulse width or duration varies with the audio signal amplitude. These signals are then applied to a switching amplifier of the class D type. High power and efficiency are achieved because of the switching action. A low-pass filter is connected to the output of the switching amplifier to average and smooth the pulses back into the original audio signal waveform. A capacitor or low-pass filter across the speaker is usually sufficient. These amplifiers are usually referred to as class D amplifiers in audio applications. They are widely used in battery-powered portable units where battery life and efficiency are paramount.

Linear Broadband Power Amplifiers

The power amplifiers described so far in this chapter are narrowband amplifiers. They provide high power output over a relatively small range of frequencies. The bandwidth of the signal to be amplified is set by the modulation method and the frequencies of the modulating signals. In so many applications, the total bandwidth is only a small percentage of the carrier frequency, making conventional *LC* resonant circuits practical. Some of the untuned push-pull power amplifiers described earlier (Figs. 8-23 and 8-24) do have a broader bandwidth up to several megahertz. However, some of the newer wireless systems now require much broader bandwidth. The best example is the code division multiple-access (CDMA) cell phone standard to be discussed in a later chapter. The CDMA system uses a modulation/multiplexing technique called spread spectrum that does, as its name implies, spread the signal out over a very wide frequency spectrum. Signal bandwidths of 1 to 20 MHz are common. Such complex modulation schemes require that the amplification be linear over a wide frequency range to ensure no amplitude, frequency, or phase distortion. Special amplification techniques have been developed to meet this need. Two common methods are discussed here.

Feedforward Amplifier. The concept behind a feedforward amplifier is that the distortion produced by the power amplifier is isolated and then subtracted from the amplified signal, producing a nearly distortion-free output signal. Figure 8-34 shows one common implementation of this idea. The wide-bandwidth signal to be amplified is fed to a power splitter that divides the signal into two equal-amplitude signals. A typical splitter may be a transformerlike device or even a resistive network. It maintains the constant impedance,

Figure 8-34   Feedforward linear power amplifier.

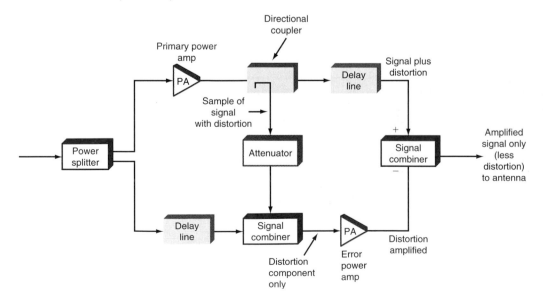

usually 50Ω, but typically also introduces some attenuation. One-half of the signal is then amplified in a linear power amplifier similar to those broadband class AB amplifiers discussed earlier. A directional coupler is used next to tap off a small portion of the amplified signal which contains the original input information as well as harmonics resulting from distortion. A directional coupler is a simple device that picks up a small amount of the signal by inductive coupling. It may be just a short copper line running adjacent to the signal line on a printed-circuit board. At microwave frequencies, a directional coupler may be a more complex device with a coaxial structure. In any case, the sample of the amplified signal is also passed through a resistive attenuator to further reduce the signal level.

The lower output of the signal splitter is sent to a delay line circuit. A delay line is a low-pass filter or a section of transmission line such as coaxial cable that introduces a specific amount of delay to the signal. It may be a few nanoseconds up to several microseconds depending upon the frequency of operation and the type of power amplifier used. The delay is used to match the delay encountered by the upper input signal in the power amplifier. This delayed signal is then fed to a signal combiner along with the attenuated signal sample from the amplifier output. Amplitude and phase controls are usually provided in both signal paths to ensure that they are of the same amplitude and phase. The combiner may be resistive or a transformerlike device. In either case, it effectively subtracts the original signal from the amplified signal, leaving only the harmonic distortion.

The harmonic distortion is now amplified by another power amplifier with a power level equal to that of the upper signal power amplifier. The upper amplifier signal is passed through the directional coupler and into a delay line that compensates for the delay introduced by the lower error signal amplifier. Again, amplitude and phase controls are usually provided to adjust the power levels of the upper and lower signal levels so they are equal. Finally, the error signal is now subtracted from the amplified combined signal in a signal coupler or combiner. This coupler, like the input splitter, is typically a transformer. The resulting output is the original amplified signal less the distortion.

Amplifiers like this are available with power levels of a few watts to over several hundred watts. The system is not perfect as the signal cancellations or subtractions are not precise because of amplitude and phase mismatches. Distortion in the lower power amplifier also contributes to overall output. Yet with close adjustments, these differences can be minimized, thereby greatly improving the linearity of the amplifier over other types. The system is also inefficient because two power amplifiers are required. But the tradeoff is wide bandwidth and very low distortion.

Adaptive Predistortion Amplification. This method of amplification uses a digital signal processing (DSP) technique to predistort the signal in such a way that when amplified the amplifier distortion will offset the predistortion characteristics, leaving a distortion-free output signal. The amplified output signal is continuously monitored and used as feedback to the DSP so that the predistorion calculations can be changed on the fly to provide an inverse predistortion that perfectly matches the amplifier's distortion.

Refer to Figure 8-35 that shows a representative system. The digitized information signal, usually voice, in serial format is fed to a digital correction algorithm in a DSP. Inside the DSP are computing algorithms that feed corrective logic signals to the digital correction algorithm to modify the signal in such a way as to be an inverse match to the distortion produced by the power amplifier.

Once the corrective action has been taken on the baseband digital signal, it is sent to a modulator that produces the signal to be transmitted. The modulation is handled by the DSP chip itself rather than a separate modulator circuit. This modulated signal is then fed to a digital-to-analog converter (DAC) where it produces the desired analog signal to be transmitted.

Next, the DAC output is sent to a mixer along with a sine wave signal from an oscillator or frequency synthesizer. A mixer is similar to a low-level amplitude modulator or analog multiplier. Its output consists of the sum and difference of the DAC and synthesizer signals. In this application, the sum signal is selected by a filter making the mixer an

Figure 8-35   Concept of adaptive predistortion amplification.

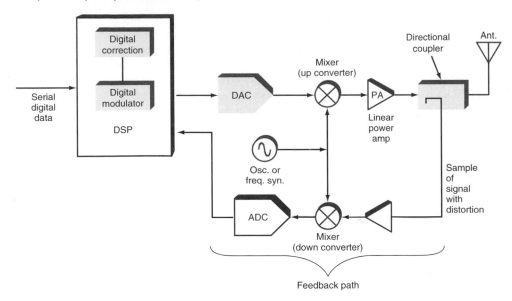

upconverter. The synthesizer frequency is chosen so that the mixer output is at the desired operating frequency. Any modulation present is also contained on the mixer output.

The predistorted signal is then amplified in a highly linear class AB power amplifier (PA) and fed to the output antenna. Note in Fig. 8-35 that the output signal is sampled in a directional coupler, amplified, and sent to another mixer used as a down converter. The synthesizer also provides the second input to this mixer. The difference frequency is selected by a filter, and the result is sent to an analog-to-digital converter (ADC). The digital output of the ADC represents the amplified signal plus any distortion produced by the PA. The DSP uses this digital input to modify its algorithm to properly correct for the actual distortion. The signal is then modified by the digital correction algorithm in such a way that most of the distortion is canceled.

While the adaptive-predistortion method of broadband amplification is complex, it provides nearly distortion-free output. Only a single power amplifier is needed, making it more efficient than the feedforward method. Several semiconductor manufacturers make the predistortion circuits needed to make this work.

8-4 Impedance–Matching Networks

Matching networks that connect one stage to another are very important parts of any transmitter. In a typical transmitter, the oscillator generates the basic carrier signal, which is then amplified, usually by multiple stages, before it reaches the antenna. Since the idea is to increase the power of the signal, the interstage coupling circuits must permit an efficient transfer of power from one stage to the next. Finally, some means must be provided to connect the final amplifier stage to the antenna, again for the purpose of transferring the maximum possible amount of power. The circuits used to connect one stage to another are known as *impedance-matching networks*. In most cases, they are *LC* circuits, transformers, or some combination. The basic function of a matching network is to provide for an optimum transfer of power through impedance-matching techniques. Matching networks also provide filtering and selectivity. Transmitters are designed to operate on a single frequency or selectable narrow ranges of frequencies. The various amplifier stages in the transmitter must confine the RF generated to these frequencies. In class C, D, and E amplifiers, a considerable number of high-amplitude harmonics are generated. These must be eliminated to prevent spurious

Impedance-matching network

Figure 8-36 Impedance matching in RF circuits.

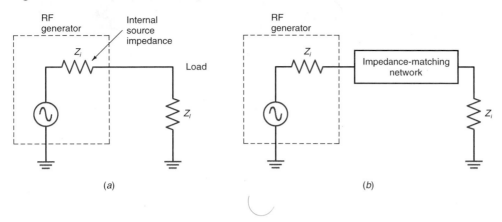

(a)   (b)

radiation from the transmitter. The impedance-matching networks used for interstage coupling accomplish this.

The basic problem of coupling is illustrated in Fig. 8-36(a). The driving stage appears as a signal source with an internal impedance of Z_i. The stage being driven represents a load to the generator with its internal resistance of Z_l. Ideally, Z_i and Z_l are resistive. Recall that maximum power transfer in dc circuits takes place when Z_i equals Z_l. This basic relationship is essentially true also in RF circuits, but it is a much more complex relationship. In RF circuits, Z_i and Z_l are seldom purely resistive and, in fact, usually include a reactive component of some type. Further, it is not always necessary to transfer maximum power from one stage to the next. The goal is to transfer a sufficient amount of power to the next stage so that it can provide the maximum output of which it is capable.

In most cases, the two impedances involved are considerably different from each other, and therefore a very inefficient transfer of power takes place. To overcome this problem, an impedance-matching network is introduced between the two, as illustrated in Fig. 8-36(b). There are three basic types of *LC* impedance-matching networks: the L network, the T network, and the π network.

Networks

L networks consist of an inductor and a capacitor connected in various L-shaped configurations, as shown in Fig. 8-37. The circuits in Fig. 8-37(a) and (b) are low-pass filters; those in Fig. 8-37(c) and (d) are high-pass filters. Typically, low-pass networks are preferred so that harmonic frequencies are filtered out.

L network

The L-matching network is designed so that the load impedance is matched to the source impedance. For example, the network in Fig. 8-37(a) causes the load resistance to appear larger than it actually is. The load resistance Z_L appears in series with the inductor of the L network. The inductor and the capacitor are chosen to resonate at the transmitter frequency. When the circuit is at resonance, X_L equals X_C. To the generator impedance Z_i, the complete circuit appears as a parallel resonant circuit. At resonance, the impedance represented by the circuit is very high. The actual value of the impedance depends upon the L and C values and the Q of the circuit. The higher the Q, the higher the impedance. The Q in this circuit is basically determined by the value of the load impedance. By proper selection of the circuit values, the load impedance can be made to appear as any desired value to the source impedance as long as Z_i is greater than Z_L.

By using the L network shown in Fig. 8-37(b), the impedance can be stepped down, or made to appear much smaller than it actually is. With this arrangement, the capacitor is connected in parallel to the load impedance. The parallel combination of C and Z_L has an equivalent series RC combination. Both C and Z_L appear as equivalent series values C_{eq} and Z_{eq}. The result is that the overall network appears as a series resonant circuit, with C_{eq} and L resonant. Recall that a series resonant circuit has a very low impedance

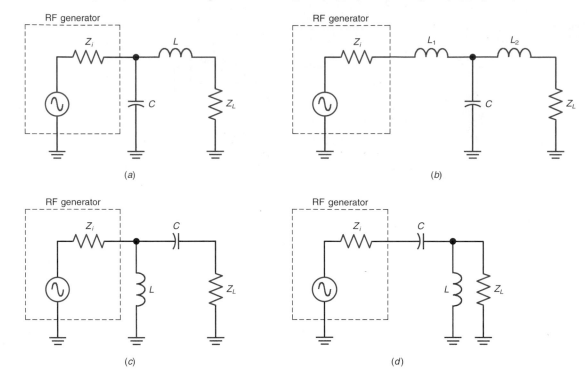

at resonance. The impedance is, in fact, the equivalent load impedance Z_{eq}, which is resistive.

The design equations for L networks are given in Fig. 8-38. Assuming that the internal source and load impedances are resistive, $Z_i = R_i$ and $Z_L = R_L$. The network in Fig. 8-38(a) assumes $R_L < R_i$, and the network in Fig. 8-38(b) assumes $R_i < R_L$.

Suppose we wish to match a 6-Ω transistor amplifier impedance to a 50-Ω antenna load at 155 MHz. In this case, $R_i < R_L$, so we use the formulas in Fig. 8-37(b).

$$X_L = \sqrt{R_i R_L - (R_i)^2} = \sqrt{6(50) - (6)^2} = \sqrt{300 - 36} = \sqrt{264} = 16.25\ \Omega$$

$$Q = \sqrt{\frac{R_L}{R_i} - 1} = \sqrt{\frac{50}{6} - 1} = 2.7$$

$$X_C = \frac{R_L R_i}{X_L} = \frac{50(6)}{16.25} = 18.46\ \Omega$$

To find the values of L and C at 155 MHz, we rearrange the basic reactance formulas as follows:

$$X_L = 2\pi f L$$

$$L = \frac{X_L}{2\pi f} = \frac{16.25}{6.28 \times 155 \times 10^6} = 16.7\ \text{nH}$$

$$X_C = \frac{1}{2\pi f C}$$

$$C = \frac{1}{2\pi f X_C} = \frac{1}{6.28 \times 155 \times 10^6 \times 18.46} = 55.65\ \text{pF}$$

In most cases, internal and stray reactances make the internal impedance and load impedances complex, rather than purely resistive. Figure 8-39 shows an example using the figures given above. Here the internal resistance is 6 Ω, but it includes an internal inductance L_i of 8 nH. There is a stray capacitance C_L of 8.65 pF across the load. The way to deal with these reactances is simply to combine them with the L network values.

Figure 8-38   The L network design equations.

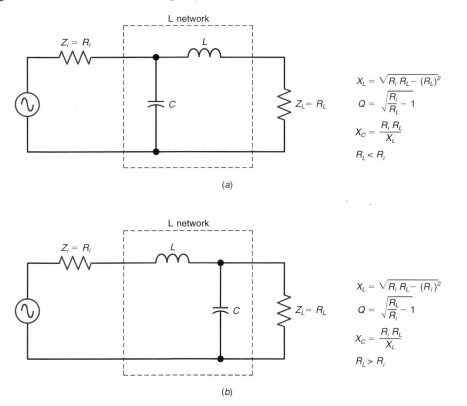

(a)

(b)

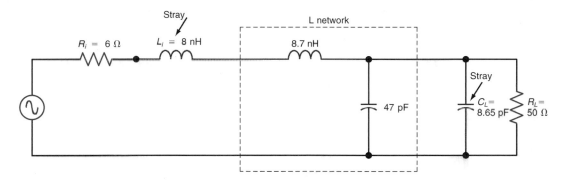

In the example above, the calculation calls for an inductance of 16.7 nH. Since the stray inductance is in series with the L network inductance in Fig. 8-39, the values will add. As a result, the L network inductance must be less than the computed value by an amount equal to the stray inductance of 8 nH, or $L = 16.7 - 8 = 8.7$ nH. If the L network inductance is made to be 8.7 nH, the total circuit inductance will be correct when it adds to the stray inductance.

A similar thing occurs with capacitance. The circuit calculations above call for a total of 55.65 pF. The L network capacitance and the stray capacitance add, for they are in parallel. Therefore, the L network capacitance can be less than the calculated value by the amount of the stray capacitance, or $C = 55.65 - 8.65 = 47$ pF. Making the L network capacitance 47 pF gives the total correct capacitance when it adds to the stray capacitance.

T and π Networks

When one is designing L networks, there is very little control over the Q of the circuit, which is determined by the values of the internal and load impedances and may not always be what is needed to achieve the desired selectivity. To overcome this problem, matching networks using three reactive elements can be used. The three most widely used impedance-matching networks containing three reactive components are illustrated in Fig. 8-40. The network in Fig. 8-40(a) is known as a π *network* because its configuration resembles the Greek letter π. The circuit in Fig. 8-40(b) is known as a *T network* because the circuit elements resemble the letter T. The circuit in Fig. 8-40(c) is also a T network, but it uses two capacitors. Note that all are low-pass filters that provide

Figure 8-40 Three-element matching networks. (a) π network. (b) T network. (c) Two-capacitor T network.

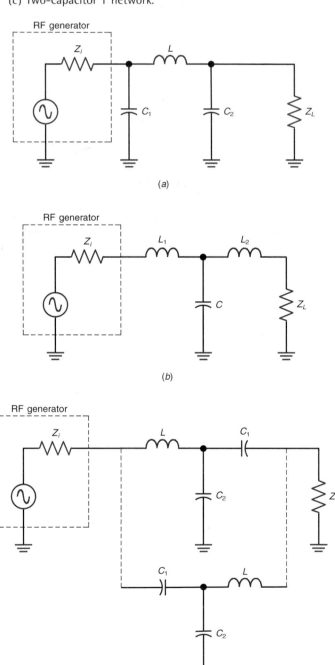

(a)

(b)

(c)

Figure 8-41 Design equations for an *LCC* T network.

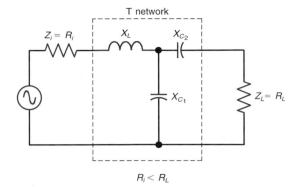

T network

$Z_i = R_i$ X_L X_{C_2}

X_{C_1} $Z_L = R_L$

$R_i < R_L$

Design Procedure:
1. Select a desired circuit Q
2. Calculate $X_L = QR_i$
3. Calculate X_{C_1}:

$$X_{C_2} = R_L \sqrt{\frac{R_i (Q^2 + 1)}{R_L} - 1}$$

4. Calculate X_{C_2}:

$$X_{C_1} = \frac{R_i (Q^2 + 1)}{Q} \times \frac{1}{1 - \dfrac{X_{C_2}}{QR_L}}$$

5. Compute final L and C values:

$$L = \frac{X_L}{2\pi f}$$

$$C = \frac{1}{2\pi f X_C}$$

maximum harmonic attenuation. The π and T networks can be designed to either step up or step down the impedance as required by the circuit. The capacitors are usually made variable so that the circuit can be tuned to resonance and adjusted for maximum power output.

The most widely used of these circuits is the T network of Fig. 8-40(*c*). Often called an *LCC* network, it is used to match the low output impedance of a transistor power amplifier to the higher impedance of another amplifier or an antenna. The design procedure and formulas are given in Fig. 8-41. Suppose once again that a 6-Ω source R_i is to be matched to a 50-Ω load R_L at 155 MHz. Assume a Q of 10. (For class C operation, where many harmonics must be attenuated, it has been determined in practice that a Q of 10 is the absolute minimum needed for satisfactory suppression of the harmonics.) To configure the *LCC* network, the inductance is calculated first:

$$X_L = QR_i$$
$$X_L = 10(6) = 60 \ \Omega$$
$$L = \frac{X_L}{2\pi f} = \frac{50}{6.28 \times 155 \times 10^6} = 51.4 \ \text{nH}$$

Next, C_2 is calculated:

$$X_{C_2} = 50 \sqrt{\frac{6(101)}{50} - 1} = 50(3.33) = 166.73 \ \Omega$$

$$C_2 = \frac{1}{2\pi f X_C} = \frac{1}{6.28 \times 155 \times 10^6 \times 166.73} = 6.16 \times 10^{-12} = 6.16 \ \text{pF}$$

Finally, C_1 is calculated

$$X_{C_1} = \frac{6(10^2 + 1)}{10} \frac{1}{1 - 166.73/(10 \times 50)} = 60.6(1.5) = 91 \ \Omega$$

$$C_1 = \frac{1}{2\pi f X_C} = \frac{1}{6.28 \times 155 \times 10^6 \times 91} = 11.3 \ \text{pF}$$

Transformers and Baluns

One of the best impedance-matching components is the *transformer*. Recall that iron-core transformers are widely used at lower frequencies to match one impedance to another. Any load impedance can be made to look like a desired load impedance by selecting the correct value of the transformer turns ratio. In addition, transformers can be connected in unique combinations called *baluns* to match impedances.

Transformer

Baluns

Figure 8-42   Impedance matching with an iron-core transformer.

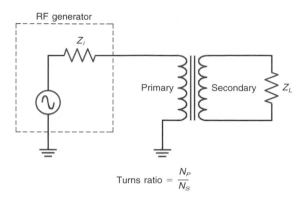

Turns ratio $= \dfrac{N_P}{N_S}$

Transformer Impedance Matching. Refer to Fig. 8-42. The relationship between the turns ratio and the input and output impedances is

$$\frac{Z_i}{Z_L} = \left(\frac{N_P}{N_S}\right)^2 \qquad \frac{N_P}{N_S} = \sqrt{\frac{Z_i}{Z_L}}$$

That is, the ratio of the input impedance Z_i to the load impedance Z_L is equal to the square of the ratio of the number of turns on the primary N_P to the number of turns on the secondary N_S. For example, to match a generator impedance of 6 Ω to a 50-Ω load impedance, the turns ratio should be as follows:

$$\frac{N_P}{N_S} = \sqrt{\frac{Z_i}{Z_L}} = \sqrt{\frac{6}{50}} = \sqrt{0.12} = 0.3464$$

$$\frac{N_S}{N_P} = \frac{1}{N_P/N_S} = \frac{1}{0.3464} = 2.887$$

This means that there are 2.89 times as many turns on the secondary as on the primary.

The relationship given above holds true only on iron-core transformers. When air-core transformers are used, the coupling between primary and secondary windings is not complete and therefore the impedance ratio is not as indicated. Although air-core transformers are widely used at Fs and can be used for impedance matching, they are less efficient than iron-core transformers.

Ferrite (magnetic ceramic) and powdered iron can be used as core materials to provide close coupling at very high frequencies. Both the primary and the secondary windings are wound on a core of the chosen material.

The most widely used type of core for RF transformers is the toroid. A *toroid* is a circular, doughnut-shaped core, usually made of a special type of powdered iron. Copper wire is wound on the toroid to create the primary and secondary windings. A typical arrangement is shown in Fig. 8-43. Single-winding tapped coils called *autotransformers* are also used for impedance matching between RF stages. Figure 8-44 shows impedance step-down and step-up arrangements. Toroids are commonly used in autotransformers.

Unlike air-core transformers, toroid transformers cause the magnetic field produced by the primary to be completely contained within the core itself. This has two important advantages. First, a toroid does not radiate RF energy. Air-core coils radiate because the magnetic field produced around the primary is not contained. Transmitter and receiver circuits using air-core coils are usually contained with magnetic shields to prevent them from interfering with other circuits. The toroid, on the other hand, confines all the magnetic fields and does not require shields. Second, most of the magnetic field produced by the primary cuts the turns of the secondary winding. Thus the basic turns ratio, input-output voltage, and impedance formulas for standard low-frequency transformers apply to high-frequency toroid transformers.

Toroid

Autotransformer

Figure 8-43 A toroid transformer.

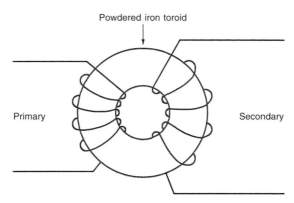

Powdered iron toroid

Primary   Secondary

Figure 8-44 Impedance matching in an autotransformer. (*a*) Step down. (*b*) Step up.

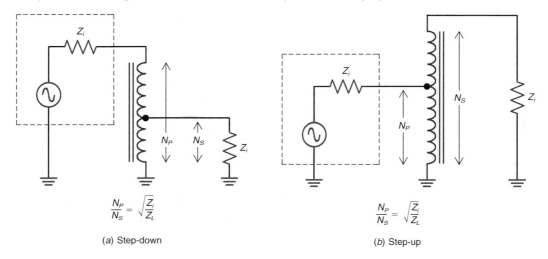

$$\frac{N_P}{N_S} = \sqrt{\frac{Z_i}{Z_L}}$$

(*a*) Step-down

$$\frac{N_P}{N_S} = \sqrt{\frac{Z_i}{Z_L}}$$

(*b*) Step-up

In most new RF designs, toroid transformers are used for RF impedance matching between stages. Further, the primary and secondary windings are sometimes used as inductors in tuned circuits. Alternatively, toroid inductors can be built. Powdered iron-core toroid inductors have an advantage over air-core inductors for RF applications because the high permeability of the core causes the inductance to be high. Recall that whenever an iron core is inserted into a coil of wire, the inductance increases dramatically. For RF applications, this means that the desired values of inductance can be created by using fewer turns of wire, and thus the inductor itself can be smaller. Further, fewer turns have less resistance, giving the coil a higher Q than that obtainable with air-core coils.

Powdered iron toroids are so effective that they have virtually replaced air-core coils in most modern transmitter designs. They are available in sizes from a fraction of an inch to several inches in diameter. In most applications, a minimum number of turns are required to create the desired inductance.

Figure 8-45 shows a toroid transformer T_1 used for interstage coupling between two class C driver amplifiers. The primary of transformer T_1 is tuned to resonance by capacitor C_1. The capacitor is adjustable, so the exact frequency of operation can be set. The relatively high output impedance of the transistor is coupled with the low input impedance of the next class C stage by a step-down transformer that provides the desired impedance-matching effects. Usually, the secondary winding is only a few turns of wire and is not resonated. The circuit in Fig. 8-45 also shows a similar transformer T_2 used for output coupling to the antenna.

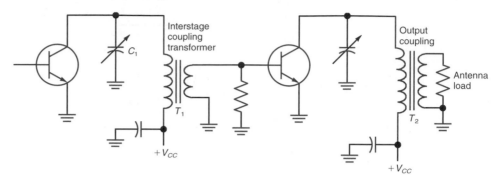

Transmission Line Transformers and Baluns

Transmission line transformer

A *transmission line* or broadband *transformer* is a unique type of transformer that is widely used in power amplifiers for coupling between stages and impedance matching. Such a transformer is usually constructed by winding two parallel wires (or a twisted pair) on a toroid, as shown in Fig. 8-46. The length of the winding is typically less than one-eighth wavelength at the lowest operating frequency. This type of transformer acts as a 1:1 transformer at the lower frequencies but more as a transmission line at the highest operating frequency.

Transformers can be connected in unique ways to provide fixed impedance-matching characteristics over a wide range of frequencies. One of the most widely used configurations is shown in Fig. 8-47. With this configuration, a transformer is usually wound on a toroid, and the numbers of primary and secondary turns are equal, giving the transformer a 1:1 turns ratio and a 1:1 impedance-matching ratio. The dots indicate the phasing of the windings. Note the unusual way in which the windings are connected. A transformer connected in this

Balun

way is generally known as a *balun* (from *bal*anced-*un*balanced) because such transformers are normally used to connect a balanced source to an unbalanced load or vice versa. In the circuit of Fig. 8-47(a), a balanced generator is connected to an unbalanced (grounded) load. In Fig. 8-47(b), an unbalanced (grounded) generator is connected to a balanced load.

Figure 8-48 shows two ways in which a 1:1 turns ratio balun can be used for impedance matching. With the arrangement shown in Fig. 8-47(a), an impedance step up is obtained. A load impedance of 4 times the source impedance Z_i provides a correct match. The balun makes the load of $4Z_i$ look like Z_i. In Fig. 8-48(b), an impedance step down is obtained. The balun makes the load Z_L look like $Z_i/4$.

Figure 8-46 A transmission line transformer.

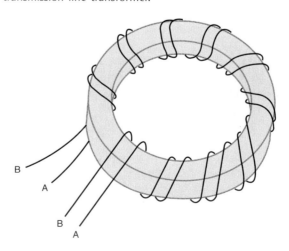

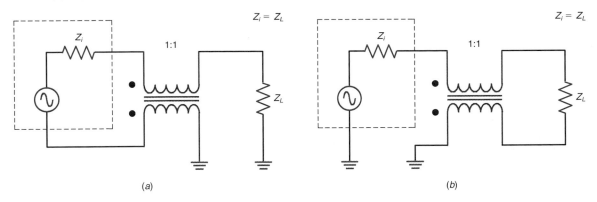

Many other balun configurations, offering different impedance ratios, are possible. Several common 1:1 baluns can be interconnected for both 9:1 and 16:1 impedance transformation ratios. In addition, baluns can be cascaded so that the output of one appears as the input to the other, and so on. Cascading baluns allow impedances to be stepped up or stepped down by wider ratios.

Note that the windings in a balun are not resonated with capacitors to a particular frequency. The winding inductances are made such that the coil reactances are 4 or more times that of the highest impedance being matched. This design allows the transformer to provide the designated impedance matching over a tremendous range of frequencies. This broadband characteristic of balun transformers allows designers to construct broadband RF power amplifiers. Such amplifiers provide a specific amount of power amplification over a wide bandwidth and are thus particularly useful in communication equipment that must operate in more than one frequency range. Rather than have a separate transmitter for each desired band, a single transmitter with no tuning circuits can be used.

When conventional tuned amplifiers are used, some method of switching the correct tuned circuit into the circuit must be provided. Such switching networks are complex and expensive. Further, they introduce problems, particularly at high frequencies. For them to perform effectively, the switches must be located very close to the tuned circuits so that stray inductances and capacitances are not introduced by the switch and the interconnecting leads. One way to overcome the switching problem is to use a broadband amplifier, which does not require switching or tuning. The broadband amplifier provides the necessary amplification as well as impedance matching. However, broadband amplifiers do

Figure 8-48 Using a balun for impedance matching. (*a*) Impedance step up. (*b*) Impedance step down.

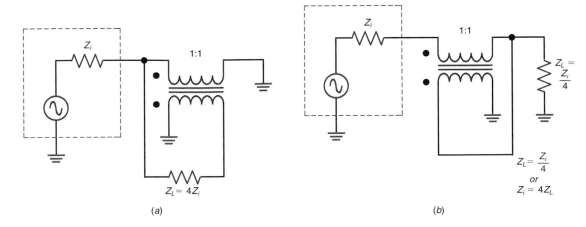

Figure 8-49 A broadband class A linear power amplifier.

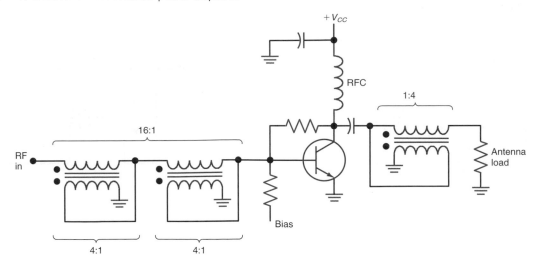

not provide the filtering necessary to get rid of harmonics. One way to overcome this problem is to generate the desired frequency at a lower power level, allowing tuned circuits to filter out the harmonics, and then provide final power amplification with the broadband circuit. The broadband power amplifier operates as a linear class A or class B push-pull circuit so that the inherent harmonic content of the output is very low.

Figure 8-49 shows a typical broadband linear amplifier. Note that two 4:1 balun transformers are cascaded at the input so that the low base input impedance is made to look like an impedance 16 times higher than it is. The output uses a 1:4 balun that steps up the very low output impedance of the final amplifier to an impedance 4 times higher to equal the antenna load impedance. In some transmitters, broadband amplifiers are followed by low-pass filters, which are used to eliminate undesirable harmonics in the output.

8-5 Typical Transmitter Circuits

Many transmitters used in recent equipment designs are a combination of ICs and discrete component circuits. Here are two examples.

Low-Power FM Transmitter. The transmitter shown in Fig. 8-50 incorporates the most up-to-date techniques. This low-power FM transmitter, which is designed to operate in the 30-MHz range, has an input power of about 3 W and a frequency deviation of 5 kHz for narrowband operation. The circuit is made up of a Freescale MC2833 single-chip FM transmitter IC, a digital shaping circuit, and a pair of power MOSFETs connected in parallel as a class E amplifier. An IC regulator provides a constant dc supply voltage from a battery pack.

The heart of the circuit is the transmitter chip. A more detailed look at this chip is given in Fig. 8-51. It is housed in a standard 16-pin DIP, and it contains a microphone amplifier with clipping diodes; an RF oscillator, which is usually crystal-controlled with an external crystal; and a buffer amplifier. Frequency modulation is produced by a variable reactance circuit connected to the oscillator. Also on the chip are two free transistors that can be connected with external components as buffer amplifiers or as multipliers and low-level power amplifiers. This chip is useful up to about 60 to 70 MHz, and at higher ranges if external multipliers are used. The chip is widely used in cordless telephones, which operate in the 46- to 49-MHz band with FM.

As shown in Fig. 8-50, the signal starts with a microphone whose signal is sent to the audio amplifier in the IC at pin 5 through C_{39}. The gain of the amplifier is set with resistor R_{11} on pins 4 and 5. The output of the amplifier is connected to the reactance

Figure 8-50 Schematic showing the modulator and tripler, class E amplifier, limiter, driver, and low-voltage regulator sections of the E-Comm transceiver. The key device is IC_1, the FM transmitter chip. (*Courtesy Electronics Now, October 1992.*)

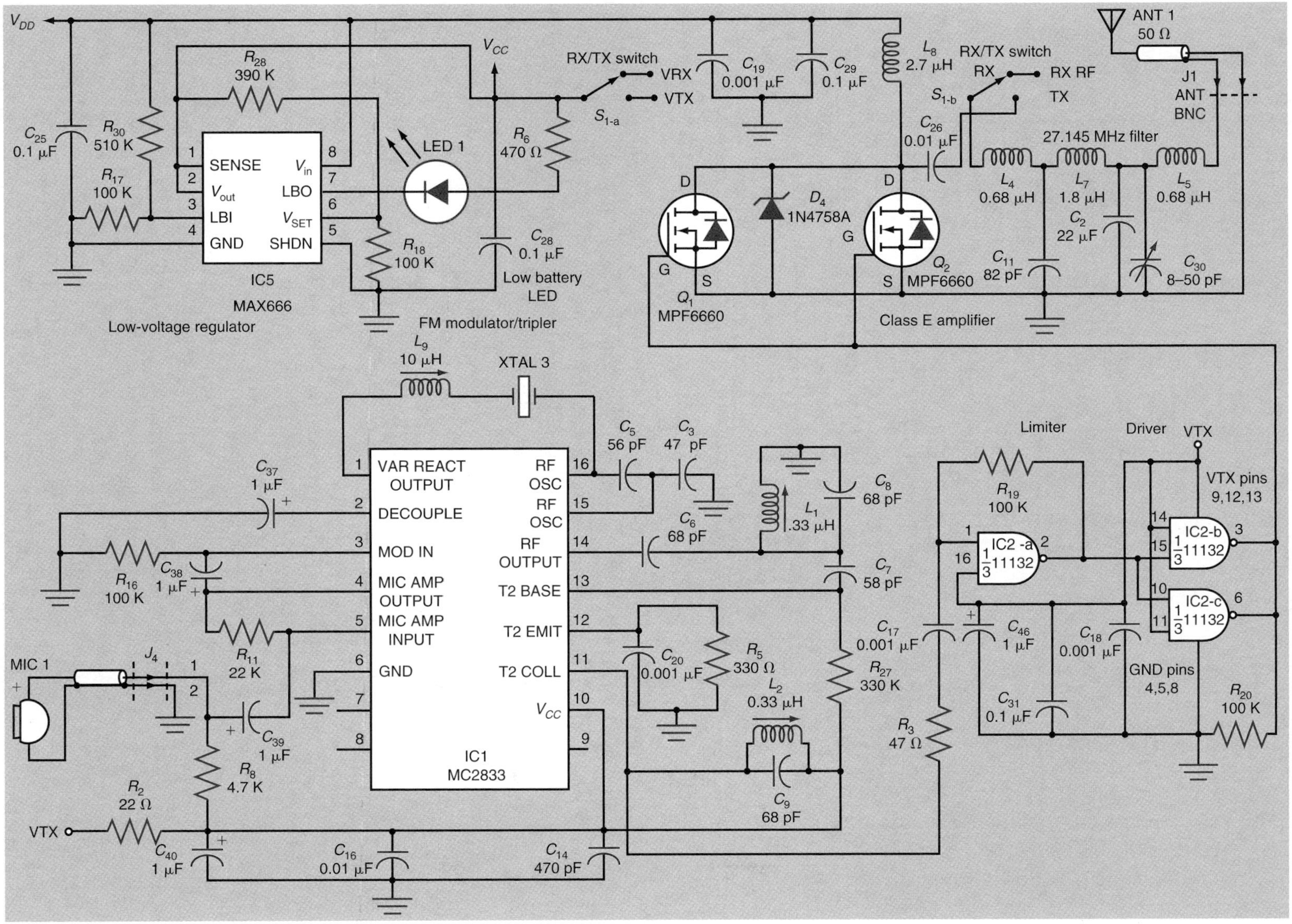

Figure 8-51 Freescale MC 2833 IC FM VHF transmitter chip.

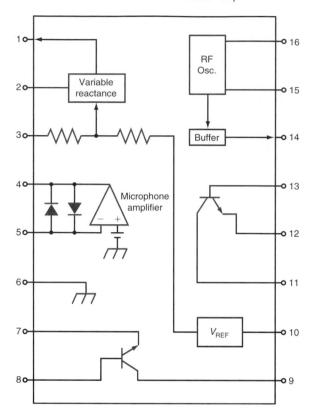

modulator via C_{38} on pin 3. This circuit connects to the oscillator whose frequency is set by the external crystal between pins 1 and 16. Assume a 10-MHz crystal. The reactance modulator pulls the crystal frequency by a small amount during modulation to produce a frequency variation.

The output of the oscillator is buffered and amplified and appears at pin 14 on the IC. The buffer amplifier has a resonant circuit (L_1 and C_8) in its output tuned to the third harmonic of the crystal or 30 MHz. In addition to multiplying the carrier frequency, the multiplier multiplies the frequency deviation by a factor of 3 to achieve the desired 5-kHz deviation. The resulting FM signal is then applied to a linear amplifier made up of one of the transistors in the IC (pins 11, 12, and 13). Its output is tuned by L_2 and C_9. Next, the FM sine wave signal is applied to one gate of a high-speed CMOS NAND-gate Schmitt trigger. IC_{2-a} shapes the sine wave into a square wave. Two additional CMOS gates, IC_{2-b} and IC_{2-c} are connected in parallel to provide a high-power square wave drive signal to the final amplifier.

The final power amplifier uses two MPF6660 RF power enhancement mode MOSFETs connected in parallel as a class E switching amplifier. Zener diode D_4 provides protection from overloads caused by mismatched antenna impedances. The π network output made up of L_4, L_7, L_5, C_{11}, C_2, and C_{30} couples the signal to the antenna, providing impedance matching and low-pass filtering to eliminate the harmonics associated with the square wave output. The input power is about 3 W, and the circuit gives about 90 percent efficiency, which means that the output power is about $0.9 \times 3 = 2.7$ W. The antenna is a "rubber ducky" vertical.

Finally, the unit is powered by a battery pack that supplies 9.6 V to IC5, a MAX666 voltage regulator. The regulator gives a constant voltage of 8 V to the transmitter circuits despite the gradual dropping of the battery voltage during operation. The IC also contains a component that senses when the battery is too discharged for proper operation and turns on the LED.

This transmitter is part of a handheld unit with a matching receiver. Receiver circuits are covered in Chap 9.

Short-Range Wireless Transmitter

There are many short-range wireless applications that require a transmitter to send data or control signals to a nearby receiver. Some examples are the small transmitters in *remote keyless entry (RKE)* devices used to open car doors, tire pressure sensors, remote-control lights and ceiling fans in homes, garage door openers, and temperature sensors. These unlicensed transmitters use very low power and operate in the FCC's industrial-scientific-medical (ISM) bands. These are frequencies set aside for unlicensed operation as defined in Part 15 of the FCC rules and regulations. The most common frequencies are 315, 433.92, 868 (Europe), and 915 MHz.

Remote keyless entry (RKE)

Figure 8-52 shows a typical transmitter IC, the Freescale MC33493/D. This CMOS device is designed to operate anywhere in the 315- to 434-MHz and 868- to 928-MHz ranges with the frequency set by an external crystal. It features OOK or FSK modulation and can handle a serial data rate up to 10 kbps. Output power is adjustable with an external resistor.

The basic transmitter circuit is simply a phase-locked loop (PLL) used as a frequency multiplier with an output power amplifier. The internal oscillator XCO uses an external crystal. The PLL multiplies an external crystal frequency by a factor of 32 or 64 to develop a PLL VCO signal at the desired operating frequency. For example, if the desired output frequency is 315 MHz, the crystal must have a frequency of 315/32 = 9.84375 MHz. For an output of 433.92 MHz, a 13.56-MHz crystal is needed. The XCO output is applied to the phase detector along with the feedback signal from the frequency dividers driven by the PLL VCO output. The divide-by-2 divider may be switched in when divide-by-64 is needed. The BAND input signal selects the divide-by-2 divider or takes it out. If the BAND signal is low, the divide-by-2 circuit is bypassed and the overall PLL multiply factor is 32. If BAND is high, the divide-by-2 circuit is inserted and the overall multiplication factor is 64. Using a 13.56-MHz crystal gives an output 867.84 MHz when the divide-by-64 factor is used.

The PLL VCO output drives a class C power amplifier. An external resistor may be inserted in the line between REXT and the dc power source to lower the power to the desired level. Maximum output with no external resistor is 5 dBm (3.1 mW). The dc supply voltage may be anything in the 1.9- to 3.6-V range usually supplied by a battery.

Figure 8-52 The Freescale MC 33493D UHF ISM transmitter IC.

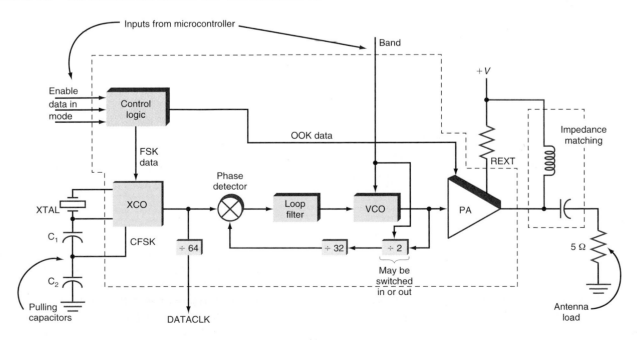

Modulation is selected by the MODE pin. If MODE is low, OOK modulation is selected. The serial binary input at the DATA pin then goes to turn the class C PA dc power off and on, on for binary 1 and off for binary 0. If MODE is high, FSK is selected. The DATA input line is then used to pull the crystal frequency between the two desired shift frequencies. A 45-kHz shift is typical. Two external capacitors C_1 and C_2 are used to pull the crystal to the desired frequencies. Either serial or parallel pulling may be used depending upon the type of crystal.

The PA output is fed to an external *LC* impedance-matching network as needed to match the 50-Ω output to the selected antenna, as Fig. 8-52 shows. Usually the antenna is a loop of copper on the printed-circuit board holding the transmitter IC.

One other feature of this chip is the data clock DATACLK output line. This output is the crystal frequency divided by 64. For a 9.84375-MHz crystal, the DATACLK output is 153.8 kHz. With a 13.56-MHz crystal, the DATACLK output is 212 kHz. This clock can be used with an external embedded microcontroller to synchronize the data stream.

This transmitter chip is designed to be used with an external microcontroller. It gets its BAND, MODE, and ENABLE signals from the microcontroller.

CHAPTER REVIEW

Summary

The starting point for all transmitters is carrier generation, which is almost always accomplished by crystal oscillators. To achieve the frequencies necessary in the VHF, UHF, and even microwave ranges, transmitters use frequency multiplier circuits.

Frequency synthesizers are variable-frequency generators using a phase-locked loop that provide the frequency stability of a crystal oscillator and the convenience of incremental tuning over a broad frequency range. The output of a frequency synthesizer is varied by using switching techniques. The two major types of frequency synthesizers are phase-locked loop (PLL) and direct digital synthesis (DDS).

The three basic types of power amplifiers used in transmitters are linear (class A, AB, or B), class C, and switching (class D, class E, class F, and class S). The class of an amplifier is determined by how it is biased. Class A amplifiers are biased so that they conduct continuously, class B amplifiers are biased at cutoff so no collector current flows with zero

input, and class C amplifiers conduct for less than one-half of the sine wave input cycle.

Both linear and class C amplifiers face the problem of self-oscillation, which is the result of positive feedback that occurs because of the interelement capacitance of the amplifying device. The process of compensating for this effect is called neutralization.

Switching amplifiers (classes D, E, and F) are more efficient than linear and class C amplifiers. This is so because they dissipate little or no power in their off and on states. Two types of broadband linear amplifier are the feedforward and adaptive predistortion amplifiers used in broadband services such as cell phone base stations.

The circuits used to connect one transmitter stage to another, called impedance-matching networks, provide for an optimum transfer of power as well as filtering and selectivity functions. The basic types of *LC* impedance-matching networks are the L network, the T network, and the π network. Transformers and baluns are widely used for impedance matching.

Questions

1. What circuits are typically part of every radio transmitter?
2. Which type of transmitter does not use class C amplifiers?
3. For how many degrees of an input sine wave does a class B amplifier conduct?
4. What is the name given to the bias for a class C amplifier produced by an input *RC* network?
5. Why are crystal oscillators used instead of *LC* oscillators to set transmitter frequency?

6. Name two ways to select crystals with switches. Which is preferred at the higher frequencies?
7. How is the output frequency of a frequency of a PLL synthesizer changed?
8. What are prescalers and why are they used in VHF and UHF synthesizers?
9. What is the purpose of the loop filter in a PLL?
10. What circuit in a direct digital synthesizer (DDS) actually generates the output waveform?
11. In a DDS, what is stored in ROM?
12. How is the output for frequency of a DDS changed?
13. What is the most efficient class of RF power amplifier?
14. What is the approximate maximum power of typical transistor RF power amplifiers?
15. What are parasitics and how are they eliminated in a power amplifier?
16. What is the main reason that switching amplifiers are used?
17. What is the difference between a class D and a class E amplifier?
18. What is a major disadvantage of a switching power amplifier?
19. Explain how a feedforward power amplifier reduces distortion.
20. In a predistortion power amplifier, what is the feedback signal?
21. Maximum power transfer occurs when what relationship exists between generator impedance Z_i and load impedance Z_L?
22. What is a toroid and how is it used? What components are made from it?
23. What are the advantages of a toroid RF inductor?
24. In addition to impedance matching, what other important function do LC networks perform?
25. What is the name given to a single winding transformer?
26. What is the name given to an RF transformer with a 1:1 turns ratio connected so that it provides a 1:4 or 4:1 impedance matching? Give a common application.
27. Why are untuned RF transformers used in power amplifiers?
28. How is impedance matching handled in a broadband linear RF amplifier?
29. What are the common impedance-matching ratios of transmission line transformers used as baluns?
30. Why are π and T networks preferred over L networks?

Problems

1. An FM transmitter has an 8.6-MHz crystal carrier oscillator and frequency multipliers of 2, 3, and 4. What is the output frequency? ◆
2. A crystal has a tolerance of 0.003 percent. What is the tolerance in ppm?
3. A 25-MHz crystal has a tolerance of ± 200 ppm. If the frequency drifts upward to the maximum tolerance, what is the frequency of the crystal? ◆
4. A PLL frequency synthesizer has a reference frequency of 25 kHz. The frequency divider is set to a factor of 345. What is the output frequency?
5. A PLL frequency synthesizer has an output frequency of 162.7 MHz. The reference is a 1-MHz crystal oscillator followed by a divider of 10. What is the main frequency divider ratio? ◆
6. A PLL frequency synthesizer has an output frequency of 470 MHz. A divide-by-10 prescaler is used. The reference frequency is 10 kHz. What is the frequency step increment?
7. A PLL frequency synthesizer has a variable-modulus prescaler of $M = 10/11$ and divider ratios of the A and N counters of 40 and 260. The reference frequency is 50 kHz. What are the VCO output frequency and the minimum frequency step increment? ◆
8. In a DDS, the ROM contains 4096 storage locations holding one cycle of sine wave values. What is the phase step increment?
9. A DDS synthesizer has a 200-MHz clock and a constant value of 16. The ROM address register has 16 bits. What is the output frequency?
10. A PLL multiplier transmitter is to operate with a 915 MHz output. With a divide factor of 64, what value of crystal is needed?
11. A class C amplifier has a supply voltage of 36 V and a collector current of 2.5 A. Its efficiency is 80 percent. What is the RF output power? ◆
12. Calculate the L and C values of an L network that is to match a 9-Ω transistor power amplifier to a 75-Ω antenna at 122 MHz.
13. Calculate what L network components will match a 4-Ω internal resistance in series with an internal inductance of 9 nH to a 72-Ω load impedance in parallel with a stray capacitance of 24 pF at a frequency of 46 MHz.
14. Design an LCC T network that will match 5-Ω internal resistance to a 52-Ω load at 54 MHz. Assume a Q of 12.
15. A transformer has 6 turns on the primary and 18 turns on the secondary. If the generator (source) impedance is 50 Ω, what should the load impedance be?
16. A transformer must match a 2500-Ω generator to a 50-Ω load. What must the turns ratio be?

◆ *Answers to Selected Problems follow Chap. 22.*

Critical Thinking

1. Name the five main parts of a PLL frequency synthesizer. Draw the circuit from memory. From which circuit is the output taken?
2. Looking at a sine wave, describe how the ROM in a DDS could use only one-fourth of the memory locations to store the sine look-up table.
3. Design an *LCC* network like that in Fig. 8-40 to match a 5.5-Ω transistor amplifier with an inductor of 7 nH to an antenna with an impedance of 50 Ω and a shunt capacitance of 22 pf.
4. To match a 6-Ω amplifier impedance to a 72-Ω antenna load, what turns ratio N_P/N_S must a transformer have?

Most crystal oscillators are packaged in metal cans like this one, designed for surface mounting.

Communication Receivers

In radio communication systems, the transmitted signal is very weak when it reaches the receiver, particularly when it has traveled over a long distance. The signal, which has shared the free-space transmission media with thousands of other radio signals, has also picked up noise of various kinds. Radio receivers must provide the sensitivity and selectivity that permit full recovery of the original intelligence signal. The radio receiver best suited to this task is known as the *superheterodyne* receiver. Invented in the early 1900s, the superheterodyne is used today in most electronic communication systems. This chapter reviews the basic principles of signal reception and discusses various superheterodyne circuits including direct conversion.

Superheterodyne

Objectives

After completing this chapter, you will be able to:

- List the benefits of a superheterodyne over a TRF receiver and identify the function of each component of a superheterodyne, including all selectivity functions.
- Express the relationship between the IF, local oscillator, and signal frequencies mathematically and calculate any one of them, given the other two.
- Explain how the design of dual-conversion receivers allows them to enhance selectivity and eliminate image problems.
- Describe the operation of the most common types of mixer circuits.
- Explain the architecture and operation of direct conversion and software-defined radios.
- List the major types of external and internal noise and explain how each interferes with signals both before and after they reach the receiver.
- Calculate the noise factor, noise figure, and noise temperature of a receiver.
- Describe the operation and purpose of the AGC circuit in a receiver.
- Explain the operation of squelch circuits.

9-1 Basic Principles of Signal Reproduction

A communication receiver must be able to identify and select a desired signal from thousands of others present in the frequency spectrum (selectivity) and to provide sufficient amplification to recover the modulating signal (sensitivity). A receiver with good selectivity will isolate the desired signal in the RF spectrum and eliminate or at least greatly attenuate all other signals. A receiver with good sensitivity involves high circuit gain.

Selectivity

Selectivity

Selectivity in a receiver is obtained by using tuned circuits and/or filters. The *LC* tuned circuits provide initial selectivity; filters, which are used later in the process, provide additional selectivity.

Bandwidth

Q **and Bandwidth.** Initial selectivity in a receiver is normally obtained by using *LC* tuned circuits. By carefully controlling the *Q* of the resonant circuit, you can set the desired selectivity. The optimum *bandwidth* is wide enough to pass the signal and its sidebands but also narrow enough to eliminate or greatly attenuate signals on adjacent frequencies. As Fig. 9-1 shows, the rate of attenuation or roll-off of an *LC* tuned circuit is gradual. Adjacent signals will be attenuated, but in some cases not enough to completely eliminate interference. Increasing the *Q* will further narrow the bandwidth and improve the steepness of attenuation, but narrowing the bandwidth in this way can be taken only so far. At some point, the circuit bandwidth may become so narrow that it starts to attenuate the sidebands, resulting in loss of information.

The ideal receiver selectivity curve would have perfectly vertical sides, as in Fig. 9-2(*a*). Such a curve cannot be obtained with tuned circuits. Improved selectivity is achieved by cascading tuned circuits or by using crystal, ceramic, or SAW filters. At lower frequencies, digital signal processing (DSP) can provide almost ideal response curves. All these methods are used in communication receivers.

> **GOOD TO KNOW**
>
> If you know the *Q* and the resonant frequency of a tuned circuit, you can calculate the bandwidth by using the equation $BW = f_r/Q$.

Figure 9-1 Selectivity curve of a tuned circuit.

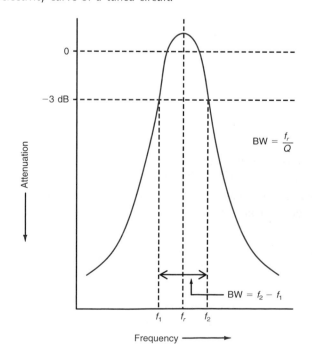

$$BW = \frac{f_r}{Q}$$

$$BW = f_2 - f_1$$

Figure 9-2 Receiver selectivity response curves. (*a*) Ideal response curve. (*b*) Practical response curve showing shape factor.

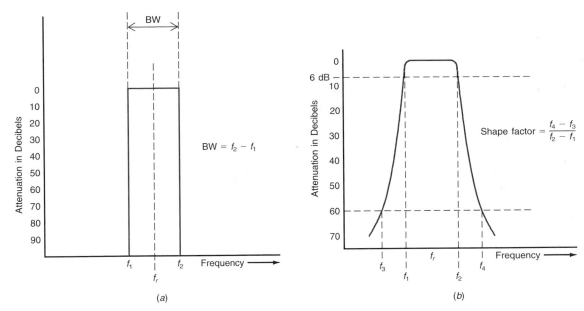

(a)

(b)

Shape Factor. The sides of a tuned circuit response curve are known as *skirts*. The steepness of the skirts, or the *skirt selectivity*, of a receiver is expressed as the *shape factor*, the ratio of the 60-dB down bandwidth to the 6-dB down bandwidth. This is illustrated in Fig. 9-2(*b*). The bandwidth at the 60-dB down points is $f_4 - f_3$; the bandwidth of the 6-dB down points is $f_2 - f_1$. Thus the shape factor is $(f_4 - f_3)/(f_2 - f_1)$. Assume, for example, that the 60-dB bandwidth is 8 kHz and the 6-dB bandwidth is 3 kHz. The shape factor is 8/3 = 2.67, or 2.67:1.

Skirt selectivity

Shape factor

The lower the shape factor, the steeper the skirts and the better the selectivity. The ideal, shown in Fig. 9-2(*a*), is 1. Shape factors approaching 1 can be achieved with DSP filters.

Sensitivity

A communication receiver's *sensitivity*, or ability to pick up weak signals, is mainly a function of overall gain, the factor by which an input signal is multiplied to produce the output signal. In general, the higher the gain of a receiver, the better its sensitivity. The greater gain that a receiver has, the smaller the input signal necessary to produce a desired level of output. High gain in communication receivers is obtained by using multiple amplification stages.

Sensitivity

Another factor that affects the sensitivity of a receiver is the signal-to-noise (*S/N*) ratio (SNR). Noise is the small random voltage variations from external sources and from noise variations generated within the receiver's circuits. This noise can sometimes be so high (many microvolts) that it masks or obliterates the desired signal. Figure 9-3 shows what a spectrum analyzer display would show as it monitored two input signals and the background noise. The noise is small, but it has random voltage variations and frequency components that are spread over a wide spectrum. The large signal is well above the noise and so is easily recognized, amplified, and demodulated. The smaller signal is barely larger than the noise and so may not be successfully received.

One method of expressing the sensitivity of a receiver is to establish the *minimum discernible signal (MDS)*. The MDS is the input signal level that is approximately equal to the average *internally* generated noise value. This noise value is called the *noise floor* of the receiver. MDS is the amount of signal that would produce the same audio power output as the noise floor signal. The MDS is usually expressed in dBm.

Another often used measure of receiver sensitivity is microvolts or decibels above 1 μV or decibels above 1 μV and decibels above 1 mW (0 dBm).

Figure 9-3   Illustrating noise, MDS, and receiver sensitivity.

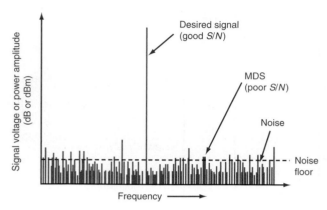

Most receivers have an antenna input impedance of 50 Ω. So a 1-μV signal produces a power P across the 50 Ω of

$$P = \frac{V^2}{R} = \frac{1 \times 10^{-6}}{50} = 2 \times 10^{-14} \text{ W}$$

Expressing this in dBm (power referenced to 1 mW) gives

$$\text{dBm} = 10 \log \frac{P}{1 \text{ mW}} = 10 \log \frac{2 \times 10^{-14}}{0.001} = -107 \text{ dBM}$$

Now if a receiver has a stated sensitivity of 10 μV, then expressing this in decibels yields

$$\text{dB} = 20 \log 10 = 20 \text{ dB}$$

The sensitivity above 1 mW then is

$$\text{dBm} = 10 - 107 = -87 \text{ dBm}$$

An input sensitivity of 0.5 μV translates to a sensitivity of

$$\text{dB} = 20 \log 0.5 = -6$$

$$\text{dBm} = -6 - 107 = -113 \text{ dBm}$$

Bit error rate (BER)

There is no one fixed way to define sensitivity. For analog signals, the signal-to-noise ratio is the main consideration in analog signals. For digital signal transmission, the *bit error rate* (*BER*) is the main consideration. BER is the number of errors made in the transmission of many serial data bits. For example, one measure is that the sensitivity is such that the BER is 10^{-10} or 1 bit error in every 10 billion bits transmitted.

Several methods for stating and measuring sensitivity have been defined in various communications standards depending upon the type of modulation used and other factors.

For example, the sensitivity of a high-frequency communication receiver is usually expressed as the minimum amount of signal voltage input that will produce an output signal that is 10 dB higher than the receiver background noise. Some specifications state a 20-dB *S/N* ratio. A typical sensitivity figure might be 1-μV input. The lower this figure, the better the sensitivity. Good communication receivers typically have a sensitivity of 0.2 to 1 μV. Consumer AM and FM receivers designed for receiving strong local stations have much lower sensitivity. Typical FM receivers have sensitivities of 5 to 10 μV; AM receivers can have sensitivities of 100 μV or higher.

Basic Receiver Configuration

Figure 9-4 shows the simplest radio receiver: a crystal set consisting of a tuned circuit, a diode (crystal) detector, and earphones. The tuned circuit provides the selectivity, the

Figure 9-4 The simplest receiver—a crystal set.

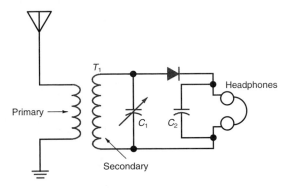

diode and C_2 serve as an AM demodulator, and the earphones reproduce the recovered audio signal.

The crystal receiver in Fig. 9-4 does not provide the kind of selectivity and sensitivity necessary for modern communication. Only the strongest signals can produce an output, and selectivity is often insufficient to separate incoming signals. The use of earphones is also inconvenient. However, a demodulator is the basic circuit in any receiver. All other circuits in a receiver are designed to improve sensitivity and selectivity, so that the demodulator can perform better.

TRF Receivers

In the *tuned-radio frequency (TRF) receiver* shown in Fig. 9-5 sensitivity has been improved by adding three stages of RF amplification between the antenna and the detector, followed by two stages of audio amplification. The RF amplifier stages increase the gain of the received signal tremendously before it is applied to the detector. The recovered signal is amplified further by the audio amplifiers, which provide sufficient gain to operate a loudspeaker.

Another design improvement is that the RF amplifiers use multiple tuned circuits. Whenever resonant *LC* circuits tuned to the same frequency are cascaded, overall selectivity is improved. The greater the number of tuned stages cascaded, the narrower the bandwidth and the steeper the skirts, as shown in Fig. 9-6. A signal above or below the center resonant frequency is attenuated by one tuned circuit, further attenuated by a second tuned circuit, and still further by a third and fourth tuned circuit. The effect is to steepen the skirts.

Tuned radio-frequency (TRF) receiver

Figure 9-5 Tuned radio-frequency (TRF) receiver.

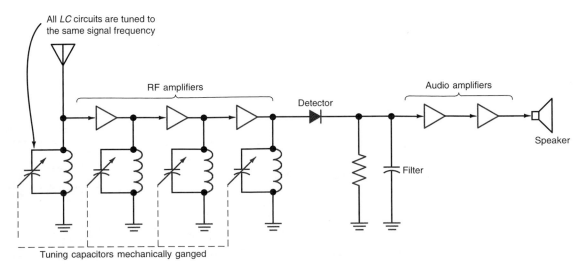

Figure 9-6   The effect of cascading tuned circuits on selectivity.

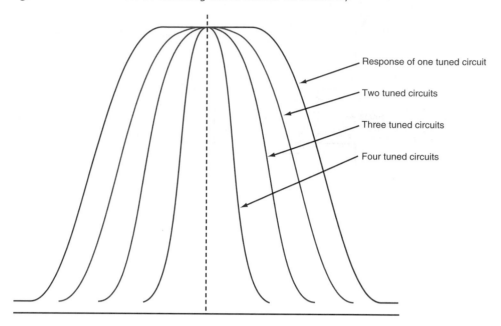

Response of one tuned circuit

Two tuned circuits

Three tuned circuits

Four tuned circuits

The main problem with TRF receivers lies in tracking the tuned circuits. In a receiver, the tuned circuits must be made variable so that they can be set to the frequency of the desired signal. In early receivers, each tuned circuit had a separate capacitor, and multiple dials had to be adjusted to tune in a signal. The solution was to gang the tuning capacitors (Fig. 9-5) so that all would be changed simultaneously when the tuning knob was rotated.

Another problem with TRF receivers is that selectivity varies with frequency. As discussed, the bandwidth of a tuned circuit increases with its resonant frequency, since $BW = f_r/Q$. (Q tends to remain nearly constant because the effective coil resistance increases slightly with frequency as a result of the skin effect.) Selectivity was good (narrow) at the low frequencies but poor (broader) at the higher frequencies. Not until the development of the superheterodyne receiver, or superhet, were these basic problems solved.

Although TRF receivers are no longer used in general communication applications, they are still found in some simple, low-cost single-frequency designs.

Figure 9-7 shows an example of a single-chip UHF receiver using multiple untuned RF amplifiers to obtain the desired gain and an external SAW filter. By operating on a fixed frequency, the SAW filter can provide the desired selectivity without the problems mentioned earlier.

Figure 9-7   A single-IC UHF TRF receiver using a SAW filter.

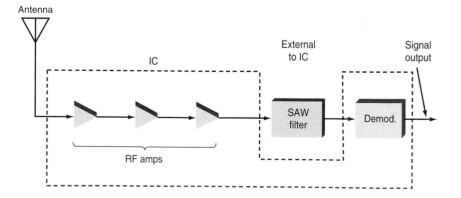

Antenna

IC

External to IC

Signal output

SAW filter

Demod.

RF amps

9-2 Superheterodyne Receivers

Superheterodyne receivers convert all incoming signals to a lower frequency, known as the *intermediate frequency (IF)*, at which a single set of amplifiers is used to provide a fixed level of sensitivity and selectivity. Most of the gain and selectivity in a superheterodyne receiver are obtained in the IF amplifiers. The key circuit is the mixer, which acts as a simple amplitude modulator to produce sum and difference frequencies. The incoming signal is mixed with a local oscillator signal to produce this conversion. Figure 9-8 shows a general block diagram of a superhetrodyne receiver. In the following sections, the basic function of each circuit is examined.

Intermediate frequency (IF)

RF Amplifiers

The antenna picks up the weak radio signal and feeds it to the *RF amplifier,* also called a *low-noise amplifier (LNA)*. Because RF amplifiers provide some initial gain and selectivity, they are sometimes referred to as *preselectors*. Tuned circuits help select the desired signal or at least the frequency range in which the signal resides. The tuned circuits in fixed tuned receivers can be given a very high Q, so that excellent selectivity can be obtained. However, in receivers that must tune over a broad frequency range, selectivity is somewhat more difficult to obtain. The tuned circuits must resonate over a wide frequency range. Therefore, the Q, bandwidth, and selectivity of the amplifier change with frequency.

RF amplifier (low-noise amplifier or preselector)

In communication receivers that do not use an RF amplifier, the antenna is connected directly to a tuned circuit, at the input to the mixer, which provides the desired initial selectivity. This configuration is practical in low-frequency applications where extra gain is simply not needed. (Most of the receiver gain is in the IF amplifier section, and even if relatively strong signals are to be received, additional RF gain is not necessary.) Further, omitting the RF amplifier may reduce the noise contributed by such a circuit. In general, however, it is preferable to use an RF amplifier. RF amplifiers improve sensitivity, because of the extra gain; improve selectivity, because of the added tuned circuits; and improve the S/N ratio. Further, spurious signals are more effectively rejected, minimizing unwanted signal generation in the mixer.

RF amplifiers also minimize oscillator radiation. The local oscillator signal is relatively strong, and some of it can leak through and appear at the input to the mixer. If the mixer input is connected directly to the antenna, some of the local oscillator signal radiates, possibly causing interference to other nearby receivers. The RF amplifier between the mixer and the antenna isolates the two, significantly reducing any local oscillator radiation.

> **GOOD TO KNOW**
>
> The key circuit in superheterodyne receivers is the mixer, which acts like a simple amplitude modulator to produce sum and difference frequencies.

Figure 9-8 Block diagram of a superheterodyne receiver.

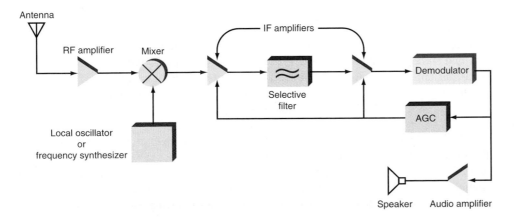

Both bipolar and field-effect transistors, made with silicon, GaAs, or SiGe, can be used as RF amplifiers. The selection is made based upon frequency, cost, integrated versus discrete, and desired noise performance.

Mixers and Local Oscillators

Mixer

Local oscillator

The output of the RF amplifier is applied to the input of the *mixer*. The mixer also receives an input from a local oscillator or frequency synthesizer. The mixer output is the input signal, the local oscillator signal, and the sum and difference frequencies of these signals. Usually a tuned circuit at the output of the mixer selects the difference frequency, or intermediate frequency (IF). The sum frequency may also be selected as the IF in some applications. The mixer may be a diode, a balanced modulator, or a transistor. MOSFETs and hot carrier diodes are preferred as mixers because of their low-noise characteristics.

The *local oscillator* is made tunable so that its frequency can be adjusted over a relatively wide range. As the local-oscillator frequency is changed, the mixer translates a wide range of input frequencies to the IF.

GOOD TO KNOW

MOSFETs and hot carrier diodes are preferred as mixers because of their low noise characteristics.

IF Amplifiers

The output of the mixer is an IF signal containing the same modulation that appeared on the input RF signal. This signal is amplified by one or more IF amplifier stages, and most of the receiver gain is obtained in these stages. Selective tuned circuits provide fixed selectivity. Since the intermediate frequency is usually much lower than the input signal frequency, IF amplifiers are easier to design and good selectivity is easier to obtain. Crystal, ceramic, or SAW filters are used in most IF sections to obtain good selectivity.

Demodulators

Demodulator (or detector)

The highly amplified IF signal is finally applied to the *demodulator*, or *detector*, which recovers the original modulating information. The demodulator may be a diode detector (for AM), a quadrature detector (for FM), or a product detector (for SSB). In modern digital superheterodyne radios, the IF signal is first digitized by an analog-to-digital converter (ADC) and then sent to a digital signal processor (DSP) where the demodulation is carried out by a programmed algorithm. The recovered signal in digital form is then converted back to analog by a digital-to-analog converter (DAC). The output of the demodulator or DAC is then usually fed to an audio amplifier with sufficient voltage and power gain to operate a speaker. For nonvoice signals, the detector output may be sent elsewhere, to a TV picture tube, e.g., or a computer.

Automatic Gain Control

Automatic gain control (AGC)

The output of a demodulator is usually the original modulating signal, the amplitude of which is directly proportional to the amplitude of the received signal. The recovered signal, which is usually ac, is rectified and filtered into a dc voltage by a circuit known as the *automatic gain control (AGC)* circuit. This dc voltage is fed back to the IF amplifiers, and sometimes the RF amplifier, to control receiver gain. AGC circuits help maintain a constant output voltage level over a wide range of RF input signal levels; they also help the receiver to function over a wide range, so that strong signals do not produce performance-degrading distortion. Virtually all superheterodyne receivers use some form of AGC.

The amplitude of the RF signal at the antenna of a receiver can range from a fraction of a microvolt to thousands of microvolts; this wide signal range is known as the *dynamic range*. Typically, receivers are designed with very high gain so that weak signals can be reliably received. However, applying a very high-amplitude signal to a receiver causes the circuits to be overdriven, producing distortion and reducing intelligibility.

With AGC, the overall gain of the receiver is automatically adjusted depending on the input signal level. The signal amplitude at the output of the detector is proportional to the amplitude of the input signal; if it is very high, the AGC circuit produces a high dc output voltage, thereby reducing the gain of the IF amplifiers. This reduction in gain eliminates the distortion normally produced by a high-voltage input signal. When the incoming signal is weak, the detector output is low. The output of the AGC is then a smaller dc voltage. This causes the gain of the IF amplifiers to remain high, providing maximum amplification.

9-3 Frequency Conversion

As discussed in earlier chapters, *frequency conversion* is the process of translating a modulated signal to a higher or lower frequency while retaining all the originally transmitted information. In radio receivers, high-frequency radio signals are regularly converted to a lower, intermediate frequency, where improved gain and selectivity can be obtained. This is called *down conversion*. In satellite communications, the original signal is generated at a lower frequency and then converted to a higher frequency for transmission. This is called *up conversion*.

Frequency conversion

Down conversion

Up conversion

Mixing Principles

Frequency conversion is a form of amplitude modulation carried out by a mixer circuit or converter. The function performed by the mixer is called *heterodyning*.

Heterodyning

Figure 9-9 is a schematic diagram of a mixer circuit. Mixers accept two inputs. The signal f_s, which is to be translated to another frequency, is applied to one input, and the sine wave from a local oscillator f_o is applied to the other input. The signal to be translated can be a simple sine wave or any complex modulated signal containing sidebands. Like an amplitude modulator, a mixer essentially performs a mathematical multiplication of its two input signals according to the principles discussed in Chaps. 2 and 3. The oscillator is the carrier, and the signal to be translated is the modulating signal. The output contains not only the carrier signal but also sidebands formed when the local oscillator and input signal are mixed. The output of the mixer, therefore, consists of signals $f_s, f_o, f_o + f_s$, and $f_o - f_s$ or $f_s - f_o$.

The local oscillator signal f_o usually appears in the mixer output, as does the original input signal f_s in some types of mixer circuits. These are not needed in the output and are therefore filtered out. Either the sum or difference frequency in the output is the desired signal. For example, to translate the input signal to a lower frequency, the lower sideband or difference signal $f_o - f_s$ is chosen. The local oscillator frequency will be chosen such that when the information signal is subtracted from it, a signal with the

Figure 9-9   Concept of a mixer.

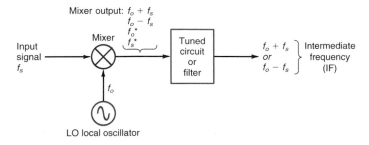

* May or may not be in the output depending upon the type of mixer.

Communication Receivers

305

desired lower frequency is obtained. When translating to a higher frequency, the upper sideband or sum signal $f_o + f_s$ is chosen. Again, the local oscillator frequency determines what the new higher frequency will be. A tuned circuit or filter is used at the output of the mixer to select the desired signal and reject all the others.

For example, for an FM radio receiver to translate an FM signal at 107.1 MHz to an intermediate frequency of 10.7 MHz for amplification and detection, a local oscillator frequency of 96.4 MHz is used. The mixer output signals are $f_s = 107.1$ MHz, $f_o = 96.4$ MHz, $f_o + f_s = 96.4 + 107.1 = 203.5$ MHz, and $f_s - f_o = 107.1 - 96.4 = 10.7$ MHz. Then a filter selects the 10.7-MHz signal (the IF, or f_{IF}) and rejects the others.

As another example, suppose a local oscillator frequency is needed that will produce an IF of 70 MHz for a signal frequency of 880 MHz. Since the IF is the difference between the input signal and local oscillator frequencies, there are two possibilities:

$$f_o = f_s + f_{IF} = 880 + 70 = 950 \text{ MHz}$$
$$f_o = f_s - f_{IF} = 880 - 70 = 810 \text{ MHz}$$

There are no set rules for deciding which of these to choose. However, at lower frequencies, say, those less than about 100 MHz, the local oscillator frequency is traditionally higher than the incoming signal's frequency, and at higher frequencies, those above 100 MHz, the local oscillator frequency is lower than the input signal frequency.

Keep in mind that the mixing process takes place on the whole spectrum of the input signal, whether it contains only a single-frequency carrier or multiple carriers and many complex sidebands. In the above example, the 10.7-MHz output signal contains the original frequency modulation. The result is as if the carrier frequency of the input signal is changed, as are all the sideband frequencies. The frequency conversion process makes it possible to shift a signal from one part of the spectrum to another, as required by the application.

Mixer and Converter Circuits

Any diode or transistor can be used to create a mixer circuit, but most modern mixers are sophisticated ICs. This section covers some of the more common and widely used types.

Diode Mixers. The primary characteristic of mixer circuits is nonlinearity. Any device or circuit whose output does not vary linearly with the input can be used as a mixer. For example, one of the most widely used types of mixer is the simple but effective diode modulator described in Chap. 3. *Diode mixers* like this are the most common type found in microwave applications.

A diode mixer circuit using a single diode is shown in Fig. 9-10. The input signal, which comes from an RF amplifier or, in some receivers, directly from the antenna, is

Diode mixer

Figure 9-10   A simple diode mixer.

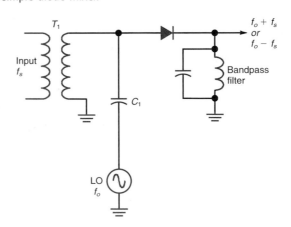

applied to the primary winding of transformer T_1. The signal is coupled to the secondary winding and applied to the diode mixer, and the local oscillator signal is coupled to the diode by way of capacitor C_1. The input and local oscillator signals are linearly added in this way and applied to the diode, which performs its nonlinear magic to produce the sum and difference frequencies. The output signals, including both inputs, are developed across the tuned circuit, which acts as a bandpass filter, selecting either the sum or difference frequency and eliminating the others.

Singly Balanced Mixers. A popular mixer circuit using two diodes is the *singly balanced mixer* illustrated in Fig. 9-11. The input signal is applied to the primary of the transformer, and the local oscillator is applied to the center tap of the secondary. The output signal is developed across an RFC and then applied to a bandpass filter or tuned circuit that selects the sum or difference frequency. In the single- or two-diode mixer, the local oscillator signal is much larger than the input signal, and so the diodes are turned off and on as switches are by the local oscillator signal.

Singly balanced mixer

In small-signal applications, germanium diodes, because of their relatively low turn-on voltage, are used in mixers. Silicon diodes also make excellent RF mixers. The best diode mixers at VHF, UHF, and microwave frequencies are hot carrier or Schottky barrier diodes.

Doubly Balanced Mixer. Balanced modulators are also widely used as mixers. These circuits eliminate the carrier from the output, making the job of filtering much easier. Any of the balanced modulators described previously can be used in mixing applications. Both the diode lattice balanced modulator and the integrated differential amplifier-type balanced modulator are quite effective in mixing applications. A version of the diode balanced modulator shown in Fig. 4-25, known as a *doubly balanced mixer* and illustrated in Fig. 9-12, is probably the single best mixer available, especially for VHF, UHF, and microwave frequencies. The transformers are precision-wound and the diodes matched in characteristics so that a high degree of carrier or local oscillator suppression occurs. In commercial products the local oscillator attenuation is 50 to 60 dB or more.

Doubly balanced mixer

FET Mixers. FETs make good mixers because they provide gain, have low noise, and offer a nearly perfect square-law response. An example is shown in Fig. 9-13. The *FET mixer* is biased so that it operates in the nonlinear portion of its range. The input signal is applied to the gate, and the local oscillator signal is coupled to the source. Again, the tuned circuit in the drain selects the difference frequency.

FET mixer

Figure 9-11   A singly balanced diode mixer.

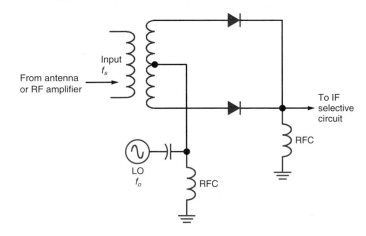

Figure 9-12 A doubly balanced mixer very popular at high frequencies.

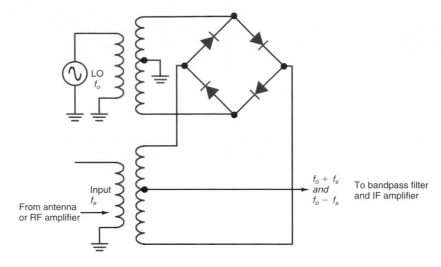

Figure 9-13 A JFET mixer.

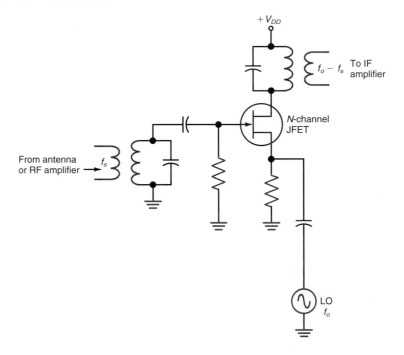

Another popular FET mixer, one with a dual-gate MOSFET, is shown in Fig. 9-14. Here the input signal is applied to one gate, and the local oscillator is coupled to the other gate. Dual-gate MOSFETs provide superior performance in mixing applications because their drain current I_D is directly proportional to the product of the two gate voltages. In receivers built for VHF, UHF, and microwave applications, junction FETs and dual-gate MOSFETs are widely used as mixers because of their high gain and low noise. Gallium arsenide FETs are preferred over silicon FETs at the higher frequencies because of their lower noise contribution and higher gain. IC mixers use MOSFETs.

One of the best reasons for using a FET mixer is that its characteristic drain current versus gate voltage curve is a perfect square-law function. (Recall that square-law formulas show how upper and lower sidebands and sum and difference frequencies are produced.) With a perfect square-law mixer response, only second-order harmonics are generated in addition to the sum and difference frequencies. Other mixers, such as diodes

Figure 9-14   A dual-gate MOSFET mixer.

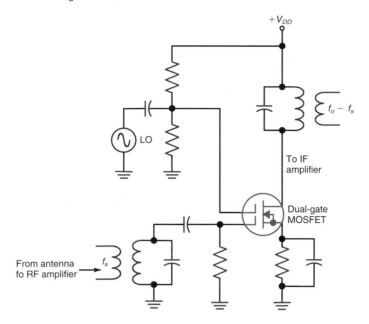

and bipolar transistors, approximate a square-law function; however, they are nonlinear, so that AM or heterodyning does occur. The nonlinearity is such that higher-order products such as the third, fourth, fifth, and higher harmonics are generated. Most of these can be eliminated by a bandpass filter that selects out the difference or sum frequency for the IF amplifier. However, the presence of higher-order products can cause unwanted low-level signals to appear in the receiver. These signals produce birdlike chirping sounds known as *birdies,* which, despite their low amplitude, can interfere with low-level input signals from the antenna or RF amplifier. FETs do not have this problem, and so FETs are the preferred mixer in most receivers, except for high-frequency applications, where diode mixers are used.

Birdies

IC Mixers. A typical IC mixer, the *NE602 mixer,* is shown in Fig. 9-15(*a*). The NE602, also known as a *Gilbert transconductance cell,* or *Gilbert cell,* consists of a doubly balanced mixer circuit made up of two cross-connected differential amplifiers. Although most doubly balanced mixers are passive devices with diodes, as described earlier, the NE602 uses bipolar transistors. Also on the chip is an NPN transistor that can be connected as a stable oscillator circuit and a dc voltage regulator. The device is housed in an 8-pin DIP. It operates from a single dc power supply voltage of 4.5 to 8 V. The circuit can be used at frequencies up to 500 MHz, making it useful in HF, VHF, and low-frequency UHF applications. The oscillator, which operates up to about 200 MHz, is internally connected to one input of the mixer. An external *LC* tuned circuit or a crystal is required to set the operating frequency.

NE602 mixer (Gilbert cell)

Figure 9-15(*b*) shows the circuit details of the mixer itself. Bipolar transistors Q_1 and Q_2 form a differential amplifier with current source Q_3, and Q_4 and Q_5 form another differential amplifier with current source Q_6. Note that the inputs are connected in parallel. The collectors are cross-connected; i.e., the collector of Q_1 is connected to the collector of Q_4 instead of Q_3, as would be the case for a parallel connection, and the collector of Q_2 is connected to the collector of Q_3. This connection results in a circuit that is like a balanced modulator in that the internal oscillator signal and the input signal are suppressed, leaving only the sum and difference signals in the output. The output may be balanced or single-ended, as required. A filter or tuned circuit must be connected to the output to select the desired sum or difference signal.

Figure 9-15 NE602 IC mixer. (*a*) Block diagram and pinout. (*b*) Simplified schematic.

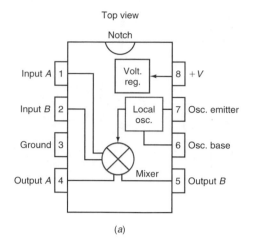

(*a*)

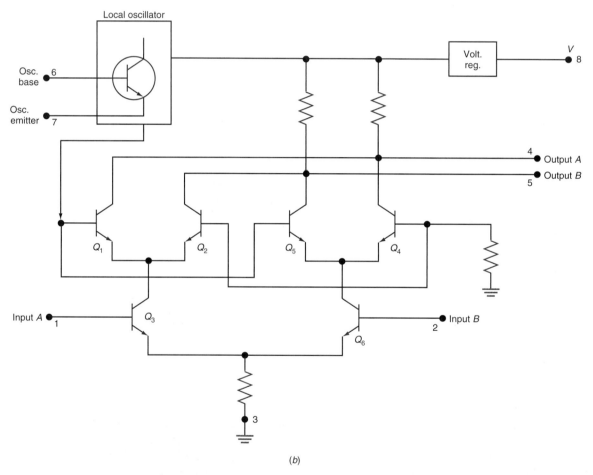

(*b*)

A typical circuit using the NE602 IC mixer is shown in Fig. 9-16. Both R_1 and C_1 are used for decoupling, and a resonant transformer T_1 couples the 72-MHz input signal to the mixer. Capacitor C_2 resonates with the transformer secondary at the input frequency, and C_3 is an ac bypass connecting pin 2 to ground. External components C_4 and L_1 form a tuned circuit that sets the oscillator to 82 MHz. Capacitors C_5 and C_6 form a capacitive voltage divider that connects the on-chip NPN transistor as a Colpitts oscillator circuit. Capacitor C_7 is an ac coupling and blocking capacitor. The output is taken from pin 4 and connected to a ceramic bandpass filter, which provides selectivity. The output, in this

Figure 9-16 NE602 mixer used for frequency translation.

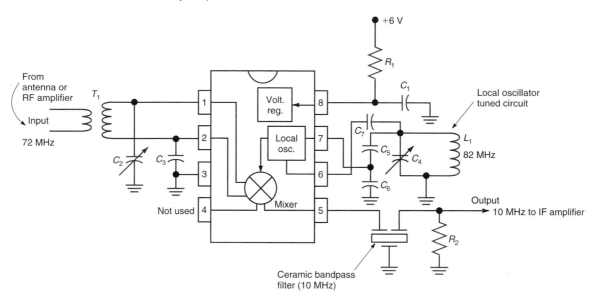

case the difference signal, or $82 - 72 = 10$ MHz, appears across R_2. The balanced mixer circuit suppresses the 82-MHz oscillator signal, and the sum signal of 154 MHz is filtered out. The output IF signal plus any modulation that appeared on the input is passed to IF amplifiers for an additional boost in gain prior to demodulation.

Image Reject Mixer. A special type of mixer is used in designs in which images cannot be tolerated. All superheterodyne receivers suffer from images (see Sec. 9-4), but some more than others because of the frequency of operation, chosen IF, and interfering signal frequency. When proper choice of IF and front-end selectivity cannot eliminate the images, an *image reject mixer* can be used. It uses Gilbert cell mixers in a configuration like that used in a phasing-type SSB generator. Referring to Chap. 4 and Fig. 4-30, you can see how this circuit can be used as a mixer. A balanced modulator is also a mixer. With this technique, the desired signal can be passed, but the image will be canceled by the phasing technique. Such circuits are sensitive to adjust, but they result in superior image performance in critical applications. This approach is widely used in modern UHF and microwave IC receivers like those in cell phones.

Image reject mixer

Local Oscillators and Frequency Synthesizers

The local oscillator signal for the mixer comes from either a conventional *LC* tuned oscillator such as a Colpitts or Clapp circuit or a frequency synthesizer. The simpler continuously tuned receivers use an *LC* oscillator. Channelized receivers use frequency synthesizers.

LC **Oscillators.** A representative local oscillator for frequencies up to 100 MHz is shown in Fig. 9-17. This type of circuit, which is sometimes referred to as a variable-frequency oscillator, or VFO, uses a JFET Q_1 connected as a Colpitts oscillator. Feedback is developed by the voltage divider, which is made up of C_5 and C_6. The frequency is set by the parallel tuned circuit made up of L_1 in parallel with C_1, which is also in parallel with the series combination of C_2 and C_3. The oscillator is set to the center of its desired operating range by a coarse adjustment of trimmer capacitor C_1. Coarse tuning can also be accomplished by making L_1 variable. An adjustable slug-tuned ferrite core moved in and out of L_1 can set the desired frequency range. The main tuning is accomplished with variable capacitor C_3, which is connected mechanically to some kind of dial mechanism that has been calibrated in frequency.

LC oscillator

Figure 9-17   A VFO for receiver local oscillator service.

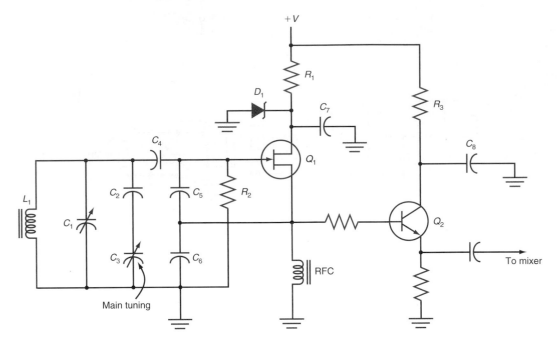

Main tuning can also be accomplished with varactors. For example, C_3 in Fig. 9-17 can be replaced by a varactor, reverse-biased to make it act as a capacitor. Then a potentiometer applies a variable dc bias to change the capacitance and thus the frequency.

The output of the oscillator is taken from across the RFC in the source lead of Q_1 and applied to a direct-coupled emitter follower. The emitter follower buffers the output, isolating the oscillator from load variations that can change its frequency. The emitter-follower buffer provides a low-impedance source to connect to the mixer circuit, which often has a low input impedance. If frequency changes after a desired station is tuned in, which can occur as a result of outside influences such as changes in temperature, voltage, and load, the signal will drift off and no longer be centered in the passband of the IF amplifier. One of the key features of local oscillators is their stability, i.e., their ability to resist frequency changes. The emitter follower essentially eliminates the effects of load changes. The zener diode gets its input from the power supply, which is regulated, providing a regulated dc to the circuit and ensuring maximum stability of the supply voltage to Q_1.

Most drift comes from the *LC* circuit components themselves. Even inductors, which are relatively stable, have slight positive temperature coefficients, and special capacitors that change little with temperature are essential. Usually, negative temperature coefficient (NPO) ceramic capacitors are selected to offset the positive temperature coefficient of the inductor. Capacitors with a mica dielectric are also used.

Frequency synthesizer

Frequency Synthesizers. Most new receiver designs incorporate *frequency synthesizers* for the local oscillator, which provides some important benefits over the simple VFO designs. First, since the synthesizer is usually of the phase-locked loop (PLL) design, the output is locked to a crystal oscillator reference, providing a high degree of stability. Second, tuning is accomplished by changing the frequency division factor in the PLL, resulting in incremental rather than continuous frequency changes. Most communication is channelized; i.e., stations operate on assigned frequencies that are a known frequency increment apart, and setting the PLL step frequency to the channel spacing allows every channel in the desired spectrum to be selected simply by changing the frequency division factor.

The former disadvantages of frequency synthesizers—higher cost and greater circuit complexity—have been offset by the availability of low-cost PLL synthesizer ICs, which make local oscillator design simple and cheap. Most modern receivers, from AM/FM car radios, stereos, and TV sets to military receivers and commercial transceivers, use frequency synthesis.

The frequency synthesizers used in receivers are largely identical to those described in Chap. 8. However, some additional techniques, such as the use of a mixer in the feedback loop, are employed. The circuit in Fig. 9-18 is a traditional PLL configuration, with the addition of a mixer connected between the VFO output and the frequency divider. A crystal reference oscillator provides one input to a phase detector, which is compared to the output of the frequency divider. Tuning is accomplished by adjusting the frequency division ratio by changing the binary number input to the divider circuit. This binary number can come from a switch, a counter, a ROM, or a microprocessor. The output of the phase detector is filtered by the loop filter into a dc control voltage to vary the frequency of the variable-frequency oscillator, which generates the final output that is applied to the mixer in the receiver.

As stated previously, one of the disadvantages of very high-frequency PLL synthesizers is that the VFO output frequency is often higher in frequency than the upper operating limit of the variable-modulus frequency divider ICs commonly available. One approach to this problem is to use a prescaler to reduce the VFO frequency before it is applied to the variable-frequency divider. Another is to reduce the VFO output frequency to a lower value within the range of the dividers by down-converting it with a mixer, as illustrated in Fig. 9-18. The VFO output is mixed with the signal from another crystal oscillator, and the difference frequency is selected. With some UHF and microwave receivers, it is necessary to generate the local oscillator signal at a lower frequency and then use a PLL frequency multiplier to increase the frequency to the desired higher level. The position of this optional multiplier is shown by the dotted line in Fig. 9-18.

As an example, assume that a receiver must tune to 190.04 MHz and that the IF is 45 MHz. The local oscillator frequency can be either 45 MHz lower or higher than the input signal. Using the lower frequency, we have $190.04 - 45 = 145.04$ MHz. Now, when the incoming 190.04-MHz signal is mixed with the 145.04-MHz signal to be generated by the synthesizer in Fig. 9-18, its IF will be the difference frequency of $190.04 - 145.04 = 45$ MHz.

Figure 9-18 A frequency synthesizer used as a receiver local oscillator.

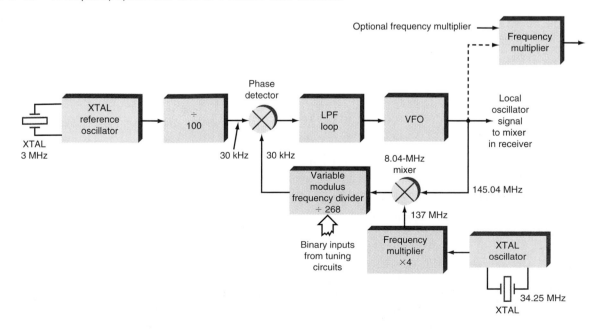

The output of the VFO in Fig. 9-18 is 145.04 MHz. It is mixed with the signal from a crystal oscillator whose frequency is 137 MHz. The crystal oscillator, set to 34.25 MHz, is applied to a frequency multiplier that multiplies by a factor of 4. The 145.04-MHz signal from the VFO is mixed with the 137-MHz signal, and the sum and difference frequencies are generated. The difference frequency is selected, for $145.04 - 137 = 8.04$ MHz. This frequency is well within the range of programmable-modulus IC frequency dividers.

The frequency divider is set to divide by a factor of 268, and so the output of the divider is $8,040,000/268 = 30,000$ Hz, or 30 kHz. This signal is applied to the phase detector. It is the same as the other input to the phase detector, as it should be for a locked condition. The reference input to the phase detector is derived from a 3-MHz crystal oscillator divided down to 30 kHz by a divide-by-100 divider. This means that the synthesizer is stepped in 30-kHz increments.

Now, suppose that the divider factor is changed from 268 to 269 to tune the receiver. To ensure that the PLL stays locked, the VFO output frequency must change. To achieve 30 kHz at the output of the divider with a ratio of 269, the divider input has to be 269×30 kHz $= 8070$ kHz $= 8.07$ MHz. This 8.07-MHz signal comes from the mixer, whose inputs are the VFO and the crystal oscillator. The crystal oscillator input remains at 137 MHz, so the VFO frequency must be 137 MHz higher than the 8.07-MHz output of the mixer, or $137 + 8.07 = 145.07$ MHz. This is the output of the VFO and local oscillator of the receiver. With a fixed IF of 45 MHz, the receiver will now be tuned to the IF plus the local oscillator input, or $145.07 + 45 = 190.07$ MHz.

Note that the change of the divider factor by one increment, from 268 to 269, changes the frequency by one 30-kHz increment, as desired. The addition of the mixer to the circuit does not affect the step increment, which is still controlled by the frequency of the reference input frequency.

9-4 Intermediate Frequency and Images

The choice of IF is usually a design compromise. The primary objective is to obtain good selectivity. Narrowband selectivity is best obtained at lower frequencies, particularly when conventional LC tuned circuits are used. Even active RC filters can be used when IFs of 500 KHz or less are used. There are various design benefits of using a low IF. At low frequencies, the circuits are far more stable with high gain. At higher frequencies, circuit layouts must take into account stray inductances and capacitances, as well as the need for shielding, if undesired feedback paths are to be avoided. With very high circuit gain, some of the signal can be fed back in phase and cause oscillation. Oscillation is not as much of a problem at lower frequencies. However, when low IFs are selected, a different sort of problem is faced, particularly if the signal to be received is very high in frequency. This is the problem of images. An *image* is a potentially interfering RF signal that is spaced from the desired incoming signal by a frequency that is 2 times the intermediate frequency above or below the incoming frequency, or

$$f_i = f_s + 2f_{IF} \qquad \text{and} \qquad f_i = f_s - 2f_{IF}$$

where f_i = image frequency
f_s = desired signal frequency
f_{IF} = intermediate frequency

This is illustrated graphically in Fig. 9-19. Note that which of the images occurs depends on whether the local oscillator frequency f_o is above or below the signal frequency.

Frequency Relationships and Images

As stated previously, the mixer in a superhetodyne receiver produces the sum and difference frequencies of the incoming signal and the local oscillator signal. Normally, the difference frequency is selected as the IF. The frequency of the local oscillator is usually

GOOD TO KNOW

Crowding of the RF spectrum increases the chance that a signal on the image frequency will cause image interference. To help remedy this, high-Q tuned circuits should be used ahead of the mixer or RF amplifier.

Figure 9-19   Relationship of the signal and image frequencies.

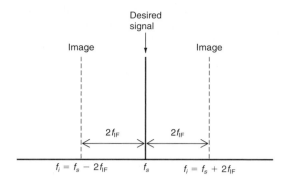

chosen to be higher in frequency than the incoming signal by the IF. However, the local oscillator frequency could also be made lower than the incoming signal frequency by an amount equal to the IF. Either choice will produce the desired difference frequency. For the following example, assume that the local oscillator frequency is higher than the incoming signal frequency.

Now, if an image signal appears at the input of the mixer, the mixer will, of course, produce the sum and difference frequencies regardless of the inputs. Therefore, the mixer output will again be the difference frequency at the IF value. Assume, e.g., a desired signal frequency of 90 MHz and a local oscillator frequency of 100 MHz. The IF is the difference $100 - 90 = 10$ MHz. The image frequency is $f_i = f_s + 2f_{IF} = 90 + 2(10) = 90 + 20 = 110$ MHz.

If an undesired signal, the image, appears at the mixer input, the output will be the difference $110 - 100 = 10$ MHz. The IF amplifier will pass it. Now look at Fig. 9-20, which shows the relationships between the signal, local oscillator, and image frequencies. The mixer produces the difference between the local oscillator frequency and the desired signal frequency, or the difference between the local oscillator frequency and the image frequency. In both cases, the IF is 10 MHz. This means that a signal spaced from the desired signal by 2 times the IF can also be picked up by the receiver and converted to the IF. When this occurs, the image signal interferes with the desired signal. In today's crowded RF spectrum, chances are high that there will be a signal on the image frequency, and image interference can even make the desired signal unintelligible. Superheterodyne design must, therefore, find a way to solve the image problem.

Figure 9-20   Signal, local oscillator, and image frequencies in a superheterodyne.

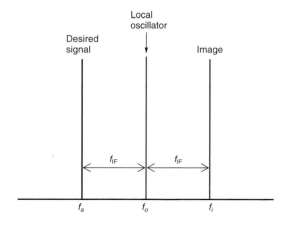

Solving the Image Problem

Image interference occurs only when the image signal is allowed to appear at the mixer input. This is the reason for using high-Q tuned circuits ahead of the mixer, or a selective RF amplifier. If the selectivity of the RF amplifier and tuned circuits is good enough, the image will be rejected. In a fixed tuned receiver designed for a specific frequency, it is possible to optimize the receiver front end for the good selectivity necessary to eliminate images. But many receivers have broadband RF amplifiers that allow many frequencies within a specific band to pass. Other receivers must be made tunable over a wide frequency range. In such cases, selectivity becomes a problem.

Assume, e.g., that a receiver is designed to pick up a signal at 25 MHz. The IF is 500 kHz, or 0.5 MHz. The local oscillator is adjusted to a frequency right above the incoming signal by an amount equal to the IF, or $25 + 0.5 = 25.5$ MHz. When the local oscillator and signal frequencies are mixed, the difference is 0.5 MHz, as desired. The image frequency is $f_i = f_s + 2f_{IF} = 25 + 2(0.5) = 26$ MHz. An image frequency of 26 MHz will cause interference to the desired signal at 25 MHz unless it is rejected. The signal, local oscillator, and image frequencies for this situation are shown in Fig. 9-21.

Now, assume that a tuned circuit ahead of the mixer has a Q of 10. Given this and the resonant frequency, the bandwidth of the resonant circuit can be calculated as $BW = f_r/Q = 25/10 = 2.5$ MHz. The response curve for this tuned circuit is shown in Fig. 9-21. As shown, the bandwidth of the resonant circuit is relatively wide. The bandwidth is centered on the signal frequency of 25 MHz. The upper cutoff frequency is $f_2 = 26.25$ MHz, the lower cutoff frequency is $f_1 = 23.75$ MHz, and the bandwidth is $BW = f_2 - f_1 = 26.25 - 23.75 = 2.5$ MHz. (Remember that the bandwidth is measured at the 3-dB down points on the tuned circuit response curve.)

The fact that the upper cutoff frequency is higher than the image frequency, 26 MHz, means that the image frequency appears in the passband; it would thus be passed relatively unattenuated by the tuned circuit, causing interference.

It is clear how cascading tuned circuits and making them with higher Qs can help solve the problem. For example, assume a Q of 20, instead of the previously given value of 10. The bandwidth at the center frequency of 25 MHz is then $f_s/Q = 25/20 = 1.25$ MHz.

The resulting response curve is shown by the darker line in Fig. 9-21. The image is now outside of the passband and is thus attenuated. Using a Q of 20 would not solve the image problem completely, but still higher Qs would further narrow the bandwidth, attenuating the image even more.

Figure 9-21 A low IF compared to the signal frequency with low-Q tuned circuits causes images to pass and interfere.

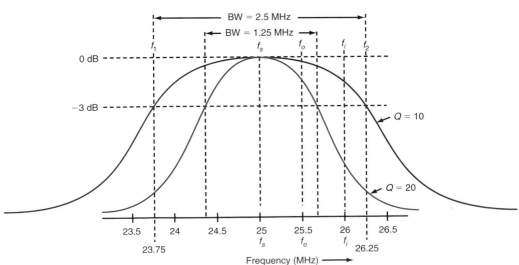

Higher Qs are, however, difficult to achieve and often complicate the design of receivers that must be tuned over a wide range of frequencies. The usual solution to this problem is to choose a higher IF. Assume, e.g., that an intermediate frequency of 9 MHz is chosen (the Q is still 10). Now the image frequency is $f_i = 25 + 2(9) = 43$ MHz. A signal at a 43-MHz frequency would interfere with the desired 25-MHz signal if it were allowed to pass into the mixer. But 43 MHz is well out of the tuned circuit bandpass; the relatively low-Q selectivity of 10 is sufficient to adequately reject the image. Of course, choosing the higher intermediate frequency causes some design difficulties, as indicated earlier.

To summarize, the IF is made as high as possible for effective elimination of the image problem, yet low enough to prevent design problems. In most receivers the IF varies in proportion to the frequencies that must be covered. At low frequencies, low values of IF are used. A value of 455 kHz is common for AM broadcast band receivers and for others covering that general frequency range. At frequencies up to about 30 MHz, 3385 kHz and 9 MHz are common IF frequencies. In FM radios that receive 88 to 108 MHz, 10.7 MHz is a standard IF. In TV receivers, an IF in the 40- to 50-MHz range is common. In the microwave region, radar receivers typically use an IF in the 60-MHz range, and satellite communication equipment uses 70- and 140-MHz IFs.

Dual–Conversion Receivers

Another way to obtain selectivity while eliminating the image problem is to use a *dual-conversion* superheterodyne *receiver*. See Fig. 9-22. The receiver shown in the figure uses two mixers and local oscillators, and so it has two IFs. The first mixer converts the incoming signal to a relatively high intermediate frequency for the purpose of eliminating the images; the second mixer converts that IF down to a much lower frequency, where good selectivity is easier to obtain.

Figure 9-22 shows how the different frequencies are obtained. Each mixer produces the difference frequency. The first local oscillator is variable and provides the tuning for the receiver. The second local oscillator is fixed in frequency. Since it need convert only one fixed IF to a lower IF, this local oscillator does not have to be tunable. In most cases, its frequency is set by a quartz crystal. In some receivers, the first mixer is driven by the fixed-frequency local oscillator, and tuning is done with the second local oscillator. Dual-conversion receivers are relatively common. Most shortwave receivers and many at VHF, UHF, and microwave frequencies use dual conversion. For example, a CB receiver operating in the 27-MHz range typically uses a 10.7-MHz first IF and a 455-kHz second IF. For some critical applications, triple-conversion receivers are used to further minimize the image problem, although their use is not common. A triple-conversion receiver uses three mixers and three different intermediate-frequency values.

Dual-conversion receiver

Figure 9-22 A dual-conversion superheterodyne.

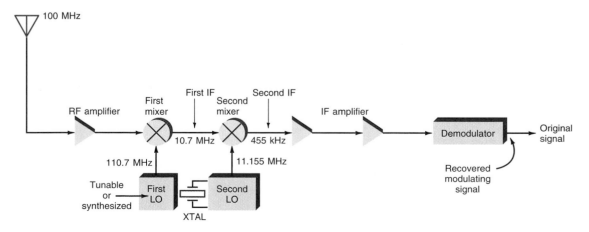

Example 9-1

A superheterodyne receiver must cover the range from 220 to 224 MHz. The first IF is 10.7 MHz; the second is 1.5 MHz. Find (*a*) the local oscillator tuning range, (*b*) the frequency of the second local oscillator, and (*c*) the first IF image frequency range. (Assume a local oscillator frequency higher than the input by the IF.)

a. $220 + 10.7 = 230.7$ MHz

$224 + 10.7 = 234.7$ MHz

The tuning range is 230.7 to 234.7 MHz

b. The second local oscillator frequency is 1.5 MHz higher than the first IF.

$$10.7 + 1.5 = 12.2 \text{ MHz}$$

c. The first IF image range is 241.4 to 245.4 MHz.

$$230.7 + 10.7 = 241.4 \text{ MHz}$$

$$234.7 + 10.7 = 245.4 \text{ MHz}$$

Direct Conversion Receivers

Direct conversion (DC) or zero-IF (ZIF) receiver

A special version of the superheterodyne is known as the *direct conversion (DC) or zero-IF (ZIF) receiver*. Instead of translating the incoming signal to another (usually lower) intermediate frequency, dc receivers convert the incoming signal directly to baseband. In other words, they perform the demodulation of the signal as part of the translation.

Figure 9-23 shows the basic ZIF receiver architecture. The low-noise amplifier (LNA) boosts the signal level before the mixer. The local oscillator (LO) frequency, usually from a PLL frequency synthesizer f_{LO}, is set to the frequency of the incoming signal f_s.

$$f_{LO} = f_s$$

The sum and difference frequencies as a result of mixing are

$$f_{LO} - f_s = 0$$

$$f_{LO} + f_s = 2f_{LO} = 2f_s$$

The difference frequency is zero. Without modulation, there is no output. With AM, the sidebands mix with the LO to reproduce the original modulation baseband signal. In this

Figure 9-23 A direct-conversion (zero-IF) receiver.

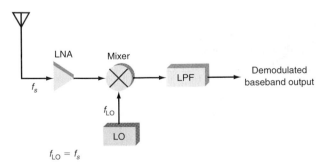

instance, the mixer is also the demodulator. The sum is twice the LO frequency that is removed by the low-pass filter (LPF).

Assume a carrier of 21 MHz and a voice modulation signal from 300 to 3000 Hz and AM. The sidebands extend from 20,997,000 to 21,003,000 Hz. At the receiver, the LO is set to 21 MHz. The mixer produces

$$21,000,000 - 20,997,000 = 3000 \text{ Hz}$$
$$21,003,000 - 21,000,000 = 3000 \text{ Hz}$$
$$21,000,000 + 21,003,000 = 42,003,000 \text{ Hz}$$
$$21,000,000 + 20,997,000 = 41,997,000 \text{ Hz}$$

A low-pass filter at the mixer output whose cutoff frequency is set to 3 kHz easily filters out the 42-MHz components.

The DC receiver has several key benefits. First, no separate IF filter is needed. This is usually a crystal, ceramic, or SAW filter that is expensive and takes up valuable printed-circuit board space in compact designs. An inexpensive *RC*, *LC*, or active low-pass filter at the mixer output supplies the needed selectivity. Second, no separate detector circuit is needed, because demodulation is inherent in the technique. Third, in transceivers that use half duplex and in which the transmitter and receiver are on the same frequency, only one PLL frequency synthesizer voltage-controlled oscillator is needed. All these benefits result in simplicity and its attendant lower cost. Fourth, there is no image problem.

The disadvantages of this receiver are subtle. In designs with no RF amplifier (LNA), the LO signal can leak through the mixer to the antenna and radiate. A LNA reduces this likelihood, but even so, careful design is required to minimize the radiation. Second, an undesired dc offset can develop in the output. Unless all circuits are perfectly balanced, dc offset can upset bias arrangements in later circuits as well as cause circuit saturation that will prevent amplification and other operations. Finally, the ZIF receiver can be used only with CW, AM, SSB, or DSB. It cannot recognize phase or frequency variations.

To use this type of receiver with FM, FSK, PM, or PSK, or any form of digital modulation, two mixers are required along with a quadrature LO arrangement. Such designs are used in most cell phones and other wireless receivers.

Figure 9-24 shows a direct conversion receiver that is typical of those using digital modulation. The incoming signal is sent to a SAW filter that provides some initial selectivity. The LNA provides amplification. The LNA output is fed to two mixers. The local oscillator (LO) signal, usually from a synthesizer, is fed directly to the upper mixer (sin θ) and to a 90° phase shifter that, in turn, supplies the lower mixer (cos θ). Remember, the LO frequency is equal to the incoming signal frequency. The mixers provide baseband signals at their outputs. The double LO frequency signals resulting from mixing are

Figure 9-24 A direct conversion receiver for FM, FSK, PSK, and digital modulation.

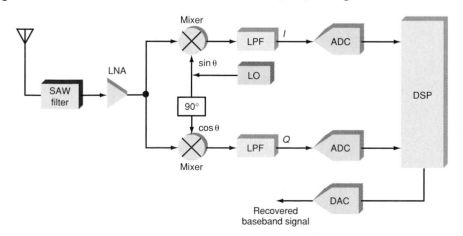

Recovered baseband signal

removed with the low-pass filters (LPFs). The two baseband signals are separated in phase by 90°. The upper signal is usually referred to as the in-phase I signal while the lower signal is referred to as the quadrature Q signal. (Quadrature means a 90° phase difference.) The I and Q signals are then sent to analog-to-digital converters (ADCs) where they are converted to binary signals. The binary signals are then sent to a digital signal processor (DSP). The DSP contains a prestored subroutine that performs the demodulation. This algorithm requires the two quadrature signals in order to have enough data to distinguish phase and frequency changes in the original signal resulting from modulation. The output from the demodulation subroutine is fed to an external digital-to-analog converter (DAC) where the original modulating signal is reproduced. This I-Q direct conversion architecture is now one of the most common receiver architectures used in cell phones and wireless networking ICs.

Software-Defined Radio

A *software-defined radio (SDR)* is a receiver in which most of the functions are performed by a digital signal processor. Figure 9-25 is a general block diagram of an SDR. While only one mixer and ADC are shown, keep in mind that the I and Q architecture of Fig. 9-24 is normally used. As in most receivers, a LNA provides initial amplification and a mixer down-converts the signal to an IF or to baseband in a DC receiver. The IF or baseband signal is then digitized by an analog-to-digital (A/D) converter. The binary words representing the IF signal with its modulation are stored in RAM. A DSP chip then performs additional filtering, demodulation, and baseband operations (voice decoding, companding, etc.).

The fastest A/D converters available today can digitize at a rate of up to 300 MHz. To meet the Nyquist requirement, this means that the highest frequency that can be digitized is less than 150 MHz. This is why the SDR must down-convert the incoming signal to an IF of less than 150 MHz. Further, the DSP must be fast enough to perform the DSP math in real time. Although DSP chips can operate at clock rates up to 1 GHz, the time it takes to execute even at these speeds limits the IF to a lower value. A practical value is in the 40- to 90-MHz range, where the A/D converter and the DSP can handle the computing chores.

An alternative is to use a dual-conversion SDR. The first mixer converts the signal to an IF that is then fed to the A/D converter, where it digitizes the data. The DSP chip is then used to down-convert the signal to an even lower IF. This mixing or down-converstion is done digitally. It is similar to the process of aliasing, in which a lower difference frequency is generated. From there the DSP performs the filtering and demodulation duties.

Figure 9-25 A software-defined radio (SDR).

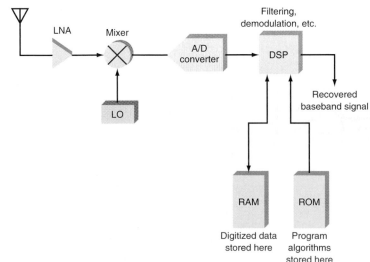

Chapter 9

The filtering, demodulation, and other processes are, of course, defined by mathematical algorithms that are in turn programmed with a computer language. The resulting programs are stored in the DSP ROM.

SDR techniques have been known for many years, but only since the early to mid-1990s have A/D converter circuits and DSP chips become fast enough to perform the desired operations at radio frequencies. SDRs have already been widely adopted in military receivers, cell phones, and cell phone base stations. As prices continue to drop and as A/D converters become even faster, these methods will become even more widely used in other communication equipment.

The benefits of SDRs are improved performance and flexibility. DSP filtering and other processes are typically superior to equivalent analog techniques. Furthermore, the receiver characteristics (type of modulation, selectivity, etc.) can be easily changed by running a different program. SDRs can be changed by downloading or switching to a new processing program that the DSP can execute. No hardware changes are required.

As A/D converters and DSPs get faster, it is expected that more receiver functions will become software-defined. The ultimate SDR is a LNA connected to an antenna and whose output goes directly to a fast A/D converter. All mixing, filtering, demodulation, and other operations are performed in DSP software.

COGNITIVE RADIO

Cognitive radio is the term describing an advanced form of SDR that is designed to help alleviate the frequency spectrum shortage. While most of the usable frequency spectrum has already been assigned by the various government regulating agencies, at any given time much of that spectrum goes unused, at least for part of the time. The question is, How can that spectrum be more efficiently assigned and used? A good example is the frequency spectrum assigned to the UHF TV stations. This spectrum in the 500- to 800-MHz range is essentially vacant except for the occasional UHF TV station. Few people watch UHF TV directly by radio. Instead, most watch by cable TV which retransmits the UHF TV station on cable. This is a huge waste of precious spectrum space, yet broadcasters are reluctant to give it up. The cell phone companies which are perpetually short of spectrum for new subscribers covet this unused but unattainable space.

A cognitive radio is designed to seek out unused spectrum space and then reconfigure itself to receive and transmit on unused portions of the spectrum that it finds. Such radios are now possible because of the availability of very agile and wide-ranging frequency synthesizers and DSP techniques. The radio could easily change frequency as well as modulation/multiplexing methods on the fly to establish communications. Look for future military and government service radios to take advantage of these techniques.

Cognitive radio

9-5 Noise

Noise is an electronic signal that is a mixture of many random frequencies at many amplitudes that gets added to a radio or information signal as it is transmitted from one place to another or as it is processed. Noise is *not* the same as interference from other information signals.

Noise

When you turn on any AM or FM receiver and tune it to some position between stations, the hiss or static that you hear in the speaker is noise. Noise also shows up on a black-and-white TV screen as snow or on a color screen as confetti. If the noise level is high enough and/or the signal is weak enough, the noise can completely obliterate the original signal. Noise that occurs in transmitting digital data causes *bit errors* and can result in information being garbled or lost.

The noise level in a system is proportional to temperature and bandwidth, and to the amount of current flowing in a component, the gain of the circuit, and the resistance of the circuit. Increasing any of these factors increases noise. Therefore, low noise is best obtained by using low-gain circuits, low direct current, low resistance values, and narrow bandwidths. Keeping the temperature low can also help.

Noise is a problem in communication systems whenever the received signals are very low in amplitude. When the transmission is over short distances or high-power transmitters are being used, noise is not usually a problem. But in most communication systems, weak signals are normal, and noise must be taken into account at the design stage. It is in the receiver that noise is the most detrimental because the receiver must amplify the weak signal and recover the information reliably.

Noise can be external to the receiver or originate within the receiver itself. Both types are found in all receivers, and both affect the signal-to-noise ratio.

Signal-to-Noise Ratio

The *signal-to-noise (S/N) ratio,* also designated SNR, indicates the relative strengths of the signal and the noise in a communication system. The stronger the signal and the weaker the noise, the higher the *S/N* ratio. If the signal is weak and the noise is strong, the *S/N* ratio will be low and reception will be unreliable. Communication equipment is designed to produce the highest feasible *S/N* ratio.

Signals can be expressed in terms of voltage or power. The *S/N* ratio is computed by using either voltage or power values:

$$\frac{S}{N} = \frac{V_s}{V_n} \qquad \text{or} \qquad \frac{S}{N} = \frac{P_s}{P_n}$$

where V_s = signal voltage
V_n = noise voltage
P_s = signal power
P_n = noise power

Assume, e.g., that the signal voltage is 1.2 μV and the noise is 0.3 μV. The *S/N* ratio is 1.2/0.3 = 4. Most *S/N* ratios are expressed in terms of power rather than voltage. For example, if the signal power is 5 μW and the power is 125 nW, the *S/N* ratio is $5 \times 10^{-6}/125 \times 10^{-9} = 40$.

The preceding *S/N* values can be converted to decibels as follows:

For voltage: dB $= 20 \log \dfrac{S}{N} = 20 \log 4 = 20(0.602) = 12$ dB

For power: dB $= 10 \log \dfrac{S}{N} = 10 \log 40 = 10(1.602) = 16$ dB

However it is expressed, if the *S/N* ratio is less than 1, the dB value will be negative and the noise will be stronger than the signal.

External Noise

External noise comes from sources over which we have little or no control—industrial, atmospheric, or space. Regardless of its source, noise shows up as a random ac voltage and can be seen on an oscilloscope. The amplitude varies over a wide range, as does the frequency. One can say that noise in general contains all frequencies, varying randomly. This is generally known as *white noise*.

Atmospheric noise and space noise are a fact of life and simply cannot be eliminated. Some industrial noise can be controlled at the source, but because there are so many sources of this type of noise, there is no way to eliminate it. The key to reliable communication, then, is simply to generate signals at a high enough power to overcome external noise. In some cases, shielding sensitive circuits in metallic enclosures can aid in noise control.

Industrial Noise. *Industrial noise* is produced by manufactured equipment, such as automotive ignition systems, electric motors, and generators. Any electrical equipment that causes high voltages or currents to be switched produces transients that create noise. Noise pulses of large amplitude occur whenever a motor or other inductive device is turned on or off. The resulting transients are extremely large in amplitude and rich in random harmonics. Fluorescent and other forms of gas-filled lights are another common source of industrial noise.

Atmospheric Noise. The electrical disturbances that occur naturally in the earth's atmosphere are another source of noise. *Atmospheric noise* is often referred to as *static*. Static usually comes from lightning, the electric discharges that occur between clouds or between the earth and clouds. Huge static charges build up on the clouds, and when the potential difference is great enough, an arc is created and electricity literally flows through the air. Lightning is very much like the static charges that we experience during a dry spell in the winter. The voltages involved are, however, enormous, and these transient electric signals of megawatt power generate harmonic energy that can travel over extremely long distances.

Like industrial noise, atmospheric noise shows up primarily as amplitude variations that add to a signal and interfere with it. Atmospheric noise has its greatest impact on signals at frequencies below 30 MHz.

Extraterrestrial Noise. *Extraterrestrial noise,* solar and cosmic, comes from sources in space. One of the primary sources of extraterrestrial noise is the sun, which radiates a wide range of signals in a broad noise spectrum. The noise intensity produced by the sun varies with time. In fact, the sun has a repeatable 11-year noise cycle. During the peak of the cycle, the sun produces an awesome amount of noise that causes tremendous radio signal interference and makes many frequencies unusable for communication. During other years, the noise is at a lower level.

Noise generated by stars outside our solar system is generally known as *cosmic noise.* Although its level is not as great as that of noise produced by the sun, because of the great distances between those stars and earth, it is nevertheless an important source of noise that must be considered. It shows up primarily in the 10-MHz to 1.5-GHz range, but causes the greatest disruptions in the 15- to 150-MHz range.

Internal Noise

Electronic components in a receiver such as resistors, diodes, and transistors are major sources of *internal noise.* Internal noise, although it is low level, is often great enough to interfere with weak signals. The main sources of internal noise in a receiver are thermal noise, semiconductor noise, and intermodulation distortion. Since the sources of internal noise are well known, there is some design control over this type of noise.

Thermal Noise. Most internal noise is caused by a phenomenon known as *thermal agitation,* the random motion of free electrons in a conductor caused by heat. Increasing the temperature causes this atomic motion to increase. Since the components are conductors, the movement of electrons constitutes a current flow that causes a small voltage to be produced across that component. Electrons traversing a conductor as current flows experience fleeting impediments in their path as they encounter the thermally agitated atoms. The apparent resistance of the conductor thus fluctuates, causing the thermally produced random voltage we call noise.

You can actually observe this noise by simply connecting a high-value (megohm) resistor to a very high-gain oscilloscope. The motion of the electrons due to room temperature in the resistor causes a voltage to appear across it. The voltage variation is completely random and at a very low level. The noise developed across a resistor is proportional to the temperature to which it is exposed.

Industrial noise

Atmospheric noise

Static

Extraterrestrial noise

GOOD TO KNOW

Low noise is best obtained by using low-gain circuits, low direct current, low resistance values, and narrow bandwidths. Keeping temperature low can also help.

PIONEERS
OF ELECTRONICS

British physicist Lord Kelvin (1824–1908) was born William Thomson. He became Lord Kelvin when he was knighted by Queen Victoria for his scientific achievements. Perhaps his best-known accomplishment is the development of the Kelvin temperature scale. This scale begins its measurement of temperature at absolute zero. This means that zero kelvins (0 K) is the total absence of thermal energy. The amount of temperature measured by 1 K is equal to that measured by 1 degree Celsius (1°C). The difference between the scales, however, is that absolute zero is 0 K and −273.15°C. One of the uses of the Kelvin scale is for the measurement of noise temperature. Other contributions of Lord Kelvin include the Kelvin absolute electrometer and the Kelvin balance. Lord Kelvin is also known for his research relating to the concept of an ideal gas which can be used to approximate the properties of real gases.

Thermal agitation is often referred to as *white noise* or *Johnson noise,* after J. B. Johnson, who discovered it in 1928. Just as white light contains all other light frequencies, white noise contains all frequencies randomly occurring at random amplitudes. A white noise signal therefore occupies, theoretically at least, infinite bandwidth. Filtered or band-limited noise is referred to as *pink noise.*

In a relatively large resistor at room temperature or higher, the noise voltage across it can be as high as several microvolts. This is the same order of magnitude as or higher than that of many weak RF signals. Weaker-amplitude signals will be totally masked by this noise.

Since noise is a very broadband signal containing a tremendous range of random frequencies, its level can be reduced by limiting the bandwidth. If a noise signal is fed into a selective tuned circuit, many of the noise frequencies are rejected and the overall noise level goes down. The noise power is proportional to the bandwidth of any circuit to which it is applied. Filtering can reduce the noise level, but does not eliminate it entirely.

The amount of open-circuit noise voltage appearing across a resistor or the input impedance to a receiver can be calculated according to Johnson's formula

$$v_n = \sqrt{4kTBR}$$

where v_n = rms noise voltage
k = Boltzman's constant (1.38×10^{-23} J/K)
T = temperature, K ($^\circ$C + 273)
B = bandwidth, Hz
R = resistance, Ω

The resistor is acting as a voltage generator with an internal resistance equal to the resistor value. See Fig. 9-26. Naturally, if a load is connected across the resistor generator, the voltage will decrease as a result of voltage divider action.

Example 9-2

What is the open-circuit noise voltage across a 100-kΩ resistor over the frequency range of direct current to 20 kHz at room temperature (25°C)?

$$v_n = \sqrt{4kTBR}$$
$$= \sqrt{4(1.38 \times 10^{-23})(25 + 273)(20 \times 10^3)(100 \times 10^3)}$$
$$v_n = 5.74 \; \mu\text{V}$$

Example 9-3

The bandwidth of a receiver with a 75-Ω input resistance is 6 MHz. The temperature is 29°C. What is the input thermal noise voltage?

$$T = 29 + 273 = 302 \text{ K}$$
$$v_n = \sqrt{4kTBR}$$
$$v_n = \sqrt{4(1.38 \times 10^{-23})(302)(6 \times 10^6)(75)} = 2.74 \; \mu\text{V}$$

Figure 9-26 A resistor acts as a tiny generator of noise voltage.

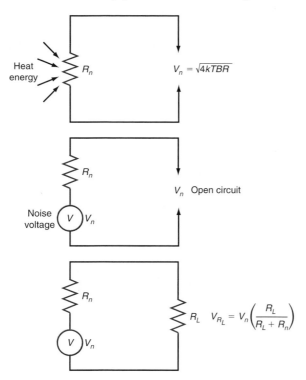

Since noise voltage is proportional to resistance value, temperature, and bandwidth, noise voltage can be reduced by reducing resistance, temperature, and bandwidth or any combination to the minimum level acceptable for the given application. In many cases, of course, the values of resistance and bandwidth cannot be changed. One thing, however, that is always controllable to some extent is temperature. Anything that can be done to cool the circuits will greatly reduce the noise. Heat sinks, cooling fans, and good ventilation can help lower noise. Many low-noise receivers for weak microwave signals from spacecraft and in radio telescopes are supercooled; i.e., their temperature is reduced to very low (cryogenic) levels with liquid nitrogen or liquid helium.

Thermal noise can also be computed as a power level. Johnson's formula is then

$$P_n = kTB$$

where P_n is the average noise power in watts.

Note that when you are dealing with power, the value of resistance does not enter into the equation.

Example 9-4

What is the average noise power of a device operating at a temperature of 90°F with a bandwidth of 30 kHz?

$$T_C = 5(T_F - 32)/9 = 5(90 - 32)/9 = 5(58)/9 = 290/9 = 32.2°C$$

$$T_K = T_C + 273 = 32.2 + 273 = 305.2 \text{ K}$$

$$P_n = (1.38 \times 10^{-23})(305.2)(30 \times 10^3) = 1.26 \times 10^{-16} \text{ W}$$

Semiconductor Noise. Electronic components such as diodes and transistors are major contributors of noise. In addition to thermal noise, semiconductors produce shot noise, transit-time noise, and flicker noise.

Semiconductor noise

Shot noise

The most common type of *semiconductor noise* is *shot noise*. Current flow in any device is not direct and linear. The current carriers, electrons or holes, sometimes take random paths from source to destination, whether the destination is an output element, tube plate, or collector or drain in a transistor. It is this random movement that produces the shot effect. Shot noise is also produced by the random movement of electrons or holes across a PN junction. Even though current flow is established by external bias voltages, some random movement of electrons or holes will occur as a result of discontinuities in the device. For example, the interface between the copper lead and the semiconductor material forms a discontinuity that causes random movement of the current carriers.

Shot noise is also white noise in that it contains all frequencies and amplitudes over a very wide range. The amplitude of the noise voltage is unpredictable, but it does follow a Gaussian distribution curve that is a plot of the probability that specific amplitudes will occur. The amount of shot noise is directly proportional to the amount of dc bias flowing in a device. The bandwidth of the device or circuit is also important. The rms noise current in a device I_n is calculated with the formula

$$I_n = \sqrt{2qIB}$$

where q = charge on an electron, 1.6×10^{-19} C

I = direct current, A

B = bandwidth, Hz

As an example, assume a dc bias of 0.1 mA and a bandwidth of 12.5 kHz. The noise current is

$$I_n = \sqrt{2(1.6 \times 10^{-19})(0.0001)(12,500)} = \sqrt{4 \times 10^{-19}} = 0.632 \times 10^{-19} \text{ A}$$
$$I_n = 0.632 \text{ nA}$$

Now assume that the current is flowing across the emitter-base junction of a bipolar transistor. The dynamic resistance of this junction r_e' can be calculated with the expression $r_e' = 0.025/I_e$, where I_e is the emitter current. Assuming an emitter current of 1 mA, we have $r_e' = 0.025/0.001 = 25\ \Omega$. The noise voltage across the junction is found with Ohm's law:

$$v_n = I_n r_e' = 0.623 \times 10^{-9} \times 25 = 15.8 \times 10^{-9}\ \text{V} = 15.8\ \text{nV}$$

This amount of voltage may seem negligible, but keep in mind that the transistor has gain and will therefore amplify this variation, making it larger in the output. Shot noise is normally lowered by keeping the transistor currents low since the noise current is proportional to the actual current. This is not true of MOSFETs, in which shot noise is relatively constant despite the current level.

Another kind of noise that occurs in transistors is called *transit-time noise*. The term *transit time* refers to how long it takes for a current carrier such as a hole or electron to move from the input to the output. The devices themselves are very tiny so the distances involved are minimal, yet the time it takes for the current carriers to move even a short distance is finite. At low frequencies, this time is negligible; but when the frequency of operation is high and the period of the signal being processed is the same order of magnitude as the transit time, problems can occur. Transit-time noise shows up as a kind of random variation of current carriers within a device, occurring near the upper cutoff frequency. Transit-time noise is directly proportional to the frequency of operation. Since most circuits are designed to operate at a frequency much less than the transistor's upper limit, transit-time noise is rarely a problem.

Transit-time noise

A third type of semiconductor noise, *flicker noise* or excess noise, also occurs in resistors and conductors. This disturbance is the result of minute random variations of resistance in the semiconductor material. It is directly proportional to current and temperature. However, it is inversely proportional to frequency, and for this reason it is sometimes referred to as $1/f$ noise. Flicker noise is highest at the lower frequencies and thus is not pure white noise. Because of the dearth of high-frequency components, $1/f$ noise is also called *pink noise*.

Flicker noise

At some low frequency, flicker noise begins to exceed thermal and shot noise. In some transistors this transition frequency is as low as several hundred hertz; in others the noise may begin to rise at a frequency as high as 100 kHz. This information is listed on the transistor data sheet, the best source of noise data.

The amount of flicker noise present in resistors depends on the type of resistor. Figure 9-27 shows the range of noise voltages produced by the various types of popular resistor types. The figures assume a common resistance, temperature, and bandwidth. Because carbon-composition resistors exhibit an enormous amount of flicker noise—an order of magnitude more than that of the other types—they are avoided in low-noise amplifiers and other circuits. Carbon and metal film resistors are much better, but metal film resistors may be more expensive. Wire-wound resistors have the least flicker noise, but are rarely used because they contribute a large inductance to the circuit, which is unacceptable in RF circuits.

Figure 9-28 shows the total noise voltage variation in a transistor, which is a composite of the various noise sources. At low frequencies, noise voltage is high, because of $1/f$ noise. At very high frequencies, the rise in noise is due to transit-time effects near

Figure 9-27 Flicker noise in resistors.

Type of Resistor	Noise Voltage Range, μV
Carbon-composition	0.1–3.0
Carbon film	0.05–0.3
Metal film	0.02–0.2
Wire-wound	0.01–0.2

Figure 9-28   Noise in a transistor with respect to frequency.

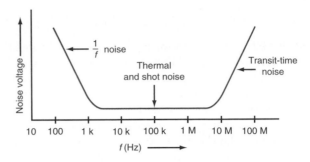

the upper cutoff frequency of the device. Noise is lowest in the midrange, where most devices operate. The noise in this range is due to thermal and shot effects, with shot noise sometimes contributing more than thermal noise.

Intermodulation distortion

Intermodulation Distortion. *Intermodulation distortion* results from the generation of new signals and harmonics caused by circuit nonlinearities. As stated previously, circuits can never be perfectly linear, and if bias voltages are incorrect in a given circuit, it is likely to be more nonlinear than intended.

Nonlinearities produce modulation or heterodyne effects. Any frequencies in the circuit mix together, forming sum and difference frequencies. When many frequencies are involved, or with pulses or rectangular waves, the large number of harmonics produces an even larger number of sum and difference frequencies. The resulting products are small in amplitude, but can be large enough to constitute a disturbance that can be regarded as a type of noise. This noise, which is not white or pink, can be predicted because the frequencies involved in generating the intermodulation products are known.

Correlated noise

Because of the predictable correlation between the known frequencies and the noise, intermodulation distortion is also called *correlated noise*. Correlated noise is produced only when signals are present. The types of noise discussed earlier are sometimes referred

Uncorrelated noise

to as *uncorrelated noise*. Correlated noise is manifested as the low-level signals called *birdies*. It can be minimized by good design.

Expressing Noise Levels

The noise quality of a receiver can be expressed as in terms of noise figure, noise factor, noise temperature, and SINAD.

Noise factor

Noise Factor and Noise Figure. The *noise factor* is the ratio of the *S/N* power at the input to the *S/N* power at the output. The device under consideration can be the entire receiver or a single amplifier stage. The noise factor or noise ratio (NR) is computed with the expression

Noise figure (NF)

$$NR = \frac{S/N \text{ input}}{S/N \text{ output}}$$

When the noise factor is expressed in decibels, it is called the *noise figure (NF)*:

$$NF = 10 \log NR \qquad dB$$

Amplifiers and receivers always have more noise at the output than at the input because of the internal noise which is added to the signal. And even as the signal is being amplified along the way, the noise generated in the process is amplified along with it.

The *S/N* ratio at the output will be less than the *S/N* ratio of the input, and so the noise figure will always be greater than 1. A receiver that contributed no noise to the signal would have a noise figure of 1, or 0 dB, which is not attainable in practice. A transistor amplifier in a communication receiver usually has a noise figure of several decibels. The lower the noise figure, the better the amplifier or receiver. Noise figures of less than about 2 dB are excellent.

Example 9-5

An RF amplifier has an *S/N* ratio of 8 at the input and an *S/N* ratio of 6 at the output. What are the noise factor and noise figure?

$$NR = \frac{8}{6} = 1.333$$

$$NF = 10 \log 1.3333 = 10(0.125) = 1.25 \text{ dB}$$

Noise Temperature. Most of the noise produced in a device is thermal noise, which is directly proportional to temperature. Therefore, another way to express the noise in an amplifier or receiver is in terms of *noise temperature* T_N. Noise temperature is expressed in kelvins. Remember that the Kelvin temperature scale is related to the Celsius scale by the relationship $T_K = T_C + 273$. The relationship between noise temperature and NR is given by

Noise temperature

$$T_N = 290(NR - 1)$$

For example, if the noise ratio is 1.5, the equivalent noise temperature is $T_N = 290(1.5 - 1) = 290(0.5) = 145$ K. Clearly, if the amplifier or receiver contributes no noise, then NR will be 1, as indicated before. Plugging this value into the expression above gives an equivalent noise temperature of 0 K:

$$T_N = 290(1 - 1) = 290(0) = 0 \text{ K}$$

If the noise ratio is greater than 1, an *equivalent noise temperature* will be produced. The equivalent noise temperature is the temperature to which a resistor equal in value to Z_o of the device would have to be raised to generate the same V_n as the device generates.

Noise temperature is used only in circuits or equipment that operates at VHF, UHF, or microwave frequencies. The noise factor or noise figure is used at lower frequencies. A good low-noise transistor or amplifier stage typically has a noise temperature of less than 100 K. The lower the noise temperature, the better the device. Often you will see the noise temperature of a transistor given in the data sheet.

SINAD. Another way of expressing the quality of communication receivers is *SINAD*—the composite *si*gnal plus the *n*oise *a*nd *d*istortion divided by noise and distortion contributed by the receiver. In symbolic form,

SINAD

$$SINAD = \frac{S + N + D \text{ (composite signal)}}{N + D \text{ (receiver)}}$$

Example 9-6

A receiver with a 75-Ω input resistance operates at a temperature of 31°C. The received signal is at 89 MHz with a bandwidth of 6 MHz. The received signal voltage of 8.3 μV is applied to an amplifier with a noise figure of 2.8 dB. Find (a) the input noise power, (b) the input signal power, (c) S/N, in decibels, (d) the noise factor and S/N of the amplifier, and (e) the noise temperature of the amplifier.

a. $T_C = 273 + 31 = 304$ K

$$v_n = \sqrt{4kTBR}$$

$$v_n = \sqrt{4(1.38 \times 10^{-23})(304)(6 \times 10^6)(75)} = 2.75 \ \mu V$$

$$P_n = \frac{(v_n)^2}{R} = \frac{(2.75 \times 10^{-6})^2}{75} = 0.1 \ \text{pW}$$

b. $P_s = \dfrac{(v_s)^2}{R} = \dfrac{(8.3 \times 10^{-6})^2}{75} = 0.918 \ \text{pW}$

c. $\dfrac{S}{N} = \dfrac{P_s}{P_n} = \dfrac{0.918}{0.1} = 9.18$

$$dB = 10 \log \frac{S}{N} = 10 \log 9.18$$

$$\frac{S}{N} = 9.63 \ \text{dB}$$

d. $NF = 10 \log NR$

$$NR = \text{antilog} \frac{NF}{10} = 10^{NF/10}$$

$$NF = 2.8 \ \text{dB}$$

$$NR = 10^{2.8/10} = 10^{0.28} = 1.9$$

$$NR = \frac{S/N \ \text{input}}{S/N \ \text{output}}$$

$$\frac{S}{N} \ \text{output} = \frac{S/N \ \text{input}}{NR} = \frac{9.18}{1.9} = 4.83$$

$$\frac{S}{N} \ (\text{output}) = \frac{S}{N} \ (\text{amplifier})$$

e. $T_N = 290(NR - 1) = 290(1.9 - 1) = 290(0.9) = 261$ K

Distortion refers to the harmonics present in a signal caused by nonlinearities.

The SINAD ratio is also used to express the sensitivity of a receiver. Note that the SINAD ratio makes no attempt to discriminate between or to separate noise and distortion signals.

To obtain the SINAD ratio, an RF signal modulated by an audio signal (usually of 400 Hz or 1 kHz) is applied to the input of an amplifier or a receiver. The composite output is then measured, giving the $S + N + D$ figure. Next, a highly selective notch (band-reject) filter is used to eliminate the modulating audio signal from the output, leaving the noise and distortion, or $N + D$.

The SINAD is a power ratio, and it is almost always expressed in decibels:

$$\text{SINAD} = 10 \log \frac{S + N + D}{N + D} \qquad \text{dB}$$

Noise in the Microwave Region

Noise is an important consideration at all communication frequencies, but it is particularly critical in the microwave region because noise increases with bandwidth and affects high-frequency signals more than low-frequency signals. The limiting factor in most microwave communication systems, such as satellites and radar, is internal noise. In some special microwave receivers, the noise level is reduced by cooling the input stages to the receiver, as mentioned earlier. This technique is called *operating with cryogenic conditions,* the term *cryogenic* referring to very cold conditions approaching absolute zero.

Noise in Cascaded Stages

Noise has its greatest effect at the input to a receiver simply because that is the point at which the signal level is lowest. The noise performance of a receiver is invariably determined in the very first stage of the receiver, usually an RF amplifier or mixer. Design of these circuits must ensure the use of very low-noise components, taking into consideration current, resistance, bandwidth, and gain figures in the circuit. Beyond the first and second stages, noise is basically no longer a problem.

The formula used to calculate the overall noise performance of a receiver or of multiple stages of RF amplification, called *Friis' formula,* is

Friis' formula

$$NR = NR_1 + \frac{NR_2 - 1}{A_1} + \frac{NR_3 - 1}{A_1 A_2} + \cdots + \frac{NR_n - 1}{A_1, A_2, \cdots, A_{n-1}}$$

where NR = noise ratio
NR_1 = noise ratio of input or first amplifier to receive the signal
NR_2 = noise ratio of second amplifier
NR_3 = noise ratio of third amplifier, and so on
A_1 = power gain of first amplifier
A_2 = power gain of second amplifier
A_3 = power gain of third amplifier, and so on

Note that the noise ratio is used, rather than the noise figure, and so the gains are given in power ratios rather than in decibels.

As an example, consider the circuit shown in Fig. 9-29. The overall noise ratio for the combination is calculated as follows:

$$NR = 1.6 + \frac{4 - 1}{7} + \frac{8.5 - 1}{(7)(12)} = 1.6 + 0.4286 + 0.0893 = 2.12$$

The noise figure is

$$NF = 10 \log NR = 10 \log 2.12 = 10(0.326) = 3.26 \text{ dB}$$

What this calculation means is that *the first stage controls the noise performance for the whole amplifier chain.* This is true even though stage 1 has the lowest NR, because after the first stage, the signal is large enough to overpower the noise. This result is true for almost all receivers and other equipment incorporating multistage amplifiers.

Figure 9-29   Noise in cascaded stages of amplification.

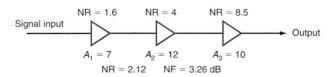

9-6 Typical Receiver Circuits

This section focuses on RF and IF amplifiers, AGC and AFC circuits, and other special circuits found in receivers.

RF Input Amplifiers

The most critical part of a communication receiver is the front end, which usually consists of the RF amplifier, mixer, and related tuned circuits and is sometimes simply referred to as the tuner. The RF amplifier, also called a low-noise amplifier (LNA), processes the very weak input signals, increasing their amplitude prior to mixing. It is essential that low-noise components be used to ensure a sufficiently high S/N ratio. Further, the selectivity should be such that it effectively eliminates images.

In some communication receivers, an RF amplifier is not used, e.g., in receivers designed for frequencies lower than about 30 MHz, where the extra gain of an amplifier is not necessary and its only contribution is more noise. In such receivers, the RF amplifier is eliminated, and the antenna is connected directly to the mixer input through one or more tuned circuits. The tuned circuits must provide the input selectivity necessary for image rejection. In a receiver of this kind, the mixer must also be of the low-noise variety. Many mixers are MOSFETs, which provide the lowest noise contribution. Low-noise bipolar transistor mixers are used in IC mixers.

Most LNAs use a single transistor and provide a voltage gain in the 10- to 30-dB range. Bipolar transistors are used at the lower frequencies, whereas at VHF, UHF, and microwave frequencies FETs are preferred.

The *RF amplifier* is usually a simple class A circuit. A typical FET circuit is shown in Fig. 9-30. FET circuits are particularly effective because their high input impedance minimizes loading on tuned circuits, permitting the Q of the circuit to be higher and selectivity to be sharper. Most FETs also have a lower noise figure than bipolars.

At microwave frequencies (those above 1 GHz), metal-semiconductor FETs, or *MESFETs*, are used. Also known as *GASFETs*, these devices are junction field-effect transistors made with gallium arsenide (GaAs). A cross section of a typical MESFET is shown in Fig. 9-31. The gate junction is a metal-to-semiconductor interface because it is in a Schottky or hot carrier diode. As in other junction FET circuits, the gate-to-source is reverse-biased, and the signal voltage between the source and the gate controls the conduction of current between the source and drain. The transit time of electrons through gallium arsenide is far shorter than that through silicon, allowing the

RF amplifier

MESFET

GASFET

GOOD TO KNOW

MESFETs provide high gain at high frequencies because electrons travel through gallium arsenide more quickly than through silicon. They are also the lowest-noise transistor available.

Figure 9-30 A typical RF amplifier used in receiver front ends.

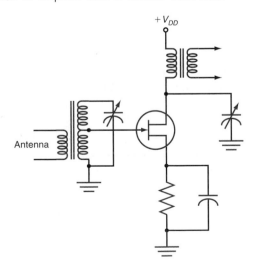

Figure 9-31 The MESFET configuration and symbol.

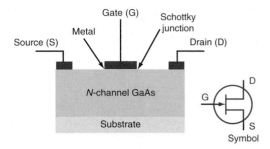

MESFET to provide high gain at very high frequencies. MESFETs also have an extremely low noise figure, typically 2 dB or less. Most MESFETs have a noise temperature of less than 200 K.

As semiconductor processing techniques have made transistors smaller and smaller, both bipolar and CMOS LNAs are widely used at frequencies up to 10 GHz. Silicon germanium (SiGe) is widely used to make bipolar LNAs, and BiCMOS (a mixture of bipolar and CMOS circuits) designs in silicon are also popular. Silicon is preferred because no special processing is required as with GaAs and SiGe.

Although single-stage RF amplifiers are popular, some small-signal applications require a bit more amplification before the mixer. This can be accomplished with a cascode amplifier, as shown in Fig. 9-32. This LNA uses two transistors to achieve not only low noise but also gains of 40 dB or more.

Figure 9-32 A cascode LNA.

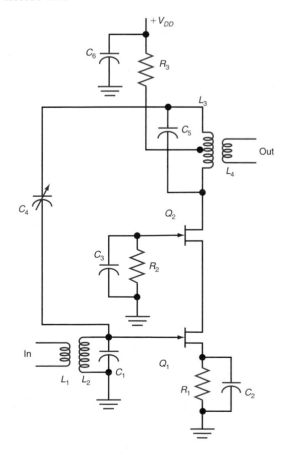

Transistor Q_1 operates as a normal common source stage. It is directly coupled to the second-stage Q_2, which is a common gate amplifier. The gate is at ac ground through C_3. The frequency range is set by the input tuned circuit L_2–C_1 and output tuned circuit L_3–C_5. Bias is provided by R_1.

One of the key benefits of the cascode circuit is that it effectively minimizes the effect of the Miller capacitance problem associated with single-stage RF amplifiers. The transistors used to implement these amplifiers, BJT, JFET, or MOSFET, all exhibit some form of interelectrode capacitance between collector and base (C_{cb}) in the BJT and between the drain and gate in FETs (C_{dg}). This capacitance introduces some feedback that makes it appear that a larger equivalent capacitance, called the Miller capacitance, appears between the base or gate to ground. This equivalent Miller capacitance C_m is equal to the interelectrode capacitance multiplied by the gain of the amplifier A less 1.

$$C_m = C_{dg}(A - 1) \qquad \text{or} \qquad C_m = C_{cb}(A - 1)$$

This capacitance forms a low-pass filter with the output impedance of the circuit driving the amplifier. The result is that the upper frequency limit of the amplifier is limited.

The cascode circuit of Fig. 9-32 effectively eliminates this problem because the output signal at the drain of Q_2 cannot get back to the gate of Q_1 to introduce the Miller capacitance. As a result, the cascode amplifier has a much wider upper frequency range.

Many RF amplifiers become unstable especially at VHF, UHF, and microwave frequencies because of positive feedback that occurs in the interelectrode capacitances of the transistors. This feedback can cause oscillation. To eliminate this problem, some kind of neutralization is normally used, as in RF power amplifiers. In Fig. 9-32, some of the output is feedback through neutralization capacitor C_4. This negative feedback cancels the positive feedback and provides the necessary stability.

Although this circuit is shown with JFETs, it can also be built with BJTs or MOSFETs. This circuit is popular in integrated-circuit cell phones and other wireless receivers and is implemented with silicon CMOS, BiCMOS, or SiGE bipolar transistors.

IF Amplifiers

As stated previously, most of the gain and selectivity in a superheterodyne receiver are obtained in the IF amplifier, and choosing the right IF is critical to good design.

Traditional IF Amplifier Circuits

IF amplifier

IF amplifiers, like RF amplifiers, are tuned class A amplifiers capable of providing a gain in the 10- to 30-dB range. Usually two or more IF amplifiers are used to provide adequate overall receiver gain. Figure 9-33 shows a two-stage IF amplifier. The amplifiers may be single-stage BJT, JFET, or MOSFET transistors or a differential amplifier. Older receivers used discrete component circuits, but today most IF amplifiers are integrated-circuit differential amplifiers, usually MOSFETs.

Figure 9-33   A two-stage IF amplifier using double-tuned transformer coupling for selectivity.

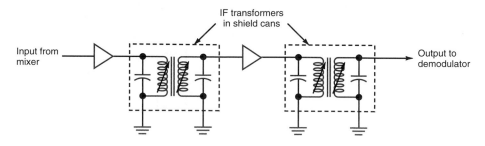

Ferrite-core transformers T_1 and T_2 are used for coupling between stages. Because they are resonant circuits, these transformers also provide the desired selectivity. The dashed lines around the transformers represent the metal can or enclosure that surrounds the transformer components to protect against radiation and undesired feedback. Older transformers were tuned with trimmer capacitors, but newer transformers are tuned with fixed capacitors and variable inductors. The ferrite cores are threaded, allowing their position to be adjusted within the coil, thus varying its inductance.

The selectivity in the IF amplifier is provided by the tuned circuits. As indicated earlier, cascading tuned circuits causes the overall circuit bandwidth to be considerably narrowed. High-Q tuned circuits are used, but with multiple tuned circuits, the bandwidth is even narrower. IF amplifiers should be designed so that the selectivity is not too sharp. A too-narrow IF bandwidth will cause sideband cutting, greatly reducing the amplitude of the higher modulating frequencies and thus distorting the received signal. The exact nature of the kinds of signals to be received must be well known so that the bandwidth of the IF amplifier can be appropriately set.

Coupled Circuit Selectivity.

When very broadband signals are received, it is sometimes necessary to widen the bandwidth of an IF amplifier. There are several ways of doing this. One technique is to connect resistors across the parallel tuned circuits, thereby lowering their Q to a value that will produce the appropriate bandwidth.

Another technique is to use overcoupled tuned circuits. The output voltage versus frequency curve for double-tuned circuits like those in Fig. 9-34 is strictly dependent on the amount of coupling or mutual inductance between the primary and secondary windings. That is, the spacing between the windings determines how much of the magnetic field produced by the primary will cut the turns of the secondary. This affects not only the amplitude of the output voltage, but also the bandwidth.

Changing the amount of coupling between the primary and secondary windings in IF coupling transformers allows the desired amount of bandwidth to be obtained. Figure 9-34 shows the response curves of a double-tuned transformer for different settings. When the windings are spaced far apart, the coils are said to be *undercoupled*. With this configuration, the amplitude is low and the bandwidth relatively narrow. At some particular degree of coupling, known as *critical coupling,* the output reaches a peak value. In most IF designs, critical coupling provides the best gain if the bandwidth provided is adequate. Moving the coils closer together and increasing the coupling widen

Critical coupling

Figure 9-34 Response curves of a double-tuned air-core transformer for various degrees of coupling.

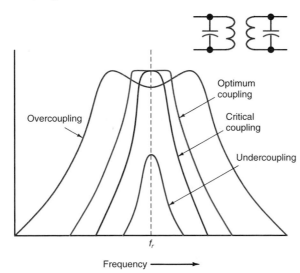

Communication Receivers

the bandwidth further. The output signal amplitude is maximum and will not increase beyond that obtained at critical coupling. This point is usually known as *optimum coupling*. Increasing the amount of coupling still further produces an effect known as *overcoupling*. The result is a double peak output response curve with a considerably wider bandwidth.

The transformer coupling design of Fig. 9-33 is rarely used today. Instead, IF amplifiers use *crystal, ceramic,* or *SAW filters* for selectivity. They are typically smaller than *LC* tuned circuits, provide higher selectivity, and require no tuning or adjustment. Today, high-performance communication receivers also use DSP filters to routinely achieve selectivity.

Limiters. In FM receivers, one or more of the IF amplifier stages is used as a *limiter,* to remove any amplitude variations on the FM signal before the signal is applied to the demodulator. Typically, limiters are simply conventional class A IF amplifiers. In fact, any amplifier will act as a limiter if the input signal is high enough. When a very large input signal is applied to a single transistor stage, the transistor is alternately driven between saturation and cutoff. For example, in an NPN bipolar class A amplifier, applying a very large positive input signal to the amplifier causes the base bias to increase, thereby increasing the collector current. When a sufficient amount of input voltage is supplied, the transistor reaches maximum conduction, where both the emitter-base and the base-collector junctions become forward-biased. At this point, the transistor is saturated, and the voltage between the emitter and collector drops to some very small value, typically less than 0.1 V. At this time, the amplifier output is approximately equal to the dc voltage drop across any emitter resistor that may be used in the circuit.

When a very large negative-going signal is applied to the base, the transistor can be driven into cutoff. The collector current drops to zero, and the voltage seen at the collector is simply the supply voltage. Figure 9-35 shows the collector current and voltage for both extremes.

Driving the transistor between saturation and cutoff effectively flattens or clips off the positive and negative peaks of the input signal, removing any amplitude variations. The output signal at the collector is, therefore, a square wave. The most critical part of the limiter design is to set the initial base bias level to that point at which *symmetric clipping*—i.e., equal amounts of clipping on the positive and negative peaks—will occur. Differential amplifiers are preferred for limiters because they produce the most symmetric clipping. The square wave at the collector, which is made up of many undesirable harmonics, is effectively filtered back into a sine wave by the tuned circuit in the collector or the output filter.

Crystal filter

SAW filter

Ceramic filter

Limiter

GOOD TO KNOW

Any amplifier will act as a limiter if the input signal is high enough.

Figure 9–35 Collector current and voltage in a bipolar limiter IF amplifier circuit.

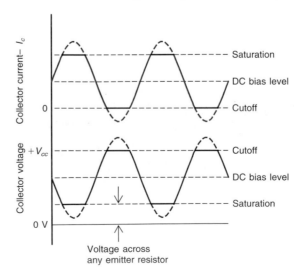

Automatic Gain Control Circuits

The overall gain of a communication receiver is usually selected on the basis of the weakest signal to be received. In most modern communication receivers, the voltage gain between the antenna and the demodulator is in excess of 100 dB. The RF amplifier usually has a gain in the 5- to 15-dB range. The mixer gain is in the 2- to 10-dB range, although diode mixers, if used, introduce a loss of several decibels. IF amplifiers have individual stage gains of 20 to 30 dB. Detectors of the passive diode type may introduce a loss, typically from -2 to -5 dB. The gain of the audio amplifier stage is in the 20- to 40-dB range. Assume, e.g., a circuit with the following gains:

RF amplifier	10 dB	
Mixer	-2 dB	
IF amplifiers (three stages)	27 dB	$(3 \times 27 = 81 \text{ total})$
Demodulator	-3 dB	
Audio amplifier	32 dB	

The total gain is simply the algebraic sum of the individual stage gains, or $10 - 2 + 27 + 27 + 27 - 3 + 32 = 118$ dB.

In many cases, gain is far greater than that required for adequate reception. Excessive gain usually causes the received signal to be distorted and the transmitted information to be less intelligible. One solution to this problem is to provide gain controls in the receiver. For example, a potentiometer can be connected at some point in an RF or IF amplifier stage to control the RF gain manually. In addition, all receivers include a volume control in the audio circuit.

The gain controls cited above are used, in part, so that the overall receiver gain does not interfere with the receiver's ability to handle large signals. A more effective way of dealing with large signals, however, is to include AGC circuits. As discussed earlier, the use of AGC gives the receiver a very wide dynamic range, which is the ratio of the largest signal that can be handled to the lowest expressed in decibels. The dynamic range of a typical communication receiver with AGC is usually in the 60- to 100-dB range.

Controlling Circuit Gain. If the IF and RF amplifiers are simple common-emitter amplifiers as used in older receivers, AGC can be implemented by controlling the collector current of the transistors. The gain of a bipolar transistor amplifier is proportional to the amount of collector current flowing. Increasing the collector current from some very low level causes the gain to increase proportionately. At some point, the gain flattens over a narrow collector current range and then begins to decrease as the current increases further. Figure 9-36 shows an approximation of the relationship between the gain variation and the collector current of a typical bipolar transistor. The gain peaks at 30 dB over the 6- to 15-mA range.

The amount of collector current in the transistor is, of course, a function of the base bias applied. A small amount of base current produces a small amount of collector current, and vice versa. In IF amplifiers the bias level is not usually fixed by a voltage divider, but controlled by the AGC circuit. In some circuits, a combination of fixed voltage divider bias plus a dc input from the AGC circuit controls overall gain.

1. The gain can be decreased by decreasing the collector current. An AGC circuit that decreases the current flowing in the amplifier to decrease the gain is called *reverse AGC*.

   Reverse AGC

2. The gain can be reduced by increasing the collector current. As the signal gets stronger, the AGC voltage increases; this increases the base current and, in turn, increases the collector current, reducing the gain. This method of gain control is known as *forward AGC*.

   Forward AGC

Figure 9-36 Approximate voltage gain of a bipolar transistor amplifier versus collector current.

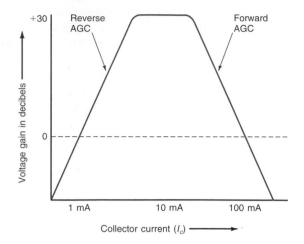

In general, reverse AGC is more common in communication receivers. Forward AGC, which is widely used in TV sets, typically requires special transistors for optimum operation.

Integrated-circuit differential amplifiers are widely used as IF amplifiers. The gain of a differential amplifier is directly proportional to the amount of emitter current flowing. Because of this, the AGC voltage can be conveniently applied to the constant-current source transistor in a differential amplifier. A typical circuit is shown in Fig. 9-37. The bias on constant-current source Q_3 is adjusted by R_1, R_2, and R_3 to provide a fixed level of emitter current I_E to differential transistors Q_1 and Q_2. Normally, the emitter-current

Figure 9-37 An IF differential amplifier with AGC.

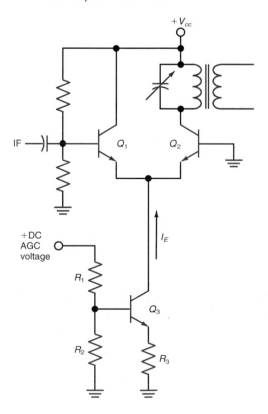

Figure 9-38 A dual-gate MOSFET's gain can be controlled with a dc voltage on the second gate.

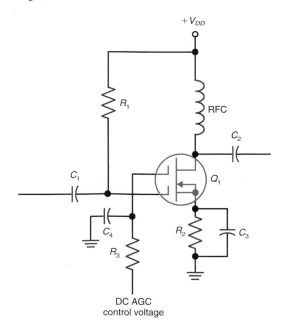

value in a constant-gain stage is fixed, and the current divides between Q_1 and Q_2. The gain is easy to control by varying the bias on Q_3. In the circuit shown, increasing the positive AGC voltage increases the emitter current and increases the gain. Decreasing the AGC voltage decreases the gain.

Figure 9-38 shows another way to control the gain of an amplifier. Here Q_1 is a dual-gate depletion mode MOSFET connected as a class A amplifier. It may be the RF amplifier or an IF amplifier. The dual-gate MOSFET actually implements a cascode circuit arrangement which is common in RF amplifiers. Normal bias is applied via R_1 to the lower gate. Additional bias is derived from the source resistor R_2. The input signal is applied to the lower gate through C_1. The signal is amplified and appears at the drain where it is coupled through R_2 to the next stage. If this circuit is the RF amplifier used ahead of the mixer, LC tuned circuits are normally used at the input and output to provide some initial selectivity and impedance matching. In IF amplifiers, several stages such as this may be cascaded to provide the gain with the selectivity coming from a single crystal, ceramic, SAW, or mechanical filter at the output of the last stage.

The dc AGC control voltage is applied to the second gate through R_3. Capacitor C_4 is a filter and decoupling capacitor. Since both gates control the drain current, the AGC voltage varies the drain current which, in turn, controls the transistor gain.

In most modern receivers, the AGC circuits are simply integrated along with the IF amplifier stages inside an IC. Some of these ICs may also have an integrated mixer or integrated demodulator. The AGC is controlled by an input voltage derived from an external circuit. Others incorporate the circuits that develop the AGC control voltage.

Deriving the Control Voltage. The dc voltage used to control the gain is usually derived by rectifying either the IF signal or the recovered information signal after the demodulator. One of the simplest and most widely used methods of AGC voltage generation in an AM receiver is to use the output from the diode detector, as shown in Fig. 9-39. The diode detector recovers the original AM information. The voltage developed across R_1 is a negative dc voltage. Capacitor C_1 filters out the IF signal, leaving the original modulating signal. The time constant of R_1 and C_1 is adjusted to eliminate the IF ripple and yet retain the highest-frequency modulating signal. The recovered signal is passed through C_2 to remove the direct current. The resulting ac

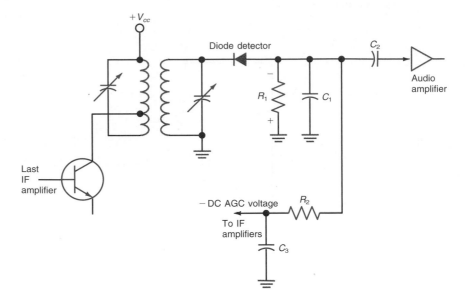

Figure 9-39 Deriving the AGC voltage from the diode detector in an AM receiver.

signal is further amplified and applied to a loudspeaker. The dc voltage across R_1 and C_1 must be further filtered to provide a pure dc voltage. This is done with R_2 and C_3. The time constant of these components is chosen to be very large so that the voltage at the output is pure direct current. The dc level varies, of course, with the amplitude of the received signal. The resulting negative voltage is then applied to one or more IF amplifier stages.

In an FM receiver, the dc voltage can usually be derived directly from the demodulator. Both Foster-Seeley discriminator and ratio detector circuits provide convenient starting points for obtaining a dc voltage proportional to the signal amplitude. With additional RC filtering, a dc level proportional to the signal amplitude is derived for use in controlling the IF amplifier gain. As mentioned previously, some FM receivers do not even use AGC because the limiters provide a crude form of gain control by clipping off signal levels higher than a specific amplitude.

In many receivers, a special rectifier circuit devoted strictly to deriving the AGC voltage is used. Figure 9-40 shows a typical circuit of this type. The input, which can be either the recovered modulating signal or the IF signal, is applied to an AGC amplifier. A voltage-doubler rectifier circuit made up of D_1, D_2, and C_1 is used to increase the voltage level high enough for control purposes. The RC filter R_1-C_2 removes any signal variations and produces a pure dc voltage. In some circuits, further amplification of the dc control voltage is necessary; a simple IC op amp such as that shown in Fig. 9-40 can be used for this purpose. The connection of the rectifier and any phase inversion in the op amp will determine the polarity of the AGC voltage, which can be either

Figure 9-40 An AGC rectifier and amplifier.

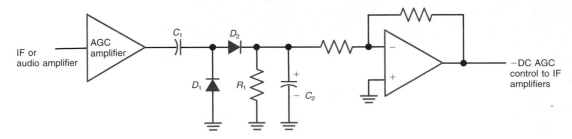

positive or negative depending upon the types of transistors used in the IF and their bias connections.

Squelch Circuits

Another circuit found in most communications receivers is a *squelch circuit.* Also called a *muting circuit,* the squelch is used to keep the receiver audio turned off until an RF signal appears at the receiver input. Most two-way communication consists of short conversations that do not take place continuously. In most cases, the receiver is left on so that if a call is received, it can be heard. When there is no RF signal at the receiver input, the audio output is simply background noise. With no input signal, the AGC sets the receiver to maximum gain, amplifying the noise to a high level. In AM systems such as CB radios, the noise level is relatively high and can be very annoying. The noise level in FM systems can also be high; in some cases listeners may turn down the audio volume to avoid listening to the noise and possibly miss a desired signal. Squelch circuits provide a means of keeping the audio amplifier turned off during the time that noise is received in the background and enabling it when an RF signal appears at the input.

Noise–Derived Squelch. *Noise-derived* squelch circuits, typically used in FM receivers, amplify the high-frequency background noise when no signal is present and use it to keep the audio turned off. When a signal is received, the noise circuit is overridden and the audio amplifier is turned on.

Figure 9-41 shows a noise-derived squelch circuit used in many communication receivers. The background noise with no signal is taken from the demodulator output and passed through C_1 and potentiometer R_1, which form a high-pass filter. Only frequencies above 6 kHz are passed (most noise is of the high-frequency variety). Also R_1 serves as a squelch level or muting threshold control. The noise is further amplified by two transistor stages and then rectified into a dc control voltage by a voltage-doubling rectifier circuit made up of C_2, C_3, D_1, and D_2. The rectifier output causes squelch gate Q_1 to saturate when no signal is present and the receiver is picking up noise only.

The Q_1 operates the squelch gate, which is made up of D_3 and D_4 and related components. Voltage dividers R_2–R_3 and R_4–R_5 provide a reverse bias to the diodes. With no signal, the noise is amplified and Q_1 saturates as described above. This places the anodes of the diodes at a voltage level below the bias voltage; both diodes are cut off, and so no signal from the demodulator reaches the audio amplifier. When a signal occurs, the audio is not passed by the 6-kHz filter, so Q_1 cuts off. The voltage at the anodes of diodes D_3 and D_4 rises to a level more positive than the bias from the voltage dividers, so the diodes conduct, providing a low-resistance path from the demodulator to the audio amplifier.

Continuous Tone–Control Squelch System. A more sophisticated form of squelch used in some systems is known as the *continuous tone-coded squelch system (CTCSS).* This system is activated by a low-frequency tone transmitted along with the audio. The purpose of CTCSS is to provide some communication privacy on a particular channel. Other types of squelch circuits keep the speaker quiet when no input signal is received; however, in communication systems in which a particular frequency channel is extremely busy, it may be desirable to activate the squelch only when the desired signal is received. This is done by having the transmitter send a very low-frequency sine wave, usually in the 60- to 254-Hz range, that is linearly mixed with the audio before being applied to the modulator. The low-frequency tone appears at the output of the demodulator in the receiver. It is not usually heard in the speaker, since the audio response of most communication systems rolls off beginning at about 300 Hz, but can be used to activate the squelch circuit.

Figure 9-41   A noise-derived squelch circuit.

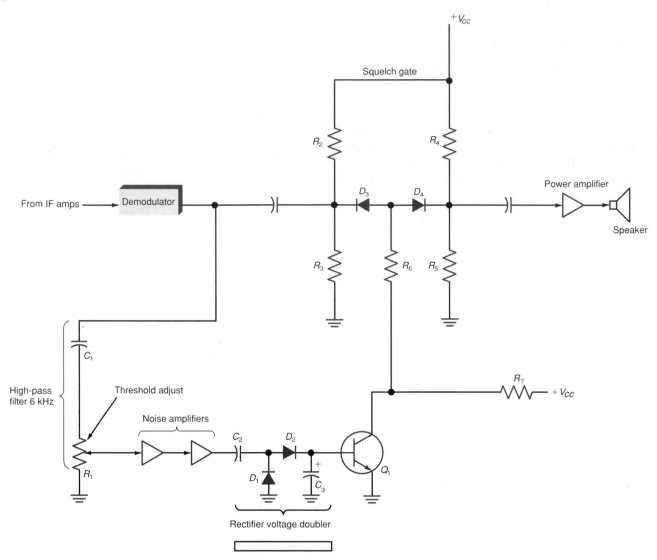

Most modern transmitters using this system have a choice of multiple tone frequencies, so that different remote receivers can be addressed or keyed up independently, providing a nearly private communication channel. The 52 most commonly used tone frequencies (given in hertz) are listed here:

60.0	100.0	151.4	192.8
67.0	103.5	156.7	196.6
69.3	107.2	159.8	199.5
71.9	110.9	162.2	203.5
74.4	114.8	165.5	206.5
77.0	118.8	167.9	210.7
79.7	120.0	171.3	218.1
82.5	123.0	173.8	225.7
85.4	127.3	177.3	229.1
88.5	131.8	179.9	133.6
91.5	136.5	183.5	241.8
94.8	141.3	186.2	250.3
97.4	146.2	189.9	254.1

At the receiver, a highly selective bandpass filter tuned to the desired tone selects the tone at the output of the demodulator and applies it to a rectifier and *RC* filter to generate a dc voltage that operates the squelch circuit.

Signals that do not transmit the desired tone will not trigger the squelch. When the desired signal comes along, the low-frequency tone is received and converted to a dc voltage that operates the squelch and turns on the receiver audio.

Digitally controlled squelch systems, known as *digital coded squelch (DCS),* are available in some modern receivers. These systems transmit a serial binary code along with the audio. There are 106 different codes used. At the receiver, the code is shifted into a shift register and decoded. If the decode AND gate recognizes the code, the squelch gate is enabled and passes the audio.

SSB and Continuous-Wave Reception

Communication receivers designed for receiving SSB or continuous-wave signals have a built-in oscillator that permits recovery of the transmitted information. This circuit, called the *beat frequency oscillator (BFO),* is usually designed to operate near the IF and is applied to the demodulator along with the IF signal containing the modulation.

Beat frequency oscillator (BFO)

Recall that the basic demodulator is a balanced modulator (see Fig. 9-42). A balanced modulator has two inputs, the incoming SSB signal at the intermediate frequency and a carrier which mixes with the incoming signal to produce the sum and difference frequencies, the difference being the original audio. The BFO supplies the carrier signal at the IF to the balanced modulator. The term *beat* refers to the difference frequency output. The BFO is set to a value above or below the SSB signal frequency by an amount equal to the frequency of the modulating signal. It is usually made variable so that its frequency can be adjusted for optimum reception. Varying the BFO over a narrow frequency range allows the pitch of the received audio to change from low to high. It is typically adjusted for most natural voice sounds. BFOs are also used in receiving CW code. When dots and dashes are transmitted, the carrier is turned off and on for short and long periods of time. The amplitude of the carrier does not vary, nor does its frequency; however, the on/off nature of the carrier is, in essence, a form of amplitude modulation.

Consider for a moment what would happen if a CW signal were applied to a diode detector or other demodulator. The output of the diode detector would be pulses of dc voltage representing the dots and dashes. When applied to the audio amplifier, the dots and dashes would blank the noise but would not be discernible. To make the dots and dashes audible, the IF signal is mixed with the signal from a BFO. The BFO signal is usually injected directly into the balanced modulator, as shown in Fig. 9-42, where the CW signal at the IF signal is mixed, or heterodyned, with the BFO signal. Since the BFO is variable, the difference frequency can be adjusted to any desired audio tone,

Figure 9-42 The use of a BFO.

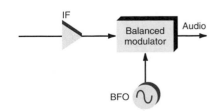

usually in the 400- to 900-Hz range. Now the dots and dashes appear as an audio tone that is amplified and heard in a speaker or earphones. Of course, the BFO is turned off for standard AM signal reception.

Integrated Circuits in Receivers

In new designs, virtually all receiver circuits are ICs. A complete receiver usually consists of three or four ICs at most, plus those discrete components that cannot be easily integrated on a chip. These include coils, transformers, high-capitance and variable capacitors, and crystal and ceramic filters. The most modern receivers are contained on a single IC.

IC receivers are typically broken down into three major sections: (1) the tuner, with RF amplifier, mixer, and local oscillator; (2) the IF section, with amplifiers, demodulator, and AGC and muting circuits; and (3) the audio power amplifier. The second and third sections are entirely implemented with ICs. The tuner may or may not be, for often the LNA is separate. For low-frequency receivers, say, those below about 200 MHz, the tuner can be in IC form also. Higher-frequency receivers require special mixer and local oscillator circuits that are routinely implemented on a single chip, including microwave receivers.

3089 IF System. Originally developed by RCA and now second-sourced by several other semiconductor manufacturers, the *3089 IF system* is one of the oldest and most widely used receiver ICs. The 3089 contains a three-stage IF amplifier, an FM demodulator, and AGC and muting circuits, and it is packaged in a standard 16-pin DIP. A block diagram of the system is shown in Fig. 9-43.

The input to this chip comes from the tuner, which consists of the RF amplifier, mixer, and local oscillator. A ceramic filter is normally used to provide the required selectivity.

In the 3089, all three IF amplifier stages use differential amplifiers that perform as amplifiers and as symmetric limiters at the higher signal levels. Note that each IF amplifier has a level detector circuit associated with it. These are AGC circuits that derive a dc control signal from the signal amplitude. The first level detector is used to provide AGC to the RF amplifier in the tuner. It is a delayed AGC; i.e., it is not fast-acting. The AGC circuit cannot respond instantaneously to a significant signal-level change because of the *RC* filter usually associated with it. It takes time for the circuit capacitance to charge or discharge to the new level, whether it is higher or lower. This AGC response delay is desirable because it prevents noise and interference from making fast, unexpected changes in receiver gain, which might distort the signal or temporarily desensitize the receiver to a weak signal.

The level detectors all drive an IF-level meter driver circuit, which can be used to operate a dc panel meter. These meters, called *S meters,* provide a way to visually display signal strength. They also function as tuning aids. When you tune a receiver, you are seeking to maximize the signal to the IF amplifier; and when a peak output is tuned for on the meter, the meter indicates that the signal is in the center of the passband and is producing maximum signal level.

The demodulator is a standard quadrature detector in which an external 22-μH coil and parallel tuned circuit provide the 90° phase shift required. The quadrature detector also has a level detector, which can be used for squelch or muting. As shown in Fig. 9-43, an external potentiometer is used for muting sensitivity. The output of the potentiometer goes to the audio mute control amplifier, which operates the internal audio amplifier. The 3089 also has an AFC output derived from the demodulator, which is used to control the local oscillator frequency to prevent drift. The audio amplifier receives its signal from the quadrature detector output, and the audio output goes to an IC audio power amplifier, which drives the speaker.

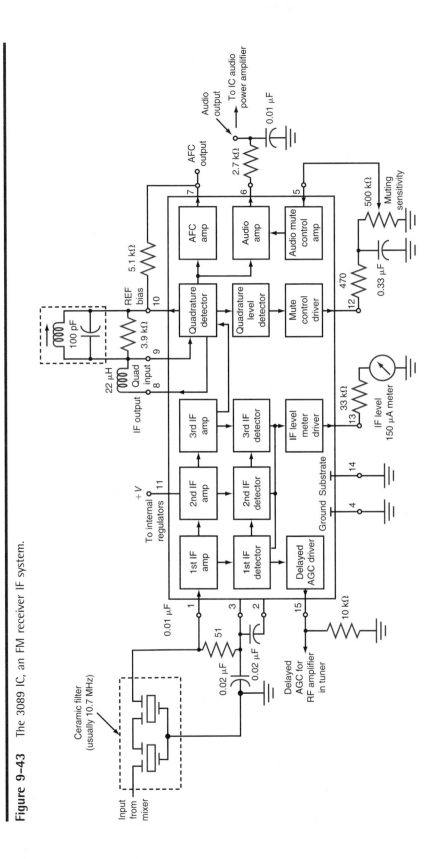

Figure 9-43 The 3089 IC, an FM receiver IF system.

9-7 Receivers and Transceivers

VHF Aircraft Communication Circuit

The typical *VHF receiver* circuit shown in Fig. 9-44 is designed to receive two-way aircraft communication between planes and airport controllers, which takes place in the VHF range of 118 to 135 MHz. Amplitude modulation is used. Like most modern receivers, the circuit is a combination of discrete components and ICs.

The signal is picked up by an antenna and fed through a transmission line to input jack J_1. The signal is coupled through C_1 to a tuned filter consisting of the series and parallel tuned circuits made up of L_1–L_5 and C_2–C_6. This broad bandpass filter passes the entire 118- to 135-MHz range.

The output of the filter is connected to an RF amplifier through C_7, which is made up of transistor Q_1 and its bias resistor R_4 and collector load R_5. The signal is then applied to the NE602 IC, U_1–C_8, which contains a balanced mixer and a local oscillator. The local oscillator frequency is set by the circuit made up of inductor L_6 and the related components. And C_{14} and D_1 in parallel form the capacitor that resonates with L_6 to set the frequency of the local oscillator. Tuning of the oscillator is accomplished by varying the dc bias on a varactor D_1. Potentiometer R_1 sets the reverse bias, which in turn varies the capacitance to tune the oscillator.

A superheterodyne receiver is tuned by varying the local oscillator frequency, which is set to a frequency above the incoming signal by the amount of the IF. In this receiver, the IF is 10.7 MHz, a standard value for many VHF receivers. To tune the 118- to 135-MHz range, the local oscillator is varied from 128.7 to 145.7 MHz.

The output of the mixer, which is the difference between the incoming signal frequency and the local oscillator frequency, appears at pin 4 of the NE602 and is fed to a ceramic bandpass filter set to the IF of 10.7 MHz. This filter provides most of the receiver's selectivity. The insertion loss of the filter is made up by an amplifier made of Q_2, its bias resistor R_{10}, and collector load R_{11}. The output of this amplifier drives an MC1350 IC through C_{16}. An integrated IF amplifier, U_2 provides extra gain and selectivity. The selectivity comes from the tuned circuit made up of IF transformer T_1. The MC1350 also contains all the AGC circuitry.

The signal at the secondary of T_1 is then fed to a simple AM diode detector consisting of D_2, R_{12}, and C_{30}. The demodulated audio signal appears across R_{12} and is then fed to op amp U_{3b}, a noninverting circuit biased by R_{13} and R_{14} that provides extra amplification for the demodulated audio and the average direct current at the detector output. This amplifier feeds the volume control, potentiometer R_2. The audio signal goes from there through C_{25} and R_{24} to another op amp, U_{3c}. Here the signal is further amplified and fed to the 386 IC power amplifier U_4. This circuit drives the speaker, which is connected via jack J_2.

The audio signal from the diode detector contains the dc level resulting from detection (rectification). Both the audio and the direct current are amplified by U_{3b} and further filtered into a nearly pure direct current by the low-pass filter made of R_{15} and C_{22}. This dc signal is applied to op amp U_{3a}, where it is amplified into a dc control voltage. This direct current at the output pin 1 of U_3 is fed back to pin 5 on the MC1350 IC to provide AGC control, ensuring a constant comfortable listening level despite wide variations in signal strength.

The AGC voltage from U_{3a} is also fed to an op amp comparator circuit made from amplifier U_{3d}. The other input to this comparator is a dc voltage from potentiometer R_3, which is used as a squelch control. Since the AGC voltage from U_{3a} is directly proportional to the signal strength, it is used as the basis for setting the squelch to a level that will blank the receiver until a signal of a predetermined strength comes along.

If the signal strength is very low or no signal is tuned in, the AGC voltage will be very low or nonexistent. This causes D_3 to conduct, effectively disabling amplifier U_{3c} and preventing the audio signal from the volume control from passing through to the

Figure 9-44 The aviation receiver–a superheterodyne unit built around four ICs–is designed to receive AM signals in the 118- to 135-MHz frequency range. (*Popular Electronics*, January 1991, Gernsback Publications, Inc.)

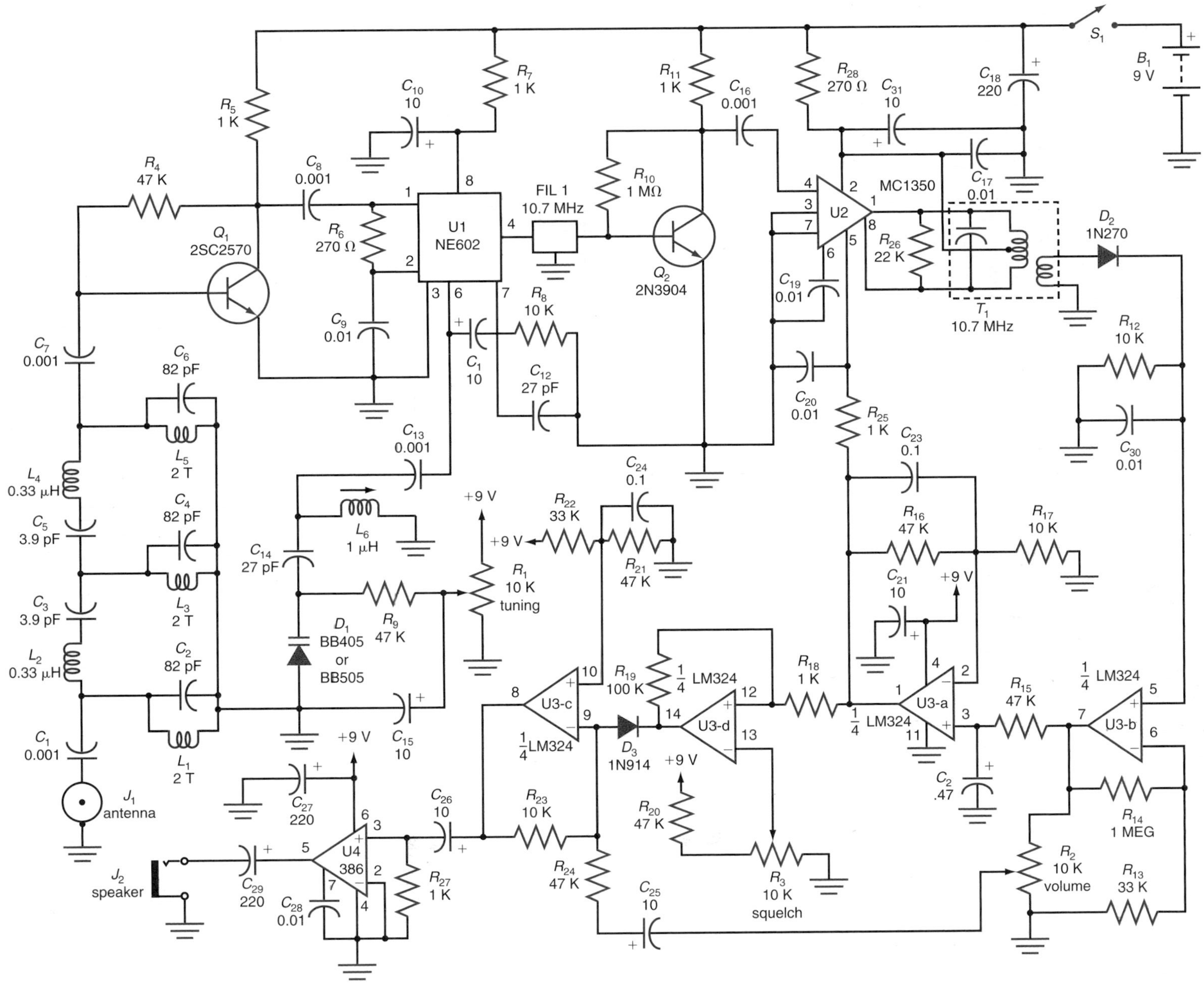

power amplifier. If a strong signal exists, D_3 will be reverse-biased and thus will not interfere with amplifier U_{3c}. As a result, the signal from the volume control passes to the power amplifier and is heard in the speaker.

Single-IC FM Receiver

The FM receiver IC chip shown in Fig. 9-45, the popular Motorola MC3363, contains all receiver circuits except for the audio power amplifier, which is a separate chip. Designed to operate at frequencies up to about 200 MHz, this chip is widely used in cordless telephones, paging receivers, and other portable applications such as remote-controlled toys and monitors and short-distance walkie-talkies. The chip is housed in a 28-pin DIP, as shown in Fig. 9-46. This dual-conversion receiver contains two mixers, two local oscillators, a limiter, a quadrature detector, and squelch circuits. The first local oscillator has a built-in varactor that allows it to be controlled by an external frequency synthesizer.

A complete receiver using the MC3363 is shown in Fig. 9-46. The receiver operates in the 30-MHz range and is the companion to the transmitter unit described in Chap. 8 and shown in Fig. 8-50. It uses the transmitter's tuned output filters for input selectivity. The output of the tuned circuit in Fig. 8-50 appears at the input to Fig. 9-46, where it feeds into an impedance-matching section made up of L_3 and C_{10}. Diodes D_1 and D_2 provide overload protection for the receiver front end. The signal goes to the transistor internal to the MC3363 on pins 2, 3, and 4, which is the RF amplifier.

The output of the RF amplifier is coupled to the first mixer through R_7 and C_{23}. The receiver is tuned to a single channel, and this frequency is fixed by an external third-

Figure 9-45 The Motorola MC3363 dual-conversion receiver IC.

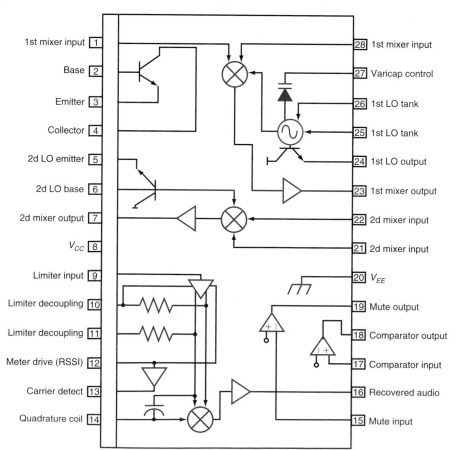

Figure 9-46 A dual-conversion IC receiver. (*Electronics Now*, October 1992.)

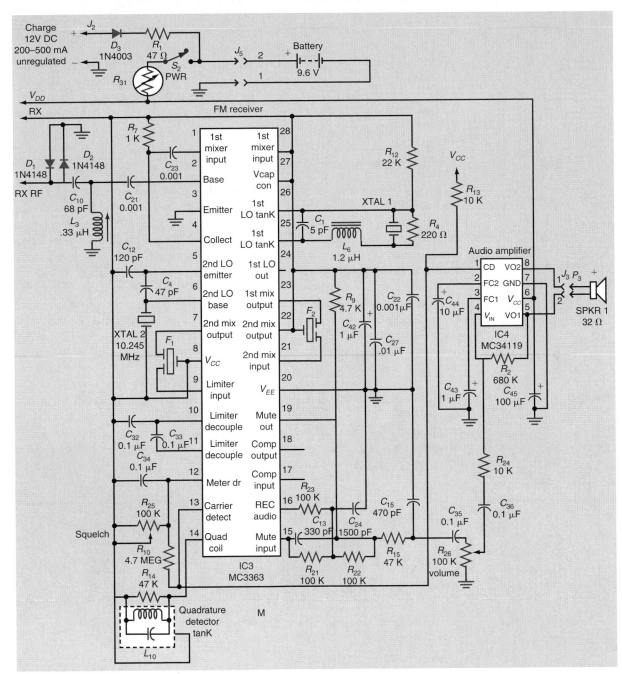

overtone-tyne crystal (XTAL1) set to a frequency 10.7 MHz greater than the incoming signal. (For example, for an input signal of 27.125 Mhz, the crystal would have a frequency of 10.7 + 27.125 = 37.825 MHz.) The crystal is connected to the receiver's first local oscillator on pins 25 and 26.

The output of the first mixer is a 10.7-MHz IF signal at pin 23. It is connected to a 10.7-MHz ceramic filter designated F_2 in Fig. 9-46. The filter output feeds into pin 21, the input to the second mixer. The second mixer is fed by a local oscillator made up of an internal transistor and related components on pins 5 and 6. It too is crystal-controlled. A 10.245-MHz crystal (XTAL2) sets the frequency. The first IF and this oscillator produce the second IF, which is the difference between 10.7 and 10.245, which is 0.455 MHz or 455 kHz. The second mixer output at pin 7 feeds a 455-kHz ceramic filter, providing additional selectivity. The filter output goes to the limiter input at pin 9, and the limiter output drives the quadrature detector. The quadrature tank coil is L10 in the diagram. The quadrature detector output (the recovered audio) is first filtered by an active low-pass filter made up of the internal op amp on pins 15 and 19 and the related resistors and capacitors. This filter cuts off frequencies above 3 kHz.

Finally, the audio signal goes to the audio power amplifier IC4, the MC34119. The squelch circuit in the MC3363 generates a carrier detect signal at pin 13, which is used to mute the IC power amplifier. The carrier detect input is pin 1 on the MC34119, and R_{26} is the volume control.

Transceivers

In the past, communication equipment was individually packaged in units based on function, and so transmitters and receivers were almost always separate units. Today, most two-way radio communication equipment is packaged so that both transmitter and receiver are in a unit known as a *transceiver*. Transceivers range from large, high-power desktop units to small, pocket-sized, handheld units. Cell phones are transceivers as are the wireless local-area networking units on PCs.

Transceivers provide many advantages. In addition to having a common housing and power supply, the transmitters and receivers can share circuits, thereby achieving cost savings and, in some cases, smaller size. Some of the circuits that can perform a dual function are antennas, oscillators, frequency synthesizers, power supplies, tuned circuits, filters, and various kinds of amplifiers. Thanks to modern semiconductor technology, most transceivers are a single silicon chip.

SSB Transceivers. Figure 9-47 is a general block diagram of a high-frequency transceiver capable of CW and SSB operation. Both the receiver and the transmitter make use of heterodyning techniques for generating the IF and final transmission frequencies, and proper selection of these intermediate frequencies allows the transmitter and receiver

Transceiver (margin)

SSB transceiver (margin)

Figure 9-47 An SSB transceiver showing circuit sharing.

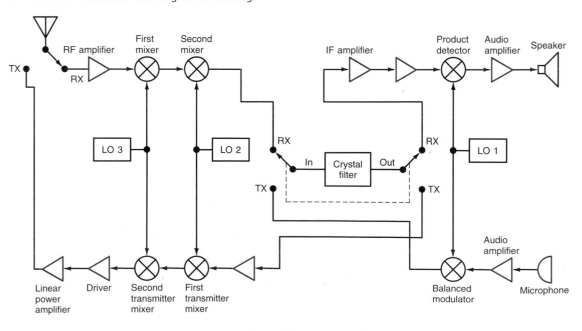

RX = receiver
TX = transmitter

Figure 9-48 A frequency synthesizer for a CB transceiver.

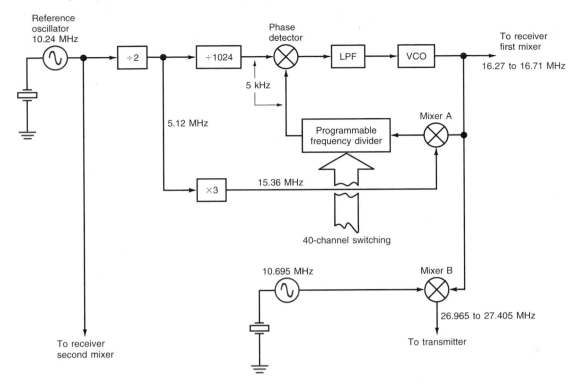

to share local oscillators. Local oscillator 1 is the BFO for the receiver product detector and the carrier for the balanced modulator for producing DSB. Later, crystal local oscillator 2 drives the second mixer in the receiver and the first transmitter mixer used for up conversion. Local oscillator 3 supplies the receiver first mixer and the second transmitter mixer.

In transmission mode, the crystal filter (another shared circuit) provides sideband selection after the balanced modulator. In the receive mode, the filter provides selectivity for the IF section of the receiver. Tuned circuits may be shared. A tuned circuit can be the tuned input for the receiver or the tuned output for the transmitter. Circuit switching may be manual, but is often done using relays or diode electronic switches. In most newer designs, the transmitter and receiver share a frequency synthesizer.

CB Synthesizers. Figure 9-48 shows a PLL synthesizer for a 40-channel CB transceiver. Using two crystal oscillators for reference and a single-loop PLL, it synthesizes the transmitter frequency and the two local oscillator frequencies for a dual-conversion receiver for all 40 CB channels. The reference crystal oscillator, which operates at 10.24 MHz, is divided by 2 with a flip-flop and then a binary frequency divider that divides by 1024 to produce a 5-kHz frequency (10.24 MHz/2 = 5.12 MHz/1024 = 5 kHz) that is then applied to the phase detector. The channel spacing is, therefore, 5 kHz.

The phase detector drives the low-pass filter and a VCO that generates a signal in the 16.27- to 16.71-MHz range. This is the local oscillator frequency for the first receiver mixer. Assume, e.g., that it is desired to receive on CB channel 1, or 26.965 MHz. The programmable divider is set to the correct ratio to produce a 5-kHz output when the VCO is 16.27. The first receiver mixer produces the difference frequency of 26.965 − 16.27 = 10.695 MHz. This is the first intermediate frequency. The 16.27-MHz VCO signal is also applied to mixer A; the other input to this mixer is 15.36 MHz, which is derived from the 10.24-MHz reference oscillator and a frequency tripler (×3). The output of mixer A drives the programmable divider, which feeds the phase detector. The 10.24-MHz reference output is also used as the local oscillator signal for the second

CB synthesizer

receiver mixer. With a 10.695 first IF, the second IF is, then, the difference, or $10.695 - 10.24$ MHz = 0.455 MHz or 455 kHz.

The VCO output is also applied to mixer B along with a 10.695-MHz signal from a second crystal oscillator. The sum output is selected, producing the transmit frequency $10.695 + 16.27 = 26.965$ MHz. This signal drives the class C drivers and power amplifiers.

Channel selection is achieved by changing the frequency-division ratio on the programmable divider, usually with a rotary switch or a digital keypad that controls a microprocessor. The circuitry in the synthesizer is usually contained on a single IC chip.

Figure 9-49 shows an integrated transceiver typical of those used in wireless local-area networks (WLANs). The transceiver is built into PCs, laptops, and other devices to wirelessly communicate with another computer and a network. This particular transceiver meets the specifications set by the IEEEs 802.11b standard for WLANs. Also known as Wi-Fi for wireless fidelity, it operates in the industrial-scientific-medical (ISM) band of 2.4 to 2.483 GHz. It is made entirely of silicon CMOS and BiCMOS and contains most of the circuitry needed to form a complete two-way radio for transmitting and receiving digital data. The device which is housed in a tiny 7×7 mm package is accompanied by two other ICs, the power amplifier (PA) for the transmitter, and a baseband processor plus the support circuits such as the A/D and D/A converters.

The transceiver can transmit at a speed of up to 11 Mb/s but can also transmit at lower rates of 5.5, 2, and 1 Mb/s depending upon the noise level, range, and operating environment. The modulation is some form of phase-shift keying. The maximum range is about 100 m although that varies with the conditions of operation. The speed of transmission is adjusted automatically depending upon the environmental situation. All these operations are handled automatically by the external processor.

The antenna feeds a bidirectional filter such as a SAW filter to provide some selectivity in both transmit and receive operations. The filter output is fed to a transmit/receive switch (T/R) that is implemented with GaAs transistors or PIN diodes and provides automatic switchover from send to receive. A balun external to the chip provides some impedance matching to the receiver input. The upper part of the transceiver chip is the receiver. The first stage is a low-noise *variable-gain amplifier (VGA)*. This amplifier drives a pair of mixers that are also fed with local oscillator signals that are 90° out of phase. Since this is a zero-IF (ZIF) radio, the local oscillator (LO) signal is at the operating frequency. This LO signal is derived from an internal PLL frequency synthesizer. The reference crystal oscillator and loop filter capacitor are off-chip. The I and Q signals generated by the mixers are filtered by on-chip active low-pass RC filters and amplified by variable-gain amplifiers. The amplified I and Q signals are sent to analog-to-digital converters (ADCs) whose outputs feed the baseband processor where demodulation and other operations are performed on the received data.

There are several other features of the receiver to consider. First, note that all transmission lines between circuits are two-wire. This means that the signals are balanced to ground and sent and received differentially. This arrangement helps minimize noise. Second, note that a digital output from the baseband processor feeds a DAC whose output is used to control the gain of the variable gain amplifiers. This is a form of AGC. Third, note that the frequency synthesizer is controlled by a serial input signal from the baseband processor. The operating channel is selected by the processor, and a serial control world is generated and sent via the serial interface to the programming interface which sets the synthesizer frequency of operation.

The transmitter is located at the bottom of the chip diagram in Fig. 9-49. Binary signals representing the modulated data to be transmitted are developed by the processor and sent to DACs in an I and Q configuration. The signals are all differential. The DAC output feed amplifiers and filters in the transceiver chip where they are sent to two I and Q mixers. Here the baseband signals are up-converted to the final transmit frequency. The on-chip synthesizer generates the carrier frequency for the mixers while a phase shifter produces the quadrature signals. The mixer output are added and then fed

Figure 9-49 Single-chip 2.4-GHz transceiver.

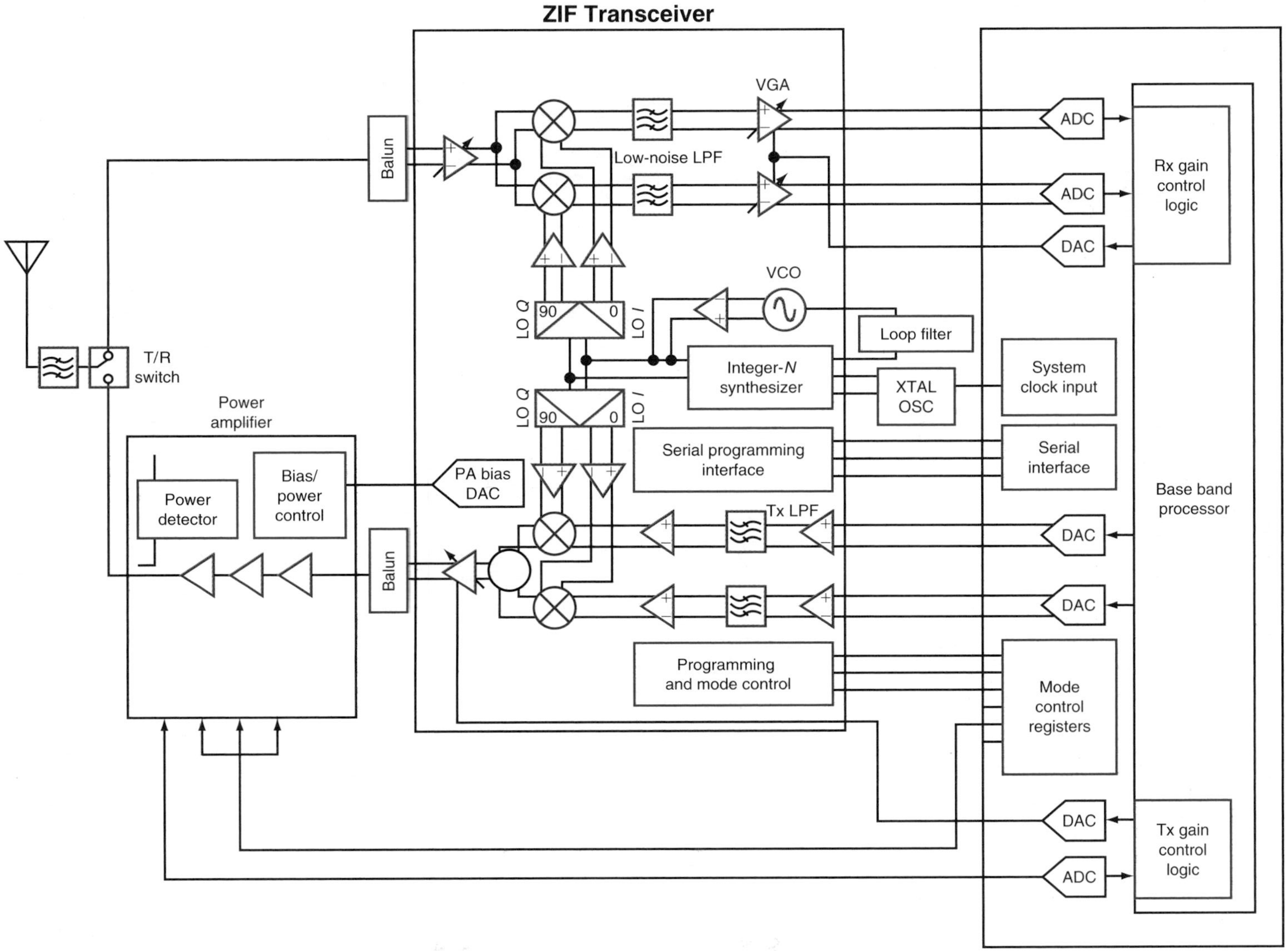

to a low-power amplifier with variable gain. This amplifier feeds an external balun that provides a match to the power amplifier (PA). The power amplifier is a class AB power amplifier with an output power of 2 dBm. The power amplifier is a separate chip and not integrated onto the transceiver chip because it generates too much heat and requires a separate package. The power amplifier output is fed to the transmit/receive switch and then the antenna via the filter.

Note that the PA has a power monitoring detector whose output is digitized by a separate ADC, and this output is used in a feedback scheme to control the transmitter power. The processor computes the desired power and sends a binary signal to a DAC that, in turn, drives the variable-gain amplifier at the transceiver output. Various mode control registers and circuits are also provided to implement the various operations dictated by the standard.

CHAPTER REVIEW

Summary

The two primary requirements for any communications receiver are *selectivity*—the ability to pick out the desired signal from all others in the frequency spectrum—and *sensitivity*—the ability to provide sufficient amplification to recover the modulating signal. Superheterodyne receivers convert all incoming signals to a lower frequency known as the intermediate frequency (IF), at which amplifiers provide optimal sensitivity and selectivity. A direct-conversion or zero-IF receiver converts directly to baseband.

IC receivers have three major sections: (1) the tuner, with RF amplifier, mixer, and local oscillator; (2) the IF section, with amplifiers, demodulator, and AGC and muting circuits; and (3) the audio power amplifier. In modern receivers all receiver functions are on one or two chips.

Mixers, like amplitude modulators, perform a mathematical multiplication of the two input signals. The output of the mixer is a signal at the IF containing the same modulation that appeared on the input RF signal; this signal is amplified by one or more IF amplifier stages. Popular types of mixer circuits are diode mixers, singly balanced mixers, doubly balanced mixers, bipolar transistor mixers, and FET mixers. Two important circuits in superheterodyne receivers are frequency synthesizers for the local oscillator and automatic gain control (AGC) circuits. AGC is a feedback system that automatically adjusts the gain of the receiver based on the amplitude of the received signal.

Two recent advances in receiver design are direct conversion (or zero IF) and software-defined radio. In zero-IF receivers, the mixer also serves as the demodulator for AM signals. For frequency and phase demodulation, direct conversion receivers use quadrature mixers (I&Q). In software-defined radios, most receiver chores such as filtering and demodulation are performed by a DSP.

Noise is random energy that interferes with the desired signal. An important indicator of noise is the *S/N* ratio, which indicates the relative strengths of the signal and the noise. The stronger the signal and the weaker the noise, the higher the *S/N* ratio. The two major sources of noise are external (industrial, atmospheric, and extraterrestrial) and internal (thermal and semiconductor). Three figures used for a receiver's noise rating are the noise factor, the ratio of the *S/N* power at the input to the *S/N* power at the output; the noise temperature; and SINAD, signal plus noise and distortion.

Squelch, or muting, circuits keep the audio amplifier turned off during the time that noise is received in the background. When an RF signal appears at the input, the audio amplifier is enabled. Two other noise control circuits are continuous tone-coded squelch systems and digital coded squelch systems.

Questions

1. How does decreasing the Q of a resonant circuit affect its bandwidth?
2. How does cascading tuned circuits affect selectivity?
3. What can happen to a modulated signal if the selectivity of a tuned circuit is too sharp?
4. How must the coil resistance be changed to narrow the bandwidth of a tuned circuit?
5. A choice is to be made between two 10.7-MHz IF filters. One has a shape factor of 2.3; the other, 1.8. Which has the better selectivity?

6. What type of receiver uses only amplifiers and a detector?
7. What type of receiver uses a mixer to convert the received signal to a lower frequency?
8. What two circuits are used to generate the IF?
9. In what stage are most of the gain and selectivity in a superheterodyne receiver obtained?
10. What circuit in a receiver compensates for a wide range of input signal levels?
11. The mixer output is usually the difference between what two input frequencies?
12. The AGC voltage controls the gain of what two stages of a receiver?
13. What do you call an interfering signal that is spaced from the desired signal by twice the IF?
14. What is the primary cause of images appearing at the mixer input?
15. What advantage does a dual-conversion superhet offer over a single-conversion superhet?
16. How can the image problem best be solved during the design of a receiver?
17. Give the expressions for the outputs of a mixer whose inputs are f_1 and f_2.
18. Name the best type of passive mixer.
19. Name the best type of transistor mixer.
20. The process of mixing is similar to what kind of modulation?
21. What is an image reject mixer?
22. What is the primary specification of a VFO used for local oscillator duty?
23. What type of local oscillator is used in most modern receivers?
24. Why are mixers sometimes used in frequency synthesizers, as in Fig. 9-18?
25. Name the three primary sources of external noise.
26. What is a direct-conversion receiver? What are its main advantages?
27. Explain the architecture and operation of a software-defined radio. What computer technology makes it possible?
28. Why is it necessary to use quadrature I and Q mixers in a (Zero-IF) receiver?
29. Name the five main types of internal noise that occur in a receiver.
30. What is the primary source of atmospheric noise?
31. List four common sources of industrial noise.
32. What is the main source of internal noise in a receiver?
33. In what units is the signal-to-noise (S/N) ratio usually expressed?
34. How does increasing the temperature of a component affect its noise power?
35. How does narrowing the bandwidth of a circuit affect the noise level?
36. Name the three types of semiconductor noise.
37. True or false? The noise at the output of a receiver is less than the noise at the input.

38. What are the three components that make up SINAD?
39. What stages of a receiver contribute the most noise?
40. What are the advantages and disadvantages of using an RF amplifier at the front end of a receiver?
41. What is the name of the low-noise transistor preferred in RF amplifiers at microwave frequencies?
42. What types of mixer have a loss?
43. How is the selectivity usually obtained in a modern IF amplifier?
44. In a double-tuned circuit, maximum bandwidth is obtained with what type of coupling?
45. What is the name given to an IF amplifier that clips the positive and negative peaks of a signal?
46. Why is clipping allowed to occur in an IF stage?
47. The gain of a bipolar class A amplifier can be varied by changing what parameter?
48. What is the overall RF–IF gain range of a receiver?
49. What is the process of using the amplitude of an incoming signal to control the gain of a receiver?
50. What is the difference between forward AGC and reverse AGC?
51. How is the gain of a differential amplifier varied to produce AGC?
52. What are two names for the circuit that blocks the audio until a signal is received?
53. Name the type of signal used to operate the circuit described in Question 52.
54. Describe the purpose and operation of a CTCSS in a receiver.
55. A BFO is required to receive what two types of signals?
56. What is the source of the signal required at the input to the 3089 receiver IC?
57. Name the three main sources of selectivity for receivers implemented with ICs.
58. In a single chip transceiver, how is the frequency of operation changed?
59. In an FM transceiver, what is the only circuitry commonly shared by the receiver and the transmitter?
60. What are the circuits shared by the transmitter and receiver in an SSB transceiver?
61. The phase-locked loop is often combined with what circuit to produce multiple frequencies in a transceiver?
62. The synthesizer in a transceiver usually generates what three frequencies?
63. Can a ZIF receiver demodulate FM or PM? How?
64. Name four functions that can be performed by DSP in an SDR.

The following questions refer to the receiver in Fig. 9-44.

65. What components or circuits determine the bandwidth of the receiver?
66. As R_1 is varied so that the voltage on the arm of the potentiometer increases toward $+9$ V, how does the frequency of the local oscillator vary?
67. What component provides most of the gain in this receiver?

68. Is the squelch signal- or noise-derived?

69. Does this receiver contain a BFO?

70. Could this circuit receive CW or SSB signals?

71. If the dc AGC voltage on pin 5 of U_2 is decreased, what happens to the gain of U_2?

72. Where would an audio signal be injected to test the complete audio section of this receiver?

73. What frequency signal would be used to test the IF section of this receiver, and where would it be connected?

74. What component would be inoperable if C_{31} became shorted?

Problems

1. A tuned circuit has a Q of 80 at its resonant frequency of 480 kHz. What is its bandwidth? ◆

2. A parallel LC tuned circuit has a coil of 4 μH and a capacitance of 68 pF. The coil resistance is 9 Ω. What is the circuit bandwidth?

3. A tuned circuit has a resonant frequency of 18 MHz and a bandwidth of 120 kHz. What are the upper and lower cutoff frequencies? ◆

4. What value of Q is needed to achieve a bandwidth of 4 kHz at 3.6 MHz?

5. A filter has a 6-dB bandwidth of 3500 Hz and a 60-dB bandwidth of 8400 Hz. What is the shape factor? ◆

6. A superhet has an input signal of 14.5 MHz. The local oscillator is tuned to 19 MHz. What is the IF?

7. A desired signal at 29 MHz is mixed with a local oscillator of 37.5 MHz. What is the image frequency? ◆

8. A dual-conversion superhet has an input frequency of 62 MHz and local oscillators of 71 and 8.6 MHz. What are the two IFs?

9. What are the outputs of a mixer with inputs of 162 and 189 MHz? ◆

10. What is the most likely IF for a mixer with inputs of 162 and 189 MHz?

11. A frequency synthesizer like the one in Fig. 9-18 has a reference frequency of 100 kHz. The crystal oscillator and the multiplier supply a signal of 240 MHz to the mixer. The frequency divider is set to 1500. What is the VCO output frequency?

12. A frequency synthesizer has a phase detector input reference of 12.5 kHz. The divider ratio is 295. What are the output frequency and the frequency change increment?

13. The signal input power to a receiver is 6.2 nW. The noise power is 1.8 nW. What is the S/N ratio? What is the S/N ratio in decibels?

14. What is the noise voltage produced across a 50-Ω input resistance at a temperature of 25°C with a bandwidth of 2.5 MHz?

15. At what frequencies is noise temperature used to express the noise in a system?

16. The noise ratio of an amplifier is 1.8. What is the noise temperature in kelvins?

◆ *Answers to Selected Problems follow Chap. 22.*

Critical Thinking

1. Why is a noise temperature of 155 K a better rating than a noise temperature of 210 K?

2. An FM transceiver operates on a frequency of 470.6 MHz. The first IF is 45 MHz and the second IF is 500 kHz. The transmitter has a frequency-multiplier chain of $2 \times 2 \times 3$. What three signals must the frequency synthesizer generate for the transmitter and two mixers in the receiver?

3. Explain how a digital counter could be connected to the receiver in Fig. 9-44 so that it would read the frequency of the signal to which it was tuned.

4. What is the effect on receiver selectivity if a resistor is connected in parallel with the tuned transformer in Fig. 9-33?

5. The circuits in a superheterodyne receiver have the following gains: RF amplifier, 8 dB; mixer, −2.5 dB; IF amplifier, 80 dB; demodulator, −0.8 dB; audio amplifier, 23 dB. What is the total gain?

6. A superheterodyne receiver receives a signal on 10.8 MHz. It is amplitude-modulated by a 700-Hz tone. The local oscillator is set to the signal frequency. What is the IF? What is the output of a diode detector demodulator?

7. A software-defined radio operates at 1900 MHz. The local oscillator is set to 1750 MHz. What is the minimum sampling frequency required in the A/D converter?

chapter 10

Multiplexing and Demultiplexing

A communication channel, or link, between two points is established whenever a cable is connected or a radio transmitter and receiver are set up between the two points. When there is only one link, only one function—whether it involves signal transmission or control operations—can be performed at a time. For two-way communication, a half duplex process is set up: Both ends of the communication link can send and receive but not at the same time.

Transmitting two or more signals simultaneously can be accomplished by running multiple cables or setting up one transmitter/receiver pair for each channel, but this is an expensive approach. In fact, a single cable or radio link can handle multiple signals simultaneously by using a technique known as *multiplexing*, which permits hundreds or even thousands of signals to be combined and transmitted over a single medium. Multiplexing has made simultaneous communication more practical and economically feasible, helped conserve spectrum space, and allowed new, sophisticated applications to be implemented.

Objectives

After completing this chapter, you will be able to:

- Explain why multiplexing techniques are necessary in telemetry, telephone systems, radio and TV broadcasting, and Internet access.
- Compare frequency division multiplexing with time-division multiplexing.
- Trace the steps in the transmission and reception of multiplexed signals.
- List the major subtypes of time-division multiplexing.
- Define pulse-code modulation, draw the diagram of a typical PCM multiplexer, and state the primary benefit of PCM over other forms of pulse modulation.
- List the characteristics of the T-carrier system.
- Explain the difference between time and frequency duplexing.

10-1 Multiplexing Principles

Multiplexing is the process of simultaneously transmitting two or more individual signals over a single communication channel, cable or wireless. In effect, it increases the number of communication channels so that more information can be transmitted. Often in communication it is necessary or desirable to transmit more than one voice or data signal simultaneously. An application may require multiple signals, or cost savings can be gained by using a single communication channel to send multiple information signals. Four applications that would be prohibitively expensive or impossible without multiplexing are telephone systems, telemetry, satellites, and modern radio and TV broadcasting.

The greatest use of multiplexing is in the telephone system, where millions of calls are multiplexed on cables, long-distance fiber-optic lines, satellites, and wireless paths. Multiplexing increases the telephone carrier's ability to handle more calls while minimizing system costs and spectrum usage.

In *telemetry,* the physical characteristics of a given application are monitored by sensitive transducers which generate electric signals that vary in response to changes in the status of the various physical characteristics. The sensor-generated information can be sent to a central location for monitoring, or can be used as feedback in a closed-loop control system. Most spacecraft and many chemical plants, e.g., use telemetry systems to monitor characteristics such as temperature, pressure, speed, light level, flow rate, and liquid level.

The use of a single communication channel for each characteristic being measured in a telemetry system would not be practical, because of both the multiple possibilities for signal degradation and the high cost. Consider, e.g., monitoring a space shuttle flight. Wire cables are obviously out of the question, and multiple transmitters impractical. If a deep-space probe were used, it would be necessary to use multiple transducers, and many transmitters would be required to send the signals back to earth. Because of cost, complexity, and equipment size and weight, this approach would not be feasible. Clearly, telemetry is an ideal application for multiplexing, with which the various information signals can all be sent over a single channel.

Finally, modern FM stereo broadcasting requires multiplexing techniques, as does the transmission of stereo sound and color in TV. Digital TV is multiplexed.

Multiplexing is accomplished by an electronic circuit known as a *multiplexer*. A simple multiplexer is illustrated in Fig. 10-1. Multiple input signals are combined by the multiplexer into a single composite signal that is transmitted over the communication medium. Alternatively, multiplexed signals can modulate a carrier before transmission.

Figure 10-1 Concept of multiplexing.

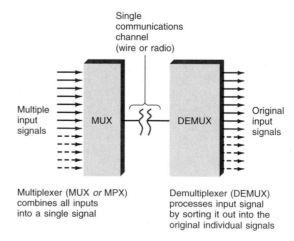

Multiplexer (MUX *or* MPX)
combines all inputs
into a single signal

Demultiplexer (DEMUX)
processes input signal
by sorting it out into the
original individual signals

At the other end of the communication link, a demultiplexer is used to process the composite signal to recover the individual signals.

The two most common types of multiplexing are frequency-division multiplexing (FDM) and time-division multiplexing (TDM). Two variations of these basic methods are frequency-division multiple access (FDMA) and time-division multiple access (TDMA). In general, FDM systems are used for analog information and TDM systems are used for digital information. Of course, TDM techniques are also found in many analog applications because the processes of A/D and D/A conversion are so common. The primary difference between these techniques is that in FDM, individual signals to be transmitted are assigned a different frequency within a common bandwidth. In TDM, the multiple signals are transmitted in different time slots on a single channel.

Another form of multiple access is known as *code-division multiple access (CDMA)*. It is widely used in cell phone systems to allow many cell phone subscribers to use a common bandwidth at the same time. This system uses special codes assigned to each user that can be identified. CDMA uses a technique called spread spectrum to make this type of multiplexing possible. Spread spectrum is covered in Chap. 12.

Code-division multiple access (CDMA)

SPATIAL MULTIPLEXING

Spatial multiplexing is the term used to describe the transmission of multiple wireless signals on a common frequency in such a way that they do not interfere with one another. One way of doing this is to use low-power transmissions so that the signals do not interfere with one another. When very low power is used, the signals do not travel very far. The transmission distance is a function of the power level, frequency, and antenna height. For example, these factors may be used to ensure that the signals do not travel more than, say, 3 mi. Beyond 3 mi, these same frequencies may be used again to carry different signals.

Another technique is to use carefully controlled antenna radiation patterns to direct the signals to different locations in such a way that signals sharing the same frequency channel do not interfere with one another. Special antennas using multiple transmit and receive elements and phase-shifting circuits form the beams of radio energy in such a way as to minimize or in some cases completely eliminate interference from nearby signals on a common channel.

Spatial multiplexing is sometimes referred to as *frequency reuse*. This technique is widely used in satellite and cellular telephone systems.

Spatial multiplexing (frequency reuse)

10-2 Frequency Division Multiplexing

In *frequency-division multiplexing (FDM)*, multiple signals share the bandwidth of a common communication channel. Remember that all channels have specific bandwidths, and some are relatively wide. A coaxial cable, e.g., has a bandwidth of about 1 GHz. The bandwidths of radio channels vary, and are usually determined by FCC regulations and the type of radio service involved. Regardless of the type of channel, a wide bandwidth can be shared for the purpose of transmitting many signals at the same time.

Frequency-division multiplexing (FDM)

Transmitter-Multiplexers

Figure 10-2 shows a general block diagram of an FDM system. Each signal to be transmitted feeds a modulator circuit. The carrier for each modulator (f_c) is on a different frequency. The carrier frequencies are usually equally spaced from one another over a specific frequency range. These carriers are referred to as *subcarriers*. Each input signal

Subcarrier

Figure 10-2 The transmitting end of an FDM system.

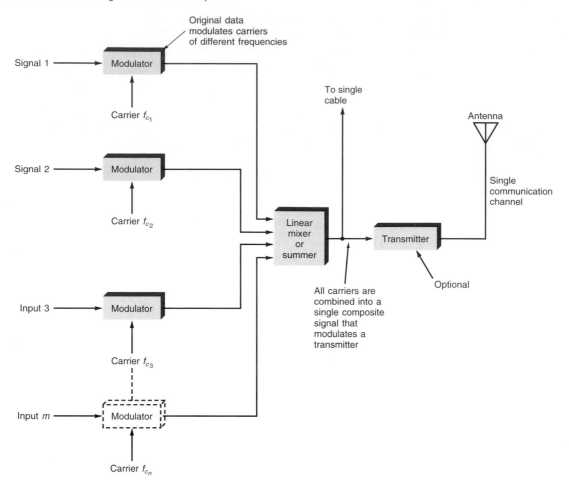

is given a portion of the bandwidth. The resulting spectrum is illustrated in Fig. 10-3. Any of the standard kinds of modulation can be used, including AM, SSB, FM, PM, or any of the various digital modulation methods. The FDM process divides up the bandwidth of the single channel into smaller, equally spaced channels, each capable of carrying information in sidebands.

The modulator outputs containing the sideband information are added algebraically in a linear mixer; no modulation or generation of sidebands takes place. The resulting output signal is a composite of all the modulated subcarriers. This signal can be used to modulate a radio transmitter or can itself be transmitted over the single communication channel. Alternatively, the composite signal can become one input to another multiplexed system.

Figure 10-3 Spectrum of an FDM signal. The bandwidth of a single channel is divided into smaller channels.

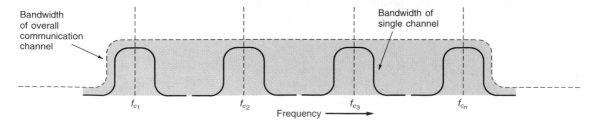

Figure 10-4 The receiving end of an FDM system.

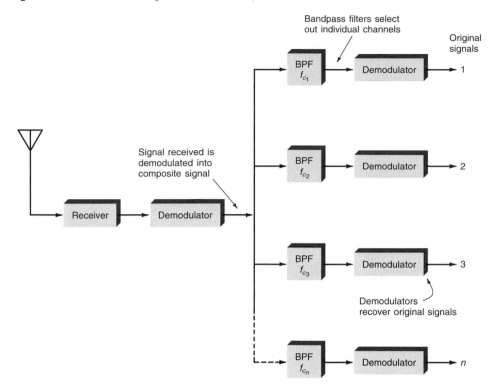

Receiver-Demultiplexers

The receiving portion of an FDM system is shown in Fig. 10-4. A receiver picks up the signal and demodulates it, recovering the composite signal. This is sent to a group of bandpass filters, each centered on one of the carrier frequencies. Each filter passes only its channel and rejects all others. A channel demodulator then recovers each original input signal.

FDM Applications

Telemetry. As indicated earlier, sensors in telemetry systems generate electric signals that change in some way in response to changes in physical characteristics. An example of a sensor is a *thermistor,* a device used to measure temperature. A thermistor's resistance varies inversely with temperature: As the temperature increases, the resistance decreases. The thermistor is usually connected into some kind of a resistive network, such as a voltage divider or bridge, and to a dc voltage source. The result is a dc output voltage, which varies in accordance with temperature and which is transmitted to a remote receiver for measurement, readout, and recording. The thermistor becomes one channel of an FDM system.

Thermistor

Other sensors have different kinds of outputs. Many have varying dc outputs, and others have ac output. Each of these signals is typically amplified, filtered, and otherwise conditioned before being used to modulate a carrier. All the carriers are then added to form a single multiplexed channel.

The conditioned transducer outputs are normally used to frequency-modulate a subcarrier. The varying direct or alternating current changes the frequency of an oscillator operating at the carrier frequency. Such a circuit is generally referred to as a *voltage-controlled oscillator (VCO)* or a *subcarrier oscillator (SCO)*. To produce FDM, each VCO operates at a different center or carrier frequency. The outputs of the subcarrier oscillators are added. A diagram of such a system is shown in Fig. 10-5.

Voltage-controlled oscillator (VCO)

Subcarrier oscillator (SCO)

Figure 10–5   An FDM telemetry transmitting system.

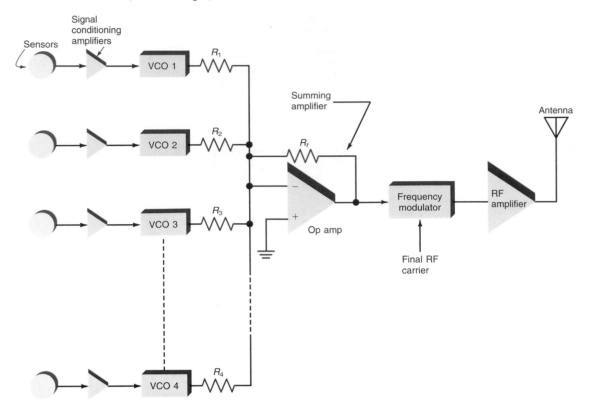

Figure 10-6(a) shows a block diagram of a typical VCO circuit. The VCOs are available as single IC chips. The popular 566 VCO IC consists of a dual-polarity current source that linearly charges and discharges an external capacitor C. The current value is set by an external resistor R_1. Together R_1 and C set the operating or center carrier frequency, which can be any value up to about 1 MHz.

The current source can be varied by an external signal, either dc or ac, which is the modulating signal from a transducer or other source. The input signal varies the charging and discharging current, thereby varying the carrier frequency. The result is direct FM.

The current source output is a linear triangular waveform that is buffered by an amplifier for external use and fed to an internal Schmitt trigger. The Schmitt trigger generates a rectangular pulse at the operating frequency that is fed to a buffer amplifier for external use.

The Schmitt trigger output is also fed back to the current source, where it controls whether the capacitor is charged or discharged. For example, the VCO may begin by charging the capacitor. When the Schmitt trigger senses a specific level on the triangular wave, it switches the current source. Discharging then occurs. The waveforms in Fig. 10-6(b) show this feedback action, which is what creates a free-running, astable oscillator.

Most VCOs are astable multivibrators whose frequency is controlled by the input from the signal conditioning circuits. The frequency of the VCO changes linearly in proportion to the input voltage. Increasing the input voltage causes the VCO frequency to increase. The rectangular or triangular output of the VCO is usually filtered into a sine wave by a bandpass filter centered on the unmodulated VCO center frequency. This can be either a conventional LC filter or an active filter made with an op amp and RC input and feedback networks. The resulting sinusoidal output is applied to the linear mixer.

The linear mixing process in an FDM system can be accomplished with a simple resistor network. However, such networks greatly attenuate the signal, and some voltage amplification is usually required for practical systems. A way to achieve the mixing and

Figure 10-6 (*a*) Typical IC VCO circuit. (*b*) Waveforms.

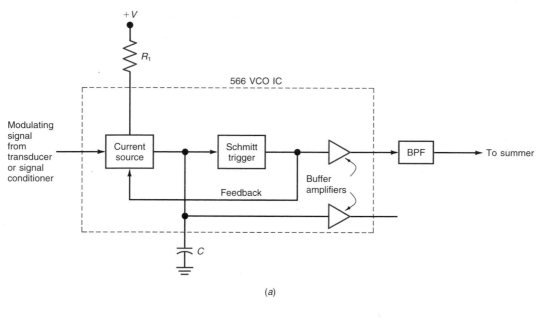

(a)

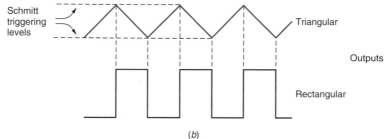

(b)

amplification at the same time is to use an op amp summer, such as that shown in Fig. 10-5. Recall that the gain of each input is a function of the ratio of the feedback resistor R_f to the input resistor value $(R_1, R_2,$ etc.). The output is given by the expression $V_{\text{out}} = -[V_1(R_f/R_1) + V_2(R_f/R_2) + V_3(R_f/R_3) + \cdots + V_n(R_f/R_n)]$.

In most cases, the VCO FM output levels are the same, and all input resistors on the summer amplifier are therefore equal. If variations do exist, amplitude corrections can be accomplished by making the summer input resistors adjustable. The output of the summer amplifier does invert the signal; however, this has no effect on the content.

The composite output signal is then typically used to modulate a radio transmitter. Again, most telemetry systems use FM, although it is possible to use other kinds of modulation schemes. A system that uses FM of the VCO subcarriers as well as FM of the final carrier is usually called an *FM/FM system*.

FM/FM system

The receiving end of a telemetry system is shown in Fig. 10-7. A standard superheterodyne receiver tuned to the RF carrier frequency is used to pick up the signal. An FM demodulator reproduces the original composite multiplexed signal, which is then fed to a demultiplexer. The demultiplexer divides the signals and reproduces the original inputs.

The output of the first FM demodulator is fed simultaneously to multiple bandpass filters, each of which is tuned to the center frequency of one of the specified subchannels. Each filter passes only its subcarrier and related sidebands and rejects all the others. The demultiplexing process is, then, essentially one of using filters to sort the composite multiplex signal back into its original components. The output of each filter is the subcarrier oscillator frequency with its modulation.

Multiplexing and Demultiplexing

Figure 10-7 An FM/FM telemetry receiver.

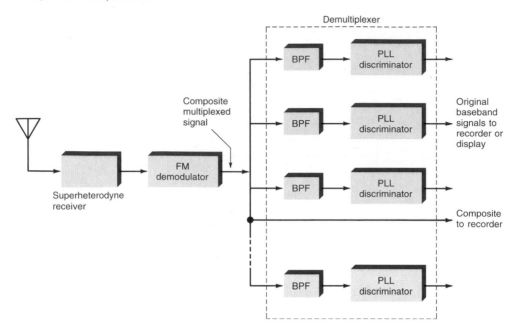

Discriminator

These signals are then applied to FM demodulators. Also known as *discriminators,* these circuits take the FM signal and recreate the original dc or ac signal produced by the transducer. The original signals are then measured or processed to provide the desired information from the remote transmitting source. In most systems, the multiplexed signal is sent to a data recorder where it is stored for possible future use. The original telemetry output signals can be graphically displayed on a strip chart recorder or otherwise converted to usable outputs.

The demodulator circuits used in typical FM demultiplexers are of either the phase-locked loop (PLL) or the pulse-averaging type. PLL circuits have superior noise performance over the simpler pulse-averaging types. A PLL discriminator is also used to demodulate the receiver output.

FDM telemetry systems, which are inexpensive and highly reliable, are still widely used in aircraft and missile instrumentation and for monitoring of medical devices, such as pacemakers.

Telephone Systems. For decades, telephone companies used FDM to send multiple telephone conversations over a minimum number of cables. In this application, the original voice signal, in the 300- to 3000-Hz range, is used to modulate a subcarrier. Lower sideband (LSB) SSB AM was used. Each subcarrier is on a different frequency, and these subcarriers are then added to form a single channel. This multiplexing process was repeated at several levels, so that up to 10,800 phone calls could be carried over a single communication channel, assuming a wide enough bandwidth.

This elaborate FDM system is no longer used. It was replaced by an all-digital time-division multiplexing (TDM) system that is described later in this chapter.

Cable TV. One of the best examples of FDM is cable TV, in which multiple TV signals, each in its own 6-MHz channel, are multiplexed on a common coaxial or fiber-optic cable and sent to nearby homes. TV signals include video and audio to modulate carriers using analog methods. Each channel uses a separate set of carrier frequencies, which can be added to produce FDM. The cable box acts as a tunable filter to select the desired channel. Figure 10-8 shows the spectrum on the cable. Each 6-MHz channel carries the video and voice of the TV signal. Coaxial and fiber-optic cables have an enormous

Figure 10-8 The spectrum on coaxial cable in a cable TV system with 6-MHz-wide channels.

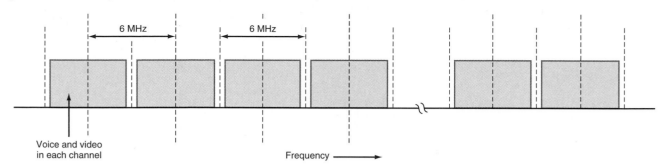

bandwidth and can carry more than one hundred TV channels. Many cable TV companies also use their cable system for Internet access. A special modem (modulator-demodulator) permits computer data to be transmitted and received at very high data speeds. You will learn more about cable modems in Chap. 11 and about cable TV in Chap. 23.

FM Stereo Broadcasting. In recording original stereo, two microphones are used to generate two separate audio signals. The two microphones pick up sound from a common source, such as a voice or an orchestra, but from different directions. The separation of the two microphones provides sufficient differences in the two audio signals to provide more realistic reproduction of the original sound. When stereo is reproduced, the two signals can come from a cassette tape, a CD, or some other source. These two independent signals must somehow be transmitted by a single transmitter. This is done through FDM techniques.

Figure 10-9 is a general block diagram of a stereo FM multiplex modulator. The two audio signals, generally called the left (L) and right (R) signals, originate at the two microphones shown in the figure. These two signals are fed to a combining circuit, where they are used to form sum ($L + R$) and difference ($L - R$) signals. The $L + R$ signal is a linear algebraic combination of the left and right channels. The composite signal it produces is the same as if a single microphone were used to pick up the sound. It is this signal that a monaural receiver will hear. The frequency response is 50 Hz to 15 kHz.

The combining circuit inverts the right channel signal, thereby subtracting it from the left channel to produce the $L - R$ signal. These two signals, the $L + R$ and $L - R$, are transmitted independently and recombined later in the receiver to produce the individual right- and left-hand channels.

The $L - R$ signal is used to amplitude-modulate a 38-kHz carrier in a balanced modulator. The balanced modulator suppresses the carrier but generates upper and lower sidebands. The resulting spectrum of the composite modulating signal is shown in Fig. 10-10. As shown, the frequency range of the $L + R$ signal is from 50 Hz to 15 kHz. Since the frequency response of an FM signal is 50 to 15 kHz, the sidebands of the $L - R$ signal are in the frequency range of 38 kHz $\pm$ 15 kHz or 23 to 53 kHz. This DSB suppressed carrier signal is algebraically added to, and transmitted along with, the standard $L + R$ audio signal.

Also transmitted with the $L + R$ and $L - R$ signals is a 19-kHz pilot carrier, which is generated by an oscillator whose output also modulates the main transmitter. Note that the 19-kHz oscillator drives a frequency doubler to generate the 38-kHz carrier for the balanced modulator.

Some FM stations also broadcast one or more additional signals, referred to as *subsidiary communications authorization (SCA) signals*. The basic SCA signal is a separate subcarrier of 67 kHz which is frequency-modulated by audio signals, usually music. SCA signals are also used to transmit weather, sports, and financial information. Special SCA FM receivers can pick up these signals. The SCA portion of the system is generally used for broadcasting background music for elevators, stores, offices, and restaurants. If an

Subsidiary communications authorization (SCA) signal

Figure 10-9 General block diagram of an FM stereo multiplex modulator, multiplexer, and transmitter.

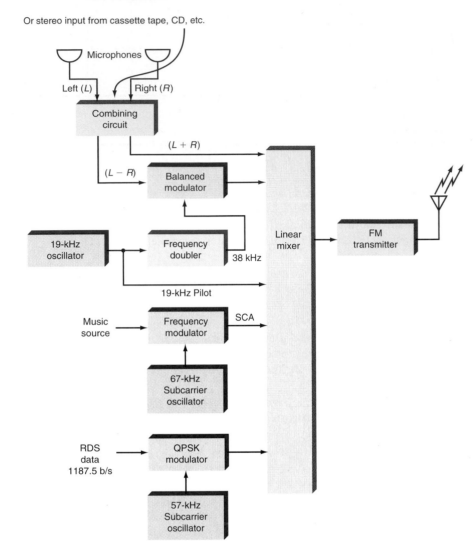

Figure 10-10 Spectrum of FM stereo multiplex broadcast signal. This signal frequency-modulates the RF carrier.

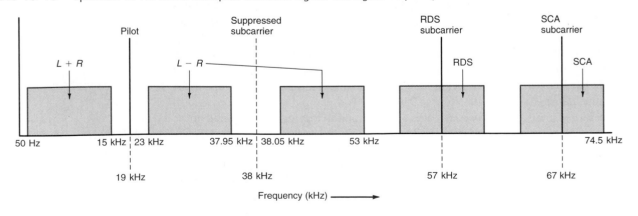

SCA system is being used, the 67-kHz subcarrier with its music modulation will also be added to the $L - R$ and $L + R$ signals to modulate the FM transmitter. Not all stations transmit SCA, but some transmit several channels, using additional, higher-frequency subcarriers.

Another alternative service provided by some FM stations is called the *Radio Data System (RDS)*. It is widely used in car radios and some home stereo receivers. It allows digital data to be transmitted to the FM receiver. Some examples of the types of data transmitted include station call letters and location, travel and weather data, and short news announcements. A popular use of RDS is to transmit the name and artist of music selections being played by the station. The transmitted data is displayed on a liquid-crystal display (LCD) in the receiver.

The data to be transmitted is used to modulate another subcarrier at 57 kHz. This is the third harmonic of the 19-kHz pilot carrier and so helps prevent interaction with the stereo signals. A form of phase modulation called *quadrature phase-shift keying (QPSK)* is used to modulate the subcarrier. The serial data rate is 1187.5 bits per second (bps).

Quadrature phase-shift keying

As in other FDM systems, all the subcarriers are added with a linear mixer to form a single signal (Fig. 10-10). This signal is used to frequency-modulate the carrier of the broadcast transmitter. Again, note that FDM simply takes a portion of the frequency spectrum. There is sufficient spacing between adjacent FM stations so that the additional information can be accommodated. Keep in mind that each additional subcarrier reduces the amount by which the main $L + R$ signal can modulate the carrier, since the maximum total modulation voltage is determined by the legal channel width.

At the receiving end, the demodulation is accomplished with a circuit similar to that illustrated in Fig. 10-11. The FM superheterodyne receiver picks up the signal, amplifies it, and translates it to an intermediate frequency, usually 10.7 MHz. It is then demodulated. The output of the demodulator is the original multiplexed signal. The additional circuits now sort out the various signals and reproduce them in their original form.

The original audio $L + R$ signal is extracted simply by passing the multiplex signal through a low-pass filter. Only the 50- to 15-kHz original audio is passed. This signal is fully compatible with monaural FM receivers without stereo capability. In a stereo

Figure 10-11 Demultiplexing and recovering the FM stereo and SCA signals.

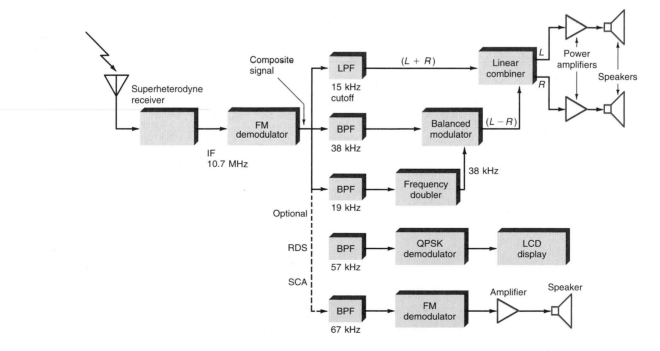

system, the $L + R$ audio signal is fed to a linear matrix or combiner where it is mixed with the $L - R$ signal to create the two separate L and R channels.

The multiplexed signal is also applied to a bandpass filter that passes the 38-kHz suppressed subcarrier with its sidebands. This is the $L - R$ signal that modulates the 38-kHz carrier. This signal is fed to a balanced modulator for demodulation.

The 19-kHz pilot carrier is extracted by passing the multiplexed signal through a narrow bandpass filter. This 19-kHz subcarrier is then fed to an amplifier and frequency doubler circuit which produces a 38-kHz carrier signal that is fed to the balanced modulator. The output of the balanced modulator, of course, is the $L - R$ audio signal. This is fed to the linear resistive combiner along with the $L + R$ signal.

The linear combiner both adds and subtracts these two signals. Addition produces the left-hand channel: $(L + R) + (L - R) = 2L$. Subtraction produces the right-hand channel: $(L + R) - (L - R) = 2R$. The left- and right-hand audio signals are then sent to separate audio amplifiers and ultimately the speaker.

If an SCA signal is being used, a separate bandpass filter centered on the 67-kHz subcarrier will extract the signal and feed it to an FM demodulator. The demodulator output is then sent to a separate audio amplifier and speaker.

If the RDS signal is used, a 57-kHz bandpass filter selects this signal and sends it to a QPSK demodulation. The recovered digital data is then displayed on the receiver's LCD. Typically, the recovered data is sent to the receiver's embedded control microprocessor that also controls the LCD display where the data is conditioned prior to display.

All the circuitry used in the demultiplexing process is usually contained in a single IC. In fact, most FM receivers contain a single chip that includes the IF, demodulator, and demultiplexer. Note that the multiplexing and demultiplexing of FM stereo in a TV set are exactly as described above, but with a different IF.

Example 10-1

A cable TV service uses a single coaxial cable with a bandwidth of 860 MHz to transmit multiple TV signals to subscribers. Each TV signal is 6 MHz wide. How many channels can be carried?

$$\text{Total channels} = 860/6 = 143.33 \text{ or } 143$$

10-3 Time-Division Multiplexing

Time-division multiplexing

In FDM, multiple signals are transmitted over a single channel, each signal being allocated a portion of the spectrum within that bandwidth. In *time-division multiplexing (TDM)*, each signal occupies the entire bandwidth of the channel. However, each signal is transmitted for only a brief time. In other words, multiple signals take turns transmitting over the single channel, as diagrammed in Fig. 10-12. Here, each of the four signals being transmitted over a single channel is allowed to use the channel for a fixed time, one after another. Once all four have been transmitted, the cycle repeats.

TDM can be used with both digital and analog signals. For example, if the data consists of sequential bytes, 1 byte of data from each source can be transmitted during the time interval assigned to a particular channel. Each of the time slots shown in Fig. 10-12 might contain a byte from each of four sources. One channel would transmit 8 bits and then halt, while the next channel transmitted 8 bits. The third channel would then transmit its data word, and so on. The cycle would repeat itself at a high rate of speed. By

Figure 10-12   The basic TDM concept.

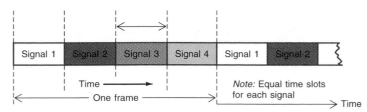

using this technique, the data bytes of individual channels can be interleaved. The resulting single-channel signal is a digital bit stream that is deciphered and reassembled at the receiving end.

The transmission of digital data by TDM is straightforward in that the incremental digital data is already broken up into chunks which can easily be assigned to different time slots. TDM can also be used to transmit continuous analog signals, whether they are voice, video, or telemetry-derived. This is accomplished by sampling the analog signal repeatedly at a high rate and then converting the samples to proportional binary numbers and transmitting them serially.

Sampling an analog signal creates pulse-amplitude modulation (PAM). As shown in Fig. 10-13, the analog signal is converted to a series of constant-width pulses whose amplitude follows the shape of the analog signal. The original analog signal is recovered by passing it through a low-pass filter. In TDM using PAM, a circuit called a multiplexer (MUX or MPX) samples multiple analog signal sources; the resulting pulses are interleaved and then transmitted over a single channel.

PAM Multiplexers

The simplest time multiplexer operates as a single-pole multiple-position mechanical or electronic switch that sequentially samples the multiple analog inputs at a high rate of speed. A basic mechanical rotary switch is shown in Fig. 10-14. The switch arm is rotated by a moter and dwells momentarily on each contact, allowing the input signal to be passed through to the output. It then switches quickly to the next channel, allowing that channel to pass for a fixed duration. The remaining channels are sampled in the same way. After each signal has been sampled, the cycle repeats. The result is that four analog signals are sampled, creating pulse-amplitude-modulated signals that are interleaved with one another. The speed of sampling is directly related to the speed of rotation, and the dwell time of the switch arm on each contact depends on the speed of rotation and the duration of contact.

Figure 10-13   Sampling an analog signal to produce pulse-amplitude modulation.

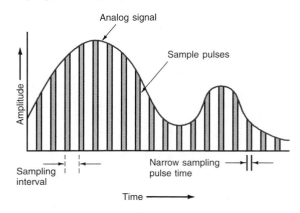

Multiplexing and Demultiplexing   **369**

Figure 10-14 Simple rotary-switch multiplexer.

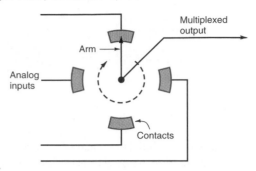

Figure 10-15 illustrates how four different analog signals are sampled by this technique. Signals A and C are continuously varying analog signals, signal B is a positive-going linear ramp, and signal D is a constant dc voltage.

Commutator Switches. Multiplexers in early TDM/PAM telemetry systems used a form of rotary switch known as a *commutator*. Multiple switch segments were attached to the various incoming signals while a high-speed brush rotated by a dc motor rapidly sampled the signals as it passed over the contacts. (Commutators have now been totally replaced by electronic circuits, which are discussed in the next section.)

In practice, the duration of the sample pulses is shorter than the time that is allocated to each channel. For example, assume that it takes the commutator or multiplexed switch 1 ms to move from one contact to another. The contacts can be set up so that each sample is 1 ms long. Typically, the duration of the sample is set to be about one-half the channel period value, in this example 0.5 ms.

Commutator

GOOD TO KNOW

The commutation rate or multiplex rate is derived by multiplying the number of samples per frame by the frame rate.

Figure 10-15 Four-channel PAM time division multiplexer.

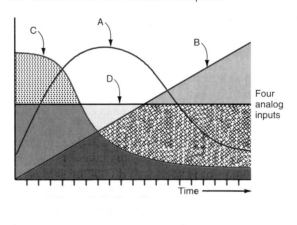

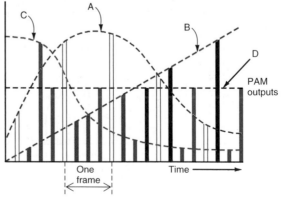

One complete revolution of the commutator switch is referred to as a *frame*. In other words, during one frame, each input channel is sampled one time. The number of contacts on the multiplexed switch or commutator sets the number of samples per frame. The number of frames completed in 1 s is called the *frame rate*. Multiplying the number of samples per frame by the frame rate yields the commutation rate or multiplex rate, which is the basic frequency of the composite signal, the final multiplexed signal that is transmitted over the communication channel.

Frame rate

In Fig. 10-15, the number of samples per frame is 4. Assume that the frame rate is 100 frames per second. The period for one frame, therefore, is $1/100 = 0.01$ s $= 10$ ms. During that 10-ms frame period, each of the four channels is sampled once. Assuming equal sample durations, each channel is thus allotted $10/4 = 2.5$ ms. (As indicated earlier, the full 2.5-ms period would not be used. The sample duration during that interval might be, e.g., only 1 ms). Since there are four samples taken per frame, the commutation rate is 4×100 or 400 pulses per second.

Electronic Multiplexers. In practical TDM/PAM systems, electronic circuits are used instead of mechanical switches or commutators. The multiplexer itself is usually implemented with FETs, which are nearly ideal on/off switches and can turn off and on at very high speeds. A complete four-channel TDM/PAM circuit is illustrated in Fig. 10-16.

The multiplexer is an op amp summer circuit with MOSFETs on each input resistor. When the MOSFET is conducting, it has a very low on resistance and therefore acts as a closed switch. When the transistor is off, no current flows through it, and it therefore acts as an open switch. A digital pulse applied to the gate of the MOSFET turns the transistor on. The absence of a pulse means that the transistor is off. The control pulses to the MOSFET switches are such that only one MOSFET is turned on at a time. These MOSFETs are turned on in sequence by the digital circuitry illustrated.

Figure 10-16 A time-division multiplexer used to produce pulse-amplitude modulation.

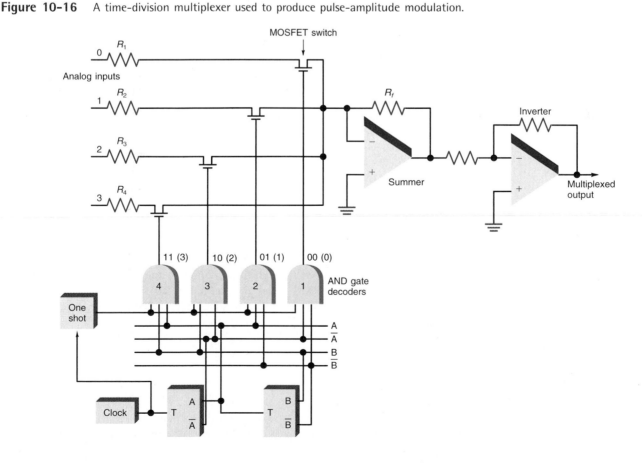

Figure 10-17   Waveforms for a PAM multiplexer.

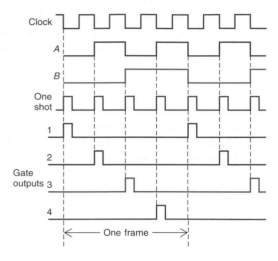

All the MOSFET switches are connected in series with resistors $(R_1\text{–}R_4)$; this, in combination with the feedback resistor (R_f) on the op amp circuit, determines the gain. For the purposes of this example, assume that the input and feedback resistors are all equal in value; in other words, the op amp circuit has a gain of 1. Since this op amp summing circuit inverts the polarity of the analog signals, it is followed by another op amp inverter which again inverts, restoring the proper polarity.

All the circuitry shown in Fig. 10-16 is usually contained on a single IC chip. MOSFET multiplexers are available with 4, 8, and 16 inputs, and these may be grouped to handle an even larger number of analog inputs.

The digital control pulses are developed by the counter and decoder circuit shown in Fig. 10-16. Since there are four channels, four counter states are needed. Such a counter can be implemented with two flip-flops, representing four discrete states—00, 01, 10, and 11—which are the binary equivalents of the decimal numbers 0, 1, 2, and 3. The four channels can therefore be labeled 0, 1, 2, and 3.

A clock oscillator circuit triggers the two flip-flop counters. The clock and flip-flop waveforms are illustrated in Fig. 10-17. The flip-flop outputs are applied to decoder AND gates that are connected to recognize the four binary combinations 00, 01, 10, and 11. The output of each decoder gate is applied to one of the multiplexed FET gates.

The *one-shot multivibrator* diagrammed in Fig. 10-16 is used to trigger all the decoder AND gates at the clock frequency. It produces an output pulse whose duration has been set to the desired sampling interval, in this case 1 ms.

Each time the clock pulse occurs, the one shot generates its pulse, which is applied simultaneously to all four AND decoder gates. At any given time, only one of the gates is enabled. The output of the enabled gate is a pulse whose duration is the same as that of the one shot.

When the pulse occurs, it turns on the associated MOSFET and allows the analog signal to be sampled and passed through the op amps to the output. The output of the final op amp is a multiplexed PAM signal like that in Fig. 10-15. The PAM output is used to modulate a carrier for transmission to a receiver. FM and PM are common modulation methods.

Demultiplexer Circuits

Once the composite PAM signal is recovered at the receiver, it is applied to a *demultiplexer (DEMUX)*. The demultiplexer is, of course, the reverse of a multiplexer. It has a single input and multiple outputs, one for each original input signal. Typical DEMUX circuitry is shown in Fig. 10-18. A four-channel demultiplexer has a single input and four

One-shot multivibrator

Demultiplexer (DEMUX)

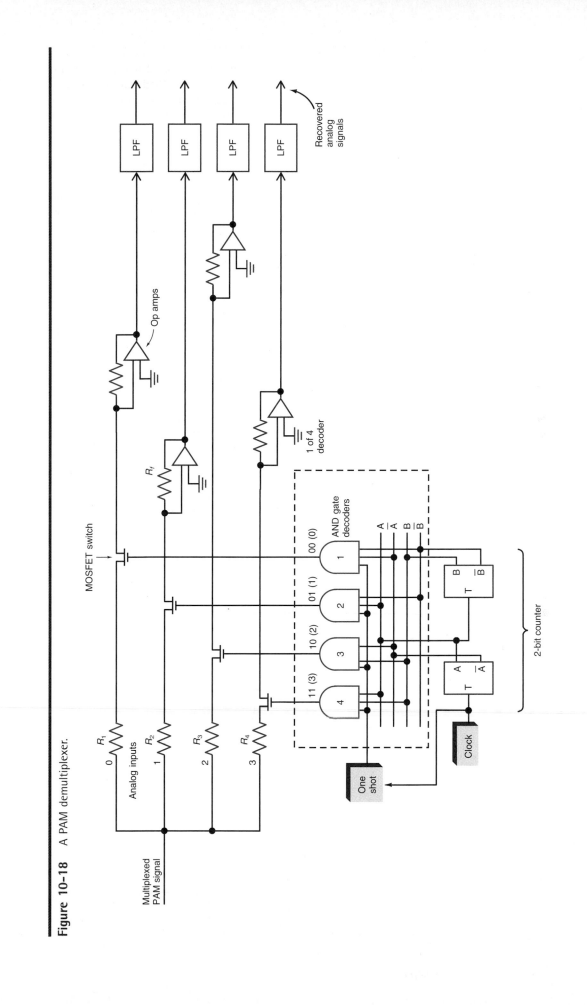

Figure 10–18 A PAM demultiplexer.

outputs. Most demultiplexers use FETs driven by a counter-decoder. The individual PAM signals are sent to op amps, where they are buffered and possibly amplified. They are then sent to low-pass filters, where they are smoothed into the original analog signals.

The main problem encountered in demultiplexing is synchronization. That is, for the PAM signal to be accurately demultiplexed into the original sampled signals, the clock frequency used at the receiver demultiplexer must be identical to that used at the transmitting multiplexer. In addition, the sequence of the demultiplexer must be identical to that of the multiplexer so that when channel 1 is being sampled at the transmitter, channel 1 is turned on in the receiver demultiplexer at the same time. Such synchronization is usually carried out by a special synchronizing pulse included as a part of each frame. Some of the circuits used for clock frequency and frame synchronization are discussed in the following sections.

Clock Recovery. Instead of using a free-running clock oscillator set to the identical frequency of the transmitter system clock, the clock for the demultiplexer is derived from the received PAM signal itself. The circuits shown in Fig. 10-19, called *clock recovery circuits,* are typical of those used to generate the demultiplexer clock pulses.

In Fig. 10-19(*a*), the PAM signal has been applied to an amplifier/limiter circuit, which first amplifies all received pulses to a high level and then clips them off at a fixed level. The output of the limiter is thus a constant-amplitude rectangular wave whose output frequency is equal to the commutation rate. This is the frequency at which the PAM pulses occur and is determined by the transmitting multiplexer clock.

The rectangular pulses at the output of the limiter are applied to a bandpass filter, which eliminates the upper harmonics, creating a sine wave signal at the transmitting clock frequency. This signal is applied to the phase detector circuit in a PLL along with the input from a voltage-controlled oscillator (VCO). The VCO is set to operate at the frequency of the PAM pulses. However, the VCO frequency is controlled by a dc error voltage applied to its input. This input is derived from the phase detector output, which is filtered by a low-pass filter into a dc voltage.

Figure 10-19 Two PAM clock recovery circuits. (*a*) Closed loop. (*b*) Open loop.

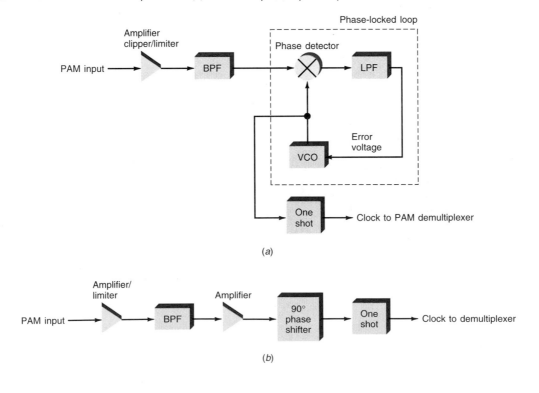

(*a*)

(*b*)

The phase detector compares the phase of the incoming PAM sine wave to the VCO sine wave. If a phase error exists, the phase detector produces an output voltage which is filtered to provide a varying direct current. The system is stabilized or locked when the VCO output frequency is identical to that of the sine wave frequency derived from the PAM input. The difference is that the two are shifted in phase by 90°.

If the PAM signal frequency changes for some reason, the phase detector picks up the variation and generates an error signal which is used to change the frequency of the VCO to match. Because of the closed-loop feature of the system, the VCO automatically tracks even minute frequency changes in the PAM signal, ensuring that the clock frequency used in the demultiplexer will always perfectly match that of the original PAM signal.

The output signal of the VCO is applied to a one-shot pulse generator which creates rectangular pulses at the proper frequency. These are used to step the counter in the demultiplexer; the counter generates the gating pulses for the FET demultiplexer switches.

A simpler, open-loop clock pulse circuit is shown in Fig. 10-19(b). Again, the PAM signal is applied to an amplifier/limiter and then a bandpass filter. The sine wave output of the bandpass filter is amplified and applied to a phase-shift circuit which produces a 90° phase shift at the frequency of operation. This phase-shifted sine wave is then applied to a pulse generator, which, in turn, creates the clock pulses for the demultiplexer. One disadvantage of this technique is that the phase-shift circuit is fixed to create a 90° shift at only one frequency, and so minor shifts in input frequency produce clock pulses whose timing is not perfectly accurate. However, in most systems in which frequency variations are not great, the circuit operates reliably.

Frame Synchronization. After clock pulses of the proper frequency have been obtained, it is necessary to synchronize the multiplexed channels. This is usually done with a special synchronizing *(sync) pulse* applied to one of the input channels at the transmitter. In the four-channel system discussed previously, only three actual signals are transmitted. The fourth channel is used to transmit a special pulse whose characteristics are unique in some way so that it can be easily recognized. The amplitude of the pulse can be higher than the highest-amplitude data pulse, or the width of the pulse can be wider than those pulses derived by sampling the input signals. Special circuits are then used to detect the sync pulse.

Figure 10-20 shows an example of a sync pulse that is higher in amplitude than the maximum pulse value of any data signal. The sync pulse is also the last to occur in the frame. At the receiver, a comparator circuit is used to detect the sync pulse. One input to the comparator is set to a dc reference voltage that is slightly higher than the maximum amplitude possible for the data pulses. When a pulse greater than the reference amplitude occurs, i.e., the sync pulse, the comparator immediately generates an output pulse, which can then be used for synchronization. Alternatively, it is possible not to transmit a pulse during one channel interval, leaving a blank space in each frame that can then be detected for the purposes of synchronization.

Frame synchronization

Sync pulse

Figure 10-20 Frame sync pulse and comparator detector.

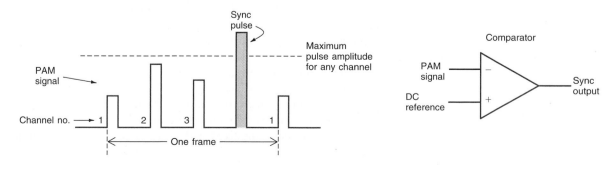

Figure 10-21 Complete PAM demultiplexer.

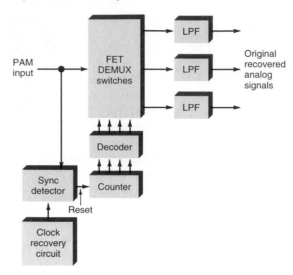

When the sync pulse is detected at the receiver, it acts as a reset pulse for the counter in the demultiplexer circuit. At the end of each frame, the counter is reset to zero (channel 0 is selected). When the next PAM pulse occurs, the demultiplexer will be set to the proper channel. Clock pulses then step the counter in the proper sequence for demultiplexing.

Finally, at the output of the demultiplexer, separate low-pass filters are applied to each channel to recover the original analog signals. Figure 10-21 shows the complete PAM demultiplexer.

10-4 Pulse-Code Modulation

Pulse-code modulation (PCM)

The most popular form of TDM uses *pulse-code modulation (PCM)* (see Sec. 7-4), in which multiple channels of digital data are transmitted in serial form. Each channel is assigned a time slot in which to transmit one binary word of data. The data streams from the various channels are interleaved and transmitted sequentially.

PCM Multiplexers

When PCM is used to transmit analog signals, the signals are sampled with a multiplexer as described previously for PAM, and then converted by an A/D converter into a series of binary numbers, where each number is proportional to the amplitude of the analog signal at the various sampling points. These binary words are converted from parallel to serial format and then transmitted.

At the receiving end, the various channels are demultiplexed and the original sequential binary numbers recovered, stored in a digital memory, and then transferred to a D/A converter that reconstructs the analog signal. (When the original data is strictly digital, D/A conversion is not required, of course.)

Any binary data, multiplexed or not, can be transmitted by PCM. Most long-distance space probes have on-board video cameras whose output signals are digitized and transmitted back to earth in binary format. Such PCM video systems make possible the transmission of graphic images over incredible distances. In computer multimedia presentations, video data is often digitized and transmitted by PCM techniques to a remote source.

Figure 10-22 A PCM system.

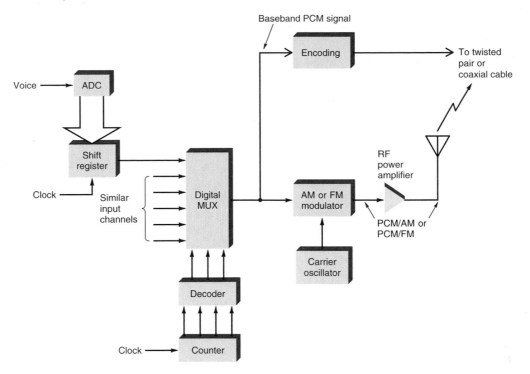

Figure 10-22 shows a general block diagram of the major components in a PCM system, where analog voice signals are the initial inputs. The voice signals are applied to A/D converters, which generate an 8-bit parallel binary word (byte) each time a sample is taken. Since the digital data must be transmitted serially, the A/D converter output is fed to a shift register, which produces a serial data output from the parallel input. In telephone systems, a codec takes care of the A/D parallel-to-serial conversion. The clock oscillator circuit driving the shift register operates at the desired bit rate.

The multiplexing is done with a simple digital MUX. Since all the signals to be transmitted are binary, a multiplexer constructed of standard logic gates can be used. A binary counter drives a decoder that selects the desired input channel.

The multiplexed output is a serial data waveform of the interleaved binary words. This baseband digital signal can then be encoded and transmitted directly over a twisted pair of cables, a coaxial cable, or a fiber-optic cable. Alternatively, the PCM binary signal can be used to modulate a carrier. A form of phase modulation known as phase-C shift keying (PSK) is the most commonly used.

Figure 10-23 shows the details of a four-input PCM multiplexer. Typically, the inputs to such a multiplexer come from an A/D converter. The binary inputs are applied to a shift register, which can be loaded from a parallel source such as the A/D converter or another serial source. In most PCM systems, the shift registers are part of a codec.

The multiplexer itself is the familiar digital circuit known as a data selector. It is made up of gates 1 through 5. The serial data is applied to gates 1 to 4; only one gate at a time is enabled, by a 1-of-4 decoder. The serial data from the enabled gate is passed through to OR gate 5 and appears in the output.

Now, assume that all shift registers are loaded with the bytes to be transmitted. The 2-bit counter AB is reset, sending the 00 code to the decoder. This turns on the 00 output, enabling gate 1. Note that the decoder output also enables gate 6. The clock pulses begin to trigger shift register 1, and the data is shifted out 1 bit at a time. The serial bits pass through gates 1 and 5 to the output. At the same time, the bit counter, which is a

Multiplexing and Demultiplexing   **377**

Figure 10-23 PCM multiplexer.

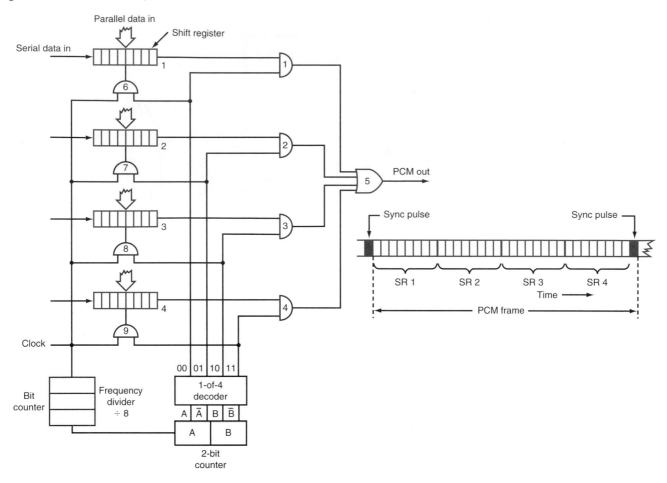

divide-by-8 circuit, keeps track of the number of bits shifted out. When 8 pulses have occurred, all 8 bits in shift register 1 have been transmitted. After counting 8 clock pulses, the bit counter recycles and triggers the 2-bit counter. Its code is now 01. This enables gate 2 and gate 7. The clock pulses continue, and now the contents of shift register 2 are shifted out 1 bit at a time. The serial data passes through gates 2 and 5 to the output. Again, the bit counter counts 8 bits and then recycles after all bits in shift register 2 have been sent. This again triggers the 2-bit counter, whose code is now 10. This enables gates 3 and 8. The contents of shift register 3 are now shifted out. The process repeats for the contents of shift register 4.

When the contents of all four shift registers have been transmitted once, one PCM frame has been formed (see Fig. 10-23). The data in the shift registers is then updated (the next sample of the analog signal converted), and the cycle is repeated.

If the clock in Fig. 10-23 is 64 kHz, the bit rate is 64 kilobits per second (kbps) and the bit interval is $1/64,000 = 15.625$ μs. With 8 bits per word, it takes $8 \times 15.625 = 125$ μs to transmit one word. This means that the word rate is $1/125 \times 10^{-6} = 8$ kbytes/s. If the shift registers get their data from an A/D converter, the sampling rate is 125 μs or 8 kHz. This is the rate used in telephone systems for sampling voice signals. Assuming a maximum voice frequency of 3 kHz, the minimum sampling rate is twice that, or 6 kHz, and so a sampling rate of 8 kHz is more than adequate to accurately represent and reproduce the analog voice signal. (Serial data rates are further explained in Chap. 11.)

As shown in Fig. 10-23, a sync pulse is added to the end of the frame. This signals the receiver that one frame of four signals has been transmitted and that another is about

Figure 10-24 A PCM receiver-demultiplexer.

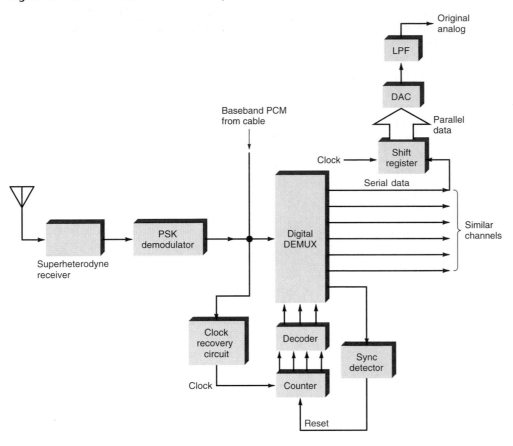

to begin. The receiver uses the sync pulse to keep all its circuits in step so that each original signal can be accurately recovered.

PCM Demultiplexers

PCM demultiplexer

At the receiving end of the communication link, the PCM signal is demultiplexed and converted back to the original data (see Fig. 10-24). The PCM baseband signal may come in over a cable, in which case the signal is regenerated and reshaped prior to being applied to the demultiplexer. Alternatively, if the PCM signal has modulated a carrier and is being transmitted by radio, the RF signal will be picked up by a receiver and then demodulated. The original serial PCM binary waveform is recovered and fed to a shaping circuit to clean up and rejuvenate the binary pulses. The original signal is then demultiplexed by means of a digital demultiplexer using AND or NAND gates. The binary counter and decoder driving the demultiplexer are kept in step with the receiver through a combination of clock recovery and sync pulse detector circuits similar to those used in PAM systems. The sync pulse is usually generated and sent at the end of each frame. The demultiplexed serial output signals are fed to a shift register for conversion to parallel data and sent to a D/A converter and then a low-pass filter. The shift register and D/A converter are usually part of a codec. The result is a highly accurate reproduction of the original voice signal.

Keep in mind that all the multiplexing and demultiplexing circuits are usually in integrated form. In fact, both MUX and DEMUX circuits are combined on a single chip to form a TDM transceiver that is used at both ends of the communications link. The individual circuits are not accessible; however, you do have access to all the inputs and outputs that allow you to perform tests, measurements, and troubleshooting as required.

Benefits of PCM

PCM is reliable, inexpensive, and highly resistant to noise. In PCM, the transmitted binary pulses all have the same amplitude and, like FM signals, can be clipped to reduce noise. Further, even when signals have been degraded because of noise, attenuation, or distortion, all the receiver has to do is to determine whether a pulse was transmitted. Amplitude, width, frequency, phase shape, and so on do not affect reception. Thus PCM signals are easily recovered and rejuvenated, no matter what the circumstances. PCM is so superior to other forms of pulse modulation and multiplexing for transmission of data that it has virtually replaced them all in communication applications.

Example 10–2

A special PCM system uses 16 channels of data, one whose purpose is identification (ID) and synchronization. The sampling rate is 3.5 kHz. The word length is 6 bits. Find (*a*) the number of available data channels, (*b*) the number of bits per frame, and (*c*) the serial data rate.

a. 16 (total no. of channels) − 1 (channel used for ID) = 15 (for data)

b. Bits per frame = 6 × 16 = 96

c. Serial data rate = sampling rate × no. bits/frame = 3.5 kHz × 96 = 336 kHz

Digital Carrier Systems

Customer premises equipment (CPE)

T carrier system

The most widespread use of TDM is in the telephone system. All modern telephone systems use digital transmission via PCM and TDM. The only place where analog signals are still used is in the local loop—the connection between a telephone company's central office (CO) and the subscriber's telephone, known as the *customer premises equipment (CPE)*. All local and long-distance connections are digital. Years ago, the telephone companies developed a complete digital transmission system called the *T carrier system*. It is used throughout the United States for all telephone calls and for the transmission of computer data including Internet access. Similar systems are used in Japan and Europe.

The T carrier system defines a range of PCM TDM systems with progressively faster data rates. The physical implementations of these systems are referred to as T-1, T-2, T-3, and T-4. The digital signals they carry are defined by the terms DS1, DS2, DS3, and DS4. It begins with the T-1 system, which multiplexes 24 basic DS1 digital voice signals that are then multiplexed into larger and faster DS2, DS3, and DS4 signals for transmission. Usually T-1 transmission is by way of a dual twisted-pair cable or coaxial cable. Wireless transmission is also common today. The T-2, T-3, and T-4 systems use coaxial cable, microwave radio, or fiber-optic cable transmission.

T–1 Systems

T-1 system

The most commonly used PCM system is the *T-1 system* developed by Bell Telephone for transmitting telephone conversations by high-speed digital links. The T-1 system multiplexes 24 voice channels onto a single line by using TDM techniques. Each serial digital word (8-bit words, 7 bits of magnitude, and 1 bit representing polarity) from the 24 channels is then transmitted sequentially. Each frame is sampled at an 8-kHz rate, producing a 125-μs sampling interval. During the 125-μs interval between analog

Figure 10-25 The T-1 frame format, serial data.

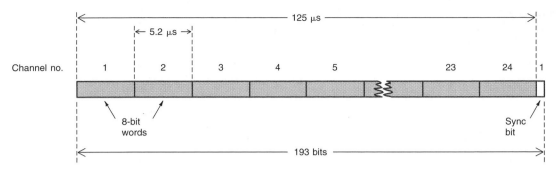

samples on each channel, 24 words of 8 bits, each representing one sample from each of the inputs, are transmitted. The channel sampling interval is 125 μs/24 = 5.2 μs, which corresponds to a rate of 192 kHz. This represents a total of 24 × 8 = 192 bits. An additional bit—a frame sync pulse—is added to this stream to keep the transmitting and receiving signals in synchronization with each other. The 24 words of 8 bits and the synchronizing bit form one frame of 193 bits. This sequence is carried out repeatedly. The total bit rate for the multiplexed signal is 193 × 8 kHz = 1544 kHz or 1.544 MHz. Figure 10-25 shows one frame of a T-1 signal.

The T-1 signal can be transmitted via cable, coaxial cable, twisted-pair cable, or fiber-optic cable; or it can be used to modulate a carrier for radio transmission. For example, for long-distance telephone calls, T-1 signals are sent to microwave relay stations, where they frequency-modulate a carrier for transmission over long distances. Also T-1 signals are transmitted via satellite or fiber-optic cable.

T-Carrier Systems

The T-1 systems transmit each voice signal at a 64-kbps rate. But they are also frequently used to transmit fewer than 24 inputs at a faster rate. For instance, a T-1 line can transmit a single source of computer data at a 1.544-Mbps rate. It can also transmit two data sources at a 722-kbps rate or four sources at a 386-kbps rate and so on. These are known as fractional T-1 lines.

T-2, T-3, and T-4 Systems

To produce greater capacity for voice traffic as well as computer data traffic, the DS1 signals may be further multiplexed into faster signals that carry even more channels. Figure 10-26 shows how four DS1 signals are multiplexed to form a DS2 signal. The result is a 6.312-Mbps serial digital signal containing 4 × 24 = 96 voice channels.

The T-2 systems are not widely used except as a steppingstone to form DS3 signals. As Fig. 10-26 shows, seven DS2 outputs are combined in a T-3 multiplexer to generate a DS3 signal. This signal contains 7 × 96 = 672 voice channels at a data rate of 44.736 Mbps. Four DS3 signals may further be multiplexed to form a DS4 signal. The T-4 multiplexer output data rate is 274.176 Mbps.

The T-1 and T-3 lines are widely used by business and industry for telephone service as well as for digital data transmission. These are dedicated circuits leased from the telephone company and used only by the subscriber so that the full data rate is available. These lines are also used in various unmultiplexed forms to achieve fast Internet access or digital data transmission other than voice traffic. The T-2 and T-4 lines are rarely used by subscribers, but they are used within the telephone system itself.

GOOD TO KNOW

Even though electronic multiplexers do not use mechanical parts to sample inputs, a complete cycle of input is still called a *frame*.

Figure 10-26   The T-carrier system.

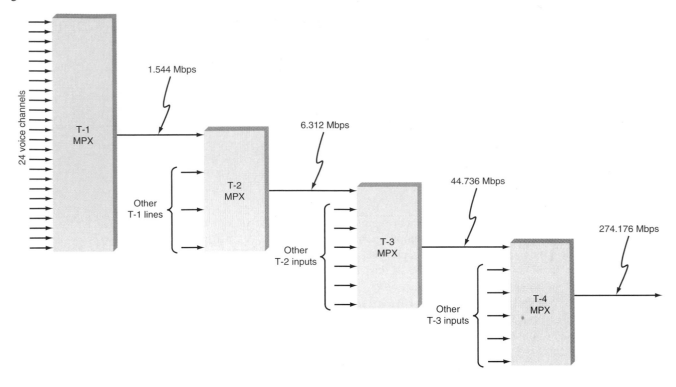

10-5 Duplexing

Duplexing is the method by which two-way communications are handled. Remember that half duplexing means that the two stations communicating take turns transmitting and receiving. Mobile, marine, and aircraft radios use half duplexing. Full duplexing means that the two stations can send and receive simultaneously. Full duplex is certainly preferred, as in phone calls. But not all systems require a simultaneous send/receive capability.

As with multiplexing, there are two ways to provide duplexing—frequency-division duplexing (FDD) and time-division duplexing (TDD). The simplest and perhaps best way to provide full duplex is to use FDD, which utilizes two separate channels, one for send and another for receive. Figure 10-27 shows the concept. The communicating parties are called station 1 and station 2. Station 1 uses the channel around f_1 for receiving only and the channel around f_2 for transmitting. Station 2 uses f_1 from transmitting and f_2 for receiving. By spacing the two channels far enough apart, the transmitter will not interfere with the receiver. Selective filters keep the signals separated. The big disadvantage of this method is the extra spectrum space required. Spectrum space is scarce and expensive. Yet most cell phone systems use this method because it is the easiest to implement and the most reliable.

Figure 10-27   Frequency-division duplexing (FDD).

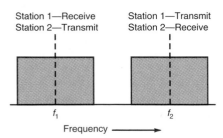

Figure 10-28   Time-division duplexing (TDD).

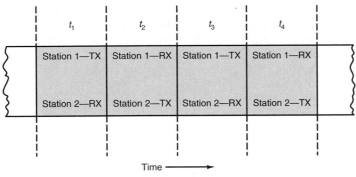

TX = transmit
RX = receive

Time-division duplexing (TDD) means that signals are transmitted simultaneously on a single channel by interleaving them in different time slots. For example, alternating time slots are devoted to transmitting and receiving. This is illustrated in Fig. 10-28. During time slot t_1, station 1 is transmitting (TX) while station 2 is receiving (RX). Then during time slot t_2, station 1 is receiving while station 2 is transmitting. Each time slot may contain one data word, such as 1 byte from an A/D converter or a D/A converter. As long as the serial data rate is high enough, a user will never know the difference.

The primary benefit of TDD is that only one channel is needed. It saves spectrum space and cost. On the other hand, the TDD method is harder to implement. The key to making it work is precise timing and synchronization between transmitter and receiver. Special synchronizing pulses or frame sequences are needed to constantly ensure that timing will not result in collisions between transmit and receive. Several of the newer third-generation cell phone systems may use TDD.

CHAPTER REVIEW

Summary

Multiplexing is the process of transmitting multiple signals simultaneously over a single communication channel. Applications that would be prohibitively expensive or impossible without multiplexing are telemetry, telephone systems, satellites, radio and TV broadcasting, and cable TV.

The two basic types of multiplexing are frequency-division multiplexing (FDM) and time-division multiplexing (TDM). In FDM, multiple signals are assigned a different frequency within a common bandwidth. In TDM, multiple signals are transmitted in different time slots.

FDM requires both mixing and amplification, which are accomplished by using a circuit known as an operational amplifier summer. The composite output signal is then typically

used to modulate a radio transmitter. Some telemetry systems use FM modulation. In cable TV systems, the FDM signal is transmitted directly on the cable. FM and TV broadcasting use FDM.

TDM can be used with both digital and analog signals. In TDM transmission of analog signals, the analog signal is repeatedly sampled at a high rate, creating pulse-amplitude modulation (PAM), and then converted to proportional binary numbers and transmitted serially. PAM can be used to time-multiplex several analog signals. The PAM signal is then usually used to modulate an RF carrier. Once the composite PAM signal is received, it is applied to a demultiplexer, which recovers the original signals.

Today virtually all TDM systems use the digital technique known as pulse-code modulation (PCM), a popular example of which is the T carrier system developed for transmitting telephone conversations by high-speed digital links. PCM is highly immune to noise, reliable, and, thanks to low-cost ICs, relatively inexpensive to implement.

There are two basic ways to implement full duplex operation—frequency-division duplexing (FDD) and time-division duplexing (TDD). FDD is the more widely used but requires more spectrum space. TDD is more economical of spectrum space but is harder to implement as precise timing and synchronization is required.

Questions

1. Is multiplexing the process of transmitting multiple signals over multiple channels?
2. State the primary benefit of multiplexing.
3. What is the name of the circuit used at the receiver to recover multiplexed signals?
4. What is the basic principle of FDM?
5. What circuit combines the multiple signals in an FDM system?
6. What is the name of the circuit to which the signals to be frequency-multiplexed are applied?
7. What kind of modulation do most telemetry multiplex systems use?
8. Name three places where telemetry is used.
9. Define spatial multiplexing. Where is it used?
10. What is the mathematical designation of the monaural signal in FM stereo multiplexing?
11. Name the four signals transmitted in FM stereo multiplex.
12. What type of modulation is used on the $L - R$ channel?
13. What type of modulation is used by the SCA signals in stereo systems?
14. What is the basic circuit used for demultiplexing in FDM systems?
15. In TDM, how do multiple signals share a channel?
16. What type of modulation is produced when an analog signal is sampled at high speed?
17. What type of circuit is used to recreate the clock pulses at the receiver in a PAM system?
18. In a PAM system, how are the multiplexer and demultiplexer kept in step with each other?
19. What is the time period called in a PAM or PCM system when all channels are sampled once?
20. What circuit is used to demodulate a PAM signal?
21. What type of switch is used in an electronic multiplexer?
22. How is the desired input to a time-division multiplexer selected?
23. How is a PAM signal transmitted?
24. What type of clock recovery circuit tracks PAM frequency variations?
25. How are analog signals transmitted in a PCM system?
26. What is the name of the IC that converts A/D and D/A signals in a PCM telephone system?
27. What is the standard audio sampling rate in a PCM telephone system?
28. What is the standard word size in a PCM telephone system?
29. What is the main benefit of PCM over PAM, FM, and other modulation techniques?
30. What is the total number of bits in one T-1 frame?
31. Does a T-1 system use baseband or broadband techniques? Explain.
32. Define half and full duplexing.
33. Explain the difference between FDD and TDD.
34. Discuss the pros and cons of FDD versus TDD.

Problems

1. How many 6-MHz-wide TV channels can be multiplexed on an 800-MHz coaxial cable? ◆
2. What is the minimum sampling rate for a signal with a 14-kHz bandwidth?
3. State the bit rate and maximum number of channels in T-1 and T-3 telephone TDM systems. ◆

◆ *Answers to Selected Problems follow Chap. 22.*

Critical Thinking

1. How long does it take to transmit one PCM frame with 16 bytes, no synchronization, and a clock rate of 46 MHz?
2. A special PCM system uses 12-bit words and 32 channels. Data is transmitted serially with no sync pulse. The time duration for 1 bit is 488.28 ns. How many bits per frame is transmitted and at what rate?
3. Can FDM be used with binary information signals? Explain.

The Transmission of Binary Data in Communication Systems

Since the mid-1970s, the proliferation of applications that send digital data over communication channels has resulted in the need for efficient methods of transmission, conversion, and reception of digital data. As with analog data, the amount of digital information that can be transmitted is proportional to the bandwidth of the communication channel and the time of transmission. This chapter further develops some of the basic concepts and equipment used in the transmission of digital data, including some mathematical principles that apply to transmission efficiency, and discusses the development of digital modulation, standards, error detection and correction methods, and spread spectrum techniques.

Objectives

After completing this chapter, you will be able to:

- Explain the difference between asynchronous and synchronous data transmission.
- State the relationship between communication channel bandwidth and data rate in bits per second.
- Name the four basic types of encoding used in the serial transmission of data.
- Describe the generation of FSK, PSK, QAM, and OFDM.
- Name three types of data modems and explain how they operate.
- Explain the need for and types of communication protocols.
- Compare and contrast redundancy, parity, block-check sequences, cyclical redundancy checks, and forward error correction.
- Explain the operation and benefits of spread spectrum systems.

11-1 Digital Codes

Data processed and stored by computers can be numerical (e.g., accounting records, spreadsheets, and stock quotations) or text (e.g., letters, memos, reports, and books). As previously discussed, the signals used to represent computerized data are digital, rather than analog. Even before the advent of computers, digital codes were used to represent data.

Early Digital Codes

Morse code

The first digital code was created by the inventor of the telegraph, Samuel Morse. The *Morse code* was originally designed for wired telegraph communication but was later adapted for radio communication. It consists of a series of "dots" and "dashes" that represent letters of the alphabet, numbers, and punctuation marks. This on/off code is shown in Fig. 11-1. A dot is a short burst of RF energy, and a dash is a burst of RF that is 3 times longer than a dot. The dot and dash on periods are separated by dot-length spaces or off periods. With special training, people can easily send and receive messages at speeds ranging from 15 to 20 words per minute to 70 to 80 words per minute.

The earliest radio communication was also carried out by using the Morse code of dots and dashes to send messages. A hand-operated telegraph key turned the carrier of a transmitter off and on to produce the dots and dashes. These were detected at the receiver and mentally converted by an operator back to the letters and numbers making up the message. This type of radio communication is known as *continuous-wave (CW) transmission*.

Continuous-wave (CW)

Baudot code

Another early binary data code was the *Baudot* (pronounced *baw-dough*) *code* used in the early teletype machine, a device for sending and receiving coded signals over a communication link. With teletype machines, it was no longer necessary for operators to learn Morse code. Whenever a key on the typewriter keyboard is pressed, a unique code is generated and transmitted to the receiving machine, which recognizes and then prints the corresponding letter, number, or symbol.

The 5-bit Baudot code is shown in Fig. 11-2. With 5 bits, $2^5 = 32$ different symbols can be represented. The different 5-bit combinations, together with two shift codes, can generate the 26 letters of the alphabet, 10 numbers, various punctuation marks, and control functions. If the message is preceded by the letter shift code (11011), all the following codes are interpreted as letters of the alphabet. Sending the figure shift code (11111) causes all the following characters to be interpreted as numbers or punctuation marks. The Baudot code is rarely used today, having been supplanted by codes that can represent more characters and symbols.

Figure 11-1 The Morse code. A dot (.) is a short click; a dash (—) is a long click.

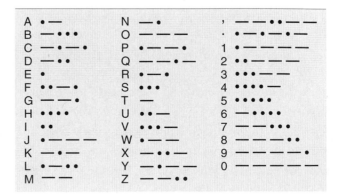

Figure 11-2 The Baudot code.

Character Shift		Binary Code					Character Shift		Binary Code				
Letter	Figure	Bit: 4	3	2	1	0	Letter	Figure	Bit: 4	3	2	1	0
A	–	1	1	0	0	0	Q	1	1	1	1	0	1
B	?	1	0	0	1	1	R	4	0	1	0	1	0
C	:	0	1	1	1	0	S	bel	1	0	1	0	0
D	$	1	0	0	1	0	T	5	0	0	0	0	1
E	3	1	0	0	0	0	U	7	1	1	1	0	0
F	!	1	0	1	1	0	V	;	0	1	1	1	1
G	&	0	1	0	1	1	W	2	1	1	0	0	1
H	#	0	0	1	0	1	X	/	1	0	1	1	1
I	8	0	1	1	0	0	Y	6	1	0	1	0	1
J	'	1	1	0	1	0	Z	"	1	0	0	0	1
K	(	1	1	1	1	0	Figure shift		1	1	1	1	1
L	)	0	1	0	0	1	Letter shift		1	1	0	1	1
M	.	0	0	1	1	1	Space		0	0	1	0	0
N	,	0	0	1	1	0	Line feed (LF)		0	1	0	0	0
O	9	0	0	0	1	1	Blank (null)		0	0	0	0	0
P	0	0	1	1	0	1							

Modern Binary Codes

Binary code

For modern data communication, information is transmitted by using a system in which the numbers and letters to be represented are coded, usually by way of a keyboard, and the binary word representing each character is stored in a computer memory. The message can also be stored on magnetic tape or disk. The following sections describe some widely used codes for transmission of digitized data.

American Standard Code for Information Interchange. The most widely used data communication code is the 7-bit binary code known as the *American Standard Code for Information Interchange* (abbreviated ASCII and pronounced *ass-key*), which can represent 128 numbers, letters, punctuation marks, and other symbols (see Fig. 11-3). With ASCII, a sufficient number of code combinations is available to represent both uppercase and lowercase letters of the alphabet.

American Standard Code for Information Interchange (ASCII)

The first ASCII codes listed in Fig. 11-3 have two- and three-letter designations. These codes initiate operations or provide responses for inquiries. For example, BEL or 0000111 will ring a bell or a buzzer; CR is carriage return; SP is a space, such as that between words in a sentence; ACK means "acknowledge that a transmission was received"; STX and ETX are start and end of text, respectively; and SYN is a synchronization word that provides a way to get transmission and reception in step with each other. The meanings of all the letter codes are given at the bottom of the table.

Hexadecimal Values. Binary codes are often expressed by using their *hexadecimal,* rather than decimal, values. To convert a binary code to its hexadecimal equivalent, first divide the code into 4-bit groups, starting at the least significant bit on the right and working to the left. (Assume a leading 0 on each of the codes.) The hexadecimal equivalents for the binary codes for the decimal numbers 0 through 9 and the letters A through F are given in Fig. 11-4.

Hexadecimal

Two examples of ASCII-to-hexadecimal conversion are as follows:

1. The ASCII code for the number 4 is 0110100. Add a leading 0 to make 8 bits and then divide into 4-bit groups: 00110100 = 0011 0100 = hex 34.
2. The letter w in ASCII is 1110111. Add a leading 0 to get 01110111; 01110111 = 0111 0111 = hex 77.

Figure 11-3   The popular ASCII code.

Bit:	6	5	4	3	2	1	0
NUL	0	0	0	0	0	0	0
SOH	0	0	0	0	0	0	1
STX	0	0	0	0	0	1	0
ETX	0	0	0	0	0	1	1
EOT	0	0	0	0	1	0	0
ENQ	0	0	0	0	1	0	1
ACK	0	0	0	0	1	1	0
BEL	0	0	0	0	1	1	1
BS	0	0	0	1	0	0	0
HT	0	0	0	1	0	0	1
NL	0	0	0	1	0	1	0
VT	0	0	0	1	0	1	1
FF	0	0	0	1	1	0	0
CR	0	0	0	1	1	0	1
SO	0	0	0	1	1	1	0
SI	0	0	0	1	1	1	1
DLE	0	0	1	0	0	0	0
DC1	0	0	1	0	0	0	1
DC2	0	0	1	0	0	1	0
DC3	0	0	1	0	0	1	1
DC4	0	0	1	0	1	0	0
NAK	0	0	1	0	1	0	1
SYN	0	0	1	0	1	1	0
ETB	0	0	1	0	1	1	1
CAN	0	0	1	1	0	0	0
EM	0	0	1	1	0	0	1
SUB	0	0	1	1	0	1	0
ESC	0	0	1	1	0	1	1
FS	0	0	1	1	1	0	0
GS	0	0	1	1	1	0	1
RS	0	0	1	1	1	1	0
US	0	0	1	1	1	1	1
SP	0	1	0	0	0	0	0
!	0	1	0	0	0	0	1
"	0	1	0	0	0	1	0
#	0	1	0	0	0	1	1
$	0	1	0	0	1	0	0
%	0	1	0	0	1	0	1
&	0	1	0	0	1	1	0
'	0	1	0	0	1	1	1
(	0	1	0	1	0	0	0
)	0	1	0	1	0	0	1
*	0	1	0	1	0	1	0

Bit:	6	5	4	3	2	1	0
+	0	1	0	1	0	1	1
,	0	1	0	1	1	0	0
-	0	1	0	1	1	0	1
.	0	1	0	1	1	1	0
/	0	1	0	1	1	1	1
0	0	1	1	0	0	0	0
1	0	1	1	0	0	0	1
2	0	1	1	0	0	1	0
3	0	1	1	0	0	1	1
4	0	1	1	0	1	0	0
5	0	1	1	0	1	0	1
6	0	1	1	0	1	1	0
7	0	1	1	0	1	1	1
8	0	1	1	1	0	0	0
9	0	1	1	1	0	0	1
:	0	1	1	1	0	1	0
;	0	1	1	1	0	1	1
<	0	1	1	1	1	0	0
=	0	1	1	1	1	0	1
>	0	1	1	1	1	1	0
?	0	1	1	1	1	1	1
@	1	0	0	0	0	0	0
A	1	0	0	0	0	0	1
B	1	0	0	0	0	1	0
C	1	0	0	0	0	1	1
D	1	0	0	0	1	0	0
E	1	0	0	0	1	0	1
F	1	0	0	0	1	1	0
G	1	0	0	0	1	1	1
H	1	0	0	1	0	0	0
I	1	0	0	1	0	0	1
J	1	0	0	1	0	1	0
K	1	0	0	1	0	1	1
L	1	0	0	1	1	0	0
M	1	0	0	1	1	0	1
N	1	0	0	1	1	1	0
O	1	0	0	1	1	1	1
P	1	0	1	0	0	0	0
Q	1	0	1	0	0	0	1
R	1	0	1	0	0	1	0
S	1	0	1	0	0	1	1
T	1	0	1	0	1	0	0
U	1	0	1	0	1	0	1

Bit:	6	5	4	3	2	1	0
V	1	0	1	0	1	1	0
W	1	0	1	0	1	1	1
X	1	0	1	1	0	0	0
Y	1	0	1	1	0	0	1
Z	1	0	1	1	0	1	0
[	1	0	1	1	0	1	1
E	1	0	1	1	1	0	0
]	1	0	1	1	1	0	1
∧	1	0	1	1	1	1	0
—	1	0	1	1	1	1	1
i	1	1	0	0	0	0	0
a	1	1	0	0	0	0	1
b	1	1	0	0	0	1	0
c	1	1	0	0	0	1	1
d	1	1	0	0	1	0	0
e	1	1	0	0	1	0	1
f	1	1	0	0	1	1	0
g	1	1	0	0	1	1	1
h	1	1	0	1	0	0	0
i	1	1	0	1	0	0	1
j	1	1	0	1	0	1	0
k	1	1	0	1	0	1	1
l	1	1	0	1	1	0	0
m	1	1	0	1	1	0	1
n	1	1	0	1	1	1	0
o	1	1	0	1	1	1	1
p	1	1	1	0	0	0	0
q	1	1	1	0	0	0	1
r	1	1	1	0	0	1	0
s	1	1	1	0	0	1	1
t	1	1	1	0	1	0	0
u	1	1	1	0	1	0	1
v	1	1	1	0	1	1	0
w	1	1	1	0	1	1	1
x	1	1	1	1	0	0	0
y	1	1	1	1	0	0	1
z	1	1	1	1	0	1	0
{	1	1	1	1	0	1	1
¦	1	1	1	1	1	0	0
}	1	1	1	1	1	0	1
	1	1	1	1	1	1	0
DEL	1	1	1	1	1	1	1

NUL = null
SOH = start of heading
STX = start of text
ETX = end of text
EOT = end of transmission
ENQ = enquiry
ACK = acknowledge
BEL = bell
BS = back space

HT = horizontal tab
NL = new line
VT = vertical tab
FF = form feed
CR = carriage return
SO = shift-out
SI = shift-in
DLE = data link escape
DC1 = device control 1

DC2 = device control 2
DC3 = device control 3
DC4 = device control 4
NAK = negative acknowledge
SYN = synchronous
ETB = end of transmission block
CAN = cancel
SUB = substitute
ESC = escape

FS = field separator
GS = group separator
RS = record separator
US = unit separator
SP = space
DEL = delete

Figure 11-4   Serial transmission of the ASCII letter M.

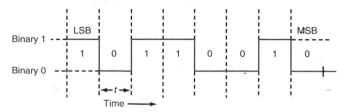

Extended Binary Coded Decimal Interchange Code. The *Extended Binary Coded Decimal Interchange Code* (EBCDIC, pronounced *ebb-see-dick*), developed by IBM, is an 8-bit code similar to ASCII allowing a maximum of 256 characters to be represented. Its primary use is in IBM and IBM-compatible computing systems and equipment. It is not as widely used as ASCII.

Extended Binary Coded Decimal
Interchange Code (EBCDIC)

11-2 Principles of Digital Transmission

As indicated in Chap. 7, data can be transmitted in two ways: parallel and serial.

Serial Transmission

Parallel data transmission is not practical for long-distance communication. Data transfers in long-distance communication systems are made serially; each bit of a word is transmitted one after another (see Fig. 11-4). The figure shows the ASCII form for the letter M (1001101) being transmitted 1 bit at a time. The LSB is transmitted first, and the MSB last. The MSB is on the right, indicating that it was transmitted later in time. Each bit is transmitted for a fixed interval of time *t*. The voltage levels representing each bit appear on a single data line one after another until the entire word has been transmitted. For example, the bit interval may be 10 μs, which means that the voltage level for each bit in the word appears for 10 μs. It would therefore take 70 μs to transmit a 7-bit ASCII word.

Expressing the Serial Data Rate. The speed of data transfer is usually indicated as number of bits per second (bps or b/s). Some data rates take place at relatively slow speeds, usually a few hundred or several thousand bits per second. Personal computers, communicate at rates up to 53,000 bps with a conventional modem or up to 8 Mbps with special digital modems over the telephone lines. However, in some data communication systems such as local-area networks, bit rates as high as tens of billions bits per second are used.

The speed of serial transmission is, of course, related to the bit time of the serial data. The speed in bits per second, denoted by bps, is the reciprocal of the bit time *t*, or bps = $1/t$. For example, assume a bit time of 104.17 μs. The speed is bps = $1/104.17 \times 10^{-6}$ = 9600 bps.

If the speed in bits per second is known, the bit time can be found by rearranging the formula: $t = 1/\text{bps}$. For example, the bit time at 230.4 kbps (230,400 bps) is $t = 1/230,400 = 4.34 \times 10^{-6} = 4.34$ μs.

Another term used to express the data speed in digital communication systems is *baud rate*. Baud rate is the number of signaling elements or symbols that occur in a given unit of time, such as 1 s. A *signaling element* is simply some change in the binary signal transmitted. In many cases it is a binary logic voltage level change, either 0 or 1, in which case, the baud rate is equal to the data rate in bits per second. In summary, the baud rate is the reciprocal of the smallest signaling interval.

Baud rate

$$\text{Bit rate} = \text{baud rate} \times \text{bit per symbol}$$

$$\text{Bit rate} = \text{baud rate} \times \log_2 S$$

where S = number of states per symbol

The symbol or signaling element can also be one of several discrete signal amplitudes, frequencies, or phase shifts, each of which represents 2 data bits or more. Several unique modulation schemes have been developed so that each symbol or baud can represent multiple bits. The number of symbol changes per unit of time is no higher than the straight binary bit rate, but more bits per unit time are transmitted. Multiple symbol changes can be combined to further increase transmission speed. For example, a popular form of modulation known as *quadrature amplitude modulation (QAM)* combines

Quadrature amplitude modulation
(QAM)

multiple amplitude levels with multiple phase shifts to produce many bits per baud. (QAM is discussed later in this chapter.) Each symbol is a unique amplitude level combined with a unique phase shift that corresponds to a group of bits. Multiple amplitude levels may also be combined with different frequencies in FSK to produce higher bit rates. As a result, higher bit rates can be transmitted over telephone lines or other severely bandwidth-limited communication channels that would not ordinarily be able to handle them. Several of these modulation methods are discussed later.

Assume, e.g., a system that represents 2 bits of data as different voltage levels. With 2 bits there are $2^2 = 4$ possible levels, and a discrete voltage is assigned to each.

00	0 V
01	1 V
10	2 V
11	3 V

In this system, sometimes called pulse-amplitude modulation (PAM), each of the four symbols is one of four different voltage levels. Each level is a different symbol representing 2 bits. Assume, e.g., that it is desired to transmit the decimal number 201, which is binary 11001001. The number can be transmitted serially as a sequence of equally spaced pulses that are either on or off [see Fig. 11-5(a)]. If each bit interval is 1 μs, the bit rate is $1/1 \times 10^{-6} = 1,000,000$ bps (1 Mbps). The baud rate is also 1 million bps.

Using the four-level system, we could also divide the word to be transmitted into 2-bit groups and transmit the appropriate voltage level representing each. The number 11001001 would be divided into these groups: 11 00 10 01. Thus the transmitted signal would be voltage levels of 3, 0, 2 and 1 V, each occurring for a fixed interval of, say, 1 μs [see Fig. 11-5(b)]. The baud rate is still 1 million because there is only one symbol or level per time interval (1 μs). However, the bit rate is 2 million bps—double the baud rate—because each symbol represents 2 bits. We are transmitting 2 bits per baud. You will sometimes see this referred to as bits per hertz or bits/Hz. The total transmission time is also shorter. It would take 8 μs to transmit the 8-bit binary word but only 4 μs to transmit the four-level signal. The bit rate is greater than the baud rate because multiple levels are used.

The PAM signal in Fig. 11-5(b) can be transmitted over a cable. In wireless applications, this signal would first modulate a carrier before being transmitted. Any type of modulation can be used, but PSK is the most common.

Figure 11-5   The bit rate can be higher than the baud rate when each symbol represents 2 or more bits. (a) Bit rate = baud rate = 1 bit/μs = 1,00,000 bps = 1 Mbps. (b) Baud rate = 1 symbol/μs = 1,000,000 baud rate; bit rate = 2 bits/Baud = 2,000,000 bps.

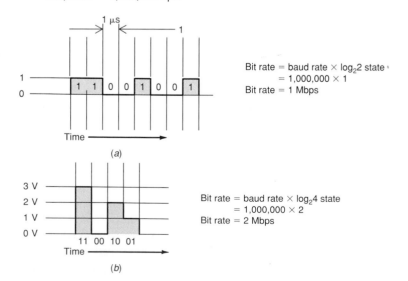

Figure 11-6 Asynchronous transmission with start and stop bits.

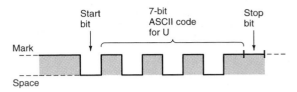

Because of the sequential nature of serial data transmission, naturally it takes longer to send data in this way than it does to transmit it by parallel means. However, with a high-speed logic circuit, even serial data transfers can take place at very high speeds. Currently that data rate is as high as 10 billion bits per second (10 Gbps) on copper wire cable and up to 100 billion bits per second (100 Gbps) on fiber-optic cable. So although serial data transfers are slower than parallel transfers, they are fast enough for most communication applications.

Asynchronous Transmission

In *asynchronous transmission* each data word is accompanied by *start and stop bits* that indicate the beginning and ending of the word. Asynchronous transmission of an ASCII character is illustrated in Fig. 11-6. When no information is being transmitted, the communication line is usually high, or binary 1. In data communication terminology, this is referred to as a *mark*. To signal the beginning of a word, a start bit, a binary 0 or space, as shown in the figure, is transmitted. The start bit has the same duration as all other bits in the data word. The transmission from mark to space indicates the beginning of the word and allows the receiving circuits to prepare themselves for the reception of the remainder of the bits.

After the start bit, the individual bits of the word are transmitted. In this case, the 7-bit ASCII code for the letter U, 1010101, is transmitted. Once the last code bit has been transmitted, a stop bit is included. The stop bit may be the same duration as all other bits and again is a binary 1 or mark. In some systems, 2 stop bits are transmitted, one after the other, to signal the end of the word.

Most low-speed digital transmission (the 1200- to 56,000-bps range) is asynchronous. This technique is extremely reliable, and the start and stop bits ensure that the sending and receiving circuits remain in step with each other. The minimum separation between character words is 1 stop plus 1 start bit, as Fig. 11-7 shows. There can also be time gaps between characters or groups of characters, as the illustration shows, and thus the stop "bit" may be of some indefinite length.

The primary disadvantage of asynchronous communication is that the extra start and stop bits effectively slow down data transmission. This is not a problem in low-speed applications such as those involving certain printers and plotters. But when huge volumes of information must be transmitted, the start and stop bits represent a significant percentage of the bits transmitted. We call that percentage the *overhead* of transmission. A 7-bit ASCII character plus start and stop bits is 9 bits. Of the 9 bits, 2 bits are not data. This represents $2/9 = 0.222$, or 22.2 percent inefficiency or overhead. Removing the start and stop bits and stringing the ASCII characters end to end allow many more data words to be transmitted per second.

Figure 11-7 Sequential words transmitted asynchronously.

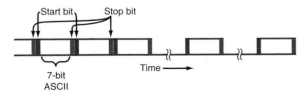

Asynchronous transmission

Start and stop bits

Figure 11-8 Synchronous data transmission.

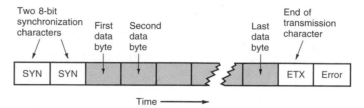

Time ⟶

Synchronous Transmission

Synchronous data transmission

The technique of transmitting each data word one after another without start and stop bits, usually in multiword blocks, is referred to as *synchronous data transmission*. To maintain synchronization between transmitter and receiver, a group of synchronization bits is placed at the beginning of the block and the end of the block. Figure 11-8 shows one arrangement. Each block of data can represent hundreds or even thousands of 1-byte characters. At the beginning of each block is a unique series of bits that identifies the beginning of the block. In Fig. 11-8, two 8-bit synchronous (SYN) codes signal the start of a transmission. Once the receiving equipment finds these characters, it begins to receive the continuous data, the block of sequential 8-bit words or bytes. At the end of the block, another special ASCII code character, ETX, signals the end of transmission. The receiving equipment looks for the ETX code; detection of this code is how the receiving circuit recognizes the end of the transmission. An error detection code usually appears at the very end of the transmission.

The special synchronization codes at the beginning and end of a block represent a very small percentage of the total number of bits being transmitted, especially in relation to the number of start and stop bits used in asynchronous transmission. Synchronous transmission is therefore much faster than asynchronous transmission because of the lower overhead.

An important consideration in synchronous transmission is how the receiving station keeps track of the individual bits and bytes, especially when the signal is noisy, since there is no clear separation between them. This is done by transmitting the data at a fixed, known, precise clock rate. Then the number of bits can be counted to keep track of the number of bytes or characters transmitted. For every 8 bits counted, 1 byte is received. The number of received bytes is also counted.

Synchronous transmission assumes that the receiver knows or has a clock frequency identical to that of the transmitter clock. Usually, the clock at the receiver is derived from the received signal, so that it is precisely the same frequency as, and in synchronization with, the transmitter clock.

Example 11-1

A block of 256 sequential 12-bit data words is transmitted serially in 0.016 s. Calculate (*a*) the time duration of 1 word, (*b*) the time duration of 1 bit, and (*c*) the speed of transmission in bits per second.

a. $t_{\text{word}} = \dfrac{0.016}{256} = 0.000625 = 625 \ \mu\text{s}$

b. $t_{\text{bit}} = \dfrac{625 \ \mu\text{s}}{12 \ \text{bits}} = 52.0833 \ \mu\text{s}$

c. $\text{bps} = \dfrac{1}{t} = \dfrac{1}{52.0833} \times 10^{-6} = 19{,}200 \ \text{bps} \quad \text{or} \quad 19.2 \ \text{kbps}$

Chapter 11

Encoding Methods

Whether digital signals are being transmitted by baseband methods or broadband methods (see Sec. 1-4), before the data is put on the medium, it is usually encoded in some way to make it compatible with the medium or to facilitate some desired operation connected with the transmission. The primary encoding methods used in data communication are summarized below.

Nonreturn to Zero. In the *nonreturn to zero (NRZ) method of encoding* the signal remains at the binary level assigned to it for the entire bit time. Figure 11-9(*a*), which shows unipolar NRZ, is a slightly different version of Fig. 11-4. The logic levels are 0 and +5 V. When a binary 1 is to be transmitted, the signal stays at +5 V for the entire bit interval. When a binary 0 is to be sent, the signal stays at 0 V for the total bit time. In other words, the voltage does not return to zero *during* the binary 1 interval.

Nonreturn to zero (NRZ) encoding

In *unipolar* NRZ, the signal has only a positive polarity. A *bipolar* NRZ signal has two polarities, positive and negative, as shown in Fig. 11-9(*b*). The voltage levels are +12 and −12 V. The popular RS-232 serial computer interface uses bipolar NRZ, where a binary 1 is a negative voltage between −3 and −25 V and a binary 0 is a voltage between +3 and +25 V.

The NRZ method is normally generated inside computers, at low speeds, when asynchronous transmission is being used. It is not popular for synchronous transmission because there is no voltage or level change when there are long strings of sequential binary 1s or 0s. If there is no signal change, it is difficult for the receiver to determine just where one bit ends and the next one begins. If the clock is to be recovered from the transmitted data in a synchronous system, there must be more frequent changes, preferably one per bit. NRZ is usually converted to another format, such as RZ or Manchester, for synchronous transmissions.

Figure 11-9 Serial binary encoding methods. (*a*) Unipolar NRZ. (*b*) Bipolar NRZ. (*c*) Unipolar RZ. (*d*) Bipolar RZ. (*e*) Bipolar RZ-AMI. (*f*) Manchester or biphase.

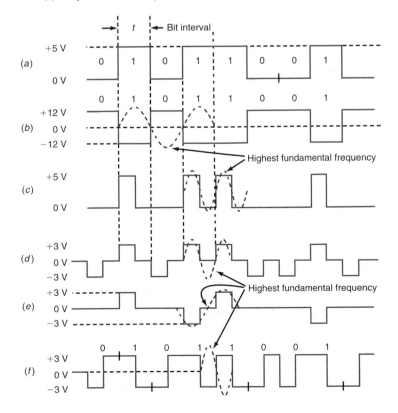

Return to Zero. In *return to zero (RZ) encoding* [see Fig. 11-9(c) and (d)] the voltage level assigned to a binary 1 level returns to zero during the bit period. *Unipolar RZ* is illustrated in Fig. 11-9(c). The binary 1 level occurs for 50 percent of the bit interval, and the remaining bit interval is zero. Only one polarity level is used. Pulses occur only when a binary 1 is transmitted; no pulse is transmitted for a binary 0.

Bipolar RZ is illustrated in Fig. 11-9(d). A 50 percent bit interval +3-V pulse is transmitted during a binary 1, and a −3-V pulse is transmitted for a binary 0. Because there is one clearly discernible pulse per bit, it is extremely easy to derive the clock from the transmitted data. For that reason, bipolar RZ is preferred over unipolar RZ.

A popular variation of the bipolar RZ format is called *alternative mark inversion (AMI)* [see Fig. 11-9(e)]. During the bit interval, binary 0s are transmitted as no pulse. Binary 1s, also called *marks,* are transmitted as alternating positive and negative pulses. One binary 1 is sent as a positive pulse, the next binary 1 as a negative pulse, the next as a positive pulse, and so on.

Manchester. *Manchester encoding,* also referred to as *biphase encoding,* can be unipolar or bipolar. It is widely used in LANs. In the Manchester system a binary 1 is transmitted first as a positive pulse, for one half of the bit interval, and then as a negative pulse, for the remaining part of the bit interval. A binary 0 is transmitted as a negative pulse for the first half of the bit interval and a positive pulse for the second half of the bit interval [see Fig. 11-9(f)]. The fact that there is a transition at the center of each 0 or 1 bit makes clock recovery very easy. However, because of the transition in the center of each bit, the frequency of a Manchester-encoded signal is 2 times an NRZ signal, doubling the bandwidth requirement.

Choosing a Coding Method. The choice of an encoding method depends on the application. For synchronous transmission, RZ and Manchester are preferred because the clock is easier to recover. Another consideration is average dc voltage buildup on the transmission line. When unipolar modes are used, a potentially undesirable average dc voltage builds up on the line because of the charging of the line capacitance. To eliminate this problem, bipolar methods are used, where the positive pulses cancel the negative pulses and the dc voltage is averaged to zero. Bipolar RZ or Manchester is preferred if dc buildup is a problem.

DC buildup is not always a problem. In some applications the average dc value is used for signaling purposes. An example is an Ethernet LAN, which uses the direct current to detect when two or more stations are trying to transmit at the same time.

Other encoding methods are also used. The encoding schemes used to encode the serial data recorded on magnetic floppy and hard disks are an example. Other schemes used in networking will be discussed later.

11–3 Transmission Efficiency

Transmission efficiency—i.e., the accuracy and speed with which information, whether it is voice or video, analog or digital, is sent and received over communication media—is the basic subject matter of a field known as *information theory.* Information theorists seek to determine mathematically the likelihood that a given amount of data being transmitted under a given set of circumstances (e.g., medium, bandwidth, speed of transmission, noise, and distortion) will be transmitted accurately.

Hartley's Law

The amount of information that can be sent in a given transmission is dependent on the bandwidth of the communication channel and the duration of transmission. A bandwidth of only about 3 kHz is required to transmit voice so that it is intelligible and recognizable. However, because of the high frequencies and harmonics produced by musical instruments, a bandwidth of 15 to 20 kHz is required to transmit music with full fidelity. Music inherently contains more information than voice, and so requires greater bandwidth. A picture signal contains more information than a voice or music signal. Therefore, greater bandwidth is required to transmit it. A typical TV signal contains both voice and picture; therefore, it is allocated 6 MHz of spectrum space.

Stated mathematically, Hartley's law is

$$C = 2B$$

Here C is the channel capacity expressed in bits per second and B is the channel bandwidth. It is also assumed that there is a total absence of noise in the channel. When noise becomes an issue, Hartley's law is expressed as

$$C = (B \log 2)\left(1 + \frac{S}{N}\right)$$

where S/N is the signal-to-noise ratio in power. These concepts are expanded upon in the coming sections.

The greater the bandwidth of a channel, the greater the amount of information that can be transmitted in a given time. It is possible to transmit the same amount of information over a narrower channel, but it must be done over a longer time. This general concept is known as *Hartley's law,* and the principles of Hartley's law also apply to the transmission of binary data. The greater the number of bits transmitted in a given time, the greater the amount of information that is conveyed. But the higher the bit rate, the wider the bandwidth needed to pass the signal with minimum distortion. Narrowing the bandwidth of a channel causes the harmonics in the binary pulses to be filtered out, degrading the quality of the transmitted signal and making error-free transmission more difficult.

Hartley's law

Transmission Media and Bandwidth

The two most common types of media used in data communication are wire cable and radio. Two types of wire cable are used, coaxial and twisted-pair (see Fig. 11-10). The coaxial cable shown in Fig. 11-10(*a*) has a center conductor sur-

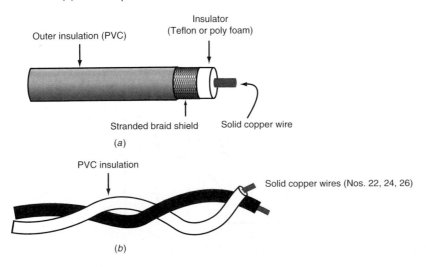

Figure 11-10   Types of cable used for digital data transmission. (*a*) Coaxial cable. (*b*) Twisted-pair cable, unshielded (UTP).

GOOD TO KNOW

The insulator in coaxial cable can sometimes break if the cable is bent sharply. This can cause loss of signal from contact between the grounded shield and the copper conductor.

GOOD TO KNOW

The rectangular shape of binary data should be preserved as much as possible. Although the data can usually be recovered if it is degraded to a sine wave, recovery is far more reliable if the original waveshape is maintained.

rounded by an insulator over which is a braided shield. The entire cable is covered with a plastic insulation.

A twisted-pair cable is two insulated wires twisted together. The one shown in Fig. 11-10(*b*) is an *unshielded twisted-pair (UTP)* cable, but a shielded version is also available. Coaxial cable and shielded twisted-pair cables are usually preferred, as they provide some protection from noise and cross talk. Cross talk is the undesired transfer of signals from one unshielded cable to another adjacent one caused by inductive or capacitive coupling.

The bandwidth of any cable is determined by its physical characteristics. All wire cables act as low-pass filters because they are made up of the wire that has inductance, capacitance, and resistance. The upper cutoff frequency of a cable depends upon the cable type, its inductance and capacitance per foot, its length, the sizes of the conductor, and the type of insulation.

Coaxial cables have a wide usable bandwidth, ranging from 200 to 300 MHz for smaller cables to 500 MHz to 3 GHz for larger cables. The bandwidth decreases drastically with length. Twisted-pair cable has a narrower bandwidth, from a few kilohertz to over 100 MHz. Again, the actual bandwidth depends on the length and other physical characteristics. Special processing techniques have extended that speed to 10 GHz over short distances (<100 m).

The bandwidth of a radio channel is determined by how much spectrum space is allotted to the application by the FCC. At the lower frequencies, limited bandwidth is available, usually several kilohertz. At higher frequencies, wider bandwidths are available, from hundreds of kilohertz to many megahertz.

As discussed in Chap. 2, binary pulses are rectangular waves made up of a fundamental sine wave plus many harmonics. The channel bandwidth must be wide enough to pass all the harmonics and preserve the waveshape. Most communication channels or media act as low-pass filters. Voice-grade telephone lines, e.g., act as a low-pass filter with an upper cutoff frequency of about 3000 Hz. Harmonics higher in frequency than the cutoff are filtered out, resulting in signal distortion. Eliminating the harmonics rounds the signal off (see Fig. 11-11).

If the filtering is particularly severe, the binary signal is essentially converted to its fundamental sine wave. If the cutoff frequency of the cable or channel is equal to or less than the fundamental sine wave frequency of the binary signal, the signal at the receiving end of the cable or radio channel will be a greatly attenuated sine wave at the signal fundamental frequency. However, the data is not lost, assuming that the *S/N* ratio is high enough. The information is still transmitted reliably, but in the minimum possible bandwidth. The sine wave signal shape can easily be restored to a

Figure 11-11 Bandwidth limitations filter out the higher harmonics in a binary signal, distorting it. (*a*) Original binary signal. (*b*) Distortion or rounding due to filtering (bandwidth limiting). (*c*) Severe distortion.

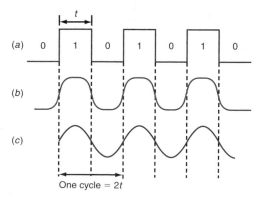

rectangular wave at the receiver by amplifying it to offset the attenuation of the transmission medium and then squaring it with a Schmitt-trigger comparator or other wave-shaping circuit.

The upper cutoff frequency of any communication medium is approximately equal to the channel bandwidth. It is this bandwidth that determines the information capacity of the channel according to Hartley's law. The channel capacity C, expressed in bits per second, is twice the channel bandwidth B, in hertz:

$$C = 2B$$

The bandwidth B is usually the same as the upper cutoff (3-dB down) frequency of the channel. This is the maximum theoretical limit, and it assumes that no noise is present.

For example, the maximum theoretical bit capacity for a 10-kHz bandwidth channel is $C = 2B = 2(10,000) = 20,000$ bps.

You can see why this is so by considering the bit time in comparison to the period of the fundamental sine wave. A 20,000-bps (20-kbps) binary signal has a bit period of $t = 1/20,000 = 50 \times 10^{-6} = 50 \ \mu s$.

It takes 2 bit intervals to represent a full sine wave with alternating binary 0s and 1s, one for the positive alternation and one for the negative alternation (see Fig. 11-11). The 2 bit intervals make a period of $50 + 50 = 100 \ \mu s$. This sine wave period translates to a sine wave frequency of $f = 1/t = 1/100 \ \mu s = 1/100 \times 10^{-6} = 10,000$ Hz (10 kHz), which is exactly the cutoff frequency or bandwidth of the channel.

Ideally, the shape of the binary data should be preserved as much as possible. Although the data can usually be recovered if it is degraded to a sine wave, recovery is far more reliable if the rectangular waveshape is maintained. This means that the channel must be able to pass at least some of the lower harmonics contained in the signal. As a general rule of thumb, if the bandwidth is roughly 5 to 10 times the data rate, the binary signal is passed with little distortion. For example, to transmit a 230.4-kbps serial data signal, the bandwidth should be at least 5×230.4 kHz (1.152 MHz) to 10×230.4 kHz (2.304 MHz).

The encoding method used also affects the required bandwidth for a given signal. For NRZ encoding, the bandwidth required is as described above. However, the bandwidth requirement for RZ is twice that for NRZ. This is so because the fundamental frequency contained in a rectangular waveform is the reciprocal of the duration of one cycle of the highest-frequency pulse, regardless of the duty cycle. The dashed lines in Fig. 11-9 show the highest fundamental frequencies for NRZ, RZ, RZ-AMI, and Manchester. AMI has a lower fundamental frequency than does RZ. The rate for Manchester encoding is twice that for NRZ and AMI.

As an example, assume an NRZ bit interval of 100 ns, which results in a bits per second-data rate of $1/t = 1/100$ ns $= 1/100 \times 10^{-9} = 10$ Mbps.

Alternating binary 1s and 0s produces a fundamental sine wave period of twice the bit time, or 200 ns, for a bandwidth of $1/t = 1/200 \times 10^{-9} = 5$ MHz. This is the same as that computed with the formula

$$B = \frac{C}{2} = \frac{10 \text{ Mbps}}{2} = 5 \text{ MHz}$$

Looking at Fig. 11-9, you can see that the RZ and Manchester pulses occur at a faster rate, the cycle time being 100 ns. The RZ and Manchester pulses are one-half the bit time of 100 ns, or 50 ns. The bit rate or channel capacity associated with this time is $1/50$ ns $= 1/50 \times 10^{-9} = 20$ Mbps. The bandwidth for this bit rate is $C/2 = 20$ Mbps$/2 = 10$ MHz.

Thus the RZ and Manchester encoding schemes require twice the bandwidth. This tradeoff of bandwidth for some benefit, such as ease of clock recovery, may be desirable in certain applications.

Multiple Coding Levels

Channel capacity can be modified by using multiple-level encoding schemes that permit more bits per symbol to be transmitted. Remember that it is possible to transmit data using symbols that represent more than just 1 bit. Multiple voltage levels can be used, as illustrated earlier in Fig. 11-5(b). Other schemes, such as using different phase shifts for each symbol, are used. Consider the equation

$$C = 2B \log_2 N$$

where N is the number of different encoding levels per time interval. The implication is that for a given bandwidth, the channel capacity, in bits per second, will be greater if more than two encoding levels are used per time interval.

Refer to Fig. 11-5, where two levels or symbols (0 or 1) were used in transmitting the binary signal. The bit or symbol time is 1 μs. The bandwidth needed to transmit this 1,000,000-bps signal can be computed from $C = 2B$, or $B = C/2$. Thus a minimum bandwidth of 1,000,000 bps/2 = 500,000 Hz (500 kHz) is needed.

The same result is obtained with the new expression. If $C = 2B \log_2 N$, then $B = C/(2 \log_2 N)$.

The logarithm of a number to the base 2 can be computed with the expression

$$\log_2 N = \frac{\log_{10} N}{\log_{10} 2}$$

$$= \frac{\log_{10} N}{0.301}$$

$$\log_2 N = 3.32 \log_{10} N$$

where N is the number whose logarithm is to be calculated. The base-10 or common logarithm can be computed on any scientific calculator. With two coding levels (binary 0 and 1 voltage levels), the bandwidth is

$$B = \frac{C}{2 \log_2 N} = \frac{1,000,000 \text{ bps}}{2(1)} = 500,000 \text{ Hz}$$

Note that $\log_2 2$ for a binary signal is simply 1.

Now we continue, using $C = 2B \log_2 N$. Since $\log_2 N = \log_2 2 = 1$,

$$C = 2B(1) = 2B$$

Now let's see what a multilevel coding scheme does. Again, we start with $B = C/(2 \log_2 N)$. The channel capacity is 2,000,000 bps, as shown in Fig. 11-5(b), because each symbol (level) interval is 1 μs long. But here the number of levels $N = 4$. Therefore 2 bits is transmitted per symbol. The bandwidth is then $(2,000,000 \text{ bps})/2 \log_2 4$. Since $\log_2 4 = 3.32 \log_{10} 4 = 3.32/(0.602) = 2$,

$$B = \frac{2,000,000}{2(2)} = \frac{2,000,000}{4} = 500,000 \text{ Hz} = 500 \text{ kHz}$$

By using a multilevel (four-level) coding scheme, we can transmit at twice the speed in the same bandwidth. The data rate is 2 Mbps with four levels of coding in a 500-kHz bandwidth compared to 1 Mbps with only two symbols (binary).

To transmit even higher rates within a given bandwidth, more voltages levels can be used, where each level represents 3, 4, or even more bits per symbol. The multilevel approach need not be limited to voltage changes; frequency changes and phase changes can also be used. Even greater increases in speed can be achieved if changes in voltage levels are combined with changes in phase or frequency.

Impact of Noise in the Channel

Another important aspect of information theory is the impact of noise on a signal. As discussed in earlier chapters, increasing bandwidth increases the rate of transmission but also allows more noise to pass, and so the choice of a bandwidth is a tradeoff.

The relationship between channel capacity, bandwidth, and noise is summarized in what is known as the *Shannon-Hartley theorem:*

Shannon-Hartley theorem

$$C = B \log_2\left(1 + \frac{S}{N}\right)$$

where C = channel capacity, bps
   B = bandwidth, Hz
   S/N = signal-to-noise ratio

Assume, e.g., that the maximum channel capacity of a voice-grade telephone line with a bandwidth of 3100 Hz and an S/N of 30 dB is to be calculated.

First, 30 dB is converted to a power ratio. If dB $= 10 \log P$, where P is the power ratio, then $P =$ antilog (dB/10). Antilogs are easily computed on a scientific calculator. A 30-dB S/N ratio translates to a power ratio of

$$P = \log^{-1}\frac{30}{10} = \log^{-1} 3 = 1000$$

The channel capacity is then

$$C = B \log_2\left(1 + \frac{S}{N}\right) = 3100 \log_2 (1 + 1000) = 3100 \log_2 1001$$

The base-2 logarithm of 1001 is

$$\log_2 1001 = 3.32 \log_{10} 1001 = 3.32(3) = 9.97 \text{ or about } 10$$

Therefore, the channel capacity is

$$C = 3100(10) = 31{,}000 \text{ bps}$$

A bit rate of 31,000 bps is surprisingly high for such a narrow bandwidth. In fact, it appears to conflict with what we learned earlier, i.e., that maximum channel capacity is twice the channel bandwidth. If the bandwidth of the voice-grade line is 3100 Hz, then the channel capacity is $C = 2B = 2(3100) = 6200$ bps. That rate is for a binary (two-level) system only, and it assumes no noise. How, then, can the Shannon-Hartley theorem predict a channel capacity of 31,000 bps when noise is present?

The Shannon-Hartley expression says that it is *theoretically* possible to achieve a 31,000-bps channel capacity on a 3100-Hz bandwidth line. What it doesn't say is that multilevel encoding is needed to do so. Going back to the basic channel capacity expression $C = 2B \log_2 N$, we have a C of 31,000 bps and a B of 3100 Hz. The number of coding or symbol levels has not been specified. Rearranging the formula, we have

$$\log_2 N = \frac{C}{2B} = \frac{31{,}000}{2(3100)} = \frac{31{,}000}{6200} = 5$$

Therefore,

$$N = \text{antilog}_2 5$$

The antilog of a number is simply the value of the base raised to the number, in this case, 2^5, or 32.

Thus a channel capacity of 31,000 can be achieved by using a multilevel encoding scheme, one that uses 32 different levels or symbols per interval, instead of a two-level (binary) system. The baud rate of the channel is still $C = 2B = 2(3100) = 6200$ Bd. But because a 32-level encoding scheme has been used, the bit rate is 31,000 bps. As it turns out, the maximum channel capacity is very difficult to achieve in practice. Typical systems limit the channel capacity to one-third to one-half the maximum to ensure more reliable transmission in the presence of noise.

Example 11 – 2

The bandwidth of a communication channel is 12.5 kHz. The *S/N* ratio is 25 dB. Calculate (*a*) the maximum theoretical data rate in bits per second, (*b*) the maximum theoretical channel capacity, and (*c*) the number of coding levels *N* needed to achieve the maximum speed. [For part (*c*), use the $\boxed{y^x}$ key on a scientific calculator.]

a. $C = 2B = 2(12.5 \text{ kHz}) = 25 \text{ kbps}$

b. $C = B \log_2 (1 + S/N) = B(3.32) \log_{10} (1 + S/N)$

$25 \text{ dB} = 10 \log P$ where $P = S/N$ power ratio

$$\log P = \frac{25}{10} = 2.5$$

$P = \text{antilog } 2.5 = \log^{-1} 2.5 = 316.2$ or $P = 10^{2.5} = 316.2$

$C = 12,500(3.32) \log_{10} (316.2 + 1)$

$= 41,500 \log_{10} 317.2$

$= 41,500(2.5)$

$C = 103,805.3 \text{ bps or } 103.8 \text{ kbps}$

c. $C = 2B \log_2 N$

$\log_2 N = C/(2B)$

$N = \text{antilog}_2 \, C/2(B)$

$N = \text{antilog}_2 \, (103,805.3)/2(12,500) = \text{antilog}_2 \, 4.152$

$N = 2^{4.152} = 17.78 \text{ or } 17 \text{ levels or symbols}$

11-4 Modem Concepts and Methods

Telephone networks, originally designed to carry voice signals, are now widely used to carry digital information as well, linking computers and computer networks across the globe. Cable TV networks designed to carry analog TV signals now also routinely carry digital data for Internet access.

Binary signals are switched dc pulses, whereas telephone and cable TV lines are designed to carry only ac analog signals. If a binary signal were applied directly to the telephone or cable TV network, it simply would not pass. The transformers, capacitors, and other ac circuitry virtually ensure that no dc signals get through in a recognizable form. Furthermore, high-speed data would be filtered out by the limited-bandwidth media.

The question is, then, How does digital data get transmitted over the telephone network? The answer is by using broadband communication techniques involving modulation, which are implemented by a *modem,* a device containing both a *mo*dulator and a *dem*odulator. Modems convert binary signals to analog signals capable of being transmitted over telephone and cable TV lines and by radio and then demodulate such analog signals, reconstructing the equivalent binary output. Figure 11-12 shows two ways that modems are commonly used in digital data transmission. In Fig. 11-12(*a*), two computers exchange data by speaking through modems. While one modem is transmitting, the other is receiving. Full duplex operation is also possible. In Fig. 11-12(*b*), a remote video terminal or personal computer is using a modem to communicate with a large server computer. Modems are also the interface between the millions of personal computers and servers that make up the Internet.

Modem

GOOD TO KNOW

If a binary signal were applied directly to a telephone network, it simply would not pass.

Figure 11-12 How modems permit digital data transmission on the telephone network.

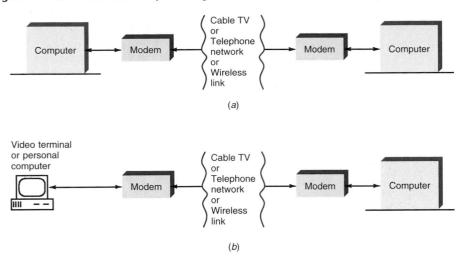

(a)

(b)

There are four widely used modem types: conventional analog dial-up modems, digital subscriber line (DSL) modems, cable TV modems, and wireless modems. The first three are discussed in the following sections.

Modulation for Data Communication

Four main types of modulation are used in modern modems: frequency-shift keying (FSK), phase-shift keying (PSK), quadrature amplitude modulation (QAM), and orthogonal frequency-division multiplexing (OFDM). FSK is used primarily in lower-speed (<500 kbps) modems in a noisy environment. PSK operates in narrower bandwidths over a wide range of speeds. QAM is a combination of both amplitude modulation and PSK. It can produce very high data rates in narrow bandwidths. OFDM operates over a very wide bandwidth and can achieve very high rates in a noisy environment.

FSK. The oldest and simplest form of modulation used in modems is *frequency-shift keying (FSK)*. In FSK, two sine wave frequencies are used to represent binary 0s and 1s. For example, a binary 0, usually called a *space* in data communication jargon, has a frequency of 1070 Hz. A binary 1, referred to as a *mark,* is 1270 Hz. These two frequencies are alternately transmitted to create the serial binary data. The resulting signal looks something like that shown in Fig. 11-13. Both of the frequencies are well within

Frequency-shift keying (FSK)

Space

Mark

Figure 11-13 Frequency-shift keying. (a) Binary signal. (b) FSK signal.

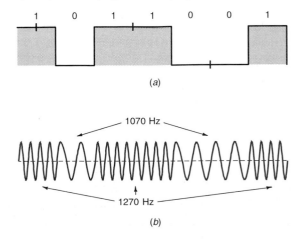

(a)

(b)

The Transmission of Binary Data in Communication Systems

Figure 11-14 The FSK signals within the telephone audio bandpass.

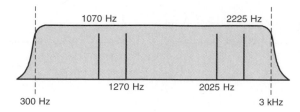

the 300- to 3000-Hz bandwidth normally associated with the telephone system, as illustrated in Fig. 11-14.

The simultaneous transmit and receive operations that are carried out by a modem, known as *full duplex operation,* require that another set of frequencies be defined. These are also indicated in Fig. 11-14. A binary 0 or space is 2025 Hz; a binary 1 or mark is 2225 Hz. These tones are also within the telephone bandwidth but are spaced far enough from the other frequencies so that selective filters can be used to distinguish between the two. The 1070- and 1270-Hz tones are used for transmitting (originate), and the 2025- and 2225-Hz tones are used for receiving (answer).

Figure 11-15 is a block diagram of the modulator and demodulator sections of an FSK modem. Each modem contains an FSK modulator and an FSK demodulator so that both send and receive operations can be achieved. Bandpass filters at the inputs to each modem separate the two tones. For example, in the upper modem, a bandpass filter allows frequencies between 1950 and 2300 Hz to pass. This means that 2025- and 2225-Hz tones will be passed, but the 1070- and 1270-Hz tones generated by the

Figure 11-15 Block diagram of the modulator-demodulator of an FSK modem.

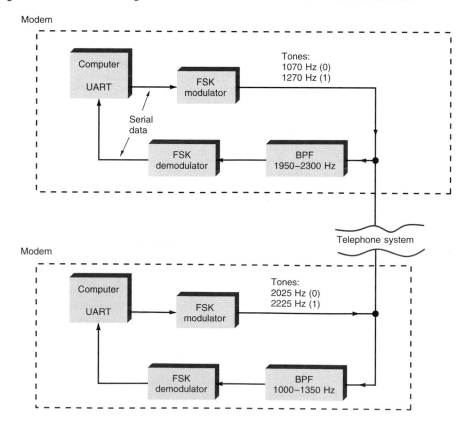

internal modulator will be rejected. The lower modem has a bandpass filter that accepts the lower-frequency tones while rejecting the upper-frequency tones generated internally.

A wide variety of modulator and demodulator circuits are used to produce and recover FSK. Virtually all the circuits described in Chap. 6 have been or could be used. A typical FSK modulator is simply an oscillator whose frequency can be switched between two frequencies. A typical demodulator is a PLL. Most modems now use digital techniques because they are simpler and more adaptable to IC implementations. A large portion of modem operations if not all operations are now implemented with DSP.

FSK signals typically occupy a wide bandwidth because of the multiple sidebands produced by the FM process. Higher orders of sidebands are also generated by the harmonics contained in the fast binary modulating signal. Any abrupt signal changes further aggravate the problem. Several techniques have been developed to improve the spectral efficiency of FSK. The term *spectral efficiency* refers to how well a specific modulation technique produces a maximum data rate in a minimal bandwidth.

When the mark and space frequencies are arbitrarily chosen, they will not be phase-coherent. That is, there will be abrupt signal changes during 0-to-1 or 1-to-0 transitions. This is illustrated in Fig. 11-16(*a*). The "glitches" or phase discontinuities produce even more harmonics and wider bandwidth. In addition, such discontinuities make demodulation more difficult and produce more bit errors.

To overcome this problem, the mark and space frequencies can be chosen so that the periods of the sine waves both cross zero at the mark-to-space and space-to-mark transitions. This is illustrated in Fig. 11-16(*b*). No phase discontinuities exist, so the

Figure 11-16   (*a*) FSK with frequencies that produce phase discontinuities. (*b*) Continuous-phase FSK with smooth transitions at the zero crossings.

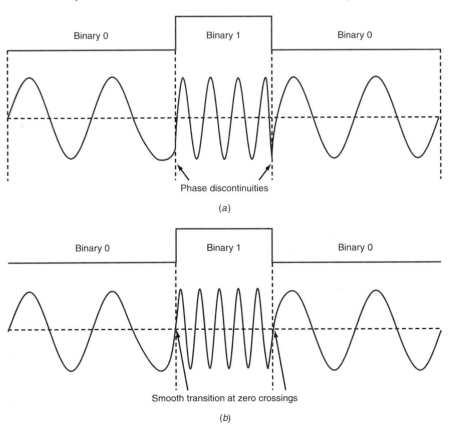

resulting bandwidth is less. This type of modulation is called *continuous-phase frequency-shift keying (CPFSK)*.

Another term for signals that start and stop at the zero crossing points is coherent. You could call this form of modulation coherent FSK. You can also have coherent ASK or OOK. Coherent versions use less bandwidth and perform better in the presence of noise.

An improved variation of CPFSK is *minimum shift keying (MSK)*. As in CPFSK, the mark and space frequencies are some integer multiple of the bit clock frequency. This ensures that the signals are fully synchronized with one another and that no phase discontinuities occur.

MSK further improves spectral efficiency by using a low modulation index. Recall from Chap. 5 that the number of pairs of sidebands produced (and therefore the wider the bandwidth) is proportional to the modulation index. With analog FM, the modulation index is

$$m_f = \frac{f_d}{f_m}$$

where f_d is the frequency deviation and f_m is the modulating signal frequency. With FSK, the modulation index m is

$$m = \Delta f(T)$$

where Δf is the deviation or frequency shift between the mark frequency f_M and the space frequency f_S.

$$\Delta f = f_S - f_M$$

Also T is the bit time or the reciprocal of the data rate.

$$T = \frac{1}{\text{bps}}$$

MSK generally specifies that m must be 0.5. However, other values (0.3) are used.

For example, assume a MSK modem with $f_M = 1200$ and $f_S = 1800$ Hz. The bit rate is 1200 bps. The modulation index is

$$\Delta f = f_S - f_M = 1800 - 1200 = 600 \text{ Hz}$$

$$T = \frac{1}{\text{bps}} = \frac{1}{1200} = 0.0008333 \text{ s}$$

$$m = \Delta f(T) = 600(0.0008333) = 0.5$$

MSK is a very spectrally efficient form of FSK. But the MSK signal bandwidth can be further reduced by prefiltering the binary modulating signal. This filter removes some of the higher-level harmonics that are responsible for the added sidebands and wider bandwidth. One of the best prefilters is called a *Gaussian low-pass filter*. It rounds the edges and somewhat lengthens the rise and fall times. This in turn reduces harmonic content. And that decreases overall signal bandwidth. *Gaussian filtered MSK* is referred to as GMSK. It is widely used in data communication and is the basis of the popular GSM digital cell phones.

PSK. In *phase-shift keying (PSK)*, the binary signal to be transmitted changes the phase shift of a sine wave character depending upon whether a binary 0 or binary 1 is to be transmitted. (Recall that phase shift is a time difference between two sine waves of the same frequency.) Figure 11-17 illustrates several examples of phase shift. A phase shift of 180°, the maximum phase difference that can occur, is known as a *phase reversal*, or *phase inversion*.

Figure 11-18 illustrates the simplest form of PSK, *binary phase-shift keying (BPSK)*. During the time that a binary 0 occurs, the carrier signal is transmitted with one phase; when a binary 1 occurs, the carrier is transmitted with a 180° phase shift.

Figure 11–17   Examples of phase shift.

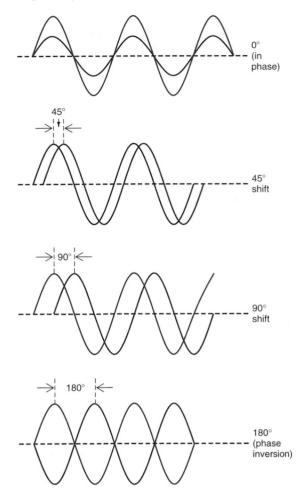

Figure 11–18   Binary phase-shift keying.

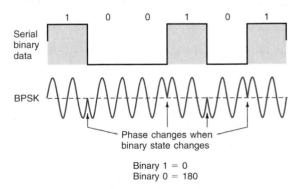

Figure 11-19 shows one kind of circuit used for generating BPSK, a standard lattice ring modulator or balanced modulator used for generating DSB signals. The carrier sine wave is applied to the input transformer T_1 while the binary signal is applied to the transformer center taps. The binary signal provides a switching signal for the diodes. When a binary 0 appears at the input, A is + and B is −, so diodes D_1 and D_4 conduct. They act as closed switches, connecting the secondary of T_1 to the primary of T_2. The windings are phased so that the BPSK output is in phase with the carrier input.

Figure 11-19 A BPSK modulator.

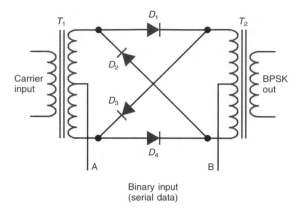

Binary input
(serial data)

When a binary 1 appears at the input, A is − and B is +, so diodes D_1 and D_4 are cut off while diodes D_2 and D_3 conduct. This causes the secondary of T_1 to be connected to the primary of T_2 but with the interconnections reversed. This introduces a 180° phase-shift carrier at the output.

Demodulation of a BPSK signal is also done with a balanced modulator. A version of the diode ring or lattice modulator can be used, as shown in Fig. 11-20. This is actually the same circuit as that in Fig. 11-19, but the output is taken from the center taps. The BPSK and carrier signals are applied to the transformers. IC balanced modulators can also be used at the lower frequencies. The modulator and demodulator circuits are identical to the doubly balanced modulators used for mixers. They are available as fully wired and tested components for frequencies up to about 1 GHz.

The key to demodulating BPSK is that a carrier with the correct frequency and phase relationship must be applied to the balanced modulator along with the BPSK signal. Typically the carrier is derived from the BPSK signal itself, using a carrier recovery circuit like that shown in Fig. 11-21. A bandpass filter ensures that only the desired BPSK signal is passed. The signal is then squared or multiplied by itself by a balanced modulator or analog multiplier by applying the same signal to both inputs. Squaring removes all the 180° phase shifts, resulting in an output that is twice the input signal frequency $(2f)$. A bandpass filter set at twice the carrier frequency passes this signal only. The resulting signal is applied to the phase detector of a PLL. Note that a ×2 frequency

Figure 11-20 A BPSK demodulator.

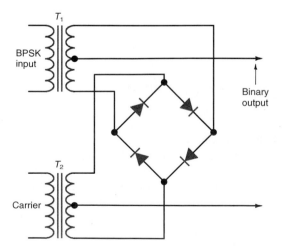

Chapter 11

Figure 11-21   A BPSK carrier recovery circuit.

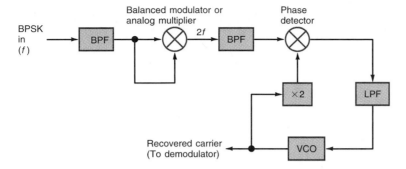

multiplier is used between the VCO and phase detector, ensuring that the VCO frequency is at the carrier frequency. Use of the PLL means that the VCO will track any carrier frequency shifts. The result is a signal with the correct frequency and phase relationship for proper demodulation. The carrier is applied to the balanced modulator-demodulator along with the BPSK signal. The output is the recovered binary data stream.

DPSK. To simplify the demodulation process, a version of binary PSK called *differential phase-shift keying (DPSK)* can be used. In DPSK, there is no absolute carrier phase reference. Instead, the transmitted signal itself becomes the phase reference. In demodulating DPSK, the phase of the received bit is compared to the phase of the previously received bit.

For DPSK to work, the original binary bit stream must undergo a process known as *differential phase coding,* in which the serial bit stream passes through an inverted exclusive-NOR circuit (XNOR), as shown in Fig. 11-22(*a*). Note that the XNOR output is applied to a 1-bit delay circuit before being applied back to the input. The delay can simply be a clocked flip-flop or a delay line. The resulting bit pattern permits the signal to be recovered because the current bit phase can be compared with the previously received bit phase.

In Fig. 11-22(*b*), the input binary word to be transmitted is shown along with the output of the XNOR. An XNOR circuit is a 1-bit comparator that produces a binary 1 output when both inputs are alike and a binary 0 output when the two bits are different. The output of the circuit is delayed a 1-bit interval by being stored in a flip-flop. Therefore, the XNOR inputs are the current bit plus the previous bit. The XNOR signal is then applied to the balanced modulator along with the carrier to produce a BPSK signal.

Demodulation is accomplished with the circuit shown in Fig. 11-23. The DPSK signal is applied to one input of the balanced modulator and a 1-bit delay circuit, either a flip-flop or a delay line. The output of the delay circuit is used as the carrier. The resulting output is filtered by a low-pass filter to recover the binary data. Typically the

Differential phase-shift keying (DPSK)

Figure 11-22   The DPSK process. (*a*) DPSK modulator. (*b*) Differential phase encoding.

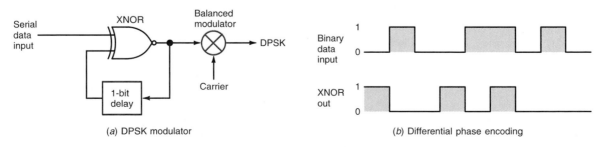

(*a*) DPSK modulator   (*b*) Differential phase encoding

Figure 11-23   A DPSK demodulator.

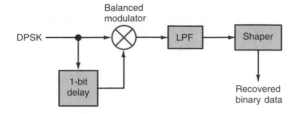

low-pass filter output is shaped with a Schmitt trigger or comparator to produce clean, high-speed binary levels.

QPSK. The main problem with BPSK and DPSK is that the speed of data transmission in a given bandwidth is limited. One way to increase the binary data rate while not increasing the bandwidth required for the signal transmission is to encode more than 1 bit per phase change. There is a symbol change for each bit change with BPSK and DPSK, so the baud (symbol) rate is the same as the bit rate. In BPSK and DPSK, each binary bit produces a specific phase change. An alternative approach is to use combinations of two or more bits to specify a particular phase shift, so that a symbol change (phase shift) represents multiple bits. Because more bits per baud are encoded, the bit rate of data transfer can be higher than the baud rate, yet the signal will not take up additional bandwidth.

Quadrature (quarternary or quadra) phase PSK (QPSK or 4-PSK)

One commonly used system for doing this is known as *quadrature, quarternary,* or *quadra phase PSK (QPSK* or *4-PSK).* In QPSK, each pair of successive digital bits in the transmitted word is assigned a particular phase, as indicated in Fig. 11-24(*a*). Each pair of serial bits, called a *dibit,* is represented by a specific phase. A 90° phase shift exists between each pair of bits. Other phase angles can also be used as long as they have a 90° separation. For example, it is common to use phase shifts of 45°, 135°, 225°, and 315°, as shown in Fig. 11-24(*b*).

The diagram in Fig. 11-24(*b*) is called a constellation diagram. It shows the modulation signal in the form of phasors. The length of the arrow or phasor indicates the peak voltage level of the signal while its angle to the axis is the phase shift. Sometimes the constellation diagram is simplified by just showing dots on the axis indicating the location of the phasor arrowhead, as in Fig. 11-24(*c*). This simplifies the diagram, but you should always imagine a phasor drawn from the origin of the axis to each dot. Constellation diagrams are widely used to show phase-amplitude modulation schemes.

Figure 11-24   Quadrature PSK modulation. (*a*) Phase angle of carrier for different pairs of bits. (*b*) Phasor representation of carrier sine wave. (c) Constellation diagram of QPSK.

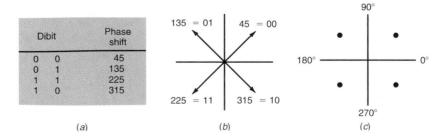

(a)	(b)	(c)

Figure 11-25 A QPSK modulator.

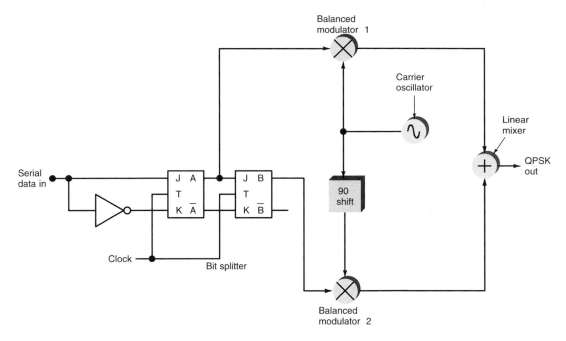

You will often hear the term *M-ary* used in discussing higher coding levels of PSK. It is derived from the word *binary* where in binary $M = 2$ so 2-ary indicates two phase shifts. QPSK is the same as 4-PSK or 4-ary PSK; 8-PSK would be called 8-ary PSK with 8 phase positions.

A circuit for producing QPSK is shown in Fig. 11-25. It consists of a 2-bit shift register implemented with flip-flops, commonly known as a *bit splitter.* The serial binary data train is shifted through this register, and the bits from the two flip-flops are applied to balanced modulators. The carrier oscillator is applied to balanced modulator 1 and through a 90° phase shifter to balanced modulator 2. The outputs of the balanced modulators are linearly mixed to produce the QPSK signal.

The output from each balanced modulator is a BPSK signal. With a binary 0 input, the balanced modulator produces one phase of the carrier. With a binary 1 input, the carrier phase is shifted 180°. The output of balanced modulator 2 also has two phase states, 180° out of phase with each other. The 90° carrier phase shift at the input causes the outputs from balanced modulator 2 to be shifted 90° from those of balanced modulator 1. The result is four different carrier phases, which are combined two at a time in the linear mixer. The result is four unique output phase states.

Figure 11-26 shows the outputs of one possible set of phase shifts. Note that the carrier outputs from the two balanced modulators are shifted 90°. When the two carriers are algebraically summed in the mixer, the result is an output sine wave that has a phase shift of 225°, which is halfway between the phase shifts of the two balanced modulator signals.

A demodulator for QPSK is illustrated in Fig. 11-27. The carrier recovery circuit is similar to the one described previously. The carrier is applied to balanced modulator 1 and is shifted 90° before being applied to balanced modulator 2. The outputs of the two balanced modulators are filtered and shaped into bits. The 2 bits are combined in a shift register and shifted out to produce the originally transmitted binary signal.

Encoding still more bits per phase change produces higher data rates. In 8-PSK, for example, 3 serial bits are used to produce a total of 8 different phase changes. In 16-PSK, 4 serial input bits produce 16 different phase changes, for an even higher data rate.

Figure 11-26 How the modulator produces the correct phase by adding two signals.

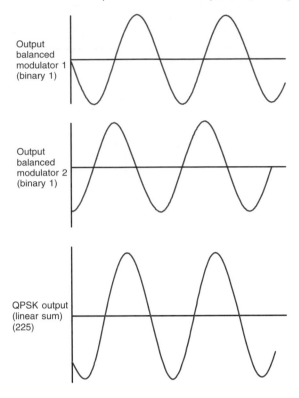

Output balanced modulator 1 (binary 1)

Output balanced modulator 2 (binary 1)

QPSK output (linear sum) (225)

Quadrature amplitude modulation (QAM)

Figure 11-28 shows the constellation diagram of a 16-PSK signal. The phase increment is 22.5°. Each phasor or dot on the diagram represents a 4-bit number. Note that since all the dots fall on a circle, the amplitude of the 16-PSK signal remains constant while only the phase changes. The radius of the circle is the peak amplitude of the signal. It is said that the signal has a constant "envelope," where the envelope is simply the line or curve connecting the peaks of the carrier sine waves. It is flat or constant, as is an FM signal.

QAM. One of the most popular modulation techniques used in modems for increasing the number of bits per baud is *quadrature amplitude modulation (QAM)*. QAM uses both amplitude and phase modulation of a carrier; not only are different phase shifts produced, but also the amplitude of the carrier is varied.

Figure 11-27 A QPSK demodulator.

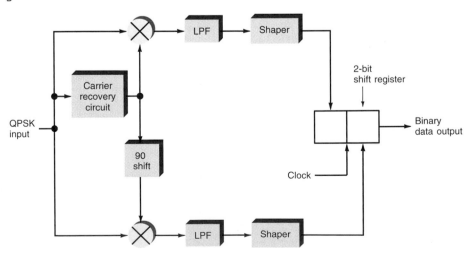

Figure 11-28 Constellation diagram of a 16-PSK signal.

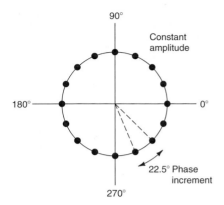

In 8-QAM, there are four possible phase shifts, as in QPSK, and two different carrier amplitudes, so that eight different states can be transmitted. With eight states, 3 bits can be encoded for each baud or symbol transmitted. Each 3-bit binary word transmitted uses a different phase-amplitude combination.

Figure 11-29 is a *constellation diagram* of an 8-QAM signal showing all possible phase and amplitude combinations. The points in the diagram indicate the eight possible phase-amplitude combinations. Note that there are two amplitude levels for each phase position. Point A shows a low carrier amplitude with a phase shift of 225°. It represents 100. Point B shows a higher amplitude and a phase shift of 315°. This sine wave represents 011.

A block diagram of an 8-QAM modulator is shown in Fig. 11-30. The binary data to be transmitted is shifted serially into the 3-bit shift register. These bits are applied in pairs to two 2-to-4 level converters. A 2-to-4 level converter circuit, basically a simple D/A converter, translates a pair of binary inputs into one of four possible dc output voltage levels. The idea is to produce four voltage levels corresponding to the different combinations of 2 input bits, i.e., four equally spaced voltage levels. These are applied to the two balanced modulators fed by the carrier oscillator and a 90° phase shifter, as in a QPSK modulator. Each balanced modulator produces four different output phase-amplitude combinations. When these are combined in the linear mixer, eight different phase-amplitude combinations are produced. The most critical part of the circuit is the 2-to-4 level converters; these must have very precise output amplitudes so that when they are combined in the linear summer, the correct output and phase combinations are produced.

Constellation diagram

Figure 11-29 A constellation diagram of a QAM signal.

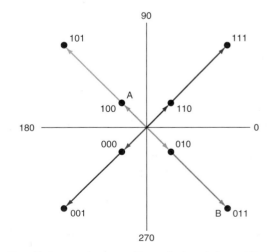

Note: Each vector has a specific amplitude and phase shift and represents one 3-bit word.

The Transmission of Binary Data in Communication Systems

Figure 11-30 An 8-QAM modulator.

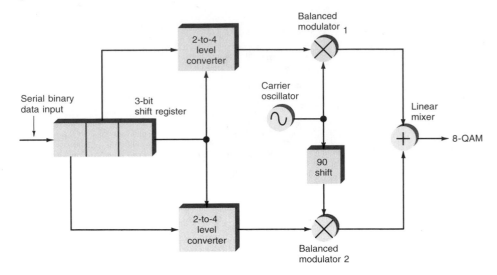

A 16-QAM signal can also be generated by encoding 4 input bits at a time. The result is 8 different phase shifts and 2 amplitude levels, producing a total of 16 different phase-amplitude combinations. Even higher data rates can be achieved with 64-QAM and 256-QAM. These signals are used in cable TV modems, satellites, and high-speed fixed broadband wireless applications.

Spectral Efficiency and Noise

As indicated earlier in this section, spectral efficiency is a measure of how fast data can be transmitted in a given bandwidth. The measure is bits per second per hertz (bps/Hz). As you have seen, different modulation methods give different efficiencies. The table shows the common efficiencies for several common types of modulation.

Modulation	Spectral efficiency, bps/Hz
FSK	<1
GMSK	1.35
BPSK	1
QPSK	2
8-PSK	3
16-QAM	4

With a modulation method like BPSK where the efficiency is 1 bps/Hz, you can actually transmit data at a rate equal to the bandwidth or

$$\text{BW} = f_b = \frac{1}{t_b}$$

where f_b is the data rate in bits per second and t_b is the bit time.

Another factor that clearly influences the spectral efficiency is the noise in the channel or the signal-to-noise *(S/N)* ratio. Obviously the greater the noise, the greater the number of bit errors. The number of errors that occur in a given time is called the *bit error rate (BER)*. The BER is simply the ratio of the number of errors that occur in 1 s of a 1-s interval of data transmission. For example, if 5 errors occur in 1 s in a 10 Mbps transmission, the BER is

$$\text{BER} = \frac{5}{10} \times 10^6 = 0.5 \times 10^{-6} \quad \text{or} \quad 5 \times 10^{-7}$$

Some modulation schemes are more immune to the noise than others.

Bit error rate (BER)

Figure 11-31 BER versus *C/N* for popular digital modulation methods.

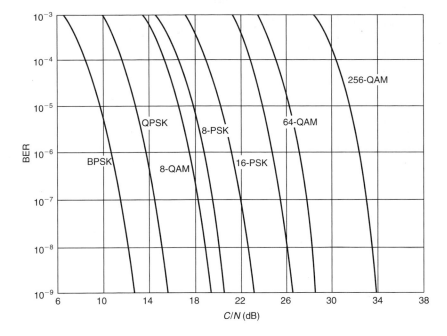

The *S/N* covered in previous chapters was the ratio of signal rms voltage to noise rms voltage. You can also use the ratio of the average signal power of the carrier plus the sidebands to the noise power, usually the thermal noise. This is called the *carrier-to-noise (C/N)* ratio. Generally *C/N* is expressed in decibels.

Carrier-to-noise (C/N)

Figure 11-31 shows the relationship between the *C/N* and BER for different modulation methods. What this graph shows is that for a given BER, the modulation methods with the fewest symbol changes or the smaller bits per hertz give the best performance at the lower *C/N* ratios. For a BER of 10^{-6}, BPSK needs only a *C/N* of 11 dB, while for 16-QAM a *C/N* of 20 dB is needed. Amplitude-modulated signals are always more susceptible to noise than are constant-envelope forms of modulation such as PSK and FSK; so you need more signal power to overcome the noise to get the desired BER. Such a graph lets you compare and evaluate different modulation schemes. Just keep in mind that bandwidth does not enter the picture here. When comparing methods, you must remember that noise increases with bandwidth.

A better measure of signal-to-noise ratio for digital data is the ratio of energy per bit transmitted to the noise power density, or E_b/N_0, usually pronounced "*E* sub *b* over *N* sub zero." Remember that energy is expressed in joules (J), where one joule per second (J/s) is equal to one watt (1 W), or 1 W = 1 J/s. Therefore, E_b is the power in 1 bit *P* multiplied by the bit time t_b, or $E_b = Pt_b$.

The noise power density in watts per hertz (W/Hz), which we call N_0, is the thermal noise power *N* divided by the bandwidth of the channel *B*. Recall that thermal noise power is $N = kTB$, where *k* is Boltzmann's constant 1.38×10^{-23}, *T* is the temperature in kelvins, and *B* is the bandwidth in hertz. Room temperature is about 290 K.

$$N_0 = \frac{kTB}{B} = kT$$

The overall result is

$$\frac{E_b}{N_0} = \frac{Pt_b}{kT}$$

The Transmission of Binary Data in Communication Systems

Figure 11-32 BER vs. E_b/N_0 for different modulation methods.

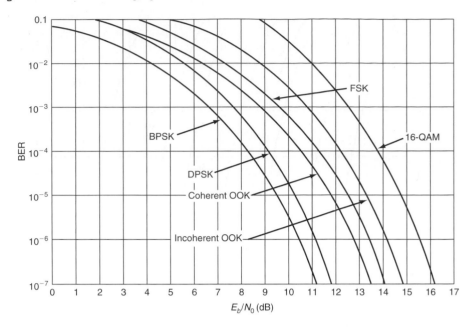

This relationship can be further manipulated to show E_b/N_0 in terms of C/N. This relationship is

$$\frac{E_b}{N_0} = \left(\frac{C}{N}\right)\left(\frac{B}{f_b}\right)$$

Here B is the bandwidth in hertz and f_b is the bit rate or bit frequency (f_b) where $f_b = 1/t_b$. Given C/N and the other factors, you can calculate E_b/N_0. What E_b/N_0 does is to take bandwidth out of the comparison. It normalizes all the different multiphase/amplitude schemes to a noise bandwidth of 1 Hz, giving you a better way to compare and contrast the various modulation methods for a given BER. You will often see curves like that in Fig. 11-31 plotted with BER on the vertical axis and E_b/N_0 rather than C/N on the horizontal axis. Figure 11-32 is an example. Note the role that coherency plays (carrier sine waves start and stop on the zero crossing). Coherent OOK needs a lower signal-to-noise ratio than incoherent OOK for a given BER.

11–5 Wideband Modulation

Most modulation methods are designed to be spectrally efficient, i.e., to transmit as many bits per hertz as possible. The goal is to minimize the use of spectrum space and to transmit the highest speed possible in the given bandwidth. However, there is another class of modulation methods that do just the opposite. These methods are designed to use more bandwidth. The transmitted signal occupies a bandwidth many times greater than the information bandwidth. Special benefits derive from such wideband modulation techniques. The two most widely used wideband modulation methods are spread spectrum and orthogonal frequency-division multiplexing.

Spread Spectrum

Spread spectrum (SS)

Spread spectrum (SS) is a modulation and multiplexing technique that distributes a signal and its sidebands over a very wide bandwidth. Traditionally, the efficiency of a modulation or multiplexing technique is determined by how little bandwidth it uses. The continued growth of all types of radio communication, the resulting crowding, and the finite bounds of usable spectrum space have made everyone in the world of data communication sensitive to how much bandwidth a given signal occupies. Designers of communication systems and equipment typically do all in their power to minimize the amount of bandwidth a signal takes. How, then, can a scheme that spreads a signal over a very wide piece of the spectrum be of value? The answer to this question is the subject of this section.

After World War II, spread spectrum was developed primarily by the military because it is a secure communication technique that is essentially immune to jamming. In the mid-1980s, the FCC authorized use of spread spectrum in civilian applications. Currently, unlicensed operation is permitted in the 902- to 928-MHz, 2.4- to 2.483-GHz, and 5.725- to 5.85-GHz ranges, with 1 W of power. Spread spectrum on these frequencies is being widely incorporated into a variety of commercial communication systems. One of the most important of these new applications is wireless data communication. Numerous LANs and portable personal computer modems use SS techniques, as does a class of cordless telephones in the 900-MHz, 2.4-, and 5.8-GHz ranges. The most widespread use of SS is in cellular telephones in the 800- to 900-MHz and 1800- to 1900-MHz ranges. It is referred to as *code-division multiple access (CDMA)*.

Code-division multiple access (CDMA)

There are two basic types of spread spectrum: *frequency-hopping (FH)* and *direct-sequence (DS)*. In frequency-hopping SS, the frequency of the carrier of the transmitter is changed according to a predetermined sequence, called pseudorandom, at a rate higher than that of the serial binary data modulating the carrier. In direct-sequence SS, the serial binary data is mixed with a higher-frequency pseudorandom binary code at a faster rate, and the result is used to phase-modulate a carrier.

Frequency–Hopping Spread Spectrum

Figure 11-33 shows a block diagram of a *frequency-hopping SS transmitter*. The serial binary data to be transmitted is applied to a conventional two-tone FSK modulator, and the modulator output is applied to a mixer. Also driving the mixer is a frequency synthesizer. The output signal from the bandpass filter after the mixer is the difference between one of the two FSK sine waves and the frequency of the frequency synthesizer. As the figure shows, the synthesizer is driven by a pseudorandom code generator, which is either a special digital circuit or the output of a microprocessor.

Frequency-hopping (FH) SS

The pseudorandom code is a serial pattern of binary 0s and 1s that changes in a random fashion. The randomness of the 1s and 0s makes the serial output of this circuit appear as digital noise. Sometimes the output of this generator is called *pseudorandom noise (PSN)*. The binary sequence is actually predictable, since it does repeat after many bit changes (hence "pseudo"). The randomness is sufficient to minimize the possibility of someone accidentally duplicating the code, but the predictability allows the code to be duplicated at the receiver.

Pseudorandom noise (PSN)

PSN sequences are usually generated by a shift register circuit similar to that shown in Fig. 11-34. In the figure, eight flip-flops in the shift register are clocked by an external clock oscillator. The input to the shift register is derived by X-ORing two or more of

Figure 11-33 A frequency-hopping SS transmitter.

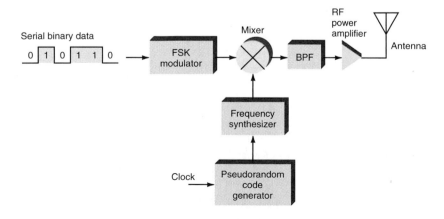

Figure 11-34   A typical PSN code generator.

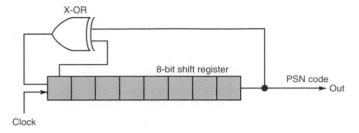

the flip-flop outputs. It is this connection that produces the pseudorandom sequence. The output is taken from the last flip-flop in the register. Changing the number of flip-flops in the register and/or which outputs are X-ORed and fed back changes the code sequence. Alternatively, a microprocessor can be programmed to generate pseudorandom sequences.

In a frequency-hopping SS system, the rate of synthesizer frequency change is higher than the data rate. This means that although the data bit and the FSK tone it produces remain constant for one data interval, the frequency synthesizer switches frequencies many times during this period. See Fig. 11-35, where the frequency synthesizer changes frequencies 4 times for each bit time of the serial binary data. The time that the synthesizer remains on a single frequency is called the *dwell time*. The frequency synthesizer puts out a random sine wave frequency to the mixer, and the mixer creates a new carrier frequency for each dwell interval. The resulting signal, whose frequency rapidly jumps around, effectively scatters pieces of the signal all over the band. Specifically, the carrier randomly switches between dozens or even hundreds of frequencies over a given bandwidth. The actual dwell time on any frequency varies with the application and data rate, but it can be as short as 10 ms. Currently, FCC regulations specify that there be a minimum of 75 hopping frequencies and that the dwell time not exceed 400 μs.

Figure 11-36 shows a random frequency-hop sequence. The horizontal axis is divided into dwell time increments. The vertical axis is the transmitter output frequency, divided into step increments of the PLL frequency synthesizer. As shown, the signal is spread out over a very wide bandwidth. Thus a signal that occupies only a few kilohertz of spectrum can be spread out over a range that is 10 to 10,000 times that wide. Because an SS signal does not remain on any one frequency for a long time, but jumps around randomly, it does not interfere with a traditional signal on any of the hopping frequencies. An SS signal actually appears to be more like background noise to a conventional narrow-bandwidth receiver. A conventional receiver picking up such a signal will not even respond to a signal of tens of milliseconds duration. In addition, a conventional receiver cannot receive an SS signal because it does not have a wide enough bandwidth and it cannot follow or track its random-frequency changes. Therefore, the SS signal is as secure as if it were scrambled.

Two or more SS transmitters operating over the same bandwidth but with different pseudorandom codes hop to different frequencies at different times, and do not typically

Figure 11-35   Serial data and the PSN code rate.

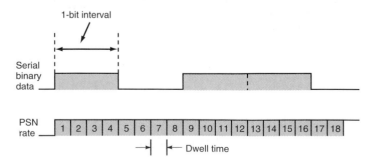

Figure 11–36 A pseudorandom frequency-hop sequence.

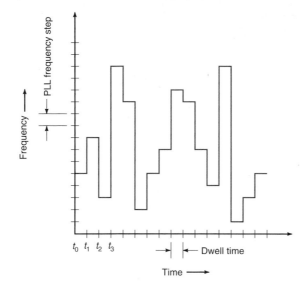

occupy a given frequency simultaneously. Thus SS is also a kind of multiplexing, as it permits two or more signals to use a given bandwidth concurrently without interference. In effect, SS permits more signals to be packed in a given band than any other type of modulation or multiplexing.

A frequency-hopping receiver is shown in Fig. 11-37. The very wideband signal picked up by the antenna is applied to a broadband RF amplifier and then to a conventional mixer. The mixer, like that in any superheterodyne receiver, is driven by a local oscillator. In this case the local oscillator is a frequency synthesizer like the one used at the transmitter. The local oscillator at the receiving end must have the same pseudorandom code sequence as that generated by the transmitter so that it can receive the signal on the correct frequency. The signal is thus reconstructed as an IF signal that contains the original FSK data. The signal is then applied to an FSK demodulator, which reproduces the original binary data train.

One of the most important parts of the receiver is the circuit that is used to acquire and synchronize the transmitted signal with the internally generated pseudorandom code. The problem of getting the two codes into step with each other is solved by a preamble signal and code at the beginning of the transmission. Once synchronization has been established, the code sequences occur in step. This characteristic makes the SS technique extremely secure and reliable. Any receiver not having the correct code cannot receive the signal.

Figure 11–37 A frequency-hopping receiver.

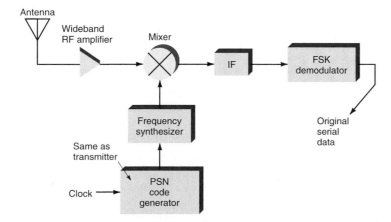

The Transmission of Binary Data in Communication Systems **417**

Figure 11-38 A direct-sequence SS transmitter.

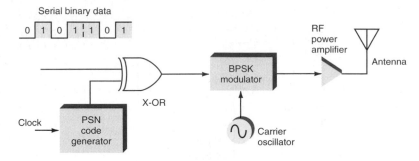

Many stations can share a common band if, instead of assigning each station a single frequency to operate on, each is given a different pseudorandom code within the same band. This permits a transmitter to selectively transmit to a single receiver without other receivers in the band being able to pick up the signal.

Direct–Sequence Spread Spectrum

Direct-sequence SS

Chipping rate

A block diagram of a *direct-sequence SS* (*DSSS*) transmitter is shown in Fig. 11-38. The serial binary data is applied to an X-OR gate along with a serial pseudorandom code that occurs faster than the binary data. Figure 11-39 shows typical waveforms. One bit time for the pseudorandom code is called a *chip,* and the rate of the code is called the *chipping rate.* The chipping rate is faster than the data rate.

The signal developed at the output of the X-OR gate is then applied to a PSK modulator, typically a BPSK device. The carrier phase is switched between 0 and 180° by the 1s and 0s of the X-OR output. QPSK and other forms of PSK can also be used. The PSK modulator is generally some form of balanced modulator. The signal phase modulating the carrier, being much higher in frequency than the data signal, causes the modulator to produce multiple, widely spaced sidebands whose strength is such that the complete signal takes up a great deal of the spectrum. Thus the resulting signal is spread. Because of its randomness, the signal looks like wideband noise to a conventional narrowband receiver.

Figure 11-39 Data signals in direct-sequence SS.

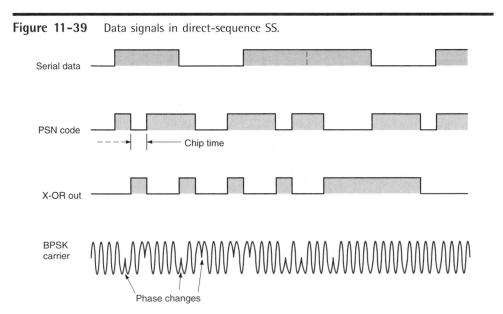

Figure 11-40 Comparison between narrowband and spread spectrum signals.

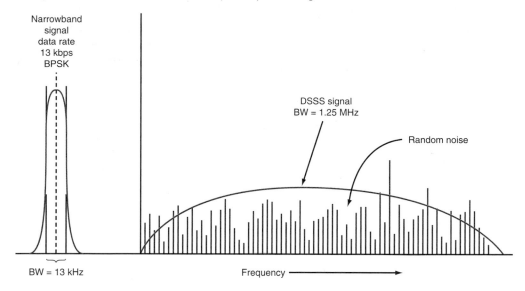

Figure 11-40 shows a standard narrowband signal and a spread spectrum signal. Assume a binary information signal that is occurring at a rate of 13 kbps. If we use BPSK with its 1 bit/Hz efficiency, we could transmit this signal in a bandwidth of about 13 kHz. Now, if we use DSSS with a chipping signal of 1.25 Mbps, the resulting signal will be spread over about 1.25 MHz of bandwidth if we use BPSK. The spread signal has the same power as the narrowband signal but far more sidebands, so the amplitudes of the carrier and sidebands are very low and just above the random noise level. To a narrowband receiver, the signal just looks like a part of the noise level.

The effect of spreading the signal is to provide a type of gain called processing gain to the signal. This gain helps to improve the overall signal-to-noise ratio. The higher the gain, the greater the ability of the system to fight interference. This processing gain G is

$$G = \frac{\text{BW}}{f_b}$$

where BW is the channel bandwidth and f_b is the data rate. For the example in Fig. 11-40, the process gain is

$$G = \frac{1.25 \text{ MHz}}{13 \text{ kbps}} = 96.15$$

In terms of decibels, this is a power gain of 19.83 dB.

One type of direct-sequence receiver is shown in Fig. 11-41. The broadband SS signal is amplified, mixed with a local oscillator, and then translated down to a lower IF in mixer 1. For example, the SS signal at an original carrier of 902 MHz might be translated down to another IF of 70 MHz. The IF signal is then compared to another IF signal that is produced in mixer 3 using a PSN sequence that is similar to that transmitted. The output of mixer 3 should be identical to the output of mixer 1 but shifted in time. This comparison process, called *correlation,* takes place in mixer 2. If the two signals are identical, the correlation is 100 percent. If the two signals are not alike in any way, the correlation is 0. The correlation process in the mixer produces a signal that is averaged in the low-pass filter at the output of mixer 2. The output signal will be a high average value if the transmitted and received PSN codes are alike.

The signal out of mixer 2 is fed to a synchronization circuit, which must recreate the exact frequency and phase of the carrier so that demodulation can take place. The synchronization circuit varies the clock frequency so that the PSN code output frequency varies, seeking the same chip rate as the incoming signal. The clock drives a PSN code

Figure 11-41   A direct-sequence receiver.

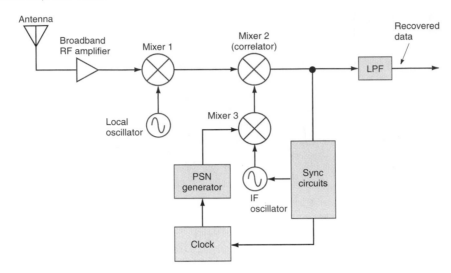

generator containing the exact code used at the transmitter. The PSN code in the receiver is the same as that of the received signal, but the two are out of sync with each other. Adjusting the clock by speeding it up or slowing it down eventually causes the two to come into synchronization.

The PSN code produced in the receiver is used to phase-modulate a carrier at the IF in mixer 3. Like all the other mixers, this one is usually of the double balanced diode ring type. The output of mixer 2 is a BPSK signal similar to that being received. It is compared to the received signal in mixer 3, which acts as a correlator. The output of mixer 3 is then filtered to recover the original serial binary data. The received signal is said to be *despread*.

Direct-sequence SS is also called *code-division multiple access (CDMA)*, or SS multiple access. The term *multiple access* applies to any technique that is used for multiplexing many signals on a single communication channel. CDMA is used in satellite systems so that many signals can use the same transponder. It is also widely used in cellular telephone systems, for it permits more users to occupy a given band than other methods.

Despread

Code-division multiple access (CDMA)

Multiple access

Benefits of Spread Spectrum

Spread spectrum is being used in more and more applications in data communication as its benefits are discovered and as new components and equipment become available to implement it.

- *Security.* SS prevents unauthorized listening. Unless a receiver has a very wide bandwidth and the exact pseudorandom code and type of modulation, it cannot intercept an SS signal.

- *Resistance to jamming and interference.* Jamming signals are typically restricted to a single frequency, and jamming one frequency does not interfere with an SS signal. Similarly, unintentional interference from a signal occupying the same band is greatly minimized and in most cases virtually eliminated.

- *Band sharing.* Many users can share a single band with little or no interference. (As more and more signals use a band, the background noise produced by the switching of many signals increases, but not enough to prevent highly reliable communication.)

- *Resistance to fading and multipath propagation.* Frequency-selective fading occurs during signal propagation because signals of different frequencies arrive at a receiver at slightly different times due to reflections from other objects. SS virtually eliminates wide variations of signal strength due to reflections and other phenomena during propagation.

- *Precise timing.* Use of the pseudorandom code in SS provides a way to precisely determine the start and end of a transmission. Thus SS is a superior method for radar and other applications that rely on accurate knowledge of transmission time to determine distance.

Orthogonal Frequency Division Multiplexing (OFDM)

Another wideband modulation method growing in popularity is called *OFDM*. Also known as *multicarrier modulation (MCM)*, this relatively new form of modulation was first proposed in the 1950s but was not seriously considered until the 1980s and early 1990s. OFDM was not widely implemented until the late 1990s because of its complexity and cost. Today, fast DSP chips make OFDM practical.

Orthogonal frequency-division multiplexing (OFDM)

Although OFDM is known as a modulation method as opposed to a multiplexing method, the term *frequency-division multiplexing* is appropriate because the method transmits data by simultaneously modulating segments of the high-speed serial bit stream onto multiple carriers spaced throughout the channel bandwidth. The carriers are frequency-multiplexed in the channel. The data rate on each channel is very low, making the symbol time much longer than predicted transmission delays. This technique spreads the signals over a wide bandwidth, making them less sensitive to the noise, fading, reflections, and multipath transmission effects common in microwave communication. Because of the very wideband nature of OFDM, it is considered to be a hybrid of spread spectrum.

Figure 11-42 shows the concept of an OFDM modem. The single serial data stream is divided into multiple slower but parallel data paths, each of which modulates a separate subcarrier. For example, a 10-Mbps data signal could be split into 1000 data signals of 10 kbps transmitted in parallel. A common format is to space the subcarriers equally across the channel by a frequency that is the reciprocal of the subcarrier symbol rate. In the situation described here, the spacing would be 10 kHz. This is what makes the carriers orthogonal. Orthogonal means that each carrier has an integer number of sine wave cycles in one bit period.

A plot of the bandwidth of each modulated carrier is the familiar $(\sin x)/x$ curve discussed in Chap. 2. (See Fig. 11-43.) Nulls occur at those points equal to the symbol rate. With this arrangement, all carriers lie at the null frequencies of the adjacent carriers.

Figure 11-42 Concept of OFDM.

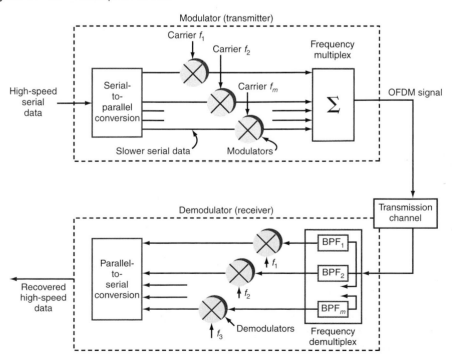

Figure 11-43   Subcarrier spectrum in OFDM.

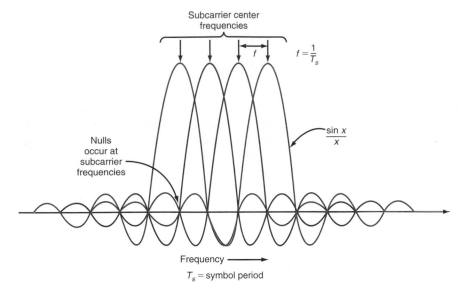

This permits simplified demultiplexing. Typically, BPSK, QPSK, or some form of QAM is used as the modulation method. In using QPSK or QAM, multiple bits per symbol are transmitted to permit a higher overall data rate. The subcarriers are algebraically added, and the resulting composite transmitted. Again referring to Fig. 11-42, note that the demodulator or receiver uses filters to separate out the individual subcarriers, and demodulators to recover the individual bit streams, which are then reassembled into the original serial data.

When tens, hundreds, or even thousands of subchannels are used, as is the case in modern systems, obviously traditional modulator, demodulator, and filter circuits are impractical because of size, complexity, and cost. However, all of these functions can be readily programmed into a fast DSP chip.

A simplified version of the process is shown in Fig. 11-44. At the transmitter or modulator, the serial data is modulated; then a serial-to-parallel conversion is performed. The inverse fast Fourier transform (IFFT) is then implemented. This process produces

Figure 11-44   Simplified processing scheme for OFDM in DSP.

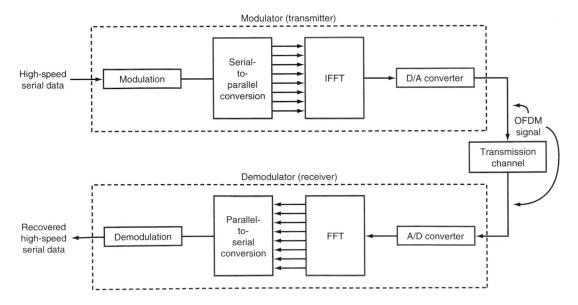

all the orthogonal subcarriers. The D/A converter converts the OFDM signal to analog form and transmits it over the communication medium.

At the receiver or demodulator, the OFDM signal is digitized by the A/D converter, and then an FFT is performed. Recall that an FFT essentially does a spectrum analysis of a time-domain signal. A sampled time-domain analog signal is translated to a frequency-domain plot of spectral content by the FFT. The receiver FFT DSP sorts out the subcarriers and demodulates the original data, which is then reassembled into the original high-speed data stream.

Like most other modems, the OFDM modem is a fast DSP chip programmed with all the mathematical algorithms that produce the functions defined by the blocks in Figs. 11-42 and 11-44.

Today, OFDM is widely used in wireless local-area networks (LANs). A version of OFDM for wired communication systems, known as *discrete multitone (DMT)*, used in ADSL modems is discussed later in this chapter. It is also the method chosen for transmitting high-quality audio in digital satellite radio broadcasting systems. More recently, it has been proposed as an alternative to the 8-VSB AM used in digital high-definition TV systems. OFDM is also used in the high speed version of wireless LANs (802.11 or Wi-Fi) and in the new broadband wireless system called WiMAX. It is also being considered for future cell phone systems. When the digital data to be transmitted is accompanied by some form of forward error correction (FEC) scheme (Trellis code, etc.), the method is called *coded OFDM* or *COFDM*.

11-6 Broadband Modem Techniques

A modem or modulator-demodulator is the circuitry used to translate a baseband signal, usually digital, to a higher transmission frequency that is better suited to the transmission medium. A good example is that digital data is not that compatible with the twisted-pair cable used in telephone systems. The bandwidth is too limited. However, modulating the data onto a carrier provides a way to transmit the data over a system originally designed for analog voice. Modems are used with all types of cables such as the telephone lines and the coaxial cable of cable TV. And modems can be of the radio variety where they are used to transmit data wirelessly. This section provides an overview of several types of popular modems.

Analog Telephone Modem

The most commonly used modem is one that connects personal computers to the telephone line. Figure 11-44 is a complete block diagram of a typical dial-up analog modem. It consists of both transmitter and receiver sections. Most modern modems are implemented by using digital signal processing techniques, and thus consist of only one or two chips.

Physically, modems were originally packaged on a single small printed-circuit board and designed to plug into the PC bus. The modem takes its power from the PC power supply, and an RJ-11 modular connector attaches the modem to the telephone line. No additional interfacing is needed.

Most modern PCs and laptop computers have a single IC modem built right into the motherboard of the computer. Earlier modems plugged into portable computers by way of a special interface called the PCMCIA or card bus.

Modem Operation. The data to be transmitted is stored in the computer's RAM. It is formatted there by the communication software installed with the computer. It is then sent 1 byte at a time to the modem. Most data transfers and operations inside a computer are parallel, but since long-distance data communication is done by using serial binary data, the modem's first job is to convert parallel data to serial data. This is done

Figure 11-45 General block diagram of a UART.

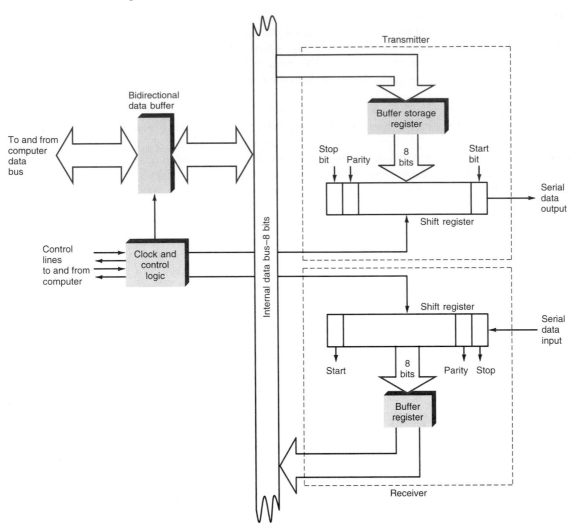

Universal asynchronous
receiver/transmitter (UART)

GOOD TO KNOW

Data compression circuits are
used in some modems because
the compressed message can be
transmitted more quickly.

with shift registers, as described earlier. It is usually carried out by a special large-scale
IC called a *universal asynchronous receiver/transmitter (UART),* a digital IC that performs parallel-to-serial conversion for transmission and serial-to-parallel conversion for
reception.

Figure 11-45 shows a general block diagram of a UART. Parallel data from the computer data bus is fed into and out of a bidirectional data buffer, which is usually a storage register with appropriate level-shifting circuits. The data, usually parallel 8 bits, is
then put on an internal data bus. Before transmission, this data is first stored in a buffer
storage register and then sent to a shift register. The internal circuitry adds start and stop
bits, which signal the beginning and end of the word, making modem operation strictly
asynchronous. The parity bit is used for error detection. A clock signal shifts the data
out serially, 1 bit at a time.

The receive section of the UART is at the bottom of Fig. 11-45. Serial data is shifted
into a shift register, where the start, stop, and parity bits are stripped off. The remaining
data is transferred to a buffer register, to the internal data bus, and through the bidirectional data buffer to the computer in parallel form. The clock and control logic circuits
in the UART control all internal shifting and data transfer operations under the direction
of control signals from the computer.

Referring again to Fig. 11-44, you can see that the serial data from the UART is passed through a scrambler circuit. This is not to encrypt the data for security, but simply to ensure that the data is random, with plenty of 0-to-1 and 1-to-0 transitions. This ensures that sufficient transitions occur so that clocking operations are reliable when long strings of binary 0s or 1s occur. Some circuits, especially adaptive equalizers and echo suppressors, rely on the randomness of the binary data for proper operation. The scrambler is a shift register with feedback that takes the serial data and scrambles it into a random binary pulse train.

The random serial data is sent to the modulator. The serial data, usually in NRZ form, modulates a low-frequency sine wave carrier inside the telephone bandwidth that extends from about 300 to 3000 Hz. Typical carrier frequencies are in the 1700-Hz range.

The output of the modulator is filtered to band-limit it and then fed to an *equalizer* circuit, which precompensates for the attenuation and distortion that the signal will receive as it is passed through the telephone system. Since the characteristics of the telephone are predictable (within limits), the equalizer processes the modulated signal in such a way to overcome attenuation and distortion so that it will appear normal at the receiving end. (The equalizer used in modems is what is called a *compromise equalizer;* it only partially corrects the problems because the exact extent of the distortion is not known.) The equalizer output is sent to the interface circuits that connect the signal to the telephone line.

During receive operations, the signal is picked off the telephone line, passed through the interface circuits, and fed to the receiver section. It first encounters an adaptive equalizer. The adaptive equalizer adjusts itself automatically to compensate for the amplitude attenuation and distortion of the signal. AGC circuits are normally used in the receiver to keep the gain constant over what can be a wide range of received signal amplitudes.

The signal is then demodulated, resulting in an NRZ serial digital signal. This is passed through a descrambler, which produces the opposite effect of the transmit scrambler. The descrambler output is the original serial data signal. This is sent to the UART, where it is translated to a parallel byte that the computer can store and use.

Although not shown in Fig. 11-44, data compression and decompression circuits are now being used in some modems. A digital signal processing (DSP) circuit takes the binary data and compresses it prior to modulation so that the transmitted signal is many bits less than the actual message length. Because fewer bits are transmitted in a given time, the overall transmission speed is higher. At the receiver, a DSP decompressor restores the shorter signal to its original length after it is demodulated.

Another feature not depicted in Fig. 11-44 is error detection and correction. All the newer modem types incorporate circuitry that can detect bit transmission errors and correct them as they occur. Such schemes, although complex and expensive, greatly increase the reliability of transmission, especially high-speed transmission in a noisy environment.

A modem is typically controlled by its own internal microprocessor-based controller. This miniature single-chip computer executes commands given through the communications software in the computer, which arrive over the computer bus. In external modems, the commands arrive serially, via the UART. The microcontroller lets the user tell the modem exactly what to do. It also implements automatic functions such as auto-dial and auto-answer operations, which most modems are capable of.

Modem Classification and Standards. The International Telecommunications Union (ITU), formerly called the CCITT, sponsors, negotiates, and maintains modem and many other communication standards. Modem standards are designated by a special V.xx symbol. The more common ones are described in Fig. 11-46.

Modems are usually capable of operating in several different V.xx modes. The modem will automatically adjust itself to the highest speed possible but will drop back to a lower speed or different mode if the receiving modem cannot handle the highest speed. Most modems in use today are the V.90 or V.92 type and are capable of speeds up to 56 kbps. In practice, the maximum speed is 53 kbps, but this is rarely achieved. Long telephone lines and noise force the modem to lower speeds to minimize errors. Typical speeds are in the 24-kbps to 50-kbps range.

Equalizer

GOOD TO KNOW

Modems are usually capable of operating in several different modes. The modem automatically adjusts itself to the highest speed possible but will drop back to a lower speed if the receiving modem cannot handle the higher speed.

Figure 11-46 Popular modem standards.

Bell 103 and 212A — Original Bell Telephone modems. The 103 used FSK at 300 bps. The 212A used PSK at 1200 bps.

V.22bis — First to use QAM. 2400 bps with fallback to 1200 bps.

V.24 — Same as EIA RS-232C/E serial interface standard.

V.32bis — 14,400 bps using TCQAM.

V.34 — Very popular and long-lived standard. TCQAM. 28,800 bps originally but extended to 33,600 bps. Automatic fallback to 31,200, 28,800, 26,400, 19,200, or 14,400 bps.

V.42 — An error correction standard used with dial-up modems.

V.42bis — An extension of V.42 that adds data compression that can further speed up transmission.

V.44 — An even better data compression standard.

V.90 — Current standard. Uses pulse-code modulation (PCM). Extends download speed to 53,000 bps (*not* 56,000 bps). Upload speed is 33,600 bps maximum.

V.92 — Most recent standard. Download speed remains at 53,000 bps maximum, but upload speed is increased to 48,000 bps. Shorter initial connection time during dial-up. Adds the ability to answer the telephone without losing the data connection.

xDSL Modems

Although the twisted-pair telephone line to the central office is normally said to have a maximum bandwidth of 4 kHz, the truth is that the bandwidth of this line varies with its length, and it can handle higher frequencies than expected. Because of the line characteristics, the higher frequencies are greatly attenuated. However, by transmitting the higher frequencies at higher voltage levels and using line compensation techniques, it is possible to achieve higher data rates than the 53 kbps of a standard analog dial-up modem. New modulation methods also permit previously unachievable line rates. The digital subscriber line (DSL) describes a set of standards set by the International Telecommunications Union that greatly extend the speed potential of the common twisted-pair telephone lines. In the term *xDSL*, the *x* designates one of several letters that define a specific DSL standard.

> xDSL modem

The most widely used form of DSL is called *asymmetric digital subscriber line (ADSL)*. This system permits downstream data rates up to 8 Mbps and upstream rates up to 640 kbps using the existing telephone lines. (*Asymmetric* means unequal upstream and downstream rates.)

> Asymmetric digital subscriber line (ADSL)
>
> Asymmetric

The connection between a telephone subscriber and the nearest telephone central office is twisted-pair cable using size 24 or 26 copper wire. Its length is usually anywhere between 9000 and 18,000 ft (2.7 to 5.5 km). This cable acts as a low-pass filter. Its attenuation to very high frequencies is enormous. Digital signals are seriously delayed and distorted by such a line. For this reason, only the lower 0- to 4-kHz bandwidth is used for voice. Traditional modems operate in the center of this voice band at 1700 to 1800 Hz using QAM.

ADSL employs some special techniques so that more of the line bandwidth can be used to increase data rates. Even though a 1-MHz signal may have an attenuation of up to 90 dB on an 18,000-ft line, special amplifiers and frequency compensation techniques make the line usable.

> Discrete multitone (DMT)

The modulation scheme used with ADSL modems is called *discrete multitone (DMT)*, another name for OFDM, discussed earlier in this chapter. It divides the upper frequency

Figure 11-47 Spectrum of telephone line used by ADSL.

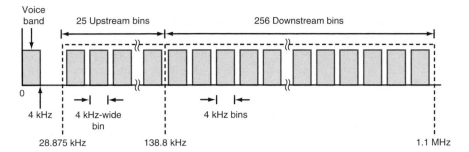

spectrum of the telephone line into 256 channels, each 4 kHz wide. See Fig. 11-47. Each channel, called a *bin,* is designed to transmit at speeds up to 15 kbps/Bd or 60 kbps.   **Bin**

Each channel contains a carrier that is simultaneously phase-amplitude-modulated (QAM) by some of the bits to be transmitted. The serial data stream is divided up so that each carrier transmits some of the bits. All bits are transmitted simultaneously. Also, all the carriers are frequency-multiplexed into the line bandwidth above the normal voice telephone channel, as Fig. 11-47 shows.

The upstream signal uses the 4-kHz bins from 25.875 to 138.8 kHz, and the down-stream signal uses bins in the 138-kHz to 1.1-MHz range.

The number of bits per baud and the data rate per bin vary according to the noise on the line. The less noise there is in each bin, the higher the data rate. Very noisy bins will carry few or no bits, whereas quiet bins can accommodate the maximum 15 kbps/Bd or 60 kbps.

This system is very complex and is implemented with a digital signal processor. The DSP chip handles all modulation and demodulation functions by simulating them digitally.

Figure 11-48 shows an ADSL modem. All DMT/OFDM modulation/demodulation is handled by the DSP chip. The digital output of the DSP is converted to analog by the D/A converter. The resulting signal is amplified, filtered, and sent to a line driver that applies a high-level signal to the line. The hybrid is a circuit or transformer that permits simultaneous transmit and receive operations on the telephone line. The transformer matches the circuit impedance to the line.

In the receive mode, the incoming DMT analog signal is amplified, filtered, and applied to a PGA for AGC. The signal is digitized by the A/D converter and applied to the DSP for recovery of the digital data.

Figure 11-48 ADSL modem—block diagram.

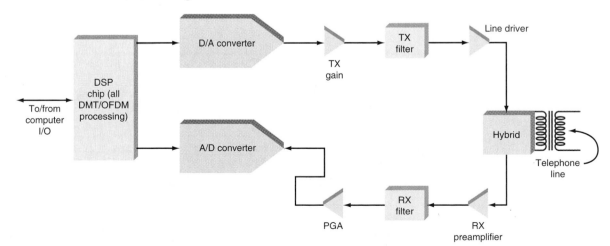

Several different levels of ADSL are available. The data rate for each depends upon the length of the subscriber twisted-pair cable. The shorter the cable, the higher the data rate. The highest standard rate is 6.144 Mbps downstream and 576 kbps upstream at a line distance not to exceed 9000 ft. The minimum rate is 1.536 Mbps downstream and 384 kbps upstream at line distances up to 18,000 ft. This is the most common form of ADSL.

ADSL is available in most large cities. Small and midsize cities many not have it yet. Many areas in large cities cannot support ADSL because of line lengths over 18,000 ft and high noise conditions. Most PC users who have access to ADSL are thrilled with the higher performance over conventional dial-up modems. ADSL is the widely used form of high-speed Internet access throughout the world. ADSL is in second place to cable TV modem broadband access.

Other forms of DSL have also been defined. Few of these are actually in use, but development continues for future applications. The most popular new versions are G.shdsl and VDSL. G.shdsl, or SHDSL, has fully symmetric up and down speeds from 192 kbps to 6 Mbps on twisted pair. It uses trellis-coded pulse-amplitude modulation (TC-PAM). The speed progressively degrades with distance down to 192 kbps at 20,000 ft. This version of DSL is expected to be popular with small businesses.

Two of the most recent versions of ADSL are ADSL2 and ADSL2+. ADSL2 extends the upper download speed to the 8- to 12-Mbps range at a distance of about 8000 ft. ADSL2+ further boosts speeds to 20 Mbps at a distance to about 4000 ft. Some newer standards referred to as bonding standards make use of two twisted pairs in the telephone cable to carry parallel data streams that effectively double the data rate for a given distance.

VDSL, or very high-speed DSL, offers a data rate of up to 52 Mbps one way (download) or 26 Mbps fully symmetrical using QAM. VDSL permits digital video to be transmitted and thus offers an alternative to cable TV systems. However, to get this speed, the twisted-pair length is limited to 1000 ft or less at 52 Mbps and less than 3500 ft at 26 Mbps. VDSL is expected to be popular in multitenant units (MTUs) such as office and apartment buildings and hotels.

Cable Modems

Many cable TV systems are set up to handle high-speed digital data transmission. The digital data is used to modulate a high-frequency carrier that is frequency-multiplexed onto the cable that also carries the TV signal.

Cable TV systems use a hybrid fiber coaxial (HFC) cable with a bandwidth of approximately 750 MHz. This spectrum is divided into 6-MHz-wide channels for TV signals. The standard VHFs and UHFs normally assigned to wireless TV are used on the cable, along with some special cable frequencies. The TV signals are therefore frequency-division-multiplexed onto the cable.

Figure 11-49 shows the spectrum of the cable. Television channels extend from 50 MHz (Channel 2) up to 550 MHz. In this 500 MHz of bandwidth, up to 83 channels of 6 MHz can be accommodated.

Figure 11-49   Cable TV spectrum showing upstream and downstream data channels on the cable.

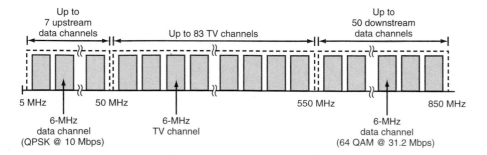

The spectrum above the TV channels, from 550 to 850 MHz, is available for digital data transmission. Standard 6-MHz channels are used, giving approximately 50 channels. These channels are used for downstream data transmission (from the remote computer down to the user).

The spectrum from 5 to 50 MHz, as you can see from Fig. 11-49, is divided into seven 6-MHz channels that are used for upstream data transmission (from the user up to the server). This frequency range may be 5 to 42 MHz in some systems or 5 to 65 MHz in other systems.

Cable modems use 64-QAM for downstream data. Using 64-QAM in a 6-MHz channel provides a data rate up to 31.2 Mbps. This method of modulation uses 64 different phase-amplitude combinations (symbols) to represent multiple bits. Because each channel is shared by multiple users, the 31.2-Mbps rate is not achieved. Typical rates are in the 500-kbps to 2-Mbps range for downloads. In some systems, 256-QAM is available to provide a maximum data rate of 41.6 Mbps in a 6-MHz channel. Higher subscriber download speeds can then be achieved.

Standard QPSK is used in the upstream channels to achieve a data rate of about 10 Mbps. With multiple users, an upstream rate of 300 to 400 kbps is normal.

Figure 11-50 shows a typical cable modem. It is basically a VHF/UHF receiver connected to the cable for downloads and a modulator/transmitter for uploads. The signal from the cable passes through the diplexer, which is a filter circuit that permits simultaneous transmit and receive operations. The signal is amplified and mixed with a local oscillator signal from the frequency synthesizer to produce an IF signal. The frequency synthesizer selects the cable channel. The IF signal is demodulated to recover the data. Reed Solomon error detection circuitry (see Sec. 11-7) finds and corrects any bit errors. The digital data then goes to an Ethernet interface to the PC. Ethernet is a popular networking system to be discussed in Chap. 12.

For transmission, the data from the computer is passed through the interface, where it is encoded for error detection. The data then modulates a carrier that is up-converted by the mixer to the selected upstream channel before being amplified and passed through the diplexer to the cable.

Cable modems provide significantly higher data rates than can be achieved over the standard telephone system. The primary limitation is the existence or availability of a cable TV system that offers such data transmission services.

Cable modem standards are set by an industry consortium called Cable Labs. The specification is referred to as the Data over Cable Service Interface Specification (DOCSIS).

Figure 11-50 Cable modem block diagram

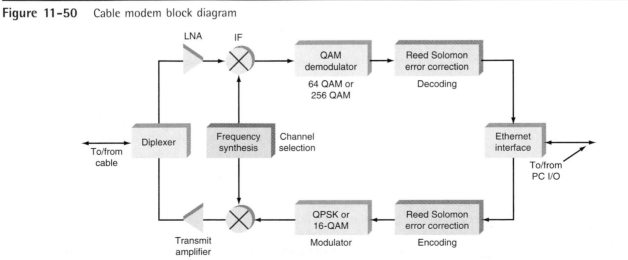

11-7 Error Detection and Correction

When high-speed binary data is transmitted over a communication link, whether it is a cable or radio, errors will occur. These errors are changes in the bit pattern caused by interference, noise, or equipment malfunctions. Such errors will cause incorrect data to be received. To ensure reliable communication, schemes have been developed to detect and correct bit errors.

Bit error rate (BER)

The number of bit errors that occur for a given number of bits transmitted is referred to as the *bit error rate (BER)*. The bit error rate is similar to a probability in that it is the ratio of the number of bit errors to the total number of bits transmitted. If there is 1 error for 100,000 bits transmitted, the BER is $1:100,000 = 10^{-5}$. The bit error rate depends on the equipment, the environment, and other considerations. The BER is an average over a very large number of bits. The BER for a given transmission depends on specific conditions. When high speeds of data transmission are used in a noisy environment, bit errors are inevitable. However, if the *S/N* ratio is favorable, the number of errors will be extremely small. The main objective in error detection and correction is to maximize the probability of 100 percent accuracy.

Example 11-3

Data is transmitted in 512-byte blocks or packets. Eight sequential packets are transmitted. The system BER is $2:10,000$ or 2×10^{-4}. On average, how many errors can be expected in this transmission?

$$8 \text{ packets} \times 512 \text{ bytes} = 4096 \text{ bytes}$$
$$4096 \text{ bytes} \times 8 \text{ bits} = 32,768 \text{ bits}$$
$$32,768/10,000 = 3.2768 \text{ sets of 10,000 bits}$$
$$\text{Average number of errors} = 2 \times 3.2768 = 6.5536$$

The process of error detection and correction involves adding extra bits to the data characters to be transmitted. This process is generally referred to as channel encoding. The data to be transmitted is processed in a way that creates the extra bits and adds them to the original data. At the receiver, these extra bits help in identifying any errors that occur in transmission caused by noise or other channel effects.

A key point about channel encoding is that it takes more time to transmit the data because of the extra bits. For example, to transmit 1 byte of data, the encoding process may add 3 extra bits for a total of 11 bits. These extra bits are called *overhead* in that they extend the time of transmission. If the bit time is 100 ns, then it takes 800 ns to send the data bits but 1100 ns to send the encoded data. The extra overhead bits add 37.5 percent more time to the transmission. As a result, to maintain a desired data rate, the overall clock speed must be increased or the lower data rate must be accepted. While the extra bits decrease the overall efficiency of transmission, remember that the benefit is more reliable data transmission because errors can be detected and/or corrected. Speed is traded off for higher-quality data transmission.

Channel encoding methods fall into to two separate categories, error detection codes and error correction codes. Error detection codes just detect the errors but do not take any corrective action. They simply let the system know that an error has occurred. Typically, these codes simply initiate retransmission until the data is received correctly. The other form of channel coding is error correction or forward error correction (FEC). These coding schemes eliminate the time-wasting retransmission and actually initiate self-correcting action.

Error Detection

Many different methods have been used to ensure reliable error detection, including redundancy, special coding and encoding schemes, parity checks, block checks, and cyclical redundancy check.

Redundancy. The simplest way to ensure error-free transmission is to send each character or each message multiple times until it is properly received. This is known as *redundancy*. For example, a system may specify that each character be transmitted twice in succession. Entire blocks or messages can be treated in the same way. These retransmission techniques are referred to as automatic repeat request (ARQ).

Redundancy

Encoding Methods. Another approach is to use an encoding scheme like the RZ-AMI described earlier, whereby successive binary 1 bits in the bit stream are transmitted with alternating polarity. If an error occurs somewhere in the bit stream, then 2 or more binary 1 bits with the same polarity are likely to be transmitted successively. If the receiving circuits are set to recognize this characteristic, single bit errors can be detected.

The turbo codes and trellis codes are another example of the use of special coding to detect errors. These codes develop unique bit patterns from the data. Since many bit patterns are invalid in trellis and turbo codes, if a bit error occurs, one of the invalid codes will appear, signaling an error that can then be corrected. These codes are covered later in this section.

Parity. One of the most widely used systems of error detection is known as *parity*, in which each character transmitted contains one additional bit, known as a *parity bit*. The bit may be a binary 0 or binary 1, depending upon the number of 1s and 0s in the character itself.

Parity

Two systems of parity are normally used, odd and even. *Odd parity* means that the total number of binary 1 bits in the character, including the parity bit, is odd. *Even parity* means that the number of binary 1 bits in the character, including the parity bit, is even. Examples of odd and even parity are indicated below. The seven left-hand bits are the ASCII character, and the right-hand bit is the parity bit.

Odd parity:	10110011
	00101001
Even parity:	10110010
	00101000

The parity of each character to be transmitted is generated by a parity generator circuit. The parity generator is made up of several levels of exclusive OR (X-OR) circuits, as shown in Fig. 11-51. Normally the parity generator circuit monitors the shift

Figure 11-51 A parity generator circuit.

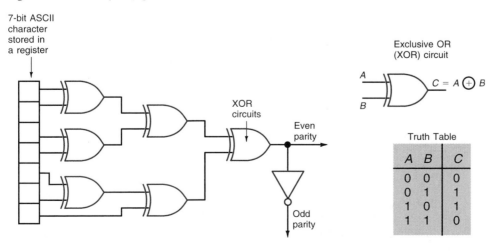

Figure 11-52 How parity is transmitted.

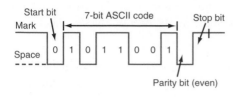

Figure 11-53 Parity checking at the receiver.

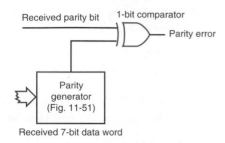

register in a UART in the computer or modem. Just before transmitting the data in the register by shifting it out, the parity generator circuit generates the correct parity value, inserting it as the last bit in the character. In an asynchronous system, the start bit comes first, followed by the character bits, the parity bit, and finally one or more stop bits (see Fig. 11-52).

At the receiving modem or computer, the serial data word is transferred into a shift register in a UART. A parity generator in the receiving UART produces the parity on the received character. It is then compared to the received parity bit in an XOR circuit, as shown in Fig. 11-53. If the internally generated bit matches the transmitted and received parity bit, it is assumed that the character was transmitted correctly. The output of the XOR will be 0, indicating no error. If the received bit does not match the parity bit generated from the received data word, the XOR output will be 1, indicating an error. The system signals the detection of a parity error to the computer. The action taken will depend on the result desired: The character may be retransmitted, an entire block of data may be transmitted, or the error may simply be ignored.

Vertical redundancy check (VRC)

The individual-character parity method of error detection is sometimes referred to as the *vertical redundancy check (VRC)*. To display characters transmitted in a data communication system, the bits are written vertically (see Fig. 11-54). The bit at the bottom is the parity, or VRC, bit for each vertical word. Horizontal redundancy checks are discussed later.

Parity checking is useful only for detecting single-bit errors. If two or more bit errors occur, the parity circuit may not detect it. If an even number of bit changes occur, the parity circuit will not give a correct indication.

Cyclical redundancy check (CRC)

Cyclical Redundancy Check. The *cyclical redundancy check (CRC)* is a mathematical technique used in synchronous data transmission that effectively catches 99.9 percent or more of transmission errors. The mathematical process implemented by CRC is essentially a division. The entire string of bits in a block of data is considered to be one giant

Figure 11-54 Vertical and horizontal redundancy checks.

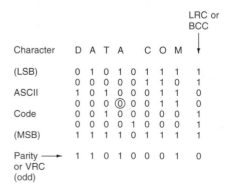

Character	D	A	T	A		C	O	M	LRC or BCC
(LSB)	0	1	0	1	0	1	1	1	1
	0	0	0	0	0	1	1	0	1
ASCII	1	0	1	0	0	0	1	1	0
	0	0	0	0	0	0	1	1	0
Code	0	0	1	0	0	0	0	0	1
	0	0	0	0	1	0	0	0	1
(MSB)	1	1	1	1	0	1	1	1	1
Parity → or VRC (odd)	1	1	0	1	0	0	0	1	0

binary number which is divided by some preselected constant. CRC is expressed by the equation

$$\frac{M(x)}{G(x)} = Q(x) + R(x)$$

where $M(x)$ is the binary block of data, called the *message function,* and $G(x)$ is the generating function. The *generating function* is a special code that is divided into the binary message string. The outcome of the division is a quotient function $Q(x)$ and a remainder function $R(x)$. The quotient resulting from the division is ignored; the remainder is known as the CRC character and is transmitted along with the data.

For convenience of calculation, the message and generating functions are usually expressed as an algebraic polynomial. For example, assume an 8-bit generating function of 10000101. The bits are numbered such that the LSB is 0 and the MSB is 7.

$$
\begin{array}{cccccccc}
7 & 6 & 5 & 4 & 3 & 2 & 1 & 0 \\
1 & 0 & 0 & 0 & 0 & 1 & 0 & 1 \\
\end{array}
$$

The polynomial is derived by expressing each bit position as a power of x, where the power is the number of the bit position. Only those terms in which binary 1s appear in the generating function are included in the polynomial. The polynomial resulting from the above number is

$$G(x) = x^7 + x^2 + x^0 \qquad \text{or} \qquad G(x) = x^7 + x^2 + 1$$

The CRC mathematical process can be programmed by using a computer's instruction set. It can also be computed by a special CRC hardware circuit consisting of several shift registers into which XOR gates have been inserted at specific points (see Fig. 11-55). The data to be checked is fed into the registers serially. There is no output, since no output is retained. The data is simply shifted in 1 bit at a time; when the data has all been transmitted, the contents will be the remainder of the division, or the desired CRC character. Since a total of 16 flip-flops are used in the shift register, the CRC is 16 bits long and can be transmitted as two sequential 8-bit bytes. The CRC is computed as the data is transmitted, and the resulting CRC is appended to the end of the block. Because CRC is used in synchronous data transmission, no start and stop bits are involved.

At the receiving end, the CRC is computed by the receiving computer and compared to the received CRC characters. If the two are alike, the message has been correctly received. Any difference indicates an error, which triggers retransmission or some other form of corrective action. CRC is probably the most widely used error detection scheme in synchronous systems. Both 16- and 32-bit CRCs are used. Parity methods are used primarily in asynchronous systems.

Figure 11-55 A CRC error detection circuit made with a 16-bit shift register and XOR gates.

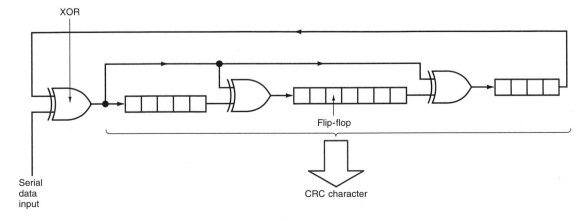

Error Correction

Forward error correction (FEC)

As stated previously, the easiest way to correct transmission errors—to retransmit any character or block of data that has an error in it—is time-consuming and wasteful. A number of efficient error correction schemes have been devised to complement the parity and BCC methods described above. The process of detecting and correcting errors at the receiver so that retransmission is not necessary is called *forward error correction (FEC)*. There are two basic types of FEC: block codes and convolutional codes.

Longitudinal redundancy check

Block–Check Character. The block-check character (BCC) is also known as a *horizontal* or *longitudinal redundancy check (LRC)*. It is the process of logically adding, by exclusive-ORing, all the characters in a specific block of transmitted data. Refer to Fig. 11-54. To add the characters, the top bit of the first vertical word is exclusive-ORed with the top bit of the second word. The result of this operation is exclusive-ORed with the top bit of the third word, and so on until all the bits in a particular horizontal row have been added. There are no carries to the next bit position. The final bit value for each horizontal row then becomes one bit in a character known as the *block-check character (BCC)*, or the *block-check sequence (BCS)*. Each row of bits is done in the same way to produce the BCC. All the characters transmitted in the text, as well as any control or other characters, are included as part of the BCC. Exclusive-ORing all bits of all characters is the same as binary addition without a carry of the codes.

Block-check character (BCC)

Block-check sequence (BCS)

The BCC is computed by circuits in the computer or modem as the data is transmitted, and its length is usually limited to 8 bits, so that carries from one bit position to the next are ignored. It is appended to the end of a series of bytes that make up the message to be transmitted. At the receiving end, the computer computes its own version of the BCC on the received data and compares it to the received BCC. Again, the two should be the same.

When both the parity on each character and the BCC are known, the exact location of a faulty bit can be determined. The individual character parity bits and the BCC bits provide a form of coordinate system that allows a particular bit error in one character to be identified. Once it is identified, the bit is simply complemented to correct it. The VRC identifies the character containing the bit error, and the LRC identifies the bit that contains the error.

Assume, e.g., that a bit error occurs in the fourth vertical character from the left in Fig. 11-54. The fourth bit down from the top should be 0, but because of noise, it is received as a 1. This causes a parity error. With odd parity and a 1 in the fourth bit, the parity bit should be 0, but it is a 1.

Next, the logical sum of the bits in the fourth horizontal row from the top will be incorrect because of the bit error. Instead of 0 it will be 1. All the other bits in the BCC will be correct. It is now possible to pinpoint the location of the error because both the vertical column where the parity error occurred and the horizontal row where the BCC error occurred are known. The error can be corrected by simply complementing (inverting) the bit from 1 to 0. This operation can be programmed in software or implemented in hardware. One important characteristic of BCC is that multiple errors may not be detected. Therefore, more sophisticated techniques are needed.

Hamming bits

Hamming code

Hamming Code. A popular FEC is the Hamming code. Hamming was a researcher at Bell Labs who discovered that if extra bits were added to a transmitted word, these extra bits could be processed in such a way that bit errors could be identified and corrected. These extra bits, like several types of parity bits, are known as *Hamming bits* and together they form a *Hamming code*. To determine exactly where the error is, a sufficient number of bits must be added. The minimum number of Hamming bits is computed with the expression

$$2^n \geq m + n + 1$$

where m = number of bits in data word
n = number of bits in Hamming code

Assume, e.g., an 8-bit character word and some smaller number of Hamming bits (say, 2). Then

$$2^n \geq m + n + 1$$
$$2^2 \geq 8 + 2 + 1$$
$$4 \geq 11$$

Two Hamming bits are insufficient, and so are three. When $n = 4$,

$$2^4 \geq 8 + 4 + 1$$
$$16 \geq 13$$

Thus 4 Hamming bits must be transmitted along with the 8-bit character. Each character requires $8 + 4 = 12$ bits. These Hamming bits can be placed anywhere within the data string. Assume the placement shown below, where the data bits are shown as a 0 or 1 and the Hamming bits are designated with an H. The data word is 01101010.

12	11	10	9	8	7	6	5	4	3	2	1
H	0	1	H	1	0	H	1	0	H	1	0

One way to look at Hamming codes is simply as a more sophisticated parity system, where the Hamming bits are parity bits derived from some of but not all the data bits. Each Hamming bit is derived from different groups of the data bits. (Recall that parity bits are derived from the data by XOR circuits.) One technique used to determine the Hamming bits is discussed below.

At the transmitter, a circuit is used to determine the Hamming bits. This is done by first expressing the bit positions in the data word containing binary 1s as a 4-bit binary number ($n = 4$ is the number of Hamming bits). For example, the first binary 1 data bit appears at position 2, so its position code is just the binary code for 2, or 0010. The other data bit positions with a binary 1 are $5 = 0101$, $8 = 1000$, and $10 = 1010$.

Next, the transmitter circuitry logically adds (XORs) these codes.

Position code 2	0010
Position code 5	0101
XOR sum	0111
Position code 8	1000
XOR sum	1111
Position code 10	1010
XOR sum	0101

This final sum is the Hamming code bits from left to right. Position code 12 is 0, position code 9 is 1, position code 6 is 0, and position code 3 is 1. These bits are interested in their proper position. The complete 12-bit transmitted word is

12	11	10	9	8	7	6	5	4	3	2	1
H	0	1	H	1	0	H	1	0	H	1	0

12	11	10	**9**	8	7	**6**	5	4	**3**	2	1
0	0	1	**1**	1	0	**0**	1	0	**1**	1	0

The Hamming bits are shown in boldface type.

Now assume that an error occurs in bit position 10. The binary 1 is received as a binary 0. The received word is

12	11	10	**9**	8	7	**6**	5	4	**3**	2	1
0	0	1	**1**	1	0	**0**	1	0	**1**	1	0

The receiver recognizes the Hamming bits and treats them as a code word, in this case 0101. The circuitry then adds (XORs) this code with the bit number of each position in the word containing a binary 1, positions 2, 5, and 8.

The Hamming code is then added to the binary numbers representing each position with a 1.

Hamming code	0101
Position code 2	0010
XOR sum	0111
Position code 5	0101
XOR sum	0010
Position code 8	1000
XOR sum	1010

This final sum is a code that identifies the bit position of the error, in this case bit 10 (1010). To correct the bit, it is simply complemented from 0 to 1. If there are no bits in error, then the XOR sum at the receiver will be zero.

Note that the Hamming code method does not work if an error occurs in one of the Hamming bits itself.

For the Hamming code method of error detection and correction to work when 2 or more bit errors occur, more Hamming bits must be added. This increases overall transmission time, storage requirements at the transmitter and receiver, and the complexity of the circuitry. The benefit, of course, is that errors are reliably detected at the sending end. The transmitter never has to resend the data, which may in fact be impossible in some applications. Not all applications require such rigid data correction practices.

Reed Solomon (RS) code

Reed Solomon Code. One of the most widely used forward error correction codes is the *Reed Solomon (RS) code*. Like Hamming codes, it adds extra parity bits to the block of data being transmitted. It uses a complex mathematical algorithm beyond the scope of this book to determine the codes. The beauty of the RS code is that it permits multiple errors to be detected and corrected. For example, a popular form of the RS code is designated RS (255,223). A block of binary data contains a total of 255 bytes; 223 bytes is the actual data and 32 bytes is parity bits computed by the RS algorithm. With this arrangement, the RS code can detect and correct errors in up to 16 corrupted bytes. An RS encoder is used on the data to be transmitted. At the receiver, the recovered data is sent to an RS decoder that corrects any errors. The encoders and decoders can be implemented with software, but hardware ICs are also available. Some common applications of the RS FEC are in music and data compact disks (CDs), cell phones, digital TV, satellite communication, and xDSL and cable TV modems.

Interleaving. Interleaving is a method used in wireless systems to reduce the effects of burst errors. Most errors in wireless transmission are caused by bursts of noise that destroy a single bit or multiple sequential bits. If we take the bits and interleave them, we have a better chance of recognizing and recovering the lost bits.

One common way of doing this is to first use an error-correcting scheme such as the Hamming code to encode the data. The data and the Hamming bits are then stored in consecutive memory locations. For example, assume four 8-bit words consisting of the data and the Hamming bits. The data words if transmitted sequentially would look like this.

12345678   12345678   12345678   12345678

Then instead of transmitting the encoded words one at a time, all the first bits of each word are transmitted, followed by all the second bits, followed by all the third bits, and so on. The result would look like this.

1111   2222   3333   4444   5555   6666   7777   8888

Now if a burst error occurs, the result may look like this.

1111   2222   3333   4444   5555   4218   7777   8888

436   *Chapter 11*

At the receiver, the de-interleaving circuits would attempt to recreate the original data, producing

$$12345478 \qquad 12345278 \qquad 12345178 \qquad 12345878$$

Now with only 1 bit in each word in error, a Hamming decoder would detect and correct the bit.

Convolutional Codes

Convolutional encoding creates additional bits from the data as do Hamming and Reed Solomon codes, but the encoded output is a function of not only the current data bits but also previously occurring data bits. Like other forms of FEC, the encoding process adds extra bits that are derived from the data itself. This is a form of redundancy that leads to greater reliability in the transmission of the data. Even if errors occur, the redundant bits allow the errors to be corrected.

Convolutional codes are beyond the scope of this text, but essentially what they do is to pass the data to be transmitted through a special shift register like that shown in Fig. 11-56. As the serial data is shifted through the shift register flip-flops, some of the flip-flop outputs are XORed together to form two outputs. These two outputs are the convolutional code, and this is what is transmitted. There are numerous variations of this scheme, but note in every case the original data itself is not transmitted. Instead, two separate streams of continuously encoded data are sent. Since each output code is different, the original data can more likely be recovered at the receiver by an inverse process. One of the more popular convolutional codes is the trellis code which is widely used in dial-up computer modems. The Viterbi code is another that is widely used in high-speed data access via satellites.

Another type of convolutional code uses feedback. These are called recursive codes because the output of the shift register is combined with the input code to produce the output streams. Figure 11-57 is an example. Recursion means taking the output from a

Figure 11-56 Convolutional encoding uses a shift register with exclusive-OR gates to create the output.

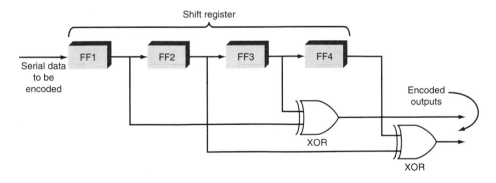

Figure 11-57 A convolutional encoder using recursion.

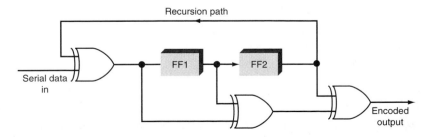

Figure 11-58 One form of turbo encoding.

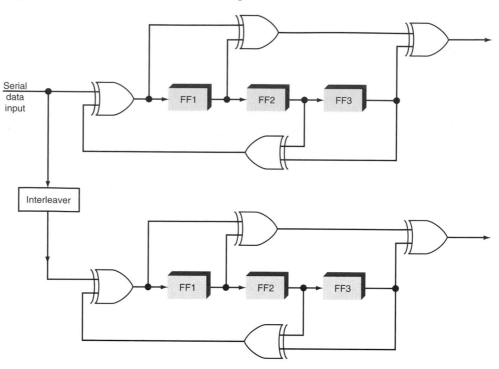

process and applying it back to the input. This procedure has been further developed to create a new class of convolutional codes called turbo codes. The turbo code is a combination of two concurrent recursive coding processes where one is derived directly from the data and the other is derived from the data that has been interleaved first. See Fig. 11-58. The result is a far more robust FEC that catches virtually all errors. Most forms of wireless data transmission today use some form of convolutional coding to ensure the robustness of the transmission path.

11-8 Protocols

Protocol

Protocols are rules and procedures used to ensure compatibility between the sender and receiver of serial digital data regardless of the hardware and software being used. They are used to identify the start and end of a message, identify the sender and the receiver, state the number of bytes to be transmitted, state a method of error detection, and for other functions. Various protocols, and various levels of protocols, are used in data communication.

The simplest form of protocol is the asynchronous transmission of data using a start bit and a stop bit (refer to Fig. 11-6) framing a single character, with a parity bit between the character bit and the stop bit. The parity bit is part of the protocol, but may or may not be used. In data communication, however, a message is more than one character. As discussed previously, it is composed of blocks, groups of letters of the alphabet, numbers, punctuation marks, and other symbols in the desired sequence. In synchronous data communication applications, the block is the basic transmission unit.

To identify a block, one or more special characters are transmitted prior to the block and after the block. These additional characters, which are usually represented by 7- or 8-bit codes, perform a number of functions. Like the start and stop bits on a character, they signal the beginning and end of the transmission. But they are also used to identify a specific block of data, and they provide a means for error checking and detection.

Figure 11-59 The handshaking process in data communication.

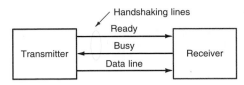

Some of the characters at the beginning and end of each block are used for handshaking purposes. These characters give the transmitter and receiver status information. Figure 11-59 illustrates the basic *handshaking* process. For example, a transmitter may send a character indicating that it is ready to send data to a receiver. Once the receiver has identified that character, it responds by indicating its status, e.g., by sending a character representing "busy" back to the transmitter. The transmitter will continue to send its ready signal until the receiver signals back that it is not busy, or is ready to receive.

Handshaking

At that point, data transmission takes place. Once the transmission is complete, some additional handshaking takes place. The receiver acknowledges that it has received the information. The transmitter then sends a character indicating that the transmission is complete, which is usually acknowledged by the receiver.

A common example of the use of such control characters is the XON and XOFF protocol used between a printer and a computer. XON is usually the ASCII character DC1, and XOFF is the ASCII character DC3. A printer that is ready and able to receive data will send XON to the computer. If the printer is not able to receive data, it sends XOFF. When the computer detects XOFF, it immediately stops sending data until XON is again received.

Asynchronous Protocols

Three popular protocols used for asynchronous ASCII-coded data transmission between personal computers, via modem, are Xmodem, Kermit, and MPN.

Xmodem. In *Xmodem,* the data transmission procedure begins with the receiving computer transmitting a negative acknowledge (NAK) character to the transmitter. NAK is a 7-bit ASCII character that is transmitted serially back to the transmitter every 10 s until the transmitter recognizes it. Once the transmitter recognizes the NAK character, it begins sending a 128-byte block of data, known as a *frame* (packet) of information (see Fig. 11-60). The frame begins with a *start-of-header (SOH) character,* which is another ASCII character meaning that the transmission is beginning. This is followed by a header, which usually consists of two or more characters preceding the actual data block which give auxiliary information. In Xmodem, the header consists of 2 bytes designating the block number. In most messages, several blocks of data are transmitted and each is numbered sequentially. The first byte is the block number in binary code. The second byte is the complement of the block number; i.e., all bits have been inverted. Then the 128-byte block is transmitted. At the end of the block, the transmitting computer sends a check-sum byte, which is the BCC, or binary sum of all the binary information sent in the block. (Keep in mind that each character is sent along with its start and stop bits, since Xmodem is an asynchronous protocol.)

Xmodem protocol

Frame

Figure 11-60 Xmodem protocol frame.

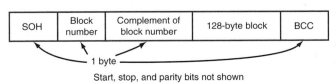

Start, stop, and parity bits not shown

Figure 11–61 Kermit asynchronous protocol.

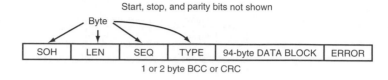

The receiving computer looks at the block data and also computes the check sum. If the check sum of the received block is the same as that transmitted, it is assumed that the block was received correctly. If the block was received correctly, the receiving computer sends an *acknowledge (ACK) character*—another ASCII code—back to the transmitter. Once ACK is received by the transmitter, the next block of data is sent. When a block has been received incorrectly because of interference or equipment problems, the check sums will not match and the receiving computer will send a NAK code back to the transmitter. A transmitter that has received NAK automatically responds by sending the block again. This process is repeated until each block, and the entire message, has been sent without errors.

When the entire message has been sent, the transmitting computer sends an *end-of-transmission (EOT) character*. The receiving computer replies with an ACK character, terminating the communication.

Kermit. Another popular asynchronous protocol is *Kermit* (see Fig. 11-61). The transmission begins with an SOH character followed by a length (LEN) character, which tells how long the block of data is. A block can be up to 94 bytes long. Next is a packet sequence number (SEQ). There can be up to 63 blocks, and these are given a sequence number so that both transmitter and receiver can keep track of long messages.

Next in the packet is a data-type (TYPE) designator. This byte may contain an ASCII control code such as ACK, NAK, EOT, EOF, or any one of a number of special codes used in the send-receive handshaking process.

The data block, which can be up to 94 bytes long, comes next. At the end of the packet is an error detection code. It can be a 1-byte check sum, a 2-byte check sum, or a 16-bit CRC.

Kermit is a very reliable protocol, since it requires that every packet sent be acknowledged by the receiver as being read correctly.

MNP. *Microcom Networking Protocols (MNPs)* are a series of protocols developed by the manufacturer Microcom to be used with asynchronous modems. They specify ways to handle error detection and correction and how to specify whether or not data compression is used. There are 10 classes of protocols. Not all modems support these protocols, but in recent years, most manufacturers have adopted them. They are reasonably easy to implement because they can be programmed into the control microcomputer used in most modems. The Microcom protocols are subdivided into classes.

The MNP class 3 protocol strips the start and stop bits from the characters in the message, which speeds up the transmission by a factor of about 8. The receiving modem puts the start and stop bits back in.

The MNP class 4 protocol uses data compression to speed up transmission. It also checks for a noisy line, and if the noise is low, sends larger blocks of data. If the noise is high, smaller blocks are sent.

The MNP class 5 protocol uses data compression.

The MNP class 6 protocol offers automatic speed increases if the line can handle it without error.

The MNP class 9 protocol offers an improved error correction method.

Acknowledge (ACK) character

End-of-transmission (EOT) character

Kermit protocol

GOOD TO KNOW

Kermit is a very reliable protocol because it requires that every packet which is sent be acknowledged by the receiver as read correctly.

Microcom Networking Protocol (MNP)

Figure 11-62   Bisync synchronous protocol.

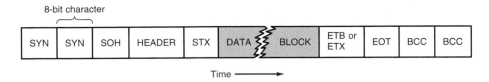

Synchronous Protocols

Protocols used for synchronous data communication are more complex than asynchronous protocols. However, like the asynchronous Xmodem and Kermit systems, they use various control characters for signaling purposes at the beginning and ending of the block of data to be transmitted.

Bisync. IBM's *Bisync protocol,* which is widely used in computer communication, usually begins with the transmission of two or more ASCII sync (SYN) characters (see Fig. 11-62). These characters signal the beginning of the transmission and are also used to initialize the clock timing circuits in the receiving modem. This ensures proper synchronization of the data transmitted a bit at a time.

    After the SYNC characters, a *start-of-header (SOH) character* is transmitted. The header is a group of characters that typically identifies the type of message to be sent, the number of characters in a block (usually up to 256), and a priority code or some specific routing destination. The end of the header is signaled by a *start-of-text (STX) character.* At this point, the desired message is transmitted, 1 byte at a time. No start and stop bits are sent. The 7- or 8-bit words are simply strung together one after another, and the receiver must sort them into individual binary words which are handled on a parallel basis farther along on the receiving circuit in the computer.

    At the end of a block an *end-of-transmission-block (ETB) character* is transmitted. If the block is the last one in a complete message, an end-of-text (ETX) character is transmitted. An *end-of-transmission (EOT) character,* which signals the end of the transmission, is followed by an error detection code, usually a 1- or 2-byte BCC.

SDLC. One of the most flexible and widely used synchronous protocols is the *synchronous data link control (SDLC) protocol* (see Fig. 11-63). SDLC is used in networks that are interconnections of multiple computers. All frames begin and end with a flag byte with the code 01111110 or hex 7E, which is recognized by the receiving computer. A sequence of binary 1s starts the clock synchronous process. Next comes an address byte that specifies a specific receiving station. Each station on the network is assigned an address number. The address hex FF indicates that the message to follow is to be sent to all stations on the network.

    A control byte following the address allows the programmer or user to specify how the data will be sent and how it will be dealt with at the receiving end. It allows the user to specify the number of frames, how the data will be received, and so on.

Bisync protocol

Start-of-header (SOH) character

Start-of-text (STX) character

End-of-transmission-block (ETB) character

End-of-transmission (EOT) character

Synchronous data link control (SDLC) protocol

Figure 11-63   The SDLC and HDLC frame formats.

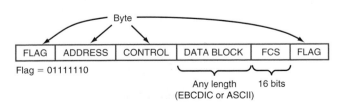

The data block (all codes are EBCDIC, not ASCII) comes next. It can be any length, but 256 bytes is typical. The data is followed by a frame-check sequence (FCS), a 16-bit CRC. A flag ends the frame.

A variation of the SDLC system, which permits interface between a larger number of different software and hardware configurations, is called *high-level data link control (HDLC)*. Its format is similar to that shown in Fig. 11-63. It may also use ASCII data and often has a 32-bit CRC/FCS.

The Open Systems Interconnection Model

As you have seen, there are many types and variations of protocols. If there is to be widespread compatibility between different systems, then some industrywide standardization is necessary. The ability of one hardware-software configuration to communicate with another, different system is known as *interoperability*. Only if all manufacturers and users adopt the same standards can true interoperability be achieved. One organization that has attempted to standardize data communication procedures is the *International Organization for Standardization*. It has come up with a framework, or hierarchy, that defines how data can be communicated. This hierarchy, known as the *open systems interconnection (OSI)* model, is designed to establish general interoperability guidelines for developers of communication systems and protocols. Even though the OSI model is not implemented by all manufacturers, the functions of each level in the OSI model outline must be accomplished by each protocol. In addition the OSI model serves as a common reference for all protocols.

The OSI hierarchy is made up of seven levels, or layers (see Fig. 11-64). Each layer is defined by software (or, in one case, hardware) and is clearly distinct from the other layers. These layers are not really protocols themselves, but they provide a way to define and partition protocols to make data transfers in a standardized way. Each layer is designed to handle messages that it receives from a lower layer or an upper layer. Each layer also sends messages to the layer above or below it according to specific guidelines. Different protocols accomplish each layer and refer to the tasks they perform by referencing the OSI model. Tasks are referred to as a layer 1, layer 2, etc. task.

As shown in the figure, the highest level is the application layer, which interfaces with the user's application program. The lowest level is the physical layer, where the electronic hardware, interfaces, and transmission media are defined.

> *Layer 1: Physical layer.* The physical connections and electrical standards for the communication system are defined here. This layer specifies interface characteristics such as binary voltage levels, encoding methods, data transfer rates, and the like.
>
> *Layer 2: Data link.* This layer defines the framing information for the block of data. It identifies any error detection and correction methods as well as any

Figure 11-64   The seven OSI layers.

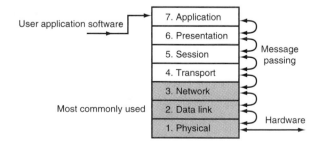

synchronizing and control codes relevant to communication. The data link layer includes basic protocols, such as HDLC and SDLC.

Layer 3: Network. This layer determines network configuration and the route the transmission can take. In some systems, there may be several paths for the data to traverse. The network layer determines the specific data routing and switching methods that can occur in the system, e.g., selection of a dial-up line, a private leased line, or some other dedicated path.

Layer 4: Transport. Included in this layer are multiplexing, if any; error recovery; partitioning of data into smaller units so that they can be handled more efficiently; and addressing and flow control operations.

Layer 5: Session. This layer handles such things as management and synchronization of the data transmission. It typically includes network log-on and log-off procedures, as well as user authorization, and determines the availability of the network for processing and storing the data to be transmitted.

Layer 6: Presentation. This layer deals with the form and syntax of the message. It defines data formatting, encoding and decoding, encryption and decryption, synchronization, and other characteristics. It defines any code translations required, and sets the parameters for any graphics operations.

Layer 7: Applications. This layer is the overall general manager of the network or the communication process. Its primary function is to format and transfer files between the communication message and the user's applications software.

The basic process is that information is added or removed as data is transmitted from one layer to another (see Fig. 11-65). Assume, e.g., that the applications program you are using contains some data that you wish to send to another computer. That data will be transmitted in the form of some kind of serial packet or frame, or a sequence of packets. The applications layer attaches the packet to some kind of header or preamble before sending it to the next level. At the presentation level, more headers and other information are added. At each of the lower levels, headers and related information are appended until the data message is almost completely encased in a much larger packet. Finally, at the physical level, the data is transmitted to the other system. As Fig. 11-65 shows, the message that is actually sent may contain more header information than actual data.

At the receiving end, the header information gets stripped off at the various levels as the data is transferred to successive levels. The headers tell the data where to go and what to do next. The data comes in at the physical level and goes up through the various layers until it gets to the applications layer, where it is finally used.

Note that it is not necessary to use all seven layers. Many modern data communication applications need only the first two or three layers to fully define a given data exchange.

The primary benefit of the OSI standard is that if it is incorporated into data communication equipment and software, compatibility between systems and equipment is more likely to be achieved. As more and more computers and networks are interconnected, e.g., on the Internet, true interoperability becomes more and more important.

Figure 11-65 The data as transmitted at the physical layer.

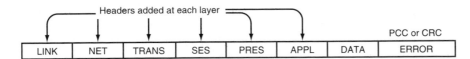

CHAPTER REVIEW

Summary

Data stored and processed by computers is encoded in binary digital form according to some standardized system, such as the American Standard Code for Information Interchange (ASCII). The signaling elements (bits, bauds, or symbols) that comprise digitally stored information can be transmitted synchronously or asynchronously. In asynchronous communication, each data word is accompanied by stop and start bits that identify the beginning and ending of the word. In synchronous transmission, start and stop bits are not used; data is transferred in multiword blocks in step with a master clock oscillator.

The speed of data transfer ranges from less than 10,000 bps (for the slower personal computers) to billions of bits per second. Data speed in digital communication systems can also be expressed in terms of baud rate. Several unique modulation schemes have been developed so that multiple bits can be included in a given symbol.

As with analog data, the amount of digital information that can be transmitted is proportional to the bandwidth of the communication channel and the time of transmission. Information theorists seek to determine, mathematically, the likelihood that a given amount of data being transmitted under a given set of circumstances (e.g., medium, bandwidth, speed of transmission, noise, and distortion) will be transmitted accurately.

Before digital serial data is sent, it is usually encoded according to the system that will facilitate the desired result. Some common encoding techniques are nonreturn to zero, return to zero, and Manchester.

Digital data is transmitted over the analog telephone network by using a modem, which both converts binary signals to analog signals and demodulates analog signals to reconstruct the equivalent binary output. The three most common modulation and demodulation techniques used in today's modems are frequency-shift keying (FSK); phase-shift keying (PSK), including binary phase-shift keying, differential phase-shift keying, and quadrature phase-shift keying; and quadrature amplitude modulation (QAM), including 8-QAM, 16-QAM and 64-QAM. One way to deal with bandwidth limitations is spread spectrum (SS), a technique that distributes a signal and its sidebands over a very wide bandwidth and offers the benefits of improved security, resistance to jamming, efficient band sharing, and resistance to fading. The two basic SS types are frequency-hopping and direct-sequence. Direct-sequence SS is the more common of the two. It is also called code-division multiple access (CDMA) and is widely used in cell phones. Orthogonal frequency division multiplexing (OFDM) is another widely used broadband technique. The three most widely used modems are analog dial-up, ADSL, and cable.

To reach a goal of 100 percent accuracy in transmission, both error detection and error correction methods must be used. Common detection techniques are redundancy, various special encoding schemes, parity, and cyclical redundancy checks. Forward error correction is the process of detecting and correcting errors at the receiver without retransmission; the most popular FEC methods are block-check sequences, Hamming codes, and Reed Solomon codes. Convolutional codes ensure nearly error-free transmission in the presence of noise.

Protocols are designed to ensure hardware and software compatibility between the sender and receiver of serial digital data. All asynchronous protocols use start and stop bits, and some include procedures such as hardware handshaking and the transmission of ASCII control characters. Three popular asynchronous protocols are Xmodem, Kermit, and MNP. Synchronous protocols, e.g., IBM's Bisync and synchronous data link control (SDLC), do not use start and stop bits. The open systems interconnection (OSI) model has been developed to provide a way to define and partition protocols so that data transfer can be truly standardized.

Questions

1. Name the earliest form of binary data communication.
2. What do you call dot-dash code transmission by radio?
3. How do you distinguish between uppercase and lowercase letters in Morse code?
4. What is Morse code for the characters C, 7, and ??
5. What 5-bit code was once used in teletype systems?
6. If the Baudot character 11011 is sent, and then 00110, what character is received?
7. What is the most widely used binary data code that uses 7 bits to represent characters?
8. What ASCII character is transmitted to ring a bell?
9. Name the two ways that bits are transmitted from one place to another.
10. What is used to signal the beginning and end of the transmission of a character in asynchronous transmissions?
11. What is the name given to the number of symbols occurring per second in a data transmission?
12. How can there be more than 1 bit per baud in data transmissions?

13. How many bits per baud (symbol) can be transmitted by a four-level FSK signal?
14. Which is faster, asynchronous transmission or synchronous transmission? Explain.
15. How is a message sent using synchronous data transmission?
16. In serial data transmission, what are the special names given to a binary 0 and a binary 1?
17. What encoding method is used in most standard digital logic signals?
18. What occurrence in serial data transmission makes it difficult to detect the clock rate when NRZ is used?
19. What two encoding methods are best for clock recovery?
20. What is a benefit of the RZ-AMI method of encoding?
21. Give two names for the encoding system that makes use of a transition at the center of each bit.
22. Explain how GMSK permits higher data rates in a smaller bandwidth.
23. What two factors does the amount of information that can be sent in a given transmission depend on?
24. True or false? Multilevel or multisymbol binary encoding schemes permit more data to be transmitted in less time, assuming a constant symbol interval.
25. For a given bandwidth system, what is the advantage of using a multisymbol encoding scheme?
26. Name the major components and circuits of a modem.
27. What is the function of a modem?
28. What is the maximum download speed of a V.90 dial-up modem?
29. What kind of modulation is used on an ADSL modem? What is its most common download speed?
30. What factor determines the maximum speed of an ADSL modem?
31. Why is a scrambler needed in a modem?
32. What is the name of the IC used to perform serial-to-parallel and parallel-to-serial conversions and other operations in a modem?
33. What is the bandwidth of a cable TV data channel?
34. What kind of modulation is used for downloads and uploads in cable TV, and what are the maximum theoretical data rates?
35. What basic circuit is used to produce BPSK?
36. What circuit is used to demodulate BPSK?
37. What circuit is used to generate the carrier to be used in demodulating a BPSK signal?
38. What type of PSK does not require a carrier recovery circuit for demodulation?
39. Name the key circuit used in a DPSK modulator.
40. How many different phase shifts are used in QPSK?
41. In QPSK, how many bits are represented by each phase shift?
42. How many bits are represented by each phase shift in 16-PSK?
43. Is carrier recovery required in a QPSK demodulator?
44. When QPSK is used, is the bit rate faster than the baud (symbol) rate?

45. What circuit is used to create a dibit in a QPSK modulator?
46. What do you call a circuit that converts a 2-bit binary code into one of four dc voltage levels?
47. QAM is a combination of what two types of modulation?
48. With QAM, can a 53-kbps signal be transmitted within a 3000-Hz bandwidth?
49. What is trellis code modulation? Why is it used?
50. What organization establishes and maintains modem standards?
51. What are the two key specifications given by modem standards?
52. State the maximum bit and baud rate and modulation method used by a modem with a V.32bis standard.
53. What is spread spectrum?
54. Name the two main types of spread spectrum.
55. What circuit generates the transmitter frequency in a frequency-hopping SS system?
56. In a frequency-hopping SS system, what circuit selects the frequency produced by the synthesizer?
57. True or false? The hop rate is slower than the bit rate of the digital data.
58. How does an SS signal appear to a narrowband receiver?
59. How are two or more stations using spread spectrum and sharing a common band identified and distinguished from one another?
60. What do you call the length of time that a frequency-hopping SS transmitter stays on one frequency?
61. What circuit is used to generate a PSN signal?
62. What is the purpose of a PSN signal?
63. In a direct-sequence SS transmitter, the data signal is mixed with a PSN signal in what kind of circuit?
64. True or false? In a direct-sequence SS system, the chip rate is faster than the data rate.
65. What type of modulation is used with direct-sequence SS?
66. What is the most difficult part of spread spectrum communication?
67. What do you call the process of comparing one signal to another in an effort to obtain a match?
68. Name the two main benefits of spread spectrum.
69. How is OFDM generated and demodulated?
70. Name four applications of OFDM.
71. Name the most common application for spread spectrum.
72. How are voice signals transmitted via spread spectrum?
73. What is another name for direct-sequence SS?
74. List two methods of error detection.
75. Name one simple but time-consuming way to ensure an error-free transmission.
76. What is the most common cause of error in data transmission?
77. What is the name given to the ratio of the number of bit errors to the total number of bits transmitted?

78. What serial bit encoding method makes it possible to detect single-bit errors?
79. What is the bit added to a transmitted character to help indicate an error called?
80. What is the name of a number added to the end of a data block to assist in detecting errors, and how is it derived?
81. What is another name for parity?
82. What logic circuit is the basic building block of a parity generator circuit?
83. What error detection system uses a BCC at the end of a data block?
84. Describe the process of generating a CRC.
85. What basic circuit generates a CRC?
86. True or false? If a bit error can be identified, it can be corrected.
87. Describe the procedure for checking the accuracy of a block transmission at the receiver using a CRC.
88. What is the name of the most popular error-correcting code?
89. What will be the outcome of X-ORing Hamming bits if no error occurs in the data word during transmission?

90. Name three types of convolutional codes.
91. What is the name given to the rules and procedures that describe how data will be transmitted and received?
92. What is the process of exchanging signals between transmitter and receiver to indicate status or availability?
93. What is a popular protocol used with PCs and printers? What control characters are used?
94. Name two popular asynchronous protocols used in PC communication via modem.
95. What do MNP protocols define, and where are they used?
96. What is the string of characters making up a message or part of a message to be transmitted called?
97. What do synchronous protocols usually begin with? Why?
98. Name two common synchronous protocols.
99. What appears at the very end of most protocol packets?
100. What does interoperability mean?
101. Name one way to achieve interoperability.
102. Name the seven OSI levels in order from highest to lowest.
103. What are the three or four most used OSI levels?

Problems

1. What is the name of the 8-bit character code used in IBM systems? ◆
2. A serial pulse train has a 70-μs bit time. What is the data rate in bits per second?
3. The data rate of a serial bit stream is 14,400 bps. What is the bit interval? ◆
4. The speed of a serial data transmission is 2.5 Mbps. What is the actual number of bits that occur in 1 s?
5. What is the channel capacity, in bits per second, of a 30-kHz-bandwidth binary system, assuming no noise? ◆
6. If an eight-level encoding scheme is used in a 30-kHz-bandwidth system, what is the channel capacity in bits per second?
7. What is the channel capacity, in bits per second, of a 15-MHz channel with an *S/N* ratio of 28 dB? (Assume a noiseless channel.) ◆

8. What is the minimum allowable bandwidth that can transmit a binary signal with a bit rate of 350 bps?
9. What is the BER if 4 errors occur in the transmission of 500,000 bits? ◆
10. Write the correct parity bits for each number.
    a. Odd 1011000 __
    b. Even 1011101 __
    c. Odd 0111101 __
    d. Even 1001110 __
11. Determine the 4 Hamming bits for the 8-bit number 11010110. Use the format illustrated in the text.
12. What is the processing gain in a DSS system with a 20-MHz channel and a data rate of 11 Mbps?

◆ *Answers to Selected Problems follow Chap. 22.*

Critical Thinking

1. Describe in detail a simple data communication system that will monitor the temperature of a remote inaccessible location and display temperature on a personal computer.
2. Suggest a future application for spread spectrum techniques and explain why SS would be appropriate for that application.

3. Name three or more common data communication applications that you may use.
4. Why is DSP necessary to implement OFDM in an ADSL modem?
5. Are spread spectrum and OFDM spectrally efficient?

Introduction to Networking and Local-Area Networks

The first computers were stand-alone machines that one person used at a time. Later, timesharing computers were invented that allowed more than one person to use the machine concurrently. When personal computers (PCs) came along, the one computer–one user paradigm came back with a vengeance. But all that is in the past. Today, most computers are networked, i.e., connected to one another so that they can communicate with one another, share resources, and access the Internet. Virtually 100 percent of business and industrial computers are networked. It is estimated that more than 70 percent of all home and personal computers are also networked. This chapter is an introduction to networking and local-area networks (LANs).

Objectives

After completing this chapter, you will be able to:

- Define the terms LAN, MAN, and WAN.
- Explain the basic purposes of LANs and describe some specific LAN applications.
- Draw diagrams of LANs arranged according to the three basic topologies: star, ring, and bus.
- Describe how repeaters, transceivers, hubs, bridges, routers, and gateways are used in LANs.
- Explain the basic difference between client-server and peer-to-peer LANs and give the advantages and disadvantages of each.
- Calculate transmission speed given the length of the data to be transmitted and the characteristics of the LAN over which the data is to be sent.

12–1 Network Fundamentals

A *network* is a communication system with two or more stations that can communicate with one another. When it is desired to have each computer communicate with two or more additional computers, the interconnections can become complex. As Fig. 12-1 indicates, if four computers are to be interconnected, there must be three links to each PC.

The number of links L required between N PCs (nodes) is determined by using the formula

$$L = \frac{N(N-1)}{2}$$

Assume, e.g., there are six PCs. The number of links is

$$L = 6(6-1)/2 = 6(5)/2 - 30/2 = 15.$$

The number of links or cables increases in proportion to the number of nodes involved. The type of arrangement shown in Fig. 12-1 is obviously expensive and impractical. Some special type of network wiring must be used, a combination of hardware and software that permits multiple computers to be connected inexpensively and simply with the minimum number of links necessary for communication.

Example 12–1

An office with 20 PCs is to be wired so that any computer can communicate with any other. How many interconnecting wires (or links, L) are needed?

$$L = \frac{N(N-1)}{2} = \frac{20(20-1)}{2} = \frac{20(19)}{2} = 190$$

Types of Networks

Node

Each computer or user in a network is referred to as a *node*. The interconnection between the nodes is referred to as the *communication link*. In most computer networks, each node is a personal computer, but in some cases a peripheral device such as a laser printer

Figure 12-1   A network of four PCs.

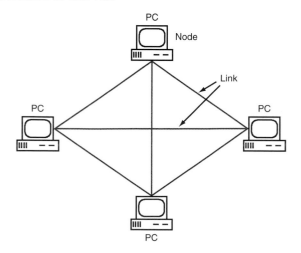

or an embedded controller built into another piece of equipment can be a node. There are four basic types of electronic networks in common use: wide-area networks (WANs), metropolitan-area networks (MANs), local-area networks (LANs), and personal-area networks (PANs). Let's take a brief look at each.

Wide-Area Networks (WANs). A WAN covers a significant geographic area. Local telephone systems are WANs, as are the many long-distance telephone systems linked together across the country and to WANs in other countries. Each telephone set is, in effect, a node in a network that links local offices and central offices. Any node can contact any other node in the system. Telephone systems use twisted-pair wire and coaxial cable, as well as microwave relay networks, satellites, and fiber-optic cabling.

There are also WANs that are not part of the public telephone networks, e.g., corporate LANs set up to permit independent intercompany communication regardless of where the various subsidiaries and company divisions, sales offices, and manufacturing plants may be. The special communication, command, and control networks set up by the military are also WANs.

The nationwide and worldwide fiber-optic networks set up since the mid-1990s to carry Internet traffic are also WANs. Known as the *Internet core* or *backbone*, these high-speed interconnections are configured as either direct point-to-point links or large rings with multiple access points. WANs make it possible for any PC or other Internet-enabled device such as a cell phone to access the World Wide Web or any entity connected to the Internet.

Metropolitan-Area Networks (MANs). MANs are smaller networks that generally cover a city, town, or village. Cable TV systems are MANs. The cable TV company receives signals from multiple sources, including local TV stations, as well as special programming from satellites, and it assembles all these signals into a single composite signal that is placed on fiber-optic and coaxial cables. The cables are then channeled to each subscriber home. The cable TV channel selector boxes are all nodes in the system. Most existing cable systems are simplex or one-way transmission systems; however, many cable companies now incorporate two-way communication capability.

Another type of MAN carries computer data. MANs are usually fiber-optic rings encircling a city that provide local access to users. Businesses, governments, schools, hospitals, and others connect their internal LANs to them. MANs also connect to local and long-distance telephone companies. The MANs, or *metro networks* as they are typically called, also provide fast and convenient connections to WANs for global Internet connectivity.

Local-Area Networks (LANs). A LAN is the smallest type of network in general use. It consists primarily of personal computers interconnected within an office or building. LANs can have as few as 3 to 5 users, although most systems connect 10 to several thousand users.

Small LANs can be used by a company to interconnect several offices in the same building; in such cases wiring can be run between different floors of the building to make the connection. Larger LANs can interconnect several buildings within a complex, e.g., large companies with multiple buildings, military installations, and college campuses.

Some LANs consist of multiple PCs that are linked both to each other and to a minicomputer or mainframe. This allows each user on the LAN to have access to the big computer as well as continue to operate independently.

Home networks of two or more PCs are also LANs and today most home LANs are fully wireless or incorporate wireless segments.

Personal-Area Networks (PANs). A PAN is a short-range wireless network that is set up automatically between two or more devices such as laptop computers, personal digital assistants (PDAs), peripheral devices, or cell phones. The distance between the

Wide-area network (WAN)

Internet core or backbone

Metropolitan-area network (MAN)

Metro networks

devices is very short, no more than about 10 m and usually much less. PANs are referred to as ad hoc networks that are set up for a specific single purpose, such as the transfer of data between the devices as required by some application. For example, a laptop computer may link up with a printer, or a PDA may need to download data from a PC. Most PANs just involve two nodes, but some have been set up to handle up to eight nodes and sometimes more.

Storage–Area Networks (SANs). SANs are an outgrowth of the massive data storage requirements developed over the years thanks to the Internet. These networks usually attach to a LAN or Internet server and are designed to store and protect huge data files. The SAN also provides users on the network access to massive data files stored in mass memory units, called *redundant arrays of independent disks (RAIDs)*. RAIDs use many hard drives interconnected to the network. RAIDs have been available for years, but they had to be located close to the computers they served to provide adequate access speed. Today, with high-speed fiber-optic links, the RAIDs may literally be located anywhere, even across the country, since access can be via the Internet or a fiber-optic WAN or MAN.

Network Hierarchy. Figure 12-2 shows a highly simplified view of how LANs, MANs, and WANs are interconnected. LANs inside a building are usually connected to a MAN that may be a local telephone central office or a special MAN set up by the organization itself or one managed by a company that leases lines to the organization. The MANs connect to the WANs, which may be a long-distance telephone network or a special optical WAN set up for Internet access or other data transmission applications. Some WANs are hierarchies of rings and direct connection points. MANs and WANs are virtually all fiber-optic networks. The interconnection points of the networks may be special computers called *servers, routers,* or switching equipment such as an add-drop multiplexer (ADM) that allows data to be added to or extracted from a ring network.

Servers, routers

Figure 12–2   Networking hierarchy.

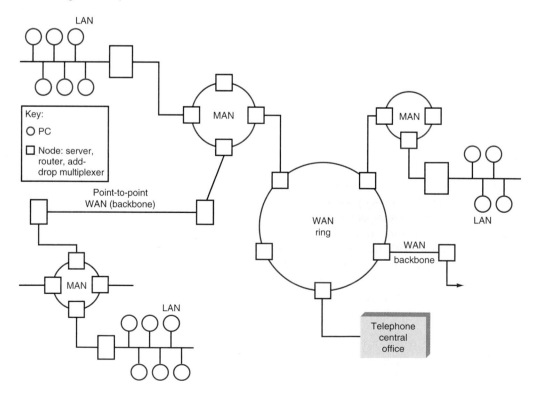

Network Topologies

The *topology* of a network describes the basic communication paths between, and methods used to connect, the nodes on a network. The three most common topologies used are star, ring, and bus. These topologies apply to the LAN, MAN, and WAN.

Topology

Star Topology. A basic *star* configuration consists of a central controller node and multiple individual stations connected to it (Fig. 12-3). The resulting system resembles a multipointed star. The central or controlling PC, often referred to as the *server,* is typically larger and faster than the other PCs and contains a large hard drive where shared data and programs are stored. Any communication between two PCs passes through the server. The server and its software manage the linkup of individual computers and the transfer of data between them.

Star topology

A star-type LAN is extremely simple and straightforward. New nodes can be quickly and easily added to the system. Further, the failure of one node does not disable the entire system. Of course, if the server node goes down, the network is disabled, although the individual PCs will continue to operate independently. Star networks generally require more cable than other network topologies, and the fact that all communication must pass through the node does place some speed restrictions on the transfer of data.

Ring Topology. In a *ring* configuration, the server or main control computer and all the computers are simply linked in a single closed loop (Fig. 12-4). Usually, data is transferred around the ring in only one direction, passing through each node. Therefore, there is some amplification and regeneration of the data at each node, permitting long transmission distances between nodes.

Ring topology

The ring topology is easily implemented and is low in cost. Expansion is generally simple, since a new node can be inserted in the ring at almost any point. The computer that has a message to send identifies the message with a code or address identifying the destination node, places it on the ring, and sends it to the next computer in the loop. Essentially, each PC in the ring receives and retransmits the message until it reaches the target PC. At that point, the receiving node accepts the message.

Figure 12-3 A star LAN configuration with a server as the controlling computer.

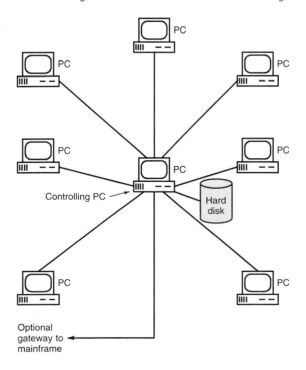

Figure 12-4 A ring LAN configuration.

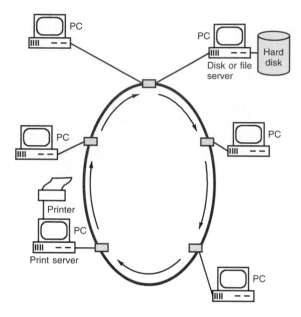

The downside of a ring network is that a failure in a single node generally causes the entire network to go down. It is also somewhat difficult to diagnose problems on a ring. Despite these limitations, the ring configuration is widely used.

Bus Topology. A *bus* is simply a common cable to which all the nodes are attached. Figure 12-5 shows the bus configuration. The bus is bidirectional in that signals can be transmitted in either direction between any two of the nodes; however, only one of the nodes can transmit at a given time. A signal to be transmitted can be destined for a single node or broadcast to all nodes simultaneously.

The primary advantage of the bus is that it is faster than any of the other topologies. The wiring is simple, and the bus can be easily expanded.

Mesh Topology. A *mesh* network is one in which each node is connected to all other nodes. You saw this in Fig. 12-1. This is called a full mesh in that every node can talk directly to any other node. Of course this leads to major costs and complications as the number of nodes increases. This problem is somewhat mitigated by the use of wireless interconnections between nodes where there is no expensive wiring and the attendant routing and maintenance problems. A variation of the full mesh is the partial mesh in which all nodes can communicate with two or more other nodes. This reduces the number of interconnections, making it more practical.

The primary value of the mesh network is that there are multiple paths for data to take from one node to another. This provides for redundancy that can provide a continuous connection when one or more of the links are broken. The lack of one link does not keep the data from reaching its destination by another path. This redundancy provides increased network reliability. Most partial mesh networks are implemented with wireless links.

Other Topologies. There are many variations and combinations of the basic topologies discussed above. For example, the *daisy chain topology* is simply a ring that has been broken. Another variation is called the *tree topology*. This topology is simply a bus design in which each node has multiple interconnections to other nodes through a star interconnection. In others, the network consists of branches from one node to two or more other nodes.

Bus topology

Mesh

Daisy chain topology

Tree topology

LOGICAL VERSUS PHYSICAL TOPOLOGY

The topologies illustrated in Figures 12-1, 12-3, 12-4, and 12-5 are what we call logical topologies as they show the connections in a formal logical sense. The connections are made in a theoretical sense that describes how the data flows and how the nodes are connected. However, in the real world, networks do not appear physically as their logical connections. Because of the random locations of the computers and other nodes and the fact that cabling must be run in patterns restricted and dictated by the buildings, the physical appearance of a network usually bears no relationship to the logical structure.

For example, logical bus-connected computers are usually of the physical star form because cables from each PC go to a central location and are terminated at a hub or switch. The bus is actually implemented inside the hub. But the physical arrangement is more of an irregular star.

The same is true for ring networks. The cable to/from each PC terminates at some central connection point, but the ring logical format is still implemented because the wires in the cables take the signals into and out of each PC to the central connection point. The ring is still there, but it looks like a star physically. Keep this key point in mind when you are working on LANs and other networks.

Figure 12-5 A bus LAN configuration.

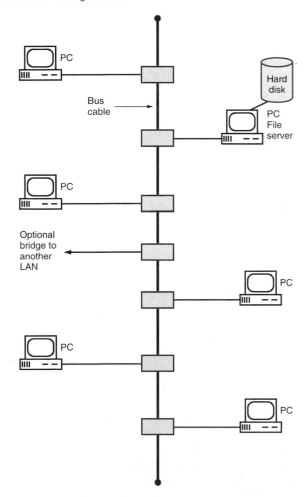

LAN Applications

The common denominator of all LANs is the communication of information. Nodes are linked to transmit and receive information, software, databases, personal messages, or anything else that can be put in binary form.

The earliest LANs were developed primarily as a way to control costs. For example, in an office with multiple PCs, the cost of permitting each computer to have its own printer was prohibitive. Networking is an inexpensive way to link all the computers in the office to a single computer to which the printer is attached. Users share the printer by transmitting information to be printed to the server, which controls print operations for all the office computers.

Another example is the sharing of large files and databases. Networks can be used to connect all the PCs in an office to a central server with a large hard drive containing the data to be shared, so that any computer can access it. Today, networks are used for many applications other than centralizing and sharing expensive peripherals such as printers and for permitting many users access to a large database.

E-mail. By far the most popular networks application is *e-mail,* the sending and receiving of electronic messages and memos. Users simply key memos on their computers and send them to other users over the network. In many organizations e-mail has all but replaced interoffice paper and telephone messages. In others, e-mail supplements the normal interoffice communication procedures.

Internet Access. PC users in organizations with LANs also access the Internet via the LAN. Rather than connect each individual PC separately to a phone line or digital subscriber line (DSL) or cable TV of Internet connection, the organization provides a LAN server that lets anyone connect to the Internet.

Groupware. *Groupware* is a set of programs that facilitate the efforts of two or more individuals working on a single project. Groupware provides ways to share common databases, exchange messages related to a particular group task, manage work flow (e.g., for project management), and maintain a common calendar. The most popular and widely used groupware program is Lotus Notes.

Client–Server and Peer-to-Peer LANs

Most LANs conform to one of two general configurations: client-server or peer-to-peer. In the *client-server* configuration one of the computers in the network, the server, essentially runs the LAN and determines how the system operates. The server, which is the largest and fastest computer in the network, manages printing operations of a central printer and controls access to a very large hard drive or bank of hard drives containing databases, files, and other information that the clients—the other computers on the network—can access. The server also provides Internet access.

The server controls clients' access to the printer, the hard drive, the files and databases, other nodes, and various other services. In many client-server systems, the server is a minicomputer such as an IBM AS/400. The server can also be a large, fast server PC.

In a *peer-to-peer* configuration, any PC can serve as either client or server. Any PC on a peer-to-peer network can have access to any other PC's files and connected peripherals. For example, a laser printer connected to any one of the PCs can be used by any of the others on the network.

Peer-to-peer LANs are smaller and less expensive than the client-server variety, and they provide a simple way to provide network communication in a small office or home. The total number of interconnected PCs is generally small, usually less than 20 machines, and the system is relatively easy to set up. Peer-to-peer networks also have some disadvantages. They usually have lower performance, meaning lower-speed

E-mail

Groupware

Client-server configuration

Peer-to-peer configuration

transmission capability. There are also manageability and security problems related to the ability of any user to access any other user's files. Protecting individual users' software and private files from access is more difficult with peer-to-peer networks than it is with client-server systems.

12-2 LAN Hardware

All LANs are a combination of hardware and software. The primary hardware devices are the computers themselves and the cables and connectors that link them. Additional pieces of hardware unique to networks include network interface cards, repeaters, hubs and concentrators, bridges, routers, gateways, and many other special interfacing devices. This section provides an overview of the specific types of hardware involved in networking.

Cables

Most LANs use some type of copper wire cable to carry data from one computer to another via baseband transmission. The digital data stored in the computer is converted to serial binary data, and voltages representing binary 1s and 0s are transmitted directly over the cable from one computer to another. The three basic cable types used in LANs are coaxial cable, twisted-pair, and fiber-optic cable. Most local-area networks started out using coaxial cable, but today twisted-pair cable dominates. Fiber-optic cable is used in higher-speed, secure networks, which are not as widespread.

Coaxial Cable. *Coaxial cable* is far superior to twisted-pair cable as a communication medium. Its extremely wide bandwidth permits very high-speed bit rates. Although loss is generally high, attenuation is usually offset by using repeaters that boost the signal level and regenerate the signal waveshape. The major benefit of coaxial cable is that it is completely shielded, so that external noise has little or no effect on it.

Coaxial cable
Twisted-pair cable
Unshielded twisted-pair (UTP)
Shielded twisted-pair (STP)

Coaxial cable is shown in Fig. 12-6. It consists of a thin center conductor surrounded by an insulating material that is, in turn, completely encircled by a shield. The shield can be crisscrossed wire braid or solid metal foil. Surrounding the shield is an outer sheath, usually made of PVC.

Coaxial cable comes in a wide variety of sizes and shapes, which are usually designated by a letter-number combination such as RG-8/U. (Chapter 13 provides more details on coaxial cable.)

Twisted–Pair Cable. *Twisted-pair cable,* as the name implies, is two insulated copper wires twisted together loosely to form a cable [Fig. 12-7(a)]. Telephone companies use twisted-pair cable to connect individual telephones to the central office. The wire is solid copper, 22-, 24-, or 26-gauge. The insulation is usually PVC. Twisted-pair cable by itself has a characteristic impedance of about 100 Ω, but the actual impedance depends on how tightly or how loosely the cable is twisted and can be anywhere from about 70 to 150 Ω.

There are two basic types of twisted-pair cables in use in LANs: *unshielded (UTP)* [Fig. 12-7(a)] and *shielded (STP)* [Fig. 12-7(b)]. UTP cables are highly susceptible to noise, particularly over long cable runs. Most twisted pairs are contained within a common cable sheath along with several other twisted pairs, and crosstalk, the coupling between

Fiber-optic cables must be tested to make sure that they meet specifications.

Figure 12-6 Coaxial cable.

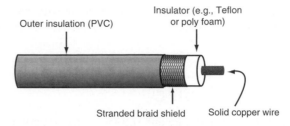

Figure 12-7 Types of twisted-pair cable. (*a*) Twisted-pair unshielded (UTP) cable. (*b*) Multiple shielded twisted-pair (STP) cable.

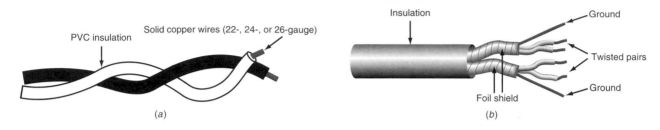

(a)　　　　　　　　　　　　　　　　　　　　(b)

adjacent cables, can also be a problem; again, this is especially true when transmission distance is great and switching speeds are high.

STP cables, which are more expensive than UTP cables, have a metal foil or braid shield around them, forming a third conductor. The shield is usually connected to ground and, therefore, provides protection from external noise and crosstalk. STP cables are, therefore, routinely used for long cable runs.

STP cables are also typically grouped together, and two, four, or more pairs can be contained within a single cable housing [Fig. 12-7(*b*)]. When multiple lines must be run between computers and connecting devices, multiple cables are the linking medium of choice.

Twisted-pair cable is available in standard types and sizes that are specified and regulated by the standards organizations the American National Standards Institute (ANSI), the Electronic Industries Alliance (EIA), and the Telecommunications Industry Association (TIA). The most often used standard is TIA 568A/568B. This standard defines several categories of twisted-pair cable. These are summarized in Fig. 12-8. Categories 2 and 4 are rarely used.

The most widely used UTP is category 5 (CAT5). It can carry baseband data at rates up to 100 Mbps at a range up to 100 m. It contains four twisted pairs within the cable and is usually terminated in RJ-45 modular connectors. A newer enhanced

Figure 12-8 Common types of twisted-pair cable.

CATEGORY	MAXIMUM DATA RATE*	IMPEDANCE (Ω)
1	Voice only	75–100
2	4 Mbps	100
3	10–16 Mbps	100
4	16–20 Mbps	100
5	100 Mbps	100
5E	155 Mbps	100
6	200–250 Mbps	100
7	600 Mbps	100

*Cable length is 100 m.

version (CAT5E) is also available. It has improved performance and a data rate up to 155 Mbps.

The category 6 and category 7 cables are now widely available. Maximum cable lengths of CAT6 and CAT7 are usually less than 100 m, and a different kind of connector will be used to handle the very high frequencies involved.

Twisted-pair cable specifications also include attenuation and near-end cross talk figures. *Attenuation* means the amount by which the cable attenuates the signal. The value for 100 m of CAT5 cable is -6.5 dB at 10 MHz and -22 dB at 100 MHz. The longer the cable, the greater the amount of loss in the cable and the smaller the output. The cable acts as a low-pass filter and also distorts digital signals. See Chap. 13 for more details.

The other key specification is *near-end cross talk (NEXT)*. *Cross talk* refers to the signal that is transferred from one twisted pair in a cable to another by way of capacitive and inductive coupling. Near-end cross talk is the signal appearing at the input to the receiving end of the cable. Like noise, NEXT can interfere with the received signal on the cable. The NEXT specification indicates the level of cross talk signal attenuation. In 100 m of CAT5 cable, it can range from -62 dB   at 1 MHz to -32 dB at 100 MHz.

Near-end cross talk (NEXT)

Many newer office buildings are constructed with special vertical channels or chambers, called *plenums,* through which cables are run between floors or across ceilings. Cable used this way, called *plenum cable,* must be made of fireproof material that will not emit toxic fumes if it catches fire. Plenum cable can be either coaxial or twisted-pair.

Plenums

Plenum cable

Fiber–Optic Cable. *Fiber-optic cable* is a nonconducting cable consisting of a glass or plastic center cable surrounded by a plastic cladding encased in a plastic outer sheath (Fig. 12-9). Most fiber-optic cables are extremely thin glass, and many are usually bundled together. Special fiber-optic connectors are required to attach them to the network equipment. Fiber-optic cables are covered in greater detail in Chap. 19. Speeds of up to 1 Tbps (terabits per second) are achievable by using fiber optics.

Fiber-optic cable

Connectors

All cables used in networks have special terminating connectors that provide a fast and easy way to connect and disconnect the equipment from the cabling and maintain the characteristics of the cable through the connection.

Coaxial Cable Connectors. Coaxial cables in networks use two types of connectors, *N connectors* and *BNC connectors,* as Fig. 12-10 shows. The N connectors are widely used in RF applications, and BNC connectors are commonly used for attaching test leads to measuring instruments such as oscilloscopes. N connectors are used on larger cables, such as RG-8/U. BNC connectors are used on smaller cables, such as RG-58/U and RG-59/U. The BNC plug attaches to a mating jack with a simple twist of the outer housing; the N connector has continuous threads.

Coaxial connector

N connector

BNC connector

A variety of connector accessories are also available to facilitate connections. One of the most popular is the BNC *T connector,* which is used to interconnect two cables to the network hardware [Fig. 12-11(a)]. T connectors provide a convenient way to break into an existing coaxial cable for attaching an additional node.

T connector

Figure 12-9 Fiber-optic cable.

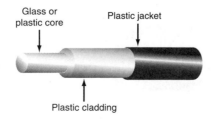

Glass or plastic core

Plastic jacket

Plastic cladding

Figure 12–10 Common coaxial connectors.

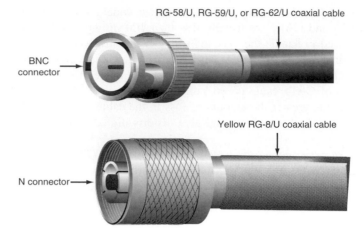

RG-58/U, RG-59/U, or RG-62/U coaxial cable

BNC connector →

Yellow RG-8/U coaxial cable

N connector →

Barrel connector

The *barrel connector* shown in Fig. 12-11(*b*) also provides a convenient way to connect two coaxial cables. This is usually done to extend a cable's length or to interconnect two existing cables end to end.

Terminator

A *terminator* is a special connector containing a resistor whose value is equal to the characteristic impedance of the coaxial cable. In LANs, the characteristic impedance of the coaxial cable is about 50 Ω. Thus the terminator is a 50-Ω resistor. Providing the correct termination value prevents reflection on the cable.

RJ-11 connector

Twisted–Pair Connectors. Most telephones attach to an outlet by way of an *RJ-11 connector* or modular plug [Fig. 12-12(*a*)]. RJ-11 connectors are used to connect PC modems to the phone line but are not used in LAN connections.

Figure 12–11 BNC connector accessories and adapters. (*a*) T connector. (*b*) Barrel connector.

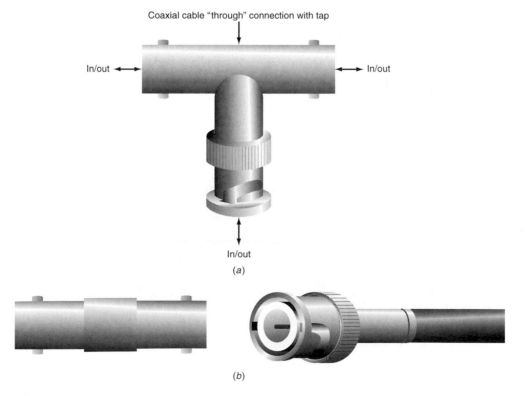

Coaxial cable "through" connection with tap

In/out ←→   ←→ In/out

In/out

(*a*)

(*b*)

Figure 12-12 Modular (telephone) connectors used with twisted-pair cable. (*a*) RJ-11. (*b*) RJ-45.

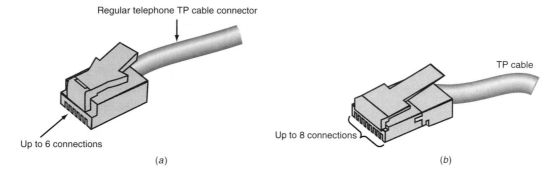

Regular telephone TP cable connector

Up to 6 connections

TP cable

Up to 8 connections

(*a*)

(*b*)

A larger modular connector known as the *RJ-45 connector* is widely used in terminating twisted pairs [see Fig. 12-12(*b*)]. The RJ-45 contains eight connectors, so it can be used to terminate four twisted pairs. Matching jacks on the equipment or wall outlets are used with these connectors. Most LANs today use RJ-45 connectors.

RJ-45 connector

Fiber–Optic Connectors. A wide range of connectors are available to terminate fiber-optic cables. Like electrical connectors, these are designed to provide a fast and easy way to attach or remove cables.

Fiber-optic connector

Network Interface Cards and Chips

A *network interface card (NIC)* provides the I/O interface between each node on a network and the network wiring. These cards can plug into the PC bus or PCMCIA slot, or are often integrated into the motherboard of the PC and provide connectors at the rear of the computer for attaching the cable connectors (see Fig. 12-13). NICs perform a

Network interface card (NIC)

Figure 12-13 A network interface card.

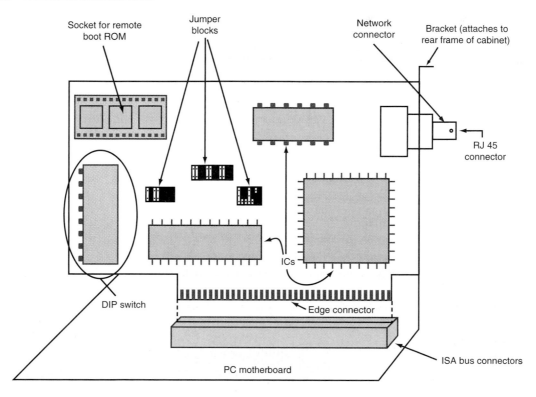

Socket for remote boot ROM

Jumper blocks

Network connector

Bracket (attaches to rear frame of cabinet)

RJ 45 connector

ICs

DIP switch

Edge connector

ISA bus connectors

PC motherboard

variety of tasks. For example, when the PC wishes to transmit information over the network, it takes data stored in RAM to be transmitted and converts it to a serial data format. This serial information is usually stored within RAM on the NIC. Logic circuitry in the NIC then groups the information into frames or packets, the format of which is defined by the communication protocol used by the LAN. Once the packet or frame has been formed, the binary data is encoded, normally by using the Manchester code, and then sent to a logic-level converter which generates the proper binary 0 and binary 1 voltage levels that are sent over the coaxial or twisted-pair cables.

Upon reception, the destination NIC recognizes when it is being addressed, i.e., when data is being sent to it. The NIC performs logic-level conversion and decoding, recovering the serial frame or packet of information; performs housekeeping functions such as error detection and correction; and places the recovered data in a buffer storage memory. The data is then converted from serial to parallel, where it is transferred to the computer RAM and used by the software.

The NIC is the key hardware component in any LAN. It completely defines the protocols and performance characteristics of the LAN. Despite their sophistication, NICs are relatively low in price and available from many manufacturers. Today, most NICs have been compressed into a single chip thanks to advanced semiconductor processing. The exception is the transformer and selected discrete components. Because most PCs are networked, NIC chips are built into all PC and laptop motherboards.

Repeaters

When signals from a NIC must travel a long distance over coaxial cables or twisted-pair cables, the binary signal is greatly attenuated by the resistance of the wires and distorted by the capacitance of the cable. In addition, the cable can pick up noise along the way. As a result, the signal can be too distorted and noisy to be received reliably.

A common solution to this problem is to use one or more repeaters along the way (see Fig. 12-14). A *repeater* is an electronic circuit that takes a partially degraded signal, boosts its level, shapes it up, and sends it on its way. Over long transmission distances, several repeaters may be required.

Repeaters are small, inexpensive devices that can be inserted into a line with appropriate connectors or built into other LAN equipment. Most repeaters are really transceivers—bidirectional circuits that can both send and receive data. Transceiver repeaters can receive signals from either direction and transmit them in the opposite direction.

Repeater

Hubs

Hub

A *hub* is a LAN accessory that facilitates the interconnections of the cables to the nodes. Whether the network topology used by a LAN is bus, ring, or star, the wiring usually resembles a star. This is so because the cabling for most networks today is permanently installed in walls, ceilings, and plenums. Bus and ring topologies, where cables logically run between individual PCs, are not convenient for plenum wiring, as they do not provide an easy way to modify the network to add or remove nodes in different parts of the office or building.

The device that facilitates such wiring is the hub, a central connecting box designed to receive the cable inputs from the various PC nodes and to connect them to the server (see Fig. 12-15). In most cases, hub wiring physically resembles a star because all the cabling comes into a central point, or hub. However, the hub wiring is such that it can

Figure 12-14 Concept of a repeater.

Figure 12-15 A hub facilitates interconnections to the server.

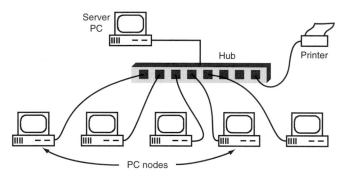

logically implement either the bus or the ring configuration. That is, inside the hub the wiring connects the nodes into a miniature ring or bus.

Hubs are usually active devices containing repeaters. Hubs amplify and reshape the signal and transmit it to all connection parts. Hubs are available with 8, 12, 16, 24, 32, and 48 parts. All signals received at the hub are repeated to all nodes connected to the hub.

Bridges

A *bridge* is a network device that is connected as a node on the network and performs bidirectional communication between two LANs (see Fig. 12-16).

A bridge can also be used when one LAN becomes too big. Most LANs are designed for a maximum upper limit of nodes. The reason for this is that the greater the number of nodes, the longer and more complex the wiring. Further, when many individuals attempt to use a LAN simultaneously, performance deteriorates greatly, leading to network delays. One way to deal with this problem is to break a large LAN into two or more smaller LANs. First it is determined which nodes communicate with other nodes the most, and then a logical breakdown into individual LANs is made. Communication between all users is maintained by interconnecting the separate LANs with bridges. The result is improved overall performance.

A bridge is generally designed to interconnect two LANs with the same protocol, e.g., two Ethernet networks. However, there are bridges that are able to accomplish protocol conversion so that two LANs with different protocols can converse.

Remote bridges are special bridges used to connect two LANs that are separated by a long distance. A bridge can use the telephone network to connect LANs in two different parts of the country, or can connect two LANs on a large campus or the grounds of a big military base through a fiber-optic cable or wireless connection.

Bridge

GOOD TO KNOW

Large LANs are often broken into two or more smaller LANs, with bridges connecting them in order to compensate for the problem of poor network performance caused by many simultaneous users.

Remote bridge

Figure 12-16 A bridge connects two LANs.

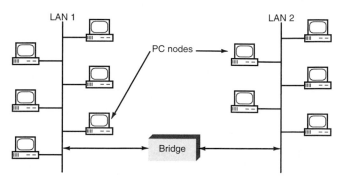

Switches

A *switch* is a hublike device used to connect individual PC nodes to the network wiring. Unlike a hub, a switch provides a means to connect or disconnect a PC from the network wiring. Switches have largely replaced hubs in most large LANs because switches greatly expand the number of possible nodes and improve performance.

As a network grows, more and more nodes are connected to the wiring. This has the effect of slowing down data transfers as all nodes must share the media. The longer cables restrict the data rate. And the greater number of users vying for the cable makes access times longer. These problems are overcome with a switch. The switch can be used to divide the LAN into smaller segments. This immediately improves performance.

LAN switches recognize individual node addresses. When transmitting data from one PC to another, the switch detects the address of the receiving PC and connects it to the wiring. Otherwise, a PC is disconnected from the wiring by the switch until it is ready to send or receive data. By reducing the loads imposed by all the unused PCs, the switch allows the network to be significantly faster.

Routers

Like bridges, *routers* are designed to connect two networks. The main difference between bridges and routers is that routers are intelligent devices that have decision-making and switching capabilities.

The basic function of a router is to expedite traffic flow on both networks and maintain maximum performance. When many users access a network at the same time, conflicts occur and speed performance is degraded. Routers are designed to recognize traffic buildup and provide automatic switching to reroute transmissions in a different direction, if possible. If transmission is blocked in one direction, the router can switch transmission through other nodes or other paths in the network.

Some routers are a combination of a bridge and a router. There are many different types of routers for the wide variety of networks in use. They can switch, perform protocol conversion, and serve as communication managers between two LANs or between a LAN and the Internet.

Gateways

A *gateway* is another internetwork device that acts as an interface between two LANs or between a LAN and a larger computer system. The primary benefit of a gateway is that it can connect networks with incompatible protocols and configurations. The gateway acts as a two-way translator that allows systems of different types to communicate.

Figure 12-17 shows a typical gateway system, one designed to interconnect one or more PC-based LANs to a mainframe. There are many different types of gateways available depending upon the equipment and protocols involved. Most gateways are computers and are sometimes referred to as *gateway servers*.

Figure 12-17   A gateway commonly connects a LAN to a larger host computer.

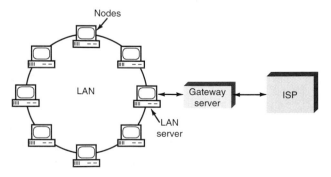

Chapter 12

As the number of companies that provide hardware increases, the functions of each device vary and devices labeled as routers may perform the functions of the switch, router, and gateway.

Modems

As discussed in Chap. 11, *modems* are interfaces between PCs and standard telephone systems. They convert the binary signals of the computer to audio-frequency analog signals compatible with the telephone system and, at the other end, convert the analog signals back to digital signals.

    Modems are widely used in home networking. The most common application is one in which remote PCs use modems to connect to an *Internet service provider (ISP)* which provides services such as Internet access and e-mail. Modems are also used in DSL and cable boxes.

Modem

Internet service provider (ISP)

Wireless LANs

One of the most complex and expensive parts of any LAN is the cabling. The cables themselves, as well as their installation and maintenance, are expensive, especially when the LAN is being installed in an existing building. In addition, in large, growing organizations, LAN needs change regularly. New users must be added, and the network must be reconfigured during expansion, reorganizations, and moves. One way to avoid the expense and headache of running and maintaining LAN cabling is to use *wireless LANs,* which communicate via radio.

    Each PC in a wireless LAN must contain a wireless modem or transceiver. This device is like a NIC that plugs in the computer, an external device that communicates through the standard USB I/O ports on a computer, or one or more chips built into the motherboard. In any case, the radio modem transceiver converts the serial binary data from the computer to radio signals for transmission and converts the received radio signals back to binary data. Wireless LANs operate as cable-connected LANs in that any node can communicate with any other node. Most wireless LANs have a top speed of 11 to 54 Mbps. See Chap. 21 for details.

Wireless LAN

12-3 Ethernet LANs

One of the oldest and by far the most widely used of all LANs is *Ethernet.* Ethernet, which was developed by Xerox Corporation at Palo Alto Research Center in the 1970s, was based on the Aloha wide-area satellite network implemented at the University of Hawaii in the late 1960s.

    In 1980, Xerox joined with Digital Equipment Corporation (now part of Hewlett-Packard) and Intel to sponsor a joint standard for Ethernet. The collaboration resulted in a definition that became the basis for the IEEE 802.3 standard. [The Institute of Electrical and Electronics Engineers (IEEE) establishes and maintains a wide range of electrical, electronic, and computing standards. The 802.X series relates to LANs.]

    Today, there are numerous variants of Ethernet. More than 95 percent of all LANs use some form of Ethernet.

Ethernet LAN

Topology

The original versions of Ethernet used a bus topology. Today, most use a physical star configuration (see Fig. 12-18). Each node uses a twisted-pair cable to connect into a centrally located hub or switch. The switch is then connected to a router or gateway which provides access to the services needed by each user. Information to be transmitted from one user to another can move in either direction on the bus, but only one node can transmit at any

Figure 12-18 The Ethernet bus.

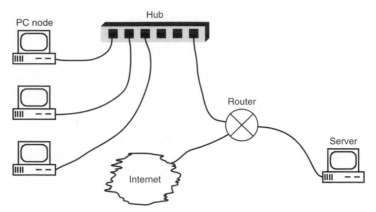

given time. The bus coaxial cable has a special terminating connector at each end containing a resistor whose value is equal to its characteristic impedance. This prevents signal reflections, which cause signal loss and significant data errors.

Encoding

Ethernet uses baseband data transmission methods. This means that the serial data to be transmitted is placed directly on the bus media. Before transmission, however, the binary data is encoded into a unique variation of the binary code known as the *Manchester code* (Fig. 12-19).

Figure 12-19(*a*) shows normal serial binary data made up of standard dc binary pulses. This serial binary data, which is developed internally by the computer, is known as *unipolar NRZ*. This data is encoded by using the Manchester format by circuits on the NIC. Manchester coding can be alternating or direct current. In ac Manchester, shown in Fig. 12-19(*b*), the signal voltage is switched between a minus level and a plus level. Each bit, whether it is a binary 0 or binary 1, is transmitted as a combination of a positive pulse followed by a negative pulse or vice versa. For example, a binary 1 is a positive pulse followed by a negative pulse. A binary 0 is a negative pulse followed by a positive pulse. During each bit interval a positive-to-negative or negative-to-positive voltage transition takes place.

This method of encoding prevents the dc voltage level on the transmission cable from building up to an unacceptable level. With standard binary pulses, an average dc voltage will appear on the cable, the value of which depends on the nature of the binary data on the cable, especially the number of binary 1s or 0s occurring in sequence. Large numbers of binary 1 bits will cause the average dc level to be very high. Long strings of binary 0s will cause the average voltage to be low. The resulting binary signal rides up and down on the average dc level, which can cause transmission errors and signal interpretation problems. With Manchester encoding, the positive and negative switching for each bit interval causes the direct current to be averaged out.

Ethernet uses dc Manchester [Fig. 12-19(*c*)]. The average dc level is about -1 V, and the maximum swing is about 0 to -2 V. The average direct current is used to detect the presence of two or more signals transmitting simultaneously.

Another benefit of Manchester encoding is that the transition in the middle of each bit time provides a means of detecting and recovering the clock signal from the transmitted data.

Speed

The standard transmission speed for Ethernet LANs is 10 Mbps. The time for each bit interval is the reciprocal of the speed, or $1/f$, where f is the transmission speed or frequency. With a 10-Mbps speed, the bit time is $1/10 \times 10^6 = 0.1 \times 10^{-6} = 0.1$ μs (100 ns) [see Fig. 12-19(*a*)].

Manchester code

Unipolar NRZ

Figure 12-19 Manchester encoding eliminates dc shift on baseband lines and provides a format from which a clock signal can be extracted. (*a*) Standard dc binary coding (unipolar NRZ). (*b*) Manchester ac encoding. (*c*) DC Manchester.

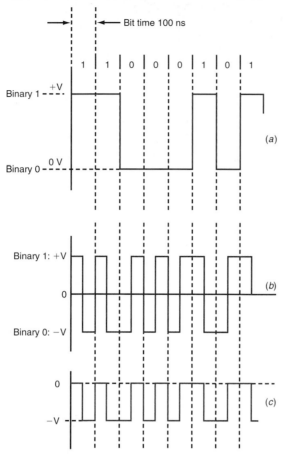

The most widely used version of Ethernet is called *Fast Ethernet.* It has a speed of 100 Mbps. Other versions of Ethernet run at speeds of 1 Gbps or 10 Gbps, typically over fiber-optic cable but also on shorter lengths of coaxial or twisted-pair cable.

Fast Ethernet

Transmission Medium

The original transmission medium for Ethernet was coaxial cable. However, today twisted-pair versions of Ethernet are more popular.

Coaxial Cable. The two main types of coaxial cable used in Ethernet networks are *RG-8/U* and *RG-58/U*. RG-8/U cable has a characteristic impedance of 53 Ω and has approximately a 0.405-in diameter. This large coaxial cable, referred to as *thick cable,* was originally designed for antenna transmission lines in RF systems but is now widely used in Ethernet LANs. It is usually bright yellow. Large type-N coaxial connectors are used to make the interconnections.

RG-8/U coaxial cable

RG-58/U coaxial cable

Figure 12-20 shows an Ethernet LAN using thick coaxial cable. Transceivers connect the nodes to the bus cable by way of NICs. A repeater extends the bus cable. Note the terminators at the end of the cable.

When thick network cable is used, it does not attach directly to the NICs. Instead, the inputs and outputs from the NIC terminate in an *attachment user interface (AUI).* This 15-pin connector, known as a *DB-15* or *DIX connector,* plugs into a matching connector on the NIC. The other end of the cable attaches to a transceiver unit, which is

Attachment user interface (AUI)

Figure 12-20 The Ethernet (10Base-5) bus.

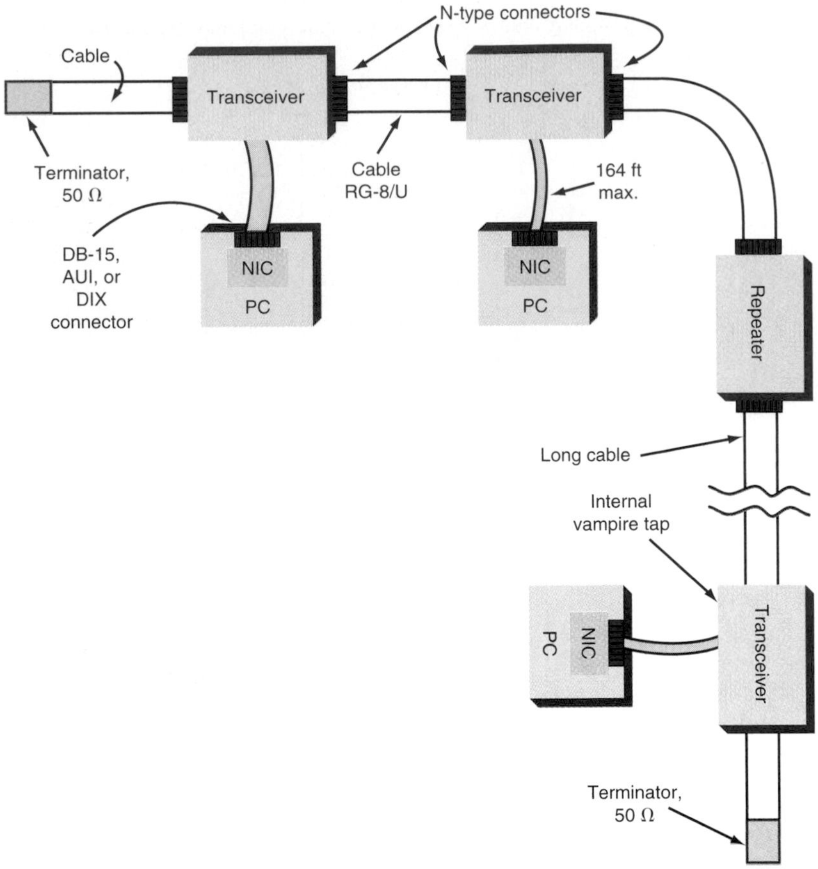

connected to the bus. The transceiver takes care of amplification and signal shaping in both transmit and receive operations.

Ethernet systems using thick coaxial cable are generally referred to as *10Base-5* systems, where *10* means a 10-Mbps speed, *Base* means baseband operation, and the *5* designates a 500-m maximum distance between nodes, transceivers, or repeaters. Ethernet LANs using thick cable are also referred to as *Thicknet.*

10Base-5 Ethernet LAN

In 10Base-5 Ethernet LANs, the cable is generally one long continuous bus that is routed from node to node. The availability of a wide variety of coaxial connectors permits shorter cables to be interconnected from computer to computer. The cable between the PC node and the transceiver can be up to 164 ft long, making node placement and wiring easy to implement. Depending on the distance between nodes, repeaters may have to be inserted to boost and rejuvenate the signal along the way. The bus is relatively easy to reconfigure; adding new nodes is not even difficult. The bus cable can be cut and new connectors installed, allowing the bus to pass through while at the same time connecting to the new NICs. Special connectors referred to as "vampire" taps can pierce the coaxial cable without cutting it.

10Base-2 Ethernet LAN

Ethernet systems implemented with thinner coaxial cable are known as *10Base-2,* or *Thinnet* systems; here the *2* indicates the maximum 200-m (actually, 185-m) run between nodes or repeaters. The most widely used thin cable is RG-58/U. It is much more flexible and easier to work with than RG-8/U cable.

Figure 12-21 shows a 10Base-2 LAN with RG-58/U. Transceivers are not used. The cable is attached directly to the NIC in each node by way of a BNC T connector, which lets the bus pass through while making the connection to the NIC. Repeaters can be used over long distances. The cable is terminated at each end with resistor terminators.

Figure 12-21 10Base-2 coaxial Ethernet bus.

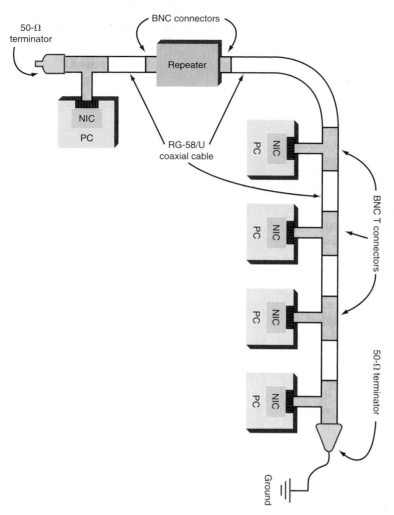

Coaxial versions of Ethernet are no longer widely used. Most LANs today use twisted-pair cable because it is simpler and less expensive. Some older legacy coaxial systems may still exist, and some special industrial applications may use them because of the interference rejection provided by the coaxial cable.

Twisted-Pair Cable. More recent versions of Ethernet use twisted-pair cable. The twisted-pair version of Ethernet is referred to as a *10Base-T network,* where the *T* stands for twisted-pair. The twisted-pair cable used in 10Base-T systems is standard 22-, 24-, or 26-gauge solid copper wire with RJ-45 modular connectors. The PC nodes connect to a hub, as shown in Fig. 12-22, which provides a convenient way to connect all nodes. Each port contains a repeater that rejuvenates the signal and buffers it for retransmission. Physically, a 10-Base-T LAN looks like a star, but the bus is implemented inside the hub itself. It is usually easier and cheaper to install 10Base-T LANs than it is to install coaxial Ethernet systems, but the transmission distances are often limited to less than 100 m.

10Base-T network

100-Mbps Ethernet. Several 100-Mbps versions have been developed including 100Base-T or 100Base-TX, also called Fast Ethernet, 100VG-AnyLAN, 100Base-T4, and 100Base-FX. All use twisted-pair cable except the FX version, which uses fiber-optic cable. The TX and FX methods use the carrier sense multiple access with collision detection (CSMA/CD) access method (described later in this chapter) and have the same packet size. The 100VG-AnyLAN and -T4 versions use unique access methods. The 100VG-AnyLAN

Figure 12-22 Hub for Ethernet 10Base-T.

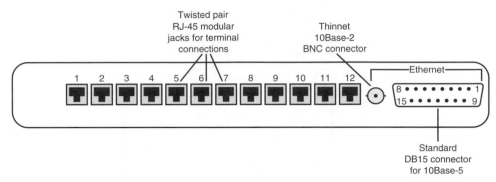

version has the IEEE standard number 802.12, and the other versions are subsets of the IEEE 802.3 standards. Most, if not all, NICs support both 10- and 100-Mbps rates.

100Base-TX (Fast Ethernet)

By far the most popular version of 100-Mbps Ethernet is *100Base-TX*, or *Fast Ethernet*. It uses two unshielded twisted pairs instead of the single pair used in standard 10Base-T. One pair is used for transmitting, and the other is used for receiving, permitting full duplex operation, which is not possible with standard Ethernet. To achieve such high speeds on UTP, several important technical changes were implemented in Fast Ethernet. First, the cable length is restricted to 100 m of category 5 UTP. This ensures minimal inductance, capacitance, and resistance, which distort and attenuate the digital data.

MLT-3 encoding

Second, a new type of encoding method is used. Called *MLT-3,* this coding method is illustrated in Fig. 12-23. The standard NRZ binary signal is shown at (*a*), and the MLT-3 signal is illustrated at (*b*). Note that three voltage levels are used: $+1$, 0, and -1 V. If the binary data is a binary 1, the MLT-3 signal changes from the current level to the next level. If the binary input is 0, no transition takes place. If the signal is currently at $+1$ and a 1111 bit sequence occurs, the MLT-3 signal changes from $+1$ to 0 to -1 and then to 0 and then to $+1$ again. What this coding method does is to greatly reduce the frequency of the transmitted signal (to one-quarter of its original signal), making higher bit rates possible over UTP.

As for access method, 100Base-T uses CSMA/CD. Most 100-Mbps Ethernet systems use the 100Base-T format described here.

100Base-T4 is a less-used form of 100Base-T. It uses category 3 or 5 UTP up to 100 m from PC to hub. This version of Ethernet uses four twisted pairs. Three pairs are used to transmit the data in a parallel format, and the fourth pair is used for access determination. A version of CSMA/CD is used. With this arrangement, three pairs essentially carry 33 Mbps, each in parallel. Full duplex is not supported.

The 100VG-AnyLAN version of Ethernet was developed by Hewlett-Packard. It is considerably different from all other versions of Ethernet. It uses category 3 UTP up to 100 m long. Most versions use all four pairs in the cable, but some use just two pairs.

The topology of 100VG-AnyLAN is a star. All cables from the PC go into a central hub that is controlled by a server computer. The access method is called *demand priority*. A proprietary encoding scheme is also used.

Figure 12-23 MLT-3 encoding used with 100Base-T Ethernet. (*a*) NRZ. (*b*) MLT-3.

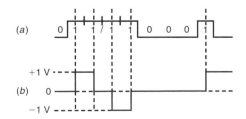

100Base-FX is a fiber-optic cable version of Ethernet. It uses two multimode fiber strands to achieve the 100-Mbps rate. The access method is CSMA/CD. This version of Ethernet is used primarily to interconnect individual LANs to one another over long distances. Full duplex operation can be achieved at distances up to 2 km.

Gigabit Ethernet. *Gigabit Ethernet* (1 GE) is capable of achieving 1000 Mbps or 1 Gbps over category 5 UTP or fiber-optic cable. The 1-Gbps speed is more readily achieved with fiber-optic cable, but it is more expensive. The UTP version of Gigabit Ethernet is defined by the IEEE standard 802.3ab and is generally referred to as 1000Base-T.

Gigabit Ethernet

The 1-Gbps rate is achieved by transmitting 1 byte of data at a time as if in a parallel data transfer system. This is done by using four UTP cables plus a coding scheme that transmits 2 bits per baud (b/Bd). The basic arrangement is shown in Fig. 12-24. One byte of data is divided into 2-bit groups. Each 2-bit group is sent to a D/A converter and an encoder that generates a five-level line code. The five levels are $+2, +1, 0, -1$, and -2 V. Each 2-bit group needs four coding levels. For example, 00 might be $+2, 01 = +1, 10 = -1$, and $11 = -2$ V. The 0-V level is not used in coding but is used in clock synchronization and in an error detection scheme. You may hear this encoding scheme referred to as PAM5. The resulting five-level code is fed to each twisted pair at a 125-MBd rate. Since 2 bits is transmitted for each baud or symbol level, the data rate on each twisted pair is 250 Mbps. Together, the four pairs produce a composite data rate of $4 \times 250 = 1000$ Mbps or 1 Gbps. Maximum distance is 100 m. Better and more reliable performance is achieved by restricting the distance from PC to hub to 25 m or less. The CSMA/CD access method is used, and communication can be half or full duplex.

The fiber-optic cable versions of Gigabit Ethernet are defined under the IEEE 802.3z standard. All data transmissions are serial, unlike transmissions with the 1000Base-T UTP version. The 1000Base-LX version uses a *single-mode fiber (SMF)* cable with a diameter of 9 μm and a transmitting laser operating at an infrared wavelength of 1310 nm. It can achieve 1 Gbps on a cable length to 10 km. When the larger, less expensive *multimode fiber (MMF)* is used, the maximum operating distances are 550 m with 50-μm-diameter fiber and 500 m with 62.5-μm-diameter fiber.

Single-mode fiber (SMF)

Multimode fiber (MMF)

A less expensive option is 1000Base-SX, which uses a lower-cost 780-μm laser and MMF cable. The maximum cable length is 550 m with 50-μm-diameter fiber and 275 m with 62.5-μm-diameter fiber. You will learn the details of fiber-optic communication in Chap. 19.

Figure 12-24 Gigabit Ethernet.

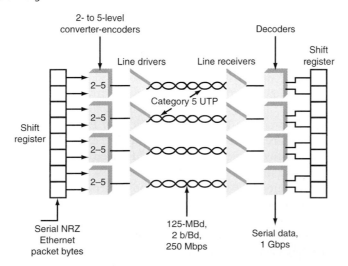

All the fiber-optic versions of Gigabit Ethernet also use a different data encoding scheme called *8B/10B*. This method takes each 8-bit byte (also called an *octet*) of data and translates it into a 10-bit word. The purpose is to ensure that an equal number of binary 0s and 1s are always transmitted. This makes clock recovery and data synchronization easier. The 8B/10B encoding scheme also allows a simple way to implement error detection for 1-bit errors. In addition, it eliminates any dc bias buildup that may occur if the signal is transmitted through a transformer. Finally, the 8B/10B code also implements frame delimiting, providing a way to clearly mark the beginning and ending of a frame of data with unique 10-bit codes.

10-Gbit Ethernet (10GE)

10-Gbit Ethernet. The newest version of Ethernet is *10-Gbit Ethernet (10 GE)*, which permits data speeds up to 10 Gbps over fiber-optic cable. The defining IEEE standard is 802.3ae. As with Gigabit Ethernet, there are several versions as indicated in the table below. All use 8B/10B coding.

Laser wavelength λ	Cable type/size (μm)	Maximum cable length
850-nm serial	MMF/50	65 m
1310-nm WWDM	MMF 62.5	300 m
1310-nm WWDM	SMF/9	10 km
1310-nm serial	SMF/9	10 km
1550-nm serial	SMF/9	40 km

Wide-wavelength division multiplexing (WWDM)

Three of the five variations use serial data transmission. The other two use what is called *wide-wavelength division multiplexing (WWDM)*. Also known as *coarse wavelength division multiplexing (CWDM)*, these versions divide the data into four channels and transmit it simultaneously over four different wavelengths of infrared light near 1310 nm. WWDM is similar to frequency-division multiplexing. This technique is described in greater detail in Chap. 19.

It is hard to believe that a 10-Gbps pulse signal could be carried over a copper cable, given the huge attenuation and distortion that the cable capacitance, inductance, and resistance can cause. Yet, today there is one copper version of 10GE now available. Called 10GBase-CX4, this version of Ethernet is standardized by the IEEE standard 802.3ak. The cable is a special twin axial coaxial cable (called twin-ax) that contains two conductors inside the outer shield. Four of these coaxial cable assemblies are combined to make a cable. The data to be transmitted is divided into four parallel paths that transmit at a 3.125-Gbps rate. The encoding is 8B/10B, meaning that only 80 percent of the bits transmitted is actual data. This gives an actual data rate of 0.8×3.125, or 2.5, Gbps. Four paths give an aggregate of 10 Gbps. The range is limited to roughly 15 m or about 50 ft. This is sufficient for connecting several servers, routers, Ethernet switches, and other equipment in wiring closets, data centers, or server farms, such as those in Internet companies where the various pieces of equipment are located close to one another.

Another copper cable version of 10GE has been in development for several years. Designated 802.3an or 10GBase-T, it is designed to use the four pairs of conductors in CAT5e or CAT6 UTP so that existing cabling can be used. The range is 100 m as with most other versions of Ethernet. Because of the severe cross talk that occurs in UTP at 10 Gbps, extensive DSP filtering and equalization is employed.

One final point is in order. Copper cable versions of Ethernet are attractive despite their complexity simply because they are significantly less expensive than fiber-optic versions. While the cost of fiber-optic equipment has declined over the years, it is still 3 to 10 times more expensive than a copper solution.

The primary application of 1-Gbps and 10-Gbps Ethernet is still in LANs. Not only can LANs be faster and handle more users, but also they can increase in size with these forms of Ethernet. Distances up to several kilometers are easily achieved with some versions. The 1-Gbps and 10-Gbps Ethernet technology is used to create a fast LAN backbone that links and aggregates slower LANs.

GOOD TO KNOW

The fast 10-Gbps Ethernet can be used as a fast LAN backbone to connect slower LANs, even over distances of up to several kilometers.

The speed and distance capabilities also make 1-Gbps and 10-Gbps Ethernet attractive for MAN applications. The 40-km version even makes 10-Gbps Ethernet appropriate for some WAN applications. It may eventually replace the more complex and expensive synchronous optical network (SONET) fiber-optic equipment now common in most MANs and WANs. SONET is covered in Chap. 19.

Wireless Ethernet. Several versions of Ethernet for transmission by radio have been developed. This permits wireless LANs to be created. The cost and complexity of buying and installing cables are eliminated, and nodes can be relocated at any time without regard to where the cable is. Each PC is equipped with a NIC that incorporates a wireless transceiver. Wireless LANs using Ethernet are described in detail in Chap. 21.

Ethernet in the First Mile. Also known as *Ethernet Passive Optical Network (EPON), Ethernet in the First Mile (EFM)* is the IEEE standard 802.3ah. It is a version of Ethernet designed to be used in fiber-optic networks that connect homes and businesses to high-speed Internet services. The *first mile,* also called the *last mile,* is a term used to describe the relatively short connection from a home or office to a local terminal or connection point that distributes data via a fiber-optic link. The EFM system uses the standard Ethernet protocols at a speed of 1.25 Gbps. It permits up to 32 users per connection, and the maximum range is about 20 km. More details are given in Chap. 19 on fiber optics.

Ethernet Passive Optical Network (EPON), Ethernet in the First Mile (EFM)

Power over Ethernet. *Power over Ethernet (PoE)* is an addition to Ethernet LANs that is used to deliver dc power to remote devices connected to the network. Specifically, it supplies about 48 V of unregulated direct current over two of the twisted pairs in a CAT5 UTP cable. This eliminates the need for some devices on the LAN to have their own power supply, and it eliminates the need for that remote device to be near a 120- or 240-V ac outlet. Some examples are wireless access points used to extend the LAN and *Voice over Internet Protocol (VoIP)* telephones which are rapidly replacing standard switched analog phones. There are numerous industrial applications as well.

Power over Ethernet (PoE)

Voice over Internet Protocol (VoIP)

Figure 12-25 shows a common PoE arrangement. On the left, a 48-V dc power supply is connected to the center taps of the I/O transformers in the Ethernet NIC. These transformers carry the serial Ethernet data. Both wires in each pair carry the

Figure 12-25 Power over Ethernet supplies dc power over the LAN cable to eliminate the need for external power.

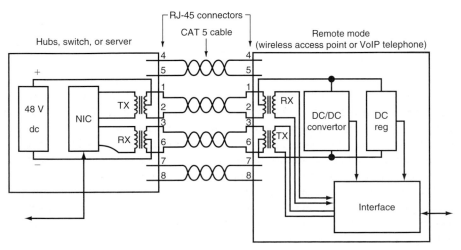

direct current. The wires in the twisted pairs are effectively in parallel for direct current. The direct current does not interfere with the data.

On the receiving end, transformers accept the signal and pass it along to the NIC circuitry in that device as usual. The dc voltage is captured from the center taps. This dc voltage is then translated to another dc level by a dc-dc converter or a voltage regulator. Voltages of 24, 12, 6, 5, and 3.3 V are common. This voltage powers the interface circuits at that end of the cable, thereby eliminating the need for a separate ac line or power supply.

The choice of 48 V was based on the fact that the wires in a CAT5 cable are very small, usually no. 28. Smaller wires have higher dc resistance and so can produce a rather large voltage drop along the cable. By keeping the voltage high, the line current is less for a given amount of power consumption in the load, thereby producing much less of a voltage drop. In practice, the maximum range is 100 m, and the voltage can usually be anything from 44 to 57 V as the dc supply is unregulated.

The maximum allowed current is 550 mA, although the current is usually held to a value of 350 mA or less. At 48 V, this translates to a maximum current consumption of 16.8 W. The standard states that the maximum desirable load is 15.4 W. Most loads consume much less than that.

Power over Ethernet is designed to work with all UTP versions of Ethernet including 10-, 100-, and 1000-Mbps systems. Only two pairs are used. The dc power is applied to the cable with a separate piece of equipment called an injector. Sometimes the direct current is supplied inside a hub or switch. Different versions of the standard vary with the pairs defined to carry the direct current and which pins on the RJ-45 connectors are used. Some companies offer variations that supply 12 V instead of 48 V.

Access Method

Access method

Carrier sense multiple access with collision detection (CSMA/CD)

Packet (frame)

Access method refers to the protocol used for transmitting and receiving information on a bus. Ethernet uses an access method known as *carrier sense multiple access with collision detection (CSMA/CD)*, and the IEEE 802.3 standard is primarily devoted to a description of CSMA/CD.

Whenever one of the nodes on an Ethernet system has data to transmit to another node, the software sends the data to the NIC in the PC. The NIC builds a *packet*, a unit of data formed with the information to be transmitted. The completed packet is stored in RAM on the NIC while the sending node monitors the bus. Since all the nodes or PCs in a network monitor activity on the bus, whenever a PC is transmitting information on the bus, all the PCs detect (sense) what is known as the carrier. The carrier in this case is the Ethernet data being transmitted at a 10-Mbps (or 100-Mbps) rate. If a carrier is present, none of the nodes will attempt to transmit.

When the bus is free, the sending station initiates transmission. The transmitting node sends one complete packet of information. In a sense, the transmitting node "broadcasts" the data on the bus so that any of the nodes can receive it. However, the packet contains a specific binary address defining the destination or receiving node. When the packet is received, it is decoded by the NIC, and the recovered data is stored in the computer's RAM.

Although a node will not transmit if a carrier is sensed, at times two or more nodes may attempt to transmit at the same time or almost the same time. When this happens, a *collision* occurs. If the stations that are attempting to transmit sense the presence of more than one carrier on the bus, both will terminate transmission. The CSMA/CD algorithm calls for the sending stations to wait a brief time and then attempt to transmit again. The waiting interval is determined randomly, making it statistically unlikely that both will attempt retransmission at the same time. Once a transmitting node gains control of the bus, no other station will attempt to transmit until the transmission is complete. This method is called a *contention system* because the nodes compete, or *contend,* for use of the bus.

Figure 12-26 Ethernet frame formats.

Ethernet (original)

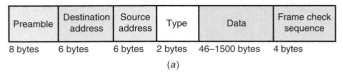

Preamble	Destination address	Source address	Type	Data	Frame check sequence
8 bytes	6 bytes	6 bytes	2 bytes	46–1500 bytes	4 bytes

(a)

IEEE 802.3

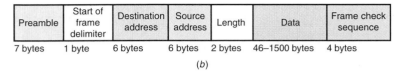

Preamble	Start of frame delimiter	Destination address	Source address	Length	Data	Frame check sequence
7 bytes	1 byte	6 bytes	6 bytes	2 bytes	46–1500 bytes	4 bytes

(b)

In Ethernet LANs with few nodes and little activity, gaining access to the bus is not a problem. However, the greater the number of nodes and the heavier the traffic, the greater the number of message transmissions and potential collisions. The contention process takes time; when many nodes attempt to use a bus simultaneously, delays in transmission are inevitable. Although the initial delay might be only tens or hundreds of microseconds, delay times increase when there are many active users. Delay would become an insurmountable problem in busy networks were it not for the packet system, which allows users to transmit in short bursts. The contention process is worked out completely by the logic in the NICs.

Packet Protocols

Figure 12-26(*a*) shows the packet (frame) protocol for the original Ethernet system, and Fig. 12-26(*b*) shows the packet protocol defined by IEEE standard 802.3. The 802.3 protocol is described here.

The packet is made up of two basic parts: (1) the frame, which contains the data plus addressing and error detection codes, and (2) an additional 8 bytes (64 bits) at the beginning, which contains the *preamble* and the *start frame delimiter (SFD)*. The preamble consists of 7 bytes of alternating 0s and 1s that help to establish clock synchronization within the NIC, and the SFD announces the beginning of the packet itself with the code 10101011.

The *destination address* is a 6-byte, 48-bit code that designates the receiving node. This is followed by a 6-byte source address that identifies the sending node. Next is a 2-byte field that specifies how many bytes will be sent in the data field. Finally, the data itself is transmitted. Any integer number of bytes in the range from 46 to 1500 bytes can be sent in one packet. Longer messages are divided up into as many separate packets as required to send the data.

Finally, the packet and frame end in a 4-byte frame check sequence generated by putting the entire transmitted data block through a *cyclical redundancy check (CRC)*. The resulting 32-bit word is a unique number designating the exact combination of bits used in the data block. At the receiving end, the NIC again generates the CRC from the data block. If the received CRC is the same as the transmitted CRC, no transmission error has occurred. If a transmission error occurs, the software at the receiving end will be notified. Sometimes when the data length is short, 1 or more padding bytes (octets) are added between the data and CRC.

It is important to point out that data transmission in Ethernet systems is synchronous: The bytes of data are transmitted end to end without start and stop bits. This speeds up data transmission but puts the burden of sorting the data on the receiving equipment. The clocking signals to be used by the digital circuits at the receiving end are derived from the transmitted data itself to ensure proper synchronization and counting of bits, bytes, fields, and frames.

Preamble

Start frame delimiter (SFD)

Cyclical redundancy check (CRC)

GOOD TO KNOW

When you want to determine how long a packet transmission will take, first find out how fast the bit rate is; then find out how large the packet will be. The frame size is the total of the data and the header size. The header size is different for each different type of packet (10-Mbps Ethernet, Token-Ring, etc.).

Example 12-2

How fast can a 1500-byte block of data be transmitted on (*a*) a 10-Mbps Ethernet (IEEE 802.3) packet and (*b*) a 16-Mbps Token-Ring packet? [*Note:* For the Ethernet IEEE 802.3 format, use Fig. 12-26(*b*); for the Token-Ring format, use Fig. 12-30(*a*)].

a. Time for 1 bit $= \dfrac{1}{10} \times 10^6 = 100$ ns

Time for 1 byte $= 8 \times 100 = 800$ ns

Time for 1526 bytes $= 1526 \times 800$ ns $= 1220.800$ ns (1.2208 ms)

b. Time for 1 bit $= \dfrac{1}{16} \times 10^6 = 62.5$ ns

Time for 1 byte $= 8 \times 62.5 = 500$ ns

Time for 1521 bytes $= 1521 \times 500 = 760{,}500$ ns (0.7605 ms)

Ethernet and the OSI Layers

It is helpful to understand how Ethernet relates to the OSI model discussed in Chap. 11. Ethernet actually defines the first two layers; layer 1 is the physical layer (PHY) and layer 2 is the data link layer. The physical layer defines all the hardware that handles the transfer of bits from one place to another. It identifies the medium type, connectors, encoding and decoding, and all the related signaling functions. Any network interface card or chip includes the physical layer components. Repeaters and hubs are also layer 1 devices.

The data link layer is actually divided into to sublayers called the *media access control (MAC)* sublayer and the *logical link control (LLC)* sublayer. The MAC layer handles all the encapsulation of the data to be transmitted, which includes the wrapping of the preamble, SOF delimiter, destination and source addresses, and CRC around the data. Hardware within the interface card or chip handles these chores.

The LLC assists the MAC sublayer in dealing with the transmitted or received data in the upper layers of the OSI stack. In the data field shown in Fig. 12-26 (*b*), several additional fields are appended to the beginning of the data. These fields control the handling of the data, identify services available, and define the protocol to be used in the upper layers. There are several variations of these additional fields which are beyond the scope of this text.

12-4 Token–Ring LAN

Another widely used LAN configuration is the *Token Ring,* originally developed by IBM. IBM's Token-Ring LAN was announced in 1985; since then, the Token-Ring configuration has been established as a standard. Its IEEE standard number is 802.5. Subsequently Token Ring has been revised and updated over the years.

Topology

In the Token-Ring configuration, all the nodes or PCs in the network are connected end to end in a continuous circle or loop (see Fig. 12-27). The data in the network travels only in one direction on the ring. The transmitted information passes through the NICs

Figure 12-27 A Token-Ring LAN.

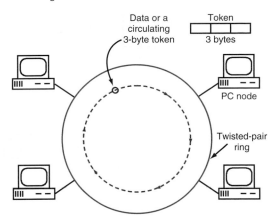

of each PC in the loop. Like Ethernet and most other standard networks, Token Ring uses baseband transmission; the binary data is placed directly on the cable. When the ring is not transmitting data, a 3-byte token code, the access key, is passed continuously around the ring.

Encoding

A modified version of the Manchester coding scheme is used.

Speed

The original Token Ring introduced by IBM ran at a speed of 4 Mbps, which is still fast enough to provide excellent communication in small- to medium-sized LANs. In 1989 IBM announced and began shipping a 16-Mbps Token-Ring system. It accommodates more nodes at higher speeds and thus provides better performance for large, active networks. Today both the 4- and 16-Mbps versions are in use. Many NICs contain circuitry that can accommodate either speed. More recent versions of Token Ring run at 100 Mbps and 1 Gbps.

Transmission Medium

Token-Ring LANs use twisted-pair cable. In general, the smaller, lower-speed systems are unshielded twisted-pair cable and the larger, faster systems use shielded twisted-pair cable. However, either type of cable can be used with either system. Connections are usually made by using RJ-45 modular connectors.

Because the data transmitted around the ring must pass through the NIC of all PCs in the loop, two twisted-pair cables are required for the connection at each node (see Fig. 12-28). One twisted pair, the *ring in (RI)*, comes into the card; another twisted pair, the *ring out (RO)*, carries the data out to the next node. Both twisted pairs are contained within a single cable and attach to the NIC by way of a DB-25 or DB-9 connector. The cable on this connector is usually terminated with the RJ-45 modular connector, which plugs into a wall jack that attaches to the ring cabling.

It is extremely difficult to run twisted-pair cable around an office area or building in a single continuous loop as required by the ring topology. To make wiring simpler, two twisted pairs in a single cable, as described earlier, are used. These cables terminate at a wiring hub referred to as a *multistation access unit (MAU)*. The input and the output of each NIC of each node appear at the MAU. The wiring inside the MAU connects the PCs in a logical ring, as shown in Fig. 12-29. The cable from each PC contains two twisted pairs, one for the RI connections and the other for the RO, which terminate at the MAU. The RI and RO connections are made inside the MAU, creating a daisy chain ring.

Multistation access unit (MAU)

Figure 12-28 Token-Ring wiring.

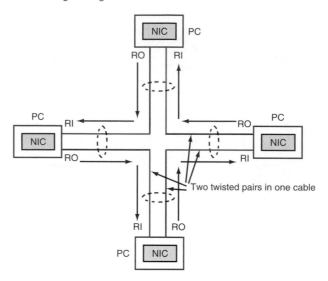

Each MAU hub is designed to accommodate eight nodes. In LANs with more nodes, multiple MAUs can be interconnected to form a larger ring. When MAUs are used as wiring hubs, the Token-Ring LAN takes on the physical configuration of a star, despite the fact that it is wired as a ring.

The MAU also contains circuitry that detects when a PC node is on. When a PC is on, the NIC sends a signal to the MAU. Inside the MAU, a relay is activated that connects the PC into the ring. If the PC node is off, the MAU does not get a signal. Thus the relay in the MAU short-circuits the connections to the PC and closes the loop for all the PCs that are turned on.

Because each node is, in effect, a repeater, there are no specific distance limitations on wiring between nodes or on total ring length. The signals coming in are read and repeated at each node; thus no dedicated repeaters are required. Despite this, there are some general guidelines about wiring distances. IBM, e.g., recommends a cable length between nodes and the MAU of less than 50 m (150 ft). Other companies have used cable lengths between node and MAU of 100 to 150 m (350 to 500 ft). The maximum practical distance between hubs and MAUs seems to be 150 m (500 ft).

IBM also recommends that the number of nodes for the 4-Mbps Token Ring be limited to 72, and that the number of nodes for the 16-Mbps system not exceed 250. Higher-speed versions have more severe distance restrictions.

Figure 12-29 Wiring of the Token Ring through the MAU.

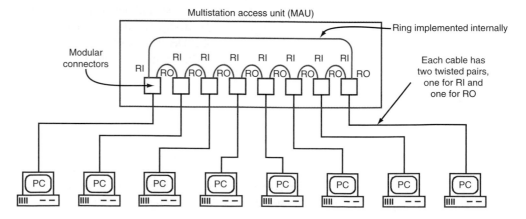

Access Method

The access method used by Token-Ring systems is *token passing*. A *token* is a unique binary word passed from one node to another around the ring. When no message is being sent, the token simply rotates around the ring from node to node. Whenever a node desires to transmit information, it captures the token. The NIC builds a packet or frame of information and transmits the data on the ring. This information passes through the NIC of each node on the ring. The receiving station designated by an address in the transmitted information recognizes its address and captures the transmitted data. After the packet of information has been received, it continues on around the ring until it comes back to the transmitting node, which takes the data off the ring. (Messages are not permitted to make the loop more than once.) The sending station then releases the token, which begins to circulate again.

The token-passing method of access is completely different from CSMA/CD. There is no contention. The station desiring to transmit captures the token and then transmits. Once it has sent one packet of data, the token passes to the next station in sequence. If the next station does not have data to transmit, the token continues to pass on around the ring until a station wanting to transmit captures the token and sends its message.

Token passing

Packet Protocol

Figure 12-30 shows the packet protocol for a Token-Ring LAN. The token frame format [Fig. 12-30(*a*)] consists of 3 bytes: a *starting delimiter (SD)*, an *access control (AC)* byte, and an *ending delimiter (ED)*. Each byte contains unique patterns of bits that the NIC recognizes. The actual token is 1 bit in the AC byte.

A typical complete data frame is shown in Fig. 12-30(*b*). It begins with the SD and AC bytes, followed by a *frame control (FC)* byte, a 6-byte destination address, and then a 6-byte source address. Some frames permit a variable-length block containing routing information if the LAN calls for it.

The data (message information) is transmitted next in a synchronous format. The block of data is limited to the number of bytes that can be transmitted in the maximum length of time allotted to the node that has the token, or 10 ms, a relatively long time within the context of data transmission rates of 4 or 16 Mbps.

The data is followed by a 4-byte (32-bit) CRC frame check sequence that is generated to catch transmission errors. The packet ends with the ending delimiter and *frame status (FS)* bytes. If the message is too long to be accommodated in one packet, the NIC assembles additional frames to be transmitted sequentially until all the data is sent. Remember, if another node captures the token between packets, there will be a wait before the next packet can be sent.

Figure 12–30 Token-Ring packet format. (*a*) Token. (*b*) Data frame.

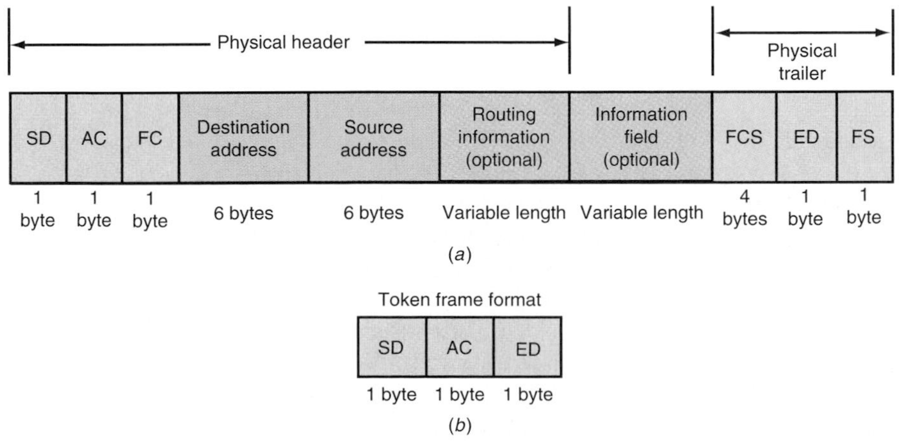

Token-Ring Status

For the most part, the majority of LANs today are implemented with Ethernet. It is estimated that more than 90 percent of all LANs are Ethernet. However, some Token-Ring installations still exist in older legacy systems and in IBM-equipped organizations. Over the years, Token Ring has been updated. In 1998, a 100-Mbps version of Token Ring (802.5t) was ratified, giving new life to this type of LAN. Today, a 1-Gbps version (802.5v) is also available. Overall, these upgrades were too little too late since Ethernet has virtually captured all the market for LAN equipment and software.

CHAPTER REVIEW

Summary

Networks are formed by computers linked by a communication medium for the purpose of information exchange. The four basic types of networks are wide-area networks (WANs), metropolitan-area networks (MANs), local-area networks (LANs), and personal-area networks (PANs). WANs and MANs are usually fiber-optic networks that connect a wide area (country, world) or a regional area (city). MANs are usually the entry point for WANs. LANs are the entry point for MANs. LANs are used inside an organization for communication with e-mail, filesharing, and Internet access.

Two other types of networks are the personal-area network (PAN) and the storage-area network (SAN). PANs connect portable devices to one another over short distances wirelessly. SANs connect servers to large-scale disk storage systems by fiber optics or via the Internet.

There are three basic LAN topologies: star, ring, and bus. The bus is the most popular. It is the topology of choice for the most widely implemented LAN, Ethernet. Ethernet can use large coaxial, small coaxial, twisted-pair, or fiber-optic cable as the transmission medium. Twisted-pair cable, especially category 5 (CAT5), is by far the most widely used. Ethernet transmission speeds are 10 and 100 Mbps and 1 Gbps over twisted-pair or 1 or 10 Gbps over fiber-optic cable. LANs can also be wireless by using radio.

There are two special versions of Ethernet, Ethernet in the First Mile (EFM) and Power over Ethernet (PoE). EFM uses the version of Ethernet for broadband access by passive optical networks to homes and small businesses. PoE is a method of distributing dc power over UTP cable to remote nodes on the network that need power, such as wireless access points and VoIP telephones.

An effective but less popular LAN is the Token Ring. It uses the ring topology, and different versions run at 4, 16, and 100 Mbps and at 1 Gbps.

Questions

1. What is the main purpose of a LAN?
2. Name an example of a WAN other than the telephone system.
3. What is a common example of a MAN?
4. What is the upper limit on the number of users on a LAN?
5. What is the name given to each PC in a network?
6. Name the four common LAN topologies.
7. Which two topologies are the most popular?
8. What is the name given to the main controlling PC in a LAN?
9. Give two examples of a WAN.
10. What transmission medium is used in most MANs and WANs?
11. What is probably the most common application on a LAN?
12. What is a SAN?
13. What transmission medium is used in a PAN?
14. What is the main advantage of a mesh networks?
15. What is the main advantage of coaxial cable over twisted-pair cable?
16. Name the two types of twisted-pair cable.
17. What is one of the primary specifications of coaxial cable?
18. Name the connector most widely used with twisted-pair cable in LANs.
19. What is the controller board used to connect a PC to a network?

20. What accessory is added to an Ethernet network if the distance between nodes is long or if the overall cable length is long and the signal is overly attenuated or distorted?
21. What piece of equipment is used to connect two LANs using the same formats or protocols?
22. What is an Ethernet switch, and what benefit does it offer?
23. What is a hub?
24. What is PoE and why is it used?
25. List the four Ethernet transmission speeds.
26. Name the basic topology of Ethernet.
27. What is the name of the line encoding method used with Ethernet, and why is it used?
28. Name the two main types of cables used with Ethernet.
29. What is the access method used by Ethernet called? (Give the full name and the abbreviation.)
30. Explain briefly how a station gains access to the LAN when Ethernet is used.
31. What is the maximum length of data that can be transmitted in one Ethernet packet?

32. What is the speed of Fast Ethernet?
33. In which layers of the OSI model does Ethernet work?
34. Explain the process by which data can be transmitted at 1 Gbps over copper cable.
35. What is the maximum transmission distance of 1-Gbps and 10-Gbps Ethernet? What two factors determine this distance?
36. What are the main applications of 1-Gbps and 10-Gbps Ethernet?
37. What data encoding scheme is used in 1-Gbps and 10-Gbps Ethernet? Why is it used?
38. What transmission medium is used in EFM?
39. What topology does a Token-Ring LAN physically resemble?
40. What are the speeds of the Token Ring?
41. What is the maximum size of the data block in a Token-Ring frame?
42. What method of error detection is used in Token Ring?
43. What types of connectors are used in Token Ring?

Problems

1. What are the basic data transmission rate of Fast Ethernet and its bit rate interval? ◆
2. What is the fastest Ethernet speed?
3. How many bytes can be transmitted in the 10 ms allowed by a LAN at the Token-Ring speed of 16 Mbps? ◆

4. Using 8B/10B encoding, what is the actual line speed of a network if the actual data rate is 10 Gbps?

◆ *Answers to Selected Problems follow Chap. 22.*

Critical Thinking

1. Networks are usually thought of in terms of general-purpose PCs in a LAN. However, other types of devices and computers are networked. Give one example.
2. Other than speed of transmission, what three key factors influence how fast two nodes in a LAN can communicate?

3. Explain why 1-Gbps/10-Gbps Ethernet is slower with the 8B/10B FEC.

USB connectors.

chapter **13**

Transmission Lines

Transmission lines in communication carry telephone signals, computer data in LANs, TV signals in cable TV systems, and signals from a transmitter to an antenna or from an antenna to a receiver. Transmission lines are critical links in any communication system. They are more than pieces of wire or cable. Their electrical characteristics are critical and must be matched to the equipment for successful communication to take place. Transmission lines are also circuits. At very high frequencies where wavelengths are short, transmission lines act as resonant circuits and even reactive components. At VHF, UHF, and microwave frequencies, most tuned circuits and filters are implemented with transmission lines. This chapter covers basic transmission line principles—theory, behavior, and applications.

Objectives

After completing this chapter, you will be able to:

- Name the different types of transmission lines and list some specific applications of each.
- Explain the circumstances under which transmission lines can be used as tuned circuits and reactive components.
- Define characteristic impedance and calculate the characteristic impedance of a transmission line by using several different methods.
- Compute the length of a transmission line in wavelengths.
- Define standing wave ratio (SWR), explain its significance for transmission line design, and calculate SWR by using impedance values or the reflection coefficient.
- State the criterion for a perfectly matched line and describe conditions that produce an improperly matched line.
- Use the Smith chart to make transmission line calculations.
- Define stripline and microstrip and state where and how they are used.

13-1 Transmission Line Basics

The two primary requirements of a *transmission line* are that (1) the line introduce minimum attenuation to the signal and (2) the line not radiate any of the signal as radio energy. All transmission lines and connectors are designed with these requirements in mind.

Types of Transmission Lines

Parallel-Wire Lines. *Parallel-wire line* is made of two parallel conductors separated by a space of $\frac{1}{2}$ in to several inches. Figure 13-1(a) shows a two-wire balanced line in which insulating spacers have been used to keep the wires separated. Such lines are rarely used today. A variation of parallel line is the 300-Ω twin-lead type shown in Fig. 13-1(b), where the spacing between the wires is maintained by a continuous plastic insulator. Parallel-wire lines are rarely used today.

Coaxial Cable. The most widely used type of transmission line is *coaxial cable,* which consists of a solid center conductor surrounded by a dielectric material, usually a plastic insulator such as Teflon [see Fig. 13-1(c)]. An air or gas dielectric, in which the center conductor is held in place by periodic insulating spacers, can also be used. Over the insulator is a second conductor, a tubular braid or shield made of fine wires. An outer plastic sheath protects and insulates the braid. Coaxial cable comes in a variety of sizes, from approximately $\frac{1}{4}$ in to several inches in diameter.

Twisted-Pair Cable. Twisted-pair cable, as the name implies, uses two insulated solid copper wires covered with insulation and loosely twisted together. See Fig. 13-1 (d). This type of cable was originally used in telephone wiring and is still used for that today.

> **GOOD TO KNOW**
>
> The major benefit of coaxial cable is that it is completely shielded so that external noise has little or no effect on it.

Figure 13-1 Common types of transmission lines. (a) Open-wire line. (b) Open-wire line called 300-Ω twin lead. (c) Coaxial cable (d) Twisted-pair cable.

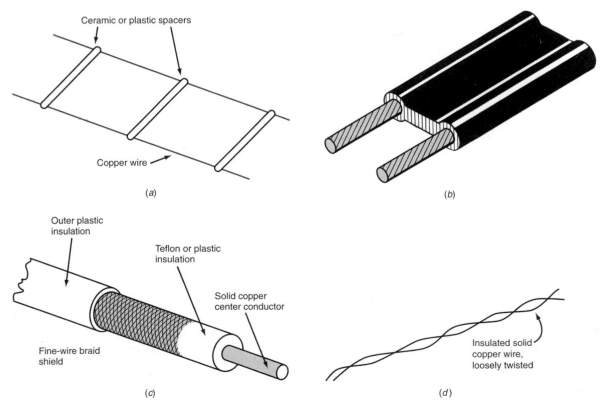

Ceramic or plastic spacers

Copper wire

(a)

(b)

Outer plastic insulation

Teflon or plastic insulation

Solid copper center conductor

Fine-wire braid shield

(c)

Insulated solid copper wire, loosely twisted

(d)

But it is also used for security system wiring of sensors and other equipment. And twisted-pair cable, as you saw in Chap. 12, is one of the most widely used types of wiring in local-area networks (LANs). It is generally known as unshielded twisted-pair (UTP) cable. There are many grades of twisted-pair cable for handling low-frequency audio or high-frequency pulses. The size of wire, type of insulation, and tightness of the twist (twists per inch) determine its characteristics. It is available with an overall braid shield and is called shielded twisted-pair (STP) cable. The most common version contains four pairs within a common insulated tubing.

Balanced Versus Unbalanced Lines

Balanced transmission line

Transmission lines can be balanced or unbalanced. A *balanced line* is one in which neither wire is connected to ground. Instead, the signal on each wire is referenced to ground. The same current flows in each wire with respect to ground, although the direction of current in one wire is 180° out of phase with the current in the other wire. In an *unbalanced line*, one conductor is connected to ground. The twisted-pair line [Fig. 13-2(*d*)] may be used in a balanced or an unbalanced arrangement, although the balanced form is more common.

Unbalanced transmission line

Open-wire line has a balanced configuration. A typical feed arrangement is shown in Fig. 13-2(*a*). The driving generator and the receiving circuit are center-tapped transformers in which the center taps are grounded. Balanced-line wires offer significant protection from noise pickup and cross talk. Because of the identical polarities of the signals on balanced lines, any external signal induced into the cable appears on both wires simultaneously but cancels at the receiver. This is called *common-mode rejection*, and noise reduction can be as great as 60 to 70 dB.

Common-mode rejection

Figure 13-2(*b*) shows an unbalanced line. Coaxial cables are unbalanced lines; the current in the center conductor is referenced to the braid, which is connected to ground. Coaxial cable and shielded twisted-pair cable provide significant but not complete protection from noise pickup or cross talk from inductive or capacitive coupling due to external signals. Unshielded lines may pick up signals and cross talk and can even radiate energy, resulting in an undesirable loss of signal.

Balun

It is sometimes necessary or desirable to convert from balanced to unbalanced operation or vice versa. This is done with a device called a *balun*, from "*bal*anced-*un*balanced."

Wavelength of Cables

The two-wire cables that carry 60-Hz power line signals into homes are transmission lines, as are the wires connecting the audio output of stereo receivers to stereo speakers.

Figure 13-2   (*a*) Balanced line. (*b*) Unbalanced line.

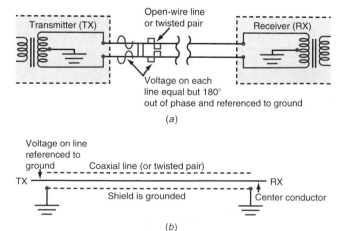

At these low frequencies, the transmission line acts as a carrier of the ac voltage. For these applications, the only characteristic of the cable of interest is resistive loss. The size and electrical characteristics of low-frequency lines can vary widely without affecting performance. An exception is conductor size, which determines current-carrying capability and voltage drop over long distances. The electrical length of conductors is typically short compared to 1 wavelength of the frequency they carry. A pair of current-carrying conductors is not considered to be a transmission line unless it is at least 0.1 λ long at the signal frequency.

Cables used to carry RF energy are not simply resistive conductors but are complex equivalents of inductors, capacitors, and resistors. Furthermore, whenever the length of a transmission line is the same order of magnitude as or greater than the wavelength of the transmitted signal, the line takes on special characteristics and requires a more complex analysis.

As discussed earlier, wavelength is the length or distance of one cycle of an ac wave or the distance that an ac wave travels in the time required for one cycle of that signal. Mathematically, wavelength λ is the ratio of the speed of light to the frequency of the signal f: $\lambda = 300,000,000/f$, where 300 million is the speed of light, in meters per second, in free space or air (300,000,000 m/s ≈ 186,400 mi/s) and f is in hertz. This is also the speed of a radio signal.

The wavelength of a 60-Hz power line signal is then

$$\lambda = \frac{300,000,000}{60} = 5 \times 10^6 \text{ m}$$

That's an incredibly long distance—several thousand miles. Practical transmission line distances at such frequencies are, of course, far smaller. At radio frequencies, however, say 3 MHz or more, the wavelength becomes considerably shorter. The wavelength at 3 MHz is $\lambda = 300,000,000/3,000,000 = 100$ m, a distance of a little more than 300 ft, or the length of a football field. That is a very practical distance. As frequency gets higher, wavelength gets shorter. At higher frequencies, the wavelength formula is simplified to $\lambda = 300/f$, where frequency is in megahertz. A 50-MHz signal has a wavelength of 6 m. By using feet instead of meters, the wavelength formula becomes $\lambda = 984/f$, where f is in megahertz (λ is now expressed in feet).

If the wavelength is known, frequency can be computed as follows:

$$f\,(\text{MHz}) = \frac{300}{\lambda\,(\text{m})} \qquad \text{or} \qquad f\,(\text{MHz}) = \frac{984}{\lambda\,(\text{ft})}$$

The distance represented by a wavelength in a given cable depends on the type of cable. The speed in a cable can be anywhere from 0.5 to 0.95 times the speed of light waves (radio waves) in space, and the signal wavelength in a cable will be proportionally

Example 13–1

For an operating frequency of 450 MHz, what length of a pair of conductors is considered to be a transmission line? (A pair of conductors does not act as a transmission line unless it is at least 0.1 λ long.)

$$\lambda = \frac{984}{450} = 2.19 \text{ ft}$$

$$0.1\,\lambda = 2.19(0.1) = 0.219 \text{ ft } (2.628 \text{ in})$$

less than the wavelength of that signal in space. Thus the calculated length of cables is shorter than wavelengths in free space. This is discussed later.

Example 13-2

Calculate the physical length of the transmission line in Example 13-1 a $\frac{3}{8}$ λ long.

$$\frac{3}{8}\lambda = \frac{2.19(3)}{8} = 0.82 \text{ ft } (9.84 \text{ in})$$

Connectors

Connector

Most transmission lines terminate in some kind of *connector,* a device that connects the cable to a piece of equipment or to another cable. An ordinary ac power plug and outlet are basic types of connectors. Special connectors are used with parallel lines and coaxial cable. Connectors, ubiquitous in communication equipment, are often taken for granted. This is unfortunate, because they are a common failure point in many applications.

Coaxial Cable Connectors. Coaxial cable requires special connectors that will maintain the characteristics of the cable. Although the inner conductor and shield braid could theoretically be secured with screws as parallel lines, the result would be a drastic change in electrical attributes, resulting in signal attenuation, distortion, and other

Coaxial connector

problems. Thus *coaxial connectors* are designed not only to provide a convenient way to attach and disconnect equipment and cables but also to maintain the physical integrity and electrical properties of the cable.

The choice of a coaxial connector depends on the type and size of cable, the frequency of operation, and the application. The most common types are the PL-259 or UHF, BNC, F, SMA, and N-type connectors.

PL-259 connector

The *PL-259 connector* is shown in Fig. 13-3(*a*); the internal construction and connection principles for the PL-259 are shown in Fig. 13-3(*b*). The body of the connector is designed to fit around the end of a coaxial cable and to provide convenient ways to attach the shield braid and the inner conductor. The inner conductor is soldered to a male pin that is insulated from the body of the connector, which is soldered or crimped to the braid. A coupler fits over the body; it has inner threads that permit the connector to attach to matching screw threads on a connector called the SO-239, which is mounted on a female chassis.

The PL-259, which is also referred to as a *UHF connector,* can be used up to low UHF values (less than 500 MHz), although it is more widely used at HF and VHF. It can accommodate both large (up to 0.5-in) and small (0.25-in) coaxial cable.

BNC connector

Another very popular connector is the *BNC connector* (Fig. 13-4). BNC connectors are widely used on 0.25-in coaxial cables for attaching test instruments, such as oscilloscopes, frequency counters, and spectrum analyzers, to the equipment being tested. BNC connectors are also widely used on 0.25-in coaxial cables in LANs and some UHF radios.

In BNC connectors, the center conductor of the cable is soldered or crimped to a male pin, and the shield braid is attached to the body of the connector. An outer shell or coupler rotates and physically attaches the connector to a mating female connector by way of a pin and cam channel on the rotating coupling [see Fig. 13-4(*b*)].

One of the many variations of BNC connectors is the *barrel connector,* which allows two cables to be attached to each other end to end, and the T coupler, which permits

SMA connector

taps on cables [see Fig. 13-4(*a*) and (*d*)]. Another variation is the *SMA connector,* which

Figure 13-3 UHF connectors. (*a*) PL-259 male connector. (*b*) Internal construction and connections for the PL-259. (*c*) SO-239 female chassis connector.

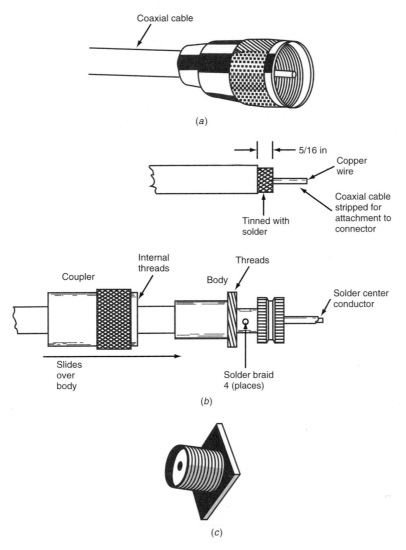

(*a*)

(*b*)

(*c*)

Figure 13-4 BNC connectors. (*a*) Male. (*b*) Female. (*c*) Barrel connector. (*d*) T connector.

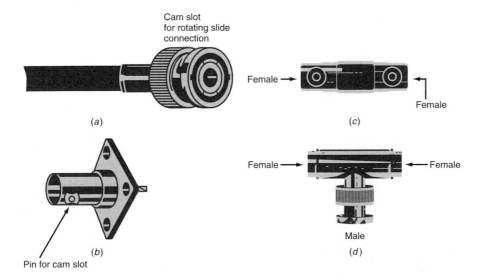

(*a*)

(*b*)

(*c*)

(*d*)

Figure 13-5 SMA connector.

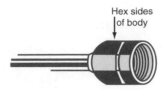

uses screw threads instead of the cam slot and pin (Fig. 13-5). The SMA connector is characterized by the hexagonal shape of the body of the male connector. Like the BNC connector, it is used with smaller coaxial cable.

F-type connector

The least expensive coaxial cable connector is the *F-type connector*, which is widely used for TV sets, VCRs, DVD players, and cable TV. The cable plug and its matching chassis jack are shown in Fig. 13-6. The shield of the coaxial cable is crimped to the connector, and the solid wire center conductor of the cable, rather than a separate pin, is used as the connection. A hex-shaped outer ring is threaded to attach the plug to the mating jack.

RCA phonograph connector

Another inexpensive coaxial connector is the well-known *RCA phonograph connector* (Fig. 13-7), which is used primarily in audio equipment. Originally designed over 60 years ago to connect phonograph pick-up arms from turntables to amplifiers, these versatile and low-cost devices can be used at radio frequencies and have been used for TV set connections in the low VHF range.

N-type connector

The best-performing coaxial connector is the *N-type connector* (Fig. 13-8), which is used mainly on large coaxial cable at the higher frequencies, both UHF and microwave. N-type connectors are complex and expensive, but do a better job than other connectors in maintaining the electrical characteristics of the cable through the interconnections.

Figure 13-6 The F connector used on TV sets, VCRs, and cable TV boxes.

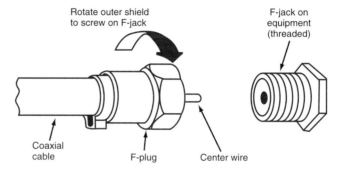

Figure 13-7 RCA phonograph connectors are sometimes used for RF connectors up to VHF.

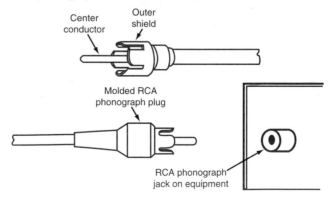

Figure 13-8 N-type coaxial connector.

Characteristic Impedance

When the length of a transmission line is longer than several wavelengths at the signal frequency, the two parallel conductors of the transmission line appear as a complex impedance. The wires exhibit considerable series inductance whose reactance is significant at high frequencies. In series with this inductance is the resistance of the wire or braid making up the conductors, which includes inherent ohmic resistance plus any resistance due to skin effect. Further, the parallel conductors form a distributed capacitance with the insulation, which acts as the dielectric. In addition, there is a shunt or leakage resistance or conductance (G) across the cable as the result of imperfections in the insulation between the conductors. The result is that to a high-frequency signal, the transmission line appears as a distributed low-pass filter consisting of series inductors and resistors and shunt capacitors and resistors [Fig. 13-9(a)]. This is called a *lumped* model of a distributed line.

In the simplified equivalent circuit in Fig. 13-9(b), the inductance, resistance, and capacitance have been combined into larger equivalent lumps. The shunt leakage resistance is very high and has negligible effect, so it is ignored. In short segments of the line, the series resistance of the conductors can sometimes be ignored because it is so low as to be insignificant. Over longer lengths, however, this resistance is responsible for considerable signal attenuation. The effects of the inductance and capacitance are considerable, and in fact they determine the characteristics of the line.

An RF generator connected to such a transmission line sees an impedance that is a function of the inductance, resistance, and capacitance in the circuit—the *characteristic* or *surge impedance* Z_0. If we assume that the length of the line is infinite, this impedance is resistive. The characteristic impedance is also purely resistive for a finite length of line if a resistive load equal to the characteristic impedance is connected to the end of the line.

Characteristic (surge) impedance

Figure 13-9 A transmission line appears as a distributed low-pass filter to any driving generator. (a) A distributed line with lumped components. (b) Simplified equivalent circuit.

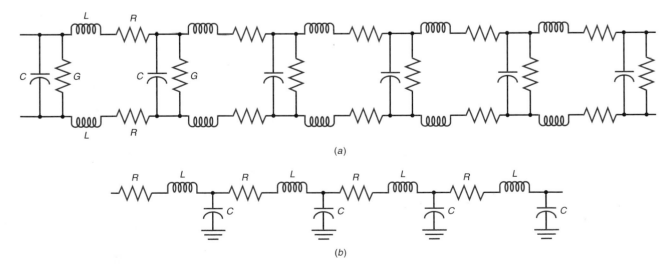

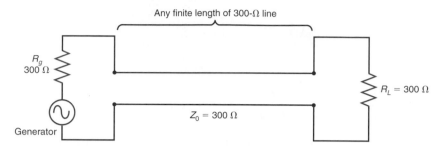

Any finite length of 300-Ω line

R_g
300 Ω

$R_L = 300$ Ω

$Z_0 = 300$ Ω

Generator

Determining Z_0 from Inductance and Capacitance. For an infinitely long transmission line, the characteristic impedance Z_0 is given by the formula $Z_0 = \sqrt{L/C}$, where Z_0 is in ohms, L is the inductance of the transmission line for a given length, and C is the capacitance for that same length. The formula is valid even for finite lengths if the transmission line is terminated with a load resistor equal to the characteristic impedance (see Fig. 13-10). This is the normal connection for a transmission line in any application. In equation form,

$$R_L = Z_0$$

If the line, load, and generator impedances are made equal, as is the case with matched generator and load resistances, the criterion for maximum power transfer is met.

An impedance meter or bridge can be used to measure the inductance and capacitance of a section of parallel line or coaxial cable to obtain the values needed to calculate the impedance. Assume, e.g., that a capacitance of 0.0022 μF (2200 pF) is measured for 100 ft. The inductance of each conductor is measured separately and then added, for a total of 5.5 μH. (Resistance is ignored because it does not enter into the calculation of characteristic impedance; however, it will cause signal attenuation over long distances.) The surge impedance is then

$$Z_0 = \sqrt{\frac{L}{C}} = \sqrt{\frac{5.5 \times 10^{-6}}{2200 \times 10^{-12}}} = \sqrt{2500} = 50 \ \Omega$$

In practice, it is unnecessary to make these calculations because cable manufacturers always specify impedance.

The characteristic impedance of a cable is independent of length. We calculated it by using a value of L and C for 100 ft, but 50 Ω is the correct value for 1 ft or 1000 ft. Note that the actual impedance approaches the calculated impedance only if the cable is several wavelengths or more in length as terminated in its characteristic impedance. For line lengths less than 1.0 λ, characteristic impedance does not matter.

Most transmission lines come with standard fixed values of characteristic impedance. For example, the widely used twin-lead balanced line [Fig. 13-1(b)] has a characteristic impedance of 300 Ω. Open-wire line [Fig. 13-1(a)], which is no longer widely used, was made with impedances of 450 and 600 Ω. The common characteristic impedances of coaxial cable are 52, 53.5, 75, 93, and 125 Ω.

• Velocity Factor

An important consideration in transmission line applications is that the speed of the signal in the transmission line is slower than the speed of a signal in free space. The velocity of propagation of a signal in a cable is less than the velocity of propagation of light in free space by a fraction called the *velocity factor* (VF), which is the ratio of the velocity in the transmission line V_p to the velocity in free space V_c:

$$VF = \frac{V_p}{V_c} \quad \text{or} \quad VF = \frac{V_p}{c}$$

where $V_c = c = 300,000,000$ m/s.

Velocity factors in transmission lines vary from approximately 0.5 to 0.9. The velocity factor of a coaxial cable is typically 0.6 to 0.8. Open-wire line has a VF of about 0.9, and 300-Ω twin-lead line has a velocity factor of about 0.8.

Calculating Velocity Factor.

The velocity factor in a line can be computed with the expression $VF = 1/\sqrt{\epsilon}$, where ϵ is the dielectric constant of the insulating material. For example, if the dielectric in a coaxial cable is Teflon, the dielectric constant is 2.1 and the velocity factor is $1/\sqrt{2.1} = 1/1.45 = 0.69$. That is, the speed of the signal in the coaxial cable is 0.69 times the speed of light, or $0.69 \times 300,000,000 = 207,000,000$ m/s (128,616 mi/s).

If a lossless (zero-resistance) line is assumed, an approximation of the velocity of propagation can be computed with the expression

$$V_p = \frac{1}{\sqrt{LC}} \quad \text{ft/s}$$

where l is the length or total distance of travel of the signal in feet or some other unit of length and L and C are given in that same unit. Assume, e.g., a coaxial cable with a characteristic impedance of 50 Ω and a capacitance of 30 pF/ft. The inductance per foot is 0.075 μH or 75 nH. The velocity of propagation per foot in this cable is

$$V_p = \frac{1}{\sqrt{75 \times 10^{-9} \times 30 \times 10^{-12}}} = 6.7 \times 10^8 \text{ ft/s}$$

or 126,262 mi/s, or 204×10^6 m/s.

The velocity factor is then

$$VF = \frac{V_p}{V_c} = \frac{204 \times 10^6}{300 \times 10^6} = 0.68$$

Calculating Transmission Line Length.

The velocity factor must be taken into consideration in computing the length of a transmission line in wavelengths. It is sometimes necessary to use a one-half or one-quarter wavelength of a specific type of transmission line for a specific purpose, e.g., impedance matching, filtering, and tuning.

The formula given earlier for one wavelength of a signal in free space is $\lambda = 984/f$. This expression, however, must be modified by the velocity factor to arrive at the true length of a transmission line. The new formula is

$$\lambda(\text{ft}) = 984 \frac{VF}{f(\text{MHz})}$$

Suppose, e.g., that we want to find the actual length in feet of a quarter-wavelength segment of coaxial cable with a VF of 0.65 at 30 MHz. Using the formula gives $\lambda = 984(VF/f) = 984(0.65/30) = 21.32$ ft. The length in feet is one-quarter of this, or $21.32/4 = 5.33$ ft.

The correct velocity factor for calculating the correct length of a given transmission line can be obtained from manufacturers' literature and various handbooks.

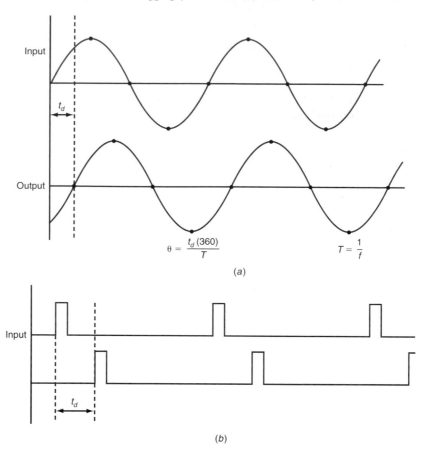

Figure 13-11 The effect of the time delay of a transmission line on signals. (*a*) Sine wave delay causes a lagging phase shift. (*b*) Pulse delay.

Input

Output

$$\theta = \frac{t_d(360)}{T} \qquad\qquad T = \frac{1}{f}$$

(*a*)

Input

t_d

(*b*)

Time delay (transit time)

Time Delay

Because the velocity of propagation of a transmission line is less than the velocity of propagation in free space, it is logical to assume that any line will slow down or delay any signal applied to it. A signal applied at one end of a line appears some time later at the other end of the line. This is called the *time delay* or *transit time* for the line. A transmission line used specifically for the purpose of achieving delay is called a *delay line.*

Figure 13-11 shows the effect of time delay on a sine wave signal and a pulse train. The output sine wave appears later in time than the input, so it is shifted in phase. The effect is the same as if a lagging phase shift were introduced by a reactive circuit. In the case of the pulse train, the pulse delay is determined by a factor that depends on the type and length of the delay line.

The amount of delay time is a function of a line's inductance and capacitance. The opposition to changes in current offered by the inductance plus the charge and discharge time of the capacitance leads to a finite delay. This delay time is computed with the expression

$$t_d = \sqrt{LC}$$

where t_d is in seconds and L and C are the inductance and capacitance, respectively, per unit length of line. If L and C are given in terms of feet, the delay time will be per foot. For example, if the capacitance of a particular line is 30 pF/ft and its inductance is 0.075 μH/ft, the delay time is

$$t_d = \sqrt{0.075 \times 10^{-6} \times 30 \times 10^{-12}} = 1.5 \times 10^{-9} \text{ or } 1.5 \text{ ns/ft}$$

A 50-ft length of this line would introduce $1.5 \times 50 = 75$ ns of delay.

Time delay introduced by a coaxial cable can also be calculated by using the formula

$$t_d = 1.016\sqrt{\epsilon} \qquad \text{ns/ft}$$

where t_d is the time delay in nanoseconds per foot and ϵ is the dielectric constant of the cable.

For example, the total time delay introduced by a 75-ft cable with a dielectric constant of 2.3 is

$$t_d = 1.016\sqrt{\epsilon} = 1.016\sqrt{2.3} = 1.016(1.517) = 1.54 \text{ ns/ft}$$

for a total delay of $1.54(75) = 115.6$ ns.

To determine the phase shift represented by the delay, the frequency and period of the sine wave must be known. The period or time T for one cycle can be determined with the well-known formula $T = 1/f$, where f is the frequency of the sine wave. Assume a frequency of 4 MHz. The period is

$$T = \frac{1}{4 \times 10^6} = 250 \times 10^{-9} = 250 \text{ ns}$$

The phase shift of the previously described 50-ft line with a delay of 75 ns is given by

$$\theta = \frac{360\, t_d}{T} = \frac{360(75)}{250} = 108°$$

Transmission line delay is usually ignored in RF applications, and it is virtually irrelevant in radio communication. However, in high-frequency applications where timing is important, transmission line delay can be significant. For example, in LANs, the time of transition of the binary pulses on a coaxial cable is often the determining factor in calculating the maximum allowed cable length.

Some applications require exact timing and sequencing of signals, especially pulses. A coaxial delay line can be used for this purpose. Obviously, a large roll of coaxial cable is not a convenient component in modern electronic equipment. As a result, artificial delay lines have been developed. These are made up of individual inductors and capacitors connected as a low-pass filter to simulate a distributed transmission line. Alternatively, a more compact distributed delay line can be constructed consisting of a coil of insulated wire wound over a metallic form. The coil of wire provides the distributed inductance and at the same time acts as one plate of a distributed capacitor. The metallic form is the other plate. Such artificial delay lines are widely used in TV sets, oscilloscopes, radar units, and many other pieces of electronic equipment.

Transmission Line Specifications

Figure 13-12 summarizes the specifications of several popular types of coaxial cable. Many coaxial cables are designated by an alphanumeric code beginning with the letters RG or a manufacturer's part number. The primary specifications are characteristic impedance and attenuation. Other important specifications are maximum breakdown voltage rating, capacitance per foot, velocity factor, and outside diameter in inches. The attenuation is the amount of power lost per 100 ft of cable expressed in decibels at 100 MHz. Attenuation is directly proportional to cable length and increases with frequency. Detailed charts and graphs of attenuation versus frequency are also available, so that users can predict the losses for their applications. Attenuation versus frequency for four coaxial cable types is plotted in Fig. 13-13. The loss is significant at very high frequencies. However, the larger the cable, the lower the loss. For purposes of comparison, look at the characteristics of 300-Ω twin-lead cable listed in Fig. 13-12. Note the low loss compared to coaxial cable.

Figure 13-12   Table of common transmission line characteristics.

Type of cable	Z_0, Ω	VF, %	C, pF/ft	Outside diameter, in	V_{max}, rms	Attenuation, dB/100 ft*
RG-8/U	52	66	29.5	0.405	4000	2.5
RG-8/U foam	50	80	25.4	0.405	1500	1.6
RG-11/U	75	66	20.6	0.405	4000	2.5
RG-11/U foam	75	80	16.9	0.405	1600	1.6
RG-58A/U	53.5	66	28.5	0.195	1900	5.3
RG-59/U	73	66	21.0	0.242	2300	3.4
RG-62A/U	93	86	13.5	0.242	750	2.8
RG-214/U	50	66	30.8	0.425	5000	2.5
9913	50	84	24.0	0.405	—	1.3
Twin-lead (open-line)	300	82	5.8	—	—	0.55

*At 100 MHz.

Example 13-3

A 165-ft section of RG-58A/U at 100 MHz is being used to connect a transmitter to an antenna. Its attenuation for 100 ft at 100 MHz is 5.3 dB. Its input power from a transmitter is 100 W. What are the total attenuation and the output power to the antenna?

$$\text{Cable attenuation} = \frac{5.3 \text{ dB}}{100 \text{ ft}} = 0.053 \text{ dB/ft}$$

$$\text{Total attenuation} = 0.053 \times 165 = 8.745 \text{ dB or } -8.745$$

$$\text{dB} = 10 \log \frac{P_{out}}{P_{in}}$$

$$\frac{P_{out}}{P_{in}} = \log^{-1} \frac{\text{dB}}{10} \qquad \text{and} \qquad P_{out} = P_{in}\left(\log^{-1} \frac{\text{dB}}{10}\right)$$

$$P_{out} = 100 \log^{-1}\left(\frac{-8.745}{10}\right) = 100 \log^{-1}(-0.8745)$$

$$P_{out} = 100(0.1335) = 13.35 \text{ W}$$

The loss in a cable can be significant, especially at the higher frequencies. In Example 13-3, the transmitter put 100 W into the line, but at the end of the line, the output power—the level of the signal that is applied to the antenna—was only 13.35 W. A major loss of 86.65 W dissipated as heat in the transmission line.

Several things can be done to minimize loss. First, every attempt should be made to find a way to shorten the distance between the transmitter and the antenna. If that is not feasible, it may be possible to use a larger cable. For the application in Example 13-3, the RG-58A/U cable was used. This cable has a characteristic impedance of 53.5 Ω, so any value near that would be satisfactory. One possibility would be the RG-8/U, with an impedance of 52 Ω and an attenuation of only 2.5 dB/100 ft. An even better choice would be the 9913 cable, with an impedance of 50 Ω and an attenuation of 1.3 dB/100 ft.

When you are considering the relationship between cable length and attenuation, remember that a transmission line is a low-pass filter whose *cutoff frequency* depends

Cutoff frequency

Figure 13-13 Attenuation versus frequency for common coaxial cables. Note that both scales on the graph are logarithmic.

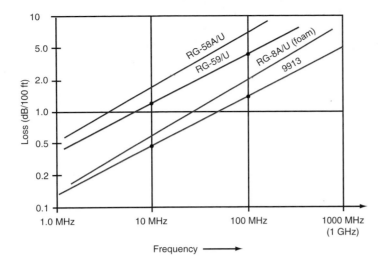

both on distributed inductance and capacitance along the line and on length. The longer the line, the lower its cutoff frequency. This means that higher-frequency signals beyond the cutoff frequency are rolled off at a rapid rate.

This is illustrated in Fig. 13-14, which shows attenuation curves for four lengths of a popular type of coaxial cable. Remember that the cutoff frequency is the 3-dB down point on a frequency-response curve. If we assume that an attenuation of 3 dB is the same as a 3-dB loss, then we can estimate the cutoff frequency of different lengths of cable. The 3-dB down level is marked on the graph. Now, note the cutoff frequency for different lengths of cable. The shorter cable (100 ft) has the highest cutoff frequency,

Figure 13-14 Attenuation versus length for RG-58A/U coaxial cable. Note that both scales on the graph are logarithmic.

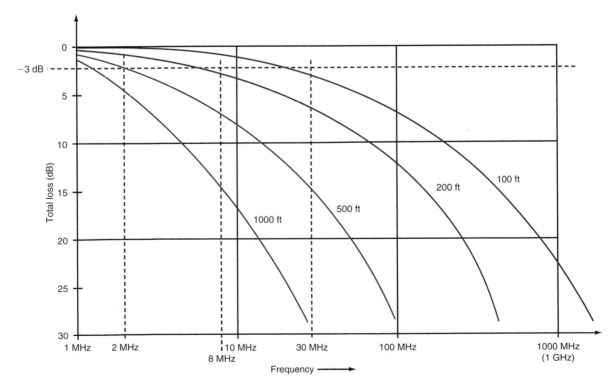

making the bandwidth about 30 MHz. The 200-ft cable has a cutoff of about 8 MHz, the 500-ft cable has a cutoff of approximately 2 MHz, and the 1000-ft cable has a cutoff of approximately 1 MHz. The higher frequencies are passed but are severely attenuated by the cable as it gets longer. It should be clear why it is important to use the larger, lower-loss cables for longer runs despite cost and handling inconvenience.

Finally, a gain antenna can be used to offset cable loss. These antennas are discussed in Chap. 14.

Example 13-4

A 150-ft length of RG-62A/U coaxial cable is used as a transmission line. Find (a) the load impedance that must be used to terminate the line to avoid reflections, (b) the equivalent inductance per foot, (c) the time delay introduced by the cable, (d) the phase shift that occurs on a 2.5-MHz sine wave, and (e) the total attenuation in decibels. (Refer to Fig. 13-12.)

a. The characteristic impedance is 93 Ω; therefore the load must offer a resistance of 93 Ω to avoid reflections.

b. $Z_0 = \sqrt{\dfrac{L}{C}}$ $\qquad Z_0 = 93\ \Omega$ $\qquad C = 13.5\ \text{pF/ft}$

$L = CZ_0^2 = 13.5 \times 10^{-12} \times (93)^2 = 116.76\ \text{nH/ft}$

c. $t_d = \sqrt{LC} = \sqrt{116.76 \times 10^{-9} \times 13.5 \times 10^{-12}} = 1.256\ \text{ns/ft}$

$150\ \text{ft} \times 1.256\ \text{ns/ft} = 188.3\ \text{ns}$

d. $T = \dfrac{1}{f} = \dfrac{1}{2.5 \times 10^{-6}} = 400\ \text{ns}$

$\theta = \dfrac{188.3(360)}{400} = 169.47°$

e. $\text{Attenuation} = \dfrac{2.8\ \text{dB}}{100\ \text{ft}} = 0.028\ \text{dB/ft}$

$150\ \text{ft} \times 0.028\ \text{dB/ft} = 4.2\ \text{dB}$

13-2 Standing Waves

When a signal is applied to a transmission line, it appears at the other end of the line some time later because of the propagation delay. If a resistive load equal to the characteristic impedance of a line is connected at the end of the line, the signal is absorbed by the load and power is dissipated as heat. If the load is an antenna, the signal is converted to electromagnetic energy and radiated into space.

If the load at the end of a line is an open circuit or a short circuit or has an impedance other than the characteristic impedance of the line, the signal is not fully absorbed by the load. When a line is not terminated properly, some of the energy is reflected from the end of the line and actually moves back up the line, toward the generator. This *reflected* voltage adds to the forward or incident generator voltage and forms a composite voltage that is distributed along the line. This pattern of voltage and its related current constitute what is called a *standing wave*.

Standing wave

Standing waves are not desirable. The reflection indicates that the power produced by the generator is not totally absorbed by the load. In some cases, e.g., a short-circuited or open line, no power gets to the load because all the power is reflected back to the generator. The following sections examine in detail how standing waves are generated.

The Relationship Between Reflections and Standing Waves

Figure 13-15 will be used to illustrate how reflections are generated and how they contribute to the formation of standing waves. Part (a) shows how a dc pulse propagates along a transmission line made up of identical LC sections. A battery (generator) is used as the input signal along with a switch to create an on/off dc pulse.

The transmission line is open at the end rather than being terminated in the characteristic impedance of the line. An open transmission line will, of course, produce a reflection and standing waves. Note that the generator has an internal impedance of R_g, which is equal to the characteristic impedance of the transmission line. Assume a transmission line impedance of 75 Ω and an internal generator resistance of 75 Ω. The 10 V supplied by the generator is, therefore, distributed equally across the impedance of the line and the internal resistance.

Now assume that the switch is closed to connect the generator to the line. As you know, connecting a dc source to reactive components such as inductors and capacitors produces transient signals as the inductors oppose changes in current while the capacitors oppose changes in voltages. Capacitor C_1 initially acts as a short circuit when the switch is closed, but soon begins to charge toward the battery voltage through L_1. As soon as the voltage at point A begins to rise, it applies a voltage to the next section of the transmission line made up of C_2 and L_2. Therefore, C_2 begins to charge through L_2. The process continues on down the line until C_4 charges through L_4, and so on. The signal moves down the line from left to right as the capacitors charge. Lossless (zero-resistance) components are assumed, and so the last capacitor C_4 eventually charges to the supply voltage.

Figure 13-15 How a pulse propagates along a transmission line.

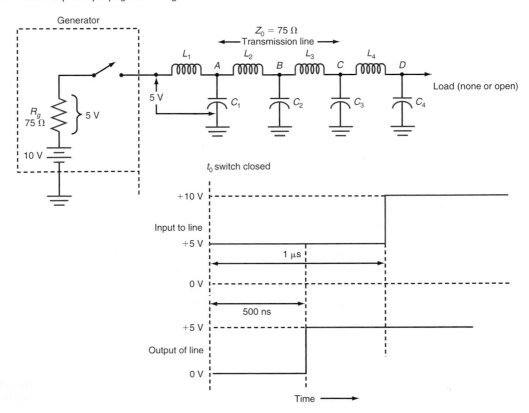

For the purposes of this illustration, assume that the length of the line and its other characteristics are such that the time delay is 500 ns: 500 ns after the switch is closed, an output pulse will occur at the end of the line. At this time, the voltage across the output capacitance C_4 is equal to 5 V or one-half of the supply voltage.

The instant that the output capacitance charges to its final value of 5 V, all current flow in the line ceases, causing any magnetic field around the inductors to collapse. The energy stored in the magnetic field of L_4 is equal to the energy stored in the output capacitance C_4. Therefore, a voltage of 5 V is induced into the inductor. The polarity of this voltage will be in such a direction that it adds to the charge already on the capacitor. Thus the capacitor will charge to 2 times the applied 5-V voltage, or 10 V.

A similar effect then takes place in L_3. The magnetic field across L_3 collapses, doubling the voltage charge on C_3. Next, the magnetic field around L_2 collapses, charging C_2 to 10 V. The same effect occurs in L_1 and C_1. Once the signal reaches the right end of the line, a reverse charging effect takes place on the capacitors from right to left. The effect is as if a signal were moving from output to input. This moving charge from right to left is the reflection, or *reflected wave*, and the input wave from the generator to the end of the line is the *incident wave*.

Reflected wave

Incident wave

It takes another 500 ns for the reflected wave to get back to the generator. At the end of 1 μs, the input to the transmission line goes more positive by 5 V, for a total of 10 V.

Figure 13-15(*b*) shows the waveforms for the input, output, and reflected voltages with respect to time. Observing the waveforms, follow the previously described action with the closing of the switch at time t_0. Since both the characteristic impedance of the line and the internal generator resistance are 75 Ω, one-half of the battery voltage appears at the input to the line at point *A*. This voltage propagates down the line, charging the line capacitors as it goes, until it reaches the end of the line and fully charges the output capacitance. At that time the current in the inductance begins to cease, with magnetic fields collapsing and inducing voltages that double the output voltage at the end of the line. Thus after 500 ns the output across the open end of the line is 10 V.

The reflection begins and now moves back down the line from right to left; after another 500 ns, it reaches the line input, the input to the line to jump to 10 V. Once the reflection stops, the entire line capacitance is fully charged to 10 V, as might be expected.

The preceding description concerned what is known as an *open-circuit load*. Another extreme condition is a *short-circuit load*. For this situation, assume a short across C_4 in Fig. 13-15(*a*).

When the switch is closed, again 5 V is applied to the input of the line, which is then propagated down the line as the line capacitors charge. Since the end of the line is short-circuited, inductor L_4 is, in effect, the load for this line. The voltage on C_3 is then applied to L_4. At this point, the reflection begins. The current in L_4 collapses, inducing a voltage which is then propagated down the line in the opposite direction. The voltage induced in L_4 is equal and opposite to the voltage propagated down the line. Therefore, this voltage is equal and opposite to the voltage on C_3, which causes C_3 to be discharged. As the reflection works its way back down the line from right to left, the line capacitance is continually discharged until it reaches the generator. It takes 500 ns for the charge to reach the end of the line and another 500 ns for the reflection to move back to the generator. Thus in a total of 1 μs, the input voltage switches from 5 to 0 V. Of course, the voltage across the output short remains zero during the entire time.

Open and shorted transmission lines are sometimes used to create special effects. In practice, however, the load on a transmission line is neither infinite nor 0 Ω; rather, it is typically some value in between. The load may be resistive or may have a reactive component. Antennas typically do not have a perfect resistance value. Instead they frequently have a small capacitive or inductive reactance. Thus the load impedance is equivalent to a series *RC* or *RL* circuit with an impedance of the form $R \pm jX$. If the load is not exactly resistive and is equal to the characteristic impedance of the line, a reflection is produced, the exact voltage levels depending on the complex impedance of the load. Usually some of the power is absorbed by the resistive part of the line; the mismatch still produces a reflection, but the reflection is not equal to the original signal, as in the case of a shorted or open load.

In most communication applications, the signal applied to a transmission line is an ac signal. This situation can be analyzed by assuming the signal to be a sine wave. The effect of the line on a sine wave is like that described above in the discussion based on an analysis of Fig. 13-15. If the line is terminated in a resistive load equal to the characteristic impedance of the line, the sine wave signal is fully absorbed by the load and no reflection occurs.

• Matched Lines

Ideally, a transmission line should be terminated in a load that has a resistive impedance equal to the characteristic impedance of the line. This is called a *matched line.* For example, a 50-Ω coaxial cable should be terminated with a 50-Ω resistance, as shown in Fig. 13-16. If the load is an antenna, then that antenna should look like a resistance of 50 Ω. When the load impedance and the characteristic impedance of the line match, the transmission goes smoothly and maximum power transfer—less any resistive losses in the line—takes place. The line can be any length. One of the key objectives in designing antenna and transmission line systems is to ensure this match.

Alternating-current voltage (or current) at any point on a matched line is a constant value (disregarding losses). A correctly terminated transmission line, therefore, is said to be *flat.* For example, if a voltmeter is moved down a matched line from generator to load and the rms voltage values are plotted, the resulting wavelength versus voltage line will be flat (see Fig. 13-17). Resistive losses in the line would, of course, produce a small voltage drop along the line, giving it a downward tilt for greater lengths.

Matched line

GOOD TO KNOW

When the mismatch between load resistive impedance and line impedance is great, the reflected power can be high enough to damage the transmitter or the line itself.

Figure 13-16 A transmission line must be terminated in its characteristic impedance for proper operation.

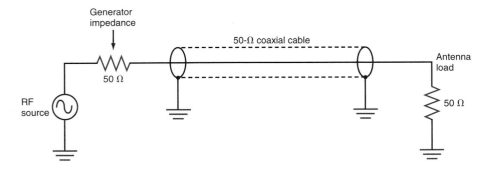

Figure 13-17 The voltage on a matched line is constant over length.

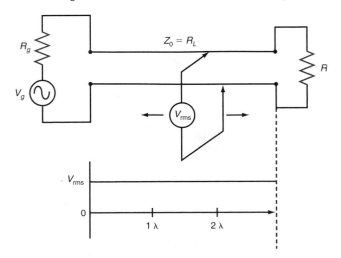

If the load impedance is different from the line characteristic impedance, not all the power transmitted is absorbed by the load. The power not absorbed by the load is reflected back toward the source. The power sent down the line toward the load is called *forward* or *incident power;* the power not absorbed by the load is called *reflected power.* The signal actually on a line is simply the algebraic sum of the forward and reflected signals.

Reflected power can represent a significant loss. If a line has a 3-dB loss end to end, only a 3-dB attenuation of the reflected wave will occur by the time the signal reaches the generator. What happens there depends upon the relative impedance of the generator and the line. Only part of the reflected energy is dissipated in the line. In cases in which the mismatch between load resistive impedance and line impedance is great, the reflected power can be high enough to actually cause damage to the transmitter or the line itself.

Shorted Lines

The condition for a short circuit is illustrated in Fig. 13-18. The graph below the transmission line shows the plot of voltage and current at each point on the line that would be generated by using the values given by a voltmeter and ammeter moved along the line. As expected in the case of a short at the end of a line, the voltage is zero when the current is maximum. All the power is reflected back toward the generator. Looking at the plot, you can see that the voltage and current variations distribute themselves according to the signal wavelength. The fixed pattern, which is the result of a composite of the forward and reflected signals, repeats every half wavelength. The voltage and current levels at the generator are dependent on signal wavelength and the line length.

The phase of the reflected voltage at the generator end of the line depends upon the length of the line. If the line is some multiple of one-quarter wavelength, the reflected wave will be in phase with the incident wave and the two will add, producing a signal at the generator that is twice the generator voltage. If the line length is some multiple of one-half wavelength, the reflected wave will be exactly 180° out of phase with the incident wave and the two will cancel, giving a zero voltage at the generator. In other words, the effects of the reflected wave can simulate an open or short circuit at the generator.

Open Lines

Figure 13-19 shows the standing waves on an open-circuited line. With an infinite impedance load, the voltage at the end of the line is maximum when the current is zero. All the energy is reflected, setting up the stationary pattern of voltage and current standing waves shown.

Figure 13-18   Standing waves on a shorted transmission line.

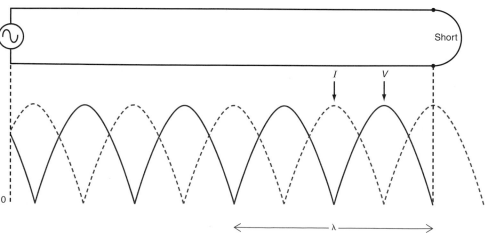

Figure 13-19   Standing waves on an open-circuit transmission line.

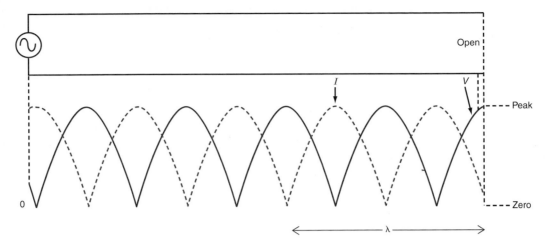

Mismatched (Resonant) Lines

Most often, lines do not terminate in a short or open circuit. Rather, the load impedance does not exactly match the transmission line impedance. Further, the load, usually an antenna, will probably have a reactive component, either inductive or capacitive, in addition to its resistance. Under these conditions, the line is said to be *resonant*. Such a mismatch produces standing waves, but the amplitude of these waves is lower than that of the standing waves resulting from short or open circuits. The distribution of these standing waves looks like that shown in Fig. 13-20. Note that the voltage or current never goes to zero, as it does with an open or shorted line.

Resonant circuit

Figure 13-20   Transmission line with mismatched load and the resulting standing waves.

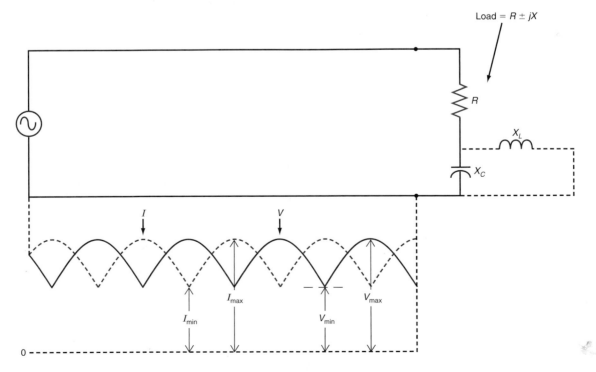

Calculating the Standing Wave Ratio

The magnitude of the standing waves on a transmission line is determined by the ratio of the maximum current to the minimum current, or the ratio of the maximum voltage to the minimum voltage, along the line. These ratios are referred to as the *standing wave ratio (SWR)*.

$$\text{SWR} = \frac{I_{max}}{I_{min}} = \frac{V_{max}}{V_{min}}$$

Under the shorted and open conditions described earlier, the current or voltage minima are zero. This produces an SWR of infinity. It means that no power is dissipated in the load; all the power is reflected.

In the ideal case, there are no standing waves. The voltage and current are constant along the line, so there are no maxima or minima (or the maximum and minima are the same). Therefore, the SWR is 1.

Measuring the maximum and minimum values of voltage and current on a line is not practical in the real world, so other ways of computing the SWR have been devised. For example, the SWR can be computed if the impedance of the transmission line and the actual impedance of the load are known. The SWR is the ratio of the load impedance Z_l to the characteristic impedance Z_0, or vice versa.

If $Z_l > Z_0$:

$$\text{SWR} = \frac{Z_l}{Z_0}$$

If $Z_0 > Z_l$:

$$\text{SWR} = \frac{Z_0}{Z_l}$$

For example, if a 75-Ω antenna load is connected to a 50-Ω transmission line, the SWR is $75/50 = 1.5$. Since the standing wave is really the composite of the original incident wave added to the reflected wave, the SWR can also be defined in terms of those waves. The ratio of the reflected voltage wave V_r to the incident voltage wave V_i is called the reflection coefficient

$$\Gamma = \frac{V_r}{V_i}$$

The *reflection coefficient* provides information on current and voltage along the line. Also, Γ = reflected power/incident power.

If a line is terminated in its characteristic impedance, then there is no reflected voltage, so $V_r = 0$ and $\Gamma = 0$. If the line is open or shorted, then total reflection occurs. This means that V_r and V_i are the same, so $\Gamma = 1$. The reflection coefficient really expresses the percentage of reflected voltage to incident voltage. If Γ is 0.5, for example, the reflected voltage is 50 percent of the incident voltage, and the reflected power is 25 percent of the incident power $[\Gamma^2 = (0.5)^2 = 0.25]$.

If the load is not matched but also is not an open or short, the line will have voltage minima and maxima, as described previously. These can be used to obtain the reflection coefficient by using the formula

$$\Gamma = \frac{V_{max} - V_{min}}{V_{max} + V_{min}} = \frac{\text{SWR} - 1}{\text{SWR} + 1}$$

The SWR is obtained from the reflection coefficient according to the equation

$$\text{SWR} = \frac{1 + \Gamma}{1 - \Gamma} = \frac{1 + \sqrt{p_r/p_i}}{1 - \sqrt{p_r/p_i}}$$

Example 13-5

An RG-11/U foam coaxial cable has a maximum voltage standing wave of 52 V and a minimum voltage of 17 V. Find (*a*) the SWR, (*b*) the reflection coefficient, and (*c*) the value of a resistive load.

a. $\text{SWR} = \dfrac{V_{max}}{V_{min}} = \dfrac{52}{17} = 3.05$

b. $\Gamma = \dfrac{V_{max} - V_{min}}{V_{max} + V_{min}} = \dfrac{52 - 17}{52 + 17} = \dfrac{35}{69}$

$\Gamma = 0.51$

or

$\text{SWR} = \dfrac{1 + \Gamma}{1 - \Gamma}$

$\Gamma = \dfrac{\text{SWR} - 1}{\text{SWR} + 1} = \dfrac{3.05 - 1}{3.05 + 1} = \dfrac{2.05}{4.05} = 0.51$

c. $\text{SWR} = 3.05 \qquad Z_0 = 75\ \Omega \qquad \text{SWR} = \dfrac{Z_l}{Z_0} = \dfrac{Z_0}{Z_l}$

$Z_l = Z_0\ (\text{SWR}) = 75(3.05) = 228.75\ \Omega$

or

$Z_l = \dfrac{Z_0}{\text{SWR}} = \dfrac{75}{3.05} = 24.59\ \Omega$

If the load matches the line impedance, then $\Gamma = 0$. The preceding formula gives an SWR of 1, as expected. With an open or shorted load, $\Gamma = 1$. This produces an SWR of infinity.

The reflection coefficient can also be determined from the line and load impedances:

$$\Gamma = \dfrac{Z_l - Z_0}{Z_l + Z_0}$$

For the example of an antenna load of 75 Ω and a coaxial cable of 50 Ω, the reflection coefficient is $\Gamma = (75 - 50)/(75 + 50) = 25/125 = 0.2$.

The importance of the SWR is that it gives a relative indication of just how much power is lost in the transmission line and the generator. This assumes that none of the reflected power is re-reflected by the generator. In a typical transmitter, some power is reflected and sent to the load again.

The curve in Fig. 13-21 shows the relationship between the percentage of reflected power and the SWR. The percentage of reflected power is also expressed by the term *return loss* and is given directly in watts or decibels (dB). Naturally, when the standing wave ratio is 1, the percentage of reflected power is 0. But as a line and load mismatch grows, reflected power increases. When the SWR is 1.5, the percentage of reflected power is 4 percent. This is still not too bad, as 96 percent of the power gets to the load.

Return loss

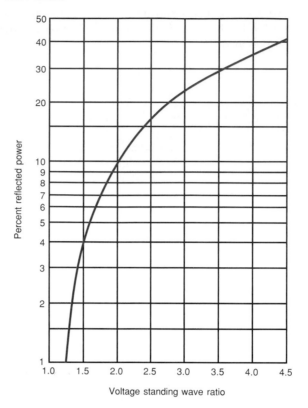

It is possible to compute reflected power P_r if given the SWR and the incident power P_i. Since $\Gamma^2 = P_r/P_i$, then $P_r = \Gamma^2 P_i$. Knowing the SWR, you can compute Γ and then solve by using the preceding equation.

$$\frac{P_r}{P_i} = \Gamma^2 = \left(\frac{\text{SWR} - 1}{\text{SWR} + 1}\right)^2$$

$$= \left(\frac{1.5 - 1}{1.5 + 1}\right)^2$$

$$= (0.2)^2$$

$$= 0.04$$

$$\frac{P_r}{P_i} = 4\%$$

One of the best and most practical ways to compute the SWR is to measure the forward power P_f and reflected power P_r and then use the formula

$$\text{SWR} = \frac{1 + \sqrt{P_r/P_i}}{1 - \sqrt{P_r/P_i}}$$

Several good circuits have been invented to measure forward and reflected power. And commercial test instruments are also available that can be inserted into a transmission line and make these measurements. The data is read from a front panel meter or digital display. Then the figures are plugged into the formula above. Some sophisticated test instruments have a built-in embedded computer to make this calculation automatically and display the SWR value.

GOOD TO KNOW

To calculate the SWR, first determine the square root of the ratio of the reflected power to the forward power $\sqrt{P_r/P_f}$. Then

$$\text{SWR} = \frac{1 + \sqrt{P_r/P_f}}{1 - \sqrt{P_r/P_f}}$$

For example, assume that you measure a forward power of 35 W and a reflected power of 7 W. The SWR is

$$\text{SWR} = \frac{1 + \sqrt{7/35}}{1 - \sqrt{7/35}} = \frac{1 + \sqrt{0.2}}{1 + \sqrt{0.2}}$$

$$\text{SWR} = \frac{1 + 0.4472}{1 - 0.4472} = \frac{1.4472}{0.5528}$$

$$\text{SWR} = 2.618$$

You will learn about the most common power-measuring circuits and equipment in Chap. 22.

For SWR values of 2 or less, the reflected power is less than 10 percent, which means that 90 percent gets to the load. For most applications this is acceptable. For SWR values higher than 2, the percentage of reflected power increases dramatically, and measures must be taken to reduce the SWR to prevent potential damage. Some solid-state systems shut down automatically if the SWR is greater than 2. The most common approach to reducing the SWR is to add or include a π, L, or T LC network to offset antenna reactance and other resistive components and to produce an impedance match. Antenna length can also be adjusted to improve the impedance match.

Example 13-6

The line input to the cable in Example 13-5 is 30 W. What is the output power? (See Fig. 13-21; disregard attenuation due to length.)

The percentage of reflected power with the SWR of 3.05 is about 25.62.

$$P_r = 0.2562(30 \text{ W}) = 7.686 \text{ W}$$

Alternatively,

$$P_r = P_i \left(\frac{\text{SWR} - 1}{\text{SWR} + 1} \right)^2$$

$$= 30 \left(\frac{3.05 - 1}{3.05 + 1} \right)^2$$

$$= 7.686 \text{ W}$$

$$P_{\text{out}} = P_i - P_i = 30 - 7.686 = 22.314 \text{ W}$$

13-3 Transmission Lines as Circuit Elements

The standing wave conditions resulting from open- and short-circuited loads must usually be avoided in working with transmission lines. However, with one-quarter and one-half wavelength transmissions, these open- and short-circuited loads can be used as resonant or reactive circuits.

Resonant Circuits and Reactive Components

Consider the shorted one-quarter wavelength ($\lambda/4$) line shown in Fig. 13-22. At the load end, voltage is zero and current is maximum. But one-quarter wavelength back, at the

Figure 13-22 A shorted one-quarter wavelength line acts as a parallel resonant circuit.

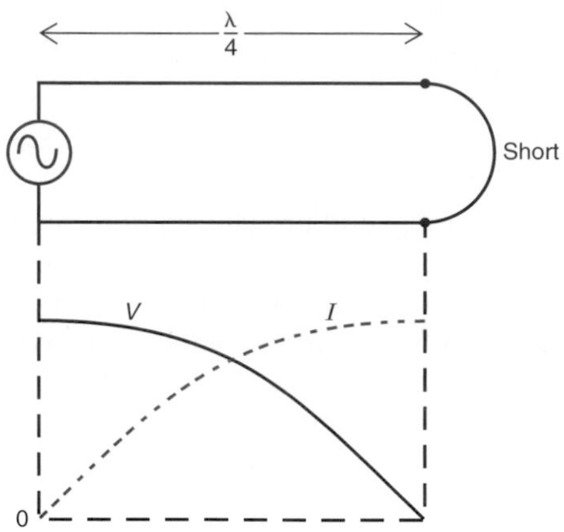

Figure 13-23 A shorted one-half wavelength line acts as a series resonant circuit.

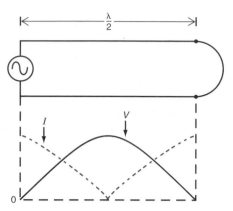

generator, the voltage is maximum and the current is zero. To the generator, the line appears as an open circuit, or at least a very high impedance. The key point here is that this condition exists at only one frequency, the frequency at which the line is exactly one-quarter wavelength. Because of this frequency sensitivity, the line acts as an *LC* tuned or resonant circuit, in this case, a parallel resonant circuit because of its very high impedance at the reference frequency.

With a shorted one-half wavelength line, the standing wave pattern is like that shown in Fig. 13-23. The generator sees the same conditions as at the end of the line, i.e., zero voltage and maximum current. This represents a short, or very low impedance. That condition occurs only if the line is exactly one-half wavelength long at the generator frequency. In this case, the line looks like a series resonant circuit to the generator.

If the line length is less than one-quarter wavelength at the operating frequency, the shorted line looks like an inductor to the generator. If the shorted line is between one-quarter and one-half wavelength, it looks like a capacitor to the generator. All these conditions repeat with multiple one-quarter or one-half wavelengths of shorted line.

Similar results are obtained with an open line, as shown in Fig. 13-24. To the generator, a one-quarter wavelength line looks like a series resonant circuit and a one-half

Figure 13-24 One-quarter and one-half wavelength open lines look like resonant circuits to a generator.

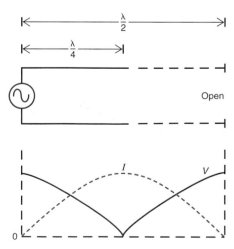

wavelength line looks like a parallel resonant circuit, just the opposite of a shorted line. If the line is less than one-quarter wavelength, the generator sees a capacitance. If the line is between one-quarter and one-half wavelength, the generator sees an inductance. These characteristics repeat for lines that are some multiple of one-quarter or one-half wavelengths.

Figure 13-25 is a summary of the conditions represented by open and shorted lines of lengths up to one wavelength. The horizontal axis is length, in wavelengths, and the vertical axis is the reactance of the line, in ohms, expressed in terms of the line's characteristic impedance. The solid curves are shorted lines and the dashed curves are open-circuit lines.

If the line acts as a series resonant circuit, its impedance is zero. If the line is of such a length that it acts as a parallel resonant circuit, its impedance is near infinity. If the line is some intermediate length, it is reactive. For example, consider a shorted one-eighth wavelength line. The horizontal divisions represent one-sixteenth wavelength, so two of these represent one-eighth wavelength. Assume that the line has a characteristic impedance of 50 Ω. At the one-eighth wavelength point on the left-hand solid curve is a reading of 1. This means that the line acts as an inductive reactance of $1 \times Z_0$, or $1 \times 50 = 50 \Omega$. An open line about three-eighths wavelength would have the same effect, as the leftmost dashed curve in Fig. 13-24 indicates.

How could a capacitive reactance of 150 Ω be created with the same 50-Ω line? First, locate the 150-Ω point on the capacitive reactance scale in Fig. 13-25. Since $150/50 = 3$, the 150-Ω point is at $X_C = 3$. Next, draw a line from this point to the right until it intersects with two of the curves. Then read the wavelength from the horizontal scale. A capacitive reactance of 150 Ω with a 50-Ω line can be achieved with an openline somewhat longer than $\frac{1}{32}$ wavelength or a shorted line a bit longer than $\frac{9}{32}$ wavelength.

Figure 13-25 Summary of impedance and reactance variations of shorted and open lines for lengths up to one wavelength.

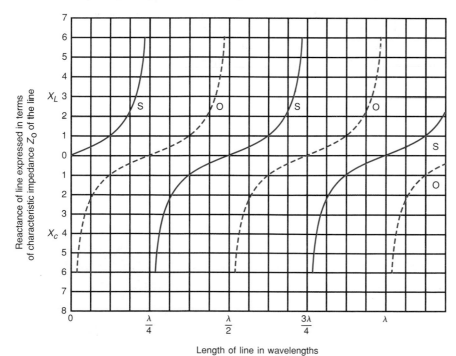

Stripline and Microstrip

At low frequencies (below about 300 MHz), the characteristics of open and shorted lines discussed in the previous sections have little significance. At low frequencies the lines are just too long to be used as reactive components or as filters and tuned circuits. However, at UHF (300 to 3000 MHz) and microwave (1 GHz and greater) frequencies the length of one-half wavelength is less than 1 ft; the values of inductance and capacitance become so small that it is difficult to realize them physically with standard coils and capacitors. Special transmission lines constructed with copper patterns on a *printed-circuit board (PCB)*, called *microstrip* or *stripline,* can be used as tuned circuits, filters, phase shifters, reactive components, and impedance-matching circuits at these high frequencies.

Printed-circuit board (PCB)

Microstrip

Stripline

A PCB is a flat insulating base made of fiberglass or some other insulating base material to which is bonded copper on one or both sides and sometimes in several layers. Teflon or ceramic is used as the base for some PCBs in microwave applications. In microwave ICs, the base is often alumina or even sapphire. The copper is etched away in patterns to form the interconnections for transistors, ICs, resistors, and other components. Thus point-to-point connections with wire are eliminated. Diodes, transistors, and other components are mounted right on the PCB and connected directly to the formed microstrip or stripline.

Microstrip. Microstrip is a flat conductor separated by an insulating dielectric from a large conducting ground plane [Fig. 13-26(*a*)]. The microstrip is usually one-quarter or one-half wavelength long. The ground plane is the circuit common. This type of microstrip is equivalent to an unbalanced line. Shorted lines are usually preferred over open lines. Microstrip can also be made in a two-line balanced version [Fig. 13-26(*b*)].

The characteristic impedance of microstrip, as with any transmission line, is dependent on its physical characteristics. It can be calculated by using the formula

$$Z = \frac{87}{\sqrt{\epsilon + 1.41}} \ln \frac{5.98h}{0.8w + t}$$

where Z = characteristic impedance
ϵ = dielectric constant
w = width of copper trace
t = thickness of copper trace
h = distance between trace and ground plane (dielectric thickness) of dielectric

Any units of measurement can be used (e.g., inches or millimeters), as long as all dimensions are in the same units. (See Fig. 13-27.) The dielectric constant of the popular FR-4 fiberglass PC board material is 4.5. The value of ϵ for Teflon is 3.

Figure 13-26 Microstrip. (*a*) Unbalanced. (*b*) Balanced.

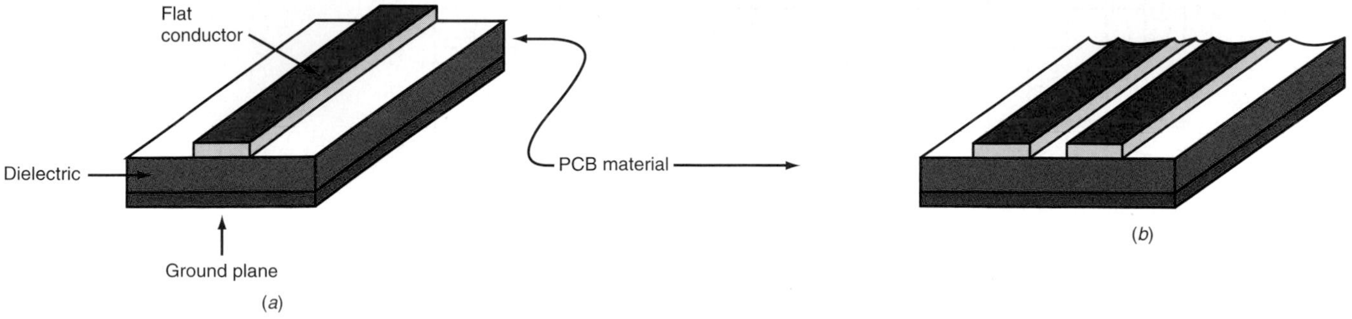

Figure 13-27   Dimensions for calculating characteristic impedance.

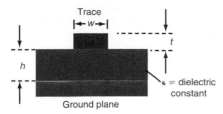

Trace

Ground plane

ϵ = dielectric constant

Figure 13-28   Stripline.

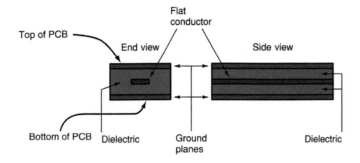

Top of PCB

End view

Flat conductor

Side view

Bottom of PCB   Dielectric

Ground planes

Dielectric

The characteristic impedance of microstrip with the dimensions $h = 0.0625$ in, $w = 0.1$ in, $t = 0.003$ in, and $\epsilon = 4.5$ is

$$Z = \frac{87}{\sqrt{4.5 + 1.41}} \ln \frac{5.98(.0625)}{0.8(0.1) + 0.003}$$

$$= 35.8 \ln 4.5 = 35.8(1.5)$$

$$Z = 53.9 \ \Omega$$

Stripline. Stripline is a flat conductor sandwiched between two ground planes (Fig. 13-28). It is more difficult to make than microstrip; however, it does not radiate as microstrip does. Radiation produces losses. The length is one-quarter or one-half wavelength at the desired operating frequency, and shorted lines are more commonly used than open lines.

The characteristic impedance of stripline is given by the formula

$$Z = \frac{60}{\epsilon} \ln \frac{4d}{0.67\pi w(0.8 + t/h)}$$

Figure 13-29 shows the dimensions required to make the calculations.

Even tinier microstrip and striplines can be made by using monolithic, thin-film, and hybrid IC techniques. When these are combined with diodes, transistors, and other components, *microwave integrated circuits (MICs)* are formed.

Microwave integrated circuits (MICs)

Figure 13-29   Dimensions for calculating stripline impedance.

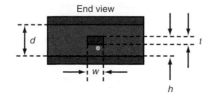

End view

Example 13-7

A microstrip transmission line is to be used as a capacitor of 4 pF at 800 MHz. The PCB dielectric is 3.6. The microstrip dimensions are $h = 0.0625$ in, $w = 0.13$ in, and $t = 0.002$ in. What are (a) the characteristic impedance of the line and (b) the reactance of the capacitor?

a. $Z_0 = \dfrac{87}{\sqrt{\epsilon + 1.41}} \ln \dfrac{5.98h}{0.8w + t}$

$\qquad = \dfrac{87}{\sqrt{3.6 + 1.41}} \ln \dfrac{5.98(0.0625)}{0.8(0.13) + 0.002}$

$\qquad = \dfrac{87}{2.24} \ln \dfrac{0.374}{0.106}$

$\qquad = (38.8)(1.26) = 48.9 \ \Omega$

b. $X_C = \dfrac{1}{2\pi f C} = \dfrac{1}{2\pi(800 \times 10^6)(4 \times 10^{-12})} = 49.74 \ \Omega$

Example 13-8

What is the length of the transmission line in Example 13-7?

Refer to Fig. 13-25. Take the ratio of X_C to Z_0.

$$\frac{X_C}{Z_0} = \frac{49.76}{48.9} = 1.02 \approx 1$$

Locate 1 on the X_C vertical region of the graph. Move to the right to encounter the dashed curve for an open line. Read the wavelength on the lower horizontal axis of $^2/_{16}$ or $\frac{1}{8} \lambda$.

$$\lambda = \frac{984}{800} = 1.23 \text{ ft} \qquad \frac{\lambda}{8} = \frac{1.23}{8} = 0.15375 \text{ ft}$$

$$0.15375 \text{ ft} \times 12 \text{ in/ft} = 1.845 \text{ in}$$

The velocity of propagation is

$$V_p = \frac{1}{\sqrt{\epsilon}} = \frac{1}{\sqrt{3.6}} = 0.527$$

$$\frac{\lambda}{8} \times V_p = 1.845 \text{ in} \times V_p = 1.845 \times 0.527 = 0.9723 \text{ in}$$

13-4 The Smith Chart

The mathematics required to design and analyze transmission lines is complex, whether the line is a physical cable connecting a transceiver to an antenna or is being used as a filter or impedance-matching network. This is so because the impedances involved are complex ones, involving both resistive and reactive elements. The impedances are in the familiar rectangular form $R + jX$. Computations with complex numbers such as this are long and

For every antenna you see here, there is a transmission line. Coaxial cable is the most common, but waveguides are used for microwaves.

time-consuming. Further, many calculations involve trigonometric relationships. Although no individual calculation is difficult, the sheer volume of the calculations can lead to errors.

In the 1930s, one clever engineer decided to do something to reduce the chance of error in transmission line calculations. The engineer's name was Philip H. Smith, and in January 1939 he published the *Smith chart*, a sophisticated graph that permits visual solutions to transmission line calculations.

Smith chart

Today, of course, the mathematics of transmission line calculations is not a problem, because of the wide availability of electronic computing options. Transmission line equations can be easily programmed into a scientific and engineering calculator for fast, easy solutions. Personal computers provide an ideal way to make these calculations, either by using special mathematics software packages or by using BASIC, Fortran, C, or another language to write the specific programs. The math software packages commonly available today for engineering and scientific computation also provide graphical outputs, if desired.

Despite the availability of all the computing options today, the Smith chart is still used. Its unique format provides a more or less standardized way of viewing and solving transmission line and related problems. Further, a graphical representation of an equation

Transmission Lines   **509**

Figure 13-30 The Smith chart.

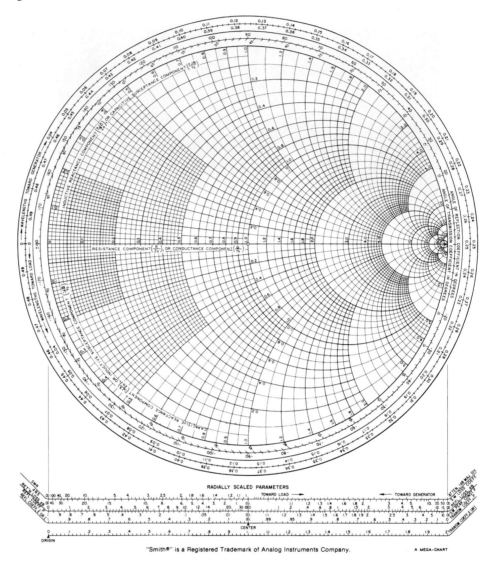

"Smith®" is a Registered Trademark of Analog Instruments Company.   A MEGA-CHART

conveys more information than is gained by a simple inspection of the equation. For these reasons, it is desirable to become familiar with the Smith chart.

Figure 13-30 is a Smith chart. This imposing graph is created by plotting two sets of orthogonal (at right angles) circles on a third circle. Smith chart graph paper is available from the original publisher, Analog Instruments Company, and from the American Radio Relay League. Smith chart graph paper is also available in some college and university bookstores.

The first step in creating a Smith chart is to plot a set of eccentric circles along a horizontal line, as shown in Fig. 13-31. The horizontal axis is the pure resistance or zero-reactance line. The point at the far left end of the line represents zero resistance, and the point at the far right represents infinite resistance. The resistance circles are centered on and pass through this pure resistance line. The circles are all tangent to one another at the infinite resistance point, and the centers of all the circles fall on the resistance line.

Each circle represents all the points of some fixed resistance value. Any point on the outer circle represents a resistance of $0\ \Omega$. The other circles have other values of resistance. The $R = 1$ circle passes through the exact center of the resistance line and is known as the *prime center*. Values of pure resistance and the characteristic impedance of transmission line are plotted on this line.

The most common transmission line impedance in use today is $50\ \Omega$. For that reason, most of the impedances and reactances are in the 50-Ω range. It is convenient, then,

Prime center

Figure 13-31 Resistance circles on a Smith chart.

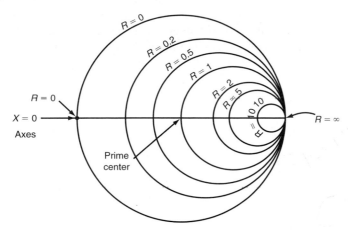

to have the value of 50 Ω located at the prime center of the chart. This means that all points on the $R = 1$ circle represent 50 Ω, all points on the $R = 0.5$ circle represent 25 Ω, all points on the $R = 2$ circle represent 100 Ω, and so on.

Smith charts are in what is called *normalized* form, with $R = 1$ at the prime center. Users customize the Smith chart for specific applications by assigning a different value to the prime center.

Normalized

The remainder of the Smith chart is completed by adding reactance circles, as shown in Fig. 13-32. Like the resistive circles, these are eccentric, with all circles

Figure 13-32 Reactance circles on the Smith chart.

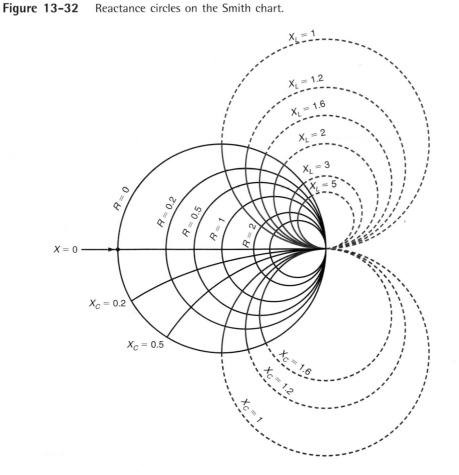

meeting at the infinite resistance point. Each circle represents a constant reactance point, with the inductive-reactance circles at the top and the capacitive-reactance circles at the bottom. Note that the reactance circles on the chart are incomplete. Only those segments of the circles within the $R = 0$ line are included on the chart. Like the resistive circles, the reactive circles are presented in normalized form. Compare Figs. 13-31 and 13-32 to the complete Smith chart in Fig. 13-30 before you proceed.

Plotting and Reading Impedance Values

The Smith chart in Fig. 13-33 shows several examples of plotted impedance values:

$$Z_1 = 1.5 + j0.5$$
$$Z_2 = 5 - j1.6$$
$$Z_3 = 0.2 + j3$$
$$Z_4 = 0.4 - j0.36$$

Locate each of those points on the Smith chart, and be sure that you understand how each was obtained.

Figure 13-33   Smith chart with four impedance values plotted.

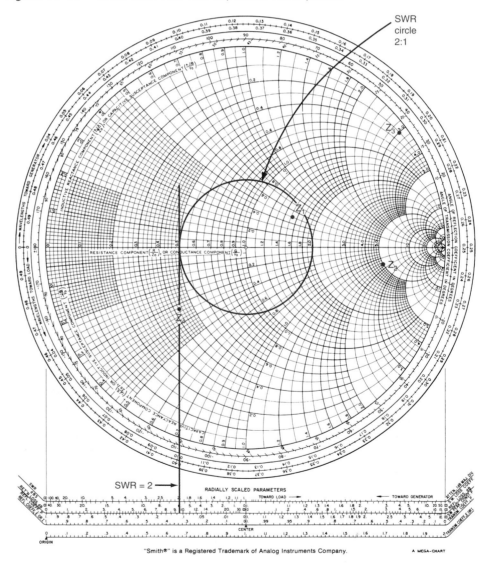

"Smith®" is a Registered Trademark of Analog Instruments Company.   A MEGA-CHART

The impedance values plotted in Fig. 13-33 are normalized values. Scaling the Smith chart to a particular impedance range requires multiplying those values by some common factor. For example, many Smith charts are plotted with 50 Ω at the prime center. The above-listed impedances for a prime center value of 50 Ω would be as follows:

$$Z_1 = 75 + j25$$
$$Z_2 = 250 - j80$$
$$Z_3 = 25 + j150$$
$$Z_4 = 20 - j18$$

To solve problems with real impedance values, put them into normalized form by dividing the resistive and reactive numbers by a factor equal to the resistance value at prime center. Then plot the numbers.

When you are reading values from a normalized Smith chart, convert them to standard impedance form by multiplying the resistance and reactance by a factor equal to the resistance of the prime center.

Wavelength Scales

The three scales on the outer perimeter of the Smith chart in Fig. 13-30 are wavelength toward the generator, wavelength toward the load, and angle of reflection coefficient in degrees. The scale labeled "toward generator" begins at the zero-resistance and zero-reactance line and moves clockwise around to the infinite resistance position. One-half of a circular rotation is 90°, and on the Smith chart scale, it represents one-quarter wavelength. One full rotation is one-half wavelength. The transmission line patterns of voltage and current distributed along a line repeat every one-half wavelength.

The scale labeled "toward load" also begins at the zero-resistance, zero-reactance point and goes in a counterclockwise direction for one complete rotation, or one-half wavelength. The infinite resistance point is the one-quarter wavelength mark.

The reflection coefficient (the ratio of the reflected voltage to the incident voltage) has a range of 0 to 1, but it can also be expressed as an angle from 0 to 360°, from 0 to positive 180°, or from 0 to minus 180°. The zero marker is at the infinite resistance point on the right hand of the resistance line.

SWR Circle

The SWR of a transmission line plotted on the Smith chart is a circle. If the load is resistive and matched to the characteristic impedance of the line, the standing wave ratio is 1. This is plotted as a single point at the prime center of the Smith chart. The impedance of the line is flat at 50 Ω or any other normalized value. However, if the load is not perfectly matched to the prime impedance, standing waves will exist. The SWR in such cases is represented by a circle whose center is the prime center.

To draw an SWR circle, first calculate the SWR by using one of the previously given formulas. For this example, assume an SWR of 2. Starting at the prime center, move to the right on the resistance line until the value 2 is encountered. Then, using a drawing compass, place the point at the center and draw a circle through the 2 mark to the right of prime center. The circle should also pass through the 0.5 mark to the left of prime center. The red circle drawn in Fig. 13-33 represents a plot of the impedance variations along an unmatched or resonant transmission line. The variations in the voltage and current standing waves mean that there is a continuous variation in the impedance along the line. In other words, the impedance at one point on the unmatched line is different from the impedance at all other points on the line. All the impedance values appear on the SWR circle.

The SWR circle can also be used to determine the maximum and minimum voltage points along the line. For example, the point at which the SWR circle crosses the resistance line to the right of the prime center indicates the point of maximum or peak voltage of the standing wave in wavelengths from the load. This is also the maximum impedance point on the line. The point at which the SWR circle passes through the resistance line to the left of prime center indicates the minimum voltage and impedance points.

The linear scales printed at the bottom of Smith charts are used to find the SWR, dB loss, and reflection coefficient. For example, to use the linear SWR scale, simply draw a straight line tangent to the SWR circle and perpendicular to the resistance line on the left side of the Smith chart. Make the line long enough that it intersects the SWR scale at the bottom of the chart. This has been done in Fig. 13-33. Note that the SWR circle for a value of 2 can be read from the linear SWR scale.

Using the Smith Chart: Examples

As discussed above, when the load does not match the characteristic impedance in a given application, the length of the line becomes a part of the total impedance seen by the generator. The Smith chart provides a way to find this impedance. Once the impedance is known, an impedance-matching circuit can be added to compensate for the conditions and make the line flat and the SWR as close to 1 as possible.

Example 1 for Fig. 13-34. The operating frequency for a 24-ft piece of RG-58A/U coaxial cable is 140 MHz. The load is resistive, with a resistance of 93 Ω. What is the impedance seen by a transmitter?

The first step is to find the number of wavelengths represented by 24 ft of cable. A 140-MHz signal has a wavelength of

$$\lambda = \frac{984}{f} = \frac{984}{140} = 7.02 \text{ ft}$$

Figure 13-34 Finding the impedance of a mismatched pair. (*a*) Actual circuit. (*b*) Equivalent circuit.

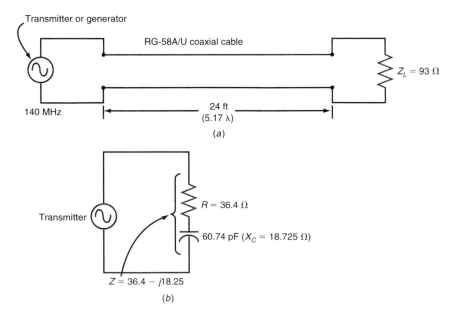

Remember that coaxial cable has a velocity factor less than 1 due to the slowdown of RF signals in a cable. The velocity factor of RG-58A/U is 0.66. Therefore, one wavelength at 140 MHz is

$$\lambda = 7.02 \times 0.66 = 4.64 \text{ ft}$$

The number of wavelengths represented by the 24-ft cable is $24/4.64 = 5.17 \ \lambda$.

As indicated earlier, the impedance variations along a line repeat every one-half wavelength and therefore every full wavelength; thus for the purposes of calculation, we need only the $0.17 \ \lambda$ of the above value.

Next, we normalize the Smith chart to the characteristic impedance of the coaxial cable, which is $53.5 \ \Omega$. The value of the prime center is $53.5 \ \Omega$. The SWR is then computed:

$$\text{SWR} = \frac{Z_l}{Z_0} = \frac{93}{53.5} = 1.74$$

This SWR is now plotted on the Smith chart. See Fig. 13-35, where point X represents the resistive load of $93 \ \Omega$.

Figure 13-35   Finding the generator load with a resistive load.

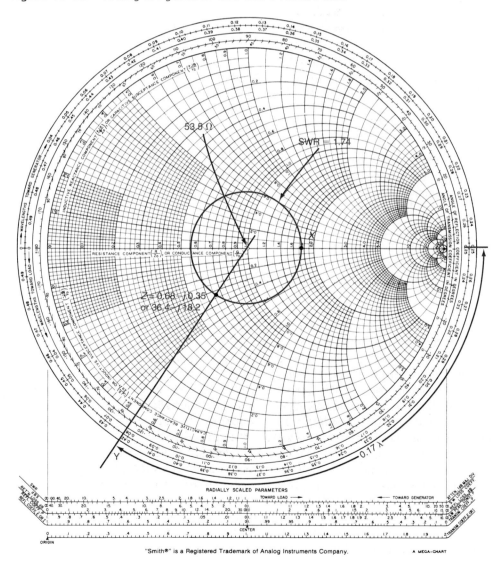

"Smith®" is a Registered Trademark of Analog Instruments Company.   A MEGA-CHART

To find the impedance at the transmitter end of the coaxial cable, we move along the line 5.17 λ from the load back to the transmitter or generator. Remember that one full rotation around the Smith chart is one-half wavelength, since the values repeat every one-half wavelength. Starting at point X, then, move in a clockwise direction (toward the generator) around the SWR circle for 10 complete revolutions, which represents 5 λ. This brings you back to point X.

Continue the clockwise rotation for an additional 0.17 λ, stop there, and mark that point on the SWR circle. Draw a line from the prime center to the location that represents 0.17 λ. The marking on the right-hand side of point X is the 0.25-λ mark. You need to go 0.17 λ from that, or 0.25 + 0.17 = 0.42. Stop at the 0.42 mark on the lower part of the circle at Y. Draw a line from there to the prime center. Refer again to Fig. 13-35.

The point at which the line cuts the SWR circle is the impedance that the generator sees. Reading the values from the chart, you have $R \approx 0.68$ and $X_C \approx 0.35$. Since the point is in the lower half of the chart, the reactance is capacitive. Thus the impedance at this point is $0.68 - j0.35$.

To find the actual value, convert from the normalized value by multiplying by the impedance of the prime center, or 53.5:

$$Z = 53.5(0.68 - j0.35) = 36.4 - j18.725 \; \Omega$$

The transmitter is seeing what appears to be a 36.4-Ω resistor in series with a capacitor with a reactance of 18.725 Ω. The equivalent capacitance value is [using $X_C = 1/(2\pi f C)$ or $C = 1/(2\pi f X_C)$]

$$C = \frac{1}{6.28(140 \times 10^6)(18.725)} = 60.74 \text{ pF}$$

The equivalent circuit is shown in Fig. 13-34(b).

Example 2 for Fig. 13-36. An antenna is connected to the 24-ft 53.5-Ω RG-58A/U line described in Example 1. The load is $40 + j30 \; \Omega$. What impedance does the transmitter see?

The prime center is 53.5 Ω, as before. Before plotting the load impedance on the chart, you must normalize it by dividing by 53.5:

$$Z_l = \frac{40 + j30}{53.5} = 0.75 + j0.56 \; \Omega$$

The plot for this value is shown in Fig. 13-36. Remember that all impedances fall on the SWR circle. You can draw the SWR circle for this example simply by placing the compass on the prime center with a radius out to the load impedance point and rotating 360°.

The SWR is obtained from the chart at the bottom of the figure by extending a line perpendicular to the resistance axis down to the SWR line. The SWR is about 2:1.

The next step is to draw a line from the prime center through the plotted load impedance so that it intersects the wavelength scales on the perimeter of the curve. Refer again to Fig. 13-36. Since our starting point of reference is the load impedance, we use the "toward generator" scale to determine the impedance at the input to the line. At point X is the wavelength value of 0.116.

Now, since the generator is 5.17λ away from the load and since the readings repeat every one-half wavelength, you move toward the generator in the clockwise direction 0.17λ: 0.116 and 0.17 = 0.286. This point is marked Y in Fig. 13-36.

Draw a line from point Y through the prime center. The point at which it intersects the SWR circle represents the impedance seen by the generator. The normalized value is $1.75 - j0.55$. Correcting this for the 53.5 Ω prime center, we get

$$Z = 53.5(1.75 - j0.55) = 93.6 - j29.4 \; \Omega$$

This is the impedance that the generator sees.

Figure 13-36 Finding the generator load with a complex load.

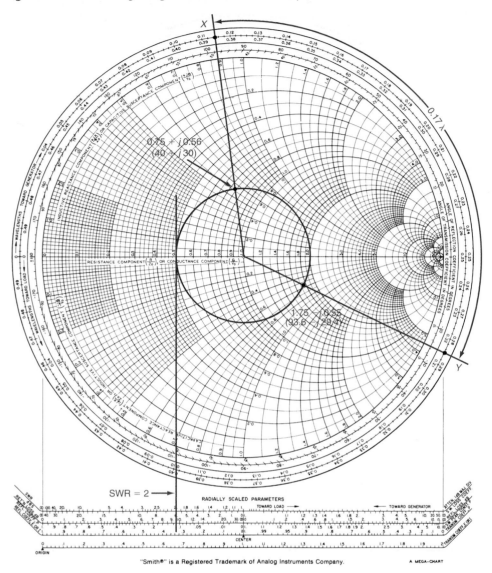

"Smith®" is a Registered Trademark of Analog Instruments Company.   A MEGA-CHART

Example 3 for Fig. 13-37. In many cases the antenna or other load impedance is not known. If it is not matched to the line, the line modifies this impedance so that the transmitter sees a different impedance. One way to find the overall impedance as well as the antenna or other load impedance is to measure the combined impedance of the load and the transmission line at the transmitter end, using an impedance bridge. Then the Smith chart can be used to find the individual impedance values.

A 50-ft RG-11/U foam dielectric coaxial cable with a characteristic impedance of 75 Ω and a velocity factor of 0.8 has an operating frequency of 72 MHz. The load is an antenna whose actual impedance is unknown. A measurement at the transmitter end of the cable gives a complex impedance of 82 + *j*43. What is the impedance of the antenna?

One wavelength at 72 MHz is 984/72 = 13.67 ft. Taking the velocity factor into account yields

$$\lambda = 13.67 \times 0.8 = 10.93 \text{ ft}$$

The length of the 50-ft cable, in wavelengths, is 50/10.93 = 4.57. As before, it is the 0.57 λ value that is of practical use.

Figure 13-37 Determining load impedance.

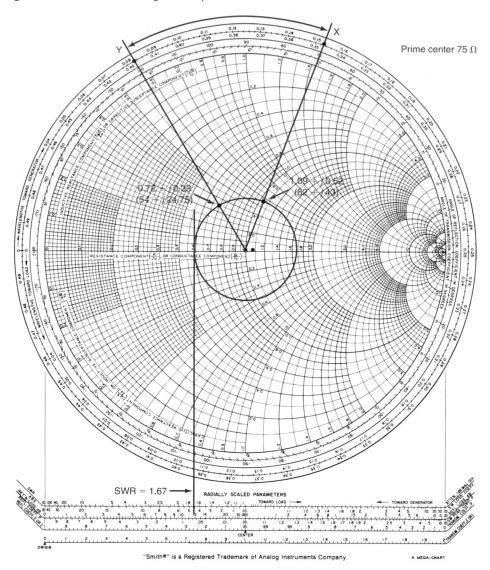

First, the measured impedance is normalized and plotted. The prime center is 75 Ω. In Fig. 13-37 $Z = (82 + j43)/75 = 1.09 + j0.52$ is plotted on the Smith chart. The SWR circle is plotted through this point from the prime center. A tangent line is then drawn from the circle to the linear SWR chart on the bottom. The SWR is 1.67.

Next, a line is drawn from prime center to the normalized impedance point and through the wavelength plots on the perimeter of the graphs. It intersects the "toward load" scale at point *X*, or 0.346λ. Refer to the figure.

The plotted impedance is what is seen at the generator end, so it is necessary to move around the graph to the load in the counterclockwise direction to find the load impedance. We move a distance equal to the length of the cable, which is 4.57λ. Nine full rotations from point *X* represents 4.5λ, which brings us back to point *X*. We then rotate 0.07λ more, counterclockwise, to complete the length. This puts us at the 0.346 + 0.07 = 0.416λ point, which is designated point *Y* in Fig. 13-37. We then draw a line from *Y* to the prime center. The point at which this line intersects the SWR circle is the actual antenna impedance. The normalized value is 0.72 + *j*0.33. Multiplying by 75 gives the actual value of the antenna impedance:

$$Z = 75(0.72 + j0.33) = 54 + j24.75 \ \Omega$$

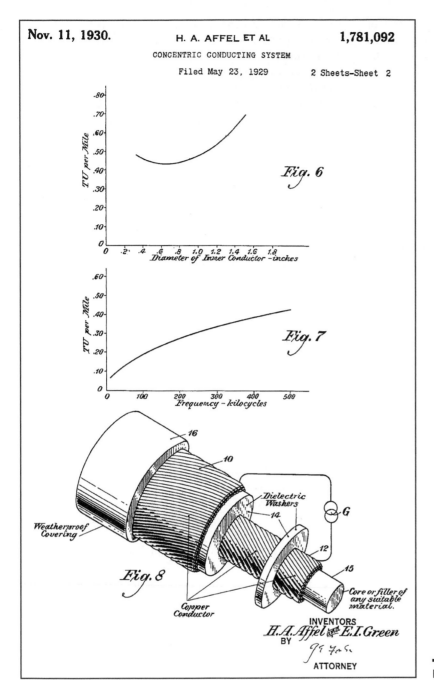

Nov. 11, 1930.   H. A. AFFEL ET AL   1,781,092
CONCENTRIC CONDUCTING SYSTEM
Filed May 23, 1929   2 Sheets-Sheet 2

Fig. 6

TU per Mile / *Diameter of Inner Conductor -inches*

Fig. 7

TU per Mile / *Frequency – kilocycles*

Fig. 8

16
10
Dielectric Washers
14
G
12
15
Weatherproof Covering
Core or filler of any suitable material.
Copper Conductor

INVENTORS
H.A.Affel and E.I.Green
BY
ATTORNEY

Patent drawing for coaxial cable, 1930.

CHAPTER REVIEW

Summary

Transmission lines are two-wire cables that connect two ends of a communication system, a transmitter to an antenna, or an antenna to a receiver. At high frequencies, they also act as resonant circuits and even reactive components.

The two basic types of wire transmission line are parallel lines, such as twin-lead wire, unshielded twisted pair, and shielded twisted pair, and coaxial cables. Each type is designed to be used with specific types of connectors. Transmission lines can be balanced or unbalanced.

When cables are used to carry RF energy, calculation of wavelength becomes important, and the distance represented by a wavelength in a given transmission depends upon the type of cable used. When the length of a transmission line is no longer than several wavelengths at the signal frequency, an RF generator connected to the transmission line sees a characteristic impedance, which is a function of the inductance, resistance, and capacitance in the circuit. Most transmission lines come with standard fixed values of characteristic impedance and attenuation. Some applications require exact timing and sequencing of signals, especially pulses. A coaxial delay line can be used for this purpose.

A transmission line that is terminated with a load that has a resistive impedance equal in value to the characteristic impedance of the line is called a matched line. The forward and reflected signals on an incorrectly terminated transmis-sion line produce standing waves, which interfere with signal transmission. The standing wave ratio (SWR) is the ratio of the maximum current or voltage along a line to the minimum current or voltage along the line. The ideal-case SWR for transmission is 1.

Standing wave conditions can be useful. For example, one-quarter or one-half wavelength segments of transmission line with an open- or short-circuit load can be used in place of resonant or reactive circuits.

Calculating impedances for transmission line applications can be complex, as they involve both resistive and reactive elements. The Smith chart offers a graphical method for obtaining the necessary impedance values. Once the values have been established, some component, such as an impedance-matching circuit, can be added to compensate for the conditions and bring the SWR as close to 1 as possible.

Questions

1. Name the two basic types of transmission lines. Which is the more widely used?
2. What is the name for a transmission line that has one of its two conductors connected to ground?
3. Name a popular type of balanced line.
4. What is the name given to the distance that a signal travels during one cycle?
5. Name a popular UHF connector used for coaxial cable.
6. What is the coaxial cable connector commonly used on test equipment and in LANs?
7. What is the best coaxial cable connector for UHF and microwave applications?
8. What type of coaxial connector is widely used for cable TV and VCR connections?
9. Describe the equivalent circuit of a transmission line.
10. What determines the characteristic impedance of a transmission line?
11. What is another name for characteristic impedance?
12. What is the importance of the velocity factor in determining cable lengths?
13. How does the cutoff frequency of a coaxial cable vary with its length?
14. Describe the pattern of the current and voltage along a properly matched transmission line.
15. Describe what happens if a transmission line is not terminated in its characteristic impedance.
16. What is the effect on transmitted power when load and transmission line impedances are mismatched?
17. Under which two conditions will all incident power on a line be reflected?
18. What is the ratio of the reflected voltage to the incident voltage on a transmission line called?
19. What do you call a mismatched transmission line?
20. How does the length of a matched transmission line affect the SWR?
21. Name two ways to implement a series resonant circuit with a transmission line.
22. How do transmission lines that are less than $\lambda/4$ or between $\lambda/4$ and $\lambda/2$ at the operating frequency act?
23. One complete revolution around a Smith chart represents how many wavelengths?

Problems

1. Compute the wavelength in meters at a frequency of 350 MHz. ◆
2. At what frequency does a line 3.5 in long represent one-half wavelength?
3. A coaxial cable has a capacitance of 30 pF/ft and an inductance of 78 nH/ft. What is the characteristic impedance? ◆
4. A coaxial cable has a capacitance of 17 pF/ft. The characteristic impedance is 75 Ω. What is the equivalent inductance per foot?
5. What is the velocity factor of a coaxial cable with a dielectric constant of 2.5? ◆
6. A manufacturer's data sheet states that the velocity factor for a specific coaxial cable is 0.7. What is the dielectric constant?
7. Determine the length in feet of a one-quarter wavelength coaxial cable with a velocity factor of 0.8 at a frequency of 49 MHz. ◆
8. What is the time delay introduced by a coaxial cable with the specifications given in Prob. 7 if the length is 65 ft?

9. What is the time delay of 120 ft of a coaxial cable with a dielectric constant of 0.7? ◆

10. How much phase shift is introduced by a coaxial cable 15 ft long to a 10-MHz sine wave signal? The dielectric constant is 2.9.

11. What is the attenuation of 350 ft of RG-11 U coaxial cable at 100 MHz?

12. What is the approximate attenuation of 270 ft of type 9913 coaxial cable at 928 MHz?

13. A transmitter has an output power of 3 W that is fed to an antenna through an RG-58A/U cable 20 ft long. How much power reaches the antenna?

14. What should be the transmission line impedance for optimum transfer of power from a generator to a 52-Ω load?

15. A 52-Ω coaxial cable has a 36-Ω antenna load. What is the SWR?

16. If the load and line impedances of the cable in Prob. 15 are matched, what is the SWR?

17. The maximum voltage along a transmission line is 170 V, and the minimum voltage is 80 V. Calculate the SWR and the reflection coefficient.

18. The reflection coefficient of a transmission line is 0.75. What is the SWR?

19. What are the reflection coefficient and SWR of an open or shorted transmission line?

20. A transmission line has an SWR of 1.65. The power applied to the line is 50 W. What is the amount of reflected power?

21. A power meter inserted into a transmission line reads 1150 W of forward power and 50 W of reflected power. What is the SWR?

22. What must be the value of the load impedance on a transmission line if the voltage variations along the line go to zero at regular intervals?

23. An open-wire transmission line 6 in long acts as a parallel resonant circuit at which frequency? (The dielectric is air.)

24. A coaxial cable has a velocity factor of 0.68. How long is one-half wavelength of this cable at 133 MHz?

25. Calculate the impedance of a microstrip line whose dimensions are $h = 0.05$ in, $w = 0.125$ in, and $t = 0.002$ in. The dielectric constant is 4.5.

26. What is the length of one-quarter wavelength of the microstrip in Prob. 25 at a frequency of 915 MHz?

27. Plot, in normalized form, the following impedances on a Smith chart: $Z_1 = 80 + j25$ and $Z_2 = 35 - j98$, assuming a prime center value of 52 Ω.

28. Plot the load impedance $104 - j58$ on a coaxial line on a Smith chart. Assume a characteristic impedance value of 75 Ω. Then find the SWR.

29. If the operating frequency is 230 MHz and the cable length is 30 ft for Prob. 28, what is the impedance at the generator if the velocity factor is 0.66? Use the Smith chart and show all work.

30. What does the SWR circle on a Smith chart look like if a 52-Ω load is connected to a 52-Ω coaxial cable?

◆ *Answers to Selected Problems follow Chap. 22.*

Critical Thinking

1. Describe the operation of a transmission line, one-quarter or one-half wavelength at the operating frequency, if the line is shorted or open. How can such a line be used?

2. At what frequencies do transmission lines implemented on a printed-circuit board become practical?

3. Compare the characteristics of microstrip and stripline. Which is preferred and why?

4. A 22-ft coaxial cable with an impedance of 50 Ω has a velocity factor of 0.78. The operating frequency is 400 MHz. The impedance of the antenna load combined with the cable impedance as measured at the generator end of the cable is $74 + j66$. What is the antenna impedance as determined on the Smith chart? What is the SWR?

5. Assume that a square wave of 10 MHz is applied to a 200-ft-long RG-58A/U coaxial cable. It is properly terminated. Describe the signal at the load, and explain why the signal appears as it does. (See Fig. 13-13.)

6. Explain how a coaxial transmission line (RG-59/U) could be used as a filter to eliminate interference at the input to a receiver. Assume a frequency of 102.3 MHz. Design the filter and calculate size.

Antennas and Wave Propagation

In wireless communication systems, an RF signal generated by a transmitter is sent into free space and eventually picked up by a receiver. The interface between the transmitter and free space and between free space and the receiver is the antenna.

The incredible variety of antenna types used in radio communication are all based on a few key concepts. This chapter introduces all the most popular and widely used antennas in HF, VHF, and UHF applications. Microwave antennas are covered in Chap. 16. Also discussed are the characteristics of free space and its ability to propagate signals over long distances. A study of wave propagation—how radio signals are affected by the earth and space in moving from transmitting antenna to receiving antenna—is critical to understanding how to ensure reliable communication over the desired distance at specific frequencies.

Objectives

After completing this chapter, you will be able to:

- Describe the characteristics of a radio wave.
- Compute the length of one-quarter wavelength and one-half wavelength antennas, given frequency of operation.
- Name the basic antenna types and give the characteristics of each.
- Explain how arrays are used to create directivity and gain.
- Describe ways in which antenna design can be modified to produce an optimal match between the impedances of a transmitter and an antenna.
- Describe the characteristics of ground waves, sky waves, and space waves.
- Compute signal strength.
- Define fading and diversity reception.

14-1 Antenna Fundamentals

Radio Waves

A radio signal is called an *electromagnetic wave* because it is made up of both electric and magnetic fields. Whenever voltage is applied to the antenna, an electric field is set up. At the same time, this voltage causes current to flow in the antenna, producing a magnetic field. The electric and magnetic fields are at right angles to each other. These electric and magnetic fields are emitted from the antenna and propagate through space over very long distances at the speed of light.

Electromagnetic wave

Magnetic Fields. A *magnetic field* is an invisible force field created by a magnet. An antenna is a type of electromagnet. A magnetic field is generated around a conductor when current flows through it. Figure 14-1 shows the magnetic field, or flux, around a wire carrying a current. Although the magnetic field is a continuous force field, for calculation and measurement purposes it is represented as individual lines of force. This is how the magnetic field appears in most antennas. The strength and direction of the magnetic field depend upon the magnitude and direction of the current flow.

The strength of a magnetic field H produced by a wire antenna is expressed by

$$H = \frac{I}{2\pi d}$$

where I = current, A
d = distance from wire, m

The SI unit for magnetic field strength is ampere-turns per meter.

Magnetic field

Electric Field. An *electric field* is also an invisible force field produced by the presence of a potential difference between two conductors. A common example in electronics is the electric field produced between the plates of a charged capacitor (Fig. 14-2). Of course, an electric field exists between any two points across which a potential difference exists.

The strength of an electric field E is expressed by

$$E = \frac{q}{4\pi \varepsilon d^2}$$

where q = charge between the two points, C
ε = permittivity
d = distance between conductors, m

Electric field

Figure 14-1   Magnetic field around a current-carrying conductor. Magnetic field strength H in ampere-turns per meter = $H = I/(2\pi d)$.

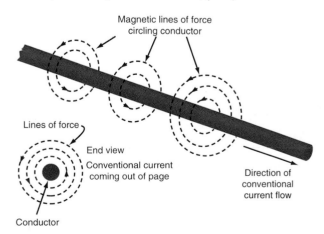

Magnetic lines of force
circling conductor

Lines of force

End view
Conventional current
coming out of page

Direction of
conventional
current flow

Conductor

Figure 14-2   Electric field across the plates of a capacitor.

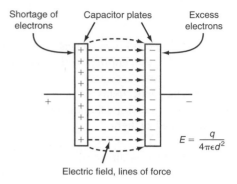

The SI unit for electric field strength is volts per meter.

Permittivity

Permittivity is the dielectric constant of the material between the two conductors. The dielectric is usually air or free space, which has an ε value of approximately $8.85 \times 10^{-12}\ \varepsilon_r$, where ε_r is the dielectric constant of the medium.

Magnetic and Electric Fields in a Transmission Line. Figure 14-3(*a*) shows the electric and magnetic fields around a two-wire transmission line. Note that at any given time, the wires have opposite polarities. During one-half cycle of the ac input, one wire is positive and the other is negative. During the negative half-cycle, the polarity reverses. This means that the direction of the electric field between the wires reverses once per cycle. Figure 14-3(*b*) is a detail of an electric field around conductors.

Note also that the direction of current flow in one wire is always opposite that in the other wire. Therefore, the magnetic fields combine, as shown in Fig. 14-3(*c*). The magnetic field lines aid one another directly between the conductors, but as the lines of force spread out, the direction of the magnetic field from one conductor is

Figure 14-3   (*a*) Magnetic and electric fields around a transmission line. (*b*) Electric field. (*c*) Magnetic fields.

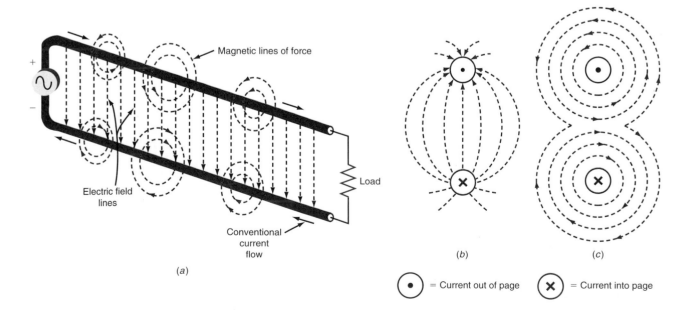

Figure 14-4 Electric and magnetic fields in a coaxial cable (cross-sectional end view). (a) Electric field. (b) Magnetic fields.

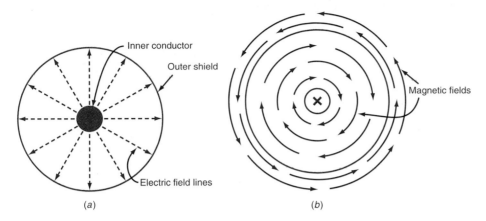

(a)   (b)

opposite that of the other conductor, so the fields tend to cancel one another. The cancellation is not complete, but the resulting magnetic field strength is extremely small. Although the magnetic and electric fields are shown separately in Fig. 14-3(b) and (c) for clarity, remember that they occur simultaneously and at right angles to one another.

A transmission line, like an antenna, is made up of a conductor or conductors. However, transmission lines, unlike antennas, do not radiate radio signals efficiently. The configuration of the conductors in a transmission line is such that the electric and magnetic fields are contained. The closeness of the conductors keeps the electric field primarily concentrated in the transmission line dielectric. The magnetic fields mostly cancel one another. The electric and magnetic fields do extend outward from the transmission line, but the small amount of radiation that does occur is extremely inefficient.

Figure 14-4(a) shows the electric fields associated with a coaxial cable, and Fig. 14-4(b) shows the magnetic fields associated with a coaxial cable. The electric field lines are fully contained by the outer shield of the cable, so none are radiated. The direction of the electric field lines reverses once per cycle.

The magnetic field around the center conductor passes through the outer shield. However, note that the magnetic field produced by the outer conductor is in the opposite direction of the field produced by the inner conductor. Since the amplitude of the current in both conductors is the same, the magnetic field strengths are equal. The inner and outer magnetic fields cancel one another, and so a coaxial cable does not radiate any electromagnetic energy. That is why coaxial cable is the preferred transmission line for most applications.

Antenna Operation

As stated above, an antenna acts as the interface between a transmitter or receiver and free space. It either radiates or senses an electromagnetic field. But the question is, What exactly is an antenna, and what is the relationship between an antenna and a transmission line? Further, how are the electric and magnetic fields produced?

The Nature of an Antenna.

If a parallel-wire transmission line is left open, the electric and magnetic fields escape from the end of the line and radiate into space [Fig. 14-5(a)]. This radiation, however, is inefficient and unsuitable for reliable transmission or reception.

The radiation from a transmission line can be greatly improved by bending the transmission line conductors so they are at a right angle to the transmission line, as shown

Figure 14-5 Converting a transmission line into an antenna. (*a*) An open transmission line radiates a little. (*b*) Bending the open transmission line at right angles creates an efficient radiation pattern.

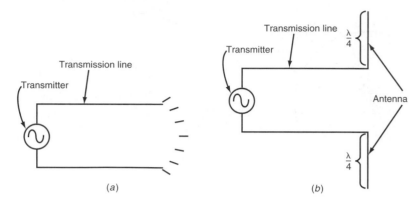

(*a*)   (*b*)

in Fig. 14-5(*b*). The magnetic fields no longer cancel and, in fact, aid one another. The electric field spreads out from conductor to conductor (Fig. 14-6). The result is an antenna. Optimum radiation occurs if the segment of transmission wire converted to an antenna is one-quarter wavelength long at the operating frequency. This makes an antenna that is one-half wavelength long.

An antenna, then, is a conductor or pair of conductors to which is applied the ac voltage at the desired frequency. In Fig. 14-5, the antenna is connected to the transmitter by the transmission line that was used to form the antenna. In most practical applications, the antenna is remote from the transmitter and receiver, and a transmission line is used to transfer the energy between antenna and transmitter or receiver. It is sometimes useful, however, to analyze an antenna as if the conductors were connected directly to the generator or transmitter, as in Fig. 14-7. The voltage creates an electric field and the current creates a magnetic field. Figure 14-7(*a*) shows the magnetic field for one polarity of the generator, and Fig. 14-7(*b*) shows the accompanying electric field. Figure 14-7(*c*) and (*d*) shows the magnetic and electric fields, respectively, for the opposite polarity of the generator.

Figure 14-6 The electric and magnetic fields around the transmission line conductors when an antenna is formed.

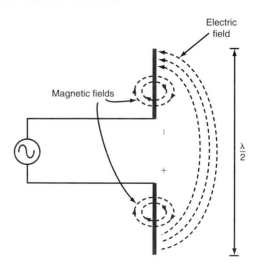

Figure 14-7 Electric and magnetic fields around an antenna.

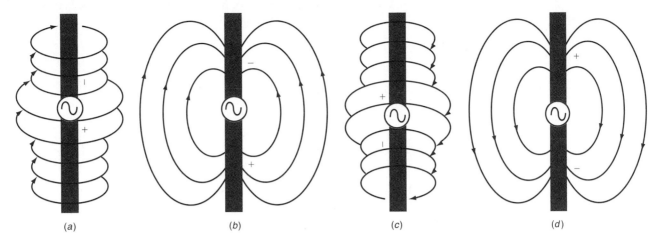

(a)   (b)   (c)   (d)

The magnetic fields vary in accordance with the applied signal from the generator, which is usually a modulated sine wave carrier. The sinusoidal electric field changing over time is similar to a current that causes the generation of the sinusoidal magnetic field. A sinusoidally varying magnetic field produces an electric field. Thus the two fields support and sustain each other. The ratio of the electric field strength of a radiated wave to the magnetic field strength is a constant. It is called the *impedance of space,* or the *wave impedance,* and is 377 Ω. The resulting fields are radiated into space at the speed of light (3×10^8 m/s or 186,400 mi/s).

Impedance of space
Wave impedance

The antenna that is radiating electromagnetic energy appears to the generator as an ideally resistive electrical load so that the applied power is consumed as radiated energy. In addition to the resistive component, an antenna can have a reactive component. The resistive component is called the *antenna radiation resistance*. This resistance does not dissipate power in the form of heat, as in electronic circuits. Instead, the power is dissipated as radiated electromagnetic energy.

Antenna radiation resistance

The Electromagnetic Field. The electric and magnetic fields produced by the antenna are at right angles to each other, and both are perpendicular to the direction of propagation of the wave. This is illustrated in several ways in Fig. 14-8. Figure 14-8(a) shows the basic right-angle relationship. Now, assume that you are looking at a small area of the space around an antenna and that the signal is moving either directly out of the page toward you or into the page away from you. Figure 14-8(b) is a view of the field lines in Fig. 14-8(a), but the perspective is shifted 90° so that you are getting an edge view. Figure 14-8(c) shows the variation in strength of the electric and magnetic fields as they move outward from the antenna. Note that the amplitude and direction of the magnetic and electric fields vary in a sinusoidal manner depending upon the frequency of the signal being radiated.

Electromagnetic field

Near Field and Far Field. Antennas actually produce two sets of fields, the near field and the far field. The *near field* describes the region directly around the antenna where the electric and magnetic fields are distinct. These fields are not the radio wave, but they do indeed contain any information transmitted. These fields weaken with the distance from the antenna, approximately by the quadruple power of the distance. The near field is also referred to as the Fresnel zone.

Near field

The *far field* which is approximately 10 wavelengths from the antenna is the radio wave with the composite electric and magnetic fields. For example, at 2.4 GHz, one wavelength is 984/2400 = 0.41 feet. The far field is 10 times that, or 4.1 ft or beyond.

Far field

Figure 14-8 Viewing the electromagnetic wave emitted by an antenna.

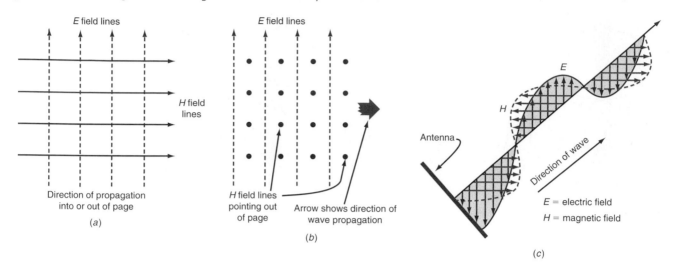

Inside that 4.1 ft lies the near field. The combined fields actually detach themselves from the antenna and radiate into space as previously described. Its strength also diminishes with distance but only at the square of the distance. The far field is also called the Fraunhofer zone.

Most wireless applications use the far field wave. And any antenna radiation patterns are valid only if measurements are taken on the far field. The near field is rarely used, but applications such as *radio-frequency identification* (*RFID*) and near field communication (NFC) make use of the near field. Some cell phone manufacturers also build in a short-range near field radio for applications such as wireless building access, ticket purchases, or automotive functions.

Polarization. *Polarization* refers to the orientation of magnetic and electric fields with respect to the earth. If an electric field is parallel to the earth, the electromagnetic wave is said to be *horizontally polarized;* if the electric field is perpendicular to the earth, the wave is *vertically polarized.* Antennas that are horizontal to the earth produce horizontal polarization, and antennas that are vertical to the earth produce vertical polarization.

Some antennas produce *circular polarization,* in which the electric and magnetic fields rotate as they leave the antenna. There can be *right-hand circular polarization (RHCP)* and *left-hand circular polarization (LHCP);* the type depends on the direction of rotation as the signal leaves the antenna. An electric field can be visualized as rotating as if the antenna were connected to a large fan blade. The electric field and the accompanying magnetic field rotate at the frequency of the transmitter, with one full rotation occurring in one cycle of the wave. Looking from the transmitter to the distant receiver, RHCP gives a clockwise rotation to the electric field and LHCP gives a counterclockwise rotation.

For optimal transmission and reception, the transmitting and receiving antennas must both be of the same polarization. Theoretically, a vertically polarized wave will produce 0 V in a horizontal antenna and vice versa. But during transmission over long distances, the polarization of waves changes slightly because of the various propagation effects in free space. Thus even when the polarization of the transmitting and receiving antennas is not matched, a signal is usually received.

A vertical or horizontal antenna can receive circular polarized signals, but the signal strength is reduced. When circular polarization is used at both transmitter and receiver, both must use either left- or right-hand polarization if the signal is to be received.

MAXWELL'S EQUATIONS

A radio wave is an electromagnetic wave, i.e., one made up of both an electric field and a magnetic field. In your earlier electronic studies, you learned that a magnetic field is created when electrons flow through a conductor. You also learned that an electric field exists between two oppositely charged bodies. The big question is, How do these two fields get together to form a radio wave, and specifically how are they propagated through space?

In the 1870s, Scottish physicist James Clerk Maxwell wrote a book predicting the existence of electromagnetic waves. Using the basic electrical theories of Faraday, Ohm, Ampère, and other electrical researchers at the time, he postulated that a rapidly changing field of one type would produce the other type of field and vice versa over time. Since one type of field produces the other, the two coexist in a self-sustaining relationship. Maxwell's equations express this relationship mathematically.

Later, in the 1880s, German physicist Heinrich Hertz proved Maxwell's theories by generating radio waves, radiating and manipulating them, and detecting them at a distance. Thus the "wireless" or radio was discovered.

Maxwell defined four basic mathematical relationships between the strength and densities of the electric and magnetic fields as they vary over time. These relationships involve partial differential equations and are, therefore, beyond the scope of this text. However, Maxwell's equations are regularly taught as part of most electrical engineering degree curricula.

Maxwell's equations tell us that an electric field changing with time acts as charges in motion or current flow that, in turn, set up a magnetic field. As the magnetic field changes over time, it sets up an electric field. The electric and magnetic fields interact with each other and sustain each other as they propagate through space at the speed of light. This explains how the electromagnetic wave can exist and move through space after it leaves the antenna or other component or apparatus that initially generates it.

Antenna Reciprocity

The term *antenna reciprocity* means that the characteristics and performance of an antenna are the same whether the antenna is radiating or intercepting an electromagnetic signal. A transmitting antenna takes a voltage from the transmitter and converts it to an electromagnetic signal. A receiving antenna has a voltage induced into it by the electromagnetic signal that passes across it. The voltage is then connected to the receiver. In both cases, the properties of the antenna—gain, directivity, frequency of operation, etc.— are the same. However, an antenna used for transmitting high power, such as in a radio or TV broadcast station, must be constructed of materials that can withstand the high voltages and currents involved. A receiving antenna, no matter what the design, can be made of wire. But a transmitting antenna for high-power applications might, e.g., be designed in the same way but be made of larger, heavier material, such as metal tubing.

In most communication systems, the same antenna is used for both transmitting and receiving, and these events can occur at different times or can be simultaneous. An antenna can transmit and receive at the same time as long as some means is provided for keeping the transmitter energy out of the front end of the receiver. A device called a *diplexer* is used for this purpose.

Antenna reciprocity

GOOD TO KNOW

An antenna can transmit and receive at the same time only if a device such as a diplexer is used to keep the transmitter energy out of the receiver.

Diplexer

The Basic Antenna

An antenna can be a length of wire, a metal rod, or a piece of tubing. Many different sizes and shapes are used. The length of the conductor is dependent on the frequency

Antennas and Wave Propagation

529

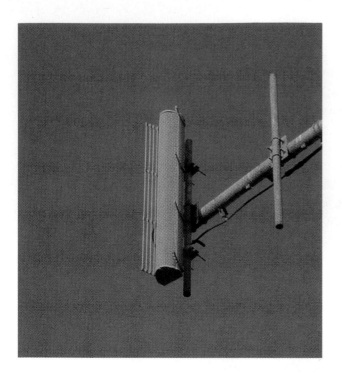

Cell phones, telephones, and other forms of data transmission are commonly accomplished by antenna. The design of the antenna can be varied to make the best match between it and the impedances of its transmitter. Microstrip technology is used in these antennas by CSA Wireless. *(Courtesy of CSA Wireless.)*

of operation. Antennas radiate most effectively when their length is directly related to the wavelength of the transmitted signal. Most antennas have a length that is some fraction of a wavelength. One-half and one-quarter wavelengths are most common.

An important criterion for radiation is that the length of the conductor be approximately one-half or one-quarter wavelength of the ac signal. A 60-Hz sine wave signal has a wavelength of $\lambda = 300{,}000{,}000/60 = 5{,}000{,}000$ m. Since 1 mi $\approx$ 1609.34 m, one wavelength of a 60-Hz signal is about $5{,}000{,}000/1609.34 = 3106.86$ mi. One-half wavelength is 1553.43 mi. Very little radiation of an electromagnetic field occurs if antenna wires are less than this length.

The same is true of wires carrying audio signals. A 3-kHz audio signal has a wavelength of $300{,}000{,}000/3000 = 100{,}000$ m, or 62.14 mi. This wavelength is so long compared to the length of the wire normally carrying such signals that little radiation occurs.

However, as frequency is increased, wavelength decreases. At frequencies from about 1 MHz up to 100 GHz, the wavelength is within the range of practical conductors and wires. It is within this range that long-distance radiation occurs. For example, a 300-MHz UHF signal has a wavelength of 1 m, a very practical length.

The other factor that determines how much energy is radiated is the arrangement of the conductors carrying the signal. If they are in the form of a cable such as a transmission line with a generator at one end and a load at the other, as shown in Fig. 14-3, very little radiation occurs at any frequency.

As seen in Figs. 14-5 and 14-6, an open transmission line can be made into an antenna simply by bending the conductors out at a right angle with the transmission line. This concept is illustrated again in Fig. 14-9. Such a line has a standing wave such that the voltage is maximum at the end of the line and the current is minimum. One-quarter wave back from the open end are a voltage minimum and a current maximum, as shown in the figure. By bending the conductors at a right angle to the transmission line at the quarter-wave point, an antenna is formed. The total length of the antenna is one-half wavelength at the frequency of operation. Note the distribution of the voltage and current standing waves on the antenna. At the center the voltage is minimum and the current is maximum.

Figure 14-9 Standing waves on an open transmission line and an antenna.

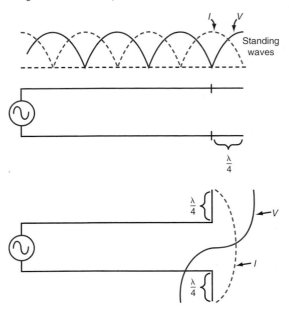

14-2 Common Antenna Types

All the most common types of antennas used in the communication industry are based on a basic dipole, and most are some modified form of the one-half wavelength dipole discussed in the last section.

The Dipole Antenna

One of the most widely used antenna types is the half-wave dipole shown in Fig. 14-10. This antenna is also formally known as the *Hertz antenna* after Heinrich Hertz, who first demonstrated the existence of electromagnetic waves. Also called a *doublet,* a dipole antenna is two pieces of wire, rod, or tubing that are one-quarter wavelength long at the operating resonant frequency. Wire dipoles are supported with glass, ceramic, or plastic insulators at the ends and middle, as shown in Fig. 14-11. Self-supporting dipoles are made from a stiff metal rod or tubing.

Radiation Resistance. The transmission line is connected at the center. The dipole has an impedance of 73 Ω at its center, which is the *radiation resistance.* At the resonant frequency, the antenna appears to be a pure resistance of 73 Ω. For maximum power transfer it is important that the impedance of the transmission line match the load. A 73-Ω coaxial cable like RG-59/U is a perfect transmission line for a dipole antenna.

Dipole antenna

Hertz antenna
Doublet

Radiation resistance

Figure 14-10 The dipole antenna.

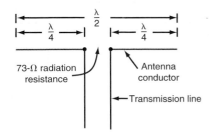

Figure 14-11   A half-wave dipole of 18 MHz.

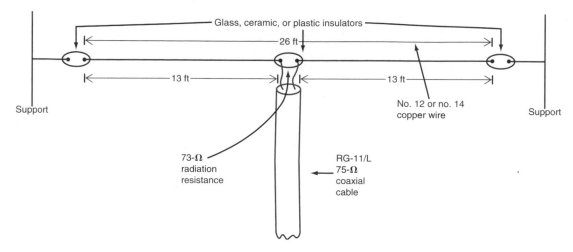

RG-11/U coaxial cable with an impedance of 75 Ω also provides an excellent match. When the radiation resistance of the antenna matches the characteristic impedance of the transmission line, the SWR is minimum and maximum power reaches the antenna.

The radiation resistance of the dipole is ideally 73 Ω when the conductor is infinitely thin and the antenna is in free space. Its actual impedance varies depending on the conductor thickness, the ratio of diameter to length, and the proximity of the dipole to other objects, especially the earth.

As the conductor thickness increases with respect to the length of the antenna, the radiation resistance decreases. A typical length-to-diameter ratio is about 10,000 for a wire antenna, for a radiation resistance of about 65 Ω instead of 73 Ω. The resistance drops gradually as the diameter increases. With a large tubing conductor, the resistance can drop to as low as 55 Ω.

The graph in Fig. 14-12 shows how radiation resistance is affected by the height of a dipole above the ground. Curves for both horizontally and vertically mounted antennas are given. The resistance varies above and below an average of about 73 Ω, depending on height in wavelengths. The higher the antenna, the less effect the earth and surrounding objects have on it, and the closer the radiation resistance is to the theoretical ideal. Given that radiation resistance is affected by several factors, it often varies from

Figure 14-12   The effect of dipole height above ground on radiation resistance.

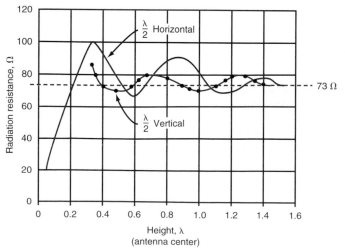

73 Ω. Nevertheless, a 75-Ω coaxial cable provides a good match, as does 50-Ω cable at lower heights with thicker conductors.

Dipole Length. An antenna is a frequency-sensitive device. In Chap. 13, you learned that the formula $\lambda = 984/f$ can be used to calculate one wavelength at a specific frequency, and $\lambda = 492/f$ can thus be used to calculate one-half wavelength. For example, one-half wavelength at 122 MHz is $492/122 = 4.033$ ft.

As it turns out, to get the dipole to resonate at the frequency of operation, the physical length must be somewhat shorter than the one-half wavelength computed with the expression given above, since actual length is related to the ratio of length to diameter, conductor shape, Q, the dielectric (when the material is other than air), and a condition known as end effect. *End effect* is a phenomenon caused by any support insulators used at the ends of the wire antenna and has the effect of adding a capacitance to the end of each wire. At frequencies up to about 30 MHz, end effect shortens the antenna by about 5 percent. Thus the actual antenna length is only about 95 percent of the computed length. The formula must be modified as follows:

$$L = \frac{492 \times 0.95}{f} = \frac{468}{f}$$

where L = length of half-wave dipole antenna.

For half-wave dipole wire antennas used below 30 MHz, the formula $L = 468/f$ provides a ballpark figure, so to speak. Minor adjustments in length can then be made to fine-tune the antenna to the center of the desired frequency range.

For example, an antenna for a frequency of 27 MHz would have a length of $468/27 = 17.333$ ft. To create a half-wave dipole, two 8.666-ft lengths of wire would be cut, probably 12- or 14-gauge copper wire. Physically, the antenna would be suspended between two points as high as possible off the ground (see Fig. 14-11). The wire conductors themselves would be connected to glass or ceramic insulators at each end and in the middle to provide good insulation between the antenna and its supports. The transmission line would be attached to the two conductors at the center insulator. The transmission line should leave the antenna at a right angle so that it does not interfere with the antenna's radiation.

At frequencies above 30 MHz, the conductor is usually thicker because thicker rods or tubing, rather than wire, is used. Using thicker materials also shortens the length by a factor of about 2 or 3 percent. Assuming a shortening factor of 3 percent, one-half wavelength would be $492 \times 0.97/f$.

Antenna Resonance. Because the antenna is one-half wavelength at only one frequency, it acts as a resonant circuit. To the generator, the antenna looks like a series resonant circuit (see Fig. 14-13). The inductance represents the magnetic field, and the capacitance represents the electric field. The resistance is the radiation resistance. As always, this resistance varies depending on antenna conductor thickness and height.

If the signal applied to the antenna is such that the antenna is exactly one-half wavelength long, the equivalent circuit will be resonant and the inductive reactance will cancel the capacitive reactance. Only the effect of the radiation resistance will be present, and the signal will radiate.

Figure 14-13   The equivalent circuit of a dipole.

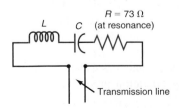

Antennas and Wave Propagation

Antenna Q

Antenna bandwidth

If the frequency of operation and the antenna length do not match, the equivalent circuit will not be resonant. Instead, like any resonant circuit, it will have a complex impedance made up of resistive and reactive components. If the frequency of operation is too low, the antenna will be too short and the equivalent impedance will be capacitive because the capacitive reactance is higher at the lower frequency. If the frequency of operation is too high, the antenna will be too long and the equivalent impedance will be inductive because the inductive reactance is higher at the higher frequency.

If the dipole is used at a frequency different from its design frequency, the antenna impedance no longer matches the transmission line impedance, so the SWR rises and power is lost. However, if the frequency of operation is close to that for which the antenna was designed, the mismatch will not be great and the antenna will work satisfactorily despite the higher SWR.

Antenna Q and Bandwidth. The *bandwidth* of an *antenna* is determined by the frequency of operation and the Q of the antenna according to the familiar relationship $BW = f_r/Q$. Although it is difficult to calculate the exact Q for an antenna, as the above relationship shows, the higher the Q, the narrower the bandwidth BW. Lowering Q widens bandwidth. In resonant circuits a high Q (>10) is usually desirable, because it makes the circuit more selective. For an antenna, low Q, and hence wider bandwidth, is desirable so that the antenna can operate over a wider range of frequencies with reasonable SWR. As a rule of thumb, any SWR below 2:1 is considered good in practical antenna work. Modern communication transceivers rarely operate at just one frequency; typically they operate on a selected channel inside a broader band of frequencies. Further, the transmitter is modulated, so there are sidebands. If the antenna has too high a Q and its bandwidth is too narrow, the SWR will be higher than 2:1 and sideband clipping can occur.

The Q and thus the bandwidth of an antenna are determined primarily by the ratio of the length of the conductor to the diameter of the conductor. When thin wire is used as the conductor, this ratio is very high, usually in the 10,000 to 30,000 range, resulting in high Q and narrow bandwidth. A length-to-diameter ratio of 25,000 results in a Q of about 14.

If the antenna conductors are made of larger-diameter wire or tubing, the length-to-diameter ratio and Q decrease, resulting in a wider bandwidth. A ratio of 1200 results in a Q of about 8.

When larger-diameter conductors are used to construct an antenna, the larger plate area causes the inductance of the conductor to decrease and the capacitance to increase. The L/C ratio is reduced for a given resonant frequency. Lowering L lowers the inductive reactance, which directly affects Q. Since $Q = X_L/R$ and $BW = f_r/Q$, lowering X_L reduces Q and increases bandwidth.

At UHF and microwave frequencies, antennas are typically made of short, fat conductors, such as tubing. It is not uncommon to see conductors with diameters as large as 0.5 in. The result is wider bandwidth.

Bandwidth is sometimes expressed as a percentage of the resonant frequency of the antenna. A small percentage means a higher Q, and a narrower bandwidth means a lower percentage. A typical wire antenna has a bandwidth in the range of 3 to 6 percent of the resonant frequency. If thicker conductors are used, this percentage can be increased to the 7 to 10 percent range, which gives lower Q and wider bandwidth.

For example, if the bandwidth of a 24-MHz dipole antenna is given as 4 percent, the bandwidth can be calculated as $0.04 \times 24 = 0.96$ MHz (960 kHz). The operating range of this antenna, then, is the 960-kHz bandwidth centered on 24 MHz. This gives upper and lower frequency limits of 24 MHz ±480 kHz or one-half the bandwidth. The operating range is 23.52 to 24.48 MHz, where the antenna is still close to resonance.

The Q and bandwidth of an antenna are also affected by other factors. In array-type antennas with many conductors, Q is affected by the number of conductors used and their spacing to the dipole. These antennas usually have high Q's and, thus, narrow bandwidths, and so off-frequency operation produces greater changes in SWR than it does with lower-Q antennas.

Figure 14-14 The conical dipole and its variation. (*a*) Conical antenna. (*b*) Broadside view of conical dipole antenna (bow tie antenna) showing dimensions. (*c*) Open-grill bow tie antenna.

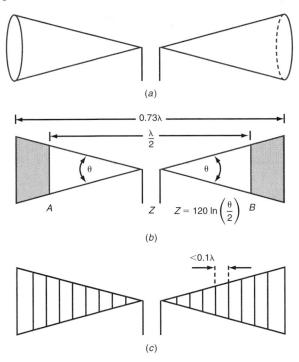

Conical Antennas. Another common way to increase bandwidth is to use a version of the dipole antenna known as the *conical antenna* [Fig. 14-14(*a*)]. Figure 14-14(*b*) shows a flat, broadside view of a conical antenna. The overall length of the antenna is 0.73λ or $0.73(984)/f = 718.32/f$. This is longer than the traditional one-half wavelength of a dipole antenna, but the physical shape changes the necessary dimensions for resonance. The cones are shaped such that the shaded area is equal to the unshaded area. When this is done, the distance between the boundaries marked by *A* and *B* in Fig. 14-14(*b*) is one-half wavelength, less about 5 percent, or approximately $468/f$, where *f* is in megahertz.

The center radiation resistance of a conical antenna is much higher than the 73 Ω usually found when straight-wire or tubing conductors are used. This center impedance is given by $Z = 120 \ln (\theta/2)$, where *Z* is the center radiation resistance at resonance and θ is the angle associated with the cone [see Fig. 14-14(*b*)]. For an angle of 30°, the center impedance is thus $Z = 120 \ln (30/2) = 120 \ln 15 = 120(2.7) = 325 \Omega$. This is a reasonably good match to 300-Ω twin-lead transmission line.

To use coaxial cable, which is usually desirable, some kind of impedance-matching network is needed to transform the high center impedance to the 50 or 75 Ω characteristic of most coaxial cables.

Cones are difficult to make and expensive, and a popular and equally effective variation of the conical antenna is the *bow tie antenna*. A two-dimensional cone is a triangle; therefore, a flat version of the conical antenna looks like two triangles or a bow tie. One bow tie version of the conical antenna in Fig. 14-14(*b*) would have the same shape and dimensions, but it would be made of flat aluminum. Bow tie antennas can also be made of a grillwork of conductors, instead of a flat plate, as shown in Fig. 14-14(*c*); this configuration reduces wind resistance. If the spacing between the conductors is made less than 0.1 wavelength at the highest operating frequency, the antenna appears to be a solid conductor to the transmitter or receiver.

The primary advantage of conical antennas is their tremendous bandwidth: they can maintain a constant impedance and gain over a 4:1 frequency range. The length of the

Conical antenna

Bow tie antenna

antenna is computed by using the center frequency of the range to be covered. For example, an antenna to cover the 4:1 range at 250 MHz to 1 GHz (1000 MHz) would be cut for a center frequency of $(1000 + 250)/2 = 1250/2 = 625$ MHz.

Dipole Polarization. Most half-wave dipole antennas are mounted horizontally to the earth. This makes the electric field horizontal to the earth; therefore, the antenna is horizontally polarized. Horizontal mounting is preferred at the lower frequencies (<30 MHz) because the physical construction, mounting, and support are easier. This type of mounting also makes it easier to attach the transmission line and route it to the transmitter or receiver.

A dipole antenna can also be mounted vertically, in which case the electric field will be perpendicular to the earth, making the polarization vertical. Vertical mounting is common at the higher frequencies (VHF and UHF), where the antennas are shorter and made of self-supporting tubing.

Radiation Pattern and Directivity. The *radiation pattern* of any antenna is the shape of the electromagnetic energy radiated from or received by that antenna. Most antennas have directional characteristics that cause them to radiate or receive energy in a specific direction. Typically that radiation is concentrated in a pattern that has a recognizable geometric shape.

The radiation pattern of a half-wave dipole has the shape of a doughnut. Figure 14-15 shows the pattern with one-half the doughnut cut away. The dipole is at the center hole of the doughnut, and the doughnut itself represents the radiated energy. To an observer looking down on the top of the dipole, the radiation pattern would appear to be a figure 8, as shown in Fig. 14-16. This horizontal radiation pattern is plotted on a polar coordinate graph in the figure. The center of the antenna is assumed to be at the center of the graph. The dipole is assumed to be aligned with the 90° to 270° axis. As shown, the maximum amount of energy is radiated at right angles to the dipole, at 0° and 180°. For that reason, a dipole is what is known as a *directional antenna*. For optimum transmission and reception, the antenna should be aligned broadside to the signal destination or source. For optimal signal transmission, the transmitting and receiving antennas must be parallel to each other.

Whenever a dipole receiving antenna is pointed toward a transmitter, or vice versa, it must be broadside to the direction of the transmitter. If the antenna is at some other angle, the maximum signal will not be received. As can be seen from the radiation pattern in Fig. 14-16, if the end of the receiving antenna is pointed directly at the transmitting antenna, no signal is received. As indicated earlier, this zero-signal condition could not occur in practice, because the radiated wave would undergo some

Figure 14–15   Three-dimensional pattern of a half-wave dipole.

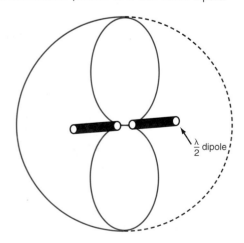

Figure 14-16 Horizontal radiation pattern of a half-wave dipole.

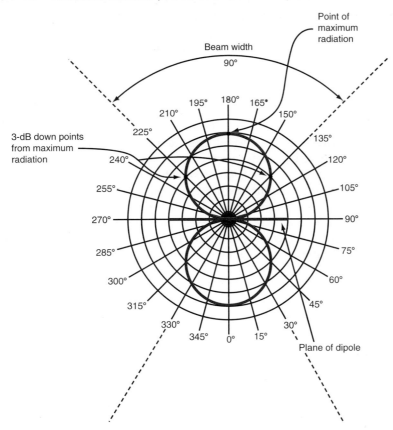

shifts during propagation and, therefore, some minimal signal would be received from the ends of the antenna.

The measure of an antenna's directivity is *beam width,* the angle of the radiation pattern over which a transmitter's energy is directed or received. Beam width is measured on an antenna's radiation pattern. The concentric circles extending outward from the pattern in Fig. 14-16 indicate the relative strength of the signal as it moves away from the antenna. The beam width is measured between the points on the radiation curve that are 3 dB down from the maximum amplitude of the curve. As stated previously, the maximum amplitude of the pattern occurs at 0° and 180°. The 3-dB down points are 70.7 percent of the maximum. The angle formed with two lines extending from the center of the curve to these 3-dB points is the beam width. In this example, the beam width is 90°. The smaller the beam width angle, the more directional the antenna.

Beam width

Antenna Gain. *Gain* was previously defined as the output of an electronic circuit or device divided by the input. Obviously, passive devices such as antennas cannot have gain in this sense. The power radiated by an antenna can never be greater than the input power. However, a directional antenna can radiate more power *in a given direction* than a nondirectional antenna, and in this "favored" direction, it acts as if it had gain. Antenna gain of this type is expressed as the ratio of the *effective radiated* output power P_{out} to the input power P_{in}. Effective radiated power is the actual power that would have to be radiated by a reference antenna (usually a nondirectional or dipole antenna) to produce the same signal strength at the receiver as the actual antenna produces. Antenna gain is usually expressed in decibels:

Gain

$$dB = 10 \log \frac{P_{out}}{P_{in}}$$

The power radiated by an antenna with directivity and therefore gain is called the *effective radiated power (ERP)*. The ERP is calculated by multiplying the transmitter power fed to the antenna P_t by the power gain A_p of the antenna:

$$\text{ERP} = A_p P_t$$

To calculate ERP, you must convert from decibels to the power ratio or gain.

The gain of an antenna is usually expressed in reference to either the dipole or an isotropic radiator. An *isotropic radiator* is a theoretical point source of electromagnetic energy. The E and H fields radiate out in all directions from the point source, and at any given distance from the point source, the fields form a sphere. To visualize this, think of a lightbulb at the center of a large world globe and the light that illuminates the inside of the sphere as the electromagnetic energy.

In what is known as the *near field* of the antenna, defined as the part of the field less than 10 wavelengths from the antenna at the operating frequency, a portion of the surface area on the sphere looks something like that shown in Fig. 14-17. In the *far field,* 10 or more wavelengths *distant* from the source, the sphere is so large that a small area appears to be flat rather than curved, much as a small area of the earth appears flat. Most far-field analysis of antennas is done by assuming a flat surface area of radiation with the electric and magnetic fields at right angles to each other.

In reality, no practical antennas radiate isotropically; instead, the radiation is concentrated into a specific pattern, as seen in Figs. 14-15 and 14-16. This concentration of electromagnetic energy has the effect of increasing the radiation power over a surface area of a given size. In other words, antenna directivity gives the antenna gain over an isotropic radiator. The mathematics involved in determining the power gain of a dipole over an isotropic source is beyond the scope of this book. As it turns out, this power gain is 1.64; in decibels, $10 \log 1.64 = 2.15$ dB.

Most formulas for antenna gain are expressed in terms of gain in decibels over a dipole (dBd). If the antenna gain is said to be 4.5 dB, this means gain as compared to a dipole. To compute the gain of an antenna with respect to an isotropic radiator (dBi), add 2.15 dB to the gain over the dipole (dBi = dBd + 2.15). In general, the more concentrated an antenna's energy, the higher the gain. The effect is as if the transmitter power were actually increased by the antenna gain and applied to a dipole.

Figure 14-17   An isotropic source and its spherical wave front.

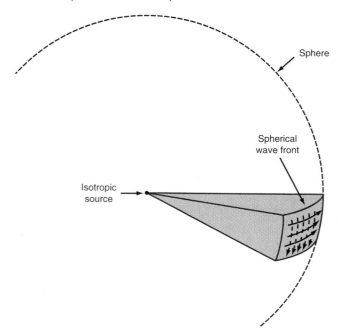

Chapter 14

Figure 14-18 Folded dipole. (*a*) Basic configuration. (*b*) Construction with twin lead.

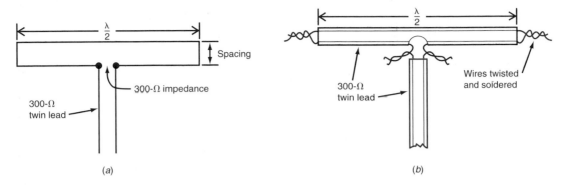

(a)   (b)

Folded Dipoles. A popular variation of the half-wave dipole is the *folded dipole,* shown in Fig. 14-18(*a*). Like the standard dipole, it is one-half wavelength long. However, it consists of two parallel conductors connected at the ends with one side open at the center for connection to the transmission line. The impedance of this popular antenna is 300 Ω, making it an excellent match for the widely available 300-Ω twin lead. The spacing between the two parallel conductors is not critical, although it is typically inversely proportional to the frequency. For very high-frequency antennas, the spacing is less than 1 in; for low-frequency antennas, the spacing may be 2 or 3 in.

The radiation pattern and gain of a folded dipole are the same as those of a standard dipole. However, folded dipoles usually offer greater bandwidth. The radiation resistance impedance can be changed by varying the size of the conductors and the spacing.

An easy way to make a folded dipole antenna is to construct it entirely of 300-Ω twin-lead cable. A piece of twin-lead cable is cut to a length of one-half wavelength, and the two ends are soldered together [Fig. 14-18(*b*)]. When one-half wavelength is calculated, the velocity factor of twin-lead cable (0.8) must be included in the formula. For example, the length of an antenna cut for 100 MHz (the approximate center of the FM broadcast band) is $\lambda/2 = 492/f = 492/100 = 4.92$ ft. By taking into account the velocity factor of twin-lead cables the final length is $4.92 \times 0.8 = 3.936$ ft.

The center of one conductor in the twin-lead cable is then cut open, and a 300-Ω twin-lead transmission line is soldered to the two wires. The result is an effective, low-cost antenna that can be used for both transmitting and receiving purposes. Such antennas are commonly used for both TV and FM radio reception.

Folded dipole

The Marconi or Ground–Plane Vertical Antenna

Another widely used antenna is the *one-quarter wavelength vertical antenna,* also called a *Marconi antenna.* It is similar in operation to a vertically mounted dipole antenna. However, it offers major advantages because it is one-half the length of a dipole antenna.

One-quarter wavelength vertical antenna

Radiation Pattern. Most half-wave dipole antennas are mounted horizontally, but they can also be mounted vertically. The radiation pattern of a vertically polarized dipole antenna is still doughnut-shaped, but the radiation pattern, as seen from above the antenna, is a perfect circle. Such antennas, which transmit an equal amount of energy in the horizontal direction, are called *omnidirectional antennas.* Because of the doughnut shape, the vertical radiation is zero and the radiation at any angle from the horizontal is greatly diminished.

Omnidirectional antenna

The same effect can be achieved with a one-quarter wavelength antenna. Figure 14-19 shows a vertical dipole with the doughnut-shaped radiation pattern. One-half of the pattern is below the surface of the earth. This is called a *vertical radiation pattern.*

Vertical radiation pattern

Vertical polarization and omnidirectional characteristics can also be achieved by using a one-quarter wavelength vertical radiator; see Fig. 14-20(*a*). This antenna is known

Figure 14-19 Side view of the radiation pattern of a vertical dipole.

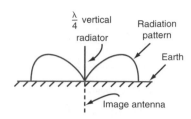

Ground-plane antenna

as a Marconi or *ground-plane antenna*. It is usually fed with coaxial cable; the center conductor is connected to the vertical radiator, and the shield is connected to earth ground. With this arrangement, the earth acts as a type of electrical "mirror," effectively providing the other one-quarter wavelength of the antenna and making it the equivalent of a vertical dipole. The result is a vertically polarized omnidirectional antenna.

Ground plane

Radial

Counterpoise

Ground Plane, Radials, and Counterpoise. The effectiveness of a vertically polarized omnidirectional antenna depends on making good electrical contact with the earth. This can be tricky. Sometimes, a reasonable ground can be obtained by driving a copper rod 5 to 15 ft long into the earth. If the earth is dry and has high resistance because of its content, however, even a ground rod is insufficient. Once a good electrical connection to the earth has been made, the earth becomes what is known as a *ground plane*. If a good electrical connection (low resistance) cannot be made to the earth, then an artificial ground plane can be constructed of several one-quarter wavelength wires laid horizontally on the ground or buried in the earth [Fig. 14-20(*b*)]. Four wires are usually sufficient, but in some antenna systems, more are used. These horizontal wires at the base of the antenna are referred to as *radials*. The greater the number of radials, the better the ground and the better the radiation. The entire ground-plane collection of radials is often referred to as a *counterpoise*.

At very high frequencies, when antennas are short, any large, flat metallic surface can serve as an effective ground plane. For example, vertical antennas are widely used on cars, trucks, boats, and other vehicles. The metallic roof of a car makes a superior ground plane for VHF and UHF antennas. In any case, the ground plane must be large enough that it has a radius of greater than one-quarter wavelength at the lowest frequency of operation.

Figure 14-20 Ground-plane antenna. (*a*) One-quarter wavelength vertical antenna. (*b*) Using radials as a ground plane.

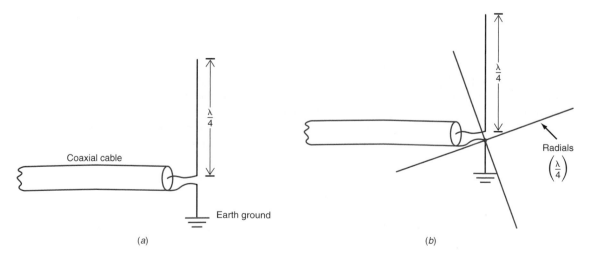

Each telecommunication dish collects electromagnetic energy from a satellite.

Radiation Resistance. The impedance of a vertical ground-plane antenna is exactly one-half the impedance of the dipole, or approximately 36.5 Ω. Of course, this impedance varies depending on the height above ground, the length/diameter ratio of the conductor, and the presence of surrounding objects. The actual impedance can drop to less than 20 Ω for a thick conductor very close to the ground.

Since there is no such thing as 36.5-Ω coaxial cable, 50-Ω coaxial cable is commonly used to feed power to the ground-plane antenna. This represents a mismatch. However, the SWR of 50/36.5 = 1.39 is relatively low and does not cause any significant power loss.

One way to adjust the antenna's impedance is to use "drooping" radials, as shown in Fig. 14-21. At some angle depending upon the height of the antenna above the ground, the antenna's impedance will be near 50 Ω, making it a nearly perfect match for most 50-Ω coaxial cable.

Antenna Length. In addition to its vertical polarization and omnidirectional characteristics, another major benefit of the one-quarter wavelength vertical antenna is its length. It is one-half the length of a standard dipole, which represents a significant savings at lower radio frequencies. For example, a one-half wavelength antenna for a frequency of 2 MHz would have to be $468/f = 468/2 = 234$ ft. Constructing a 234-ft vertical antenna would present a major structural problem, as it would require a support at least that long. An alternative is to use a one-quarter wavelength vertical antenna, the length of which would only have to be $234/f = 117$ ft. Most low-frequency transmitting antennas use the one-quarter wavelength of vertical configuration for this reason. AM broadcast stations in the 535- to 1635-kHz range use one-quarter wavelength vertical antennas because they are short, inexpensive, and not visually obtrusive. Additionally, they provide an equal amount of radiation in all directions, which is usually ideal for broadcasting.

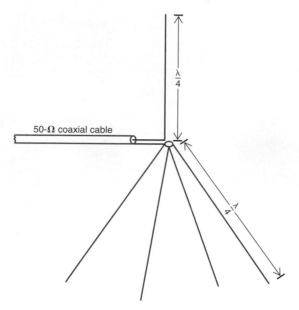

50-Ω coaxial cable

$\frac{\lambda}{4}$

$\frac{\lambda}{4}$

For many applications, e.g., with portable or mobile equipment, it is not possible to make the antenna a full one-quarter wavelength long. A cordless telephone operating in the 46- to 49-MHz range would have a one-quarter wavelength of $234/f = 234/46 = 5.1$ ft. A 5-ft whip antenna, or even a 5-ft telescoping antenna, would be impractical for a device you have to hold up to your ear. And a one-quarter wavelength vertical antenna for a 27-MHz CB walkie-talkie would have to be an absurdly long $234/27 = 8.7$ ft!

To overcome this problem, much shorter antennas are used, and lumped electrical components are added to compensate for the shortening. When a vertical antenna is made less than one-quarter wavelength, the practical effect is a decreased inductance. The antenna no longer resonates at the desired operating frequency, but at a higher frequency. To compensate for this, a series inductor, called a *loading coil,* is connected in series with the antenna coil (Fig. 14-22). The loading coil brings the antenna back into resonance at the desired frequency. The coil is sometimes mounted external to the equipment at the base of the antenna so that it can radiate along with the vertical rod. The coil can also be contained inside a handheld unit, as in a cordless telephone.

Loading coil

In some cases, the loading coil is placed at the center of the vertical conductor (see Fig. 14-23). CB antennas and cellular telephone antennas use this method. The CB antennas have a large coil usually enclosed inside a protective housing. The shorter cellular telephone antennas use a built-in self-supporting coil that looks like and serves the dual purpose of a flexible spring.

Figure 14-22 Using a base leading coil to increase effective antenna length.

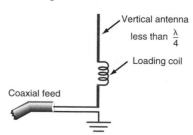

Vertical antenna less than $\frac{\lambda}{4}$

Loading coil

Coaxial feed

Figure 14-23   A center-loaded shortened vertical.

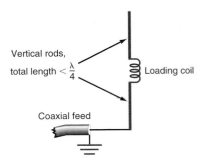

Vertical rods,
total length $< \frac{\lambda}{4}$

Loading coil

Coaxial feed

In both types of inductive vertical antennas, the inductor can be made variable, so that the antenna can be tuned; or the inductor can be made larger than needed and a capacitor connected in series to reduce the overall capacitance and tune the coil to resonance.

Another approach to using a shortened antenna is to increase the effective capacitance represented by the antenna. One way to do this is to add conductors at the top of the antenna, as shown in Fig. 14-24. Sometimes referred to as a *top hat,* this structure increases the capacitance to surrounding items, bringing the antenna back into resonance. Obviously, such an arrangement is too top-heavy and inconvenient for portable and mobile antennas. However, it is sometimes used in larger fixed antennas at lower frequencies.

Top hat

One of the most popular variations of the one-quarter wavelength vertical antenna is the $\frac{5}{8}\lambda$ vertical antenna ($5\lambda/8$, or 0.625λ). Like a one-quarter wavelength ground-plane antenna, the $\frac{5}{8}$-wavelength vertical antenna is fed at the base with coaxial cable and has four or more one-quarter wavelength radials or the equivalent (i.e., body of a car).

A $\frac{5}{8}\lambda$ vertical antenna is one-eighth wavelength longer than one-half wavelength. This additional length gives such an antenna about a 3-dB gain over a basic dipole and one-quarter wavelength vertical antenna. The gain comes from concentrating the radiation into a narrower vertical radiation pattern at a lower angle to the horizon. Thus $\frac{5}{8}\lambda$ vertical antennas are ideal for long-distance communication.

Since a $\frac{5}{8}\lambda$ antenna is not some integer multiple of a one-quarter wavelength, it appears too long to the transmission line; i.e., it looks like a capacitive circuit. To compensate for this, a series inductor is connected between the antenna and the transmission line, making the antenna a very close match to a 50-Ω coaxial transmission line. The arrangement is similar to that shown in Fig. 14-22.

The $\frac{5}{8}\lambda$ antenna is widely used in the VHF and UHF bands, where its length is not a problem. It is very useful in mobile radio installations, where omnidirectional antennas are necessary but additional gain is needed for reliable transmission over longer distances.

Figure 14-24   Using a top hat capacitive load to lower the resonant frequency of a shortened vertical antenna.

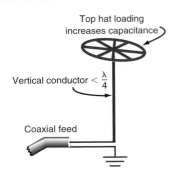

Top hat loading
increases capacitance

Vertical conductor $< \frac{\lambda}{4}$

Coaxial feed

Example 14-1

Calculate the length of the following antennas and state their radiation resistance at 310 MHz: (a) dipole; (b) folded dipole (twin lead; $Z = 300\ \Omega$, velocity factor $= 0.8$); (c) bow tie ($\theta = 35°, 0.73\lambda$); (d) ground plane (vertical).

a. $\dfrac{\lambda}{2} = \dfrac{492 \times 0.97}{f} = \dfrac{477.25}{310} = 1.54\ \text{ft}$

Dipole radiation resistance $= 73\ \Omega$

b. $\dfrac{\lambda}{2} = \dfrac{492}{f} = \dfrac{492}{310} = 1.587\ \text{ft}$

Taking the velocity factor into account gives $0.8 \times 1.587 = 1.27\ \text{ft}$.

c. $\lambda = \dfrac{984}{f} = \dfrac{984}{310} = 3.16\ \text{ft}$

$0.73\lambda = 3.16(0.73) = 2.3\ \text{ft}$

$$Z = 120 \ln \dfrac{\theta}{2} = 120 \ln \dfrac{35}{2}$$

$$= 120 \ln 17.5 = 120(2.862)$$

$$= 343.5\ \Omega$$

d. $\dfrac{\lambda}{4} = \dfrac{234}{f} = 0.755\ \text{ft}$

$Z = 36.5\ \Omega$

The impedance of the ground-plane antenna is one-half the impedance of the dipole ($73 \div 2 = 36.5\ \Omega$).

Directivity

In many types of communication systems, it is desirable to use antennas with omnidirectional characteristics, i.e., antennas which can send messages in any direction and receive them from any direction. In others, it is more advantageous to restrict the direction in which signals are sent or received. This requires an antenna with directivity.

Directivity refers to the ability of an antenna to send or receive signals over a narrow horizontal directional range. In other words, the physical orientation of the antenna gives it a highly directional response or directivity curve. A directional antenna eliminates interference from other signals being received from all directions other than the direction of the desired signal. A highly directional antenna acts as a type of filter to provide selectivity based on the direction of the signal. The receiving antenna is pointed directly at the station to be received, thereby effectively rejecting signals from transmitters in all other directions.

Directional antennas provide greater efficiency of power transmission. With omnidirectional antennas, the transmitted power radiates out in all directions. Only a small portion of the power is received by the desired station; the rest is, in effect, wasted. When the antenna is made directional, the transmitter power can be focused into a narrow beam directed toward the station of interest.

The conventional half-wave dipole has some directivity in that it sends or receives signals in directions perpendicular to the line of the antenna. (This is illustrated in Fig. 14-16, which shows the figure-8 response curve of a half-wave dipole.) The half-wave dipole antenna is directional in that no signal is radiated from or picked up from its ends. Such an antenna is referred to as *bidirectional*, since it receives signals best in two directions.

Directivity

GOOD TO KNOW

Directional antennas provide great efficiency of power transmission because the transmitter power can be focused into a narrow beam directed toward the station of interest.

Bidirectional antenna

Figure 14-25 Radiation pattern of a highly directional antenna with gain. (*a*) Horizontal radiation pattern. (*b*) Three-dimensional radiation pattern.

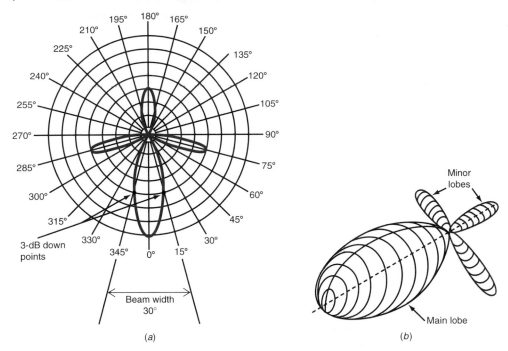

(a)

(b)

Antennas can also be designed to be *unidirectional;* unidirectional antennas send or receive signals in one direction only. Figure 14-25(*a*) shows the directivity pattern of a highly directional antenna. The larger loop represents the main response curve for the antenna. Maximum radiation or reception is in the direction of 0°. The three smaller patterns or loops going off in different directions from the main larger pattern are called *minor lobes.* A three-dimensional version of the horizontal radiation pattern shown in Fig. 14-25(*a*) is given in Fig. 14-25(*b*).

Few antennas are perfectly unidirectional. Because of various imperfections, some power is radiated (or received) in other directions (the minor lobes). The goal is to eliminate or at least minimize the minor lobes through various antenna adjustments and enhancements designed to put more power into the main lobe.

The beam width on a standard half-wave dipole is approximately 90°. This is not a highly directional antenna. The narrower the beam width, of course, the better the directivity and the more highly focused the signal. The antenna whose pattern is shown in Fig. 14-25 has a beam width of 30°. At microwave frequencies, antennas with beam widths of less than 1° have been built; these provide pinpoint communication accuracy.

Unidirectional antenna

Minor lobe

Gain

When a highly directive antenna is used, all the transmitted power is focused in one direction. Because the power is concentrated into a small beam, the effect is as if the antenna had amplified the transmitted signal. Directivity, because it focuses the power, causes the antenna to exhibit gain, which is one form of amplification. An antenna cannot, of course, actually amplify a signal; however, because it can focus the energy in a single direction, the effect is *as if* the amount of radiated power were substantially higher than the power output of the transmitter. The antenna has power gain.

The power gain of an antenna can be expressed as the ratio of the power transmitted P_{trans} to the input power of the antenna P_{in}. Usually, however, power gain is expressed in decibels:

$$dB = 10 \log \frac{P_{trans}}{P_{in}}$$

Antennas and Wave Propagation

545

The total amount of power radiated by the antenna, the ERP, is, as stated previously, the power applied to the antenna multiplied by the antenna gain. Power gains of 10 or more are easily achieved, especially at the higher RFs. This means that a 100-W transmitter can be made to perform as a 1000-W transmitter when applied to an antenna with gain.

Relationship Between Directivity and Gain

The relationship between the gain and the directivity of an antenna is expressed mathematically by the formula

$$B = \frac{203}{(\sqrt{10})^x}$$

where B = beam width of antenna, deg
x = antenna power gain in decibels divided by 10 (x = dB/10)

The beam width is measured at the 3-dB down points on the radiation pattern. It assumes a symmetric major lobe.

For example, the beam width of an antenna with a gain of 15 dB over a dipole is calculated as follows (x = dB/10 = 15/10 = 1.5):

$$B = \frac{203}{(\sqrt{10})^{1.5}} = \frac{203}{3.162^{1.5}} = \frac{203}{5.62} = 36.1°$$

It is possible to solve for the gain, given the beam width, by rearranging the formula and using logarithms:

$$x = 2 \log \frac{203}{B}$$

The beam width of an antenna of unknown gain can be measured in the field, and then the gain can be calculated. Assume, e.g., a measured −3-dB beam width of 7°. The gain is

$$x = 2 \log \frac{203}{7} = 2 \log 29 = 2(1.462) = 2.925$$

Since x = dB/10, dB = 10x. Therefore the gain in decibels is 2.925 × 10 = 29.25 dB.

To create an antenna with directivity and gain, two or more antenna elements are combined to form an *array*. Two basic types of antenna arrays are used to achieve gain and directivity: parasitic arrays and driven arrays.

Antenna array

Parasitic Arrays

Parasitic array

Parasitic elements
Driven element

A *parasitic array* consists of a basic antenna connected to a transmission line plus one or more additional conductors that are not connected to the transmission line. These extra conductors are referred to as *parasitic elements,* and the antenna itself is referred to as the *driven element.* Typically the driven element is a half-wave dipole or some variation. The parasitic elements are slightly longer than and slightly less than one-half wavelength long. These parasitic elements are placed in parallel with and near the driven elements. A common arrangement is illustrated in Fig. 14-26. The elements of the antenna are all mounted on a common boom. The boom does not have to be an insulator. Because there is a voltage null at the center of a one-half wavelength conductor at the resonant frequency, there is no potential difference between the elements and so they can all be connected to a conducting boom with no undesirable effect. In other words, the elements are not "shorted together."

Reflector

The *reflector,* a parasitic element which is typically about 5 percent longer than the half-wave dipole-driven element, is spaced from the driven element by a distance of 0.15λ to 0.25λ. When the signal radiated from the dipole reaches the reflector, it induces

Figure 14-26   A parasitic array known as a Yagi antenna.

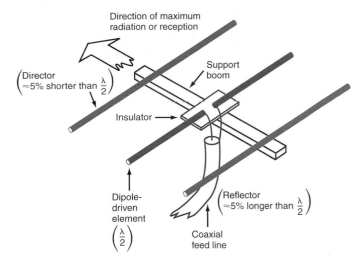

a voltage into the reflector and the reflector produces some radiation of its own. Because of the spacing, the reflector's radiation is mostly in phase with the radiation of the driven element. As a result, the reflected signal is added to the dipole signal, creating a stronger, more highly focused beam in the direction of the driven element. The reflector minimizes the radiation to the right of the driven element and reinforces the radiation to the left of the driven element (see Fig. 14-26).

Another kind of parasitic element is a director. A *director* is approximately 5 percent shorter than the half-wave dipole driven element and is mounted in front of the driven element. The directors are placed in front of the driven element and spaced by some distance between approximately one-tenth and two-tenths of a wavelength from the driven element. The signal from the driven element causes a voltage to be induced into the director. The signal radiated by the director then adds in phase to that from the driven element. The result is increased focusing of the signal, a narrower beam width, and a higher antenna gain in the direction of the director. The overall radiation pattern of the antenna in Fig. 14-16 is very similar to that shown in Fig. 14-25.

Director

An antenna made up of a driven element and one or more parasitic elements is generally referred to as a *Yagi antenna,* after one of its inventors. The antenna elements are usually made of aluminum tubing and mounted on an aluminum cross member or boom. Since the centers of the parasitic elements are neutral electrically, these elements can be connected directly to the boom. For the best lightning protection, the boom can then be connected to a metal mast and electrical ground. An antenna configured in this way is often referred to as a *beam antenna* because it is highly directional and has very high gain. Gains of 3 to 15 dB are possible with beam angles of 20° to 40°. The three-element Yagi antenna shown in Fig. 14-26 has a gain of about 8 dB when compared to a half-wave dipole. The simplest Yagi is a driven element and a reflector with a gain of about 3 dB over a dipole. Most Yagis have a driven element, a reflector, and from 1 to 20 directors. The greater the number of directors, the higher the gain and the narrower the beam angle.

Yagi antenna

Beam antenna

Additional gain and directivity can be obtained by combining two or more Yagis to form an array. Two examples are shown in Fig. 14-27. In Fig. 14-27(*a*), two nine-element Yagis are stacked. The spacing determines overall gain and directivity. In Fig. 14-27(*b*), two nine-element Yagis are mounted side by side in the same plane. The gain and beam width depend on the spacing between the two arrays and on how the transmission line is connected.

The drive impedance of a Yagi varies widely with the number of elements and the spacing. The parasitic elements greatly lower the impedance of the driven element, making it less than 10 Ω in some arrangements. Typically, some kind of impedance-matching

Figure 14-27 Yagi arrays. (*a*) Stacked. (*b*) Side by side.

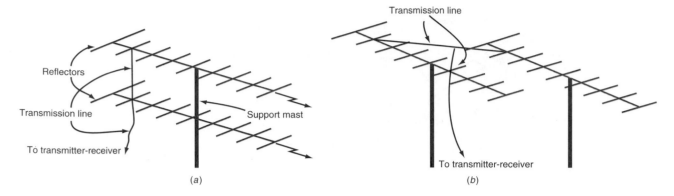

(*a*)   (*b*)

circuit or mechanism must be used to attain a reasonable match to 50-Ω coaxial cable, which is the most commonly used feed line.

Despite the focusing action of the reflector and director in a Yagi, so that most power is radiated in the forward direction, a small amount is lost to the rear, making the Yagi a less than perfect directional antenna. Thus, in addition to the gain and beam width, another specification of a Yagi is the ratio of the power radiated in the forward direction to the power radiated in the backward direction, or the *front-to-back (F/B) ratio:*

Front-to-back (F/B) ratio

$$\text{F/B} = 10 \log \frac{P_f}{P_b} \quad \text{dB}$$

where P_f = forward power

P_b = backward power

Relative values of forward and backward power are determined by estimating the sizes of the loops in the radiation pattern for the antenna of interest. When the radiation patterns are plotted in decibels rather than in terms of power, the F/B ratio is simply the difference between the maximum forward value and the maximum rearward value, in decibels.

By varying the number of parasitic elements and their spacing, it is possible to maximize the F/B ratio. Of course, varying the number of elements and their spacing also affects the forward gain. However, maximum gain does not occur with the same conditions required to achieve the maximum F/B ratio. Most Yagis are designed to maximize F/B ratio rather than gain, thus minimizing the radiation and reception from the rear of the antenna.

Yagis are widely used communication antennas because of their directivity and gain. At one time they were widely used for TV reception, but since they are tuned to only one frequency, they are not good for reception or transmission over a wide frequency range. Amateur radio operators are major users of beam antennas. And many other communication services use them because of their excellent performance and low cost.

Beam antennas such as Yagis are used mainly at VHF and UHF. For example, at a frequency of 450 MHz, the elements of a Yagi are only about 1 ft long, making the antenna relatively small and easy to handle. The lower the frequency, the larger the elements and the longer the boom. In general, such antennas are only practical above frequencies of about 15 MHz. At 15 MHz, the elements are in excess of 35 ft long. Although antennas this long are difficult to work with, they are still widely used in some communication services.

Driven Arrays

Driven array

The other major type of directional antenna is the *driven array,* an antenna that has two or more driven elements. Each element receives RF energy from the transmission line, and different arrangements of the elements produce different degrees of directivity and

Figure 14-28   Types of collinear antennas.

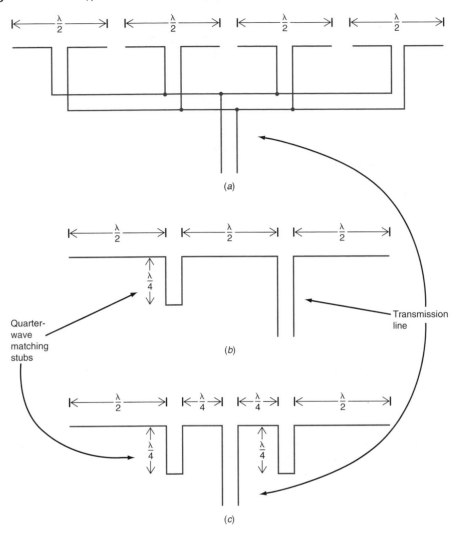

gain. The three basic types of driven arrays are the collinear, the broadside, and the end-fire. A fourth type is the wide-bandwidth log-periodic antenna.

Collinear Antennas. *Collinear antennas* usually consist of two or more half-wave dipoles mounted end to end. (See Fig. 14-28.) The lengths of the transmission lines connecting the various driven elements are carefully selected so that the energy reaching each antenna is in phase with all other antennas. With this configuration, the individual antenna signals combine, producing a more focused beam. Three different transmission line connections are shown in the figure. Like dipole antennas, collinear antennas have a bidirectional radiation pattern, but the two beam widths are much narrower, providing greater directivity and gain. With four or more half-wave elements, minor lobes begin to appear. A typical collinear pattern is shown in Fig. 14-29.

The collinear antennas in Fig. 14-28(*b*) and (*c*) use half-wave sections separated by shorted quarter-wave matching stubs which ensure that the signals radiated by each half-wave section are in phase. The greater the number of half-wave sections used, the higher the gain and the narrower the beam width.

Collinear antennas are generally used only on VHF and UHF bands because their length becomes prohibitive at the lower frequencies. At high frequencies, collinear antennas are usually mounted vertically to provide an omnidirectional antenna with gain.

Collinear antenna

Figure 14-29   Radiation pattern of a four-element collinear antenna.

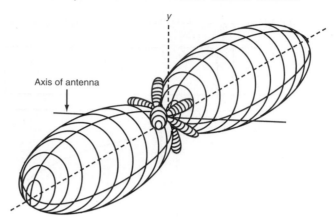

Axis of antenna

Broadside array

Broadside Antennas.

A *broadside array* is, essentially, a stacked collinear antenna consisting of half-wave dipoles spaced from one another by one-half wavelengths, as shown in Fig. 14-30. Two or more elements can be combined. Each is connected to the other and to the transmission line. The crossover transmission line ensures the correct signal phasing. The resulting antenna produces a highly directional radiation pattern, not in the line of the elements, as in a Yagi, but broadside or perpendicular to the plane of the array. Like the collinear antenna, the broadside is bidirectional in radiation, but the radiation pattern has a very narrow beam width and high gain. Its radiation pattern also resembles that of the collinear antenna, as shown in Fig. 14-29. The radiation pattern is at a right angle to the plane of the driven elements.

End-fire array

End-Fire Antennas.

The *end-fire array,* shown in Fig. 14-31(*a*), uses two half-wave dipoles spaced one-half wavelength apart. Both elements are driven by the transmission line. The antenna has a bidirectional radiation pattern, but with narrower beam widths and lower gain. The radiation is in the plane of the driven elements as in a Yagi. The

Figure 14-30   A broadside array.

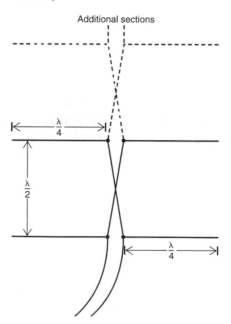

Additional sections

$\frac{\lambda}{4}$

$\frac{\lambda}{2}$

$\frac{\lambda}{4}$

Figure 14-31 End-fire antennas. (*a*) Bidirectional. (*b*) Unidirectional.

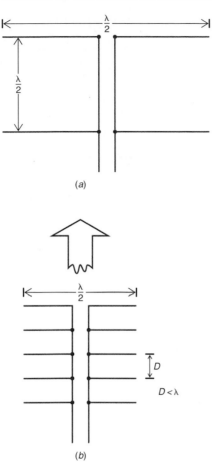

(a)

(b)

end of the fire array in Fig. 14-31(*b*) uses five driven elements spaced some fraction of a wavelength *D* apart. By careful selection of the optimal number of elements with the appropriately related spacing, a highly unidirectional antenna is created. The spacing causes the lobe in one direction to be canceled so that it adds to the other lobe, creating high gain and directivity in one direction.

Log–Periodic Antennas. A special type of driven array is the *wide-bandwidth log-periodic antenna* (see Fig. 14-32). The lengths of the driven elements vary from long to short and are related logarithmically. The longest element has a length of one-half wavelength at the lowest frequency to be covered, and the shortest element is one-half wavelength at the higher frequency. The spacing is also variable. Each element is fed with a special short transmission line segment to properly phase the signal. The transmission line is attached at the smallest element. The driving impedance ranges from about 200 to 800 Ω and depends on the length-to-diameter ratio of the driven element. The result is a highly directional antenna with excellent gain.

The great advantage of the log-periodic antenna over a Yagi or other array is its very wide bandwidth. Most Yagis and other driven arrays are designed for specific frequency or a narrow band of frequencies. The lengths of the elements set the operating frequency. When more than one frequency is to be used, of course, multiple antennas can be used. The log-periodic antenna can achieve a 4:1 frequency range, giving it a very wide bandwidth. The driving impedance remains constant over this range. Most TV antennas in use today are of the log-periodic variety so that they can provide high gain and directivity on both VHF and UHF TV channels. Log-periodic antennas are

Log-periodic array

Antennas and Wave Propagation

Figure 14-32   Log-periodic antenna.

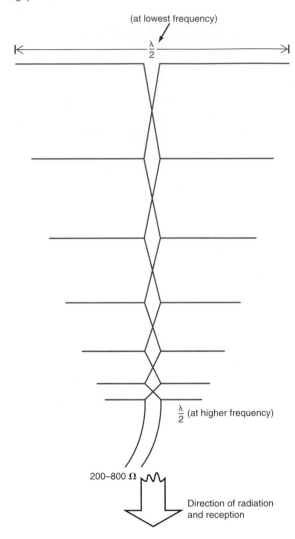

also used in other two-way communication systems where multiple frequencies must be covered.

Impedance Matching

One of the most critical aspects of any antenna system is to ensure maximum power transfer from the transmitter to the antenna. An important part of this, of course, is the transmission line itself. When the characteristic impedance of the transmission line matches the output impedance of the transmitter and the impedance of the antenna itself, the SWR will be 1:1 and maximum power transfer will take place.

The best way to prevent a mismatch between antenna and transmission line is through correct design. In practice, when mismatches do occur, some corrections are possible. One solution is to tune the antenna, usually by adjusting its length to minimize the SWR. Another is to insert an impedance-matching circuit or antenna tuner between the transmitter and the transmission line, such as the balun or *LC* L, T, or π networks described previously. These circuits can make the impedances equal so that no mismatch occurs, or at least the mismatch is minimized. The ideal is an SWR of 1, but any SWR value below 2 is usually acceptable.

Today, most transmitters and receivers are designed to have an antenna impedance of 50 Ω. Receivers must see an antenna system including transmission line that looks like a

Impedance matching

Example 14-2

An antenna has a gain of 14 dB. It is fed by an RG-8/U transmission line 250 ft long whose attenuation is 3.6 dB/100 ft at 220 MHz. The transmitter output is 50 W. Calculate (a) the transmission line loss and (b) the effective radiated power.

a. Total line attenuation: 3.6 dB/100 ft

$$\frac{250}{100} = 2.5 \text{ hundreds}$$

$$2.5 \times 3.6 = -9 \text{ dB} \qquad \text{(minus dB indicates a loss)}$$

$$dB = 10 \log P$$

$$P = \log^{-1} \frac{dB}{10} = \log^{-1}(-0.9) = 0.126$$

$$P = \frac{P_{out}}{P_{in}}$$

where P_{in} = power into transmission line
P_{out} = power out of transmission line to antenna

$$\frac{P_{out}}{P_{in}} = 0.126$$

$$P_{in} = 50 \text{ W}$$

$$P_{out} = 0.126(50) = 6.3 \text{ W}$$

b. Gain = 14 dB

$$\text{Power ratio} = \log^{-1} \frac{dB}{10}$$

$$= \log^{-1} 1.4$$

$$= 25.1$$

$$\text{ERP} = \text{power to antenna} \times \text{power ratio} = 6.3(25.1) = 158.2 \text{ W}$$

generator with a 50-Ω resistive impedance. Transmitters must see an antenna including transmission line with a 50-Ω resistive impedance over the desired operating frequency range if the SWR is to be close to 1 and maximum power is to be transferred to the antenna.

For low-frequency applications (<30 MHz) using wire antennas high above the ground, the 50-Ω coaxial cable is a reasonable match to the antenna, and no further action need be taken. However, if other antenna designs are used, or if the feed point impedance is greatly different from that of a 50-Ω coaxial cable, some means must be used to ensure that the impedances are matched. For example, most Yagis and other multiple-element antennas have an impedance that is not even close to 50 Ω. Further, an antenna that is not the correct length for the desired frequency will have a large reactive component that can severely affect the SWR and power output.

In situations in which a perfect match between antenna, transmission line, and transmitter is not possible, special techniques collectively referred to as *antenna tuning* or *antenna matching* are used to maximize power output and input. Most of these techniques are aimed at impedance matching, i.e., making one impedance look like another through the use of tuned circuits or other devices. Several of the most popular methods are described in the following sections.

Q Sections. A *Q section,* or *matching stub,* is a one-quarter wavelength of coaxial or balanced transmission line of a specific impedance that is connected between a load and

Q section (matching stub)

Figure 14-33 A one-quarter wavelength matching stub or Q section.

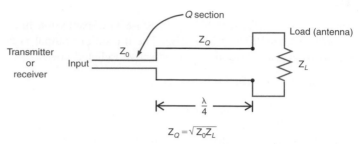

a source for the purpose of matching impedances (see Fig. 14-33). A one-quarter wavelength transmission line can be used to make one impedance look like another according to the relationship

$$Z_Q = \sqrt{Z_0 Z_L}$$

where Z_Q = characteristic impedance of quarter-wave matching stub or Q section
Z_0 = characteristic impedance of transmission line or transmitter at input of Q section
Z_L = impedance of load

Usually Z_L is the antenna feed point impedance.

For example, suppose it is desired to use a Q section to match a standard 73-Ω coaxial transmission line to the 36.5-Ω impedance of a quarter-wave vertical antenna. Using the equation, we have

$$Z_Q = \sqrt{36.5 \times 73} = \sqrt{2664.5} = 51.6$$

This tells us that a one-quarter wavelength section of 50-Ω coaxial cable will make the 73-Ω transmission line look like the 36.5-Ω antenna or vice versa. In this way maximum power is transferred.

When open-wire balanced transmission lines are used, a special balanced line can be designed and built to achieve the desired impedance. If it is desirable to use standard coaxial cable, then standard available impedances must be used. This means that the Q section will have to be 50, 75, or 93 Ω, since these are the only three standard values commonly available.

In using Q section techniques, it is important to take into account the velocity factor of the cable used to make the Q section. For example, if a Q section of 93 Ω is needed, then RG-62/U coaxial cable can be used. If the operating frequency is 24 MHz, the length of a quarter-wave is $246/f = 246/24 = 10.25$ ft. The velocity factor for this cable is 0.86, and so the correct length for the cable is $10.25 \times 0.86 = 8.815$ ft.

Two or more Q sections can be used in series to achieve the desired match, with each section performing an impedance match between its input and output impedances.

Baluns. Another commonly used impedance-matching technique makes use of a *balun,* a type of transformer used to match impedances. Most baluns are made of a ferrite core, either a toroid or rod, and windings of copper wire. Baluns have a very wide bandwidth and, therefore, are essentially independent of frequency. Baluns can be created for producing impedance-matching ratios of $4:1$, $9:1$, and $16:1$. Some baluns have a $1:1$ impedance ratio; the sole job of this type of balun is to convert between a balanced and unbalanced condition with no phase reversal.

Figure 14-34 shows a $4:1$ balun made with bifilar windings on a toroid core. *Bifilar windings,* in which two parallel wires are wound together as one around the core, provide maximum coupling between wires and core. The connections of the wires are such as to

Balun

Bifilar windings

Figure 14-34 A bifilar toroidal balun for impedance matching.

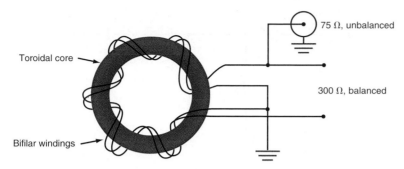

provide a balanced-to-unbalanced transformation. A common example of such a balun is one that provides a 75-Ω unbalanced impedance on one end and a 300-Ω balanced impedance on the other. Baluns are fully bidirectional; i.e., either end can be used as the input or the output. Such baluns have excellent broadband capabilities and work over a very wide frequency range.

Baluns can also be constructed from coaxial cable. A one-half wavelength coaxial cable is connected between the antenna and the feed line, shown in Fig. 14-35. When the length of the half-wave section for such a configuration is calculated, the velocity factor of the coaxial cable must be taken into account. Baluns of this type provide a 4:1 impedance transformation. For example, they can easily convert a 300-Ω antenna to a 75-Ω load and vice versa.

Antenna Tuners. When baluns and matching sections cannot do the job, antenna tuners are used. An *antenna tuner* is a variable inductor, one or more variable capacitors, or a combination of these components connected in various configurations. L, T, and π networks are all widely used. The inductor and capacitor values are adjusted until the SWR indicates that the impedances match.

It is important to mention that using an antenna tuner at the transmitter only "tricks" the transmitter into seeing a low SWR. In reality, the SWR on the line between the tuner and the antenna is still high.

One popular variation of the antenna tuner is a *transmatch circuit,* which uses a coil and three capacitors to tune the antenna for optimal SWR (see Fig. 14-36). Capacitors

Antenna tuner

Transmatch circuit

Figure 14-35 A coaxial balun with a 4:1 impedance ratio.

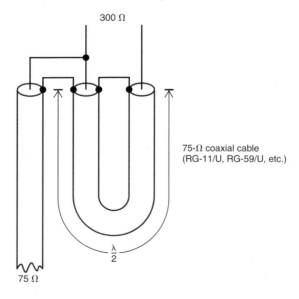

300 Ω

75-Ω coaxial cable
(RG-11/U, RG-59/U, etc.)

$\dfrac{\lambda}{2}$

75 Ω

Figure 14-36 An antenna tuner.

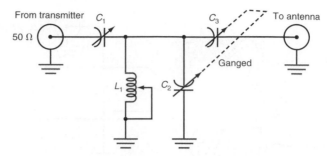

C_2 and C_3 are ganged together and tuned simultaneously. Transmatch circuits can be made to work over a wide frequency range, typically from about 2 to 30 MHz. A similar circuit can be made by using lower values of inductance and capacitance for matching VHF antennas. At UHF and microwave frequencies, other forms of impedance matching are used.

To use this circuit, an SWR meter is usually connected in series with the antenna transmission line at the transmatch output. Then L_1 is adjusted for minimum SWR, with the transmitter feeding power to the antenna. Then C_1 and C_2–C_3 are adjusted for minimum SWR. This procedure is repeated several times, alternately adjusting the coil and capacitors until the lowest SWR is reached.

A major development in antenna tuners is the automatic tuner. This may be an L, T, π, or other configuration that is self-adjusting. For example, the inductor in Fig. 14-36 would be provided with multiple taps along its length. By selecting the desired tap, the inductance could be changed. In place of each capacitor would be banks of parallel connected capacitors that could be switched in or out. All the inductor taps and capacitors are switched in or out by relay contacts. The relays in turn are operated by an embedded microcontroller.

The microcontroller gets its inputs from feedback provided by the outputs of an RF power meter. Today's power meters measure both forward and reflected power and provide a dc output proportional to those values. These dc values are converted to binary by an analog-to-digital converter and sent to the microcontroller. In the microcontroller resides a program that implements an algorithm that automatically adjusts the inductor and capacitors until the reflected power is reduced to zero or at least minimized. The modern antenna tuner can produce an excellent match in a few seconds or less.

Example 14-3

Calculate the length of the impedance-matching section needed for a Q section to match a 50-Ω transmitter output to a Yagi with a feed impedance of 172 Ω. The operating frequency is 460 MHz.

$$Z_Q = \sqrt{Z_0 Z_L} = \sqrt{50(172)} = \sqrt{8600} = 92.74\ \Omega$$

The amount 92.74 Ω is a close match to RG-62 A/U coaxial cable with an impedance of 93 Ω (see Fig. 13-14). The velocity factor is 0.86.

$$\frac{\lambda}{4} = \frac{246}{f} \times \text{VF} = \frac{246}{460} \times 0.86 = 0.5347 \times 0.86 = 0.46\ \text{ft}\ (5.5\ \text{in})$$

Example 14-4

Calculate the length of the impedance-matching section needed for a $\lambda/4$ balun to convert 75 Ω to 300 Ω. The operating frequency is 460 MHz. RG-59 A/U coaxial cable is the line of choice; the velocity factor of RG-59 A/U cable is 0.66.

$$\frac{\lambda}{4} = 0.5347 \times 0.66 = 0.353 \text{ ft (4.23 in)}$$

14-3 Radio Wave Propagation

Once a radio signal has been radiated by an antenna, it travels or propagates through space and ultimately reaches the receiving antenna. The energy level of the signal decreases rapidly with distance from the transmitting antenna. The electromagnetic wave is also affected by objects that it encounters along the way such as trees, buildings, and other large structures. In addition, the path that an electromagnetic signal takes to a receiving antenna depends upon many factors, including the frequency of the signal, atmospheric conditions, and time of day. All these factors can be taken into account to predict the propagation of radio waves from transmitter to receiver.

Optical Characteristics of Radio Waves

Radio waves act very much as light waves do. Light waves can be reflected, refracted, diffracted, and focused by other objects. The focusing of waves by antennas to make them more concentrated in a desired direction is comparable to a lens focusing light waves into a narrower beam. Understanding the optical nature of radio waves gives a better feel for how they are propagated over long distances.

Radio wave

Reflection. Light waves are reflected by a mirror. Any conducting surface looks like a mirror to a radio wave, and so radio waves are reflected by any conducting surface they encounter along a propagation path. All metallic objects reflect radio waves, especially if the metallic object is at least one-half wavelength at the frequency of operation. Any metallic object on a transmission path, such as building parts, water towers, automobiles, airplanes, and even power lines, causes some reflection. Reflection is also produced by other partially conductive surfaces, such as the earth and bodies of water.

Reflection

Radio wave reflection follows the principles of light wave reflection. That is, the angle of reflection is equal to the angle of incidence, as shown in Fig. 14-37. The radio wave

Figure 14-37 How a conductive surface reflects a radio wave.

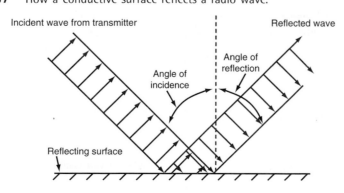

is shown as a wave front. To simplify the drawing, only the electrical lines of force, designated by arrows, are shown. The angle of incidence is the angle between the incoming line of the wave and a perpendicular line to the reflecting surface. The angle of reflection is the angle between the reflected wave and the perpendicular line.

A perfect conductor would cause total reflection: All the wave energy striking the surface would be reflected. Since there are no perfect conductors in the real world, the reflection is never complete. But if the reflecting surface is a good conductor, such as copper or aluminum, and is large enough, most of the wave is reflected. Poorer conductors simply absorb some of the wave energy. In some cases, the wave penetrates the reflecting surface completely.

As Fig. 14-37 shows, the direction of the electric field approaching the reflecting surface is reversed from that leaving the surface. The reflection process reverses the polarity of a wave. This is equivalent to a 180° phase shift.

Refraction

Refraction. *Refraction* is the bending of a wave due to the physical makeup of the medium through which the wave passes. The speed of a radio wave, like the speed of light, is approximately 300,000,000 m/s (186,400 mi/s) in free space, i.e., in a vacuum or air. When light passes through another medium, such as water or glass, it slows down. The slowing down as light enters or exits a different medium causes the light waves to bend.

The same thing happens to a radio wave. As a radio wave travels through free space, it encounters air of different densities, the density depending on the degree of ionization (caused by an overall gain or loss of electrons). This change of air density causes the wave to be bent.

Index of refraction

The degree of bending depends on the *index of refraction* of a medium n, obtained by dividing the speed of a light (or radio) wave in a vacuum and the speed of a light (or radio) wave in the medium that causes the wave to be bent. Since the speed of a wave in a vacuum is almost the same as the speed of a wave in air, the index of refraction for air is very close to 1. The index of refraction for any other medium will be greater than 1, with how much greater depending upon how much the wave speed is slowed.

Figure 14-38 shows how a wave is refracted. The incident wave from a transmitter travels through air, where it meets a region of ionized air which causes the speed of propagation to slow. The incident wave has an angle of θ_1 to a perpendicular to the boundary line between air and the ionized air. The bent (refracted) wave passes through the ionized air; it now takes a different direction, however, which has an angle of θ_2 with respect to the perpendicular.

Snell's law

The relationship between the angles and the indices of refraction is given by a formula known as *Snell's law:*

$$n_1 \sin \theta_1 = n_2 \sin \theta_2$$

where n_1 = index of refraction of initial medium
n_2 = index of refraction of medium into which wave passes
θ_1 = angle of incidence
θ_2 = angle of refraction

Figure 14-38 How a change in the index of refraction causes bending of a radio wave.

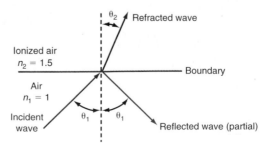

Figure 14-39 Diffraction causes waves to bend around obstacles.

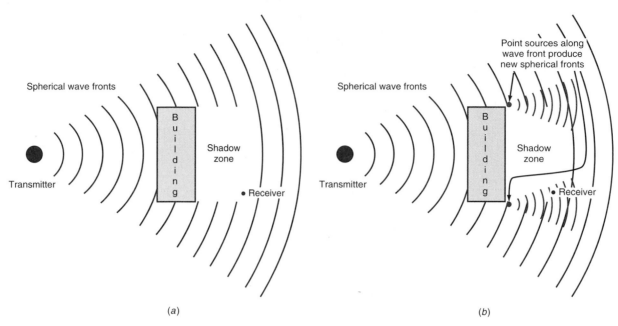

(a)   (b)

Note that there will also be some reflection from the boundary between the two media because the ionization causes the air to be a partial conductor. However, this reflection is not total; a great deal of the energy passes into the ionized region.

Diffraction. Remember that light and radio waves travel in a straight line. If an obstacle appears between a transmitter and receiver, some of the signal is blocked, creating what is known as a *shadow zone* [Fig. 14-39(a)]. A receiver located in the shadow zone cannot receive a complete signal. However, some signal usually gets through due to the phenomenon of *diffraction,* the bending of waves around an object. Diffraction is explained by what is known as *Huygens' principle.* Huygens' principle is based on the assumption that all electromagnetic waves, light as well as radio waves, radiate as spherical wave fronts from a source. Each point on a wave front at any given time can be considered as a point source for additional spherical waves. When the waves encounter an obstacle, they pass around it, above it, and on either side. As the wave front passes the object, the point sources of waves at the edge of the obstacle create additional spherical waves that penetrate and fill in the shadow zone. This phenomenon, sometimes called *knife-edge diffraction,* is illustrated in Fig. 14-39(b).

Diffraction

Huygens' principle

Radio Wave Propagation Through Space

The three basic paths that a radio signal can take through space are the ground wave, the sky wave, and the space wave.

Ground Waves. *Ground* or *surface waves* leave an antenna and remain close to the earth (see Fig. 14-40). Ground waves actually follow the curvature of the earth and can, therefore, travel at distances beyond the horizon. Ground waves must have vertical polarization to be propagated from an antenna. Horizontally polarized waves are absorbed or shorted by the earth.

Ground wave

Ground wave propagation is strongest at the low- and medium-frequency ranges. That is, ground waves are the main signal path for radio signals in the 30-kHz to 3-MHz

Figure 14-40   Ground or surface wave radiation from an antenna.

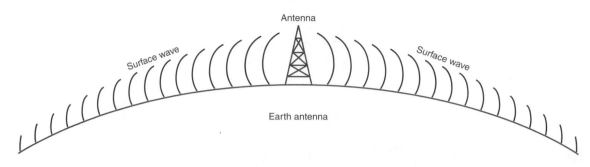

range. The signals can propagate for hundreds and sometimes thousands of miles at these low frequencies. AM broadcast signals are propagated primarily by ground waves during the day and by sky waves at night.

The conductivity of the earth determines how well ground waves are propagated. The better the conductivity, the less the attenuation and the greater the distance the waves can travel. The best propagation of ground waves occurs over salt water because the water is an excellent conductor. Conductivity is usually lowest in low-moisture areas such as deserts.

At frequencies beyond 3 MHz, the earth begins to attenuate radio signals. Objects on the earth and features of the terrain become the same order of magnitude in size as the wavelength of the signal and thus absorb or adversely affect the signal. For this reason, the ground wave propagation of signals above 3 MHz is insignificant except within several miles of the transmitting antenna.

Sky wave

Ionosphere

D layer

E layer

F layer

Sky Waves. Sky wave signals are radiated by the antenna into the upper atmosphere, where they are bent back to earth. This bending of the signal is caused by refraction in a region of the upper atmosphere known as the *ionosphere* (see Fig. 14-41). Ultraviolet radiation from the sun causes the upper atmosphere to *ionize,* i.e., to become electrically charged. The atoms take on or lose electrons, becoming positive or negative ions. Free electrons are also present. At its lowest point, the ionosphere is approximately 30 mi (50 km) above the earth and extends as far as 250 mi (400 km) from the earth. The ionosphere is generally considered to be divided into three layers, the *D layer,* the *E layer,* and the *F layer;* the F layer is subdivided into the F_1 and F_2 layers.

The D and E layers, the farthest from the sun, are weakly ionized. They exist only during daylight hours, during which they tend to absorb radio signals in the medium-frequency range from 300 kHz to 3 MHz.

The F_1 and F_2 layers, the closest to the sun, are the most highly ionized and have the greatest effect on radio signals. The F layers exist during both day and night.

The primary effect of the F layer is to cause refraction of radio signals when they cross the boundaries between layers of the ionosphere with different levels of ionization. When a radio signal goes into the ionosphere, the different levels of ionization cause the radio waves to be gradually bent. The direction of bending depends on the angle at which the radio wave enters the ionosphere and the different degrees of ionization of the layers, as determined by Snell's law.

Figure 14-41 shows the effects of refraction with different angles of radio signals entering the ionosphere. When the angle is large with respect to the earth, the radio signals are bent slightly, pass on through the ionosphere, and are lost in space. Radiation directly vertical from the antenna, or 90° with respect to the earth, passes through the ionosphere. As the angle of radiation decreases from the vertical, some signals continue to pass through the ionosphere. But at some critical angle, which varies with signal frequency, the waves begin to be refracted back to the earth. The smaller the angle with respect to the earth, the more likely it is that the waves will be refracted and sent back to earth. This effect is so pronounced that it actually appears as though the radio wave has been reflected by the ionosphere.

Figure 14-41 Sky wave propagation.

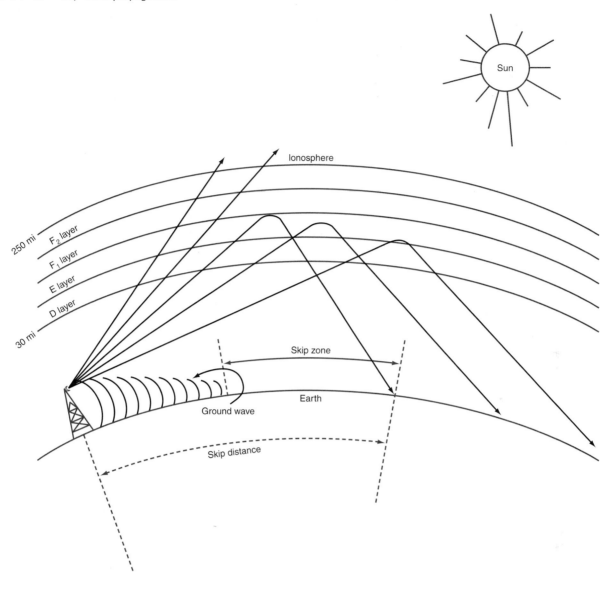

In general, the higher the frequency, the smaller the radiation angle required for refraction to occur. At very high frequencies, essentially those above about 50 MHz, refraction seldom occurs regardless of the angle. VHF, UHF, and microwave signals usually pass through the ionosphere without bending. However, during a period of sunspot activity, or other unusual electromagnetic phenomena, VHF and even UHF waves may be refracted by the ionosphere.

Reflected radio waves are sent back to earth with minimum signal loss. The result is that the signal is propagated over an extremely long distance. This effect is most pronounced in the 3- to 30-MHz or shortwave range, which permits extremely long distance communication.

In some cases, the signal reflected back from the ionosphere strikes the earth, is reflected back up to the ionosphere, and is re-reflected back to earth. This phenomenon is known as *multiple-skip* or *multiple-hop* transmission. For strong signals and ideal ionospheric conditions, as many as 20 hops are possible. Multiple-hop transmission can extend the communication range by many thousands of miles. The maximum distance of a single hop is about 2000 mi, but with multiple hops, transmissions all the way around the world are possible.

Figure 14-42 Line-of-sight communication by direct or space waves.

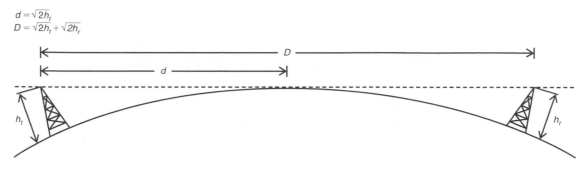

Skip distance

Skip zone

Direct waves
Space wave

Line-of-sight communication

The distance from the transmitting antenna to the point on earth where the first refracted signal strikes the earth to be reflected is referred to as the *skip distance* (see Fig. 14-41). If a receiver lies in that area between the place where the ground wave is fully attenuated and the point of first reflection from the earth, no signal will be received. This area is called the *skip zone.*

Space Waves. The third method of radio signal propagation is by *direct waves,* or *space waves.* A direct wave travels in a straight line directly from the transmitting antenna to the receiving antenna. Direct wave radio signaling is often referred to as *line-of-sight communication.* Direct or space waves are not refracted, nor do they follow the curvature of the earth.

Because of their straight-line nature, direct wave signals travel horizontally from the transmitting antenna until they reach the horizon, at which point they are blocked, as shown in Fig. 14-42. If a direct wave signal is to be received beyond the horizon, the receiving must be high enough to intercept it.

Obviously, the practical transmitting distance with direct waves is a function of the height of the transmitting and receiving antennas. The formula for computing the distance between a transmitting antenna and the horizon is

$$d = \sqrt{2h_t}$$

where h_t = height of transmitting antenna, ft
d = distance from transmitter to horizon, mi

Radio horizon

This is called the *radio horizon.*

To find the practical transmission distance D for straight-line wave transmissions, the height of the receiving antenna must be included in the calculations:

$$D = \sqrt{2h_t} + \sqrt{2h_r}$$

where h_r = height of receiving antenna, ft. For example, if a transmitting antenna is 350 ft high and the receiving antenna is 25 ft high, the longest practical transmission distance is

$$D = \sqrt{2(350)} + \sqrt{2(25)} = \sqrt{700} + \sqrt{50} = 26.46 + 7.07 = 33.53 \text{ mi}$$

Line-of-sight communication is characteristic of most radio signals with a frequency above approximately 30 MHz, particularly VHF, UHF, and microwave signals. Such signals pass through the ionosphere and are not bent. Transmission distances at those frequencies are extremely limited, and it is obvious why very high transmitting antennas must be used for FM and TV broadcasts. The antennas for transmitters and receivers operating at the very high frequencies are typically located on top of tall buildings or on mountains, which greatly increases the range of transmission and reception.

To extend the communication distance at VHF, UHF, and microwave frequencies, special techniques have been adopted. The most important of these is the use of repeater stations (see Fig. 14-43). A *repeater* is a combination of a receiver and a

Repeater

Figure 14-43 How a repeater extends the communication distance of mobile radio units at VHF and UHF.

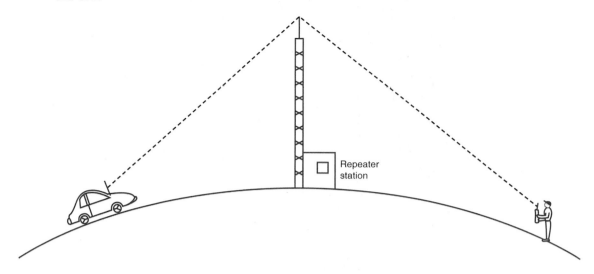

Repeater station

transmitter operating on separate frequencies. The receiver picks up a signal from a remote transmitter, amplifies it, and retransmits it (on another frequency) to a remote receiver. Usually the repeater is located between the transmitting and receiving stations, and therefore it extends the communication distance. Repeaters have extremely sensitive receivers and high-power transmitters, and their antennas are located at high points.

Repeaters are widely used to increase the communication range for mobile and handheld radio units, the antennas for which are naturally not very high off the ground. The limited transmitting and receiving range of such units can be extended considerably by operating them through a repeater located at some high point.

In high-activity areas, a repeater used for mobile units will become overloaded when too many users try to access it at the same time. When that happens, some users have to wait until free time becomes available, continuing to call until they get through. Such access delays are only a nuisance in some cases, but are clearly not acceptable when emergency services are unable to get through.

Although multiple repeaters can be used to ease overcrowding, they are often inadequate because communication activity is not equally distributed among them. This problem is solved by using *trunked repeater systems* in which two or more repeaters are under the control of a computer system that can transfer a user from an assigned but busy repeater to another, available repeater. Thus the communication load is spread around between several repeaters.

Trunked repeater systems

Repeaters can also be used in series, as shown in Fig. 14-44. Each repeater contains a receiver and a transmitter. The original signal is picked up, amplified, and retransmitted on a different frequency to a second repeater, which repeats the process. Typically, such relay stations are located 20 to 60 mi apart, mostly at high elevations to ensure reliable communication over very long distances. Microwave relay stations are used by many telephone companies for long-distance communication.

The "ultimate" repeater is, of course, a *communication satellite*. Most communication satellites are located in a geostationary orbit 22,500 mi above the equator. Since at that distance it takes exactly 24 h to rotate around the earth, communication satellites appear stationary. They act as fixed repeater stations. Signals sent to a satellite are amplified and retransmitted to earth long distances away. The receiver-transmitter combination within the satellite is known as a *transponder*. Most satellites have many transponders, so that multiple signals can be relayed, making possible worldwide communication at microwave frequencies. (See Chap. 17.)

Communication satellite

Transponder

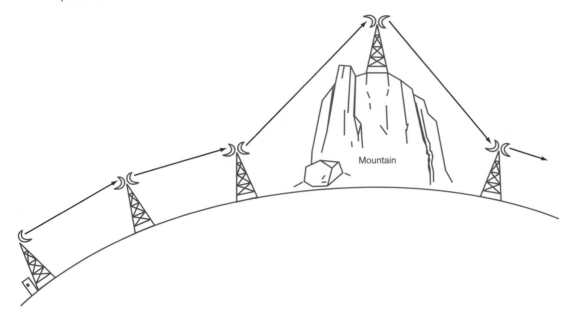

Calculating Received Power

A transmitted signal is radiated at a specific power level. The output power of a transmitter can be accurately determined by calculation or measurement. That power level is increased if the antenna has gain because of improved directivity. As a signal leaves an antenna, it immediately begins to become attenuated. Basically, the degree of attenuation is proportional to the square of the distance between the transmitter and receiver. As discussed previously, other factors also affect attenuation. Ground wave signals are greatly attenuated by objects on the earth, which block the signals and reduce their level at the receiver. In sky wave propagation, the ionospheric conditions and the number of hops determine the signal level at the receiver, with each hop further reducing the signal level. Space wave signals are simply absorbed and attenuated by objects in their path such as trees or walls.

Despite these factors, it is possible to predict the approximate power level at a receiver, and such calculations are quite accurate for the short distances characteristic of direct or space wave transmission.

In analyzing the transmission of radio waves, it is often useful to start with the concept of an isotropic radiator, or point source of radio waves. That is, the signal radiates spherically in all directions. The power density at a given distance from an isotropic radiator is predicted by the formula

$$P_d = \frac{P_t}{4\pi d^2}$$

where P_d = power density of signal, W/m^2
  d = distance from point source, m
  P_t = total transmitted power, W

The distance d is really the radius of an imaginary sphere enclosing the source, and $4\pi d^2$ is the area of that sphere at any given distance. However, since practical antennas are not purely isotropic sources, this formula must be modified somewhat. For example, if the transmitting antenna is a dipole, the dipole has a gain of 1.64 (or 2.15 dB) over an isotropic source, so the result must be multiplied by 1.64.

Suppose a transmitter puts a 50-W signal into a dipole antenna. The power density of the signal at a distance of 30 mi (48.3 km, or 48,300 m) is

$$P_d = \frac{1.64P_t}{4\pi d^2} = \frac{1.64(50)}{4(3.1416)(48,300)^2}$$

$$= 3 \times 10^{-9} = 3 \text{ nW/m}^2$$

Knowing the power density at a given distance is not a particularly useful thing. However, the formula for power at a given distance can be expanded to derive a general formula for computing the actual power value of a signal at a receiving antenna:

$$P_r = \frac{P_t G_t G_r \lambda^2}{16\pi^2 d^2}$$

where λ = signal wavelength, m
$\quad d$ = distance from transmitter, m
$\quad P_r, P_t$ = received and transmitted power, respectively
$\quad G_r, G_t$ = receiver and transmitter antenna gains expressed as a power ratio and referenced to an isotropic source

If the gains are those in reference to a dipole, each must be converted to a power ratio and multiplied by 1.64 before being used in the formula.

This formula is normally used only for ground wave, direct wave, or space wave calculations. It is not used for sky wave signal predictions because the refraction and reflection that occur make predictions highly inaccurate.

As an example, assume that a transmitter is operating at 150 MHz with a power of 3 W into a one-quarter wavelength vertical antenna. The receiver, which is 20 mi (32.2 km, or 32,200 m) away, has an antenna with a gain of 8 dB. What is the received power?

The wavelength at 150 MHz is $\lambda = 300/f = 300/150 = 2$ m. The gain of the quarter-wave vertical transmitting antenna is the same as that for a dipole. With a dipole gain of 1, we must multiply this by 1.64 to get the gain over an isotropic source.

The gain of the receiving antenna is 8 dB. The gain is usually expressed as the gain over a dipole. To convert to gain with respect to an isotropic source, we add 2.15 dB. This is the same as multiplying the power ratio represented by the gain in decibels by 1.64. The result is $8 + 2.15 = 10.15$ dB.

This must now be converted to an actual power ratio. Since dB = 10 log (P_{out}/P_{in}), where P_{in} and P_{out} are the input and output power of the antenna, respectively,

$$\frac{P_{out}}{P_{in}} = \log^{-1}\left(\frac{dB}{10}\right) = \log^{-1}\left(\frac{10.15}{10}\right) = \log^{-1}(1.015) = 10.35$$

The received power can now be calculated.

$$P_r = \frac{3(1.64)(10.35)(2)^2}{16(9.87)(32,200)^2} = \frac{203.7}{16.37 \times 10^{+10}}$$

$$= 1.24 \times 10^{-9} \text{ W or } 1.24 \text{ nW}$$

The expansive use of telephone networks has resulted in integratively designed transmission of telephone and data communication. The antennas are covered with a fiberglass material to protect them from the weather. The covering material does not impede the RF signals.

If the receiver antenna, transmission line, and front-end input impedance are 50 Ω, we can calculate the input voltage, given this input power. Since $P = V^2/R$, $V = \sqrt{PR}$. Substituting into $V = \sqrt{PR}$ gives

$$V = \sqrt{(1.24 \times 10^{-9})(50)} = 2.5 \times 10^{-4} = 250 \ \mu\text{V}$$

This is a relatively strong signal; most good narrowband receivers can generate full intelligible output with 1 μV or less.

Path attenuation

Path Attenuation. Another way to predict received power is to estimate the total power attenuation over a transmission path. This attenuation in decibels is given by

$$\text{dB loss} = 37 \text{ dB} + 20 \log f + 20 \log d$$

where f = frequency of operation, MHz
$\quad\quad d$ = distance traveled, mi

The distance can also be given in kilometers, in which case the 37-dB figure must be changed to 32.4 dB. Isotropic antennas are assumed.

The attenuation over a 20-mi path at a frequency of 150 MHz is

$$\text{dB loss} = 37 + 20 \log 150 + 20 \log 20 = 37 + 43.52 + 26 = 106.52$$

The dB loss formula, then, tells us that for every doubling of the distance between transmitter and receiver, the attenuation increases about 6 dB.

Common Propagation Problems

Fading

Although radio waves pass right through most objects on their way from transmitter to receiver, they are negatively influenced by these objects. The result is a common problem called *fading*. Good design of a communication system can minimize but not completely eliminate fading. One way to overcome fading is to use a diversity system.

Fading. One of the primary effects of radio wave propagation is called fading. Fading is the variation in signal amplitude at the receiver caused by the characteristics of the signal path and changes in it. Fading causes the received signal to vary in amplitude, typically making it smaller. Under some conditions the received signal may actually be larger than a direct path signal depending upon the specific communication situation. Fading is caused by four factors: variation in distance between transmitter and receiver, changes in the environmental characteristics of the signal path, the presence of multiple signal paths, and relative motion between the transmitter and receiver.

As a receiver gets farther away from a transmitter, the signal gets weaker just because the path length is increasing. If the receiver moves closer to the transmitter, the signal strength increases. Both types of situations occur when one or perhaps both of the transceivers are moving with respect to the other. It is especially noticeable in airplanes and in cars. This type of fading is generally gradual and does not result in severe or rapid swings in signal amplitude.

Shadow fading

Fading is also caused by objects coming between the transmitter and receiver. Known as *shadow fading,* this occurs if a vehicle containing a transceiver moves in such a way that a large building or a mountain comes between it and a base station transceiver. The obstacle causes the signal to be attenuated, resulting in fading. When a car enters a tunnel, the signal may be greatly attenuated so that fading occurs. Even the movement of a rainstorm or snowstorm between transmitter and receiver can cause fading. Weather-related effects are especially pronounced at the higher microwave frequencies, where the signal wavelengths are in the same size range as the raindrops or snowflakes that produce massive signal scattering by reflection.

Multipath interference
Rayleigh fading

One of the worst sources of fading is *multipath interference*. Sometimes called *Rayleigh fading*, this type of fading occurs when a transmitted signal takes multiple paths to the receiver because of reflections. The term *Rayleigh* refers to a particular type of statistical response curve that mathematically describes the variation of the received signal. As you saw

Example 14-5

A 275-ft-high transmitting antenna has a gain of 12 dB over a dipole. The receiving antenna, which is 60 ft high, has a gain of 3 dB. The transmitter power is 100 W at 224 MHz. Calculate (a) the maximum transmitting distance and (b) the received power at the distance calculated in part (a). (There is 1.61 km/mi.)

a. $D = \sqrt{2h_t} + \sqrt{2h_r}$

$\quad = \sqrt{2(275)} + \sqrt{2(60)} = \sqrt{550} + \sqrt{120}$

$\quad = 23.45 + 10.95 = 34.4 \text{ mi}$

$34.4 \text{ mi} \times 1.61 \text{ km/mi} = 55.4 \text{ km}$

$55.4 \text{ km} \times 1000 = 55,400 \text{ m}$

b. $P_r = \dfrac{P_t G_t G_r \lambda^2}{16\pi^2 d^2}$

$\quad = 100 \text{ W}$

$G_t = 12 \text{ dB over dipole}$

$\quad = 12 \text{ dB} + 2.15 \text{ dB}$

$\quad = 14.15 \text{ dB}$

$\text{dB} = 10 \log \dfrac{P_{\text{out}}}{P_{\text{in}}}$

$\dfrac{P_{\text{out}}}{P_{\text{in}}} = \log^{-1}\left(\dfrac{\text{dB}}{10}\right) = \log^{-1}\left(\dfrac{14.15}{10}\right) = \log^{-1}(1.415)$

$\quad = 26$

$G_t = 26$

$G_r = 3 \text{ dB} + 2.15 \text{ dB} = 5.15 \text{ dB}$

$\dfrac{P_{\text{out}}}{P_{\text{in}}} = \log^{-1}\left(\dfrac{5.15}{10}\right) = \log^{-1}(0.515) = 3.27$

$G_r = 3.27$

$\lambda = \dfrac{300}{f} = \dfrac{300}{224} = 1.34 \text{ m}$

$\lambda^2 = (1.34)^2 = 1.8$

$P_r = \dfrac{100(26)(3.27)(1.8)}{16(3.14)^2(55,400)^2}$

$\quad = \dfrac{15,303.6}{4.85 \times 10^{11}} = 3.16 \times 10^{-8} = 31.6 \text{ nW}$

earlier in this chapter, radio signals are easily reflected by conducting objects. The signal is usually radiated by a nondirectional antenna over a wide horizontal range in such a way that it will strike the receiver antenna directly by way of the direct line-of-sight space wave, but it may also strike many obstacles along the way. Buildings, water towers, hills and mountains, and even moving vehicles all have reflected surfaces that will direct a signal along a separate path to the receiving antenna. The signal may also be reflected from the ground or water. The result is that multiple signals reach the receiver antenna at different times.

Reflected signals take a longer path than a direct signal, so they are delayed and arrive at the antenna later than the direct signal. This time delay is seen as a phase shift whose magnitude is a function of the total signal path distance and the wavelength (frequency) of the signal. Keep in mind that as you saw earlier in the chapter, reflections produce a 180° phase shift that worsens the problem. The receiver sees a composite of all the received signals. The phase shifts are usually such that they cancel the direct path signal, resulting in an overall weaker signal. But the delays could also be such that the reflected signals arrive in phase with the direct signal, thereby causing the signals to add in phase and actually produce a higher-level signal.

Another type of fading is caused by movement of either the transmitter or the receiver. When the transmitter is in a car, plane, or other vehicle, rapid movement toward or away from the receiver introduces a signal frequency change called a *Doppler shift* (to be described in Chap. 16). Movement that causes the transmitter and receiver to get closer to each other causes the signal frequency to increase. Movement that increases the distance causes a frequency decrease. Large signal-frequency changes produce lower-level signals because the signals are partially out of the passband of the receiver's selective filters. In digital systems that predominantly use some form of phase-shift modulation, the Doppler shift confuses the demodulator and produces bit errors.

In most cases, several types of fading occur simultaneously. Multipath fading and shadow fading are the worst offenders. If you have ever used a cell phone from a moving car in a changing environment, you know that fading can cause significant signal variations, including no service at all. Using a cell phone or radio in a large city with many tall buildings produces extreme multipath interference and shadow fading. Using cell phones or other wireless equipment inside a building essentially does the same.

When digital communication is involved, multipath fading can cause intersymbol interference. If high-speed data transmission is used, the symbols are short, and multipath delays may be of the same order of magnitude. A symbol received directly may be different from a symbol received from a reflected source. The result is severe bit errors.

Although fading can occur on signals of any frequency, it is most pronounced in UHF and microwave communication, where the signal wavelengths are very short compared to the path distances and size of reflective surfaces. Fading is a significant problem with cell phones and other radio equipment especially when one or more of the transceivers is in motion. Fading is also a problem in long-distance shortwave communication when the signal can take several bounces off the ionosphere and the earth to produce canceling or reinforcing signals. Fading signal variations can be only a few decibels or as much as 20 to 30 dB. A significant amount of fading can make communication unreliable.

To overcome fading, most communication systems have a built-in fading margin. That is, they have a high enough transmitter power and sufficient receiver sensitivity to ensure that the weaker reflective signals do not degrade the direct signal as much. A fade margin of at least 5 dB is built into most systems.

Multipath fading can also be greatly minimized by using highly directive antennas, either at the transmitter or at the receiver or at both. Narrow transmit and receive beams virtually eliminate multiple paths and the related fading. However, in most communication systems, nondirectional (i.e., omnidirectional) antennas are the norm. Cell phones, two-way radios, and even base stations must have broad azimuth coverage to receive or transmit signals over a wide area.

Broadband signals are much less sensitive to multipath fading than narrowband signals are. If broadband methods such as spread spectrum (CDMA) or OFDM are used, the signals are spread over a very wide frequency range. Multiple reflections of signals over a wide range of frequencies are received in such a way that less cancellation or intersymbol interference occurs. This is a major factor in designing new communication systems where fading is expected.

Diversity System. Fading can also be minimized by using what is called a *diversity system*. A diversity system uses multiple transmitters, receivers, or antennas to mitigate the problems caused by multipath signals. Two common types of diversity are frequency and spatial. With frequency diversity, two separate sets of transmitters and receivers operating on different frequencies are used to transmit the same information simultaneously. The theory is that signals on different frequencies will react differently to the various

Diversity system

fading mechanisms, thereby resulting in a least one reliably received signal. To be effective, the frequencies should be widely spaced from one another. Of course, such systems are very expensive since they require two transmitters, receivers, and antennas, all on different frequencies. The scarcity of frequency spectrum also makes this type of system impractical. It is rarely used except in cases where extreme reliability is a must.

Another more widely used form of diversity is called *space* or *spatial diversity*. It uses two receiver antennas spaced as far apart as possible to receive the signals. Diversity systems are used mainly at base stations rather than in portable or handheld units. The basic idea is that antennas at slightly different locations will receive different variations of the signals, with one being better than another. The spacing may be horizontal or vertical, whichever is the more convenient. However, in some cases one arrangement will be superior to the other.

Diversity reception is particularly difficult at shortwave frequencies, where the spacing will typically be many hundreds or even thousands of feet. Only horizontal spacing is used. At UHF and microwave frequencies, wide antenna spacing is relatively simple because the wavelengths are short. In general, the wider the spacing (10 or 20 wavelengths or more), the better. Many systems use the relationship $h/d = 11$ to determine a minimum and, as it turns out, optimum spacing for antennas. In this relationship, h is the height of the antenna and d is the spacing distance. For antennas that are 100 ft high, the minimum spacing would be

$$\frac{100}{d} = 11$$

$$d = \frac{100}{11} = 9.09 \text{ ft}$$

Figure 14-45(*a*) shows a typical spatial diversity system. The two antennas feed a combiner network, where the two signals are linearly added. The result is a larger signal and

Spatial diversity

Figure 14-45 Common diversity reception systems. (*a*) Signals are combined for maximum output. (*b*) Only the strongest signal is switched to the receiver.

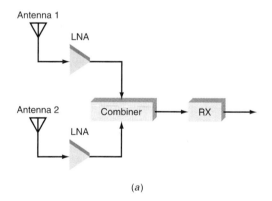

(*a*)

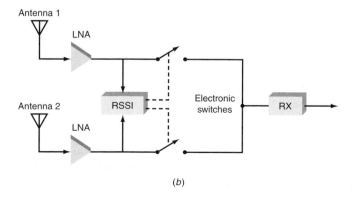

(*b*)

minimized fading effects. The signals may be combined at the antenna or almost anywhere in the receiver. Some systems combine after the LNAs. Others combine after the IF. In still other systems, two completely separate receivers are used and the signals are combined after demodulation. Experimentation is required to determine the best results.

Another form of spatial diversity is selective or switched diversity. In this system, shown in Fig. 14-45(*b*), the two antennas feed separate LNAs whose outputs are monitored by circuits that measure received signal strength. In cell phone systems, these circuits are called *received signal strength indicators (RSSIs)*. They determine the signal with the greatest strength and switch that signal to the remainder of the receiver circuits. All this is done automatically at high speed, ensuring that the receiver always has the strongest signal.

Received signal strength indicators (RSSIs)

Diversity systems are widely used in the newer cell phone systems and in wireless LANs that work indoors and, in some cases, with mobile wireless units (laptop computers, PDAs, etc.) that are frequently in motion. New techniques such as multiple-input, multiple-output (MIMO) and smart (adaptive) antennas are now being used to further improve transmission in multipath environments. These techniques will be covered in Chap. 16.

CHAPTER REVIEW

Summary

Radio signals are electromagnetic waves. Signal voltage applied to an antenna creates an electric field and also causes current to flow, producing a magnetic field. The electric and magnetic fields propagate through space over very long distances at the speed of light. The electric and magnetic fields are perpendicular to each other and perpendicular to the direction of propagation of the wave.

If the electric field of a signal is parallel to the earth, the electromagnetic wave is said to be horizontally polarized; if it is perpendicular to the earth, the wave is vertically polarized. For optimal transmission and reception, both the transmitting and the receiving antennas must be of the same polarization.

Antennas are some fraction of the signal wavelength, most commonly one-half or one-quarter wavelength. A one-half wavelength antenna fed at its center by the transmission line is called a dipole. Most antennas are some modified form of a basic one-half wavelength dipole. Two common variations are the folded dipole antenna and the Marconi or ground-plane vertical antenna. Each type offers different impedance values, directional radiation patterns, and gain.

To create an antenna with directivity and gain, two or more antenna elements are combined to form an array. The two basic array types are parasitic arrays, consisting of parasitic elements such as reflectors and directors, and driven arrays. Driven arrays are subdivided into broadside, collinear, end-fire, and log-periodic arrays. Yagi antennas are made up of a driven element and one or more parasitic elements.

For a given application, the characteristic impedance of a transmission line must match the output impedance of the transmitter and the impedance of the antenna itself. Mismatches cause an unfavorable standing wave ratio and a loss of power. Antenna length, bandwidth, and Q are two characteristics that can be adjusted, through design, to produce the best match. Specialized components, such as Q sections, baluns, and tuners, can also be used.

The three basic types of radio wave propagation paths are ground waves, sky waves, and space waves (direct waves). Ground waves, which are vertically polarized, follow the curvature of the earth and are strongest at medium and low frequencies. Sky wave signals are radiated by an antenna into the upper atmosphere, where they are refracted in the ionosphere and sent back to earth. Direct waves travel in a straight line from the transmitting antenna to the receiving antenna, and they are limited to what is known as line-of-sight transmission. To extend communication distances at VHF, UHF, and microwave frequencies, repeater stations can be used.

For direct and space wave transmissions over relatively short distances, it is possible to use simple formulas to predict attenuation, given distance between transmitter and receiver, gain, and power figures.

A major problem in many communication systems is fading, the varying received signal strength caused by distance variation, motion between transmitter and receiver, obstacles between transmitter and receiver, and multipath reception caused by unwanted reflections. Fading can be minimized by using diversity reception techniques.

Questions

1. What is the basic makeup of a radio wave?
2. What is the name of the mathematical expressions that describe the behavior of electromagnetic waves? Give a general statement regarding what these expressions state.
3. Describe the relationship between the types of fields that make up a radio wave. What is the orientation of these fields with respect to the transmission direction of the wave?
4. What part of a radio wave determines its polarization with respect to the earth's surface?
5. What is the polarization of a radio wave with a magnetic field that is horizontal to the earth?
6. Define what is meant by antenna reciprocity.
7. What is the name of the antenna type upon which most other antennas are based? What is the length of this antenna in terms of wavelength at the operating frequency?
8. What is the most commonly used medium for connecting an antenna to a transmitter or receiver?
9. What is the length of a dipole?
10. State the theoretical radiation resistance or drive impedance of a dipole.
11. What factors affect the impedance of a dipole and how?
12. What is the equivalent circuit of a dipole at resonance, above resonance, and below resonance?
13. What factors influence antenna bandwidth?
14. Describe the horizontal radiation pattern of a dipole. What is the three-dimensional shape of a dipole radiation pattern?
15. Define what is meant by directivity in an antenna.
16. What unit is used to state and measure an antenna's directivity?
17. Define antenna gain. What gives an antenna gain?
18. What is an isotropic radiator?
19. What gain does a dipole have over an isotropic radiator?
20. State two ways to express antenna gain.
21. Define ERP.
22. What is a folded dipole? What are its advantages over a standard dipole?
23. What two names are used to refer to a one-quarter wavelength vertical antenna?
24. What is the horizontal radiation pattern of a one-quarter wavelength vertical antenna?
25. What term describes an antenna that radiates equally well in all horizontal directions?
26. What must a one-quarter wavelength vertical antenna have for proper operation, in addition to a vertical conductor radiator?
27. What is the name of a group of conductors connected to the transmission line at the base of some vertical antennas?
28. Name the most commonly used type of transmission line for one-quarter wavelength vertical antennas.
29. What term describes the directivity of an antenna that transmits or receives equally well in two opposite directions?
30. What is a unidirectional antenna?
31. Why does a directional antenna amplify a signal?
32. What physical attributes must an antenna have to exhibit gain and directivity?
33. Name the two basic classes of antenna arrays.
34. Name the three basic elements in a Yagi antenna.
35. Name the two parasitic elements in a Yagi beam antenna.
36. What is the upper limit on the number of directors that a Yagi can have?
37. What is the typical beam width range of a Yagi antenna?
38. State two factors that affect the beam width of a Yagi.
39. Define the term *front-to-back ratio* as it applies to a Yagi.
40. True or false? A Yagi can be operated with either horizontal or vertical polarization.
41. Which polarization is the more common for a Yagi?
42. List three kinds of driven arrays.
43. Describe the radiation pattern of the three most common types of driven arrays.
44. Do driven arrays have gain over a dipole?
45. What is the name of a popular wideband driven array with a unidirectional radiation pattern?
46. State one major benefit of a log-periodic antenna over other types of driven arrays.
47. What is the most common value of antenna input/output impedance for modern transmitters and receivers?
48. Why is impedance matching necessary in some antenna installations?
49. What is the name of a type of transformer used for impedance matching in antenna installations?
50. What do you call a one-quarter wavelength section of transmission line that is used for impedance matching between antenna and transmission line?
51. What is the impedance-matching ratio of a coaxial balun?
52. Describe how an antenna tuner works.
53. What is the usual procedure for adjusting an antenna tuner?
54. What are the requirements for the reflection of a radio signal?
55. Will a vertically polarized radio wave be received by a horizontally polarized antenna?
56. What is circular polarization? State the two types of circular polarization.
57. Will horizontally and vertically polarized antennas pick up circularly polarized signals?
58. What is refraction as it refers to radio waves? What causes refraction of radio waves?

59. What is diffraction as it refers to radio waves? Is diffraction harmful or advantageous in radio communication?
60. Name the three paths that a radio signal can take through space.
61. What two names are used to refer to a radio wave that propagates near the surface of the earth? Are they vertically or horizontally polarized?
62. What is the frequency range of signals that propagate best over the surface of the earth?
63. State the name of the radio wave that is refracted by the ionosphere.
64. What is the frequency range of signals that propagate best by ionospheric refraction?
65. What is the ionosphere? How is it different from atmospheric layers that are closer to the earth? Why does it differ?
66. What layer of the ionosphere has the greatest effect on a radio signal?
67. True or false? The ionosphere acts as a mirror, reflecting radio waves back to earth.

68. What phenomenon of propagation makes worldwide radio communication possible?
69. What factors determine whether a radio wave is refracted by the ionosphere or passes through to outer space?
70. What do you call a radio wave that propagates only over line-of-sight distances?
71. At what frequencies is line-of-sight transmission the only method of propagation?
72. How can the transmission distance of a UHF signal be increased?
73. What technique is used to increase transmission distances beyond the line-of-sight limit at VHF and above?
74. What is a repeater station and how does it work?
75. Which conditions cause the worst kind of fading?
76. What is the name given to multipath fading?
77. Why is frequency diversity undesirable?
78. How does spatial diversity minimize fading?
79. Which produces the fewest fading effects, narrowband or wideband signals?

Problems

1. What is the approximate impedance of a dipole that is 0.1λ aboveground? 0.3λ aboveground? ◆
2. What is the length of a wire dipole antenna at 16 MHz?
3. A wire dipole antenna has length of 27 ft. What is its frequency of operation? What is its approximate bandwidth? ◆
4. At low frequencies (less than 30 MHz), what is the normal polarization of a dipole?
5. What is the most common transmission line used with low-frequency dipoles? ◆
6. The power applied to an antenna with a gain of 4 dB is 5 W. What is the ERP?
7. What is the length of a folded dipole made with a 300-Ω twin lead for a frequency of 216 MHz? ◆
8. Calculate the length of a one-quarter wavelength vertical antenna at 450 MHz.
9. What is the length of the driven element in a Yagi at 290 MHz? ◆
10. Calculate the length of the coaxial loop used in a coaxial balun for a frequency of 227 MHz. Assume a velocity factor of 0.8.

11. What is the length in inches of a one-quarter wavelength impedance-matching section of coaxial cable with a velocity factor of 0.66 at 162 MHz?
12. How far away is the radio horizon of an antenna 100 ft high?
13. What is the maximum line-of-sight distance between a paging antenna 250 ft high and a pager receiver 3.5 ft off the ground? ◆
14. What is the path attenuation between transmitter and receiver at a frequency of 1.2 GHz and a distance of 11,000 mi?
15. An antenna has a dBd gain of 6. What is its gain with respect to an isotropic radiator?
16. A cell phone antenna tower 240 ft high uses spatial diversity. What is the minimum desirable antenna separation?

◆ *Answers to Selected Problems follow Chap. 22.*

Critical Thinking

1. Of two dipoles, one using wire conductors and one using thin tubing conductors, which has a wider bandwidth?
2. The sun is at an angle of 30° to the horizon and causes a tall one-quarter wavelength vertical antenna to cast a shadow 400 ft long. At what frequency does the antenna resonate?

3. Where are the two conductors of a coaxial transmission line connected in a one-quarter wavelength vertical antenna installation?
4. Compare and contrast the operation of a one-quarter wavelength vertical antenna to that of a one-half wavelength dipole.

5. What is the feed impedance of a one-quarter wavelength vertical antenna? What factors influence this impedance?

6. If a vertical antenna is too short, what can be done to bring it into resonance?

7. If a vertical antenna is too long for the desired frequency of operation, what can be done to make it resonant?

8. See Fig. 14-46. What is the beam width of this antenna? Calculate its gain. Calculate the front-to-back ratio of this antenna in decibels. *Note:* This radiation pattern uses a logarithmic scale and is calibrated in decibels. The maximum radiation point on the major lobe at 0° is given the value of 0 dB. The -3-dB down points are clearly identified so that beam width can be determined. The center of the plot is -100 dB.

9. A transmitter has a directional antenna with a gain of 4 dB. The frequency is 72 MHz. The power into the antenna is 2 W. The receiver uses a one-quarter wavelength ground plane. The distance between the transmitting and receiving antennas is 3 mi. What is the power at the receiver? What is the receiver input voltage with a 50-Ω input impedance?

10. A 915-MHz transmitter sends a signal to a receiver. The transmitter output power is 1 W. Both transmitter and receiver use one-quarter wavelength vertical antennas. The receiver input impedance is 50 Ω. What is the maximum distance that can be achieved and still deliver 1 μV into the receiver front end?

11. What is the length of a $^5/_8\lambda$ vertical antenna at 902 MHz?

12. A very low-power battery-operated transmitter at an inaccessible location is to send telemetry data to a data collection center. The signal is often too weak for reliable data recovery. List some practical steps that can be taken to improve transmission.

Figure 14-46 Radiation pattern for Critical Thinking question 8.

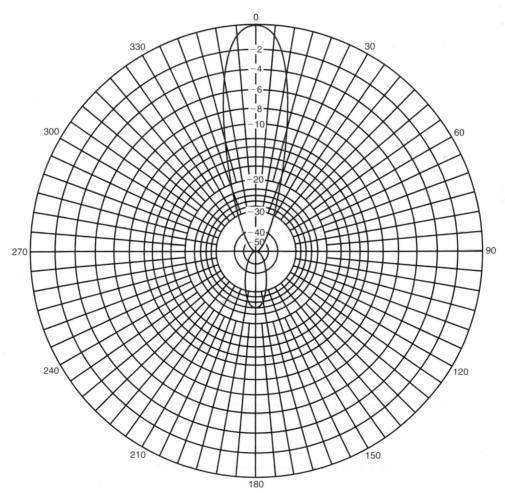

Internet Technologies

The *Internet* is a worldwide interconnection of computers by means of many complex networks. The Internet was established in the late 1960s under the sponsorship of the Department of Defense and later through the National Science Foundation. It provided a way for universities and companies doing military and government research to communicate and to share computer files and software. In the early 1990s, the Internet was privatized and opened to anyone. Today it is a system with hundreds of millions of users across the world.

This chapter discusses the technology used in transporting data in packets through large numbers of computers called servers and Internet appliances called routers and the extensive fiber-optic Internet backbone. Technologies such as Frame Relay, asynchronous transfer mode (ATM), and SONET are discussed. The transmission protocol TCP/IP is discussed and the operation of a router is explained. This chapter also includes an introduction to storage-area networks (SANs). Finally, the subject of Internet data security is defined, and methods of encryption and authentication are considered.

Objectives

After completing this chapter, you will be able to:

- Define the Internet.
- List the main applications of the Internet.
- Explain how data is transmitted over the Internet.
- Name and explain the three main types of transmission systems used in the Internet including ATM, SONET, and Frame Relay.
- Explain the details of the TCP/IP protocol.
- Explain the operation of a router.
- Identify and explain the major types of storage-area networks and their transmission technologies including Fibre Channel and iSCSI.
- Name the main types of security systems used with the Internet.
- Name the most commonly used encryption systems on the Internet.

15-1 Internet Applications

The Internet is a communication system that accomplishes one of three broad uses: (1) share resources; (2) share files or data; and (3) communication. The primary applications of the Internet are e-mail, file transfer, the World Wide Web, e-commerce, searches, Voice over Internet Protocol, and video.

E–Mail. *E-mail* is the exchange of notes, letters, memos, and other personal communication by way of e-mail software and service companies. You write a message to a person and send it over the Internet via your e-mail provider, which, in turn, transfers it to the e-mail provider of the receiving person. That person retrieves the message at his or her convenience. E-mail is one of the most common communication methods in use today and the number one application of the Internet.

File Transfer. *File transfer* refers to the ability to transfer files of data or software from one computer to another. The file may be text, digitized photographic data, a computer program, and so on. A file transfer program (FTP) allows you to access files in remote computers and download them to your computer, where they may be used. Files can also be "attached" to e-mail messages. File transfer is an excellent research tool because it allows you to access massive amounts of data in the form of books, articles, newspapers, brochures, data sheets, and hundreds of other sources. Apple Computer's iTunes allows purchase and download of music.

File transfer

World Wide Web. Whenever most people refer to the Internet, they are really talking about the *World Wide Web (WWW)*, or the Web for short. The Web is a specialized part of the Internet where companies, organizations, the government, or even individuals can post information for others to access and use. To do this, you establish a website, which is a computer that stores the information you wish to dispense. The information can be on your own computer or in the computer of a Web service provider. The information is presented in the form of pages, a logic subunit which may contain text, graphics, animations, sound, and even video.

World Wide Web

You access these websites through the Internet by way of a special piece of software known as a *browser*. The browser allows you to call up a desired website by name or to search the Internet websites containing information of interest to you. The browser is the software that lets you navigate and explore the Web and access and display the information. The two most widely used browsers are Netscape Navigator and Microsoft Internet Explorer.

Browser

A key feature of the Web is *hypertext*. Hypertext is a method that allows different pages or websites to be linked. When Web pages are created, usually with a language called *hypertext markup language (HTML)*, programmers can insert links to other pages on the website, to other parts of the same page, or even to different websites. For example, a page may contain a highlighted word (usually in blue letters), which means that it is a link to some related topic on a different page or different site. Clicking with the mouse on that word automatically takes you to the related information. This ability to link to related or relevant information is one of the most powerful and useful features of the Web.

Hypertext

Hypertext markup language (HTML)

E-commerce. *E-commerce* or electronic commerce refers to doing business over the Internet, usually buying and selling goods and services by way of the Web. Companies selling products or services set up websites describing their wares and offering them for sale. Individuals may buy these items by giving a credit card number or making other forms of payment. The product is then shipped by mail or overnight carrier. Online shopping is expected to grow significantly in the future. Two of the larger e-commerce retailers are Amazon, a book reseller, and eBay, a site where some items are sold by auction.

E-Commerce

Searches. An Internet search allows a person to look for information on any given topic. Several companies offer the use of free search "engines," which are specialized software that can look for websites related to the desired search topic. The most commonly used search sites are Google, Yahoo, and Microsoft Network (MSN). You simply access the website through your browser and type in one or more keywords that direct the search engine. Within seconds, the search engine returns a listing of all those sites it finds containing the keywords. Most searches return hundreds and even thousands of references such as websites, advertisements, press releases, magazine and newspaper articles, and many other resources.

Voice over Internet Protocol (VoIP). VoIP is the technique of replacing standard telephone service with a digital voice version with calls taking place over the Internet. Voice is digitized, compressed, and transmitted over the Internet in packets to the called location. There it is decompressed and reconverted to analog. VoIP has largely replaced standard telephone services in companies and governments and has made great inroads into homes across the United States and other countries. Its growth continues.

Video over Internet Protocol. Video or TV over the Internet (IPTV) is also becoming more common. The video (and accompanying audio) is digitized, compressed, and sent via the Internet. It is expected to gradually replace some video transmitted over the air and by cable television systems. Both IPTV and VoIP make use of the standard Internet connections used for data.

How the Internet Works

The Internet is the ultimate data communication network. It uses virtually every conceivable type of data communication equipment and technique. All the concepts you learned earlier apply to the Internet. Just keep in mind that the information is transmitted as serial binary pulses, usually grouped as bytes (8-bit chunks) of data within larger groups called *packets*. All the different types of communication media are used including twisted-pair cable, coaxial cable, fiber-optic cable, satellites, and other wireless connections.

Packets

Internet Addresses. Each individual or computer on the Internet must have some kind of identifier or address. An addressing system for the Internet uses a simplified name-address scheme that defines a particular hierarchy. The upper level of the hierarchy is called a *top-level domain (TLD)*. A domain is a specific type of organization using the Internet. Such domains are assigned a part of the Internet address. The most common domains and their address segments are as follows:

Domain	Address Segment
Commercial companies	.com
Educational institutions	.edu
Nonprofit organizations	.org
Military	.mil
Government	.gov
Internet service providers	.net
Air transportation	.aero
Business	.biz
Cooperatives	.coop
Information sites	.info
International organizations	.int
Mobile	.mobi
Museums	.museum
Families and individuals	.name
Professions	.pro
Travel-related companies	.travel

| Adult sites | .xxx |
| Country | .us, .uk, .fr, .jp, .de (United States, United Kingdom, France, Japan, Germany) |

Another part of the address is the host name. The *host* refers to the particular computer connected to the Internet. A host is a computer, device, or user on the network. A server provides services such as e-mail, Web pages, and DNS. The host name is often the name of the company, organization, or department sponsoring the computer. For example, IBM's host name is ibm.

Host

E-mail Address.
The first part of the address is the user's name or some abbreviation, concatenation, or nickname. You might use your name for an e-mail address or some made-up name that you would recognize. The complete address might look like this: <billbob@xyz.net>.

The user name is separated from the host by the @ symbol. The host, in this case xyz, is the e-mail service provider. Note the dot between the host name and the domain name. This address gets converted to a series of numbers used by computers on the Internet to identify and locate one another.

WWW Addresses.
To locate sites on the Web, you use a special address called a *uniform resource locator (URL)*. A typical URL is <http://www.abs.com/newinfo>.

Uniform resource locator (URL)

The first part of the URL specifies the communication protocol to be used, in this case *hypertext transfer protocol (http)*. The www, of course, designates the World Wide Web. The abs.com part is the domain or the computer on which the website exists. The item after the slash (/) indicates a directory within the website software. Most websites have multiple directories, which are also usually further subdivided into pages.

Hypertext transfer protocol (http)

Initial Connections.
A PC is connected to the Internet in a variety of ways. The most common method is through a modem that connects to the telephone system. There are traditional dial-up modems such as those described earlier and the newer *asynchronous digital subscriber line (ADSL)* and cable TV modems. Wireless modems are also available.

A common way of connecting to the Internet is to use a LAN to which your PC may be connected. Most company and organization PCs are almost always connected to a LAN. The LAN has a server that handles the Internet connection, which may be by way of a T1 line to the telephone system or a fiber-optic connection.

Role of the Telephone System.
The familiar telephone system is often the first link to the Internet. Because it is so large and so convenient, it is a logical way to make connections to remote computers. A standard telephone line is used with dial-up modems and DSL modems. In many applications the telephone system is the only connection between computers. Many corporate computers are connected in this way, and you can connect directly to an online service provider in this way. Its primary function is to connect you to a facility known as an *Internet service provider (ISP)*.

Internet service provider (ISP)

Broadband Connections.
Although most connections to the Internet occur with a modem over the telephone lines, more and more individuals are acquiring broadband connections. A *broadband connection* is a fast Internet connection provided by a local telephone company or some other organization. The term *fast* here means a data transmission speed greater than that which can be obtained with a telephone modem, typically 53 kbps. A broadband connection can provide speeds up to many megabits per second.

Broadband connection

If you have ever accessed the Internet or World Wide Web by way of a common telephone modem, you know that it takes a long time to receive information. Long waits are common. Although most modems are capable of receiving at up to 53 kbps, few actually achieve this speed because their distance to the central office is too long or the noise level on the line is high. More commonly, the connection speed is 19.2 to 40 kbps. If you are using e-mail regularly and downloading large files from the Web, you are spending a great deal of time waiting. A broadband connection speeds things up significantly.

The most widely used broadband connection is a cable TV modem. Cable TV companies often act as Internet service providers. The cable TV system is ideal for supporting fast data transfers. A special cable modem is provided that connects to a PC by way of an Ethernet interface. Data transfer rates vary depending upon the amount of traffic on the line, but they can reach as high as 6 Mbps.

The second most widely used broadband connection is the *digital subscriber line (DSL)*. It is most often offered by a local telephone company but may be delivered by an independent supplier. It gives a data rate from 1.5 to 6 Mbps. Other fast but less used broadband services are offered by electrical utilities (broadband over power line) and satellite TV companies. Internet access is also available through cell phones.

Internet Service Provider. An ISP is a company set up especially to tap into the network known as the Internet. It can be an independent company or a local telephone company or a cable TV company. The ISP has one or more servers to which are connected dozens, hundreds, or even thousands of modems, DSLs, or cable connections from subscribers. It is usually the ISP that provides the software you use in communication over the Internet. The ISP also usually provides e-mail service.

There are thousands of ISPs across the United States. Most are set up in medium-size to large cities. You access your ISP by dialing its local telephone number. If you live in a small town or a rural area, you can connect to an ISP in a nearby city, but you may have to make a long-distance call to do this. The ISP is connected to the Internet backbone by way of a fast digital interface such as T1 or T3 lines or by a faster method via a MAN.

15-2 Internet Transmission Systems

The Internet is made up of a huge number of individual elements. These elements include the servers, routers, and transmission media that carry the signals. A huge number of components and interconnections are involved. This section shows the overall operation of the Internet, the transmission protocols used, as well as the most widely used transmission media.

Frame Relay

Frame Relay (FR) is a packet-switching protocol standardized by the ITU-T. It packages data to be transmitted into FR frames that have the structure shown in Fig. 15-1. The 8-bit flags signal the beginning and ending of a packet. A two-octet (byte) address field contains all the details regarding the exact destination of the packet through the network. The data field is variable and may contain up to 4096 octets. A two-octet *frame check sequence (FCS)* is an error detection code that is compared to the FCS calculated from the received data. If any transmission error occurs, the receiving unit asks for a retransmission.

Frame check sequence (FCS)

Figure 15-1 A Frame Relay (FR) frame or packet.

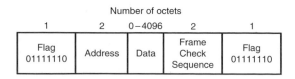

FR is protocol-independent in that it can carry the data from any other transmission method such as Ethernet. Ethernet packets are put into the data field and carried unmodified to the destination. Frame Relay uses existing dedicated digital circuits leased by telecommunications companies. Both T1 and T3 lines are widely used. The equipment used in handling FR is frame relay switches that examine the destination address and forward the packets through one switch after another until the destination is reached. The most common use of FR is in LAN-to-LAN connections where the LANs are widely separated, as in two different company locations. The FR service is usually offered by most local telecommunications carriers.

Asynchronous Transfer Mode

Asynchronous transfer mode (ATM) is also a packet-switching system for transmitting data. It uses very short 53-byte packets with a 48-byte data payload and a 5-byte header that designates the destination as well as the type of data to be handled. Any kind of data may be transmitted in this way including voice, video, and computer data. The packet format is shown in Fig. 15-2.

ATM is transmitted by way of switches that pass the packets from one to the other over the network. An ATM network sets up a virtual circuit that gives the appearance of a single continuous path for the data from source to destination. In reality, the packets most likely take several different paths through the network and may even arrive out of order. The ATM system handles these occurrences such that all data arrives in the desired order.

The ATM network can use almost any physical layer, but most often it is some fiber-optic system. Early ATM networks operated at 155 and 622 Mbps. Today, ATM uses 2.5- and 10-Gbps optical networks.

Asynchronous transfer mode (ATM)

SONET*

The *Synchronous Optical Network (SONET)* was developed to transmit digitized telephone calls in T1 format over fiber-optic cable at high speeds. Its primary use is to send time-multiplexed voice or data over switched networks. SONET is used between telephone central offices, between central offices and long-distance carrier facilities, and for long-distance transmission. Most Internet backbones are SONET point-to-point connections or rings.

SONET is by far the most widely used optical data transmission network in the United States. It is an American National Standards Institute (ANSI) standard as well as a subset of the broader international standard known as the *Synchronous Digital Hierarchy (SDH)*. The latter was developed and sanctioned by the International Telecommunications Union (ITU) and then used throughout the rest of the world.

SONET is a physical (PHY) layer standard that defines the method for formatting and transmitting data in sync with a master timing system keyed to an atomic clock. The standard defines different optical-carrier (OC) speeds, from 51.84 Mbps to 39.812 Gbps. The electric signal to be transmitted is called the *synchronous transport signal (STS)* and

Synchronous Optical Network (SONET)

Synchronous Digital Hierarchy (SDH)

Synchronous Transport Signal (STS)

*Portions of this section are reprinted with permission from *Electronic Design,* November 20, 2000. Copyright 2000. Penton Media Inc.

Figure 15-2 An ATM packet.

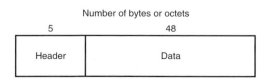

Number of bytes or octets

5	48
Header	Data

Figure 15-3   SONET data rates.

Sonet level	STM level	Data rate
STS-1/OC-1	—	51.84 Mbps
STS-3/OC-3	STM-1	155.52 Mbps
STS-12/OC-12	STM-4	622.08 Mbps
STS-48/OC-48	STM-16	2.488 Gbps*
STS-192/OC-192	STM-64	9.953 Gbps*
STS-768/OC-768	STM-256	39.812 Gbps*

*Note that when referencing these data rates, they are typically rounded to 2.5, 10, and 40 Gbps, respectively.

Synchronous transport mode

Add-drop multiplexor

can exist at several levels or speeds, although the base speed (STS-1) is 51.84 Mbps. In the SDH system, the signal is also called the *synchronous transport mode,* or *STM.* Figure 15-3 lists the most common channel speeds.

Earlier systems used OC-3 and OC-12 levels, but today most SONET systems have been upgraded to operate at OC-48. The OC-192 and OC-768 levels are gradually being deployed as technology and economics permit.

SONET is a time-division multiplexing (TDM) transmission scheme that sends time-interleaved data in fixed-length frames of 810 bytes. The frame format consists of nine 90-byte rows. The bytes are transmitted consecutively from left to right and from top to bottom. See Fig. 15-4. In each row 4 bytes is for overhead, and 86 bytes per row is for data payload. The overhead bytes contain framing, control, parity, and pointer information for managing the payload.

Although SONET can operate in a point-to-point link, the most common topology is a ring. Multiple nodes make up the ring, with an *add-drop multiplexer (ADM)* at each node. One of the key features and benefits of a SONET system is that the ADM permits data from multiple sources to be added to or extracted from the data stream as required. The ring topology also offers data transmission redundancy, which ensures survivability when a cable has been cut.

Figure 15-4   SONET frame format.

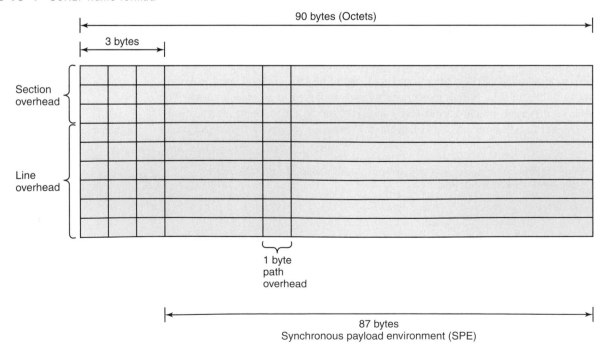

The basic SONET transmission speed is OC-1, or 51.84 Mbps. This is the rate achieved when one SONET frame is transmitted in 125 μs. If higher-speed transmission is desired, as is usually the case, multiple frames are transmitted in this time period by a system of byte interleaving.

Most of today's long-distance telephone service, as well as the Internet backbone, is made up of a hierarchy of large and small SONET rings. Even though SONET was designed to carry synchronous voice traffic in a circuit-switched environment, its primary function is data transmission with no regard to its content. As a result, SONET can carry asynchronous, packet-switched data from asynchronous transfer-mode (ATM), Frame Relay, Ethernet, or Transport Control Protocol/Internet Protocol (TCP/IP) equipment.

Routers

The *router* is the single most important piece of equipment in the Internet. You have probably heard of the world's largest router manufacturer, Cisco. Routers interconnect the various segments of the WAN backbones. Routers are also used to connect MANs to WANs, LANs to others LANs, and LANs to MANs. The routers connect to one another and to the various servers to form a large mesh network connected usually by fiber-optic cable.

A router is an Intelligent computerlike device that looks at all packets transmitted to it and examines their Internet protocol (IP) destination addresses. It then determines the best next path for the data to take to its destination. The router stores information about the other routers and networks to which it is connected and about any nearby networks. This information is stored in a routing table that is compared to the destination address on all incoming packets. Routing algorithms determine the best (closest, fastest) connection and then retransmit the packet. In any given Internet transmission, most packets pass through several routers before they reach their final target computer.

Physically, a router can have several configurations. Any high-speed device can be used that has serial inputs and outputs and a way to examine packet addresses and make decisions regarding routing and switching. For example, a computer with a fast processor, adequate memory, serial I/O, and the appropriate software can serve as a router. Today, most routers are specialized pieces of equipment optimized for the routing function.

The basic configuration of a modern router is shown in Fig. 15-5. It consists of a group of line cards that plug into connectors on a printed-circuit board back plane. The back plane contains the copper interconnecting lines to allow the line cards to transmit and receive data

GOOD TO KNOW

Researchers Partha Mitra and Jason Stark determined that the maximum theoretical capacity of a single optical fiber is 100 terabits (Tbits) equivalent to 20 billion simultaneous one-page e-mails.

Router

Figure 15-5 General block diagram of a router.

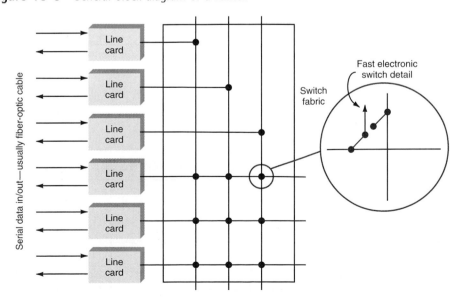

from one another. Transfer speeds are typically many gigabits per second. The line cards communicate with one another through a switch fabric. The switch fabric is a large digital equivalent of a cross bar switch. A cross bar switch is a matrix of switches in a row-column format that allows any one of multiple input lines to be connected to any one of multiple output lines. Superfast electronic switches connect the line cards. See Fig. 15-5.

Serializer/deserializer (SERDES)

Each line card has a serial input port and a serial output port. These are usually optical fiber lines, but they could also be RJ-45 Ethernet ports. At the receive input, the optical data is converted to serial electric signals that are then deserialized by a circuit called a *serializer/deserializer (SERDES)* that performs both serial-to-parallel and parallel-to-serial data conversion. The parallel data is then stored in a random-access memory (RAM). This data is then examined by a special processor to determine its destination address. The processor uses a routing lookup table and algorithms that decide which output line is to be used. Any input may be routed to any output in this way. The processor then passes the packet to the switch fabric where it is sent to an output port. The packet is serialized by a SERDES circuit and then sent to the output port by using SONET, ATM, Ethernet, or other protocol.

The Internet Backbone

Internet backbone

The *Internet backbone* is a collection of companies that install, service, and maintain large nationwide and even worldwide networks of high-speed fiber-optic cable. The companies own the equipment and operate it to provide universal access to the Internet. In many ways, these companies *are* the Internet, for they provide the basic communication medium, equipment, and software that permit any computer to access any other.

Although each of the backbone providers has its own nationwide network, the providers are usually connected to one another to provide many different paths from one computer to another. There are more than 50 of these interconnection points in the United States. Known as *network access points (NAPs)*, these facilities provide the links between backbones.

Network access points (NAPs)

Figure 15-6 shows a diagram of the main components of the Internet. On the left are three ways in which a PC is connected for Internet access, to the telephone local

Figure 15-6 Simplified diagram of the Internet.

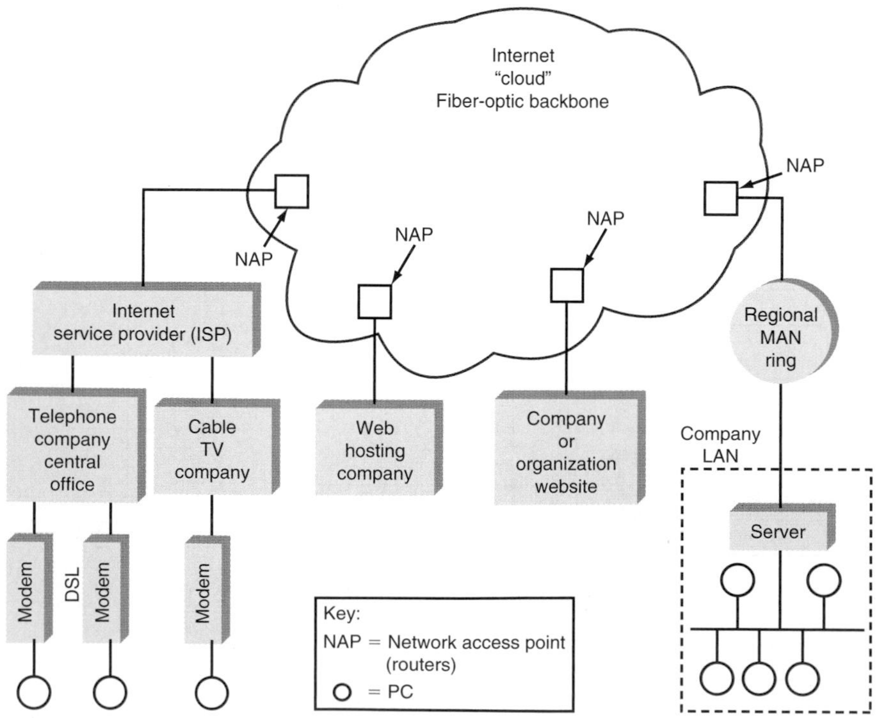

office via a conventional or DSL modem or by way of a cable modem and cable TV company. The telephone company central office and cable TV company connect to an Internet service provider by way of T1 or T3 lines or a local MAN. (*Note:* The telephone company or cable TV company may also be the ISP.) The ISP contains multiple servers that handle the traffic. Each connects to the Internet by way of a router that attaches to the Internet backbone at one of the many network access points. The primary equipment at network access points is routers that determine the destination of packets. The "cloud" in Fig. 15-6 represents the Internet backbone. This is one of several large fiber-optic networks, either point-to-point or a ring, that carry the Internet data and interconnect with one another to exchange data.

On the right in Fig. 15-6 are additional connections. Here a company LAN accesses the backbone by way of a regional MAN. A company server contains a website that is regularly accessed by others. Also shown is a Web hosting company that stores the websites and Web pages of others.

The Packet–Switching Transmission System

The Internet is a packet-switching system. Data to be sent is divided up into short chunks called *packets* and transmitted one at a time. The term *datagram* is also used to describe a packet. Packets are typically less than 1500 octets long. (*Note:* In data communication, a byte is more often called an *octet,* an 8-bit word.) Each packet finds its own way through the complex maze that is the Internet by using addressing information stored in the packet. Not all packets take the same path through the system, and in fact the packets may arrive at the receiving end out of the order in which they were sent.

Figure 15-7 shows a simplified version of the packet-switching concept. The packet-switching network consists of multiple exchanges with high-speed packet switches that connect to multiple inputs and outputs. The multiple interexchange links produce many possible paths through the network.

Several popular packet-switching systems are in use in Internet backbones. These include Frame Relay and asynchronous transfer mode (ATM) systems. ATM systems are the fastest and most widely used. They break down all data into 53-byte packets and transmit them over fiber-optic networks at speeds of up to 10 Gbps (and soon 40 Gbps). ATM is also used to transmit voice and video. In some instances, the packets are packaged into

Packet
Datagram

Figure 15–7   The packet-switching concept showing nodes in the backbone.

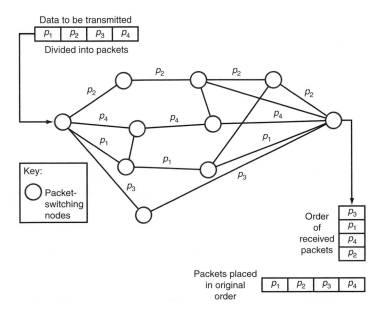

Figure 15–8 Comparing the OSI and TCP/IP layers.

OSI layers	TCP/IP layers
7: Application	Application
6: Presentation	
5: Session	Host-to-host (TCP)
4: Transport	
3: Network	IP
2: Data link	Network access
1: Physical	

Packet over SONET (PoS)

TCP/IP protocol

long continuous synchronous data streams and transmitted by way of the Synchronous Optical Network (SONET). Such transmissions are called *packet over SONET (PoS)*.

Packet switching requires a set of software protocols that make sure that the data is properly partitioned, transmitted, received, and reassembled. On the Internet, these protocols are called *TCP/IP*. TCP means Transmission Control Protocol, and IP means Internet Protocol. Both protocols are used to send and receive data over the Internet. TCP/IP was invented in the 1960s, when the Advanced Research Project Agency (ARPA) of the Department of Defense (DoD) tested and established the first packet network. It was first implemented on large computers running the UNIX operating system. Today, TCP/IP is by far the most widely implemented data communication protocol, and it is at the heart of the Internet.

TCP/IP is a layered protocol similar to the OSI seven-layer model discussed in Chap. 11. TCP/IP does not implement all seven layers, although the effect is the same. Some operations of the individual seven layers are combined to form four layers, as shown in Fig. 15-8. The upper, or applications, layer works with other protocols that implement the desired application. The most widely used are TELNET, which permits a remote PC to connect via the telephone system to the Internet; the *file transfer protocol (FTP)*, which facilitates the transmission of long files of data; the *simple mail transfer protocol (SMTP)*, which implements e-mail; and the *hypertext transfer protocol, (http)*, which provides access to the World Wide Web.

File transfer protocol (FTP)

Simple mail transfer protocol (SMTP)

Hypertext transfer protocol (http)

Transmission Control Protocol. The host-to-host layer is really TCP. It divides the data into packets to be sent. To each packet is attached a comprehensive header containing source and destination addresses, a sequence number, an acknowledgment number, a check sum for error detection and correction, and other information. The TCP header is shown in Fig. 15-9. It consists of 20 bytes or octets, minimum, depending upon what options are added. At the receiving end, TCP reassembles the packets in the proper order and sends them to the application. It also asks for retransmission if a packet is received in error.

TCP is used only to prepare the packets for transmission and reassemble the packets when received. It does not implement the actual packet transmission over the Internet. As you will see, that is the job of IP. TCP's job is to offer some assurance that the packets arrive in a reliable and high-quality form by providing the error detecting and correcting check sums, sequence numbers, and other features.

The TCP header is a single serial bit stream in which each of the fields is transmitted one after another. This header is difficult to draw and print as a long single line,

GOOD TO KNOW

When a packet is sent from a host to another host over the Internet, it is first assembled into packets with TCP format; then the IP layer adds a new header to send the packet over the Internet.

Figure 15-9 TCP headers.

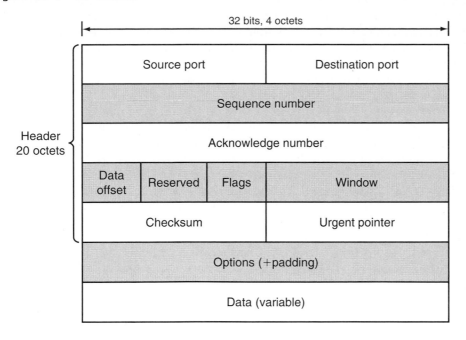

so the presentation format in Fig. 15-9 is used. This representation is widely used to illustrate other protocols and headers. In reality, data is transmitted serially from left to right and from top to bottom, row after row, with 4 octets per row.

Internet Protocol. The TCP packet is then sent to the IP layer, where the IP protocol is used. The IP layer ensures that the packet gets to its destination over the Internet. In Fig. 15-10, an IP header is attached before the resulting new packet is transmitted.

Figure 15-10 IP header (IPv4).

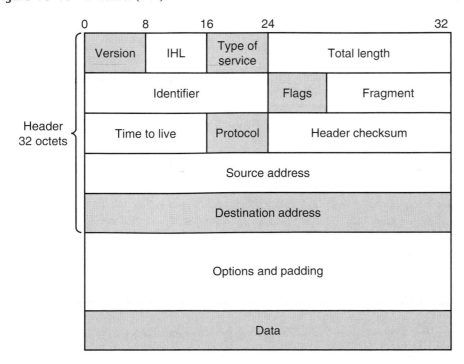

Note in Fig. 15-10 that the destination address has 32 bits. Every computer has one of these addresses assigned to it. The addresses are usually expressed in what is called the *dotted decimal form*. The 32-bit address is divided into four 8-bit segments. Each segment can represent 256 numbers from 0 through 255. A typical address would look like this: 125.63.208.7. Dots separate the four decimal segments. These addresses are assigned by an organization known as the *Internet Assigned Numbers Authority (IANA)*. The addresses are subdivided into classes (A through E) to represent different types and sizes of networks, organizations, and individual users.

The most widely used IP protocol is called *IPv4*, or *Internet Protocol version 4*. Its primary limitation is that its destination address size is only 32 bits; i.e., it is limited to 2^{32}, or 4,294,967,296, locations or users. Although more than 4 billion addresses would seem to be enough, as the Internet has grown, this has become a severe limitation. The newer version is IPv6, which has a 128-bit address field that should provide more than enough destinations (2^{128}) for massive future expansion. Also IPv6 provides for authentication and multi-

casting. *Multicasting* is the ability of IP to move fast audio and video data over the Internet from a single source to multiple destinations. IPv6 is implemented on many computers, but not all. It is slowly being phased in as software and hardware are updated in new systems.

Finally, the network access layer that contains the physical layer connection transports the data. As explained earlier, there are several different ways that TCP/IP information is transmitted. For example, the packets may get packaged into Ethernet packets in a LAN and then delivered to a MAN for ultimate connection to the packet-switched WAN backbone. Or the TCP/IP packets may be delivered over the telephone line, as with a dial-up modem. Alternatively, the TCP/IP data may travel via a cable TV modem to an ISP that connects to a MAN via a T1 or T3 line and then to the WAN. The WAN may carry the data via ATM or SONET through the backbone.

TCP/IP*

Several protocols or sets of rules are used to communicate over a computer network. Each protocol must accomplish several tasks, such as encapsulation, fragmentation and reassembly, connection control, ordered delivery, flow control, error control, addressing, multiplexing, and transmission services. Perhaps the most common protocol used to accomplish these tasks is the Transmission Control Protocol/Internet Protocol or TCP/IP. TCP/IP is the communications protocol that hosts use to communicate over the Internet, and it establishes a virtual connection between a destination and source host. TCP/IP uses two protocols to accomplish this task, TCP and IP.

TCP enables two hosts to establish a connection and exchange data. TCP will guarantee both the delivery of data and that the packets will be delivered in the same order in which they were sent. Remember, packets are sent through a network according to the best path. This "best path" choice does not guarantee that all packets will take the same path, nor will they arrive in the same order they were sent. TCP has the job of ensuring that all the packets are received and put back into the correct order before they are passed up the protocol stack.

IP determines the format of the packets. The IP packet format is not discussed in detail, but there are a total of 20 octets in IP version 4 packets. These bits are used to select the type of service, length of datagram, identification number, flags, time to live, next-higher protocol, header checksum, and various addresses. The packet IP provides a function similar to the address on a postal letter. You write the address on the letter and put it in the mailbox. You and the receiver know where it is sent from and to whom it is being sent, but the path is determined by someone else. That someone else is the routers in the network between the destination and the source. It is TCP/IP that establishes the connection between the destination and the source. TCP steps in and cuts the letter up into smaller pieces or packets and then sends them, ensuring all the packets are received and put back into the proper order.

*Courtesy of Walter Currier, Assistant Professor of Computer and Electrical Engineering at Liberty University, Lynchburg, VA.

UDP

The *User Datagram Protocol (UDP)* is another protocol used at the transport level. UDP provides a connectionless service for applications. UDP provides few error recovery services, unlike TCP. However, like TCP, UDP uses IP to route its packets throughout the Internet. UDP is used when the arrival of a message is not absolutely critical. You may recall that from time to time you receive letters for Current Resident in your mailbox. The senders of this "junk mail" are not concerned that everyone receive the package they send. UDP is similar to the Current Resident mail and is often used to send broadcast messages over a network. A broadcast message is a message that is sent periodically to all hosts on the network in order to locate users and collect other data on the network. UDP messages are also used to request responses from nodes or to disseminate information. Another application of UDP is in the use of real-time applications. With real-time applications, retransmitting and waiting for arrival of packets are not possible so TCP is not used for these applications. When real-time data (voice or video) is routed, a connectionless UDP protocol is used. If packets get dropped or fail to arrive, the overall message is usually not corrupted beyond recognition.

Internet and Addressing

You have seen the components that make up a network and how information travels across a network, but how does a packet find its intended destination? The Internet is organized in a hierarchical structure. The entire network is often referred to as the Internet or the World Wide Web. The Internet is subdivided into several smaller networks which are all interconnected by routers.

The Internet connects several separate segments or networks together by using routers. Routers need some way to identify the destination network to which a packet is bound. Routers accomplish this by using the network IP address. All devices on that network share the same network address, but have unique host addresses. Packets get routed from network to network until they arrive at the network that contains the host to whom the packet has been sent.

A good example of a hierarchal structure is how large military units are organized. The largest structure is the brigade. The brigade is separated into two regiments with three battalions per regiment. Each battalion has five companies. Each company then has platoons, and each platoon has squads. An individual soldier or sailor is in a squad, platoon, company, battalion, and so on. If you want to contact all the soldiers or sailors in a particular company, you can send a message to just that company. The same is true for a battalion, or platoon. In a computer network, the ability to send messages to an individual host on a particular network is also important. Each network is then connected into the entire Internet or the so-called Internet cloud.

We can break each connection to the cloud into its own network, and each network would be connected to the cloud by using a router. Every computer connected off the router is considered to be on the same network.

This arrangement is similar to a family. The router would represent a single family, say, the Jones family; and all the segments represent children in the Jones family. We can easily identify who is in the Jones family by looking at the last name. A router can recognize who is in its network by using a set of numbers called an IP address.

When a computer receives a packet from the router, the computer first checks the destination MAC address of the packet at the data link layer. If it matches, it's then passed on to the network layer. At the network layer, it checks the packet to see if the destination IP address matches the computer's IP address. From there, the packet is processed as required by the upper layers. On the other hand, the computer may be generating a packet to send to the router. Then as the packet travels down the OSI model and reaches the network layer, the destination and source IP address of this packet are added in the IP header.

The format of an IP address is called dotted decimal, and it consists of four numbers from 0 to 255 separated by periods or dots. Each number between the periods is considered an octet because it represents 8 binary bits.

Example: 35.75.123.250

The dotted decimal format is convenient for people to use, but in reality the router will convert this number to binary, and it sees the above dotted decimal number as a continuous string of 32 bits. The example below shows an IP address in decimal notation. This IP address (35.75.123.250) is then converted to binary, which is what the computer understands. It's easier for us to remember four different numbers than thirty-two 0s or 1s.

35   .75   .123   .250
00100011.01001011.01111011.11111010

The above IP address would look like 00100011010010110111101111111010 to the computer. The hexadecimal version is 234B7BFA.

To provide flexibility, the early designers of the IP address standard sat down to sort out the range of numbers that were going to be used by all computers. They organized the IP address into five classes, and we normally use three of these classes. When people apply for IP addresses, they are given a certain range within a specific class depending on the size of their network.

In Table 15-1, you can see the five classes. The first three classes (A, B, and C) are used to identify workstations, routers, switches, and other devices, whereas the last two classes (D and E) are reserved for special use. The IP addresses listed above are not all usable by hosts!

An IP address consists of 32 bits, which means it is 4 bytes long. The first octet (first 8 bits or first byte) of an IP address is enough for us to determine the class to which it belongs. And depending on the class to which the IP address belongs, we can determine which portion of the IP address is the network ID and which is the host ID. For example, if you were told that the first octet of an IP address was 168, then, using the above table, you would notice that it falls within the 128 to 191 range, which makes it a class B IP address.

Earlier we said that companies are assigned different IP ranges within these classes, depending on the size of their networks. For instance, if a company required 1000 IP addresses, it would probably be assigned a range that falls within a class B network rather than a class A or C.

To get the information to the correct host, the IP address is divided into two parts, the network ID and the host ID. These two parts contain two pieces of valuable information:

1. It tells us which network the device is part of (network ID).
2. It identifies that unique device within the network (host ID).

Table 15-1	Five Different Classes of IP Addresses		
Class	Range of IP Addresses		
A	1.0.0.0	to	127.255.255.255
B	128.0.0.0	to	191.255.255.255
C	192.0.0.0	to	223.255.255.255
D	224.0.0.0	to	239.255.255.255
E	240.0.0.0	to	255.255.255.255

Figure 15-11 Example showing network and host IDs.

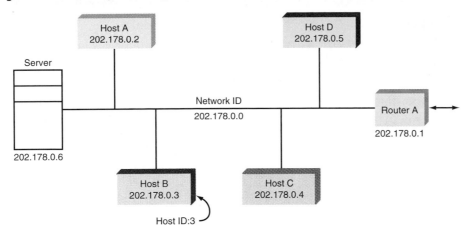

Figure 15-11 gives you an example to help you understand the concept.

Routers will look at the first number or octet to determine the class of the IP address. The class indicates how many bits are used to represent the network ID and how many bits are used to represent the host ID. In the above picture, you can see a small network. We have assigned a class C IP range for this network. Remember that class C IP addresses are for small networks. Looking now at host B, you will see that its IP address is 202.178.0.3. The network ID is 202.178, and the host ID is 0.3.

Table 15-2 contains the range of numbers used to determine the class of the network and the number of bits available to assign to a network and the hosts on that network. For example, 140.179.220.200 is a class B address. The 140 falls within the 128 to 191 range, which makes it a class B IP address. So, by default, the network part of the address (also known as the *network address*) is defined by the first 2 octets (140.179.x.x), and the node part is defined by the last 2 octets (x.x.220.200).

Now we can see how the class determines, by default, which part of the IP address belongs to the network (N) and which part belongs to the host (h).

- Class A—0NNNNNNN.hhhhhhhh.hhhhhhhh.hhhhhhhh
- Class B—10NNNNNN.NNNNNNNN.hhhhhhhh.hhhhhhhh
- Class C—110NNNNN.NNNNNNNN.NNNNNNNN.hhhhhhhh

Take a class A IP address as an example to understand exactly what is happening. Any class A network has a total of 7 bits for the network ID (bit 8 is always set to 0) and 24 bits for the host ID. Now all we need to do is to calculate how much is 7 bits: $2^7 = 128$ networks, while $2^{24} = 16,777,216$ hosts in each network. Of the 16,777,216 hosts

Table 15-2	Identifying Network and Host ID		
Class	Range of First Octet	Number of Network ID Bits	Number of Host ID Bits
A	1–127	8	24
B	128–191	16	16
C	192–223	24	8

Table 15-3 — IP Addresses Unusable by Hosts

IP Address	Function
Default—Network 0.0.0.0	Refers to the default route. This route simplifies routing tables used by IP.
Loopback—Network 127.0.0.1	Reserved for loopback. The address 127.0.0.1 is often used to refer to the local host. Using this address, applications can address a local host as if it were a remote host.
Network address—IP address with all host bits set to 0 (for example, 192.168.0.0)	Refers to the actual network itself. For example, network 202.178.0.0 can be used to identify network 202.178. This type of notation is often used within routing tables.
Subnet/network broadcast—IP address with all node bits set to 1 (for example, 202.178.255.255)	IP addresses with all node bits set to 1 are local network broadcast addresses and must *not* be used. Some examples: 125.255.255.255 (class A), 190.30.255.255 (class B), 203.31.218.255 (class C).
Network broadcast—IP address with all bits set to 1 (for example, 255.255.255.255)	The IP address with all bits set to 1 is a broadcast address and must *not* be used. These are destined for all nodes on a network, no matter what IP address they might have.

in each network, 2 cannot be used. One is the network address, and the other is the network broadcast address (see Table 15-3). Therefore when we calculate the "valid" hosts in a network, we always subtract 2.

If asked how many valid hosts you can have on a class A network, you should answer 16,777,214, *not* 16,777,216. The same story applies for the other two classes we use, class B and class C. The only difference is that the number of networks and hosts changes because the bits assigned to them are different for each class. So if asked how many valid hosts you can have on a class B network, you should answer 65,534, *not* 65,536. And if asked how many valid hosts you can have on a class C network, you should answer 254, *not* 256.

Now you've learned that even though we have three classes of IP addresses that we can use, some IP addresses have been reserved for special use. This doesn't mean you can't assign them to a workstation; but if you did, it would create serious problems within your network. For this reason it's best to avoid using these IP addresses. Table 15-3 shows the IP addresses that you should avoid using.

It is imperative that every network, regardless of class and size, have a network address (first IP address, for example, 202.178.0.0 for class C network) and a broadcast address (last IP address, for example, 202.178.0.255 for class C network), as mentioned in Table 15-3 and the diagrams above, which cannot be used. So when you calculate available IP addresses in a network, always remember to subtract 2 from the number of IP addresses within that network.

Reserved Host ID Numbers

When you design a network, you will be given a network ID from a controlling authority. The network ID portion of your Internet address cannot change, but the host ID

Table 15-4	Network Classes with Network Masks	
Class Type	Network Range	Network Masks
A	1.0.0.0 to 127.255.255.255	255.0.0.0
B	128.0.0.0 to 191.255.255.255	255.255.0.0
C	192.0.0.0 to 223.255.255.255	255.255.255.0

portion of the IP address is the bits you own, and you can assign any IP within the host ID portion to the computers on your network, other than the all-0s or all-1s host ID mentioned above. Typically designers will assign the first assignable IP to the first host, but there is no restriction and any number can be assigned.

Just as the name *Jones* identifies the family members, the network ID identifies your network, but how does the router figure out that the network ID is a match? Information travels in packets, and each packet has a header. The header will contain the IP address of the computer to which the packet is being sent. The router will use a special sequence of bits called the *network mask* to determine if the packet is being sent to its network. **Network mask** The network mask has all 1s in the network ID and all 0s in the host ID. This mask is then logically ANDed to the packet, and the router will see if the destination host is on its network. In our 35.0.0.0 network, the network mask would be 255.0.0.0. If the network were a class B, the network mask would be 255.255.0.0, and for a class C the network mask would be 255.255.255.0.

Network Mask

Table 15-4 shows our three network classes with their respective network masks.

An IP address consists of two parts, the network ID and the host ID. We can see this once again shown below, where the IP address is analyzed in binary, because this is the way you should work when dealing with network masks:

Class C IP Address

IP address:	192.	178.	0.	4
Network mask:	255.	255.	255.	0

Conversion to Binary

IP address:	11001010.	1011 0010	0000 0000.	0000 0100
Network mask:	1111 1111.	1111 1111.	1111 1111.	0000 0000
ANDed result	11001010	1011 0010		0000 0000
		Network ID		Host ID

GOOD TO KNOW

All class C IP addresses have a 24-bit network mask (255.255.255.0).
All class B IP addresses have a 16-bit network mask (255.255.0.0).
All class A IP addresses have an 8-bit network mask (255.0.0.0).

This class C network uses 21 bits for the network ID (remember, the first 3 bits in the first octet is set) and 8 bits for the host ID. The network mask is what splits the network ID and host ID.

We are looking at an IP address with its network mask for the first time. What we have done is to take the decimal network mask and convert it to binary, along with the IP address. It is essential to work in binary because it makes things clearer and we can avoid making mistakes. The 1s in the network mask are ANDed with the IP address, and the result defines the network ID. If we change any bit within the network ID of the IP address, then we immediately move to a different network. So in this example, we have a 24-bit network mask (twenty-four 1s, counting from left to right).

Subnet Mask

Subnetting

As the network grows, it becomes increasingly difficult to efficiently route all the traffic, since the router needs to keep track of all the hosts on its network. Let's say that we have a simple MAN of two routers. Whenever a packet is sent from one host to another on our network, the router will route the packet to the proper host. When the packet is sent to a host connected to another router, the network mask will be used to determine that the packet is not for our network, and the router will send the packet over the communication link to the other network. If each router has only a few hosts connected to it, then the number of packets that the router has to route is relatively small. As the network grows, this number of hosts gets quite large. For a class A network there could be 2^{24}, or close to 17 million, hosts. As a means to help the routers more efficiently route packets and manage the size of their router tables, a technique called *subnetting* is used.

When we subnet a network, we basically split it into smaller networks. By subnetting the network, we can partition it into as many smaller networks as we need. By default, all types of classes (A, B, and C) have a network mask. However, a network mask other than the default can be used. This is called a subnet mask. The use of an IP address with a subnet mask other than the network default results in the standard host bits (the bits used to identify the host ID) being divided into two parts: a subnet ID and host ID. Take the same IP address as above, and divide it further into a subnet ID and host ID, thus changing the default network mask. Then by using the ANDing technique described earlier, smaller subnets can be created and identified as required.

MAC Address Versus IP Address

Media Access Control (MAC)

Recall the *media access control (MAC)* address discussed earlier. The MAC address is a unique address assigned to the physical device. The IP address is a logical address used to determine where in the network a host is located. As in the postal example, the state could be considered the network, with the city the subnet, and the individual person as the host in the network. The MAC address is like the Social Security number (SSN) that each person has. The SSN gives no information about where the person is located, but does uniquely identify that person. The MAC identifies the manufacturer and has a unique number associated with it. The IP address is used to find out where the MAC is so that packets can be routed to the host.

An Example Transmission

Assume an e-mail application in which a message packaged by the SMTP is sent to the TCP layer for formatting. The TCP is then sent to the IP layer, which further packages the message for transmission. Also assume that the IP packets are encapsulated into Ethernet LAN packets and sent to a WAN. At the first WAN node, the router strips off the Ethernet packets, recovers the IP header, and reveals the destination. The router makes its routing decision and then reencapsulates the packets into the protocol of the WAN for this application ATM. The ATM packets then travel from router to router, where the same thing occurs. The packets finally reach the final destination, where the router again recovers the IP packets.

At the receiving end, the IP layer verifies that the packet has come to the right place and then strips off the header and passes it to the TCP layer. Here any errors are detected and retransmission is requested if necessary. Then all the packets are reassembled in the correct order, the header is removed, and the final data is sent to the applications layer, where it is sent to the SMTP protocol that delivers the e-mail.

15-3 Storage–Area Networks

Storage-area networks (SANs) are one of the faster-growing segments of data communications. SANs and similar storage systems provide a way to meet the Internet's nearly insatiable need for data storage. Besides the need for mass storage of data for websites and e-mail, there is a growing demand for storage at large companies; also the government needs storage for backup storage, disaster recovery, and retention of files for government regulations such as Sarbanes-Oxley. Add to that the growing use of databases for every conceivable need, and the result is a need clearly beyond even the very large disk drives in the servers. Special storage systems have been created to hold these massive data resources, and special networks and communications systems have been developed to ensure rapid access to this data.

Early storage systems were made up of external disk drives that were not inside the PCs or servers. These external drives were connected to the PC or server by way of a fast parallel transmission bus referred to as the *Small Computer Storage Interface (SCSI)*, nicknamed the "skuzzy" interface. A formalized set of binary commands allowed the computer to control the disk drives to store or access data. Over the years, the parallel bus became faster and faster, but the distance over which data can be transmitted in parallel format declined to several feet with the increased speed. This form of mass storage was referred to as *direct attached storage (DAS)*. While DAS is still used today, larger, more flexible systems using fast serial data transfer are available.

Small Computer Storage Interface (SCSI)

Direct Attached Storage (DAS)

One of these new systems is called *network attached storage (NAS)*. These systems are made up of a *redundant array of independent disks (RAID)* or *just a bunch of disks (JBOD)*. These large boxes of disk drives are typically connected to a PC or server by way of the installed Ethernet LAN. They are assigned an IP address so that data can be accessed in a file format. Anyone connected to the LAN can access the data on the disks if authorization is provided.

Network Attached Storage (NAS)

Redundant Array of Independent Disks (RAID)

Just a Bunch of Disks (JBOD)

Very large storage needs are met by storage area networks. These use the RAIDs and JBODs that are connected via the SAN to the various network servers, database servers, or other computers designed to provide access to the data. Figure 15-12 shows a typical arrangement. All the various user PCs or workstations are connected to a central LAN using 100-MHz or 1-GHz Ethernet over twisted-pair cable to which all the specialized servers (mail, data, application-specific, etc.) are attached. The servers are also connected to the SAN which, in turn, is attached to each disk storage unit. Fiber-optic connections are typically used in connecting the SAN to the servers, but in some cases a 1-GHz twisted-pair Ethernet connection may be used. With this arrangement, any individual PC or workstation may access the data in any disk system or establish backup files via the LAN and the SAN. That access may actually be via the Internet. SANs use block transfers of data instead of file transfers, where fixed-size blocks of data are transferred rather than complete files.

The connection between the servers and the SAN is made usually by a fiber-optic network known as *Fibre Channel (FC)*. A newer connection system called *iSCSI* or *Internet SCSI* ("I skuzzy") uses the installed Ethernet LAN plus Ethernet switches.

Fibre Channel (FC)

Fibre Channel

Fibre Channel is an optical fiber transmission standard established by the American National Standards Institute in the late 1980s. The standard defines a protocol and a

Figure 15-12 The basic architecture of a SAN.

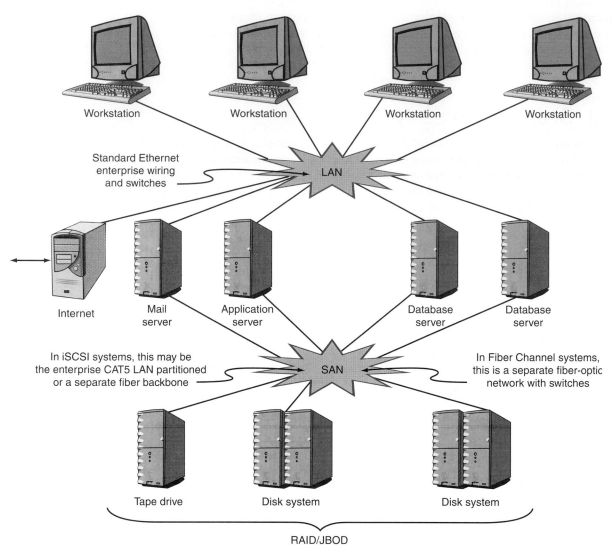

fiber-optic physical layer (PHY) that can be used to connect computers and storage systems in a loop or ring, point-to-point or through switches. Early FC systems transmitted at a rate of only a few hundred megabits per second, but today systems transmit at 1, 2, 4, or 10 Gbps. Very high data rate is essential in a SAN if any large block of data is going to be accessed by a user in a reasonable time.

The FC PHY defines the use of a pair of fiber-optic cables, one for transmit and one for receive. Most systems use 50- or 62.5-μm-diameter multimode fiber (see Chap. 19) for relatively short connection distances up to several hundred meters. Single-mode fiber can also be used for longer-range connections from 10 to 40 km. The short-range systems use 850-μm infrared (IR) laser transmitters while the longer-range systems use either 1310- or 1550-μm IR laser transmitters.

The protocol defines a 2148-byte packet or frame that is transmitted with 8B/10B coding. A 4-byte (32-bit) CRC is used for error detection. The actual transmission speeds are 1.0625, 2.125, 4.25, or 10.5 Gbps. The connection to the fiber-optic cable is made through an interface card known as a *host bus adapter (HBA)*. Each HBA plugs into the bus of the server or the storage control unit of the RAID or JBOD. The connections may be a direct, point-to-point link or a ring. Today, most connections are made through a very high-speed electronic cross point switching arrangement known as a switch fabric. The switch fabric is packaged into a box with the fiber-optic cable connectors, and the

Host Bus Adapter (HBA)

entire unit is called an FC switch. It appears inside the SAN "cloud" shown in Fig. 15-12. This arrangement essentially lets any server or disk drive node on the network connect to any other. Most FC switches permit up to 2^{24} devices to be connected.

One of the primary advantages of the FC SAN is that it is inherently secure. Since it is not connected to the LAN or the Internet, it is essentially immune to outside hacking, virus, spam, or other attacks normally associated with the Internet. The FC SAN is completely separate from any other network connections.

FC systems can also communicate over longer distances via the Internet by using a new protocol called *Fibre Channel over Internet Protocol (FCIP)* developed by the Internet Engineering Task Force (IETF). This protocol encapsulates the FC frames into packets and transmits them via TCP/IP. With this arrangement, multiple FC SANs can be interconnected and managed over an IP network. Another arrangement called *Internet Fibre Channel Protocol (iFCP)* allows FC SANs to be linked by using TCP/IP with standard Ethernet switches or routers. The switch or router becomes a gateway that takes the place of an FC switch.

Fibre Channel over Internet Protocol (FCIP)

Internet Fibre Channel Protocol (iFCP)

Internet SCSI

While FC is used in more than 90 percent of all SANs because of its speed, flexibility, and reliability, its main downside is high cost. Recently, a lower-cost SAN connection system called *Internet SCSI (iSCSI)* has been developed. It uses standard off-the-shelf Ethernet components and TCP/IP software so widely available. This system is also an IETF standard. It uses the same command and control protocol developed for parallel bus SCSI DAS systems except it uses serial data transfers over Ethernet.

Internet SCSI (iSCSI)

Figure 15-13 shows the data flow in a data access operation in an iSCSI SAN. The PC requesting the data notifies the server, and the server operating system then produces an appropriate SCSI command. The iSCSI protocol that is implemented in the Ethernet *network interface card (NIC)* encapsulates the SCSI command into TCP, then into IP packets that are transmitted by using Ethernet on available LAN connections. In Fig. 15-12, the SAN "cloud" is usually just the installed LAN wiring with Ethernet routers and switches. The connection may also be through a MAN or WAN using the Internet. On the receiving end, which is the target RAID or JBOD, the process is reversed, as Fig. 15-13 shows.

Network Interface Card (NIC)

Figure 15-13 The sequence of operations for accessing data by using the iSCSI protocol.

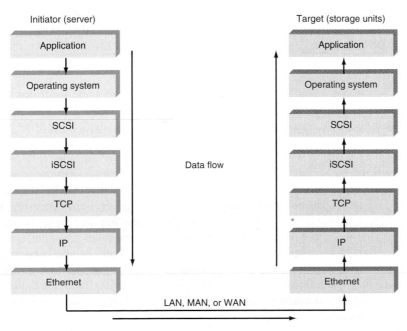

The NICs used in the servers and the RAID/JBOD systems are standard Ethernet interfaces but are iSCSI-enabled. They may even incorporate what is called a TCP/IP off-load engine. This is a special processor designed to handle the TCP/IP operations in hardware rather than in software on the server as usual. This greatly speeds up all operations.

The primary benefit of an iSCSI SAN is its lower cost and use of existing LAN wiring or the Internet. The main disadvantage is that such systems are at risk to hacking, viruses, and other such security problems. This can be taken care of by using security software and data encryption methods, but these increase the cost and greatly slow down all data transmission operations.

15-4 Internet Security

One of the most important aspects of the Internet is security of the data being transmitted. *Security* refers to protecting the data from interception and protecting the sending and receiving parties from unwanted threats such as viruses and spam. And it means protecting the equipment and software used in the networks. The Internet or any network-connected computer is subject to threats by hackers, individuals who deliberately try to steal data or damage computer systems and software just for the challenge.

Prior to the Internet, computer security was primarily limited to sensitive government and military data transmissions. Some large companies also used security measures to transmit critical data. Today the Internet has forced all users and organizations to employ security measures to protect their computer and data. The Internet greatly expanded development in security measures. Wireless systems are also very vulnerable to hacker attacks simply because radio waves are easily picked up and used by anyone with an appropriate receiver. Over the past years, security for wireless systems has been developed and widely deployed.

Most security measures are implemented in software. Some security techniques can also be implemented in hardware such as data encryption chips.

Types of Security Threats

The most common form of threat is the ability of a hacker to link to an existing network and literally read the data being transmitted. Some types of connections permit disk files to be accessed, e-mail files to be read, data to be modified, and new unwanted data to be added. There are a huge number of specific ways in which data can be read, stolen, compromised, or corrupted. The other common forms of security threats are explained below.

Viruses. A virus is a small program designed to implement some nefarious action in a computer. A virus typically rides along with some other piece of information or program so that it can be surreptitiously inserted into the computer's hard drive or RAM. The virus program is then executed by the processor to do its damage. Any number of viruses have been created over the years and transmitted by e-mails to unsuspecting computers. Sometimes viruses arrive by way of a Trojan horse, a seemingly useful and innocent program that hides the virus. Viruses typically interfere with the operating system, causing it to do unwanted things or not to perform certain functions. Viruses can affect the executable programs on the computer, the file directory, the data files themselves, and the boot programs. Besides making the computer unusable, a virus can erase or corrupt files, cause unknown e-mails to be transmitted, or take other malicious actions. Like a real virus, computer viruses are designed to spread themselves within the computer or to be retransmitted to others in e-mails. These viruses are called *worms* as they automatically duplicate and transmit themselves from network to network and computer to computer.

Spam. A more recent threat, while not actually damaging, is unwanted ads and solicitations via e-mail called spam. Spam clogs up the e-mail system with huge quantities of unwanted data and uses transmission time and bandwidth that could be used in a more productive way. Spam is not illegal, but you must remove the spam yourself, thereby using up valuable time, not to mention memory space in your e-mail system.

Spyware. Spyware is a kind of software that monitors a computer and its user while he or she accesses the Internet or e-mail. It then collects data about how that user uses the Internet such as Internet website access, shopping, etc. It uses this information to send unsolicited ads and spam. Some examples of dangerous practices are the capture of credit card numbers, delivery of unsolicited pop-up ads, and capture of Web-browsing activity and transmission to a person or company for use in unauthorized promotions.

Denial-of-Service (DoS) Attacks. This is a process that transmits errors in the communications protocol and causes the computer to crash or hang up. This type of vandalism doesn't steal information, but it does prevent the user from accessing the operating system, programs, data files, applications programs, or communications links. It is the easiest form of attack that serves no useful purpose other than to hurt others.

One special type of DoS attack is called *smurfing*. A smurf attack usually overwhelms ISP servers with a huge number of worthless packets, thereby preventing other ISP subscribers from using the system. Smurfing makes use of a technique called *pinging*. A ping is the transmission of an inquiry by way of the Internet Control Message Protocol (ICMP) that is a part of TCP/IP to see if a particular computer is connected to the Internet and active. In response to the ping, the computer sends back a message confirming that it is connected. Hackers substitute the ISP's own address for the return message so it gets repeatedly transmitted, thus tying up the system.

Smurfing

Pinging

Security Measures

To protect data and prevent the kinds of malicious hacking described, special software or hardware is used. Here is a brief summary of some of the techniques used to secure a computer system or network.

Encryption and Decryption. Encryption is the process of obscuring information so that it cannot be read by someone else. It involves converting a message to some other form that makes it useless to the reader. Decryption is the reverse process that translates the encrypted message back to readable form.

Encryption has been used for centuries by governments and the military, mainly to protect sensitive material from enemies. Today it is still heavily used by the government and the military but also by companies and individuals as they strive to protect their private information. The Internet has made encryption more important than ever as individuals and organizations send information to one another. For example, encryption ensures that a customer's credit card number is protected in e-commerce transactions (buying items over the Internet). Other instances are automated teller machine (ATM) accesses and sending private financial information. Even digitized voice in a cell phone network can be protected by encryption.

Figure 15-14 shows the basic encryption process. The information or message to be transmitted is called *plaintext*. In binary form, the plaintext is encrypted by using some predetermined computer algorithm. The output of the algorithm is called *ciphertext*. The ciphertext is the secret code that is transmitted. At the receiving end, the reverse algorithm is performed on the ciphertext to generate the original plaintext.

Plaintext

Ciphertext

Most encryption processes combine the plaintext with another binary number called a secret key. The key is used as part of the algorithmic computing process. To translate the ciphertext back to plaintext, the receiving computer must also know the secret key. The strength of the encryption (meaning how secure the data is from deliberate attempts at decryption by brute computing force) is determined by the number of bits in the key. The greater the number of bits, the more difficult the key is to discover.

Figure 15-14 The basic encryption/decryption process.

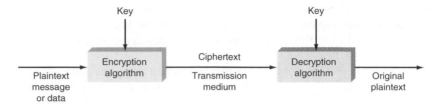

There are two basic types of encryption: *secret key encryption (SKE)*, also called private key encryption, and *public key encryption (PKE)*. SKE is said to be symmetric because both sending and receiving parties must have the same key. The problem with this method lies in sharing the key. How do you transmit or distribute the secret key in a secure manner? This problem led to the development of PKE.

PKE is known as *asymmetric encryption.* It uses two keys, a public key and a private key, in the encryption process. The public key can be openly shared in public. In fact, the public key is sent by the receiver to the transmitting party, and it is used in the encryption process. A secret key is also needed in the decryption process. The original PKE method used two factors of a large prime number for the public and private keys. Other methods are used today.

There are literally dozens of different types of encryption methods. Two of the most common SKEs are the *Data Encryption Standard (DES)* and the *Advanced Encryption Standard (AES)* developed by the National Institute of Standards and Technology (NIST) for the U.S. government. DES uses a 56-bit key for encryption. The key is actually 64 bits or 8 bytes long, where 1 bit of each byte is a parity bit. The remaining eight 7-bit bytes make up the key. The plaintext is encoded or encrypted in 64-bit blocks.

DES was found to be insufficiently secure as the key could actually be discovered by a very fast computer, simply by trying all the related key combinations. This led to the development of 3DES (pronounced triple-dez) which puts the plaintext through three separate sequential DES encryptions, creating a virtually unbreakable code.

The AES algorithm was developed by NIST to replace DES with a method better suited to network use and hardware as well as software implementation. The resulting cipher is known as the Rijndael algorithm, named after one of its creators. It uses 128-, 192-, or 256-bit keys, making it ultrasecure and essentially impossible to break.

There are numerous private key methods. Another one that is often used is Rivest Cipher #4 (RC4), developed by Ron Rivest at the Massachusetts Institute of Technology. The key length can vary from about 40 to 128 bits.

The original public key encryption concept is known as the Diffie-Hellman key exchange, named for its inventors Whitfield Diffie and Martin Hellman. The Diffie-Hellman algorithm uses random number generation and logarithms to create the keys. One of the most commonly used is the RSA method developed by Rivest, Shamir, and Adleman. It uses large prime numbers to generate 512-, 1024-, or 2048-bit keys to ensure maximum protection.

A newer PKE algorithm is the *elliptic curve cryptosystem (ECC).* Its prime advantage is that is uses a smaller 160-bit key that provides equivalent protection of a 1024-bit RSA key, but computation is significantly faster. Large keys make it more secure.

Here is a step-by-step sequence to show how PKE is used.

1. Two parties X and Y wish to communicate. Party X will transmit information to Y.

2. Both X and Y have encryption software that generates both public and private keys.

3. The receiving party Y first transmits the public key to X. This can be done by using nonsecure or unencrypted methods.

4. The transmitting party X then uses the public key to encrypt the message which is sent to Y.

5. Then Y decrypts the message by applying the private key that matches the public key.

Even though a message is encrypted by whatever method, there is no way to be sure that it arrives unmodified. It could be changed during transmission due to some form of interception and attack or just muddled by noise. To prevent this problem, methods of ensuring data integrity have been created. These methods are referred to as *hash functions.* Hash functions are a kind of one-way encryption. They allow you to determine if the original message has been changed in any way during transmission.

Hash functions

What a hashing function or algorithm does is to compress the plaintext of any length into a fixed-length binary number. The hash process may be like a checksum where the bytes of the message are added or XORed together to create a single byte or longer word. Generating a CRC is another similar example. Modern hash functions are more complex to be more secure. The hash function takes plaintext of an arbitry length and maps it to fixed-length blocks. The algorithm is performed on the blocks of data. The result is a digest of the message. The two most commonly used hash functions are designated Message Digest 5 (MD5), which produces a 128-bit hash value, and Secure Hash Algorithm-1 (SHA-1), which generates a 160-bit hash value. SHA-1 is generally more secure, but both are available in just about all security software. A newer SHA-2 further improves security.

Authentication. *Authentication* is the process of verifying that you are who you say you are. It is a way that you or someone you are communicating with really can confirm a true identity. Authentication ensures that the transmitting and receiving parties are really who they say they are and that their identities have not been stolen or simulated. Digital authentication allows computer users to confidently access the Internet, other networks, computers, software, or other resources such as bank accounts if they can verify their identities. Authentication is widely used in most Internet transactions such as e-commerce as it provides a way to control access, keep out unauthorized users, and keep track of those who are using the resources.

Authentication

The most common methods of authentication are the use of passwords or personal identification numbers (PIN). Coded ID cards are another way. More recently, biometric methods of identification are being used as security tightens with more and more transactions. Some common biometric ID methods are fingerprint scans, retinal eye scans, voiceprints, or video facial recognition.

Passwords and PINs are often encrypted before transmission so they cannot be stolen. This is done with the one-way encryption method known as hashing just described. When you key in your password, a hashing algorithm encrypts it and sends it to the authorizing computer, which decrypts it and then compares it to the password you originally created. The process cannot be reversed and so is secure.

The most commonly used process of authentication in network communications is the use of digital certificates. Also known as certificate-based authentication, this method uses hashing and public key encryption to verify identity in various transactions. A digital certificate is a message or document created by a computer that has been "signed" by some trusted authority or third party. Several companies have been established to provide public and private keys for this purpose. Known as a *certification authority (CA),* these organizations issue public keys to those individuals or organizations and vouches for their identity. The digital signature binds the person or company to that public key.

Certification Authority (CA)

Here is the general process for creating a digital signature.

1. The message to be sent is first put through the hash process to produce a digest.
2. The digest is then encrypted by using the sender's private key. The encrypted hash is the signature.
3. That signature is appended as a header (or trailer) to the message to be transmitted. The combination is transmitted.
4. At the receiving end, the signature is decrypted by using the sender' public key. The result is the original hash of the message.
5. The message itself is then put through the same hash function. The decrypted hash and the recreated hash are compared. If the two are the same, authentication is ensured.

Secure Socket Layer (SSL). The processes of encryption/decryption and authentication are used together to ensure secure transactions over the Internet. All these processes are combined into a protocol known as the *Secure Socket Layer* (*SSL*). The resulting process renders the exchange of private information such as credit card numbers safe and secure. Without such a system, use of the Internet would be more limited. E-commerce simply would not exist without the safeguards of SSL. SSL or a more advanced version called *Transport Layer Security* (*TLS*) usually resides in layers 5, 6, or 7 of the OSI model. The SSL or TLS protocol is implemented in the browser software such as Microsoft Internet Explorer (IE) or Netscape. It uses public key and private key encryption and authentication via hashing. SSL was originally developed by Netscape, which invented the browser.

To show how all these techniques are used, here is an example of an e-commerce transaction. This transaction is handled through the browser software as a person we call client *X* accesses the desired website-designated server *Y* via the Internet.

1. Server *X* transmits its public key to client *X*. It is signed by a digital signature as described earlier.
2. The client then generates a secret key.
3. Client *X* uses the public key to encrypt the secret key, which is sent to server *Y*.
4. Client *X* encrypts the message, using private key encryption methods, and sends the message to server *Y*.
5. Server *Y* decrypts the private key previously sent and then uses it to decrypt the message.
6. Hashing and digital signatures are used throughout the process to ensure identity.

The encryption, decryption, and authentication processes are computationally intense. Even with very high-speed computers, the encryption, decryption, and hashing algorithms take considerable time to execute. The longer the key, e.g., the longer the calculation time. This adds to the time of transmission via the Internet. The result can be a significant delay. This is the price to be paid for secure transmissions. Since most of these processes are software, using a faster computer will speed up the process. In more recent systems, hardware is used to speed up the process. Most security algorithms can be implemented in hardware. This hardware can be special processors or logic chips dedicated to the purpose. All significantly speed up the security measures.

Firewalls. A *firewall* is a piece of software that monitors transmissions on a network and inspects the incoming information to see if it conforms to a set of guidelines established by the software or the organization or person owning the network. The firewall controls the flow of traffic from the Internet to a LAN or PC or between LANs or other networks.

The most common type of firewall operates at the network layer in the OSI model. It examines TCP/IP packets and acts as a filter to block access from inputs that do not match a set of rules set up in the firewall. The firewall screens packets for specific IP sources or destinations, packet attributes, domain name, or other factors. Firewalls are the first line of defense against intrusions by unwanted sources. Today, any computer connected to the Internet should have a firewall. These are available as a software program loaded into a PC that screens according to the guidelines set up by the software producer. Some operating systems such as Microsoft Windows now come with a built-in firewall. More sophisticated firewalls are available for LANs and other networks. These usually can be configured by the network administrator to filter on special rules as needed by the organization.

Antivirus, Antispam, and Antispyware Software. There are commercial programs designed to be installed on a computer to find and eliminate these security problems. The antivirus and antispyware programs scan all files on the hard drive either automatically or on command, to look for viruses. The antivirus software looks for a pattern of code unique to each virus, and when it is identified, the software can remove the virus or in some cases quarantine and isolate the infected file so that it does no harm. Antispyware works the same way by scanning all files, searching for patterns that designate a spyware program. It then removes the program.

Antispam software is typically set up to monitor incoming e-mail traffic and look for clues to whether it is legitimate e-mail or spam. It then blocks the spam from the e-mail inbox and places it in a special bulk e-mail file. You will never see the spam unless your e-mail system allows you to look in a special bulk file normally furnished by the e-mail provider. Antispam software is not perfect, and because of its rules for blocking spam it can also affect desired e-mail. It is worthwhile to examine the bulk files occasionally to be sure legitimate e-mails are getting through. Most antispam programs allow you to change the filtering rules to ensure you get all desired mail while the real spam is rejected.

Virtual Private Network (VPN). One way to achieve security on a LAN is to use software measures to block off segments of a network or create a subnetwork using software to assign access only to authorized users. This is referred to as a *virtual LAN* or *VLAN*. Security can also be achieved when you are connecting two remote LANs by using a leased line. The lease line, such as a T1 or T3 connection, is totally dedicated to just the connection between the LANs. No one else has access. While this works well, it is very expensive. A popular alternative is to create a secure connection through the Internet by using a *virtual private network* (*VPN*). In a VPN, the data to be transmitted is encrypted, encapsulated in a special packet, and then sent over the Internet.

Virtual LAN

Virtual private network (VPN)

VPNs use one of two special protocols for the encapsulation and encryption process. One of these is IPsec (Internet Protocol security), a protocol created and supported by the Internet Engineering Task Force. IPsec encrypts the data along with the TCP header and then adds another header that identifies the kind of encryption used plus a trailer that contains the authentication. An IP header is added to form the datagram or packet to be transmitted.

Next, this datagram is encrypted and encapsulated in one additional IP datagram which is also encrypted. The combined packet is transmitted. This process is referred to as *tunneling*. The packet containing the message is encrypted, which in turn is wrapped in a second IP packet and encrypted again. This in effect forms a secure tunnel through the otherwise insecure Internet. The routers at the sending and receiving ends of the VPN sort out all the source and destination addresses for the proper delivery of the data.

Tunneling

While IPsec has been widely used for VPNs, it is gradually being replaced by SSL, which is also an IETF standard. Both IPsec and SSL are usually implemented in software, but hardware versions are available that greatly speed up transactions.

Wireless Security. Security in wireless systems is important because it is relatively easy to capture a radio signal containing important information. A directional antenna and sensitive receiver designed for the specific wireless service, such as a wireless LAN and a computer, are all you need. Wireless data can be protected by encryption, and a number of special methods have been developed especially for wireless systems. These are discussed later in the wireless chapters, Chaps. 20 and 21.

Internet security is a very broad and complex subject that is far beyond the scope of this chapter. It is one of the most critical and fastest-growing segments of the networking industry.

CHAPTER REVIEW

Summary

The Internet is a global network made up of multiple networks interconnected. This "open" network permits anyone to communicate with any one else by way of email, websites and file transfers. Today, business is conducted over the Internet and the search capability has become one of the most useful features. Links and connections are made by using Internet addresses called uniform resource locators (URLs) and email addresses. URLs use a suffix called a domain name to indicate the type of website.

Data transmission over the Internet is via packet switching. Most initial links are by way of some broadband connection such as that provided by an Internet service provider (ISP). The most typical connection is cable TV or a DSL line although some links are made via dial-up modems over the phone lines. The data is eventually transmitted over the Internet backbone, a complex interconnection of high speed fiber optic networks. The most common protocols used are the asynchronous transfer mode (ATM) and the synchronous optical network (SONET). Frame relay is also used. Typical back bone speeds are 155,622 Mbps and 2.5 or 10 Gbps.

The key piece of equipment of the Internet is the router. This device examines the destination addresses of the messages and compares them to routing tables of addresses in the router to determine the optimum path for the message. The protocol used is the transmission control protocol and the Internet protocol (TCP/IP) both packet-based. Dedicated computers known as servers store, forward and receive the data via the routers. Internet protocol addresses are based on a 32-bit word and it is formatted into what is called dotted decimal where each 8-bit segment is expressed as a decimal number from 0 to 255 creating a four number address such as 168.42.93.57.

Storage area networks (SANs) are a relatively recent addition to computer systems as the need for massive storage created by the Internet information has grown. These large arrays of disk drives are usually connected to servers and the Internet by way of Fibre Channel (FC), a fast fiber optical network designed for SANs. Another system is an extension of the older Small Computer Systems Interface (SCSI) parallel interface for storage. The newer version is called iSCSI and uses faster serial data transfers and standard Ethernet interfaces.

Internet security is a real problem today and there are many threats to the data and messages sent. Viruses, worms, spam, spyware and denial of service attacks are common. Special hardware and software have been developed to deal with these problems and provide security. Some of these include data encryption, authentication, passwords and PIN numbers, and the secure socket layer (SSL). Firewalls and antivirus, antispam, and antispyware software also provide an added level of security. Virtual private networks (VPNs) let you use the Internet securely as well.

Questions

1. List five common applications of the Internet.
2. In the Internet address *www.qxrj.net,* what is the type of domain and what is the host?
3. What symbol or character designates an e-mail address?
4. Name the three most common ways that a person accesses the Internet from home.
5. What do you call the company that provides an Internet connection to subscribers?
6. What piece of software is needed to access the World Wide Web?
7. What is the name of the transmission system that sends data in packets as large as 5002 octets by using dedicated telephone lines?
8. What is the size of the packet used in ATM systems?
9. What piece of equipment is used in the transmission of ATM packets?
10. What is the transmission medium of most ATM systems?
11. What does SONET mean?
12. Name the two most common applications for SONET.
13. What network topologies are common with SONET?
14. What are the designation and speed of the fastest SONET connection?
15. What controls the data transfer in a SONET system?
16. Name the equipment used to get data into and out of a SONET system.
17. What is the total number of bytes or octets in a SONET frame?
18. True or false? SONET can be used to carry packetized data.
19. What specifically does a router look for during packet transmission?
20. How does a router know where to send a packet?
21. What is the name of the basic input/output interfaces used in routers?
22. What is the name of the circuit that lets any input be connected to any output in a router?
23. True or false? A PC with software can be used as a router.
24. What is the name of the circuit used for serial-to-parallel conversion?
25. What is the name given to the fiber-optic network making up the Internet?
26. What WAN transmission media and protocol are used in the Internet?
27. Explain the process of packet switching.
28. What is the name of the equipment used to inspect every packet transmitted over the Internet?
29. What software protocol prepares packets for transmission over the Internet?
30. What software protocol actually transmits the packets?
31. What software makes sure that any packets transmitted out of sequence get reassembled correctly?
32. Is TCP actually called into play during a packet's transmission from source to destination? Explain.
33. Name the two basic speeds of the Internet backbone.
34. What name is given to the format of an IP address?

35. How many different classes of IP addresses are there?
36. What devices are identified by IP addresses in classes A, B, and C?
37. In the IP address 133.46.182.9, identify the class, network ID, and host ID.
38. What is a mask?
39. What logical process is used with the mask and the IP address?
40. What is a subnet?
41. What are the two names given to clusters of disk drives used for massive storage?
42. What is the name given to external disk drives connected to a PC or server via the SCSI?
43. What is SCSI? What is its nickname?
44. Name the two kinds of large-scale storage system networked to PCs. What is the main difference between them?
45. In a SAN, what is the name of the interface and protocol used in connecting the disk drive systems to the servers?
46. What is the physical medium used in an FC system?
47. What are the speeds of transmission of an FC system?
48. Name the interface used to attach a server or disk system to the SAN.
49. How are servers and disk systems connected to one another in a SAN?
50. What is the name of the SAN interface used in place of FC?
51. What is the advantage of the SAN interface in Question 43 over FC? Its disadvantage?

52. What protocol and medium are commonly used with iSCSI?
53. List four common Internet security threats.
54. What is the name given to a virus that replicates itself?
55. True or false? Spam and spyware are illegal.
56. What is spyware?
57. How do you detect and get rid of spam, spyware, and viruses?
58. Name the two broad types of encryption and decryption.
59. What are the names given to the input and output of an encryption algorithm?
60. What is a key?
61. What is the primary problem with secret key encryption?
62. How does public key encryption work?
63. Name four specific methods of secret key encryption.
64. Name two specific methods of public key encryption.
65. What approach is used to make encryption stronger and more resistant to brute-force computing efforts to break it?
66. What is a hash function? Name the two most common types.
67. What is the purpose of a digital signature?
68. What protocol is used for authentication and encryption in e-commerce transactions?
69. What is a firewall?
70. Explain the operation of a VPN.
71. What two protocols are used to establish a VPN?
72. True or false? Security procedures may also be implemented in hardware.

Problems

1. What is the shortest time for transmitting 50 ATM packets at 2.5 Gbps?
2. How many bits does it take to represent an IP address of 124.76.190.38?
3. Convert the IP address 222.155.8.17 to binary and hexadecimal.
4. Logically AND the number 125 with 128, 0, 15, and 240. Express the result in decimal.

Critical Thinking

1. Besides its use in SANs, where else could Fibre Channel be used?
2. How many keys are needed in secret key encryption? Public key encryption?
3. Can a RAID be accessed by a PC over the Internet? Explain.
4. Suggest some ways to avoid wireless security problems without encryption.

chapter 16

Microwave Communication

As the use of electronic communication has increased over the years, the frequency spectrum normally used for radio signals has become extremely crowded. In addition, there has been an increasing need for more spectrum space to carry wider-bandwidth video and digital information. One of the primary solutions to this problem has been to move more radio communication higher in the spectrum, specifically the microwave range, the 1- to 300-GHz range.

In the past, because of the difficulty of generating, transmitting, and receiving microwave signals, only those who truly needed and could afford the special equipment used this part of the spectrum. Today, thanks to advances in semiconductor technology, microwaves are now being more widely used. This has opened the microwave spectrum to all sorts of new services which include wireless local-area networks, cellular and cordless phones, digital satellite radio, and wireless broadband, to name a few. In this chapter we take a look at some of the components and techniques used in modern microwave radios.

Objectives

After completing this chapter, you will be able to:

- Explain the reasons for the growing use of microwaves in communications.
- Identify the circuits that require the use of special microwave components.
- Define the term *waveguide,* explain how a waveguide works, and calculate the cutoff frequency of a waveguide.
- Explain the purpose and operation of direction couplers, circulators, isolators, T sections, cavity resonators, and microwave vacuum tubes.
- Describe the operation of the major types of microwave diodes.
- Name five common types of microwave antennas and calculate the gain and beam width of horn and parabolic dish antennas.
- Explain the basic concepts and operation of pulsed and Doppler radar.

16-1 Microwave Concepts

Microwaves are the ultrahigh, superhigh, and extremely high frequencies directly above the lower frequency ranges where most radio communication now takes place and below the optical frequencies that cover infrared, visible, and ultraviolet light. The outstanding benefits for radio communication of these extremely high frequencies and accompanying short wavelengths more than offset any problems connected with their use. Today, most new communication services and equipment use microwaves.

Microwave

Microwave Frequencies and Bands

The practical microwave region is generally considered to extend from 1 to 30 GHz, although some definitions include frequencies up to 300 GHz. Microwave signals of 1 to 30 GHz have wavelengths of 30 cm (about 1 ft) to 1 cm (or about 0.4 in).

The microwave frequency spectrum is divided up into groups of frequencies, or bands, as shown in Fig. 16-1. Frequencies above 40 GHz are referred to as *millimeter waves* because their wavelength is only millimeters (mm). Note that parts of the L and S bands overlap part of the UHF band, which is 300 to 3000 MHz. Frequencies above 300 GHz are in the *submillimeter band*. Currently the only communication in either the millimeter or submillimeter ranges is for research and experimental activities.

Millimeter wave

Submillimeter wave

Benefits of Microwaves

Every electronic signal used in communication has a finite bandwidth. When a carrier is modulated by an information signal, sidebands are produced. The resulting signal occupies a certain amount of bandwidth, called a *channel*, in the radio-frequency (RF) spectrum. Channel center frequencies are assigned in such a way that the signals using

Channel

Figure 16-1 Microwave frequency bands.

Band designation	Frequency range
L band	1 to 2 GHz
S band	2 to 4 GHz
C band	4 to 8 GHz
X band	8 to 12 GHz
K_u band	12 to 18 GHz
K band	18 to 26.5 GHz
K_a band	26.5 to 40 GHz
Q band	30 to 50 GHz
U band	40 to 60 GHz
V band	50 to 75 GHz
E band	60 to 90 GHz
W band	75 to 110 GHz
F band	90 to 140 GHz
D band	110 to 170 GHz
Submillimeter	>300 GHz

Millimeter waves

each channel do not overlap and interfere with signals in adjacent channels. As the number of communication signals and channels increases, more and more of the spectrum space is used up. Over the years as the need for electronic communication has increased, the number of radio communication stations has increased dramatically. As a result, the radio spectrum has become extremely crowded.

Use of the radio-frequency spectrum is regulated by the federal government. In the United States, this job is assigned to the Federal Communications Commission (FCC). The FCC establishes various classes of radio communication and regulates the assignment of spectrum space. For example, in radio and TV broadcasting, certain areas of the spectrum are set aside, and frequency assignments are given to stations. For two-way radio communication, other portions of the spectrum are used. The various classes of radio communication are assigned specific areas in the spectrum within which they can operate. Over the years, the available spectrum space, especially below 300 MHz, has essentially been used up. In many cases, communication services must share frequency assignments. In some areas, new licenses are no longer being granted because the spectrum space for that service is completely full. In spite of this, the demand for new electronic communication channels continues. The FCC must, on an ongoing basis, evaluate users' needs and demands and reassign frequencies as necessary. Many compromises have been necessary.

Technological advances have helped solve some problems connected with overcrowding. For example, the selectivity of receivers has been improved so that adjacent channel interference is not as great. This permits stations to operate on more closely spaced frequencies.

On the transmitting side, new techniques have helped squeeze more signals into the same frequency spectrum. A classic example is the use of SSB, where only one sideband is used rather than two, thereby cutting the spectrum usage in half. Limiting the deviation of FM signals also helps to reduce bandwidth. In data communication, new modulation techniques such as PSK and QAM have been used to narrow the required bandwidth of transmitted information or to transmit at higher speeds in narrower bandwidths. Digital compression methods also transmit more information through a narrow channel. Multiplexing techniques help put more signals or information into a given bandwidth. Broadband schemes such as spread spectrum and *orthogonal frequency-division multiplexing (OFDM)* allow many radios to share a single bandwidth.

The other major approach to solving the problem of spectrum crowding has been to move into the higher frequency ranges. Initially, the VHF and UHF bands were tapped. Today, most new communication services are assigned to the microwave region.

To give you some idea why more bandwidth is available at the higher frequencies, let's take an example. Consider a standard AM broadcast station operating on 1000 kHz. The station is permitted to use modulating frequencies up to 5 kHz, thus producing upper and lower sidebands 5 kHz above and below the carrier frequency, or 995 and 1005 kHz. This gives a maximum channel bandwidth of $1005 - 995 = 10$ kHz. This bandwidth represents $10/1000 = 0.01$ or 1 percent of the spectrum space at that frequency.

Now consider a microwave carrier frequency of 4 GHz. One percent of 4 GHz is $0.01 \times 4{,}000{,}000{,}000 = 40{,}000{,}000$ or 40 MHz. A bandwidth of 40 MHz is incredibly wide. In fact, it represents all the low-frequency, medium-frequency, and high-frequency portions of the spectrum plus 10 MHz. This is the space that might be occupied by a 4-GHz carrier modulated by a 20-MHz information signal. Obviously, most information signals do not require that kind of bandwidth. A voice signal, e.g., would take up only a tiny fraction of that. A 10-kHz AM signal represents only $10{,}000/4{,}000{,}000{,}000 = 0.00025$ percent of 4 GHz. Up to 4000 AM broadcast stations with 10-kHz bandwidths could be accommodated within the 40-MHz (1 percent) bandwidth.

Obviously, then, the higher the frequency, the greater the bandwidth available for the transmission of information. This not only gives more space for individual stations, but also allows wide-bandwidth information signals such as video and high-speed digital data to be accommodated. The average TV signal has a bandwidth of approximately 6 MHz. It is impractical to transmit video signals on low frequencies because they use up entirely too much spectrum space. That is why most TV transmission is in the VHF and UHF ranges. There is even more space for video in the microwave region.

Wide bandwidth also makes it possible to use various multiplexing techniques to transmit more information. Multiplexed signals generally have wide bandwidths, but these can be easily handled in the microwave region. Finally, transmission of high-speed binary information often requires relatively wide bandwidths, and these are also easily transmitted on microwave frequencies.

Disadvantages of Microwaves

The higher the frequency, the more difficult it becomes to analyze electronic circuits. The analysis of electronic circuits at lower frequencies, say, those below 30 MHz, is based upon current-voltage relationships (circuit analysis). Such relationships are simply not usable at microwave frequencies. Instead, most components and circuits are analyzed in terms of electric and magnetic fields (wave analysis). Thus techniques commonly used for analyzing antennas and transmission lines can also be used in designing microwave circuits. Measuring techniques are, of course, also different. In low-frequency electronics, currents and voltages are calculated. In microwave circuits, measurements are of electric and magnetic fields. Power measurements are more common than voltage and current measurements.

Another problem is that at microwave frequencies, conventional components become difficult to implement. For example, a common resistor that looks like pure resistance at low frequencies does not exhibit the same characteristics at microwave frequencies. The short leads of a resistor, although they may be less than an inch, represent a significant amount of inductive reactance at very high frequencies. A small capacitance also exists between the leads. These small stray and distributed reactances are sometimes called *residuals*. Because of these effects, at microwave frequencies a simple resistor looks like a complex *RLC* circuit. This is also true of inductors and capacitors. Figure 16-2 shows equivalent circuits of components at microwave frequencies.

To physically realize resonant circuits at microwave frequencies, the values of inductance and capacitance must be smaller and smaller. Physical limits become a problem. Even a 0.5-in piece of wire represents a significant amount of inductance at microwave frequencies. Tiny surface-mounted chip resistors, capacitors, and inductors have partially solved this problem. Furthermore, as integrated-circuit dimensions have continued to decrease, smaller and smaller on-chip inductors and capacitors have been made successfully.

Another solution is to use distributed circuit elements, such as transmission lines, rather than lumped components, at microwave frequencies. When transmission lines are cut to the appropriate length, they act as inductors, capacitors, and resonant circuits. Special versions of transmission lines known as striplines, microstrips, waveguides, and cavity resonators are widely used to implement tuned circuits and reactances.

In addition, because of inherent capacitances and inductances, conventional semiconductor devices such as diodes and transistors simply will not function as amplifiers, oscillators, or switches at microwave frequencies.

Figure 16-2 Equivalent circuits of components at microwave frequencies. (*a*) Resistor. (*b*) Capacitor. (*c*) Inductor.

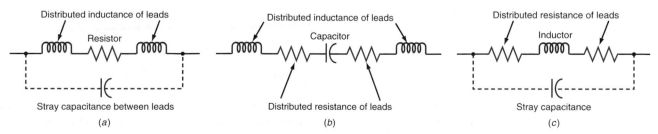

(*a*)   (*b*)   (*c*)

Another serious problem is transistor *transit time*—the amount of time it takes for the current carriers (holes or electrons) to move through a device. At low frequencies, transit times can be neglected; but at microwave frequencies, they are a high percentage of the actual signal period.

This problem has been solved by designing smaller and smaller microwave diodes, transistors, and ICs and using special materials such as gallium arsenide (GaAs), indium phosphide (InP), and silicon germanium (SiGe) in which transit time is significantly less than in silicon. In addition, specialized components have been designed for microwave applications. This is particularly true for power amplification, where special vacuum tubes known as klystrons, magnetrons, and traveling-wave tubes are the primary components used for power amplification.

Another problem is that microwave signals, as do light waves, travel in perfectly straight lines. This means that the communication distance is usually limited to line-of-sight range. Antennas must be very high for long-distance transmission. Microwave signals penetrate the ionosphere, so multiple-hop communication is not possible.

Microwave Communication Systems

Like any other communication system, a microwave communication system uses transmitters, receivers, and antennas. The same modulation and multiplexing techniques used at lower frequencies are also used in the microwave range. But the RF part of the equipment is physically different because of the special circuits and components that are used to implement the components.

Microwave transmitter

Transmitters. Like any other transmitter, a *microwave transmitter* starts with a carrier generator and a series of amplifiers. It also includes a modulator followed by more stages of power amplification. The final power amplifier applies the signal to the transmission line and antenna. The carrier generation and modulation stages of a microwave application are similar to those of lower-frequency transmitters. Only in the later power amplification stages are special components used.

Figure 16-3 shows several ways that microwave transmitters are implemented. The special microwave stages and components are shaded. In the transmitter circuit shown in Fig. 16-3(a) a microwave frequency is first generated in the last multiplier stage. The operating frequency is 1680 MHz, where special microwave components and techniques must be used. Instead of tuned circuits made of loops of wire for inductors and discrete capacitors, microstrip transmission lines are used as tuned circuits and as impedance-matching circuits. SAW filters are the most commonly used filters in low-power circuits. One or more additional power amplifiers are then used to boost the signal to the desired power level. Both bipolar and MOSFET microwave power transistors are available that give power levels up to about 100 W or more. When FM is used, the remaining power amplifiers can also be class C, which provides maximum efficiency. If more power is desired, several transistor power amplifiers can be paralleled, as in Fig. 16-3(a).

If AM is used in a circuit like that in Fig. 16-3(a), an amplitude modulator can be used to modulate one of the lower-power amplifier stages after the multiplier chain. When this is done, the remaining power amplifier stages must be linear amplifiers to preserve the signal modulation.

For very high-power output levels—beyond several hundred watts—a special amplifier must be used, e.g., the klystron.

Figure 16-3(b) shows another possible transmitter arrangement, in which a mixer is used to up-convert an initial carrier signal with or without modulation to the final microwave frequency.

The synthesizer output and a microwave local oscillator signal are applied to the mixer. The mixer then translates the signal up to the desired final microwave frequency. A conventional crystal oscillator using fifth-overtone VHF crystals followed by a chain of frequency multipliers can be used to develop the local oscillator frequency. Alternatively, one

Figure 16-3 Microwave transmitters. (*a*) Microwave transmitter using frequency multipliers to reach the microwave frequency. The shaded stages operate in the microwave region. (*b*) Microwave transmitter using up-conversion with a mixer to achieve an output in the microwave range.

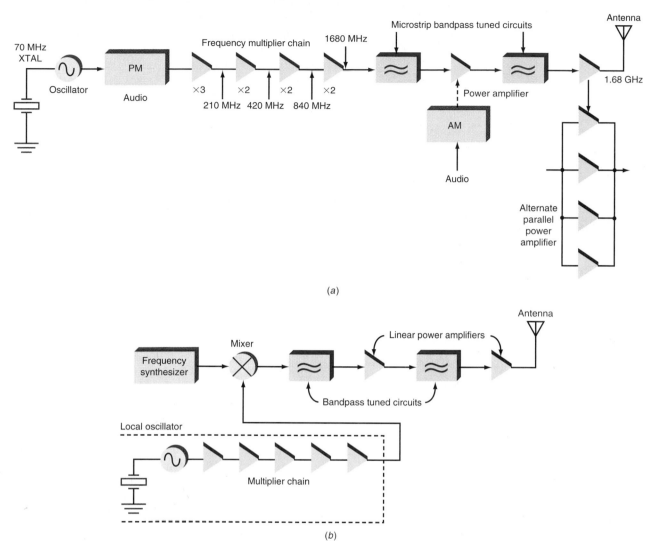

(*a*)

(*b*)

of several special microwave oscillators could be used, e.g., a Gunn diode, a microwave semiconductor in a cavity resonator, or a dielectric resonator oscillator.

The output of the mixer is the desired final frequency at a relatively low power level, usually tens or hundreds of milliwatts at most. Linear power amplifiers are used to boost the signal to its final power level. At frequencies less than about 10 GHz, a microwave transistor can be used. At the higher frequencies, special microwave power tubes are used.

The tuned bandpass circuits shown in Fig. 16-3(*b*) can be microstrip transmission lines when transistor circuits are used, or cavity resonators when the special microwave tubes are used.

Modulation could occur at several places in the circuit in Fig. 16-3(b). An indirect FM phase modulator might be used at the output of the frequency synthesizer; for some applications, a PSK modulator would be appropriate.

Receivers. *Microwave receivers,* like low-frequency receivers, are the superheterodyne type. Their front ends are made up of microwave components. Most receivers use double conversion. A first down-conversion gets the signal into the UHF or VHF range, where it can be more easily processed by standard methods. A second conversion

Microwave receivers

Each of these microwave antennas located in Stephen Butte, Washington, is specifically designed for long-distance telephone signal transmission. They are covered to reduce weather-related maintenance problems.

Low-noise amplifier (LNA)

reduces the frequency to an IF appropriate for the desired selectivity.

Figure 16-4 is a general block diagram of a double-conversion microwave receiver. The antenna is connected to a tuned circuit, which could be a cavity resonator or a microstrip or stripline tuned circuit. The signal is then applied to a special RF amplifier known as a *low-noise amplifier (LNA)*. Special low-noise transistors, usually gallium arsenide FET amplifiers, must be used to provide some initial amplification. Another tuned circuit connects the amplified input signal to the mixer. Most mixers are of the doubly balanced diode type, although some simple single-diode mixers are also used.

The local oscillator signal is applied to the mixer. The mixer output is usually within the UHF or VHF range. The 700- to 800-MHz range is typical. A SAW filter selects out the difference signal, which in Fig. 16-4 is 12 GHz − 11.2 GHz = 0.8 GHz, or 800 MHz.

The remainder of the receiver is typical of other superheterodynes. Note that the desired selectivity is obtained with a SAW filter, which is sometimes used to provide a specially shaped IF response. Many of the newer microwave cell phone and LAN receivers are of the direct conversion type, and selectivity is obtained with *RC* and DSP filters.

In more recent microwave equipment such as cell phones and wireless networking interfaces, the microwave frequencies are generated by a phase-locked loop (PLL) operating as a multiplier. See Fig. 16-5. The VCO produces the desired local oscillator (LO) or final transmitting frequency directly. The VCO frequency is controlled by the phase detector and its low-pass loop filter. The frequency divider and input crystal determine the output frequency. Recent advances in quartz crystal design permit input oscillator frequencies up to 200 MHz. In Fig. 16-5, the 155-MHz crystal combined with a frequency divider of 20 produces a VCO output at $155 \times 20 = 3100$ MHz, or 3.1 GHz. The ÷20 divider reduces the 3.1-GHz output to 155 MHz to match the input crystal signal at the phase detector, as required for closed-loop control. Of course, the divider can also be part of a microcontroller-based frequency synthesizer that is designed to permit setting the output to multiple channel frequencies as required by the application.

Transmission Lines. The transmission line most commonly used in lower-frequency radio communication is coaxial cable. However, coaxial cable has very high attenuation at microwave frequencies, and conventional cable is unsuitable for

Figure 16-4 A microwave receiver. The shaded areas denote microwave circuits.

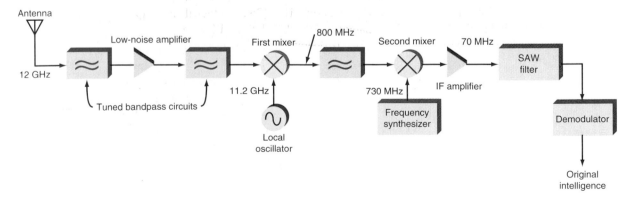

Figure 16-5 A phase-locked loop (PLL) multiplier is the primary signal source in modern microwave transceivers.

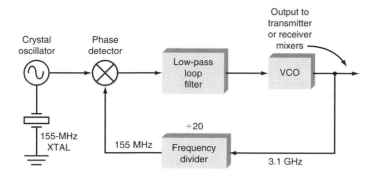

LNAs make use of these transistors from Agilent. These transistors, called PHEMTs, are enhancement-mode "pseudomorphic high-electron-mobility" transistors.

carrying microwave signals except for very short runs, usually several feet or less. Newer types of coaxial cables permit lengths of up to 100 ft at frequencies to 10 GHz.

Special microwave coaxial cable that can be used on the lower microwave bands—L, S, and C—is made of hard tubing rather than wire with an insulating cover and a flexible braid shield. The stiff inner conductor is separated from the outer tubing with spacers or washers, forming low-loss coaxial cable known as *hard line* cable. The insulation between the inner conductor and the outer tubing can be air; in some cases a gas such as nitrogen is pumped into the cable to minimize moisture buildup, which causes excessive power loss. This type of cable is used for long runs of transmission lines to an antenna on a tower. At higher microwave frequencies, C band and upward, a special hollow rectangular or circular pipe called *waveguide* is used for the transmission line (see Sec. 16-3).

Antennas. At low microwave frequencies, standard antenna types, including the simple dipole and the one-quarter wavelength vertical antenna, are still used. At these frequencies antenna sizes are very small; e.g., the length of a half-wave dipole at 2 GHz is only about 3 in. A one-quarter wavelength vertical antenna for the center of the C band is only about 0.6 in long. At the higher frequencies, special antennas are generally used (see Sec. 16-6).

Hard line

Waveguide

Microwave transistor amplifier

Monolithic microwave integrated circuits (MMICs)

Microstrip tuned circuit

16-2 Microwave Lines and Devices

Today, although vacuum tubes and microwave tubes such as the klystron and magnetron are still used, especially for higher-power applications, most microwave systems use *transistor amplifiers*. Over the years, semiconductor manufacturers have learned to make transistors work at these higher frequencies. Special geometries are used to make bipolar transistors that provide both voltage and power gain at frequencies up to 10 GHz. Microwave FET transistors have also been created such as the MESFET described in Chap. 9. The use of gallium arsenide (GaAs) and silicon germanium (SiGe) rather than pure silicon has further increased the frequency capabilities of MOSFETs and bipolars. Both small-signal and power MOSFETs are available to operate at frequencies up to about 10 GHz. Since most microwave communication activity takes place in the lower-frequency ranges (L, S, and C bands), transistors are the primary active components used.

In the following sections, we discuss both discrete transistor types with microstrip tuned circuits and *monolithic microwave integrated circuits (MMICs)*.

Microstrip Tuned Circuits

Before specific amplifier types are introduced, it is important to examine the method by which tuned circuits are implemented in microwave amplifiers. Lumped components such

Microwave Communication   **611**

Figure 16-6 Microstrip transmission line used for reactive circuits. (*a*) Perspective view. (*b*) Edge or end view. (*c*) Side view (open line). (*d*) Side view (shorted line).

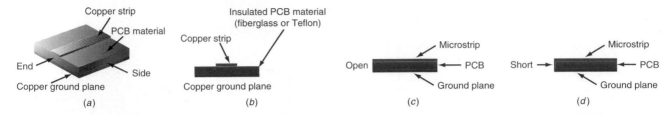

Microstrip

as coils and capacitors are still used in some cases at the high UHF and low microwave frequencies (below about 2 GHz) to create resonant circuits, filters, or impedance-matching circuits. However, at higher frequencies, standard techniques for realizing such components become increasingly harder to implement. Instead, transmission lines, specifically *microstrip*, are used.

Chapter 13 described how transmission lines can be used as inductors and capacitors as well as series and parallel resonant circuits. These are readily implemented at the microwave frequencies because one-half or one-quarter wavelength transmission lines are only inches or some fraction thereof at those frequencies. Microstrip is preferred for reactive circuits at the higher frequencies because it is simpler and less expensive than stripline, but stripline is used where shielding is necessary to minimize noise and cross talk. The tuned circuits are created by using a copper pattern *printed-circuit board (PCB)* upon which are mounted the transistors, ICs, and other components of the circuit.

Figure 16-6 shows several views of microstrip transmission line used for a reactive circuit. The PCB is usually made of G-10 or FR-4 fiberglass or a combination of fiberglass and Teflon. The bottom of the PCB is a thin solid copper sheet which serves as a ground plane and one side of the transmission line. The copper strip is the other conductor of the transmission line.

Figure 16-6(*a*) is a perspective view of a microstrip line; Fig. 16-6(*b*) is an end view; and Fig. 16-6(*c*) and (*d*) are side views. Both open and shorted segments of line can be used, although shorted segments are preferred because they do not radiate as much as open segments. One-quarter wavelength sections are preferred because they are shorter and take up less space on the PCB. Figure 16-7 summarizes the open- and shorted-line possibilities for microstrip.

An important characteristic of microstrip is its impedance. As discussed in Chap. 13, the characteristic impedance of a transmission line depends on its physical characteristics, in this case, e.g., on the width of the strip and the spacing between the strip and the copper ground plane, which is the thickness of the PCB material. The dielectric constant of the insulating material is also a factor. Most characteristic impedances are

Figure 16-7 Equivalent circuits of open and shorted microstrip lines.

LENGTH	SHORTED LINES	OPEN LINES
Less than $\lambda/4$	Inductor	Capacitor
$\lambda/4$	Parallel resonant or open circuit	Series resonant or short circuit
$>\lambda/4, <\lambda/2$	Capacitor	Inductor
$\lambda/2$	Series resonant circuit or short repeater	Parallel resonant circuit or open repeater

Figure 16-8 How a one-quarter wavelength microstrip can transform impedances and reactances.

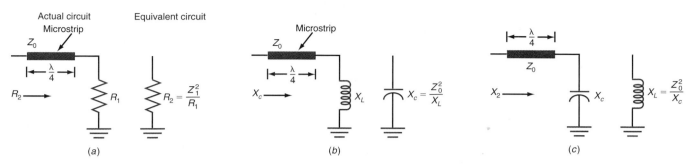

(a)   (b)   (c)

less than 100 Ω; 50 Ω is the most common, followed closely by 75 Ω. Values higher than 100 Ω are used for cases when impedance-matching requirements demand it.

One-quarter wavelength transmission line can be used to make one type of component look like another. For example, in Fig. 16-8(a), the $\lambda/4$ microstrip line can make a resistor at one end look like a resistance of another value; specifically,

$$R_2 = \frac{Z_0^2}{R_1}$$

Here R_1 is the resistance value of the resistor connected to one end of the line, and Z_0 is the characteristic impedance of the microstrip, such as 50 Ω. If R_1 is the characteristic impedance of the microstrip, such as 50 Ω, the line is matched and the generator sees 50 Ω. If R_1 is 150 Ω, then the other end of the line will have a value of R_2, or

$$\frac{50^2}{150} = \frac{2500}{150} = 16.67 \ \Omega$$

This application is the same as the one-quarter wavelength matching transformer or Q section discussed in Chap. 14. Recall that two impedances can be matched by using a line of length $\lambda/4$ according to the relationship $Z_0 = \sqrt{R_1 R_2}$, where Z_0 is the characteristic impedance of the one-quarter wavelength line and R_1 and R_2 are the input and output impedances to be matched.

A quarter-wavelength line can also make a capacitor look like an inductor or an inductance look like a capacitance [see Fig. 16-8(b) and (c)]. For example, a 75-Ω microstrip $\lambda/4$ long will make an inductive reactance of 30 Ω look like a capacitive reactance of

$$X_c = \frac{Z_0^2}{X_l} = \frac{75^2}{30} = \frac{5635}{30} = 187.5 \ \Omega$$

Figure 16-9 shows the physical configurations for equivalent coils and capacitors in microstrip form. The thin segment of microstrip shown in Fig. 16-9(a) acts like a series inductor. Figure 16-9(b) shows a short right-angle segment whose end is grounded; this microstrip acts as a parallel inductor. When series capacitance or capacitive coupling is needed, a small capacitor can be created by using the ends of microstrip lines as tiny capacitor plates separated by an air dielectric [Fig. 16-9(c)]. A shunt capacitor can be created by using a short, fat segment of microstrip as in Fig. 16-9(d). As these general forms demonstrate, it is often possible to visualize or even draw the equivalent circuit of a microstrip amplifier by observing the pattern on the PCB.

Microstrip can also be used to realize coupling from one circuit, as illustrated in Fig. 16-10. One microstrip line is simply placed parallel to another segment of microstrip. The degree of coupling between the two depends on the distance of separation and the

Figure 16-9 Common microstrip patterns and their equivalents.

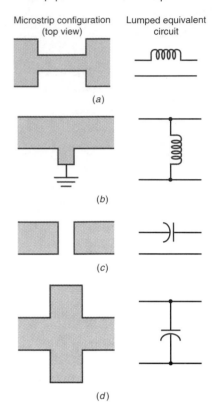

Microstrip configuration (top view)   Lumped equivalent circuit

(a)

(b)

(c)

(d)

length of the parallel segment. The closer the spacing and the longer the parallel run, the greater the coupling. There is always signal loss by such a coupling method, but it can be accurately controlled.

Although microstrip performs best when it is a straight line, 90° turns are often necessary on a PCB. When turns must be used, a straight right-angle turn, like that shown in Fig. 16-11(a), is forbidden because it acts as a low-pass filter across the line. A gradually curved line, like that shown in Fig. 16-11(b) (or Fig. 16-10), is preferred when the turn radius is much greater than the width of the line. An acceptable alternative method is shown in Fig. 16-11(c). The cut on the corner is critical. Note that the dimensions must be held to $\lambda/4$, which is one-quarter the width of the microstrip.

A special form of microstrip is the *hybrid ring* shown in Fig. 16-12. The total length of the microstrip ring is 1.5λ. There are four taps or ports on the line, spaced at one-quarter wavelength ($\lambda/4$) intervals, which can be used as inputs or outputs.

Hybrid ring

GOOD TO KNOW

Microstrip patterns are made directly onto printed-circuit boards.

Figure 16-10 Coupling between microstrips.

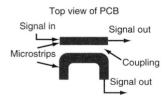

Top view of PCB

Signal in   Signal out

Microstrips   Coupling

Signal out

Figure 16-11 Microstrip 90° turns. (*a*) Right angles such as this should be avoided. (*b*) Gradual curves or turns are preferred. (*c*) This arrangement is also acceptable.

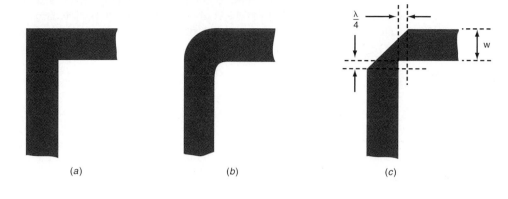

(*a*) (*b*) (*c*)

Figure 16-12 A microstrip hybrid ring.

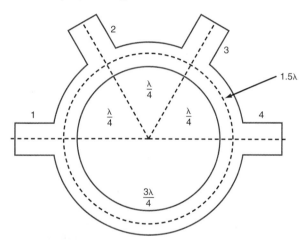

Now, a signal is applied to port 1, and some interesting things happen. Output signals appear at ports 2 and 4, but their levels are at one-half the power of the input. Thus the circuit acts as a power divider to supply two signals of equal level to other circuits. There is no output at port 3. The effect of applying a signal at port 4 is similar. Equal half-power outputs appear at ports 1 and 3, but no signal appears at port 2.

If individual signals are applied simultaneously to ports 1 and 3, the output at port 2 will be their sum and the output at port 4 will be their difference.

The unique operation of the hybrid ring makes it very useful for splitting signals or combining them.

Microstrip can be used to create almost any tuned circuit necessary in an amplifier, including resonant circuits, filters, and impedance-matching networks. Figure 16-13(*a*) shows how a low-pass filter is implemented with microstrip sections. The component shown is a highly selective low-pass filter for use in the 1- to 3-GHz range depending on the exact dimensions. The lumped constant equivalent circuit is shown in Fig. 16-13(*b*). The transmission line segments are formed on the PCB itself and connected end to end as required. The transistors or ICs are then soldered to the PCB along with any resistors or larger discrete components that may be needed.

Example 16-1

A one-quarter wave Q-matching section made of microstrip is designed to match a source of 50 Ω to a load of 136 Ω at 5.8 GHz. The PCB dielectric constant ϵ_r is 2.4. Calculate (a) the required impedance of the microstrip and (b) its length.

a. $Z_Q = \sqrt{Z_{source} \times Z_{load}}$

$= \sqrt{50(136)} = 82.46\ \Omega$

b. $\lambda = \dfrac{300}{f}$ MHz

5.8 GHz = 5800 MHz

$\lambda = \dfrac{300}{5800} = 0.0517$ m

$\dfrac{\lambda}{4} = \dfrac{0.0517}{4} = 0.012931$ m

1 m = 38.37 in

$\dfrac{\lambda}{4} = 0.012931(39.37) = 0.51$ in

Velocity of propagation $= \dfrac{1}{\sqrt{\epsilon_r}} = \dfrac{1}{\sqrt{2.4}} = 0.645$

$\dfrac{\lambda}{4} = (0.51\text{ in})\,(0.645) = 0.3286$ in

Microwave Transistors

Microwave transistor

Microwave transistors, whether they are bipolar or FET types, operate just as other transistors do. The primary differences between standard lower-frequency transistors and microwave types are internal geometry and packaging.

Figure 16-13   A microstrip filter. (a) Microstrip low-pass filter. (b) Lumped constant equivalent circuit.

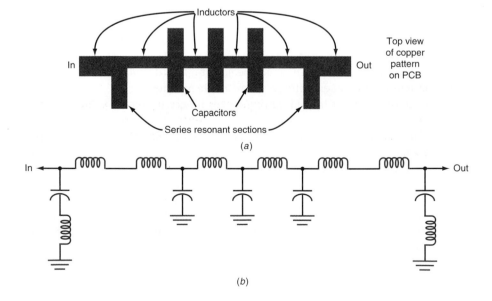

(a)

(b)

Figure 16-14 Microwave transistors. (*a*) and (*b*) Low-power small signal. (*c*) FET power. (*d*) NPN bipolar power.

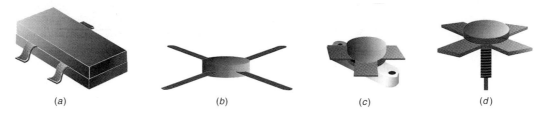

(*a*)　　　　　(*b*)　　　　　(*c*)　　　　　(*d*)

To reduce internal inductances and capacitances of transistor elements, special chip configurations known as *geometries* are used that permit the transistors to operate at higher power levels and at the same time minimize distributed and stray inductances and capacitances.

Figure 16-14 shows several types of microwave transistors. Figure 16-14(*a*) and (*b*) is low-power small-signal microwave transistors; these types are either NPN silicon or gallium arsenide FET. Note the very short leads. Both packages are designed for surface mounting directly to microstrip on the PCB. The transistor in Fig. 16-14(*b*) has four leads, usually two emitter (or source) leads plus the base (or gate) and the collector (or drain). The two emitter leads in parallel ensure low inductance. Some transistors of this type have two base or two collector leads instead of two emitter leads.

Figure 16-14(*c*) shows an enhancement-mode power MOSFET. The short, fat leads are thick strips of copper that are soldered directly to the microstrip circuitry on the PCB. These wide loads also help transfer heat away from the transistor. Figure 16-14(*d*) is an NPN power transistor with two emitter leads. The short, fat leads ensure low inductance and also permit high currents to be accommodated. The fat copper strips help to dissipate heat. The devices in Fig. 16-14(*c*) and (*d*) can handle power levels up to several hundred watts.

Transistors for small-signal amplification and oscillators are available for frequency up to about 100 GHz. For power amplification, transistors are available for frequencies up to 20 GHz.

Most microwave transistors continue to be made of silicon. As transistor geometry sizes have decreased below 0.09 μm (90 nm), switching speeds and amplification frequencies have increased well into the microwave region. CMOS digital integrated circuits, which are made with MOSFETs, can operate up to 10 GHz. RF and linear/analog circuits made with CMOS and used in low-power microwave radios as well as in optical fiber transmission circuits can achieve operation up to 10 GHz. But beyond 10 GHz, special devices are necessary.

You have already seen how GaAs MESFETs, a type of JFET using a Schottky barrier junction, can operate at frequencies in excess of 5 GHz. A variant of the MESFET called a *high electron mobility transistor (HEMT)* extends the frequency range beyond 20 GHz by adding an extra layer of semiconductor material such as AlGaAs.

High electron mobility transistor (HEMT)

A highly popular device known as a *heterojunction bipolar transistor (HBT)* is making even higher-frequency amplification possible in both discrete form and integrated circuits. A heterojunction is formed with two different types of semiconductor materials. Some popular combinations are indium-phosphide (InP) and silicon-germanium (SiGe). Other combinations include AlGaAs/GaAs and InGaAsP. The InP HBTs operate at frequencies up to 50 GHz, and SiGe HBTs have been developed to operate up to 200 GHz. Both small-signal and power amplification versions are available.

Heterojunction bipolar transistor (HBT)

Small-Signal Amplifiers

A *small-signal microwave amplifier* can be made up of a single transistor or multiple transistors combined with a biasing circuit and any microstrip circuits or components as required. Most microwave amplifiers are of the tuned variety. That is, their bandwidth is set by the application and implemented by microstrip series or parallel tuned circuits,

Small-signal microwave amplifier

Figure 16-15 A single-stage class A RF microwave amplifier.

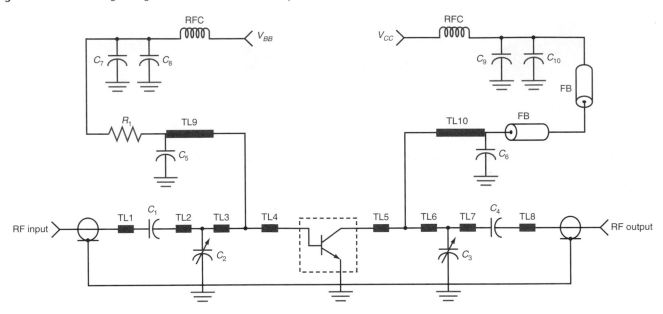

and then microstrip lines are used to perform the various impedance-matching duties required to get the amplifier to work.

Another type of small-signal microwave amplifier is a *multistage integrated circuit,* a variety of MMIC. Besides amplifiers, other MMICs available include mixers, switches, and phase shifters.

Multistage integrated-circuit amplifier

Transistor amplifier

Transistor Amplifiers. Figure 16-15 shows a microwave amplifier for small signals. This type of amplifier is commonly used in the front end of a microwave receiver to provide initial amplification for the mixer. A low-noise transistor is used. The typical gain range is 10 to 25 dB.

Most microwave amplifiers are designed to have input and output impedances of 50 Ω. In the circuit shown in Fig. 16-15, the input can come from an antenna or another microwave circuit. The blocks labeled TL are microstrip sections that act as tuned circuits, inductors, or capacitors. An input microstrip TL1 acts as an impedance-matching section, and TL2 and TL3 form a tuned circuit. TL4 is another impedance-matching section that matches the tuned circuit to the complex impedance of the base input. The tuned circuits set the bandwidth of the amplifier input.

A similar sequence of impedance-matching sections and tuned circuits is used in the collector to set the bandwidth and to match the transistor collector impedance to the output. And C_2 and C_3 are variable capacitors that allow some tuning of the bandwidth.

All the other components are of the surface-mounted chip type to keep lead inductances short. The microstrip segments TL9 and TL10 act as inductors, forming part of the substantial decoupling networks used on the base bias supply V_{BB} and the collector supply V_{CC}. The base supply voltage and the value of R_1 set the base bias, thus biasing the transistor into the linear region for class A amplification. RFCs are used in the supply leads to keep the RF out of the supply and to prevent feedback paths through the supply that can cause oscillation and instability in multistage circuits. Ferrite beads (FB) are used in the collector supply lead for further decoupling.

GOOD TO KNOW

Most microwave amplifiers are designed to have input and output impedances of 50 Ω.

Monolithic microwave integrated circuit (MMIC)

MMIC Amplifiers. A common *monolithic microwave integrated-circuit (MMIC)* amplifier is one that incorporates two or more stages of FET or bipolar transistors made on a common chip to form a multistage amplifier. The chip also incorporates resistors for biasing and small bypass capacitors. Physically, these devices look like the transistors in Fig. 16-14(*a*) and (*b*). They are soldered to a PCB containing microstrip circuits for impedance matching and tuning.

Figure 16-16 A class A microwave power amplifier.

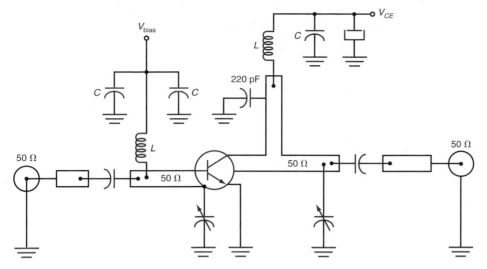

Another popular form of MMIC is the *hybrid circuit,* which combines an amplifier
IC connected to microstrip circuits and discrete components of various types. All the
components are formed on a tiny alumina substrate which serves as both a base and a
place to form microstrip lines. Surface-mounted chip resistors, capacitors, transistors, and
MMIC amplifiers are connected. The entire unit is packaged into a housing, usually metal
for shielding, and connected to additional circuits on a PCB.

Hybrid circuit

Power Amplifiers. A typical *class A microwave power amplifier* is shown in
Fig. 16-16. The microstrip lines are used for impedance matching and tuning. Most
microstrip circuits simulate L-type matching networks with a low-pass configuration.
Small wire loop inductors and capacitors are used to form the decoupling networks to
prevent feedback through the power supply, which would cause oscillation. The input
and output impedances are 50 Ω. This particular stage operates at 1.2 GHz and provides
an output power of 1.5 W. Typical power supply voltages are 12, 24, and 28 V, but can
go to 36 or 48 V for high-power applications.

Class A microwave power amplifier

Note that bias is not supplied by a resistive voltage divider. Instead, it usually comes
from a separate bias-current circuit like that shown in Fig. 16-17, which is an ordinary

Figure 16-17 A constant-current bias supply for a linear power amplifier.

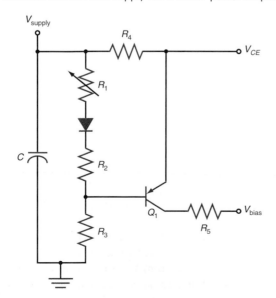

Figure 16-18 An FET power amplifier.

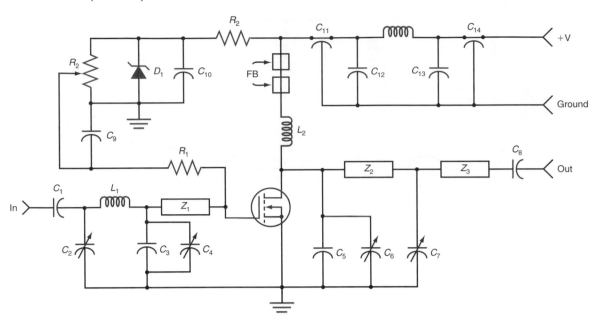

constant-current source. Resistors R_1, R_2, and R_3 form a voltage divider to set the base voltage on Q_1. A voltage is developed across emitter resistor R_4. This voltage divided by R_4 gives the value of the current supplied by Q_1 to the transistor in the microwave amplifier. A diode in series with the voltage divider provides some temperature compensation for variations in the emitter-base voltage that occur in Q_1. Resistor R_1 is adjustable to set the bias current to the precise value for optimum power and minimum distortion. Most power amplifiers obtain their bias from constant-current sources; this provides a bias current that is relatively temperature-independent and thus provides superior protection against damage.

The bias is applied to the base of the amplifier in Fig. 16-16 through a small inductor. This and the bypass capacitors keep the microwave energy out of the bias circuit.

The single-stage FET power amplifier shown in Fig. 16-18 can achieve a power output of 100 W in the high UHF and low microwave region. The transistor is an enhancement-mode FET; that means that the FET does not conduct with drain voltage applied. A positive gate signal of 3 V or more must be applied to achieve conduction. The gate signal is supplied by a lower-power driver stage, and the gate bias is supplied by a zener diode. The gate voltage is made adjustable with potentiometer R_2. This allows the bias to be set for linear class B or class C operation. For linear class A operation, a quiescent value of drain current is set with the bias. This bias circuit can also be used to adjust the power output over a small range.

In the circuit shown in Fig. 16-18, a combination of LC tuned circuits and microstrips is used for tuning and impedance matching. Here L_1 is a tiny hairpin loop of heavy wire, and L_2 is a larger inductor made of multiple turns of wire to form a high impedance for decoupling.

16-3 Waveguides and Cavity Resonators

Long parallel transmission lines, such as 300-Ω twin-lead, radiate electromagnetic energy while transporting it from one place to another. As the frequency of operation gets higher, the amount of radiation from the line increases. At microwave frequencies, virtually all the energy is radiated; almost no energy ever reaches the end of the transmission line. Coaxial cable eliminates the radiation problem, but it has significant loss at the higher frequencies. For the lower microwave frequencies, special coaxial cables have been developed that can be used up to approximately 6 GHz if the length is kept short (less than 100 ft). Above this frequency, coaxial cable loss is too great. Short lengths of coaxial cable, several feet or less, can be used to interconnect pieces of equipment that are close together, but for longer runs other methods of transmission must be used. To minimize loss, special types of coaxial cables with large inner conductors and shields have been developed; however, in most cases these cables are rigid rather than flexible, which makes them difficult to use.

Another problem is the power-handling limit of coaxial cables. The larger cables can carry power levels up to about 1 kW. Higher power levels needed in radar and satellite applications require another solution.

Waveguides

Most microwave energy transmission above about 6 GHz is handled by *waveguides,* which are hollow metal conducting pipes designed to carry and constrain the electromagnetic waves of a microwave signal. Most waveguides are rectangular. Waveguides can be used to carry energy between pieces of equipment or over longer distances to carry transmitter power to an antenna or microwave signals from an antenna to a receiver.

Waveguide

Waveguides are made of copper, aluminum, or brass. These metals are extruded into long rectangular or circular pipes. Often the insides of waveguides are plated with silver to reduce resistance, keeping transmission losses to a very low level.

Signal Injection and Extraction. A microwave signal to be carried by a waveguide is introduced into one end of the waveguide with an antennalike probe that creates an electromagnetic wave which propagates through the waveguide. The electric and magnetic fields associated with the signal bounce off the inside walls back and forth as the signal progresses down the waveguide. The waveguide totally contains the signal so that none escapes by radiation.

The probe shown in Fig. 16-19(*a*) is a one-quarter wavelength vertical antenna at the signal frequency that is inserted in the waveguide one-quarter wavelength from the end, which is closed. The signal is usually coupled to the probe through a short coaxial cable and a connector. The probe sets up a vertically polarized electromagnetic wave in the waveguide, which is then propagated down the line. Because the probe is located one-quarter wavelength from the closed end of the waveguide, the signal from the probe is reflected from the closed end of the line back toward the open end. Over a one-quarter wavelength distance, the reflected signal appears back at the probe in phase

Figure 16-19   Injecting a sine wave into a waveguide and extracting a signal.

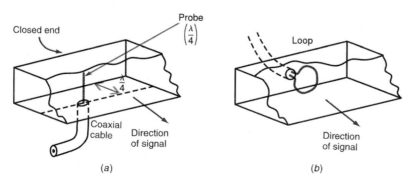

to aid the signal going in the opposite direction. Remember that a radio signal consists of both electric and magnetic fields at right angles to each other. The electric and magnetic fields established by the probe propagate down the waveguide at a right angle to the two fields of the radio signal. The position of the probe determines whether the signal is horizontally or vertically polarized. In Fig. 16-19(a), the electric field is vertical, so the polarization is vertical. The electric field begins propagating down the line and sets up charges on the line that cause current to flow. This in turn generates a companion magnetic field at right angles to the electric field and the direction of propagation.

A loop can also be used to introduce a magnetic field into a waveguide. The loop in Fig. 16-19(b) is mounted in the closed end of the waveguide. Microwave energy applied through a short piece of coaxial cable causes a magnetic field to be set up in the loop. The magnetic field also establishes an electric field, which is then propagated down the waveguide.

Probes and loops can also be used to extract a signal from a waveguide. When the signal strikes a probe or a loop, a signal is induced which can then be fed to other circuitry through a short coaxial cable.

Waveguide Size and Frequency. Figure 16-20 shows the most important dimensions of a rectangular waveguide: the width a and the height b. Note that these are the inside dimensions of the waveguide. The frequency of operation of a waveguide is determined by the size of a. This dimension is usually made equal to one-half wavelength, a bit below the lowest frequency of operation. This frequency is known as the *waveguide cutoff frequency*. At its cutoff frequency and below, a waveguide will not transmit energy. At frequencies above the cutoff frequency, a waveguide will propagate electromagnetic energy. A waveguide is essentially a high-pass filter with a cutoff frequency equal to

$$f_{co} = \frac{300}{2a}$$

where f_{co} is in megahertz and a is in meters.

For example, suppose it is desired to determine the cutoff frequency of a waveguide with an a dimension of 0.7 in. To convert 0.7 in to meters, multiply by 2.54 to get centimeters and divide by 100 to get meters: 0.7 in = 0.01778 m. Thus,

$$f_{co} = \frac{300}{2(0.01778)} = 8436 \text{ MHz} = 8.436 \text{ GHz}$$

Normally, the height of a waveguide is made equal to approximately one-half the a dimension, or 0.35 in. The actual size might be 0.4 in.

Waveguide cutoff frequency

Figure 16-20 The dimensions of a waveguide determine its operating frequency range.

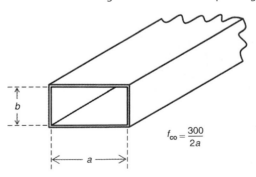

$$f_{co} = \frac{300}{2a}$$

Example 16-2

A rectangular waveguide has a width of 0.65 in and a height of 0.38 in. (*a*) What is the cutoff frequency? (*b*) What is a typical operating frequency for this waveguide?

a. 0.65 in × 2.54 = 1.651 cm

$$\frac{1.651}{100} = 0.01651 \text{ m}$$

$$f_{co} = \frac{300}{2a} = \frac{300}{2(0.01651)} = 9085 \text{ MHz} = 9.085 \text{ GHz}$$

b. $f_{co} = 0.7f$

$$f = \frac{f_{co}}{0.7} = 1.42f_{co}$$

$$= 1.42(9.085) = 12.98 \text{ GHz and above}$$

Example 16-3

Would the rectangular waveguide in Example 16-2 operate in the C band?

The C band is approximately 4 to 6 GHz. Since a waveguide acts as a high-pass filter with a cutoff of 9.085 GHz, it will not pass a C band signal.

Signal Propagation. When a probe or loop launches energy into a waveguide, the signal enters the waveguide at an angle so that the electromagnetic fields bounce off the sidewalls of the waveguides as the signal propagates along the line. In Fig. 16-21(*a*), a vertical probe is generating a vertically polarized wave with a vertical electric field and a magnetic field at a right angle to the electric field. The electric field is at a right angle to the direction of wave propagation, so it is called a *transverse electric (TE) field*. Figure 16-21(*b*) shows how a loop would set up the signal. In this case, the magnetic field is transverse to the direction of propagation, so it is called a *transverse magnetic (TM) field*.

Signal propagation

Transverse electric (TE) field

Transverse magnetic (TM) field

Figure 16-21 (*a*) Transverse electric field. (*b*) Transverse magnetic field.

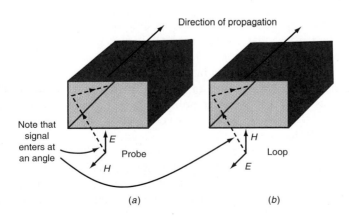

Figure 16-22 Wave paths in a waveguide at various frequencies. (*a*) High frequency. (*b*) Medium frequency. (*c*) Low frequency. (*d*) Cutoff frequency.

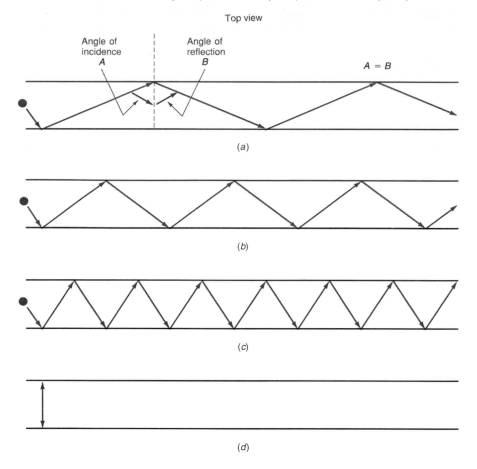

The angles of incidence and reflection depend on the operating frequency (see Fig. 16-22). At high frequencies, the angle is large and therefore the path between the opposite walls is relatively long, as shown in Fig. 16-22(*a*). As the operating frequency decreases, the angle also decreases and the path between the sides shortens. When the operating frequency reaches the cutoff frequency of the waveguide, the signal simply bounces back and forth between the sidewalls of the waveguide. No energy is propagated.

Whenever a microwave signal is launched into a waveguide by a probe or loop, electric and magnetic fields are created in various patterns depending upon the method of energy coupling, frequency of operation, and size of waveguide. Figure 16-23 shows typical fields in a waveguide. In the end view, the lines represent the electric field E lines. The dots represent the magnetic field H lines. In the top view the dashed lines represent the H field, and the ✕s and dots represent the E field. The ✕ means that the line is going into the page; the • means that the line is coming out of the page.

The pattern of the electromagnetic fields within a waveguide takes many forms. Each form is called an *operating mode*. As indicated earlier, the magnetic or the electric field must be perpendicular to the direction of propagation of the wave. In the TE mode, the electric field exists across the guide and no E lines extend lengthwise along the guide. In the TM mode, the H lines form loops in planes perpendicular to the walls of the guide, and no part of an H line is lengthwise along the guide.

Subscript numbers are used along with the TE and TM designations to further describe the E and H field patterns. A typical designation is $TE_{0,1}$. The first number indicates the number of half-wavelength patterns of transverse lines that exist along the short dimension of the guide through the center of the cross section. Transverse lines are those

Figure 16-23 Electric (*E*) and magnetic (*H*) fields in a rectangular waveguide.

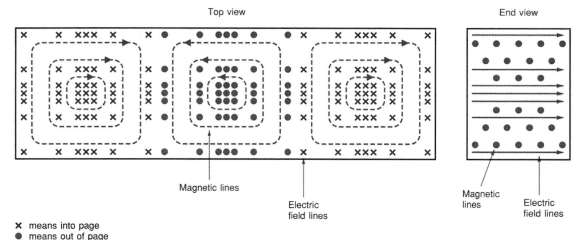

Top view

End view

Magnetic lines

Electric field lines

Magnetic lines

Electric field lines

✕ means into page
● means out of page

perpendicular to the walls of the guide. The second number indicates the number of transverse half-wavelength patterns that exist along the long dimension of the guide through the center of the cross section. If there is no change in the field intensity of one dimension, a zero is used.

The waveguide in Fig. 16-23 is TE because the *E* lines are perpendicular (transverse) to the sides of the guide. Looking at the end view in Fig. 16-23 along the short length, we see no field intensity change, so the first subscript is 0. Along the long dimension of the end view, the *E* lines spread out at the top and bottom but are close together in the center. The actual field intensity is sinusoidal—zero at the ends, maximum in the center. This is one-half of a sine wave variation, and so the second subscript is 1. The mode of the line in Fig. 16-23 is therefore $TE_{0,1}$. This, by the way, is the main or dominant mode of most rectangular waveguides. Many other patterns are possible, such as the two shown in Fig. 16-24.

Figure 16-24 Other waveguide operating modes.

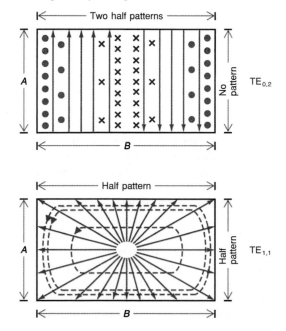

Two half patterns

A

No pattern

B

$TE_{0,2}$

Half pattern

A

Half pattern

B

$TE_{1,1}$

Figure 16-25 A choke joint permits sections of waveguide to be interconnected with minimum loss and radiation.

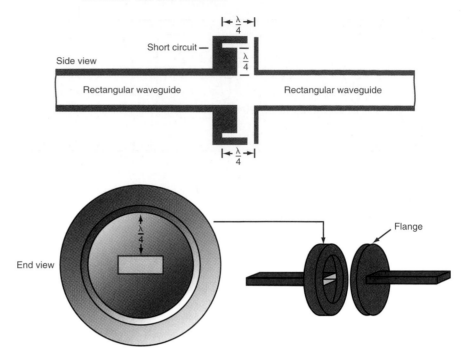

Side view

Short circuit

$\frac{\lambda}{4}$

$\frac{\lambda}{4}$

Rectangular waveguide

Rectangular waveguide

$\frac{\lambda}{4}$

End view

$\frac{\lambda}{4}$

Flange

Waveguide Hardware and Accessories

In some ways, waveguides have more in common with plumbing equipment than they do with the standard transmission lines used in radio communication. And, like plumbing, waveguides have a variety of special parts, such as couplers, turns, joints, rotary connections, and terminations. Most waveguides and their fittings are precision-made so that the dimensions match perfectly. Any mismatch in dimensions or misalignment of pieces that fit together will introduce significant losses and reflections. Waveguides are available in a variety of standard lengths which are interconnected to form a path between a microwave generator and its ultimate destination.

Choke joint

Connection Joints. Figure 16-25 shows a *choke joint* which is used to interconnect two sections of waveguide. It consists of two flanges connected to the waveguide at the center. The right-hand flange is flat, and the one at the left is slotted one-quarter wavelength deep at a distance of one-quarter wavelength from the point at which the walls of the guide are joined. The one-quarter waves together become a half wave and reflect a short circuit at the place where the walls are joined. Electrically, this creates a short circuit at the junction of the two waveguides. The two sections can actually be separated as much as one-tenth wavelength without excessive loss of energy at the joint. This separation allows room to seal the interior of the waveguide with a rubber gasket for pressurization. Some long waveguides are pressurized with nitrogen gas to reduce moisture buildups. The choke joint effectively keeps the RF inside the waveguide. And it introduces minimum loss, 0.03 dB or less.

Curved Sections. Special curved waveguide sections are available for making 90° bends. Curved sections introduce reflections and power loss, but these are kept small by proper design. When the radius of the curved section is greater than 2λ at the signal frequency, losses are minimized. Figure 16-26 shows several different configurations. Flanges are used to connect curve sections with straight runs.

T Sections. It is occasionally necessary to split or combine two or more sources of microwave power. This is done with *T sections* or *T junctions* (Fig. 16-27). The T can

T section (T junction)

Figure 16-26 Curved waveguide sections. (*a*) Right-angle bends. (*b*) Twisted section.

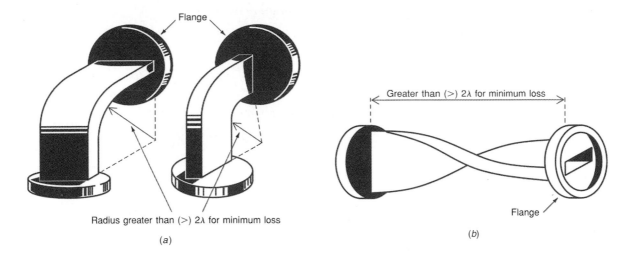

Flange

Greater than (>) 2λ for minimum loss

Radius greater than (>) 2λ for minimum loss

Flange

(*a*)

(*b*)

Figure 16-27 Waveguide T sections. (*a*) Shunt T. (*b*) Series T.

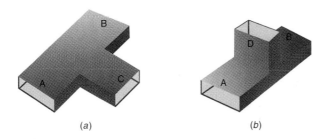

(*a*)

(*b*)

be formed on the short or long side of the waveguide. If the junction is formed on the short side, it is called a *shunt T.* If the junction is formed on the long side, it is called a *series T.* Each T section has three ports which can be used as inputs or outputs.

If a signal is propagated along a waveguide connected to the C port of a shunt T like that in Fig. 16-27(*a*), equal amounts of the signal will appear in phase at output ports A and B. The power level of the signals at A and B is one-half the input power. Such a device is called a *power divider.*

The shunt T can also combine signals. If the signal input to A is of the same phase as the signal input to B, they will be combined at the output port C. The output power is the sum of the individual powers. This is called a *power combiner.*

In the series T shown in Fig. 16-27(*b*), a signal entering port D will be split into two half-power signals which appear at ports A and B but which are 180° out of phase with each other.

Hybrid T's. A special device called a *hybrid T* can be formed by combining the series and shunt T sections (see Fig. 16-28). Sometimes referred to as a *magic T,* this device is used as a duplexer to permit simultaneous use of a single antenna by both a transmitter and a receiver. The antenna is connected to port B. Any received signal is passed to port D, which is connected to the receiver front end. Port A is terminated and not used. The received signal will not enter port C, which is connected to the transmitter.

A transmitted signal will pass through to the antenna at port B, but will not enter port D to the receiver. If a transmitter and receiver operate on the same frequency or near the same frequencies, and the transmitter output signal is at a high power level, some means must be used to prevent the power from entering the receiver and causing damage. The hybrid T can be used for this purpose.

Power divider

Power combiner

Hybrid T

Figure 16-28 A hybrid or magic T used as a duplexer.

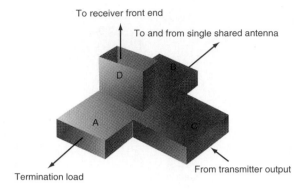

To receiver front end

To and from single shared antenna

Termination load

From transmitter output

Termination

Terminations. In many cases it is necessary to terminate the end of an unused port of a waveguide. For example, port A in Fig. 16-28 would be connected to a load of the correct impedance to prevent high reflections, unfavorable SWR, and loss. If the length of the line is some multiple of one-quarter or one-half wavelength, it may be possible to open or short the waveguide. In most cases this is not possible, and other means of termination are used.

One approach to termination is to insert a pyramid-shaped metallic section in the end of the line, as shown in Fig. 16-29(a). The taper provides a correct match. Usually the tapered section is movable so that the termination can be adjusted for minimum SWR.

It is also possible simply to fill the end of the line with a powdered graphite resistive material, as shown in Fig. 16-29(b). This absorbs the signal and dissipates it as heat so that no reflections occur.

Termination can be accomplished by using a resistive material shaped as a triangle or wedge at the end of a closed line [see Fig. 16-29(c)]. The tapered resistive element is oriented to match the orientation of the electric field in the guide. The magnetic component of the wave induces a voltage into the tapered resistive material, and current flows. Thus the signal is absorbed and dissipated as heat.

Directional Couplers. One of the most commonly used waveguide components is the directional coupler. *Directional couplers* are used to facilitate the measurement of microwave power in a waveguide and the SWR. They can also be used to tap off a small portion of a high-power microwave signal to be sent to another circuit or piece of equipment.

The directional coupler in Fig. 16-30 is simply a short segment of waveguide with coupling joints that are designed to be inserted into a longer run of waveguide between a transmitter and an antenna or between some source and a load. A similar section of waveguide is physically attached to this short segment of line. It is terminated at one end, and the other end is bent away at a 90° angle. The bent section is designed to attach to a microwave power meter or an SWR meter. The bent section is coupled to the straight section by two

Directional coupler

Figure 16-29 Terminating matched loads for waveguides.

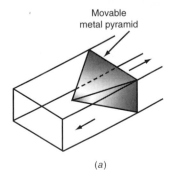

Movable metal pyramid

(a)

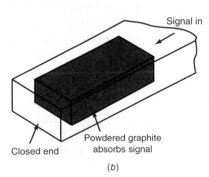

Signal in

Closed end

Powdered graphite absorbs signal

(b)

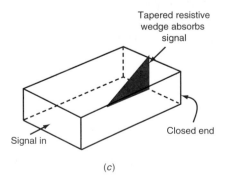

Tapered resistive wedge absorbs signal

Signal in

Closed end

(c)

Figure 16-30 Directional coupler.

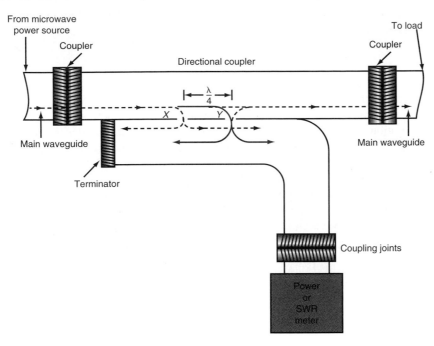

holes (X and Y) which are one-quarter wavelength apart at the frequency of operation. Some of the microwave energy passes through the holes in the straight section into the bent section. The amount of coupling between the two sections depends on the size of the holes.

If energy from the source at the left moves through the waveguide from left to right, microwave energy passes through hole X into the bent section. The energy splits in half. Part of the energy goes to the left, where it is absorbed by the terminator; the other half moves to the right toward hole Y. This is indicated by the dashed lines.

Additional energy from the straight section enters hole Y. It, too, splits into two equal components, one going to the left and the other going to the right. These components are indicated by the solid lines. The energy moving from hole Y to the left toward hole X in the curved section cancels the energy moving from hole X to Y. The signal moving from hole Y to hole X travels a total distance of one-half wavelength or 180°, so it is out of phase with the signal at hole X. But between Y and X the signals are exactly out of phase and equal in amplitude, so they cancel. Any small residual signal moving to the left is absorbed by the terminator. The remaining signal moves on to the power or SWR meter.

The term *directional coupler* derives from the operation of the device. A portion of the energy of signals moving from left to right will be sampled and measured. Any signal entering from the right and moving to the left will simply be absorbed by the terminator; none will pass on to the meter.

Directional coupler

The amount of signal energy coupled into the bend section depends on the size of the holes. Usually only a very small portion of the signal, less than 1 percent, is extracted or sampled. Thus the primary signal is not materially attenuated. However, a sufficient amount of signal is present to be measured. The exact amount of signal extracted is determined by the coupling factor C, which is determined by the familiar formula for power ratio:

$$C = 10 \log \frac{P_{in}}{P_{out}}$$
$$= \text{coupling factor, dB}$$

where P_{in} = amount of power applied to straight section

P_{out} = signal power going to power meter

Most directional couplers are available with a fixed coupling factor, usually 10, 20, or 30 dB.

The amount of power that is actually measured can be calculated by rearranging the power ratio formula:

$$P_{out} = \frac{P_{in}}{\log^{-1}(C/10)} = \frac{P_{in}}{10^{C/10}}$$

For example, if the input power is 2000 W to a directional coupler with a coupling factor of 30, the actual power to the meter is $2000/10^3 = 2000/1000 = 2$ W.

Bidirectional coupler

A variation of the directional coupler is the *bidirectional coupler.* Using two sections of line to sample the energy in both directions allows determination of the SWR. One segment of line samples the incident or forward power, and the other is set up to sample the reflected energy in the opposite direction.

One of the most widely used forms of directional coupler today is that made with microstrip transmission line, such as shown in Fig. 16-10. Directional couplers can be made right on the printed-circuit board holding the other circuitry. Directional couplers are also available as separate units with coaxial connectors for inputs and outputs.

Cavity Resonators

Cavity resonator

A *cavity resonator* is a waveguidelike device that acts as a high-Q parallel resonant circuit. A simple cavity resonator can be formed with a short piece of waveguide one-half wavelength long, as illustrated in Fig. 16-31(a). The ends are closed. Energy is coupled into the cavity with a coaxial probe at the center, as shown in the side view in Fig. 16-31(b). When microwave energy is injected into the cavity, the signal bounces off the shorted ends of the waveguide and reflects back toward the probe. Because the probe is located one-quarter wavelength from each shorted end, the reflected signal reinforces the signal at the probe. The result is that the signal bounces back and forth off the shorted ends. If the signal is removed, the wave continues to bounce back and forth until losses cause it to die out.

This effect is pronounced at a frequency where the length of the waveguide is exactly one-half wavelength. At that frequency, the cavity is said to resonate and acts as a parallel resonant circuit. A brief burst of energy applied to the probe will make the cavity oscillate; the oscillation continues until losses cause it to die out. Cavities such as this have extremely high Q, as high as 30,000. For this reason, they are commonly used to create resonant circuits and filters at microwave frequencies.

A cavity can also be formed by using a short section of circular waveguide such as that shown in Fig. 16-32. With this shape, the diameter should be one-half wavelength at the operating frequency. Other cavity shapes are also possible.

Figure 16-31 Cavity resonator made with waveguide. (*a*) A λ/2 section of rectangular waveguide used as a cavity resonator. (*b*) Side view of cavity resonator showing coupling of energy by a probe.

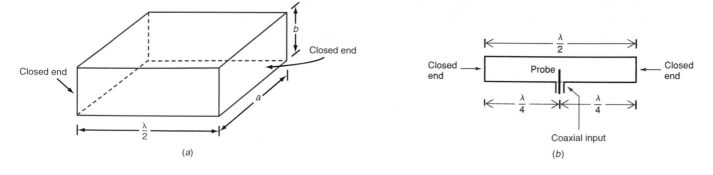

(a)

(b)

Figure 16-32 Circular resonant cavity with input-output loops.

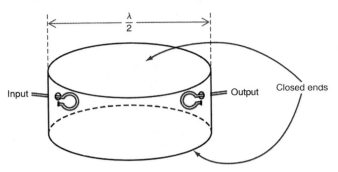

Typically, cavities are specially designed components. Often, they are hollowed out sections in a block of metal that have been machined to very precise dimensions for specific frequencies. The internal walls of the cavity are often plated with silver or some other low-loss material to ensure minimum loss and maximum Q.

Some cavities are also tunable. One wall of the cavity is made movable, as shown in Fig. 16-33(a). An adjustment screw moves the end wall in and out to adjust the resonant frequency. The smaller the cavity, the higher the operating frequency. Cavities can also be tuned with adjustable plugs in the side of the cavity, as in Fig. 16-33(b). As the plug is screwed in, more of it intrudes into the cavity and the operating frequency goes up.

Circulators

A *circulator* is a three-port microwave device used for coupling energy in only one direction around a closed loop. A schematic diagram of a circulator is shown in Fig. 16-34(a). Microwave energy applied to port 1 is passed to port 2 with only minor attenuation; however, the signal will be greatly attenuated on its way to port 3. The loss from port 1 to port 3 is usually 20 dB or more. A signal applied to port 2 will be passed with little attenuation to port 3, but little or none will reach port 1.

The primary application of a circulator is as a *diplexer,* which allows a single antenna to be shared by a transmitter and receiver. As shown in Fig. 16-34(a), the signal from the transmitter is applied to port 1 and passed to the antenna, which is connected to port 2.

Circulator

Diplexer

Figure 16-33 Adjustable or tunable cavities. (a) Cylindrical cavity with adjustable disk. (b) Cavity with adjustable plugs.

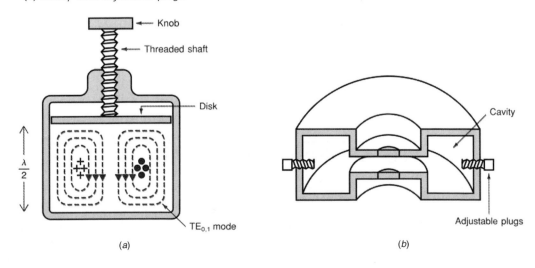

(a)

(b)

Figure 16-34 Circulator. (*a*) Circulator used as a diplexer. (*b*) Construction of a circulator.

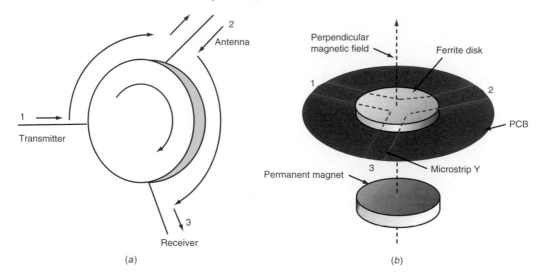

(*a*) (*b*)

The transmitted signal does not reach the receiver connected to port 3. The received signal from the antenna comes into port 2 and is sent to port 3, but is not fed into the transmitter on port 1.

A common way of making a circulator is to create a microstrip pattern that looks like a Y, as shown in Fig. 16-34(*b*). The three ports are 120° apart. These ports, which are formed on PCBs, can be connected to coaxial connectors.

Ferrite disk

On top of the Y junction, and sometimes on the bottom, is a *ferrite disk*. Ferrites are ceramics made of compounds such as $BaFe_2O_3$. They have magnetic properties similar to those of iron or steel, so they support magnetic fields, but they do not cause the induction of eddy currents. Iron is often mixed with zinc or manganese. A widely used type of ferrite is yttrium-iron-garnet (YIG). A permanent magnet is then placed such that it forces magnetic lines of force perpendicular to the device through the ferrite. This strong magnetic field interacts with the magnetic fields produced by any input signals, creating the characteristic action of the circulator.

The type of circulator described here is for low and medium power levels. For high-power applications, a circulator made with a waveguidelike structure containing ferrite material is used.

Isolators

Isolator

Isolators are variations of circulators, but they have one input and one output. That is, they are configured like the circulator diagrammed in Fig. 16-34, but only ports 1 and 2 are used. An input to port 1 is passed to port 2 with little attenuation. Any reflected energy, as would occur with an SWR higher than 1, is not coupled back into port 1. Isolators are often used in situations where a mismatch, or the lack of a proper load, could cause reflection so large as to damage the source.

16-4 Microwave Semiconductor Diodes

Small-Signal Diodes

Small-signal diode

Diodes used for signal detection and mixing are the most common microwave semiconductor devices. Two types are widely used: the point-contact diode and the Schottky barrier or hot carrier diode.

The typical *semiconductor diode* is a junction formed of P- and N-type semiconductor materials. Because of the relatively large surface area of the junction, diodes exhibit a high capacitance which prevents normal operation at microwave frequencies. For this reason, standard PN junction diodes are not used in the microwave region.

Semiconductor diode

Point–Contact Diodes. Perhaps the oldest microwave semiconductor device is the *point-contact diode,* also called a *crystal diode.* A point-contact diode comprises a piece of semiconductor material and a fine wire which makes contact with the semiconductor material. Because the wire makes contact with the semiconductor over a very small surface area, the capacitance is extremely low. Current flows easily from the cathode, the fine wire, to the anode, the semiconductor material. However, current does not flow easily in the opposite direction.

Most early point-contact diodes used germanium as the semiconductor material, but today these devices are made of P-type silicon with a fine tungsten wire as the cathode. (See Fig. 16-35.) The forward threshold voltage is extremely low.

Point-contact diodes are ideal for small-signal applications. They are widely used in microwave mixers and detectors and in microwave power measurement equipment. They are extremely delicate and cannot withstand high power. They are also easily damaged and therefore must be used in such a way to minimize shock and vibration.

Point-contact diode (crystal diode)

GOOD TO KNOW

Although they are widely used in microwave mixers and detectors and in microwave power measurement equipment, point-contact diodes are extremely delicate and cannot withstand high power.

Hot Carrier Diodes. For the most part, point-contact diodes have been replaced by *Schottky diodes,* sometimes referred to as *hot carrier diodes.* Most Schottky diodes are made with N-type silicon on which has been deposited a thin metal layer. Gallium arsenide is also used. The semiconductor forms the cathode, and the metal forms the anode. The structure and schematic symbol for a Schottky diode are shown in Fig. 16-36. Typical anode materials are nickel chromium and aluminum, although other metals, e.g., gold and platinum, are also used.

Like the point-contact diode, the Schottky diode is extremely small and therefore has a tiny junction capacitance. It also has a low bias threshold voltage. It conducts at a forward bias of 0.2 to 0.3 V, whereas the silicon junction diode conducts at 0.6 V. Higher voltage drops across PN silicon diodes result from bulk resistance effects. Schottky diodes are ideal for mixing, signal detection, and other low-level signal operations. They are widely used in balanced modulators and mixers. Because of their very high-frequency response, Schottky diodes are used as fast switches at microwave frequencies.

The most important use of microwave diodes is as *mixers.* Microwave diodes are usually installed as part of a waveguide or cavity resonator tuned to the incoming signal

Schottky diode

Hot carrier diode

Mixer

Figure 16-35 A point-contact diode.

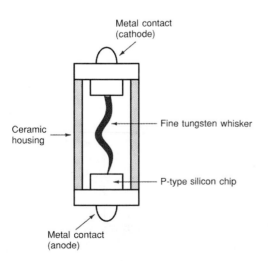

Metal contact (cathode)

Ceramic housing

Fine tungsten whisker

P-type silicon chip

Metal contact (anode)

Figure 16-36 Hot carrier or Schottky diode.

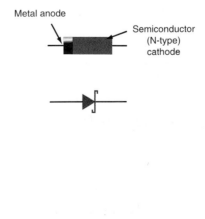

Metal anode

Semiconductor (N-type) cathode

frequency. A local-oscillator signal is injected with a probe or loop. The diode mixes the two signals and produces the sum and difference output frequencies. The difference frequency is usually selected with another cavity resonator or with a low-frequency LC tuned circuit. In a microwave circuit, the mixer is usually the input circuit. This is so because it is desirable to convert the microwave signal down to a lower-frequency level as early as possible, at a point where amplification and demodulation can take place with simpler, more conventional electronic circuits.

Frequency–Multiplier Diodes

Microwave diodes designed primarily for frequency-multiplier service include varactor diodes and step-recovery diodes.

Varactor Diodes. A *varactor diode* is basically a voltage variable capacitor. When a reverse bias is applied to the diode, it acts as a capacitor. Its capacitance depends upon the value of the reverse bias. Varactor diodes made with gallium arsenide are optimized for use at microwave frequencies. Their main application in microwave circuits is as frequency multipliers.

Figure 16-37 shows a varactor frequency-multiplier circuit. When an input signal is applied across the diode, it alternately conducts and cuts off. The result is a nonlinear or distorted output containing many harmonics. When a tuned circuit is used in the output, the desired harmonic is selected and others are rejected. Since the lower harmonics produce the greatest amount of energy, varactor multipliers are usually used only for doubling and tripling operations. In Fig. 16-37, the input tuned circuit L_1–C_2 resonates at the input frequency f_{in}, and the output tuned circuit L_2–C_3 resonates at 2 or 3 times the input frequency, as desired. In practice, the tuned circuits are not actually made up of individual inductors or capacitors. Instead, they are microstrip, stripline, or cavity resonators. Capacitors C_1 and C_4 are used for impedance matching.

A varactor frequency multiplier does not have gain like that of a class C amplifier used as a multiplier. In fact, a varactor introduces a signal power loss. However, it is a relatively efficient circuit, and the output can be as high as 80 percent of the input. Typical efficiencies are in the 50 to 80 percent range. No external source of power is required for this circuit; only the RF input power is required for proper operation. Outputs up to 50 W are obtainable with special high-power varactors.

Varactors are used in applications in which it is difficult to generate microwave signals. Usually it is a lot easier to generate a VHF or UHF signal and then use a series of frequency multipliers to put it into the desired microwave region. Varactor diodes are available for producing relatively high-power outputs at frequencies up to 100 GHz.

Step–Recovery Diodes. Another diode used in microwave frequency-multiplier circuits like that in Fig. 16-37 is the *step-recovery diode* or *snap-off varactor*. It is a PN-junction diode made with gallium arsenide or silicon. When it is forward-biased, it conducts as any diode, but a charge is stored in the depletion layer. When reverse bias

Figure 16-37 A varactor frequency multiplier.

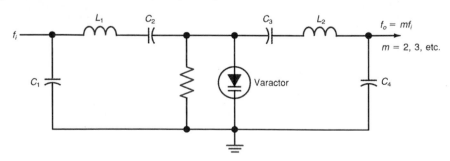

is applied, the charge keeps the diode on momentarily. Then the diode turns off abruptly. This snap-off produces an extremely high-intensity reverse current pulse with a duration of about 10 ps (1 ps = 10^{-12} s) or less. It is extremely rich in harmonics. Even the higher harmonics are of relatively high amplitude.

Step-recovery diodes can also be used circuits like that in Fig. 16-37 to produce multipliers with power ratings up to 5 and 10 W. Power ratings of 50 W can be obtained. Operating frequencies up to 100 GHz are possible with an efficiency of 80 percent or better.

Oscillator Diodes

Oscillator diode

Three types of diodes other than the tunnel diode that can oscillate due to negative-resistance characteristics are the Gunn diodes and IMPATT and TRAPATT diodes.

Gunn Diodes. *Gunn diodes,* also called *transferred-electron devices (TEDs),* are not diodes in the usual sense because they do not have junctions. A Gunn diode is a thin piece of N-type gallium arsenide (GaAs) or indium phosphide (InP) semiconductor which forms a special resistor when voltage is applied to it. This device exhibits a negative-resistance characteristic. That is, over some voltage range, an increase in voltage results in a decrease in current and vice versa, just the opposite of Ohm's law. When it is so biased, the time it takes for electrons to flow across the material is such that the current is 180° out of phase with the applied voltage. If a Gunn diode so biased is connected to a cavity resonant near the frequency determined by the electron transit time, the resulting combination will oscillate. The Gunn diode, therefore, is used primarily as a microwave oscillator.

Gunn diodes are available that oscillate at frequencies up to about 150 GHz. In the lower microwave range, power outputs from milliwats up to several watts are possible. The thickness of the semiconductor determines the frequency of oscillation. However, if the cavity is made variable, the Gunn oscillator frequency can be adjusted over a narrow range.

IMPATT and TRAPATT Diodes. Two other microwave diodes widely used as oscillators are the *IMPATT* and *TRAPATT diodes.* Both are PN-junction diodes made of silicon, GaAs, or InP. They are designed to operate with a high reverse bias that causes them to avalanche or break down. A high current flows. Over a narrow range, a negative-resistance characteristic is produced that causes oscillation when the diode is mounted in a cavity and properly biased. IMPATT diodes are available with power ratings up to about 25 W to frequencies as high as about 300 GHz. Pulsed power of several hundred watts is possible. IMPATT diodes are preferred over Gunn diodes if higher power is required. Their primary disadvantages are their higher noise level and higher operating voltages.

PIN Diodes

A *PIN diode* is a special PN-junction diode with an I (intrinsic) layer between the P and N sections, as shown in Fig. 16-38(*a*). The P and N layers are usually silicon, although GaAs is sometimes used. In practice, the I layer is a very lightly doped N-type semiconductor.

At frequencies less than about 100 MHz, the PIN diode acts just like any other PN junction diode. At higher frequencies, it acts like a variable resistor or like a switch. When the bias is zero or reverse, the diode acts like a high value of resistance, 5k Ω and higher. If a forward bias is applied, the diode resistance drops to a very low level, typically a few ohms or less. When the amount of forward bias is varied, the value of the resistance can be varied over a linear range. The characteristic curve for a PIN diode is shown in Fig. 16-38.

Gunn diode or transferred-electron device (TED)

GOOD TO KNOW

Police radar guns use Gunn diode oscillators to generate microwave signals, and radar detector receivers use Gunn diodes as local oscillators.

IMPATT diode

TRAPATT diode

PIN diode

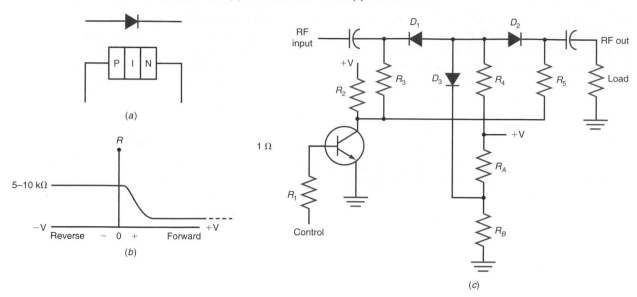

PIN diodes are used as switches in microwave circuits. A typical application is to connect a PIN diode across the output of a microwave transmission line like microstrip or stripline. When the diode is reverse-biased, it acts as a very high resistance and has little effect on the normally much lower characteristic impedance of the transmission line. When the diode is forward-biased, it shorts the line, creating almost total reflection. PIN diodes are widely used to switch sections of quarter- or half-wavelength transmission lines to provide varying phase shifts in a circuit.

Tee switching configuration

A popular switching circuit is the *tee configuration* shown in Fig. 16-38(*c*), which is a combination of two series switches and a shunt switch. It has excellent isolation between input and output. A simple saturated transistor switch is used for control. When the control input is binary 0, the transistor is off, so positive voltage is applied to the cathodes of diodes D_1 and D_2. Diode D_3 conducts; because of the voltage divider, which is made up of R_A and R_B, the cathode of D_3 is at a lower positive voltage level than its anode. Under this condition, the switch is off, so no signal reaches the load.

When the control input is positive, or binary 1, the transistor conducts, pulling the cathodes of D_1 and D_2 to ground through resistors R_3 and R_5. The anodes of D_1 and D_2 are at positive voltage through resistor R_4, so these diodes conduct. Diode D_3 is cut off. Thus the signal passes to the load.

PIN diodes are also sometimes used for their variable-resistance characteristics. Varying their bias to change resistance permits variable-voltage attenuator circuits to be created. PIN diodes can also be used as amplitude modulators (see Chap. 4).

16-5 Microwave Tubes

Microwave tube

Vacuum tube

Before transistors were invented, all electronic circuits were implemented with *vacuum tubes*. Tubes are devices used for controlling a large current with a small voltage to produce amplification, oscillation, switching, and other operations. Today, vacuum tubes are used only for special applications. The *cathode-ray tube (CRT)* used in TV sets, computer monitors, oscilloscopes, spectrum analyzers, and other display devices is a special form of vacuum tube that shows no signs of obsolescence. LCDs offer the only alternative to the CRT, and their quality has increased significantly over the years.

Vacuum tubes are also still found in microwave equipment. This is particularly true in microwave transmitters used for producing high output power. Bipolar field-effect transistors can produce power in the microwave region up to approximately several hundred watts, but many applications require more power. Radio transmitters in the UHF and low microwave bands use standard vacuum tubes designed for power amplification. These can produce power levels of up to several thousand watts. At higher microwave frequencies (above 2 GHz) special tubes are used. In satellites and their earth stations, TV stations, and in some military equipment such as radar, very high output powers are needed. Special microwave tubes developed during World War II— the klystron, the magnetron, and the traveling-wave tube—are still widely used for such microwave power amplification.

Klystrons

A *klystron* is a microwave vacuum tube using cavity resonators to produce velocity modulation of an electron beam which produces amplification. Figure 16-39 is a schematic diagram of a two-cavity klystron amplifier. The vacuum tube itself consists of a cathode which is heated by a filament. At a very high temperature, the cathode emits electrons. These negative electrons are attracted by a plate or collector, which is biased with a high positive voltage. Thus current flow is established between the cathode and the collector inside the evacuated tube.

    The electrons emitted by the cathode are focused into a very narrow stream by using electrostatic and electromagnetic focusing techniques. In *electrostatic focusing,* special elements called *focusing plates* to which have been applied high voltages force the electrons into a narrow beam. Electromagnetic focusing makes use of coils around the tube

Klystron

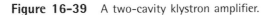

Figure 16-39 A two-cavity klystron amplifier.

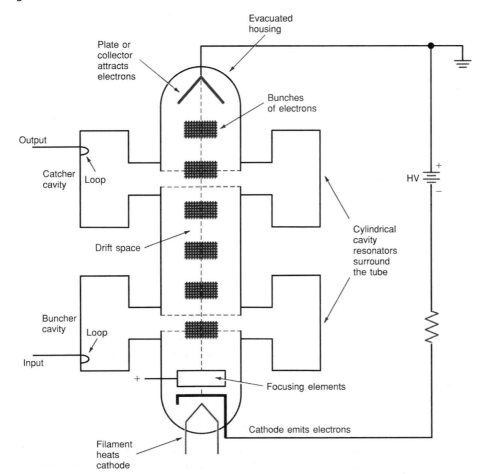

through which current is passed to produce a magnetic field. This magnetic field helps focus the electrons into a narrow beam.

The sharply focused beam of electrons is then forced to pass through the centers of two cavity resonators that surround the open center cavity. The microwave signal to be amplified is applied to the lower cavity through a coupling loop. This sets up electric and magnetic fields in the cavity which cause the electrons to speed up and slow down as they pass through the cavity. On one half-cycle, the electrons are speeded up; on the next half cycle of the input, they are slowed down. The effect is to create bunches of electrons that are one-half wavelength apart in the drift space between the cavities. This speeding up and slowing down of the electron beam is known as *velocity modulation*. Since the input cavity produces bunches of electrons, it is commonly referred to as the *buncher cavity.*

Since the bunched electrons are attracted by the positive collector, they move on through the tube, eventually passing through the center of another cavity known as the catcher cavity. Because the bunched electrons move toward the collector in clouds of alternately dense and sparse areas, the electron beam can be referred to as a *density-modulated* beam.

As the bunches of electrons pass through the catcher cavity, the cavity is excited into oscillation at the resonant frequency. Thus the dc energy in the electron beam is converted to RF energy at the cavity frequency, and amplification occurs. The output is extracted from the catcher cavity with a loop.

Klystrons are also constructed with additional cavities between the buncher and catcher cavities. These intermediate cavities produce further bunching which causes increased amplification of the signal. If the buncher cavities are tuned off center frequency from the input and output cavities, the effect is to broaden the bandwidth of the tube. The frequency of operation of a klystron is set by the sizes of the input and output cavities. Since cavities typically have high Qs, their bandwidth is limited. By lowering the Qs of the cavities and by introducing intermediate cavities, wider-bandwidth operation can be achieved.

Klystrons are no longer widely used in most microwave equipment. Gunn diodes have replaced the smaller reflex klystrons in signal-generating applications because they are smaller and lower in cost and do not require high dc supply voltages. The larger multicavity klystrons are being replaced by traveling-wave tubes in high-power applications.

Magnetrons

Another widely used microwave tube is the *magnetron,* a combination of a simple diode vacuum tube with built-in cavity resonators and an extremely powerful permanent magnet. The typical magnetron assembly shown in Fig. 16-40 consists of a circular anode into which has been machined with an even number of resonant cavities. The diameter of each cavity is equal to a one-half wavelength at the desired operating frequency. The anode is usually made of copper and is connected to a high-voltage positive direct current.

In the center of the anode, called the *interaction chamber,* is a circular cathode that emits electrons when heated. In a normal diode vacuum tube, the electrons would flow directly from the cathode straight to the anode, causing a high current to flow. In a magnetron tube, however, the direction of the electrons is modified because the tube is surrounded by a strong magnetic field. The field is usually supplied by a C-shaped permanent magnet centered over the interaction chamber. In the figure, the field is labeled *perpendicular to the page;* this means that the lines of force could be coming out of the page or going into the page depending upon the construction of the tube.

The magnetic fields of the moving electrons interact with the strong field supplied by the magnet. The result is that the path for electron flow from the cathode is not directly to the anode, but instead is curved. By properly adjusting the anode voltage and the strength of the magnetic field, the electrons can be made to bend such that they rarely reach the anode and cause current flow. The path becomes circular loops, as illustrated

Figure 16-40 A magnetron tube used as an oscillator.

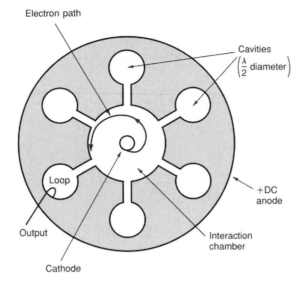

Electron path

Cavities
$\left(\frac{\lambda}{2} \text{ diameter}\right)$

Loop

+DC anode

Output

Interaction chamber

Cathode

Note: Magnetic field is perpendicular to the page.

in Fig. 16-40. Eventually, the electrons do reach the anode and cause current flow. By adjusting the dc anode voltage and the strength of the magnetic field, the electron path is made circular. In making their circular passes in the interaction chamber, electrons excite the resonant cavities into oscillation. A magnetron, therefore, is an oscillator, not an amplifier. A takeoff loop in one cavity provides the output.

Magnetrons are capable of developing extremely high levels of microwave power. Thousands and even millions of watts of power can be produced by a magnetron. When operated in a pulsed mode, magnetrons can generate several megawatts of power in the microwave region. Pulsed magnetrons are commonly used in radar systems. Continuous-wave magnetrons are also used and can generate hundreds and even thousands of watts of power. A typical application for a continuous-wave magnetron is for heating purposes in microwave ovens.

Traveling-Wave Tubes

One of the most versatile microwave RF power amplifiers is the *traveling-wave tube* (*TWT*), which can generate hundreds and even thousands of watts of microwave power. The main virtue of the TWT is an extremely wide bandwidth. It is not resonant at a single frequency.

Figure 16-41 shows the basic structure of a traveling-wave tube. It consists of a cathode and filament heater plus an anode that is biased positively to accelerate the electron beam forward and to focus it into a narrow beam. The electrons are attracted by a positive plate called the *collector* to which is applied a very high dc voltage. Traveling-wave tubes can be anywhere from 1 ft to several feet. In any case, the length of the tube is usually many wavelengths at the operating frequency. Permanent magnets or electromagnets surround the tube, keeping the electrons tightly focused into a narrow beam.

Encircling the length of a traveling-wave tube is a helix or coil. The electron beam passes through the axis of the helix. The microwave signal to be amplified is applied to the end of the helix near the cathode, and the output is taken from the end of the helix near the collector. The purpose of the helix is to provide a path for the RF signal that will slow down its propagation. The propagation of the RF signal along the helix is made approximately equal to the velocity of the electron beam from cathode to collector. The helix is configured such that the wave traveling along it is slightly slower than that of the electron beam.

Traveling-wave tube (TWT)

Figure 16-41   A traveling-wave tube (TWT).

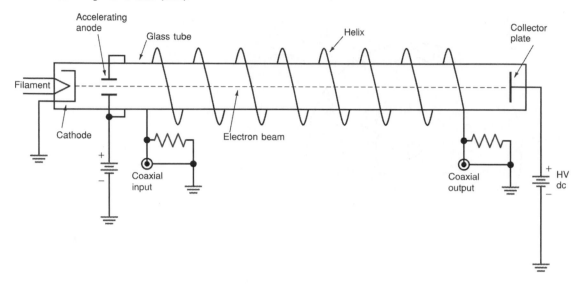

The passage of the microwave signal down the helix produces electric and magnetic fields that interact with the electron beam. The effect on the electron beam is similar to that in a klystron. The electromagnetic field produced by the helix causes the electrons to be speeded up and slowed down. This produces velocity modulation of the beam, which in turn produces density modulation. Density modulation, of course, causes bunches of electrons to group one wavelength apart. These bunches of electrons travel down the length of the tube toward the collector. Since the density-modulated electron beam is essentially in step with the electromagnetic wave traveling down the helix, the electron bunches induce voltages into the helix which reinforce the existing voltage. The result is that the strength of the electromagnetic field on the helix increases as the wave travels down the tube toward the collector. At the end of the helix, the signal is considerably amplified. Coaxial cable of waveguide structures is used to extract energy from the helix.

Traveling-wave tubes can be made to amplify signals in a range from UHF to hundreds of gigahertz. Most TWTs have a frequency range of approximately 2:1 in the desired segment of the microwave region to be amplified. TWTs can be used in both continuous and pulsed modes of operation. One of the most common applications of TWTs is as power amplifiers in satellite transponders.

Miscellaneous Microwave Tubes

Backward wave oscillator (BWO)

In a *backward wave oscillator (BWO)*, a variation of the TWT, the wave travels from the anode end of the tube back toward the electron gun, where it is extracted. BWOs can generate up to hundreds of watts of microwave power in the 20- to 80-GHz range. The operating frequency of the BWO can easily be tuned by varying the collector voltage.

Gyrotron

Gyrotrons, which are built and operated as klystrons, are used for amplification at microwave frequencies above 30 GHz into the millimeter wave range. They can also be connected to operate as oscillators. Gyrotrons are the only devices currently available for power amplification and signal generation in the millimeter wave range.

Crossed-field amplifier (CFA)

A *crossed-field amplifier (CFA)* is similar to a TWT. Its gain is lower but is somewhat more efficient. For a given power level, the operating voltage of a CFA is usually lower than that of a TWT. The bandwidth is about 20 to 60 percent of the design frequency. Power levels up to several megawatts in a pulse mode can be achieved.

16-6 Microwave Antennas

All the antennas we discussed in Chap. 15 can also be used at microwave frequencies. However, these antennas will be extremely small. At 5 GHz a half-wave dipole is slightly less than 1 in long, and a one-quarter wavelength vertical is slightly less than 0.5 in long. These antennas can, of course, radiate microwave signals, but inefficiently. Because of the line-of-sight transmission of microwave signals, highly directive antennas are preferred because they do not waste the radiated energy and because they provide an increase in gain, which helps offset the noise problems at microwave frequencies. For these important reasons, special high-gain, highly directive antennas are normally used in microwave applications.

Low-Frequency Antennas

At low microwave frequencies, less than 2 GHz, standard antennas are commonly used, including the dipole and its variations such as the bow tie, the Yagi, and the ground-plane antenna. Another variation is the *corner reflector* shown in Fig. 16-42. This antenna is a fat, wide-bandwidth, half-wave dipole fed with low-loss coaxial cable. Behind the dipole is a reflector made of solid sheet metal, a group of closely spaced horizontal rods, or a fine-mesh screen material to reduce wind resistance. This arrangement gives better reflection than a simple rod reflector as used in a Yagi, so the gain is higher.

The angle of the reflector is usually 45°, 60°, or 90°. The spacing between the dipole and the corner of the reflector is usually in the range from about 0.25λ to 0.75λ. Within that spacing range, the gain varies only about 1.5 dB. However, the feed point impedance of the dipole varies considerably with spacing. The spacing is usually adjusted for the best impedance match. It is common to use 50- or 75-Ω coaxial cable, which is relatively easy to match. A folded dipole can also be used if the antenna is operated in the UHF or VHF ranges. The overall gain of a corner reflector antenna is 10 to 15 dB. Higher gains can be obtained with a parabolic reflector, but corner reflectors are easier to make and much less expensive.

Horn Antennas

As discussed previously, waveguides are the most predominant type of transmission line used with microwave signals. Below approximately 6 GHz, special coaxial cables can be used effectively if the distances between the antenna and the receiver or transmitter are less than 50 ft. In most microwave systems, waveguides are preferred because of their low loss. Microwave antennas, therefore, must be some extension of or compatible with a waveguide. Waveguides are, of course, inefficient radiators if simply left open at the end. The problem with using a waveguide as a radiator is that it provides a poor impedance match with free space, and the mismatch results in standing waves and reflected power. The result is tremendous power loss of the radiated signal. This mismatch can be offset by simply flaring the end of the waveguide to create a *horn antenna,* as shown in Fig. 16-43. The longer and more gradual the flair, the better the impedance match and the lower the loss. Horn antennas have excellent gain and directivity. The longer the horn, the greater its gain and directivity.

Figure 16-42 A corner reflector used with a dipole for low microwave frequencies.

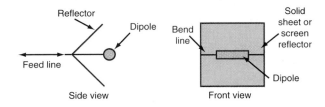

Figure 16-43 Basic horn antenna.

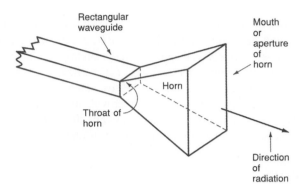

Different kinds of horn antennas can be created by flaring the end of the waveguide in different ways. For example, flaring the waveguide in only one dimension creates a sectoral horn, as shown in Fig. 16-44(*a*) and (*b*). Two of the sides of the horn remain parallel with the sides of the waveguide, and the other dimension is flared. Flaring both dimensions of the horn produces a pyramidal horn, as shown in Fig. 16-44(*c*). If a circular waveguide is used, the flare produces a conical horn, as in Fig. 16-44(*d*).

The gain and directivity of a horn are a direct function of its dimensions; the most important dimensions are horn length, aperture area, and flare angle (Fig. 16-45).

The length of a typical horn is usually 2λ to 15λ at the operating frequency. Assuming an operating frequency of 10 GHz, the length of λ is $300/f = 300/10,000 = 0.03$ m, where f is in megahertz. A length of 0.03 m equals 3 cm; 1 in equals 2.54 cm, so a wavelength at 10 GHz is $3/2.54 = 1.18$ in. Thus a typical horn at 10 GHz could be anywhere from about 2½ to 18 in long. Longer horns are, of course, more difficult to mount and work with, but provide higher gain and better directivity.

Figure 16-44 Types of horn antennas. (*a*) Sectoral horn. (*b*) Sectoral horn. (*c*) Pyramidal horn. (*d*) Conical horn.

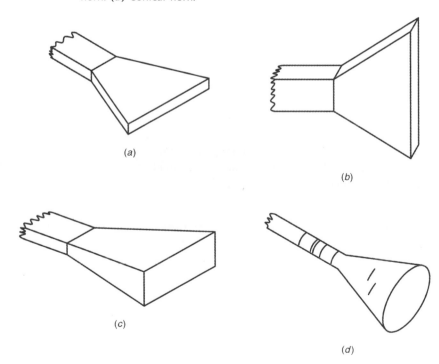

(a)

(b)

(c)

(d)

Figure 16-45 Dimensions of a horn.

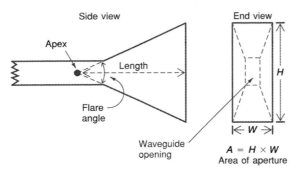

The aperture is the area of the rectangle formed by the opening of the horn, and it is the product of the height and width of the horn, as shown in Fig. 16-45. The greater this area, the higher the gain and directivity. The flare angle also affects gain and directivity. Typical flare angles vary from about 20° to 60°. Obviously, all these dimensions are interrelated. For example, increasing the flare angle increases the aperture area, and for a given aperture area, decreasing the length increases the flare angle. Any of these dimensions can be adjusted to achieve a desired design objective.

Beam Width. Remember that the directivity of an antenna is measured in terms of beam width, the angle formed by extending lines from the center of the antenna response curve to the 3-dB down points. In the example in Fig. 16-46, the beam width is approximately 30°. Horn antennas typically have a beam angle somewhere in the 10° to 60° range.

The signal radiated from an antenna is three-dimensional. The directivity patterns indicate the horizontal radiation pattern of the antenna. The antenna also has a vertical radiation pattern. Figure 16-47 is a typical plot of the vertical radiation pattern of a horn antenna. On pyramidal and circular horns, the vertical beam width is usually about the same angle as the horizontal beam width. This is not true for sectoral horns.

The horizontal beam width B of a pyramidal horn is computed by using the expression

$$B = \frac{80}{w/\lambda}$$

where w = horn width
λ = wavelength of operating frequency

Figure 16-46 Directivity of an antenna as measured by beam width.

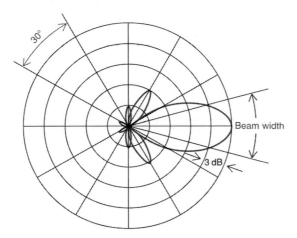

Figure 16-47 Vertical directivity of an antenna.

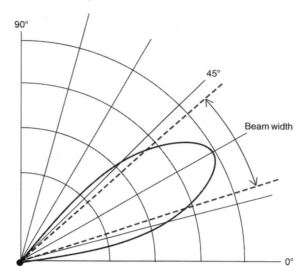

As an example, assume an operating frequency of 10 GHz (10,000 MHz), which gives a wavelength of 0.03 m, as computed earlier. If the pyramidal horn is 10 cm high and 12 cm wide, the beam width is $80/(0.12/0.03) = 80/4 = 20°$.

Gain. The gain of a pyramidal horn can also be computed from its dimensions. The approximate power gain of a pyramidal horn antenna is

$$G = 4\pi\frac{KA}{\lambda^2}$$

where A = aperture of horn, m^2
λ = wavelength, m
K = constant derived from how uniformly the phase and amplitude of electromagnetic fields are distributed across aperture

The typical values of K are 0.5 to 0.6.

Let us take as an example the previously described horn (height = 10 cm; width = 12 cm). The aperture area is

$$A = \text{height} \times \text{width} = 10 \times 12 = 120 \text{ cm}^2 = 0.012 \text{ m}^2$$

The operating frequency is 10 GHz, so $\lambda = 0.03$ m or 3 cm. The gain is

$$G = \frac{4(3.14)(0.5)(0.012)}{(0.03)^2} = \frac{0.07536}{0.0009} = 83.7$$

This is the power ratio P. To find the gain in decibels, the standard power formula is used:

$$dB = 10 \log P$$

where P is the power ratio or gain. Here,

$$dB = 10 \log 83.7$$
$$= 10(1.923)$$
$$= 19.23$$

This is the power gain of the horn over a standard half-wave dipole or quarter-wave vertical.

Bandwidth. Most antennas have a narrow bandwidth because they are resonant at only a single frequency. Their dimensions determine the frequency of operation. Bandwidth is an important consideration at microwave frequencies because the spectrum transmitted on the microwave carrier is usually very wide so that a considerable amount of information can be carried. Horns are essentially nonresonant or aperiodic, which means that they can operate over a wide frequency range. The bandwidth of a typical horn antenna is approximately 10 percent of the operating frequency. The bandwidth of a horn at 10 GHz is approximately 1 GHz. This is an enormous bandwidth—plenty wide enough to accommodate almost any kind of complex modulating signal.

Parabolic Antennas

Horn antennas are used by themselves in many microwave applications. When higher gain and directivity is desirable, it can easily be obtained by using a horn in conjunction with a parabolic reflector. A *parabolic reflector* is a large dish-shaped structure made of metal or screen mesh. The energy radiated by the horn is pointed at the reflector, which focuses the radiated energy into a narrow beam and reflects it toward its destination. Because of the unique parabolic shape, the electromagnetic waves are narrowed into an extremely small beam. Beam widths of only a few degrees are typical with parabolic reflectors. Of course, such narrow beam widths also represent extremely high gains.

Parabolic reflector

A parabola is a common geometric figure. See Fig. 16-48. A key dimension of a parabola is a line drawn from its center at point Z to a point on the axis labeled F, which is the focal point. The ends of a parabola theoretically extend outward for an infinite distance, but for practical applications they are limited. In the figure, the limits are shown by the dashed vertical line; the endpoints are labeled X and Y.

The distance between a parabola's focal point and any point on the parabola and then to the vertical dashed line is a constant value. For example, the sum of lines FA and AB is equal to the sum of lines FC and CD. This effect causes a parabolic surface to collimate electromagnetic waves into a narrow beam of energy. An antenna placed at the focal point F will radiate waves from the parabola in parallel lines. If used as a receiver, the parabola will pick up the electromagnetic waves and reflect them to the antenna located at the focal point.

Figure 16-48 Cross-sectional view of a parabolic dish antenna.

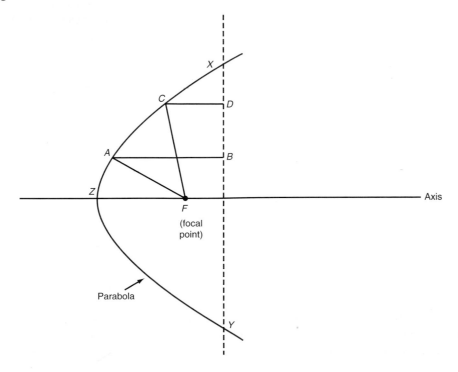

Figure 16-49 Sending and receiving with a parabolic reflector antenna.

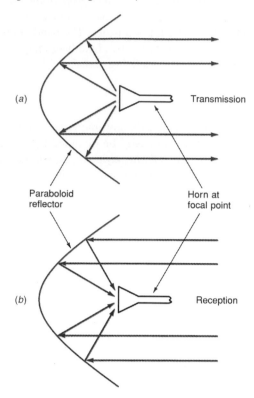

The key thing to remember about a parabolic reflector is that it is not two-dimensional. If a parabola is rotated about its axis, a three-dimensional dish-shaped structure results. This is called a paraboloid.

Figure 16-49 shows how a parabolic reflector is used in conjunction with a conical horn antenna for both transmission and reception. The horn antenna is placed at the focal point. In transmitting, the horn radiates the signal toward the reflector, which bounces the waves off the reflector and collimates them into a narrow parallel beam. When used for receiving, the reflector picks up the electromagnetic signal and bounces the waves toward the antenna at the focal point. The result is an extremely high-gain, narrow-beamwidth antenna.

Note that any common antenna type (e.g., a dipole) can be used with a parabolic reflector to achieve the effects just described.

Gain. The gain of a parabolic antenna is directly proportional to the aperture of the parabola. The aperture, the area of the outer circle of the parabola, is

$$A = \pi R^2$$

The gain of a parabolic antenna is given by the simple expression

$$G = 6\left(\frac{D}{\lambda}\right)^2$$

where G = gain, expressed as a power ratio
  D = diameter of dish, m
  λ = wavelength, m

Most parabolic reflectors are designed so that the diameter is no less than λ at the lowest operating frequency. However, the diameter can be as much as 10λ if greater gain and directivity is required.

For example, the power gain of a 5-m-diameter dish at 10 GHz ($\lambda = 0.03$, as computer earlier) is

$$G = 6\left(\frac{5}{0.03}\right)^2$$
$$= 6(166.67)^2$$
$$= 6(27,778.9) = 166,673$$

Expressed in decibels, this is

$$dB = 10 \log 166,673 = 10(5.22) = 52.2$$

Beam Width. The beam width of a parabolic reflector is inversely proportional to the diameter. It is given by the expression

$$B = \frac{70}{D/\lambda}$$

The beam width of our 5-m, 10-GHz antenna is $70(5/0.03) = 70/166.67 = 0.42°$.

With a beam width of less than $0.5°$, the signal radiated from a parabolic reflector is a pencil-thin beam that must be pointed with great accuracy in order for the signal to be picked up. Usually both the transmitting and receiving antennas are of a parabolic design and have extremely narrow beam widths. For that reason, they must be accurately pointed if contact is to be made. Despite the fact that the beam from a parabolic reflector spreads out and grows in size with distance, directivity is good and gain is high. The precise directivity helps to prevent interference from signals coming in at angles outside of the beam width.

Feed Methods. Keep in mind that the parabolic dish is not the antenna, only a part of it. The antenna is the horn at the focal point. There are many physical arrangements used in positioning the horn. One of the most common is the seemingly awkward configuration shown in Fig. 16-50. The waveguide feeds through the center of the parabolic dish and is curved around so that the horn is positioned exactly at the focal point.

Another popular method of feeding a parabolic antenna is shown in Fig. 16-51. Here the horn antenna is positioned at the center of the parabolic reflector. At the focal point is another small reflector with either a parabolic or a hyperbolic shape. The electromagnetic radiation from the horn strikes the small reflector, which then reflects the energy toward the large dish which in turn radiates the signal in parallel beams. This arrangement is known as a *Cassegrain feed*.

The Cassegrain feed has several advantages over the feed arrangement in Fig. 16-50. The first is that the waveguide transmission line is shorter. In addition, the radical bends

Cassegrain feed

Figure 16-50 Standard waveguide and horn feed.

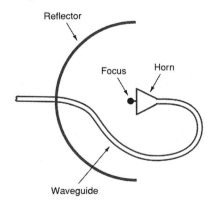

Figure 16-51 Cassegrain feed.

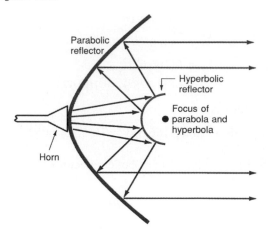

in the waveguide are eliminated. Both add up to less signal attenuation. The noise figure is also improved somewhat. Most large earth station antennas use a Cassegrain feed arrangement.

Many other feed arrangements have been developed for parabolic reflectors. In antennas used for satellite TV reception, a waveguide is not used. Instead, a horn antenna mounted at the focal point is usually fed with large microwave coaxial cable. In other large antenna systems, various mechanical arrangements are used to permit the antenna to be rotated or its position otherwise physically changed. Many earth station antennas must be set up so that their azimuth and elevation can be changed to ensure proper orientation for the receiving antenna. This is particularly true of antennas used in satellite communication systems.

Example 16-4

A parabolic reflector antenna has a diameter of 5 ft. Calculate (*a*) the lowest possible operating frequency, (*b*) the gain at 15 GHz, and (*c*) the beam width at 15 GHz. (The lowest operating frequency occurs where the dish diameter is λ.)

a. $\lambda = \dfrac{984}{f_{MHz}}$

$= 5$ ft

$f_{MHz} = \dfrac{984}{5} = 196.8$ MHz

b. There is 3.28 ft/m, so the diameter is $5/3.28 = 1.524$ m.

15 GHz = 15,000 MHz

$\lambda = \dfrac{300}{f_{MHz}} = \dfrac{300}{15,000} = 0.02$ m

$G = 6\left(\dfrac{D}{\lambda}\right)^2 = 6\left(\dfrac{1.524}{0.02}\right)^2 = 34,838.6 \qquad$ (power density ratio)

$dB = 10 \log 34,838.6 = 45.42$

c. $B = \dfrac{70}{D/\lambda} = \dfrac{70\lambda}{D} = \dfrac{70(0.02)}{1.524} = 0.92°$

Figure 16-52 · The helical antenna.

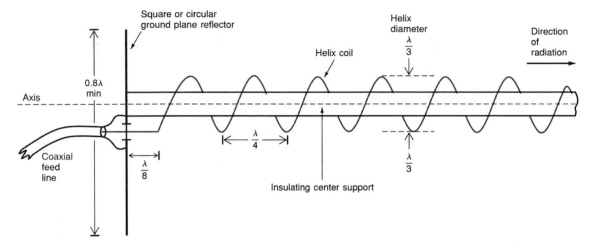

Some sophisticated systems use multiple horns on a single reflector. Such multiple feeds permit several signals on different frequencies to be either radiated or received with a single large reflecting structure.

Helical Antennas

A *helical antenna,* as its name suggests, is a wire helix (Fig. 16-52). A center insulating support is used to hold heavy wire or tubing formed into a circular coil or helix. The diameter of the helix is typically one-third wavelength, and the spacing between turns is approximately one-quarter wavelength. Most helical antennas use from 6 to 8 turns. A circular or square ground-plane antenna or reflector is used behind the helix. Figure 16-52 shows a coaxial feed line. Helical antennas are widely used at VHF and UHF ranges.

Helical antenna

The gain of a helical antenna is typically in the 12- to 20-dB range, and beam widths vary from approximately 12° to 45°. Although these values do not compare favorably with those obtainable with horns and parabolic reflectors, helical antennas are favored in many applications because of their simplicity and low cost.

Most antennas transmit either a vertically or a horizontally polarized electromagnetic field. With a helical antenna, however, the electromagnetic field is caused to rotate. This is known as *circular polarization.* Either right-hand (clockwise) or left-hand (counterclockwise) circular polarization can be produced, depending on the direction of winding of the helix. Because of the rotating nature of the magnetic field, a circularly polarized signal can easily be received by either a horizontally or a vertically polarized receiving antenna. A helical receiving antenna can also easily receive horizontally or vertically polarized signals. Note, however, that a right-hand circularly polarized signal will not be picked up by a left-hand circularly polarized antenna, and vice versa. Therefore, helical antennas used at both transmitting and receiving ends of a communication link must both have the same polarization.

Circular polarization

To obtain greater gain and narrower beam width, several helical antennas can be used in an array with a common reflector. A popular arrangement is a group of four helical antennas.

Bicone Antennas

Most microwave antennas are highly directional. But in some applications an omnidirectional antenna may be required. One of the most widely used omnidirectional

Figure 16-53 The omnidirectional bicone antenna.

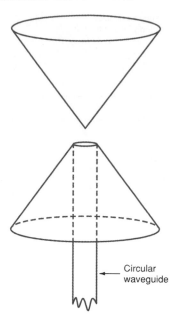

Circular
waveguide

microwave antennas is the *bicone* (Fig. 16-53). The signals are fed into bicone antennas through a circular waveguide ending in a flared cone. The upper cone acts as a reflector, causing the signal to be radiated equally in all directions with a very narrow vertical beam width. A version of the bicone replaces the upper cone with a flat horizontal disk that performs the same function.

Slot Antennas

A *slot antenna* is a radiator made by cutting a one-half wavelength slot in a conducting sheet of metal or into the side or top of a waveguide. The basic slot antenna is made by cutting a one-half wavelength slot in a large metal sheet. It has the same characteristics as a standard dipole antenna, as long as the metal sheet is very large compared to λ at the operating frequency. A more common way of making a slot antenna is shown in Fig. 16-54. The slot must be one-half wavelength long at the operating frequency. Figure 16-54(*a*) shows how the slots must be positioned on the waveguide to radiate. If the slots are positioned on the center lines of the waveguide sides, as in Fig. 16-54(*b*), they will not radiate.

Figure 16-54 Slot antennas on a waveguide. (*a*) Radiating slots. (*b*) Nonradiating slots.

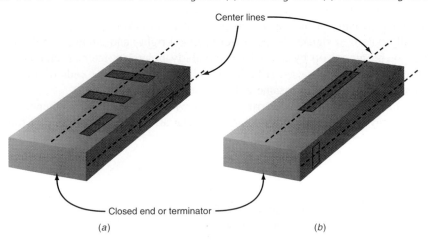

Center lines

Closed end or terminator

(*a*) (*b*)

Figure 16-55 A slot antenna array.

Several slots can be cut into the same waveguide to create a slot antenna array (Fig. 16-55). Slot arrays, which are equivalent to driven arrays with many elements, have better gain and better directivity than single-slot antennas.

Slot antennas are widely used on high-speed aircraft. External antennas would be torn off at such high speeds or would slow the aircraft. The slot antenna can be integrated into the metallic skin of the aircraft. The slot itself is filled in with an insulating material to create a smooth skin surface.

Dielectric (Lens) Antennas

As discussed previously, radio waves, similar to light waves, can be reflected, refracted, diffracted, and otherwise manipulated. This is especially true of microwaves, which are close in frequency to light. Thus a microwave antenna can be created by constructing a device that serves as a lens for microwaves just as glass or plastic can serve as a lens for light waves. These *dielectric* or *lens antennas* use a special dielectric material to collimate or focus the microwaves from a source into a narrow beam. Figure 16-56 shows how a lens concentrates light rays from a source into a focused narrow beam. A dielectric lens antenna operates in a similar way.

An example of a lens antenna is one used in the millimeter wave range. The microwave energy is coupled to a horn antenna through a waveguide. A dielectric lens is placed over the end of the horn, which focuses the waves into a narrower beam with greater gain and directivity. In technical terms, the lens takes the microwaves from a source with a spherical wave front (e.g., a horn antenna) and concentrates them into a plane wave front. A lens like that shown in Fig. 16-57(*a*) can be used. The shape of the lens ensures that all the entering waves with a spherical wave front are put into phase at the output to create the concentrated plane wave front. However, the lens in Fig. 16-57(*a*) will work only when it is very thick at the center. This creates great signal loss, especially at the lower microwave frequencies. To get around this problem, a stepped or zoned lens, such as the one shown in Fig. 16-57(*b*), can be used. The spherical wave front is still converted to a focused plane wave front, but the thinner lens causes less attenuation.

Lens antennas are usually made of polystyrene or some other plastic, although other types of dielectric can be used. They are rarely used at the lower microwave frequencies. Their main use is in the millimeter range above 40 GHz.

Dielectric (lens) antenna

Figure 16-56 How a lens focuses light rays.

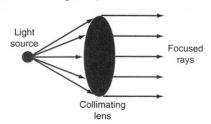

Figure 16-57   Lens antenna operations. (*a*) Dielectric lens. (*b*) Zoned lens.

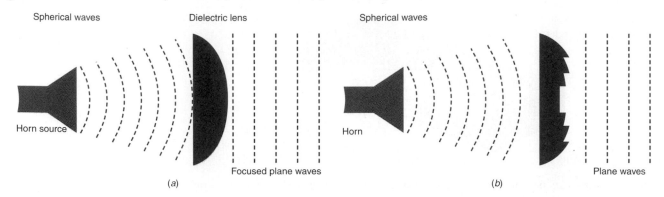

Spherical waves   Dielectric lens

Horn source

Focused plane waves

(*a*)

Spherical waves

Horn

Plane waves

(*b*)

Patch Antennas

Patch antenna

Patch antennas are made with microstrip on PCBs. The antenna is a circular or rectangular area of copper separated from the ground plane on the bottom of the board by the thickness of the PCB's insulating material (see Fig. 16-58). The width of the rectangular antenna is approximately one-half wavelength, and the diameter of the circular antenna

Figure 16-58   Patch or microstrip antennas. (*a*) Coaxial feed. (*b*) Edge feed.

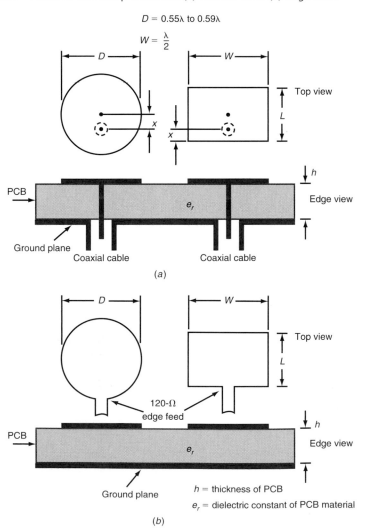

$D = 0.55\lambda$ to 0.59λ

$W = \dfrac{\lambda}{2}$

D W

Top view

x x L

h

PCB e_r Edge view

Ground plane   Coaxial cable   Coaxial cable

(*a*)

D W

Top view

L

120-Ω edge feed

h

PCB e_r Edge view

Ground plane

h = thickness of PCB

e_r = dielectric constant of PCB material

(*b*)

652   *Chapter 16*

is about 0.55λ to 0.59λ. In both cases the exact dimensions depend upon the dielectric constant and the thickness of the PCB material. The most commonly used PC board material for patch antennas is a Teflon-fiberglass combination.

The feed method for patch antennas can be either coaxial or edge. With the coaxial method, the center conductor of a coaxial cable is attached somewhere between the center and the edge of the patch, and the coaxial shield is attached to the ground plane [Fig. 16-58(a)]. If the antenna is fed at the edge, a length of microstrip is connected from the source to the edge, as shown in Fig. 16-58(b). The impedance of the edge feed is about 120 Ω. A quarter-wave Q section can be used to match this impedance to the 50-Ω impedance that is characteristic of most circuits. When coaxial feed is used, the impedance is zero at the center of the antenna and increases to 120 Ω at the edge. Correctly positioning the coaxial cable center on the patch [dimension x in Fig. 16-58(a)] allows extremely accurate impedance matching.

Patch antennas are small, inexpensive, and easy to construct. In many applications they can simply be integrated on the PCB with the transmitter or receiver. A disadvantage of patch antennas is their narrow bandwidth, which is usually no more than about 5 percent of the resonant frequency with circular patches and up to 10 percent with rectangular patches. The bandwidth is directly related to the thickness of the PCB material, i.e., the distance between the antenna and the ground plane [h in Fig. 16-58(b)]. The greater the thickness of the PCB dielectric, the greater the bandwidth.

The radiation pattern of a patch antenna is approximately circular in the direction opposite to that of the ground plane.

Phased Arrays

A *phased array* is an antenna system made up of a large group of similar antennas on a common plane. Patch antennas on a common PCB can be used, or separate antennas such as dipoles can be physically mounted together in a plane. See Fig. 16-59. Slot antennas are also used. The antennas are driven by transmission lines that incorporate impedance-matching, power-splitting, and phase-shift circuits. The basic purpose of an array is to improve gain and directivity. Arrays also offer better control of directivity, since individual antennas in an array can be turned off or on, or driven through different phase shifters. The result is that the array can be "steered"—i.e., its radiation pattern can be pointed over a wide range of different directions—without physically moving the antenna, as is necessary with Yagis or parabolic dish antennas.

Phased array

Figure 16-59 An 8 × 8 phase array using patch antennas. (Feed lines are not shown.)

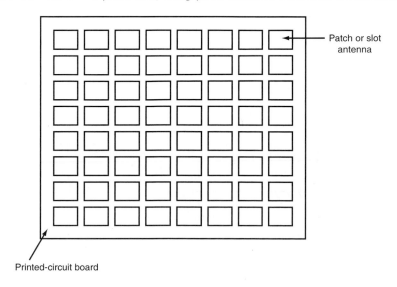

Patch or slot antenna

Printed-circuit board

There are two common arrangements for phased arrays. In one configuration, the multiple antennas are driven by a common transmitter or feed a common receiver. A second approach is to have a low-power transmitter amplifier or low-noise receiver amplifier associated with each dipole or patch in the array. In both cases, the switching and phase shifting are under the control of a microprocessor or computer. Different programs in the processor select the gain, directivity, and other factors as required by the application.

Most phased arrays are used in radar systems, but they are finding applications in cell phone systems and in satellites.

Printed–Circuit Antennas

Because antennas are so small at microwave frequencies, they can be conveniently made right on a printed-circuit board that also holds the transmitter and/or receiver ICs and related circuits. No separate antenna structure, feed line, or connectors are needed. The patch and slot antennas discussed previously are examples. But there are a few other types that are widely used. These are the loop, the inverted-F, and the meander line antennas shown in Fig. 16-60.

Figure 16-60 Popular PCB antennas. (*a*) Loop (*b*) Inverted-F. (*c*) Meander line.

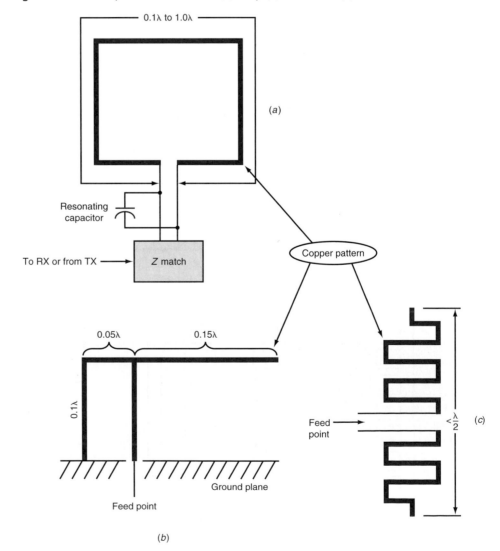

The Loop. A loop antenna is just as its name implies, a single closed loop that is usually rectangular but could be round as well. See Fig. 16-60(a). The length of the loop is usually in the 0.1λ to λ at the operating frequency. The loop is usually resonated with a parallel capacitor. Because the characteristic impedance of the loop is very low, about $10\ \Omega$ at 0.5λ but as high as $120\ \Omega$ for λ, some form of microstrip transmission line is used to match the impedance to the receiver or transmitter. Loops are relatively inefficient but are effective in short-range applications such as garage door openers and remote keyless entry radios in cars and in pagers.

Inverted–F. This unique antenna is a variation of the ground plane as it is designed to work over a conducting ground plane. See Fig. 16-60(b). Note the lengths of the various segments. These are experimented with to get the desired performance as well as the impedance match to the transmission line. A desirable feature is that the radiation pattern is effectively omnidirectional.

Meander Line. The meander line in Fig. 16-60(c) is an attempt to shorten an antenna by bending the conductors back on themselves to save space. The design is essentially a half-wave dipole and performs as one. All sorts of variations have been created by using curved sections and cross patterns in addition to the bow tie design discussed earlier.

Dielectric Antennas. A dielectric antenna is one whose copper pattern elements are formed on some type of resonant dielectric material such as ceramic or some derivative thereof. The dielectric is designed to be resonant to the frequency of operation and actually contributes to the radiation. Dielectric antennas often use the inverted-F, dipole, or meander line configurations. No ground plane is necessary for operation. Dielectric antennas are small and have a wide bandwidth. These antennas are often available as a component for mounting on a printed-circuit board.

Intelligent Antenna Technology

Intelligent antennas or smart antennas are antennas that work in conjunction with electronic decision-making circuits to modify antenna performance to fit changing situations. They adapt to the signals being received and the environment in which they transmit. Also called adaptive antennas, these new designs greatly improve transmission and reception in multipath environments and can also multiply the number of users of a wireless system. Some popular adaptive antennas today use diversity, multiple-input multiple-output, and automatic beam forming.

Diversity. Diversity was discussed in Chap. 14. It uses two or more antennas that receive the signal from different physical positions. In this way, they get different signals. The antenna with the strongest signal is selected, or the signals are combined to produce a stronger overall signal. More and more, microwave equipment is using diversity simply because of the degrading effects of multiple signal paths, reflections, diffractions, and other conditions that weaken the signal in complex environments. Some systems use three or four antennas with different modes of signal selection to optimize reception.

MIMO. *Multiple-input, multiple-output (MIMO)* takes the idea of diversity to a whole new level. It uses two or more antennas for receiving but also uses two or more antennas for transmission. One possible arrangement is shown in Fig. 16-61. The data to be transmitted is divided into two separate streams of bits that are transmitted simultaneously. Since there are two separate data paths, the effect is to double the actual data rate of transmission. For example, in one type of wireless LAN, the maximum data rate is 54 Mbps. With two data streams, the composite throughtput is 108 Mbps. The transmit antennas are physically separated by a wavelength or more, so that they truly generate different paths to the receiver. The modulation is usually some form of OFDM, and the data is transmitted in the same bandwidth.

Figure 16-61 In MIMO, two transmitters send parallel data, doubling the data rate, to multiple receivers that process the signals to improve gain and reliability.

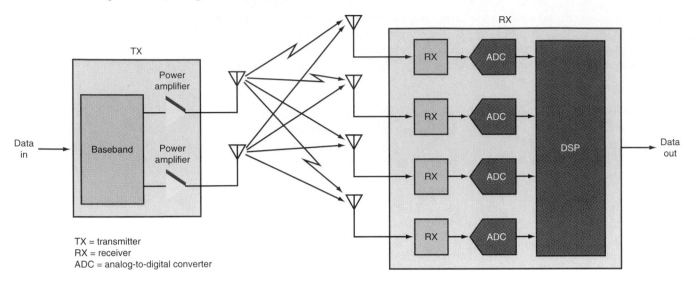

TX = transmitter
RX = receiver
ADC = analog-to-digital converter

At the receiving end, three or more antennas are used. The two transmitted signals take different paths to the four receiver antennas shown in Fig. 16-61. The receiver antennas are separated by a wavelength or so, providing multiple paths for each of the two transmitted signals. The signals may experience multipath reflections and other anomalies along the way. The receiver antennas pick up everything. The outputs of the four receivers are then digitized with analog-to-digital converters (ADCs), and their outputs are combined in a DSP. Special algorithms programmed into the DSP manipulate and combine the signals in different ways to minimize the multipath effects and to create usable signals where none were available with just one transmitter and receiver. MIMO provides an amazing increase in signal gain and reliability. And despite the seemingly high cost and complexity of such a scheme, in reality the very small low-cost IC receivers and transmitters make this technique very simple and affordable from the hardware point of view. The real complexity and "magic" is in the DSP. MIMO makes a previously impossible wireless application work by providing multiple signals that can be combined and processed to produce a usable signal.

Adaptive Beam Forming

Adaptive antennas are systems that automatically adjust their characteristics to the environment. They use beam-forming and beam-pointing techniques to zero in on signals to be received and to ensure transmission under noisy conditions with interference from other sources.

Beam-forming antennas use multiple antennas such as the phase arrays discussed earlier. By using many antennas, the transmit/receive pattern can be adjusted as required by the situation. The beam may be narrowed or widened, and the direction of the beam may be electronically adjusted on the fly thanks to electronic controls. Such directional antennas can pinpoint a specific signal while tuning out interfering signals on the same frequency at nearby locations. Beam-forming antennas also have high gain, which helps boost the signal strength of a desired signal and improves the link reliability since noise and interference have been minimized.

There are two kinds of adaptive antennas, switched beam arrays and adaptive arrays. The radiation pattern of a switched beam antenna looks something like that in Fig. 16-62. The antenna itself is usually multiple phased arrays. For example, there may be four phased arrays each capable of covering 90° to 100° of azimuth. Multiple beams are formed in that 90° range. The electronics controlling the arrays steers the beam by some predetermined algorithm. The most common arrangement is for the antenna to scan the

Figure 16-62 A switched beam intelligent antenna.

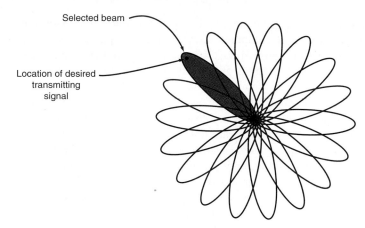

entire 360° range in seeking a signal. As each beam is switched in, the signal strength is monitored. The beam with the strongest signal is then selected. Signals outside the beam will not be received at all, or at most only a small amount of signal will be present. Such antennas can provide a gain of 20 to 50 dB.

An adaptive antenna can also cover the entire 360° range but uses more sophisticated control algorithms. The adaptive array not only seeks out the stongest signal and adjusts the beam width to enhance it, but also recognizes interfering signals and adjusts the antenna to null out the interfering signal. See Fig. 16-63. Adaptive arrays track signals and then fine-tune themselves for best reception. All this takes place automatically at electronic speeds.

Both switched beam arrays and adaptive arrays are already being employed in some cell phone systems and in newer wireless LANs. They are particularly beneficial to cell phone systems because they can actually boost the system capacity since they can reuse the same frequencies multiple times and allow the antenna to keep signals on the same frequency from interfering with one another. This concept is known as *spatial division multiplexing* or *spatial division multiple access (SDMA)*.

Figure 16-63 An adaptive array zeros in on the desired signal while mulling out interfering signals.

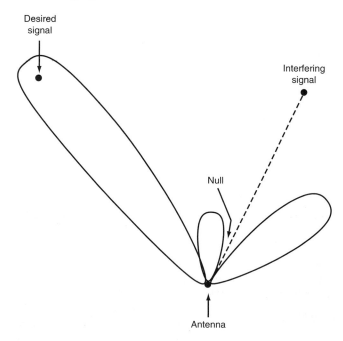

16-7 Microwave Applications

The communication applications in which microwaves are most widely used today are telephone communication, computer networking, cell phones, satellites, and radar. However, there are many other significant uses of microwave frequencies in communication. For example, TV stations use microwave relay links instead of coaxial cables to transmit TV signals over long distances, and cable TV networks use satellite communication to transmit programs from one location to another. Communication with satellites, deep-space probes, and other spacecraft is usually done by microwave transmission because microwave signals are not reflected or absorbed by the ionosphere, as are many lower-frequency signals. Electromagnetic radiation from the stars is also primarily in the microwave region; only sensitive radio receivers and large antennas operating in the microwave region are used to map outer space with far greater precision than could be achieved with optical telescopes. Finally, microwaves are also used for heating—in the kitchen (microwave ovens), in medical practice (diathermy machines used to heat muscles and tissues without causing skin damage), and in industry for melting material and heat treating.

Figure 16-64 summarizes the major applications of microwaves. The military uses microwave for multichannel communication with ranges from 1 km to 160 km for line-of-sight and troposcatter communications. The multiple channels can carry a variety of communication signals.

Figure 16-64 Major applications of microwave radio.

1. Radar
   a. Aircraft and marine navigation
   b. Military threat detection
   c. Altimeters
   d. Weather plotting
   e. Traffic speed enforcement
   f. Automotive collision avoidance and speed control

2. Satellite
   a. Telephone communication
   b. TV transmission (cable, short-range, direct broadcast)
   c. Surveillance
   d. Weather plotting
   e. Navigation (GPS, etc.)

3. Wireless local-area and personal-area networks
   a. IEEE 802.11b/g   Ethernet, 2.4 GHz, rates of 11 to 54 Mbps
   b. IEEE 802.11a   Ethernet, 5 GHz, rate of 54 Mbps
   c. 802.11n   Ethernet, 2.4 GHz, rate to 250 Mbps
   d. Bluetooth   2.4 GHz, rate to 3 Mbps
   e. HomeRF   2.4 GHz, rate to 10 Mbps
   f. Ultrawideband   Rate to 1 Gbps

4. Wireless broadband access to the Internet
   a. MMDS
   b. LMDS
   c. WiMAX

5. Cell phones (allocations in the 1.8-, 1.9-, and 2.3-GHz ranges)

6. Heating
   a. Microwave ovens (domestic)
   b. Microwave heating (industrial)

7. Radio telescopes

The rest of this chapter is devoted to a discussion of radar. Satellites are covered in detail in Chap. 17. Cell phones, wireless computer networks, and other special microwave applications are covered in Chaps. 21 and 22.

Radar

The electronic communication system known as *radar* (*ra*dio *d*etection *a*nd *r*anging) is based on the principle that high-frequency RF signals are reflected by conductive targets. The usual targets are airplanes, missiles, ships, and automobiles. In a radar system, a signal is transmitted toward the target. The reflected signal is picked up by a receiver in the radar unit. The reflected or return radio signal is called an *echo*. The radar unit can then determine the distance to the target (range), its direction (azimuth), and in some cases its elevation (distance above the horizon).

The ability of radar to determine the distance between a remote object and the radar unit is dependent upon knowing the exact speed of radio signal transmission. In most radar applications, nautical miles are used instead of statute miles to express transmission speeds. One nautical mile is equal to 6076 ft. The speed of a radio signal is 162,000 nautical miles per second. (Sometimes, a special unit known as a radar mile is used. One radar mile is equal to 6000 ft.) It takes a radio signal 5.375 µs to travel 1 mi and 6.18 µs to travel 1 nautical mile.

A radar signal must travel twice the distance between the radar unit and the remote target. The signal is transmitted, a finite time passes before the signal reaches the target and is reflected, and the signal then travels an equal distance back to the source. If an object is exactly 1 nautical mile away, the signal takes 6.18 µs to reach the target and 6.18 µs to return. The total elapsed time from the instant of initial transmission to the reception of the echo is 12.36 µs.

The distance to a remote target is calculated by using the expression

$$D = \frac{T}{12.36}$$

where D = distance between radar unit and remote object, nautical miles
T = total time between transmission and reception of signal, µs

In short-distance applications, the yard is the common unit of distance measurement. A radio signal travels 328 yd/µs, so the distance to an object in yards is computed as

$$D = \frac{328T}{2} = 164T$$

A measured time of 5.6 µs corresponds to a distance of 164(5.6) = 918.4 yd.

To obtain a strong reflection or echo from a distant object, the wavelength of the radar signal should be small compared to the size of the object being observed. If the wavelength of the radar signal is long with respect to the distant object, only a small amount of energy will be reflected. At higher frequencies, the wavelength is shorter and therefore the reflected energy is greater. For optimal reflection, the size of the target should be one-quarter wavelength or more at the transmitted frequency.

The shorter the wavelength of the signal compared to the observed object, the higher the resolution or definition of the remote object. In most cases, it is necessary only to detect the presence of a remote object. But if very short wavelengths are used, in many cases the actual shape of an object can be clearly determined.

The term *cross section* is often used with reference to a radar target. If a target is at least 10 times larger than λ of the radar signal detecting it, the cross section is constant. The cross section of a target, a measure of the area of the target "illuminated" by the radar signal, is given in square meters. A target's cross section is determined by the size of the object, the unique geometry of the target, the viewing angle, and the position.

The larger the cross section, the greater the reflected signal power, the greater the distance of detection, and the higher the probability of the signal being greater than the noise. Another factor influencing the return signal strength is the material of the target. Metal returns the greatest signal; other materials can also reflect radar waves, but not as effectively. Some objects will absorb radar waves, making the reflected signal very small.

The F117, the U.S. Air Force's stealth fighter, has a large physical cross section (about 1 m^2), but it is designed to deflect and absorb any radar signal aimed at it. All the plane's surfaces are at an angle such that radio signals striking them are not reflected directly back to the radar unit. Instead, they are reflected off at an angle, and little if any reflected energy is received. The surface of the aircraft is also coated with a material that absorbs radio waves. This combination gives the F117 an effective cross section the size of a small bird.

All the most important factors affecting the amount of received signal reflected from a target are summed up in what is called the *radar equation:*

$$P_r = \frac{P_t G \sigma A_e}{(4\pi)^2 R^4}$$

where P_r = received power
G = antenna gain (product of transmitting and receiving gains)
σ = cross section of target
A_e = effective area of receiving antenna (dish area)
R = range or distance to target

Most of the relationships between the variables are obvious. However, given the fact that the received power is inversely proportional to the fourth power of the distance to the target, it is not surprising that radar range is generally so limited. Keep in mind that the radar equation does not take into account the *S/N* ratio. Very low-noise receiver front ends are essential for target acquisition.

Since radar uses microwave frequencies, line-of-sight communication results. In other words, radar cannot detect objects beyond the horizon. Objects do not have to be physically visible, but must be within line-of-sight radio distance in order for detection to occur.

The relationship between range, azimuth, and elevation can be expressed by a right triangle, as shown in Fig. 16-65. Assume that the radar is land-based and used to detect aircraft. The distance between the radar unit and the remote airplane is the hypotenuse of the right triangle. The angle of elevation is the angle between the hypotenuse and the baseline, which is a line tangent to the surface of the earth at the radar location.

Figure 16-65 The trigonometry of radar.

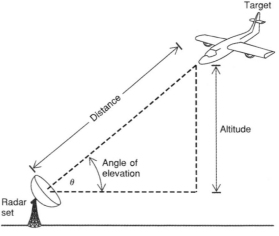

Altitude = distance × sin θ

The altitude is defined by the angle of elevation. The greater the angle of elevation, the greater the altitude. Knowing the range and the angle of elevation allows the altitude to be computed by using standard trigonometric techniques.

Another important factor in locating a distant object is knowledge of its direction with respect to the radar set. If the radar station is fixed and land-based, the direction (bearing or azimuth) of the remote object is usually given as a compass direction in degrees. Recall that true north is 0 or 360°, east is 90°, south is 180°, and west is 270°. If the radar unit is located in a moving vehicle, such as an airplane or a ship, the azimuth is given as a relative bearing with respect to the forward direction of the vehicle. Straight ahead is 0° or 360°, directly to the right is 90°, directly behind is 180°, and directly to the left is 270°.

The ability of a radar unit to determine the direction of a remote object requires the use of a highly directional antenna. An antenna with an extremely narrow beam width will receive signals only over a narrow angle. The narrower the beam width of the antenna, the more precisely the actual bearing can be determined.

Since most radar systems operate in the microwave region, highly directional antennas are easily obtained. Horns with parabolic reflectors are the most common, and beam widths of less than 1° are readily attainable. These highly directional antennas are continuously rotated 360°. The same antenna is used for transmitting the original signal and receiving the reflected signal.

Circuits within the radar unit are calibrated so that the direction in which the antenna is pointing is accurately known. When the echo is received, it is compared to the calibrated values and the precise direction determined.

The ability of a radar unit to determine the altitude of a remote target depends on the vertical beam width of the radar antenna. The radar antenna may scan vertically while measuring the distance of the object during the scan. When the object is detected, the vertical elevation of the antenna is noted and the actual altitude is then computed.

Example 16-5

A radar set detects the presence of an aircraft. The time between the radiated and received pulses is 9.2 μs. The antenna is set to an angle of elevation of 20°. Determine (a) the line-of-sight distance to the aircraft in statute miles and (b) the altitude of the aircraft.

a. $D \text{ (nautical miles)} = \dfrac{T}{12.36} = \dfrac{9.2}{12.36} = 0.744$

1 statute mile = 5280 ft

1 nautical mile = 6076 ft

$\dfrac{5280}{6076} = 0.87$ statute mile/nautical mile

$D \text{ (statute miles)} = 0.744(0.87) = 0.647$

b. $A = D \sin \theta$

where θ = angle of elevation

$A = 0.647 \sin 20$

$= 0.647(0.342) = 0.22$ mi

$= 0.22(5280 \text{ ft}) = 1161.6$ ft

Figure 16-66   Pulsed radar signals.

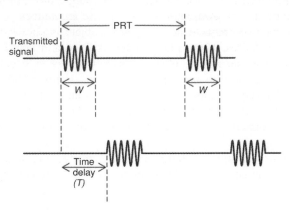

Pulsed Radar. There are two basic types of radar systems: *pulsed* and *continuous-wave (CW)*. There are also numerous variations of each. By far the most commonly used radar system is the pulsed type. Signals are transmitted in short bursts or pulses, as shown in Fig. 16-66. The duration or width W of the pulse is very short and, depending upon the application, can be anywhere from less than 1 μs to several microseconds. The time between transmitted pulses is known as the *pulse repetition time (PRT)*. If the PRT is known, the *pulse repetition frequency (PRF)* can be determined by using the formula

$$PRF = \frac{1}{PRT}$$

For example, if the pulse repetition time is 150 μs, the PRF is $1/150 \times 10^{-6} =$ 666.7 kHz.

The ratio of the pulse width to the PRT is known as the *duty cycle*. The duty cycle is normally expressed as a percentage:

$$\text{Duty cycle} = W \times \frac{100}{PRT}$$

For example, a pulse width of 7 μs with a PRT of 280 μs produces a duty cycle of $7 \times 100/280 = 2.5$ percent.

It is during the interval between the end of the transmitted pulse and the beginning of the next pulse in sequence in Fig. 16-66 that the echo is received.

The duration of the transmitted pulse and the PRT are extremely critical in determining the performance of a radar system. Very short-range radars have narrow pulses and short pulse repetition times. If the target is only a short distance away, the echo travel time will be relatively short. In short-range radars, the pulse width is made narrow to ensure that the pulse is terminated before the echo of the target is received. If the pulse is too long, the return signal may be masked or blanked by the transmitted pulse. Long-range radars typically have a longer pulse repetition time because it takes longer for the echo to return. This also permits a longer burst of energy to be transmitted, ensuring a stronger return.

If the PRT is too short relative to the distance of the target, the echo may not return during the time interval between two successive pulses, but after the second transmitted pulse. This is known as a *double range* or *second return echo*. Naturally, such echoes lead to imprecise distance measurements.

Continuous–Wave (CW) Radar. In *continuous-wave (CW) radar,* a constant-amplitude continuous microwave sine wave is transmitted. The echo is also a constant-amplitude microwave sine wave of the same frequency, but of lower amplitude and obviously shifted in phase. The question is, How is such a signal used to determine target characteristics?

The answer has to do with the object itself. In most cases, the target is moving with respect to the radar unit. The reflected signal from a moving airplane, ship, missile, or automobile undergoes a frequency change. It is this frequency change between the transmitted signal and the returned signal that is used to determine the speed of the target.

The frequency shift that occurs when there is relative motion between the transmitting station and a remote target is known as the *Doppler effect*. A familiar example of the Doppler effect is the fixed-frequency sound waves emitted by an automobile horn. If the horn sounds when the car is stationery, you will hear a single tone. However, if the car is moving toward you while the horn is on, you will experience a tone of continuously increasing frequency. As the car moves closer to you, the sound waves are compressed, creating the effect of a higher-frequency signal. If the car is moving away from you with its horn on, you will experience a continually decreasing frequency. As the car moves away, the sound waves are stretched out, creating the effect of a lower-frequency signal. This same effect works on both radio and light waves.

Doppler effect

In a Doppler system, the transmitter sends out a continuous-frequency signal. If the frequency difference between a transmitted signal and a reflected signal is known, the relative speed between the radar unit and the observed object can be determined by using the formula

$$V = \frac{f\lambda}{1.03}$$

where f = frequency difference between transmitted and reflected signals, Hz
λ = wavelength of transmitted signal, m
V = relative velocity between the two objects, mi/h

Assume, e.g., a frequency shift of 1500 Hz at a frequency of 10 GHz. A frequency of 10 GHz represents a wavelength of $300/f$ (in MHz) $= 300/10,000 = 0.03$ m. The speed is therefore $(1500)(0.03)/1.03 = 43.7$ mi/h.

In CW radar, it is the Doppler effect that provides frequency modulation of the reflected carrier. For there to be a frequency change, the observed object must be moving toward or away from the radar unit. If the observed object moves parallel to the radar unit, there is no relative motion between the two and no frequency modulation occurs.

The greatest value of CW radar is its ability to measure the speed of distant objects. Police radar units use CW Doppler radar for measuring the speed of cars and trucks.

Some radar systems combine both pulse and Doppler techniques to improve performance and measurement capabilities. One such system evaluates successive echoes to determine phase shifts that indicate when a target is moving. Such radars are said to incorporate *moving target indication (MTI)*. Through a variety of special signal processing techniques, multiple moving targets can be distinguished not only from one another but from fixed targets as well.

Moving target indication (MTI)

Block Diagram Analysis. Figure 16-67 is a block diagram of a typical pulsed radar unit. There are four basic subsystems: the antenna, the transmitter, the receiver, and the display unit.

The transmitter in a pulsed radar system invariably uses a magnetron. Recall that a magnetron is a special high-power vacuum tube oscillator that operates in the microwave region. The cavity size of the magnetron sets the operating frequency. A master timing generator develops the basic pulses used for triggering the magnetron. The timing generator sets the pulse duration, the PRT, and the duty cycle. The pulses from the timing network trigger the magnetron into oscillation, and it emits short bursts of microwave energy. Magnetrons are capable of extremely high power, especially when operated on a pulse basis. Continuous average power may be low, but when pulsed, magnetrons can produce many megawatts of power for the short duration required by the application. This helps ensure a large reflection. Klystrons and TWTs are commonly used in CW Doppler radars. Low-power radars such as those used by the police for speed detection use Gunn diodes.

Figure 16-67 General block diagram of a pulsed radar system.

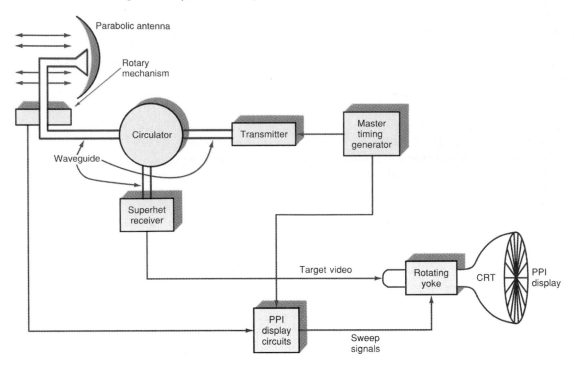

In Fig. 16-67, you can see that the transmitter output is passed through a circulator and then applied to the antenna. The circulator is a type of duplexer that allows the transmitter and receiver to share a single antenna and prevents the high-power transmitted signal from getting into the receiver and damaging it.

A radar duplexer is a waveguide assembly containing special devices that prevent interference between the transmitter and receiver. The most commonly used device is a spark gap tube. Spark gap tubes are either *TR (transmit-receive)* or *anti-transmit-receive (ATR)* types. TR tubes prevent transmitter power from reaching the receiver. When RF energy from the transmitter is detected, the spark gap breaks down, creating a short circuit for the RF energy. ATR tubes effectively disconnect the transmitter from the circuit during the receive interval. The TR and ATR tubes, when combined with the appropriate one-quarter and one-half wavelength waveguides, provide effective isolation between transmitter and receiver. In low-power radars, PIN diodes or a standard circulator can be used for this purpose.

The antenna system is typically a horn with a parabolic reflector that produces a very narrow beam width. A special waveguide assembly with a rotating joint allows the waveguide and horn antenna to be rotated continuously over 360°.

The same antenna is also used for reception. During the pulse off time, the received signal passes through the antenna, the associated waveguide, and either the duplexer or the circulator to the receiver. The receiver is a standard high-gain superheterodyne type. The signal is usually fed directly to a mixer in most systems, although some radars use an RF amplifier. The mixer is typically a single diode or diode bridge assembly. The local oscillator feeds a signal to the mixer at the appropriate frequency so that an IF is developed. Both klystrons and Gunn diodes are used in local-oscillator applications. The IF amplifiers provide very high gain prior to demodulation. Some radar receivers use double conversion.

The demodulator in a pulsed radar system is typically a diode detector, since only the pulses must be detected. In Doppler and other more complex radars, some form of frequency- and phase-sensitive demodulator is used. Phase-locked loops are common in this application. The output of the demodulator is fed to a video amplifier which creates signals that are ultimately displayed.

The display in most radar systems is a CRT. Various display formats can be used. The most common type of CRT display is known as a *type P display* or *plan position indicator (PPI)*. PPI displays show both the range and the azimuth of a target. The center of the display is assumed to be the location of the radar unit. Concentric circles indicate the range. The azimuth or direction is indicated by the position of the reflected target on the screen with respect to the vertical radius line. The targets show up as lighted blips on the screen.

Type P display

Plan position indicator (PPI)

The PPI display is developed by a sophisticated scanning system tied into antenna rotation. As the antenna is rotated by a motor, an encoder mechanism sends signals to the PPI control circuits that designate the azimuth or direction of the antenna. At the same time, the horizontal and vertical deflection coils (the *yoke*) around the neck of the CRT rotate in synchronism with the antenna. The electron beam in the CRT is swept from the center of the screen out to the edge. The sweep begins at the instant of pulse transmission and sweeps outward from the center. The beginning of each transmitted pulse begins another sweep of the electron beam. As the deflection yoke rotates, the beam moves in such a way that it appears that a radius from the center to the edge is continuously rotating. Target reflections appear as lighted blips on the screen. Calibrations on the screen in the form of graticule markings or superimposed electron beam patterns permit distance and azimuth values to be read directly.

Some PPI radars scan only a narrow range of azimuth rather than the entire 360°. Airplanes with radars in the nose, e.g., only scan a 90° to 180° range forward.

An important high-tech type of radar known as *phased array radar* provides greater flexibility in scanning narrow sectors and tracking multiple targets. Instead of a single horn and parabolic reflector, multiple dipoles or patch antennas are used. Slots in a waveguide are also used. A half-wave dipole at microwave frequencies is very short. Therefore, many can be mounted together in a matrix or array. The result is a special collinear array above a reflecting surface with very high gain. By using a system of separate feed lines and a variable phase shifter for each antenna, the beam width and directivity can be controlled electronically. This permits rapid scanning and on-the-fly adjustment of directivity. Phase array radars eliminate the mechanical systems needed for conventional radars.

Phased array radar

UWB Radar. The newest form of radar is called *ultrawideband (UWB) radar*. It is a form of pulsed radar that radiates a stream of very short pulses several hundred picoseconds long rather than a burst of RF at a specific carrier frequency. The resulting spectrum as determined by a Fourier analysis is very broad, usually several gigahertz wide. This spectrum typically overlaps radio signals within the signal bandwidth, commonly in the 1- to 10-GHz range. To avoid interference with other radio signals, very low power (microwatts) is used. The very narrow pulses give this radar extreme precision and resolution of small objects and details. However, the low power restricts operation to short distances (<100 m). The circuitry used is relatively simple, so it is possible to make inexpensive, single-chip radars. These are used in short-range collision detection systems in airplanes and soon will be in automobiles for automatic braking based upon distance from the vehicle ahead.

Ultrawideband (UWB) radar

Another application of UWB radar is personnel detection on the battlefield. These radars can penetrate walls to detect the presence of human beings. Chapter 22 provides more details on UWB systems.

Radar Applications. One of the most important uses of radar is in weapons defense systems and in safety and navigation systems. Search radars are used to locate enemy missiles, planes, and ships. Tracking radars are used on missiles and planes to locate and zero in on targets. Radars are widely used on planes and ships for navigating blind in fog or bad weather. Radars help ground controllers locate and identify nearby planes. Special radars assist planes in landing in bad weather when visibility is near zero.

Radars are also used as altimeters to measure height. High-frequency radars can actually be used to plot or map the terrain in an area. Special terrain-following radars allow high-speed jets to fly very close to the ground to avoid detection by enemy radar. Aircraft collision avoidance systems also use radar.

In civilian applications, radars are used on boats of all sizes for navigation in bad weather. The police use radar to catch speeders. Small handheld Doppler radar units can also be used in sporting events—to time race cars or determine the speed of a pitched baseball or tennis serve. Some of the newer upper-scale cars such as Lexus and Mercedes Benz have built in radars that detect the distance between the vehicle ahead and automatically adjust the speed to maintain a safe distance. These short-range radars operate in the 20- to 40-GHz bands. Finally, ground and satellite based radars are widely used to track clouds, storms, and other phenomena for the purpose of weather forecasting.

CHAPTER REVIEW

Summary

Microwaves occupy the 1- to 300-GHz frequency range. The best-known microwave applications are telephone systems (microwave relay systems), radar, satellite communication, and heating. Today, most new communication services are assigned to the microwave region.

The current-voltage relationships that exist at low frequencies are not usable at microwave frequencies. At these frequencies even a 0.5-in piece of wire represents a significant amount of inductance. Instead of lumped components, distributed circuit elements are used. Microstrip can be used to create almost any tuned circuit for microwave transmission, including resonant circuits, filters, and impedance-matching networks.

Most microwave energy transmission above 6 GHz is handled by waveguides, which are hollow metal tubes designed specifically for those high frequencies. Waveguides have a variety of special parts, such as directional couplers, turns, joints, rotary connections, and terminations. Other special microwave components are cavity resonators (waveguidelike devices that act as high-Q parallel resonant circuits), circulators (diplexers that allow a single antenna to be shared by a transmitter and a receiver), and isolators.

Diodes that have been specially designed for microwave use are point-contact diodes, Schottky (hot carrier) diodes, tunnel diodes, varactor diodes, step-recovery diodes, and oscillators (Gunn, IMPATT, and TRAPATT diodes). Bipolar and field-effect transistors made with silicon operate up to a frequency of 2 GHz. By using semiconductor materials such as GaAs, InP, and SiGe, transistors can be made to operate at frequencies up to 200 GHz. CMOS circuits made of silicon are used to make radios that operate up to 10 GHz. Tubes used for microwave power amplification are the klystron, the magnetron, and the traveling-wave tube.

One widely used microwave antenna is the horn, a flared waveguide with high gain and customizable directivity. Horns are often used in combination with parabolic reflectors, the familiar dish antennas associated with TV and satellite transmissions. Other types of microwave antennas are the helical antenna, the slot, the dielectric lens antenna, and the patch. Patch antennas are made with microstrip on printed-circuit boards (PCBs). Other popular PCB antennas are the loop, inverted-F, and the meander line. A phased array is an antenna system made up of a large group of similar antennas on a common plane. Patch or slot antennas on a common PCB can be used, or separate antennas such as dipoles can be physically mounted together in a plane. Phased arrays are used in military radars and in special beam-switching and beam-forming antennas known as intelligent antennas.

Radar, a microwave-frequency communication system used to detect objects at a distance that cannot be observed visually, is based on the principle that high-frequency RF signals are reflected by conductive targets. The two basic types of radar are pulsed and continuous-wave (Doppler).

Questions

1. What is the microwave frequency range?
2. What is the primary advantage of using microwave frequencies?
3. List seven reasons why microwaves are more difficult to work with than lower-frequency signals.
4. What is the designation of the lowest-frequency microwave band?
5. What is the name for microwaves above 40 GHz?
6. Name four techniques that have helped squeeze more signals into a given spectrum space.
7. Why won't conventional transistors work at microwave frequencies?
8. Name two common methods for generating a microwave signal with conventional lower-frequency (VHF and UHF) components.
9. Name the parts of a microwave receiver and a microwave transmitter that require special microwave components.
10. Name the two types of transmission lines made of PCB material used to produce circuits and components in the microwave region. Which is the more commonly used and why?
11. List three common circuits or functions performed by microstrip in microwave equipment.
12. What is the name of the microstrip device that has four ports and allows multiple circuits and devices to be interconnected without interfering with one another?
13. How are microwave transistors different from lower-frequency transistors?
14. What three semiconductor materials are used instead of silicon to make transistors and ICs perform satisfactorily in the microwave region?
15. How are microstrip lines used in microwave amplifiers?
16. Up to what frequency are microwave transistors and ICs available?
17. How are microwave linear amplifiers biased and why?
18. What is the main disadvantage of using coaxial cable at microwave frequencies?
19. Coaxial cable is not used beyond a certain frequency. What is that frequency?
20. Name the special coaxial cable made for microwave applications. What is its main disadvantage?
21. Name the two ways to couple or extract energy from a waveguide.
22. A waveguide acts as what kind of filter?
23. True or false? A waveguide with a cutoff frequency of 6.35 GHz will pass a signal of 7.5 GHz.
24. The magnetic and electric fields in a waveguide are designated by what letters?
25. State the basic operating mode of most waveguides.
26. What do you call the coupling mechanism used to attach two sections of a waveguide?
27. Name the two types of T sections of a waveguide and briefly tell how each is used.
28. Explain how a hybrid T works and give a common application.
29. What devices are used to terminate a waveguide? Why is termination necessary?
30. State the main purpose of a directional coupler.
31. What useful device is created by a one-half wavelength section of waveguide shorted at both ends?
32. A cavity resonator acts as what type of circuit?
33. How are resonant cavities used in microwave circuits?
34. What characteristic of a cavity can be adjusted by mechanically varying the cavity's dimensions?
35. What is a circulator?
36. What is an isolator?
37. Describe the anode and cathode of a point-contact diode.
38. What are two names for a microwave diode with an N-type silicon cathode and a metal anode forming a junction?
39. Name the most common application of microwave signal diodes.
40. What is the name of a microwave diode that acts as a variable-voltage capacitor? What is its main application?
41. Name two types of diodes widely used as frequency multipliers in the microwave region. Do such multiplier circuits amplify?
42. Name two other diodes used as microwave diode oscillators.
43. Name two ways that a PIN diode is used at microwave frequencies.
44. Does an IMPATT diode operate with forward or reverse bias?
45. True or false? Klystrons can be used as amplifiers or oscillators.
46. True or false? A magnetron operates as an amplifier.
47. True or false? A TWT is an amplifier.
48. The cavities in a klystron produce what kind of modulation of the electron beam?
49. State at which points in a klystron amplifier the input is applied and the output is taken from.
50. Low-power klystrons are being replaced by what semiconductor device? High-power klystrons are being replaced by what other device?
51. What component of a magnetron causes the electrons to travel in circular paths?
52. State two major applications of magnetrons.
53. What is the main application of a TWT?
54. How is density modulation of the electron beam in a TWT achieved?
55. What is the main benefit of a TWT over a klystron?
56. What is the most commonly used microwave antenna?
57. How does increasing the length of a horn affect its gain and beam width?
58. What is the typical beam width range of horn?
59. Describe the bandwidth of horns.

60. State the geometric shape of a commonly used microwave antenna reflector.
61. For a dish reflector to work, how must the antenna be positioned?
62. What type of antenna is most commonly used with a dish reflector?
63. What do you call the feed arrangement of a dish with a horn at its center and a small reflector at the focal point? What is its purpose?
64. Name two benefits of a helical antenna.
65. What is the gain range of a helical antenna?
66. What is the typical beam width range of a helical antenna?
67. What is the type of polarization radiated by a helical antenna?
68. What does RHCP mean? LHCP?
69. True or false? A helical antenna can receive either vertically or horizontally polarized signals.
70. True or false? An RHCP antenna can receive a signal from an LHCP antenna.
71. Name a popular omnidirectional microwave antenna.
72. What is the typical length of a slot antenna?
73. Where are slot antennas most often used?
74. How is a slot antenna driven?
75. What basic function is performed by a lens antenna?
76. What is the name given to antennas made with microstrip?
77. Name the two common shapes of microstrip antennas.
78. Name two ways that microstrip antennas are fed.
79. What is the name of the antenna made of multiple patch dipoles? Where is it used?
80. Name three popular PCB antennas other than the patch. Where are they used?
81. Define MIMO. Where is it used?
82. Name the two types of intelligent antennas. Where are they used?
83. List two benefits of beam switched and beam-forming antennas.
84. Explain how spatial division multiple access can increase the subscriber capacity of a cell phone system.
85. List four popular uses for microwaves in communication.
86. Knowing the speed of radio signals allows radar to determine what characteristic of a target?
87. What characteristic of a radar antenna determines the azimuth of a target?
88. What characteristic of a target determines the amount of reflected signal?
89. Radars operate in what microwave bands?
90. True or false? Both transmitter and receiver in a radar share the same antenna.
91. Name the two main types of radar.
92. What is the name given to a frequency shift of sound, radio, or light waves that occurs as the result of the relative motion between objects?
93. True or false? Distance can be measured with CW radar.
94. What is the key component in a radar transmitter?
95. What are TR and ATR tubes? Why are they used?
96. What is the output display in most radar sets?
97. Name three major military and four civilian uses of radar.
98. Name one advantage and one disadvantage of UWB radar.

Problems

1. The TV channels 2 through 13 occupy a total bandwidth of about 72 MHz. At a frequency of 200 MHz, what is the bandwidth percentage? What is the bandwidth percentage at 20 GHz? ◆
2. A one-quarter wavelength piece of 75-Ω microstrip line has a capacitive load of $X_c = 22\ \Omega$. What impedance does a driving generator see?
3. What is the length of a one-quarter wavelength microstrip at 2.2 GHz? (The dielectric constant of the PCB is 4.5.) ◆
4. A rectangular waveguide has a width of 1.4 in and a height of 0.8 in. What is the cutoff frequency of the waveguide?
5. A 20-dB directional coupler is installed in a waveguide carrying a 5-W 6-GHz signal. What is the output of the coupler? ◆
6. What is the length of a dipole at 2.4 GHz?
7. A horn antenna is 5λ long at 7 GHz. What is that length in inches? ◆
8. A horn aperture is 6 cm wide and 4 cm high. The operating frequency is 18 GHz. Compute the beam angle and the gain. Assume $K = 0.5$.
9. A parabolic reflector antenna has a diameter of 6 m. The frequency of operation is 20 GHz. What are the gain and beam width? ◆
10. What is the approximate lowest frequency of operation of a dish with a diameter of 18 in?
11. State the speed of a radar signal in microseconds per nautical mile.
12. Convert 22 standard statute miles to nautical miles.
13. The elapsed time between radiating a radar signal and receiving its echo is 23.9 μs. What is the target distance in nautical miles?
14. How many microseconds is the time period between the transmitted signal and receipt of an echo from a target 1800 yd away?
15. For optimum reflection from a target, what should the size of the target be, in wavelengths of the radar signal?
16. A CW radar operates at a frequency of 9 GHz. A frequency shift of 35 kHz is produced by a moving target. What is its speed?

◆ *Answers to Selected Problems follow Chap. 22.*

Critical Thinking

1. If microwave power transistors are limited to, say, an upper power limit of 100 W each, explain how you could build a microwave power amplifier with a power rating of 1 kW.

2. A NASA deep-space probe operates on 8.4 GHz. Its location is past Mars, heading into outer space. Explain in detail how you would go about maintaining and/or improving signal strength at the main receiver in California.

3. Explain how the speed of an aircraft can be measured by using pulse radar techniques.

4. What semiconductor technology originally developed for computer and digital applications is now widely used to make microwave radio circuits?

chapter 17

Satellite Communication

Satellite

Orbit

A *satellite* is a physical object that *orbits,* or rotates about, some celestial body. Satellites occur in nature, and our own solar system is a perfect example. The earth and other planets are satellites rotating about the sun. The moon is a satellite to the earth. A balance between the inertia of the rotating satellite at high speed and the gravitational pull of the orbited body keeps the satellite in place.

Satellites are launched and orbited for a variety of purposes. The most common application is communication in which the satellite is used as a repeater. In this chapter, we introduce satellite concepts and discuss how satellites are identified and explained. We summarize the operation of a satellite ground station and review typical satellite applications, with particular emphasis on the Global Positioning System, a worldwide satellite-based navigational system.

Objectives

After completing this chapter, you will be able to:

- Define the terms *posigrade, retrograde, geocenter, apogee, perigee, ascending, descending, period, angle of inclination, geosynchronous, latitude, longitude,* and *meridian.*
- State the operative physical principles of launching a satellite and maintaining its orbit.
- Draw a block diagram of the communication system in a communication satellite, give its name, and explain how it works.
- List the six main subsystems of a satellite.
- Draw a block diagram of a satellite ground station, identifying the five main subsystems and explaining the operation of each.
- Name three common applications for satellites and state which is the most common.
- Explain the concept and operation of the Global Positioning System. Draw a block diagram of a GPS receiver and explain the function of each component.

17-1 Satellite Orbits

The ability to launch a satellite and keep it in orbit depends upon following well-known physical and mathematical laws that are referred to collectively as *orbital dynamics*. In this section we introduce these principles before discussing the physical components of a satellite and how it is used in various communication applications.

Principles of Satellite Orbits and Positioning

If a satellite were launched vertically from the earth and then released, it would fall back to earth because of gravity. For the satellite to go into orbit around the earth, it must have some forward motion. For that reason, when the satellite is launched, it is given both vertical and forward motion. The forward motion produces inertia, which tends to keep the satellite moving in a straight line. However, gravity tends to pull the satellite toward the earth. The inertia of the satellite is equalized by the earth's gravitational pull. The satellite constantly changes its direction from a straight line to a curved line to rotate about the earth.

If a satellite's velocity is too high, the satellite will overcome the earth's pull and go out into space. It takes an escape velocity of approximately 25,000 mi/h to cause a spacecraft to break the gravitational pull of the earth. At lower speeds, gravity constantly pulls the satellite toward the earth. The goal is to give the satellite acceleration and speed that will exactly balance the gravitational pull.

The closer the satellite is to earth, the stronger the effect of the earth's gravitational pull. So in low orbits, the satellite must travel faster to avoid falling back to earth. The lowest practical earth orbit is approximately 100 mi. At this height, the satellite's speed must be about 17,500 mi/h to keep the satellite in orbit. At this speed, the satellite orbits the earth in approximately 1½ h. Communication satellites are usually much farther from earth. A typical distance is 22,300 mi. A satellite need travel only about 6800 mi/h to stay in orbit at that distance. At this speed, the satellite rotates about the earth in approximately 24 h, the earth's own rotational time.

There is more to a satellite orbit than just the velocity and gravitational pull. The satellite is also affected by the gravitational pull of the moon and sun.

A satellite rotates about the earth in either a circular or an elliptical path, as shown in Fig. 17-1. Circles and ellipses are geometric figures that can be accurately described

Figure 17-1 Satellite orbits. (*a*) Circular orbit. (*b*) Elliptical orbit.

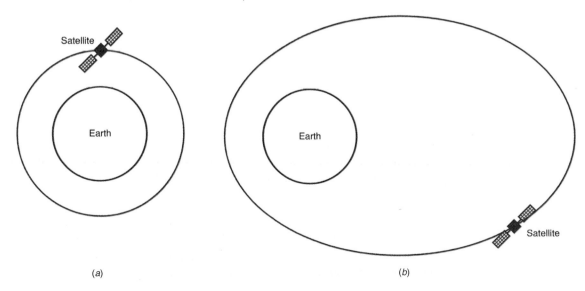

(*a*)   (*b*)

Figure 17-2 The orbital plane passes through the geocenter.

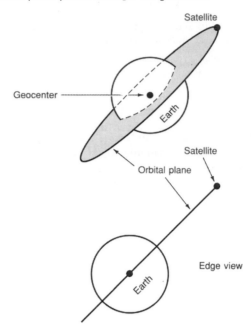

mathematically. Because the orbit is either circular or elliptical, it is possible to calculate the position of a satellite at any given time.

A satellite rotates in an orbit that forms a plane passing through the center of gravity of the earth called the *geocenter* (Fig. 17-2). In addition, the direction of satellite rotation may be either in the same direction as the earth's rotation or against the direction of earth's rotation. In the former case, the orbit is said to be *posigrade,* and in the latter case, *retrograde.* Most orbits are posigrade. In a circular orbit, the speed of rotation is constant. However, in an elliptical orbit, the speed changes depending upon the height of the satellite above the earth. Naturally the speed of the satellite is greater when it is close to the earth than when it is far away.

Satellite Height. In a circular orbit, the height is simply the distance of the satellite from the earth. However, in geometric calculations, the height is really the distance between the center of the earth and the satellite. In other words, that distance includes the radius of the earth, which is generally considered to be about 3960 mi (or 6373 km). A satellite that is 5000 mi above the earth in circular orbit is 3960 + 5000, or 8960, mi from the center of the earth (see Fig. 17-3).

When the satellite is in an elliptical orbit, the center of the earth is one of the focal points of the ellipse (refer to Fig. 17-4). In this case, the distance of the satellite from the earth varies according to its position. Typically the two points of greatest interest are the highest point above the earth—the *apogee*—and the lowest point—the *perigee.* The apogee and perigee distances typically are measured from the geocenter of the earth.

Satellite Speed. As indicated earlier, the speed varies according to the distance of the satellite from the earth. For a circular orbit the speed is constant, but for an elliptical orbit the speed varies according to the height. Low earth satellites of about 100 mi in height have a speed in the neighborhood of 17,500 mi/h. Very high satellites such as communication satellites, which are approximately 22,300 mi out, rotate much more slowly, a typical speed of such a satellite being in the neighborhood of 6800 mi/h.

Satellite Period. The *period* is the time it takes for a satellite to complete one orbit. It is also called the *sidereal period.* A sidereal orbit uses some external fixed or apparently motionless object such as the sun or star for reference in determining a sidereal

GOOD TO KNOW

Before being launched into space, satellites undergo testing for vibration, center of gravity, vacuum, and thermal shock.

Apogee

Perigee

Satellite period

Sidereal period

Figure 17-3 Satellite height.

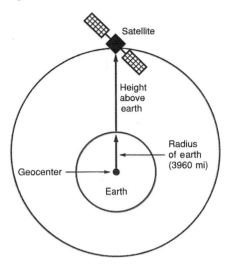

period. The reason for using a fixed reference point is that while the satellite is rotating about the earth, the earth itself is rotating.

Another method of expressing the time for one orbit is the revolution or synodic period. One *revolution* (1 r) is the period of time that elapses between the successive passes of the satellite over a given meridian of earth longitude. Naturally, the synodic and sidereal periods differ from each other because of the earth's rotation. The amount of time difference is determined by the height of the orbit, the angle of the plane of the orbit, and whether the satellite is in a posigrade or retrograde orbit. Period is generally expressed in hours. Typical rotational periods range from about 1½ h for a 100-mi height to 24 h for a 22,300-mi height.

Angle of Inclination. The *angle of inclination* of a satellite orbit is the angle formed between the line that passes through the center of the earth and the north pole and a line that passes through the center of the earth but that is also perpendicular to the orbital plane. This angle is illustrated in Fig. 17-5(*a*). Satellite orbits can have inclination angles of 0° through 90°.

Angle of inclination

Figure 17-4 Elliptical orbit showing apogee and perigee.

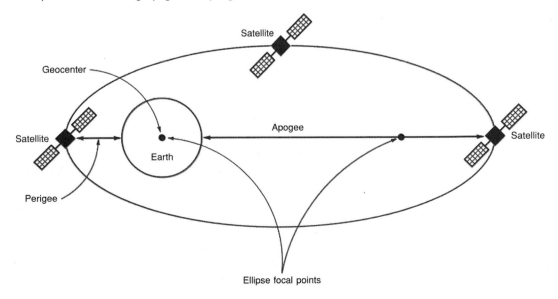

Satellite Communication **673**

Figure 17-5 (*a*) Angle of inclination. (*b*) Ascending and descending orbits.

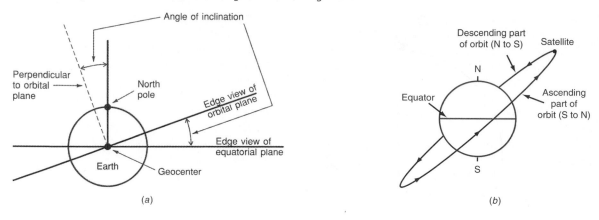

(a)   (b)

Another definition of *inclination* is the angle between the equatorial plane and the satellite orbital plane as the satellite enters the northern hemisphere. This definition may be a little bit easier to understand, but it means the same thing as the previous definition.

When the angle of inclination is 0°, the satellite is directly above the equator. When the angle of inclination is 90°, the satellite passes over both the north and the south poles once for each orbit. Orbits with 0° inclination are generally called *equatorial orbits,* and orbits with inclinations of 90° are referred to as *polar orbits.*

Equatorial orbit

Polar orbit

When the satellite has an angle of inclination, the orbit is said to be either *ascending* or *descending.* As the satellite moves from south to north and crosses the equator, the orbit is ascending. When the satellite goes from north to south across the equator, the orbit is descending [see Fig. 17-5(*b*)].

Angle of elevation

Angle of Elevation. The *angle of elevation* of a satellite is the angle that appears between the line from the earth station's antenna to the satellite and the line between the earth station's antenna and the earth's horizon (see Fig. 17-6). If the angle of elevation is too small, the signals between the earth station and the satellite have to pass through much more of the earth's atmosphere. Because of the very low powers used and the high absorption of the earth's atmosphere, it is desirable to minimize the amount of time that the signals spend in the atmosphere. Noise in the atmosphere also contributes to poor performance. The lower the angle of radiation, the more time that this signal spends in the atmosphere. The minimum practical angle of elevation for good satellite performance is generally 5°. The higher the angle of elevation, the better.

Figure 17-6 Angle of elevation.

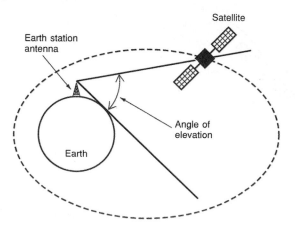

Geosynchronous Orbits. To use a satellite for communication relay or repeater purposes, the ground station antenna must be able to follow or track the satellite as it passes overhead. Depending upon the height and speed of the satellite, the earth station is able to use it only for communication purposes for that short period when it is visible. The earth station antenna tracks the satellite from horizon to horizon. But at some point, the satellite disappears around the other side of the earth. At this time, it can no longer support communication.

One solution to this problem is to launch a satellite with a very long elliptical orbit so that the earth station can "see" the apogee. In this way the satellite stays in view of the earth station for most of its orbit and is useful for communication for a longer time. It is only during that short time when the satellite disappears on the other side of the earth (perigee) that it cannot be used.

The intermittent communication caused by these orbital characteristics is highly undesirable in many communication applications. One way to reduce interruptions is to use more than one satellite. Typically three satellites, if properly spaced in the correct orbits, can provide continuous communication at all times. However, multiple tracking stations and complex signal switching or "hand-off" systems between stations are required. Maintaining these stations is expensive and inconvenient.

Despite the cost and complexity of multiple-satellite systems, they are widely deployed in global telecommunication applications. These systems use anywhere from 24 to more than 100 satellites. At any given time, multiple satellites are in view anywhere on earth, making continuous communication possible.

Multiple satellite systems are usually located in two ranges above the earth. *Low earth-orbiting satellites,* commonly referred to as *LEOs,* are placed in the range of 400 to 1000 mi above the earth. *Medium earth-orbiting satellites* or *MEOs* occupy the range of 1000 to 6000 mi above the earth.

The greater the height above the earth, the better the view and the greater the radio area coverage on the earth's surface. When the goal is broader coverage per satellite, the MEO is obviously preferred over the LEO. However, the higher the satellite, the higher the power required for reliable communication and the longer the delay. Even though radio waves travel at 186,400 mi/s, there is a noticeable delay in any voice signal from the uplink to the downlink. For MEOs, the round-trip delay averages about 100 ms. The delay in LEOs averages 10 ms.

The best solution is to launch a synchronous or geostationary satellite. In a *geosynchronous earth orbit (GEO),* the satellite orbits the earth about the equator at a distance of 22,300 mi (or 35,888 km). A satellite at that distance rotates about the earth in exactly 24 h. In other words, the satellite rotates in exact synchronism with the earth. For that reason, it appears to be fixed or stationary, thus the terms *synchronous, geosynchronous,* or *geostationary* orbit. Since the satellite remains apparently fixed, no special earth station tracking antennas are required. The antenna is simply pointed at the satellite and remains in a fixed position. With this arrangement, continuous communication is possible. Most communication satellites in use today are of the geosynchronous variety. Approximately 40 percent of the earth's surface can be "seen" or accessed from such a satellite. Users inside that area can use the satellite for communication. Higher power is required for such a great distance, and the round-trip delay is about 260 ms, which is very noticeable in voice communication.

Position Coordinates in Latitude and Longitude. To use a satellite, you must be able to locate its position in space. That position is usually predetermined by the design of the satellite and is achieved during the initial launch and subsequent position adjustments. Once the position is known, the earth station antenna can be pointed at the satellite for optimum transmission and reception. For geosynchronous satellites, the earth station antenna can be adjusted once, and it will remain in that position except for occasional minor adjustments. The positions of other satellites above the earth vary according to their orbital characteristics. To use these satellites, special tracking systems must be employed. A tracking system is essentially an antenna whose position can be changed

Synchronous (geosynchronous or geostationary) orbit

to follow the satellite across the sky. To maintain optimum transmission and reception, the antenna must be continually pointed at the satellite as it rotates. In this section, we discuss methods of locating and tracking satellites.

The location of a satellite is generally specified in terms of latitude and longitude, just as a point on earth would be described. The satellite location is specified by a point on the surface of the earth directly below the satellite. This point is known as the *subsatellite point (SSP)*. The subsatellite point is then located by using conventional latitude and longitude designations.

Latitude and longitude form a system for locating any given point on the surface of the earth. This system is widely used for navigational purposes. If you have ever studied a globe of the earth, you have seen the lines of latitude and longitude. The lines of longitude, or *meridians,* are drawn on the surface of the earth between the north and south poles. Lines of latitude are drawn on the surface of the earth from east to west, parallel to the equator. The centerline of latitude is the equator, which separates the earth into the northern and southern hemispheres.

Latitude is defined as the angle between the line drawn from a given point on the surface of the earth to the point at the center of the earth called the *geocenter* and the line between the geocenter and the equator (see Fig. 17-7). The 0° latitude is at the equator, and 90° latitude is at either the north or south pole. Usually an N or an S is added to the latitude angle to designate whether the point is in the northern or southern hemisphere.

A line drawn on the surface of the earth between the north and south poles is generally referred to as a *meridian.* A special meridian called the *prime meridian* is used as a reference point for measuring longitude. It is the line on the surface of the earth, drawn between the north and south poles, that passes through Greenwich, England. The longitude of a given point is the angle between the line connected to the geocenter of the earth to the point where the prime meridian and equator and the meridian containing the given point of interest intersect (see Fig. 17-7). The designation east or west is usually added to the longitude angle to indicate whether the angle is being measured to the east or west of the prime meridian. As an example, the location of Washington, D.C., is given by the latitude and longitude of 39° north and 77° west.

To show how latitude and longitude are used to locate a satellite, refer to Fig. 17-8. This figure shows some of the many geosynchronous communication satellites serving

Subsatellite point (SSP)

Meridians

Latitude
Geocenter

Prime meridian

Figure 17-7 Tracking and navigation by latitude and longitude.

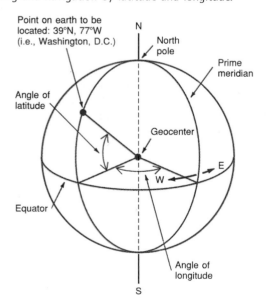

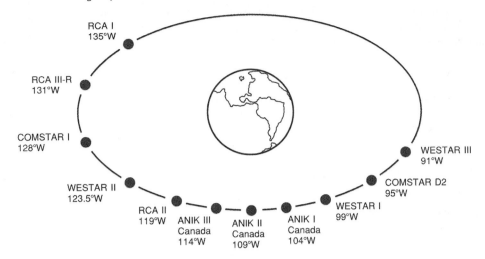

the United States and other parts of North America. Since geosynchronous satellites rotate about the equator, their subsatellite point is on the equator. For that reason, all geosynchronous satellites have a latitude of 0°.

Azimuth and Elevation. Knowing the location of the satellite is insufficient information for most earth stations that must communicate with the satellite. The earth station really needs to know the azimuth and elevation settings of its antenna to intercept the satellite. Most earth station satellite antennas are highly directional and must be accurately positioned to "hit" the satellite. The azimuth and elevation designations in degrees tell where to point the antenna (see Fig. 17-9). *Azimuth* refers to the direction where north is equal to 0°. The *azimuth angle* is measured clockwise with respect to north. The *angle of elevation* is the angle between the horizontal plane and the pointing direction of the antenna.

Once the azimuth and elevation are known, the earth station antenna can be pointed in that direction. For a geosynchronous satellite, the antenna will simply remain in that position. For any other satellite, the antenna must be moved as the satellite passes overhead.

For geosynchronous satellites, the angles of azimuth and elevation are relatively easy to determine. Because geosynchronous satellites are fixed in position over the equator, special formulas and techniques have been developed to permit easy determination of azimuth and elevation for any geosynchronous satellite for any point on earth. Example 17-1 shows how calculations are made.

Azimuth

Azimuth angle

Angle of elevation

Figure 17-9   Azimuth and elevation: Azimuth = 90°; elevation = 40°.

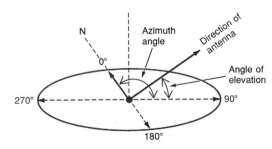

Example 17-1

A satellite earth station is to be located at longitude 95° west, latitude 30° north. The satellite is located at 121° west longitude in geosynchronous orbit. Determine the approximate azimuth and elevation settings for the antenna.

a. Difference between longitude of satellite and longitude of location:

$$121 - 95 = 26°$$

b. Locate longitude difference (relative longitude) and latitude.

c. Determine the elevation from the curves:

$$45°$$

d. Determine the radial position:

$$137°$$

e. Calculate the actual azimuth:

$$360 - 137 = 223°$$

17-2 Satellite Communication Systems

Communication satellite

Communication satellites are not originators of information to be transmitted. Although some other types of satellites generate the information to be transmitted, communication satellites do not. Instead, these satellites are relay stations for earth sources. If a transmitting station cannot communicate directly with one or more receiving stations because of line-of-sight restrictions, a satellite can be used. The transmitting station sends the information to the satellite, which in turn retransmits it to the receiving stations. The satellite in this application is what is generally known as a *repeater.*

Repeater

Repeaters and Transponders

Figure 17-10 shows the basic operation of a communication satellite. An earth station transmits information to the satellite. The satellite contains a receiver that picks up the transmitted signal, amplifies it, and translates it on another frequency. The signal on the new frequency is then retransmitted to the receiving stations on earth. The original signal being transmitted from the earth station to the satellite is called the *uplink,* and the retransmitted signal from the satellite to the receiving stations is called the *downlink.* Usually the downlink frequency is lower than the uplink frequency. A typical uplink frequency is 6 GHz, and a common downlink frequency is 4 GHz.

Uplink
Downlink
Transponder

The transmitter-receiver combination in the satellite is known as a *transponder.* The basic functions of a transponder are amplification and frequency translation (see Fig. 17-11). The reason for frequency translation is that the transponder cannot transmit and receive on the same frequency. The transmitter's strong signal would overload, or "desensitize," the receiver and block out the very small uplink signal, thereby prohibiting any communication. Widely spaced transmit and receive frequencies prevent interference.

Transponders are also wide-bandwidth units so that they can receive and retransmit more than one signal. Any earth station signal within the receiver's bandwidth will be amplified, translated, and retransmitted on a different frequency.

Although the typical transponder has a wide bandwidth, it is used with only one uplink or downlink signal to minimize interference and improve communication reliability. To be economically feasible, a satellite must be capable of handling several

Figure 17-10   Using a satellite as a microwave relay link.

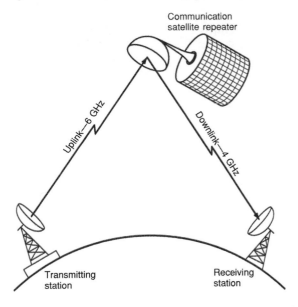

Figure 17-11   A satellite transponder.

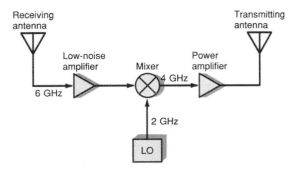

channels. As a result, most satellites contain multiple transponders, each operating at a different frequency. A typical communication satellite has 24 channels, 12 vertically polarized and 12 horizontally polarized. Each transponder represents an individual communication channel. Various multiple-access schemes are used so that each channel can carry multiple information transmissions.

Frequency Allocations. Most communication satellites operate in the microwave frequency spectrum. However, there are some exceptions. For example, many military satellites operate in the 200- to 400-VHF/UHF range. Also, the amateur radio OSCAR satellites operate in the VHF/UHF range. VHF, UHF, and microwave signals penetrate the ionosphere with little or no attenuation and are not refracted to earth, as are lower-frequency signals in the 3- to 30-MHz range.

The microwave spectrum is divided up into frequency bands that have been allocated to satellites as well as other communication services such as radar. These frequency bands are generally designated by a letter of the alphabet. Figure 17-12 shows the various frequency bands used in satellite communication.

One of the most widely used satellite communication bands is the C band. The uplink frequencies are 5.925 to 6.425 GHz. In any general discussion of the C band, the uplink is generally said to be 6 GHz. The downlink is in the 3.7- to 4.2-GHz range. But again, in any general discussion of the C band, the downlink is nominally said to be 4 GHz. Occasionally, the C band is referred to by the designation 6/4 GHz, where the uplink frequency is given first.

Satellite Communication   **679**

Figure 17-12 Frequency bands used in satellite communication.

BAND	FREQUENCY
P	225–390 MHz
J	350–530 MHz
L	1530–2700 MHz
S	2500–2700 MHz
C	3400–6425 MHz
X	7250–8400 MHz
Ku	10.95–14.5 GHz
Ka	17.7–31 GHz
Q	36–46 GHz
V	46–56 GHz
W	56–100 GHz

Over the past several years there has been a steady move toward the higher frequencies. Currently, the Ku band is receiving the most attention. The uplinks are in the 14- to 14.5-GHz range, and the downlinks are from 11.7 to 12.2 GHz. You will see the Ku band designated as 14/12 GHz. Use of the Ka band is also increasing.

C band

Most new communication satellites will operate in the Ku band. This upward shift in frequency is happening because the *C band* is overcrowded. Many communication satellites are in orbit now, most of them operating in the C band. However, there is some difficulty with interference because of the heavy usage. The only way this interference will be minimized is to shift all future satellite communication to higher frequencies. Naturally, the electronic equipment that can achieve these higher frequencies is more complex and expensive. Yet, the crowding and interference problems cannot be solved in any other way. Further, for a given antenna size, the gain is higher in the Ku band than in the C band. This can improve communication reliability while decreasing antenna size and cost.

Two other bands of interest are the X and L bands. The military uses the X band for its satellites and radar. The L band is used for navigation as well as marine and aeronautical communication and radar.

Spectrum Usage

Recall the frequencies designated for the C band uplink and downlink. These are 5925 to 6425 and 3700 to 4200 MHz, respectively. You can see that the bandwidth between the upper and lower limits is 500 MHz. This is an incredibly wide band, capable of carrying an enormous number of signals. In fact, 500 MHz covers all the radio spectrum so well known from VLF through VHF and beyond. Most communication satellites are designed to take advantage of this full bandwidth. This allows them to carry the maximum possible number of communication channels. Of course, this extremely wide bandwidth is one of the major reasons why microwave frequencies are so useful in communication. Not only can many communication channels be supported, but also very high-speed digital data requiring a wide bandwidth is supported.

The transponder receiver "looks at" the entire 500-MHz bandwidth and picks up any transmission there. However, the input is "channelized" because the earth stations operate on selected frequencies or channels. The 500-MHz bandwidth is typically divided into 12 separate transmit channels, each 36 MHz wide. There are 4-MHz guard bands between channels that are used to minimize adjacent channel interference (see Fig. 17-13). Note the center frequency for each channel. Remember that the uplink frequencies are translated by frequency conversion to the downlink channel. In both cases, the total bandwidth (500 MHz) and channel bandwidth (36 MHz) are the same. Onboard the satellite, a separate transponder is allocated to each of the 12 channels.

Although 36 MHz seems narrow compared to 500 MHz, each transponder bandwidth is capable of carrying an enormous amount of information. For example, one

Figure 17-13 Receive and transmit bandwidths in a C-band communication satellite.

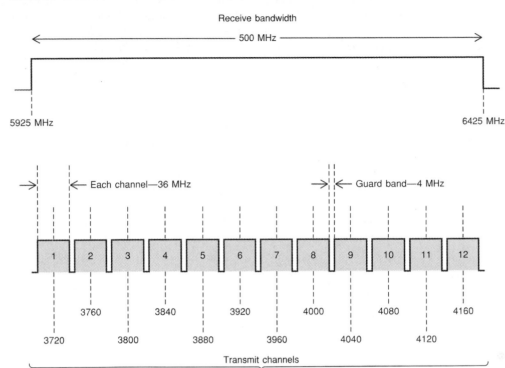

Channel center frequencies (MHz)

typical transponder can handle up to 1000 one-way analog telephone conversations as well as one full-color TV channel. Each transponder channel can also carry high-speed digital data. Using certain types of modulation, a standard 36-MHz bandwidth transponder can handle digital data at rates of up to 60 Mbps.

Example 17-2

A satellite transponder operates in the C band (see Fig. 17-13). Assume a local-oscillator frequency of 2 GHz.

a. What is the uplink receiver frequency if the downlink transmitter is on channel 4? The downlink frequency of channel 4 is 3840 MHz (Fig. 17-13). The downlink frequency is the difference between the uplink frequency f_u and the local-oscillator frequency f_{LO}:

$$f_d = f_u - f_{LO}$$

Therefore,

$$f_u = f_d + f_{LO}$$
$$= 3840 + 2000 = 5840 \text{ MHz} = 5.84 \text{ GHz}$$

b. What is the maximum theoretical data rate if one transponder is used for binary transmission?
The bandwidth of one transponder channel is 36 MHz. For binary transmission, the maximum theoretical data rate or channel capacity C for a given bandwidth B is

$$C = 2B$$
$$= 2(36) = 72 \text{ Mbps}$$

Although the transponders are quite capable, they nevertheless rapidly become overloaded with traffic. Further, at times there is more traffic than there are transponders to handle it. For that reason, numerous techniques have been developed to effectively increase the bandwidth and signal-carrying capacity of the satellite. Two of these techniques are known as *frequency reuse* and *spatial isolation*.

Frequency reuse

Frequency Reuse. One system for effectively doubling the bandwidth and information-carrying capacity of a satellite is known as *frequency reuse*. In this system, a communication satellite is provided with two identical sets of 12 transponders. The first channel in one transponder operates on the same channel as the first transponder in the other set, and so on. With this arrangement, the two sets of transponders transmit in the same frequency spectrum and, therefore, appear to interfere with each other. However, this is not the case. The two systems, although operating on exactly the same frequencies, are isolated from each other by the use of special antenna techniques.

One technique for keeping transmissions separate is to use different antenna polarizations. For example, a vertically polarized antenna will not respond to a horizontally polarized signal and vice versa. Or a *left-hand circularly polarized (LHCP)* antenna will not respond to a *right-hand circularly polarized (RHCP)* signal and vice versa.

Left-hand circularly polarized (LHCP)

Right-hand circularly polarized (RHCP)

Another technique is to use spatial isolation. By using narrow beam or spot beam antennas, the area on the earth covered by the satellite can be divided up into smaller segments. Earth stations in each segment may actually use the same frequency, but because of the very narrow beam widths of the antennas, there is no interference between adjacent segments. This technique is referred to a *spatial-division multiple access (SDMA)* in that access to the satellite depends on location and not frequency.

Frequency-division multiple access (FDMA)

Time-division multiple access (TDMA)

Access Methods. To maximize the use of the available spectrum in satellite transponders and to ensure access for as many users as possible, all satellites use some form of multiplexing. Frequency-division multiplexing (FDM), more commonly called *frequency-division multiple access (FDMA)*, was widely used in early satellites. Today, time-division multiplexing (TDM), also known as *time-division multiple access (TDMA)*, is more prevalent. This digital technique assigns each user a time slot on the full bandwidth of the transponder channel. Modulation methods are BPSK and QPSK, although multilevel QAM (16 QAM, 32 QAM, and 256 QAM) is also used to increase digital transmission speeds in a given bandwidth. Spread spectrum is used in some of the newer satellites. Also known as *code-division multiple access (CDMA)*, this digital method spreads the signals of multiple users over the full transponder channel bandwidth and sorts them by use of pseudorandom codes. CDMA also provides the security so important in today's wireless systems and today, more and more satellites use SDMA to give multiple access while conserving spectrum.

Code-division multiple access (CDMA)

17-3 Satellite Subsystems

All satellite communication systems consist of two basic parts, the satellite or spacecraft and two or more earth stations. The satellite performs the function of a radio repeater or relay station. Two or more earth stations may communicate with one another through the satellite rather than directly point-to-point on the earth.

Satellites vary is size from about 1 ft^3 for a small LEO satellite to more than 20 ft long. The largest satellites are roughly the size of the trailer on an 18-wheeler. Weight ranges from about 100 lb for the smaller satellites to nearly 10,000 lb for the largest.

The heart of a communication satellite is the communication subsystem. This is a set of transponders that receive the uplink signals and retransmit them to earth. A transponder is a repeater that implements a wideband communication channel that can carry many simultaneous communication transmissions.

The transponders are supported by a variety of additional "housekeeping" subsystems. These include the power subsystem, the telemetry tracking and command

Figure 17-14 General block diagram of a communication satellite.

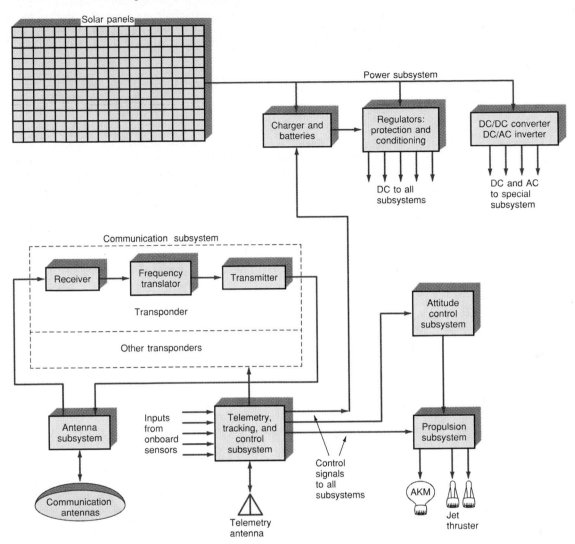

subsystems, the antennas, and the propulsion and attitude stabilization subsystems. These are essential to the self-sustaining nature of the satellite.

Figure 17-14 is a general block diagram of a satellite. All the major subsystems are illustrated. The solar panels supply the electric power for the spacecraft. They drive regulators that distribute dc power to all other subsystems. And they charge the batteries that operate the satellite during eclipse periods. And ac-to-dc converters and dc-to-ac inverters are used to supply special voltages to some subsystems. Total power capability runs from a few hundred watts in the smaller units to several kilowatts in the largest systems.

The communication subsystem consists of multiple transponders. These receive the uplink signals, amplify them, translate them in frequency, and amplify them again for retransmission as downlink signals. The transponders share an antenna subsystem for both reception and transmission.

The telemetry, tracking, and command (TT&C) subsystem monitors onboard conditions such as temperature and battery voltage and transmits this data back to a ground station for analysis. The ground station may then issue orders to the satellite by transmitting a signal to the command subsystem, which then is used to control many spacecraft functions such as firing the jet thrusters.

Satellite Communication   **683**

The jet thrusters and the *apogee kick motor (AKM)* are part of the propulsion subsystem. They are controlled by commands from the ground.

The attitude control subsystem provides stabilization in orbit and senses changes in orientation. It fires the jet thrusters to perform attitude adjustment and station-keeping maneuvers that keep the satellite in its assigned orbital position.

Communication Subsystems

The main payload on a communication satellite, of course, is the communication subsystem that performs the function of a repeater or relay station. An earth station takes the signals to be transmitted, known as *baseband signals,* and modulates a microwave carrier. The three most common baseband signals are voice, video, and computer data. These uplink signals are then amplified, translated in frequency, and retransmitted on the downlink to one or more earth stations. The component that performs this function

is known as a *transponder.* Most modern communication satellites contain at least 12 transponders. More advanced satellites contain many more. These transponders operate in the microwave frequency range.

The basic purpose of a transponder is simply to rejuvenate the uplink signal and retransmit it over the downlink. In this role the transponder performs the function of an amplifier. By the time the uplink's signal reaches the satellite, it is extremely weak. Therefore, it must be amplified before it can be retransmitted to the receiving earth station.

However, transponders are more than just amplifiers. An *amplifier* is a circuit that takes a signal and increases the voltage or power level of that signal without changing its frequency or content. Such a transponder, then, literally consists of a receiver and a transmitter that operate on the same frequency. Because of the close proximity of the transmitter and the receiver in the satellite, the high transmitter output power for the downlink is picked up by that satellite receiver. Naturally, the uplink signal is totally obliterated. Further, the transmitter output fed back into the receiver input causes oscillation.

To avoid this problem, the receiver and transmitter in the satellite transponder are designed to operate at separate frequencies. In this way, they will not interfere with each other. The frequency spacing is made as wide as practical to minimize the effect of the transmitter desensitizing the receiver. In many repeaters, even though the receive and transmit frequencies are different, the high output power of the transmitter can still affect the sensitive receiver input circuits and, in effect, desensitize them, making them less sensitive in receiving the weak uplink signals. The wider the frequency spacing between transmitter and receiver, the less of desensitizing problem this is.

In typical satellites, the input and output frequencies are separated by huge amounts. At C band frequencies, the uplink signal is in the 6-GHz range and the downlink signal is in the 4-GHz range. This 2-GHz spacing is sufficient to eliminate most problems. However, to ensure maximum sensitivity and minimum interference between uplink and downlink signals, the transponder contains numerous filters that not only provide channelization but also help to eliminate interference from external signals regardless of their source.

Three basic transponder configurations are used in communication satellites. They are all essentially minor variations of one another, but each has its advantages and disadvantages. These are the single-conversion, double-conversion, and regenerative transponders.

A *single-conversion transponder* uses a single mixer to translate the uplink signal to the downlink frequency. A *dual-conversion transponder* makes the frequency translation in two steps with two mixers. No demodulation occurs. A *regenerative repeater* demodulates the uplink signal after the frequency is translated to some lower intermediate frequency. The recovered baseband signal is then used to modulate the downlink signal.

Multichannel Configurations. Virtually all modern communication satellites contain multiple transponders. This permits many more signals to be received and transmitted. A typical commercial communication satellite contains 12 transponders, 24 if frequency reuse is incorporated. Military satellites often contain fewer transponders, whereas the newer, larger commercial satellites have provisions for up to 50 channels. Each transponder operates on a separate frequency, but its bandwidth is wide enough to carry multiple channels of voice, video, and digital information.

There are two basic multichannel architectures in use in communication satellites. One is a broadband system, and the other is a fully channelized system.

Broadband System.

Broadband system

As indicated earlier, a typical communication satellite spectrum is 500 MHz wide. This is typically divided into 12 separate channels, each with a bandwidth of 36 MHz. The center frequency spacing between adjacent channels is 40 MHz, thereby providing a 4-MHz spacing between channels to minimize adjacent channel interference. Refer to Fig. 17-13 for details. A wideband repeater (Fig. 17-15) is designed to receive any signal transmitted within the 500-MHz total bandwidth.

The receive antenna is connected to a low-noise amplifier (LNA) as in every transponder. Very wideband tuned circuits are used so that the entire 500-MHz bandwidth is received and amplified. A low-noise amplifier, usually a GaAs FET, provides gain. A mixer translates all incoming signals to their equivalent lower downlink frequencies. In a C-band communication satellite, the incoming signals are located between 5.925 and 6.425 MHz. A local oscillator operating at the frequency of 2.225 GHz is used to translate the inputs to the 3.7- to 4.2-GHz range. A wideband amplifier following the mixer amplifies this entire spectrum.

Channelization Process.

Channelization process

The *channelization process* occurs in the remainder of the transponder. For example, in a 12-channel satellite, 12 bandpass filters, each centered on one of the 12 channels, are used to separate all the various received signals. Figure 17-13 shows the 12 basic channels with their center frequencies, each having a bandwidth of 36 MHz. The bandpass filters separate out the unwanted mixer output signals and retain only the difference signals. Then individual *high-power amplifiers (HPAs)* are used to increase the signal level. These are usually *traveling-wave tubes (TWTs)*. The output of each TWT amplifier is again filtered to minimize harmonic and intermodulation distortion problems. These filters are usually part of a larger assembly known as a *multiplexer*, or *combiner*. This is a waveguide–cavity resonator assembly that filters and combines all the signals for application to a single antenna.

High-power amplifier (HPA)

Traveling-wave tube (TWT)

Figure 17-15 A broadband multiple-channel repeater.

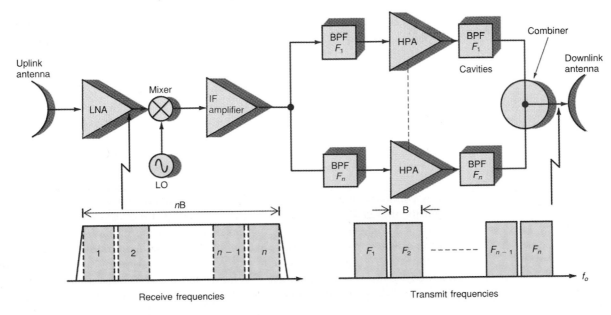

Solar panel

It is logical to assume that if the receive function can be accomplished by wideband amplifier and mixer circuits, then it must be possible to provide the transmit function in the same way. However, it is generally not possible to generate very high output power over such wide bandwidth. The fact is that no components and circuits can do this well. The high-power amplifiers in most transponders are traveling-wave tubes that inherently have limited bandwidth. They operate well over a small range but cannot deal with the entire 500-MHz bandwidth allocated to a satellite. Therefore, to achieve the high power levels, the channelization process is used.

Power Subsystem

Today virtually every satellite uses solar panels for its basic power source. *Solar panels* are large arrays of photocells connected in various series and parallel circuits to create a powerful source of direct current. Early solar panels could generate hundreds of watts of power. Today huge solar panels are capable of generating many kilowatts. A key requirement is that the solar panels always be pointed toward the sun. There are two basic satellite configurations. In cylindrical satellites, the solar cells surround the entire unit, and therefore some portion of them is always exposed to sunlight. In body-stabilized, or three-axis, satellites, individual solar panels are manipulated with various controls to ensure that they are correctly oriented with respect to the sun.

Solar panels generate a direct current that is used to operate the various components of the satellite. However, the dc power is typically used to charge secondary batteries that act as a buffer. When a satellite goes into an eclipse or when the solar panels are not properly positioned, the batteries take over temporarily and keep the satellite operating. The batteries are not large enough to power the satellite for a long time; they are used as a backup system for eclipses, initial satellite orientation and stabilization, or emergency conditions.

The basic dc voltage from the solar panels is conditioned in various ways. For example, it is typically passed through voltage regulator circuits before being used to power individual electronic circuits. Occasionally, voltages higher than those produced by the solar panels must also be generated. For example, the TWT amplifiers in most communication transponders require thousands of volts for proper operation. Special dc-to-dc converters are used to translate the lower dc voltage of the solar panels to the higher dc voltage required by the TWTs.

Telemetry, Command, and Control Subsystems

Telemetry, command, and control (TC&C) subsystem

All satellites have a *telemetry, command, and control (TC&C) subsystem* that allows a ground station to monitor and control conditions in the satellite. The telemetry system is used to report the status of the onboard subsystems to the ground station (see Fig. 17-16). The telemetry system typically consists of various electronic sensors for measuring temperatures, radiation levels, power supply voltages, and other key operating characteristics. Both analog and digital sensors may be used. The sensors are selected by a multiplexer and then converted to a digital signal, which then modulates an internal transmitter. This transmitter sends the telemetry information back to the earth station, where it is recorded and monitored. With this information, the ground station then determines the operational status of the satellite at all times.

A command and control system permits the ground station to control the satellite. Typically, the satellite contains a command receiver that receives control signals from an earth station transmitter. The control signals are made up of various digital codes that tell the satellite what to do. Various commands may initiate a telemetry sequence, activate thrusters for attitude correction, reorient an antenna, or perform other operations as required by the special equipment specific to the mission. Usually, the control signals are processed by an onboard computer.

Figure 17-16 (*a*) General block diagram of a satellite telemetry unit. (*b*) The command
receiver and controller.

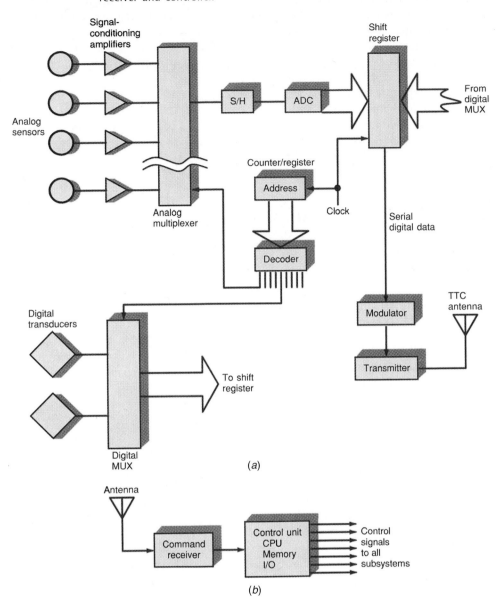

Most satellites contain a small digital computer, usually microprocessor-based, that acts as a central control unit for the entire satellite. The computer contains a built-in ROM with a master control program that operates the computer and causes all other subsystems to be operated as required. The command and control receiver typically takes the command codes that it receives from the ground station and passes them on to the computer, which then carries out the desired action.

The computer may also be used to make necessary computations and decisions. Information collected from the telemetry system may be first processed by the computer before it is sent to the ground station. The memory of the computer may also be used to store data temporarily prior to processing or prior to its being transmitted back to earth. The computer may also serve as an event timer or clock. Thus the computer is a versatile control element that can be reprogrammed via the command and control system to carry out any additional functions that may be required, particularly those that were not properly anticipated by those designing the mission.

Applications Subsystems

The applications subsystem is made up of the special components that enable the satellite to fulfill its intended purpose. For a communication satellite, this subsystem is made up of the transponders.

An observation satellite such as those used for intelligence gathering or weather monitoring may use TV cameras or infrared sensors to pick up various conditions on earth and in the atmosphere. This information is then transmitted back to earth by a special transmitter designed for this purpose. There are many variations of this subsystem depending upon the use. The *Global Positioning System (GPS)* for satellites is an example of a subsystem, the application payload for which is used for navigation. This system is discussed in detail later in this chapter.

Global Positioning System (GPS)

17-4 Ground Stations

Ground station

The *ground station,* or earth station, is the terrestrial base of the system. The ground station communicates with the satellite to carry out the designated mission. The earth station may be located at the end user's facilities or may be located remotely with ground-based intercommunication links between the earth station and the end user. In the early days of satellite systems, earth stations were typically placed in remote country locations. Because of their enormous antennas and other critical requirements, it was not practical to locate them in downtown or suburban areas. Today, earth stations are still complex, but their antennas are smaller. Many earth stations are now located on top of tall buildings or in other urban areas directly where the end user resides. This offers the advantage of eliminating complex intercommunication systems between the earth station and the end user.

Like the satellite, the earth station is made up of a number of different subsystems. The subsystems, in fact, generally correspond to those onboard the satellite but are larger and much more complex. Further, several additional subsystems exist at earth stations that would not be appropriate in a satellite.

Ground control equipment (GCE) subsystem

An earth station consists of five major subsystems: the antenna subsystem, the receive subsystem, the transmit subsystem, the *ground control equipment (GCE) subsystem,* and the power subsystem (see Fig. 17-17). Not shown here are the telemetry, control, and instrumentation subsystems.

The power subsystem furnishes all the power to the other equipment. The primary source of power is the standard ac power lines. This subsystem operates power supplies that distribute a variety of dc voltages to the other equipment. The power subsystem also consists of emergency power sources such as diesel generators, batteries, and inverters to ensure continuous operation during power failures.

Diplexer

As shown in Fig. 17-17, the antenna subsystem usually includes a *diplexer,* i.e., a waveguide assembly that permits both the transmitter and the receiver to use the same antenna. The diplexer feeds a bandpass filter (BPF) in the receiver section that ensures that only the received frequencies pass through to the sensitive receiving circuits. This bandpass filter blocks the high-power transmit signal that can occur simultaneously with reception. This prevents overload and damage to the receiver.

The output of the bandpass filter feeds a low-noise amplifier that drives a power divider. This is a waveguidelike assembly that splits the received signal into smaller but equal power signals. The power divider feeds several down converters. These are standard mixers fed by local oscillators (LOs) that translate the received signals down to an intermediate frequency, usually 70 MHz. A bandpass filter ensures the selection of the proper sidebands out of the down converter.

The IF signal containing the data is then sent to the GCE receive equipment, where it is demodulated and fed to a demultiplexer, where the original signals are finally obtained. The demultiplexer outputs are usually the baseband or original communication

Figure 17-17 General block diagram of an earth station.

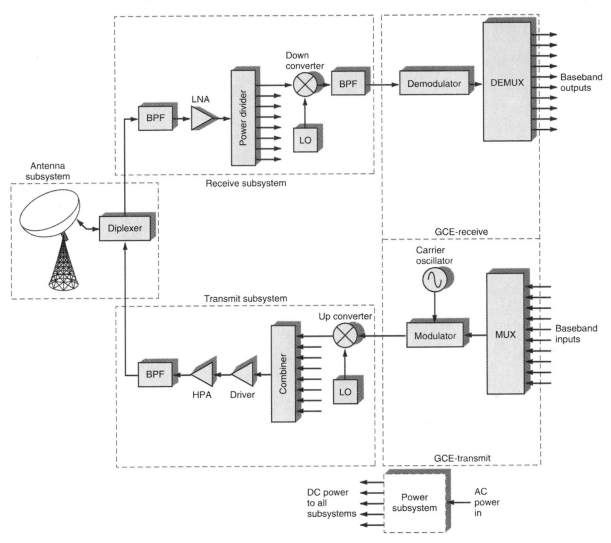

signals. In actual systems, several levels of demodulation and demultiplexing may have to take place to obtain the original signals. In the GCE transmit subsystem, the baseband signals such as telephone conversations are applied to a multiplexer that permits multiple signals to be carried on a single channel.

Antenna Subsystems

All earth stations have a relatively large parabolic dish antenna that is used for sending and receiving signals to and from the satellite. Early satellites had very low power transmitters, and so the signals received on earth were extremely small. Huge high-gain antennas were required to pick up minute signals from the satellite. The earth station dishes were 80 to 100 ft or more in diameter. Antennas of this size are still used in some satellite systems today, and even larger antennas have been used for deep-space probes.

Modern satellites now transmit with much more power. Advances have also been made in receiver components and circuitry. For that reason, smaller earth station antennas are now practical. In some applications, antennas having as small as 18-in diameter can be used.

Antenna subsystem

Typically, the same antenna is used for both transmitting and receiving. A diplexer is used to permit a single antenna to be used for multiple transmitters and/or receivers. In some applications, a separate antenna is used for telemetry and control functions.

The antenna in an earth station must also be steerable. That is, it must be possible to adjust its azimuth and elevation so that the antenna can be properly aligned with the satellite. Earth stations supporting geosynchronous satellites can generally be fixed in position, however. Azimuth and elevation adjustments are necessary to initially pinpoint the satellite and to permit minor adjustments over the satellite's life.

Receive Subsystems

Receive subsystem

The downlink is the *receive subsystem* of the earth station. It usually consists of very low-noise preamplifiers that take the small signal received from the satellite and amplify it to a level suitable for further processing. The signal is then demodulated and sent on to other parts of the communication system.

Receiver circuit

Receiver Circuits. The receive subsystem consists of the LNA, down converters, and related components. The purpose of the receive subsystem is to amplify the downlink satellite signal and translate it to a suitable intermediate frequency. From that point, the IF signal is demodulated and demultiplexed as necessary to generate the original baseband signals.

Refer to the general block diagram of the receive subsystem shown in Fig. 17-18. Figure 17-18(a) shows a typical dual-conversion down converter. The input bandpass filter passes the entire 500-MHz satellite signal. This is fed to a mixer along with a local oscillator. The output of the mixer is an IF signal, usually 770 MHz. This is passed through a bandpass filter at that frequency with a bandwidth of 36 MHz.

The signal is then applied to another mixer. When combined with the local-oscillator frequency, the mixer output is the standard 70-MHz IF value. An IF of 140 MHz is used in some systems. A 36-MHz-wide bandpass filter positioned after the mixer passes the desired channel.

In dual-conversion down converters, two different tuning or channel selection arrangements are used. One is referred to as *RF tuning*, and the other is referred to as *IF tuning*. In RF tuning, shown in Fig. 17-18(a), the first local oscillator is made adjustable. Typically, a frequency synthesizer is used in this application. The frequency

Figure 17-18 Dual-conversion down converters. (*a*) RF tuning. (*b*) IF tuning.

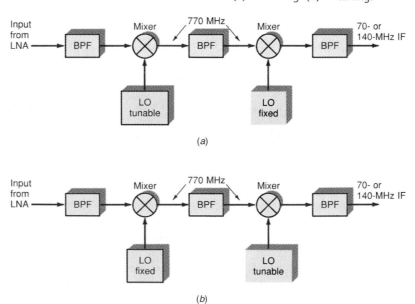

(a)

(b)

synthesizer generates a highly stable signal at selected frequency increments. The frequency synthesizer is set to a frequency that will select a desired channel. In RF tuning, the second local oscillator has a fixed frequency to achieve the final conversion.

In IF tuning, the first local oscillator is fixed in frequency and the second is made tunable. Again, a frequency synthesizer is normally used for the tunable local oscillator [see Fig. 17-18(b)].

Example 17-3

Refer to Fig. 17-18(b). If the earth station downlink signal received is at $f_s = 4.08$ GHz, what local-oscillator frequencies f_{LO} are needed to achieve IFs of 770 and 140 MHz?

Assume LO frequencies lower than the received signals. First IF:

$$f_{IF} = f_s - f_{LO}$$
$$f_{LO} = f_s - f_{IF} = 4080 - 770 = 3310 \text{ MHz}$$

Second IF:

$$f_{LO} = f_s - f_{IF}$$
$$f_s = 770 \text{ MHz}$$
$$f_{LO} = 770 - 140 = 630 \text{ MHz}$$

Receiver Ground Control Equipment. The receiver *ground control equipment (GCE)* consists of one or more racks of equipment used for demodulating and demultiplexing the received signals. The down converters provide initial channelization by transponder, and the demodulators and demultiplexing equipment process the 70-MHz IF signal into the original baseband signals. Other intermediate signals may be developed as required by the application.

Ground control equipment (GCE)

The outputs from the down converters are usually made available on a patch panel of coaxial cable connectors. These are interconnected via coaxial cables to the demodulators. The demodulators are typically packaged in a thin, narrow vertical module that plugs into a chassis in a rack. Many of these demodulators are provided. They are all identical in that they demodulate the IF signal. In FDM systems, each demodulator is an FM detector. The most commonly used type is the phase-locked loop discriminator. Equalization and deemphasis are also taken care of in the demodulator.

In systems using TDM, the demodulators are typically used to detect four-phase, or quadrature, PSK at 60 or 120 Mbps. The IF is usually at 140 MHz. Again, a patch panel between the down converters and the demodulators permits flexible interconnection to provide any desired configuration. When video is transmitted, the output of the FM demodulator is the baseband video signal, which can then be transmitted by cable or used on the premises as required.

If the received signals are telephone calls, the demodulator outputs are sent to demultiplexing circuits. Again, in many systems a patch panel between the demodulator outputs and the demultiplexer inputs is provided to make various connections as required. In FDM systems, standard frequency-division multiplexing is used. This consists of additional single-sideband (SSB) demodulators and filters. Depending upon the number of signals multiplexed, several levels of channel filters and SSB demodulators may be required to generate the original baseband voice signals. Once this is done, the signal may be transmitted over the standard telephone system network as required.

In TDM systems, time-division demultiplexing equipment is used to reassemble the original transmitted data. The original baseband digital signals may be developed in some cases, or in others these signals are used with modems as required for interconnecting the earth station with the computer that will process the data.

Transmitter Subsystems

The uplink is the transmitting subsystem of the earth station. It consists of all the electronic equipment that takes the signal to be transmitted, amplifies it, and sends it to the antenna. In a communication system, the signals to be sent to the satellite might be TV programs, multiple telephone calls, or digital data from a computer. These signals are used to modulate the carrier which is then amplified by a large traveling-wave tube or klystron amplifier. Such amplifiers usually generate many hundreds of watts of output power. This is sent to the antenna by way of microwave waveguides, combiners, and diplexers.

Transmit subsystem

The *transmit subsystem* consists of two basic parts, the up converters and the power amplifiers. The up converters translate the baseband signals modulated on to carriers up to the final uplink microwave frequencies. The power amplifiers generate the high-power signals that are applied to the antenna. The modulated carriers are created in the transmit GCE.

Transmit Ground Control Equipment.
The transmit subsystem begins with the baseband signals. These are first fed to a multiplexer, if multiple signals are to be carried by a single transponder. Telephone calls are a good example. Frequency- or time-division multiplexers are used to assemble the composite signal. The multiplexer output is then fed to a modulator. In analog systems, a wideband frequency modulator is normally used. It operates at a carrier frequency of 70 MHz with a maximum deviation of ± 18 MHz. Video signals are fed directly to the modulator; they are not multiplexed.

In digital systems, analog signals are first digitized with PCM converters. The resulting serial digital output is then used to modulate a QPSK modulator.

Transmitter circuit

Transmitter Circuits.
Once the modulated IF signals have been generated, up conversion and amplification will take place prior to transmission. Individual up converters are connected to each modulator output. Each up converter is driven by a frequency synthesizer that allows selection of the final transmitting frequency. The frequency synthesizer selects the transponder it will use in the satellite. The synthesizers are ordinarily adjustable in 1-kHz increments so that any up converter can be set to any channel frequency or transponder.

As in down converters, most modern up converters use dual conversion. Both RF tuning and IF tuning are used (Fig. 17-19). With IF tuning, a tunable carrier from a frequency synthesizer is applied to the mixer to convert the modulated signal to an IF level, usually 700 MHz. Another mixer fed by a fixed-frequency local oscillator (LO) then performs the final up conversion to the transmitted frequency.

In RF tuning, a mixer fed by a fixed-frequency local oscillator performs an initial up conversion to 700 MHz. Then a sophisticated RF frequency synthesizer applied to a second mixer provides up conversion to the final microwave frequency.

In some systems, all the IF signals at 70 or 140 MHz are combined prior to the up conversion. In this system, different carrier frequencies are used on each of the modulators to provide the desired channelization. When translated by the up converter, the carrier frequencies will translate to the individual transponder center frequencies. A special IF combiner circuit mixes all the signals linearly and applies them to a single up converter. This up converter creates the final microwave signal.

In most systems, however, individual up converters are used on each modulated channel. At the output of the up converters, all the signals are combined in a microwave combiner which produces a single output signal that is fed to the final amplifiers. This arrangement is illustrated in Fig. 17-20.

Figure 17-19   Typical up converter circuits. (*a*) IF tuning. (*b*) RF tuning.

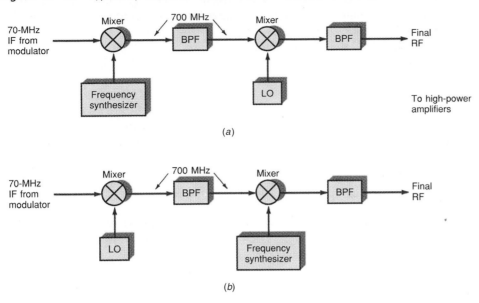

(*a*)

(*b*)

Figure 17-20   An RF combiner and power amplifiers.

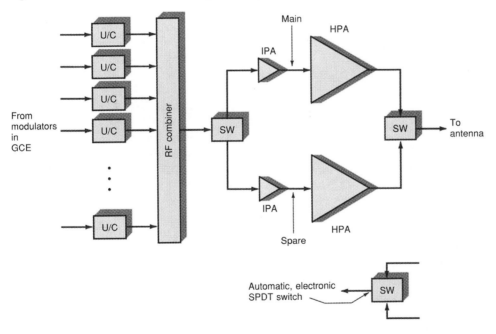

The final combined signal to be transmitted to the satellite appears at the output of the RF combiner. But it must first be amplified considerably before being sent to the antenna. This is done by the power amplifier. The power amplifier usually begins with an initial stage called the *intermediate-power amplifier (IPA)*. This provides sufficient drive to the final high-power amplifier (HPA). Note also in Fig. 17-20 that redundant amplifiers are used. The main IPA and HPA are used until a failure occurs. Switches (SW) automatically disconnect the defective main amplifier and connect the spare to ensure continuous operation. The amplified signal is then sent to the antenna via a wave-guide, the diplexer, and a filter.

Three types of power amplifiers are used in earth stations: transistor, traveling-wave tube (TWT), and klystron. Transistor power amplifiers are used in small and medium earth stations with low power. Powers up to 50 W are common. Normally, power gallium

Intermediate-power amplifier (IPA)

arsenide field-effect transistors (GaAs MESFETs) or silicon MOSFETs are used in this application. The newer GaAs MESFETs and MOSFET power devices, when operated in parallel, can achieve powers of up to approximately 100 W. Further improvements are expected.

Most medium- and high-power earth stations use either TWTs or klystrons for the power amplifiers. There are two typical power ranges, one in the 200- to 400-W range and the other in the 2- to 3-kW range. The amount of power used depends upon the location of the station and its antenna size. Satellite transponder characteristics also influence the power required by the earth station. As improvements have been made in low-noise amplifiers, and satellites have been able to carry higher-power transponders, earth station transmitter power requirements have greatly decreased. This is also true for antenna sizes.

Power Subsystems

Power subsystem

Most earth stations receive their power from the normal ac mains. Standard power supplies convert the ac power to the dc voltages required to operate all subsystems. However, most earth stations have backup power systems. Satellite systems, particularly those used for reliable communication of telephone conversations, TV programs, computer data, and so on, must not go down. The backup power system takes over if an ac power failure occurs. The backup power system may consist of a diesel engine driving an ac generator. When ac power fails, an automatic system starts the diesel engine. The generator creates the equivalent ac power, which is automatically switched to the system.

Uninterruptible power supplies (UPS)

Smaller systems may use *uninterruptible power supplies (UPS)*, which derive their main power from batteries. Large battery arrays drive dc-to-ac inverters which produce the ac voltages for the system. Uninterruptible power supplies are not suitable for long power failures and interruptions because the batteries quickly become exhausted. However, for short interruptions of power, i.e., less than an hour, they are adequate.

Telemetry and Control Subsystems

The telemetry equipment consists of a receiver and the recorders and indicators that display the telemetry signals. The signal may be received by the main antenna or a separate telemetry antenna. A separate receiver on a frequency different from that of the communication channels is used for telemetry purposes. The telemetry signals from the various sensors and transducers in the satellite are multiplexed onto a single carrier and are sent to the earth station. The earth station receiver demodulates and demultiplexes the telemetry signals into the individual outputs. These are then recorded and sent to various indicators, such as strip chart recorders, meters, and digital displays. Signals may be in digital form or converted to digital. They can be sent to a computer where they can be further processed and stored.

The control subsystem permits the ground station to control the satellite. This system usually contains a computer for entering the commands that modulate a carrier that is amplified and fed to the main antenna. The command signals can make adjustments in the satellite attitude, turn transponders off and on, actuate redundant circuits if the circuits they are backing up fail, and so on.

Instrumentation subsystem

In some satellite systems where communication is not the main function, some instrumentation may be a part of the ground station. *Instrumentation* is a general term for all the electronic equipment used to deal with the information transmitted back to the earth station. It may consist of demodulators and demultiplexers, amplifiers, filters, A/D converters, or signal processors. The *instrumentation subsystem* is in effect an extension of the telemetry system. Besides relaying information about the satellite itself, the telemetry system may be used to send back information related to various scientific experiments being conducted on the satellite. In satellites used for surveillance, the instrumentation may be such that it can deal with digital still photographs or TV signals sent back from an onboard camera. The possibilities are extensive depending upon the actual satellite mission.

P60974370

...es of electronic communication

2460268X

e: 20/04/12 23:59

Total items: 1
Total fines: £6.80
28/03/2012 19:11
Issued: 5
Overdue: 0

Thank you for using Self Service.
Please keep your receipt.

Overdue books are fined at 40p per day for
1 week loans, 10p per day for long loans.

Very Small–Aperture Terminal

A *very small-aperture terminal (VSAT)* is a miniature low-cost satellite ground station. In the past, most ground stations were large and expensive, large because of the huge dish antennas often needed and expensive because of the equipment costs. But over the years, semiconductor and other technology breakthroughs have greatly reduced the size and cost of ground stations. The VSAT is one result. These units are extremely small and mount on the top or side of a building and in some versions even fit into a suitcase. Costs range from a few thousand dollars to no more than about $6000 today. They can be installed very quickly by plugging them in and pointing the antenna.

Very small-aperture terminal (VSAT)

A VSAT is a full receive-transmit earth station. Most of those used in the United States and Europe operate in the Ku band, so the antennas are very small, typically 0.6 to 2.4 m (about 2 to 8 ft) in diameter. VSATs in Asia, Africa, and Latin America use the C band which requires much larger dishes. *Receive-only (RO)* VSATs are also available for special applications such as digital video broadcasts (DVBs). Most of the transmit-receive electronics in a VSAT is contained in a housing mounted on the dish. A box containing the feed horn antenna at the focal point of the dish contains the LNA, down converters, and demodulators in the receiver and the transmit power amplifier. In some systems these circuits may be contained in a housing near the base of the antenna to keep the coaxial connection between the feed horn antenna and transmitter-receiver unit short. A cable then connects the baseband inputs and outputs (voice, video, or data) to a computer which is connected to the system.

Receive-only (RO)

The most common application of VSATs today is in connecting many remote company or organization sites to a main computer system. For example, most gas stations and retail stores use VSATs as point-of-sale (PoS) terminals to transmit sales transaction information to the home office, check customer credit cards, and relay inventory data. Companies such as Shell, Wal-Mart, and Barnes & Noble all use VSATs. Most stores that sell lottery tickets use a VSAT as do most tollbooths using SpeedPass and other radio-frequency identification (RFID) of vehicles for tolls. A good example of a RO VSAT is the settop box receiver used by consumers for Direct Broadcast Satellite (DBS) TV reception.

17-5 Satellite Applications

Every satellite is designed to perform some specific task. Its predetermined application specifies the kind of equipment it must have onboard and its orbit. Although the emphasis on satellites in this chapter is communication, satellites are useful for observation purposes.

Communication Satellites

The main application for satellites today is in communication. Satellites used for this purpose act as relay stations in the sky. They permit reliable long-distance communication worldwide. They solve many of the growing communication needs of industry and government. Communication applications will continue to dominate this industry.

The primary use of communication satellites is in long-distance telephone service. Satellites greatly simplify long-distance calls not only within but also outside the United States.

Another major communication application is TV. For years, TV signals have been transmitted through satellites for redistribution. Because of the very high-frequency signals involved in TV transmission, other long-distance transmission methods are not technically or economically feasible. Special coaxial cables and fiber-optic cables as well as microwave relay links have been used to transmit TV signals from one place to another. However, with today's communication satellites, TV signals can be transmitted easily from one place to another. All the major TV networks and cable TV companies rely on communication satellites for TV signal distribution.

Direct Broadcast Satellite (DBS). A more recent satellite TV service is the *Direct Broadcast Satellite (DBS) service.* This is a TV signal distribution system designed to distribute signals directly to consumers. Special broad U.S. coverage satellites with high power transmit cable-TV-like services direct to homes equipped with the special DBS receivers. These newer satellites work at the higher microwave frequencies (K band) with higher power so that only very small antennas are needed. The DBS system allows dishes with diameters as small as 18 in to be used reliably. The newer low-noise GaAs FET receiver front ends make this possible.

Satellite Cell Phones. An application under development is satellite-based cellular telephone service. Current cellular telephone systems rely on many low-power, ground-based cells to act as intermediaries between the standard telephone system and the millions of roving cellular telephones. The proposed new systems use low-earth-orbit satellites to perform the relay services to the main telephone system or to make connection directly between any two cellular telephones using the system.

Although nearly a dozen different satellite cellular telephone systems have been proposed, the most advanced system is that developed by Motorola. Known as *Iridium,* it uses a constellation of 66 satellites in six polar orbits with 11 satellites per orbit 420 mi above the earth (refer to Fig. 17-21). The satellites operate in the L band over the frequency range of 1.61 to 1.6265 GHz. The satellites communicate with ground stations called *gateways* that connect the system to the public switched telephone network. The satellites also communicate among themselves. Both gateway and intersatellite communication takes place over K_a band frequencies. The system provides truly global coverage between any two handheld cellular telephones or between one of the cellular telephones and any other telephone on earth.

Each satellite in the system transmits back to earth, creating 48 spot beams that in effect form moving cell areas of coverage. The spot beams provide spatial separation to permit more than one satellite to use the same frequencies for simultaneous communication. Frequency reuse makes efficient use of a small number of frequencies.

Transmission is effected by digital techniques. A special form of QPSK is used for modulation. *Time-division multiple access (TDMA),* a special form of time-division multiplexing, is used to provide multiple voice channels per satellite.

In addition to voice communication, Iridium will be able to provide a whole spectrum of other communication services including

1. *Data communication*   E-mail and other computer communication.
2. *Fax*   Two-way facsimile.
3. *Paging*   Global paging to receivers with a two-line alphanumeric display.

Figure 17-21   The Motorola Iridium satellite cellular telephone system.

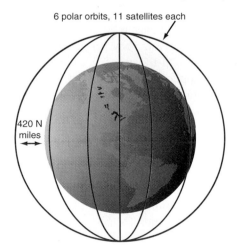

6 polar orbits, 11 satellites each

420 N miles

4. *Radio Determination Services (RDSs)*   A subsystem that permits satellites to locate transceivers on earth. Accuracy is expected to be within 3 mi.

The Iridium system became fully operational in the 1999–2000 period, and it is still operational on a limited scale. The cost of service is high but affordable by companies and governments needing global communications.

Other worldwide satellite telephone systems have experienced similar problems. Next to Iridium, Globalstar is the most widely used. This LEO system uses 48 satellites and CDMA for voice and data communication. Other, smaller worldwide satellite systems are Teledesic, G2, and INMARSAT.

Digital Satellite Radio

One of the newest satellite applications is in *digital satellite radio* or the *digital audio radio service (DARS)*. This service provides hundreds of channels of music, news, sports, and talk radio primarily to car portable and home radios. Conventional AM and FM radio stations cover only short distances and are subject to local and even national radio propagation effects. As you are traveling by car or truck, radio stations come and go every 40 mi or so. And if you are driving in the rural areas of the United States, you may not get any station. This is not so with satellite radio, which provides full continuous coverage of the station you select wherever you are in the United States. The system uses digital transmission techniques that ensure high-quality stereo sound that is immune to noise. Furthermore, the digital format allows the satellites to transmit other information such as song title and artist, type of music, and other data, which are displayed on a LCD screen.

Digital satellite radio or digital audio radio service (DARS)

These aircraft function a bit as satellites do. The aircraft, while overhead, provide broadband communication to customers below. Customers over thousands of square miles can receive services such as movies, videoconferencing, Internet, and phone. The service is not hindered by weather, buildings, trees, or terrain features. The system, called Halo, has tested well, providing 52 Mbps per second. The idea is to fly Proteus aircraft over major cities in fixed patterns at 51,000 ft or higher (farther up than planes and bad weather). The system has been developed by Raytheon and Angel Technologies.

Two such systems in the United States are XM Satellite Radio and Sirius Satellite Radio. XM Satellite Radio uses two geosynchronous satellites, named *Rock* and *Roll*, positioned to cover the continental United States. The Sirius Satellite Radio system uses three elliptical-orbit satellites. Each satellite appears over the United States for approximately 16 h/aday. One satellite is always available for coverage anywhere in the United States. Both the XM and Sirius systems operate in the 2.3-GHz S band.

XM and Sirius generate their programming as do other radio stations, and then they up-link their digitized signals to the satellites. The satellites provide the continuous coverage to vehicles on the road and homes in the United States.

The microwave receivers are mounted in cars as are other car radios, but they require a special outside or window-mounted antenna for optimum reception. Typically, the unit also contains standard AM and FM receiver capability. These satellite digital radios are available as an option in most new cars, and several manufacturers are offering after-market add-on radios for existing vehicles. Home and portable units are also available.

Surveillance Satellites

Another application of satellites is in surveillance or observation. From their vantage point high in the sky, satellites can look at the earth and transmit what they see to ground stations for a wide variety of purposes. For example, military satellites are used to perform reconnaissance. Onboard cameras take photographs that can later be ejected from the satellite and brought back to earth for recovery. TV cameras can take pictures and send them back to earth as electric signals. Infrared sensors detect heat sources. Small radars can profile earth features.

Intelligence satellites collect information about enemies and potential enemies. They permit monitoring for the purpose of proving other countries' compliance with nuclear test ban and missile stockpile treaties.

There are many different kinds of observation satellites. One special type is the meteorological, or weather, satellite. These satellites photograph cloud cover and send back to earth pictures that are used for determining and predicting the weather. Geodetic satellites photograph the earth for the purpose of creating more accurate and more detailed maps.

Navigation Satellites

A third applications area is navigation. Electronic systems have been used for years to provide accurate position information to ships, airplanes, and land-based vehicles. Loran and Omega are well-known systems used in marine navigation. *Loran* is a highly accurate system that uses coastal stations along all U.S. water borders. More than 50 stations transmit on a frequency of about 100 kHz in a range that is very limited.

The *Omega system* uses signals in the 10- to 13-kHz range, and with eight stations it can cover the world. In both cases accuracy is limited to within the 1000- to 2000-ft range. Yet this is good accuracy for general navigation.

Aircraft use the *VOR/DME Tacan system* of navigation. *VOR* means *VHF omnidirectional ranging* and *DME* means *distance-measuring equipment*. *Tacan* refers to *Tactical Air Navigation*, which is a similar but higher-frequency, higher-accuracy version of the VOR/DME used by the military. VOR uses hundreds of beacon stations transmitting a rotating narrow-beam signal that vectors planes from one VOR station to another along common air traffic routes. DME permits accurate distance measurements between stations. VOR transmits in the 108- to 118-MHz range, and Tacan operates in the 960- to 1215-MHz range. Typical accuracy is in the 200- to 600-ft range. Obviously, use is limited to the United States.

Satellite systems overcome some of the problems with land-based navigation systems. First, they can provide global coverage so that planes and ships can navigate anywhere on earth. By using multiple satellites, global coverage can be obtained, thus overcoming the limitations of earth-based systems that are confined in range by signal propagation conditions.

By using satellite systems, very high-frequency signals can be used. High-frequency signals give better resolution and accuracy in positioning. Finally, satellite systems can

Surveillance satellite

Navigation satellite

Loran system

Omega system

VHF omnidirectional ranging (VOR)

Distance-measuring equipment (DME)

Tactical Air Navigation (Tacan)

provide altitude positioning data that other systems cannot. The navigation systems built by the Navy such as Transit have all been phased out in favor of the GPS. Section 17-6 covers GPS in detail.

17-6 Global Positioning System

The *Global Positioning System (GPS)*, also known as *Navstar,* is a satellite-based navigation system that can be used by anyone with an appropriate receiver to pinpoint her or his location on earth. The array of GPS satellites transmits highly accurate, time-coded information that permits a receiver to calculate its exact location in terms of the latitude and longitude on earth as well as the altitude above sea level.

Global Positioning System (GPS) or Navstar

GPS was developed by the U.S. Air Force for the Department of Defense as a continuous global radio navigation system that all elements of the military services would use for precision navigation. Development was started in 1973, and by 1994, the system was fully operational.

The GPS Navstar system is an *open navigation system;* i.e., anyone with a GPS receiver can use it. The system is designed, however, to provide a base navigation system with an accuracy to within 100 m. A supplementary part of the system, at one time accessible only to the military and called *selective availability (SA)*, provides much greater precision, to within 10 to 20 m. This precision is now available to any GPS user. The base system is highly accurate and more so than virtually any other electronic navigation system in existence. As a result, GPS is gradually replacing older military systems and civilian land-based systems.

Open navigation system

Selective availability (SA)

The GPS is an excellent example of a modern satellite-based system and the high-technology communication techniques used to implement it. Learning about GPS is an excellent way to bring together and illustrate the complex concepts presented in this chapter and previous chapters.

The GPS consists of three major segments: the space segment, the control segment, and the user segment.

Space Segment

The *space segment* is the constellation of satellites orbiting above the earth that contain transmitters which send highly accurate timing information to GPS receivers on earth. The receivers in the user segment themselves may be used on land, sea, or air.

Space segment

GPS mapping aids are small enough to be mounted on a dashboard or stashed in a backpacker's pack. They're popular with boaters, too.

Figure 17-22   The GPS space segment.

6 orbits of 24 satellites, 55° angle of inclination
with height of 10,898 mi

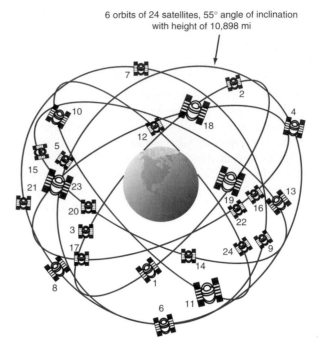

The fully implemented GPS consists of 24 main operational satellites plus 3 active spare satellites (see Fig. 17-22). The satellites are arranged in six orbits, each orbit containing 3 or 4 satellites. The orbital planes form a 55° angle with the equator. The satellites orbit at a height of 10,898 nautical miles above the earth (20,200 km). The orbital period for each satellite is approximately 12h (11 h 58 min).

Each of the orbiting satellites contains four highly accurate atomic clocks. They provide precision timing pulses used to generate a unique binary code, i.e., a pseudorandom code identifying the specific satellite in the constellations that is transmitted to earth. The satellite also transmits a set of digitally coded ephemeris data that completely defines its precise orbit. The term *ephemeris* is normally associated with specifying the location of a celestial body. Tables have been computed so that it is possible to pinpoint the location of planets and other astronomical bodies at precise locations and times. Ephemeris data can also be computed for any orbiting body such as a satellite. This data tells where the satellite is at any given time, and its location can be specified in terms of the satellite ground track in precise latitude and longitude measurements. Ephemeris information is coded and transmitted from the satellite, providing an accurate indication of the exact position of the satellite above the earth at any given time. The satellite's ephemeris data is updated once a day by the ground control station to ensure accuracy.

A GPS receiver on earth is designed to pick up signals from three, four, or more satellites simultaneously. The receiver decodes the information and, using the time and ephemeris data, calculates the exact position of the receiver. The receiver contains a high-speed, floating-point microcomputer that performs the necessary calculations. The output of the receiver is a decimal display of latitude and longitude as well as altitude. Readings from only three satellites are necessary for latitude and longitude information only. A fourth satellite reading is required to compute altitude. Most GPS receivers now include a detailed map display on a color LCD.

Each GPS satellite carries two transmitters that together transmit the timing and location signals to the earth receivers. One of the transmitters sends a signal called L1 on 1575.42 MHz. The signal transmitted is a *pseudorandom code (PRC)* called the *coarse acquisition (C/A)* code. It is transmitted at a 1-Mbps rate using BPSK. It repeats every 1023 bits. The C/A code is like the pseudorandom codes used in spread spectrum in that they are used to distinguish between transmitted signals at the receiver. All 24 GPS satellites transmit on the same frequency, but the PRC is unique to each satellite so that the receiver can tell them apart.

Each satellite also contains another transmitter on a frequency of 1227.6 MHz. This is called the L2 signal. It contains another PRC known as the P code transmitted at a 10-Mbps rate. The P code can also be encrypted, and in this form it was called the Y code. The P and Y codes were originally designed for use by only the military so that military GPS receivers would be more accurate than the civilian (or enemy) receivers which receive only L1 signal. This feature was called *selective availability (SA)*. The SA feature was abandoned on May 1, 2000. Today all receivers pick up both signals, enabling all receivers to have an accuracy that identifies the location to within 10 m.

Selective availability (SA)

The basic information contained in the L1 signal consists of almanac data, ephemeris data, and the current date and time. The almanac data effectively notifies each receiver of where each satellite is during the day. The almanac data helps the receiver to initially lock onto a signal. The ephemeris data contains the exact position and timing of each satellite. It is this data that the receiver uses in the calculations to pinpoint its location. The time and date signal comes from the atomic clocks carried by each satellite.

Like all other satellites, GPS satellites contain a TT&C unit that is used by the ground stations to transmit updated ephemeris data and to make sure the satellite is in its exact position. Small thrusters fired from the ground allow the ground stations to correct minor drift that introduces errors into the measurements.

Control Segment and Atomic Clocks

The *control segment* of the GPS refers to the various ground stations that monitor the satellites and provide control and update information. The master control station is operated by the U.S. Air Force in Colorado Springs. Additional monitoring and control stations are located in Hawaii, Kwajalein, Diego Garcia, and the Ascension Islands. These four monitoring stations are not staffed. They constantly monitor the satellites and collect range information from each. The positions of these monitoring stations are accurately known. The information is sent back to the master control station in Colorado, where all the information is collected and position data on each satellite calculated. The master control station then transmits new ephemeris and clock data to each satellite on the S-band uplink once per day. This data updates the NAV-msg or navagation message, a 50 bps signal that modulates the L1 carrier and contains bits that describe the satellite orbits, clock corrections, and other system characteristics.

Control segment

The telemetry data transmitted as part of the NAV-msg is also received by the ground control station to keep track of the health and status of each receiver. The uplink S-band control system allows the ground station to do some station keeping to correct the satellite position as needed. Positioning is accomplished with the hydrazene thrusters.

Atomic Clocks. The precision timing signals are derived from atomic clocks. Most digital systems derive their timing information from a precision crystal oscillator called a *clock*. Crystal oscillators, although precise and stable, do not have the necessary precision and stability for the GPS. Timing data must be extremely precise to provide accurate navigation information.

Atomic clocks are electronic oscillators that use the oscillating energy of a gas to provide a stable operating frequency. Certain chemicals have atoms that can oscillate between low and high energy levels. This frequency of oscillation is extremely precise and stable.

Atomic clock

Cesium and rubidium are used in atomic clocks. In gaseous form, they are irradiated by electromagnetic energy at a frequency near their oscillating point. For cesium, this is 9,192,631,770 Hz; for rubidium, it is 6,834,682,613 Hz. This signal is generated by a quartz crystal-controlled oscillator whose frequency can be adjusted by the application of a control voltage to a varactor. The oscillation signal from the cesium (or rubidium) gas is detected and converted to a control signal that operates the oscillator. An error detector determines the difference between the crystal frequency and the cesium oscillations and generates the control voltage to set the crystal oscillator precisely. The output of the crystal oscillator is then used as the accurate timing signal. Frequency dividers and phase-locked loops are provided to generate the lower-frequency signals to be used. These signals operate the C/A and P code generators.

The GPS satellite contains two cesium clocks and two rubidium clocks. Only one is used at any given time. The clocks are kept fully operational so that if one fails, another can be switched in immediately.

GPS Receivers

GPS receiver

A *GPS receiver* is a complex superheterodyne microwave receiver designed to pick up the GPS signals, decode them, and then compute the location of the receiver. The output is usually an LCD display giving latitude, longitude, and altitude information and/or a map of the area.

There are many different types of GPS receivers. More than 40 manufacturers now make some form of GPS receiver. The larger and more sophisticated units are used in military vehicles. There are also sophisticated civilian receivers for use in various kinds of precision applications such as surveying and mapmaking. Different models are available for use in aircraft, ships, and trucks. Handheld units are also available.

The most widely used GPS receiver is the popular handheld portable type, not much larger than an oversized handheld calculator. Most of the circuitry used in making a GPS receiver has been reduced to integrated-circuit form, thereby permitting an entire receiver to be contained in an extremely small, portable battery-operated unit. Keep in mind that all GPS receivers are not only superheterodyne communication receivers but also sophisticated computers. A considerable amount of high-level mathematics must be carried out to compute the receiver position from the received data.

Figure 17-23 is a general block diagram of a GPS receiver, typical of the simpler, low-cost handheld units on the market. The receiver consists of the antenna, the RF/IF section, the frequency standard clock oscillator, and a frequency synthesizer that provides local-oscillator signals as well as clock and timing signals for the rest of the receiver.

Other sections of the receiver include a digital signal processor and a control microcomputer along with its related RAM and ROM. Interface circuits provide connection to the LCD display.

Figure 17-23 A GPS receiver.

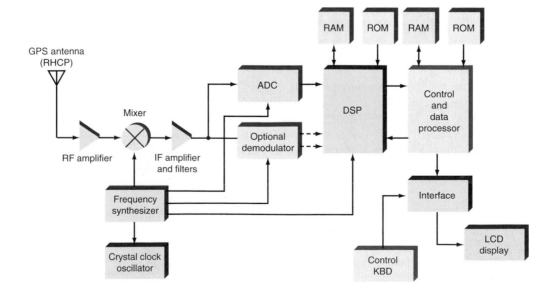

Figure 17-24 Navigating with the GPS.

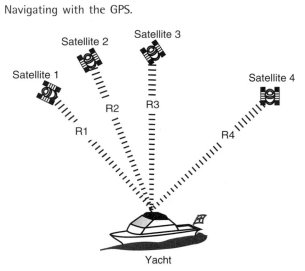

The antenna system is a type of patch antenna made on a printed-circuit board. It is designed to receive right-hand circularly polarized (RHCP) signals from the GPS satellites. In the handheld units, the antenna is part of the single physical structure and is connected directly to the receiver front end. In some larger and more complex GPS receivers, the antenna is a separate unit and may be mounted at a high clear point and connected to the receiver with coaxial cable.

The receiver can determine its exact position only by computing the position information obtained from four satellites. The receiver picks up signals from four satellites simultaneously (see Fig. 17-24). R1 through R4 are the ranges to the satellites from the yacht. Because spread spectrum is used, the receiver ignores all the signals except the one whose pseudocode has been entered and used to obtain lock. Once the receiver locks on the one satellite, all the information is extracted from the satellite. Then the C/A code is switched to another satellite within view, and the process is repeated. In other words, the receiver performs a time multiplexing operation on the four satellites within view of the receiver. The data is extracted from each of the four satellites and stored in the receiver's memory. Data from three satellites is needed to fix the receiver's position. If data from a fourth satellite is available, altitude can be calculated.

GPS TRIANGULATION

The determination of the location of a GPS receiver is based on measuring the distance between the receiver and three satellites. The distance is determined by measuring the time of arrival of the satellite signals and then computing distance based on the speed of radio waves, with correction factors. See Fig. 17-25. Assume three satellites A, B, and C. The receiver first computes the distance from the receiver to satellite A. Note that the distance from A is on a circle that falls on a wide range of locations on earth. Then the receiver calculates the distance to satellite B. That distance is defined along another circle. The two circles intersect at two points. One of those is the exact location, but we don't know which until we get a third satellite reading. The distance from satellite C intersects with the other circles at only one point. That is the location of the receiver.

Now, mentally translate each circle to a sphere. Then you can see how the signals really come together. Using a fourth satellite gives a fourth intersection point that enables the altitude to be determined.

Figure 17-25 How triangulation works to locate a GPS receiver.

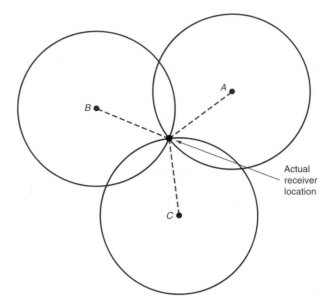

Actual
receiver
location

Once all the data has been accumulated, the high-speed control and data microprocessor in the receiver performs the final calculations. The microprocessor is typically a 16-bit unit with floating-point capability. Floating-point numbers must be used to provide the precision of calculation for accurate location.

The computations performed by the receiver are given below. First the receiver calculates the ranges R1 through R4 to each of the four satellites. These are obtained by measuring the time shift between the received pulses, which is the delay in transmission between the satellite and the receiver. These are the T_1 through T_4 times given below. Multiplying these by the speed of light c gives the range in meters between the satellite and the receiver. These four range values are used in the final calculations:

$$R1 = c \times T_1$$
$$R2 = c \times T_2$$
$$R3 = c \times T_3$$
$$R4 = c \times T_4$$

The basic ranging calculation is the solution to four simultaneous equations, as indicated in Fig. 17-26. The X, Y, and Z values are derived from the NAV-msg data transmitted by each of the satellites. The C_B is the clock bias. Since there is a difference between the clock frequency in the receiver and the clock in the satellites, there will be some difference called the *bias*. However, by factoring the clock error into the equations, it will be canceled out.

The goal of the microprocessor is to solve for the user position, designated U_X, U_Y, and U_Z in the equations. Once the calculation has been made, the microprocessor converts that information to the latitude, longitude, and altitude data, which is displayed on the LCD screen. The receiver display also shows the time of day, which is highly precise because it is derived from the atomic clocks in the satellites.

Enhanced GPS

Over the years a number of services have been created to improve on the accuracy of the GPS. That inherent accuracy is less than 10 m now that the selective availability has been disabled. Yet for many, this accuracy is insufficient. Errors caused by signal propagation speed differences in the ionosphere and troposphere, minor variations in satellite position, multipath signals, and even timing differences caused by clock drift add up to loss of position accuracy.

Figure 17-26 The equations solved by the GPS receiver.

Compute position coordinates
(four equations with four unknowns)

$$(X_1 - U_X)^2 + (Y_1 - U_Y)^2 + (Z_1 - U_Z)^2 = (R_1 - C_B)^2$$

$$(X_2 - U_X)^2 + (Y_2 - U_Y)^2 + (Z_2 - U_Z)^2 = (R_2 - C_B)^2$$

$$(X_3 - U_X)^2 + (Y_3 - U_Y)^2 + (Z_3 - U_Z)^2 = (R_3 - C_B)^2$$

$$(X_4 - U_X)^2 + (Y_4 - U_Y)^2 + (Z_4 - U_Z)^2 = (R_4 - C_B)^2$$

Solve for position coordinates
(U_X, U_Y, U_Z) and clock bias (C_B)

One of the enhanced services is *differential GPS (DGPS)*. This service is implemented by the U.S. Coast Guard and is available only in the United States, mostly on the coasts and along major waterways. DGPS uses a fixed station whose precise location is known. This station then monitors all satellites and compares location data from the satellite to its known position. It determines any errors in position that they create and transmits these errors to GPS receivers, where the error data updates the received data to give a more accurate position. The error signals are transmitted on a separate radio so a GPS receiver must also have a receiver for the DGPS signal to provide the error correction information. DGPS-enabled GPS radios are widely available but cost a bit more than a standard unit. If you want the most accurate position information available with an error of less than 5 m, this is the service to use, if it is available locally.

Differential GPS (DGPS)

Another enhanced GPS is called the *Wide-Area Augmentation System (WAAS)*. It was developed by the Federal Aviation Administration (FAA) and the Department of Transportation (DoT) so that aircraft could use GPS for blind instrument control landings. Currently the error even with DGPS is just too great to assume that a plane will be able to precisely identify the end of the runway and its extremities.

Wide-Area Augmentation System (WAAS)

The WAAS consists of about 25 ground stations around the United States with precisely known locations and two coastal stations that collect all the data from the other stations. The collected data is used to determine all errors, and then differential correction signals are transmitted up to one of two geosynchronous satellites that in turn transmit the correction signals to GPS receivers. As with DGPS, the receiver must be WAAS-enabled to receive the corrective data. Use of WAAS improves the accuracy with an error of less than 3 ft.

GPS Applications

The primary application of GPS is military and related navigation. GPS is used by all services for ships, aircraft of all sorts, and ground troops. Civilian uses have also increased dramatically because of the availability of many low-cost portable receivers. In fact, it is now possible to purchase a handheld receiver for less than $200. Most civilian applications involve navigation, which is usually marine or aviation-related. Hikers and campers and other outdoors sports enthusiasts also use GPS.

Commercial applications include surveying, mapmaking, and construction. Vehicle location is a growing application for trucking and delivery companies, taxi, bus, and train transportation. Police, fire, ambulance, and forest services also use GPS. GPS-based navigation systems are now widely available as accessories in cars to provide a continuous readout of current vehicle location.

GPS is finding new applications every day. For instance, it is used to keep track of fleets of trucks. A GPS receiver in each truck transmits its position data by way of a wireless connection, such as a wireless local-area network or cell phone. Many new cell phones contain a GPS receiver that automatically reports the location of the user if he

GOOD TO KNOW

The GPS was first widely used in Operation Desert Storm with much success. Nonmilitary applications of this system involve navigation, surveying, mapmaking, and vehicle tracking devices.

An Air Force major uses Global Positioning System satellites to cover flights and weather fronts.

or she makes a 911 call. Called Enhanced 911 service (E911), this feature is one that all cell phone companies must provide. While most location-based services will be used for 911 calls, eventually other location services may be developed for cell phones. Not all cell phones use GPS. Some use a unique triangulation method based on the cell phone being able to be in touch with at least three cell sites. Finally, GPS receivers are so inexpensive and accurate that they have led to a new hobby called *geocaching*. In this sport, one team hides an item or "treasure" and then gives the other team coordinates to follow to find the treasure within a given time.

Geocaching

CHAPTER REVIEW

Summary

A satellite is a natural or human-made physical object that orbits, or rotates, around some celestial body. Satellites are kept in orbit by programming them in accordance with physical and mathematical laws collectively referred to as orbital dynamics.

For a satellite to orbit the earth, it must have some forward motion. Thus, when it is launched, it is given both vertical and forward motion. The forward motion produces inertia, which tends to keep the satellite moving in a straight line. However, gravity tends to pull the satellite toward earth. The inertia of the satellite is equalized by the earth's gravitational pull. The satellite constantly changes its direction from a straight line to a curved line to rotate about the earth. Ground stations receive signals from the satellite that give the exact position of the satellite. Corrections to the orbital path can be made from ground stations.

Communication satellites are not originators of information; rather, they are repeaters. A transmitting station sends the information to the satellite, which in turn retransmits it to the receiving stations.

Every satellite is designed to perform some specific task. Its predetermined application specifies the kind of equipment it must have onboard and its orbit. The main application for satellites today is communication. Satellites used for this purpose act as relay stations in the sky. They permit reliable long-distance communication worldwide. The primary use of communication satellites is for long-distance telephone calls. Another major application is in TV signal transmission.

Satellites are also used for military surveillance, meteorological observations, and the monitoring of natural resources and pollution.

Navigation of ships, airplanes, and land-based vehicles is another common use of satellites. The Global Positioning System (GPS), also known as Navstar, is a satellite-based navigation system that can be used by anyone with an appropriate receiver to pinpoint his or her location on earth. The GPS is being used increasingly by civilians for navigation assistance in, e.g., hiking or boating. Other commercial uses are for mapmaking and surveying, transportation, and police and fire protection. GPS accuracy is improved by the use of enhancement systems known as differential GPS and the Wide-Area Augmentation System.

Questions

1. Is a satellite that rotates in the same direction as the earth's rotation said to be in a posigrade orbit or a retrograde orbit?
2. What is the geometric shape of a noncircular orbit?
3. What is the name of the center of gravity of the earth?
4. State the name for the time of one orbit.
5. Define *ascending orbit* and *descending orbit.*
6. State the effects on a satellite signal if the angle of elevation is too low.
7. What do you call a satellite that rotates around the equator 22,300 mi from the earth?
8. Why are small jet thrusters on a satellite fired occasionally?
9. What is the name of the point on the earth directly below a satellite?
10. Name the two angles used to point a ground station antenna toward a satellite.
11. What is the basic function and purpose of a communication satellite? Why are satellites used instead of standard earthbound radio?
12. What element of a satellite system transmits an uplink signal to the satellite?
13. What do you call the signal path from a satellite to a ground station? What is the name of the signal path from the satellite to ground?
14. State the name of the basic communication electronics unit on a satellite. Explain the purpose and operation of the four main elements.
15. What is the most common operational frequency range of most communication satellites?
16. One of the most popular satellite frequency ranges is 4 to 6 GHz. What letter name is given to that range?
17. Military satellites often operate in which frequency band?

18. What is the frequency range of the Ku band? Ka band?
19. What is the bandwidth of a typical satellite transponder?
20. A typical C-band transponder can carry how many channels? What is the bandwidth of each?
21. Name three common baseband signals handled by a satellite.
22. How is power amplification achieved in a transponder?
23. What are two common transponder intermediate frequencies?
24. What is the name of the circuit that provides channelization in a transponder?
25. What is the main power supply in a satellite?
26. What is used to power the satellite during an eclipse?
27. State the purpose of the TC&C subsystem on a satellite.
28. The signals to be communicated by the earth station to the satellite are known as _____ signals.
29. What does the receiver in a ground station do to the downlink signal before demodulation and demultiplexing?
30. What kind of circuit is often used to replace local oscillators for channel selection in earth station transmitters and receivers?
31. Name the three main types of power amplifiers used in earth stations.
32. Name the four access methods used in satellites. Which is the most widely used?
33. Name the two digital radio satellite services in the United States and state the frequency of operation.
34. What is a VSAT? Where is it used?
35. Give an example of an RO VSAT.

Problems

1. What is the angle of inclination of a satellite that orbits the equator? ◆
2. What standard navigation coordinates are used to locate a satellite in space?
3. How are satellite attitude adjustments made from the ground station? ◆
4. How do GPS receivers distinguish between the different satellite signals all transmitted on the same frequencies?
5. What GPS signal transmits position data? What is the data rate? ◆
6. What microwave band is used by GPS?
7. Explain the need for and concept behind differential GPS. ◆
8. What is the purpose of using multilevel QAM in a satellite transponder?

◆ Answers to Selected Problems follow Chap. 22.

Critical Thinking

1. How is a satellite kept in orbit? Explain the balancing forces.
2. Define the frequency and format of the C/A and P code signals. What is the Y code, and why is it needed?
3. Assume a satellite communication system based upon a highly elliptical polar orbit approximately centered over the Atlantic Ocean. Communication is to be maintained between a U.S. station and one in the United Kingdom. The frequency of operation is a 435-MHz uplink and 145-MHz downlink. Discuss the type of antennas that might be used for both uplink and downlink. What are the implications—pros and cons—of each type? Will tracking or positioning equipment be needed? If so, for which types of antennas? Will communication ever be interrupted; if so, when?
4. Explain how you can find where you are on the earth if you have a GPS receiver that gives outputs in latitude and longitude.
5. Frequency reuse in a satellite allows two transponders to share a common frequency because the signals are kept from interfering with each other by using different antenna polarizations. How else could frequency reuse be achieved in a given area on earth?
6. List and explain the tradeoffs (distance, power, cost, etc.) of a global cell phone system using LEO, MEO, and GEO.
7. How do GPS receivers tell one satellite from another?

Portable overseas tracking and data acquisition station used by the Japanese.

Telecommunication Systems

The telephone system is the largest and most complex electronic communication system in the world. It uses just about every type of electronic communication technique available including virtually all the ones described in this book. The telephone communication system is so large and widely used that no text on electronic communication would be complete without a discussion of it.

Although the primary purpose of the telephone system is to provide voice communication, it is also widely used for many other purposes including facsimile transmission and computer data transmission. Data transmission via modems and DSL was discussed in Chap. 11, so it will not be repeated here. The chapter concludes with coverage of the newest telephone technology called Voice over Internet Protocol (VoIP).

Objectives

After completing this chapter, you will be able to:

- Name and describe the components in conventional and electronic telephones.
- Describe the characteristics of the various signals used in telephone communication.
- State the general operation of a cordless telephone.
- Describe the operation of a PBX.
- Explain the hierarchy of signal transmission within the telephone system.
- Explain the operation of a facsimile machine.
- Explain the operation of a paging system.
- Describe the operation of an Internet Protocol telephone.

18-1 Telephones

The original telephone system was designed for full duplex analog communication of voice signals. Today, the telephone system is still primarily used for voice, but it employs mostly digital techniques, not only in signal transmission but also in control operations.

The telephone system permits any telephone to connect with any other telephone in the world. This means that each telephone must have a unique identification code—the 10-digit telephone number assigned to each telephone. The telephone system provides a means of recognizing each individual number and provides switching systems that can connect any two telephones.

The Local Loop

Standard telephones are connected to the telephone system by way of a two-wire, twisted-pair cable that terminates at the local exchange or central office. As many as 10,000 telephone lines can be connected to a single central office (see Fig. 18-1). The connections from the central office go to the "telephone system" represented in Fig. 18-1 by the large "cloud." This part of the system, which is mainly long distance, is described in Sec. 18-2. A call originating at telephone A will pass through the central office and then into the main system, where it is transmitted via one of many different routes to the central office connected to the desired location designated as B in Fig. 18-1. The connection between nearby local exchanges is direct rather than long distance.

The two-wire, twisted-pair connection between the telephone and the central office is referred to as the *local loop* or *subscriber loop*. You will also hear it referred to as the *last mile* or the *first mile*. The circuits in the telephone and at the central office form a complete electric circuit, or loop. This single circuit is analog and carries both dc and ac signals. The dc power for operating the telephone is generated at the central office and supplied to each telephone over the local loop. The ac voice signals are transmitted along with the dc power. Despite the fact that only two wires are involved, full duplex operation, i.e., simultaneous send and receive, is possible. All dialing and signaling operations are also carried on this single twisted-pair cable.

Telephone Set

A basic telephone or *telephone set* is an analog baseband transceiver. It has a handset which contains a microphone and a speaker, better known as a *transmitter* and a *receiver*.

Figure 18-1 The basic telephone system.

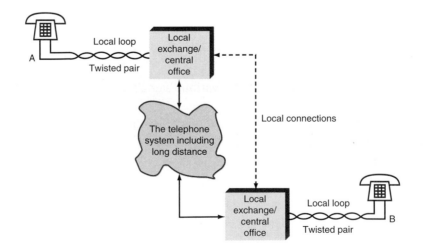

It also contains a ringer and a dialing mechanism. Overall, the telephone set fulfills the following basic functions.

The receive mode provides

1. An incoming signal that rings a bell or produces an audio tone indicating that a call is being received
2. A signal to the telephone system indicating that the signal has been answered
3. Transducers to convert voice to electric signals and electric signals to voice

The transmit mode

1. Indicates to the telephone system that a call is to be made when the handset is lifted
2. Indicates that the telephone system is ready to use by generating a signal called the *dial tone*
3. Provides a way of transmitting the telephone number to be called to the telephone system
4. Receives an indication that the call is being made by receiving a ringing tone
5. Provides a means of receiving a special tone indicating that the called line is busy
6. Provides a means of signaling the telephone system that the call is complete

All telephone sets provide these basic functions. Some of the more advanced electronic telephones have other features such as multiple line selection, hold, speaker phone, call waiting, and caller ID.

Figure 18-2 is a basic block diagram of a telephone set. The function of each block is described below. Detailed circuits for each of the blocks and their operation are described later when the standard and electronic telephones are discussed in detail.

Ringer. The *ringer* is either a bell or an electronic oscillator connected to a speaker. It is continuously connected to the twisted pair of the local loop back to the central office. When an incoming call is received, a signal from the central office causes the bell or ringer to produce a tone.

Switch Hook. A *switch hook* is a double-pole mechanical switch that is usually controlled by a mechanism actuated by the telephone handset. When the handset is "on the hook," the hook switch is open, thereby isolating all the telephone circuitry from the central office local loop. When a call is to be made or to be received, the handset is taken off the hook. This closes the switch and connects the telephone circuitry to the local loop. The direct current from the central office is then connected to the telephone, closing its circuits to operate.

Figure 18-2 Basic telephone set.

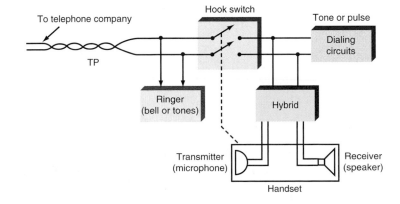

Dial tone

Ringer

Switch hook

Dialing circuit

Dual-tone multifrequency (DTMF)
system

Handset

Hybrid circuit

Dialing Circuits. The *dialing circuits* provide a way for entering the telephone number to be called. In older telephones, a pulse dialing system was used. A rotary dial connected to a switch produced a number of on/off pulses corresponding to the digit dialed. These on/off pulses formed a simple binary code for signaling the central office.

In most modern telephones, a tone dialing system is used. Known as the *dual-tone multifrequency (DTMF) system,* this dialing method uses a number of pushbuttons that generate pairs of audio tones that indicate the digits called.

Whether pulse dialing or tone dialing is used, circuits in the central office recognize the signals and make the proper connections to the dialed telephone.

Handset. This unit contains a microphone for the transmitter and a speaker or receiver. When you speak into the transmitter, it generates an electric signal representing your voice. When a received electric voice signal occurs on the line, the receiver translates it to sound waves. The transmitter and receiver are independent units, and each has two wires connecting to the telephone circuit. Both connect to a special device known as the hybrid.

Hybrid. The *hybrid circuit* is a special transformer used to convert signals from the four wires from the transmitter and receiver to a signal suitable for a single two-line pair to the local loop. The hybrid permits *full duplex,* i.e., simultaneous send and receive, analog communication on the two-wire line. The hybrid also provides a side tone from the transmitter to the receiver so that the speaker can hear her or his voice in the receiver. This feedback permits automatic voice-level adjustment.

Standard Telephone and Local Loop

Figure 18-3 is a simplified schematic diagram of a conventional telephone and the local loop connections back to the central office. The circuitry at the central office is discussed in greater detail later. For now, note that the central office applies a dc voltage over the twisted-pair line to the telephone. This dc voltage is approximately -48 V with respect to ground in the open-circuit condition. When a subscriber picks up the telephone, the

Figure 18-3 Standard telephone circuit diagram showing connection to central office.

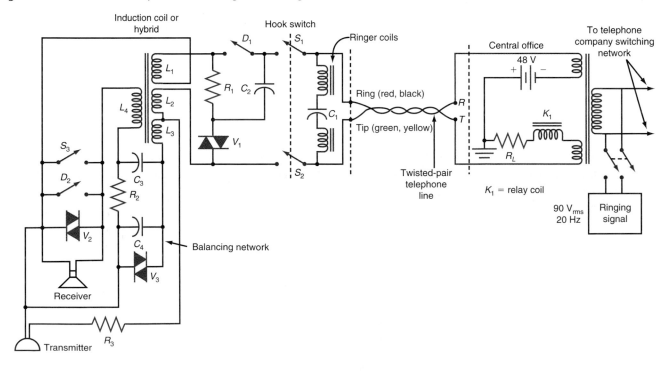

switch hook closes, connecting the circuitry to the telephone line. The load represented by the telephone circuitry causes current to flow in the local loop and the voltage inside the telephone to drop to approximately 5 to 6 V.

The amount of current flowing in the local loop depends upon a number of factors. The dc voltage supplied by the central office may not be exactly −48 V. It can, in fact, vary many volts above or below the 48-V normal value.

As Fig. 18-3 shows, the central office also inserts some resistance R_L to limit the total current flow if a short circuit occurs on the line. This resistance can range from about 350 to 800 Ω. In Fig. 18-3, the total resistance is approximately 400 Ω.

The resistance of the telephone itself also varies over a relatively wide range. It can be as low as 100 Ω and as high as 400 Ω, depending upon the circuitry. The resistance varies because of the resistance of the transmitter element and because of the variable resistors called *varistors* used in the circuit to provide automatic adjustment of line level.

The local loop resistance depends considerably on the length of the twisted pair between the telephone and the central office. Although the resistance of copper wire in the twisted pair is relatively low, the length of the wire between the telephone and the central office can be many miles long. Thus the resistance of the local loop can be anywhere from 1000 to 1800 Ω, depending upon the distance. The local loop length can vary from a few thousand feet up to about 18,000 ft.

Finally, the frequency response of the local loop is approximately 300 to 3400 Hz. This is sufficient to pass voice frequencies that produce full intelligibility. An unloaded twisted pair has an upper cutoff frequency of about 4000 Hz. But this cutoff varies considerably depending upon the overall length of the cable. When long runs of cable are used, special loading coils are inserted into the line to compensate for excessive roll-off at the higher frequencies.

The wires in Fig. 18-4 end at terminals on the telephone labeled *tip* and *ring*. These designations refer to the plug used to connect telephones to one another at the central office. At one time, large groups of telephone operators at the central office used plugs and jacks at a switchboard to connect one telephone to another manually. The jack is shown in Fig. 18-4. The tip and the ring are metallic contacts that are attached to the wires in the cable. They touch spring-loaded contacts in the jack to make the connection. Although such plugs and jacks are no longer used, the tip and ring designation is still used to refer to the two wires of the local loop.

The wires are also usually color-coded red and green. The tip wire is green and is usually connected to ground; the ring wire is red. Many telephone cables into a home or an office also contain a second twisted pair if a separate telephone line is to be installed. These wires are usually color-coded black and yellow. Black and yellow correspond to ring and tip, respectively, where yellow is ground. Other color combinations are used in telephone wiring.

Ringer. In Fig. 18-3, the circuitry connected directly to the tip and ring local loop wires is the ringer. The ringer in most older telephones is an electromechanical bell. A pair of electromagnetic coils is used to operate a small hammer that alternately strikes two small metallic bells. Each bell is made of different materials so that each produces a slightly different tone. When an incoming call is received, a voltage from the central office operates the electromagnetic coils which, in turn, operate the hammer to ring the bells. The bells make the familiar tone produced by most standard telephones.

Varistor

Tip

Ring

GOOD TO KNOW

When long runs of cable are used, special loading coils are inserted into the line to compensate for large amounts of roll-off at higher frequencies.

GOOD TO KNOW

In the United States, the telephone ringing voltage occurs for 1 s followed by a 3-s break. Telephones in other countries may use different ringing sequences. Smaller office-based systems in the United States may also use other sequences for internal calls.

Figure 18-4 Tip and ring designation on an old plug and jack.

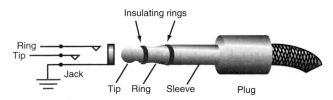

Figure 18-5 Telephone ringing sequence. (*a*) United States and Europe. (*b*) United Kingdom.

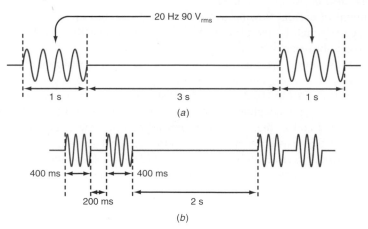

In Fig. 18-3, the ringing coils are connected in series with a capacitor C_1. This allows the ac ringing voltage to be applied to the coils but blocks the 48 V of direct current, thus minimizing the current drain on the 48 V of power supplied at the central office.

The ringing voltage supplied by the central office is a sine wave of approximately 90 V_{rms} at a frequency of about 20 Hz. These are the nominal values, because the actual ringing voltage can vary from approximately 80 to 100 V_{rms} with a frequency somewhere in the 15- to 30-Hz range. This ac signal is supplied by a generator at the central office. The ringing voltage is applied in series with the -48-V dc signal from the central office power supply. The ringing signal is connected to the local loop line by way of a transformer T_1. The transformer couples the ringing signal into its secondary winding where it appears in series with the 48-V dc supply voltage.

The standard ringing sequence is shown in Fig. 18-5. In U.S. telephones, the ringing voltage occurs for 1 s followed by a 3-s interval. Telephones in other parts of the world use different ringing sequences. For example, in the United Kingdom, the standard ring sequence is a higher-frequency tone occurring more frequently, and it consists of two ringing pulses 400 ms long, separated by 200 ms. This is followed by a 2-s interval of quiet before the tone sequence repeats.

Transmitter. The transmitter is the microphone into which you speak during a telephone call. In a standard telephone, this microphone uses a carbon element that effectively translates acoustical vibrations into resistance changes. The resistance changes, in turn, produce current variations in the local loop representing the speaker's voice. A dc voltage must be applied to the transmitter so that current flows through it during operation. The 48 V from the central office is used in this case to operate the transmitter.

Figure 18-6 is a simplified diagram showing how a telephone transmitter works. The basic transmitter element is a small module containing fine carbon granules. One side of the module is a flexible diaphragm. Whenever you speak, you create acoustic energy in the form of the movement of air. The air molecules move in accordance with your voice frequency. The acoustic energy from the voice reaches the diaphragm and causes it to vibrate in accordance with the speech. An outward acoustic pressure wave causes the carbon granules to be compressed. Pushing the carbon granules closer together causes the overall resistance of the element to decrease. When the acoustic energy moves in the opposite direction, the carbon granules expand outward. Since they are less tightly compressed, their resistance increases. The transmitter element is in series with the telephone circuit, which includes the 48-V central office battery and the speaker in the remote handset. Speaking into the transmitter causes the current flow in the circuit to vary in accordance with the voice signal. The resulting ac voice signal produced on the telephone line is approximately 1 to 2 V_{rms}.

Figure 18-6 The transmitter and receiver in a telephone.

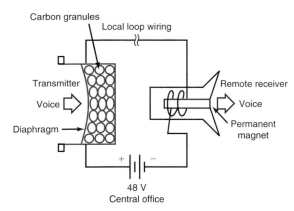

Receiver. The receiver, or earpiece, is basically a small permanent-magnet speaker. A thin metallic diaphragm is physically attached to a coil which rests inside a permanent magnet. Whenever a voice signal comes down a telephone line, it develops a current in the receiver coil. The coil produces a magnetic field that interacts with the permanent-magnet field. The result is vibration of the diaphragm in the receiver, which converts the electric signal to the acoustic energy that supplies the voice to the ear. As it comes in over the local loop lines, the voice signal has an amplitude of approximately 0.5 to 1 V_{rms}.

Hybrid. The hybrid is a transformerlike device that is used to simultaneously transmit and receive on a single pair of wires. The hybrid, which is also sometimes referred to as an *induction coil,* is really several transformers combined into a single unit. The windings on the transformers are connected in such a way that signals produced by the transmitter are put on the two-wire local loop but do not occur in the receiver. In the same way, the transformer windings permit a signal to be sent to the receiver, but the resulting voltage is not applied to the transmitter.

In practice, the hybrid windings are set up so that a small amount of the voice signal produced by the transmitter does occur in the receiver. This provides feedback to the speaker so that she or he may speak with normal loudness. The feedback from the transmitter to the receiver is referred to as the *side tone.* If the side tone were not provided, there would be no signal in the receiver and the person speaking would have the sensation that the telephone line was dead. By hearing his or her own voice in the receiver at a moderate level, the caller can speak at a normal level. Without the side tone, the speaker tends to speak more loudly, which is unnecessary.

Side tone

Automatic Voice Level Adjustment. Because of the wide variation in the different loop lengths of the two telephones connected to each other, the circuit resistances will vary considerably, thereby causing a wide variation in the transmitted and received voice signal levels. All telephones contain some type of component or circuit that provides *automatic voice level adjustment* so that the signal levels are approximately the same regardless of the loop lengths. In the standard telephone, this automatic loop length adjustment is handled by components called *varistors.* These are labeled V_1, V_2, and V_3 in Fig. 18-3.

Automatic voice level adjustment

Varistors

A varistor is a nonlinear resistance element whose resistance changes depending upon the amount of current passing through it. When the current passing through the varistor increases, its resistance decreases. A decrease in current causes the resistance to increase.

The varistors are usually connected across the line. In Fig. 18-3, varistor V_1 is connected in series with resistor R_1. This varistor automatically shunts some of the current away from the transmitter and the receiver. If the loop is long, the current will be relatively low and the voltage at the telephone will be low. This causes the resistance of the varistor to increase, thus shunting less current away from the transmitter and receiver. On short local loops, the current will be high and the voltage at the telephone will be high. This causes

Figure 18-7 Current pulses produced by dialing.

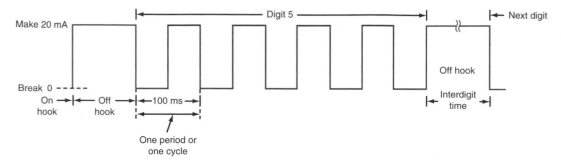

Pulse dialing

Tone dialing

Dual-tone multifrequency
(DTMF) system

the varistor resistance to decrease; thus more current is shunted away from the transmitter and receiver. The result is a relatively constant level of transmitted or received speech.

Note that a second varistor V_3 is used in the balancing network. The balancing network (C_3, C_4, R_2) works in conjunction with the hybrid to provide the side tone discussed earlier. The varistor adjusts the level of the side tone automatically.

Pulse Dialing. The term *dialing* is used to describe the process of entering a telephone number to be called. In older telephones, a rotary dial was used. In more modern telephones, pushbuttons that generate electronic tones are used for "dialing."

The use of a rotary dialing mechanism produces what is known as *pulse dialing*. Rotating the dial and releasing it cause a switch contact to open and close at a fixed rate, producing current pulses in the local loop. These current pulses are detected by the central office and used to operate the switches that connect the dialing telephone to the called telephone.

Figure 18-7 shows the process of dialing the number 5. Five on/off pulses are produced. This is followed by a variable time interval until the next digit is dialed.

Referring to Fig. 18-3, you can see that the pulse-dialed switch D_1 is connected in series with the telephone circuit. This switch contact is normally closed when the dial is not in use. It remains closed as the dial is rotated in the clockwise direction. When the dial is released, the switch opens and closes at a constant rate, as indicated earlier, opening and closing the circuit and switching the loop current off and on.

Also during the dialing process, switch D_2, which is mounted on the dialing mechanism, closes. This short-circuits the receiver to prevent the current pulses from producing loud clicks in the receiver. Although pulse dialing is no longer used in modern telephones, there are many older dial phones still in use, so telephone companies continue to support this technique.

Tone Dialing. Although some dial telephones are still in use and all central offices can accommodate them, most modern telephones use a dialing system known as *Touch-Tone*. It uses pairs of audio tones to create signals representing the numbers to be dialed. This dialing system is referred to as the *dual-tone multifrequency (DTMF) system*.

A typical DTMF keyboard on a telephone is shown in Fig. 18-8. Most telephones use a standard keypad with 12 buttons or switches for the numbers 0 through 9 and the special symbols * and #. The DTMF system also accommodates 4 additional keys for special applications.

In Fig. 18-8 numbers represent audio frequencies associated with each row and column of pushbuttons. For example, the upper horizontal row containing the keys for 1, 2, and 3 is labeled 697, which means that when any one of these three keys is depressed, a sine wave of 697 Hz is produced. Each of the four horizontal rows produces a different frequency. The horizontal rows generate what is generally known as the *low group of frequencies*.

A higher group of frequencies is associated with the vertical columns of keys. For example, the keys for the numbers 2, 5, 8, and 0 produce a frequency of 1336 Hz when depressed.

If the number 2 is depressed, two sine waves are generated simultaneously, one at 697 Hz and the other at 1336 Hz. These two tones are linearly mixed. This combination

Figure 18-8 DTMF keypad.

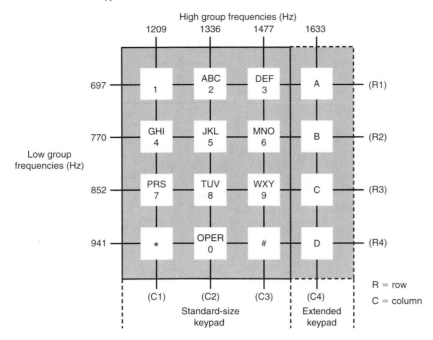

produces a unique sound and is easily detected and recognized at the central office as the signal representing the dialed digit 2. The tolerance on the generated frequencies is usually within ±1.5 percent.

The earliest TouchTone telephones used a simple single-transistor oscillator to generate two tones simultaneously. Today in electronic phones an IC generates the tones.

Electronic Telephones

When solid-state circuits came along in the late 1950s, an electronic telephone became possible and practical. Today, most new telephones are electronic, and they use integrated-circuit technology.

The development of the microprocessor has also affected telephone design. Although simple electronic telephones do not contain a microprocessor, most multiple-line and full-feature telephones do. A built-in microprocessor permits automatic control of the telephone's functions and provides features such as telephone number storage and automatic dialing and redialing that are not possible in conventional telephones.

Typical IC Electronic Telephone. The major components of an electronic telephone circuit are shown in Fig. 18-9. Most of the functions are implemented with circuits contained within a single IC.

In Fig. 18-9, note that the TouchTone keypad drives a DTMF tone generator circuit. An external crystal or ceramic resonator provides an accurate frequency reference for generating the dual dialing tones.

The tone ringer is driven by the 20-Hz ringing signal from the phone line and drives a piezoelectric sound element.

The IC also contains a built-in line voltage regulator. It takes the dc voltage from the local loop and stabilizes it to provide a constant voltage to the internal electronic circuits. An external zener diode and transistor provide bias to the electret microphone.

The internal speech network contains a number of amplifiers and related circuits that fully duplicate the function of a hybrid in a standard telephone. This IC also contains a microcomputer interface. The box labeled MPU is a single-chip *microprocessing unit*. Although it is not necessary to use a microprocessor, if automatic dialing and other functions are implemented, this circuit is capable of accommodating them.

IC electronic telephone

GOOD TO KNOW

Newer telephones with built-in microprocessors are able to provide users with features such as telephone number storage and automatic dialing.

Figure 18-9 Single-chip electronic telephone.

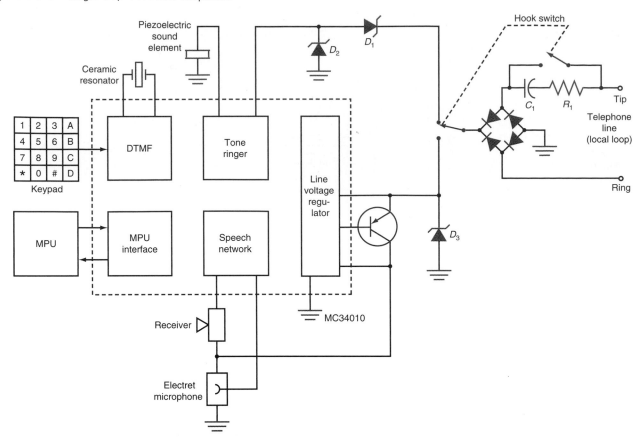

Finally, note the bridge rectifier and hook switch circuit. The twisted pair from the local loop is connected to the tip and ring. Both the 48-V dc and 20-Hz ring voltages will be applied to this bridge rectifier. For direct current, the bridge rectifier provides polarity protection for the circuit, ensuring that the bridge output voltage is always positive. When the ac ringing voltage is applied, the bridge rectifier rectifies it into a pulsating dc voltage. The hook switch is shown with the telephone on the hook or in the "hung-up" position. Thus the dc voltage is not connected to the circuit at this time. However, the ac ringing voltage will be coupled through the resistor and capacitor to the bridge, where it will be rectified and applied to the two zener diodes D_1 and D_2 that drive the tone ringer circuit.

When the telephone is taken off the hook, the hook switch closes, providing a dc path around the resistor and capacitor R_1 and C_1. The path to the tone ringer is broken, and the output of the bridge rectifier is connected to zener diode D_3 and the line voltage regulator. Thus the circuits inside the IC are powered up, and calls may be received or made.

The tone ringer circuit is shown in Fig. 18-10. This IC is the Motorola MC34010 which is representative of a typical single-chip telephone IC. When an incoming call is being received, the central office applies the 20-Hz sine wave to the local loop. This passes through capacitor C_{17}, and R_1 is applied to the bridge rectifier circuit. The output of the bridge rectifier is applied to two zener diodes D_1 and D_2. Diode D_1 provides a threshold level that must be exceeded before the tone ringer circuits are enabled. A dc voltage developed across D_2 and filtered by R_1 and C_{15} is applied to the tone ringer input circuits, powering an internal bias circuit generating 21 V dc. This is applied to an internal threshold detector circuit. When the ringing voltage reaches a specific level, it triggers and powers up the remaining circuits. Because there is a threshold to be exceeded, false ringing will not occur when stray signals appear on the line or when the dc voltage from the loop drops to a low level.

Figure 18-10 Tone ringer circuit.

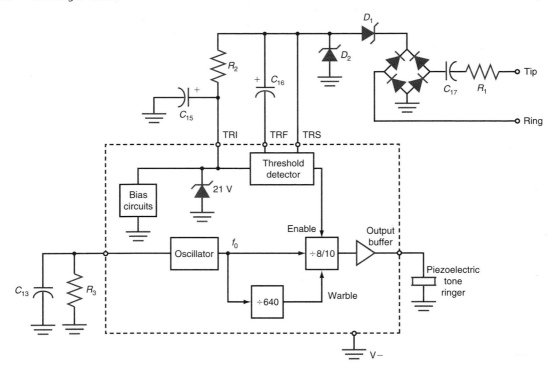

The ringing tone is generated by an internal oscillator circuit, the frequency of which is set to 8 kHz by external components C_{13} and R_3. The 8-kHz signal is applied to a frequency divider which can divide by 8 or by 10. The division ratio is determined by the warble which is generated by the oscillator and a circuit which is divided by 640. This produces a 12.5-Hz warble square wave that alternates the frequency divider between divide by 8 and divide by 10. As a result, the frequency divider output is a rectangular wave that switches between 800 and 1000 Hz at a 12.5-Hz rate. The signal is applied to an output buffer amplifier that drives an external piezoelectric tone ringer. This type of tone ringer is a transducer that reproduces the signal applied to it. It is unlike the previously described tone ringer that produces a fixed resonant tone output in the 2- to 3-kHz range.

Details of the DTMF dialing generator are shown in Fig. 18-11. The row and column connections to the TouchTone keypad are made to the keypad comparator and logic circuits. The output of the logic circuits goes to program two counter circuits, a column counter and a row counter. These counters generate 8-bit binary numbers that drive two digital-to-analog (D/A) converters. As the counters are stepped, they generate a binary number sequence that, when applied to the D/A converter, produces a stairstep approximation of the sine wave at a frequency related to the number key pressed. The column counter and its D/A converter generate one tone, and the row counter and its D/A converter generate the second tone. The two tones are combined at the output of the D/A converters and applied to the output buffer amplifier. The signal is then connected to the telephone line.

Timing for the circuitry is provided by a built-in oscillator that runs at 500 kHz. This frequency is set by an external ceramic resonator and capacitors C_1 and C_2.

Details of the speech network are shown in Fig. 18-12. This collection of amplifiers and other circuits duplicates the function of the hybrid transformer and the related level control circuits in a conventional telephone.

The dc voltage from the local loop through the bridge rectifier is applied to C_{12} and zener diode D_3, providing a stabilized dc voltage for the internal regulator circuit. The dc voltage from transistor T_1 and capacitor C_9 biases the electret microphone in the telephone handset. This bias voltage at input pin TXO also biases the transmit amplifier. The voice signal from the microphone is coupled through capacitor C_5 and R_{13}, C_4 and R_{12} to the transmit amplifier. The peak limiter circuit provides feedback to the transmit amplifier

Figure 18-11 DTMF dialing circuits.

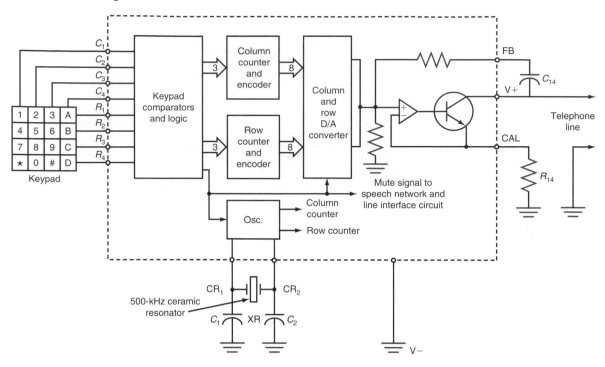

Figure 18-12 Electronic speech circuits.

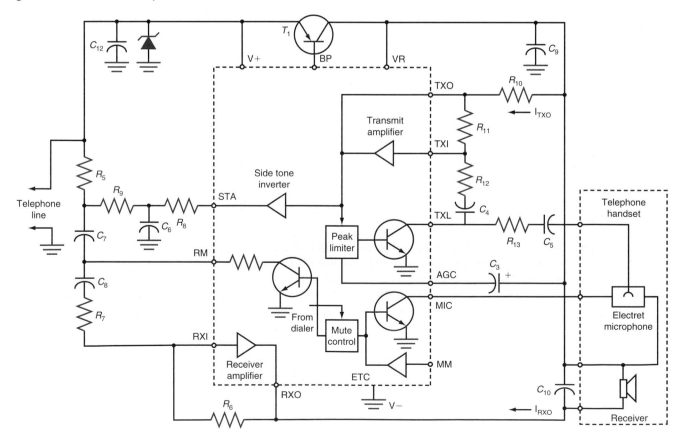

to reduce the gain and minimize distortion when very loud talkers use the telephone. The transmit amplifier output appears at the TXO pin and is applied to R_{10} and transistor T_1 to the telephone line. Thus T_1 acts as a final amplifier for the voice signal before being placed on the local loop.

The output of the transmit amplifier is inverted by a side tone amplifier (STA) and appears at the STA output. Here it is applied to an RC network and reduced in level. It is applied through C_7, C_8, and R_7 to the receive amplifier at input RXI. The receive amplifier output at RXO drives the small receiver speaker in the handset.

When an incoming voice signal is received, it is passed through R_5, C_7, C_8, and R_7 to the receive amplifier, where it again is passed to the speaker in the handset.

Finally, note that this circuit has a built-in mute control driven from the dialer circuit. Whenever dialing takes place, the mute control shuts off the transmit circuits and reduces the amplitude of the signal to the receive amplifier.

Microprocessor Control. All modern electronic telephones contain a built-in microcontroller. Like any microcontroller, it consists of the CPU, a ROM in which a control program is stored, a small amount of random access read-write memory, and I/O circuits. The microcontroller, usually a single-chip IC, may be directly connected to the telephone IC, or some type of intermediate interface circuit may be used.

The functions performed by the microcomputer include operating the keyboard and any LCD display, if present. Some other functions involve storing telephone numbers and automatically redialing. Many advanced telephones have the capability of storing 10 or more commonly called numbers. The user puts the telephone into a program mode and uses the TouchTone keypad to enter the most frequently dialed numbers. These are stored in the microcontroller's RAM. To automatically dial one of the numbers, the user depresses a pushbutton on the front of the telephone. This may be one of the TouchTone pushbuttons, or it may be a separate set of pushbuttons provided for the purpose. When one of the pushbuttons is depressed, the microcontroller supplies a preprogrammed set of binary codes to the DTMF circuitry in the telephone IC. Thus the number is automatically dialed. Other features implemented by the microcontroller are caller ID and an answering machine.

Voice Mail. Previously called an *answering machine,* this feature is implemented on most electronic phones. The microcontroller automatically answers the call after a preprogrammed number of rings and saves the voice message. In older answering machines, the message was recorded on a tape cassette. But in modern phones, the voice message is digitized, compressed, and then stored in a small flash ROM ready for replay. The outgoing message is also stored there.

Answering machine

Voice mail

Caller ID. *Caller ID,* also known as the *calling line identification service,* is a feature that is now widely implemented on most electronic telephones. To make use of this service, you must sign up and pay for it monthly. With this feature, any calling number will be displayed on an LCD readout when the phone is ringing. This allows you to identify the caller.

Caller ID (calling line identification service)

The caller ID service sends a digitized version of the calling number to your phone during the first and second rings. The data transmitted includes the date, time, and calling number. Data is transmitted by FSK, where a binary 1 (mark) is a 1200-Hz tone and a binary 0 (space) is a 2200-Hz tone. The data rate is 1200 bps.

There are two message formats in use, the *single-data message format (SDMF)* and the *multiple-data message format (MDMF).* The SDMF is illustrated in Fig. 18-13. One-half second after the first ring, 80 bytes of alternating 0s and 1s (hex 05) is transmitted for 250 ms followed by 70 ms of mark symbols. These two signals provide initialization and synchronization of the caller ID circuitry in the phone. This is followed by 1 byte describing the message type. This is usually a binary 4 (00000100), indicating the SDMF. This is followed by a byte containing the message length, usually the number of digits in the calling number. Next the data is transmitted. This is

Single-data message format (SDMF)

Multiple-data message format (MDMF)

Figure 18-13 The caller ID transmission format.

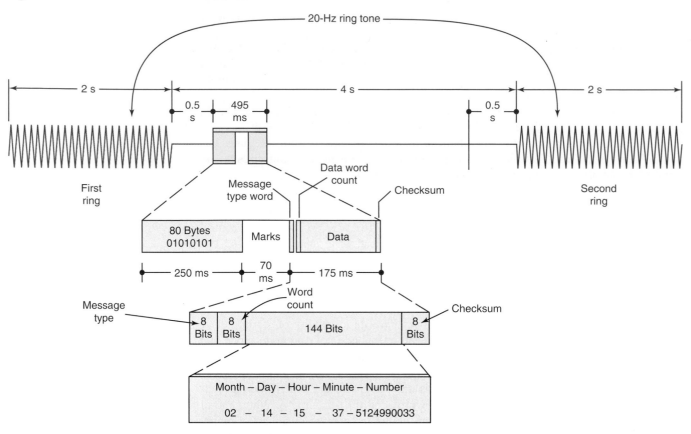

the date, time, and the 10-digit phone number transmitted as ASCII bytes with the least significant digit first. The data format is 2 digits for the month, 2 digits for the day, 2 digits for the hour (military time), 2 digits for the minutes, and up to 10 digits for the calling number. For example, if the date is February 14, the time is 3:37 p.m., and the calling number is 512-499-0033, the data sequence would be 0214153751249900033. The final byte in the message is the checksum that is used for error detection. The checksum is the 2s complement sum (XOR) of all the data bytes not including the initialization and sync signals.

If the calling number is outside the calling area, the system will display an O on the LCD rather than the calling number. Furthermore, a caller may also have his or her number blocked. This can be done by setting it up with the service provider in advance or by dialing *67 prior to making the call. This will cause a P to be displayed on the LCD instead of the calling number.

A more advanced data format is the MDMF. It is similar to the SDMF but includes an extra field for the name of the calling party plus additional identification bytes.

Line Interface. Most telephones are connected by way of a thin multiwire cable to a wall jack. A special connector on the cable, called an *RJ-11 modular connector,* plugs into the matching wall jack. Two local loops are available if needed.

The wall jack is connected by way of wiring inside the walls to a central wiring point called the *subscriber interface.* Also known as the *wiring block* or *modular interface,* this is a small plastic housing containing all the wiring that connects the line from the telephone company to all the telephone wires in the house. Many houses and apartments are wired so that there is a wall jack in every room.

Figure 18-14 is a general diagram of the modular interface. The line from the telephone company usually passes through a protector that provides lightning protection. It then

RJ-11 modular connector

Subscriber interface

Wiring block or modular interface

Figure 18-14 Subscriber interface.

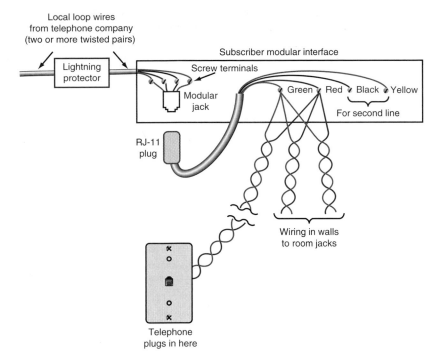

terminates at the interface box. An RJ-11 jack and plug are provided to connect to the rest of the wiring. This gives the telephone company a way to disconnect the incoming line from the rest of the house wiring and makes testing and troubleshooting easier.

All the wiring is made by way of screw terminals. For a single-line house, the green and red tip and ring connections terminate at the terminals, and all wiring to the room wall jacks is connected in parallel at these terminals.

If a second line is installed, the black and yellow wires, which are the tip and ring connections, are also terminated at screw terminals. They are then connected to the inside house wiring.

Connections on the RJ-11 connector are shown in Fig. 18-15. The red and green wires terminate at the two center connections, and the black and yellow wires terminate at the two outside connections. Most telephone wire and RJ-11 connectors have four wires and connections. Some cables have only the two inner wires. With four wires a two-line phone can be accommodated.

Figure 18-15 Connections to modular plug.

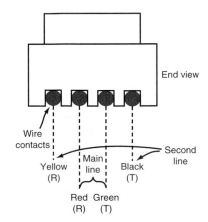

Cordless Telephones

Virtually all offices and most homes now have two or more telephones, and many homes and apartments have a standard telephone jack in every room. This permits a single phone to be moved easily from one place to another, and it permits multiple (extension) phones. However, the ultimate convenience is a *cordless telephone,* which uses two-way radio transmission and provides total portability. Today, most homes have a cordless unit.

Cordless Telephone Concepts.
A cordless telephone is a full duplex, two-way radio system made up of two units, the portable unit or handset and the base unit. The base unit is wired to the telephone line by way of a modular connector. It receives its power from the ac line. The base unit is a complete transceiver in that it contains a transmitter that sends the received audio signal to the portable unit and receives signals transmitted by the portable unit and retransmits them on the telephone line. It also contains a battery charger that rejuvenates the battery in the handheld unit.

The portable unit is also a battery-powered transceiver. This unit is designed to rest in the base unit where its battery can be recharged. Both units have an antenna.

The transceivers in both the portable and the base units use full duplex operation. To achieve this, the transmitter and receiver must operate on different frequencies.

Figure 18-16 shows simplified block diagrams of the base and portable units of a typical cordless telephone. Both the base unit and the handset contain an embedded microcontroller that controls all operations including the keyboard and display. A high percentage of cordless units also contain a caller ID function, and many contain a voice mail feature. An analog-to-digital converter translates a received voice message to digital; it is compressed by the microcontroller and then stored in a flash memory connected to the microcontroller.

Frequency Allocations.
The FCC has set aside four primary frequency bands for cordless telephones: 43 to 50 MHz, 902 to 928 MHz, 2.4 to 2.45 GHz, and 5.8 GHz. The older analog phones used 25 assigned duplex frequency pairs in the 43- to 50-MHz range. In the 902- to 928-MHz ISN band there are more channels but the number depends upon the technology used. The 2.4-GHz band has up to 100 wide channels where many spread spectrum signals can exist concurrently and channels are determined by a pseudorandom code. The 5.8-GHz band is the most recent addition with plenty of spectrum space for multiple channels. Most of the newer phones use the 900- or 2.4-GHz bands. The phones are programmed to automatically seek a channel pair with no activity and minimum noise.

GOOD TO KNOW

The base unit and receiver unit in cordless telephones each contain a separate full duplex, complete transceiver.

Figure 18-16 General block diagram of a cordless telephone.

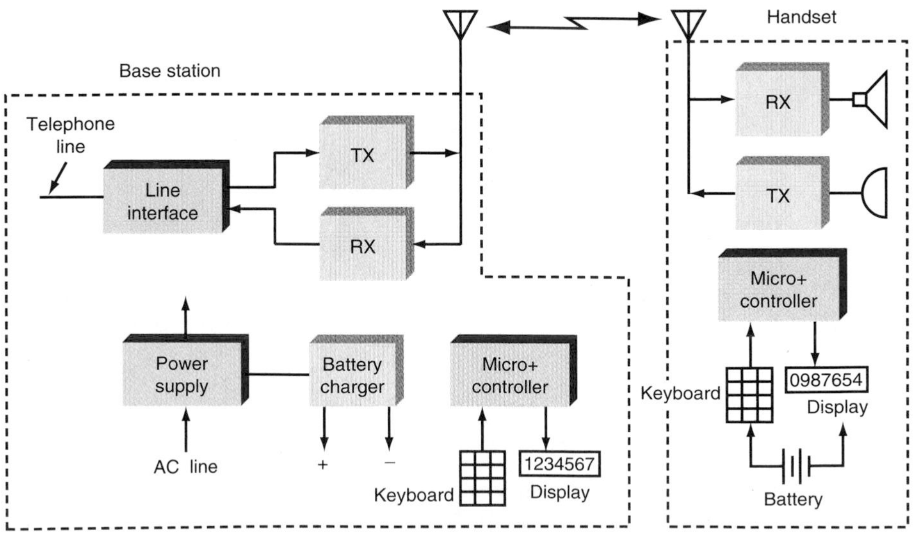

Chapter 18

Cordless Phone Features, Capabilities, and Limitations. The frequency range defines the three basic classes of cordless telephones available today, but there are other considerations. Here is a summary of the three basic types.

The simplest and least expensive cordless phones use the 43- to 50-MHz range. They are analog phones using frequency modulation. The transmitter output power is limited to 500 mW, and this, in turn, limits the transmission range to a maximum of about 1000 ft, depending upon the environment. The FCC created these limitations deliberately to reduce the amount of interference with nearby cordless telephones as well as the many wireless baby monitors and toy walkie-talkies using the same frequencies. While some 43- to 50-MHz phones are still available, for the most part they have been replaced by the newer digital phones.

Although these older phones work well enough, they are susceptible to noise and their range is limited. If higher quality and longer range are desired, phones in the 900-MHz, 2.4-GHz, or 5.8-GHz range can be used.

Three types of 900-MHz phones are available. These are analog, digital, and spread spectrum. The analog phones use FM. Although they can transmit over a longer distance, they are still susceptible to noise. A digital 900-MHz phone is also available. It uses *Gaussian FSK (GFSK)* modulation. The best 900-MHz phones use *direct-sequence spread spectrum (DSSS)*. With a power of up to 1 W, the transmission distance is a maximum of about 5000 to 7000 ft, depending on the environment and terrain. Both types of digital phones are highly immune to noise.

Gaussian FSK (GFSK)

Direct-sequence spread spectrum (DSSS)

The newer and perhaps the best cordless phones use DSSS in the 2.4-GHz or 5.8-GHz bands. Their maximum range is nearly 7000 ft, and they are virtually immune to local noise. Although these phones are far more expensive, they offer the highest-quality sound and greatest reliability.

For the most part, cordless phones in the United State have used proprietary designs rather than those conforming to a particular standard. Since the phones are only intended to work in a home or small office setting and there is no requirement that the phone interoperate with other cordless phones, any technology will work as long as it meets the FCC's frequency and operating mode guidelines. The situation is different in Europe where standards for cordless phones have existed for many years. The newest standard created by the European Telecommunications Standards Institute (ETSI), called *Digital Enhanced Cordless Telecommunications (DECT)* has now been approved for use in the United States. DECT works in the 1.8- to 1.9-GHz bands in Europe, but versions for the 900-MHz, 2.4-GHz, and 5.8-GHz bands have been developed for U.S. use.

Digital Enhanced Cordless Telecommunications (DECT)

The DECT phones are digital, using Gaussian FSK modulation. Instead of using frequency-division duplexing (FDD) with two channels, DECT uses only a single channel and time-division duplexing (TTD). In a single channel, time-division multiplexing permits 12 users per channels. Typically 10 channels are available. The raw data rate is 1.152 Mbps. It is expected that U.S. cordless phones will eventually use the DECT technology. Some DECT phones are already available.

18-2 Telephone System

Most of us take telephone service for granted, as we do other so-called utilities, e.g., electric power. In the United States telephone service is excellent. But this is certainly not the case in many other countries in the world.

When we refer to the *telephone system,* we are talking about the organizations and facilities involved in connecting your telephone to the called telephone regardless of where it might be in United States or anywhere else in the world. You will sometimes hear the telephone system referred to as the Plain Old Telephone Service (POTS). A number of different companies are involved in long-distance calls, although a single company is usually responsible for local calls in a given area. These companies make up the telephone system, and they design, build, maintain, and operate all the facilities

Telephone system

and equipment used in providing universal telephone service. A vast array of equipment and technology are employed. Practically every conceivable type of electronic technology is used to implement worldwide telephone service, and that continues to change as Internet calling known as Voice over Internet Protocol (VoIP) grows.

The *telephone,* a small but relatively complex entity, is nothing compared to the massive system that backs it up. The telephone system can connect any two telephones in the world, and most people can only speculate on the method by which this connection takes place. It takes place on many levels and involves an incredible array of systems and technology. Obviously, it is difficult to describe such a massive system here. However, in this brief section, we attempt to describe the technical complexities of interconnecting telephones, the central office and the subscriber line interface that connect each user to the telephone system, the hierarchy of interconnections within the telephone system, and the major elements and general operation of the telephone system. Long-distance operation and special telephone interconnection systems such as the PBX are also discussed. VoIP is introduced.

Subscriber Interface

Most telephones are connected to a local central office by way of the two-line, twisted-pair local loop cable. The central office contains all the equipment that operates the telephone and connects it to the telephone system that makes the connection to any other telephone.

Each telephone connected to the central office is provided with a group of basic circuits that power the telephone and provide all the basic functions such as ringing, dial tone, and dialing supervision. These circuits are collectively referred to as the *subscriber interface* or the *subscriber line interface circuit (SLIC)*. In older central office systems, the subscriber interface circuits used discrete components. Today, most functions of the subscriber line interface are implemented by one or perhaps two integrated circuits plus supporting equipment. The subscriber line interface is also referred to as the *line side interface.*

The SLIC provides seven basic functions generally referred to as *BORSCHT* (representing the first letters of the functions *battery, overvoltage protection, ringing, supervision, coding, hybrid,* and *test*). A general block diagram of the subscriber interface and BORSCHT functions is given in Fig. 18-17.

Battery. The subscriber line interface at the central office must provide a dc voltage to the subscriber to operate the telephone. In the United States, this is typically −48V dc with respect to ground. The actual voltage can be anything between approximately

Figure 18-17 BORSCHT functions in the subscriber line interface at the central office.

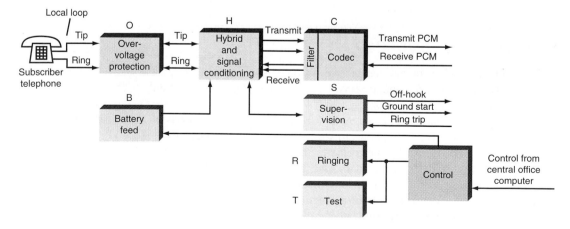

Margin glossary terms:

Telephone

Subscriber interface or subscriber line interface circuit (SLIC)

BORSCHT

Battery

−20 and −80 V when the phone is on the hook, i.e., disconnected. The voltage at the telephone drops to approximately 6 V when the phone is taken off the hook. The large difference between the on-hook and off-hook voltages has to do with the large voltage drop that occurs across the components in the telephone and the long local loop cable.

Overvoltage Protection. The circuits and components that protect the subscriber line interface circuits from electrical damage are referred to collectively as *overvoltage protection*. The phone lines are vulnerable to many types of electrical problems. Lightning is by far the worst threat, although other hazards exist, including accidental connection to an electric power line or some type of misconnection that would occur during installation. Induced disturbances from other sources of noise can also cause problems. Overvoltage protection ensures reliable telephone operation even under such conditions.

Overvoltage protection

Ringing. When a specific telephone is receiving a call, the telephone local office must provide a ringing signal. As indicated earlier, this is commonly a 90-V_{rms} ac signal at approximately 20 Hz. The SLIC must connect the ringing signal to the local loop when a call is received. This is usually done by closing relay contacts that connect the ringing signal to the line. The SLIC must also detect when the phone is picked up (off hook) so that the ringing signal can be disconnected.

Ringing

Supervision. *Supervision* refers to a group of functions within the subscriber line interface that monitor local loop conditions and provide various services. For example, the supervision circuits in the SLIC detect when a telephone is picked up to initiate a new call. A sensing circuit recognizes the off-hook condition and signals circuits within the SLIC to connect a dial tone. The caller then dials the desired number, which causes interconnection through the telephone system.

The supervision circuits continuously monitor the line during the telephone call. The circuits sense when the call is terminated and provide the connection of a busy signal if the called number is not available.

Supervision

GOOD TO KNOW

On-hook and off-hook signals differ in resistance across tip and ring conductors. On-hook minimum dc resistance is 30,000 Ω, and off-hook maximum resistance is 200 Ω.

Coding. *Coding* is another name for A/D conversion and D/A conversion. Today, many telephone transmissions are made by way of serial digital data methods. The SLIC may contain codec that converts the analog voice signals to serial PCM format or converts received digital calls back to analog signals to be placed on the local loop. Transmission over trunk lines to other central offices or toll offices or for use in long-distance transmission is typically by digital PCM signals in modern systems.

The staff of this telephone network operational control center must be well trained in electronics and computer technology to operate a complex communication system. The network staff are currently routing long-distance telephone calls.

Hybrid. Recall that in the telephone, a hybrid circuit (also known as a *two-wire to four-wire circuit*), usually a transformer, provides simultaneous two-way conversations on a single pair of wires. The hybrid combines the signal from the telephone transmitter with the received signal to the receiver on the single twisted-pair cable. It keeps the signals separate within the telephone.

A hybrid is also used at the central office. It effectively translates the two-wire line to the subscriber back into four lines, two each for the transmitted and received signals. The hybrid provides separate transmit and receive signals. Although a single pair of lines is used in the local loop to the subscriber, all other connections to the telephone system treat the transmitted and received signals separately and have independent circuits for dealing with them along the way.

Test Signals. To check the status and quality of subscriber lines, the phone company often puts special test tones on the local loop and receives resulting tones in return. These can give information about the overall performance of the local loop. The SLIC provides a way to connect the test signals to the local loop and to receive the resulting signals for measurement.

BORSCHT Functions. The basic BORSCHT functions are usually divided into two groups, high voltage and low voltage. The high-voltage parts of the system are the battery feed, overvoltage protection, ringing circuits, and test circuits. The low-voltage group includes the supervision, coding, and hybrid functions. In older systems, all the functions were implemented with discrete component circuits. Today, these functions are generally divided between two ICs, one for the high-voltage functions and the other for the low-voltage functions. However, single-chip SLIC BORSCHT ICs are now available.

Telephone Hierarchy

Whenever you make a telephone call, your voice is connected through your local exchange to the telephone system. From there it passes through at least one other local exchange, which is connected to the telephone you are calling. Several other facilities may provide switching, multiplexing, and other services required to transmit your voice. The telephone system is referred to as the *public switchod telephone network (PSTN)*. The organization of this hierarchy in the United States is discussed in the next sections.

Central Office. The *central office* or *local exchange* is the facility to which your telephone is directly connected by a twisted-pair cable. Also known as an *end office (EO),* the local exchange can serve up to 10,000 subscribers, each of whom is identified by a four-digit number from 0000 through 9999 (the last four digits of the telephone number).

The local exchange also has an exchange number. These are the three additional digits that make up a telephone number. Obviously, there can be as many as 1000 exchanges with numbers from 000 through 999. These exchanges become part of an area code region, which is defined by an additional three-digit number. Each area code is fully contained within one of the geographic areas assigned to one of the Bell regional operating companies.

RBOC. In 1984 the U.S. Justice Department broke up the AT&T telephone system to encourage competition. Seven *regional Bell operating companies (RBOCs)* were formed to provide local telephone service. The original RBOCs were Ameritech, Bell Atlantic, Bell South, NYNEX, Pacific Bell, Southwestern Bell, and U.S. West.

Mergers over the years have reduced the number of RBOCs to four. SBC Communications, for instance, is the merger of Ameritech, Pacific Bell, and Southwestern Bell. Verizon is the merger of NYNEX and Bell Atlantic. U.S. West merged with a new company, Qwest. The only remaining original RBOC is Bell South.

Some areas are not served by a RBOC. These primarily rural areas obtain service from independent phone companies. One large independent, GTE *(General Telephone and Electronics),* is now part of Verizon.

All these companies (RBOC or independent) are called *local exchange carriers,* or *local exchange companies (LECs).*

Long Distance. With the breakup mentioned earlier, AT&T became the long-distance supplier for the system. Competition from new companies such as MCI, LDDS, and Sprint developed quickly. Today, AT&T is still the leading supplier of long-distance service. WorldCom acquired LDDS and MCI, and Sprint remains independent. WorldCom went bankrupt and has now returned as MCI. Recently, additional consolidation has occurred. SBC acquired AT&T and will become known as AT&T. Verizon also acquired MCI. Sprint acquired cell phone company Nextel.

Operational Relationships. The LECs provide telephone services to designated geographic areas referred to as *local access and transport areas (LATAs)*. The United States is divided into approximately 200 LATAs. The LATAs are defined within the individual states making up the seven operating regions. The LECs provide the telephone service for the LATAs within their regions but do not provide long-distance service for the LATAs.

Local access and transport areas (LATAs)

Long-distance service is provided by long-distance carriers known as *interexchange carriers (IXCs)*. The IXCs are the familiar long-distance carriers such as AT&T (now SBC), WorldCom (now Verizon), and US Sprint. Long-distance carriers must be used for the interconnection for any inter-LATA connections. The LECs can provide telephone service within the LATAs that are part of their operating region, but links between LATAs within a region, even though they may be directly adjacent to one another, must be made through an IXC.

Interexchange carriers (IXCs)

Each LATA contains a *serving,* or *point-of-presence (POP), office* that is used to provide the interconnections to the IXCs. The local exchanges communicate with one another via individual trunks. And all local exchanges connect to an LEC central office, which provides trunks to the POP. At the POP the long-distance carriers can make their interface connections. The POPs must provide equal access for any long-distance carrier desiring to connect. Many POPs are connected to multiple IXCs, but in many areas, only one IXC serves a POP.

Point of presence (POP) office

Figure 18-18 summarizes the hierarchy just discussed. Individual telephones within a LATA connect to the local exchange or central office by way of the two-wire local loop. The central offices within an LATA are connected to one another by trunks. These trunks may be standard baseband twisted-pair cables run underground or on telephone poles, but they may also be coaxial cable, fiber-optic cable, or microwave radio links. In some areas, two or more central offices are located in the same building or physical facility. Trunk interconnections are usually made by cables.

The local exchanges are also connected to an LEC central office when a connection cannot be made between two local exchanges that are not directly trunked. The call passes from the local exchange to the LEC central office, where the connection is made to the other local exchange.

The LEC central office is also connected to the POP. Depending upon the organization of the LEC within the LATA, the LEC central office may contain the POP.

Figure 18-18   Organization of the telephone system in the United States.

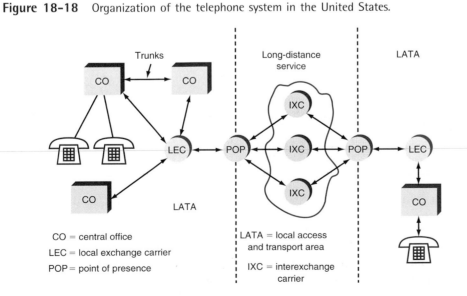

Note in Fig. 18-18 that the POP provides the connections to the long-distance carriers, or IXCs. The "cloud" represents the long-distance networks of the IXCs. The long-distance network connects to the remote POPs, which in turn are connected to other central offices and local exchanges.

Most other long-distance carriers have their own specific hierarchical arrangements. A variety of switching offices across the country are linked by trunks using fiber-optic cable or microwave relay links. Multiplexing techniques are used throughout to provide many simultaneous paths for telephone calls.

In all cases, the various central offices and routing centers provide switching services. The whole idea is to permit any one telephone to directly connect with any other specific telephone. The purpose of all the different levels in the telephone system hierarchy is to provide the interconnecting trunk lines as well as switching equipment that makes the desired interconnection.

The connections between central offices, central offices and LEC and POPs are digital and use the T1 and T3 multiplexing schemes described in Chap. 12. The transmission method in long distance is fiber-optic cable using protocols known as the asynchronous transfer mode (ATM) and the synchronous optical network (SONET). These systems are described in Chap. 12.

Private Telephone System

Telephone service provided to companies or large organizations with many employees and many telephones is considerably different from basic local loop service provided for individuals. Depending upon the size of the organization, there may be dozens, hundreds, or even thousands of telephones required. It is simply not economical to provide each telephone in the organization with its own separate local loop connection to the central office. It is also an inefficient use of expensive facilities to use a remote central office for intercompany communication. For example, an individual in one office often may need to make an intercompany call to a person in another office, which may be only a few doors down the hall or a couple of floors away. Making this connection through the local exchange is wasteful.

Private telephone system

This problem is solved by the use of *private telephone systems* within a company or organization. Private telephone systems implement telephone service among the telephones in the organization and provide one or more local loop connections to the central office. The two basic types of private telephone systems are known as *key systems* and *private branch exchanges.*

Key system

Key Systems. *Key systems* are small telephone systems designed to serve from 2 to 50 user telephones within an organization. Commercially available systems usually have provisions for 6, 10, 12, or 50 telephones.

Simple key telephone systems are made up of the individual telephone units generally referred to as *stations,* all of which are connected to a central answering station. The central answering station is connected to one or more local loop lines known as *trunks* back to the local exchange. Most systems also contain a central electronic switching unit that makes all the internal and external connections.

The telephone sets in a key system typically have a group of pushbuttons that allow each telephone to select two or more outgoing trunking lines. Phone calls are made in the usual way.

Private branch exchange (PBX)

Private Automatic branch exchange (PABX)

Private Branch Exchange. A *private branch exchange,* or *PBX,* as it is known, is a private telephone system for larger organizations. Most PBXs are set up to handle 50 or more telephone interconnections. They can handle thousands of individual telephones within an organization. These systems may also be referred to as *private automatic branch exchanges (PABXs)* or *computer branch exchanges (CPXs).* Of the three terms, the expression *PBX* is the most widely known and used.

A PBX (see Fig. 18-19) is, in effect, a miniature complete telephone system. It provides baseband interconnections to all the telephones in an organization. All the telephones connect to a central switching system which makes intercompany connections as well as external connections to multiple trunk lines to the central office.

Like the key system, the PBX offers the advantages, of efficiency and cost reduction when many telephones are required. Interoffice calls can be completed by the PBX system without accessing the local exchange. Further, it is more economical to limit the number of trunk lines to the central office, for not all telephones in the organization will be attempting to access an outside line at one time.

The modern PBX is usually fully automated by computer control. Although no operator is required, most large organizations have one or more operators who answer incoming telephone calls and route them appropriately with a control console. However, some PBXs are automated so that the individual user's telephone whose extension is the last four digits of the telephone number can be called directly from outside.

As you can see from Fig. 18-19, the PBX is made up of line circuits that are similar to the subscriber line interface circuits discussed earlier. The matrix is the electronic switch that connects any phone to any other phone in the system. It also permits conference calls. The trunk circuits interface to the local loop lines to the central office. All the circuits are under the control of a central computer dedicated to the operation of the PBX.

An alternative to the PBX is known as *Centrex*. This service, normally provided by the local telephone company, performs the function of a PBX but uses special equipment, and most of the switching is carried out by the local exchange switching equipment over special trunk lines. Its advantage over a standard PBX is that the high initial cost of PBX equipment can be avoided by leasing the Centrex equipment from the telephone company.

Centrex

Today, as more companies adopt VoIP systems, the older style PBX systems are gradually disappearing in favor of an equivalent system that uses VoIP standards. These systems attach to the company's LAN system that typically uses Ethernet to connect phones to a base or key unit for distribution and calling features such as voice mail and PBX-like answering capability. Most of these functions are implemented in software with a server dedicated to this function.

Figure 18-19 A PBX.

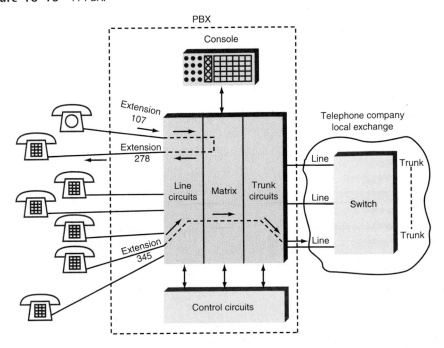

18-3 Facsimile

Facsimile (fax)

Facsimile, or *fax,* is an electronic system for transmitting graphic information by wire or radio. Facsimile is used to send printed material by scanning it and converting it to electronic signals that modulate a carrier to be transmitted over the telephone lines. Since modulation is involved, fax transmission can also take place by radio. With facsimile, documents such as letters, photographs, line drawings, or any printed information can be converted to an electric signal and transmitted with conventional communication techniques. The components of a fax system are illustrated in Fig. 18-20.

Although facsimile is used to transmit pictures, it is not TV because it does not transmit sound messages or live scenes and motion. However, it does use scanning techniques that are generally similar to those used in TV. A scanning process is used to break a printed document up into many horizontal scan lines which can be transmitted and reproduced serially.

How Facsimile Works

The early facsimile machines scanned the document to be transmitted with a light and photocell arrangement. A scanning head consists of a light source and a photocell. A light source, focused to a tiny point with a lens system, was used to scan the document. The lens was also used to focus the reflected light from on the document onto the photocell. As the light scanned the letters and numbers in a typed or printed document or the gray scale in a photograph, the photocell produced a varying electronic signal whose output amplitude was proportional to the amount of reflected light. This baseband signal was then used to amplitude- or frequency-modulate a carrier in the audio frequency range. This permitted the signal to be transmitted over the telephone lines.

Figure 18-21 shows how a printed letter might have been scanned. Assume that the letter F is black on a white background. The output of a photodetector as it scans across line *a* is shown in Fig. 18-21(*a*). The output voltage is high for white and low for black. The output of the photodetector is also shown for scan lines *b* and *c*. The output of the photodetector is used to modulate a carrier, and the resulting signal is put on the telephone line.

The resolution of the transmission is determined by the number of scan lines per vertical inch. The greater the number of lines scanned, the finer the detail transmitted and the higher the quality of reproduction. Older systems had a resolution of 96 lines per inch (LPI), and the new systems have 200 LPI.

On the receiving end, a demodulator recovered the original signal information, which was then applied to a stylus. The purpose of the stylus was to redraw the original information on a blank sheet of paper. A typical stylus converted the electric signal to heat variations that burned the image into heat-sensitive paper. Other types of printing mechanisms were used.

Today's modern fax machine is a high-tech electrooptical machine. Scanning is done electronically, and the scanned signal is converted to a binary signal. Then digital transmission with standard modem techniques is used.

Figure 18-20 Components of a facsimile system.

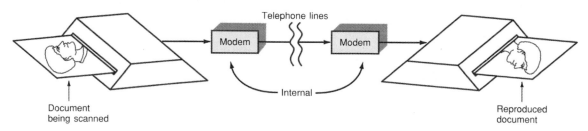

Document being scanned

Telephone lines

Modem

Modem

Internal

Reproduced document

Figure 18-21 Output of a photosensitive detector during different scans.

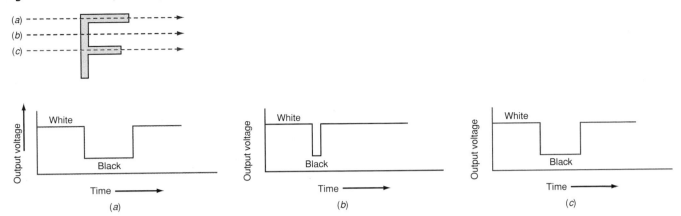

Figure 18-22 is a block diagram of a modern fax machine. The transmission process begins with an image scanner that converts the document to hundreds of horizontal scan lines. Many different techniques are used, but they all incorporate a photo- (light-) sensitive device to convert light variations along one scanned line into an electrical voltage. The resulting signal is then processed in various ways to make the data smaller and thus faster to transmit. The resulting signal is sent to a modem where it modulates a carrier set to the middle of the telephone voice spectrum bandwidth. The signal is then transmitted to the receiving fax machine over the public switched telephone network.

The receiving fax machine's modem demodulates the signal that is then processed to recover the original data. The data is decompressed and then sent to a printer, which reproduces the document. Since all fax machines can transmit as well as receive, they are referred to as *transceivers*. The transmission is half duplex because only one machine may transmit or receive at a time.

Most fax machines have a built-in telephone, and the printer can also be used as a copy machine. An embedded microcomputer handles all control and operation including paper handling.

Image Processing

Most fax machines use *charge-coupled devices (CCDs)* for scanning. A CCD is a light-sensitive semiconductor device that converts varying light amplitudes to an electric signal. The typical CCD is made up of many tiny reverse-biased diodes that act as capacitors which are manufactured in a matrix on a silicon chip (see Fig. 18-23). The base

Image processing

Charge-coupled device (CCD)

Figure 18-22 Block diagram of modem fax machine.

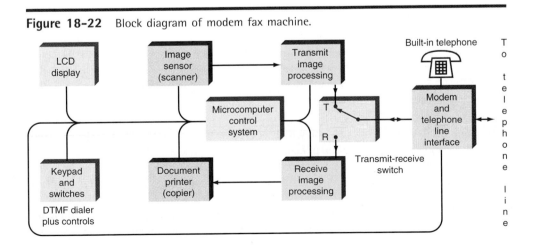

Figure 18-23 A charge-coupled device is used to scan documents in modern fax machines. (*a*) Cross section. (*b*) Detail of capacitor matrix.

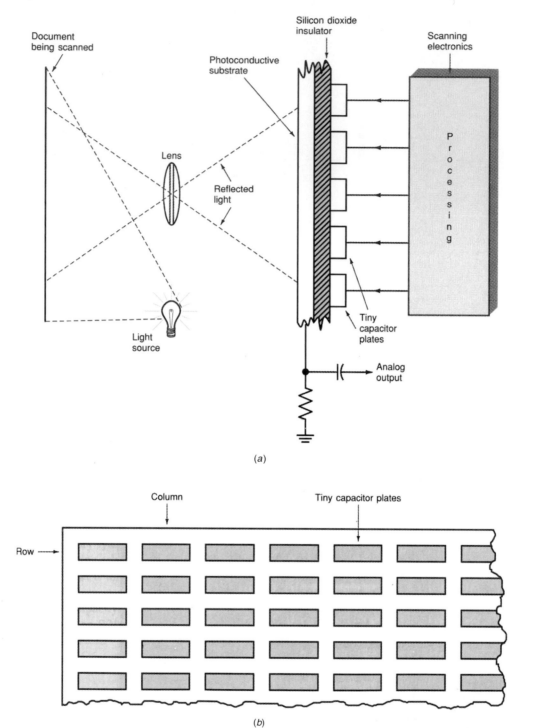

(*a*)

(*b*)

forms one large plate of a capacitor which is electrically separated by a dielectric from many thousands of tiny capacitor plates, as shown. When the CCD is exposed to light, the CCD capacitors charge to a value proportional to the light intensity. The capacitors are then scanned or sampled electronically to determine their charge. This creates an analog output signal that accurately depicts the image focused on the CCD.

A CCD is actually a device that breaks up any scene or picture into *individual picture elements,* or *pixels.* The greater the number of CCD capacitors, or pixels, the higher the

Pixels

Figure 18-24 Scanning mechanism in a fax machine.

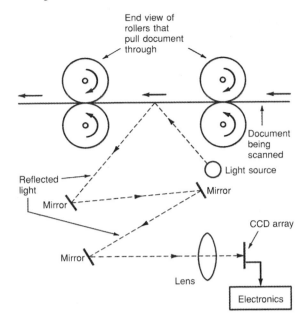

resolution and the more faithfully a scene, photograph, or document can be reproduced. CCDs are available with a matrix of many thousands of pixels, thereby permitting very high-resolution picture transmission. CCDs are widely used in modern video cameras in place of the more delicate and more expensive vidicon tubes. In the video camera (camcorder), the lens focuses the entire scene on a CCD matrix. This same approach is used in some fax machines. In one type of fax machine, the document to be transmitted is placed face down as it might be in a copy machine. The document is then illuminated with brilliant light from a xenon or fluorescent bulb. A lens system focuses the reflected light on a CCD. The CCD is then scanned, and the resulting output is an analog signal whose amplitude is proportional to the amplitude of the reflected light.

In most desktop fax machines, the entire document is not focused on a single CCD. Instead, only a narrow portion of the document is lighted and examined as it is moved through the fax machine with rollers. A complex system of mirrors is used to focus the lighted area on the CCD (see Fig. 18-24).

The more modern fax machines use another type of scanning mechanism that does not use lenses. The scanning mechanism is an assembly made up of an LED array and a CCD array. These are arranged so that the entire width of a standard $8\frac{1}{2} \times 11$ in page is scanned simultaneously one line at a time. The LED array illuminates a narrow portion of the document. The reflected light is picked up by the CCD scanner. A typical scanner has 2048 light sensors forming one scan line. Figure 18-25 shows a side view of the scanning mechanism. The 2048 pixels of light are converted to voltages proportional to the light variations on one scanned line. These voltages are converted from a parallel format to a serial voltage signal. The resulting analog signal is amplified and sent to an AGC circuit and an S/H amplifier. The signal is then sent to an A/D converter where the light signals are translated to binary data words for transmission.

Data Compression

An enormous amount of data is generated by scanning one page of a document. A typical $8\frac{1}{2} \times 11$ in page represents about 40,000 bytes of data. This can be shortened by a factor of 10 or more with *data compression techniques.* Furthermore, because of the narrow bandwidth of telephone lines, data rates are limited. That is why it takes so long to transmit one page of data. Developments in high-speed modems have helped reduce

Data compression techniques

Figure 18-25 LED/CCD scanner mechanism in a modern fax machine.

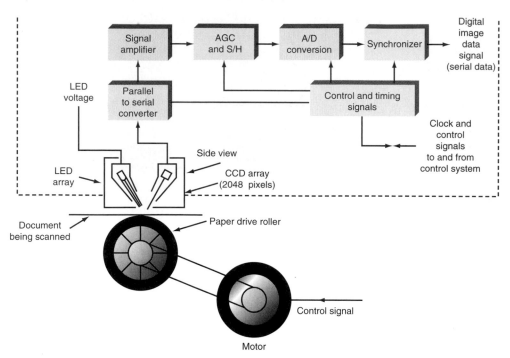

the transmission time, but the most important developments are data compression techniques that reduce the overall amount of data, which significantly decreases the transmission time and telephone charges.

Data compression is a digital data processing technique that looks for redundancy in the transmitted signal. White space or continuous segments of the page that are the same shade produce continuous strings of data words that are the same. These can be eliminated and transmitted as a special digital code that is significantly faster to transmit. Other forms of data compression use various mathematical algorithms to reduce the amount of data to be transmitted.

The data compression is carried out by a *digital signal processing (DSP) chip.* This is a high-speed microprocessor with embedded ROM containing the compression program. The digital data from the A/D converter is passed through the DSP chip, from which comes a significantly shorter string of data that represents the scanned image. This is what is transmitted, and in far less time than the original data could be transmitted.

At the receiving end, the demodulated signal is decompressed. Again, this is done through a DSP chip especially programmed for this function. The original data signal is recovered and sent to the printer.

Modems

Every fax machine contains a built-in *modem* that is similar to a conventional data modem for computers. These modems are optimized for fax transmission and reception. And they follow international standards so that any fax machine can communicate with any other fax machine.

A number of different modulation schemes are used in fax systems. Analog fax systems use AM or FM. Digital fax uses PSK or QAM. To ensure compatibility between fax machines of different manufacturers, *facsimile standards* have been developed for speed, modulation methods, and resolution by the *International Telegraph and Telephone Consultative Committee,* better known by its French abbreviation, *CCITT.* The CCITT is now known as the *ITU-T,* or *International Telecommunications Union.* The ITU-T fax standards are divided into four groups:

Modem

Facsimile standards

International Telegraph and Telephone Consultative Committee (CCITT)

International Telecommunications Union (ITU-T)

1. ***Group 1 (G1 or GI):*** Analog transmission using frequency modulation where white is 1300 Hz and black is 2100 Hz. Most North American equipment uses 1500 Hz for white and 2300 Hz for black. The scanning resolution is 96 lines per inch (LPI). Average transmission speed is 6 minutes per page ($8\frac{1}{2} \times 11$ in or A4 metric size, which is slightly longer than 11 in).

2. ***Group 2 (G2 or GII):*** Analog transmission using FM or vestigial sideband AM. The vestigial sideband AM uses a 2100-Hz carrier. The lower sideband and part of the upper sideband are transmitted. Resolution is 96 LPI. Transmission speed is 3 min or less for an $8\frac{1}{2} \times 11$ in or A4 page.

3. ***Group 3 (G3 or GIII):*** Digital transmission using PCM black and white only or up to 32 shades of gray. PSK or QAM to achieve transmission speeds of up to 9600 Bd. Resolution's 200 LPI. Transmission speed is less than 1 minute per page, with 15 to 30 s being typical.

4. ***Group 4 (G4 or GIV):*** Digital transmission, 56 kbps, resolution up to 400 LPI, and speed of transmission less than 5 s.

The older G1 and G2 machines are no longer used. The most common configuration is group 3. Most G3 machines can also read the G2 format.

The G4 machines are not yet widely used. They are designed to use digital transmission only with no modem over very wideband dedicated digital-grade telephone lines. Both G3 and G4 formats also employ digital data compression methods that shorten the binary data stream considerably, thereby speeding up page transmission. This is important because shorter transmission times cut long-distance telephone charges and reduce operating costs.

Fax Machine Operation

Figure 18-26 is a simplified block diagram of the transmitting circuits in a modern G3 fax transceiver. The analog output from the CCD array is serialized and fed to an A/D converter which translates the continuously varying light intensity into a stream of binary numbers. Sixteen gray scale values between white and black are typical. The binary data

Figure 18-26 Block diagram of a facsimile machine.

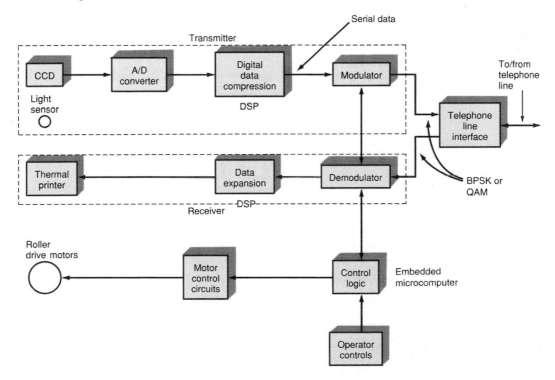

is sent to a DSP digital data compression circuit as described earlier. The binary output in serial data format is used to modulate a carrier which is transmitted over the telephone lines. The techniques are similar to those employed in modems. Speeds of 2400/4800 and 7200/9600 Bd are common. Most systems use some form of PSK or QAM to achieve very high data rates on voice-grade lines.

In the receiving portion of the fax machine, the received signal is demodulated and then sent to DSP circuits, where the data compression is removed and the binary signals are restored to their original form. The signal is then applied to a printing mechanism.

The most common fax printer today is an ink jet printer like those popularly used with PCs. In the high-priced machines, laser scanning of an electrosensitive drum, similar to the drum used in laser printers, produces output copies by using the proven techniques of xerography.

The control logic in Fig. 18-26 is usually an embedded microcomputer. Besides all the internal control functions it implements, it is used for "handshaking" between the two machines that will communicate. This ensures compatibility. Handshaking is usually carried out by exchanging different audio tones. The called machine responds with tones designating its capability. The calling machine compares this to its own standards and then either initiates the transmission or terminates it because of incompatibility. If the transmission proceeds, the calling machine sends synchronizing signals to ensure that both machines start at the same time. The called machine acknowledges the receipt of the sync signal, and transmission begins. All the protocols for establishing communication and sending and receiving the data are standardized by the ITU-T. Transmission is half duplex.

As improvements have been made in picture resolution quality, transmission speed, and cost, facsimile machines have become much more popular. The units can be easily attached with standard RJ-11 modular connectors to any telephone system. In most business applications, the fax machine is typically dedicated to a single line. Most fax machines feature fully automatic operation with microprocessor-based control. A document can be sent to a fax machine automatically. The sending machine simply dials the receiving machine and initiates the transmission. The receiving machine answers the initial call and then reproduces the document before hanging up.

Most fax machines have a built-in telephone and are designed to share a single line with conventional voice transmission. The built-in telephone usually features Touch-Tone dialing and number memory plus automatic redial and other modern telephone features. Most fax machines also have automatic send and receive features for fully unattended operation. Smaller portable fax machines are available. Fax machines may also be used with standard cellular telephone systems in automobiles.

Another popular variation is the *fax modem,* an internal modem designed to be plugged into a personal computer. It allows standard modem operation for connection to online services and any desired remote computer usage. However, this device also contains the circuits for a fax modem but without the image scanning and printing.

A document produced with a word processor is usually stored in ASCII format as a file on disk. This file can be transmitted serially to the fax modem, which compresses it and sends it over the telephone lines. The receiving computer demodulates the signal, decompresses it, and stores it in RAM and then on disk. The resulting document can be read on the video monitor or printed by the usual means.

Fax modem

18-4 Paging Systems

Paging

Paging systems

Paging is a radio communication system designed to signal individuals wherever they may be. *Paging systems* operate in the simplex mode, for they broadcast signals or messages to individuals who carry small battery-operated receivers. Millions of people carry paging receivers. Typically, they work in jobs that require maintaining constant communication with their employer and/or customers. The paging receiver operates continuously. To contact an individual with a pager, all you need to do is make a telephone call. A paging company will send a radio signal that will be received by the pager. The paging receiver

Figure 18-27 The paging process.

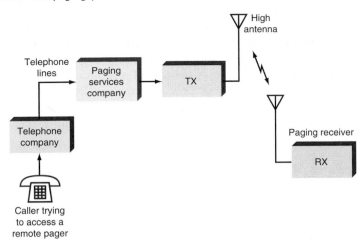

usually has a built-in audible signaling device or silent vibrator that informs the person that she or he is being paged. The signal may be as simple as an audio tone that indicates that the individual should call a telephone number to receive a message or otherwise make contact. Alternatively, the paging company can transmit a short printed message to the paging receiver. Some paging receivers have a small LCD screen on which a telephone number is displayed. This tells the paged individual which number to call. Some paging receivers have larger LCD screens that are capable of displaying several lines of alphanumeric text information. The newest pagers are two-way devices that receive data or send data in the form of numerically coded messages or short alphanumeric text.

Paging System Operation

Although the paging business is not owned by telephone companies, it is closely allied with the telephone business, because the telephone system provides the initial and final communication process. The most common paging process is described below (see Fig. 18-27).

To contact a person who has a pager, an individual dials the telephone number assigned to that person. The call is received at the office of the paging company. The paging company responds with one or more signaling tones that tell the caller to enter the telephone number the paged person should call. Once the number is entered, the caller presses the pound sign key to signal the end of the telephone entry. The calling party then hangs up.

The paging system records the telephone number in a computer and translates this number to a serial binary-coded message. A unique protocol is used. The message is transmitted as a data bit stream to the paging receiver. The serial binary-coded message modulates the carrier of a radio transmitter. Paging systems usually operate in the VHF and UHF ranges. A variety of bands have been assigned by the FCC specifically for paging purposes. The most popular are in the ranges of 149 to 175, 400 to 500, and 920 to 940 MHz.

The paging company usually has a large antenna system mounted on a tower or the top of a tall building so that its communication range is considerable. Most paging systems can locate an individual within a 30-mi radius.

When the signal is transmitted, the paging receiver picks it up on its assigned frequency. The paged individual receives the message, as described above, and responds.

As indicated earlier, modern paging systems can transmit complete messages to an individual. These messages are usually entered on a personal computer and transmitted through a modem over the telephone system to the paging company. In this case the message is received, stored in a computer, and formulated into the correct protocol for transmission to the pager. The length of the message is typically restricted to several lines of text. A typical paging receiver may be able to display four lines of 20 characters each.

Paging Protocols

Cap code

The earliest paging systems used a tone signaling system. Each paging receiver is assigned a special code called a *cap code,* which is a sequence of numbers or a combination of letters and numbers. The code is broadcast over the paging region. If the pager is within the region, it will pick up and recognize its unique code. The cap code is encoded by using audio tones. Early systems used a two-tone system.

The tones frequency-modulate the transmitter carrier in a fixed protocol or sequence. All paging receivers pick up every transmission, but they recognize only their own code. When a code is recognized, the beeper in the receiver goes off, informing the user that a call has been received.

The older analog tone systems have been replaced by digital protocols that not only increase the number of possible subscribers but also make transmission more reliable because of the inherent noise immunity of digital transmission and because it is possible to employ error correction and detection methods.

POCSAG (Post Office Code Standardisation Advisory Group) code

Dozens of different digital paging protocols have been developed over the years. The most widely used are the POCSAG and FLEX protocols. The *POCSAG (Post Office Code Standardisation Advisory Group) code* is a worldwide standard developed under the sponsorship of the British Post Office. It can support up to 2 million users on a system. POCSAG uses FSK with a deviation of ± 4.5 kHz, with the lower frequency representing a logic 1 and the upper frequency representing a logic 0. The original data rate was 512 bps, but today most transmissions are at 1200 bps. A 2400-bps version is also available. POCSAG has mostly been phased out in favor of the FLEX system.

FLEX system

Thus the most widely used digital paging format is the *FLEX system,* developed by Motorola. One of its biggest benefits is that it can accommodate up to 5 billion users on a system. It uses a sophisticated two- or four-level modulation scheme to achieve data rates of 1600, 3200, or 6400 bps. The higher the data rate, the faster a message can be sent and the greater the number of users who can be accommodated on a system. FLEX uses a two- or four-level FSK modulation. The deviations are ± 1600 and ± 4800 Hz. Each shift frequency represents 2 bits in the four-level scheme:

$$-4800 = 00$$
$$-1600 = 01$$
$$+1600 = 11$$
$$+4800 = 10$$

The FM demodulator output looks like the one shown in Fig. 18-28.

At the basic symbol rate using two levels, the data rate is 1600 bps. Using four levels, the rate is $4 \times 1600 = 6400$ bps.

The data rate automatically adjusts to the amount of traffic being handled. If traffic is light, the 1600-bps rate is used. During moderate traffic, 3200 bps is used. With maximum use, the speed steps up to 6400 bps.

Data is transmitted continuously and synchronously. The data rate is referenced to a clock derived from a highly precise atomic clock such as that used in the Global

Figure 18-28 FM demodulating output for FLEX system.

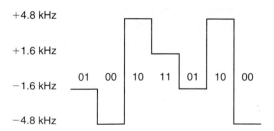

Figure 18-29 FLEX protocol for pagers. *(Courtesy Motorola)*

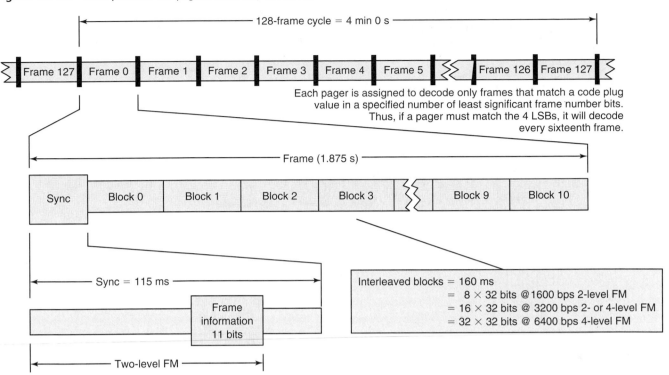

Positioning System. The basic data structure is shown in Fig. 18-29. One complete cycle consists of 128 frames of information. The cycle time is 4 min. Each frame is made up of synchronizing segments and 11 blocks of data. An 11-bit frame information segment following an initial synchronizing signal tells the pager the frame and cycle in which it resides. A following sync segment identifies the data rate. The message is then transmitted in one of the following blocks of information. Error detection and error correction are included in each block.

Today FLEX is the dominant pager technology, and Motorola is the leading pager supplier. However, Motorola has licensed the technology to many other companies. A more recent version of FLEX is REFLEX, which is a two-way pager technology. It allows the pager subscriber to transmit as well as receive, to send messages in acknowledgment of the receipt of a message, or to compose and send an original message. Two-way pagers contain a tiny alphanumeric keyboard. The newest form of FLEX is INFLEXION. This system allows the pager subscriber to send a voice message.

As cell phone use has grown and as cell phones have incorporated short messaging and e-mail capability, paging has been in a steep decline. Although pagers are still in use, the overall number of pager users continues to decrease as cell phone usage increases. At this time, the future of paging is unknown.

Paging Receiver

A *paging receiver* is a small battery-powered superheterodyne receiver. Most pagers use a single-chip IC receiver similar to the one described in Chap. 9. Both single- and double-conversion models are available. Direct conversion receivers (ZIF) are also used. Most basic paging systems use some form of frequency modulation.

A FLEX paging receiver is shown in Fig. 18-30. This is a simple single-conversion receiver. The local oscillator is crystal-controlled and therefore sets the paging receiver to a specific frequency.

The output of the FM demodulator circuit is sent to the FLEX decoder, which recognizes the receiver's unique code. The microcontroller provides the necessary message

Paging receiver

Figure 18-30 A FLEX paging receiver.

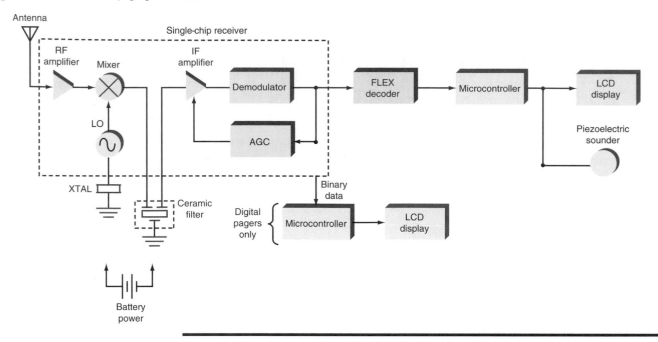

Historical PBX switchboard from Century Telephone Construction Co. of Buffalo, New York. With this model, one operator would be responsible for connecting all the office calls. Today's computerized units can handle thousands of individual telephones within an organization.

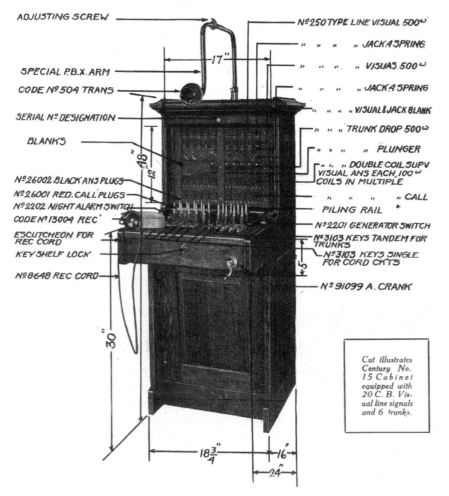

decoding and display operation. It stores the message in its memory and then displays it on the LCD. A control program handles all protocol disassembly and LCD display formatting and operation as well as the "beeper" activation.

18-5 Internet Telephony

Internet telephony, also called *Internet Protocol (IP)* telephony or *Voice over Internet Protocol (VoIP)*, uses the Internet to carry digital voice telephone calls. VoIP, in effect, for the most part, bypasses the existing telephone system, but not completely. It has been in development for over a decade, but only recently has it become practical and popular. VoIP is a highly complex digital voice system that relies on high-speed Internet connections from cable TV companies, phone companies supplying DSL, and other broadband systems including wireless. It uses the Internet's vast fiber-optic cabling network to carry phone calls without phone company charges. This new telephony system is slowly replacing traditional phones, especially in large companies. It offers the benefits of lower long-distance calling charges and reduces the amount of new equipment needed since phone service is essentially provided over the same local-area network (LAN) that interconnects the PCs in an organization. VoIP is rapidly growing in use and in the future is expected to replace standard phones in many companies and homes. While the legacy PSTN will virtually never go away, over time it will play a smaller and smaller role as VoIP is more widely adopted or as more and more individuals choose a cell phone as their main telephone service.

Internet Protocol (IP)

Voice over Internet Protocol (VoIP)

VoIP Fundamentals

There are two basic parts to an IP phone call: the "dialing" process which establishes an initial connection and the voice signal flow.

Voice Signal Flow. Figure 18-31 shows the signal flow and major operations that take place during an IP phone call. The voice signal is first amplified and digitized by

Figure 18-31 Signal flow in a VoIP system.

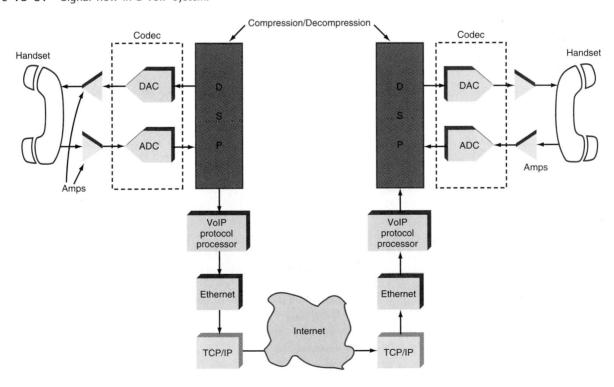

an analog-to-digital converter (ADC) which is part of a coder-decoder (codec) circuit that also includes a digital-to-analog converter (DAC). The ADC usually samples the voice signal at 8 kHz and produces an 8-bit word for each sample. These samples occur one after another serially and therefore produce a 64-kbps digital signal. A relatively wide bandwidth is needed to transmit this bit stream (64 kHz or more). To reduce the data rate and the need for bandwidth, the bit stream is processed by a voice encoder that compresses the voice signal. This compression is usually done by DSP either in a separate DSP processor chip or as hardwired logic on a larger chip. The output is at a greatly reduced serial digital data rate.

The type of compression used is determined by International Telecommunications Union standards. Various mathematical algorithms beyond the scope of this text are used. The 64-kbps digital signal is designated as standard G.711 and is better known as *pulse-code modulation (PCM)*, covered earlier in this book. Standard G.729a is probably the most common compression standard used and results in an 8-kbps digital voice signal. Another popular standard is G.723 which produces an even more highly compressed 5.3-kbps signal at the expense of some voice quality. Numerous other compression standards are used, and they are selected based upon the application. Most VoIP phones contain all the common compression standard algorithms in the DSP memory for use as called for. The signal is also processed in the DSP to provide echo cancellation, a problem in digital telephony.

The resulting serial digital signal is put into a special packet by a microcomputer processor running a VoIP protocol and then transmitted by Ethernet over a LAN or via a high-speed Internet connection such as is available from a cable TV company or on DSL. From there the signal travels over standard available Internet connections using TCP/IP through multiple servers and routers until it comes to the desired location.

At the receiving phone, the process is reversed. The Internet signal gets converted back to Ethernet, and then the VoIP processor recovers the original packet. From there, the compressed data is extracted, decompressed by a DSP, and sent to the DAC in the codec where the original voice is heard.

One of the main problems with VoIP is that it takes a relatively long time to transmit the voice data over the Internet. The packets may take different routes through the Internet. They all do eventually arrive at their intended destination, but often the packets are out of sequence. The receiving phone must put them back together in the correct sequence. This takes time.

Furthermore, even though the signals traverse the high-speed optical Internet lines at gigabit speeds, the packets pass through numerous routers and servers, each adding transit time or latency. *Latency* is the delay between the time the signal is transmitted and the time it is received. It has been determined that the maximum acceptable latency is about 150 ms. Any longer time is noticeable by the user. One party may have to wait a short time before responding to avoid talking while the signal is still be received. This annoying wait is unacceptable to most. Keeping the latency below 150 ms minimizes this problem.

Link Establishment. In the PSTN, the dialing process initiates multiple levels of switching that literally connects the calling phone to the called phone. That link is maintained for the duration of the call because the switches stay in place and the electronic paths stay dedicated to the call. In Internet telephony, no such temporary dedicated link is established because of the packetized nature of the system. Yet some method must be used to get the voice data to the desired phone. This is taken care of by a special protocol developed for this purpose. The initial protocol used was the ITU H.323. Today, however, a newer protocol established by the Internet Engineering Task Force (IETF) called the *session initiation protocol (SIP)* has been adopted as the de facto standard. In both cases the protocol sets up the call and then makes sure that the voice packets produced by the calling phone get sent to the receiving phone in a timely manner.

Latency

Session initiation protocol (SIP)

Internet Phone Systems

There are two basic types of IP phones: those used in the home and those used in larger organizations. The concepts as described above are the same for both, but the details are slightly different.

Home VoIP. To establish IP phone service in the home, the subscriber must have some form of high-speed Internet service. Cable TV provides this service in most homes, but it can also be provided over the standard POTS local loop with DSL. In addition, the subscriber must have a VoIP interface. This is called different things by the different service providers. A common example is the Analog Terminal Adapter (ATA). This device connects the standard home telephone to the existing broadband Internet modem. Another configuration is a VoIP gateway that contains the ATA circuitry as well as the broadband modem.

A general block diagram of an ATA is shown in Fig. 18-32. Notice that the ATA allows standard telephones and cordless phones to attach to the ATA via the usual RJ-11 modular plug. In fact, the input to the ATA is the phone wiring in the home. The home wiring is disconnected from the subscriber interface at the connection provided outside the home by the phone company. In this way, any of the available home phones can be used with the ATA over installed wiring. Note in the figure that since standard phones are used, they must be provided with SLIC BORSCHT functions. The SLIC circuitry is usually packaged in a single IC chip, and often the codec is also contained on this chip.

The codec inputs and outputs go to one or more processors where the H.323 or SIP protocol is implemented and where the DSP functions for compression and decompression reside. An Ethernet interface is also provided. The Ethernet signal connects to the broadband modem for cable TV service or DSL. If the cable modem is used, the POTS and last mile local loop are simply not used. However, if DSL service provides the broadband connection, the POTS connection is used for the DSL modem. The home phone wiring must be disconnected from the POTS line as described earlier.

Enterprise IP Phones. IP phones in companies or large organizations are especially designed for VoIP service. The telephone set contains all the ATA circuitry except for the SLIC and connects directly to the available Ethernet connection usually supplied to each desk. No broadband modem is needed. Since most employees will also have a PC connected to the LAN, a two-port Ethernet switch in the phone or PC provides a single Ethernet connection to the LAN that the phone and PC share.

Figure 18-32 Analog terminal adapter (ATA) or VoIP gateway.

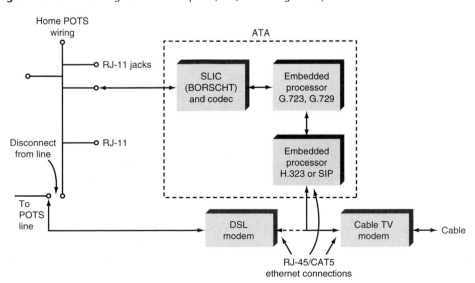

A major benefit of IP phones is that they may also use wireless Ethernet connections. Wireless Ethernet, generally called Wi-Fi or the IEEE standard designation 802.11, is widely used to extend the LANs in most companies. If the IP phone is equipped with a wireless Ethernet transceiver, then no wired connection is needed. Already some cell phone manufacturers are including Wi-Fi VoIP in some models. In this way, a person's cell phone works outside the company with the standard cell site service but also serves as the person's company phone with a wireless Ethernet connection when inside the company. Wireless systems such as this are covered in Chap. 22.

CHAPTER REVIEW

Summary

This chapter introduced the parts of the telecommunication systems most commonly in use and accessible. The basic operation of standard and electronic telephones was discussed, as were the basic parts and operations of cordless telephones, facsimile machines, and pagers. The chapter also covered telecommunication systems, including the ISDN, for telephones, facsimile machines, pagers, and Internet Protocol phones.

The standard telephone was invented in 1876, long before electronic components and circuits had been discovered. For that reason, the telephone was basically an electrical device, and all functions were handled by simple passive components. The original telephone system was designed for full duplex analog communication of voice signals. Today the telephone system is primarily digital, not only in signal transmission but also in control mechanisms. Today most new telephones are electronic, and they use IC technology. The development of the microprocessor has also affected telephone design. A built-in microprocessor permits automatic control of a multiple-line and full-featured telephone's functions. A microprocessor provides such features as number storage, automatic dialing, and redialing that are not possible with standard telephones.

The ultimate convenience in the home or office is the cordless telephone, which uses two-way radio transmission and provides total portability. An advanced version of the cordless telephone is the digital spread-spectrum cordless telephone. Because of the use of digital techniques, signal quality is excellent with virtually no noise.

The telephone itself is a small but relatively complex entity, but it is nothing compared to the massive system that backs it up. The telephone system can interconnect any two telephones in the world, and most people can only speculate on the method by which this interconnection takes place. In reality, it takes place on many levels and involves an incredible array of systems and technology managed by an equally complex array of businesses. Through the telephone system, facsimile machines can send printed information by wire or radio.

Paging systems were developed in the 1980s, and their popularity grew during the 1990s. The most widely used is Motorola's FLEX type. Currently, paging has declined significantly because more people are using cell phones, including phones with short-message service (SMS) and e-mail.

Voice over Internet Protocol phones are becoming more widely used in homes and offices. The digitized voice is compressed, packetized, and transmitted over a LAN or broadband Internet connection. The ITU and IETF establish standards that specify the type of compression and connection methods.

Questions

1. Define specifically what is meant by the *local loop*.
2. What type of power supply is used to power a standard telephone, what are its specifications, and where is it located?
3. State the characteristics of the ringing signal supplied by the telephone company.
4. What is a hybrid?
5. True or false? Most telephone companies can still accommodate pulse dial telephones.
6. What type of transmitter (microphone) is used in a standard telephone, and how does it work?
7. Define what is meant by *tip* and *ring* and state the colors used to represent them.
8. What is the name of the TouchTone dialing system?
9. What two tone frequencies are generated when you press the pound key (#) on a TouchTone phone?
10. What is the name of the building or facility to which every telephone is connected?

11. What kind of microphone is used in an electronic telephone?

12. What is the purpose of the bridge rectifier circuit at the input to the connection of the telephone to the line to the telephone company?

13. Name one type of low-cost sound device used to implement the bell or ringer in an electronic telephone.

14. True or false? In an electronic telephone, the hybrid is a special type of transformer.

15. Give two names or designations for the standard connector used on telephones.

16. State the four frequency ranges used by cordless telephones, and tell which type of modulation is used.

17. What do you call the circuits that make up the connections to each telephone at the telephone office?

18. What is a key telephone system?

19. Does your local telephone company supply long-distance service?

20. Define POTS and PSTN.

21. True or false? Fax can transmit photographs and drawings as well as printed text.

22. What is the most common transmission medium for fax signals? What other medium is commonly used?

23. True or false? Facsimile was invented before radio.

24. Who sets the standards for fax transmission?

25. Vestigial sideband AM is used in what group type of fax machines?

26. What is the name of the semiconductor photosensitive device used in most modern fax machines to convert a scanned line to an analog signal?

27. What is the group designation given to most modern fax machines?

28. To ensure compatibility between sending and receiving fax machines, the control logic carries out a procedure by using audio tones to establish communications. What is this process called?

29. What circuit in the fax machine makes the fax signal compatible with the telephone line?

30. What is the upper speed limit of a G3 fax machine over the telephone lines?

31. What is the resolution of a G3 fax machine in lines per inch?

32. Is fax transmission usually full duplex or half duplex?

33. Are fax signals representing the image to be transmitted before they are prepared for the telephone lines analog or digital?

34. True or false? Group 4 transmissions do not use the standard telephone lines.

35. What are the speed and resolution of group 4 fax transmissions?

36. What does caller ID do?

37. Explain how caller ID data is transmitted to the telephone.

38. Describe the modulation and data rate used in caller ID.

39. Name two ways that caller ID can be blocked.

40. Name the three frequency ranges used by cordless telephones.

41. What are the three primary disadvantages of the older analog cordless phones?

42. Name three types of 900-MHz cordless phone.

43. What type of modulation is used in the best digital cordless phones?

44. Why are the 2.4-GHz and 5.8-GHz cordless phones inherently secure?

An Internet-enabled, cordless SurFone. It includes a chip with 900-MHz operation.

45. What is the approximate maximum distance over which a digital cordless phone can transmit?
46. Name the major parts of a paging system, and briefly explain its operation.
47. Name the three data rates of the FLEX system.
48. Explain how the highest FLEX data speed is achieved.
49. Do VoIP phone systems use the PSTN? Explain.
50. What is the name of the circuit that does the data conversion in an IP phone?
51. What are the designations of the basic protocol standard used to implement VoIP?
52. What are some of the standards for compression and decompression used in IP phones?

53. Why is compression needed in IP phones?
54. How are compression and decompression accomplished in an IP phone?
55. What two transmission protocols are widely used in sending and receiving IP phone calls?
56. Give two names of the equipment used in a home to implement VoIP.
57. True or false? A home VoIP system can use regular analog phones.
58. True or false? An office IP phone needs an SLIC.
59. What effect greatly degrades the quality of a VoIP call and makes it annoying? What causes this problem?

Problems

1. What is an SLIC?
2. List the basic BORSCHT functions that are performed by the telephone company for every telephone. ◆
3. Explain what each of the number groups in a 10-digit telephone number mean.
4. Describe the types of possible links between telephone exchanges. Are they two-line or four-line? ◆
5. Briefly define the terms *LATA, LEC, POP,* and *IXC.* Which one means "long-distance carrier"?
6. State the name and basic specifications and benefits of the newest class of digital cordless telephone.
7. Explain the process and hardware used to convert images to be transmitted to electric signals in a fax machine. ◆
8. Describe two methods of scanning used in modern fax machines.
9. What is the most commonly used type of printer in a fax machine?

10. Describe how fax signals are processed to speed up transmission.
11. What is the symbol time for the 1600- and 6400-bps data rates in FLEX?
12. How many bits can be transmitted during one block of FLEX data (Fig. 18-29) at 1600 and 6400 bps?
13. What is the basic frequency response of the telephone local loop? Can it carry digital as well as analog signals?
14. What is the serial data speed of a G.711 VoIP signal?
15. What are the data rates of G.723 and G.729a compressed VoIP voice signals?
16. State the maximum allowable latency in an IP phone call.

◆ *Answers to Selected Problems follow Chap. 22.*

Critical Thinking

1. Discuss how it would be possible to use the standard ac power lines for telephone voice transmission. What circuits might be needed, and what would be the limitations of this system?

2. Explain the factors that limit the range of a cordless telephone.

chapter 19

Optical Communication

Optical communication systems use light to transmit information from one place to another. Light is a type of electromagnetic radiation like radio waves. Today, infrared light is being used increasingly as the carrier for information in a communication system. The transmission medium is either free space or a special light-carrying cable called a *fiber-optic cable*. Because the frequency of light is extremely high, it can accommodate very high rates of data transmission with excellent reliability. This chapter introduces the basic concepts, circuits and the most widely used optical networking technologies.

Optical communication systems

Fiber-optic cable

Objectives

After completing this chapter, you will be able to:

- Define the terms *optical* and *light*. Name the three main bands of the optical spectrum, and state their wavelength ranges.
- State eight benefits of fiber-optic cables over electrical cables for communication. Name six typical communication applications for fiber-optic cable.
- Explain how light is propagated through a fiber-optic cable. Name the three basic types of fiber-optic cables, and state the two materials from which they are made. Calculate the transmission loss in decibels of fiber-optic cable and connectors over a distance.
- Name the two types of optical transmitter components and their main operating characteristics.
- Explain the operation of an optical detector and receiver.
- State the nature and frequency range of the infrared band, and name natural and artificial sources of infrared light.

19-1 Optical Principles

Optics is a major field of study in itself and far beyond the scope of this book. Most courses in physics introduce the basic principles of light and optics. We review briefly those principles that relate directly to optical communication systems and components.

Light

Light, radio waves, and microwaves are all forms of electromagnetic radiation. Light frequencies fall between those of microwaves and X-rays, as shown in Fig. 19-1(*a*). Radio frequencies range from approximately 10 kHz to 300 GHz. Microwaves extend from 1 to 300 GHz. The range of about 30 to 300 GHz is generally defined as millimeter waves.

Further up the scale is the *optical spectrum,* made up of infrared, visible, and ultraviolet light. The frequency of the optical spectrum is in the range of 3×10^{11} to 3×10^{16} Hz. This includes both the infrared and the ultraviolet bands as well as the visible parts of the spectrum. The visible spectrum is from 4.3×10^{14} to 7.5×10^{14} Hz.

Figure 19-1 The optical spectrum. (*a*) Electromagnetic frequency spectrum showing the optical spectrum. (*b*) Optical spectrum details.

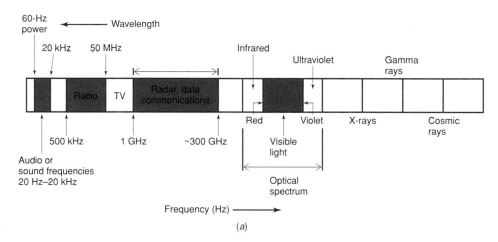

(a)

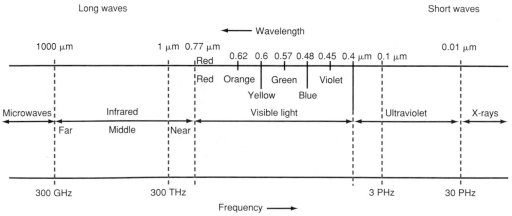

GHz = gigahertz = 10^9 Hz, THz = terahertz = 10^{12} Hz, PHz = petahertz = 10^{15} Hz

(b)

We rarely refer to the "frequency of light." Light is expressed in terms of wavelength. Recall that *wavelength* is a distance measured in meters between peaks of a wave. It is calculated with the familiar expression

$$\lambda = \frac{300{,}000{,}000}{f}$$

where λ (lambda) is the wavelength in meters, 300,000,000 is the speed of light in meters per second, and f is the frequency in hertz. The details of the optical spectrum are shown in Fig. 19-1(b).

Light waves are very short and usually expressed in *nanometers* (nm, one-billionth of a meter) or micrometers (μm, one-millionth of a meter). An older term for *micrometer* is *micron*. *Visible light* is in the 400- to 700-nm range or 0.4 to 0.7 μm depending upon the color of the light. Short-wavelength light is violet (400 nm) and red (700 nm) is long-wavelength light. Nanometer

Micrometer (μm) or micron

Visible light

Another unit of measure for light wavelength is the *angstrom*. One angstrom (Å) is equal to 10^{-10} m or 10^{-4} μm. To say it the other way, 1 μm equals 10,000 Å. Angstrom (Å)

Right below visible light is a region known as *infrared*. Its spectrum is from 0.7 to 1000 μm. Sometimes you will hear infrared referred to as *near* or *far infrared*. Near infrared is those frequencies near the optical spectrum and far infrared is lower in frequency near the upper microwave region [see Fig. 19-1(b)]. Infrared spectrum

Right above the visible spectrum is the *ultraviolet range*. The ultraviolet range is above violet visible light which has a wavelength of 400 nm or 0.4 μm up to about 10^{-8} m or 0.01 μm or 10 nm. The higher the frequency of the light, the shorter the wavelength. The primary source of ultraviolet light is the sun. Infrared and ultraviolet are included in what we call the *optical spectrum*. Ultraviolet range

Speed of Light. Light waves travel in a straight line as microwaves do. Light rays emitted by a candle, lightbulb, or other light source move out in a straight line in all directions. Light waves are assumed to have a spherical wave front as do radio waves. The *speed of light* is approximately 300,000,000 m/s, or about 186,000 mi/s, in free space. These are the values normally used in calculation, but for a more accurate outcome, the actual values are closer to 2.998×10^8 m/s, or 186,280 mi/s. Speed of light

The speed of light depends upon the medium through which the light passes. The figures given above are correct for light traveling in *free space*, i.e., for light traveling in air or a vacuum. When light passes through another material such as glass, water, or plastic, its speed is slower.

Physical Optics

Physical optics refers to the ways that light can be processed. Light can be processed or manipulated in many ways. For example, lenses are widely used to focus, enlarge, or decrease the size of light waves from some source. Physical optics

Reflection. The simplest way of manipulating light is to reflect it. When light rays strike a reflective surface, such as a mirror, the light waves are thrown back or reflected. By using mirrors, the direction of a light beam can be changed. Reflection

The reflection of light from a mirror follows a simple physical law. That is, the direction of the reflected light wave can be easily predicted if the angle of the light beam striking the mirror is known (refer to Fig. 19-2). Assume an imaginary line that is perpendicular with the flat mirror surface. A perpendicular line, of course, makes a right angle with the surface, as shown. This imaginary perpendicular line is referred to as the *normal*. The normal is usually drawn at the point where the mirror reflects the light beam.

If the light beam follows the normal, the reflection will simply go back along the same path. The reflected light ray will exactly coincide with the original light ray.

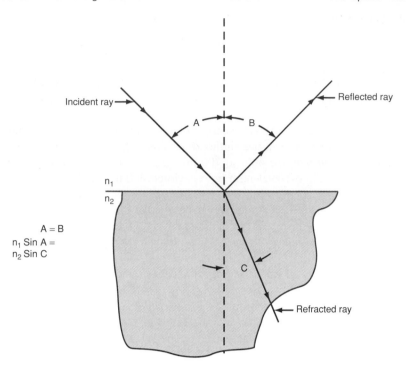

$A = B$
$n_1 \sin A =$
$n_2 \sin C$

If the light ray strikes the mirror at some angle *A* from the normal, the reflected light ray will leave the mirror at the same angle *B* to the normal. This principle is known as the *law of reflection.* It is usually expressed in the following form: *The angle of incidence is equal to the angle of reflection.*

The light ray from the light source is usually called the *incident ray.* It makes an angle *A* with the normal at the reflecting surface, called the *angle of incidence.* The reflected ray is the light wave that leaves the mirror surface. Its direction is determined by the angle of reflection *B,* which is exactly equal to the angle of incidence.

Refraction. The direction of the light ray can also be changed by *refraction,* which is the bending of a light ray that occurs when the light rays pass from one medium to another. In reflection, the light ray bounces away from the reflecting surface rather than being absorbed by or passing through the mirror. Refraction occurs only when light passes through transparent material such as air, water, and glass. Refraction takes place at the point where two different substances come together. For example, where air and water come together, refraction will occur. The dividing line between the two different substances or media is known as the *boundary,* or *interface.*

To visualize refraction, place a spoon or straw into a glass of water, as shown in Fig. 19-3(*a*). If you observe the glass of water from the side, it will look as if the spoon or straw is bent or offset at the surface of the water.

Another phenomenon caused by refraction occurs whenever you observe an object under water. You may be standing in a clear stream and observing a stone at the bottom. The rock is in a different position from where it appears to be from your observation [see Fig. 19-3(*b*)].

The refraction occurs because light travels at different speeds in different materials. The speed of light in free space is typically much higher than the speed of light in water, glass, or other materials. The amount of refraction of the light of a material is usually expressed in terms of the *index of refraction n.* This is the ratio of the speed of light in air to the speed of light in the substance. It is also a function of the light wavelength.

Naturally, the index of refraction of air is 1, simply because 1 divided by itself is 1. The refractive index of water is approximately 1.3, and that of glass is 1.5.

Law of reflection

Incident ray

Angle of incidence

Refraction

GOOD TO KNOW

The simplest way of manipulating light is to reflect it. The direction of reflected light can be predicted by applying the law of reflection: the angle of incidence is equal to the angle of reflection.

Index of refraction

Figure 19-3 Examples of the effect of refraction.

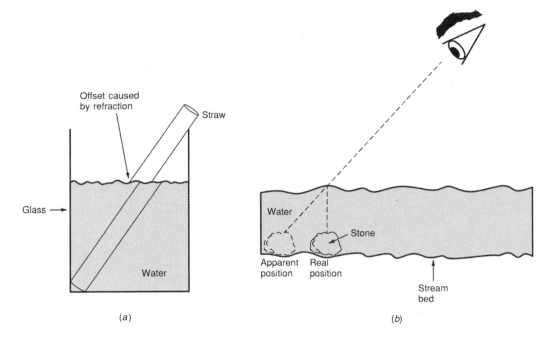

Offset caused by refraction

Straw

Glass →

Water

(a)

Water

Stone

Apparent position   Real position

Stream bed

(b)

To get a better understanding of this idea, consider a piece of glass with a refractive index of 1.5. This means that light will travel through 1.5 ft of air, but during that same time, the light will travel only 1 ft through the glass. The glass slows down the light wave considerably. The index of the refraction is important because it tells exactly how much a light wave will be bent in various substances.

When a light ray passes from one medium to another, the light wave is bent according to the index of refraction. In figure 19-2, the incident ray strikes the surface at angle A to the normal but is refracted at an angle C. The relationship between the angles and indices of refractions are

$$n_1 \sin A = n_2 \sin C$$

Now refer to figure 19-4. A light ray is passing through air. It makes an angle A with the normal. At the interface between the air and the glass, the direction of the light ray is changed. The speed of light is slower; therefore, the angle that the light beam makes to the normal is different from the incident angle. If the index of refraction is known, the exact angle can be determined with the formula given earlier.

If the light ray passes from the glass back into air, it will again change direction, as Fig. 19-4 shows. The important point to note is that the angle of the refracted ray B is not equal to the angle of incidence A.

If the angle of incidence is increased, at some point the angle of refraction will equal 90° to the normal, as shown in Fig. 19-5(a). When this happens, the refracted light ray in red travels along the interface between the air and glass. In this case, the angle of incidence A is said to be the *critical angle*. The critical angle value depends upon the index of refraction of the glass.

If you make the angle of incidence greater than the critical angle, the light ray will be reflected from the interface [see Fig. 19-5(b)]. When the light ray strikes the interface at an angle greater than the critical angle, the light ray does not pass through the interface into the glass. The effect is as if a mirror existed at the interface. When this occurs, the angle of reflection B is equal to the angle of incidence A as if a real mirror were used. This action is known as *total internal reflection,* which occurs only in materials in which the velocity of light is slower than that in air. This is the basic principle that allows a fiber-optic cable to work.

GOOD TO KNOW

The bending of light rays known as refraction occurs because light travels at different speeds in different materials. This phenomenon takes place at the point at which two different substances come together.

Critical angle

Total internal reflection

Figure 19-4 How light rays are bent when passing from one medium to another.

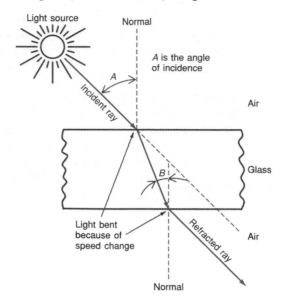

Figure 19-5 Special cases of refraction. (*a*) Along the surface. (*b*) Reflection.

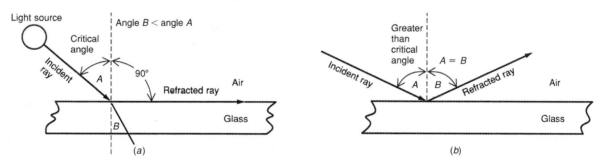

19-2 Optical Communication Systems

Optical communication systems use light as the carrier of the information to be transmitted. As indicated earlier, the medium may be free space as with radio waves or a special light "pipe" or *waveguide* known as *fiber-optic cable*. Both media are used, although the fiber-optic cable is far more practical and more widely used. This chapter focuses on fiber-optic cable.

Rationale for Light Wave Communication

The main limitation of communication systems is their restricted information-carrying capabilities. In more specific terms, this means that the communication medium can carry just so many messages. This information-handling ability is directly proportional to the bandwidth of the communication channel. Using light as the transmission medium provides vastly increased bandwidths. Instead of using an electric signal traveling over a cable or electromagnetic waves traveling through space, the information is put on a light beam and transmitted through space or through a special fiber-optic waveguide.

Light Wave Communication in Free Space

Figure 19-6 shows the elements of an optical communication system using free space. It consists of a light source modulated by the signal to be transmitted, a photodetector

Figure 19-6 Free-space optical communication system.

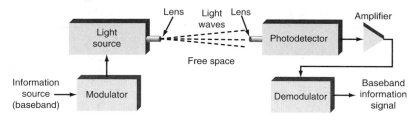

to pick up the light and convert it back into an electric signal, an amplifier, and a demodulator to recover the original information signal.

Light Sources.

A transmitter is a light source. Other common light sources are *light-emitting diodes (LEDs)* and lasers. These sources can follow electric signal changes as fast as 10 GHz or more.

Lasers generate *monochromatic,* or single-frequency, light that is fully *coherent;* i.e., all the light waves are lined up in sync with one another and as a result produce a very narrow and intense light beam.

Light-emitting diodes (LEDs)

Modulator.

A *modulator* is used to vary the intensity of the light beam in accordance with the modulating baseband signal. *Amplitude modulation,* also referred to as *intensity modulation,* is used where the information or intelligence signal controls the brightness of the light. Analog signals vary the brightness continuously over a specified range. This technique is used in some cable TV systems. Digital signals simply turn the light beam off and on at the data rate.

A modulator for analog signals can be a power transistor in series with the light source and its dc power supply (see Fig. 19-7). The voice, video, or other information signal is applied to an amplifier that drives the class A modulator transistor. As the analog signal goes positive, the base drive on the transistor increases, turning the transistor on harder and decreasing its collector-to-emitter base voltage. This applies more of the supply voltage to the light source, making it brighter. A negative-going or decreasing signal amplitude drives the transistor toward cutoff, thereby reducing its collector current and increasing the voltage drop across the transistor. This decreases the voltage to the light source.

Modulator
Amplitude modulation
Intensity modulation

GOOD TO KNOW

Most light wave communication is accomplished through pulse modulation, although amplitude modulation is used with analog signals.

Figure 19-7 A simple light transmitter with series amplitude modulator. Analog signals: transistor varies its conduction and acts as a variable resistance. Pulse signals: Transistor acts as a saturated on/off switch.

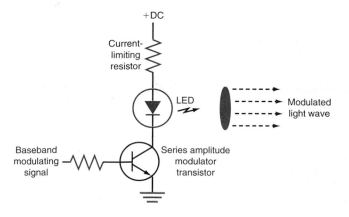

Frequency modulation is not used in light communication. There is no practical way to vary the frequency of the light source, even a monochromatic source such as an LED or a laser. Amplitude modulation is used with analog signals, but otherwise most light wave communication is accomplished by pulse modulation.

Pulse modulation refers to turning the light source off and on in accordance with some serial binary signal. The most common type of pulse modulation is pulse-code modulation (PCM), which is serial binary data organized into bytes or longer words. RZ and Manchester formats are common.

Receiver. The modulated light wave is picked up by a photodetector. This is usually a photodiode or transistor whose conduction is varied by the light. The small signal is amplified and then demodulated to recover the originally transmitted signal. Digital processing may be necessary. For example, if the original signal is voice that was digitized by an A/D converter before being transmitted as a PCM signal, then a D/A converter will be needed at the receiver to recover the voice signal.

Communication by light beam in free space is impractical over very long distances because of the great attenuation of the light due to atmospheric effects. Fog, haze, smog, rain, snow, and other conditions absorb, reflect, refract, and disperse the light, greatly attenuating it and thereby limiting the transmission distance. Artificial light beams used to carry information are obliterated during daylight hours by the sun. And they can be interfered with by any other light source that points in the direction of the receiver. Distances are normally limited to several hundred feet with low-power LEDs and lasers. If high-powered lasers are used, a distance of several miles may be possible.

Light beam communication has become far more practical with the invention of the *laser,* a special high-intensity, single-frequency light source. It produces a very narrow beam of brilliant light of a specific wavelength (color). Because of its great intensity, a laser beam can penetrate atmospheric obstacles better than other types of light can, thereby making light beam communication more reliable over longer distances. When lasers are used, the light beam is so narrow that the transmitter and receiver must be perfectly aligned with each other for communication to occur. Although this causes initial installation alignment problems, it also helps to eliminate external interfering light sources.

Fiber-Optic Communication System

Instead of free space, some type of light-carrying cable can be used. Today, fiber-optic cables have been highly refined. Cables many miles long can be constructed and then interconnected for the purpose of transmitting information. Thanks to these fiber-optic cables, a new transmission medium is now available. Its great advantage is its immense information-carrying capacity (wide bandwidth). Whereas hundreds of telephone conversations may be transmitted simultaneously at microwave frequencies, many thousands of signals can be carried on a light beam through a fiber-optic cable. When multiplexing techniques similar to those in telephone and radio systems are used, fiber-optic communication systems have an almost limitless capacity for information transfer.

The components of a typical fiber-optic communication system are shown in Fig. 19-8. The information signal to be transmitted may be voice, video, or computer data. The first step is to convert the information to a form compatible with

Pulse modulation

Receiver

Laser

Optical waveguide and laser research created a surge of new products in the communication field.

Figure 19-8 Basic elements of a fiber-optic communication system.

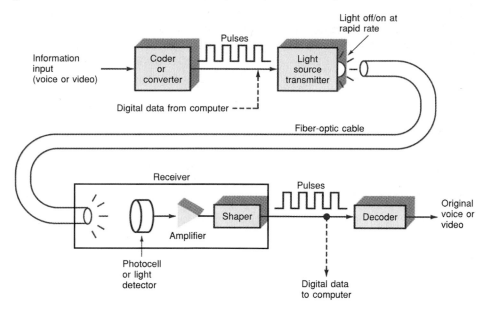

the communication medium, usually by converting continuous analog signals such as voice and video (TV) signals to a series of digital pulses. An A/D converter is used for this purpose.

These digital pulses are then used to flash a powerful light source off and on very rapidly. In simple low-cost systems that transmit over short distances, the light source is usually a light-emitting diode that emits a low-intensity infrared light beam. Infrared beams such as those used in TV remote controls are also used in transmission.

The light beam pulses are then fed into a fiber-optic cable, which can transmit them over long distances. At the receiving end, a light-sensitive device known as a *photocell,* or *light detector,* is used to detect the light pulses. It converts the light pulses to an electric signal. The electrical pulses are amplified and reshaped back into digital form. They are fed to a decoder, such as a D/A converter, where the original voice or video is recovered.

Light detector (photocell)

In very long transmission systems, repeater units must be used along the way. Since the light is greatly attenuated when it travels over long runs of cable, at some point it may be too weak to be received reliably. To overcome this problem, special relay stations are used to pick up the light beam, convert it to electrical pulses that are amplified, and then retransmit on another light beam.

Applications of Fiber Optics

Fiber-optic communication systems are being used more and more each day. Their primary use is in long-distance telephone systems and cable TV systems. Fiber-optic networks also form the core or backbone of the Internet. Fiber-optic cables are no more expensive or complex to install than standard electrical cables, yet their information-carrying capacity is many times greater.

Fiber-optic communication system

Fiber-optic communication systems are being used in other applications. For example, they are being used to interconnect computers in networks within a large building, to carry control signals in airplanes and in ships, and in TV systems because of the wide bandwidth. In all cases, the fiber-optic cables replace conventional coaxial or twisted-pair cables. Figure 19-9 lists some of the applications in which fiber-optic cables are being used.

Figure 19-9 Applications of fiber-optic cables.

1. TV studio to transmitter interconnection eliminating a microwave radio link

2. Closed-circuit TV systems used in buildings for security

3. Secure communication systems at military bases

4. Computer networks, wide area, metro, and local area

5. Shipboard communication

6. Aircraft communication/controls

7. Interconnection of measuring and monitoring instruments in plants and laboratories

8. Data acquisition and control signal communication in industrial process control systems

9. Nuclear plant instrumentation

10. College campus communication

11. Utilities (electric, gas, and so on) station communication

12. Cable TV systems replacing coaxial cable

13. The Internet

Bandwidth

GOOD TO KNOW

The small size and brittleness of fiber-optic cable along with the special tools and techniques required make it difficult to work with.

Benefits of Fiber Optics

The main benefit of fiber-optic cables is their enormous information-carrying capability. This capacity is dependent upon the bandwidth of the cable. *Bandwidth* refers to the range of frequencies that a cable will carry. Electric cables such as coaxial have a wide bandwidth (up to about 750 MHz), but the bandwidth of fiber-optic cable is much greater. Data rates in excess of 100 GHz have been achieved, and even higher rates have been achieved with multiplexing. Fiber-optic cable has many other benefits, summarized in Fig. 19-10.

There are some disadvantages to fiber-optic cable. For example, its small size and brittleness make it difficult to work with. Special, expensive tools and techniques are required.

Figure 19-10 Benefits of fiber-optic cables over conventional electrical cables.

1. *Wider bandwidth.* Fiber-optic cables have high information-carrying capability.

2. *Low loss.* Fiber-optic cables have less signal attenuation over a given distance than an equivalent length of coaxial cable.

3. *Lightweight.* Glass or plastic cables are much lighter than copper cables and offer benefits when low weight is critical (e.g., aircraft).

4. *Small size.* Practical fiber-optic cables are much smaller in diameter than electrical cables and thus can be contained in a relatively small space.

5. *Security.* Fiber-optic cables cannot be as easily "tapped" as electrical cables, and they do not radiate signals that can be picked up for eavesdropping purposes. There is less need for complex and expensive encryption techniques.

6. *Interference immunity.* Fiber-optic cables do not radiate signals, as some electrical cables do, and cause interference to other cables. They are immune to the picking up of interference from other sources.

7. *Greater safety.* Fiber-optic cables do not carry electricity. Therefore, there is no shock hazard. They are also insulators and thus not susceptible to lightning strikes as electrical cables are. They can be used in corrosive and/or explosive environments without danger of sparks.

19-3 Fiber-Optic Cables

Fiber-optic cable

A *fiber-optic cable* is thin glass or plastic cable that acts as a light "pipe." It is not really a hollow tube carrying light, but a long, thin strand of glass or plastic fiber. Fiber cables have a circular cross section with a diameter of only a fraction of an inch. Some fiber-optic cables are the size of a human hair. A light source is placed at the end of the fiber, and light passes through it and exits at the other end of the cable. How the light propagates through the fiber depends upon the laws of optics.

Principles of Fiber-Optic Cable

Fiber-optic cables operate on the optical principles of total internal reflection as described earlier in this chapter. Figure 19-11(*a*) shows a thin fiber-optic cable. A beam of light is focused on the end of the cable. It can be positioned in a number of different ways so that the light enters the fiber at different angles. For example, light ray *A* enters the cable perpendicular to the end surface. Therefore, the light beam travels straight down the fiber and exits at the other end. This is the most desirable condition.

Critical Angle. The angle of light beam *B* is such that its angle of incidence is less than the critical angle, and therefore refraction takes place. The light wave passes through the fiber and exits the edge into the air at a different angle.

The angle of incidence of light beams *C* and *D* is greater than the critical angle. Therefore, total internal reflection takes place, and the light beams are reflected off the surface of the fiber cable. The light beam bounces back and forth between the surfaces until it exits at the other end of the cable.

When the light beam reflects off the inner surface, the angle of incidence is equal to the angle of reflection. Because of this principle, light rays entering at different angles will take different paths through the cable, and some paths will be longer than others. Therefore, if multiple light rays enter the end of the cable, they will take different paths, and so some will exit sooner and some later than others.

In practice, the light source is placed so that the angle is such that the light beam passes directly down the center axis of the cable so that the reflection angles are great. This prevents the light from being lost because of refraction at the interface. Because of total internal reflection, the light beam will continue to propagate through the fiber even though it is bent. With long slow bends, the light will stay within the cable.

The core of the fiber has a higher index of refraction than the cladding surrounding the core as shown in figure 19-11(*b*). The light enters the core at an infinite number of angles but only those rays entering the core at an angle less than the critical angle actually pass down the core. These light rays are reflected off the interface between core and cladding as they pass down the cable.

Figure 19-11 (*a*) Light rays in a fiber-optic cable. (*b*) Critical angle and cone of acceptance.

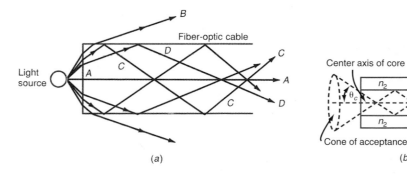

Figure 19-11(b) shows the critical angle of the cable θ_c, sometimes referred to as the *acceptance angle* θ_A. This angle is formed between the center axis line of the core and the line that defines the maximum point where light entering the cable will undergo total internal reflection. If the light beam entering the end of the cable has an angle less than the critical angle, it will be internally reflected and propagated down the cable.

Numerical Aperture. Refer to the angles in Fig. 19-11(b). External to the end of the cable is what is called a *cone of acceptance;* it is defined by the critical angle. Any light beam outside the cone will not be internally reflected and transmitted down the cable.

The cone of acceptance defines the *numerical aperture (NA)* of the cable. This is a number less than 1 that gives some indication of the range of angles over which a particular cable will work. The NA can be calculated with the expression

$$NA = \sin \theta_c$$

For example, if the critical angle is 20°, the NA is

$$NA = \sin 20° = 0.342$$

The NA can also be determined from the indices of refraction of the core and cladding:

$$NA = \sqrt{n_1{}^2 - n_2{}^2}$$

If $n_1 = 1.5$ and $n_2 = 1.4$, the numerical aperture is

$$NA = \sqrt{(1.5)^2 - (1.4)^2} = \sqrt{2.25 - 1.96} = \sqrt{0.29} = 0.5385$$

Typical numerical apertures for common cables are 0.275 and 0.29.

Example 19-1

The numerical aperture of a fiber-optic cable is 0.29. What is the critical angle?

$$NA = \sin \theta_c$$
$$0.29 = \sin \theta_c$$
$$\theta_c = \sin^{-1} 0.29 = \arcsin 0.29$$
$$\theta_c = 16.86°$$

Fiber-Optic Cable Construction

Fiber-optic cables come in a variety of sizes, shapes, and types. The simplest cable contains a single strand of fiber; a complex cable is made up of multiple fibers with different layers and other elements.

The portion of a fiber-optic cable that carries the light is made from either glass or plastic. Another name for glass is *silica*. The optical characteristics of glass are superior to those of plastic. However, glass is far more expensive and more fragile than plastic. Although plastic is less expensive and more flexible, its attenuation of light is greater. For a given intensity, light will travel a farther distance in glass than in plastic.

The construction of a fiber-optic cable is shown in Fig. 19-11(b). The glass or plastic optical fiber is contained within an outer *cladding*. The index of refraction of the outer cladding N_2 is slightly less than the index of refraction N_1 of the core. Typical values for N_1 and N_2 are 1.5 and 1.4, respectively. Over the cladding is a plastic jacket similar to the outer insulation on an electrical cable.

Cone of acceptance

Numerical aperature (NA)

Silica

Cladding

GOOD TO KNOW

Plastic fiber-optic cables are less expensive and more flexible than glass, but the optical characteristics of glass are superior.

Figure 19-12   Basic construction of a fiber-optic cable.

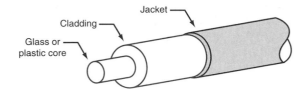

Figure 19-13   Typical layers in a fiber-optic cable.

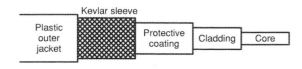

The fiber, which is called the *core,* is usually surrounded by a protective cladding (see Fig. 19-12). The cladding is also made of glass or plastic but has a lower index of refraction. This ensures that the proper interface is achieved so that the light waves remain within the core. In addition to protecting the fiber core from nicks and scratches, the cladding gives strength. Some fiber-optic cable has a glass core with a glass cladding. Other cables have a plastic core with a plastic cladding. Another arrangement, *plastic-clad silica (PCS) cable,* is a glass core with a plastic cladding.

Typically the division between the core and the cladding cannot be seen; they are usually made of the same types of material. Over the cladding is usually a plastic *jacket* similar to the outer insulation on an electrical cable.

In addition to the core, its cladding, and a jacket, fiber-optic cable usually contains one or more elements to form a complete cable. The simplest cable is the core with its cladding surrounded by a protective jacket. As in electrical cables, this outer jacket, or insulation, is made of some type of plastic, typically polyethylene, polyurethane, or polyvinyl chloride (PVC). The main purpose of this outer jacket is to protect the core and cladding from damage. Usually fiber-optic cables are buried underground or strung between supports. Therefore, the fibers must be protected from moisture, dirt, temperature variations, and other conditions. The outer jacket also helps minimize physical damage such as cuts, nicks, and crushing.

The more complex cables may contain two or more fiber-optic elements. Typical cables are available with 2, 6, 12, 18, and 24 optical fiber cores.

There are many different types of cable configurations. Many have several layers of protective jackets. Some cables incorporate a flexible strength or tension element that helps minimize damage to the fiber-optic elements when the cable is being pulled or must support its own weight. Typically, this strength member is made up of a stranded steel or a special yarn known as *Kevlar,* which is strong and preferred over steel because it is an insulator. In some cables, Kevlar forms a protective sleeve or jacket over the cladding. Most claddings are covered with a clear protective coating for added strength and resistance to moisture and damage (see Fig. 19-13).

Fiber-optic cables are also available in a flat ribbon form (see Fig. 19-14). Flat ribbon cable works well with multiple fibers. Handling and identification of individual fibers are easy. The flat cable is also more space-efficient for some applications.

Types of Fiber–Optic Cables

There are two basic ways of classifying fiber-optic cables: The first method is by the index of refraction, which varies across the cross section of the cable. The second method of classification is by *mode,* which refers to the various paths the light rays can take in passing through the fiber. Usually these two methods of classification are combined to define the types of cable.

Step Index Cable. The two ways to define the index of refraction variation across a cable are the step index and the graded index. *Step index* refers to the fact that there is a sharply defined step in the index of refraction where the fiber core and the cladding interface. It means that the core has one constant index of refraction N_1 and the cladding has another constant index of refraction N_2. When the two come together,

Core

Jacket

Mode of cable

Step index cable

Figure 19-14   Multicore flat ribbon cable.

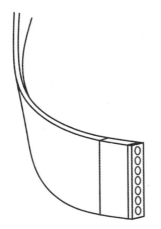

Figure 19-15 A step index cable cross section.

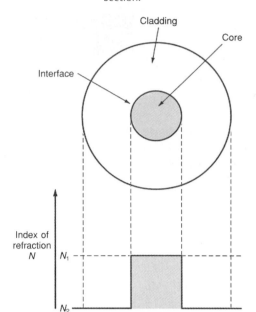

Figure 19-16 Graded index cable cross section.

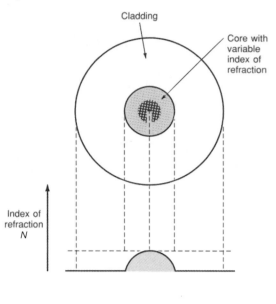

there is a distinct step (see Fig. 19-15). If you were to plot a curve showing the index of refraction as it varies vertically from left to right across the cross section of the cable, you would see a sharp increase in the index of refraction as the core is encountered and then a sharp decline in the index of refraction as the cladding is encountered.

Graded index cable

Graded Index Cable. The other type of cable has a *graded index*. Here, the index of refraction of the core is not constant. Instead, it varies smoothly and continuously over the diameter of the core (see Fig. 19-16). As you get closer to the center of the core, the index of refraction gradually increases, reaching a peak at the center and then declining as the other outer edge of the core is reached. The index of refraction of the cladding is constant.

Cable Mode. *Mode* refers to the number of paths for the light rays in the cable. There are two classifications: single mode and multimode. In *single mode,* light follows a single path through the core; in *multimode,* the light takes many paths.

Multimode

Each type of fiber-optic cable uses one of these methods of rating the index or mode. In practice, there are three commonly used types of fiber-optic cable: *multimode step index, single-mode step index,* and *multimode graded index.*

Multimode step index cable

Multimode Step Index Cable. The *multimode step index fiber cable* is probably the most common and widely used type. It is also the easiest to make and therefore the least expensive. It is widely used for short to medium distances at relatively low pulse frequencies.

The main advantage of a multimode stepped index fiber is its large size. Typical core diameters are in the 50- to 1000-μm range. Such large-diameter cores are excellent at gathering light and transmitting it efficiently. This means that an inexpensive light source such as an LED can be used to produce the light pulses. The light takes many hundreds or even thousands of paths through the core before exiting (refer to Fig. 19-11). Because of the different lengths of these paths, some of the light rays take longer to reach the other end of the cable than do others. The problem with this is that it stretches the light pulses.

Figure 19-17   A multimode step index cable.

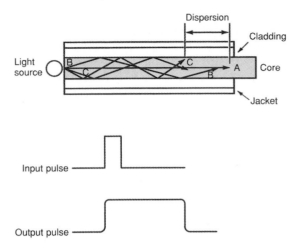

For example, in Fig. 19-17, a short light pulse is applied to the end of the cable by the source. Light rays from the source travel in multiple paths. At the end of the cable, the rays that travel the shortest distance reach the end first. Other rays begin to reach the end of the cable later, until the light ray with the longest path finally reaches the end, concluding the pulse. In Fig. 19-17, ray *A* reaches the end first, then *B*, and then *C*. The result is a pulse at the other end of the cable that is lower in amplitude because of the attenuation of the light in the cable and increased in duration because of the different arrival times of the various light rays. This stretching of the pulse is referred to as *modal dispersion*.

Modal dispersion

Because the pulse has been stretched, pulses at the input cannot occur at a rate faster than the output pulse duration permits. Otherwise, the pulses will essentially merge (see Fig. 19-18). At the output, one long pulse will occur and will be indistinguishable from the three separate pulses originally transmitted. This means that incorrect information will be received. The only cure for this problem is to reduce the pulse repetition rate or the frequency of the pulses. When this is done, proper operation occurs. But with pulses at a lower frequency, less information can be handled. The key point here is that the type of cable selected must have a modal dispersion sufficiently low to handle the desired upper frequency of operation.

Another type of dispersion is chromatic dispersion, which also causes pulse stretching. *Chromatic dispersion* occurs when multiple wavelengths of light are used, as in *dense wavelength-division multiplexing (DWDM)* systems. Since higher light frequencies travel faster than lower light frequencies, pulse stretching occurs. This type of dispersion is the most troubling at data rates above 10 Gbps. Special fiber sections or filters are used to correct it.

Chromatic dispersion

Dense wavelength-division multiplexing (DWDM)

One more type of pulse stretching is caused by *polarization mode dispersion (PMD)*. This is a phenomenon that occurs in *single-mode fiber (SMF)*. SMF essentially supports two orthogonal (at a 90° angle) polarizations along the cable. Since the cable is not a

Single-mode fiber (SMF)

Figure 19-18   The effect of modal dispersion on pulses occurring too rapidly in a multimode step index cable.

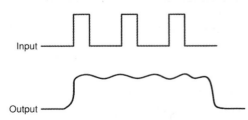

perfect cylinder and because of cable distortions caused by bending, twisting, and other stresses, pulses with different polarization orientations can travel at different velocities in different parts of the cable. The orientations of connectors and splices also introduce this effect and distort the signal. The most common effect is two pulses produced for each transmitted pulse. PMD is generally measured in picoseconds, so it does not materially affect pulses at rates under about 5 Gbps. At 10 Gbps and above, PMD becomes noticeable especially in long-haul WANs and MANs. As with chromatic dispersion, special cable and physical compensators are available to correct it.

One of the newer and better ways to deal with dispersion is to use *electronic dispersion compensation (EDC)*. EDC uses equalization techniques to adjust the received waveform to compensate for any type of dispersion. Equalizers can be made to correct the frequency response of a circuit, but also can be designed to correct for phase and amplitude differences. Most equalizers are fixed circuits and so only correct for problems of one type. However, adaptive equalizers adjust the correction for different speeds and distances. Adaptive equalizers adjust themselves to the fiber link into which they are connected after a short "training" period provides them with information about the type and degree of distortion produced. The equalizer then adds or subtracts energy to or from the signal in the preceding two or more pulse intervals to correct for the dispersion.

Finite impulse response (FIR)

An electronic dispersion compensation device is usually deployed at the serial output of the receiver. The equalizer is essentially a *finite impulse response (FIR)* DSP filter. The equalizer uses feedforward and feedback taps whose coefficients can be automatically adjusted. The received signal is sampled by a very fast A/D converter, and the samples are processed in a DSP chip. The DSP chip computes coefficients that are actively applied to the equalizer. The equalizer in effect performs the inverse function of the dispersive cable, thereby correcting the problem. Most modern 10-Gbps optical transceivers employ EDC. It essentially doubles the range that a given optical transceiver can cover.

Single-mode step index cable

Single–Mode Step Index Cable. A *single-mode* or *monomode step index fiber cable* essentially eliminates modal dispersion by making the core so small that the total number of modes or paths through the core is minimized (see Fig. 19-19). Typical core sizes are 2 to 15 μm. The only path through the core is down the center. With minimum refraction, little pulse stretching occurs. The output pulse has essentially the same duration as the input pulse.

Single-mode step index fibers are by far the best since the pulse repetition rate can be high and the maximum amount of information can be carried. For very long-distance transmission and maximum information content, single-mode step index fiber cables should be used.

The main problem with this type of cable is that it is extremely small, difficult to make, and therefore very expensive. It is also more difficult to handle. Splicing and making interconnections are more difficult. Finally, for proper operation, an expensive, super-intense light source such as a laser must be used. For long distances, however, this is the type of cable preferred.

Multimode graded index cable

Multimode Graded Index Cable. *Multimode graded index fiber cables* have several modes, or paths, of transmission through the cable, but they are much more orderly and predictable. Figure 19-20 shows the typical paths of the light beams. Because of the continuously varying index of refraction across the core, the light rays are bent smoothly and repeatedly converge at points along the cable. The light rays near the edge of the core take a longer path but travel faster because the index of refraction is lower. All the modes or light paths tend to arrive at one point simultaneously. The result is less modal dispersion. As a result, this cable can be used at very high pulse rates, and therefore a considerable amount of information can be carried. This type of cable is also much wider in diameter, with core sizes in the 50- to 100- μm range. Therefore, it is easier to splice and interconnect, and cheaper, less intense light sources can be used.

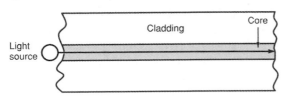

Figure 19-19 Single-mode step index cable.

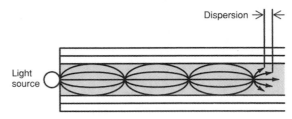

Figure 19-20 A multimode graded index cable.

Fiber-Optic Cable Specifications

The most important specifications of a fiber-optic cable are size, attenuation, and bandwidth. The NA is also sometimes given, although this specification is needed only when connectors are being designed—a rare occurrence, because standard connectors are available.

Cable Size. Fiber-optic cable comes in a variety of sizes and configurations as previously indicated. Size is normally specified as the diameter of the core, and cladding is given in micrometers (μm), where 1 micrometer is one-millionth of a meter. For example, a common size for multimode fiber is 62.5/125, where 62.5 is the diameter of the core and 125 is the diameter of the cladding, both in micrometers (see Fig. 19-21). A PVC or polyurethane plastic jacket over the outer cladding gives a total outside diameter of about 3 mm or 0.118 in. Other common cable sizes for multimode fiber are 50/125 and 100/140, although the 62.5/125 is by far the most widely used. Common sizes for single-mode fibers are 9/125 or 8.3/125 μm.

    Cables come in two common varieties, simplex and duplex. Simplex cable, as the name implies, is just a single-fiber core cable, as shown in Fig. 19-21. In a common duplex cable, as shown in Fig. 19-22, two cables are combined within a single outer cladding.

Attenuation. The most important specification of a fiber-optic cable is its attenuation. *Attenuation* refers to the loss of light energy as the light pulse travels from one end of the cable to the other. The light pulse of a specific amplitude or brilliance is applied to one end of the cable, but the light pulse output at the other end of the cable will be much lower in amplitude. The intensity of the light at the output is lower because of various losses in the cable. The main reason for the loss in light intensity over the length of the cable is light absorption, scattering, and dispersion.

Cable size

Attenuation

Figure 19-21 Fiber-optic cable dimensions.

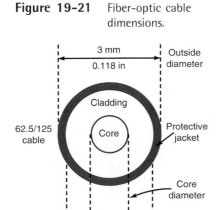

Figure 19-22 Cross section of a duplex cable.

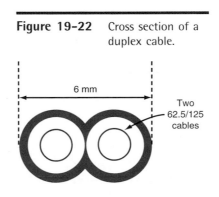

Absorption refers to how light energy is converted to heat in the core material because of the impurity of the glass or plastic. This phenomenon is similar to electrical resistance. *Scattering* refers to the light lost due to light waves entering at the wrong angle and being lost in the cladding because of refraction. *Dispersion,* as mentioned, refers to the pulse stretching caused by the many different paths through the cable. Although no light is lost as such in dispersion, the output is still lower in amplitude than the input, although the length of the light pulse has increased in duration.

The amount of attenuation varies with the type of cable and its size. Glass has less attenuation than plastic. Wider cores have less attenuation than narrower cores of the same material. Wider cores have less absorption and much more dispersion. Wider cores are also usually plastic, which has greater absorption capacity than glass.

More important, attenuation is directly proportional to the length of the cable. It is obvious that the longer the distance the light has to travel, the greater the loss due to absorption, scattering, and dispersion.

The attenuation of a fiber-optic cable is expressed in decibels (dB) per unit of length. The standard is decibels per kilometer. The standard decibel formula is used

$$dB = 10 \log \frac{P_{out}}{P_{in}}$$

where P_{out} is the power out and P_{in} is the power in. Because light intensity is a type of electromagnetic radiation, it is normally expressed and measured in power units, watts, or some fraction thereof. Figure 19-23 shows the percentage of output power for various decibel losses.

For example, 3 dB represents half power. In other words, a 3-dB loss means that only 50 percent of the input appears at the output. The other 50 percent of the power is lost in the cable. The higher the decibel figure, the greater the attenuation and loss. A 30-dB loss means that only one-thousandth of the input power appears at the end. The standard specification for fiber-optic cable is attenuation expressed in terms of decibels per kilometer (dB/km).

The attenuation ratings of fiber-optic cables vary over a considerable range. The finest single-mode step index cables have an attenuation of only 1 dB/km. However, very large core plastic fiber cables can have an attenuation of several thousand decibels per kilometer. A typical 62.5/125 cable has a loss in the 3- to 5-dB/km range. Typically, fibers with an attenuation of less than 10 dB/km are called *low-loss fibers,* and those with an attenuation between 10 and 100 dB/km are *medium-loss fibers. High-loss fibers*

Figure 19-23   Decibel (dB) loss table.

Loss, dB	Power Output, %
1	79
2	63
3	50
4	40
5	31
6	25
7	20
8	14
9	12
10	10
20	1
30	0.1
40	0.01
50	0.001

An optometer for measuring the power output of laser diodes, LEDs, and fiber-optic cables over the range 20 to 60 dBm.

have over 100-dB/km ratings. Naturally, the smaller the decibel number, the less the attenuation and the better the cable.

The total attenuation for a particular cable can be determined from the attenuation rating of the cable. For example, if a cable has an attenuation of 3.75 dB/km, a 5-km-long cable has a total attenuation of 3.75 × 15, or 56.25 dB. If two cables are spliced together and one has an attenuation of 17 dB and the other 24 dB, the total attenuation is simply the sum, or 17 + 24, or 41 dB.

Bandwidth. The *bandwidth* of a fiber-optic cable determines the maximum speed of the data pulses the cable can handle. The bandwidth is normally stated in terms of megahertz-kilometers (MHz·km). A common 62.5/125-μm cable has a bandwidth in the 100- to 300-MHz·km range. Cables with 500 and 600 MHz·km are also common. Even higher-bandwidth cables are available to carry gigahertz-range signals.

As the length of the cable is increased, the bandwidth decreases in proportion. If a 160-MHz·km cable length is doubled from 1 to 2 km, its bandwidth is halved to 80 MHz·km.

Bandwidth

Example 19–2

A fiber-optic cable has a bandwidth rating of 600 MHz·km. What is the bandwidth of a 500-ft segment of cable?

$$1 \text{ km} = 0.62 \text{ mi}$$
$$1 \text{ mi} = 5280 \text{ ft}$$
$$1 \text{ km} = 0.62(5280) = 3274 \text{ ft}$$
$$500 \text{ ft} = \frac{500}{3274} = 0.153 \text{ km}$$
$$600 \text{ MHz·km} = 600 \text{ MHz } (1 \text{ km})$$
$$600 \text{ MHz·km} = x_{\text{MHz}}(0.153 \text{ km})$$

$$x_{\text{MHz}} = \frac{600 \text{ MHz·km}}{0.153 \text{ km}} = 3928.3 \text{ MHz, or } 3.93 \text{ GHz}$$

GOOD TO KNOW

For calculating bandwidths of mile segments of fiber-optic cable when the bandwidth rating is given in units of megahertz-kilometers, the mile values must first be converted to kilometers.

Frequency Range. Most fiber-optic cable operates over a relatively wide light *frequency range*, although it is normally optimized for a narrow range of light frequencies. The most commonly used light frequencies are 850, 1310, and 1550 nm (or 0.85, 1.31, and 1.55 μm). The cable has minimum attenuation to these frequencies.

Figure 19-24 shows attenuation versus wavelength for a typical cable. Note that there is a peak in the attenuation at approximately 1.4 μm. This is caused by hydroxyl ions or undesired hydrogen oxygen ions produced during the manufacturing process. These so-called water losses are avoided by carefully selecting the frequency of light transmission. Minimal loss occurs at 1.3 and 1.55 μm. Most LED or laser light sources operate at one of these wavelengths.

Frequency range

Connectors and Splicing

When long fiber-optic cables are needed, two or more cables can be spliced together. The ends of the cable are perfectly aligned and then fused together with heat. A variety of connectors are available that provide a convenient way to splice cables and attach them to transmitters, receivers, and repeaters.

Optical Communication

767

Figure 19-24 Attenuation versus wavelength of a typical fiber-optic cable.

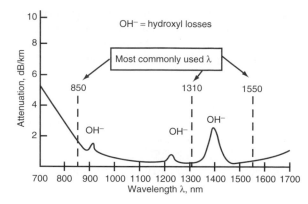

Connectors. *Connectors* are special mechanical assemblies that allow fiber-optic cables to be connected to one another. Fiber-optic connectors are the optical equivalent of electrical plugs and sockets. They are mechanical assemblies that hold the ends of a cable and cause them to be accurately aligned with the ends of another cable. Most fiber-optic connectors either snap or twist together or have threads that allow the two pieces to be screwed together.

Connectors ensure precise alignment of the cables. The ends of the cables must be aligned with precision so that maximum light from one cable is transferred to another. A poor splice or connection will introduce excessive attenuation as the light exits one cable and enters the other. Figure 19-25 shows several ways that cores can be misaligned. A connector can correct these problems.

A typical fiber-optic connector is shown in Fig. 19-26(*a*). One end of the connector, called the *ferrule,* holds the fiber securely in place. A matching fitting holds the other fiber securely in place. When the two are screwed together, the ends of the fibers touch, thereby establishing a low-loss coupling. Figure 19-26(*b*) shows in greater detail how the connector aligns the fibers.

Dozens of different kinds of connectors are available for different applications. They are used in the parts of the system where it may be desirable to occasionally disconnect the fiber-optic cable for making tests or repairs. Connectors are normally used at the end of the cable applied to the light source or the end of the cable connected to the photodetector.

Connectors are also used at the repeater units where the light is picked up, converted to an electrical pulse, amplified and reshaped, and then used to create a new pulse to continue the transmission over a long line. Connectors are used on the back of interface adapters that plug into computers.

The two most common connector designations are ST and SMA. The *ST connectors,* also referred to as *bayonet connectors,* use a half-twist cam-type arrangement like that used in BNC coaxial connectors. They are convenient for quick connect and disconnect. The *SMA connectors* are about the same size but have threaded connections.

Figure 19-25 Misalignment and rough end surfaces cause loss of light and high attenuation. (*a*) Too much end separation. (*b*) Axial misalignment. (*c*) Angular misalignment. (*d*) Rough, uneven surfaces.

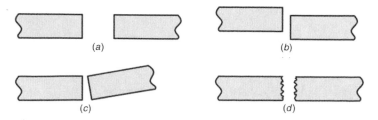

Figure 19-26 Details of a fiber cable connector.

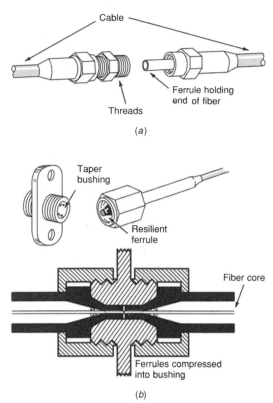

Splicing. *Splicing* fiber-optic cable means permanently attaching the end of one cable to another. This is usually done without a connector. The first step is to cut the cable, called *cleaving* the cable, so that it is perfectly square on the end. Cleaving is so important to minimizing light loss that special tools have been developed to ensure perfect cuts.

The two cables to be spliced are then permanently bonded together by heating them instantaneously to high temperatures so that they fuse or melt together. Special tools and splicing machines must be used to ensure perfect alignment.

Installing a connector begins with cleaving the fiber so that it is perfectly square. Polishing usually follows. Again, the special cleaving and polishing machines devised for this purpose must always be used. Poorly spliced cable or poorly installed connectors create an enormous loss.

19-4 Optical Transmitters and Receivers

In an optical communication system, transmission begins with the *transmitter*, which consists of a carrier generator and a modulator. The carrier is a light beam that is usually modulated by turning it on and off with digital pulses. The basic transmitter is essentially a light source.

The *receiver* is a light or photodetector that converts the received light back to an electric signal. In this section the types of light sources used in fiber-optic systems and the transmitter circuitry, as well as the various light detectors and the related receiver circuits, are discussed.

Light Sources

Conventional *light sources* such as incandescent lamps cannot be used in fiber-optic systems because they are too slow. To transmit high-speed digital pulses, a very fast light

Margin notes: Splicing · Cleaving · Transmitter · Receiver · Light sources

source must be used. The two most commonly used light sources are *light-emitting diodes (LEDs)* and *semiconductor lasers.*

Light–Emitting Diodes. A *light-emitting diode (LED)* is a PN-junction semiconductor device that emits light when forward-biased. When a free electron encounters a hole in the semiconductor structure, the two combine, and in the process they give up energy in the form of light. Semiconductors such as gallium arsenide (GaAs) are superior to silicon in light emission. Most LEDs are GaAs devices optimized for producing red light. LEDs are widely used for displays indicating whether a circuit is off or on, or for displaying decimal and binary data. However, because an LED is a fast semiconductor device, it can be turned off and on very quickly and is capable of transmitting the narrow light pulses required in a digital fiber-optics system.

LEDs can be designed to emit virtually any color light desired. The LEDs used for fiber-optic transmission are usually in the red and near-infrared ranges. Typical wavelengths of LED light commonly used are 0.85, 1.31, and 1.55 μm, more commonly designated 850, 1310, and 1550 nm where 1 micrometer (μm) equals 1000 nm. These frequencies are all in the near-infrared range just below red light, which is not visible to the naked eye. These frequencies have been chosen primarily because most fiber-optic cables have the lowest losses in these frequency ranges.

One physical arrangement of the LED is shown in Fig. 19-27(*a*). A P-type material is diffused into the N-type substrate, creating a diode. Radiation occurs from the P-type material and around the junction. Figure 19-27(*b*) shows a common light radiation pattern.

The light output from an LED is expressed in terms of power. Typical light output levels are in the 10- to 50-μW range. Sometimes the light output is expressed in dBm or dB referenced of 1 mW (milliwatt). Common levels are -15 to -30 dBm. Forward-bias current levels to achieve this power level are in the 50- to 200-mA range. High-output LEDs with output ratings in the 600- to 2500-μW range are also available.

A typical LED used for lighting is relatively slow to turn on and off. A typical turn-on/turn-off time is about 150 ns. This is too slow for most data communication applications by fiber optics. Faster LEDs capable of data rates up to 50 MHz are available. For faster data rates, a laser diode must be used.

Special LEDs are made just for fiber-optic applications. These units are made of *gallium arsenide* or *indium phosphide (GaAs or InP)* and emit light at 1.3 μm. Other LEDs with as many as six multiple layers of semiconductor material are used to optimize the device for a particular frequency and light output. Figure 19-28 shows some LED assemblies made to accept ST bayonet and SMA fiber-optic cable connectors. These are made for PC board mounting.

Laser Diodes. The other commonly used light transmitter is a *laser,* which is a light source that emits coherent monochromatic light. *Monochromatic light* is a pure single-frequency light. Although an LED emits red light, that light covers a narrow spectrum

Figure 19-27 (*a*) Typical LED construction. (*b*) Light radiation pattern.

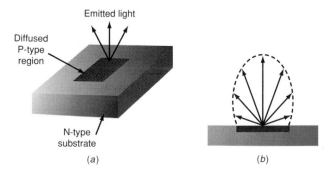

Figure 19-28   LED assemblies with connectors.

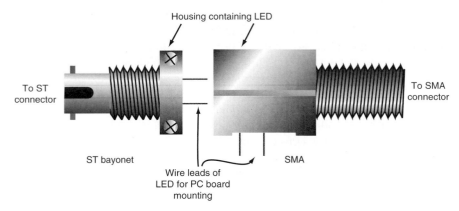

around the red frequencies. *Coherent* refers to the fact that all the light waves emitted are in phase with one another. Coherence produces a focusing effect on a beam so that it is narrow and, as a result, extremely intense. The effect is somewhat similar to that of using a highly directional antenna to focus radio waves into a narrow beam that also increases the intensity of the signal.

Coherent

The most widely used light source in fiber-optic systems is the *injection laser diode (ILD)*, also known as a *Fabry-Perot (FP) laser*. Like the LED, it is a PN junction diode usually made of GaAs. See Fig. 19-29. At some current level, it emits a brilliant light. The physical structure of the ILD is such that the semiconductor structure is cut squarely at the ends to form internal reflecting surfaces. One of the surfaces is usually coated with a reflecting material such as gold. The other surface is only partially reflective. When the diode is properly biased, the light is emitted and bounces back and forth internally between the reflecting surfaces. The distance between the reflecting surfaces has been carefully measured so that it is some multiple of a half wave at the light frequency. The bouncing back and forth of the light waves causes their intensity to reinforce and build up. The structure is like a cavity resonator for light. The result is an incredibly high-brilliance, single-frequency light beam that is emitted from the partially reflecting surface.

Injection laser diode (ILD)

Fabry-Perot (FP) laser

Injection laser diodes are capable of developing light power up to several watts. They are far more powerful than LEDs and, therefore, are capable of transmitting over much longer distances. Another advantage ILDs have over LEDs is their ability to turn off and on at a faster rate. High-speed laser diodes are capable of gigabit per second digital data rates.

Although most of the light emitted by an FP ILD is at a single frequency, light is also emitted at frequencies from slightly below to slightly above the main light frequency.

Figure 19-29   A Fabry-Perot injection laser diode.

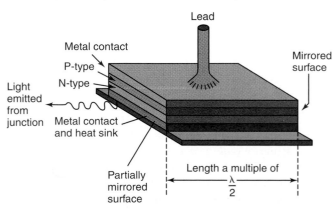

As a result, the light occupies a narrow spectrum on the fiber. In systems using multiple light wavelengths, FP lasers can interfere with one another. To overcome this problem, a *distributed feedback (DFB) laser* can be used. This laser is made with a cavity that contains an integrated grating structure that acts as a selective filter. The output from a DFB laser has a much narrower bandwidth and is nearer to a single light wavelength than any other type of laser. FP and DFB lasers are the highest-power lasers made and are used primarily in long-distance fiber transmission and in metropolitan-area networks.

Another type of laser used in fiber-optic systems is the *vertical cavity surface-emitting laser (VCSEL)*. Instead of emitting light from the edge of the diode, the VCSEL is made on the surface of silicon wafer–like transistors and integrated circuits. The cavity is made vertical to the wafer surface so that light is emitted from the surface. VCSELs are very easy to make and low in cost. Furthermore, they can be made in arrays on the wafer surface. The output power emitted by a VCSEL is greater than that of a LED but less than that of an FB or a DFB laser. Its bandwidth is less than that of a LED but wider than that of either the FB or the DFB laser. This makes VCSELs ideal for short-distance LANs or MANs. Most VCSELs are 850 nm, but recently 1310-nm VCSELs became available; 1550-nm VCSELs are not yet available.

The most popular laser frequencies are 850, 1310, and 1550 nm. Some lasers are also made to operated at 980 and 1490 nm. The 850- and 1310-nm lasers are used mainly in LANs and in some MANs. The 1550-nm lasers are used primarily in long-distance fiber-optic systems since the fiber attenuation is less near 1550 nm than it is in the 850- and 1310-nm bands.

A recent development is a *tunable laser* whose frequency can be varied by changing the dc bias on the device or mechanically adjusting an external cavity. A wavelength range of more than 100 nm above and below a center wavelength is possible. Wider-range tunable lasers are being developed. Tunable lasers are used in applications where multiple frequencies are needed, as in dense wavelength-division multiplexing (DWDM). Instead of having to inventory many expensive lasers, each on a separate wavelength, a single laser or at most several lasers are adequate to cover multiple frequencies.

The lasers are modulated by the data in two ways, direct and indirect. In direct modulation, the data turns the laser off and on to form the binary pulses. Actually the laser is never turned off completely but is kept biased on at a low light level because it takes too much time to turn the laser on. This limits the speed of operation. To generate a maximum light pulse, the bias is then increased to its peak value by a binary 1 level. This method works well up to about several gigabits per second, but for rates beyond this, an indirect modulator is used.

An indirect modulator is a semiconductor device physically attached to a laser that is fully on at all times. The modulator, sometimes called an *electroabsorption modulator (EAM)*, acts as an electrically activated shutter that is turned off and on by the data. Data rates from 10 to more than 40 Gbps can be achieved with an EAM. Sometimes the EAM is packaged with the laser, but it is also available as a separate component.

Lasers dissipate a tremendous amount of heat and, therefore, must be connected to a heat sink for proper operation. Because their operation is heat-sensitive, most lasers are used in a circuit that provides some feedback for temperature control. This not only protects the laser but also ensures proper light intensity and frequency. Many lasers use a thermoelectric cooler based upon the Peltier effect.

Light Transmitters

LED Transmitter. A *light transmitter* consists of the LED and its associated driving circuitry. A typical circuit is shown in Fig. 19-30. The binary data pulses are applied to a logic gate which, in turn, operates a transistor switch Q_1 that turns the LED off and on. A positive pulse at the NAND gate input causes the NAND output to go to zero. This turns off Q_1, so the LED is then forward-biased through R_2 and turns on. With zero input, the NAND output is high, so that Q_1 turns on and shunts current away from the LED. Very high current pulses are used to ensure very bright light. High intensity is

Figure 19-30 Optical transmitter circuit using an LED.

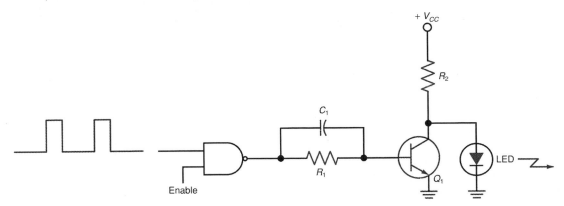

required if data is to be transmitted reliably over long distances. Most LEDs are capable of generating power levels up to approximately several thousand microwatts. With such low intensity, LED transmitters are good for only short distances. Further, the speed of the LED is limited. Turn-off and turn-on times are no faster than tens of nanoseconds, and so transmission rates are limited. Most LED-like transmitters are used for short-distance, low-speed, digital fiber-optic systems.

Laser Transmitter. A typical laser driver circuit is shown in Fig. 19-31. Most of the circuitry is contained in a single integrated circuit designated the VSC7940 and made by Vitesse Semiconductor Corporation. It operates from 3.3 or 5 V and contains special automatic power control (APC) circuitry that maintains a constant laser output. This circuit can operate at data rates up to 3.125 Gbps.

The input data is in differential form, as are most logic signals above several hundred megahertz. A multiplexer is used to either pass the data directly to the laser driver

Figure 19-31 A typical laser driver circuit. *(Courtesy Vitesse Semiconductor Corp.)*

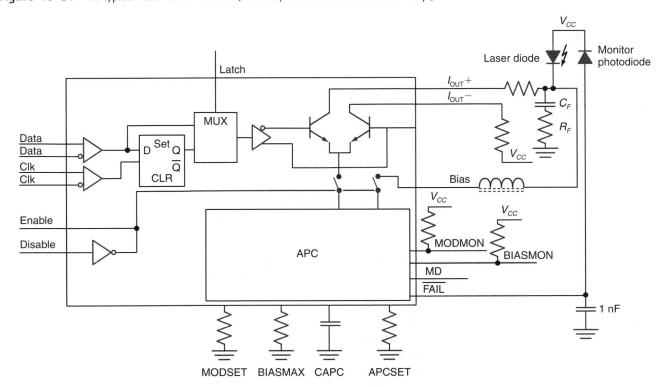

Figure 19-32 Data formats used in fiber-optic transmission. (*a*) NRZ (nonreturn-to-zero). (*b*) RZ (return-to-zero). (*c*) Biphase (Manchester).

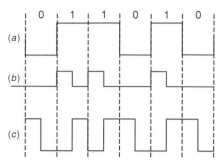

transistors or select data that is clocked via a flip-flop and an external differential clock signal. Enable/disable signals are used to turn the laser off or on as desired.

The laser diode is connected to the driver by means of several resistors and a capacitor that set the current and switching response. Most laser packages also contain a photodiode that is used to monitor the laser light output and provide feedback to the APC circuit in the chip. If laser output varies because of temperature variations or decreases because of lifetime variations, the APC automatically corrects this, ensuring a constant brightness. The output driver transistors can switch up to 100 mA and have typical rise and fall times of 60 ps. External resistors and capacitors set the laser bias, modulation bias level, average current, and control loop response time.

Data format

Data Formats. Digital data is formatted in a number of ways in fiber-optic systems. To transmit information by fiber-optic cable, data is usually converted to a serial digital data stream. A common NRZ serial format is shown in Fig. 19-32(*a*). The NRZ format at *A* was discussed in Chap. 11. Each bit occupies a separate time slot and is either a binary 1 or binary 0 during that time period.

In the RZ format of Fig. 19-32(*b*), the same time period is allotted for each bit, but each bit is transmitted as a very narrow pulse (usually 50 percent of the bit time) or as an absence of a pulse. In some systems the Manchester or biphase code is used, as shown in Fig. 19-32(*c*). Although the NRZ code can be used, the RZ and Manchester codes are easier to detect reliably at the higher data rates and are thus preferred.

Light Detectors

The receiver part of the optical communication system is relatively simple. It consists of a detector that senses the light pulses and converts them to an electric signal. This signal is amplified and shaped into the original serial digital data. The most critical component is the light sensor.

Photodiode

Photodiode. The most widely used light sensor is a *photodiode*. It is a silicon PN-junction diode that is sensitive to light. This diode is normally reverse-biased, as shown in Fig. 19-33. The only current that flows through it is an extremely small reverse leakage current. When light strikes the diode, this leakage current increases significantly. This current flows through a resistor and develops a voltage drop across it. The result is an output voltage pulse.

Phototransistor

Phototransistor. The reverse current in a diode is extremely small even when exposed to light. The resulting voltage pulse is very small and so must be amplified. The base-collector junction is exposed to light. The base leakage current produced causes a larger emitter-to-collector current to flow. Thus the transistor amplifies the small leakage

Figure 19-33 The conversion of light, by a photodiode, to voltage pulses.

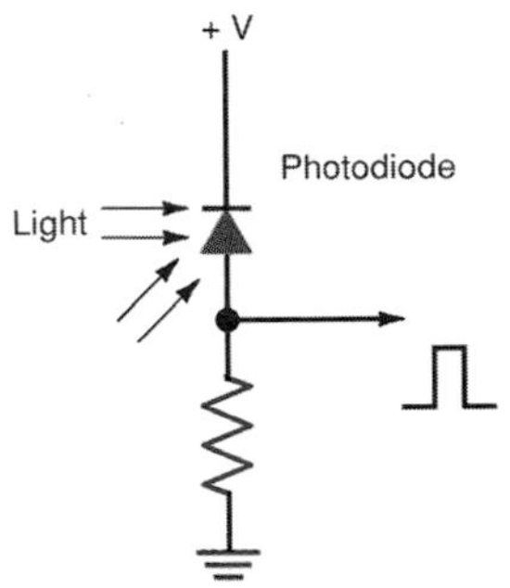

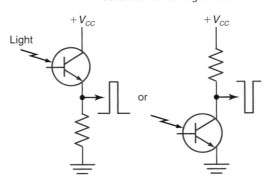

Figure 19-34 A phototransistor is more sensitive to the light level.

current into a larger, more useful output (see Fig. 19-34). Phototransistor circuits are far more sensitive to small light levels, but they are relatively slow. Thus further amplification and pulse shaping are normally used.

PIN Diode. The sensitivity of a standard PN-junction photodiode can be increased and the response time decreased by creating a new device that adds an undoped or *intrinsic (I) layer* between the P and N semiconductors. The result is a *PIN diode* (Fig. 19-35). The thin P layer is exposed to the light which penetrates to the junction, causing electron flow proportional to the amount of light. The diode is reverse-biased, and the current is very low until light strikes the diode, which significantly increases the current.

PIN diodes are significantly faster in response to rapid light pulses of high frequency. And their light sensitivity is far greater than that of an ordinary photodiode.

PIN diode

Avalanche Diode. The *avalanche photodiode (APD)* is a more widely used photosensor. It is the fastest and most sensitive photodiode available, but it is expensive and its circuitry is complex. Like the standard photodiode, the APD is reverse-biased. However, the operation is different. The APD uses the reverse breakdown mode of operation that is commonly found in zener and IMPATT microwave diodes. When a sufficient amount of reverse voltage is applied, an extremely high current flows because of the avalanche effect. Normally, several hundred volts of reverse bias, just below the avalanche threshold, are applied. When light strikes the junction, breakdown occurs and a large current flows. This high reverse current requires less amplification than the small current in a standard photodiode. Germanium APDs are also significantly faster than the other photodiodes and are capable of handling the very high gigabit-per-second data rates possible in some systems.

Avalanche photodiode (APD)

GOOD TO KNOW

The avalanche photodiode is the fastest and most sensitive photodiode available.

Figure 19-35 Structure of a PIN photodiode.

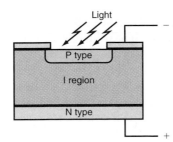

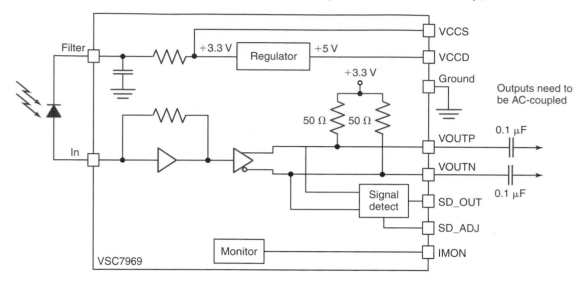

Light Receivers

Light receiver

Figure 19-36 shows a representative *light receiver* circuit. This integrated circuit, the VSC7969 by Vitesse Semiconductor Corporation, uses an external PIN or APD photodiode and can operate at rates to 3.125 Gbps. The input stage, generally known as a *transimpedance amplifier (TIA)*, converts the diode current to an output voltage and amplifies it. The following stage is a limiter that shapes up the signal and applies it to a differential driver amplifier. The output is capacitively coupled to the next stage in the system. A signal detect circuit provides a CMOS logic output that indicates the presence of an input signal if the diode current exceeds a specific lower limit. A photodiode current monitor is also provided. The circuit operates from either 3.3 or 5 V dc, and an onboard regulator operates the circuitry and provides bias to the photodiode.

Transimpedance amplifier (TIA)

Optical Transceivers

Optical transceivers or transponders are assemblies called optical modules into which both the light transmitter and light receiver are packaged together to form a single module. See Fig. 19-37. These modules form the interface between the optical transmission medium and the electrical interface to the computer or other networking equipment.

These modules are made up of the *transmit optical subassembly (TOSA)* and the *receive optical subassembly (ROSA)*. Each is provided with an optical connector to get the signals into and out of the unit. These subassemblies connect to the interface circuits that supply the transmit signals and receive the input signals. The entire unit is housed in a metal enclosure suitable for mounting on a printed-circuit board of a router line card or other interface circuit.

Figure 19-38 shows a block diagram of a typical transceiver module. The optical fiber cable connections are on the right while the electrical connections to the networking equipment are on the left. The optical input fiber connects to a PIN diode or APD IR detector and transimpedance amplifier (TIA). The photodiode and TIA are a single package called the receive optical subassembly. This is followed by additional amplification in a *postamplifier (PA)*. If electronic dispersion compensation (EDC) is used, it appears here in the signal flow. The EDC output then connects to a *clock and data recovery (CDR)* unit.

Figure 19-37 A standardized optical transponder or transceiver.

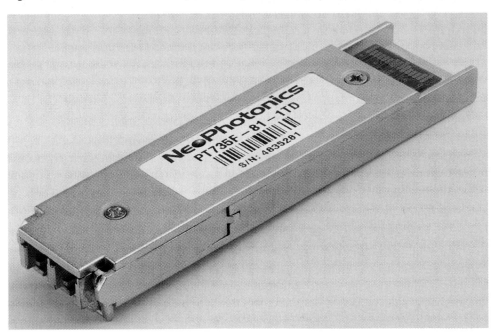

Figure 19-38 Block diagram of one type of MSA optical transceiver module.

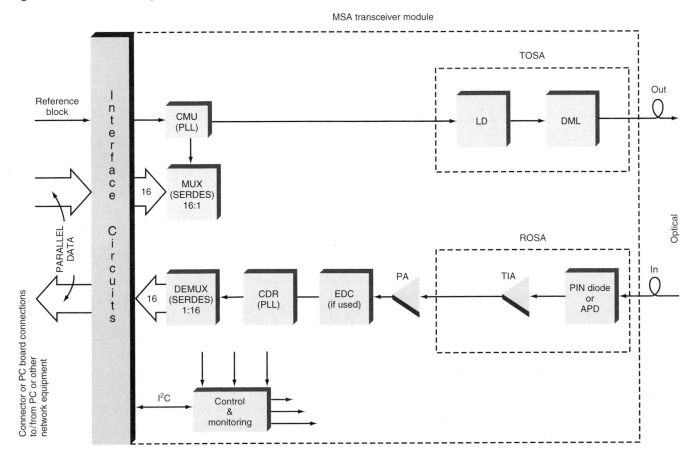

The CDR is a phase-locked loop (PLL) that extracts the clock signal from the incoming data. The recovered clock signal is then used to time the serial input data and related operations. The PLL voltage-controlled oscillator operates at the clock frequency and is locked to the incoming data frequency. The PLL serves as a filter and signal regeneration circuit to produce a clean clock signal from the degraded received signal.

A serializer/deserializer (SERDES) circuit converts the serial data to parallel and demultiplexes the individual data words and sends them to the interface, where they connect to a host computer or other equipment. The electrical interface circuits are sometimes parallel but may also be serial. A 16-bit word is common.

On the transmit side of the transceiver, the data to be transmitted is usually received from a network device in parallel form via the interface. The parallel data is converted to serial data by a serializer/deserializer circuit that serves as a multiplexer where the parallel words are put into a serial sequence for transmission. The serial data is clocked out of the multiplexer by a clock signal at the desired transmission rate.

A reference clock oscillator, usually external to the transceiver, is multiplied up to the desired clock rate by a *clock multiplier unit (CMU)*. The CMU is a PLL with a frequency divider in the feedback path used as a frequency multiplier. The serial data is then sent to the *laser driver (LD)*, where it operates the *directly modulated laser (DML)* diode. Direct modulation simply means that the data turns the laser off and on to transmit. In some modules that use higher-power, higher-frequency lasers, the serial data drives an external modulator that in turn interrupts the light path of a continuously operating laser. Note that the laser driver and modulated laser form a unit called the transmit optical subassembly.

Finally, most transceivers include a serial port called the I^2C port that is used to monitor specific conditions in the module (laser temperature, supply volage, etc.) and to control some aspect of the module.

Over the years, a variety of such subassemblies have been developed by different manufacturers. This has led to interconnection and interoperability problems, meaning that units of different manufacturers cannot be used with one another. Lack of standardization meant that no second sources of products were possible. The result is that manufacturers of optical transceivers and network equipment have come together to standardize on optical transceiver sizes, mechanical characteristics, electrical characteristics, and connectors.

A set of standards known as a *multisource agreements (MSAs)* has emerged. Referred to as optical modules, the most common transceiver types are listed below.

300-pin. The most widely used format and standard. Converts between 16-bit 622.08-Mbps electric signals and 10-Gbps optical signals. See Fig. 19-38. For SONET system, 16-bit words are supplied at the 622.08-MHz rate and converted to serial data at a rate 16 times greater, or 9.953 Gbps. This serial rate is usually rounded off and expressed as 10 Gbps. The electrical interface is a standard called SPI-4 developed by the Optical Internetworking Foundation (OIF) or one called XSBI, developed by the IEEE. The data paths are for 16-bit words, and two-wire differential connection are used.

SFF. Small form factor. A module developed for lower-speed optical applications in the 1-, 2-, and 4-Gbps range.

SFP. Small form factor pluggable. A module of 1-, 2-, or 4-Gbps applications but is "pluggable," meaning that it uses fiber-optic cable connectors rather than the short fiber links used with SFF modules. The modules are also "hot pluggable," meaning that they may be put into or taken out of the system with power on. Their circuits are protected from transients and surges that may occur during hot pluggability.

XENPAK. A standard for a 10-Gbps optical system using the standard Ethernet 10-Gbps attachment unit interface (XAUI) electrical interface. (*Note:* The X is a roman numeral for 10). The XAUI interface uses four 3.125-Gbps serial electrical channels to achieve a maximum data rate of $4 \times 3.125 = 12.5$ Gbps. This is the gross serial optical rate in 10-Gbps Ethernet systems. The net rate is 10 Gbps, but because of the 8B/10B error correction feature of this protocol, the additional bits cause the gross rate to be higher by a factor of 10/8: $10 \times 1.25 = 12.5$ Gbps.

X2. This standard is similar to XENPAK but is a smaller and less complex package since it does not have to deal with the high heat dissipation common in XENPAK modules. X2 is used in the shorter-reach applications.

XFP. A 10-Gbps small form factor pluggable. It does not use serial-to-parallel or parallel-to-serial conversion. The electrical interface is a standard 10-Gbps serial interface referred to as XFI. The input and output frequency range is 9.95 to 10.7 Gbps.

XPAK. The newest format is a smaller version of a XENPAK module for reaches to 10 km.

Most of these modules focus on 10-Gbps applications which include SONET, Ethernet, and Fibre Channel. Each has a slightly different data rate depending upon the standard, use of FEC or not, type of FEC, and other factors. However, the modules are essentially "protocol-agnostic," meaning they can handle any standard or protocol. The modules are also generally classified by the range or reach of the optical signals. The basic categories are

Very short reach (VSR)—300 to 600 m or less

Short reach (SR)—2 km

Intermediate reach (IR)—10 to 40 km

Long reach (LR)—40 to 80 km

Very long reach (VLR)—120 km

The VSR and SR modules use 850-nm lasers, some SR and LR modules use 1310-nm lasers. The LR and VLR modules use 1550-nm lasers.

Performance Considerations

The most important specification in a fiber-optic communication system is the *data rate,* i.e., the speed of the optical pulses. The very best systems use high-power injection laser diodes and APD detectors. This combination can produce data rates of several billion (giga) bits per second (bps). The rate is known as a *gigabit rate.* Depending upon the application, data rates can be anywhere from about 20 Mbps to 40 Gbps.

Data rate

The performance of a fiber-optic cable system is usually indicated by the *bit rate–distance product.* This rating tells the fastest bit rate that can be achieved over a 1-km cable. Assume a system with a 100-Mbps·km rating. This is a constant figure and is a product of the megabits per second and the kilometer values. If the distance increases, the bit rate decreases in proportion. In the above system, at 2 km the rate drops to 50 Mbps. At 4 km, the rate is 25 Mbps and so on.

The upper data rate is also limited by the dispersion factor. The rise and fall times of the received pulse are increased by an amount equal to the dispersion value. If the dispersion factor is 10 ns/km, over a 2-km distance the rise and fall times are increased by 20 ns each. The data rate can never be more than the frequency corresponding to the sum of the rise and fall times at the receiver.

A handy formula for determining the maximum data rate R in megabits per second (Mbps) for a given distance D in kilometers of cable with a dispersion factor of d, given in microseconds per kilometer (μs/km), is

$$R = \frac{1}{5dD}$$

Assume a cable length of 8 km and a dispersion factor of 10 ns/km, or 0.01 μs/km:

$$R = \frac{1}{5(0.01)(8)} = \frac{1}{0.4} = 2.5 \text{ Mbps}$$

This relationship is only an approximation, but it is a handy way to predict system limitations.

Example 19-3

A measurement is made on a fiber-optic cable 1200 ft long. Its upper frequency limit is determined to be 43 Mbps. What is the dispersion factor d?

$$1 \text{ km} = 3274 \text{ ft}$$

$$D = 1200 \text{ ft} = \frac{1200}{3274} = 0.367 \text{ km}$$

$$R = \frac{1}{5dD}$$

$$d = \frac{1}{5RD}$$

$$= \frac{1}{5(43 \times 10^6)(0.367)} = 12.7 \text{ ns/km}$$

Power Budget

Power (flux) budget

A *power budget,* sometimes called a *flux budget,* is an accounting of all the attenuation and gains in a fiber-optic system. Gains must be greater than losses for the system to work. A designer must determine whether an adequate amount of light power reaches the receiver. Does the receiver get as much input light as its sensitivity dictates, given the power output of the transmitter and all the cable and other losses?

Losses in fiber-optic cable system

Losses. There are numerous sources of *losses* in a fiber-optic cable system:

1. *Cable losses.* These vary with the type of cable and its length. The range is from less than 1 dB/km up to tens of decibels per kilometer.

2. *Connections between cable and light source and photodetector.* These vary widely depending upon how they are made. Today, the attachment is made by the manufacturer, who specifies a loss factor. The resulting assembly has a connector to be attached to the cable. Such terminations can produce attenuations of 1 to 6 dB.

3. *Connectors.* Despite their precision, connectors still introduce losses. These typically run from 0.5 to 2 dB each.

4. *Splices.* If done properly, the splice may introduce an attenuation of only a few tenths of a decibel. But the amount can be much more, up to several decibels if splicing is done incorrectly.

5. *Cable bends.* If a fiber-optic cable is bent at too great an angle, the light rays will come under a radically changed set of internal conditions. The total internal reflection will no longer be effective, for angles have changed due to the bend. The result is that some of the light will be lost by refraction in the cladding. If bend radii are made 1000 times more than the diameter of the cable, the losses are minimal. Decreasing the bend radius to less than about 100 times the cable diameter will increase the attenuation. A bend radius approaching 100 to 200 times the cable diameter could cause cable breakage or internal damage.

While fiber-optic cable does carry visible light as shown here, in practice, networking technologies use infrared (IR) light which is invisible to the human eye.

All the above losses will vary widely depending upon the hardware used. Check the manufacturer's specifications for every component to be sure that you have the correct information. Then, as a safety factor, add 5 to 10 dB of loss. This contingency factor will cover incidental losses that cannot be predicted.

Calculating the Budget. The losses work against the light generated by the LED or ILD. The idea is to use a light power sufficient to give an amount of received power in excess of the minimum receiver sensitivity with the losses in the system.

Assume a system with the following specifications:

1. Light transmitter LED output power: 30 μW
2. Light receiver sensitivity: 1 μW
3. Cable length: 6 km
4. Cable attenuation: 3 dB/km, $3 \times 6 = 18$ dB total
5. Four connectors: attenuation 0.8 dB each, $4 \times 0.8 = 3.2$ dB total
6. LED-to-connector loss: 2 dB
7. Connector-to-photodetector loss: 2 dB
8. Cable dispersion: 8 ns/km
9. Data rate: 3 Mbps

First, calculate all the losses; add all the decibel loss factors.

$$\text{Total loss, dB} = 18 + 3.2 + 2 + 2 = 25.2 \text{ dB}$$

Also add a 4-dB contingency factor, making the total loss 25.2 + 4, or 29.2, dB. What power gain is needed to overcome this loss?

$$dB = 10 \log \frac{P_t}{P_r}$$

where P_t is the transmitted power and P_r is the received power

$$29.2 \text{ dB} = 10 \log \frac{P_t}{P_r}$$

Therefore,

$$\frac{P_t}{P_r} = 10^{dB/10} = 10^{2.92} = 831.8$$

If P_t is 30 μW, then

$$P_r = \frac{30}{831.8} = 0.036 \text{ μW}$$

Accordingly, the received power will be 0.036 μW. The sensitivity of the receiver is only 1 μW. The received signal is below the threshold of the receiver. This problem may be solved in one of three ways:

1. Increase transmitter power.
2. Get a more sensitive receiver.
3. Add a repeater.

In an initial design, the problem would be solved by increasing the transmitter power and/or increasing receiver sensitivity. Theoretically a lower-loss cable could also be used. Over short distances, a repeater is an unnecessary expense; therefore using a repeater is not a good option.

Assume that the transmitter output power is increased to 1 mW or 1000 μW. The new received power then is

$$P_r = \frac{1000}{831.8} = 1.2 \text{ μW}$$

This is just over the threshold of the receiver sensitivity. Now, we can determine the upper frequency or data rate.

$$R = \frac{1}{5dD} = \frac{1}{5(0.008)(6)} = 4.1666 \text{ Mbps}$$

This is higher than the proposed data rate of 3 Mbps, so the system should work.

Amplification

Regeneration

Optical-electrical-optical (OEO) conversion

Regeneration and Amplification. There are several ways to overcome the attenuation experienced by a signal as it travels over fiber-optic cable. First is to use newer types of cable that inherently have lower losses and fewer dispersion effects. The second method is to use regeneration. *Regeneration* is the process of converting the weak optical signal to its electrical equivalent, then amplifying and reshaping it electronically, and retransmitting it on another laser. This process is generally known as *optical-electrical-optical (OEO) conversion*. It is an expensive process since in most systems regeneration is necessary about every 40 km of distance, even with the newer lower-attenuation cables. This process is especially expensive in multifrequency systems such

Figure 19-39 An erbium-doped fiber amplifier.

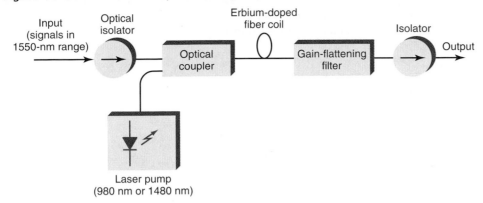

as dense wavelength-division multiplexing (DWDM), in which many signals must undergo OEO conversion.

A third method, and the best, is to use an optical amplifier. Optical amplifiers boost signal level without OEO conversion. A typical optical amplifier is the *erbium-doped fiber amplifier (EDFA)* shown in Fig. 19-39. The weak optical input signal is applied to an optical isolator to prevent signal reflections. This signal is then applied to an optical combiner that linearly mixes the signal to be amplified with one generated by an internal laser diode pump. The laser pump operates at a higher frequency (lower wavelength) than the wavelength of the signals to be amplified. If 1550-nm signals are being amplified, the pump laser operates at 980 or 1480 nm.

The combined signals are then fed into a coiled length of optical fiber that has been heavily doped with erbium ions. (Erbium is one of the rare earth elements.) The laser pump signal excites the erbium atoms to a higher-energy state. When the photons produced by the input signal pass through the erbium fiber, they interact with the excited erbium atoms, causing them to relax to their normal state. During this process, additional photons are released at the same wavelength of the input signals, resulting in amplification. A single EDFA produces a gain in the range of 15 to 20 dB. When two amplifiers with independent pump lasers and doped fiber coils are cascaded, a composite gain of up to 35 dB is possible. With such amplification, the signals can be transmitted over a distance of up to almost 200 km without OEO regeneration.

Erbium-doped fiber amplifier (EDFA)

GOOD TO KNOW

A laser pump uses a coil of erbium-doped fiber to release additional photons at the same frequency as the input photons.

19-5 Wavelength–Division Multiplexing

Data is most easily multiplexed on fiber-optic cable by using time-division multiplexing (TDM), as in the T1 system or in the SONET system described later in this chapter. However, developments in optical components make it possible to use frequency-division multiplexing (FDM) on fiber-optic cable (called *wavelength-division multiplexing*, or *WDM*), which permits multiple channels of data to operate over the cable's light wave bandwidth.

Wavelength-division multiplexing, another name for frequency-division multiplexing, has been widely used in radio, TV, and telephone systems. The best example today is the multiplexing of dozens of TV signals on a common coaxial cable coming into the home.

In WDM, different frequencies or "colors" of infrared light are employed to carry individual data streams. These are combined and carried on a single fiber. Although frequency as a parameter is more widely used to distinguish the location of wireless signals below 300 GHz, at light frequencies the wavelength parameter is the preferred measure.

Wavelength-division multiplexing (WDM)

Remember that the relationship between wavelength λ in meters and frequency f is $f = c/\lambda$, where c is the speed of light in a vacuum or 2.998×10^8 m/s. The speed of light in fiber cable is a bit less than that, or about 2.99×10^8 m/s. Optical wavelength is usually expressed in nanometers or micrometers. Optical frequencies are expressed in terahertz (THz), or 10^{12} Hz.

Data to be transmitted in a fiber-optic network is used to modulate (by OOK or ASK) a laser-generated infrared light. Infrared signals best match the light-carrying characteristics of fiber-optic cable, which has an attenuation response to infrared light such that the lowest attenuation (about 0.2 dB/km) occurs in two narrow bands of frequencies, one centered at 1310 nm and the other at 1550 nm.

Coarse Wavelength–Division Multiplexing

The first *coarse WDM (CWDM) systems* used two channels operating on 1310 and 1550 nm. Later, four channels of data were multiplexed. Figure 19-40 illustrates a CWDM system. A separate serial data source controls each laser. The data source may be a single data source or a multiple TDM source. Current systems use light in the 1550-nm range. A typical four-channel system uses laser wavelengths of 1534, 1543, 1550, and 1557.4 nm. Each laser is switched off and on by the input data. The laser beams are then optically combined and transmitted over a single-fiber cable. At the receiving end of the cable, special optical filters are used to separate the light beams into individual channels. Each light beam is detected with an optical sensor and then filtered into the four data streams.

Dense Wavelength–Division Multiplexing

Dense wavelength-division multiplexing (DWDM) refers to the use of 8, 16, 32, 64, or more data channels on a single fiber. Standard channel wavelengths have been defined by the International Telecommunications Union (ITU) as between 1525 and 1565 nm with a 100-GHz (approximately 0.8-nm) channel spacing.

The block of channels between about 1525 and 1565 nm is called the *C or conventional band*. Most DWDM activity currently occurs in the C band. Another block of wavelengths from 1570 to 1610 nm is referred to as the *long-wavelength band,* or *L band.* Wavelengths in the 1525- to 1538-nm range make up the *S band.*

Current DWDM systems allow more than 160 individual data channels to be carried simultaneously on a single fiber at data rates up to 40 Gbps, giving an overall capacity of 160×40, or 6400, Gbps (6.4 Tbps). The potential for future systems is over 200 channels per fiber at a data rate of 40 Gbps. Even more channels can be transmitted on a

Figure 19-40 A CWDM fiber-optic system.

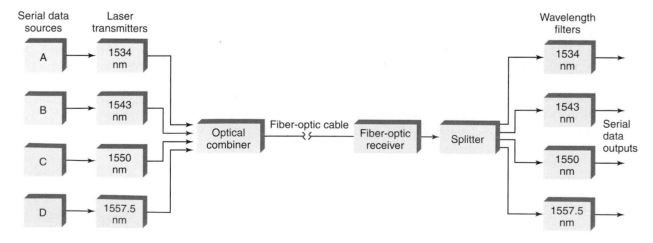

Figure 19-41 General concept of an array waveguide grating.

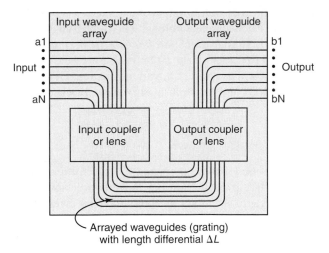

Arrayed waveguides (grating)
with length differential ΔL

single fiber as better filters and optical components called *splitters* become available to permit 50-, 25-, or even 12.5-GHz channel spacing.

Splitters

The key components in a DWDM system are the multiplexer and demultiplexer and the multiwavelength lasers discussed elsewhere in this chapter. Numerous methods have been developed over the years to add and separate optical signals. Optical couplers can be used for multiplexing, and optical filters such as fiber Bragg gratings or thin films can be implemented for demultiplexing. But the one method that appears to be emerging as the most popular is the *arrayed waveguide grating (AWG)*. This device is an array of optical waveguides of different lengths made with silica (SiO_2) on a silicon chip, and it can be used for both multiplexing and demultiplexing.

Arrayed waveguide grating (AWG)

Figure 19-41 shows the concept of an AWG. The multiple inputs are fed into a cavity or coupler region that acts as a lens to equally divide the inputs to each of the waveguides. Every waveguide in the grating has a length L that differs from its neighbor by ΔL. This produces a phase difference in the beams coming out of the gratings into an output cavity.

The output cavity acts as a lens to refocus the beams from all the grating waveguides onto the output waveguide array. Any output contains the multiplexed inputs. For demultiplexing, the single multiwavelength input signal is applied to any of the inputs where the signals of different wavelengths propagate through the grating. The grating acts as multiple filters to separate the signals into individual paths that appear at the multiple outputs.

AWGs are popular because they are relatively easy to produce with standard semiconductor processes, making them very inexpensive. They have very low insertion loss and low cross talk. A typical unit has 32 or 40 inputs and 32 or 40 outputs with 100-GHz spacing on ITU grid wavelengths. Some products have up to 64 or 80 input and output channels, and devices with higher channel counts are under development. AWG insertion loss is in the 2- to 4-dB range with 20 to 30 dB of adjacent channel cross talk attenuation.

19-6 Passive Optical Networks

The primary applications for fiber-optic networks are in wide-area networks such as long-distance telephone service and the Internet backbone. As speeds have increased and prices have declined, fiber-optic technology has been adopted into metropolitan-area networks, storage-area networks (SANs), and local-area networks. The most widely used of these technologies—SONET, Ethernet, and Fibre Channel—have been discussed in previous chapters.

A newer and growing fiber-optic system is the *passive optical network (PON)*, a type of metropolitan-area network technology. This technology is also referred to as *fiber to the home (FTTH)*. Similar terms are *fiber to the premises* or *fiber to the curb*, designated as *FTTP* or *FTTC*. The term *FTTx* is used, where *x* represents the destination. PONs are already widely used in Japan and Korea. Here in the United States, both AT&T (formerly SBC) and Verizon already have PONs in place with more on the way.

The PON Concept

Most optical networking uses active components to perform optical-to-electrical and electrical-to-optical (OEO) conversions during transmission and reception. These conversions are expensive because each requires a pair of transceivers and the related power supply. Over long distances, usually 10 to 40 km or more, repeaters or optical amplifiers are necessary to overcome the attenuation, restore signal strength, and reshape the signal. These OEO repeaters and amplifiers are a nuisance as well as expensive and power-hungry. One solution to this problem is to use a passive optical network. The term *passive* implies no OEO repeaters, amplifiers, or any other device that uses power. Instead, the transmitter sends the signal out over the network cable, and a receiver at the destination picks it up. There are no intervening repeaters or amplifiers. Only passive optical devices such as splitters and combiners are used. By using low-attenuation fiber-optic cable, powerful lasers, and sensitive receivers, it is possible to achieve distances of up to about 20 km without intervening active equipment. This makes PONs ideal for metropolitan-area networks.

The PON method has been adopted by telecommunications carriers as the medium of choice for their very high-speed broadband Internet connections to consumers and businesses. These metropolitan networks cover a portion of a city or a similar-size area. PONs will be competitive with cable TV and DSL connections but faster than both. Using optical techniques, the consumer can have an Internet connection speed of 100 Mbps or higher. This is much faster than the typical 1- to 6-Mbps cable TV and DSL connections. And PONs make digital TV distribution more practical. Furthermore, they provide the extra bandwidth to carry Internet phone calls (VoIP).

PON Technologies

There are several different types of PONs and standards. The earliest standard, called APON, was based upon ATM packets and featured speeds of 115.52 and 622.08 Mbps. A more advanced version, called BPON, features a data rate up to 1.25 Gbps. The most recent version is a superset of BPON, called GPON, for Gigabit PON. It provides download speeds up to 2.5 Gbps and upload speeds up to 1.25 Gbps. One of the key features of GPON is that it uses encapsulation, a technique that makes it protocol-agnostic. Any type of data protocol including TDM (such as T1 or SONET) or Ethernet can be transmitted.

Figure 19-42 shows a basic block diagram of a BPON/GPON network. The carrier central office (CO) serves as the Internet service provider (ISP), TV supplier or telephone carrier as the case may be. The equipment at the carrier central office (CO) is referred to as the optical line terminal (OLT). It develops a signal on 1490 nm for transmission (download) to the remote terminals. This signal carries all Internet data and any voice signals as in Voice over IP (VoIP). If TV is transmitted, it is modulated on to a 1550 nm laser. The 1550 and 1490 nm outputs are mixed or added in a passive combiner to create a coarse wavelength division multiplexed (CWDM) signal. This master signal is then sent to a passive splitter that divides the signal into four equal power levels for transmission over the first part of the network. In BPON the data rate is 622 Mbps or 1.25 Gbps but with GPON it is 2.5 Gbps.

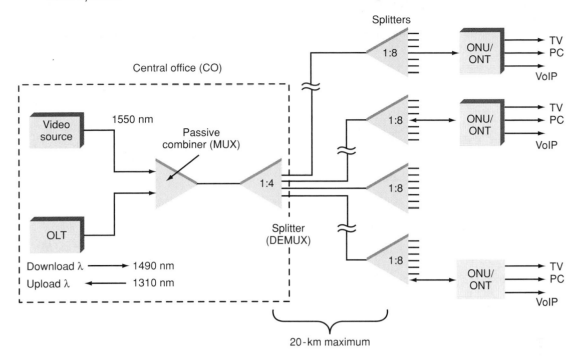

Additional splitters are used along the way to further split the signals for distribution to multiple homes. Splitters are available to divide the power by 2:1, 4:1, 8:1, 16:1 and 64:1. One OLT can transmit to up to 64 destinations up to about 20 km. The upper limit may be 16 devices depending upon the ranges involved. Just remember that each time the signal is split, its power is decreased by the split ratio. The power out of each port on a 4:1 splitter is only one-fourth of the input power. Splitters are passive demultiplexers (DEMUX).

Note also, because the splitters are optical devices made of glass or silicon, they are bidirectional. They also serve as combiners in the opposite direction. In this capacity they serve as multiplexers (MUX).

At the receiving end, each subscriber has an *optical networking unit (ONU)* or *optical networking terminal (ONT)*. These boxes connect to your PC, TV set, and/or VoIP telephone. The ONU/ONT is a two-way device, meaning that it can transmit as well as receive. In VoIP or Internet applications, the subscriber needs to transmit voice and dialing data back to the OLT. This is done over a separate 1310-nm laser using the same fiber-optic path. The splitters are bidirectional and also work as combiners or multiplexers. The upload speed is 155 Mbps in BPON and 1.25 Gbps in GPON.

> Optical networking unit (ONU) or optical networking terminal (ONT)

Another widely used standard is EPON, or Ethernet PON. While BPON and GPON have been adopted as the North American PON standard, EPON is the de facto PON standard in Japan, Korea, and some European nations. EPON is one part of the popular IEEE Ethernet standard and is designated 802.3ah. You will also hear it referred to as *Ethernet in the first mile (EFM)*. The term *first mile* refers to that distance between the subscriber and any central office. Sometimes it is also called the *last mile*.

> Ethernet in the first mile (EFM)

The topology of EPON is similar to that of GPON, but the downstream and upstream data rates are symmetric at 1.25 Gbps. The downstream is on 1490 nm while the upstream is on 1310 nm. Standard Ethernet packets are transmitted with a data payload to 1518 bytes. A real plus for EPON is that since it is Ethernet, it is fully compatible with any other Ethernet LAN.

While PONs provide the ultimate in bandwidth and data rate for home broadband connections, they are expensive. Carriers must invest in a huge infrastructure that requires rewiring the area served, again. Fiber-optic cables are mostly laid underground, but that process is expensive because right-of-way must be bought and trenches dug. Cables can be carried overhead on existing poles at less expense, but the effect is less aesthetically pleasing.

While some PON services such as Verizon's Fois service take the fiber directly to the home, other systems rely on the standard POTS twisted-pair cable that is already in place virtually everywhere. In such cases, the fiber cable is run to a neighborhood terminal or gateway. Such a gateway may also serve an apartment complex or multiunit dwelling such as a condominium building. Then the signals are distributed over the standard twisted-pair telephone cable that is already in place. An advanced form of digital subscriber line called ADSL2 or ADSL2+ is used with a data rate to 24 Mbps. A more advanced version called *VDSL,* for *video DSL,* is also used in some cases. It supports a data rate to 50 Mbps over shorter runs of twisted-pair cable.

Video DSL

CHAPTER REVIEW

Summary

Today, light is being used increasingly as the carrier for information in a communication system. Although visible light is used to some extent, most optical communication systems are infrared. The transmission medium is either free space or a special light-carrying vacuum called a *fiber-optic cable.* The light is modulated by the information to be transmitted. Because the frequency of light is extremely high, it can accommodate very wide bandwidths of baseband information and/or extremely high rates of data transmission with excellent reliability.

Fiber optic cable is available in two basic forms: plastic and glass. The plastic cable is more flexible and less expensive but has greater attenuation. Glass cable is more expensive and fragile but has less attenuation. The most common forms of fiber cable are *multimode fiber (MMF)* and *single-mode fiber (SMF).* Common sizes are 62.5/125 and 50/125 μm, where the first number is the diameter of the inner fiber and the second number is the cladding diameter. These sizes are for MMF. The typical size of SMF is 9/125 μm. Besides attenuation, fiber cable also distorts the light pulse, causing it to lengthen and reduce the potential data rate. This is called dispersion. It is usually corrected electronically today.

The most common light transmitters are lasers. The main types are the Fabry-Perot, distributed feedback (DFB), and vertical cavity surface-emitting lasers (VCSELs). The most common light frequencies are 850, 1310, and 1550 nm. The two widely used light receivers are the PIN photodiode and the avalanche photodiode (APD). A special amplifier called a transimpedance amplifier (TIA) is used to amplify the photodiode signal.

The most dramatic development in fiber-optic communication systems in recent years is dense wavelength-division multiplexing (DWDM). This technique, which is simply frequency-division multiplexing (FDM) for optical fiber, permits multiple wavelengths of light to be transmitted simultaneously over a single fiber. The potential is more than 160 wavelengths or channels per fiber, with each carrying data at a rate up to 40 Gbps. DWDM dramatically increases the data-carrying capacity of any fiber. A device called an array waveguide grating (AWG) is used to multiplex and demultiplex DWDM signals.

The light transmitter and receiver components are usually packaged into a single unit called a transceiver, transponder, or module. The package is metal with optical cable connectors on one side and electrical interface connections on the other. Various sizes and protocols are accommodated by several standards called multisource agreements (MSAs). The various MSA formats are designated SFF, SFP, X2, XENPAK, XFP, XPAK, and 300-pin. Most are for 10-Gbps systems.

Optical fiber techniques are used mainly in WANs, MANs, and SANs. SONET and Fibre Channel are the most common protocols. Fiber is also used in the faster Ethernet LANs. A newer form of fiber netowork is the passive optical network (PON). A PON is a metropolitan network that uses no optical-electrical-optical (OEO) conversions, repeaters, or optical amplifiers. Only passive optical devices called combiners and splitters are used. PONs are being deployed by telecommunications companies for high-speed Internet service as well as VoIP and TV distribution. The most common protocols used are GPON (2.5-Gbps download, 1.25-Gbps upload) and EPON (1.25-Gbps symmetric).

Questions

1. True or false? Light is electromagnetic radiation.
2. The optical spectrum is made up of three parts. Name them.
3. Which segment of the optical spectrum has the highest frequencies? The lowest frequencies?
4. Light waves travel in a
   a. Circle
   b. Straight line
   c. Curve
   d. Random path
5. What units are used to express the wavelength of light?
6. State the lowest wavelength and highest wavelength of visible light, and state the color of each.
7. What is the wavelength range of infrared light?
8. True or false? The speed of light is faster in glass or plastic than it is in air.
9. What is the name of the number that tells how fast light travels in a medium compared to air?
10. What device can be used to bounce or change the direction of a light wave?
11. What is the term used to describe the bending of light rays due to speed changes in moving from one medium to another?
12. When the angle of refraction is 90° to the normal, describe how the ray travels in relationship to the two media involved.
13. What special effect occurs when the incident ray strikes the interface between two media at an angle greater than the critical angle?
14. What factor determines the critical angle in a medium?
15. Of what value are ultraviolet rays in communications?
16. Name the two most common transmission media in an optical communication system.
17. Describe how a light source is modulated. What type of modulation is the most common?
18. Describe how both analog and digital signals may modulate a light transmitter.
19. What are the two main types of light sources used in optical communication systems? Which is preferred and why?
20. What factors limit the free-space transmission of information on a light beam?
21. What type of signals do fiber-optic cables carry?
22. What optical principle makes fiber-optic cable possible?
23. What two materials are used to make fiber-optic cable?
24. Name the three main types of information carried by fiber-optic cables.
25. What is the major application of fiber-optic cable?
26. State the main benefit of fiber-optic cable over electrical cable.
27. True or false? Fiber-optic cable has greater loss than electrical cable over long distances.
28. True or false? Fiber-optic cable is smaller, lighter, and stronger than electrical cable.
29. State the two main disadvantages of fiber-optic cable.
30. What is the name of the device that converts light pulses to an electric signal?
31. What is the name given to the regenerative units used to compensate for signal attenuation over long distances?
32. Which material has the better optical characteristics and lower loss?
   a. Plastic
   b. Glass
   c. They are equal.
33. What covers and protects the core in a fiber-optic cable?
34. In PCS-type cable, what materials are used to make the core and the cladding?
35. The index of refraction is highest in the
   a. Core
   b. Cladding
36. List the three main types of fiber-optic cable.
37. What is the name given to the phenomenon of stretching of the light pulse by the fiber-optic cable? What actually causes it?
38. Light pulse stretching occurs in what two types of cable? What type of fiber-optic cable does not cause any significant degree of pulse stretching?
39. What type of cable is the best to use for very high-frequency pulses?
40. Pulse stretching by modal dispersion causes the information capacity of a cable to
   a. Increase
   b. Decrease
41. Define chromatic dispersion.
42. What is electronic dispersion compensation?
43. What is the typical core diameter range of a single-mode step index cable?
44. What types of covering are applied over the cladding in a fiber-optic cable to protect against moisture and damage?
45. True or false? Fiber-optic cables are available with multiple cores.
46. What term is used to refer to light loss in a cable?
47. What three major factors cause light loss in a fiber-optic cable?
48. How is the amount of light loss in a fiber-optic cable expressed and measured?
49. True or false? Fiber-optic cables may be spliced.
50. State how fiber-optic cables are conveniently linked and attached to one another and to related equipment.
51. Name the designations of the two most common types of fiber-optic cable connectors, and state the primary difference between them.

52. Name three common and popular sizes of fiber-optic cable.
53. Define what is meant by *bandwidth* as it applies to fiber-optic cable. What units are used to express bandwidth?
54. What causes the greatest loss peaks in a fiber-optic cable? At what frequencies do these peak losses occur?
55. What are the two most common light sources used in fiber-optic transmitters?
56. True or false? Visible light is the most common type of light used in fiber-optic systems.
57. Name the three most common light frequencies used in fiber-optic cable systems. Why are these frequencies used?
58. True or false? The light from a 1.55-μm LED is visible.
59. What semiconductor material are LEDs usually made of?
60. What term is used to describe a single light frequency?
61. What do you call the condition of all emitted light waves being in phase?
62. What special structure is created when reflective surfaces are added to a laser diode? How does it affect the light waves generated and emitted?
63. For normal operation, LEDs and ILDs are
    a. Reverse-biased
    b. Forward-biased
64. Which is faster?
    a. LED
    b. ILD
65. Which produces the brighter light?
    a. LED
    b. ILD
66. List three common laser types. Which is preferred for long-distance applications?
67. What is the main benefit of a tunable laser?
68. During normal operation, all photodiodes are
    a. Reverse-biased
    b. Forward-biased
69. Name the two most sensitive and fastest light detectors.
70. Name the two main circuits in a fiber-optic receiver.
71. Aside from the reason stated in the text, name one key reason why Manchester encoding might be preferred.
72. What popular local-area network system uses fiber-optic cable?
73. What electrical unit is used to state the output of a light transmitter and the sensitivity of a light receiver?
74. List three common laser operating frequencies in fiber-optic systems. Which is preferred for long-distance operation, and why?
75. State why the laser is never turned off completely in direct laser modulation.
76. What is the name given to the external modulator used with some lasers? Why is it preferred over direct modulation?
77. Explain the operation of the automatic power control circuit used with most laser driver circuits.

78. What is the name given to the input amplifier used in optical receiver circuits?
79. What is regeneration in a fiber-optic network, and why is it needed?
80. Explain the process of regeneration and give another name for it.
81. Describe the main components in an optical amplifier.
82. What is the normal gain range of an EDFA?
83. State the primary benefit of an optical amplifier.
84. What is an MSA? What are its benefits?
85. What does the X mean in the abbreviations of the different MSAs?
86. What is the name of the device to which the MSAs apply?
87. Describe a ROSA.
88. Describe a TOSA.
89. What is a CDR and how does it work?
90. What is a CMU and how does it work?
91. What is the abbreviation used to describe a circuit that converts from serial to parallel or parallel to serial?
92. List the most widely used MSA formats.
93. What is the parallel electrical interface on a 300-pin MSA?
94. Describe the electrical interface on a XENPAX module. What does XAUI mean?
95. What is the electrical interface on an XFP module?
96. What does *reach* mean? Define *long reach*.
97. What is wavelength-division multiplexing?
98. How many channels or wavelengths are commonly used in CDWM systems?
99. What is the maximum number of channels possible in DWDM systems?
100. What is the name of the device commonly used to multiplex and demultiplex IR signals in a DWDM system?
101. Give the wavelength ranges of the C, L, and S IR bands.
102. Name the device used to multiplex and demultiplex DWDM signals.
103. Categorize passive optical networks as one of the following: WAN, MAN, SAN, LAN.
104. What are PONs used for?
105. What are the designations for the equipment at the carrier central office and the customer premises?
106. What is a combiner? A splitter? How are they used?
107. If the light input to 16:1 splitter is 400 μW, what is the power available at each output?
108. What is the primary U.S. PON standard called? What are its download and upload speeds?
109. What is EPON? Where is it used?
110. Give the EPON data rates.
111. Name the light wavelengths used in most PONs and tell what they are used for.
112. Explain how data signals are distributed in PONs where the fiber does not run all the way to the home or office.
113. Define FTTH.
114. What is the first/last mile?

Problems

1. State the speed of light in air in meters per second and miles per second. ◆
2. If a light ray strikes a mirror at an angle of 28.4° from the normal, at what angle is it reflected from the normal?
3. Draw a simple block diagram of an optical communication system, and explain the purpose and operation of each element. ◆
4. Explain the process by which voice and video signals are transmitted by light beam.
5. A cable has a loss of 9 dB. What percentage of its input power will appear at the output? ◆
6. Express 1.2 km in terms of miles and feet.
7. How many kilometers are there in 6 mi? ◆
8. Four cables with attenuations of 7, 16, 29, and 34 dB are spliced together. What is the total attenuation in decibels?
9. A fiber-optic cable has a bandwidth of 160 MHz·km. What is the bandwidth of a 1-mi segment? A 0.5-mi segment? ◆
10. Explain briefly how light falling on the PN junction of a photodiode causes the diode's conductance to change.
11. The bit rate–distance product of a system is 600 Mbps·km. What is the speed rating at 8 km?
12. What is the average maximum distance between repeaters in a fiber-optic system?
13. Explain how analog signals are transmitted over a fiber-optic cable.
14. A fiber-optic cable system has a dispersion factor of 33 ns/km. The length of the system is 0.8 km. What is the highest data rate that can be achieved on this link?
15. Calculate the minimum receiver sensitivity needed to reliably detect a signal transmitted with a power of 0.7 mW over a link that is 3.5 km long with attenuation of 2.8 dB/km. It has two splices, each with an attenuation of 0.3 dB, and four connectors, each with attenuation of 1 dB. The losses in the connections to the IPD and APD devices are 2 dB each. Assume a contingency factor of 5 dB. The dispersion factor is 18 ns/km. What is the maximum data rate that this system can achieve?
16. A DWDM system has 64 OC-48 50 NET channels. What is the composite or aggregate total data rate?

◆ *Answers to Selected Problems follow Chap. 22.*

Critical Thinking

1. Name three potential new applications for fiber-optic communication not on the list in Fig. 19-9 but that take advantage of the benefits listed in Fig. 19-10.
2. Explain how a single fiber-optic cable can handle two-way communication, both half and full duplex.
3. Could an incandescent light be used for a fiber-optic transmitter? Explain its possible benefits and disadvantages.
4. Compare a wireless radio system with a fiber-optic communication system for digital data communications over a distance of 1 km. Assume a desired data rate of 75 Mbps. Give pros and cons, advantages and disadvantages, of each. Which one would be better, all factors being considered?
5. Explain how byte data can be transmitted in parallel in a DWDM system.

chapter 20

Cell Phone Technologies

Wireless

Wireless was the original word for "radio." Essentially, it meant *wireless telegraphy*. The term was widely used in the early twentieth century but fell out of use in the United States and was almost entirely replaced by the word *radio* before World War II. But today the term wireless is back in a big way. It still means "radio," but it has some specific modern implications. Nowadays, wireless refers primarily to the enormous cellular telephone industry. After all, cell phones are basically sophisticated two-way radios.

With over 800 million cell phones sold in 2005, the cell phone is the largest-volume consumer electronics device. It has changed the way that we communicate. Also 2005 was the year in which cell phone subscribers numbered more than wired telephone subscribers. Furthermore, as the data speed of the newer digital cell phone transmissions increases, more cell phone applications are possible. These include cameras, Internet access, e-mails, audio, gaming, and video. This chapter provides a technical overview of cell phone standards and operation. Other short-range wireless technologies are covered in Chap. 21.

Objectives

After completing this chapter, you will be able to:

- Describe the cell phone operational concept.
- Name the three most common second-generation digital cell phone systems and describe the features of each.
- Define the cell phone terms *2G*, *2.5G*, *3G*, and *4G*.
- Describe the block diagram architecture of a modern 2.5/3G digital cell phone.
- State the features and benefits and applications of 3G cell phones.
- Explain the applications and benefits of location-based technologies in cell phones.
- Describe the architecture and operation of a cell phone base station.

20-1 Cellular Telephone Systems

A *cellular radio system* provides standard telephone service by two-way radio at remote locations. Cellular radios or telephones were originally installed in cars or trucks, but today most are handheld models. Cellular telephones permit users to link up with the standard telephone system, which permits calls to any part of the world.

The Bell Telephone Company division of AT&T developed the cellular radio system during the 1970s and fully implemented it during the early 1980s. Today, cellular radio telephone service is available worldwide. The original U.S. cell phone system, known as the *advanced mobile phone system,* or *AMPS,* was based on analog radio technologies. AMPS has gradually been phased out and replaced by second-generation (2G) and third-generation (3G) digital cell phone systems. This section provides an overview of this awesome worldwide network.

Cellular radio system

Advanced mobile phone system (AMPS)

Cellular Concepts

The basic concept behind the cellular radio system is that rather than serving a given geographic area with a single transmitter and receiver, the system divides the service area into many smaller areas known as *cells,* as shown in Fig. 20-1. The typical cell covers only several square miles and contains its own receiver and low-power transmitter. The coverage of a cell depends upon the density (number) of users in a given area. See Fig. 20-2. For a heavily populated city, many small cells are used to ensure service. In less populated rural areas, fewer cells are used. Short cell antenna towers limit the cell coverage area. Higher towers give broader coverage. The cell site is designed to reliably serve only persons and vehicles in its small cell area.

Cells

Each cell is connected by telephone lines or a microwave radio relay link to a master control center known as the *mobile telephone switching office (MTSO).* The MTSO controls all the cells and provides the interface between each cell and the main telephone office. As the person with the cell phone passes through a cell, it is served by the cell transceiver. The telephone call is routed through the MTSO and to the standard telephone system. As the person moves, the system automatically switches from one cell to the next. The receiver in each cell station continuously monitors the signal strength of the

Mobile telephone switching office (MTSO)

Figure 20-1 The area served by a cellular telephone system is divided into small areas called *cells. Note:* Cells are shown as ideal hexagons, but in reality they have circular to other geometric shapes. These areas may overlap, and the cells may be of different sizes.

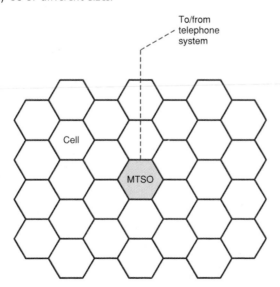

Figure 20-2 Area of cell coverage is determined by antenna height.

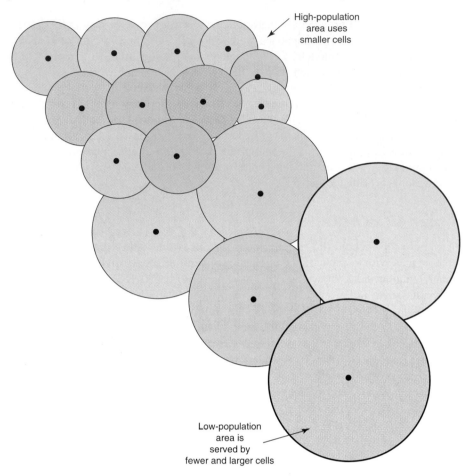

High-population area uses smaller cells

Low-population area is served by fewer and larger cells

mobile unit. When the signal strength drops below a desired level, it automatically seeks a cell where the signal from the mobile unit is stronger. The computer at the MTSO causes the transmission from the person to be switched from the weaker cell to the stronger cell. This is called a *handoff*. All this takes place in a very short time and is completely unnoticeable to the user. The result is that optimum transmission and reception are obtained.

Frequency Allocation

Cellular radio systems operate in the UHF and microwave bands as assigned by the Federal Communications Commission (FCC). The original frequency assignments were in the 800- to 900-MHz range previously occupied by the mostly unused UHF TV channels 68 through 83. Figure 20-3 shows the most widely used bands. The frequencies between 824 and 849 MHz are reserved for the uplink transmissions from the cell phone to the base station. These are also called the reverse channels. The frequencies between 869 and 894 MHz are the downlink bands from base station to cell phone. Both of these 25-MHz segments of spectrum were originally divided into 832 channels 30 kHz wide. While these are still used, the different cell phone technologies use different amounts of bandwidth, such as 30 kHz, 200 kHz, and 1.25 MHz, so this spectrum gets used in different ways by different cell phone companies in different locations.

Another commonly used block of spectrum is shown in Fig. 20-4(*a*). Again, the use of this spectrum varies depending upon the cell phone carrier and the geographic area. A more recently allocated block of spectrum is shown in Fig. 20-4(*b*). These two blocks of 60 MHz are referred to as the personal communications systems (PCS) channels.

Handoff

Figure 20-3 Standard U.S. UHF cell phone spectrum.

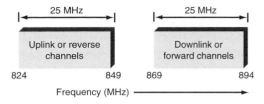

Figure 20-4 Additional U.S. cell phone spectrum. (*a*) 890 to 960 MHz and (*b*) 1850 to 1990 MHz are called the personal communication system PCS band.

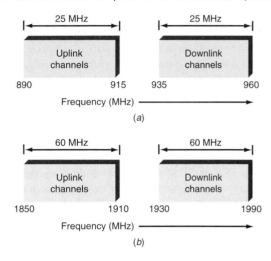

While the range at these higher microwave frequencies is somewhat less than that achievable in the UHF bands, this block of frequencies provides greater system capacity, meaning more subscribers. Also the antennas are smaller at these frequencies.

One of the major issues in the cell phone business lies in obtaining more spectrum for more subscribers. More subscribers mean greater income. Yet, spectrum is scarce and very expensive. In the future, more spectrum should become available as the remaining UHF TV channels are abandoned for digital high-definition TV in 2009. This will make more spectrum available in the 700- to 800-MHz range. More recently, the military and government have moved some of their systems, thereby freeing up space in the 1700- to 1750-MHz range. Some spectrum is also available in the 1900- to 2300-MHz range for newer third-generation systems. Also keep in mind that different countries use different spectrum blocks. For example, in Europe the most commonly used bands are 900 and 1800 MHz.

Multiple Access

Multiple access refers to how the subscribers are allocated to the assigned frequency spectrum. Access methods are the ways in which many users share a limited amount of spectrum. These are similar to multiplexing methods you learned about in previous chapters. The techniques include frequency reuse, frequency-division multiple access (FDMA), time-division multiple access (TDMA), code-division multiple access (CDMA), and spatial-division multiple access (SDMA).

Multiple access

Frequency Reuse. In frequency reuse, individual frequency bands are shared by multiple base stations and users. This is possible by ensuring that one subscriber or base station does not interfere with any others. This is achieved by controlling such factors as transmission power, base station spacing, and antenna height and radiation patterns. With low-power and lower-height antennas, the range of a signal is restricted to only a mile

Figure 20-5 Horizontal antenna radiation pattern of a common cell site showing 120° sectors that permit frequency reuse.

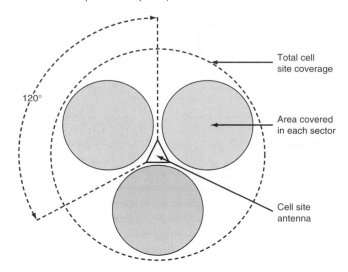

or so. Furthermore, most base stations use sectorized antennas with 120° radiation patterns that transmit and receive over only a portion of the area they cover. See Fig. 20-5. In any given city, the same frequencies are used over and over simply by keeping cell site base stations isolated from one another.

Frequency-Division Multiple Access. FDMA systems are like frequency-division multiplexing in that they allow many users to share a block of spectrum by simply dividing it up into many smaller channels. See Fig. 20-6. Each channel of a band is given an assigned number or is designated by the center frequency of the channel. One subscriber is assigned to each channel. Typical channel widths are 30 kHz, 200 kHz, 1.25 MHz, and 5 MHz. There are usually two similar bands, one for uplink and the other for downlink.

Time-Division Multiple Access. TDMA relies on digital signals and operates on a single channel. Multiple users use different time slots. Because the audio signal is sampled at a rapid rate, the data words can be interleaved into different time slots, as Fig. 20-7 shows. Of the two common TDMA systems in use, one allows three users per frequency channel and the other allows eight users per channel.

Figure 20-6 Frequency-division multiple-access (FDMA) spectrum.

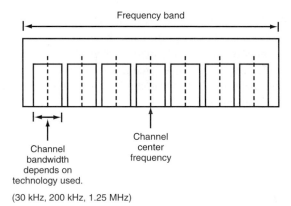

Figure 20-7 Time division multiple access (TDMA). Different callers use different time slots on the same channel.

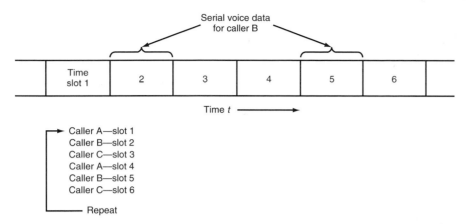

Code-Division Multiple Access. CDMA is just another name for spread spectrum. A high percentage of cell phone systems use *direct sequence spread spectrum (DSSS)*. Here the digital audio signals are encoded in a circuit called a vocoder to produce a 13-kbps serial digital compressed voice signal. It is then combined with a higher-frequency chipping signal. One system uses a 1.288-Mbps chipping signal to encode the audio, spreading the signal over a 1.25-MHz channel. See Fig. 20-8. With unique coding, up to 64 subscribers can share a 1.25-MHz channel.

Figure 20-8 Code-division multiple access (CDMA). (*a*) Spreading the signal. (*b*) Resulting bandwidth.

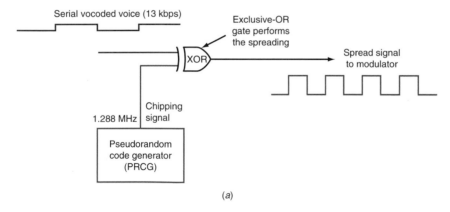

(*a*)

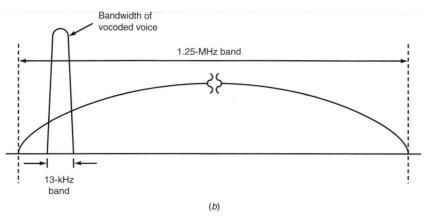

(*b*)

Figure 20-9 The concept of spatial-division multiple access (SDMA) using highly directional antennas.

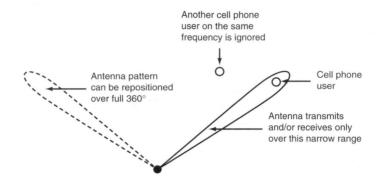

Spatial–Division Multiple Access. This form of access is actually an extension of frequency reuse. It uses highly directional antennas to pinpoint users and reject others on the same frequency. In Fig. 20-9, very narrow antenna beams at the cell site base station are able to lock in on one subscriber but block another while both subscribers are using the same frequency. Modern antenna technology using adaptive phased arrays is making this possible. Such antennas allow cell phone carriers to expand the number of subscribers by more aggressive frequency reuse because finer discrimination can be achieved with the antennas. SDMA is also widely used in wireless local-area networks (WLANs) and other broadband wireless applications.

Duplexing

Duplexing refers to the ways in which two-way radio or telephone conversations are handled. Many two-way radio applications still use *half duplex* where one party talks at a time. The communicating individuals take turns speaking and listening. Telephone communications has always been *full duplex,* where both parties can simultaneously send and receive. All cell phone systems are full duplex.

To achieve full duplex operation, however, special arrangements must be made. The most common arrangement is called *frequency-division duplexing (FDD)*. In FDD, separate frequency channels are assigned for the transmit and receive functions. The transmit and receive channels are spaced so that they do not interfere with one another inside the cell phone or base station circuits. The uplink and downlink channels in Figs. 20-3 and 20-4 are an example.

Another arrangement is *time-division duplexing (TDD)*. This is less common but is used in a few systems. The system assigns the transmit and receive data to different time slots, both on the same frequency. For example, the transmitted and received data is alternated in sequential time slots. While the transmitted and received signals do indeed occur at different times, the speed of the signals is fast enough that a human feels as though they are occurring at the same time.

FDD is far more widely used than TDD.

20-2 The Advanced Mobile Phone System (AMPS)

AMPS was the first cell phone system in the United States. It was based on traditional FM radio technology. Although analog AMPS cell phones are due to be phased out beginning in 2007, millions are still in use. Furthermore they are a good illustration of the architecture and operation of any cell phone.

Duplexing

Half duplex

Full duplex

frequency-division duplexing (FDD)

Time-division duplexing (TDD)

Figure 20-10 General block diagram of a typical AMPS unit (cellular radio).

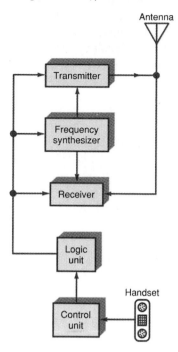

Typical AMPS Handset

Fig 20-10 is a general block diagram of a typical AMPS unit. It consists of five major sections: transmitter, receiver, synthesizer, logic unit, and control unit. Mobile radios derive their operating power from a built-in rechargeable battery. The transmitter and receiver share a single antenna. The sections are discussed below.

Transmitter. The transmitter block diagram is shown in Fig. 20-11. It is a low-power FM unit operating in the frequency range of 825 to 845 MHz. Channel 1 is 825.03 MHz, channel 2 is 825.06 MHz, and so on. The carrier furnished by a frequency synthesizer

Transmitter

Figure 20-11 AMPS transmitter.

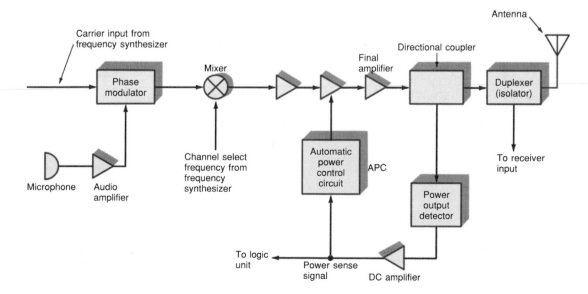

is phase-modulated by the voice signal. The phase modulator produces a deviation of ± 12 kHz. Preemphasis is used to help minimize noise. The modulator output is translated up to the final transmitter frequency by a mixer whose second input also comes from the frequency synthesizer. The mixer output is fed to class C or class E power amplifier stages where the output signal is developed. The final amplifier stage is designed to supply about 600 mW to the antenna.

A unique feature of the transmitter is that its output power is controllable by the cell site and MTSO. Special control signals picked up by the receiver are sent to an *automatic power control (APC) circuit* that sets the transmitter to one of eight power output levels. The APC circuit can introduce power attenuation in steps of 4 dB from 0 dB (600 mW) to 28 dB (6.3 mW). This is done by controlling the supply voltage to one of the intermediate-power amplifier stages.

The output power of the transmitter is monitored internally by built-in circuits. A microstrip directional coupler taps off an accurate sample of the transmitter output power and rectifies it into a proportional dc signal. This signal is used in the APC circuit and is transmitted back to the cell site, permitting the MTSO to know the current power level.

The APC feature permits optimum cell site reception with minimal power. It also helps to minimize interference from other stations in the same or adjacent cells.

The transmitter output is fed to a duplexer circuit or isolator that allows the transmitter and receiver to share the same antenna. Since cellular telephone units use full duplex operation, the transmitter and receiver operate simultaneously. The transmit and receive frequencies are spaced 45 MHz apart to minimize interference. However, an isolator is still needed to keep transmitter power out of the sensitive receiver. The duplexer consists of two very sharp bandpass filters, one for the transmitter and one for the receiver. The transmitter output passes through this filter to the antenna.

Receiver. The *receiver* is typically a dual-conversion superheterodyne (refer to Fig. 20-12). A radio-frequency (RF) amplifier boosts the level of the received cell site signal. The receiver frequency range is 870.03 to 889.98 MHz. The receive channels are spaced 30 kHz apart. The first mixer translates the incoming signal down to a first intermediate frequency (IF) of 82.2 MHz. Some receivers use a 45-MHz first IF. The local-oscillator signal for the mixer is derived from the frequency synthesizer. The local-oscillator (LO) frequency sets the receive channel. The signal passes through IF amplifiers and filters to

Figure 20-12 AMPS receiver.

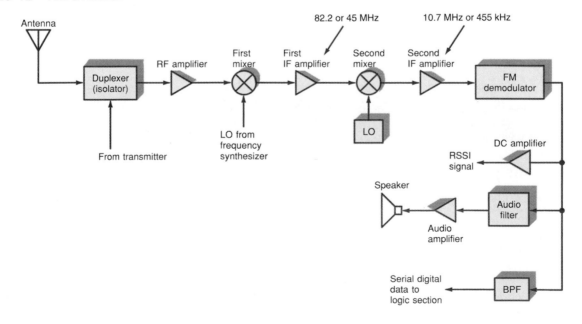

800   *Chapter 20*

Figure 20-13 AMPS frequency synthesizer.

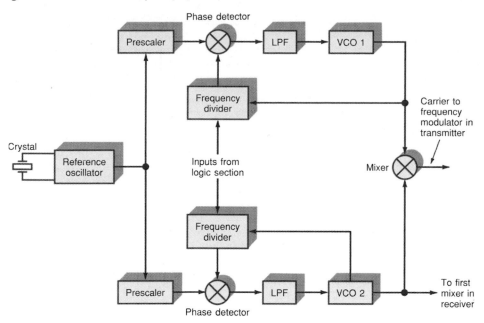

the second mixer, which is driven by a crystal-controlled local oscillator. The second IF is usually either 10.7 MHz or 455 kHz. The signal is then demodulated, deemphasized, filtered, and amplified before it is applied to the output speaker in the handset.

The output of the demodulator is also fed to other filter circuits that select out the control audio tones and digital control data stream sent by the cell site to set and control both the transmitter and the receiver. The demodulator output is also filtered into a dc level whose amplitude is proportional to the strength of the received signal. This is the *receive signal strength indicator (RSSI) signal* that is sent back to the cell site so that the MTSO can monitor the received signal from the cell and make decisions about switching to another cell.

Receive signal strength indicator (RSSI) signal

Frequency Synthesizer. The *frequency synthesizer* section develops all the signals used by the transmitter and receiver (see Fig. 20-13). It uses standard phase-locked loop (PLL) circuits and a mixer. A crystal-controlled oscillator provides the reference for the PLLs. One PLL incorporates a voltage-controlled oscillator (VCO) (number 2) whose output frequency is used as the local oscillator for the first mixer in the receiver. This signal is mixed with the output of a second PLL VCO to derive the transmitter output frequency.

Frequency synthesizer

As in other PLL circuits, the output VCO frequency is determined by the frequency-division ratio of the divider in the feedback path between the VCO and the phase detector. In a cellular radio, this frequency-division ratio is supplied by the MTSO via the cell site. When a mobile unit initiates or is to receive a call, the MTSO computer selects an unused channel. It then transmits a digitally coded signal to the receiver containing the frequency-division ratios for the transmitter and receiver PLLs. This sets the transmit and receive channel frequencies.

Logic Unit. The *logic unit* shown in Fig. 20-14 contains the master control circuitry for the cellular radio. It is made up of an embedded microcontroller with both RAM and ROM plus additional circuitry used for interpreting signals from the MTSO and cell site and generating control signals for the transmitter and receiver.

Logic unit

All cellular radios contain a programmable read-only memory (PROM) chip called the *number assignment module (NAM)*. The NAM contains the *mobile identification number (MIN)*, which is the telephone number assigned to the unit. The NAM PROM

Number assignment module (NAM)

Mobile identification number (MIN)

Figure 20-14 Logic control circuits in an AMPS.

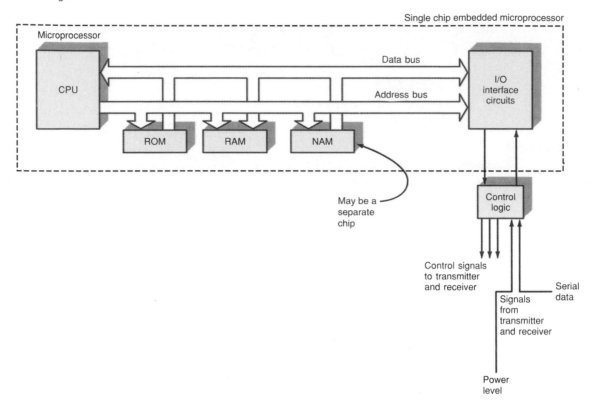

is "burned" when the cellular radio is purchased and the MIN assigned. This chip allows the radio to identify itself when a call is initiated or when the radio is interrogated by the MTSO.

All cellular mobile radios are fully under the control of the MTSO through the cell site. The MTSO sends a serial digital data stream at 10 kbps through the cell site to the radio to control the transmit and receive frequencies and transmitter power. The MTSO monitors the received cell signal strength at the cellular radio by way of the RSSI signal, and it monitors the transmitter power level. These are transmitted back to the cell site and MTSO. Audio tones are also used for signaling purposes.

Control unit

Control Unit. The *control unit* contains the handset with speaker and microphone. This may be a standard handset as used in a regular telephone on a mobile unit. However, these circuits are built into the handheld units. The main control unit contains a complete TouchTone dialing circuit (see Fig. 20-15). The control unit is operated by a separate microprocessor that drives the LCD display and other indicators. It also implements all manual control functions. The microprocessor memory permits storage of often called numbers and an autodial feature.

Operational Procedure

Described below is the sequence of operations that occur when a person initiates a cellular telephone call:

1. The operator applies power to the unit. This turns on the transmitter and receiver.
2. The receiver seeks an open control channel. Twenty-one control or paging channels are used to establish initial contact with a cell and the MTSO. When contact is made, the cell site reads the NAM data and the MTSO computer verifies that it is a valid number.
3. The operator enters the number to be called via a keyboard.

Figure 20-15 Control unit with handset.

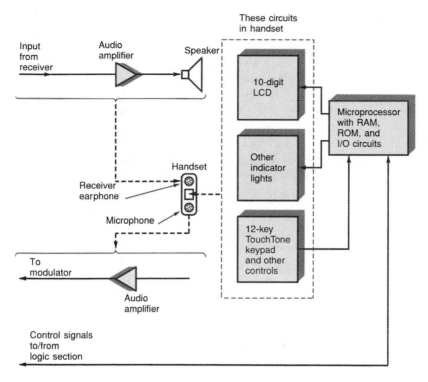

4. The operator sends the telephone number to be called by pressing a send or call button.

5. The cell and MTSO search for an open channel and send the frequency data to a cellular transceiver.

6. The RSSI signal is read to determine the optimum cell selection. The transmitter power is adjusted.

7. Handshake signals are exchanged, signifying that contact has been established.

8. The MTSO calls the designated number. Conversation takes place.

9. If the mobile unit passes from one cell to another, the MTSO senses the RSSI signal and "hands off" the mobile signal from one cell to another to maintain maximum signal strength.

10. The call is terminated.

When a mobile unit is to receive a call, the MTSO and cell site transmit a call signal containing the MIN over a control channel. The transceiver monitors the control channels, identifies its MIN, and turns on. From that point on, the sequence beginning at step 5 above is similar.

20-3 Digital Cell Phone Systems

Digital cell phone system

The original AMPS described earlier uses analog communication methods. However, all new cell phones and systems use digital methods. These all-digital systems were developed primarily to expand the capacity of the cell phone systems already in place. The rapid growth of the number of wireless subscribers forced the carriers to seek new and more efficient methods of increasing the number of users a system could handle. The main problem was that the carriers were restricted by the Federal Communications Commission to specific segments of the frequency spectrum. No additional space was available for expansion. Digital techniques provide several ways to multiplex many users into the same spectrum space.

The use of digital techniques brought several additional benefits. Digital communication systems are inherently more robust than analog systems in that they are more reliable in a noisy environment. Furthermore, digital circuits can be made smaller and more power-efficient, and therefore handsets can be more compact and can operate for longer times on a single battery charge. Finally, digital cell phones greatly facilitate the transmission of data as well as voice so that data services such as e-mail and Internet access are possible with a cell phone.

Most modern digital cell phones are referred to as *second-generation (2G) phones*. The first generation, of course, was the analog AMPS phone. Today, third-generation (3G) cell phones and systems are in use. The 2G phones have been enhanced with high-speed data capability to create an intermediate designation called *2.5G phones*. In the following sections, you will learn about the current 2G phones and the 2.5G enhancements and gain an overview of the new 3G phone designs.

2G Cell Phone Systems

Three basic *second-generation (2G) digital cell phone systems* are in wide use today. Two use time-division multiplexing, and the third uses spread spectrum (SS). The TDM systems are the Global System for Mobile Communications (GSM) and the IS-136 standard for time-division multiple access. The SS system is code-division multiple access. The three systems are widely used in the United States. However, Europe and much of the rest of the world have standardized on the GSM system. These systems are described in the next sections.

Vocoders. To use digital data transmission techniques first requires that the voice be digitized. The circuit that does this is a *vocoder*, a special type of analog-to-digital (A/D) converter and digital-to-analog (D/A) converter. With voice frequencies as high as 4 kHz, the minimum Nyquist sampling rate is 2 times the highest frequency, or 8 kHz. This means that the A/D converter in a vocoder should sample the voice signal every 125 μs and generate a proportional binary word. Assuming that it is an 8-bit value, during the 125-μs period, the 8 bits is transmitted serially. This translates to a serial data rate of $125/8 = 15.625$ μs/bit, or $1/15.625 \times 10^{-6} = 64$ kbps. This is how the T1 telephone system described in Chap. 12 works.

This serial data signal, representing the voice, is now used to modulate the carrier and the composite signal transmitted over the assigned channel. Recall that the bandwidth required to transmit a digital signal depends primarily upon the data rate. The higher the data rate, the wider the bandwidth required. As a rule of thumb, the bandwidth is roughly equal to the data rate. For example, a 64-kbps signal would require about 64 kHz of bandwidth. That represents 1 bit/Hz. Different modulation methods result in different degrees of data rate per bandwidth. Some are more spectrally efficient than others. A 1-bit/Hz rating is essentially wasteful of precious spectrum space. If the 30-kHz AMPS channels are to be used to transmit 64-kbps voice, a more efficient modulation scheme is needed, or some other technique is required.

The main function of a vocoder is data compression. Data compression techniques are used to process the digitized voice signal in such a way as to reduce the number of bits needed to represent the voice reliably. This in turn allows the speed of data transmission to be reduced to a level compatible with that of the available channel bandwidth. In modern cell phones a variety of vocoding data compression schemes are used. An A/D converter is followed by a digital signal processing (DSP) chip that does the compression in accordance with some algorithm. The vocoder then generates a serial digital voice signal at a rate of 7.4 to 13 kbps. This permits three to eight voice signals to occupy the same channel by using TDM. At the receiver, the demodulated digital data is sent to the vocoder, where a DSP chip takes the serial bits and converts them back to binary words representing the voice. A D/A converter then recreates the voice. All 2G and 3G phones contain a vocoder.

Figure 20-16 The IS-136 TDMA frame format.

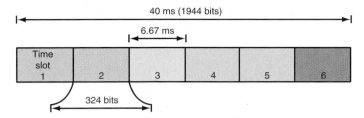

IS-136 TDMA.

IS-136 (*IS* means "interim standard") is the Telecommunications Industry Association (TIA) standard that fully describes the time-division multiple-access (TDMA) cell phone system. Also known as *digital AMPS (DAMPS), American digital cellular (ADC),* or *North American TDMA (NA-TDMA),* IS-136 TDMA is derived from an earlier standard, IS-54. The TDMA designation is used to distinguish this system from GSM, which is also a TDMA system. IS-136 operates concurrently on the same 800- to 900-MHz band channels used by AMPS. It is also used in the PCS-1900 bands. The vocoder uses *algebraic code excited linear predictive (ACELP)* speech compression, resulting in a 7.4-kbps serial voice data rate. This permits three subscribers to concurrently use a single 30-kHz channel. As in AMPS, full duplexing is achieved by using separate channels for simultaneous transmit and receive. This is referred to as frequency-division duplexing (FDD).

The IS-136 system provides for six time slots in the TDMA frame. See Fig. 20-16. Two time slots are assigned to each of three users. Time slots 1 and 4 create channel 1, slots 2 and 5 produce channel 2, and slots 3 and 6 create channel 3. Each slot is 6.67 ms long. During this time, a 324-bit frame is transmitted. It consists of the voice data plus various control signals and error detection and correction bits. The data rate in the 30-kHz channel is 48.6 kbps, giving 48.6 kbps/30 kHz = 1.62 bps/Hz. This impressive spectral efficiency is achieved with *$\pi/4$-DQPSK modulation,* which is a variation of the DPSK and QPSK methods described earlier. Like QPSK, this method encodes 2 bits per symbol and produces quadrature outputs that are added to get the final output. However, the carrier is shifted in 90° increments to produce a total of eight different carrier phases. The constellation diagram is shown in Fig. 20-17. Note the bit pairs in parentheses and the two phase positions for each bit pair. Differential encoding makes demodulation easy because the received signal is phase-compared to the previously received carrier phase and not some absolute phase value. As of this writing, $\pi/4$-DQPSK is one of the most efficient modulation methods.

IS-136 TDMA

Digital AMPS (DAMPS), American digital cellular (ADC), or North American TDMA (NA-TDMA)

GOOD TO KNOW

Using ACELP speech compression permits three subscribers to use a single 30-kHz channel.

$\pi/4$-DQPSK modulation

Figure 20-17 $\pi/4$-DQPSK modulation constellation diagram.

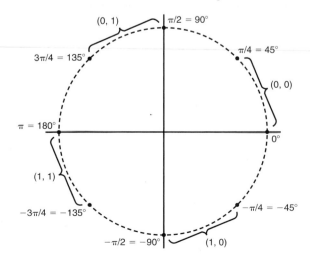

The IS-136 TDMA system is now being phased out nationwide in favor of a more improved and widely used TDMA standard referred to as GSM. The limited channel capacity of IS-136 and its lack of a high-speed data transmission process have made it obsolete. While some IS-136 phones are still in use, they are now mostly discontinued and some base stations no longer support this standard.

 GSM (Global System for Mobile Communications)

GSM. The most widely used 2G digital system is GSM. *GSM* originally stood for "Group Special Mobile" but has become known as the "Global System for Mobile Communications." It was developed in Europe under the auspices of the European Telecommunications Standardization Institute (ETSI) to replace the many incompatible analog systems used in different European countries. The GSM was designed to permit widespread roaming from country to country throughout Europe. GSM is implemented primarily in the 900-MHz band in Europe, but is also used in the 1800-MHz (1.8-GHz) range in Europe, where it is referred to as the *digital cellular system (DCS),* or *DCS-1800.* GSM is also widely implemented in the United States in both the 800- and 1900-MHz personal communication system band. It has now mostly replaced IS-136 systems in the United States. Cingular and T-Mobile are the two main carriers supporting GSM.

Digital cellular system (DCS), or DCS-1800

Like IS-136, GSM uses TDMA. The vocoder uses a compression scheme called *regular pulse excitation-linear prediction coding (RPE-LPC)* or *residual excited linear predictive (RELP) coding* that produces a 13-kbps voice bit stream. It allows eight telephone calls to be transmitted concurrently in a single 200-kHz-wide channel. The modulation method, known as *Gaussian minimum shift keying (GMSK),* is similar to frequency-shift keying (FSK) but has improved spectral properties that allow higher speeds to be transmitted in a narrower channel. A Gaussian response filter shapes the serial digital bit stream before modulation to narrow the signal bandwidth. The basic GSM data rate is 270 kbps in the 200-kHz channel, giving 270 kbps/200 kHz = 1.35 bits/Hz. Considerable error detection and correction coding is used to improve the reliability in the presence of noise, multipath fading, interference, and Doppler shifts. The basic GSM TDMA frame is shown in Fig. 20-18. Each frame is 4.615 ms long, and each voice slot is 0.577 ms long. GSM also uses a frequency-hopping scheme to minimize interchannel interference. The hop rate is 217 hops per second, or about 1200 bits per hop. FDD is used for full duplex operation.

Regular pulse excitation-linear prediction coding (RPE-LPC) or residual excited linear predictive (RELP) coding

Gaussian minimum shift keying (GMSK)

GSM continues to be the most dominant cell phone technology in the world. However, this 2G technology is slowly giving way to newer, better, faster 2.5G and 3G versions. Two key additions to GSM are general packet radio service (GPRS) and enhanced data rate for GSM evolution (EDGE). These are packet-based data services designed to permit Internet access, e-mail, and other forms of digital data transmission. These technologies are described later under the section 2.5G Cell Phone Systems.

IS-95 CDMA

IS-95 CDMA. This TIA cell phone standard is called code-division multiple access (CDMA). Also known as *cdmaOne,* it uses spread spectrum. This system was invented by Qualcomm, a company that makes the chip sets used in CDMA cell phones. The company also holds most of the patents in this field. CDMA uses direct sequence spread spectrum

Figure 20-18 A GSM TDMA frame for eight time slots.

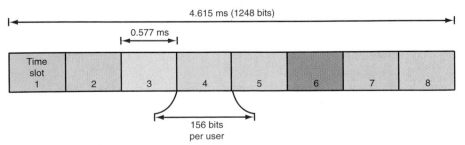

(DSSS) with a 1.2288-MHz chipping rate that spreads the signal over a 1.25-MHz channel. As many as 64 users can use this band simultaneously with little or no interference or degradation of service, although in practice typically only 10 to 40 subscribers occupy a channel at one time. This CDMA system uses FDD for duplexing.

As in other cell phone systems, CDMA takes the voice signal and digitizes it in a vocoder. The output is a 13-kbps serial voice signal that is further processed before it is used to modulate the carrier. The digitized voice is fed to an exclusive-OR (XOR) gate where it is mixed with a 64-bit pseudorandom code occurring at the chip rate of 1.2288 Mbps. This signal is then used to modulate the carrier with QPSK. The carrier may be in the regular 800- to 900-MHz band or in the PCS-1900 band. The resulting signal occupies a huge bandwidth spread over a wide spectrum. It may also coexist with up to 64 other CDMA signals that use the same carrier but have different pseudorandom codes. These special codes are known as *Walsh codes* and are chosen so that they are easily recognized and recovered at the receiver by using the correlation technique described earlier.

Walsh codes

A key part of a CDMA system is APC. All cell phones have APC, but for CDMA it is especially important. For the receivers to recover a CDMA signal, all incoming signal levels must be at the same power level. This ensures that the receiver does not confuse a higher-power signal with a lower-power signal during the decorrelation detection process. The base stations increase the power level of weak distant signals and decrease the power level of signals near the cell site.

Digital Cell Phone Circuits

Digital cell phones are quite different from the analog phones discussed earlier. Because they use digital techniques and pulse modulation methods, and since massive growth in cellular usage has caused spectrum crowding and interference problems, new architectures and circuits have been developed. Furthermore, because of the numerous standards, a variety of different circuits have been created. Three major trends dominate the cell phone evolution: increased digital processing over analog processing, increased integration of circuitry on a few chips, and multimode/multiband phones. Some new digital phones may also contain AMPS circuitry. If a subscriber roams into an area lacking a carrier that uses digital technology, the phone reverts to analog, which is still supported in most areas. In addition, most new phones contain circuitry that allows the phone to operate in several bands. A typical GSM phone, e.g., may work in the 800- to 900-MHz range as well as in the 1900-MHz PCS band.

Figure 20-19 is a block diagram of a 2G cell phone. The RF section contains the transmitter and receiver circuits including mixers, local oscillators or frequency synthesizers for channel selection, the receiver low-noise amplifier (LNA), and the transmitter power amplifier (PA). The baseband section contains the vocoder with its A/D and D/A converters plus a DSP chip that handles many processing functions typically performed by analog circuits in older systems. For example, today most baseband and intermediate-frequency filtering is done digitally, as are modulation, demodulation, and mixing.

An embedded controller handles all the digital control and signaling, handoffs, and connection and identification operations that take place transparent to the subscriber. It also takes care of running the display and keyboard and all user functions such as number storage, autodialing, and caller ID. Because of the complexity of the baseband and control functions, this embedded controller is usually a very fast (more than 100-MHz) 32-bit microprocessor with considerable RAM, ROM, and flash memory. A separate DSP chip handles the signal processing duties.

In addition to adopting digital techniques in 2G and later phones, designers have worked hard to eliminate costly components such as filters and to create circuitry that conserves power and thereby provides longer battery life. This has led to some interesting architectures, especially in the receiver section. Although superheterodyne designs are still used, several variations have emerged as dominant. These are the direct-conversion and very low-IF designs.

GOOD TO KNOW

Cellular phones equipped with a digital signal processor (DSP) chip can compress speech to fit many digital calls in the same amount of radio spectrum previously required for analog calls.

Figure 20-19 Block diagram for a 2G digital cell phone.

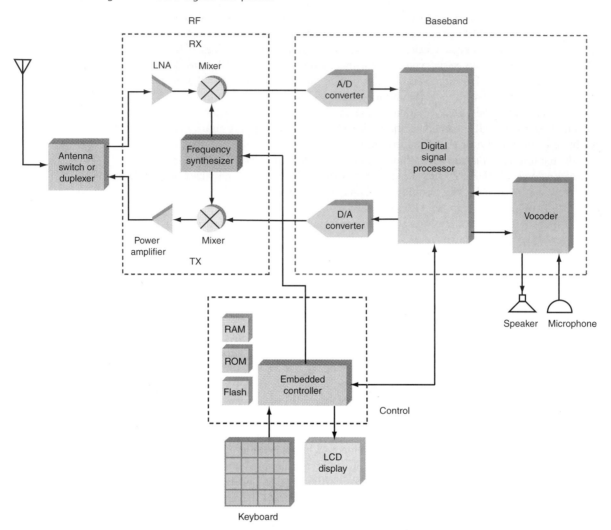

Direct Conversion. The direct-conversion or zero-IF design sets the LO frequency to the incoming signal frequency so that the translation is made directly to the baseband signal. See Fig. 20-20. Since direct conversion works only with double-sideband (DSB) suppressed AM signals, changes have been made to accommodate FSK, BPSK, QPSK, and other forms of digital modulation. Specifically, the incoming signal is applied to two mixers simultaneously. One mixer receives the LO signal directly (sine), and the other receives a signal shifted 90° (cosine). This results in down conversion to baseband as well as the generation of in-phase *(I)* and quadrature *(Q)* signals that preserve the frequency and phase information in the signal necessary for demodulation.

Direct conversion is popular because it eliminates the need for an expensive and physically large selective SAW IF filter. It also eliminates the imaging problem so common in superheterodyne designs, especially in the crowded multiband cellular spectrum. With direct conversion, baseband filtering can be accomplished by using simple low-pass *RC* filters and/or DSP filters. The *I* and *Q* signals are digitized, and a DSP chip performs additional filtering, demodulation, and voice decoding. Modern IC designs have essentially eliminated the LO leakage and dc offset problems ordinarily associated with direct-conversion designs.

Low IF. Another popular alternative is low-IF architecture. When an IF is used near the baseband frequencies, filtering is simple and very effective. A typical GSM chip set using

Figure 20-20 A direct-conversion receiver.

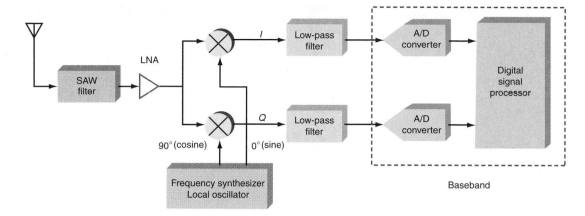

this design is shown in Fig. 20-21. Made by Silicon Laboratories' 0.18-μm CMOS, the Si4200 transceiver and Si4133T frequency synthesizer can work in any of the GSM bands in the United States or Europe. Most 2G and later phones are multiband phones that can operate in three or four bands, thereby permitting widespread roaming. The antenna connects to an electronic transmit-receive (TX/RX) switch made with GaAs PIN diodes. The incoming signal is applied to one of three SAW filters that select the desired band, the 900- or 1800-MHz European bands or the 800- or 1900-MHz PCS bands in the United States. A newer version of the chip includes the European 900-MHz band. Each SAW filter drives individual LNAs that feed a block of quadrature and image reject mixers.

An image reject mixer uses a technique similar to the phasing method of generating a single-sideband (SSB) signal described in Chap. 4. The simplified concept is shown

Figure 20-21 Silicon Labs' GSM transceiver chip set.

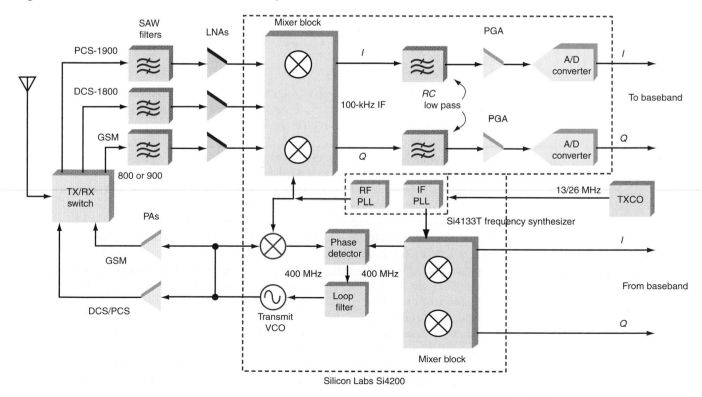

Figure 20-22 An image reject mixer.

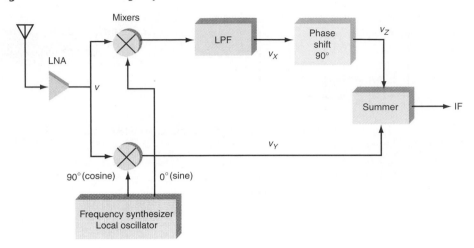

in Fig. 20-22. Assume that the incoming composite signal v consists of the desired signal $A_s \cos \omega_s t$ plus the image signal $A_i \cos \omega_i t$ or

$$v = A_s \cos \omega_s t + A_i \cos \omega_i t$$

Remember that the image frequency is twice as high as the IF. This composite signal is applied to two mixers that are also fed with quadrature (90° separation) LO signals ($\sin \omega_o t$ or $\cos \omega_o t$). Each mixer produces the sum and difference frequencies. Assuming that we keep the difference, we filter out the sum frequencies with low-pass filters. As a result, we generate signals v_X and v_Y:

$$v_X = A_s \sin/2[(\sin(\omega_o - \omega_s)t] + A_i/2[\sin(\omega_o - \omega_i)t]$$
$$v_Y = A_s \sin/2[\cos(\omega_o - \omega_s)t] + A_i/2[\cos(\omega_o - \omega_i)t]$$

Shifting the signal at X by 90° produces the signal v_Z.

$$v_Z = A_s \sin/2[\cos(\omega_s - \omega_o)t] - A_i/2[\cos(\omega_o - \omega_i)t]$$

Adding the signals v_Y and v_Z produces

$$A_s \cos(\omega_o - \omega_s)t$$

Note that the image signal components cancel and thus have been effectively rejected. The difference IF signal is preserved. As in the phasing method of SSB, the amount of the rejection depends on how well the circuits and signal levels are balanced and the precision of the 90° phase shifts.

The quadrature signal technique is also used to recover the phase and frequency content of the GMSK-modulated GSM signal. In the mixer block of Fig. 20-21, the incoming signal is applied to two mixers simultaneously. One receives the LO signal from the RF PLL directly (sine), and the other receives a signal shifted 90° (cosine). This results in down conversion as well as the generation of in-phase *(I)* and quadrature *(Q)* signals necessary for demodulation. The IF signals are the difference between the incoming signal and LO or, in this case, 100 kHz. Simple integrated *RC* low-pass filters are used to eliminate the sum signals resulting from the conversion. This step is followed by programmable gain amplifiers (PGAs) that equalize the signal levels. The low-IF signals are then applied to two delta-sigma ($\Delta\Sigma$) A/D converters running at 13 MHz.

The digitized signals go to the baseband circuit, which is usually a DSP chip, where they receive a second down conversion to baseband and filtering. The signals are then demodulated by the DSP. In this chip, the DSP circuitry is hardwired rather than being programmable, as in a typical DSP chip. The recovered digital data is then sent to the vocoder, where the voice signal is recovered.

In the transmitter section, the voice signal is digitized by the vocoder, whose output is sent to the DSP, where mixing is performed to produce the I and Q signals. Digital-to-analog converters convert the I and Q signals to analog. The signals are sent to the mixer block in the Si4200 chip. Again, see Fig. 20-21. The mixers work with the IF PLL to generate a 400-MHz intermediate signal that is used to produce the final operating frequency. The output carrier is generated by the transmit VCO, which is part of a PLL that converts the 400-MHz signal to the final frequency. The PLL acts as a bandpass filter to minimize noise, interference, and spurious signals, thereby eliminating the need for an output SAW filter.

The second chip in Fig. 20-21 is the Si4133T frequency synthesizer. A more detailed diagram of this chip is shown in Fig. 20-23. It consists of three PLL synthesizers with programmable divide-by-N frequency dividers. All the synthesizers operate from an external temperature-controlled crystal oscillator (TCXO) running at 13 or 26 MHz. The oscillator signal is divided internally by 2 if a 26-MHz signal is used, and then the 13-MHz signal is divided by either 64 or 65 and used as the reference for all three PLLs.

In Fig. 20-23 in the two upper PLLs, the RF VCOs provide a center frequency of 1900 or 1350 MHz depending on the band selected. The lower PLL VCO runs at 825 MHz and is used to operate the IF circuits. The PLL frequencies are set as usual by selecting the proper frequency-division ratio in the feedback paths of the PLLs. A 22-bit serial data word from the embedded controller in the cell phone sets the frequency of each PLL as the need for channel change is determined by the base station.

Thanks to ever-decreasing semiconductor circuit processes, the two chips described earlier are now available on a single chip. In fact, several companies such as Silicon Laboratories, Philips Semiconductor, ST Microelectronics, Texas Instruments, and others now produce single-chip cell phones that allow manufacturers to produce even smaller and cheaper cell phone models.

Figure 20-23   Simplified block diagram of Silicon Labs' SI4133T frequency synthesizer chip.

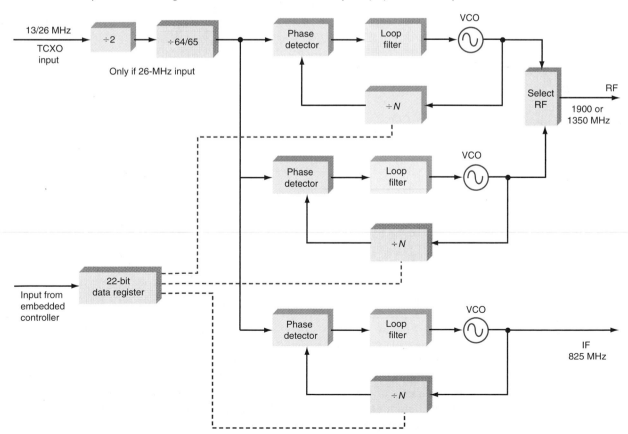

2.5G Cell Phone Systems

The designation *2.5G* refers to a generation of cell phones between the original second-generation (2G) digital phones and the newer third-generation (3G) phones. The 2.5G phones bring data transmission capability to 2G phones in addition to normal voice service. A 2.5G phone permits subscribers to exchange e-mails and access the Internet by cell phone. Because of the small screen size and a small or very restricted keyboard, data transmission capability is limited but available to those who need it.

Currently, three technologies are used in 2.5G systems: GPRS, EDGE, and CDMA2000. Although GPRS and EDGE systems have been implemented, they are generally considered to be temporary solutions to the need for data transmission capability in cell phones. True high-speed packet data capability is available with 3G phones. The CDMA2000 technology is an extension and improvement of the IS-95A/B standard.

GPRS (general packet radio service)

GPRS. One popular 2.5G technology is the *general packet radio service (GPRS).* This system is designed to work with GSM phones. It uses one or more of the eight TDMA time slots in a GSM phone system to transmit data rather than digitized voice. Depending on how many of the eight time slots are used, the data rate can vary from about 20 kbps up to a maximum of 160 kbps. A typical rate is about 40 kbps, which is more than enough for e-mail and short message service (SMS) but poor for Internet access.

Each GSM frame has eight time slots for data. Refer to Fig. 20-18. The overall bit rate is 270 kbps. In voice operation, each slot contains the compressed or vocoded voice signal. In GPRS, other types of data can be transmitted. The data rate that can be achieved is a function of the type of coding used (FEC) and the number of time slots allotted to the data. The GPRS standard, which was created by the European Telecommunications Standards Institute (ETSI), is now maintained by the 3rd Generation Partnership Project (3GPP). It defines four levels of data coding referred to a CS-1 through CS-4. The most robust coding scheme CS-1 produces fewer errors, but the maximum data speed per slot is 8 kbps. The least robust coding method is CS-4, but it produces a data rate to 20 kbps. To achieve maximum data rate, you could use all eight slots for a rate of 8×20 kbps = 160 kbps. However, this is never done. Instead, GPRS defines 12 classes that give different levels of data speed. The selection of the desired class is made by the cell phone carrier who sets just how much of the network capacity is devoted to voice and to data. The classes are shown in the table.

GPRS Class	Number of Downlink Slots	Number of Uplink Slots
1	1	1
2	2	1
3	2	2
4	3	1
5	2	2
6	3	2
7	3	3
8	4	1
9	3	2
10	4	2
11	4	3
12	4	4

The odd classes are rarely, if ever, used. Class 2 is the most used minimal configuration. Class 12 gives downlink and uplink data rates of 80 kbps maximum. The carrier usually adjusts the class to match its own mix of voice and data users and often charges the data user on a per-kbps basis. Keep in mind, too, that the GPRS method involves an automatic rate adjustment algorithm that adjusts the class and data rate to the robustness of the wireless channel. Over shorter distances with less noise and interference, the system can achieve the maximum data rate. Over longer distances with added noise, the system adjusts itself to a lower data rate to ensure accurate transmission of the data.

Virtually all modern GSM phones come with GPRS, but the user must sign up for services (instant messaging, e-mail, etc.) related to this capability.

EDGE. A faster 2.5G technology is *enhanced data rate for GSM evolution (EDGE)*. It is based upon the GPRS system but uses 8-PSK modulation instead of GMSK to achieve even higher data rates up to 384 kbps.

EDGE (enhanced data rate for GSM evolution)

EDGE is sometimes referred to as enhanced GPRS (EGPRS). It is usually implemented as a software upgrade to the base stations but also requires a linear power amplifier. Both hardware and software changes are needed in a GPRS handset. EDGE uses the GPRS class concept whereby the data rate is a function of the encoding and the number of time slots used. By using $3\pi/8$-8PSK modulation, 3 bits is coded per symbol change, thereby tripling the gross data rate. The theoretical maximum data rate is 473.6 kbps with all eight slots used. A more typical implementation is the use of four slots for a data rate of 236.8 kbps. Again, a data rate algorithm automatically backs off on the rate as channel conditions degrade due to noise or increased distance. Typical everyday rates are usually over 100 kbps but less than 200 kbps.

One of the key changes required when EDGE is implemented is the need for linear power amplifiers both at the base station and in the handset. GMSK as used in GSM and GPRS is a type of FM with a constant envelope (amplitude) carrier that changes in frequency with the modulation. FM permits more efficient class C, D, E, and F amplifiers to be used. These amplifiers clip or distort the amplitude of a signal but with FM that does not interfere with the modulation. When $3\pi/8$-8PSK is used, the envelope does change as the signal switches from one phase to another. Therefore, to retain the information content, the amplitude of the signal must be preserved through amplification. A class AB linear power amplifier must be used. Some base stations already use such amplifiers and so may simply adjust them to maximum linear operation rather than maximum efficiency.

In the handsets, efficient class C or E/F power amplifiers in the transmitter must be replaced with a class AB linear amplifier. This is a significant change in a handset as the lower efficiency produces greater heat and shortens the battery life.

Some manufactures have solved the problem by incorporating a new form of modulation called polar modulation. Refer to Fig. 20-24 which shows the basic concept of this method. In polar modulation, the normal I and Q rectangular baseband outputs from the DSP to the transmitter are first converted to a polar format with an amplitude phasor and a phase angle component. The phase signal is usually fed directly to a DAC whose output is the input to a phase-locked loop (PLL) that's used as a phase/frequency modulator. The output is an FSK signal. The PLL VCO output signal is then fed directly to a PA that operates near the saturation/clipping level. It can use any one of the more efficient nonlinear amplifiers designs such as class E.

A DAC converts the amplitude phasor to a signal that can now amplitude-modulate the PA. The method is similar to that used in older high-level AM collector modulators.

Figure 20-24 Basic concept of a polar modulated EDGE transmitter.

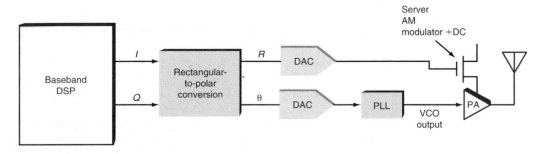

Refer to Fig. 4-13. The amplitude signal is used to control a device such as a MOSFET in series with the PA's collector or drain supply to actually perform the amplitude variation. In older AM transmitters, the final PA operated efficiently as a class C amplifier while the modulation was applied with a transformer in the collector supply line. With this polar modulation scheme you can use an efficient PA but with the necessary amplitude variation components. There are numerous other schemes to produce polar modulation.

CDMA2000. A third, different form of 2.5G digital cell phone is designated CDMA2000. This standard was developed by Qualcomm. It is an extension of the widely used IS-95 CDMA standard also known as cdmaOne. The earliest versions of this radio system were correctly designated as a 2.5G technology, but subsequent improved versions have clearly made it a 3G technology because of the high data rates it can achieve.

The basic CDMA2000 data transmission method is generally called 1×RTT (radio transmission technology). It uses the same 1.25-MHz-wide channels but also changes the modulation and coding formats to actually double the voice capacity over that in IS-95. The data capability is packet-based and permits a data rate of up to 144 kbps which is comparable to EDGE. A version designated 3×RTT uses three 1.25-MHz channels for a total bandwidth of 3.75 MHz. By using a higher chip rate, a maximum date rate roughly 3 times the 1×RTT speed (432 kbps) is possible.

The more recent version is called 1×EV-DO or Evolution-Data Optimized. It has a higher data rate approaching 3.1 Mbps downlink and an uplink rate up to 1.8 Mbps. These speeds definitely qualify this system as a 3G technology. CDMA2000 and its variations are supported by U.S. carriers Verizon and Sprint Nextel.

iDEN Technology

A unique cell phone methodology you will sometimes hear about is the integrated digital enhanced network (iDEN) developed by Motorola for the Nextel cell phone system. It is a digital TDMA system that puts six channels in a 25-kHz band. One of the unique features of the system is that it permits immediate push-to-talk (PTT) communications by those on the network. There is no need to dial. Just push the transmit button as you would in other half duplex two-way radio systems. It also supports paging and a slow data rate messaging feature. A faster version called WiDEN using four 25-kHz channels to get a data rate of 96 kbps has been developed but has not been adopted.

Nextel was recently acquired by Sprint and so has deemed iDEN an older technology for which there is no upgrade path. Users will no doubt be directed to the newer CDMA2000 methods.

3G Cell Phone Systems

3G cell phone system

International Mobile Telecommunication 2000 (IMT-2000)

Third-generation (3G) cell phones are true packet data phones. They feature enhanced digital voice and high-speed data transmission capability. Third-generation phones were originally described by the term *International Mobile Telecommunication 2000,* or *IMT-2000.* The 2000 refers to 2000 MHz, the approximate center of the frequency range defined for 3G (1800 to 2200 MHz). The goal of the International Telecommunications Union (ITU) was to define a worldwide standard for future cell phones to which all other systems could evolve, thereby providing full global roaming. An IMT-2000 phone can achieve a data rate up to 2.048 Mbps in a fixed position, 384 kbps in a slow-moving pedestrian environment, and 144 kbps in a fast mobile environment. With such high-speed capability, a 3G phone can do lots more than just transmit high-quality digital voice.

Some potential 3G applications include fast e-mail and Internet access. With larger color screens and full keyboards, cell phones can act more as small computer terminals.

The 3G phones are already being packaged with personal digital assistants (PDAs) such as the Palm Pilot. High speed also permits the transmission of video. Subscribers can watch a movie on their 3G phones, although the small screen limits viewing. In some models, a built-in CCD camera lets cell phones become picture phones and digital cameras.

UMTS 3G

The ITU did not specify a particular technology to implement 3G. However, it did recommend one worldwide version known as wideband CDMA (WCDMA). This system is also known as the *Universal Mobile Telecommunications Service (UMTS)*. While the standard is still based in the ITU, it is maintained and promoted by the Third Generation Partnership Project (3GPP).

Universal Mobile Telecommunications Service (UMTS)

WCDMA is a direct sequence spread spectrum technology. In the most popular configuration, it is designed to use a 3.84-MHz chipping rate in 5-MHz-wide bands. Duplexing is FDD requiring the matching of 5-MHz channels. The modulation is QPSK. It can achieve a packet data rate up to 2 Mbps.

A key problem in implementing 3G is the need for huge portions of spectrum. New spectrum is scarce and expensive. In Europe, paired bands in the 1900- to 2200-MHz range are available. In the United States, the 806- to 890-MHz range can be used for 3G in some areas. The Department of Defense has transferred some of its spectrum in the 1710- to 1885-MHz range to the FCC for auction to carriers. Also, some segments of the 2500- to 2690-MHz band will be available. The exact 3G spectrum varies widely depending on which part of the world you are in, making it extremely difficult to design a cell phone that is fully operable worldwide.

The UMTS 3G standard also defines a TDD version known as TD-SCDMA for time-division synchronous code-division multiple access. It is designed to use a 1.6-MHz-wide channel with a chipping rate of 1.28 MHz. Different time slots in the time-multiplexed data stream are assigned to uplink and downlink activity. The number of uplink and downlink channels may be dynamically assigned so that a carrier can adjust the system to the traffic load at any given time. The primary benefits of TD-SCDMA are that less spectrum is needed. Only a single 1.6-MHz channel is needed. Furthermore, since duplexing is TDD, there is no need for paired spectrum as in WCDMA or GSM or any other FDD system. The downside is that the system is more complex because of the extreme need for accurate timing and synchronization required for proper operation. So far, the only nation to adopt TD-SCDMA as a standard is China.

When the 3G WCDMA standard was first adopted, it was assumed that it would be put into use far faster than it has. During the past years, wireless technology has changed, making the original specifications somewhat behind the times. The maximum 2-Mbps data rate was assumed to be fast enough for any service. But today, the demand for faster data speeds is growing, especially because of the growing demand for mobile video service. Because of the need for faster systems, a new system compatible with WCDMA has been developed. Known as *high-speed downlink packet access (HSDPA)*, this so-called 3.5G technology is an add-on to WCDMA systems. HSDPA is to WCDMA as 1xEV-DO is to cdma2000. It provides a packet data rate from the base station to the handset over 5 times that of the 2-Mbps maximum rate of WCDMA.

High-speed downlink packet access (HSDPA)

HSDPA uses an adaptive coding and modulation scheme with QPSK and 16-QAM. Data is transmitted in 2-ms frames. There are 12 categories of HSDPA that define different coding and modulation schemes. The minimum is category 11,900 kbps using QPSK. Category 6 gives 3.6 Mbps using 16-QAM. The maximum data rate is 14.4 Mbps using 16-QAM in category 10. The actual rate achieved is a function of the link quality. High noise and long range give a lower rate. The rate adapts to the channel conditions automatically.

While most data needs will be served by a high-speed downlink capability, in some applications a fast uplink may be needed. This is accommodated by a companion standard known as *high-speed uplink packet access (HSUPA)*. A fast handset to base station rate is more difficult to implement so uplink rates are naturally slower. HSUPA provides a maximum data rate of 5.76 Mbps. Again, the rate adapts to the channel conditions.

High-speed uplink packet access (HSUPA)

It has taken many years for WCDMA 3G to be adopted, as it requires a complete new system infrastructure to be put in place. All new base stations require an investment of many billions of dollars. Furthermore, it requires significantly more spectrum than older technologies, and today governments sell spectrum to the highest bidder after adding additional costs. The availability (or not) of spectrum in many parts of the world is also an issue. All new, more expensive, and complex handsets are also needed. It is no wonder that cell phone carriers have delayed this investment. They did so by expanding their 2G systems to 2.5G with lower-cost upgrades to existing equipment and systems. New handset designs were also evolutions rather than the revolution required by 3G. The high-speed packet data access provided by 2.5G systems has been more than adequate in satisfying users with e-mail, messaging, Internet access, and other features such as digital cameras and even video.

The normal planned upgrade path for carriers now using GSM, GPRS, and EGDE is to adopt WCDMA. This is the preplanned path for most European carriers and the U.S. carriers Cingular and T-Mobile. Currently over one-half of the European carriers have launched 3G service. In the United States, 3G service is only available in a few major cities, but the conversion is underway. Carriers must continue to maintain their older systems as the 3G systems are phased in. In addition, handset manufacturers now offer 3G phones all of which also contain full GSM, GPRS, and EDGE capability if the subscriber happens to be in an area not served by 3G.

CDMA2000

While CDMA2000 started as a 2.5G method, its subsequent upgrades to 1xEV-DO have made it clearly a valid 3G technology. Already, both Verizon and Sprint Nextel have implemented 1xEV-DO in many U.S. cities. The downlink packet rate can be as high as 2.4 Mbps while the uplink rate is a maximum of 153.6 kbps. This technology uses the existing 1.25-MHz bands used for all CDMA2000.

A newer version known as 1xEV-DV for Evolution-Data/Voice has a maximum packet rate of 3.07 Mbps. Uplink speed is the same as that of 1xEV-DO. It has not yet been widely adopted.

CDMA2000 is by far the most widely adopted 3G technology worldwide. What makes it so successful and desirable is that it is fully backward-compatible with all other versions of CDMA2000 as well as the older cdmaOne (IS-95) phones and systems. It is expected that WCDMA usage will eventually equal and possible surpass CDMA2000 use although both will continue to be supported.

What is interesting is that other non–cell phone technologies such as wireless local-area network (WLAN) and wireless metropolitan-area network (WMAN) technologies will compete with 3G systems in some areas. Both of the wireless systems, described in Chap. 21, were developed for reasons other than voice communications. WLANs are used to connect computers into networks. WMANs are used as broadband networks to compete with cable TV and DSL lines. Yet, by using Voice over Internet Protocol (VoIP) methods, both of these wireless methods are capable of providing voice service. It is possible that future cell phones will be capable of placing calls from one of three different systems.

4G Systems

While 3G systems have yet to be fully deployed, already work is continuing on a fourth-generation cell phone system. It is expected that the 4G system will feature orthogonal frequency-division multiplexing (OFDM), MIMO, or adaptive antenna arrays and have a packet data rate of up to 20 Mbps. Such systems are not expected until well after 2010.

Advanced Cell Phones

Cell phones are no longer just two-way radios for making phone calls. Handset manufacturers have built in a wide range of features that make the cell phone the hottest, most desired consumer electronic product ever developed. Here are just a few of the neat features you can now enjoy. As you consider each, think of what electronic circuits and systems are needed to make these features work.

Color LCD Screens. Early cell phone screens were monochrome, but today color is common. It brightens the display and permits advanced features such as digital cameras and video to be implemented.

Digital Cameras. Over 50 percent of all cell phones sold today have a built-in digital camera. A lens focuses a scene on a CCD imaging device, which converts it to a serial color bit stream that is stored in memory. The picture is displayed on the cell phone's color LCD screen. It may then be transmitted to any other phone or PC via e-mail. While these cameras do not have the resolution of standalone digital cameras, their quality is excellent with as many a 2 megapixels. Many are also capable of storing a small amount of real-time video imaging as a camcorder does.

E-mail. While many cell phones are capable of sending and receiving e-mail via a Wireless Internet connection, special phones designed especially to deliver and accept e-mail have become enormously popular. The most widespread is the Research In Motion (RIM) product known as BlackBerry. This device is a fully functional cell phone, but it has a full alphanumeric QWERTY keyboard that subscribers use to thumb-type messages. The accompanying e-mail service automatically pushes any new e-mails out to the device through special software and services implemented by the cell phone carrier and RIM. Similar phones such as the Palm Treo and models from Motorola and Nokia are also available.

Games. Many cell phones incorporate individual user games with color graphics and sound. New games may be downloaded, and in some cases gaming is interactive online via the Internet.

GPS. A few cell phones contain a complete Global Positioning System satellite navigation receiver for pinpointing the exact location.

Internet Access. Virtually all cell phones today allow access to the Internet. A special operating system reformats the images for the small screen.

MP3 Players. Many of the newer phones are now available with MP3 music players that allow a user to store hundreds of favorite songs in flash memory for replay when the cell phone is not being used. Some phones offer a miniature hard drive to store even more songs.

Push-to-Talk. This feature is common on all half duplex two-way radios. This special service allows users within a group or company to stay in touch instantly without dialing.

FM Radio. The availability of a single-chip FM radio lets cell phone manufacturers build a regular 88- to 108-MHz stereo FM radio into a phone. The headset cable does double duty as the antenna.

Wireless Headsets. Headsets are widely available for cell phones. Most attach to your ear and include a single earphone and a microphone. It makes hands-free cell phone operation possible. A key irritation is the cable between the phone and headset. The problem in now solved with the common availability of a wireless headset. Both the headset and the phone have built-in short-range transceivers that talk to one another, eliminating the cable. The technology used is Bluetooth, described in Chap. 21.

Video. Thanks to high-speed data transmission in cell phones and video compression technology, TV on a cell phone is now possible. Although watching movies or long programs is impractical, many short video programs are being developed and new video services tested.

Location-Based Technology. Another feature included in modern phones is *enhanced 911 (E911)* capability. This system is mandated by the U.S. government. All cell phone carriers must have a system that makes it possible to locate any cell phone position automatically. This permits emergency medical services or automobile towing crews to find the cell phone used to make the 911 emergency call. Several different systems have been adopted by the various carriers. Most CDMA phones contain a GPS receiver that transmits its coordinates digitally to the carrier, from which they can be forwarded to emergency services. GSM, GPRS, and EDGE phones use a system called Uplink—Time Difference of Arrival (U-TDOA), a method of triangulation based on cell phone signals being received at three different cell sites. Although many subscribers like the idea of having their location known, others believe that this is an invasion of privacy. Carriers are attempting to devise various "location services" that they might sell to subscribers. For example, such services could provide information or advertisements about nearby restaurants or parking lots.

Figure 20-25 shows a generic block diagram of an advanced cell phone. Called a feature phone or smart phone, this device contains many of the features listed above. No phone has all features. Users choose from hundreds of offerings that provide just the features they want.

Note in Fig. 20-25 that the average cell phone today contains multiple radios. This one contains a UMTS/WCDMA 3G radio but also a complete GSM/GPRS/EDGE radio

Figure 20-25 General block diagram of an advanced cell phone.

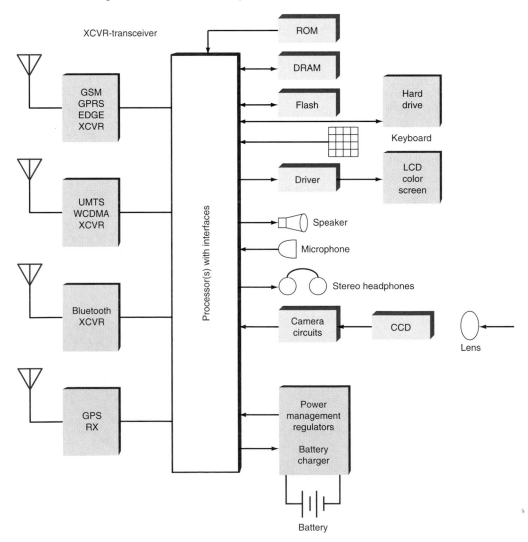

when WCDMA service is not available. This phone also has a Bluetooth transceiver for wireless headset operation, but it can also be used to interface to a laptop computer for Internet access via the cell phone service. A GPS receiver (RX) is included for navigation.

Most of the special and advanced features are implemented by one or more processors. The processor may be a single powerful 32-bit microcontroller or a 32-bit microcontroller plus one or more DSPs. Since cell phones are essentially software-defined radios today, virtually all modulation, demodulation, encoding, decoding, and voice or video compression and decompression are handled by one or more processors. The processor runs an operating system that manages the DRAM and flash memory and all the inputs, outputs, and applications. Software is stored in a ROM.

A key element of a cell phone is its battery supply and power management hardware. The battery operates a whole array of multiple voltage regulators and dc-dc converters that supply all the different operating voltages to the various ICs, interfaces, and peripheral devices. The power management system also shuts down unneeded circuits and devices when they are not in use, to extend battery life and time between charges. The battery charger is built in. The tradeoff in any cell phone today is features versus talk and charge time. For every new feature added, more current is drawn from the battery so battery life shortens and charge intervals decrease.

Base Stations

The most complex and expensive part of any cellular telephone system is the network of *base stations* that carriers must have to make it all work. Over the years, carriers have added many more base stations to handle the constantly growing number of subscribers. In addition, each base station has expanded and become more complex because of the growing number and variety of radio standards it must handle. Base stations must continue to support AMPS through 2007 as well as 2G technologies, 2.5G enhancements, and now 3G systems. Support for multiple standards has led to considerably more equipment. An effort has been made by base station manufacturers to consolidate the equipment by using software-defined radio techniques and DSP. These methods permit the base station receivers to accommodate existing multiple standards and to be able to work with new standards by reprogramming rather than replacing equipment.

Base stations consist of multiple receivers and transmitters so that many calls can be handled on many different channels simultaneously. The transmitters in the cell site are much more powerful than those in the handsets. Power levels up to 40 W are typical. These power levels are achieved with highly linear broadband class A or class AB power amplifiers. Superior linearity is critical, especially in CDMA systems that cover a broad spectrum. Nonlinearities produce intermodulation and spurious signals that can make a system inoperable. The power amplifiers incorporate special feedback and cancellation techniques that eliminate distortion and thus ensure linear operation.

The most visible feature of a base station, of course, is its antenna on a tower. The antennas used by base stations must serve many transmitters and receivers by means of isolators, combiners, and splitters. Base station antennas have become directional rather than omnidirectional, as the cell patterns suggest. This "sectorization" of the cell site has helped to increase subscriber capacity with minimal cost. Most base stations use a triangular antenna array that looks like the one shown in Fig. 20-26. On each side of the triangular frame is an array of three vertical antennas forming a broadside or collinear array that may also use reflectors. Each of the three arrays produces a gain of about 8 dB and an antenna pattern with a beam width of 120°. This divides the cell coverage into three equal sectors. See Fig. 20-27. This directional capability provides excellent isolation of the three sectors, which in turn permits the same channel frequencies to be used in each of three 120° sectors. Some sites also use two supplementary antennas that provide spatial diversity, which greatly improves the reception of weak signals from the handset. Carriers are beginning to use more sophisticated "smart" beam-forming antennas with more elements, arrays, and sectors to further improve capacity as well as to meet the forthcoming E911 location requirement of the U.S. government.

Base station

Figure 20-26 Typical triangular cell site antenna.

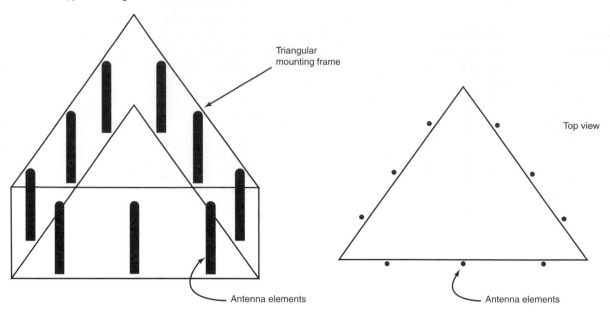

Figure 20-27 Horizontal radiation and reception pattern of a typical cell site antenna.

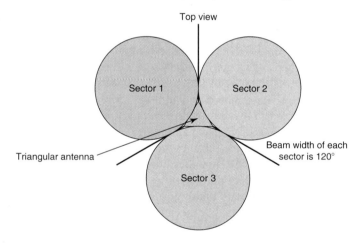

CHAPTER REVIEW

Summary

The cellular radio system was developed during the 1970s and came into wide use during the 1980s. Basically, a cellular radio system provides standard telephone service by two-way radio. The original cellular phone system was called the advanced mobile phone system, or AMPS. It was based on analog radio technologies.

Cell phone systems divide a geographic area into cells, each of which covers a few square miles. Each cell is connected to a master control center known as the mobile telephone switching office (MTSO).

The five sections of a typical handset are the transmitter, receiver, frequency synthesizer, logic unit, and control unit.

The use of analog cell phones expanded rapidly, so much so that new and more efficient ways to increase the number of users a system could handle were needed. Digital systems have provided the solution to this problem. Analog systems will remain in use for some time, but digital techniques offer many benefits and will no doubt replace the analog systems in time.

Second-generation (2G) digital cell phones are now widely used (the first generation was the analog AMPS). Two 2G systems use time-division multiplexing (TDM). One is the Global System for Mobile Communications (GSM), and the other is the IS-136 standard for time-division multiple access (TDMA). The third 2G system is the IS-95 standard for spread spectrum (SS) and is called code-division multiple access (CDMA).

An intermediate system, called 2.5G, is the generation of cell phones between 2G and the new third-generation (3G) phones. The 2.5G phones provide data transmission capability in addition to voice service. Three 2.5G technologies have emerged: general packet radio service (GPRS), enhanced data rate for GSM evolution (EDGE), and CDMA2000.

Third-generation (3G) cell phones are true packet data phones, featuring enhanced digital voice and high-speed data transmission. They were originally described by the term *International Mobile Telecommunication 2000*, or *IMT-2000*. Applications of 3G cell phones include fast e-mail and Internet access and even video. In some models a CCD camera turns 3G phones into picture phones and cameras. Another feature is E911 capability, which makes it possible to locate any cell phone position.

It is expected that two 3G systems will eventually coexist: wideband CDMA (WCDMA) and CDMA2000 (an enhancement by Qualcomm of IS-95). Fourth-generation (4G) systems using OFDM are already being developed.

Questions

1. What key communication technology is referred to when the term *wireless* is used?
2. What is the name of the original North American analog cell phone system?
3. Name the two main frequency bands used for cell phones in the United States.
4. What primary characteristic determines cell site coverage?
5. What does *PCS* mean, and what does it refer to?
6. Name four access methods used in cellular systems.
7. True or false? Cellular telephone radios operate full duplex.
8. What type of modulation is used in an analog cellular radio?
9. What is the maximum power output of an AMPS mobile handset?
10. What is the transmit frequency range?
11. What is the receive frequency range?
12. What is the channel spacing between AMPS cellular stations?
13. What is the separation between send and receive frequencies?
14. Which type of duplexing is the most widely used?
15. What circuit in the cellular telephone allows a transmitter and receiver to share an antenna?
16. What is the source of the frequency divider ratios in the frequency synthesizer PLLs in the cellular telephone?
17. Name two signals that are transmitted back to the cell site and monitored by the MTSO.
18. Name three conditions in the transceiver controlled by the MTSO.
19. What is the name of the section of the cellular transceiver that interprets the serial digital data from the cell site and MTSO?
20. What is the NAM, and where is it kept?
21. What type of IC implements the control section of a cell phone?
22. How are the transmit and receive frequencies used by a cellular telephone determined?
23. What is the purpose of the APC circuit? What operates this circuit?
24. What technique allows multiple base stations to share a common channel?
25. Name the major benefits of digital cellular radio over analog.
26. What is a vocoder?
27. List the three primary 2G digital cell phone systems used worldwide. Which is the most widely used?
28. Name the two primary functions of a vocoder.
29. In IS-136 TDMA, how many users can use a single 30-kHz channel?
30. What is the modulation used in IS-136 TDMA?
31. What is the data rate in an IS-136 channel?
32. What organization maintains the standards for IS-136 and IS-95?
33. What unit of measure is used to determine the spectral efficiency of a modulation scheme?
34. What is the bandwidth of a GSM channel?
35. How many users can share a channel in GSM?
36. True or false? GSM uses TDMA.
37. What modulation is used in GSM?
38. What type of multiuser access is available in IS-95 CDMA?
39. What is the bandwidth of a typical CDMA channel? State the maximum number of subscribers that can use the channel.
40. What feature of CDMA is critical to its success in reception?
41. By what method is a CDMA channelized?
42. What two receiver architectures are common in most modern 2G cell phones? State why they are preferred over the older superheterodyne architectures.
43. Are low-IF receivers dual-conversion?
44. Why is the *I/Q* circuit arrangement used in cell phones?

45. What is an image reject mixer? Explain briefly how it works.
46. List four functions typically performed by a DSP chip in the baseband part of a cell phone.
47. At what point in most receivers are the A/D converters placed?
48. In the GSM transceiver described in the text, how is the channel frequency selected?
49. Describe the basic antenna structure of a typical cell site base station.
50. Describe the antenna radiation pattern and explain how it permits sectorization.
51. How does the antenna permit frequency reuse?
52. What types of amplifiers are used in base stations? What is their power level?
53. Why is power amplifier linearity so important?
54. Name the three 2.5G technologies widely used to transmit data via cell phone.
55. What makes EDGE faster than GPRS?
56. What name did the ITU give to the 3G cell phone effort?
57. State the basic characteristics and features of a 3G cell phone.
58. Name four data applications of 2.5G and 3G phones.
59. What is E911? Name two ways in which it is implemented.
60. Name the most popular e-mail phone.

61. Is CDMA2000 a 2G or 3G technology? Explain. Who developed it?
62. What is the typical average data rate of a phone with GPRS? EDGE?
63. What are the names of the fastest versions of CDMA2000? What are their maximum data rates?
64. What 3G technology did China adopt?
65. Name the technology that makes WCDMA faster. What modulation methods are used?
66. List the primary bands to be used for 3G phones in the United States. What are two potential future bands?
67. What is the name of the modulation method adopted by some EDGE handsets?
68. Why do EDGE cell phones and base stations require linear power amplifiers?
69. True or false? A WCDMA 3G handset can communicate with a CDMA2000 base station.
70. True of false? A WCDMA 3G handset can communicate with a GSM/GPRS/EDGE base station.
71. Explain the push-to-talk feature on some cell phones.
72. Name the U.S. carriers that use GSM/GPRS/EDGE and WCDMA.
73. Name the U.S. carriers that use CDMA2000.
74. What antenna technology permits increased subscriber growth on existing base stations?
75. Briefly describe what a 4G phone may be like.

Problems

1. Compute the spectral efficiency of modulation that gives a data rate of 2.4 Mbps in a bandwidth of 1.5 MHz. ◆
2. Without any special modulation methods, approximately what bandwidth is needed to transmit a data rate of 1.2 Mbps?
3. How many phase positions does π/4-DQPSK have? ◆
4. A cell phone receiver has a first IF of 82 MHz and a second IF of 456 kHz. What is the LO frequency on the second mixer?

5. In a low-IF cell phone receiver, what is the IF? ◆
6. How many GSM channels can exist in the spectrum of Fig. 20-4(b)?

◆ *Answers to Selected Problems follow Chap. 22.*

chapter **21**

Wireless Technologies

Introduction

The most prevalent wireless technology is that associated with cell phones as described in Chap. 20. However, there are many more wireless systems and applications in common use today. These are primarily short-range systems that have a range of a few inches up to several miles depending upon the application. This chapter describes the more widely used of these popular systems. Each is generally defined by a specific industry standard and is identified with one or at least a few well-known applications. These popular technologies and their main applications are summarized in the table below. Note that each technology is identified by its common name or an IEEE standard number. In Fig. 21-1, the technologies are compared by both data rate and range. As you can see, some of the technologies overlap in their coverage. There is a wireless system that fits virtually every need.

Figure 21-1 Range versus data rate: common wireless technologies.

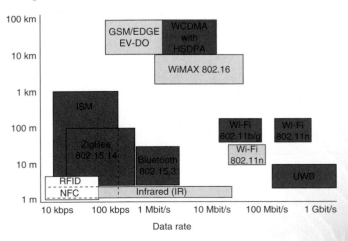

Wireless Technology	Primary Applications
Wi-Fi 802.11	Wireless Ethernet LAN
Bluetooth	Short-range wireless primarily for audio
ZigBee 802.15.4	Home/industrial monitoring and control
Ultrawideband (UWB)	High speed and short range for computer peripherals and video connections
WiMAX 802.16	MAN broadband wireless
Radio-frequency ID (RFID)	Wireless bar code for tracking and ID
Near-field communications (NFC)	Ultrashort-range for smart cards and ID
Industrial scientific medical (ISM)	Mixed-use unlicensed short-range telemetry and control
Infrared (IR)	Optical wireless for control and data

The following sections describe each of these technologies in greater detail.

Objectives

After completing this chapter, you will be able to:

- Define personal-area network (PAN).
- Name three popular PAN technologies.
- Identify the features, benefits, applications, and operation of the wireless technologies Wi-Fi, Bluetooth, ZigBee, WiMAX, and ultrawideband (UWB).
- Explain the operation and applications of the wireless technologies RFID, NFC, and IR.

21-1 Wireless LAN

Local-area networks (LANs) within a company, government agency, hospital, or other organization typically use CAT5 or CAT6 unshielded twisted pair as the transport medium. However, more and more, wireless extensions to these LANs are becoming popular as are entirely wireless LANs. Low-cost wireless modems installed in personal computers and laptops make this possible. Two common configurations are shown in Fig. 21-2(*a*). Here a wireless access point (AP) is connected to an existing wired LAN, usually through an Ethernet switch. This AP contains a transceiver that can cover a specific geographic area, usually inside a building. This area usually extends out to no more than about 100 m, but generally the range is less due to the great signal attenuation of the walls, ceilings, floors, and other obstructions. PCs or laptops within that range and containing a radio modem can link up with the AP which, in turn, connects the PC or laptop to the main LAN and any services generally available via that LAN such as e-mail and Internet access.

Another popular configuration is shown in Fig. 21-2(*b*). Here the AP is connected to the main LAN or more commonly to an Internet service provider (ISP) by way of a

Figure 21-2 Types of WLANs. (*a*) Access point extension to a wired LAN. (*b*) Public access point via an Internet service provider (ISP).

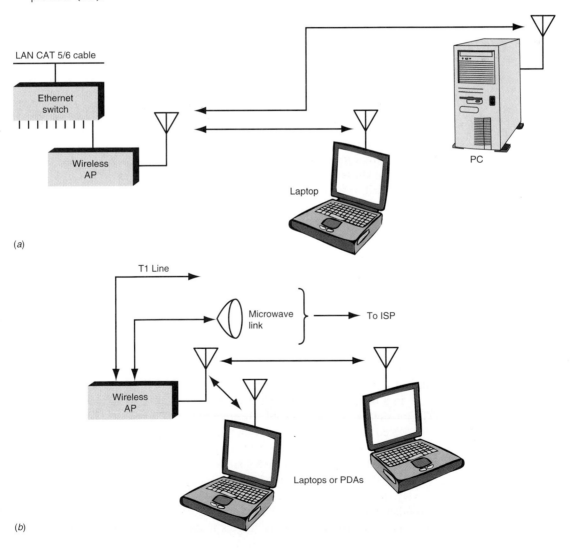

long-range interconnection such as a hardwired T1 or T3 line or a microwave relay link such as WiMAX, as described later in this chapter. The AP is usually installed in a restaurant, coffee shop, airport, hotel, convention center, or other public place. It is more commonly known as a "hot spot." Some cities are also installing municipal hot spots. Anyone with a laptop equipped with a LAN modem interface can link up to the AP and access his or her e-mail or the Internet. There are hundreds of thousands of hot spots around the world.

What makes the wireless LAN so appealing is that it offers flexibility, convenience, and lower costs. To add a node to an existing wired LAN, the main problem is the new wiring. If such wiring is not in place already, it is time-consuming and expensive to pull cables through walls and ceilings and to install connectors. Moving computers within a building because of office reconfiguration is a huge problem and expense unless existing wiring can be reused. By using a wireless extension such problems essentially disappear. Any computer can be located at any new point quickly and easily at no additional cost. As long as the computer is within the range of the AP, the connection is automatic. Wireless is a great way to expand an existing network.

Wireless LANs also serve our continuing need to be more mobile in our jobs and activities. The cell phone has given us freedom to maintain communications anywhere, at any time, and virtually in any place. The wireless LAN also gives us that same portability for our computers, mainly laptops which have essentially become the de facto PC form factor. Within an organization, a user can take her or his laptop to the conference room for a meeting, to a colleague's office, or to the cafeteria for lunch. And with all the available hot spots, we can use our laptops almost anywhere, especially while we are traveling.

Another growing use of wireless LANs is in the implementation of home networks. As more and more families become multi-PC users, there is a need to interconnect each PC to a broadband Internet connection such as a DSL or cable TV line. It allows each user to access e-mail or the Internet or to share a common peripheral such as a printer. Most homeowners do not want to wire their homes with CAT5/6 cable at great expense. Installing a wireless LAN is fast, easy, and very inexpensive these days. A special box called a residential gateway or wireless router connects to the cable TV or DSL and serves as the access point. This gateway or router uses a software approach called *network address translation (NAT)* to make it appear as if each networked PC has its own Internet address, when in reality only the one associated with the incoming broadband line is used.

Network address translation (NAT)

Hardware of Wireless LANs

The hardware devices in a wireless LAN are the access point or the gateway/router and the radio modems in the PCs. The access point is just a box containing a transceiver that interfaces to an existing LAN by way of CAT5/6 wiring. It typically gets its dc operating power via the twisted-pair cabling since the dc supply voltage is superimposed on the data. The IEEE 802.3af standard related to furnishing dc power over the network cable is referred to as *Power over Ethernet (PoE)*. The AP is usually mounted high on a wall or ceiling to give good coverage to a specific area. The antenna may be built into the box or may be a separate array that gives directionality to the AP to ensure coverage of a desired area and minimum interference to other nearby WLANs.

Power over ethernet (PoE)

In a home network, the gateway or router is designed to attach to the DSL or cable TV modem with CAT5/6 cable. It often attaches to one of the PCs in the home network by cable. The other PCs link to the gateway/router wirelessly.

The radio modems for each PC take many forms. All are transceivers with an accompanying antenna. The transceivers are usually a single chip in most of the newer systems. In the older systems, the modem is contained on a plug-in card for the PC/PCI bus. Today, it is more common to have the radio modem built into the PC motherboard. Another form is an external modem that connects to the PC by way of a USB port common on all PCs. The modem gets its dc power via the USB port.

For laptops, the modem is most often built in, so no special installation is needed. Another popular format is the card bus modem that plugs into the PCMCIA slot available on most laptops.

Wireless LAN Standards

Over the years, a number of wireless LAN methods have been developed, tested, and abandoned. One standard has emerged as the most flexible, affordable, and reliable. Known as the IEEE 802.11 standard, it is available in multiple forms for different needs. The table shows the different versions of the standard and some technical details.

IEEE Standard	Frequency, GHz	Access	Max. Data Rate, Mbps	Max. Range, m
802.11a	5	OFDM	54	50
802.11b	2.4	DSSS	11	100
802.11g	2.4	OFDM	54	100
802.11n	2.4	OFDM/MIMO	600	100

IEEE 802.11b. The earliest useful and most widely adopted version of the 802.11 standard is 802.11b. It operates in 11 channels in the 2.4-GHz unlicensed ISM band. This band extends from 2.4 to 2.4835 MHz for a total bandwidth of 83.5 MHz. The center frequencies of each channel are given below.

Channel	Center Frequency, GHz
1	2.412
2	2.417
3	2.422
4	2.427
5	2.432
6	2.437
7	2.442
8	2.447
9	2.452
10	2.457
11	2.462

Note that the channels are spaced 5 MHz apart over the spectrum. However, each channel is 22 MHz wide so the channels overlap. Any given AP uses one of these channels. The access method is direct sequence spread spectrum (DSSS) so that multiple signals may share the same band. Channel assignments are critical in facilities where multiple WLANs exist, so that interference is minimized.

The 802.11b standard specifies a maximum data rate to 11 Mbps. This rate is achieved only under the most favorable path conditions such as minimum range and minimum noise. Increasing range or noise causes the rate to automatically drop off to 5.5, 2, or 1 Mbps. This helps ensure a reliable connection despite the lower speed. At the 1- and 2-Mbps rates, the serial data signal is XORed with an 11-bit code called the Barker code to produce the DSSS signal. This particular bit sequence has unique properties that make it easy to receive and decode. The Barker sequence is 10110011000. Each serial data bit is XORed with this code. For modulation, 1 Mbps is achieved with DBPSK. For 2 Mbps, the modulation is DQPSK.

To achieve its faster rates of 5.5 and 11 Mbps, a different form of coding called *complementary code keying (CCK)* is used. The serial data signal is then modified by

Complementary code keying (CCK)

using one of 64 eight-bit codes to represent 6 bits of the serial data signal. The bit coded bits are the chips. The modulation is *differential quadrature phase-shift keying (DQPSK)*. Four chips per 6-bit sequence are used to achieve the 5.5-Mbps rate. The use of CCK greatly improves the performance of the signal under noise and multipath conditions because the unique codes have properties that make them easier to identify and decode under adverse conditions.

As conditions degrade between the AP and the wireless node due to increased distance, noise, or number of obstacles, the transceiver automatically readjusts to the changing conditions by adjusting the data rate downward, first to 5.5 Mbps also using DQPSK/CCK, then to 2 Mbps using DQPSK alone, and then to 1 Mbps using DBPSK. The maximum allowed *equivalent isotropic radiated power (EIRP)* is 1 W. Most IC transceivers produce an output of 100 mW. Gain antennas may be used as long as the output power plus the antenna gain is less than the 1-W EIRP allowed. A variety of power amplifier accessories and antennas are commercially available to customize each LAN.

The overall range depends upon environmental conditions. Indoors the range is typically less than 100 ft at 11 Mbps. It drops to 1 Mbps at about 300 ft. Outdoors with a clear line of sight and maximum EIRP, a range up to 8 km (about 12 mi) can be achieved.

IEEE 802.11a. The 802.11a standard was developed next. It uses the unlicensed 5-GHz band. There are three authorized segments: 5.15 to 5.25 GHz with 50-mW maximum power, 5.25 to 5.35 GHz with 250-mW maximum power, and 5.725 to 5.825 GHz at a maximum of 1 W of power. Each of these bands is divided into multiple nonoverlapping 20-MHz-wide channels. Each channel is designed to carry an OFDM signal made up of 52 subcarriers, 48 for data and the other 4 for error correction codes. Each of the subcarriers is about 300 kHz wide.

As with the 802.11b standard, the 802.11a version supports a wide range of data rates. The fastest is 54 Mbps. Other backoff rates usually include 48, 36, 24, 18, 12, 9, and 6 Mbps. Each uses a different modulation scheme. For 6 Mbps, BPSK is used. For 12 Mbps, QPSK is used. For the higher rates QAM is used; 16-QAM gives 24 Mbps while 64-QAM is used to achieve 54 Mbps. The standard provides for backoff data rates as the link conditions deteriorate due to increased range, noise, or multipath interference.

The key advantage of the 802.11a standard is that the frequency band is much less used than the busy 2.4-GHz band which contains microwave ovens, cordless phones, Bluetooth wireless, and a number of other services, all of which can cause interference at one time or another, thereby producing interference that can block communications or at least decrease the range and data rate. With fewer interfering signals in the 5-GHz band, there is less interference and greater reliability.

The downside of this standard is its shorter range. As frequency of operation increases, the given transmission range typically decreases. Indoor operation greatly reduces range because 5-GHz signals are more easily absorbed and reflected than 2.4-GHz signals. With 802.11a, the maximum range is about 50 m at the maximum data rate.

IEEE 802.11g. The 802.11g standard was an attempt to extend the data rate within the popular 2.4-GHz band. Using OFDM, this standard provides for a maximum data rate of 54 Mbps at 100 ft indoors. As with the 802.11a standard, there are lower backoff rates, as described earlier, as the communications path degrades. The 802.11g standard also accommodates the 802.11b standards and so is fully backward-compatible. An 802.11b transceiver can talk to an 802.11g AP but at the lower data rate. An 802.11g transceiver can also talk to an 802.11b AP but also at the lower data rate.

IEEE 802.11n. The newest standard is the 802.11n version which was developed to further increase the data rate. It also uses the 2.4-GHz band and OFDM. A primary feature of this standard is the use of multiple-input multiple-output (MIMO) antenna systems to improve reliability of the link. APs for 802.11n use two or more transmit antennas

Differential quadrature phase-shift keying (DQPSK)

Equivalent isotropic radiated power (EIRP)

828   *Chapter 21*

and three or more receive antennas. The wireless nodes use a similar arrangement. In each case multiple transceivers are required for the AP and the node. This arrangement permits a data rate in the 100- to 600-Mbps range at a distance up to 100 m. MIMO systems greatly mitigate multipath problems and help extend the range and reliability of the wireless link.

In all these standards, the carrier sense multiple access with collision avoidance (CSMA/CA) access method is used to minimize conflicts among those wireless nodes seeking access to the AP. Each transceiver listens before it transmits on a channel. If the channel is occupied, the transceiver waits a random period before attempting to transmit again. This process continues until the channel is free for transmission.

Wireless Security. The 802.11 standard also includes provision for encryption to protect the privacy of wireless users. Since radio signals can literally be picked up by anyone with an appropriate receiver, those concerned about privacy and security should use the encryption feature built into the system. The basic security protocol is called *Wired Equivalent Privacy (WEP)* and uses the RC4 encryption standard and authentication. WEP may be turned off or on by the user. It does provide a basic level of security; however, WEP has been cracked by hackers and is not totally secure from the most high-tech data thieves. Two stronger encryption standards called *Wi-Fi Protected Access (WPA)* and WPA2 are also available in several forms to further boost the encryption process. The IEEE also has a security standard called 802.11i that provides the ultimate in protection. It removes all doubt about the safety of data transmission for even the toughest applications. Another standard, 802.11x, provides a secure method of authentication for wireless transactions.

Equivalent privacy (WEP)

Wi-Fi protected access (WPA)

While the IEEE standards attempt to establish the technical specifications of the transceivers and their interfaces, different implementations of the standards by different semiconductor manufacturers have created interoperability problems. What this means is that even though each manufacturer has met the standard, some minor variation prevents the transceiver from communicating with the transceiver of another manufacturer. Such interoperability problems greatly limited the adoption during the early stages of 802.11b WLANs. It was so bad that manufacturers banded together to form the Wi-Fi Alliance. Wi-Fi is short for *wireless fidelity,* and it has been adopted as the trademark of 802.11 products. The Wi-Fi Alliance set up testing and certification standards that all vendors had to meet to ensure full interoperability of all products. Any Wi-Fi certified AP or wireless node will talk to any other Wi-Fi certified products. The full interoperability brought about by the testing and certification process has made this wireless standard popular and very widely used.

21-2 PANs and Bluetooth

A *personal-area network (PAN)* is a very small network that is created informally or on an ad hoc basis. It typically involves only two or three nodes, but some systems permit many nodes to be connected in a small area. Although PANs can be wired, today all PANs are wireless.

Personal-area network (PAN)

The most popular wireless PAN system is *Bluetooth,* a standard developed by the cell phone company Ericsson for use as a cable replacement. The objective was to provide hands-free cell phone operation by eliminating the cable connecting a cell phone to a headset. Today, this is one of the main applications of Bluetooth, but it also has other cable replacement applications.

Bluetooth

Bluetooth is a digital radio standard that uses frequency-hopping spread spectrum (FHSS) in the unlicensed 2.4-GHz ISM band. It hops over 79 frequencies spaced 1 MHz apart from 2.402 to 2.480 GHz. The hop rate is 1600 hops per second. The dwell time on each frequency, therefore, is $1/1600 = 625$ μs. During this time, digital data is transmitted. The total data rate is 1 Mbps, but some of that is overhead (headers, error detection and correction, etc.). The actual data rate is 723.2 kbps simplex or 433.9 kbps duplex. The data, which may be voice or any other digitized information, is put into packets and transmitted sequentially in as many as five time slots. The serial data signal is Gaussian-filtered, and then FSK is used for modulation. The frequency shift between binary 0 and 1 is ±160 kHz.

Three levels of transmission power have been defined, depending upon the application. For short distances up to 10 m, class 3 power at 0 dBm (1 mW) is used. For longer distances or more robust operation in an environment with obstacles and noise, the higher-power class 2 can be used with 4 dBm or 2.5 mW. Maximum Bluetooth range is about 100 m and is achieved with class 1 power of 20 dBm or 100 mW.

Bluetooth transceivers are available as single-chip transceivers that interface to the device to be part of a PAN. These devices invariably contain some kind of embedded controller that handles the application. If voice is used, a vocoder is needed.

Bluetooth is set up so that the wireless transceiver constantly sends out a search signal and then listens for other nearby, similarly equipped Bluetooth devices. If another device comes into range, the two Bluetooth devices automatically interconnect and exchange data. These devices form what is called a *piconet*, the linking of one Bluetooth device that serves as a master controller to up to seven other Bluetooth slave devices. Once the PAN has been established, the nodes can exchange information with one another. Bluetooth devices can also link to other piconets to establish larger *scatternets*. See Fig. 21-3.

A newer version 2.0 of Bluetooth is called Enhanced Data Rate (EDR). It has all the features described earlier but increases the overall data rate to 3 Mbps. The 3-Mbps rate includes all the headers and other overhead. The raw data rate is 3 times the 723 kbps rate mentioned earlier for a net rate of more than 2.1 Mbps. The new protocol still transmits at 1 Mbps using GFSK for accessing and recognizing inputs to establish a link and for the protocol headers. However, it uses a different modulation method to achieve the higher data rate in the data payload.

A gross data rate of 2.1 Mbps is achieved by using a form of QPSK called *π/4-differential QPSK*. It uses 90° spaced phase shifts of +135°, +45°, −45°, and −135°. It transmits 2 bits per symbol with a symbol rate of 1 Mbps.

Piconet

Scatternet

Figure 21-3 Bluetooth piconet with scatternet link. Up to seven devices can be actively connected.

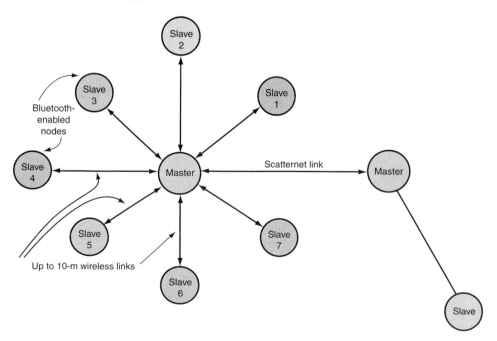

Staying in touch is as easy as using this small PC card by 3com. The card uses Bluetooth technology, which enables you to surf on a laptop without being plugged into a phone line.

To reach the 3-Mbps rate, an eight-phase differential phase-shift keying (8DPSK) modulation scheme is used. It transmits 3 bits per symbol. Otherwise all other characteristics of the Bluetooth standard are the same.

The most recent version is one that uses ultrawideband (UWB) for transmission. UWB is discussed later in this chapter. Bluetooth uses the OFDM version standardized by the WiMedia Alliance. It permits Bluetooth to achieve data rates from 100 Mbps at the typical 10-m Bluetooth range to 480 Mbps over 2 to 4 m.

The main applications for Bluetooth are cordless headsets for cell phones, wireless connections between PCs, or laptop computers and PDAs. A popular configuration is a PAN that is formed by a number of laptops used by people attending a meeting. They can exchange electronic business cards and other documents electronically. Other Bluetooth applications are wireless printer–PC connections, laptop–cell phone connections, wireless audio headsets, and wireless digital camera–TV set connections. Any wireless connection over a short distance that is within the data rate capability of Bluetooth is a potential application.

The Bluetooth standard is maintained by the Bluetooth Special Interest Group (SIG) and supported by more than 2000 manufacturers. Chip set prices have dropped below $4 for bulk purchases, making Bluetooth usable in many low-cost consumer items such as toys.

21-3 ZigBee and Mesh Wireless Networks

ZigBee is the commercial name for another PAN network technology based on the IEEE 802.15.4 wireless standard. Like Bluetooth, it is a short-range technology with networking capability. It was designed primarily for commercial, industrial, and home monitoring and control applications. The 802.15.4 standard defines the so-called air interface, which is the physical layer (PHY or layer 1 of the OSI standard) and the media access control (MAC or layer 2) of the system. The ZigBee Alliance, an organization of chip, software, and equipment vendors of ZigBee products, specifies additional higher levels of layers including networking and security.

ZigBee is designed to operate in the license-free spectrum available in the world. This is defined by the FCC Part 15 in the United States. There are three basic bands and versions.

Frequency Band	Number of Channels	Modulation	Max. Data Rate, Kbps
868 MHz (Europe)	1	DSSS/BPSK	20
915 MHz	1	DSSS/BPSK	40
2.4 GHz	16	DSSS/O-QPSK	250

While the data rates are low, this is not usually an issue since most applications are simply transmitting sensor data or making simple on/off operations. Transmission is by packets with a maximum size of 128 bytes, 104 of which is data. Both 16- and 64-bit addressing modes are available, although the maximum is considered to be up to 65,536 total nodes. The access method as with Wi-Fi is CSMA/CA. The most widely used version is the one operating in the 2.4-GHz band.

As for range, it varies considerably with the application and the environment. Using 2.4 GHz, the typical maximum indoor range is about 30 m. That can extend to 400 m outdoors with a clear line of sight. Maximum range is obtained at 868 or 900 MHz and can be as much as 1000 m with a line-of-sight path.

ZigBee's virtue is its versatile networking capability. The standard supports three topologies: star, mesh, and cluster tree. The most commonly used are the star and mesh, illustrated in Fig. 21-4. These network topologies are made up of three types of Zig-Bee nodes: a ZigBee coordinator (ZC), a ZigBee router (ZR), and ZigBee end device (ZED). The ZC initiates a network formation. There is only one ZC per network. The ZR serves as monitor or control device that observes a sensor or initiates off/on operations on some end device. It also serves as a router as it can receive data from other nodes and retransmit it to other nodes. The ZED is simply an end monitor or control device that only receives data or transmits it. It does not repeat or route. The ZC and ZR nodes are called *full-function devices (FFDs)*, and the ZED is known as a *reduced-function device (RFD)*.

The star configuration in Fig. 21-4(*a*) is the most common, where a centrally located ZR accepts data from or distributes control data to other ZRs or ZEDs. The central ZR then communications back to the ZC which serves as the master controller for the system.

In the mesh topology, most of the nodes are ZRs which can serve as monitor and control points but also can repeat or route data to and from other nodes. The value of the mesh topology is that it can greatly extend the range of the network. If a node lacks the power or position to reach the desired node, it can transmit its data through adjacent nodes that pass along the data until the desired location is reached. While the maximum range between nodes may be only 30 m or less, the range is multiplied by passing data from node to node over a much longer range and wider area.

An additional feature of the mesh topology is network reliability or robustness. If one node is disabled, data can still be routed through other nodes over alternate paths. With redundant paths back to the ZC, a ZigBee mesh ensures that data reaches its destination regardless of unfavorable conditions. Many critical applications require this level of reliability.

As for applications, ZigBee can address a wide range of wireless needs. It was designed primarily for monitoring and control. Monitoring refers to looking at a wide range of physical conditions, especially temperature, humidity, pressure, the presence of light, speed, and position information. Sensors generate an analog signal representing the physical variable that is amplified and otherwise conditioned and then converted to digital data that is transmitted back to the central monitoring location where decisions are made. This characteristic makes ZigBee a superior short-range telemetry system in what is being called wireless sensor networks.

Control refers to the sending of command signals to initiate some action. Typically commands are used to turn things off and on. Some examples are lights, motors, solenoids, relays, and other devices that perform some type of function.

Full-function devices (FFDs)

Reduced-function device (RFD)

Figure 21-4 Most common ZigBee network topologies. (*a*) Star. (*b*) Mesh.

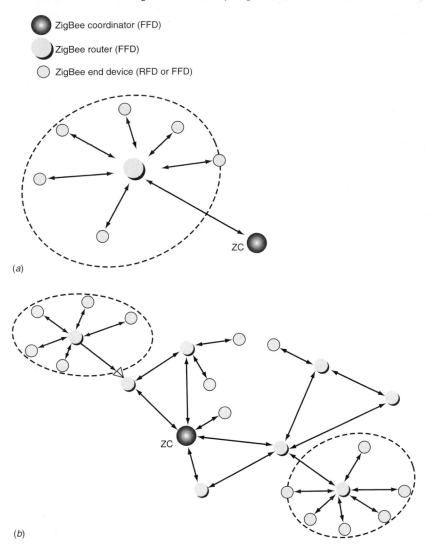

- ● ZigBee coordinator (FFD)
- ◗ ZigBee router (FFD)
- ○ ZigBee end device (RFD or FFD)

(*a*)

(*b*)

Some popular applications include monitoring and controlling lights; heating, ventilating, and air conditioning (HVAC) systems in large buildings; and industrial monitoring and control in factories, chemical plants, and manufacturing operations. Automatic electric and gas meter reading is a major new application. Other applications include medical uses such as wireless patient monitoring, automotive sensor systems, military battlefield monitoring, and a whole host of consumer applications such as home monitoring and control, remote control of other objects, and security. Because ZigBee is so low-cost and battery-operated, it can be used in a wide range of situations, most of which probably have not been discovered yet.

21-4 WiMAX and Wireless Metropolitan-Area Networks

Metropolitan-area networks (MANs) are primarily fiber-optic networks, most often SONET rings, that connect enterprise LANs to WANs or the Internet backbone. Another typical MAN is a local cable TV network. Now there is wireless contender for metropolitan-area networking. Known as WiMAX, it is the wireless system defined by the IEEE 802.16

standard. It was developed to provide a wireless alternative to consumers for broadband Internet connections. These connections are now dominated by cable TV and DSL, but with the new WiMAX standard, *wireless Internet service providers (WISPs)* may soon be offering wireless broadband connections.

The primary standard is known as IEEE 802.16-2004 or 802.16d. Its primary applications will fit into two basic categories: *point-to-point (P2P)* or *point-to-multipoint (PMP)*. The P2P mode is for applications requiring the transfer of data between two points. Common examples are cell site backhaul from a base station to the switching office or Wi-Fi hot spot interconnections to the ISP. Both of these applications typically rely on hardwired T1 or T3 connections which are very expensive. A wireless backhaul link is far less expensive, not to mention easier to install.

The PMP mode is a broadcast mode from a central base station to multiple surrounding nodes. In this mode WiMAX serves as a WISP for homes or businesses. In both modes, the service is assumed to be fixed; i.e., none of the nodes are mobile.

WiMAX was designed to operate anywhere in the 2- to 6-GHz range wherever appropriate spectrum is available. The spectrum may be licensed or unlicensed depending upon its location and the host country. The most common bands are 2.3, 2.5 and 5.8 GHz in the United States and 3.5 GHz in Europe, Asia, and Canada.

The maximum data rate is 75 Mbps, but that is usually divided up among a large number of users. Speed is set by bandwidth which can be anything from 1.75 to 20 MHz. The WISP will allocate bandwidth and speed to users based on their needs and charge accordingly. The maximum range of a single base station is about 30 mi although in a practical system one base station will usually only cover a range of 2 to 10 km (3.2 to 6 mi). Full duplex using time-division duplexing (TDD) or frequency-division duplexing (FDD) is supported by the standard. One or the other is chosen based on the application and the spectrum and service available.

WiMAX uses a 256-carrier OFDM system with adaptive modulation. OFDM is a superior method that mitigates the line-of-sight (LOS) problems that occur in serving a large area. Reflections from buildings and other structures cause multipath problems that can stop a transmission. Trees and houses can absorb the signal, making reception poor or nonexistent. Yet OFDM helps to lessen these problems. The modulation method of each OFDM channel is selected automatically depending upon the range, noise, and data rate. The standard supports BPSK, QPSK, 16-QAM, 64-QAM, and 256-QAM. BPSK would be selected for longer range and lower speeds; 64-QAM or 256-QAM would be selected for shorter ranges to give higher speeds.

A mobile version of WiMAX is now available. The standard is IEEE 802.16e 2005. It is designed to permit nodes to be mobile while maintaining contact with a base station. An example is a WiMAX-enabled laptop on a commuter train or in a car or truck. The maximum range is about 3 mi, but that will depend upon the environment and terrain. Speeds up to about 75 mi/h can be accommodated. Data rates are adaptable to the environment and range but can hit a maximum of 15 Mbps under ideal conditions. IEEE 802.16e uses 2048-channel OFDM to achieve these characteristics. The access method is OFDMA (access) where the 2048 channel OFDM signal is divided among multiple users. With 64 channels assigned to each subscriber, one OFDMA bandwidth can accommodate up to 32 users.

The initial application of WiMAX has been backhaul where it makes remote cell sites and hot spots fast, easy, and inexpensive to connect back to the service provider. WISP applications have been slow to develop, but it is expected that some areas, such as those not now served by cable TV or DSL, will adopt WiMAX as the broadband Internet service as well as subscription TV and VoIP. WiMAX is especially attractive to developing countries where a wireless infrastructure is far easier and less expensive to install than a traditional wired telephone or cable TV system.

It is also said that the mobile version of WiMAX could compete with 3G cell phone data systems. A WiMAX plug-in card in a laptop could accomplish the same thing as a 3G cell phone card in the same laptop. Like many wireless technologies, WiMAX is new and will find its optimum applications over time.

Wireless Internet service providers (WISPs)

Point-to-point (P2P)

Point-to-multipoint (PMP)

21-5 Infrared Wireless

Perhaps the most widespread wireless system uses infrared light for short-distance data communication. The most ubiquitous example is the wireless remote control on virtually all TV sets, VCRs, and DVD players and on most audio CD stereo systems. This standard feature on most consumer electronic products is so common that it is taken for granted. Infrared has also been used for wireless PANs. Because light travels in straight lines, there must be a clear path from transmitter to receiver for the system to work. Furthermore, light does not penetrate walls, floors, and ceilings, as wireless radio does, so IR LANs are not practical. Wireless PANs, however, are widely used to link nearby laptops, PCs, and PDAs such as Palm Pilots. This section explores the most popular IR wireless systems.

TV Remote Control

Almost every TV set sold these days, regardless of size or cost, has a wireless remote control. Other consumer electronic products have remote controls including VCRs, cable TV converters, CD and DVD players, stereo audio systems, and some ordinary radios. Generic remote controls are available to hook up to any device that you wish to control remotely.

All these devices work on the same basic principle. A small handheld battery-powered unit transmits a serial digital code via an IR beam to a receiver that decodes it and carries out the specific action defined by the code. A TV remote control is one of the more sophisticated of these controls, for it requires many codes to perform volume control, channel selection, and other functions.

Figure 21-5 is a general block diagram of a remote control transmitter. In most modern units, all the circuitry, except perhaps for the IR LED driver transistors, is contained within a single IC. The purpose of the transmitter is to convert a keyboard entry to a serial binary code that is transmitted by IR to the receiver.

The keyboard is a matrix of momentary-contact single-pole single-throw (SPST) pushbuttons. The arrangement shown is organized as eight rows and four columns. The row and column connections are made to a keyboard encoder circuit inside the IC. Pulses generated internally are applied to the column lines. When a key is depressed, the pulses from one of the column outputs are connected to one of the row inputs. The encoder circuit converts this input to a unique binary code representing a number for channel selection or some function such as volume control. Some encoders generate as few as 6 bits, and others generate up to 32-bit codes. Also 9- and 10-bit codes are very common.

The serial output is generated by the shift register as data is shifted out. A standard *nonreturn to zero (NRZ)* serial code is generated. This is usually applied to a serial encoder to generate a standard biphase or Manchester code. Recall that the biphase code provides more reliable transmission and reception because there is a signal change for every 0-to-1 or 1-to-0 transition. The actual bit rate is usually in the 30- to 70-kbps range.

The serial bit stream turns a higher-frequency pulse source off and on according to the code's binary 1s and 0s. The transmitter IC contains a clock oscillator that runs at a frequency in the 445- to 510-kHz range. A typical unit runs at 455 kHz, using an external ceramic resonator to set the frequency. The serial data turns the 455-kHz pulses off and on. For example, a binary 1 generates a burst of 16 pulses of 455 kHz, as shown in Fig. 21-6. When a binary 0 occurs in the data train, no pulses are transmitted. The figure shows a 6-bit code (011001) with a start pulse. The period T of the 455-kHz pulses is 2.2 µs. The pulse width is set for a duty cycle of about 25 percent, or $T/4$.

If 16 pulses make up a binary 1 interval, that duration is 16×2.2, or 35.2, µs. This translates to a code bit rate of $1/35.2 \times 10^{-6}$, or 28.410 kHz.

The 455-kHz pulses modulate the IR light source by turning it off and on. The IR source is usually one or more IR LEDs. These are driven by a Darlington transistor pair external to the IC, as shown in Fig. 21-5. Two or more LEDs are used to ensure a sufficient level of IR radiation to the receiver in the TV set. The LED current is usually very high, giving high IR output levels for reliable transmission to the receiver. Some

Figure 21-5 IR TV remote control transmitter.

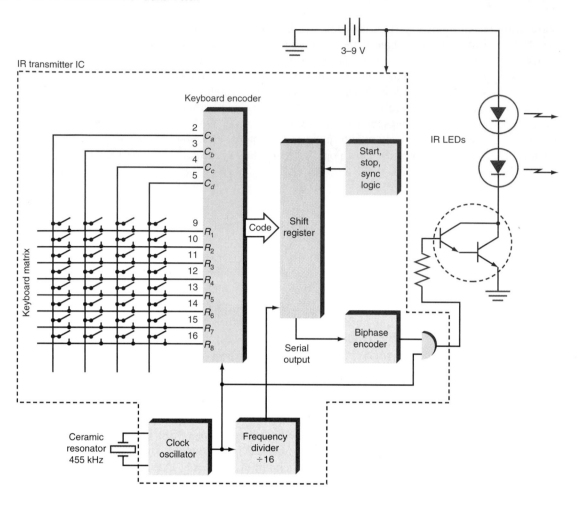

Figure 21-6 Pulse waveforms in the IR remote control transmitter.

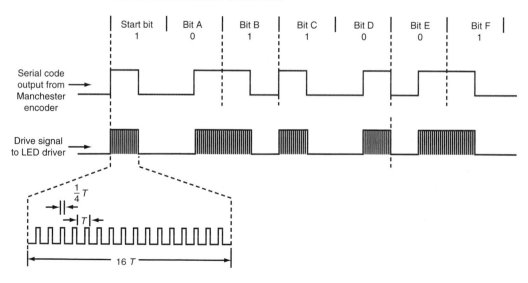

Figure 21-7 The IR receiver and control microprocessor.

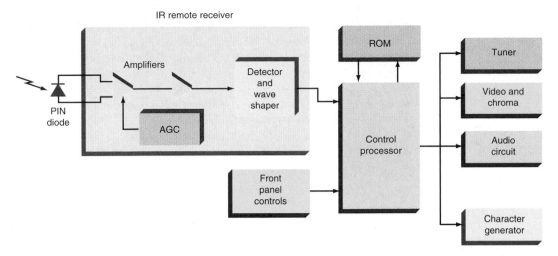

remote units use three LEDs for a wide-angle transmission signal so that a high-amplitude signal will be received regardless of the direction in which the remote is pointed.

An IR receiver is shown in Fig. 21-7. The PIN IR photodiode is mounted on the front of the TV set, where it picks up the IR signal from the transmitter. The received signal is very small despite the fact that the distance between the transmitter and receiver is only 6 to 15 ft on average. Two or more high-gain amplifiers boost the signal level. Most circuits have some form of automatic gain control (AGC). The incoming pulses are detected, shaped, and converted to the original serial data train. This serial data is then read by the control microcomputer that is usually part of the TV receiver.

The microcontroller is a dedicated microcomputer built into every TV set. A master control program is stored in a ROM. The microcomputer converts inputs from the remote control and front panel controls to output signals that control the various functions in a TV set, such as channel selection and volume control.

The microcontroller inputs and decodes the incoming signal and then issues output control signals to all other circuits: the PLL frequency synthesizer that controls the TV tuner, the volume control circuits in the audio section, and in the more advanced receivers, chroma and video such as hue, saturation, brightness, and contrast. The microcontroller also generates, sometimes with the help of an external IC, the characters and simple graphics that can be displayed on the screen. Most microcontrollers also contain a built-in clock.

IR PANs

Besides remote control, the primary application for IR data communication is in short-distance links between computers, computers and printers, or ad hoc PANs. For IR *short distance* typically means up to 1 m. Under some special conditions, the distance can be extended up to 9 m maximum. And there must be a clear line of sight between transmitter and receiver. Some of the more common applications for IR are given in Fig. 21-8.

Figure 21-9 shows a block diagram of an IR transceiver. It connects to interface circuitry in the PC or PDA. The interface is typically a small embedded controller inside the computer or PDA. It can also be a part of the PDA controller. The encoder puts the serial digital data from the PC or PDA into the proper format for transmission. A high-current bipolar transistor or MOSFET drives one or more IR LEDs. The receiver consists of the PIN diode that picks up the IR light from a nearby transmitter. The signal is amplified and shaped and then sent to the decoder, which recovers the original data. Figure 21-10 shows the physical arrangement of the transceiver module. It contains the LED and PIN diodes plus the related circuitry. The LED and PIN diode are positioned next to each other, about 0.5 to 1 in apart.

Figure 21-8 Common applications for IR data communication.

◆ Wireless printer and/or scanner connection from desktop PC, laptops, PDAs, and calculators.
◆ PC-to-PC connections for electronic exchange of business cards and short data files.
◆ Laptop-to-laptop connections in an ad hoc PAN in meetings.
◆ PC-to-PDA, laptop-to-PDA, or PDA-to-PDA communication.
◆ Wireless mouse or keyboard for a desktop PC.
◆ Wireless computers, TV game controllers, and toys.
◆ Digital camera to TV or computer connection.
◆ Cell phone camera to laptop or PC.

Figure 21-9 IR wireless LAN transceiver.

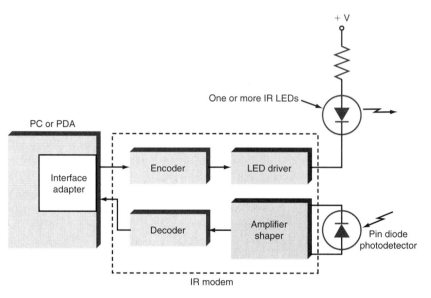

Figure 21-10 IR transceiver module.

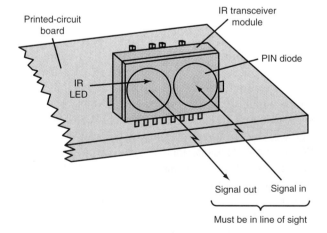

Although all this circuitry is very small and can be mounted into the PC, laptop, cell phone, or PDA, it must be positioned so that it can be directly pointed at the device to which it will communicate. The angle of radiation is relatively small, from about 15° to 30°. Therefore, alignment between transmitter and receiver is critical for maximum signal reception. And, of course, the path must be completely unobstructed. Although many laptops and PDAs have built-in transceivers, their use is often restricted by this need for line of sight. Awkward physical arrangements are usually needed for proper operation. A better arrangement is a transceiver *dongle,* which consists of a cable attached to the interface in the PC or PDA and to the movable dongle containing the LED and PIN diode. The dongle can then be positioned where it has a clear view of the device to which it will communicate.

dongle

IrDA System

The most widely used IR data communication system was developed by Hewlett-Packard in the early 1990s. It has since become an international standard that is maintained by the Infrared Data Association (IrDA). The complete interface and system are referred to as *IrDA.* The systems are designed for short range. The low-power version has a range of 20 to 30 cm. A more typical range is about 1 m. This range can be extended to 1.6 m by boosting the LED current to 250 mA. Using a higher-current LED with 500 mA gives a range of up to 2.2 m. Further range extensions can be achieved by using multiple parallel modules with 500-mA LEDs. The maximum usable range is 8.9 m. Most IR LEDs operate in the 850- to 900-nm range.

IrDA system

As for data speed, the original standard supported speeds up to 115.2 kbps. Most systems used speeds as low as 2.4 kbps, although 9600 bps was very common. Different versions support speeds of 576 kbps and 1.152 Mbps up to 4 Mbps, which is the most commonly used version of IrDA. A 16-Mbps version is also available.

Although TV remote controls use a modulated IR beam, IrDA does not. It uses baseband transmission that requires encoding and decoding. The standard NRZ serial data is converted to pulses especially encoded for IR operation. In the original 115.2-kbps system, the encoding is called 3/16 *modulation.* This scheme uses return-to-zero inverted (RZI) data and is shown in Fig. 21-11. The NRZ data is converted to pulses 3/16 of a bit period long, one for each binary 0 transmitted. This ensures that for very long strings of continuous 0s, short pulses are transmitted. These short pulses lower the overall power requirements and permit higher peak LED power to be transmitted.

The 4-Mbps version uses another encoding scheme, called *4 PPM (pulse position modulation).* It uses a symbol rate of 2-Mbps symbols per second with an interval of 500 ns.

Figure 21-11 The 3/16 data encoding used in 115.2-kbps IrDA. *(Courtesy Agilent Technologies)*

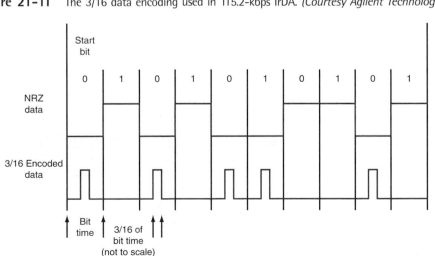

Figure 21-12 The 4 PPM encoding needed to achieve a 4-Mbps data rate with IrDA. *(Courtesy Agilent Technologies)*

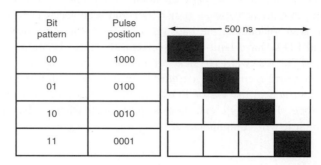

Bit pattern	Pulse position
00	1000
01	0100
10	0010
11	0001

In each 500-ns symbol time, 2 bits is transmitted. Each of the four 2-bit groups is represented by one of four unique time positions within that time frame. See Fig. 21-12. Most IrDA interfaces contain the hardware and software to encode or decode either format.

21-6 Radio-Frequency Identification and Near-Field Communications

Radio-frequency identification (RFID)

Another growing wireless technique is *radio-frequency identification (RFID)*. You can think of it as the wireless version of bar codes. This technology uses thin, inexpensive tags or labels containing passive radio circuits that can be queried by a remote wireless interrogation unit. The tags are attached to any item that is to be monitored, tracked, accessed, located, or otherwise identified. RFID tags are widely used in inventory control, container and parcel shipping, capital equipment and other asset management, baggage handling, and manufacturing and production line tracking. They are also widely used for automatic toll collection and parking access for vehicles. Other applications for RFID tags are personnel security checking and access, animal tracking, and theft prevention. As the technology has developed, prices have dropped and many new applications have been discovered.

The basic concept of RFID is illustrated in Fig. 21-13. The tag is a very thin label-like device into which is embedded a simple passive single-chip radio transceiver and antenna. The chip also contains a memory that stores a digital ID code unique to the tagged item. For the item to be identified, it must pass by the interrogation or reader unit, or the reader unit must physically go to a location near the item. Longer-range systems cover a complete building or area. The reader unit sends out a radio signal that

Figure 21-13 Basic concept and components of an RFID system.

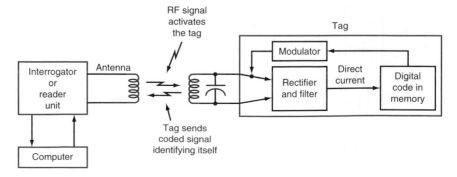

Figure 21-14 Block diagram of RFID interrogator (reader).

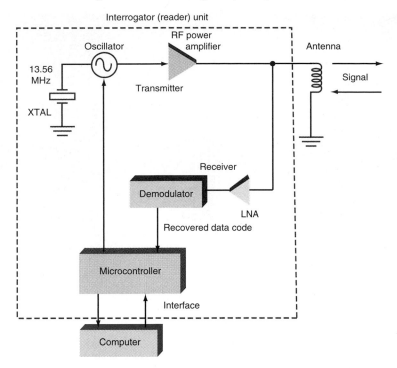

may travel from a few inches up to no more than 100 ft or so. The radio signal is strong enough to activate the tag. The tag rectifies and filters the RF signal into direct current that operates the transceiver. This activates a low-power transmitter that sends a signal back to the interrogator unit along with its embedded ID code. The reader then checks its attached computer, where it notes the presence of the item and may perform other processing tasks associated with the application.

RFID systems operate over the full radio spectrum. Commercial systems have been built to operate from 50 kHz to 2.4 GHz. The most popular ranges are 125 kHz, 13.56 MHz, 902 to 928 MHz, and 2.45 GHz. The 125-kHz and 13.56-MHz units operate only over short distances up to several feet, whereas the 902- to 928-MHz and 2.45-GHz units can operate up to about 100 ft. Most of the tags are passive; i.e., they have no power source of their own. They rely upon the interrogator unit to supply a large enough RF signal to rectify for dc power. However, some active tages containing small flat batteries are available, and they can operate over a much larger range.

Figure 21-14 shows a block diagram of a typical 13.56-MHz RFID interrogator unit. A 13.56-MHz crystal oscillator generates the basic RF signal, which is amplified and sent to the antenna. A microcontroller gates the oscillator on for a short time, and then the receiver waits for a response from the tag. The antenna picks up the weak tag signal. The receiver amplifies and demodulates it and then recovers the serial data code. The microcontroller communicates with the attached computer to do whatever processing is needed in the ID process.

Some RFID tag configurations are shown in Fig. 21-15. They consist of a flat spiral inductor and a capacitor that make up a 13.56-MHz tuned circuit that serves as the antenna. The transceiver chip is contained in the black dot on the tag. A block diagram of the circuitry is given in Fig. 21-16. A typical tag is the model MCRF 355/360, a product of Microchip Technology Inc. The resonant circuit picks up the interrogator signal as if it were the induced signal in a transformer secondary rather than an actual received electromagnetic radio wave. When the voltage reaches about 4 V_{p-p}, the power circuits are activated. The RF is rectified in a voltage multiplier circuit, filtered, and regulated into the direct current that operates the remaining circuits.

Figure 21-15 RFID tag configurations.

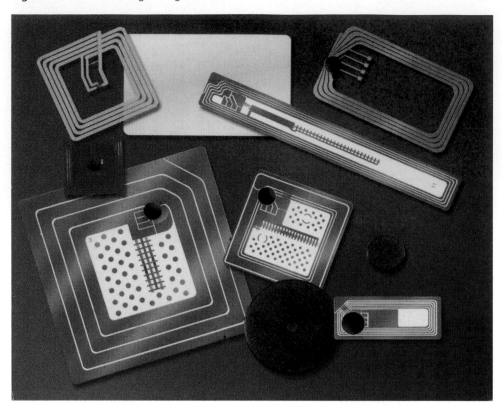

Figure 21-16 Block diagram of RFID tag.

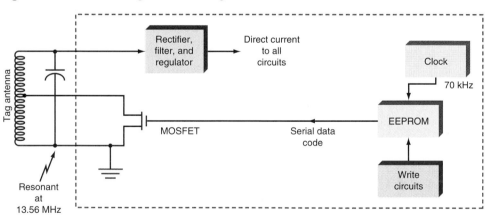

The unique ID code is stored in an electrically erasable programmable read-only memory (EEPROM) in the tag chip. In this device, the code is 154 bits long. Figure 21-17 shows the format of the packet sent to the reader. The 9-bit header initiates synchronization at the reader receiver for clock recovery. The customer's unique ID number is encoded with 13 bytes. The checksum provides for error checking at the reader. The code is stored in the chip by the manufacturer with a contact transmitter that activates the chip and writes the code into memory. The tag chip contains the EEPROM write circuitry that stores the code.

The ID code in EEPROM is read out serially in NRZ data format, which is then converted to a Manchester or biphase signal that is used to modulate the carrier sent back to the reader. See Fig. 21-18. The Manchester code is used so that the clock can be easily recovered from the data in the reader. The data rate is typically 70 kbps. With a bit

Figure 21-17 Format of coded packets transmitted to the reader by the tag. *(Courtesy Microchip Technology)*

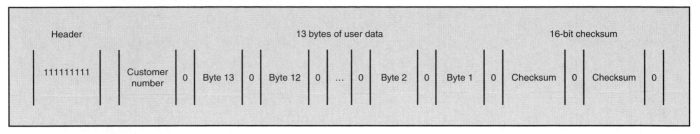

9 bit header
8 bit customer number
104 bits (13 × 8) of user data
17 bits of zeros between each byte, header, and checksum
16 bits of checksum
Total: 154 bits

Figure 21-18 Data waveforms in the tag. The Manchester-coded signal operates the cloaking transistor. *(Courtesy Microchip Technology)*

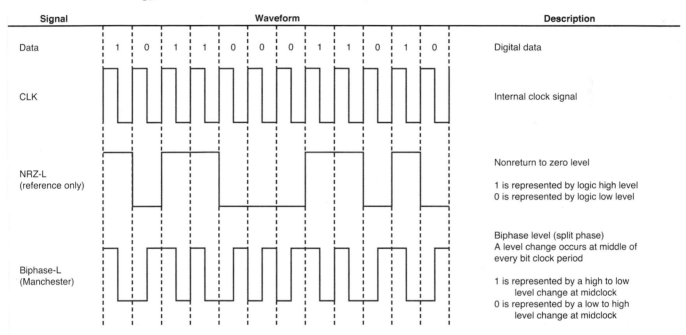

time of $1/70 \times 10^3 = 14.28$ μs, it takes 154×14.28 μs $= 2200$ μs, or 2.2 ms, to transmit the 154-bit code.

The modulation used is a form of amplitude modulation called *backscatter modulation*. The circuitry was shown in Fig. 21-16, and the process is shown in Fig. 21-19. The coil in the external tuned circuit/antenna is tapped. A MOSFET switching transistor is connected to that tap inside the tag. The data to be transmitted is applied to this transistor. When the transistor is off, the carrier is passed to the tuned circuit and sent to the reader, which reads the signal as a binary 1. When the transistor is turned on, a portion of the coil is shorted, making the external tuned circuit resonant at a frequency of 3 to 6 MHz higher than its 13.56-MHz design frequency. This signal is out of the frequency range of the reader, so it receives a much lower-level signal that is interpreted as a binary 0. During the time the transistor is on, the signal is said to be *cloaking*. With the transistor off, the signal is uncloaked. The cloaking and uncloaking process produces amplitude shift keying (ASK) at the reader receiver. See Fig. 21-19.

Backscatter modulation

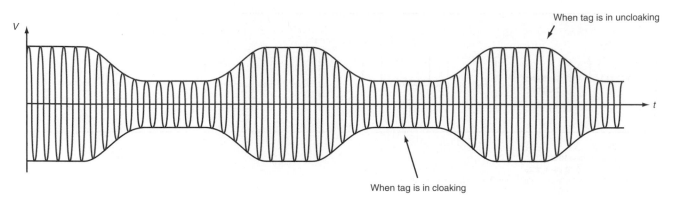

When tag is in uncloaking

When tag is in cloaking

In some systems, the tag tuned circuit serves as the secondary winding of a transformer where the reader antenna is the primary winding. The binary pulses to be transmitted modify the impedance of the tag antenna, and this, in turn, causes an amplitude shift in the reader. As the tag's data is read out, the process loads and unloads the secondary winding tag antenna, causing a reflected impedance back into the reader antenna. A typical carrier may have a 100-V peak-to-peak amplitude which will shift by several hundred millivolts as data is being transmitted by the tag. A peak detector recovers the signal.

The most recent new RFID standard is called Gen 2 for second generation. It is a standard developed by more than 60 companies worldwide. The standard is under the auspices of EPCGlobal, the organization that also standardizes the Electronic Product Code (EPC) to be used on all tagged items. The Gen 2 standard operates in the 900-MHz region with different frequencies being used in different countries depending upon the local regulations. The 868-MHz frequency is common in Europe while 915 MHz is common in the United States. The standard uses ASK backscatter and supports a 96-bit EPC plus a 32-bit error correction code and a kill command that deactivates a tag after it is read.

A key benefit of the new standard is that it is designed to read multiple tags faster. Tag read rates as high as 1500 tags per second are possible, although in most situations the read rate will be no more than 500 to 1000 tags per second, which is still much faster than the 100 tag per second maximum imposed by the older standards. The Gen 2 tags are also more robust and can operate reliably in an environment with multiple readers transmitting and receiving simultaneously. The RFID field is still new but growing as tag prices drop and as greater security measures are developed.

Near-Field Communications

One of the newest forms of wireless is a version of RFID called *near-field communications (NFC)*. It is an ultrashort-range wireless whose range is rarely more than a few inches. It is a technology used in smart cards and cell phones to pay for purchases or gain admittance to some facilities.

Near field means the near field of a radio wave. As discussed earlier, a radio wave is made up of both electric and magnetic fields. At a distance of about 10 wavelengths or more at the operating frequency, the radio wave behaves just as Maxwell's equations describe it. The two fields exchange energy and reinforce each other as it passes from transmitting antenna to receiving antenna. This is the so-called far field. At a distance of less than 10 wavelengths from the transmitting antenna is the near field, where the individual electric and magnetic fields exist. The electric field is not useful, but the magnetic field is used for short-range communications. The way to imagine NFC is as the magnetic field between the windings of a transformer. The coefficient of coupling is very low because of a large distance between the primary winding (the transmitting antenna) and secondary winding (the receiving antenna). The primary limitation of the near field is that the magnetic field strength drops off at a rate of about $1/d^6$, where d is the distance. With only low power the range is very limited. The far field only drops off at a $1/d^2$ rate.

NFC is standardized internationally. The technology is similar to that used in RFID. It is similar to and compatible with the technology used in smart cards, those credit cards with an internal chip that allow you to pay for something by just passing the card over a point-of-sale (POS) terminal reader.

The standard specifies an operating frequency of 13.56 MHz, the international no-license band and one of the ISM band Part 15/18 frequencies in the United States. The transfer data rate is 106, 212, or 424 kbps. The speed depends upon the range, which is up to maximum of 20 cm or about 8 in. In most cases the actual range will be only a few inches or no more than 10 cm.

The standard also specifies an active and a passive mode of operation. In the active mode, both parties have powered transceivers. This means that each node has a battery or some other power supply. Either unit may initiate a transmission, which is half duplex with a "listen before transmit" protocol. One of the devices is the initiator, and the other device becomes the target.

In the passive mode, the target is a passive device such as an RFID tag. The tag gets its operational power from the field transmitted by the initiator. It then transmits data back to the initiator by modulating the magnetic field, using backscatter AM.

There are several intended applications for this ultrashort-range technology. The most frequent use is an automatic payment tool such as a smart card. But instead of using a smart card, the NFC transceiver is built into your cell phone. To buy something, you just tap your cell phone on the reader or pass it within an inch or so, and your credit card account is automatically billed. You could use it to buy theater tickets or even to pay for a plane, train, or hotel charge.

The second most useful application is automatic gated entry. Passing your cell phone near the reader allows you entry into a building, parking lot, or other controlled area. NFC chips are also expected to be incorporated into the next-generation passports.

Another proposed use is to set up and initiate other forms of wireless. Some short-range wireless modes such as Wi-Fi or Bluetooth require that the two parties desiring a peer-to-peer link first exchange information to set up the correct protocol. It is sometimes called pairing. Putting your cell phone, laptop, or other device next to the device to be connected to allows all this protocol setup to be exchanged automatically. After that the two devices then automatically begin talking in the new wireless mode faster and at a longer distance. NFC chips are small and inexpensive, so look for them to be more widely used as further applications are discovered.

21-7 Ultrawideband Wireless

Perhaps the newest and most unusual form of wireless is known as *ultrawideband (UWB) wireless*. There are two basic forms of UWB, the original version based on very narrow impulses and the newer kind based on OFDM. Both spread the signal over a very wide range of spectrum but at a very low signal level so it does not interfere with other signals operating over those frequencies. Both methods are used, but the newer OFDM version appears to have captured the greatest number of manufacturing companies and the applications. Both types of UWB are covered in this section.

Ultrawideband (UWB) wireless

The original UWB discovered in the 1960s is known as impulse, baseband, or carrierless wireless. This form of UWB transmits data in the form of very short pulses, typically less than 1 ns. From the Fourier theory discussed in Chap. 12, you know that a fundamental frequency sine wave and many harmonics can represent any pulse train. A UWB signal using very short pulses with a low duty cycle occupies a very wide bandwidth. A UWB signal is defined as having a bandwidth at least 25 percent of the center frequency, or 1.5 GHz minimum. Another definition specifies UWB as occupying more than 500 MHz of spectrum.

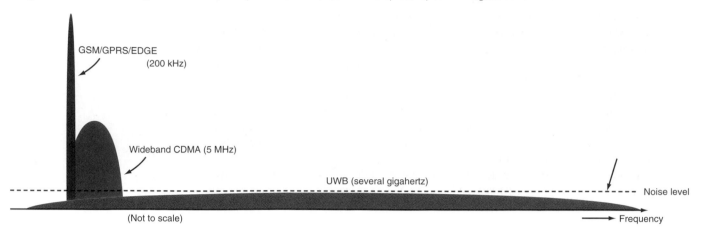

Figure 21-20 shows a UWB signal spectrum compared to a standard 30-kHz cell phone channel and a 5-MHz wideband CDMA (spread spectrum) cell phone channel.

The FCC permits UWB in the 3.1- to 10.6-GHz range. The only other services in this region are satellites, radars, broadband wireless, and wireless networks. UWB equipment spreads its signals over much of that range, but the power level is so low that there is essentially no interference to other services. UWB is like spread spectrum in that many users can share a single wide bandwidth simultaneously.

A UWB signal starts as a very low duty cycle (<1 percent) rectangular pulse stream at some pulse repetition interval (PRI). The pulses are then Gaussian-filtered and differentiated to produce the final pulses to be transmitted. The pulses are applied directly to the antenna (see Fig. 21-21). Known as *monocycles,* these pulses are not just one cycle of a sine wave. They are shaped by a Gaussian filter. The pulse width sets the center frequency of the signal and the half-power bandwidth. The center frequency is approximately the reciprocal of the pulse width. For a pulse width of 500 ps, the center frequency is $1/500 \times 10^{-12} = 2$ GHz.

The serial data to be transmitted is then encoded with a unique pseudorandom code like that used in CDMA. This method effectively "channelizes" the system so that multiple users can share the spectrum but still be individually identified. The coded signal then modulates the pulse train by either PPM or BPSK. Both methods are illustrated in Fig. 21-22. In PPM, the position of the pulse may occur sooner or later in time than a pulse with no modulation. A binary 0 may be represented by an earlier pulse, and a binary 1 as a later pulse, or vice versa. The time shift is small compared to the pulse width. Because of the very small time differences, the timing

Figure 21-21 Basic UWB waveform of repetitive monocycles.

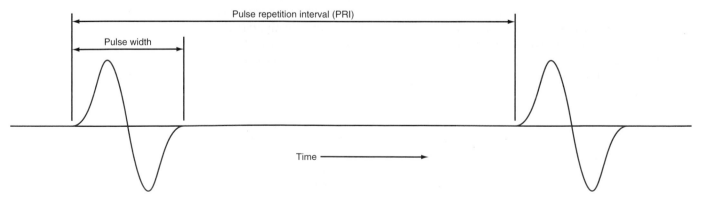

Fig. 21-22 Types of UWB modulation. (*a*) Pulse position modulation (PPM). (*b*) Binary phase-shift keying (BPSK).

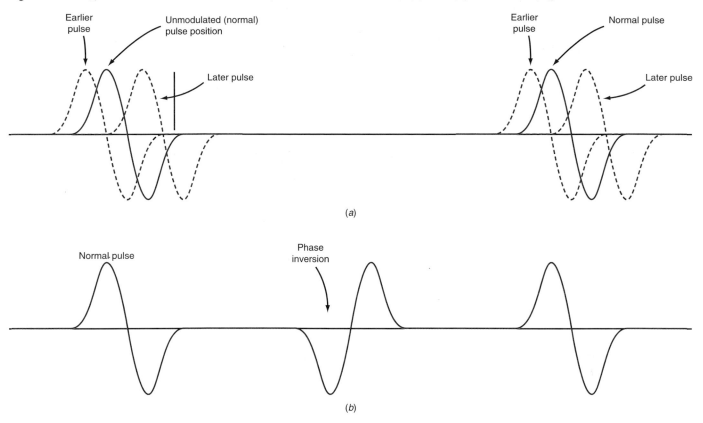

(*a*)

(*b*)

clock generating the pulses must be very precise and stable, with minimal jitter, to ensure recovery.

In BPSK, the PRI is constant, and the data bits produce a normal pulse for a binary 0 and a phase-inverted pulse for a binary 1.

Impulse UWB Hardware

UWB circuitry is mostly patented or proprietary, and details about it are generally unavailable. However, some basic facts are known.

Transmitter. Early UWB transmitters used a pulse-forming *LC* low-pass network or transmission line along with a snap (step) recovery diode (SRD) to generate the pulses in a PPM system. Newer circuits using BPSK can generate the pulses with standard CMOS integrated circuits. The pulses are applied directly to the antenna.

Receiver. Refer to Fig. 21-23. The receiver amplifies the incoming signal and then applies it to a correlator consisting of a multiplier, where it is multiplied by a stream of coded pulses similar to those transmitted. The multiplier output is integrated to produce a correlated output. If the multiplier output exceeds a specific level, it is considered to be detected and recovered. The recognized signal is then demodulated into the original data.

Antenna. Most communication antennas are resonant to a frequency set by their length or other physical dimension. They also have a narrow bandwidth. New types of broadband antennas have been developed for UWB. They take many forms and include microwave horns as well as uniquely patterned patch antennas that significantly increase the bandwidth.

Fig. 21-23   Basic concept of a UWB receiver.

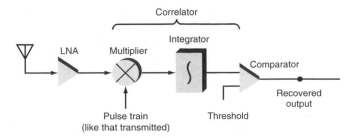

Multiband OFDM UWB

The newest form of UWB is called multiband OFDM or MB-OFDM UWB. The term *multiband* is derived from the fact that many OFDM carriers make up the signal. This form of UWB divides the lower end of the assigned spectrum into three 528-MHz-wide channels, as shown in Fig. 21-24. These bands extend from 3.168 to 4.952 GHz. Note the center frequencies of the three bands. Each band is designed to hold an OFDM data signal. There are 128 carriers per band, and each carrier has a bandwidth of 4.125 MHz. Of the bands 100 actually carry the data while 12 are used as pilot carriers to aid in establishing communications with nearby nodes. The remaining carriers serve as guard bands on either side to prevent interference between the three portions of spectrum. The signal to be transmitted is divided up among the carriers, and each is modulated by BPSK or QPSK depending on the data speed selected. In Fig. 21-24, you will see a dashed line that designates the maximum allowed operating power specified by the FCC. It is a very low −41 dBm/MHz which generally prevents any interference to other services.

The system is designed to permit a wide range of data rates from about 53 to 480 Mbps. The most often mentioned speed is 110 Mbps at a range up to 10 m. A speed of up to 480 Mbps is possible but only at a range of 2 to 3 m.

Implementation of an OFDM UWB transceiver is just like that of any OFDM device. DSP chips are used to create the transmit carriers with the inverse fast Fourier transform (IFFT), and a DSP chip in the receiver uses the FFT for recovery of the data. MB-OFDM UWB radios are usually a single-chip IC containing all functions.

Figure 21-24   Operating spectrum for multiband OFDM UWB.

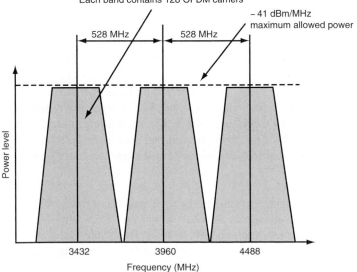

There is no UWB standard. Companies worked for years in an IEEE Task Group to create a single standard to be designated 802.15.3a. No consensus could be reached, so companies went their different ways. The largest group of companies banded together in the WiMedia Alliance to create a standard that most could agree upon. Today the MB-OFDM form of UWB is the defacto standard. The WiMedia Alliance maintains this standard. The impulse form of UWB championed by Freescale Semiconductor is supported by the UWB Forum, another industry consortium.

Advantages and Disadvantages of UWB

UWB offers many benefits to radar, imaging, and communication applications:

1. Superior resolution in radar and imaging.
2. Immunity to multipath propagation effects.
3. Higher data rates than are possible with other wireless technologies.
4. License-free operation.
5. No interference to other signals using the same frequency band. UWB signals appear as random noise to conventional radios.
6. Power-efficient, extremely low-power operation. Peak power levels are in the milliwatt region, and average power is in microwatts.
7. Simple circuitry, most of which can be integrated in standard CMOS.
8. Potentially low cost.

The primary disadvantage, which is also an advantage, is low power. It severely limits the range of operation. The range can be extended in military radar with higher power levels, but the power level in commercial and consumer applications is severely restricted by the FCC. Typical ranges are from a few inches up to no more than about 100 ft.

Primary Application of UWB

The primary application of impulse UWB to date has been in military radar. The very short pulse widths of electromagnetic energy permit very fine resolution of target distance and detail. Short pulses also give UWB the ability to penetrate surfaces to see what is behind them. For that reason, UWB is an excellent electronic imaging technique. It is especially effective in seeing through leaves, trees, and foliage. UWB radars can even see underground to detect mines, pipes, and so on. UWB radar is used by fire, emergency, and police personnel to see through walls and doors. Medical versions permit body imaging for diagnosis.

Under development are low-cost, short-range UWB radars that can be used in cars and trucks for collision avoidance, automatic braking, improved air bag deployment, and suspension systems. These systems will operate in the 24- to 29-GHz range and are expected to appear in cars before 2010.

There are several target markets for UWB. First is computer peripherals. UWB, when married to the popular PC and laptop USB (Universal Serial Bus) interface, permits a wireless USB connection. USB is used almost exclusively today for connecting devices to PCs and laptops such as printers, mice, external drives, and networking equipment. By making the interface wireless, cables and the hassle of connecting them are eliminated.

UWB is also attractive for wirelessly connecting video equipment. Because of the very high-speed nature of digital TV and video, a superhigh-speed wireless technology is needed for transport. While 802.11a/g/n WLAN gear can handle it, it is more expensive and consumes greater power. Using a UWB link gives even higher speed with very low power consumption. The range is limited, but video links are usually short from TV set to cable box or DVD player or to a camcorder or wireless speakers. Wireless ports for USB as well as the popular FireWire or IEEE1394 serial interface are now used to connect video equipment. The multiband OFDM form of UWB is also the newest physical layer air interface for Bluetooth, as described earlier.

CHAPTER REVIEW

Summary

- There are many short-range wireless technologies used to replace cables, set up networks, and transmit monitor and control information. The technology varies widely depending upon the range, data rate, and other factors. The most common technologies are Bluetooth, Wi-Fi, ZigBee, WiMAX, infrared, ultrawideband, RFID, and NFC.

- Wireless LANs provide a fast, easy way to expand an existing Ethernet LAN without wires. It offers wireless services to laptops in public locations by way of hot spot access points. WLANs are standardized by the IEEE under the 802.11 designation. Different versions offer different data rates. IEEE 802.11b, g, and n operate in the 2.4-GHz band and have maximum data rates of 11, 54, and 600 Mbps, respectively. IEEE 802.11a operates in the 5-GHz band and offers rates to 54 Mbps.

- Bluetooth is a personal-area network (PAN) that automatically forms ad hoc networks between nearby transceivers. It is also used for short-distance cable replacement such as in cell phone headsets. Bluetooth operates in the 2.4-GHz band using FHSS and FSK. Also 1- and 3-Mbps versions are available plus a UWB version.

- ZigBee is an expanded version of the IEEE 802.15.4 PAN standard. It is a low-speed wireless technology used primarily for sensor monitoring and control in buildings, homes, and factories. It operates in the 868- and 902- to 928-MHz and 2.4-GHz bands. Data rates are 20, 40, and 250 kbps, respectively. ZigBee transceivers can form ad hoc mesh networks to extend range and reliability.

- WiMAX is the commercial name for the IEEE 802.16d/e standard for metropolitan-area networks (MANs). It can provide very high-speed connections for fixed broadband services such as cable TV and DSL. It is also useful in backhaul operations. A mobile version permits in-motion services. The service can operate in available bands from 2 to 6 MHz. The modulation/access method is OFDM, and data rates to 75 Mbps at a range of up to 30 mi are possible.

- Infrared (IR) is high-frequency light (800 to 1000 nm) used for wireless applications such as TV remote controls. It can also be used in short-range (less than a few feet) data applications such as connections to laptops, cell phones, or PDAs. One popular standard is IrDA which permits rates from 115.2 kbps up to 4 Mbps.

- Radio-frequency identification (RFID) is an electronic equivalent of bar coding. Small flat tags attached to containers, products, or objects allow the items to be tracked and accounted for wirelessly. Passive tags use power from an interrogator unit to activate the tag that transmits a special electronic product code, using ASK backscatter modulation. Read range is only a few feet, but active tags containing a small battery can be read at distances up to many feet. Popular operating frequency ranges are 125 kHz, 13.56 MHz, 868 or 915 MHz, and 2.4 GHz. A version of RFID called near-field communications (NFC) is being used in smart cards and cell phones to facilitate purchases.

- Ultrawideband (UWB) transmits data as short pulses that spread the data over a very wide spectrum at least 500 MHz wide. It was originally used for radar but is now adapted to fast data transmission such as digitized video. A newer version used OFDM in three 528-MHz-wide bands to send data at rates to 480 Mbps.

Questions

1. What is the primary benefit of a wireless LAN?
2. Name the IEEE standard for WLANs.
3. What is the commercial nickname for this standard?
4. List the four main version of the IEEE WLAN standard, and state the operating frequency range and maximum data rate.
5. What is an access point? A hot spot?
6. What is the name of the unit that forms the base unit of a home WLAN?
7. What is the value of a home network?
8. Describe how interoperability is achieved with IEEE standard WLAN equipment.
9. What is a PAN?
10. Name the two most popular PAN standards.
11. Describe the access mode and modulation type used in the original version of Bluetooth.
12. Describe the access mode and modulation type used in the newest version of Bluetooth.
13. True or false? The fastest version of Bluetooth uses UWB.
14. What is the approximate maximum communications range of class 1 and class 3 Bluetooth power categories?
15. What is a piconet? A scatternet?
16. List five popular applications for Bluetooth. Name the most popular application.
17. What is the IEEE standard for ZigBee?
18. State the main application for ZigBee.
19. List the operating frequency ranges of ZigBee and the maximum data rates for each.

20. Which version of ZigBee is the most popular?
21. What are the access method and modulation used in ZigBee?
22. Name the two primary network topologies used with ZigBee.
23. Define ZC, ZR, and ZED.
24. Explain how a mesh network increases the transmission range and reliability.
25. List five popular applications for ZigBee.
26. What is the popular name for the IEEE 802.16d/e standard?
27. Explain why IEEE 802.16 is a MAN technology.
28. What are the two main topologies of the 802.16 standard?
29. Name the general operating frequency range of 802.16.
30. What two broadband services might get competition from 802.16?
31. Name the access and modulation methods used in 802.16.
32. List the maximum operating speeds and range of 802.16.
33. What is 802.16e?
34. True or false? VoIP can be used with 802.16.
35. State the major use for IR wireless.
36. List the transmission format, baseband or modulation for each IR wireless device.
    a. Remote control
    b. IrDA
37. State the maximum data rate of the IrDA standard.
38. List three popular applications for IR.
39. What physical requirement is necessary for IR systems to work?
40. What is RFID?
41. RFID is the electronic equivalent of what popular coding and ID scheme?
42. Where does the circuitry in a RFID tag get the dc voltage to operate its circuits?

43. Name the three primary operating frequencies for RFID tags.
44. What is the approximate maximum operating range of a 13.56-MHz tag?
45. Name the type of modulation used in most passive tags.
46. List five potential applications of RFID.
47. What is the most recent version of RFID standards? State the
    a. Operating frequency range
    b. Advantages and benefits
48. Where is the Electronic Product Code stored in a tag?
49. What is NFC?
50. State the operating frequency and data rates of the NFC standard.
51. What are the main two applications for NFC?
52. Describe the basic concept of UWB. Is a carrier used?
53. What is the name of the type of pulse transmitted by UWB?
54. State the primary operating frequency range of UWB. What frequency range is used in automotive applications?
55. State the two types of modulation used with UWB.
56. List the major advantages of UWB over other wireless methods.
57. What is the primary disadvantage of UWB?
58. Give five major applications for UWB. Which has been the most widespread to date?
59. Explain the access mode, modulation, and frequency bands for the newest form of UWB.
60. What is the maximum data rate of this newest form of UWB?
61. Name two popular applications of the newest form of UWB.

Problems

1. Refer to Fig. 21-1.
   a. What low-speed wireless technology gives the greatest range?
   b. What technology gives the maximum speed over the greatest range?

2. The pulse width of a UWB monocycle is 60 ps. What is the approximate center frequency? Answers to Selected Problems follow Chap. 22.

Critical Thinking

1. Name three wireless applications not covered in this chapter.
2. Think up three new wireless applications that would be useful and potentially practical.

3. In selecting a LAN or PAN, what three critical factors should be considered for each application?

chapter

22

Communication Tests and Measurements

This book was written primarily to educate those of you seeking to become specialists in communication electronics. If you are employed in communication electronics, your work will involve some form of testing and measurement. The work may be installation, operation, servicing, repair, and maintenance or testing to specifications or standards.

The purpose of this chapter is to introduce you to the wide range of special testing equipment and measurement procedures used in communication electronics. The chapter concludes with a special section on troubleshooting techniques and EMI reduction.

Because of the wide range of communication equipment available and the many differences from manufacturer to manufacturer, it is difficult to be specific. Only general test and measurement procedures are given. Just keep in mind that when you are performing real tests and measurements, you must familiarize yourself with not only the test equipment used but also the specific transmitter, receiver, or other communication device being tested. Manuals for both the test instruments and the communication equipment should be available for reference.

Objectives

After completing this chapter, you will be able to:

- List 10 common test instruments used in testing communication equipment and describe the basic operation of each.
- Name common communication equipment tests carried out on transmitters, receivers, and antennas including frequency measurements, power measurements, SWR measurements, sensitivity, and spectrum analyses.
- Describe the basic troubleshooting procedures used for locating problems in transmitters and receivers.
- Define electromagnetic interference (EMI), list its sources, and describe the measures to control it.

22-1 Communication Test Equipment

This section gives a broad overview of the many different types of test instruments available for use with communication equipment. It is assumed that you already know basic test and measurement techniques used with conventional low-frequency test equipment such as multimeters, signal generators, and oscilloscopes. In your communication work, you will continue to use standard oscilloscopes and multimeters for measuring voltages, currents, and resistance. Coverage of these basic test instruments will not be repeated here. We will, however, draw upon your knowledge of the principles of those instruments as they apply to the test equipment discussed in this section.

Voltage Measurements

The most common measurement obtained for most electronic equipment is voltage. This is particularly true for dc and low-frequency ac applications. In RF applications, voltage measurements may be important under some conditions, but power measurements are far more common at higher frequencies, particularly microwave. In testing and troubleshooting communication equipment, you will still use a dc voltmeter to check power supplies and other dc conditions. There are also occasions when measurement of RF, that is, ac, voltage must be made.

There are two basic ways to make ac voltage measurements in electronic equipment. One is to use an ac voltmeter. Most conventional voltmeters can measure ac voltages from a few millivolts to several hundred volts. Typical bench or portable ac multimeters are restricted in their frequency range to a maximum of several thousand kilohertz. Higher-frequency ac voltmeters are available for measuring audio voltages up to several hundred thousand kilohertz. For higher frequencies, special RF voltmeters must be used. The second method is to use an oscilloscope as described below.

RF Voltmeters. An *RF voltmeter* is a special piece of test equipment designed to measure the voltage of high-frequency signals. Typical units are available for making measurements up to 10 MHz. Special units capable of measuring voltages from microvolts to hundreds of volts at frequencies up to 1 to 2 GHz are also available.

RF voltmeters are made to measure sine wave voltages, with the readout given in *root mean square (rms)*. Most RF voltmeters are of the analog variety with a moving pointer on a background scale. Measurement accuracy is within the 1 to 5 percent range depending upon the specific instrument. Accuracy is usually quoted as a percentage of the reading or as a percentage of the full-scale value of the voltage range selected. RF voltmeters with digital readout probes are also available with somewhat improved measurement accuracy.

RF voltmeter

Root mean square (rms)

RF Probes. One way to measure RF voltage is to use an *RF probe* with a standard dc multimeter. RF probes are sometimes referred to as *detector probes*. An RF probe is basically a rectifier (germanium or hot carrier) with a filter capacitor that stores the peak value of the sine wave RF voltage. The external dc voltmeter reads the capacitor voltage. The result is a peak value that can easily be converted to root mean square by multiplying it by 0.707.

Most RF probes are good for RF voltage measurements to about 250 MHz. The accuracy is about 5 percent, but that is usually very good for RF measurements.

RF probe (detector probe)

Oscilloscopes. Two basic types of oscilloscopes are used in RF measurements: the *analog oscilloscope* and the *digital storage oscilloscope (DSO)*.

Analog oscilloscopes amplify the signal to be measured and display it on the face of a CRT at a specific sweep rate. They are available for displaying and measuring RF voltages to about 500 MHz. As a rule, an analog oscilloscope should have a bandwidth of 3 or more times the highest-frequency component (a carrier, a harmonic, or a sideband) to be displayed.

Digital storage oscilloscopes, also known as *digital,* or *sampling, oscilloscopes,* are growing in popularity and rapidly replacing analog oscilloscopes. DSOs use high-speed

Analog oscilloscope

Digital storage oscilloscope (DSO; digital, or sampling, oscilloscope)

sampling or A/D techniques to convert the signal to be measured to a series of digital words that are stored in an internal memory. Sampling rates vary depending upon the oscilloscope but can range from approximately 20 million samples per second to more than 8 billion samples per second. Each measurement sample is usually converted to an 8- or 10-bit parallel binary number that is stored in an internal memory. Oscilloscopes with approximately 512 kbytes of memory or more are available depending upon the product.

Today, more than 50 percent of oscilloscope sales are of the digital sampling type. DSOs are very popular for high-frequency measurements because they provide the means to display signals with frequencies up to about 30 GHz. This means that complex modulated microwave signals can be readily viewed, measured, and analyzed.

Power Meters

Power meter

As indicated earlier, it is far more common to measure RF power than it is to measure RF voltage or current. This is particularly true in testing and adjusting transmitters that typically develop significant output power. One of the most commonly used RF test instruments is the *power meter*.

Power meters come in a variety of sizes and configurations. One of the most popular is a small in-line power meter designed to be inserted into the coaxial cable between a transmitter and an antenna. The meter is used to measure the transmitter output power supply to the antenna. A short coaxial cable connects the transmitter output to the power meter, and the output of the power meter is connected to the antenna or dummy load.

A more sophisticated power meter is the bench unit designed for laboratory or production line testing. The output of the transmitter or other device whose power is to be measured is connected by a short coaxial cable to the power meter input.

Power meters may have either an analog readout meter or a digital display. The dial or display is calibrated in milliwatts, watts, or kilowatts. The dial can also be calibrated in terms of dBm. This is the decibel power reference to 1 milliwatt (mW). In the smaller, handheld type of power meter, an SWR measurement capability is usually included.

The operation of a power meter is generally based on converting signal power to heat. Whenever current flows through a resistance, power is dissipated in the form of heat. If the heat can be accurately measured, it can usually be converted to an electric signal that can be displayed on a meter.

Example 22-1

An RF voltmeter with a detector probe is used to measure the voltage across a 75-Ω resistive load. The frequency is 137.5 MHz. The measured voltage is 8 V. What power is dissipated in the load?

An RF power meter with a detector probe produces a peak reading of voltage V_P:

$$V_P = 8 \text{ V}$$

$$V_{rms} = 0.707 V_P$$

$$= 0.707(8) = 5.656 \text{ V}$$

$$\text{Power} = \frac{V^2}{R} = \frac{(5.656)^2}{75}$$

$$= \frac{32}{75} = 0.4265 \text{ W}$$

$$= 426.5 \text{ mW}$$

Figure 22-1 Monomatch power/SWR meter.

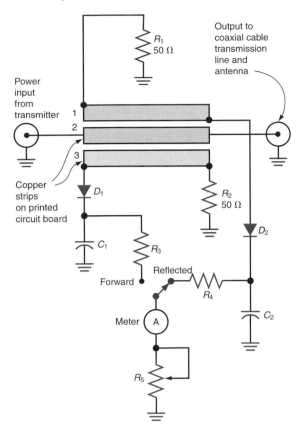

Power can also be measured indirectly. If the load impedance is known and resistive, you can measure the voltage across the load and then calculate the power with the formula $P = V^2/R$.

Power Measurement Circuits

Relatively simple circuits can be used to measure power in transmitters and RF power circuits. An example is the monomatch circuit shown in Fig. 22-1. It uses a 50-Ω transmission line made with a microstrip on a small printed-circuit board (PCB). The center conductor is the segment labeled 2 in the schematic. On each side of the center conductor are narrower pickup loops labeled 1 and 3. An RF voltage proportional to the forward and reverse (reflected) power is produced as the result of capacitive and inductive coupling with the center conductor. The voltage in segment 3 represents the forward power. It is rectified by diode D_1 and filtered by C_1 into a proportional dc voltage. This voltage is applied through multiplier resistors R_3 and R_5 to a meter whose scale is calibrated in watts of power. Note the 50-Ω resistors that terminate the pickup loops for impedance matching.

The voltage induced in pickup loop 1 is proportional to the reflected power. It is rectified by D_2 and filtered into a proportional direct current by C_2. A switch is used to select the display of either the forward or the reflected power. Resistor R_5 is used to calibrate the meter circuit, using an accurate power meter as a standard.

Another popular power measurement circuit is the directional coupler shown in Fig. 22-2. A short piece of 50-Ω coaxial cable serves as the single-turn primary winding on a transformer made with a toroid core and a secondary winding of many turns of fine wire. When RF power is passed through the coaxial section, a stepped-up voltage is induced into the secondary winding. Equal-value resistors R_1 and R_2 divide the voltage

Figure 22-2   Directional coupler power measurement.

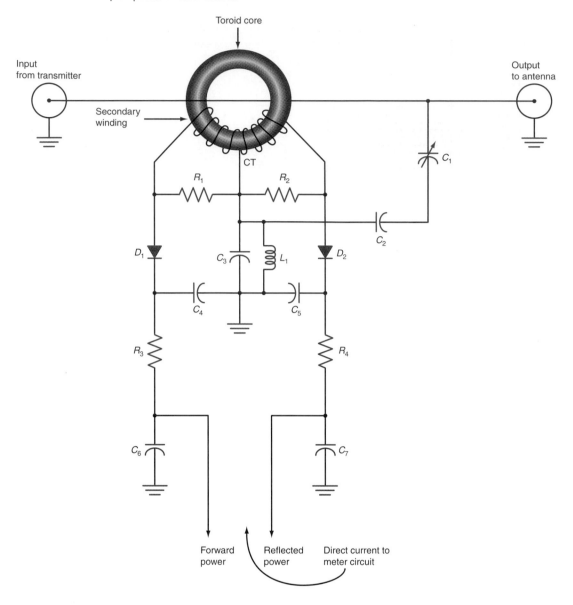

equally between two diode rectifier circuits made up of D_1 and D_2 and the related components.

A voltage divider made up of C_1, C_2, C_3, and L_1 samples the voltage at the output of the circuit. This voltage is applied to both diode rectifiers along with the voltages from the transformer secondary. When these voltages are combined, the rectified outputs are proportional to the forward and reflected voltages on the line. Low-pass filters R_3-C_6 and R_4-C_7 smooth the rectified signals into direct current. A meter arrangement like that in Fig. 22-1 is used to display either forward or reflected power.

Both circuits can be designed to handle power levels from a few milliwatts to many kilowatts. When low-level signals are used, the diodes must be of the germanium or hot carrier type with low bias threshold voltages (0.2 to 0.4 V) to provide sufficient accuracy of measurement. With careful design and adjustment these circuits can give an accuracy of 90 percent or better. Because the circuits are so small, they are often built into the transmitter or other circuit along with the meter and switch.

Example 22-2

If the forward power and reflected power in a circuit are known, the SWR can be calculated. If the forward power is 380 W and the reflected power is 40 W, what is the SWR?

$$\text{SWR} = \frac{1 + \sqrt{P_R/P_F}}{1 - \sqrt{P_R/P_F}}$$

$$= \frac{1 + \sqrt{40/380}}{1 - \sqrt{40/380}}$$

$$= \frac{1 + 0.324}{1 - 0.324} = \frac{1.324}{0.675} = 1.96$$

Dummy Loads

A *dummy load* is a resistor that is connected to the transmission line in place of the antenna to absorb the transmitter output power. When power is measured or other transmitter tests are done, it is usually desirable to disconnect the antenna so that the transmitter does not radiate and interfere with other stations on the same frequency. In addition, it is best that no radiation be released if the transmitter has a problem or does not meet frequency or emission standards. The dummy load meets this requirement. The dummy load may be connected directly to the transmitter coaxial output connector, or it may be connected to it by a short piece of coaxial cable.

Dummy load

The load is a resistor whose value is equal to the output impedance of the transmitter and that has sufficient power rating. For example, a CB transmitter has an output impedance of 50 Ω and a power rating of about 4 W. The resistor dummy load must be capable of dissipating that amount of power or more. For example, you could use three 150-Ω, 2-W resistors in parallel to give a load of 150/3, or 50, Ω and 3 $\times$ 2, or 6, W. Standard composition carbon resistors can be used. The tolerance is not critical, and resistors with 5 or 10 percent tolerance will work well.

For low-power transmitters such as CBs and amateur radios, an incandescent light bulb makes a reasonably good load. A type 47 pilot light is widely used for transmitter outputs of several watts. An ordinary lightbulb of 75 or 100 W, or higher, can also be used for higher-power transmitters.

The best dummy load is a commercial unit designed for that purpose. These units are usually designed for some upper power limit such as 200 W or 1 kW. The higher-power units are made with a resistor immersed in oil to improve its heat dissipation capability without burning up. A typical unit is a resistor installed in a 1-gal can filled with insulating oil. A coaxial connector on top is used to attach the unit to the transmitter. Other units are mounted in an aluminum housing with heat fins to improve heat dissipation. The resistors are noncritical, but they must be noninductive. The resistor should be as close to pure resistance at the operating frequency as possible.

Standing Wave Ratio Meters

Standing wave ratio meters

The SWR can be determined by calculation if the forward and reflected power values are known. Some SWR meters use the monomatch or directional coupler circuits described above and then implement the SWR calculation given in Example 22-2 with op amps and analog multiplier ICs. But you can also determine SWR directly.

Figure 22-3 Bridge SWR meter.

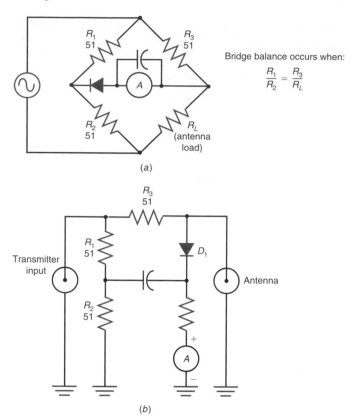

Bridge balance occurs when:

$$\frac{R_1}{R_2} = \frac{R_3}{R_L}$$

(a)

(b)

Figure 22-3(*a*) shows a bridge SWR meter. A bridge is formed of precision, noninductive resistors and the antenna radiation resistance. In some SWR meters, resistors are replaced with a capacitive voltage divider. The meter is connected to measure the unbalance of the bridge. The transmitter is the ac power source.

Figure 22-3(*b*) shows the circuit rearranged so that the meter and one side of the bridge are grounded, thereby creating a better match to unbalanced coaxial transmission lines. Note the use of coaxial connectors for the transmitter input and the antenna and transmission line. The meter is a basic dc microammeter. Diode D_1 rectifies the RF signal into a proportional direct current. If the radiation resistance of the antenna is 50 Ω, the bridge will be balanced and the meter reading will be zero. The meter is calibrated to display an SWR of 1. If the antenna radiation resistance is not 50 Ω, the bridge will be unbalanced and the meter will display a reading that is proportional to the degree of unbalance. The meter is calibrated in SWR values.

Signal Generators

Signal generator

A *signal generator* is one of the most often needed pieces of equipment in communication equipment servicing. As its name implies, a signal generator is a device that produces an output signal of a specific shape at a specific frequency and, in communication applications, usually with some form of modulation. The heart of all signal generators is a variable-frequency oscillator that generates a signal that is usually a sine wave. Audio-frequency sine waves are required for the testing of audio circuits in communication equipment, and sine waves in the RF range from approximately 500 kHz to 30 GHz are required to test all types of RF amplifiers, filters, and other circuits. This section provides a general overview of the most common types of signal generators used in communication testing and servicing.

Function Generators. A *function generator* is a signal generator designed to generate sine waves, square waves, and triangular waves over a frequency range of approximately 0.001 Hz to about 3 MHz. By changing capacitor values and varying the charging current with a variable resistance, a wide range of frequencies can be developed. The sine, square, and triangular waves are also available simultaneously at individual output jacks.

A function generator is one of the most flexible signal generators available. It covers all the frequencies needed for audio testing and provides signals in the low-RF range. The precision of the frequency setting is accurate for most testing purposes. A frequency counter can be used for precise frequency measurement if needed.

The output impedance of a function generator is typically 50 Ω. The output jacks are BNC connectors which are used with 50- or 75-Ω coaxial cable. The output amplitude is continuously adjustable with a potentiometer. Some function generators include a switched resistive attenuator that allows the output voltage to be reduced to the millivolt and microvolt levels.

Because of the very low cost and flexibility of a function generator, it is the most popular bench instrument in use for general testing of radio amplifiers, filters, and low-frequency RF circuits. The square wave output signals also make it useful in testing digital circuits.

RF Signal Generators. Two basic types of RF signal generators are in use. The first is a simple, inexpensive type that uses a variable-frequency oscillator to generate RF signals in the 100-kHz to 500-MHz range. The second type is frequency-synthesized.

These simple RF signal generators contain an output level control that can be used to adjust the signal to the desired level, from a few volts down to several millivolts. Some units contain built-in resistive step attenuators to reduce the signal level even further. The more sophisticated generators have built-in level control or automatic gain control (AGC). This ensures that the output signal remains constant while it is tuned over a broad frequency range.

Most low-cost signal generators allow the RF signal being generated to be amplitude-modulated. Normally, a built-in audio oscillator with a fixed frequency somewhere in the 400- to 1000-Hz range is included. A modulation level control is provided to adjust the modulation from 0 to 100 percent. Some RF generators have a built-in frequency modulator.

Such low-cost signal generators are useful in testing and troubleshooting communication receivers. They can provide an RF signal at the signal frequency for injection into the antenna terminals of the receiver. The generator can produce signals that can substitute for local oscillators or can be set to the intermediate frequencies for testing IF amplifiers.

The output frequency is usually set by a large calibrated dial. The precision of calibration is only a few percent, but more precise settings can be obtained by monitoring the signal output on a frequency counter.

When any type of generator based on *LC* or *RC* oscillators is being used, it is best to turn the generator on and let it warm up for several hours before it is used. When a generator is first turned on, its output frequency will drift because of changes in capacitance, inductance, and resistance values. Once the circuit has warmed up to its operating temperature, these variations cease or drop to a negligible amount.

Newer generators use *frequency synthesis* techniques. These generators include one or more mixer circuits that allow the generator to cover an extremely wide range of frequencies. The great value of a frequency-synthesized signal generator is its excellent frequency stability and precision of frequency setting.

Most frequency-synthesized signal generators use a front panel keyboard. The desired frequency is entered with the keypad and displayed on a digital readout. As with other signal generators, the output level is fully variable. The output impedance is typically 50 Ω, with both BNC- and N-type coaxial connectors being required.

Frequency-synthesized generators are available for frequencies into the 20- to 30-GHz range. Such generators are extremely expensive, but they may be required if precision measurement and testing is necessary.

GOOD TO KNOW

A function generator is one of the most flexible generators available. It covers all the frequencies needed for audio testing and provides signals in the low-RF range.

Careers are available in the repair and testing of communication equipment. Such equipment varies widely from one manufacturer to another, so manuals are needed for both the test equipment and the products to be tested. Here, a technician uses RF equipment for vector network analysis, spectrum analysis, and instrument controller testing.

Sweep generator

Sweep Generators. A *sweep generator* is a signal generator whose output frequency can be linearly varied over some specific range. Sweep generators have oscillators that can be frequency-modulated, where a linear sawtooth voltage is used as a modulating signal. The resulting output waveform is a constant-amplitude sine wave whose frequency increases from some lower limit to some upper limit (see Fig. 22-4).

Sweep generators are normally used to provide a means of automatically varying the frequency over a narrow range to plot the frequency response of a filter or amplifier or to show the bandpass response curve of the tuned circuits in equipment such as a receiver. The sweep generator is connected to the input of the circuit, and the upper and lower frequencies are determined by adjustments on the generator. The generator then automatically sweeps over the desired frequency range.

At the same time, the output of the circuit being tested is monitored. The amplitude of the output will vary according to the frequency, depending on the type of circuit being tested. The output of the circuit is connected to an RF detector probe. The resulting signal is the envelope of the RF signal as determined by the output variation of the circuit being tested. The signal displayed on the oscilloscope is an amplitude plot of the frequency response curve. The horizontal axis represents the frequency being varied with time, and the output represents the amplitude of the circuit output at each of the frequencies.

Figure 22-4 Sweep generator output.

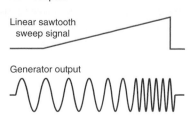

Linear sawtooth sweep signal

Generator output

Figure 22-5 Testing frequency response with a sweep generator.

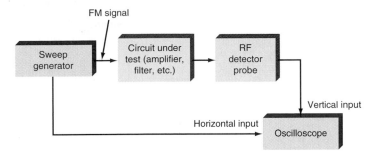

Figure 22-5 shows the general test setup. The linear sweep from the sweep generator is used in place of the oscilloscope's internal sweep so that the displayed response curve is perfectly synchronized with the generator.

Most sweep generators also have *marker capability;* i.e., one or more reference oscillators are included to provide frequency markers at selected points so that the response curve can be actively interpreted. Marker increments may be 100 kHz or 1 MHz. They are added to (linearly mixed with) the output of the RF detector probe, and the composite signal is amplified and sent to the vertical input of the oscilloscope. Sweep generators can save a considerable amount of time in testing and adjusting complex tuned circuits in receivers and other equipment.

Most function generators have built-in sweep capability. If sweep capability is not built in, often an input jack is provided so that an external sawtooth wave can be connected to the generator for sweep purposes.

Arbitrary Waveform Generators. A newer type of signal generator is the *arbitrary waveform generator.* It uses digital techniques to generate almost any waveform. An arbitrary waveform generator stores binary values of a desired waveform in a memory. These binary words are fed sequentially to a digital-to-analog converter that produces a stepped approximation of the desired wave. Most arbitrary waveform generators come with preprogrammed standard waves such as sine, rectangular, sawtooth, and triangular waves, and amplitude modulation. These generators are set up so that you can program a waveform. The arbitrary waveform generator provides a fast and easy way to generate almost any signal shape. Because digital sampling techniques are used, the upper frequency limit of the output is usually below 1 GHz.

Frequency Counters

One of the most widely used communication test instruments is the *frequency counter.* It measures the frequency of transmitters, local and carrier oscillators, frequency synthesizers, and any other signal-generating circuit or equipment. It is imperative that the frequency counter operate on its assigned frequency to ensure compliance with rules and regulations and to avoid interference with other services.

A frequency counter displays the frequency of a signal on a decimal readout. Counters are available as bench instruments or portable battery-powered units. A block diagram of a frequency counter is shown in Fig. 22-6. Almost all digital counters are made up of six basic components: input circuit, gate, decimal counter, display, control circuits, and time base. In various combinations, these circuits permit the counter to make time and frequency measurements.

Frequency Measurement. *Frequency* is a measure of the number of events or cycles of a signal that occur in a given time. The usual unit of frequency measurement is hertz (Hz), or cycles per second. The time base generates a very precise signal that

Marker capability

Arbitrary waveform generator

Frequency counter

Frequency

Figure 22-6   Block diagram of a frequency counter.

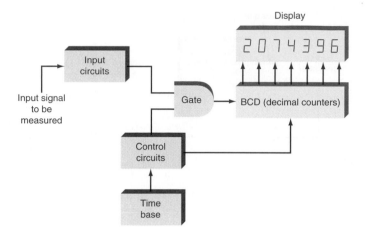

is used to open, or enable, the main gate for an accurate period of time to allow the input pulses to pass through to the counter. The time base accuracy is the most critical specification of the counter. The counter accumulates the number of input cycles that occur during that 1-s interval. The display then shows the frequency in cycles per second, or hertz.

The number of decade counters and display digits also determines the resolution of the frequency measurement. The greater the number of digits, the better the resolution will be. Most low-cost counters have at least 5 digits of display. This provides reasonably good resolution on most frequency measurements. For very high-frequency measurements, resolution is more limited. However, good resolution can still be obtained with a minimum number of digits by optimum selection of the time base frequency.

In most counters that have a selectable time base, the position of the display decimal point is automatically adjusted as the time base signal is adjusted. In this way, the display always shows the frequency in units of hertz, kilohertz, or megahertz. Some of the more sophisticated counters have an automatic time base selection feature called *autoranging*. Special autoranging circuitry in the counter automatically selects the best time base frequency for maximum measurement resolution without overranging. *Overranging* is the condition that occurs when the count capability of the counter is exceeded during the count interval. The number of counters and display digits determines the count capability and thus the over-range point for a given time base.

Prescaling. All the techniques for measuring high frequencies involve a process that converts the high frequency to a proportional lower frequency that can be measured with conventional counting circuitry. This translation of the high frequency to the lower frequency is called *down conversion*.

Prescaling is a down-conversion technique that involves the division of the input frequency by a factor that puts the resulting signal into the normal frequency range of the counter. It is important to realize that although prescaling permits the measurement of higher frequencies, it is not without its disadvantages, one of which is loss of resolution. One digit of resolution is lost for each decade of prescaling incorporated.

The prescaling technique for extending the frequency-measuring capability of a counter is widely used. It is simple to implement with modern, high-speed ICs. It is also the most economical method of extending the counting range. Prescalers can be built into the counter and switched in when necessary. Alternatively, external prescalers, which are widely available for low-cost counters, can be used. Most prescalers operate in the range of 200 MHz to 20 GHz. For frequencies beyond 20 GHz, more sophisticated down-conversion techniques must be used.

Spectrum Analyzers

The *spectrum analyzer* is one of the most useful and popular communication test instruments. Its basic function is to display received signals in the frequency domain. Oscilloscopes are used to display signals in the time domain. The sweep circuits in the oscilloscope deflect the electron beam in the CRT across the screen horizontally. This represents units of time. The input signal to be displayed is applied to deflect the electron beam vertically. Thus, electronic signals that are voltages occurring with respect to time are displayed on the oscilloscope screen.

The spectrum analyzer combines the display of an oscilloscope with circuits that convert the signal to the individual frequency components dictated by Fourier analysis of the signal. Signals applied to the input of the spectrum analyzer are shown as vertical lines or narrow pulses at their frequency of operation.

Figure 22-7 shows the display of a spectrum analyzer. The horizontal display is calibrated in frequency units, and the vertical part of the display is calibrated in voltage, power, or decibels. The spectrum analyzer display shows three signals at frequencies of 154.7, 157.8, and 160.2 MHz. The vertical height represents the relative strength of the amplitude of each signal. Each signal might represent the carrier of a radio transmitter. A spectrum analysis of any noise between the signals is shown.

The spectrum analyzer can be used to view a complex signal in terms of its frequency components. A graticule on the face of the CRT allows the frequency spacing between adjacent frequency components to be determined. The vertical amplitude of the scale is calibrated in decibels. The spectrum analyzer is extremely useful in analyzing complex signals that may be difficult to analyze or whose content may be unrecognizable.

The four basic techniques of spectrum analysis are bank of filters, swept filter, swept spectrum superheterodyne, and fast Fourier transform (FFT). All these methods do the same thing, i.e., decompose the input signal into its individual sine wave frequency components. Both analog and digital methods are used to implement each type. The superheterodyne and FFT methods are the most widely used and are discussed below.

Spectrum analyzer

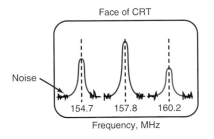

Figure 22-7 A frequency display of a spectrum analyzer.

A spectrum analyzer from Tektronix, covering frequencies to 14 GHz.

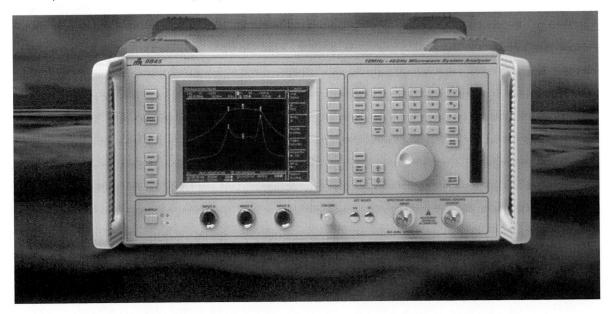

Communication Tests and Measurements

Figure 22-8 Superheterodyne RF spectrum analyzer.

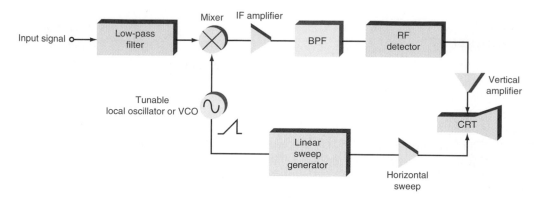

Superheterodyne RF spectrum analyzer

Intermediate frequency (IF)

Superheterodynes. Perhaps the most widely used RF spectrum analyzer is the *superheterodyne* type (see Fig. 22-8). It consists of a broadband front end, a mixer, and a tunable local oscillator. The frequency range of the input is restricted to some upper limit by a low-pass filter in the input. The mixer output is the difference between the input signal frequency component and the local-oscillator frequency. This is the *intermediate frequency (IF)*.

As the local-oscillator frequency increases, the output of the mixer stays at the IF. Each frequency component of the input signal is converted to the IF value by the varying local-oscillator signal. If the frequency components are very close to one another and the bandwidth of the IF bandpass filter (BPF) is broad, the display will be just one broad pulse. Narrowing the bandwidth of the IF BPF allows more closely spaced components to be detected. Most spectrum analyzers have several switchable selectivity ranges for the IF.

Spectrum analyzers are available in many configurations with different specifications, and they are designed to display signals from approximately 100 kHz to approximately 30 GHz. Most RF and microwave spectrum analyzers are superheterodynes. Spectrum analyzers are usually calibrated to provide relatively good measurement accuracy of the signal. Most signals are displayed as power or decibel measurements, although some analyzers provide for voltage level displays. The input is usually applied through a 50-Ω coaxial cable and connector.

Fast Fourier transform (FFT) spectrum analyzer

FFT Spectrum Analyzers. The *fast Fourier transform (FFT)* method of spectrum analysis relies on the FFT mathematical analysis. FFT spectrum analyzers give a high-resolution display and are generally superior to all other types of spectrum analyzers. However, the upper frequency of the input signal is limited to frequencies in the tens of megahertz range.

In addition to measuring the spectrum of a signal, spectrum analyzers are useful in detecting harmonics and other spurious signals generated unintentionally. Spectrum analyzers can be used to display the relative signal-to-noise ratio, and they are ideal for analyzing modulation components and displaying the harmonic spectrum of a rectangular pulse train.

Despite their extremely high prices (usually $10,000 to $50,000), spectrum analyzers are widely used. Many critical testing and measurement applications demand their use, especially in developing new RF equipment and in making final tests and measurements of manufactured units. Spectrum analyzers are also used in the field for testing cable TV systems, cellular telephone systems, fiber-optic networks, and other complex communication systems.

Network Analyzers

Network analyzer

A *network analyzer* is a test instrument designed to analyze linear circuits, especially RF circuits. It is a combination instrument that contains a wide-range sweep sine wave generator and a CRT output that displays not only frequency plots as does a spectrum analyzer but also plots of phase shift versus frequency.

Network analyzers are used by engineers to determine the specific performance characteristics of a circuit they are designing, such as a filter, an amplifier, or a mixer. They are also useful in analyzing transmission lines and even individual components. The network analyzer applies a swept-frequency sine wave and measures the circuit output. The resulting measurement data is then used to produce an output display such as an amplitude versus frequency plot, a phase shift versus frequency, or even a plot of complex impedance values on a Smith chart display.

Network analyzers completely describe the performance or characterization of a circuit. This type of information is useful not only to engineers creating the circuits but also to those in manufacturing who have to produce and test the circuit. Despite their very high cost, these instruments are widely used because of the valuable information they provide and the massive amount of design and test time they save.

Field Strength Meters

One of the least expensive pieces of RF test equipment is the *field strength meter (FSM)*, a portable device used for detecting the presence of RF signals near an antenna. The FSM is a sensitive detector for RF energy being radiated by a transmitter into an antenna. It provides a relative indication of the strength of the electromagnetic waves reaching the meter.

Field strength meter (FSM)

The field strength meter is a vertical whip antenna, usually of the telescoping type, connected to a simple diode detector. The diode detector is exactly like the circuit of a simple crystal radio or a detector probe, as described earlier, but without any tuned circuits so that the unit will pick up signals on any frequency.

The field strength meter does not give an accurate measurement of signal strength. In fact, its only purpose is to detect the presence of a nearby signal (within about 100 ft or less). Its purpose is to determine whether a given transmitter and antenna system are working. The closer the meter is moved to the transmitter and antenna, the higher the signal level.

A useful function of the meter is the determination of the radiation pattern of an antenna. The field strength meter is adjusted to give the maximum reading in the direction of the most radiation from the antenna. The meter is moved in a constant-radius circle around the antenna for 360°. Every 5° or 10°, a field strength reading is taken from the meter. The resulting set of readings can be plotted on polar graph paper to reveal the horizontal radiation pattern of the antenna.

Other types of field strength meters are available. A simple meter may be built to incorporate a resonant circuit to tune the input to a specific transmitter frequency. This makes the meter more sensitive. Some meters have a built-in amplifier to make the meter even more sensitive and useful at greater distances from the antenna.

An *absolute* (rather than *relative*) *field strength meter* is available for accurate measurements of signal strength. The strength of the radiated signal is usually measured in microvolts per meter (μV/m). This is the amount of voltage the signal will induce into an antenna that is 1 m long. An absolute field strength meter is calibrated in units of microvolts per meter. Highly accurate signal measurements can be made.

Absolute field strength meter

Other Test Instruments

There are hundreds of types of communication test instruments, most of which are very specialized. The ones described previously in this chapter, plus those listed in the above table, are the most common, but there are many others including the many special test instruments designed by equipment manufacturers for testing their production units or servicing customers' equipment.

Instrument	Purpose
Absorption wave meter	Variable tuned circuit with an indicator that tells when the tuned circuit is resonant to a signal coupled to the meter by a transmitter; provides a rough indication of frequency.
Impedance meter	An instrument, usually of the bridge type, that accurately measures the impedance of a circuit, a component, or even an antenna at RF frequencies.
Dip oscillator	A tunable oscillator used to determine the approximate resonant frequency of any deenergized LC–resonant circuit. The oscillator inductor is coupled to the tuned circuit inductor, and the oscillator is tuned until its feedback is reduced as energy being taken by the tuned circuit, which is indicated by a "dip," or reduction, in current on a built-in meter. The approximate frequency is read from a calibrated dial.
Noise bridge	Bridge circuit driven by a random noise voltage source (usually a reverse-biased zener diode that generates random "white" or pseudorandom "pink" noise) that has an antenna or coaxial cable as one leg of the bridge; used to make antenna characteristic impedance measurements and measurements of coaxial cable velocity factor and length.

Absorption wave meter

Impedance meter

Dip oscillator

Noise bridge

22-2 Common Communication Tests

Hundreds, even thousands, of different tests are made on communication equipment. However, some common tests are widely used on all types of communication equipment. This section summarizes the most common tests and measurements made in the servicing of communication equipment.

Most of the tests described here focus on standard radio communication equipment. Tests for transmitters, receivers, and antennas will be described, as will special microwave tests, fiber-optic cable tests, and tests on data communication equipment. Keep in mind that the test procedures described are general. To make specific tests, follow the test setups recommended by the test equipment manufacturer. Always have the manuals for the test equipment and the equipment being serviced on hand for reference.

Transmitter Tests

Four main tests are made on most transmitters: tests of frequency, modulation, and power, and tests for any undesired output signal component such as harmonics and parasitic radiations. These tests and measurements are made for several reasons. First, any equipment that radiates a radio signal is governed by Federal Communications Commission (FCC) rules and regulations. For a transmitter to meet its intended purpose, the FCC specifies frequency, power, and other measurements to which the equipment must comply. Second, the tests are normally made when the equipment is first installed to be sure that everything is working correctly. Third, such tests may be performed to troubleshoot equipment. If the equipment is not working properly, these tests are some of the first that should be made to help identify the trouble.

Frequency Measurement. Regardless of the method of carrier generation, the frequency of the transmitter is important. The transmitter must operate on the assigned frequency to comply with FCC regulations and to ensure that the signal can be picked up by a receiver tuned to that frequency.

Figure 22-9 Transmitter frequency measurement.

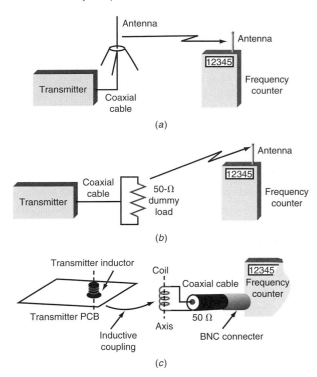

(a)

(b)

(c)

The output of a transmitter is measured directly to determine its frequency. The transmitted signal is independently picked up and its frequency measured on a frequency counter. Figure 22-9 shows several methods of picking up the signal. Many frequency counters designed for communication work come with an antenna that picks up the signal directly from the transmitter [see Fig. 22-9(a)]. The transmitter is keyed up (turned on) while it is connected to its regular antenna, and the antenna on the counter picks up the signal and translates it to one that can be measured by the counter circuitry. No modulation should be applied, especially if FM is used.

The transmitter output can be connected to a dummy load [see Fig. 22-9(b)]. This will ensure that no signal is radiated, but that there will be sufficient signal pickup to make a frequency measurement if the counter and its antenna are placed near the transmitter.

Another method of connecting the counter to the transmitter is to use a small coil, as shown in Fig. 22-9(c). A small pickup coil can be made of stiff copper wire. Enamel copper wire, size AWG 12 or 14 wire, is formed into a loop of two to four turns. The ends of the loop are connected to a coaxial cable with a BNC connector for attachment to the frequency counter. The loop can be placed near the transmitter circuits. This method of pickup is used when the transmitter has been opened and its circuits have been exposed. For most transmitters, the loop has been placed only in the general vicinity of the circuitry. Normally the loop picks up radiation from one of the inductors in the final stage of the transmitter. Maximum coupling is achieved when the axis of the turns of the loop is parallel to the axis of one of the inductors in the final output stage.

Once the signal from the transmitter is coupled to the counter, the counter sensitivity is adjusted, and the counter is set to the desired range for displaying the frequency. The greater the number of digits the counter can display, the more accurate the measurement.

Normally this test is made without modulation. If only the carrier is transmitted, any modulation effects can be ignored. Modulation must not be applied to an FM transmitter, because the carrier frequency will be varied by the modulation, resulting in an inaccurate frequency measurement.

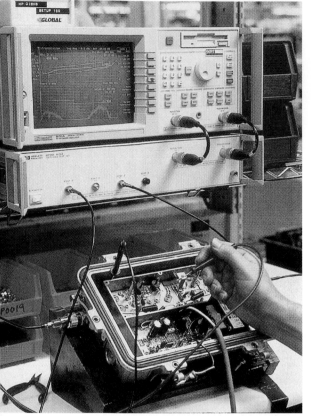

The quality of crystals today is excellent; thus, off-frequency operation is not common. If the transmitter is not within specifications, the crystal can be replaced. In some critical pieces of equipment, the crystal may be in an oven. If the oven temperature control circuits are not working correctly, the crystal may have drifted off frequency. This calls for repair of the oven circuitry or replacement of the entire unit.

If the signal source is a frequency synthesizer, the precision of the reference crystal can be checked. If it is within specifications, perhaps an off-frequency operation is being caused by a digital problem in the phase-locked loop (PLL). An incorrect frequency-division ratio, faulty phase detector, or poorly tracking voltage-controlled oscillator (VCO) may be the problem.

Modulation Tests. If AM is being used in the transmitter, you should measure the percentage of modulation. It is best to keep the percentage of modulation as close to 100 as possible to ensure maximum output power and below 100 to prevent signal distortion and harmonic radiation. In FM or PM transmitters, you should measure the frequency deviation with modulation. The goal with FM is to keep the deviation within the specific range to prevent adjacent channel interference.

The best way to measure AM is to use an oscilloscope and display the AM signal directly. To do this, you must have an oscilloscope whose vertical amplifier bandwidth is sufficient to cover the transmitter frequency. Figure 22-10(a) shows the basic test setup. An audio signal generator is used to amplitude-modulate the transmitter. An audio signal of 400 to 1000 Hz is applied in place of the microphone signal.

The transmitter is then keyed up, and the oscilloscope is attached to the output load. It is best to perform this test with a dummy load to prevent radiation of the signal. The oscilloscope is then adjusted to display the AM signal. The display will appear as shown in Fig. 22-10(b).

Power Measurements. Most transmitters have a tune-up procedure recommended by the manufacturer for adjusting each stage to produce maximum output power. In older transmitters, tuned circuits between stages have to be precisely adjusted in the correct sequence. In modern solid-state transmitters, there are fewer adjustments, but in most cases there are some adjustments in the driver and frequency multiplier stages as well as tuning adjustments for resonance at the operating frequency to the final amplifier. There may be impedance-matching adjustments in the final amplifier to ensure full coupling of the power to the antenna. The process is essentially that of adjusting the tuned circuits to resonance. These measurements are generally made while monitoring the output power of the transmitter.

The procedure for measuring the output power is to connect the transmitter output to an RF power meter and the dummy load, as shown in Fig. 22-11. The transmitter is keyed up without modulation, and adjustments are made on the transmitter circuit to tune for maximum power output. With the test arrangement shown, the power meter will display the output power reading.

Once the transmitter is properly tuned up, it can be connected to the antenna. The power into the antenna will then be indicated. If the antenna is properly matched to the transmission line, the amount of output power will be the same as that in the dummy load. If not, SWR measurements should be made. Most modern power meters measure both forward and reflected power, so SWR measurements are easier to make. It may be necessary to adjust the antenna or a matching circuit to ensure maximum output power with minimum SWR.

Harmonics and Spurious Output Measurements. A common problem in transmitters is the radiation of undesirable harmonics or spurious signals. Ideally, the output of the transmitter should be a pure signal at the carrier frequency with only those sideband components produced by the modulating signal. However, most transmitters

Figure 22-10 AM measurement. (*a*) Test setup. (*b*) Typical waveforms.

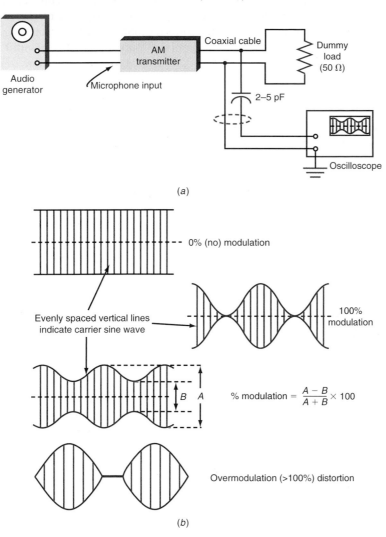

(a)

(b)

% modulation $= \dfrac{A - B}{A + B} \times 100$

will generate some harmonics and spurious signals. Transmitters that use class C, class D, and class E amplifiers generate a high harmonic content. If the tuned circuits in the transmitter are properly designed, the Q's will be high enough to reduce the harmonic content level sufficiently. However, this is not always the case.

Another problem is that other spurious signals can be generated by transmitters. In high-power transmitters particularly, parasitic oscillations can occur. These are caused by the excitation of small tuned circuits whose components are the stray inductances and capacitances in the circuits or the transistors of the tubes involved. Parasitic oscillations can reach high levels and cause radiation on undesired frequencies.

Figure 22-11 Power measurement.

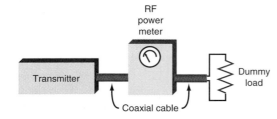

For most transmitters, the FCC specifies maximum levels of harmonic and spurious radiation. Intermodulation distortion in mixers and nonlinear circuits also produces unwanted signals. Normally these signals must be at least 30 or 40 dB down from the main carrier signal. The best way to measure harmonics and spurious signals is to use a spectrum analyzer. The transmitter output is modulated with an audio tone, and its output is monitored directly on the spectrum analyzer. It is usually best to feed the transmitter output into a dummy load for this measurement. The spectrum analyzer is then adjusted to display the normal carrier and sideband pattern. The search for high-level signals can begin for spurious outputs by tuning the spectrum analyzer above and below the operating frequency. The spectrum analyzer can be tuned to search for signals at the second, third, and higher harmonics of the carrier frequency. If signals are detected, they can be measured to ensure that they are sufficiently low in power to meet FCC regulations and/or the manufacturer's specifications. The spectrum analyzer is then tuned over a broad range to ensure that no other spurious non-harmonic signals are present.

It may be possible to reduce the harmonic and spurious output content by making a minor transmitter tuning adjustment. If not, to meet specifications, it is often necessary to use filters to eliminate unwanted harmonics or other signals.

Antenna and Transmission Line Tests

If the transmitter is working correctly and the antenna has been properly designed, about the only test that needs to be made on the transmission line and antenna is for standing waves. It will tell you whether any further adjustments are necessary. If the SWR is high, you can usually tune the antenna to reduce it.

You may also run into a transmission line problem. It may be open or short-circuited, which will show up on an SWR test as infinite SWR. But there may be other problems such as a cable that has been cut, short-circuited, or crushed between the transmitter and receiver. These kinds of problems can be located with a time-domain reflectometer test.

SWR Tests. The test procedure for SWR is shown in Fig. 22-12. The SWR meter is connected between the transmission line and the antenna. Check with the manufacturer of the SWR meter to determine whether any specific connection location is required or whether other conditions must be met. Some of the lower-cost SWR meters must be connected directly at the antenna or a specific number of half wavelengths from the antenna back to the transmitter.

Once the meter is properly connected, key up the transmitter without modulation. The transmitter should have been previously tuned and adjusted for maximum output power. The SWR can be read directly from the instrument's meter. In some cases, the meter will give measurements for the relative amount of incident or forward and reflected power or will read out in terms of the reflection coefficient, in which case you must calculate the SWR as described earlier. Other meters will read out directly in SWR. The maximum range is usually 3 : 1.

Figure 22-12 SWR measurement.

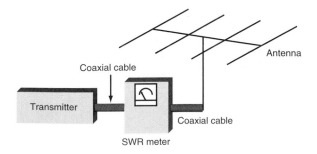

The ideal SWR is 1 or 1:1, which means that all the power generated by the transmitter is absorbed by the antenna load. Nevertheless, in even the best systems, perfect matching is rarely achieved. Any mismatch will produce reflected power and standing waves. If the SWR is less than 2:1, the amount of power that will be lost or reflected will be minimal.

The primary procedure for reducing the SWR is to make antenna adjustments, usually in the form of modifying the element lengths to more closely tune the antenna to the frequency of operation. Using many antennas makes it possible to adjust their length over a narrow range to fine-tune the SWR. Other antennas permit adjustment to provide a better match of the transmission line to the driven element of the antenna. These adjustments can be made one at a time, and the SWR monitored.

TDR Tests. *Time-domain reflectometry (TDR)* is a pulse for cables and transmission lines of all types. It is widely used in finding faults in cables used for digital data transmission, but it can be used for RF transmission lines. (Refer to the section Data Communication Tests later in this chapter for details.)

Receiver Tests

The primary tests for receivers involve sensitivity and noise level. The greater the sensitivity of the receiver, the higher its gain and the better job it does of receiving very small signals.

As part of the sensitivity testing, the signal-to-noise (*S/N*) ratio is also usually measured indirectly. The ability of a receiver to pick up weak signals is just as much a function of the receiver noise level as it is of overall receiver gain. The lower the noise level, the greater the ability of the receiver to detect weak signals.

As part of an overall sensitivity check, some receiver manufacturers specify an audio power output level. Since receiver sensitivity measurements are usually made by measuring the speaker output voltage, power output can also be checked if desired.

In this section, some of the common tests made on receivers are described. The information is generic, and the actual testing procedures often differ from one receiver manufacturer to another.

Equipment Required. To make sensitivity and noise measurements, the following equipment is necessary:

1. **Dual-Trace Oscilloscope.** The vertical frequency response is not too critical, for you will be viewing noise and audio frequency signals.

2. **RF Signal Generator.** This generator provides an RF signal at the receiver operating frequency. It should have an output attenuator so that signals as low as 1 μV or less can be set. This may indicate the need for an external attenuator if the generator does not have built-in attenuators or output level adjustments. The generator must also have modulation capability, either AM or FM, depending upon the type of receiver to be tested.

3. **RF Voltmeter.** The RF voltmeter is needed to measure the RF generator output voltage in some tests. Some higher-quality RF generators have an RF voltmeter built in to aid in setting the output attenuator and level controls.

4. **Frequency Counter.** A frequency counter capable of measuring the RF generator output frequency is needed.

5. **Multimeter.** A multimeter capable of measuring audio-frequency (AF) voltage levels is needed. Any analog or digital multimeter with ac measurement capability in the AF range can be used.

6. **Dummy Loads.** A dummy load is needed for the receiver antenna input for the noise test. This can be either a 50- or a 75-Ω resistor attached to the appropriate coaxial input connector. A dummy load is needed for the speaker. Since most noise and sensitivity tests are made with maximum receiver gain including audio gain, it

> **GOOD TO KNOW**
>
> The ability of a receiver to pick up weak signals is just as much a function of receiver noise level as it is of overall receiver gain. The lower the noise level, the greater the ability of the receiver to detect weak signals.

is not practical or desirable to leave the speaker connected. Most communication receivers have an audio output power capability of 2 to 10 W, which is sufficiently high to make the output signal level too high for comfort. A speaker dummy load of 4, 8, or 16 Ω, depending upon the speaker impedance, is needed. Be sure that the dummy load can withstand the maximum audio power output of the receiver. Do not use wire-wound resistors for this application, for they have too much inductance. Check the receiver's specifications for both the impedance and the maximum power level specification.

Noise

Noise Tests. *Noise* consists of random signal variations picked up by the receiver or caused by thermal agitation and other conditions inside the receiver circuitry. External noise cannot be controlled or eliminated. However, noise contributed by the receiver can be controlled. Every effort is made during the design to minimize internally generated noise and thus to improve the ability of the receiver to pick up weak signals.

Most noise generated by the receiver occurs in the receiver's front end, primarily the RF amplifier and the mixer. Careful attention is given to the design of both these circuits so that they contribute minimum noise.

Because noise is a totally random signal that is a composite of varying frequency and varying amplitude signals, it is somewhat difficult to measure. However, the following procedure has become a common and popular method that is easy to implement.

Refer to the test setup shown in Fig. 22-13. The antenna is removed from the receiver, and a dummy load of the correct impedance is used to replace it. A carbon composition resistor of 50 or 75 Ω can be used. The idea is to prevent the receiver from picking up any signals while maintaining the correct impedance.

At the output of the receiver, the speaker is replaced with a dummy load. Most speakers have an output impedance of 4 or 8 Ω. Check the receiver's specifications, and connect an appropriate value of resistor in place of the speaker. Be sure that the dummy load resistors can withstand the output power.

Finally, connect the dual-trace oscilloscope across the dummy speaker load. The same signal should be displayed on both channels of the oscilloscope. The displayed signal will be an amplified version of the noise produced by the receiver and amplified by all the stages between the antenna input and the speaker. Follow this step-by-step procedure:

1. Turn on the receiver, and tune it to a channel where no signal will be received.
2. Set the receiver volume control to maximum. If the receiver has any type of RF or IF gain control, it too should be set to its maximum setting.
3. Set the oscilloscope input for the lower trace (channel B) to ground. Most oscilloscopes have a switch that allows the input to be set for ac measurements, dc measurements, or ground. Grounding the channel B input will prevent any signal from being displayed. At this time you will see a straight horizontal line for the lower trace. Adjust that lower trace so that it lines up with one of the horizontal graticule lines near the bottom of the oscilloscope screen. This will provide a voltage measurement reference.

GOOD TO KNOW

To determine the noise level when the horizontal spacing on the oscilloscope and the vertical sensitivity are known, multiply the horizontal spacing by the sensitivity and divide by 2.

Figure 22-13   Noise test setup.

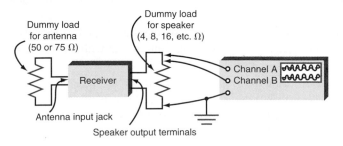

Figure 22-14 Noise measurement procedure.

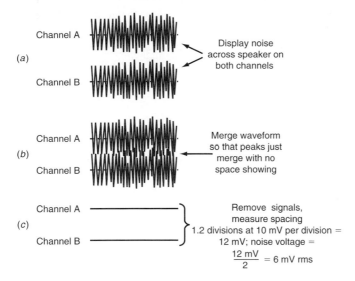

(a) Channel A ⟋ Channel B — Display noise across speaker on both channels

(b) Channel A ⟋ Channel B — Merge waveform so that peaks just merge with no space showing

(c) Channel A — Channel B — Remove signals, measure spacing
1.2 divisions at 10 mV per division = 12 mV; noise voltage =
$$\frac{12 \text{ mV}}{2} = 6 \text{ mV rms}$$

4. Set the channel B input and the channel A input to alternating current. Adjust the vertical sensitivities of channels A and B so that they are on the same range. Make the adjustments so that the signal is displayed something like that shown in Fig. 22-14(*a*). You should see exactly the same noise pattern on both channels.

5. Using the vertical position control on the upper or A channel, move the upper noise trace downward so that it begins to merge with the noise signal on channel B. The correct adjustment for the position of the channel A noise signal is such that the peaks of the upper and lower signals just barely merge. This will generally be indicated at the point where there is no blank space between upper and lower traces. The signal should look something like that shown in Fig. 22-14(*b*).

6. Now set both oscilloscope inputs to ground, thereby preventing the noise signal from being displayed. You will see two straight horizontal lines. The distance between the two lines is a measure of the noise voltage. The value indicated is 2 times the root mean square (rms) noise voltage [see Fig. 22-14(*c*)].

Assume that the adjustments described above were made and the separation between the two horizontal traces is 1.2 vertical divisions. If the vertical gains of both channels are set to the 10 mV per division range, the noise reading is 1.2 × 10 mV = 12 mV. The rms noise voltage is one-half of this figure, or 12 mV/2 = 6 mV.

Power Output Tests. Sometimes it is necessary to measure the receiver's total power output capability. This is a good general test of all the receiver circuits. If the receiver can supply the manufacturer's specified maximum output power into the speaker with a given low RF signal level input, the receiver is operating correctly.

The test setup for the power output test is shown in Fig. 22-15. An RF signal generator for the correct frequency is used as the primary signal source. It must also be possible to modulate this generator with either AM or FM depending upon the type of receiver.

It is also desirable to connect a frequency counter to the signal generator output to provide an accurate measure of the receiver input frequency. Most communication receivers operate on specific frequency channels. For the test to be valid, the generator output frequency must be set to the center of the receiver frequency channel. This will usually be known from the receiver's specifications. The signal generator is tuned, and the frequency is set by monitoring the digital readout on the frequency counter.

Be sure to replace the speaker with a resistive load of an impedance equal to that of the speaker, such as 8 Ω. The dummy speaker load should also be able to carry the

Figure 22-15 Power output test.

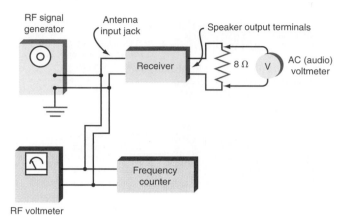

maximum rated output power of the receiver. Finally, connect an ac voltmeter across the speaker dummy load.

Turn on the receiver, and set the volume control to the maximum. If the receiver has a variable RF or IF gain control, you must set it to the maximum setting also. At this time, any other receiver features should be disabled. For example, in an FM receiver, the squelch should be turned off or disabled. In an AM receiver, if a noise limiter is used, it too should be turned off.

To begin the test, follow this procedure:

1. Set the RF generator output level to 1 mV. If the RF signal generator has a built-in RF voltmeter, use it to make this setting. Otherwise, an external RF voltmeter may be needed, as shown in Fig. 22-15.

2. Set the signal generator for modulation of the appropriate type. If AM is used, set the percentage of modulation for 30. If FM is used, set the deviation for ± 3 kHz. In most signal generators, the percentage of AM and the frequency deviation for FM are fixed. Refer to the signal generator specifications to find out what these values are.

3. With everything appropriately adjusted, measure the ac voltage across the speaker dummy load. This will be an rms reading.

4. To determine the receiver power output, use the standard power formula $P = V^2/R$.

Assume that you measure an rms voltage of 4 V across an 8-Ω speaker. The power output will be

$$P = \frac{V^2}{R} = \frac{4^2}{8} = \frac{16}{8} = 2 \text{ W}$$

An optional test is to observe the signal across the speaker load with an oscilloscope. Most RF generators modulate the input signal with a sine wave of 400 Hz or 1 kHz. If an oscilloscope is placed across the dummy speaker load, the sine wave will be seen. This will indicate whether the receiver is distorting. The oscilloscope can also be used in place of the audio voltmeter to make the voltage measurement across the dummy speaker load. Remember that oscilloscope measurements are peak to peak. The peak-to-peak value must be converted to root mean square to make the power output calculation.

20-dB Quieting Sensitivity Tests.

In most cases, the sensitivity of a receiver is expressed in terms of the minimum RF voltage at the antenna terminals that will produce a specific audio output power level. Most measurements factor in the effect of noise.

The method of sensitivity measurement is determined according to whether AM or FM is used. Since most modern radio communication equipment uses frequency modulation,

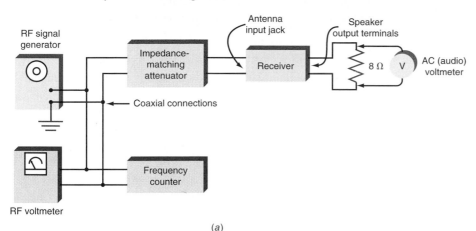

(*a*)

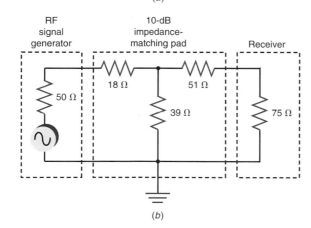

(*b*)

measuring FM receiver sensitivity is illustrated. There are two basic methods, quieting and SINAD. The *quieting method* measures the amount of signal needed to reduce the output noise to 20 dB. As the signal level increases, the noise level decreases until the limiters in the IF section begin to start their clipping action. When this happens, the receiver output "quiets"; i.e., its output is silent and blanks out the noise.

The *SINAD test* is a measure of the input signal voltage that will produce at least a 12-dB signal-to-noise ratio. The noise value includes any harmonics that are produced by the receiver circuits because of distortion.

The test setup for receiver sensitivity measurements is shown in Fig. 22-16(*a*). It consists of an RF signal generator, an RF voltmeter, a frequency counter, the receiver to be tested, and a voltmeter to measure the output across the speaker dummy load.

It is often necessary to provide an impedance-matching network between the generator and the receiver antenna input terminals. Most RF generators have a 50-Ω output impedance. This may match the receiver input impedance exactly. However, some receiver input impedances may be different. For example, if a receiver has a 75-Ω input impedance, some form of impedance matching will be required. This is usually handled by a resistive attenuator known as an *impedance-matching pad,* which is a resistive T network that provides the correct match between the receiver input and the generator. A typical impedance-matching pad is shown in Fig. 22-16(*b*). It matches the 50-Ω generator output to the 50-Ω receiver antenna input. Since resistors are used, the impedance-matching circuit is also an attenuator. With the values given in Fig. 22-16(*b*), the signal attenuation is 10 dB. This must be factored into all signal generator measurements to obtain the correct sensitivity figure. Different values of resistors can be used to create a pad with the correct impedance-matching qualities but with lower or higher values of attenuation.

Quieting method

SINAD test

Impedance-matching pad

Some manufacturers specify a special input network made up of resistors, inductors, and/or capacitors for this or other sensitivity tests. This network helps simulate the antenna accurately in equipment that uses special antennas.

Follow this procedure to make the 20-dB quieting measurement:

1. Turn on the receiver, and set it to an unused channel.
2. Leave the signal generator off so that no signal is applied.
3. Set the receiver gain to maximum with any RF or IF gain control, if available.
4. Adjust the volume control of the receiver so that you read some convenient value of noise voltage on the meter connected across the speaker. One volt rms is a good value if you can achieve it; but if not, any other convenient value will do.
5. Turn on the signal generator, but set the output level to zero or some very low value. Adjust the generator frequency to the center of the receiver's channel setting. Turn off the modulation so that the generator supplies carrier only.
6. Increase the signal generator output signal level a little at a time, and observe the voltage across the speaker. The noise voltage level will decrease as the carrier signal gets strong enough to overpower the noise. Increase the signal level until the noise voltage drops to one-tenth of its previous value.
7. Measure the generator output voltage on the generator meter or the external RF voltmeter.
8. If an attenuator pad or other impedance-matching network was used, subtract the loss it introduces. The resulting value is the voltage level that produces 20 dB of quieting in the receiver.

Assume that you measured a generator output of 5 μV that produces the 20-dB noise decrease. This is applied across a 50-Ω load producing an input power of $P^2 = V^2/R = (5 \times 10^{-6})^2/50 = 0.5$ pW. This is attenuated further by the 10-dB matching pad to a level of one-tenth, or 0.05 pW, which translates to a voltage level across 50 Ω as

$$V = \sqrt{PR} = \sqrt{0.05 \times 10^{-12} \times 50} = 1.58 \times 10^{-6} = 1.58 \; \mu\text{V}$$

This is the receiver sensitivity. It takes 1.58 μV of a signal to produce 20 dB of quieting in the receiver.

For a good communication receiver, the 20-dB quieting value should be under 1 μV. A typical value is in the 0.2- to 0.5-μV range. The lower the value, the better the sensitivity.

Blocking and Third–Order Intercept Tests. As the spectrum has become more crowded and as modulation advances have permitted higher speed per hertz of bandwidth, the potential for adjacent channel interference has significantly increased. To ensure minimum adjacent channel interference, receiver specifications have become tighter. This is especially true in cell phones. Whole suites of tests must be passed by the receiver in order to meet the specifications of a specific cell phone standard such as GSM or CDMA.

An example is the receiver blocking test that makes measurements to ensure that signals from an adjacent channel do not block or desensitize the channel being used. A very strong signal near the receiving frequency has the effect of lowering the gain of the receiver. Any small signal being received will be decreased in amplitude or even blocked completely. Some specifications call for the receiver to be able to receive a weak signal when the adjacent channel signal is 60 to 70 dB greater in level. The ability to meet this test depends upon the filtering selectivity of the receiver.

But perhaps the most difficult test is the *third-order intercept test,* designated *TOI* or *IP3.* This test is a measure of the linearity of amplifiers, mixers, and other circuits. When two signals are applied to a circuit, any nonlinearity in the circuit causes a mixing or modulation effect. The larger the input signals, the more likely the amplifier will be driven into a nonlinear region where mixing will occur. Sum and difference signals will be produced. Some of the resulting so-called intermodulation products are problematic because they occur at a frequency near or inside the receiver bandpass and

Third-order intercept test (TOI or IP3)

Figure 22-17 Third-order intermodulation products.

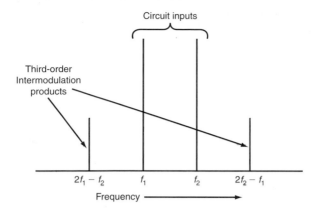

interfere with the signal being received. Such signals, because they are so close to the desired signal, cannot be filtered out. These intermodulation signals must be reduced as much as possible in the design of the receiver by selecting more linear components or giving greater attention to biasing schemes and operating points.

Figure 22-17 shows two signals f_1 and f_2 that appear at the input to an amplifier. The nonlinear action of the amplifier generates a wide range of sums and differences, including those with the second and third harmonics. Most of these undesired products will be filtered out by the receiver IF filters. The third-order products cause the most problems since they will most likely fall within the receiver IF bandpass. The third-order products are $2f_1 \pm f_2$ and $2f_2 \pm f_1$, where the terms $2f_1 - f_2$ and $2f_2 - f_1$ cause the most problems.

To measure the third-order problem, two equal power signals are applied to the amplifier or other circuit to be tested. The frequency spacing is usually small and often made equal to the normal channel spacing used. The power of the input signals is gradually increased, and measurements are made on the amplifier output to determine the levels of the test signals and the third-order products. As the power levels of the input signals increase, the third-order signal power increases as the cube of the input power change. On a logarithmic scale, the rate of increase of third-order products is 3 times that of the original signals.

Refer to Fig. 22-18. If you plot the output signal amplitudes versus input signal power increases, at some point the power levels reach their limit and flatten out. If you

Figure 22-18 Third-order intercept plot of an RF amplifier.

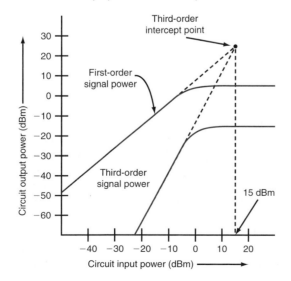

Communication Tests and Measurements

extend the linear portions of the two curves, they will meet at a theoretical point where the initially lower third-order signals equal the main input signals in amplitude. This is the third-order intercept point. The input power at that point is used as a measure of the intermodulation. In Fig. 22-18 the IP3 point is at 15 dBm. Typical IP3 values are between 0 and 35 dBm. The higher the IP3 value, the better the circuit linearity and the lower the intermodulation products.

Microwave Tests

Microwave tests are generally similar to those performed on standard transmitters and receivers. Transmitter measurements include output power, deviation, harmonics, and spurious signals as well as modulation. The techniques are similar but require the use of only those test instruments whose frequency response is in the desired microwave region. The same goes for receiver measurements and antenna transmission line tests. The procedures are generally the same, but the equipment is different. For example, with power measurements, a directional coupler is normally used as the transmitter output to reduce the signal to a proper level for measurement with the power meter.

Data Communication Tests

The tests for wireless data communication equipment are essentially the same as those for standard RF communication as described above. The only difference is the type of modulation used to apply the binary signal to the carrier. FSK and its many variants, as well as PSK and spread spectrum, are the most widely used. Special FSK and PSK deviation and modulation meters are available to make these measurements.

For data communication applications in which binary signals are baseband on coaxial and twisted-pair cables, such as in LANs, more conventional testing methods may be used. For example, binary test patterns may be initiated in the transmitting equipment, and the signal viewed on the oscilloscope at the receiving end. Tests of signal attenuation and wave shape can then be made.

Eye Diagrams. A common method of analyzing the quality of binary data transmitted on a cable is to display what is known as the *eye diagram* on a common oscilloscope. The eye diagram, or pattern, is a display of the individual bits overlapped with one another. The resulting output looks like an open eye. The shape of the pattern and the degree that the eye is "open" can be used to determine many things about the quality of transmission.

Eye diagrams are used for testing because it is difficult to display long streams of random serial bits on an oscilloscope. The randomness of the data prevents good synchronization of the oscilloscope with the data, and thus the display jitters and changes continuously. Sending the same pattern of bits such as repeating the ASCII code for the letter U (alternating 1s and 0s) may help the synchronization process, but the display of one whole word on the screen usually does not provide sufficient detail to determine the nature of the signal. The eye diagram solves these problems.

Figure 22-19(*a*) shows a serial pulse train of alternating binary 0s and 1s that is applied to a transmission line. The transmission line, either coaxial or twisted pair, is a low-pass filter, and therefore it eliminates or at least greatly attenuates the higher-frequency components in the pulse train. It also delays the signal. The result is that the signal is rounded and distorted at the end of the cable and the input to the receiver [see Fig. 22-19(*b*)].

The longer the cable and/or the higher the bit rate, the greater the distortion. The pulses tend to blur into one another, causing what is called *intersymbol interference (ISI)*. ISI makes the voltage levels for binary 1 and 0 closer together with 1 bit smearing or overlapping into the other. This makes the receiver's job of clearly distinguishing a binary 1 from a binary 0 more difficult. Further, noise is usually picked up along the transmission path, making the received signal a poor representation of the original data signal.

Figure 22-19 (*a*) Ideal signal before transmission. (*b*) Signal attenuated, distorted, and delayed by medium (coaxial or twisted pair). (*c*) Severely attenuated signal with noise.

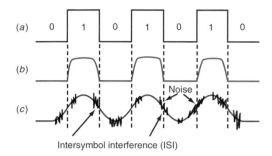

Too much signal rounding or ISI introduces bit errors. Figure 22-19(*c*) is a severely distorted, noisy data signal.

The eye pattern provides a way to view a serial data signal and to make a determination about its quality. To display the eye diagram, you need an oscilloscope with a bandwidth at least 5 times the maximum bit rate. For example, if the bit rate is 10 Mbps, the oscilloscope bandwidth should be at least 10×5 MHz, or 50 MHz. The higher the better. The oscilloscope should have triggered sweep. Either a conventional analog or a digital oscilloscope can be used.

Apply the baseband binary signal at the end of the cable and the receiver input to the vertical input. Adjust the sweep rate of the oscilloscope so that a 1-bit interval takes up the entire horizontal width of the screen. Use the variable sweep control to fine-tune the display and use the trigger controls to stabilize the display. The result is an eye diagram.

Several different eye patterns are shown in Fig. 22-20. The multiple lines represent the overlapping pulses occurring over time. Their amplitude and phase shift are slightly shifted from sweep to sweep, thereby giving the kind of pattern shown. If the signal has not been severely rounded, delayed, or distorted, it might appear as shown in Fig. 22-20(*a*). The eye is "wide open" and has a trapezoidal shape. This is a composite display of the rise and fall times of the pulses overlapping one another. The steeper the sides, the less the distortion. The eye pattern in Fig. 22-20(*a*) indicates wide bandwidth of the medium.

In Fig. 22-20(*b*), the pattern looks more like an open eye. The pulses are rounded, indicating that the bandwidth is limited. In fact, the pulses approach the shape of a sine wave. The pattern shown in Fig. 22-20(*c*) indicates more severe bandwidth limiting. This reduces the amplitude of the rounded pulses, resulting in a pattern that appears to be an eye that is closing. The more the eye closes, the narrower the bandwidth, the greater the distortion, and the greater the intersymbol interference. The difference between the binary 0 and 1 levels is less, and the chance is greater for the receiver to misinterpret the level and create a bit error.

Note further in Fig. 22-20(*c*) that the amplitudes of some of the traces are different from others. This is caused by noise varying the amplitude of the signal. The noise can further confuse the receiver, thus producing bit errors. The amount of voltage between the lowest of the upper patterns and the highest of the lower patterns as shown is called the *noise margin*. The smaller this value, the greater the noise and the greater the bit error rate. Noise margin is sometimes expressed as a percentage based upon the ratio of the noise margin level *a* in Fig. 22-20(*c*) to the maximum peak-to-peak value of the eye *A*.

The eye diagram is not a precise measurement method. But it is an excellent way to get a quick qualitative check of the signal. The eye pattern tells at a glance the degree of bandwidth limitation, signal distortion, jitter, and noise margin.

Noise margin

Figure 22-20 Eye diagrams. (*a*) Good bandwidth. (*b*) Limited bandwidth. (*c*) Severe bandwidth limitation with noise.

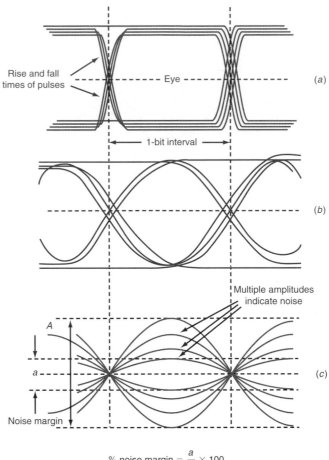

% noise margin = $\frac{a}{A} \times 100$

Pattern generator

Bit error rate (BER) analyzer

An eye diagram on an oscilloscope that is optimized for optical data communication testing.

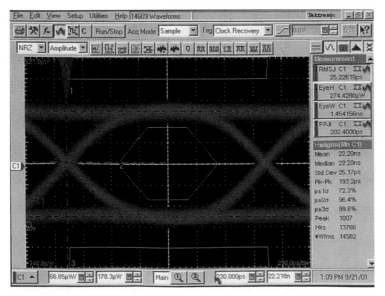

Pattern Generators. A *pattern generator* is a device that produces fixed binary bit patterns in serial form to use as test signals in data communication systems. The pattern generator may generate a repeating ASCII code or any desired stream of 1s and 0s. Pattern generators are used to replace the actual source of data such as the computer. Their output patterns can be changed to standard codes or messages or may be programmable to some desired sequence. A pattern generator may be implemented in software at the sending computer.

Bit Error Rate Tests. At the other end of the link, the pattern generator sequence is detected and compared to the known actual pattern or message sent. Any errors in the comparison indicate errors. The instrument that detects the pulse's pattern and compares it is called a *bit error rate (BER) analyzer*. It compares on a bit-by-bit basis the transmitted and received data to point out every bit error made. It keeps track of the total number of bits sent and the number of errors that occur and then computes the BER by dividing the number of errors by the number of bits sent. The BER tester must of course know the exact pattern or message sent by the pattern generator.

$$\text{Percent BER} = \frac{\text{number of errors detected}}{\text{numbers of bits sent}} \times 100$$

TDR Tests with an Analyzer. Special data communication test instruments are also available. A popular instrument is the *TDR tester*, also known as a *cable analyzer* or *LAN meter*. This instrument, often handheld, is connected to coaxial or twisted-pair cable and is able to make tests and measurements, some of which are

1. Tests for open or short circuits and impedance anomalies on coaxial or twisted-pair cables.
2. Measurements of cable length, capacitance, and loop resistance.
3. Measurements of cable attenuation.
4. Tests for cable miswiring such as so-called split pairs.

A *split pair* is a wiring error often made when a cable contains multiple twisted-pair lines. One of the wires from one twisted pair is wrongly paired with one wire from another twisted pair.

Many tests are based upon what is called *time-domain reflectometry (TDR)*. The TDR technique can be used on any cable or transmission line, such as antenna lines or LAN cables, to determine SWR, short and open circuits, and characteristic impedance mismatches between cable and load. It can even determine the distance to the short or open circuit or to any other glitch anywhere along the line. TDR testing is based upon the presence of standing waves on the line if impedances are not matched.

The basic TDR process is to apply a rectangular pulse to the cable input and to monitor the signal at the input. The test setup is shown in Fig. 22-21. If the load impedance is matched, the pulse will be absorbed by the load and no reflections will occur. However, if there is a short or open circuit or impedance mismatch, a pulse will be reflected from the point of the mismatch.

Protocol Analyzers. The most sophisticated data communications test equipment is the *protocol analyzer*. Its purpose is to capture and analyze the data transmitted in a particular system. Most data communication systems transmit data in frames or packets that include preamble information such as sync bits or frames, start of header codes, addresses of source and destination, and a finite block of data, followed by error detection codes. A protocol analyzer can capture these frames, analyze them, and tell you whether the system is operating properly. The analyzer will determine the specific protocol being captured and then indicate whether the data transmitted is following the protocol or whether there are errors in transmission or in formatting the frame.

(margin notes) TDR tester (cable analyzer or LAN meter)

Split pair

time-domain reflectometry (TDR)

Protocol analyzer

Figure 22-21   TDR test setup.

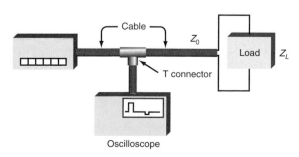

Protocol analyzers contain microcomputers programmed to recognize a wide range of data communication protocols such as Bisync, SDLC, HDLC, Ethernet, SONET, and other LAN and network protocols. These instruments normally cost tens of thousands of dollars and may, in fact, be primarily a computer containing the stored protocols and software that read and store and then compare and analyze the received data so that it can report any differences or errors. Most protocol analyzers have a video display.

Special Test Sets. As communications equipment, especially wireless, has become more complex, it has been necessary to develop special test systems for specific protocols. For example, special test sets have been developed for GSM and CDMA cell phones and Bluetooth. These test sets combine multiple instruments into a common enclosure along with a computer for control. The instruments include signal generators (synthesizers) operating on the desired frequency bands along with the proper modulation and protocol for a given standard. This part of the test set permits testing of the receiver. Another section will include a calibrated receiver set up to receive any transmitted signal. An internal computer is programmed to conduct a precise sequence of tests and then to record and analyze the data. Test sets automate the testing procedure and identify those units (cell phones, etc.) that pass or fail the tests. These automated systems usually include internal spectrum analyzers for display of results.

Fiber-Optic Test Equipment and Measurements

A variety of special instruments are available for testing and measuring fiber-optic systems. The most widely used fiber-optic instruments are the *automatic splicer* and the *optical time-domain reflectometer (OTDR)*.

Automatic Splicers. Splicing fiber-optic cable is a common occurrence in installing and maintaining fiber-optic systems. This operation can be accomplished with hand tools especially made for cutting, polishing, and splicing the cable. However, as the cable thickness has gotten finer, hand splicing has become more difficult than ever. It is very difficult to align the two cable ends perfectly before the splice is made.

To overcome this problem, a special splicer has been developed by several manufacturers. It provides a way to automatically align the cable ends and splice them. The two cables to be spliced are stripped and cleaved by hand and then placed in the unit. A special mechanism holds the two cable ends close together. Then an optical system with a light source, lenses, and light sensors detects the physical alignment of the two cables, and a servo feedback mechanism drives a motor so that the two cable ends are perfectly centered on each other. An optical viewing screen is provided so that the operator can view the alignment from two directions at 90° to each other.

Once the alignment is perfect, the splicer is activated. The splicer is a pair of probes centered over the junction of the two cable ends. Pressing the "splice" button causes the probes to generate an electric arc hot enough to fuse the two glass cable ends together.

The automatic splicer is very expensive, but it must be used because it is not possible for human beings to make good splices visually and by hand. Handmade splices have high attenuation, whereas minimum attenuation is best achieved with the automatic splicer.

Optical Time-Domain Reflectometer. Another essential instrument for fiber-optic work is the optical TDR, or OTDR. It is an oscilloscopelike device with a CRT display and a built-in microcomputer.

The OTDR works as a standard TDR in that it generates a pluse, in this case, a light pulse, and sends it down a cable to be tested. If there is a break or defect, there is a light reflection, just as there is a reflection on an electric transmission line. The reflection is detected. Internal circuitry measures the time between the transmitted and reflected pulses so that the location of the break or other fault can be calculated and displayed. The OTDR also detects splices, connectors, and other anomalies such as dents in the cable. The attenuation of each of these irregularities can be determined and displayed.

Figure 22-22   Jitter on a data signal.

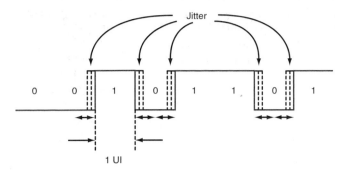

Optical Signal Analyzer.   A newer breed of versatile optical test instrument makes multiple measurements. In addition to providing the OTDR measurement, this unit is a sampling oscilloscope capable of displaying signals of more than 10 Gbps. It can also be used to show eye patterns, optical output power, and jitter.

Jitter is a type of noise that shows up as a time variation of the leading and trailing edges of a binary signal. See Fig. 22-22. It appears as a kind of phase or frequency shift in which the time period for 1 bit is lengthened or shortened at a rapid rate. Jitter shows up on an oscilloscope as a blurring of the 0-to-1 and 1-to-0 transitions of a binary signal. It is not much of a problem at lower data rates, but as data rates go above 1 GHz, jitter becomes more prevalent. On fiber-optic data systems, jitter is a major problem. Furthermore, jitter is difficult to measure. Most optical fiber communication networks such as SONET have jitter specifications that must be met. Therefore, some accurate method of measurement is needed. Some of the newer optical signal analyzers have a jitter measurement capability. Jitter is usually expressed as a percentage of the *unit interval (UI)*. The UI is the bit time of the data signal. For example, a jitter measurement may be 0.01 UI or 10 mUI.

22-3 Troubleshooting Techniques

Some of the main duties of a communication expert are to troubleshoot, service, and maintain communication equipment. Most communication equipment is relatively reliable and requires little maintenance. However, equipment does fail. Most electronic communication equipment fails because of on-the-job wear and tear. Of course, it is still possible for equipment to fail as the result of component defects. The equipment might fail eventually because of poor design, exceeding of product capabilities, or misapplication. In any case, you must locate such failures and repair them. This is where troubleshooting techniques are valuable. The goal is to find the trouble quickly, solve the problem, and put the equipment back into use as economically as possible.

General Servicing Advice

One of the main decisions you must make in dealing with any kind of electronic equipment is to repair or not to repair. Because of the nature of electronic equipment today, repairing it may not be the fastest and most economical approach.

Assume that you have a defective radio transceiver. One of your options is to send the unit out for repair. Repair rates run anywhere from $25 per hour to over $100 per hour depending upon the equipment, the manufacturer, and other factors. If the problem is a difficult one, it may take several hours to locate and repair it. Many communication transceivers are inexpensive units that may cost less to buy new than to repair.

There are two types of repair approaches: (1) replace modules or (2) troubleshoot to the component level and replace individual components. Some electronic equipment is built in sections or modules. The module is, in most cases, a separate PCB containing a portion of the circuitry inside the unit. The typical arrangement might be for the receiver to be on one PCB, the transmitter on another, and the power supply on still another, with another unit such as a tuner or frequency synthesizer also separate. A fast and easy way to troubleshoot and repair a unit is to replace the entire defective module. If you are a manufacturer repairing your own units in volume for customers, or if your organization does many repairs of a similar nature on a particular brand or model of equipment, repair at the component level is the best approach.

Common Problems

Many repairs can be made quickly and easily because they result from problems that occur on a regular basis. Some of the most common problems in communication equipment are power supply failures, cable and connector failures, and antenna troubles.

Power Supplies. All equipment is powered by some type of dc power supply. If the power supply doesn't work, the equipment is completely inoperable. Therefore, one of the first things you should do is to check that the power supply is working.

If the unit is used in a fixed location and operates from standard ac power lines, the first test should be to check for ac power and the availability of the correct dc power supply voltages. Is the unit plugged in, and if so, does ac power actually get to the outlet? If ac power is indeed available, check the power supply inside the unit next. These power supplies convert ac power to one or more dc voltages to operate the equipment. Open the equipment and, using the manufacturer's service information, determine the power supply voltages. Then use a multimeter to verify that they are at the correct levels. Most power supplies these days are regulated, and therefore the voltages should be very close to those specified, at least within ± 5 percent. Anything outside that range should be suspect. Any voltages that are obviously quite different from the specified value indicate a power supply problem.

Another common power supply problem is bad batteries. With continuous usage, batteries quickly run down. If primary batteries are used, the batteries must be replaced with new ones. If secondary or rechargeable batteries continue to fail even after short periods of use after charging, it means that they, too, should be replaced. Most rechargeable batteries can be charged and discharged only so many times before they are no longer effective.

Cables and Connectors. Perhaps the most common failure points in any electronic system or equipment are the mechanical components. Connectors and cables are mechanical and can be a weak link in electronic equipment. Once it has been confirmed that the power supplies are operating correctly in the equipment, the next step is to check the cables and connectors. Start by verifying that the connectors are correctly attached. Another common problem is for the cable attached to the connector to break internally. Most of the time the cable does not break completely, but one or more wires in the cable may be broken while others remain attached.

Occasionally connectors get dirty. Removing the connector and cleaning the connections often solve the problem. It may be necessary to replace the connectors to ensure a reliable physical connection, however.

Antennas. Another common failure in communication systems is the antenna. In most cases, antennas on portable equipment are fragile. A bad antenna is a common problem on handheld transceivers, cordless telephones, cellular telephones, and similar equipment.

Documentation

Before you begin any serious detailed troubleshooting and repair of communication equipment, be sure that you have all the necessary documentation. This includes the manufacturer's user operation manual and any technical service manuals that you can acquire. Nothing speeds up troubleshooting and repair faster than having all the technical information before you begin. Manufacturers often regularly identify common problems and suggest troubleshooting approaches. But perhaps most important, manufacturers provide specifications as well as measurement data and procedures that are critical to the operation of the equipment. By having this information, you will be able to make the necessary tests, measurements, and adjustments to ensure that the equipment complies with the specifications.

Troubleshooting Methods

There are two basic approaches to troubleshooting transmitters, receivers, and other equipment: signal tracing and signal injection. Both methods work equally well and may often be used together to isolate a difficult problem.

Signal Tracing. A commonly used technique in troubleshooting communications equipment is called *signal tracing*. The idea is to use an oscilloscope or other signal detection device to follow a signal through the various stages of the equipment. As long as the signal is present and of the correct amplitude, the circuits are good. The point at which you lose the signal in the equipment or at which the signal no longer conforms to specifications is the location of the problem.

Signal tracing

To perform signal tracing in a transmitter, you need some type of monitoring or measuring instrument: an RF voltmeter or an oscilloscope, an RF detector probe on an oscilloscope, a spectrum analyzer, and power meters and frequency counters.

Figure 22-23 is a block diagram of a generic FM communication transmitter. It is assumed that the power supply is working and that it is supplying the correct voltage to all the circuits. Further, it is assumed that the transmitter has been declared inoperable. A quick verification of this should be made by testing the transmitter to see whether it is putting out a signal. A good overall check is to connect a dummy load, key up the transmitter, and attempt to pick up the signal on a nearby frequency counter or field strength meter with antenna. If no signal is indicated, troubleshooting can begin.

To troubleshoot the transmitter, connect a dummy load to the antenna jack and turn on the power. Using the signal-tracing method, start with the carrier signal source. Locate the carrier crystal oscillator or the frequency synthesizer responsible for generating the basic carrier signal. Monitor the output of this circuit on the oscilloscope to see that it is generating a sine wave carrier of the correct amplitude and frequency. The frequency of the signal can be checked with a frequency counter. For the circuit in Fig. 22-23, you should measure and observe a 13.5-MHz sine wave.

If the carrier oscillator or frequency synthesizer is working correctly, you can go on to the next stages. Most transmitters have additional buffer amplifiers, frequency

Figure 22-23   FM transmitter.

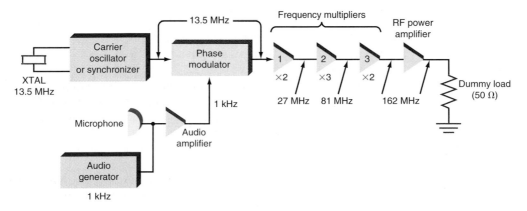

multipliers, and power amplifiers. Each should be checked in sequence from one step to the other. In Fig. 22-23, the carrier signal passes through a phase modulator to a sequence of three frequency multipliers. The output of the phase modulator should be the carrier frequency. The output of the first multiplier, a doubler, should be 27 MHz. The output of the second multiplier should be 3×27, or 81, MHz. The output of the third multiplier should be 2×81, or 162, MHz. This is also the output of the final power amplifier stage.

The output of every individual stage or circuit should be verified. The schematics or service manual will usually provide typical values of signal level at the output of each stage or at selected points in the transmitter. When discrete component circuits are used, it is easy to check each transistor stage. In modern equipment each stage is implemented with one or more ICs. In such cases you can verify the input and determine what outputs are available. Then you can determine what you should expect at the outputs.

You may lose the signal at some point as you follow it through the stages. For example, you may be tracing the signal through a sequence of frequency multipliers and power amplifiers, as shown in Fig. 22-23. Assume that the carrier oscillator is working correctly, as was the first multiplier. You measure the correct frequency at the output of the first multiplier. This signal also appears at the input to the next stage. But suppose you can measure no signal at the output of the second multiplier. If this is the case, you have effectively isolated the problem to the second frequency multiplier circuit. It has an input but no output.

The most likely problem is a bad transistor or IC. Active components such as transistors and ICs fail more often than passive components such as resistors, capacitors, and inductors or transformers. You should verify bias voltages to determine that they are correct; but if they are and the circuit still does not operate, usually there is an open or short-circuited transistor.

Other components may also fail. Capacitors fail more often than any other type of component except the semiconductor devices. Resistors are less likely to fail but can open or change in value. Inductors rarely fail. Delicate, sensitive components such as crystal, ceramic, and SAW filters can also break. Once you isolate the problem to a particular component, turn off the power, replace the suspected part, and repeat the tests. Continue testing until the transmitter delivers a signal to the dummy load.

If the carrier circuits are working but the unit is not receiving modulation, verify that the microphone is operating correctly. In portable and mobile operation, the microphone is a separate unit attached by a cable to the transmitter. Microphone cables and connectors frequently fail because of the continuous stress they receive. Microphones are also subject to physical damage because they are banged around and almost universally abused in a mobile operating environment.

If the microphone is OK, start checking the audio amplifier and modulator circuits. You can use a simple audio signal generator to inject the signal into the audio amplifiers and follow it through with an oscilloscope. A common test is to use a 1-kHz sine wave at a low level to prevent distortion. Then follow the signal through the circuitry with an oscilloscope, noting both the amplitude and any distortion that may occur. Distortion is really the result of applying too much signal to the input. Remember, microphones are very low-level voltage-generating devices. The audio circuits are designed to provide a substantial amount of gain even with signal input voice signals that are in the microvolt or millivolt range.

Using higher output voltage from a signal generator will overload the circuitry and cause slipping and distortion. If you see that the audio signals you are tracing are square or otherwise distorted, back off on the signal generator output voltage until the distortion disappears. Then continue with the signal tracing. The point at which you lose the signal or at which the signal becomes distorted or attenuated is the location of the problem. Once the transmitter is delivering power to the dummy load, you will want to run frequency, power output, and deviation tests to ensure that everything is working correctly and that the unit meets manufacturer's specifications and FCC regulations.

You can also perform signal tracing on a receiver. You will need an RF signal generator with appropriate modulation and an oscilloscope, RF voltmeter, or other signal-measuring instrument with which to trace the signal. You may prefer to use a speaker dummy load rather than the speaker itself.

Signal Injection. *Signal injection* is somewhat similar to signal tracing. It is normally used with receivers. The process is to use signal generators of the correct output frequency to inject a signal into the various stages of the receiver and to check for the appropriate output response, usually a correct signal in the speaker.

Signal injection

Signal injection is the opposite of signal tracing, for it starts at the speaker output and works backward through the receiver from speaker to antenna. The signal injection begins by testing the audio power output amplifier. You inject a 1-kHz sine wave from an audio oscillator or function generator into the input of the amplifier. Follow the receiver documentation's information with regard to how much signal should be present at the outputs of each stage. If an audio signal is heard, the speaker and power amplifier are OK. Then inject the audio signal into any other audio amplifier stages back to the demodulator circuit output.

The next injection takes place at the input to the second mixer. Set the RF signal generator to 4.5 MHz with appropriate audio modulation. You should hear the audio tone in the speaker. If you do not, check the local-oscillator and IF stages as described before. Do not overlook the demodulator. Keep in mind that you may not actually be able to get to the inputs and outputs of some circuits because they may be contained within an IC. If this is the case, you can restrict the injection to the IC input while monitoring the output. If you get no output, replace the IC.

Next, inject a signal of 45 MHz with modulation at the input to the first mixer. You should hear the tone; but if you do not, inspect and test the first IF stages and local oscillator. Finally, test the RF amplifier with a signal at the receive frequency of 478 MHz.

Modern Troubleshooting. Although signal tracing and signal injection are valid troubleshooting approaches, they are more difficult than ever when applied to a modern communications device. Examples are cell phones and wireless LAN transceivers or line cards in a router. These devices use only one or more ICs to implement the entire signal chain. It is not possible to access the internal circuits of the chips, making the procedures described previously impossible, if not more difficult. These techniques have limited use in such highly integrated devices, although the principles are still valid.

Boundary Scan and the JTAG Standard. Modern networking and communications equipment is made up primarily of digital ICs. These are large-scale devices made for surface mounting on a printed-circuit board (PCB). It is difficult and sometimes not possible to access the pins on an IC, making signal tracing or injection worthless in troubleshooting. Realizing this, IC designers eventually figured out a way not only to test the chip after it is made and packaged, but also to test the equipment built with these chips. This method is called boundary scan.

The purpose of boundary scan is to provide a way to observe test points inside the chip that are not normally accessible and to observe signals at IC pins inaccessible because of their surface mounting on a PCB. And it provides a way to apply test signals to selected points in the circuit and then to monitor the results. Figure 22-24 shows the boundary scan circuitry that is built into many complex ICs. The logic circuitry to be tested is in the center and referred to as core logic. Surrounding the core is a series of boundary scan cells (BSCs). Each BSC is made up of a pair of flip-flops and some multiplexers that can receive inputs and store them or output their contents when interrogated. The BSC flip-flops are also connected to form a large shift register, the boundary scan register. This register provides the basic storage unit for inputting and outputting data. The input and output are serial. Serial data can be input to the register via the TDI line. Data stored in the register can be read out serially by monitoring the TDO line. An external clock signal TCK and a test mode select (TMS) signal control the serial data

---

Communication Tests and Measurements

Figure 22-24 The boundary scan circuitry is built into the IC.

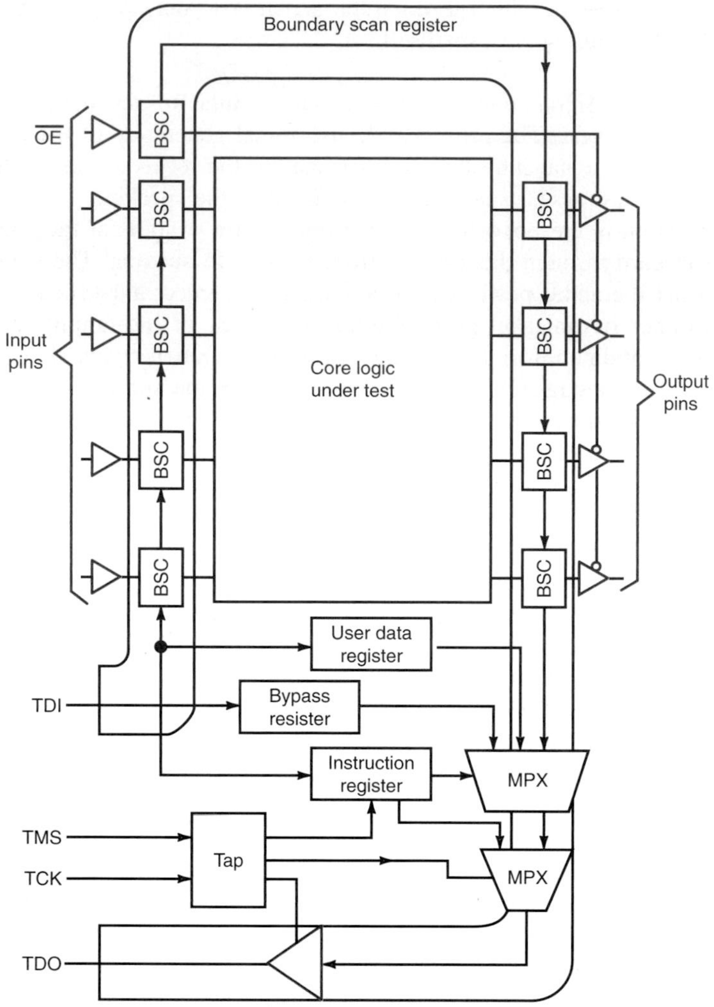

Note: The boundary scan register is shifted TDI to TDO.

speed through the test access port (TAP) controller. The function of the boundary scan circuitry is determined by a binary-coded instruction that is entered serially.

To read the data in at the internally monitored points, the boundary scan circuits are fed a serial instruction called INTEST. This is stored in the instruction register and decoded. The internal circuits are then configured by the logic to read the data from the core logic and store it in the BSCs. The data in the boundary scan register is then shifted out and read serially. This serial data is generally sent to a computer where a program uses it to determine if the results are correct.

Other instructions may also be entered to control the various functions. A SAMPLE/PRELOAD instruction is used to set the core logic to some desired state before a test is run. The CLAMP instruction can also be used to set the core logic levels to some predetermined pattern. The RUNBIST instruction causes a predetermined, built-in self-test (BIST) to be executed.

The boundary scan circuitry and its operation have been standardized by the Institute of Electrical and Electronics Engineers (IEEE) and is designated as the Joint Test Action Group (JTAG) 1149.1 standard. Most large complex ICs now contain a JTAG interface and internal circuitry. It is used to test the chip. Then later it may be used by the equipment manufacturer as a way to implement a larger test program for the equipment. Because a computer is programmed to use the JTAG interface, most testing and troubleshooting can be automated. An engineer or technician can actually sit in front of a computer, running the test software and exercise and test the equipment and monitor the results.

22-4 Electromagnetic Interference Testing

A growing problem in electronic communication is *electromagnetic interference (EMI)*. Also known in earlier years as *radio-frequency interference (RFI)* and *TV interference (TVI)*, now EMI is defined as any interference to a communication device by any other electronic device. Since all electronic circuits and equipment emit some form of EMI, they are potentially sources of interference with sensitive communication devices such as cell phones, cordless phones, radio and TV sets, pagers, and wireless LANs. As the number of computers has increased and as more and more cell phones and other wireless devices have come into use, the EMI problem has become a major one. The problem is so great that the FCC has created interference standards that must be met by all electronic devices. Before any electronic equipment can be sold and used, it must be tested and certified by the FCC to ensure that it does not emit radiation in excess of the allowed amount. Such strict rules and regulations have made design and manufacturing more difficult and have increased the cost of electronic products. But the result has been fewer interference problems between electronic products and a reduction in disruption of their use.

Today, a large part of the work of a communication technician or engineer is EMI testing and EMI minimization in products. This section briefly describes EMI, its sources, common techniques for reducing EMI, and EMI testing procedures.

Sources of Electromagnetic Interference

Any radio transmitter is a source of EMI. Although transmitters are assigned to a specific frequency or band, they can nevertheless cause interference because of the harmonics, intermodulation products, or spurious signals they produce.

Receivers are also a source of EMI. A local oscillator or frequency synthesizer generates low-level signals that, if not minimized, can interfere with nearby equipment.

Almost all other electronic devices can also generate EMI. Perhaps the worst offenders are computers. Computers contain millions of logic circuits that switch off and on at rates up to and in excess of 2 GHz. Because of the short pulses with fast rise and fall times that are generated, these circuits naturally generate a massive number of harmonics. The problem is so severe that computer manufacturers must comply with some of the FCC's toughest EMI reduction rules. Computer manufacturers use every trick in the book to get the radiation level down to the FCC specifications.

Of course, any electronic equipment that uses an embedded microcontroller, which today is almost every electronic device, is a potential source of EMI.

Another major source of EMI is switching power supplies. More than 80 percent of all power supplies in use today are of the switching variety, and this percentage is increasing. Power supplies that use switching regulators generate very high levels of pulse energy and radiate many high-amplitude harmonics. Inverters (dc-to-ac converters) in uninterruptible power supplies (UPS) and dc-to-dc converters also use switching methods and therefore produce harmonics and radiation.

The 60-Hz power line is another source of interference. Electrical transient pulses caused by high-power motors and other equipment turning off or on can disrupt computer operation or create a type of noise for communication equipment. Just the large magnetic and electric fields produced by the ubiquitous power lines can cause hum in stereo amplifiers, noise in sensitive medical equipment, and interference in communication receivers.

GOOD TO KNOW

The local oscillator of a receiver can be a source of electromagnetic interference.

Another form of EMI is *electrostatic discharge (ESD)*. This is the dissipation of a large static electric field. Lightning is the perfect example. The huge pulse of current produced by lightning generates an enormous number of harmonics that show up as noise in radio receivers. Anyplace where static buildup occurs can produce ESD, which not only can destroy integrated circuits and transistors but also can generate pulses that manifest themselves as noise in a receiver.

EMI is transmitted between electronic devices by several means. Interfering signals may travel by way of electromagnetic radiation (radio) from one unit to another. EMI may also be passed along by inductive or capacitive coupling when two units are close to each other. Cross talk on adjacent cables is an example. And interfering signals may be passed along from one piece of equipment to another by way of the ac power line that both use as the main power source.

Reduction of Electromagnetic Interference

The three basic techniques for reducing the level of EMI are grounding, shielding, and filtering. All these methods are used in the design of new equipment as well as in reducing EMI in applications in which the equipment is already deployed. Here is a brief summary of the techniques most often used.

Grounding. A poor electrical ground often causes EMI. As you know, a *ground* is the common reference point for most, if not all, voltages in a circuit. This ground shows up in many different physical forms. It may be a metal chassis or rack, the metal frame of a building, or water pipes. The best ground is an earth ground formed when a long copper rod is driven into the ground. Inside the equipment, the ground is formed on the PCBs on which the components are mounted. The ground is usually a wide copper strip or in some cases a broad copper ground plane formed on one side of the PCB.

In the design of equipment, especially RF circuits, the ground is a key consideration. Much care is taken in forming it, routing it, and connecting to it. Short, wide, and very low-resistance connections are the best. In equipment that uses both analog (linear) and digital circuits, separate grounding paths are usually formed for each type of circuit. This helps minimize interference to the sensitive analog circuits by the noisier digital circuits. The two ground systems are eventually connected, of course, but at only one point in the equipment.

Circuit grounds in equipment are in place and cannot be changed. But it is possible for the grounding connections of different pieces of equipment in a system to cause some form of EMI. Many times, the EMI can be eliminated or greatly reduced by simply experimenting with different ground arrangements. These are some useful guidelines:

1. If a piece of equipment does not have a ground, add one. A connection to a large common ground is preferred, especially one connected to earth ground.

2. Ground connections should always be kept as short as possible. Ground wires should be no longer than $\lambda/4$ at the highest frequency of operation.

3. Ground cables should be large and have low resistance. Stranded copper wire of a size greater than AWG 10 is preferred. The size should be greater if the ground carries a large current. Copper braid as wide as practical also makes a good ground connection.

4. If multiple pieces of equipment are involved and signals are passed from one unit to another, ground loops may exist. A *ground loop* is formed when multiple circuits or pieces of equipment are connected to a common ground but at different points. See Fig. 22-25. Current flow in the ground connection can produce a voltage drop across a part of the ground. That voltage then shows up in series, with the very small signals at the inputs to other circuits or equipment causing interference. Ground loops are eliminated by connecting all circuits or equipment to a single point on the common ground. See Fig. 22-26.

Figure 22-25   Grounds are not perfect short circuits, especially at high frequencies, where skin effect makes the resistance higher and permits ground loops to form.

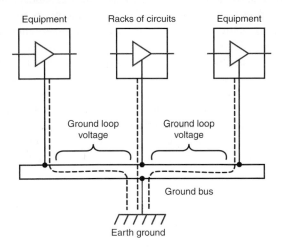

Figure 22-26   Single-point grounds are best because they minimize the formation of ground loops.

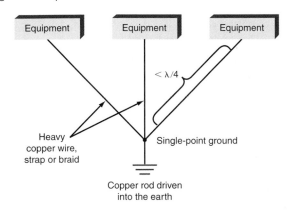

5. EMI is often caused by the incorrect connection of coaxial cable shields. The shield braid may be broken or open at one of the connectors. In some applications, grounding only one end of the shield instead of at both ends of the cable reduces EMI. This is also true of any shielded wire. Experimentation with the cable shields and grounds can often reduce EMI.

6. Remember that all ac power connections have a ground associated with them. In two-wire ac connections, the neutral wire is grounded at the entry point of the alternating current. In three-wire systems, the third wire is also a ground used primarily for safety purposes. However, the lack of a third ground can also cause interference. Adding the third ground wire often solves EMI problems.

Shielding. *Shielding* is the process of surrounding EMI-emitting circuits or sensitive receiving circuits with a metal enclosure to prevent the radiation or pickup of signals. Often, just placing a metal plate between circuits or pieces of equipment to block radiation is sufficient to reduce or eliminate EMI. The metal reflects any radiated signals and can actually absorb some of the radiated energy.

Almost all communication equipment is made with extensive use of shielding. Oscillators and frequency synthesizers are almost always packaged in a shielded can or enclosure. Individual transmitter or receiver stages are often shielded from one another

Shielding

with blocking plates or completely surrounding enclosures. Switching power supplies are always shielded in their own enclosure. Just remember that when shielding is used, ventilation holes are usually necessary to release any heat produced by the circuits being shielded. These holes must be as small as possible. If the holes have a diameter that is near $\lambda/2$ of the signals being used, the holes can act as slot antennas and radiate the signals more effectively. Good initial design prevents this problem.

In some cases, RF signals leak from shielded enclosures where the various panels of metal come together. A continuous shield is the most effective, but most shielded boxes must have a removable panel or two to permit assembly and repair access. If the panels are not securely mated with a low-resistance contact, RF will leak out of the opening. Securely attaching panels with multiple screws and making sure that the metal is not dirty or oxidized will solve this problem. If not, special flexible metal seals have been designed to attach to enclosures and panels that must mate with one another. These seals will eliminate the leakage.

Finally, radiation EMI can sometimes be reduced in a larger system by moving the equipment around. By placing the offending units farther apart, the problem can be eliminated or at least minimized. Remember that the strength of a radiated signal varies as the square of the distance between the transmitting and receiving circuits. Even a short-distance move is all that may be needed to reduce the interference to an acceptable level.

Filtering

Filtering. The third method of EMI reduction is *filtering*. Filters allow desired signals to pass and undesired signals to be significantly reduced in level. Filters are not much help in curing radiation or signal coupling problems, but they are a very effective way to deal with conducted EMI, interference that is passed from one circuit or piece of equipment to another by actual physical conduction over a cable or other connection.

Some types of filters used in reducing EMI include

1. Bypass and decoupling circuits or components used on the dc power supply lines inside the equipment. Typical decoupling circuits are shown in Fig. 22-27. Instead of using a physical inductor (usually called a *radio-frequency choke*, or *RFC*) in series with the dc line, small cylindrical ferrite beads can be placed over a wire conductor to form a small inductance. These beads are widely used in high-frequency equipment. The bypass capacitors must have a very low impedance, even for RF signals; i.e., ceramic or mica capacitors must be used. Any plastic dielectric or electrolytic capacitors used for decoupling must be accompanied by a parallel ceramic or mica to make the filtering effective at the higher frequencies.

2. High- or low-pass filters used at the inputs and outputs of the equipment. A common example is the low-pass filter that is placed at the output on most transmitters to reduce harmonics in the output.

Radio-frequency choke (RFC)

Figure 22-27 Decoupling circuits are low-pass filters that keep high-frequency signals out of the power supply.

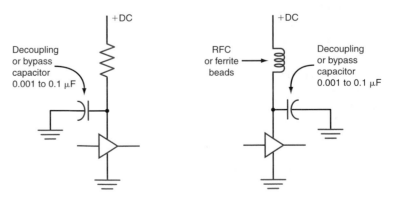

Figure 22-28 Alternating-current power line filter works both ways, removing high-frequency signals on input and output.

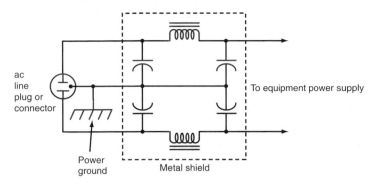

3. ac power line filters. These are low-pass filters placed at the ac input to the equipment power supplies to remove any high-frequency components that may pass into or out of equipment connected to a common power line. Figure 22-28 shows a common ac power line filter that is now built into almost all types of electronic equipment, especially computers and any equipment using sensitive high-gain amplifiers (medical, stereo sound, industrial measuring equipment, communications receivers, etc.).

4. Filters on cables. By wrapping several turns of a cable around a toroid core, as shown in Fig. 22-29, interfering signals produced by inductive or capacitive coupling can be reduced. Any common-mode signals induce voltages into the core, where they are canceled out. This approach works on ac power cords or any signal-carrying cable.

Measurement of Electromagnetic Interference

The rules and regulations pertaining to EMI are given in the Code of Federal Regulations, Title 47, Parts 15 and 18. Anyone working in the communication field should have copies of the code, which can be obtained from the Government Printing Office. Essentially these guidelines state the maximum signal strength levels permitted for certain types of equipment. The accepted levels of radiation vary considerably depending upon the type of equipment, the environment in which it is used, and the frequency range. The FCC also distinguishes between intentional radiators such as wireless LANs and other wireless units and unintentional radiators such as computers.

Radiation is measured with a field strength meter. It may be a simple device, as described earlier in this chapter, or a sensitive broadband communication receiver with a calibrated antenna. In either case, the measurement unit is microvolts per meter (μV/m). This is the amount of received signal picked up by an antenna 1 m long at some specified distance.

Figure 22-29 A toroid core cancels common-mode signals picked up by radiation or through capacitive or inductive coupling.

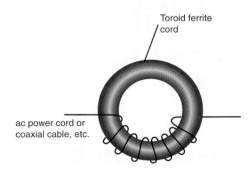

Toroid ferrite cord

ac power cord or coaxial cable, etc.

As discussed in the text, there's no need for cables when communication devices have Bluetooth. Several hundred million Bluetooth devices are sold each year. Thus there is a need to accurately and quickly test products before they come to market. This spectrum analyzer has Bluetooth wireless test capability.

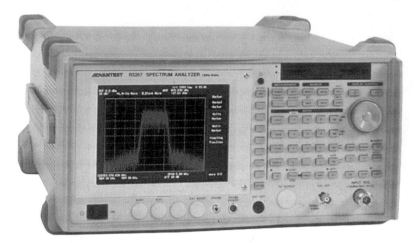

Consider the FCC regulations for a computer. Conducted EMI must not exceed 250 μV over the frequency range of 450 kHz to 30 MHz on the ac power line. Radiated emissions may not exceed a specific field strength level at a given distance for specified frequency ranges. The measurement distance is either 3 or 10 ft. At 10 ft the received signal strength may not exceed

100 μV/m between 30 and 88 MHz

150 μV/m between 88 and 216 MHz

200 μV/m between 216 and 960 MHz

500 μV/m above 960 MHz

Several manufacturers make complete EMI test systems that are sophisticated field strength meters or special receivers with matched antennas that are used to "sniff out" EMI. Some units have inductive or capacitive accessory probes that are designed to pick up magnetic or electric fields radiated from equipment. These probes are good for finding radiation leaks in shielded enclosures, connectors, or cables. The antennas are directional so that they can be scanned around an area or a piece of equipment to help pinpoint the radiation source and its frequency. Once the nature of the radiation is determined, grounding, shielding, or filtering steps can be taken to eliminate it.

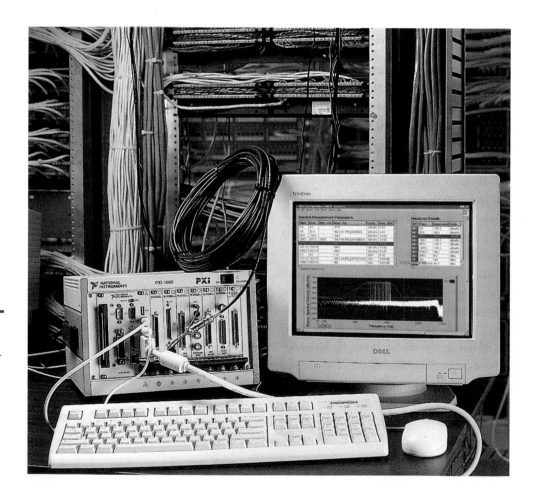

Devices such as PDAs and cell phones use flat-panel displays. To inspect these displays, technicians use software such as Labview by National Instruments. A technician can test the audio, electrical, and RF components. By using this program, it is possible to diagnose problems with alignment, pixel defect, color, and contrast.

Testing communication products with specialized computer software.

CHAPTER REVIEW

Summary

The work performed by communication electronics experts will involve some form of testing and measurement over the course of equipment installation, operation, servicing, repair, and maintenance. Engineering and manufacturing engineers and technicians will also perform a wide range of tests and measurements in their work, including verifying and making final adjustments to specifications. It is important, therefore, for them to be familiar with the types of malfunctions that can occur and the factors to consider in repair and maintenance of equipment.

There exist a wide range of choices in electronic communication testing equipment such as oscilloscopes and multimeters for measuring ac and dc voltages, current, resistance, and frequency. Measurements are routinely made of frequency, power, modulation, and SWRs. Some equipment is manufactured with its own testing microcomputers built in; other equipment must be tested with separate equipment. Because electronic communication equipment is so complex and varied in makeup, the manufacturers' own manuals are the best resource in diagnosing and repairing their equipment.

Most modern communications products are made up of complex ICs that contain all or most of the circuitry required for the application. This makes testing and troubleshooting that use older procedures difficult. Many modern digital circuits are tested by way of a built-in testing system referred to as boundary scan that permits IC pins and internal circuit points not brought out to pins to be monitored. Boundary scan permits automated testing and troubleshooting by computer.

This chapter also described the importance of electromagnetic interference (EMI) in communication work. Most electronic equipment generates signals that can potentially interfere with other electronic equipment. Computers and other digital devices are the major offenders. EMI can the minimized by a combination of good initial design plus a mix of grounding, shielding, and filtering techniques.

Questions

1. What is the most important specification of an ac voltmeter for measuring radio-frequency voltages?
2. What is a digital oscilloscope? Name the major sections of a DSO, and describe how they work.
3. What kind of resistors must be used in a dummy load?
4. Name two common types of SWR meters.
5. Name two common power-measuring circuits.
6. Show how the SWR can be determined from power measurements.
7. What is the name of the versatile generator that generates sine, square, and triangular waves?
8. How is the output level of an RF generator controlled?
9. What is the typical output impedance of an RF generator?
10. What cautions should you take and what guidelines should you follow when connecting measuring instruments to the equipment being tested? Why should you allow an RF generator to warm up before using it?
11. What is a sweep generator, and how does it work?
12. Name two applications for a sweep generator.
13. How is the output of the circuit being tested with a sweep generator detected and displayed?
14. Name the six main sections of a frequency counter.
15. What is a prescaler? Name two types.
16. What is a network analyzer? How is it used?
17. What is displayed on the CRT of a spectrum analyzer? Explain.
18. State the function of a field strength meter. What is its circuit?
19. Name the four most common transmitter tests.
20. What test instrument is used to measure the modulation of an AM transmitter?
21. What kind of test instrument is best for detecting the presence of harmonics and spurious radiation from a transmitter?
22. Generally, what level of SWR should not be exceeded in normal operation?
23. What does *SINAD* mean?
24. What does the receiver blocking test determine?
25. What does the third-order intercept test in a receiver measure?
26. What does an eye pattern show?
27. What does a "closing eye" usually indicate about the medium or circuit being tested?
28. Explain how an excessively noisy binary bit stream appears in an eye diagram.
29. What kind of instrument is used for bit error rate testing?
30. Describe the basic process of bit error rate testing.
31. What is a protocol analyzer? How does it display its results?
32. What types of components are most likely to fail in communication equipment?
33. What circuits should be tested first before any troubleshooting is initiated?
34. Besides the right test equipment, what should you have on hand before you attempt to service or troubleshoot any kind of communication equipment?
35. Name two ways in which an antenna may be defective.
36. Name the two basic types of troubleshooting methods used with communication equipment. Which is more likely to be used on receivers? Transmitters?
37. What test instruments are needed to carry out the two troubleshooting methods given above?
38. Name four measurements often made in fiber-optic systems.
39. In what type of equipment and in what frequency range is jitter a problem?
40. What is jitter?
41. What is the common unit for jitter measurement?
42. Why are large-scale ICs difficult to test by signal injection and tracing?
43. What is the name of the internal circuitry built into many large-scale ICs for test purposes?
44. By what circuit is the test data entered or accessed?
45. How is data entered and monitored with built-in test circuits?
46. What controls the function of the test circuits?
47. Name the standard of this test method.
48. Define EMI.
49. Who mandates EMI control?
50. How can receivers be a source of EMI?
51. How can a power supply be an EMI source?
52. Explain the role of the 60-Hz ac power line in EMI.
53. Name three ways by which EMI is transmitted.
54. What is the most common source of EMI?
55. List the three techniques used to reduce EMI.
56. What is considered the ground on a PCB?
57. What is the best type of ground?
58. What is a ground loop? What causes it?
59. How are ground loop problems eliminated?
60. State the conductor requirements for a good ground.
61. How does shielding minimize EMI?
62. What can be done to prevent radiation leakage from a shield?
63. What is the effect of power supply decoupling circuits on EMI?
64. What can be used in place of an RFC in a decoupling circuit?
65. Explain what ac power line filters do.
66. What type of interference does a toroid core deal with most effectively?
67. Where can rules and regulations for EMI be found?
68. Name two types of EMI radiation testing equipment.
69. What is the basic measuring unit for radiated EMI? What does it mean?
70. What components in an EMI test system allow you to pinpoint the source of radiation?

Problems

1. Explain the theory behind the operation of an arbitrary waveform generator. What is its main limitation? ◆

2. Explain the operation of a superheterodyne spectrum analyzer. What determines the resolution of the analyzer?

3. A noise test is performed on a receiver. Where is the noise measured? The spacing between the two horizontal lines on the oscillope screen is 2.3 divisions. The vertical sensitivity calibration is 20 mV per division. What is the noise level? (Use the procedure described in the text.) ◆

4. Refer to the transmitter diagram in Fig. 22-23. No modulation is applied to the circuit. The transmitter output power across the dummy load is zero. A measurement of the signal at the output of the phase modulator reveals a sine wave of 14.8 MHz. What is the problem? Explain why there is no output. Solve this problem in two ways: (*a*) Assume that the carrier frequency is as shown in Fig. 22-23. (*b*) Assume a carrier crystal of 14.8 MHz.

5. In Fig. 22-23, the output of the third multiplier is OK, but there is no output across the dummy load. What component is most likely defective? ◆

6. A small radio transmitter puts out 5 W of forward power but produces 1 W of reflected power. What is the SWR?

◆ *Answers to Selected Problems follow Chap. 22.*

Critical Thinking

1. State two benefits of DSOs over analog oscilloscopes.

2. How do you lower the SWR if it is too high?

3. A handheld cellular telephone worth approximately $175, consumer retail price, was obtained free when the cellular service was first initiated. It does not work. A call to the dealer indicates that this model can be repaired, but the labor rate is $75 per hour, and most repairs require a minimum of 2 h to fix. Parts charges are extra. Should you have the telephone fixed, or should you buy a new one? Give your reasons for the decision you make.

4. Give three examples of how household appliances could produce EMI and the electronic equipment they may interfere with.

5. State the potential for CB radios, FM family radios, and amateur radio equipment to interfere with other household electronic equipment.

6. Can the JTAG system test analog/RF circuits?

Answers to Selected Problems

CHAPTER 1

1-1. 7.5 MHz, 60 MHz, 3750 MHz, or 3.75 GHz

1-3. In radar and satellites

CHAPTER 2

2-1. 50,000

2-3. 30,357

2-5. 5.4, 0.4074

2-7. 14 dB

2-9. 37 dBm

CHAPTER 3

3-1. $m = (V_{max} - V_{min})/(V_{max} + V_{min})$

3-3. 100 percent

3-5. 80 percent

3-7. 3896 kHz, 3904 kHz; BW = 8 kHz

3-9. 800 W

CHAPTER 4

4-1. 28.8 W, 14.4 W

4-3. 200 μV

CHAPTER 5

5-1. m_f = 12 kHz/2 kHz = 6

5-5. −0.1

5-7. 8.57

5-9. 3750 Ω, 3600 or 3900 Ω (EIA)

CHAPTER 6

6-1. 1.29 MHz

6-3. 3141.4 Hz or ±1570.7 Hz at 4000-Hz modulating frequency

6-5. f_0 = 446.43 kHz, f_L = 71.43 kHz

CHAPTER 7

7-1. 7 MHz

7-3. 8 kHz − 5 kHz = 3 kHz

7-5. 92.06 dB

CHAPTER 8

8-1. 206.4 MHz

8-3. 25.005 MHz

8-5. 1627

8-7. 132 MHz, 50 kHz

8-11. 72 W

CHAPTER 9

9-1. 6 kHz

9-3. 18.06 and 17.94 MHz

9-5. 2.4

9-7. 46 MHz

9-9. 27, 162, 189, and 351 MHz

CHAPTER 10

10-1. 133

10-3. T1: Bit rate 1.544 MHz, 24 channels
T3: Bit rate 44.736 MHz, 672 channels

CHAPTER 11

11-1. EBCDIC

11-3. 69.44 μS

11-5. 60 kbps

11-7. 278.95 Mbps

11-9. 8×10^{-6}

CHAPTER 12

12-1. 10 Mbps; $t = 1/10 \times 10^6 = 0.1 \times 10^{-6} = 0.1 \mu S = $ 100 nS.

12-3. 20,000 bytes, which includes all of the beginning and ending fields and delimiters.

CHAPTER 13

13-1. 0.857 m

13-3. 51 Ω

13-5. VF = 0.6324

13-7. 4.02 ft

13-9. 1.325 nS

CHAPTER 14

14-1. 0.1 wavelength = 20 Ω; 0.3 wavelength = 90 Ω (see Fig. 14-11)

14-3. 468/27 = 17.333 mHz, 520 kHz to 1.04 MHz

14-5. 75-Ω coaxial cable

14-7. 1.822 ft

14-9. 1.65 ft

CHAPTER 15

15-1. 1.06 μS

15-3. Binary:
11011110100110110000010000
0001111
Hex: DE9B080F

CHAPTER 16

16-1. 36 percent, 0.36 percent

16-3. 0.632 in

16-5. 50 mW

16-7. 8.43 in

16-9. Gain = 960,000 or 59.82 dB; beamwidth = 0.145°

CHAPTER 17

17-1. Zero degrees (equator), 90° (polar)

17-3. The TC&C system is used to fire small hydrazine thrusters that move the satellite in one of several directions.

17-5. Satellite position information is transmitted as part of the NAV-msg 1500-bit message containing ephemeris and timing information. The data rate is 50 bps.

17-7. Differential GPS uses a precisely placed earth station that receives the L1 GPS signals and computes the error between them and its own precisely known location. The error is transmitted to special DGPS receivers that can decipher the error data and correct the output. Accuracies to within 3 to 6 ft are possible.

CHAPTER 18

18-1. Subscriber line interface circuit (SLIC) performs the BORSCHT functions.

18-2. BORSCHT functions are battery power, overvoltage protection, ringing, supervision, coding, hybrid, test.

18-4. Links between exchanges are four line rather than two line; that is, the send and receive signals are separated before transmission between exchanges. These links may be twisted pair, fiber-optic cable, or microwave radio and may be analog or digital, in either case with multiplexing.

18-7. The document to be transmitted is divided into many fine horizontal lines as it is scanned a line at a time by a transducer that converts the dark and light portions into an electrical signal that is used to modulate a carrier.

CHAPTER 19

19-1. 300,000,000 m/s, 186,000 mi/s
19-3. See Fig. 19-8 and the related text.
19-5. 12 percent
19-7. 9.6 km
19-9. 100 MHz, 200 MHz

CHAPTER 20

20-1. Spectral efficiency is data rate divided by bandwidth: 2.4 Mbps/1.5 Mhz = 1.6 bits/Hz.
20-3. π/4-DQPSK has a total of eight phase positions.
20-5. Typical low IF is 100 kHz or 120 kHz

CHAPTER 21

21-1. a. ISM band
b. Wi-Fi 802.11n
21-2. 16.67 GHz

CHAPTER 22

22-1. An arbitrary waveform generator uses digital techniques to generate a wave of any shape by storing binary samples in a memory and outputting them through a digital-to-analog converter and filter. Primary limitation is that the maximum frequency is about 100 MHz.

22-3. Noise is measured across a dummy resistive load replacing the speaker. 23 mV rms

22-5. The RF power amplifier transistor is probably defective.

Glossary

3DES triple data encryption standard Using the DES encryption three times to improve security.

2G cell phone system Second-generation digital cell phone system. The systems are GSM, TDMA, and CDMA.

2.5G cell phone system A generation of cell phone between the original second-generation (2G) digital phone and the newer third-generation (3G) phone. It brings data transmission capability to 2G phones.

3G cell phone system Third-generation cell phones, which are true packet data phones with enhanced digital voice and high-speed data transmission capability.

4G cell phone system Fourth generation of cell phones. Used to describe still-to-come advanced cell phone technologies that feature OFDM and packet data rates up to 100 Mbps in a mobile environment.

10Base-2 Ethernet LAN A type of Ethernet local-area network system implemented with thin coaxial cable.

10Base-5 Ethernet LAN An Ethernet system using thick coaxial cable.

10Base-T Ethernet LAN The twisted-pair version of Ethernet. The most widely used version.

10-gigabit Ethernet A new version of Ethernet that permits data speeds up to 10 Gbps over fiber-optic cable.

3GPP third generation partnership project An organization dedicated to developing and improving cell phone technologies that are standardized through the ITU.

100Base-TX. *See* Fast Ethernet.

1000Base-T Ethernet *See* Gigabit Ethernet.

1496/1596 circuit A typical IC balanced modulator that can work at carrier frequencies up to approximately 100 MHz and can achieve a carrier suppression of 50 to 65 dB.

3089 IF system A receiver IC originally developed by RCA.

A

Absolute field strength meter Device used for accurate measurements of signal strength. Usually measured in microvolts per meter.

Absorption In fiber-optic communication, the way light energy is converted to heat in the core material owing to the impurity of the glass or plastic.

Absorption wave meter Variable tuned circuit with an indicator that tells when the tuned circuit is resonant to a signal coupled to the meter by a transmitter; provides a rough indication of frequency.

Access method The protocol used for transmitting and receiving information on a bus.

Accumulator The combination of a register and adder.

Acknowledge character (ACK) A character that acknowledges that a transmission was received.

Advanced encryption standard (AES) An encryption method developed by the National Institute of Standards and Technology (NIST) to be more secure than the older DES standard because it uses 128, 192 or 256 bit keys.

Advanced mobile phone system (AMPS) The original cell phone system, based on analog radio technologies.

A-law compander The type of companding used in European telephone networks.

Aliasing A problem that occurs when the sampling frequency is not high enough. Aliasing causes a new signal near the original to be created.

All-pass filter A filter that passes all frequencies equally well over its design range but has a fixed or predictable phase shift characteristic.

American digital cellular (ADC) *See* IS-136 TDMA.

American Standard Code for Information Interchange (ASCII) The most widely used data communication code. This is a 7-bit binary code.

Amplification The boosting of signal voltage and power.

Amplitude modulation Varying the amplitude of a carrier signal to transmit information.

Amplitude shift keying (ASK) Amplitude modulation by square waves or rectangular binary pulses.

Analog multiplier A type of IC that can be used as a balanced modulator. It uses differential amplifiers operating in a linear mode.

Analog oscilloscope Device that amplifies the signal to be measured and displays it on the face of a CRT at a specific sweep rate.

Analog signal A smoothly and continuously varying voltage or current.

Analog-to-digital conversion (A/D conversion) The process of sampling or measuring an analog signal at regular intervals and converting it into a binary value.

Angle modulation A collective term referring to FM and PM.

Angle of elevation The angle between the line from the earth station's antenna to the satellite and the line between the same antenna and the earth's horizon.

Angle of incidence An angle A between the incident ray and the normal at the reflecting surface.

Angle of inclination The angle formed between the line that passes through the center of the earth and the north pole and the line that passes through the center of the earth and is also perpendicular to the orbital plane.

Angstrom (Å) A unit of measure for light wavelength, equal to 10^{-10} m or 10^{-4} μm.

Answering machine A feature on an electronic telephone that answers a call after a preprogrammed number of rings and saves the voice message. More formally called voice mail.

Antenna An electromechanical device used to transmit or receive radio signals.

Antenna array Two or more antenna elements combined to create an antenna with directivity and gain.

Antenna bandwidth The range of frequencies over which an antenna operates efficiently; the difference between the upper and lower cutoff frequencies of the antenna, which acts as a resonant circuit or bandpass filter.

Antenna directivity The ability of an antenna to send or receive signals over a narrow horizontal or vertical directional range.

Antenna Q The ratio of inductive reactance to resistance. Although it is difficult to calculate the exact Q for an antenna, the higher the Q, the narrower the bandwidth. Lowering Q widens bandwidth.

Antenna radiation pattern The geometric shape of the signal strength around an antenna.

Antenna reciprocity The condition that exists when the characteristics and performance of an antenna are the same whether the antenna is radiating or intercepting a signal.

Antenna resonance The frequency at which an antenna has peak voltage or current.

Antenna subsystem An assemblage of devices used for both transmitting and receiving signals.

Antenna tuner A configuration consisting of a variable inductor, one or more variable capacitors, or a combination of these components connected in various ways.

Antenna tuning A technique used to maximize power output by impedance matching.

Anti-aliasing filter A low-pass filter usually placed between the signal source and the A/D converter to ensure that no signal with a frequency greater than one-half the sample frequency is passed.

Antilog The number obtained when the base is raised to the logarithm that is the exponent.

Anti-transmit-receive (ATR) tube A type of spark gap tube. It effectively disconnects the transmitter from the circuit during the receive interval.

Apogee The highest point of a satellite above the earth.

Arbitrary waveform generator A signal generator that uses digital techniques to generate almost any waveform.

Array waveguide grating (AWG) An array of optical waveguides of different lengths made with silica on a silicon chip that can be used for both multiplexing and demultiplexing.

Asynchronous data transmission A transmission method in which each data word is accompanied by start and stop bits that indicate the beginning and ending of the word.

Asynchronous digital subscriber line (ADSL) The most widely used form of digital subscriber line (DSL). It permits data rates up to 8 Mbps and upstream rates up to 640 kbps using the existing telephone lines.

Asynchronous protocol The simplest form of protocol for ASCII-coded data transmission, using a start bit and a stop bit framing a single character, and with a parity bit between the character bit and the stop bit.

Asynchronous transfer mode (ATM) system A packet-switching system used in Internet backbones. ATM systems break data into 53-byte packets and transmit them over fiber-optic networks at speeds up to 10 Gbps.

Atmospheric noise The electrical disturbance that occurs naturally in the earth's atmosphere. Also known as static.

Atomic clock An electronic oscillator that uses the oscillating energy of a gas to provide a stable operating frequency.

Attachment user interface (AUI) Device in which inputs and outputs from the NIC terminate when thick Ethernet network cable is used.

Attenuation The reduction in signal amplitude over distance.

Audio The voice or sound portion of a communication system.

Authentication An electronic method of verifying who a person is for the secure exchange of information.

Automatic frequency control (AFC) A feedback control circuit that is used in high-frequency receivers and keeps the local oscillator on frequency.

Automatic gain control (AGC) A feature of receivers that ensures that the output signal remains constant over a broad range of input signal amplitudes.

Automatic power control (APC) circuit In the analog AMPS cellular phone system, a circuit that sets the transmitter to one of eight power output levels.

Automatic splicer A device used to achieve highly accurate alignment and splicing of fiber-optic cable.

Automatic voice level adjustment The process of having a circuit sense and adjust the gain of an amplifier used to amplify voice signals in a radio or telephone system.

Autoranging A feature of frequency counters that selects the best time-base frequency for maximum measurement resolution without overranging.

Avalanche photodiode (APD) A widely used, sensitive, and fast photosensor. It is reverse-biased.

Azimuth The compass direction in which north is equal to $0°$.

B

B − Y signal I and Q color signals are also called $R − Y$ and $B − Y$. Combining the three color signals causes Y to be subtracted from R or B signals.

Backbone LAN A central network with nodes that are other LANs. This provides a convenient way to interconnect a variety of different LANs for organizationwide communication.

Backward wave oscillator (BWO) A type of TWT in which the wave travels from the anode end back toward the electron gun, where it is extracted.

Balanced modulator A circuit that generates a DSB signal, suppressing the carrier and leaving only the sum and difference frequencies at the output.

Balanced transmission line A transmission line with neither wire connected to ground. Instead, the signal on each wire is referenced to ground.

Balun A device used to convert transmission lines from balanced to unbalanced operation.

Bandpass filter A filter that passes frequencies over a narrow range between lower and upper cutoff frequencies.

Band-reject, bandstop, or notch filter A filter that rejects or stops frequencies over a narrow range but allows frequencies above and below to pass.

Bandwidth The narrow frequency range over which the signal amplitude in a circuit is highest. The portion of the electromagnetic spectrum occupied by a signal. The range of frequencies a cable will carry. It determines the maximum speed of the data pulses the cable can handle.

Barrel connector A variation of the BNC connector that allows two cables to be attached to one another end to end.

Base station A cell site in a cellular telephone system, the most visible feature of which is its antenna on a tower.

Baseband LAN A local-area network in which the data to be transmitted is placed directly on the transmission medium, usually a twisted pair cable, as binary voltage levels.

Basic rate interface (BRI) A type of ISDN connection using one twisted pair. Two bearer (B) voice channels and one data control (D) channel are multiplexed on the line.

Baud rate In digital communication systems, the number of signaling elements or symbols that occur in a given unit of time.

Baudot code Rarely used today, this binary data code was used in early teletype machines.

Beam antenna An antenna that is highly directional and has very high gain.

Beamwidth The measure of an antenna's directivity.

Beat frequency oscillator (BFO) An oscillator that is built into receivers designed to receive SSB or CW signals.

Bessel filter A filter that provides the desired frequency response but has a constant time delay in the passband.

Bessel function A mathematical process used to solve the FM equation.

Bicone antenna A widely used omnidirectional microwave antenna. Signals are fed into the antenna through a waveguide or coaxial cable ending in a flared cone.

Bidirectional antenna An antenna that receives signals best in two directions.

Bidirectional coupler A variation of the directional coupler in which two sections of line are used to sample the energy in both directions, allowing determination of the SWR.

Binary check code, block-check character, or block-check sequence (BCC or BCS) The logical sum of the data bytes in a block used as a method of error detection.

Binary code A type of code consisting of patterns of 0s and 1s. Each pattern represents number a letter of the alphabet, or some special symbol such as punctuation or a mathematical operation.

Binary phase-shift keying (BPSK) A type of PSK in which a standard lattice ring modulator or balanced modulator is used for generating DSB signals.

Binary phase-shift keying (BPSK) or phase-shift keying (PSK) The process of modulating a carrier with binary data by changing the carrier phase $180°$.

Biphase encoding See Manchester encoding.

Bipolar transistor mixer A special mixer that consists of a single transistor biased into the nonlinear range to produce analog multiplication.

Birdies Random "chirping" or "tweeting" sounds at the audio output of some receivers due to intermodulation products.

Bisync protocol An IBM protocol, widely used in computer communication. It usually begins with the transmission of two or more ASCII sync (SYN) characters.

Bit error rate (BER) The number of bit errors that occur for a given number of bits transmitted.

Bit error rate (BER) analyzer A device that compares the transmitted and received data on a bit-by-bit basis to point out every bit error made.

Block A group of data which represent hundreds or even thousands of 1-byte characters.

Bluetooth A wireless PAN system. One of its applications is to provide hands-free cell phone operation.

BNC connector Connector that is widely used with coaxial cables to attach test instruments to equipment being tested. Often used in LANs and some UHF radios.

BORSCHT Seven basic functions provided by the subscriber line interface circuit (SLIC): **B**attery, **O**vervoltage protection, **R**inging, **S**upervision, **C**oding, **H**ybrid, and **T**est.

Bow-tie antenna A popular and effective variation of the dipole antenna which uses two-dimensional cones or triangles.

Bridge Electronic equipment connected as a node on the network that performs bidirectional communication between two LANs.

Bridge circuit An electrical network consisting of four impedances connected to form a rectangle, with one pair of diagonally opposite corners connected to an input device and the other pair to an output device.

Broadband connection A fast Internet connection provided by a local telephone company or other organization. A broadband connection can provide speeds up to several megabits per second.

Broadband LAN A type of LAN in which the binary data to be transmitted is used to modulate a carrier, which is then placed on the transmission medium.

Broadband system One of two basic multichannel architectures in use in communication satellites.

Broadband transmission The process of changing a baseband signal using a modulator circuit.

Broadside array A stacked collinear antenna consisting of half-wave dipoles separated from one another by one-half wavelengths.

Brouter A combination of a bridge and a router that can switch, perform protocol conversion, and serve as a communication manager between two LANs.

Browser The software that makes it possible to navigate and explore the Web and to access and display information.

Bus topology A topology in which all of the nodes in a LAN are connected to a common cable.

Butterworth filter Filter with maximum flatness in response in the passband and uniform attenuation with frequency.

C

C band The frequency under which older satellite systems operate. This frequency requires large dish antennas.

Caller ID (calling line identification service) A feature on many electronic telephones.

The calling number is displayed on an LCD readout when the phone is ringing, permitting the recipient of the call to identify the caller.

Cap code Special code assigned to some types of paging receivers.

Capacitor A circuit component made up of two parallel conductors separated by an insulating medium. A capacitor is capable of storing energy in the form of a charge and an electric field.

Capture effect The effect caused by two or more FM signals occurring simultaneously on the same frequency. The stronger signal captures the channel, eliminating the weaker channel.

Capture range The range of frequencies over which a PLL will capture an input signal.

Carrier A high-frequency signal usually a sine wave whose characteristics are changed by having a baseband signal impressed upon it.

Carrier frequency The frequency generated by an unmodulated radio, radar, carrier communication, etc., or the average frequency of the emitted wave when modulated by a symmetrical signal.

Carrier generator An oscillator.

Carrier recovery circuit A term applying to a number of techniques that can be used in a receiver to generate a carrier signal that has the correct frequency and phase relationship to the original transmitting carrier.

Carrier sense multiple access with collision detection (CSMA/CD) The access method used in Ethernet.

Carson's rule A method of determining the bandwidth of an FM signal.

Cassegrain feed A method of feeding a parabolic antenna in which electromagnetic radiation from the horn strikes a small reflector that reflects the energy toward a large dish which then radiates it in parallel beams.

Cathode-ray tube (CRT) A TV picture tube, or vacuum tube.

CATV (cable TV) A system of delivering a TV signal to home receivers by way of a coaxial and/or fiber-optic cable rather than by radio wave propagation.

CATV converter Known as a cable TV box, this tuner can select the special cable TV channels and convert them to a frequency that any TV set can pick up.

Cauer filter A filter producing more rapid attenuation than a Chebyshev filter. Though Cauer filters provide greater attenuation out of the passband, they do so with an even higher ripple.

Cavity resonator A waveguide-like device that acts like a high-Q parallel resonant circuit.

CB synthesizer Often used in citizens band radios, this is usually a PLL with one or more mixers that generate the carrier with the local oscillator frequencies for the receiver.

CDMA *See* Code division multiple access.

Cells In the cellular radio system, the small service areas.

Cellular radio system Standard telephone service provided by two-way radio at remote locations.

Central office (local exchange) A facility to which each telephone is directly connected by a twisted-pair cable.

Centrex An alternative to private branch exchange (PBX) that uses special equipment whereby most of the switching is carried out the local exchange switching equipment over special trunk lines.

Ceramic filter A filter made with a manufactured crystal-like compound with the same piezoelectric qualities as quartz. Like a crystal filter, the ceramic filter provides high selectivity but is less expensive.

Channel The amount of bandwidth that a signal occupies in the radio frequency spectrum.

Channelization process The process of dividing a range of frequencies into multiple bands of frequencies called *channels* over which transmissions will be made.

Characteristic (surge) impedance Impedance that is a function of the inductance, resistance, and capacitance in a transmission line.

Charged coupled device (CCD) Often used in fax machines, this is a light-sensitive semiconductor device that converts varying light amplitudes into an electrical signal.

Chebyshev filter A filter with extremely good selectivity. The attenuation rate is much higher than that of the Butterworth filter. The main problem with this filter is ripple in the passband.

Choke *See* Inductor.

Choke joint Two flanges connected to a waveguide at the center that is used to interconnect two sections of waveguide.

Chromatic dispersion In a multimode step index cable, a type of dispersion that occurs when multiple wavelengths of light are used.

Chrominance signal Color portion of a color TV signal. I and Q signals transmitted with luminance information in the bandwidth allotted to the TV signal.

Circular polarization When electric and magnetic fields rotate as they leave an antenna.

Circulator A three-port microwave device used for coupling energy in only one direction around a closed loop.

Cladding A plastic sheath surrounding a fiber-optic cable with an index of refraction that keeps light waves in the core.

Class A amplifier A linear amplifier. This is not an efficient amplifier and thus makes a poor power amplifier.

Class AB amplifier Similar to a Class B amplifier but biased with a small amount of conduction to eliminate crossover distortion.

Class B amplifier An amplifier that is biased at cutoff so that no collector current flows with zero input. This type of amplifier is used mainly as a push-pull amplifier.

Class C amplifier This amplifier is very efficient because it conducts for even less than one-half of the sine-wave input cycle.

Class D amplifier   An efficient switching amplifier that uses a pair of transistors to produce a square-wave current in a tuned circuit.

Class E amplifier   An efficient switching amplifier in which only a single transistor is used, either bipolar or MOSFET.

Class F amplifier   A variation of the class E amplifier. It contains an additional resonant network in the collector or drain circuit.

Class S amplifier   This amplifier uses switching techniques with a scheme of pulse-width modulation. It is found mainly in audio applications.

Cleaving   The first step in cutting fiber-optic cable, so that it is perfectly square on the end.

Client-server configuration   The general LAN configuration in which one of the computers in a network runs the LAN and determines how the system operates.

Clock recovery circuit   A circuit used to generate clock pulses from a received signal.

Coarse wavelength division multiplexing (CWDM)   A method of multiplexing data on fiber-optic cable. *See also* Dense wavelength division multiplexing.

Coaxial cable   A widely used type of transmission line that consists of a solid center conductor surrounded by a dielectric material. In cable TV systems, it is usually 75-Ω RG-6/U cable.

Coaxial connector   An electromechanical cable termination device designed to provide a convenient way to attach and disconnect equipment and cables and maintain the integrity and electrical properties of a cable.

Code division multiple access (CDMA)   A digital cell phone system using direct sequence spread spectrum (DSSS). *See also* Direct-sequence SS; IS-95 CDMA; Time division multiple access.

Codec   A single, large-scale IC chip that takes care of all A/D and D/A conversion and related functions.

Coding   Also known as analog-to-digital (A/D) conversion.

Coil   *See* Inductor.

Collector modulator   A high-level AM modulator that takes a low-level signal and amplifies it to a high-power level.

Collinear antenna   An antenna consisting of two or more half-wave dipoles mounted end to end.

Color burst   In color signal generation in TV, a sample of the 3.58-MHz subcarrier signal is added to the composite video signal. This synchronizes color demodulation at the receiver and is called *color burst.*

Color signal generation   The generation of color signals using three simultaneous signals (R, G, and B) during the scanning process by the light-sensitive imaging devices.

Colpitts oscillator   An oscillator in which the feedback is derived from a capacitive voltage divider.

Common mode rejection   In a balanced transmission line, any external signal induced into the cable appears on both wires simultaneously but cancels at the receiver. This is called *common mode rejection,* and it significantly reduces noise.

Communication   The process of exchanging information.

Communication satellite   Type of satellite that makes long-distance communication possible by serving as a relay station in the sky.

Commutating filter   A variation of a switched capacitor filter made of discrete resistors and capacitors with MOSFET switches driven by a counter and decoder.

Commutator   A form of rotary switch used in multiplexers in early TDM/PAM telemetry systems.

Companding   A process of signal compression and expansion that is used to overcome problems of distortion and noise in the transmission of audio signals.

Compression   A process that reduces the number of binary words needed to represent a given analog signal.

Computer branch exchange (CPX)   *See* Private branch exchange.

Conductor   A wire or cable that carries current from one place to another.

Cone of acceptance   The area external to the end of a fiber-optic cable and defined by the critical angle. Any light beam outside the cone will not be internally reflected and transmitted down the cable.

Conical antenna   A dipole antenna that uses two cone-shaped elements.

Connector   A special mechanical assembly that allows fiber-optic cables to be joined to one another or used to connect a transmission line to a piece of equipment or to another transmission line.

Constant-*k* filter   A filter that makes the product of the capacitive and inductive reactances a constant value *k*.

Continuous phase frequency-shift keying (CPFSK)   A type of frequency modulation that eliminates phase discontinuities.

Continuous tone-control squelch system (CTCS)   A system activated by a low-frequency tone transmitted along with audio to provide some communication privacy on a particular channel.

Continuous-wave (CW) radar   A type of radar using a constant-amplitude continuous microwave sine wave. The echo is also a constant-amplitude microwave sine wave of the same frequency but of lower amplitude and shifted in phase.

Continuous-wave (CW) transmission   A code transmission such as Morse code in which the signal is turned on and off to transmit a message.

Control segment   The part of the GPS system consisting of the various ground stations that monitor the satellites and provide control and update information.

Converter   *See* Mixer.

Copper digital data interface (CDDI)   Standard wire cable version of the FDDI system. It is less expensive than FDDI because twisted pair is used instead of fiber-optic cable.

Cordless telephone   Full-duplex, two-way radio system made up of two units, the portable unit or handset, and the base unit.

Core   The glass fiber in a fiber-optic cable that is usually surrounded by a protective plastic cladding.

Corner reflector   A type of high-frequency antenna in which a reflector is placed behind the dipole to improve gain and directivity.

Correlated noise   Another name for intermodulation distortion. Correlated noise is produced only when signals are present. It is manifested as the signals called *birdies.*

Cosmic noise   Noise generated by stars outside our solar system.

Counterpoise   The entire ground-plane collection of radials.

Coupled circuit   A circuit linked by way of a magnetic field through transformer action.

Critical angle   The angle between the incident light ray and the normal to the glass fiber surface. This value depends upon the index of refraction of the glass.

Critical coupling   The degree of magnetic coupling at which the output reaches a peak value.

Crossed-field amplifier (CFA)   Similar to a TWT, this amplifier has a lower gain but is somewhat more efficient.

CRT   *See* Cathode-ray tube.

Crystal diode   *See* Point-contact diode.

Crystal filter   A filter made from the same type of quartz crystal normally used in crystal oscillators. It is used in communication receivers in which superior selectivity is required.

Crystal oscillator   An oscillator in which the frequency of the AC output is determined by the mechanical properties of a piezoelectric crystal.

Crystal radio receiver   An early radio receiver that had weak reception because no active amplification was used.

Cutoff frequency   The frequency at which the output of a frequency-selective device, such as a filter or amplifier, drops to 70.7 percent of its maximum output. Also known as the half-power or 3-dB down point.

Cyclical redundancy check (CRC)   A mathematical technique used in data transmission that catches 99.9 percent or more of transmission errors.

D

D layer   The layer of the ionosphere that is farthest from the sun. This layer is weakly ionized and exists only in daylight.

Daisy chain topology   A variation of a ring topology that can be described as a ring that has been broken.

Data   Information to be communicated.

Data bus   Multiple parallel lines that carry binary data.

Data communication   A technique that can be used to transmit voice, video, and other analog signals in digital form.

Data compression techniques   A digital data processing technique that looks for redundancy in the transmitted signal to speed up transmission.

Data conversion   In digital communication, the conversion of data in analog form into digital form and vice versa.

Data encryption standard (DES)   One of the first and still widely used encryption standards created by the National Institute of Standards and Technology (NIST). It uses 56-bit keys that are no longer considered to provide the ultimate in security.

Data rate   The speed of the binary pulses in a communication system.

dBm   A power level expressed in dB referenced to one milliwatt.

Decibel   The unit of measure created as a way of expressing the hearing response of the human ear to various sound levels but now primarily used to express gain or loss in electronic circuits, cables or communication links.

Decimation in time (DIT)   The processing that takes place with the fast Fourier transform.

Decimator   A digital filter.

Decryption   Deciphering an encrypted message to recover the original data.

De-emphasis   A process used to return the frequency response of a preemphasized signal to its normal level.

De-emphasis circuit   A simple low-pass filter with a time constant of 75 μs.

Definition   The resolution of detail in a TV picture.

Delay line   A circuit that delays a signal or sample by some constant time interval.

Delta modulation   A special form of A/D conversion that results in a continuous serial data signal being transmitted.

Demodulator (detector)   A circuit that accepts modulated signals and recovers the original modulating information.

Demultiplexer   A device with a single input and multiple outputs, one for each original input signal.

Dense wavelength division multiplexing (DWDM)   A light frequency multiplexing method that uses 8, 16, 32, 64, or more data channels on a single fiber.

Depletion region   A thin area with no free carriers, holes, or electrons around a PN semiconductor junction that acts like a thin insulator that prevents current flow through the device.

Detector   *See* Demodulator.

Detector probe   *See* RF probe.

Deviation meter   A device designed to measure the amount of carrier deviation of an FM/PM transmitter.

Deviation ratio   The number of times per second the carrier frequency deviates above and below its center frequency.

Dial tone   In a telephone set, the audio note indicating that the telephone is ready to use.

Dialing circuit   A circuit that provides a way for entering the telephone number to be called. The system used is either a pulse or a tone system.

Dielectric   The insulating portion of a capacitor.

Dielectric (lens) antenna   An antenna that uses a special dielectric material to collimate or focus microwaves from a source into a narrow beam.

Differential amplifier modulator   An amplifier with high gain and good linearity that can be amplitude modulated 100 percent.

Differential phase-shift keying (DPSK)   A type of phase modulation that simplifies the demodulation process because the transmitted signal itself becomes the phase reference.

Diffraction   The bending of waves around an object.

Digital AMPS (DAMPS)   *See* IS-136 TDMA.

Digital cell phone system   The system that has superseded the original analog AMPS system.

Digital private branch exchange (DPBX)   A telephone switching system in which all switching is done by IC transistor switches that have replaced the mechanical relays in older PBXs and PABXs.

Digital satellite radio or **digital audio radio service (DARS)**   A satellite service that provides hundreds of channels of music, news, sports, and talk radio primarily to car radios.

Digital signal   A signal that changes in steps or discrete increments. Most digital signals use binary or two-state codes.

Digital signal processing (DSP)   The processing of analog signals by digital computing methods.

Digital storage oscilloscope (DSO; digital or sampling oscilloscope)   A device that uses high-speed sampling or A/D techniques to convert the signal to be measured into a series of digital words that are stored in internal memory.

Digital Subscriber Line (DSL)   The process of transmitting digital data over analog telephone lines using OFDM/DMT.

Digital-to-analog (D/A) conversion   The process of translating multiple binary numbers back into the equivalent analog voltage.

Digital TV (DTV)   *See* High-definition TV.

Diode   A unidirectional semiconductor device used for rectification, signal detection, and mixing.

Diode detector   A simple and widely used amplitude demodulator.

Diode mixer   A widely used mixer that uses a diode.

Diode modulator   A simple amplitude modulator.

Diode ring or **lattice modulator**   A popular and widely used type of balanced modulator.

Dip oscillator   A tunable oscillator used to determine the approximate resonant frequency of any de-energized *LC*-resonant circuit.

Diplexer   A type of filter or circulator that allows a single antenna to be shared by a transmitter and receiver.

Dipole antenna   An antenna consisting of two pieces of wire, rod, or tubing that together are one-half wavelength long at the operating resonant frequency. Also called Hertz antenna or doublet.

Direct attached storage (DAS)   A method of communications between a computer and external storage like hard disks using a standard parallel interface over short distances.

Direct Broadcast Satellite (DBS) service   A TV signal distribution system that uses K band signals to send cable-TV-like service to homes equipped with 18-in.-diameter satellite antennas.

Direct conversion (DC) or **zero IF (ZIF) receiver**   A special version of the superheterodyne that translates RF directly to baseband.

Direct digital synthesis (DDS)   A form of frequency synthesis in which a sine-wave output is generated digitally with a DAC.

Direct-sequence SS or **code division multiple access (CDMA)**   A type of spread spectrum in which the serial binary data is mixed with a higher-frequency pseudorandom binary code at a faster rate and then used to phase-modulate a carrier.

Directional coupler   A commonly used waveguide component that facilitates the measurement of microwave power in a waveguide and the SWR.

Directivity   The ability of an antenna to send or receive signals over a narrow horizontal and vertical directional range.

Director   A kind of parasitic element in a Yagi that is shorter than the half-wave dipole-driven element and is mounted in front of the element to increase gain and directivity.

Discrete Fourier transform (DFT)   An algorithm that can be used in a DSP processor to analyze the frequency content of an input signal.

Discrete Multitone (DMT)   The name for OFDM used in DSL.

Discriminator   A circuit in a system that takes an FM signal and re-creates the original DC or AC signal produced by the transducer.

Dish antenna   Special antenna using a parabolic reflector which selects the signal from the desired satellite and provides very high gain.

Dispersion   Pulse stretching caused by the many different paths through a fiber-optic cable.

Distance measuring equipment (DME)   Equipment that permits accurate distance measurements between stations in aircraft navigation.

Distortion   A condition that causes a signal's shape to be changed by a circuit. The introduction of harmonics into a signal.

Distributed feedback laser (DFL)   A laser made with a cavity that contains an integrated grating structure that acts like a selective filter.

Diversity system   A way to minimize fading caused by multipath signals by using multiple transmitters, receivers, or antennas.

Doppler effect   A frequency shift that occurs when there is motion between a transmitting station and a remote target.

Dotted decimal   A way of simplifying an IP address by converting each byte or octet of the address into a decimal number and separating the decimal numbers with a dot or period.

Double-conversion down converter   One of two types of down converter used in earth station receivers. It resolves the problems of image rejection and tuning or channel selection difficulties.

Double-sideband suppressed carrier (DSSC or DSB) signal   A signal containing both upper and lower sidebands, with the carrier suppressed.

Doubler   A frequency multiplier that multiplies the frequency by 2.

Doublet   *See* Dipole antenna.

Doubly balanced mixer   A balanced modulator.

Down conversion   The process used in radio receivers in which high-frequency radio signals are converted to a lower, intermediate frequency.

Driven array   A directional antenna that has two or more driven elements.

Drooping radial   A wire or other conductor that serves as a part of the ground plane of a quarter-wave vertical antenna. A conductor that is positioned at an angle greater than 90° from the vertical element of the antenna.

Dual-conversion receiver   A receiver in which the first mixer converts the incoming signal to a relatively high intermediate frequency, for the purpose of eliminating images, and the second mixer converts that IF to a much lower frequency at which good selectivity is easier to obtain.

Dual-conversion transponder   A device that translates the uplink signal to the downlink frequency in two steps with two mixers.

Dual-tone multifrequency (DTMF) system   The tone dialing system used in most modern telephones.

Dummy load   A resistor that is connected to the transmission line in place of the antenna to absorb transmitter output power.

Duplex communication   Communication that flows in two directions.

Dynamic nonhierarchical routing (DNHR)   In long-distance telephone service, a method permitting long-distance connections with two or fewer switching centers.

E

E-commerce   Business done over the Internet.

E layer   The layer of the ionosphere next to the D layer. This layer is weakly ionized and exists only in daylight.

EDGE (enhanced data for GSM evolution)   A faster 2.5G technology that used 8-PSKS modulation to achieve data rates up to 384 kbps.

Effective radiated power (ERP)   The power radiated by an antenna with directivity and therefore gain.

Electric field   An invisible force field produced by the presence of a potential difference between two conductors.

Electromagnetic field   A radio wave.

Electromagnetic interference (EMI)   Any interference to a communication device by any other electrical or electronic device.

Electromagnetic spectrum   The range of electromagnetic signals encompassing all frequencies.

Electromagnetic wave   A signal such as a radio signal made up of both electric and magnetic fields.

Electronic communication system   A group of devices including a transmitter, a channel or medium, and a receiver.

Electronic mail (E-mail)   The sending and receiving of electronic messages. Users create messages on their computers and send them to other users over a network.

Electrostatic discharge   The dissipation of a large electric field; lightning is an example.

Elevation   *See* Angle of elevation.

Encoding   A process in which binary data is transformed into a unique variation of the binary code, such as the Manchester code, prior to transmission.

Encryption   The process of encoding data in such a way as to obscure its meaning and content so that it can be transmitted securely.

End effect   A phenomenon caused by any support insulators used at the ends of a wire antenna. It causes a capacitance to be added to the end of each wire.

End-fire array   An antenna that uses two half-wave dipoles spaced one-half wavelength apart and has a bidirectional radiation pattern.

End of transmission block character (ETB)   A character transmitted to signal the end of a data block.

End of transmission character (EOT)   A character transmitted to signal the end of the data transmission.

Engineer   In the field of electronics, a person who holds a bachelor's or an advanced degree and specializes in design.

Enhanced 911 (E911) capability   A feature included with 2.5G and 3G phones that is mandated by the U.S. government. This feature makes it possible to locate any cell phone position automatically.

Enhanced data for GSM evolution   *See* EDGE.

Envelope   An imaginary line connecting the positive peaks and negative peaks of the carrier wave form. The envelope gives the exact shape of the modulating information signal. Also, the outline of the peaks of individual signals like harmonics in the frequency spectrum.

Envelope delay   The time needed for a point on a waveform to pass through a filter.

Envelope detector   Another name for diode detector.

Equatorial orbit   A satellite orbit with an inclination of 0°.

Erbium-doped fiber amplifier (EDFA)   A type of optical amplifier.

Error   One or more incorrect bits in the transmission of information that is often caused by a very high noise level. It can be detected by special circuitry.

Error signal   The output of a phase detector circuit.

Ethernet LAN   Developed by Xerox Corporation, one of the oldest and one of the most widely used LAN types. It uses bus topology and baseband data-transmission methods.

Extended Binary Coded Decimal Interchange Code (EBCDIC)   An 8-bit code used mainly in IBM and IBM-compatible computing systems and equipment. It allows a maximum of 256 characters to be represented.

External noise   Random AC voltage that comes from sources over which we have little or no control—industrial, atmospheric, or space.

Extraterrestrial noise   An electrical disturbance that is solar or cosmic (from space).

Extremely high frequency (EHF)   The frequency range from 30 to 300 GHz.

Extremely low frequency (ELF)   The frequency range from 30 to 300 Hz.

Eye diagram   A diagram or pattern that is a display of overlapping individual bits of binary data displayed on a common oscilloscope.

F

F layer   The layer of the ionosphere closest to the sun. It is highly ionized, has the greatest effect on radio signals, and exists day and night.

F-type connector   An inexpensive coaxial connector used in TVs, VCRs, and cable TV.

Facsimile   An electronic system for transmitting graphical information by wire or radio.

Facsimile standards   Standards established by the International Telecommunications Union to ensure compatibility of fax machines made by different manufacturers.

Fading   A problem that occurs when radio waves pass through objects on their way from transmitter to receiver. The radio waves are negatively affected by these objects.

Far field   The radio wave beyond about one wavelength from the antenna.

Fast Ethernet   A newer version of Ethernet, currently the most widespread version of Ethernet. It has a speed of 100 Mbps. Also called 100Base-TX.

Fast Fourier transform (FFT)   A special version of the DFT algorithm developed to speed up the calculation of signal spectrum analysis.

Fast Fourier transform (FFT) spectrum analyzer   A test instrument that gives a frequency domain display similar to other types of spectrum analyzers.

Ferrite disk   A ceramic, sometimes made of yttrium-iron-garnet (YIG), used in circulators.

FET mixer A mixer that uses the square law response of a field-effect transistor for frequency conversion.

FET phase modulator A modulator with a phase shifter made up of a capacitor and the variable resistance of a field-effect transistor.

Fiber digital data interface (FDDI) A high-speed fiber-optic cable network offering a data transmission rate of 100 Mbps.

Fiber-optic cable A nonconducting cable consisting of a glass or plastic center cable surrounded by a plastic cladding encased in a plastic outer sheath and used in optical communication systems. It acts as a light "pipe."

Fiber-optic communication system A communication system that uses electronic digital and optical multi-plexing techniques for high speed data communication.

Fiber-optic connector A connector designed to provide a fast and easy way to attach or remove fiber-optic cables.

Fiber-optic interrepeater link (FOIRL) A fiber-optic communication channel with repeaters at each end that are designed to interconnect two Ethernet networks.

Fibre Channel (FC) An optical fiber transmission standard with speeds to 2 Gbps used primarily in storage area networks (SANs).

Fibre Distributed Data Interface (FDDI) A LAN standard designed to connect up to 500 nodes (PCs, etc.) in a ring configuration over a distance of up to 60 miles.

Field In the generation of a video signal, one complete scanning of a scene, which contains $262\frac{1}{2}$ lines.

Field strength meter (FSM) A portable device used for detecting the presence of RF signals near an antenna.

File transfer The transfer of files, records, or whole databases from one place to another.

Filter A frequency-selective circuit designed to pass some frequencies and reject others.

Filtering The use of circuits to separate frequencies.

Final power amplifier The high-power amplifier that drives to an antenna or other load in an RF system.

Finite impulse response (FIR) filter (nonrecursive filter) A popular DSP filter whose output is a function of the sum of products of the current input samples.

Firewall A piece of software that monitors all incoming data to a network for the purpose of establishing whether the data meets the standards or guidelines set by an organization or individual for admission to the network and computers.

Flash converter A fast analog-to-digital converter that uses a large resistive voltage divider and multiple analog comparators.

Flicker noise An electrical disturbance that occurs in resistors and conductors and is the result of minute random variations of resistance in the semiconductor material.

FM/FM system A system that uses the FM of the VCO subcarriers as well as the FM of the final carrier in a telemetry system.

Folded dipole A variation of the half-wave dipole consisting of two parallel conductors connected at the ends with one side open at the center.

Forward AGC A method of reducing gain by increasing the collector current.

Forward error correction (FEC) The process of detecting and correcting errors at the receiver so that retransmission is not necessary.

Forward (incident) power The power sent down a transmission line toward the load.

Foster–Seeley discriminator One of the earliest FM demodulators, no longer widely used.

Fourier theory The process for accurately analyzing and expressing complex nonsinusoidal signals in terms of harmonics using calculus.

Frame The portion of a packet that contains the data to be communicated plus addressing and error detection codes.

Frame rate The number of frames transmitted in 1 second.

Frame relay A packet switching system used by the major telecommunications companies for data transfer.

Free-running frequency The normal operating frequency of a VCO, as determined by internal frequency-determining components.

Frequency In electronics, the number of cycles of a repetitive wave that occur in a given period of time.

Frequency conversion The process of translating a signal to a higher or lower frequency while retaining the originally transmitted information.

Frequency-correcting network, predistorter, or 1/f filter A low-pass filter that causes the higher modulating frequencies to be attenuated.

Frequency counter A test instrument that measures signal frequency.

Frequency deviation (f_d) The amount of change in carrier frequency produced by the modulating signal.

Frequency demodulator A circuit used to recover the original modulating signal from an FM transmission.

Frequency divider A circuit with an output frequency that is some integer submultiple of the input frequency.

Frequency division multiplexing (FDM) or frequency division multiple access (FDMA) A type of multiplexing in which multiple signals share the bandwidth of a common communication channel.

Frequency-domain display A plot of signal amplitude versus frequency.

Frequency-hopping (FH) SS A type of spread spectrum in which the frequency of the carrier of the transmitter is changed according to a predetermined sequence at a rate less than the serial binary data modulating the carrier.

Frequency modulation Modulation in which the instantaneous frequency of the modulated waves differs from the carrier frequency by an amount proportional to the instantaneous value of the modulating wave.

Frequency modulator A circuit that varies carrier frequency in accordance with the modulating signal.

Frequency multiplier A circuit with an output frequency that is some integer multiple of the input frequency.

Frequency-multiplier diode A diode designed primarily for frequency-multiplier service. Varactor diodes and step-recovery diodes are frequency-multiplier diodes.

Frequency reuse A technique used for effectively increasing the bandwidth and information-carrying capacity of a satellite. A means by which cells within a cellular telephone system can share the same frequency channel. *See also* Spatial multiplexing.

Frequency-shift keying (FSK) A type of modulation widely used in the transmission of binary data. Used primarily in low-speed modems, FSK uses two sine-wave frequencies that represent binary 0s and 1s.

Frequency spectrum The electromagnetic spectrum.

Frequency synthesizer A signal-generating circuit whose output can be changed in discrete increments by digital means.

Friis's formula The formula used to calculate the overall noise performance of a receiver or of multiple stages of RF amplification.

Front-to-back (F/B) ratio The ratio of the power radiated in the forward direction to the power radiated in the backward direction.

Function generator A signal generator that generates sine waves, square waves, and triangular waves over a frequency range of about 0.001 Hz to about 2 MHz.

G

GaAsFET A junction field-effect transistor made with gallium arsenide. *See also* Gallium arsenide.

Gain Amplification. The ratio of output to input of an amplifier, circuit or antenna.

Gallium arsenide (GaAs) semiconductor A compound semiconductor used in microwave transistors and LEDs made specifically for fiber-optic applications. It emits light at 1.3 μm.

GASFET Another name for MESFET.

Gateway An internetwork device that acts as an interface between two LANs or between a LAN and a larger computer system. This two-way translator allows different types of systems to communicate.

Gaussian filtered MSK (GMSK) A method of prefiltering MSK in which harmonic content and overall signal bandwidth are reduced.

General packet radio service *See* GPRS.

Geocenter The center of the earth.

Gigabit Ethernet The most recent version of Ethernet, capable of achieving 1000 Mbps or 1 Gbps over Category 5 UTP or fiber-optic cable. Also called 1000Base-T Ethernet.

Gilbert transconductance cell The balanced mixer made with transisters and used in most IC mixers.

Global Positioning System (GPS) Satellite-based navigation system that can be used by anyone with an appropriate receiver to pinpoint his or her location on earth.

Global System for Mobile Communications *See* GSM.

GPRS (general packet radio service) The most popular 2.5G technology, designed to work with GSM phones. It transmits data as well as digitized voice.

GPS receiver A complex superheterodyne microwave receiver designed to pick up GPS signals, decode them, and compute the location of the receiver.

Graded index cable A cable whose index of refraction for the core varies smoothly and continuously over the diameter of the core.

Grid A control element added between the plate and the cathode of a vacuum tube to control the current flow.

Ground The common reference point for most voltages in a circuit.

Ground control equipment (GCE) subsystem Equipment at a satellite earth station, which is used for demodulating and demultiplexing the received signals.

Ground loop An unintentional circuit that occurs when multiple circuits or pieces of equipment are connected to a common ground at different points.

Ground plane A conducting surface or array of conductors over $\frac{1}{4} \lambda$ long used as one element in a vertical antenna. The earth becomes known as the ground plane once a good electrical connection to earth has been made.

Ground-plane antenna. *See* Marconi vertical antenna.

Ground station Also known as an earth station, the terrestrial base of the satellite communication system.

Ground wave A wave that follows the curve of the earth and can therefore travel to distances beyond the horizon.

Groupware A set of programs that allow two or more individuals working on a single project to share databases, exchange messages, manage work flow, and maintain a calendar.

GSM (Global System for Mobile Communications or Group Special Mobile) The most widely used 2G digital phone system.

Gunn diode A thin piece of N-type gallium arsenide (GaAs) or indium phosphide (InP) semiconductor which forms a resistor when voltage is applied to it. Also called transferred-electron device (TED). Uses its negative resistance characteristics to generate a microwave signal.

Gyrotron The only device available for power amplification and signal generation in the millimeter-wave range.

H

Hamming code A popular method of forward error correction (FEC), a code that uses extra bits added to a transmitted word and processed to identify and correct bit errors.

Handoff A switching at the mobile telephone switching office (MTSO) from a weaker cell to a stronger cell, to provide optimum transmission and reception.

Handset The portion of a telephone that contains the speaker and microphone, which serve as transmitter and receiver.

Handshaking The exchange of status information between transmitter and receiver in a digital system.

Harmonic A sine wave whose frequency is some integer multiple of a fundamental sine wave.

Hartley's law A law that states that the greater the bandwidth of a channel, the greater the amount of information that can be transmitted in a given time.

Harvard architecture A type of microprocessor in which there are two memories: a program or instruction memory and a data memory. There are two data paths between the memories.

Hash function or algorithm A simple method of encryption that converts a plaintext message into a compressed binary number representing that data.

Helical antenna A wire helix in which the diameter of the helix is typically one-third wavelength. Spacing between turns is about one-quarter wavelength.

Hertz antenna *See* Dipole antenna.

Heterodyne processing A method of translating the incoming TV signal to a different frequency.

Heterodyne processor A module used by cable companies to translate receiver signals to the desired channel.

Heterodyning The function performed by a mixer circuit in frequency conversion.

Heterojunction bipolar transistor (HBT) A transistor that makes high-frequency amplification possible in both discrete form and integrated circuits. It is formed with two different types of semiconductor materials.

High-definition TV (HDTV) or digital TV (DTV) System that transmits fine picture detail and enhanced sound.

High electron mobility transistor (HEMT) A variant of the MESFET that extends the frequency range beyond 20 GHz by adding an extra layer of semiconductor material.

High frequency (HF) The frequency range from 30 to 300 MHz.

High-level AM A variation of voltage and power in the final RF amplifier stage of a transmitter.

High-level data link control (HDLC) protocol A popular modulation technique used in modems for increasing the number of bits per baud.

High-pass filter A filter that passes frequencies above the cutoff but rejects frequencies below it.

High-power amplifier (HPA) An amplifier used to increase signal level.

High-Speed Token Ring (HSTR) A new version of Token Ring that runs at 100 Mbps.

Horn antenna An antenna created by flaring the end of the waveguide. The more gradual the flair, the better the impedance match and the lower the loss.

Hot carrier diode Also known as a Schottky diode, this diode is made with N-type silicon on which is a thin metal layer.

Hub A LAN accessory that functions like a central connecting box. It is designed to receive the cable inputs from various PC nodes and connect them to the server.

Huygens' principle The principle based on the assumption that all electromagnetic waves radiate as spherical wavefronts from a source. When the waves encounter an obstacle, they pass around it, above it, and on either side of it.

Hybrid circuit A popular form of MMIC that combines all amplifier ICs connected to microstrip circuits and discrete components of various types. Also, a special transformer used to convert signals from the four wires of a transmitter and receiver into a signal suitable for a single two-line pair to the local loop.

Hybrid fiber cable (HFC) system A newer system of cable TV in which a fiber-optic cable and a coaxial cable rather than just a coaxial cable is used.

Hybrid ring A special form of microstrip with four taps or ports on the line spaced at one-quarter wavelength intervals which can be used as inputs or outputs.

Hybrid T A waveguide-like device with combined series and shunt T sections that is used to permit simultaneous use of an antenna by both a transmitter and a receiver.

Hypertext On the Web, a method that allows different pages or sites to be linked.

Hypertext transfer protocol (http) The first part of a URL. It specifies the communication protocol to be used.

I

I and Q The in-phase and quadrature signals used in most digital modulation and demodulations circuits.

I color signal A TV color signal with the following specifications: 60 percent red, 28 percent green, and -32 percent blue.

IC balanced modulator A balanced modulator circuit using differential amplifiers.

IC electronic telephone A telephone unit in which all the circuits are fully integrated on a single chip of silicon.

IC receiver An integrated circuit radio receiver. A receiver on a single chip.

IF amplifier An amplifier that uses tuned circuits with crystal, ceramic, or SAW filters in order to provide good selectivity. *See also* Intermediate frequency.

Image An interfering RF signal that is spaced from the desired incoming signal by a frequency that is two times the intermediate frequency above or below the incoming frequency.

Image processing The processing of data by algorithms that are used to improve the resolution clarity, speed, or signal-to-noise ratio of a transmitted picture.

Image reject mixer A special type of mixer used in designs in which images cannot be tolerated.

IMPATT diode A PN-junction diode that is made of silicon, GaAs, or InP and operates with a high reverse bias that causes it to avalanche, or break down. It is available with power ratings to about 25 W and to frequencies as high as 30 GHz.

Impedance The total opposition of the components produced by combining resistance, inductance and/or capacitance in series or parallel.

Impedance matching A procedure done to ensure that maximum power transfer will take place between a transmitter or receiver and an antenna with transmission line.

Impedance-matching network A circuit used to transfer maximum signal power from one circuit to another.

Impedance-matching pad A resistive T network that provides the correct match between the receiver and input and the generator.

Impedance meter An instrument (often bridge type) that accurately measures the impedance of a circuit, a component, or an antenna at RF frequencies.

Impedance of space *See* Wave impedance.

Incident power *See* Forward power.

Incident ray Light ray from a light source. The input wave from a generator to the end of a fiber-optic transmission line.

Index of refraction A figure obtained by dividing the speed of a light wave in a vacuum by the speed of a light wave in a medium that causes the wave to be bent. The ratio of the speed of light in air to the speed of light in the substance.

Indirect FM The FM produced by a phase modulator.

Indium phosphide (InP) semiconductor A compound semiconductor material made with indium and phosphorus.

Inductor An electronic component made by winding multiple turns of a wire on a form. Also called a *coil* or *choke.*

Industrial noise Electrical disturbance created by manufactured equipment such as automotive ignition systems, electric motors, and generators.

Infinite impulse response (IIR) filter (recursive filter) A recursive DSP filter that uses feedback. Each new output sample is calculated using both the current output and past samples.

Information In electronic communication systems, the message or data.

Infrared communication A form of optical communication that utilizes signals from 0.7 to 1000 μm.

Infrared LAN The least expensive form of wireless LAN. It uses infrared transceivers.

Infrared (IR) wireless The use of infrared light for short-distance data communication.

Infrared spectrum The region of the optical spectrum just below visible light, from 0.7 to 1000 μm.

Injection laser diode (ILD) A PN-junction diode usually made with GaAs. It produces a low-level light over a broad frequency range with a low-level forward-biased current.

Insertion loss The loss a filter introduces to the signals in the passband (usually in decibels). Passive filters introduce attenuation because of resistive losses in the components and the voltage divider effect.

Institute of Electrical and Electronics Engineers (IEEE) A professional society that establishes and maintains a wide range of electrical, electronic, and computing standards.

Instrumentation subsystem An extension of a satellite telemetry system; a general term for all the electronic equipment used to deal with the information transmitted back to the earth station.

Integrated circuit *See entries under* IC.

Integrated services digital network (ISDN) system A digital communication interface designed to replace the local analog loop used in the public switched network but rarely used.

Intensity modulation A type of modulation used when the information or intelligence signal controls the brightness of a laser transmitter. A form of amplitude modulation.

Interexchange carriers (IXCs) Long-distance telephone service carriers.

Interlaced scanning The meshing of two sequentially scanned fields on a TV CRT.

Intermediate frequency (IF) The lower frequency to which superheterodyne receivers convert all incoming signals. The mixer output of a superheterodyne receiver, it is the difference between the input signal frequency component and the local oscillator frequency.

Intermediate-power amplifier (IPA) An amplifier in a chain of several amplifiers used to boost the power level of a signal before transmission. The amplifiers are usually referred to as low-power, intermediate-power, and high-power amplifiers.

Intermodulation distortion Distortion resulting from the generation of new signals and harmonics arising from circuit nonlinearities.

Intermodulation products The unintentional mixing of signals and harmonics to produce undesired interfering signals.

Internal noise Noise caused by heat in electronic components in a receiver such as resistors, diodes, and transistors.

International Organization for Standardization An organization that has developed a framework, or hierarchy, known as the open systems interconnection model, that defines how data can be communicated.

International Telecommunications Union (ITU) An agency of the United Nations whose duties include setting standards for various areas within the communication field. Formerly known as the CCITT, it establishes standards to ensure compatability of telecommunication equipment of different manufacturers.

Internet A worldwide connection of computers by means of a complex network of many networks.

Internet backbone A collection of worldwide networks of high-speed fiber-optic cable that carry all Internet traffic.

Internet Engineering Task Force (IETF) An organization that develops and standardizes technologies and protocols related to Internet data transfer.

Internet service provider (ISP) A company set up especially to tap into the Internet.

Interoperability The ability of equipment from one manufacturer to work compatibly with that of another.

Intersymbol interference (ISI) Pulses that blur into one another cause this type of interference. The longer the cable or the higher the bit rate, the greater the distortion.

Ionosphere The region of the upper atmosphere where ultraviolet radiation from the sun causes the atmosphere to become electrically charged, or ionize.

IP address Internet protocol address. A 32-bit binary number assigned to a specific computer to identify its presence on the Internet to a router.

IPv4 (Internet Protocol version 4) The currently most widely used IP protocol. Its destination address size, 32 bits, limits the number of users. IPv4 will be replaced by IPv6 that uses a 128-bit address.

IrDA system The most widely used IR data communication system.

Iridium system An advanced satellite cellular telephone system that uses a constellation of 66 satellites in six polar orbits with 11 satellites per orbit 420 miles above the earth.

iSCSI Internet Small Computer Systems Interface (SCSI). A version of the popular SCSI parallel data interface used to connect hard drives and tape drives to computers that uses Ethernet for connectivity.

IS-95 CDMA A TIA cell phone standard, also known as CDMA One, which uses spread spectrum.

IS-136 TDMA The Telecommunications Industry Association (TIA) standard that describes the TDMA cell phone system.

Isolator A type of circulator with one input and one output.

Isotropic radiator A theoretical point source of electromagnetic energy.

J

Jacket The protective insulation that surrounds the core of a cable with its cladding.

JBOD Just a bunch of disks Term used to describe a collection of hard disk drives used externally for a computer or server to store massive amounts of data.

Jitter A type of noise that shows up as a time variation of the leading and trailing edges of a binary signal.

K

Kermit protocol A once popular asynchronous protocol that requires that every packet that is sent be acknowledged by the receiver as read correctly.

Key A secret binary number used in the encryption process that must be known or exchanged to decrypt the data.

Key system A small telephone system designed to serve from 2 to 50 user telephones within an organization.

L

L network An inductor and a capacitor connected in various L-shaped configurations used for impedance matching.

Laser A single-frequency light source that produces a very narrow beam of brilliant light of a specific wavelength.

Laser diode PN-junction diode usually made with a GaAs compound that produces coherent laser light. It is the most widely used light source in fiber-optic systems.

Laser transmitter The circuitry associated with the laser in a fiber-optic transceiver.

Latitude The angle between a line from a given point on the surface of the earth to the geocenter and the line between the geocenter and the equator.

Lattice modulator *See* Diode ring.

Law of reflection This law states that the angle of incidence is equal to the angle of reflection.

***LC* filter** A filter made with inductors and capacitors.

***LC* oscillator** A signal-generating circuit whose frequency is set by an inductor-capacitor combination. Colpitts, Hartley, and Clapp circuits are examples of this type of oscillator.

LED transmitter A type of transmitter employing a light-emitting diode that is used for short-distance, low-speed, digital fiber-optic systems.

Left-hand circular polarization (LHCP) The term that describes radiation which leaves an antenna with a counterclockwise rotation.

Light A type of electromagnetic radiation that occupies the part of the frequency spectrum lower in frequency than x-rays but higher in frequency than microwaves and includes infrared, visible, and ultra-violet light.

Light detector (photocell) A light-sensitive device that converts light pulses into an electrical signal.

Light-emitting diode (LED) A PN-junction semiconductor device that emits light when forward-biased.

Light receiver A type of receiver in which current through the photodiode, generated when light is sensed, produces a current that is amplified in an op amp.

Light transmitter A laser or LED and its associated driving circuitry.

Limiter A circuit used to remove any amplitude variations on the FM signal before the signal is applied to the demodulator.

Linear amplifier An amplifier that provides an output that is an identical, enlarged replica of the input. An amplifier whose output is a straight line function of its input.

Line-of-sight communication A communication signal that travels in a straight line directly from the transmitting antenna to the receiving antenna.

Loading coil A series inductor used to bring a short antenna into resonance at the desired frequency.

Local access and transport areas (LATAs) Designated geographical areas to which local exchange carriers (LECs) provide telephone services.

Local-area network (LAN) An information network that transmits data from one place to another using some form of cable or wireless communication signal. A group of PCs in an office or company that are connected in order to share data and other resources.

Local exchange *See* Central office.

Local exchange companies (or carriers; LECs) Companies to which local exchanges that provide local telephone service are connected.

Local loop (subscriber loop) A two-wire, twisted-pair connection between a telephone and the central office.

Local oscillator Either a conventional *LC* tuned or crystal oscillator or a frequency synthesizer used to produce a continuous sine wave. Usually drives a mixer in a receiver.

Lock range The range of frequencies over which a PLL can track an input signal and remain locked.

Log-periodic array An antenna in which the lengths of the driven elements vary and are related logarithmically. This type of antenna also provides a very wide bandwidth as well as gain and directivity.

Long-distance operation Service offered by interexchange carriers that provide the interconnection for any inter-LATA connections.

Longitudinal redundancy check (LRC) The process of logically adding, by exclusive ORing, all the characters in a specific block of transmitted data used for error detection.

Loran system A highly accurate marine navigation satellite system using coastal stations along U.S. water borders.

Losses Attenuation or weakening of a signal in a circuit, coaxial, or fiber-optic cable.

Low frequency (LF) The frequency range from 30 to 300 kHz.

Low-level AM Amplitude modulation at a low power level that must be amplified by a linear amplifier considerably before being transmitted.

Low-noise amplifier (LNA) An RF amplifier that uses special low-noise transistors to provide initial amplification in a receiver.

Low-pass filter A circuit that introduces no attenuation at frequencies below the cutoff frequency but greatly attenuates all signals with frequencies above the cutoff.

Luminance (*Y*) signal The signal that results when different levels of light along each scan line are transmitted as shades of gray between black and white.

M

Magnetic field The invisible force field created by a magnet.

Magnetron A microwave vacuum tube that is a diode tube with built-in cavity resonators and an extremely powerful permanent magnet.

Manchester code A type of the binary line code that prevents the DC voltage level on a transmission cable from building up to an unacceptable level. A unipolar or bipolar coding system widely used in LANs. Also called biphase encoding.

Marconi (ground-plane) vertical antenna A quarter-wavelength vertical radiator used to achieve vertical polarization and omnidirectional characteristics. *See also* Quarter-wavelength vertical antenna.

Marker capability The term describing the condition when a sweep generator has one or more reference oscillators that enable the response curve to be actively interpreted.

Matched line The ideal situation in which a transmission line terminates in a load that has a resistive impedance equal to the characteristic impedance of the line.

Matching stub *See Q* section.

Media Accesses Controller (MAC) The circuitry making up Layer 2 of an OSI type protocol.

Medium frequency (MF) The frequency range from 300 to 3000 kHz.

MESFET A junction field-effect transistor made with gallium arsenide (GaAs). Also called GASFET.

Metal-oxide-semiconductor field-effect transistor (MOSFET) A transistor often used in switching amplifier applications. It provides extremely low power dissipation even with high current. The basic building block of most CMOS digital integrated circuits.

Metal-semiconductor field-effect transistor (MESFET) *See* GaAsFET.

Metropolitan-area network (MAN) A network that covers a city, town, or village. A cable TV system is an example of a MAN.

Microcom Networking Protocol (MNP) A series of protocols used with asynchronous modems that specify ways to handle error detection and correction and how to specify if data compression is used.

Micrometer (μm) or micron One-millionth of a meter. Light waves or small dimensions in integrated circuits are expressed in terms of this unit.

Microstrip A flat conductor separated from a conducting ground plane by an insulating dielectric used as a transmission line or tuned circuit. It is preferred for reactive circuits at microwave frequencies.

Microstrip tuned circuit A section of microstrip transmission line used as a resonant circuit.

Microwaves The frequencies from 1 GHz to 30 GHz.

Microwave antenna Any antenna used at frequencies above 1 GHz.

Microwave transistor A transistor that operates like other transistors, whether it is a bipolar or an FET type. The differences between this and a lower-frequency transistor are geometry and packaging that is optimized for frequencies above 1 GHz.

Microwave transistor amplifier The transistor amplifier used in the front end of a microwave receiver to provide initial amplification for the mixer. The typical gain range is 20 to 25 dB. Also a power transistor used in microwave transmitters.

Microwave tube A type of tube that includes the klystron, magnetron, and traveling-wave tube.

Millimeter wave An electromagnetic wave at a frequency above 40 GHz. The part of the frequency spectrum lower in frequency than x-rays but higher in frequency than microwaves (about 30 to 300 GHz).

Minimum shift keying (MSK) An improved version of FSK or continuous phase frequency-shift keying in which the signals are fully synchronized with one another.

Minor lobe The portion of the directional response curve of an antenna that denotes the transmission of RF energy over a particular range of angles different from the major direction of radiation of the antenna.

Mixer (converter) Any device or circuit whose output does not vary linearly with the input. A device that translates the signal up or down to another frequency.

MLT-3 encoding The encoding method used in Fast Ethernet.

Mobile identification number (MIN) The telephone number assigned to a cellular unit.

Mobile telephone switching office (MTSO) The master control center to which the cells in a cellular phone system are connected by telephone lines or microwave radio relay links.

Modal dispersion In a multimode step index cable, an attenuation or stretching of a pulse at the end of the cable.

Modem A device that converts binary signals into analog signals capable of being transmitted over the telephone lines and demodulates them, re-creating the binary output.

Modulation The process by which a baseband voice, video, or digital signal is modified by another, higher-frequency signal called the carrier.

Modulation index The ratio of the frequency deviation to the modulating frequency.

Modulator A circuit used to vary some aspect of a radio carrier or light signal in accordance with the modulating baseband signal.

Monochrome CRT A black and white picture tube, a vacuum tube called a cathode-ray tube.

Monolithic microwave integrated circuit (MMIC) A circuit that incorporates two or more stages of FET or bipolar transistors made on a common chip to form a multi-stage amplifier.

Morse code A series of dots and dashes that represent letters of the alphabet, numbers, and punctuation marks.

MOSFET See Metal-oxide-semiconductor field-effect transistor.

Motion Picture Experts Group (MPEG) The organization that establishes technical standards for movies and video.

μ-law compander The type of companding used in telephone systems in the United States and Japan.

Multicarrier modulation (MCM) See Orthogonal frequency division multiplexing.

Multicasting The ability of IP to move fast audio and video data over the Internet from a single source to multiple destinations.

Multimode graded index cable A type of cable that has several modes, or paths, of transmission through the cable but is orderly and predictable.

Multimode step index cable A type of cable used for short to medium distances at relatively low pulse frequencies. It is a widely used type of fiber-optic cable.

Multipath interference Fading that occurs when a transmitted signal takes multiple paths to the receiver because of reflections. Also called Rayleigh fading.

Multiple data message format (MDMF) A message format used in caller ID.

Multiple input multiple output (MIMO) A wireless technique that makes use of multiple antennas at the transmitter and receiver to make use of multipath signals to improve communications range and reliability.

Multiplexer A device or circuit that time- (or frequency-) shares a single channel with multiple signals.

Multiplexing A technique which allows more than one signal to be transmitted concurrently over a single medium.

Multistage integrated circuit amplifier A type of small-signal microwave amplifier or monolithic microwave integrated circuit (MMIC).

Multistation access unit (MAU) The basic wiring hub in a Token-Ring network.

Muting circuit See Squelch circuit.

N

N connector Widely used in RF applications, this type of connector is used on larger cables such as RG-8/U.

Nanometer (nm) One-billionth of a meter. Light waves are often expressed in terms of nanometers.

Narrow-band FM (NBFM) A special case of FM in which the modulation process produces only a single pair of significant sidebands like those produced by AM.

National Telecommunications and Information Administration (NTIA) A regulatory body of the U.S. government whose function is to regulate electromagnetic emissions for the government and military services.

Navigation satellite A satellite that provides accurate position information to ships, airplanes, and land-based vehicles.

Near field The radio wave within about one wavelength of the radiating antenna.

Near field communications (NFC) A short range wireless technology similar to RFID that uses the near magnetic field of a radio wave for data transfers at distances to 8 inches. Used for payment and access when used in a cell phone.

NE566 IC VCO An IC voltage-controlled oscillator.

NE602 IC mixer A popular circuit that uses bipolar transistors connected as Gilbert transconductance cells.

Network analyzer A test instrument designed to analyze linear circuits. It is a combination instrument that contains a wide-range sweep sine-wave generator and a CRT output that displays not only frequency plots like a spectrum analyzer but also plots of phase shift versus frequency.

Network attached storage (NAS) A group of hard disks usually connected to a computer or server via standard Ethernet.

Network interface card (NIC) A device that provides the I/O interface between each node on a network and the network wiring.

Neutralization A process through which one signal, equal in amplitude to an original signal, is fed back, resulting in the two signals canceling one another.

Noise In electronics, any signal that is a mixture of many frequencies at many amplitudes that gets added to a radio or information signal as it is transmitted from one place to another or as it is processed. Random signal variations picked up by a receiver or caused by thermal agitation and other conditions inside the receiver circuitry. Random, undesirable electronic energy that enters a communication system via the communication medium and interferes with the transmitted message.

Noise bridge A bridge circuit driven by a random noise voltage source that has an antenna or coaxial cable as one leg of the bridge. Used to make antenna characteristic impedance measurements and measurements of coaxial cable velocity factor and length.

Noise factor The ratio of the S/N power at the input to the S/N power at the output.

Noise figure The noise factor as expressed in decibels.

Noise immunity Resistance to noise. Digital signals are more immune to noise than analog signals.

Noise margin The amount of voltage between the highest expected noise voltage and a threshold level above which triggering occurs.

Noise temperature Another way to express noise in an amplifier or receiver using the Kelvin scale.

Nonreturn to zero encoding (NRZ) A method of encoding in which the signal remains at the binary level assigned to it for the entire bit time.

North American TDMA (NA-TDMA) *See IS-136 TDMA.*

Notch, band-reject, or bandstop filter A type of filter that rejects or stops frequencies over a narrow range but allows frequencies above and below to pass.

Number assignment module (NAM) A programmable read-only (PROM) chip in a cellular radio.

Numerical aperture (NA) A number less than 1 that indicates the range of angles over which a particular cable will work.

Nyquist frequency The minimum sampling frequency at which the high-frequency information in an analog signal can be retained. It is equal to two times the highest signal frequency.

O

Octet An 8-bit number, word or block of data. Another name for byte.

Omnidirectional antenna An antenna that transmits equally well in all horizontal directions. It has a circular horizontal transmission pattern.

One-shot multivibrator A pulse generator circuit that produces one output pulse whose duration has been set to some desired interval.

ON–OFF keying (OOK) A form of amplitude modulation where the carrier is turned off or on by a modulating binary signal.

Open systems interconnection (OSI) protocol A framework or hierarchy designed to establish general interoperability guidelines for developers of communication systems and protocols.

Optical communication system A communication system that uses light to transmit information from one place to another.

Optical-electrical-optical (OEO) conversion *See* Regeneration.

Optical signal analyzer An optical test instrument that provides the OTDR measurement and is also a sampling oscilloscope capable of displaying signals of more than 10 Gbps.

Optical spectrum A group of signals with frequencies higher than 300 GHz. The range of frequencies that consists of infrared, visible, and ultraviolet light (range of 3×10^{11} to 3×10^{16} Hz).

Optical time domain reflectometer (OTDR) A device that generates a light pulse and sends it down a fiber-optic cable to be tested. If there is a defect in the line, light will be reflected. The time between the generated and reflected pulses is measured.

Optics The aspect of physics concerned with the behavior of light.

Orbit The path taken by a satellite as it circles the earth.

Orthogonal frequency division multiplexing (OFDM) A modulation method by which data is transmitted by simultaneously modulating segments of the high-speed bit stream onto multiple carriers spaced throughout the channel bandwidth. Also called multicarrier modulation (MCM).

Oscillator diode A special semiconductor device (Gunn diode, TRAPPAT diode, etc.) that generates a microwave signal, usually a sine wave at the resonant frequency, when used with a tuned circuit element such as a waveguide, microstrip, microstrip, or stripline.

Overmodulation In amplitude modulation, the term describing the condition when the modulating voltage is greater than the carrier voltage.

Overranging In amplitude modulation, the condition that occurs when the count capability of a counter is exceeded during the count interval.

Oversampling converter A type of converter that uses a clock or sampling frequency that is many times the minimum Nyquist rate required for other types of converters. The $\Sigma\Delta$ converter is an example of an oversampling converter.

Overtone Similar to a harmonic. An integer multiple of a fundamental oscillation frequency.

Overtone crystal A crystal that is cut in a special way so that it optimizes its oscillation at an overtone of the basic crystal frequency.

Overvoltage protection The protection afforded by circuits and components that protect the subscriber line interface circuits from electrical damage.

P

Packet A unit of data made up of the frame of data and additional bytes at the beginning and end that contain addresses and error correction codes.

Paging receiver A small, battery-operated superheterodyne receiver that enables individuals to be signaled, wherever they may be.

Paging system A popular communication system that operates in the simplex mode, for it broadcasts signals only to small battery-operated receivers.

PAM demultiplexer A circuit that recovers data multiplexed with PAM.

PAM/FM system An arrangement in which PAM signals frequency-modulate a carrier.

PAM/FM/PM system An arrangement in which PAM signals frequency-modulate a subcarrier that, in turn, phase modulates the carrier.

Parabolic antenna An antenna in which the energy radiated by the horn is pointed at a parabolic reflector that focuses the energy into a narrow beam and reflects it toward its destination. When receiving, the antenna focuses signals on the horn.

Parallel data transmission A type of data transfer in which all the bits of a code word are transferred simultaneously. This method of data transfer is not practical in long-distance communication systems.

Parallel resonant circuit A selective circuit formed when the inductor and capacitor are connected in parallel with the applied voltage.

Parallel-wire line A transmission line made up of two parallel conductors separated by a space of $\frac{1}{2}$ in. to several inches.

Parasitic array A basic antenna connected to a transmission line plus one or more additional conductors that are not connected to the transmission line. A Yagi antenna.

Parasitic oscillation A type of aberrant oscillation that occurs when a circuit oscillates at a higher frequency unrelated to the tuned frequency.

Parity A widely used system of error detection. Each character transmitted contains 1 additional bit.

Passband The frequency range over which a filter or amplifier passes signals. This is the range between cutoff frequencies or between the cutoff frequency and zero.

Patch antenna An antenna made with microstrip on PCBs. It is a circular or rectangular area of copper separated from the ground plane by the thickness of the PCB's insulating material.

Path attenuation Loss created by the distance of free space between a transmitting and receiving antenna, usually expressed as a power ratio or in decibels.

Pattern generator A device that produces fixed binary bit patterns in serial form to use as test signals in data communication systems.

PCM demultiplexer A circuit that recovers data multiplexed using PCM.

Peak envelope power (PEP) The maximum power produced on voice amplitude peaks in single sideband communication.

Peer-to-peer configuration A type of LAN that is smaller and less expensive than client-server LANs and provides a simple way to build a network within an office.

Percentage of modulation The ratio of the amplitude of the modulating signal to the amplitude of the carrier.

Perigee Lowest point of a satellite orbit above the earth.

Permittivity The dielectric constant of the material between two conductors.

Personal-area network (PAN) A very small network created informally or on an ad hoc basis. Nowadays, all PANs are wireless.

Personal communication service (PCS) The band of frequencies near 1900 MHz that is used for cellular telephones.

Phase modulation (PM) The signal that results when the amount of phase shift of a constant-frequency carrier is varied in accordance with a modulating signal.

Phase modulator A circuit often used instead of direct FM because the carrier oscillator can be optimized for frequency accuracy and stability.

Phased array An antenna system made up of a large group of similar antennas on a common plane to provide controllable gain and directivity.

Phased array radar A type of radar that provides flexibility in scanning narrow sectors and tracking multiple targets by using a phased antenna array.

Phase-locked loop (PLL) A frequency- or phase-sensitive feedback control circuit used in frequency demodulation, frequency synthesizers, and various filtering and signal-detection applications.

Phase-shift keying (PSK) A type of modulation in which the binary signal changes the phase shift of a sine-wave character depending upon whether a binary 0 or a binary 1 is to be transmitted. Also called binary phase-shift keying (BPSK).

Phasor A line or arrow whose length is proportional to the peak value of the sine wave being represented and its angle indicating phase position.

Photocell *See* Light detector.

Photodiode Silicon PN-junction diode that is sensitive to light. It is the most widely used light sensor.

Phototransistor A light-sensitive transistor used for light detection.

Physical optics A term referring to the ways that light can be processed.

PHY Abbreviation of the physical layer of a data communications system based on the OSI system.

π network Network configuration resembling the Greek letter π.

π/4-DQPSK modulation A method of modulation used in IS-136 TDMA.

Piconet The linking of one Bluetooth device that serves as a master controller to up to seven other Bluetooth devices.

Pierce oscillator An oscillator configuration using field-effect transistors or bipolar transistors with crystals.

Piezoelectric effect Vibration that occurs when a crystal is excited by an AC signal across its plates.

Pilot carrier A low-level carrier signal. It is used to help in the recovery of DSB and SSB signals.

PIN diode A device used as a switch or variable resistor in a microwave circuit. This PN-junction diodes has an I layer between the P and N sections.

PIN diode modulator A modulator using a special type of silicon junction diode designed for use at frequencies above approximately 100 MHz.

Pixel A tiny dot of light, thousands of which make up a television or computer monitor screen. Each pixel can be any of hundreds of colors.

PL-259 connector A male connector that fits on the end of a coaxial cable and provides a way to attach the shield braid and inner conductor.

Plaintext The name given to any message or data to be encrypted.

Plan-position indicator (PPI) A type of CRT display that shows both the range and the azimuth of a target on a radar set.

Plenum cable Coaxial or twisted-pair cable run between floors and across ceilings through special channels or chambers built into the structure.

PLL demodulator A circuit that recovers FM signals. It has an S/N ratio better than that of other FM detectors.

Point-contact diode A semiconductor diode made from a piece of P-type semiconductor material and a fine tungsten wire that makes contact with the semiconductor material at a small point of contact. Used in radar receivers.

Point of presence (POP) office A serving office used to provide interconnections to the IXCs.

Polar orbit A satellite orbit that passes over the north and south poles.

Polarization The orientation of the electric field in a radio wave with respect to the earth.

Pole A frequency at which there is a high impedance in the circuit.

Posigrade orbit An orbit that moves in the same direction as the earth's rotation.

Power (flux) budget An accounting of all attenuation and gain in a fiber-optic system.

Power combiner A component, circuit, or waveguide assembly that sums the power outputs of two or more circuits.

Power divider A component, circuit, or waveguide assembly that equally divides a common input amongst two or more outputs.

Power meter A commonly used RF test meter that measures the forward and reflected power output of a circuit.

Power subsystem The basic power source of a satellite system. Most satellites use solar panels for its basic power source.

Preamble Multiple bytes of data in a packet protocol that help to establish clock synchronization.

Pre-emphasis A technique that helps to offset high-frequency noise interference in FM radio.

Prescaler A special frequency divider normally used between the high-output frequency of a VCO and the programmable part of a divider in a PLL frequency synthesizer. Also a divider used ahead of a frequency counter.

Prescaling A frequency division technique that involves the division of the input frequency by a factor that puts the resulting signal into the normal frequency range of the counter.

Primary rate interface (PRI) An ISDN connection type made up of one twisted pair. This method uses 23 bearer (B) channels and one (D) channel multiplexed on the line.

Prime meridian Special meridian used as a reference point for measuring longitude.

Private automatic branch exchange (PABX) A telephone switching system that is

extremely small, fast, and efficient. Many PABXs are designed to carry digital signals and analog voice signals.

Private branch exchange (PBX) A telephone switching system, dedicated to a single customer, which provides telephone switching and links inside an organization and also connects to the standard public telephone network.

Private telephone system Telephone service provided to companies or large organizations with many employees and many telephones.

Product detector A balanced modulator used to recover SSB or DSB signals.

Progressive scanning Line-after-line scanning used in HDTV and computer monitor screens.

Protocol Rules and procedures used to ensure compatibility between the sender and receiver of serial digital data.

Protocol analyzer A test instrument used to capture and analyze the data transmitted in a data communication system.

Public key encryption (PKE) A method of encryption that uses two keys, one public and the other private.

Pulse-amplitude modulation (PAM) Modulation that produces a series of constant-width pulses whose amplitudes vary with the analog signal. The pulses are usually narrow compared to the period of sampling.

Pulse-averaging discriminator A circuit that converts FM signals into constant-amplitude pulses and filters them into the original signal.

Pulse-code modulation (PCM) A form of time division multiplexing (TDM) in which multiple channels of digitized voice signals are transmitted in serial form. Each channel is assigned a time slot in which to transmit one binary word of data.

Pulse dialing The type of dialing produced by a rotary dialing mechanism in older telephones.

Pulse-duration modulation *See* Pulse-width modulation.

Pulse modulation (PM) Modulation used to transmit analog data in the form of pulses whose amplitude, width, or time position is varied.

Pulse-position modulation (PPM) A procedure in which the pulses change position according to the amplitude of the analog signal.

Pulse spectrum The frequency distribution of the sinusoidal components of a pulse in relative amplitude and relative phase.

Pulse-width modulation (PWM) or pulse-duration modulation (PDM) A procedure in which the width or duration of the pulses varies according to the amplitude of the analog signal.

Pulsed radar A radar system in which signals are transmitted in short bursts or pulses.

Push-pull amplifier An amplifier using two tubes or transistors biased for Class AB, B, or C, connected in such an arrangement that both the positive and the negative alternations of the input are amplified.

Q

Q The quadrature signal is an I-Q modulator.

Q color signal A TV color signal with the following specifications: 21 percent red, -52 percent green, 31 percent blue.

Q section (matching stub) A quarter wavelength of coaxial or balanced transmission line of a specific impedance that is connected between a load and a source in order to match impedances.

Quadrative A 90° phase relationship.

Quadrature amplitude modulation (QAM) A modulation technique used in modems for increasing the number of bits per baud. It uses both amplitude and phase modulation of a carrier.

Quadrature detector An FM demodulator that produces a phase shift of 90° at the unmodulated carrier frequency. It is primarily used in TV audio demodulation.

Quadrature modulation A type of modulation-multiplexing that combines two carriers shifted by 90°, each modulated by a separate signal.

Quadrature (quaternary or quadra) phase PSK (QPSK or 4-PSK) A system in which each pair of successive digital bits in the transmitted word is assigned a particular phase. It is used to increase the binary data rate while maintaining the bandwidth needed to transmit the signal.

Quantizing error Error associated with the analog-to-digital conversion process.

Quarter-wavelength vertical antenna An antenna whose main radiating element is a wire, tubing, or conductor that is one-quarter wavelength at the main operating frequency. Also called a Marconi or ground-plane antenna.

Quieting method A type of sensitivity measurement that determines the amount of RF input signal needed to reduce output noise to 20 dB.

R

R − Y signal A color signal also called $B - Y$. Com-bining three color signals causes Y to be subtracted from R or B signals.

Radar (radio detection and ranging) An electronic communication system based on the principle that high-frequency RF signals are reflected by conductive targets.

Radial A quarter-wavelength wire laid horizontally on the ground or buried in the earth at the base of an antenna to create an artificial ground plane.

Radiation resistance The impedance of an antenna at its driving point.

Radio The general term applied to any form of wireless communication from one point to another.

Radio frequency identification (RFID) A growing wireless technique similar to bar coding.

Radio horizon The distance at which direct wave signals can no longer be received. It is a function of the height of the transmitting and receiving antennas.

Radio wave An electromagnetic wave that can be reflected, refracted, diffracted, and focused.

Rayleigh fading *See* Multipath interference.

RC filter A filter using combinations of resistors and capacitors to achieve the desired frequency response.

RCA phonograph connector An inexpensive coaxial connector used primarily in audio equipment.

Reactance The opposition to AC flow that is offered by coils and capacitors.

Reactance modulator A circuit that uses a transistor amplifier that acts like either a variable capacitor or an inductor to produce direct FM.

Receive signal strength indicator (RSSI) signal In the analog AMPS cellular telephone system, a signal sent back to the cell site so that the mobile telephone switching office (MTSO) can monitor the received signal from the cell and decide whether to switch to another cell.

Receiver The speaker in a telephone handset. A collection of electronic components and circuits that accepts the transmitted message from a channel and converts it back to a form understandable by people. A device in which a modulated light wave is picked up by a photodetector. The signal is amplified and then demodulated to recover the original signal.

Redundancy The simplest way to ensure error-free transmission. Each character or message is sent multiple times until it is properly received.

Redundant array of independent disks (RAID) A collection of hard disk drives used to store data remotely from a computer or server.

Reed Solomon A widely used forward error correction technique that encodes and decodes both wired and wireless data streams to improve reliability of data transfer.

Reed-Solomon (RS) code A forward error correction code that adds extra parity bits to the block of data being transmitted. It permits multiple errors to be detected and corrected.

Reference regulator A zener diode that receives the DC supply voltage as an input and translates it into a precise reference voltage.

Reflected power The power not absorbed by a load and bounced back to the transmitter.

Reflection When light rays strike a reflective surface, the light waves are thrown back, or reflected. The angle of reflection is equal to the angle of incidence. Radio waves are re-flected by any conducting surface they encounter along a path.

Reflection coefficient The ratio of the reflected voltage wave V_r to the incident voltage wave V_i. It provides information on current and voltage along a transmission line.

Reflector A parasitic element that is typically about 5 percent longer than the half-wave dipole-driven element. A parabolic dish used in microwave antennas.

Refraction The bending of a light ray that occurs when light rays pass from one medium to another. The bending of radio waves in the ionosphere.

Regeneration The process of converting a weak optical signal to its electrical equivalent, amplifying and reshaping it electronically, and then retransmitting it on another laser. Also used to rejuvenate electrical signals transmitted over long distances.

Regenerative repeater A circuit that picks up transmitted signals, amplifies them, and retransmits them sometimes on another frequency.

Remote bridge A special bridge used to connect two LANs that are separated by a long distance.

Repeater A circuit that takes a partially degraded signal, boosts its level, and shapes it up. Over long distances, several repeaters may be required. A satellite used to receive information from a transmitting station and then retransmit it to receiving stations.

Resolution The amount of detail produced by a TV signal. In D/A converters, the total number of increments the D/A converter produces over its output voltage range.

Resonant circuit A tuned circuit or filter that has a maximum response, either voltage or current, at a particular (resonant) frequency. A circuit using a combination of inductance and capacitance or materials that exhibits resonance such as ceramic or quartz crystals.

Retrograde orbit The orbit that moves in the direction opposite to the earth's rotation.

Return to zero (RZ) encoding A method of encoding in which the voltage level assigned to a binary 1 level returns to zero during the bit period.

Reverse AGC A method of reducing gain by decreasing the collector current.

RF amplifier Also known as a preselector, or low-noise amplifier that provides some initial gain and selectivity in a receiver.

RF probe (detector probe) A rectifier with a filter capacitor that stores the peak value of the sine-wave RF voltage.

RF signal generator A signal generator used in testing. A variable-frequency oscillator or a frequency-synthesizer.

RF voltmeter A special piece of test equipment optimized for measuring the voltage of high-frequency signals.

RG-8/U coaxial cable One of two types of coaxial cable used in Ethernet networks and antenna transmission lines. It has a characteristic impedance of 53 Ω, is about 0.4 in. in diameter, and is referred to as thick cable.

RG-58/U coaxial cable A widely used thin cable, this cable is used for antenna transmission lines for CB radios and cellular telephones as well as for LANs.

Right-hand circular polarization (RHCP) The term that describes radiation which leaves an antenna with a clockwise rotation.

Ring   A terminal on a telephone at which wires end.

Ring topology   A LAN configuration in which one PC acts as a server and the computers are linked together in a closed loop. Data is usually transferred around the ring in only one direction, passing through each node.

Ringer   A bell or electronic oscillator connected to a speaker that is used to indicate the presence of an incoming call.

Ringing   A 20 Hz signal provided by a local telephone office that indicates when a specific telephone is receiving a call.

Ripple   Amplitude variation with frequency in the passband, or the repetitive rise and fall of the signal level in the passband of some types of filters.

Rise time   The time it takes a pulse voltage to rise from its 10 percent value to its 90 percent value.

RJ-11 connector   A modular plug containing up to six terminals through which most telephones attach to an outlet.

RJ-45 connector   A large modular connector used in terminating twisted pairs in Ethernet LANs. It contains up to eight terminals.

Root-mean-square (rms)   An average that expresses current or voltage magnitude.

Router   An intelligent device with decision-making and switching capabilities, designed to connect two networks.

Rubbering   A term sometimes applied to the whole process of fine-tuning a crystal in an oscillator.

S

Sample-and-hold (S/H) circuit (track/store circuit)   A circuit that accepts an analog input signal and passes it through, unchanged, during its sampling mode. In the hold mode, the amplifier remembers or stores a particular voltage level at the instant of sampling.

Sampling   The process of "looking at" an analog signal for a brief time. During this interval, the amplitude of the analog signal is allowed to be passed or stored.

Satellite   A physical object that orbits, or rotates, around some celestial body.

Satellite period (sidereal period)   The time it takes for a satellite to complete one orbit.

Satellite receiver   A special subsystem designed to work with a TV set, consisting of a dish antenna, a low-noise amplifier and down converter, an IF section with appropriate demodulators for both video and sound, and a method of interconnecting to a conventional TV set.

Satellite TV   The distribution of TV signals via satellite.

SatNav receiver   A communication receiver designed to be used as part of a satellite navigation system.

SAW filter   A solid-state filter used to obtain bandpass selectivity in RF or IF amplifiers.

Scanning   A technique that divides a rectangular scene up into individual lines.

Scattering   Light lost due to light waves entering at the wrong angle and being lost in the cladding of a cable because of refraction.

Scatternet   A linking of piconets.

Schmitt trigger circuit   A level detector that switches between specific voltage levels.

Schottky diode   A diode made with N-type silicon with a thin metal layer. Also known as a hot carrier diode.

Secure socket layer (SSL)   A process involving encryption and authentication to provide for secure transactions like credit card purchases over the Internet.

Selective fading   A phenomenon that occurs when the carrier and sidebands arrive at a receiver a slightly different times, causing a phase shift that can make them cancel one another instead of adding up to the original AM signal.

Selectivity   The ability of a communication receiver to identify and select a desired signal from the thousands of others present in the frequency spectrum and minimize all others.

Self-oscillation   Output voltage that finds its way back to the input of the amplifier with the correct amplitude and phase. The amplifier sometimes oscillates at its tuned frequency, and at other times at a higher frequency.

Semiconductor diode   A unidirectional current junction formed of P- and N-type semiconductor materials.

Semiconductor noise   Noise created by electronic components such as diodes and transistors.

Sensitivity   The ability of a communication receiver to pick up weak signals. This is mainly a function of overall gain.

Serial data transfer   Data transfer in which each bit of a word is transmitted one after another.

Serial-parallel conversion   A technique for converting between serial and parallel transmission and vice versa. It is usually done by shift registers.

Serializer/deserializer (SERDES) device   A circuit used for serial-to-parallel or parallel-to-serial data conversion.

Series modulator   An amplitude-modulating scheme that eliminates the need for a large, heavy, and expensive modulation transformer.

Series resonant circuit   A circuit made up of inductance, capacitance, and resistance used in filters.

Settling time   The amount of time it takes for the output voltage of a D/A converter or S/H circuit to stabilize to within a specific voltage range after a change in binary input.

Shadow fading   Fading caused by objects coming between a transmitter and receiver, for example, when a large building comes between a vehicle containing a transceiver and a base station transceiver.

Shannon-Hartley theorem   The relationship between channel capacity, bandwidth, and noise. This is expressed as
$$C = B \log_2 (1 + S/N).$$

Shape factor   The ratio of the pass bandwidth to the stop bandwidth of a bandpass filter. The smaller the ratio, the greater the selectivity. The steepness of the skirts or the skirt selectivity of a receiver.

Shielding   The process of surrounding EMI-emitting circuits or sensitive receiving circuits with a metal enclosure to prevent the radiation of pickup of signals.

Shift register   A sequential logic circuit made up of a number of flip-flops connected in cascade.

Shot noise   Noise produced by the random movement of electrons or holes across a PN junction. Also, the most common type of semiconductor noise.

Sideband   A new signal generated as part of the modulation process. Sidebands are at sum and difference frequencies directly above and directly below the carrier frequency.

Sidereal period   *See* Satellite period.

Sigma-delta (ΣΔ) converter   A variation of the delta converter, also known as a delta-sigma of charge balance converter. This circuit provides extreme precision, wide dynamic range, and low noise in analog-to-digital conversion.

Signal bandwidth   The frequency range occupied by a signal.

Signal bias   A bias method used in Class C amplifiers involving adjusting the time constant of R_1 and C_1 so that an average DC reverse-bias voltage is established. The applied voltage causes the transistor to conduct, but only on the peaks.

Signal generator   A device that produces an output signal of a specific shape at a specific frequency and, in communication, usually with some form of modulation.

Signal injection   A common way to troubleshoot receivers by using a signal generator of the correct output frequency to test stages of a receiver or transmitter for the correct output response.

Signal plus noise and distortion   *See* SINAD.

Signal processing   The manipulation of a signal by amplification, filtering, modulation, or other mathematical operations.

Signal-to-noise ratio (S/N)   A ratio indicating the relative strengths of a signal and noise in a communication system.

Signal tracing   A common way to troubleshoot equipment by using a signal detection device to follow a signal through various stages of the equipment.

Silica   Another name for glass.

Simplex communication   Communication that flows in one direction.

SINAD (signal plus noise and distortion)   A way of expressing the quality of communication receivers. SINAD equals the composite signal plus the noise and distortion divided by noise and distortion contributed by the receiver.

SINAD test   A method of sensitivity measurement that determines the input signal voltage that will produce at least a 12-dB signal-to-noise ratio.

Single-conversion down converter A circuit used to translate the frequency of a transmitted or received signal to a lower frequency. A mixer with an output that is at a lower frequency than its input.

Single-conversion transponder A device that uses a single mixer to translate the uplink signal to the downlink frequency.

Single data message format (SDMF) A message format used in caller ID.

Single-mode step index cable A cable used for long-distance transmission and maximum content. It eliminates modal dispersion by minimizing the paths through the core.

Single-sideband suppressed carrier (SSSC) or single-sideband (SSB) modulation A form of AM in which the carrier is suppressed and one of the sidebands is eliminated.

Single-sideband transceiver A transceiver that utilizes single sideband signals. Like AM and CW transceivers, in this transceiver the transmitter and receiver can share many circuits.

Singly balanced mixer A mixer circuit using two diodes.

Skin effect The tendency of electrons to flow near and on the outer surface of a conductor at frequencies in the VHF, UHF, and microwave regions.

Skip distance The distance from the transmitting antenna to the point on earth where the first refracted signal strikes the earth and is reflected.

Skirt selectivity The steepness of the sides of a tuned circuit response curve.

Sky wave A signal that is radiated by an antenna into the upper atmosphere, where it is bent back to earth because of refraction in the upper atmosphere.

Slope detector The simplest frequency demodulator. It makes use of a tuned circuit and a diode detector to convert frequency variations into voltage variations.

Slot antenna A radiator made by cutting a half-wavelength rectangular opening in a conducting sheet of metal or into a waveguide. It is used on high-speed aircraft.

SMA connector A coaxial cable connector that is characterized by the hexagonal shape of the body of the male connector.

Small-signal diode A diode used for signal detection and mixing. Two examples are the point-contact diode and the Schottky barrier, or hot-carrier, diode.

Small-signal microwave amplifier An amplifier made of a single transistor or multiple transistors combined with a biasing circuit and any microstrip circuits or components as required. Most of these amplifiers are of the integrated circuit variety.

Smith chart A graph that permits visual solutions to transmission-line calculations.

Snap-off varactor *See* Step-recovery diode.

Snell's law A formula that gives the relationship between the angles and the indices of refraction.

Software-defined radio A receiver in which most of the functions are performed by a DSP.

Solar panel A device that converts the light energy of the sun into a voltage. It is the most common source of power used in satellites.

SONET *See* Synchronous Optical Network.

Space segment The part of a satellite communication or navigation station that is in space.

Space wave A wave that travels in a straight line directly from the transmitting antenna to the receiving antenna. It is also known as line-of-sight communication.

Spam Unwanted and unsolicited email messages.

Spatial isolation A technique used to increase the bandwidth and signal-carrying capacity of a satellite.

Spatial multiplexing (frequency reuse) The transmission of multiple wireless signals on a common frequency in such a way that the signals do not interfere with one another.

Spectrum analysis The process of examining a signal to determine its frequency content.

Spectrum analyzer An oscilloscope-like test instrument used to display received signals in the frequency domain.

Speech processing The way the voice signal used in communication is modified before being applied to the modulator.

Speed of light Approximately 300,000,000 m/s, or about 186,000 mi/s, in free space. Light waves travel in a straight line.

Splatter A type of harmonic sideband interference, so called because of the way it sounds at the receiver.

Splicing Permanently attaching the end of one fiber-optic cable to another. Also connecting electrical conductors.

Split pair A wiring error often made when a cable contains multiple twisted-pair lines.

Spread spectrum (SS) A modulation and multiplexing technique that distributes a signal and its sidebands over a very wide bandwidth.

Spyware Software attached to your computer unknowingly that monitors your activity, computer usage, and collects data and reports it back to its initiator without your knowledge or permission.

Square-law function A current variation in proportion to the square of the input signals.

Squelch (muting) circuit A circuit used to keep the receiver audio turned off until an RF signal appears at the receiver input.

SSB *See* Single-sideband.

Standards In communication systems, specifications and guidelines that companies and individuals follow to ensure compatibility between transmitting and receiving equipment.

Standing wave A composite of forward and reflected voltage distributed along a transmission line, not matched to the load, which indicates that the power produced by the generator is not totally absorbed by the load.

Standing wave ratio (SWR) The ratio of maximum current to minimum current, or the ratio of maximum voltage to minimum voltage along a transmission line.

Standing wave ratio (SWR) meter A meter that measures forward and reflected power and therefore can display SWR.

Star topology A network configuration consisting of a central controller node and multiple individual stations connected to it.

Start and stop bits Binary levels that indicate the beginning and ending of a word in asynchronous data transmission.

Start frame delimiter (SFD) The portion of a packet protocol which announces the beginning of the packet.

Start of header (SOH) character The character that begins a frame. It is an ASCII character which means that the transmission is beginning.

Start of text (STX) character An ASCII character indicating the start of text.

Static Electrical disturbances that occur naturally in the earth's atmosphere. Also called atmospheric noise.

Step index cable A cable with a sharply defined step in the index of refraction where the fiber core and the cladding interface.

Step-recovery diode A PN-junction diode made with gallium arsenide or silicon.

Stop band The range of frequencies outside the passband, that is, the range of frequencies that are greatly attenuated by the filter. Frequencies in this range are rejected.

Storage area network (SAN) A network of disk drives used to access massive amounts of data.

Straight-through processor A device that picks up TV signals from a local station and amplifies the signal before multiplexing it onto the main cable.

Subcarrier In an FDM system, each signal to be transmitted feeds a modulator circuit. The carrier for each modulator is on a different frequency and is called a subcarrier.

Subcarrier oscillator (SCO) A modulated oscillator used in a telemetry system.

Submillimeter wave An electromagnetic wave at a frequency below 30 GHz.

Subsatellite point (SSP) The point on the earth that is directly below a satellite and through which the satellite's location is specified.

Subscriber interface (subscriber line interface circuit, SLIC) A group of basic circuits that power the telephone and provide all the basic functions such as ringing, dial tone, and dialing supervision.

Subscriber loop *See* Local loop.

Subsidiary communication authorization (SCA) signal A separate subcarrier of 67 kHz that is frequency-modulated by audio signals, usually music, and transmitted with an FM broadcast signal.

Successive approximations converter A type of A to D converter in which the bits driving a DAC are turned on one at a time from MSB to USB to estimate input voltage level.

Successive approximations register (SAR) A special register that causes each bit in a register driving a DAC to be turned on one at a time from MSB to LSB until the closest binary value is stored in the register.

Superheterodyne A receiver that converts signals to the intermediate frequency and for which a single set of amplifiers provides a fixed level of sensitivity and selectivity.

Superheterodyne RF spectrum analyzer A widely used type of RF spectrum analyzer.

Superhigh frequency (SHF) The frequency range from 3 to 30 GHz.

Supervision The group of functions in the subscriber line interface that monitor local loop conditions and provide services.

Surface acoustic wave (SAW) filter A special form of crystal filter. It operates like a fixed tuned bandpass filter designed to provide the exact selectivity required by a given application.

Surge impedance See Characteristic impedance.

Surveillance satellite A satellite used for purposes such as military reconnaissance, map making, and weather forecasting.

Sweep generator A signal generator with an output frequency that can be linearly varied over some specific range.

Switch A hublike device used to connect individual PC nodes to the network wiring.

Switch hook A double-pole mechanical switch that is usually actuated by a telephone handset.

Switched capacitor filter (SCF) An active IC filter made of op amps, capacitors, and transistor switches. It can be designed to operate as a high-pass, low-pass, bandpass, or bandstop filter.

Switched integrator The basic building block of a switched capacitor filter (SCF).

Switching amplifier A transistor that is used as a switch and is either conducting or nonconducting.

Sync pulse A pulse applied to one of the input channels at a transmitter to synchronize the multiplexed channels at the receiver.

Synchronizing circuit A circuit needed to keep the sweep of a TV receiver in step with the transmitted signal.

Synchronous data link control (SDLC) protocol A flexible and widely used synchronous protocol, it is used in networks that are interconnections of multiple computers.

Synchronous data transmission The technique of transmitting each data word one after another without start and stop bits, synchronized with a clock signal, usually in multiword blocks.

Synchronous detection A method in which an internal clock signal at the carrier frequency in a receiver switches the AM signal off and on, producing rectification similar to that in standard diode detectors.

Synchronous detector A circuit that uses an internal clock signal at the carrier frequency to switch the AM signal off and on.

Synchronous (geostationary or geosynchronous) satellite A satellite that appears to remain in a fixed position because it rotates in exact synchronism with the earth.

Synchronous Optical Network (SONET) An optical network developed to transmit digitized telephone calls and data over fiber-optic cable at high speed.

T

T-carrier system A digital telephone system used throughout the United States.

T connector A coaxial connector accessory that provides a convenient way to attach an additional node to an existing coaxial cable.

T junction See T section.

T network A type of impedance-matching network. The configuration of circuit elements resembles the letter T.

T-1 system A PCM system developed by Bell Telephone for transmitting telephone conversations by high-speed digital links. It multiplexes 24 voice channels onto a single line using TDM techniques.

T section A device used to split or combine two or more sources of microwave power.

Tactical Air Navigation (Tacan) The higher-frequency system of air navigation used by the military.

Tank circuit A parallel-resonant LC circuit.

TCP/IP protocol A software protocol, at the heart of the Internet, that ensures that data is properly partitioned, transmitted, received, and reassembled.

TDD Telecommunication device for the deaf.

TDR tester (cable analyzer or LAN meter) An instrument used to make tests and measurements of items such as opens or short circuits, cable attenuation, or cable miswiring.

Technician A person working in the electronics industry with some kind of postsecondary education in electronics—usually 2 years of formal, post–high school training usually involved with equipment service, maintenance, repair, installation, or operation.

Technologist In the field of electronics, a technologist usually holds a bachelor's degree in electronics technology from a technical college or university usually employed as an engineer.

Tee switching configuration In a PIN diode, a combination of two series switches and a shunt switch.

Telemetry A communication system used for remote monitoring or measurement.

Telemetry, command, and control (TC&C) subsystem Equipment consisting of a receiver and the recorders and indicators that allow a ground station to monitor and control conditions in a satellite.

Telephone hierarchy The organization of the path in which a call travels from the person placing the call to the person receiving the call.

Telephone set An analog baseband receiver with a microphone and speaker, better known as a transmitter and receiver. It also uses a ringer and dialing mechanism.

Telephone system An analog system originally designed for full-duplex communication

of voice signals. It now employs digital techniques.

Termination A resistive load connected to a transmission line or waveguide to prevent reflections on the line.

Terminator A special connector containing a resistor whose value is equal to the characteristic impedance of the coaxial cable.

Test signal A special tone placed on the local loop by the phone company to check the status and quality of lines.

Thermal agitation Random variation of electrons in an electronic component due to heat energy.

Thermal noise Noise resulting from the random motion of free electrons in a conductor caused by heat.

Thermistor A resistor whose resistance varies inversely with temperature.

Third-order intercept test (TOI or IP3) A measure of the linearity of amplifiers, mixers, and other circuits.

Time delay (transit time) The time it takes for a signal applied at one end of a transmission line to appear at the other end of the line.

Time division multiple access (TDMA) A special form of time division multiplexing that provides multiple voice channels per satellite or telephone spectrum.

Time division multiple access (TDMA) cell phone system See IS-136 TDMA.

Time division multiplexing (TDM) A time-sharing or sampling technique that makes it possible for each signal to occupy the entire bandwidth of a channel. However, each signal is transmitted for only a brief period of time.

Time domain display A display of variations in voltage, current, or power with respect to time as on an oscilloscope screen.

Time domain reflectometry (TDR) A test for all types of cables and transmission lines. It is widely used in finding faults in cables used for digital data transmission.

Tip A terminal on a telephone at which wires end.

Token passing An access method used by Token-Ring systems in which a binary word is passed around the ring to communicate when a node desires to send data.

Token Ring A LAN configuration developed by IBM. It uses twisted pair.

Tone dialing A dialing system used in most modern telephones that uses pairs of audio tones to create signals representing the numbers to be dialed. See DMTF.

Top hat In a shortened antenna, an arrangement in which conductors are added at the top of the antenna.

Topology The physical paths used to connect the nodes on a network.

Toroid A doughnut-shaped core used in RF transformers, usually made of a special type of powdered iron.

Transceiver A communication equipment package in which both transmitter and receiver are in a single unit.

Transferred–electron device (TED) *See* Gunn diode.

Transimpedance amplifier (TIA) The input stage of a light receiver circuit during which the diode current is converted into an output voltage and amplified.

Transit system A satellite navigation system used by the U.S. Navy.

Transit time *See* Time delay.

Transit-time noise Noise caused by the time it takes for a current carrier such as a hole or electron to move from the input to the output.

Transmatch circuit A type of antenna tuner that uses a coil and three capacitors to tune the antenna for optional SWR.

Transmission line A cable that carries radio signals, telephone signals, computer data, TV signals, etc. The two requirements of any transmission line are that it introduce minimum attenuation and distortion of the signal, and not radiate any of the signal as radio energy.

Transmission line transformer A type of transformer widely used in power amplifiers for coupling between stages and impedance matching.

Transmit-receive (TR) tube A special vacuum tube used as a fast high-power switch that permits both the receiver and a high-power transmitter to share a single antenna without damage to the receiver.

Transmit subsystem In a satellite, collection of electronic equipment that takes the signal to be transmitted, amplifies it, and sends it to the antenna.

Transmitter A collection of electronic components and circuits designed to convert an electrical signal into one that can be transmitted over a given medium.

Transponder The transmitter-receiver combination in a satellite.

Transverse electric (TE) field The electric field at a right angle to the direction of wave propagation.

Transverse magnetic (TM) field The magnetic field transverse to the direction of propagation.

TRAPATT diode A PN-junction diode made of silicon, GaAs, or InP. It is designed to operate with a high reverse bias that causes it to avalanche, or break down.

Traveling-wave tube (TWT) A versatile microwave RF power amplifier. It can generate hundreds and even thousands of watts of microwave power.

Tree topology A bus topology in which each node has multiple interconnections to other nodes through a star interconnection.

Trellis code modulation (TCM) A special form of QAM that facilitates error detection and correction.

Triode A type of vacuum tube with three elements—the cathode, the grid, and the plate.

Tripler A frequency multiplier that multiplies the frequency by 3.

Trunk cable The main output cable in a modern cable TV system usually a fiber-optic cable.

Trunked repeater system A system in which two or more repeaters are under the control of a computer system that can transfer a user from an assigned but busy repeater to another, available repeater.

TTY Teletypewriter.

Tunable laser A laser used in fiber-optic systems whose frequency can be varied by changing the DC bias on the device or mechanically adjusting an external cavity.

Tuned circuit A circuit made up of inductors and capacitors that resonate at specific frequencies.

Tuned radio frequency (TRF) receiver A receiver in which sensitivity is increased by multiple stages of RF amplification followed by a demodulator. The main problem with this receiver is tracking the tuned circuits.

Tuner An RF unit with LNA, mixer and local oscillator used as the front-end in a receiver to produce an IF output.

Tuning synthesizer A local oscillator set to frequencies that will convert the RF signals to the IF.

TV remote control A small handheld battery-powered unit that transmits a serial digital code via an IR beam to a receiver that decodes it and carries out the specific action defined by the code.

TV signal Voice and video signals that occupy a channel in the spectrum with a bandwidth of 6 MHz.

TV spectrum allocation An allotted range. For example, using VHF and UHF frequencies, TV stations in the United States use the frequency range between 54 and 806 MHz. Sixty-eight 6-MHz TV channels are assigned frequencies within this range.

TVRO (TV receive-only) system A satellite receiver containing circuitry for controlling the positioning of a satellite dish antenna.

Twisted pair Two insulated copper wires twisted together loosely to form a transmission line.

Type-P display A kind of radar display using a cathode-ray tube (CRT) that shows target reflections as vertical "blips" with respect to a horizontal time sweep across the face of the CRT.

U

Ultrahigh frequency (UHF) The frequency range from 300 to 3000 MHz.

Ultraviolet range The range of frequencies above violet visible light, which has a wavelength of 400 nm, or 0.4 μm.

Ultrawideband (UWB) radar Radar in the form of pulsed radar that radiates a stream of very short pulses several hundred picoseconds long. The resulting spectrum is very broad.

Ultrawideband (UWB) wireless A type of wireless that transmits data in the form of very short pulses.

Unbalanced transmission line A transmission line in which one conductor is connected to ground.

Unidirectional antenna An antenna that sends or receives signals in one direction only.

Uniform resource locator (URL) A special address used to locate sites on the Web.

Universal asynchronous receiver transmitter (UART) A special large-scale digital IC that performs parallel-to-serial and serial-to-parallel conversion.

Up conversion The process in which the original signal is generated at a lower frequency and then converted to a higher frequency for transmission.

Uplink The original signal transmitted from an earth station to a satellite.

User datagram protocol (UDP) A protocol used with TCP/IP.

V

Varactor A diode designed to optimize the variable capacitance exhibited as reverse bias is changed.

Varactor diode A voltage variable capacitor. When a reverse bias is applied to the diode, it acts like a capacitor.

Varactor modulator A modulator that utilizes a varactor.

Varactor phase modulator A simple phase-shift circuit that can be used as a phase modulator by using a varactor.

Variable frequency oscillator (VFO) An oscillator used to provide continuous tuning over a desired range.

Varistor A nonlinear resistance element whose resistance changes depending upon the amount of voltage applied across it.

Velocity factor The ratio of the velocity of propagation of a signal in a transmission line to its velocity in free space.

Vertical-cavity surface-emitting laser (VCSEL) A laser used in fiber-optic systems that is made on the surface of silicon wafer–like transistors and integrated circuits.

Vertical radiation pattern The part of the doughnut-shaped radiation pattern of an antenna indicating its vertical response.

Very high frequency (VHF) The frequency range from 30 to 300 MHz.

Very low frequency (VLF) The frequency range from 15 to 20 kHz.

Vestigial sideband signal An AM signal where a portion of one sideband is suppressed. Used in TV.

VHF omnidirectional ranging (VOR) A system of air navigation used by nonmilitary aircraft.

Video intermediate frequency A standard 45.75-MHz frequency for TV pictures or 41.25 MHz for sound.

Video signal A voltage variation representing variations in light intensity along a scan line.

Virtual private network (VPN) A technique that uses software to establish a virtual network over a LAN or the Internet for the purpose of providing security only to authorized users.

Virus   An unwanted program that disrupts or destroys the software and hardware of a computer.

Visible light   The frequency in the 400- to 700-nm range, or 0.4 to 0.7 μm, depending upon the color of the light.

Visible spectrum   A group of signals with frequencies ranging from 0.4 to 0.7 μm.

Vocoder   A circuit that digitizes voice signals. *See also* Codec.

Voice frequency (VF)   A group of signals with frequencies ranging from 300 to 3000 Hz. This is the range of normal speech.

Voice mail   *See* Answering machine.

Voltage-controlled oscillator (VCO) or voltage-controlled crystal oscillator (VXO)   A circuit in which the varying direct or alternating current changes the frequency of an oscillator operating at the carrier frequency. An oscillator often used in applications in which voltage to frequency conversion is required. Its frequency is controlled by an external input voltage.

Von Neumann architecture   The stored-program concept that is the basis of operation of all digital computers.

Von Neumann bottleneck   Term describing the fact that only one path exists between the memory and the CPU. Therefore, only one data or instruction word can be accessed at a time. This greatly limits execution speed.

W

Waveguide   A hollow metal conducting pipe designed to carry and constrain the electromagnetic waves of a microwave signal. It has a rectangular or circular cross section and is made of copper, aluminum, or brass.

Waveguide cutoff frequency   The frequency below which a waveguide will not transmit energy.

Wavelength   The distance occupied by one cycle of a wave (usually expressed in meters).

Wavelength division multiplexing (WDM)   Frequency division multiplexing (FDM) used on fiber-optic cable. It permits multiple channels of data to operate independently on different light wavelengths in a fiber-optic cable.

WCDMA– wideband CDMA   A version of CDMA that uses 5 MHz wide channels to implement 3G cell phone systems. A standard based on the 3GPP.

White (Johnson) noise   Noise containing all frequencies randomly occurring at random amplitudes.

Wide-area network (WAN)   One of three basic types of electronic networks in common use. The long-distance telephone systems linked together are also this type of network. The Internet backbone is a WAN.

Wide-wavelength division multiplexing (WWDM)   A variation of 10-gigabit Ethernet in which data is divided into four channels and transmitted simultaneously over four different wavelengths of infrared light near 1310 nm. It is similar to frequency division multiplexing.

Wi-Fi   The trade name of the wireless local area networking technology defined by the IEEE 802.11 standard. It uses the 2.4 and 5.8 GHz bands to send high speed data at a range up to 100 meters. It is widely used to access the Internet via hot spots with a laptop computer.

WiMAX   World interoperability for Microwave Access. The trade name for a broadband wireless technology based on the IEEE 802.16 standard. Use for fixed broadband access and for mobile access at high data rates with cell sites up to 3 miles away.

Wireless   Originally, the word for radio. Nowadays, the term refers mainly to the cellular telephone industry. (Cell phones are sophisticated two-way radios and radio LANs).

Wireless LAN   A network communicating through radio and infrared techniques. In this system, each PC must contain a wireless modem or transceiver.

World Wide Web   A special part of the Internet where companies, organizations, government, or individuals can post information for others to access and use.

X

xDSL modem   A specific digital subscriber line (DSL) standard. The "x" is one of several letters that define the standard.

Xmodem protocol   A protocol used for asynchronous ASCII-coded data transmission between computers via modem. These transmissions begin with a NAK character.

Y

Yagi antenna   A directional gain antenna made up of a driven element and one or more parasitic elements.

Z

Zero   A frequency at which there is zero impedance in a circuit.

Zero IF (ZIF) receiver   *See* Direct conversion receiver.

ZigBee   The trade name of a wireless technology that is based on the IEEE standard 802.15.4. It uses low speed data in the ISM UHF and low microwave bands to implement monitor and control functions over short distances in home, commercial and industrial applications. It is based on the concept of mesh networks.

Credits

CHAPTER 1

p. 17: © CORBIS; p. 21: Courtesy Federal Communications Commission; p. 25: © Michael Newman/Photo Edit; p. 26: © Theodore Anderson/Getty Images

CHAPTER 2

p. 41: © Mark Steinmetz; p. 90: Courtesy Hewlett-Packard Company

CHAPTER 7

p. 207: Courtesy Analog Devices, Inc.

CHAPTER 8

p. 296: Courtesy Fox Electronics

CHAPTER 12

p. 455: Courtesy Agilent Technologies; p. 479 (left & right): Courtesy Kawasaki Microelectronics

CHAPTER 13

p. 509: © David Joel/Getty Images

CHAPTER 14

p. 530: Courtesy Jaybeam Wireless; p. 541: © Lonnie Duka/Getty Images; p. 565: © Peter Timmermans/Getty Images

CHAPTER 16

p. 610: © Randy Wells/Getty Images; p. 611 and 635: Courtesy Agilent Technologies

CHAPTER 17

p. 697: Courtesy Raytheon Company; p. 699: Courtesy Garmin Ltd.; p. 706: Courtesy Department of Defense

CHAPTER 18

p. 727: © Bob Krist/Getty Images; p. 742: © Mike Sandman; p. 747: © Atmel

CHAPTER 19

p. 756: © Steve Dunwell Photograph/Index Stock; p. 766: Courtesy Gigahertz Optik, Inc.; p. 777: Courtesy NeoPhotonics Corporation; p. 781: © PhotoDisc; p. 782: Courtesy Lucent Technologies/Bell Labs

CHAPTER 21

p. 831: Courtesy 3Com; p. 842: Courtesy Microchip Technology

CHAPTER 22

p. 860: Courtesy Hewlett-Packard Company; p. 863: Courtesy Tektronix; p. 868: Courtesy Hewlett-Packard Company; p. 880: Courtesy Tektronix; p. 894 (top): Courtesy Tektronix; p. 894 (bottom): Photo Courtesy of National Instruments Corporation; p. 895: Photo Courtesy of National Instruments Corporation

Index

Battery, 726–727
Baud rate, 389
Baudot code, 386–389
Bayonet connectors, 768
Beam antennas, 547–548
Beamwidth, antenna, 537, 643–645, 647
Beat frequency oscillator (BFO), 343
Bessel functions, 160–164
Bessel (Thomson) filters, 64–65
Biasing methods, 269–271
Bicone antennas, 649–650
Bidirectional antennas, 544
Bidirectional couplers, 630
Bifilar windings, 554–555
Binary data, 385–446
    modern codes, 8, 387–389
    transmission, 10, 389–400
        asynchronous, 391, 439–441
        bisynchronous, 441
        coding levels, 398
        encoding methods, 393–394, 431, 464
        modems in, 10, 400–414, 463, 736–737
        serial, 204, 389–391
        synchronous, 392, 441–442
Binary phase-shift keying (BPSK), 158, 404–407,
    846–848
Bins, 427
Biphase encoding, 394, 842–843
Birdies, 309, 328
Bisync protocol, 441
Bit error rate (BER), 300, 412, 430
    analyzer, 880–881
Bit rate-distance product, 779
Bit splitters, 409
BlackBerry, 817
Block-check character (BCC), 434
Block-check sequence (BCS), 434
Block diagram analysis, 663–665
Blocking test, 876–878
Bluetooth, 819, 845, 849, 894
    PANs and, 832–834
BNC connectors, 457–458, 484–485
BORSCHT, 726–727
Boundary scan, 887–889
Boundary scan cells (BSCs), 887
Bow-tie antenna, 535
BPSK (binary phase-shift keying), 407, 846–848
Bridge circuits, 125
Bridge-T filters, 68
Bridges, 461
Broadband connection, 577–578
Broadband transmission, 9–11, 21, 685
Broadcasting, FM stereo, 365–368
Broadside arrays, 550
Browsers, 575
Built-in self-test (BIST), 888
Bus topology, 452, 463–464
Butterworth filters, 64

C (conventional) band, 679–681, 784
Cable mode, 762
Cable modems, 428–429
Cable television (CATV), 22, 364–365,
    428–429, 577–578, 582–583
Cables. *See also* Coaxial cable; Twisted-pair cable
    fiber-optic, 457, 470, 582, 749, 758–769
    local area network, 455–457

size, 765
    troubleshooting, 884
Caller ID (calling line identification), 721–722
Cameras, digital, 817
Cap code, 740
Capacitive reactance, 41
Capacitors, 41–42
Capture effect, 169–170
Capture range, 194–195
Careers, communication, 24–27
Carrier sense multiple access with collision
        avoidance (CSMA/CA), 832, 835
Carrier sense multiple access with collision
        detection (CSMA/CD), 469, 472
Carrier-to-noise (C/N) ratio, 413
Carrier(s), 9, 94–95
    current transmission, 5
    deviation, 156
    frequency, 175
    generators, 248–265
    phase shift, 146–148
    recovery circuits, 135–136
Carson's rule, 164
Cascaded circuits, 33
    noise in, 331
Cassegrain feed, 642, 647–648
Cathode-ray tubes (CRTs), 863
Cauer (elliptical) filters, 64
Cavity resonators, 630–631
CB synthesizers, 351–352
CB transceivers, 351–352
CDMA2000, 814, 816
Cell phones, 23, 792–822
    advanced, 816–819
    satellite, 696
    single-chip, 811
    systems, 793–798
    video, 817
Cell site antennas, 820
Cells, 793–794
Cellular radio system, 793
Celsius scale, 325
Central office (local exchange), 728, 786
Centrex, 731
Ceramic filters, 71–75, 336
Certification authority (CA), 599
Channelization process, 685–686
Channels, 605–606
    bandwidth, 19
Characteristic (surge) impedance, 487–488,
    512–513
Charged coupled devices (CCDs),
    733–735, 737
Chebyshev filters, 64
Chip geometries, 617
Chipping rate, 418
Chips, 418
Choke joints, 626
Chromatic dispersion, 763
Circuit gain, 337–339
Circulators, 631–632
Citizen's radio, 23
Cladding, 760
Class A amplifiers, 265
    buffers, 266
    microwave power, 619–620
Class AB amplifiers, 265

Class B amplifiers, 265
    push-pull, 267–268
Class C amplifiers, 265–266, 269–273
Class D amplifiers, 276–277
Class E amplifiers, 277
Class F amplifiers, 277
Class S amplifiers, 277–278
Cleaving, 769
Client-server networks, 454–455
Cloaking, 843
Clock and data recovery (CDR) unit, 776
Clock multiplier unit (CMU), 778
Clock recovery circuits, 374–375
Coarse acquisition (C/A) code, 700
Coarse wavelength division multiplexing
        (CWDM), 470, 784, 786
Coaxial cable, 481, 482, 525, 610, 621
    connectors, 457–458, 484–487
    in local area networks, 454–455, 465–467
    specifications, 491–494
Code-division multiple access (CDMA), 359, 415,
        420, 682, 797, 806–807, 819, 846
Code of Federal Regulations, 893
Codecs, 234–235
Coding, 727
Cognitive radio, 321
Coherent detectors, 136
Coherent light, 771
Collector modulators, 128–130
Collectors, 639
Collinear antennas, 549–550
Collision, 472
Color LCD screens, 817
Colpitts oscillator, 250
Comb response, 78
Common mode rejection, 482
Communication, defined, 3
Communication channels, 4–5, 398–400
Communication links, 448
Communication satellites, 563, 678–682, 695–697
Communication subsystems, satellite, 684–686
Communication systems, 3–5. *See also* specific
        types of communication
Communication test equipment, 852–900
Communications Act of 1934, 20
Commutating filters, 76–78
Commutator switches, 370–371
Compact disks (CDs), 835
Companding, 232–234
Complementary code keying (CCK), 830
Complex programmable logic devices
        (CPLDs), 237
Compression circuits, 97. *See also*
        Data compression
Computer branch exchanges (CPXs), 730
Computers, 200. *See also* Internet; Local area
        networks
Cone of acceptance, 760
Conical antennas, 535–536
Connection joints, 626
Connectors, 484–487, 768
    coaxial cable, 457–458, 484–487
    fiber-optic, 457, 768–769
    troubleshooting, 884
    twisted-pair, 458–459
Constellation diagrams, 408, 411
Contention systems, 472